Springer Handbook of Enzymes
Supplement Volume S1

Dietmar Schomburg and
Ida Schomburg (Eds.)

Springer Handbook of Enzymes

Supplement Volume S1
Class 1 · Oxidoreductases
EC 1

coedited by Antje Chang

Second Edition

Professor Dietmar Schomburg
e-mail: d.schomburg@tu-bs.de

Dr. Ida Schomburg
e-mail: i.schomburg@tu-bs.de

Dr. Antje Chang
e-mail: a.chang@tu-bs.de

Technical University Braunschweig
Bioinformatics & Systems Biology
Langer Kamp 19b
38106 Braunschweig
Germany

Library of Congress Control Number: 2008942397

ISBN 978-3-540-85187-5 2nd Edition Springer Berlin Heidelberg New York

The first edition was published as the "Enzyme Handbook, edited by D. and I. Schomburg".

Springer is a part of Springer Science+Business Media
springer.com

Printed in Germany

Cover design: Erich Kirchner, Heidelberg
Typesetting: medionet Publishing Services Ltd., Berlin

Printed on acid-free paper 2/3141m-5 4 3 2 1 0

Preface

Today, as the full information about the genome is becoming available for a rapidly increasing number of organisms and transcriptome and proteome analyses are beginning to provide us with a much wider image of protein regulation and function, it is obvious that there are limitations to our ability to access functional data for the gene products – the proteins and, in particular, for enzymes. Those data are inherently very difficult to collect, interpret and standardize as they are widely distributed among journals from different fields and are often subject to experimental conditions. Nevertheless a systematic collection is essential for our interpretation of genome information and more so for applications of this knowledge in the fields of medicine, agriculture, etc. Progress on enzyme immobilisation, enzyme production, enzyme inhibition, coenzyme regeneration and enzyme engineering has opened up fascinating new fields for the potential application of enzymes in a wide range of different areas.

The development of the enzyme data information system BRENDA was started in 1987 at the German National Research Centre for Biotechnology in Braunschweig (GBF), continued at the University of Cologne from 1996 to 2007, and then returned to Braunschweig, to the Technical University, Institute of Bioinformatics & Systems Biology. The present book "Springer Handbook of Enzymes" represents the printed version of this data bank. The information system has been developed into a full metabolic database.

The enzymes in this Handbook are arranged according to the Enzyme Commission list of enzymes. Some 4,000 "different" enzymes are covered. Frequently enzymes with very different properties are included under the same EC-number. Although we intend to give a representative overview on the characteristics and variability of each enzyme, the Handbook is not a compendium. The reader will have to go to the primary literature for more detailed information. Naturally it is not possible to cover all the numerous literature references for each enzyme (for some enzymes up to 40,000) if the data representation is to be concise as is intended.

It should be mentioned here that the data have been extracted from the literature and critically evaluated by qualified scientists. On the other hand, the original authors' nomenclature for enzyme forms and subunits is retained. In order to keep the tables concise, redundant information is avoided as far as possible (e.g. if K_m values are measured in the presence of an obvious cosubstrate, only the name of the cosubstrate is given in parentheses as a commentary without reference to its specific role).

The authors are grateful to the following biologists and chemists for invaluable help in the compilation of data: Cornelia Munaretto and Dr. Antje Chang.

Braunschweig
Autumn 2008

Dietmar Schomburg, Ida Schomburg

List of Abbreviations

A	adenine
Ac	acetyl
ADP	adenosine 5'-diphosphate
Ala	alanine
All	allose
Alt	altrose
AMP	adenosine 5'-monophosphate
Ara	arabinose
Arg	arginine
Asn	asparagine
Asp	aspartic acid
ATP	adenosine 5'-triphosphate
Bicine	N,N'-bis(2-hydroxyethyl)glycine
C	cytosine
cal	calorie
CDP	cytidine 5'-diphosphate
CDTA	trans-1,2-diaminocyclohexane-N,N,N,N-tetraacetic acid
CMP	cytidine 5'-monophosphate
CoA	coenzyme A
CTP	cytidine 5'-triphosphate
Cys	cysteine
d	deoxy-
D-	(and L-) prefixes indicating configuration
DFP	diisopropyl fluorophosphate
DNA	deoxyribonucleic acid
DPN	diphosphopyridinium nucleotide (now NAD^+)
DTNB	5,5'-dithiobis(2-nitrobenzoate)
DTT	dithiothreitol (i.e. Cleland's reagent)
EC	number of enzyme in Enzyme Commission's system
E. coli	Escherichia coli
EDTA	ethylene diaminetetraacetate
EGTA	ethylene glycol bis(-aminoethyl ether) tetraacetate
ER	endoplasmic reticulum
Et	ethyl
EXAFS	extended X-ray absorption fine structure
FAD	flavin-adenine dinucleotide
FMN	flavin mononucleotide (riboflavin 5'-monophosphate)
Fru	fructose
Fuc	fucose
G	guanine
Gal	galactose

GDP	guanosine 5'-diphosphate
Glc	glucose
GlcN	glucosamine
GlcNAc	N-acetylglucosamine
Gln	glutamine
Glu	glutamic acid
Gly	glycine
GMP	guanosine 5'-monophosphate
GSH	glutathione
GSSG	oxidized glutathione
GTP	guanosine 5'-triphosphate
Gul	gulose
h	hour
H4	tetrahydro
HEPES	4-(2-hydroxyethyl)-1-piperazineethane sulfonic acid
His	histidine
HPLC	high performance liquid chromatography
Hyl	hydroxylysine
Hyp	hydroxyproline
IAA	iodoacetamide
IC 50	50% inhibitory concentration
Ig	immunoglobulin
Ile	isoleucine
Ido	idose
IDP	inosine 5'-diphosphate
IMP	inosine 5'-monophosphate
ITP	inosine 5'-triphosphate
K_m	Michaelis constant
L-	(and D-) prefixes indicating configuration
Leu	leucine
Lys	lysine
Lyx	lyxose
M	mol/l
mM	millimol/l
m-	*meta*-
Man	mannose
MES	2-(N-morpholino)ethane sulfonate
Met	methionine
min	minute
MOPS	3-(N-morpholino)propane sulfonate
Mur	muramic acid
MW	molecular weight
NAD^+	nicotinamide-adenine dinucleotide
NADH	reduced NAD
$NADP^+$	NAD phosphate
NADPH	reduced NADP
NAD(P)H	indicates either NADH or NADPH

NBS	N-bromosuccinimide
NDP	nucleoside 5'-diphosphate
NEM	N-ethylmaleimide
Neu	neuraminic acid
NMN	nicotinamide mononucleotide
NMP	nucleoside 5'-monophosphate
NTP	nucleoside 5'-triphosphate
o-	*ortho-*
Orn	ornithine
p-	*para-*
PBS	phosphate-buffered saline
PCMB	*p*-chloromercuribenzoate
PEP	phosphoenolpyruvate
pH	-log10[H^+]
Ph	phenyl
Phe	phenylalanine
PHMB	*p*-hydroxymercuribenzoate
PIXE	proton-induced X-ray emission
PMSF	phenylmethane-sulfonylfluoride
p-NPP	*p*-nitrophenyl phosphate
Pro	proline
Q_{10}	factor for the change in reaction rate for a 10°C temperature increase
Rha	rhamnose
Rib	ribose
RNA	ribonucleic acid
mRNA	messenger RNA
rRNA	ribosomal RNA
tRNA	transfer RNA
Sar	N-methylglycine (sarcosine)
SDS-PAGE	sodium dodecyl sulfate polyacrylamide gel electrophoresis
Ser	serine
T	thymine
t_H	time for half-completion of reaction
Tal	talose
TDP	thymidine 5'-diphosphate
TEA	triethanolamine
Thr	threonine
TLCK	N^{α}-*p*-tosyl-L-lysine chloromethyl ketone
T_m	melting temperature
TMP	thymidine 5'-monophosphate
Tos-	tosyl- (*p*-toluenesulfonyl-)
TPN	triphosphopyridinium nucleotide (now $NADP^+$)
Tris	tris(hydroxymethyl)-aminomethane
Trp	tryptophan
TTP	thymidine 5'-triphosphate
Tyr	tyrosine
U	uridine

U/mg	µmol/(mg*min)
UDP	uridine 5'-diphosphate
UMP	uridine 5'-monophosphate
UTP	uridine 5'-triphosphate
Val	valine
Xaa	symbol for an amino acid of unknown constitution in peptide formula
XAS	X-ray absorption spectroscopy
Xyl	xylose

Index of Recommended Enzyme Names

Description of Data Fields

All information except the nomenclature of the enzymes (which is based on the recommendations of the Nomenclature Committee of IUBMB (International Union of Biochemistry and Molecular Biology) and IUPAC (International Union of Pure and Applied Chemistry) is extracted from original literature (or reviews for very well characterized enzymes). The quality and reliability of the data depends on the method of determination, and for older literature on the techniques available at that time. This is especially true for the fields *Molecular Weight* and *Subunits*.

The general structure of the fields is: **Information – Organism – Commentary – Literature**

The information can be found in the form of numerical values (temperature, pH, K_m etc.) or as text (cofactors, inhibitors etc.).

Sometimes data are classified as *Additional Information*. Here you may find data that cannot be recalculated to the units required for a field or also general information being valid for all values. For example, for *Inhibitors*, *Additional Information* may contain a list of compounds that are not inhibitory.

The detailed structure and contents of each field is described below. If one of these fields is missing for a particular enzyme, this means that for this field, no data are available.

1 Nomenclature

EC number

The number is as given by the IUBMB, classes of enzymes and subclasses defined according to the reaction catalyzed.

Systematic name

This is the name as given by the IUBMB/IUPAC Nomenclature Committee

Recommended name

This is the name as given by the IUBMB/IUPAC Nomenclature Committee

Synonyms

Synonyms which are found in other databases or in the literature, abbreviations, names of commercially available products. If identical names are frequently used for different enzymes, these will be mentioned here, cross references are given. If another EC number has been included in this entry, it is mentioned here.

CAS registry number

The majority of enzymes have a single chemical abstract (CAS) number. Some have no number at all, some have two or more numbers. Sometimes two enzymes share a common number. When this occurs, it is mentioned in the commentary.

2 Source Organism

For listing organisms their systematic name is preferred. If these are not mentioned in the literature, the names from the respective literature are used. For example if an enzyme from yeast is described without being specified further, yeast will be the entry. This field defines the code numbers for the organisms in which the enzyme with the respective EC number is found. These code numbers (form <_>) are displayed together with each entry in all fields of BRENDA where organism-specific information is given.

3 Reaction and Specificity

Catalyzed reaction

The reaction as defined by the IUBMB. The commentary gives information on the mechanism, the stereochemistry, or on thermodynamic data of the reaction.

Reaction type

According to the enzyme class a type can be attributed. These can be oxidation, reduction, elimination, addition, or a name (e.g. Knorr reaction)

Natural substrates and products

These are substrates and products which are metabolized in vivo. A natural substrate is only given if it is mentioned in the literature. The commentary gives information on the pathways for which this enzyme is important. If the enzyme is induced by a specific compound or growth conditions, this will be included in the commentary. In *Additional information* you will find comments on the metabolic role, sometimes only assumptions can be found in the references or the natural substrates are unknown.
In the listings, each natural substrate (indicated by a bold **S**) is followed by its respective product (indicated by a bold **P**). Products are given with organisms and references included only if the respective authors were able to demonstrate the formation of the specific product. If only the disappearance of the substrate was observed, the product is included without organisms of references. In cases with unclear product formation only a ? as a dummy is given.

Substrates and products

All natural or synthetic substrates are listed (not in stoichiometric quantities). The commentary gives information on the reversibility of the reaction,

on isomers accepted as substrates and it compares the efficiency of substrates. If a specific substrate is accepted by only one of several isozymes, this will be stated here.
The field *Additional Information* summarizes compounds that are not accepted as substrates or general comments which are valid for all substrates. In the listings, each substrate (indicated by a bold **S**) is followed by its respective product (indicated by a bold **P**). Products are given with organisms and references included if the respective authors demonstrated the formation of the specific product. If only the disappearance of the substrate was observed, the product will be included without organisms or references. In cases with unclear product formation only a **?** as a dummy is given.

Inhibitors

Compounds found to be inhibitory are listed. The commentary may explain experimental conditions, the concentration yielding a specific degree of inhibition or the inhibition constant. If a substance is activating at a specific concentration but inhibiting at a higher or lower value, the commentary will explain this.

Cofactors, prosthetic groups

This field contains cofactors which participate in the reaction but are not bound to the enzyme, and prosthetic groups being tightly bound. The commentary explains the function or, if known, the stereochemistry, or whether the cofactor can be replaced by a similar compound with higher or lower efficiency.

Activating Compounds

This field lists compounds with a positive effect on the activity. The enzyme may be inactive in the absence of certain compounds or may require activating molecules like sulfhydryl compounds, chelating agents, or lipids. If a substance is activating at a specific concentration but inhibiting at a higher or lower value, the commentary will explain this.

Metals, ions

This field lists all metals or ions that have activating effects. The commentary explains the role each of the cited metal has, being either bound e.g. as Fe-S centers or being required in solution. If an ion plays a dual role, activating at a certain concentration but inhibiting at a higher or lower concentration, this will be given in the commentary.

Turnover number (min^{-1})

The k_{cat} is given in the unit min^{-1}. The commentary lists the names of the substrates, sometimes with information on the reaction conditions or the type of reaction if the enzyme is capable of catalyzing different reactions with a single substrate. For cases where it is impossible to give the turnover number in the defined unit (e.g., substrates without a defined molecular weight, or an undefined amount of protein) this is summarized in *Additional Information*.

Specific activity (U/mg)

The unit is micromol/minute/milligram of protein. The commentary may contain information on specific assay conditions or if another than the natural substrate was used in the assay. Entries in *Additional Information* are included if the units of the activity are missing in the literature or are not calculable to the obligatory unit. Information on literature with a detailed description of the assay method may also be found.

K_m-Value (mM)

The unit is mM. Each value is connected to a substrate name. The commentary gives, if available, information on specific reaction condition, isozymes or presence of activators. The references for values which cannot be expressed in mM (e.g. for macromolecular, not precisely defined substrates) are given in *Additional Information.* In this field we also cite literature with detailed kinetic analyses.

K_i-Value (mM)

The unit of the inhibition constant is mM. Each value is connected to an inhibitor name. The commentary gives, if available, the type of inhibition (e.g. competitive, non-competitive) and the reaction conditions (pH-value and the temperature). Values which cannot be expressed in the requested unit and references for detailed inhibition studies are summerized under *Additional information.*

pH-Optimum

The value is given to one decimal place. The commentary may contain information on specific assay conditions, such as temperature, presence of activators or if this optimum is valid for only one of several isozymes. If the enzyme has a second optimum, this will be mentioned here.

pH-Range

Mostly given as a range e.g. 4.0–7.0 with an added commentary explaining the activity in this range. Sometimes, not a range but a single value indicating the upper or lower limit of enzyme activity is given. In this case, the commentary is obligatory.

Temperature optimum (°C)

Sometimes, if no temperature optimum is found in the literature, the temperature of the assay is given instead. This is always mentioned in the commentary.

Temperature range (°C)

This is the range over which the enzyme is active. The commentary may give the percentage of activity at the outer limits. Also commentaries on specific assay conditions, additives etc.

4 Enzyme Structure

Molecular weight

This field gives the molecular weight of the holoenzyme. For monomeric enzymes it is identical to the value given for subunits. As the accuracy depends on the method of determination this is given in the commentary if provided in the literature. Some enzymes are only active as multienzyme complexes for which the names and/or EC numbers of all participating enzymes are given in the commentary.

Subunits

The tertiary structure of the active species is described. The enzyme can be active as a monomer a dimer, trimer and so on. The stoichiometry of subunit composition is given. Some enzymes can be active in more than one state of complexation with differing effectivities. The analytical method is included.

Posttranslational modifications

The main entries in this field may be proteolytic modification, or side-chain modification, or no modification. The commentary will give details of the modifications e.g.:

- proteolytic modification <1> (<1>, propeptide Name) [1];
- side-chain modification <2> (<2>, N-glycosylated, 12% mannose) [2];
- no modification [3]

5 Isolation / Preparation / Mutation / Application

Source / tissue

For multicellular organisms, the tissue used for isolation of the enzyme or the tissue in which the enzyme is present is given. Cell-lines may also be a source of enzymes.

Localization

The subcellular localization is described. Typical entries are: cytoplasm, nucleus, extracellular, membrane.

Purification

The field consists of an organism and a reference. Only references with a detailed description of the purification procedure are cited.

Renaturation

Commentary on denaturant or renaturation procedure.

Crystallization

The literature is cited which describes the procedure of crystallization, or the X-ray structure.

Cloning

Lists of organisms and references, sometimes a commentary about expression or gene structure.

Engineering

The properties of modified proteins are described.

Application

Actual or possible applications in the fields of pharmacology, medicine, synthesis, analysis, agriculture, nutrition are described.

6 Stability

pH-Stability

This field can either give a range in which the enzyme is stable or a single value. In the latter case the commentary is obligatory and explains the conditions and stability at this value.

Temperature stability

This field can either give a range in which the enzyme is stable or a single value. In the latter case the commentary is obligatory and explains the conditions and stability at this value.

Oxidation stability

Stability in the presence of oxidizing agents, e.g. O_2, H_2O_2, especially important for enzymes which are only active under anaerobic conditions.

Organic solvent stability

The stability in the presence of organic solvents is described.

General stability information

This field summarizes general information on stability, e.g., increased stability of immobilized enzymes, stabilization by SH-reagents, detergents, glycerol or albumins etc.

Storage stability

Storage conditions and reported stability or loss of activity during storage.

References

Authors, Title, Journal, Volume, Pages, Year.

(+)-*trans*-Carveol dehydrogenase 1.1.1.275

1 Nomenclature

EC number
1.1.1.275

Systematic name
(+)-trans-carveol:NAD^+ oxidoreductase

Recommended name
(+)-trans-carveol dehydrogenase

Synonyms
carveol dehydrogenase

2 Source Organism

<1> *Rhodococcus opacus* (no sequence specified) [2]
<2> *Carum carvi* (no sequence specified) [1]

3 Reaction and Specificity

Catalyzed reaction
(+)-trans-carveol + NAD^+ = (+)-(S)-carvone + NADH + H^+

Reaction type
oxidation
redox reaction
reduction

Natural substrates and products
S (+)-trans-carveol + NAD^+ <2> (<2> enzyme catalyses the final step of (+)-carvone biosynthesis, which is developmentally regulated, key enzyme of the pathway is limonene-6-hydroxylase [1]) (Reversibility: ?) [1]
P (+)-carvone + NADH <2> [1]

Substrates and products
S (+)-trans-carveol + NAD^+ <1, 2> (<2> best substrate [1]; <1> slow reaction [2]; <2> stereospecificity [1]; <2> enzyme catalyses the final step of (+)-carvone biosynthesis, which is developmentally regulated, key enzyme of the pathway is limonene-6-hydroxylase [1]) (Reversibility: ?) [1, 2]

P (+)-carvone + NADH <1, 2> [1, 2]
S (+)/(-)-trans-carveol + NAD^+ <2> (Reversibility: ?) [1]
P (+)-carvone + NADH <2> [1]
S (-)-cis-carveol + NAD^+ <2> (<2> 65% activity compared to (+)/(-)-trans-carveol [1]) (Reversibility: ?) [1]
P (-)-carvone + NADH <2> [1]
S (-)-trans-carveol + NAD^+ <2> (<2> 5% activity compared to (+)/(-)-trans-carveol [1]; <2> product identification not possible, below detection limit of GC-MS [1]) (Reversibility: ?) [1]
P ?
S (-)/(+)-cis-carveol + NAD^+ <2> (Reversibility: ?) [1]
P (-)-carvone + NADH <2> [1]

Cofactors/prosthetic groups

NAD^+ <1, 2> (<1> dependent on [2]; <2> absolutely dependent on, $NADP^+$ cannot substitute for NAD^+ and cannot act synergistically [1]) [1, 2]

pH-Optimum

10 <2> (<2> about [1]) [1]

Temperature optimum (°C)

30 <2> (<2> assay at [1]) [1]

5 Isolation/Preparation/Mutation/Application

Source/tissue

fruit <2> [1]
Additional information <2> (<2> developmental regulation, overview [1]) [1]

Localization

cytosol <2> [1]

Application

biotechnology <1> (<1> applicability of strains with high enzyme content or recombinant overproducing strains for production of (+)-carvone, which is used as a flavor compound [2]) [2]

References

[1] Bouwmeester, H.J.; Gershenzon, J.; Konings, M.C.J.M.; Croteau, R.: Biosynthesis of the monoterpenes limonene and carvone in the fruit of caraway. I. Demonstration of enzyme activities and their changes with development. Plant Physiol., **117**, 901-912 (1998)

[2] Duetz, W.A.; Fjallman, A.H.M.; Ren, S.; Jourdat, C.; Witholt, B.: Biotransformation of D-limonene to (+) trans-carveol by toluene-grown Rhodococcus opacus PWD4 cells. Appl. Environ. Microbiol., **67**, 2829-2832 (2001)

Serine 3-dehydrogenase 1.1.1.276

1 Nomenclature

EC number
1.1.1.276

Systematic name
L-serine:$NADP^+$ 3-oxidoreductase

Recommended name
serine 3-dehydrogenase

CAS registry number
9038-55-5

2 Source Organism

<1> *Escherichia coli* (no sequence specified) [2, 3]
<2> *Saccharomyces cerevisiae* (no sequence specified) [3]
<3> *Agrobacterium tumefaciens* (no sequence specified) [1, 2]

3 Reaction and Specificity

Catalyzed reaction
L-serine + $NADP^+$ = 2-aminomalonate semialdehyde + NADPH + H^+ (<3> sequential mechanism [1])

Reaction type
oxidation
redox reaction
reduction

Natural substrates and products
S L-serine + $NADP^+$ <1, 2, 3> (Reversibility: ?) [1, 2, 3]
P 2-aminomalonate semialdehyde + NADPH <3> [2]

Substrates and products
S 2-methyl-DL-serine + $NADP^+$ <1, 2, 3> (<1> 59% activity compared to L-serine [3]; <3> 8.7% activity compared to L-serine [1]; <2> 21% activity compared to L-serine [3]) (Reversibility: ?) [1, 3]
P 2-amino-2-methyl-3-oxopropanoic acid + NADPH

S 3-hydroxypropionate + $NADP^+$ <1, 2, 3> (<3> 1.6% activity compared to L-serine [1]; <1> 23% activity compared to L-serine [3]; <2> 13% activity compared to L-serine [3]) (Reversibility: ?) [1, 3]
P 3-oxopropanoic acid + NADPH
S D-3-hydroxybutyrate + $NADP^+$ <1, 2> (<1> 1% activity compared to L-serine [3]; <2> 12% activity compared to L-serine [3]) (Reversibility: ?) [3]
P 3-oxobutyrate + NADPH
S D-3-hydroxyisobutyrate + $NADP^+$ <1, 2, 3> (<1> 90% activity compared to L-serine [3]; <2> 78% activity compared to L-serine [3]; <3> 3.5% activity compared to L-serine [1]) (Reversibility: ?) [1, 3]
P 2-methyl-3-oxopropanoic acid + NADPH
S D-glycerate + $NADP^+$ <1, 2, 3> (<1> 48% activity compared to L-serine [3]; <2> 6% activity compared to L-serine [3]; <3> 12.9% activity compared to L-serine [1]) (Reversibility: ?) [1, 3]
P ?
S D-serine + $NADP^+$ <1, 2, 3> (<1> 83% activity compared to L-serine [3]; <3> 55.2% activity compared to L-serine [1]; <2> 71% activity compared to L-serine [3]) (Reversibility: ?) [1, 3]
P 2-aminomalonate semialdehyde + NADPH
S D-threonine + $NADP^+$ <1, 2> (<1> 156% activity compared to L-serine [3]; <2> 478% activity compared to L-serine [3]) (Reversibility: ?) [3]
P D-2-amino-3-oxobutanoic acid + NADPH
S L-3-hydroxybutyrate + $NADP^+$ <1, 2> (<2> 85% activity compared to L-serine [3]; <1> 9% activity compared to L-serine [3]) (Reversibility: ?) [3]
P 3-oxobutyrate + NADPH
S L-3-hydroxyisobutyrate + $NADP^+$ <1, 2, 3> (<1> 96% activity compared to L-serine [3]; <3> 8.5% activity compared to L-serine [1]; <2> 182% activity compared to L-serine [3]) (Reversibility: ?) [1, 3]
P 2-methyl-3-oxopropanoic acid + NADPH
S L-allo-threonine + $NADP^+$ <1, 2> (<1,2> best substrate [3]; <1> 137% activity compared to L-serine [3]; <2> 397% activity compared to L-serine [3]) (Reversibility: ?) [3]
P L-2-amino-3-ketobutyrate + NADPH <1, 2> (<1,2> L-2-amino-3-ketobutyrate spontaneously decarboxylates into aminoacetone [3]) [3]
S L-glycerate + $NADP^+$ <1, 2, 3> (<1> 28% activity compared to L-serine [3]; <3> 11.3% activity compared to L-serine [1]; <2> 5% activity compared to L-serine [3]) (Reversibility: ?) [1, 3]
P ?
S L-serine + $NADP^+$ <1, 2, 3> (<3> best substrate [1]; <3> enzyme belongs to the short-chain alcohol dehydrogenase family, similar reaction mechanism [2]) (Reversibility: ?) [1, 2, 3]
P 2-aminomalonate semialdehyde + NADPH <3> (<3> 2-aminomalonate semialdehyde spontaneously decarboxylates into 2-aminoacetaldehyde [2]) [2]
S Additional information <1, 2, 3> (<1,2> no activity with DL-homoserine, O-methyl-L-serine, N-acetyl-DL-serine, DL-isoserine, DL-threo-3-phenyl-

serine, malonate, DL-malate, DL-lactate, DL-tartrate, D-gluconate, citrate, isocitrate, glycolate, glycerol [3]; <3> no activity with 4-hydroxybutyrate, DL-2-hydroxybutyrate, DL-2-hydroxyisobutyrate, DL-2-hydroxycaproate, DL-2-hydroxyisocaproate, 3-phosphoglycerate, DL-lactate, L-malate, glycollate, glycerol, D-threonine, DL-threo-phenylserine, DL-homoserine, DL-isoserine, L-alanine, *β*-alanine, L-aspartate, L-glutamate, L-valine, L-leucine, L-methionine, L-lysine, L-phenylalanine, and glycine [1]; <1,2> L-threonine, D-allo-threonine are poor substrates [3]) (Reversibility: ?) [1, 3]

P ?

Inhibitors

2-methylmalonate <3> (<3> competitive against L-serine, noncompetitive against $NADP^+$ [1]) [1]
Co^{2+} <3> (<3> weak [1]) [1]
D-cysteine <3> (<3> competitive against L-serine, noncompetitive against $NADP^+$ [1]) [1]
Hg^{2+} <3> (<3> strong [1]) [1]
L-cysteine <3> (<3> competitive against L-serine, noncompetitive against $NADP^+$ [1]) [1]
malonate <3> (<3> competitive against L-serine, noncompetitive against $NADP^+$ [1]) [1]
N-ethylmaleimide <3> [1]
tartronate <3> (<3> competitive against L-serine, noncompetitive against $NADP^+$ [1]) [1]
p-chloromercuribenzoate <3> (<3> strong [1]) [1]
Additional information <3> (<3> no inhibition by EDTA, sodium arsenate, *o*-phenethroline, *α*,*α*-dipyridyl and very slight inhibition by sodium azide [1]) [1]

Cofactors/prosthetic groups

$NADP^+$ <1, 2, 3> (<1,2,3> dependent on [1,2,3]; <1,2,3> NAD^+ cannot replace $NADP^+$ [1,3]) [1,2,3]

Metals, ions

Additional information <3> (<3> no effect with Cu^{2+}, Mg^{2+}, Mn^{2+}, Zn^{2+}, Fe^{2+} [1]) [1]

Turnover number (min^{-1})

4.2 <1> (L-serine, <1> pH 9.0, 30°C [3]) [3]
5.76 <1> (L-allo-threonine, <1> pH 9.0, 30°C [3]) [3]
11 <2> (L-serine, <2> pH 9.0, 30°C [3]) [3]
43.8 <2> (L-allo-threonine, <2> pH 9.0, 30°C [3]) [3]
Additional information <1, 2> (<1,2> substrate specificity [3]) [3]

Specific activity (U/mg)

0.0013 <1> [2]
0.043 <3> (<3> recombinant enzyme in Escherichia coli [2]) [2]
2.4 <1> (<1> purified recombinant enzyme [3]) [3]

6 <2> (<2> purified recombinant enzyme [3]) [3]
45.3 <3> (<3> purified enzyme [1]) [1]

K_m-Value (mM)

0.029 <3> ($NADP^+$, <3> pH 9.0, 30°C [1]) [1]
0.5 <2> ($NADP^+$, <2> pH 9.0, 30°C [3]) [3]
0.54 <1> ($NADP^+$, <1> pH 9.0, 30°C [3]) [3]
3 <2> (L-allo-threonine, <2> pH 9.0, 30°C [3]) [3]
29 <1> (L-allo-threonine, <1> pH 9.0, 30°C [3]) [3]
40 <1> (L-serine, <1> pH 9.0, 30°C [3]) [3]
42 <3> (L-serine, <3> pH 9.0, 30°C [1]) [1]
44 <3> (D-serine, <3> pH 9.0, 30°C [1]) [1]
54 <3> (L-glycerate, <3> pH 9.0, 30°C [1]) [1]
56 <3> (D-glycerate, <3> pH 9.0, 30°C [1]) [1]
95 <2> (L-serine, <2> pH 9.0, 30°C [3]) [3]
Additional information <1, 2> (<1,2> substrate specificity [3]) [3]

K_i-Value (mM)

0.067 <3> (L-cysteine, <3> versus L-serine, pH 9.0, 30°C [1]) [1]
0.21 <3> (L-cysteine, <3> versus $NADP^+$, pH 9.0, 30°C [1]) [1]
0.59 <3> (D-cysteine, <3> versus L-serine, pH 9.0, 30°C [1]) [1]
9.9 <3> (D-cysteine, <3> versus $NADP^+$, pH 9.0, 30°C [1]) [1]
10 <3> (2-methylmalonate, <3> versus $NADP^+$, pH 9.0, 30°C [1]) [1]
11 <3> (malonate, <3> versus L-serine, pH 9.0, 30°C [1]) [1]
12 <3> (2-methylmalonate, <3> versus L-serine, pH 9.0, 30°C [1]) [1]
13 <3> (tartronate, <3> versus L-serine, pH 9.0, 30°C [1]) [1]
17 <3> (malonate, <3> versus $NADP^+$, pH 9.0, 30°C [1]) [1]
18 <3> (tartronate, <3> versus $NADP^+$, pH 9.0, 30°C [1]) [1]

pH-Optimum

8.5 <1, 2> (<1,2> oxidation of L-serine [3]) [3]
9.1 <3> [1]

Temperature optimum (°C)

30 <1, 3> (<1,3> assay at [1,2]) [1, 2]

4 Enzyme Structure

Molecular weight

100000 <3> (<3> gel filtration [1]) [1]
105000 <1> (<1> recombinant enzyme, gel filtration [3]) [3]
110000 <2> (<2> recombinant enzyme, gel filtration [3]) [3]

Subunits

tetramer <1, 2, 3> (<1> 4 * 27000, recombinant enzyme, SDS-PAGE [3]; <2> 4 * 29000, recombinant enzyme, SDS-PAGE [3]; <3> x * 25000-26600, recombinant enzyme, SDS-PAGE and amino acid sequence determination [2]) [2, 3]

5 Isolation/Preparation/Mutation/Application

Purification

<1> (recombinant from overproducing Escherichia coli, to homogeneity) [3]
<2> (recombinant from overproducing Escherichia coli, to homogeneity) [3]
<3> [1]
<3> (recombinant from Escherichia coli, to homogeneity) [2]

Cloning

<1> (overexpression in Escherichia coli JM109) [3]
<2> (overexpression in Escherichia coli JM109) [3]
<3> (DNA sequence determination and analysis, expression in Escherichia coli JM109) [2]

6 Stability

pH-Stability

6.5-10 <1> (<1> 10 min, 40°C, stable [3]) [3]
7-8 <3> (<3> stable at [1]) [1]
7.5-10.5 <2> (<2> 10 min, 40°C, stable [3]) [3]

Temperature stability

35 <3> (<3> stable at [1]) [1]
40 <2> (<2> 10 min, 0.1 M potassium phosphate, pH 7.4, 0.01% 2-mercaptoethanol, 10% glycerol, stable [3]) [3]
55 <1> (<1> 10 min, 0.1 M potassium phosphate, pH 7.4, 0.01% 2-mercaptoethanol, 10% glycerol, stable [3]) [3]

References

[1] Chowdhury, E.K.; Higuchi, K.; Nagata, S.; Misono, H.: A novel $NADP^+$ dependent serine dehydrogenase from Agrobacterium tumefaciens. Biosci. Biotechnol. Biochem., **61**, 152-157 (1997)
[2] Fujisawa, H.; Nagata, S.; Chowdhury, E.K.; Matsumoto, M.; Misono, H.: Cloning and sequencing of the serine dehydrogenase gene from Agrobacterium tumefaciens. Biosci. Biotechnol. Biochem., **66**, 1137-1139 (2002)
[3] Fujisawa, H.; Nagata, S.; Misono, H.: Characterization of short-chain dehydrogenase/reductase homologues of Escherichia coli (YdfG) and Saccharomyces cerevisiae (YMR226C). Biochim. Biophys. Acta, **1645**, 89-94 (2003)

3β-Hydroxy-5β-steroid dehydrogenase 1.1.1.277

1 Nomenclature

EC number
1.1.1.277

Systematic name
3β-hydroxy-5β-steroid:NADP$^+$ 3-oxidoreductase

Recommended name
3β-hydroxy-5β-steroid dehydrogenase

Synonyms
3β-hydroxysteroid 5β-dehydrogenase
3β-hydroxysteroid 5β-oxidoreductase
3β-hydroxysteroid 5β-progesterone oxidoreductase
dehydrogenase, 3β-hydroxy steroid 5β

CAS registry number
162731-81-9

2 Source Organism

<1> *Digitalis lanata* (no sequence specified) [2, 3]
<2> *Digitalis purpurea* (no sequence specified) [1]

3 Reaction and Specificity

Catalyzed reaction
3β-hydroxy-5β-pregnane-20-one + NADP$^+$ = 5β-pregnan-3,20-dione + NADPH + H$^+$

Reaction type
oxidation
redox reaction
reduction

Natural substrates and products
S 3β-hydroxy-5β-pregnane-20-one + NADP$^+$ <1, 2> (<1,2> biosynthesis of cardenolides [1,2]; <1> biosynthesis and transformation of pregnane derivates [3]) (Reversibility: r) [1, 2, 3]
P 5β-pregnane-3,20-dione + NADPH + H$^+$

Substrates and products

S 3β-hydroxy-5β-pregnane-20-one + $NADP^+$ <1, 2> (<1,2> stereospecific [1,2]; <1,2> biosynthesis of cardenolides [1,2]; <1> biosynthesis and transformation of pregnane derivates [3]) (Reversibility: r) [1, 2, 3]

P 5β-pregnane-3,20-dione + NADPH + H^+<1, 2> [1, 2, 3]

Inhibitors

Mg^{2+} <1> [2]

Cofactors/prosthetic groups

NADH <2> (<2> affinity to NADH is lower than to NADPH [1]) [1]
NADPH <1, 2> [1,2,3]

Specific activity (U/mg)

0.0003 <1> (<1> cell suspension culture [2]) [2]
0.00084 <1> (<1> leaf [2]) [2]
Additional information <1> [3]

K_m-Value (mM)

0.00139 <2> (NADH) [1]
0.0171 <2> (5β-Pregnane-3,20-dione) [1]
0.3 <2> (NADPH) [1]

5 Isolation/Preparation/Mutation/Application

Source/tissue

cell suspension culture <1> [2]
embryo <1> (<1> somatic embryo and proembryogenic mass [3]) [3]
leaf <1> [2, 3]
shoot <1, 2> (<1> light- and dark-grown [2]) [1, 2]
Additional information <2> (<2> no activity in undifferentiated cell cultures [1]) [1]

Localization

mitochondrion <1> [3]
soluble <1, 2> [1, 2]

References

[1] Seitz, H.U.; Gaertner, D.E.: Enzymes in cardenolide-accumulating shoot cultures of Digitalis purpurea. Plant Cell Tissue Organ Cult., **38**, 337-344 (1994)

[2] Stuhlemmer, U.; Kreis, W.: Cardenolide formation and activity of pregnane-modifying enzymes in cell suspension cultures, shoot cultures and leaves of Digitalis lanata. Plant Physiol. Biochem., **34**, 85-91 (1996)

[3] Lindemann, P.; Luckner, M.: Biosynthesis of pregnane derivatives in somatic embryos of Digitalis lanata. Phytochemistry, **46**, 507-513 (1997)

3β-Hydroxy-5α-steroid dehydrogenase 1.1.1.278

1 Nomenclature

EC number
1.1.1.278

Systematic name
3β-hydroxy-5α-steroid:NADP$^+$ 3-oxidoreductase

Recommended name
3β-hydroxy-5α-steroid dehydrogenase

Synonyms
3β-hydroxy steroid dehydrogenase <1> [3]
3β-hydroxysteroid oxidoreductase <5> [4]
3β-hydroxysteroid-5α-oxidoreductase <5> [1]
dehydrogenase, 3(17)β-hydroxy steroid
Additional information <5> (<5> 3β-hydroxysteroid oxidoreductase is exclusively associated with the 5α-progesterone reductase in microsomes [4]) [4]

CAS registry number
9015-81-0

2 Source Organism

<1> *Cavia porcellus* (no sequence specified) [3]
<2> *Homo sapiens* (no sequence specified) [2]
<3> *Rattus norvegicus* (no sequence specified) [3, 5, 6, 7, 8]
<4> *Oryctolagus cuniculus* (no sequence specified) [3]
<5> *Digitalis lanata* (no sequence specified) [1,4]

3 Reaction and Specificity

Catalyzed reaction
3β-hydroxy-5α-pregnane-20-one + NADP$^+$ = 5α-pregnan-3,20-dione + NADPH + H$^+$

Reaction type
oxidation
redox reaction
reduction

Natural substrates and products

S 5α-pregnan-3,20-dione + NADPH <1, 2, 3, 4, 5> (<3> low activity [7]; <3> biosynthetic pathway, overview [5,7]) (Reversibility: r) [1, 2, 3, 4, 5, 7]

P 3β-hydroxy-5α-pregnane-20-one + $NADP^+$ <1, 2, 3, 4, 5> [1, 2, 3, 4, 5, 7]

Substrates and products

S 5α-pregnan-3,20-dione + NAD(P)H <1, 2, 3, 4, 5> (<3> low activity [7]; <3> activity below detection limit of 0.1 fmol/mg protein/min [6]) (Reversibility: r) [1, 2, 3, 4, 5, 6, 7]

P 3β-hydroxy-5α-pregnan-20-one + $NAD(P)^+$ <1, 2, 3, 4, 5> (<3,5> i.e. 5α-pregnan-3β-ol-20-one [1,4,5,6,7]) [1, 2, 3, 4, 5, 6, 7]

S 5α-pregnan-3,20-dione + NADPH <1, 2, 3, 4, 5> (<3> low activity [7]; <3> biosynthetic pathway, overview [5,7]) (Reversibility: r) [1, 2, 3, 4, 5, 7, 8]

P 3β-hydroxy-5α-pregnane-20-one + $NADP^+$ <1, 2, 3, 4, 5> [1, 2, 3, 4, 5, 7]

S Additional information <2> (<2> progesterone is no substrate [2]) (Reversibility: ?) [2]

P ?

Inhibitors

2-mercaptoethanol <5> (<5> activates at 1 mM, inhibitory at 10 mM [4]) [4]
2-mercaptoethanol/cysteine <5> [4]
5α-pregnan-3,20-dione <2> [2]
ascorbate <5> [4]
Ca^{2+} <5> [4]
Co^{2+} <5> [4]
Mg^{2+} <5> [4]
Mn^{2+} <5> [4]
NADPH <5> (<5> inhibitory at concentrations above 1 mM [4]) [4]
progesterone <2> [2]
Zn^{2+} <5> [4]

Cofactors/prosthetic groups

NAD^+ <2, 5> (<2> less effective than $NADP^+$ [2]) [1,2,4]
NADH <2, 3, 5> (<2> less effective than NADPH [2]; <3> cannot substitute for NADPH [5]) [1,2,4,5]
$NADP^+$ <1, 2, 3, 5> [2, 3, 4, 5]
NADPH <1, 2, 3, 5> (<5> required [4]) [2, 3, 4, 5, 8]

Activating compounds

DTT <5> (<5> 1 mM, 1.75fold increase in activity [4]) [4]

Specific activity (U/mg)

Additional information <1, 3> (<3> activity below detection limit of 0.1 fmol/mg protein/min [6]; <1> activity is determined as progesterone-metabolizing activity including steroid 5α-reductase and 3β-hydroxysteroid dehydrogenase activity [3]) [3, 6]

K_m-Value (mM)

0.0185-0.02 <5> (5α-pregnan-3,20-dione, <5> pH 8.0, 25°C [4]) [4]
0.05-0.12 <5> (NADPH, <5> pH 8.0, 25°C [4]) [4]

K_i-Value (mM)

0.0095 <2> (5α-pregnan-3,20-dione, <2> in presence of NADH [2]) [2]
0.0125 <2> (5α-pregnan-3,20-dione, <2> in presence of NADPH [2]) [2]
0.035 <2> (progesterone, <2> in presence of NADH [2]) [2]
0.066 <2> (progesterone, <2> in presence of NADPH [2]) [2]

pH-Optimum

7 <1> (<1> assay at [3]) [3]
7.5 <5> (<5> assay at [1]) [1]
8 <5> [4]

pH-Range

6-9 <5> [4]

Temperature optimum (°C)

25 <5> [4]
37 <1, 3, 5> (<1,3,5> assay at [1,3,5]) [1, 3, 5]

Temperature range (°C)

5-35 <5> (<5> half-maximal activity at 5°C and 35°C [4]) [4]

5 Isolation/Preparation/Mutation/Application

Source/tissue

cell suspension culture <5> [4]
cerebral cortex <3> (<3> activity below detection limit of 0.1 fmol/mg protein/min [6]) [6]
colon <1> [3]
gastric mucosa <1, 3, 4> [3]
parotid gland <3> (<3> only male rats, no activity in tissues of pregnant and non-pregnant female rats [5]) [5]
pituitary gland <3> (<3> anterior, low activity [7]) [7]
salivary gland <3> (<3> only male rats, no activity in tissues of pregnant and non-pregnant female rats [5]) [5]
skin <2> (<2> pubic [2]) [2]
small intestine <1> [3]
somatic embryo <5> (<5> light and dark-grown [1]) [1]
stomach <1> [3]
Additional information <5> (<5> proembryonic mass [1]) [1]

Localization

cytosol <2, 3, 5> (<3> no activity [5]; <5> major part [4]) [2, 4, 5]
endoplasmic reticulum <5> (<5> minor part [4]) [4]
microsome <1, 2, 3, 4, 5> [2, 3, 4, 5]

References

[1] Lindemann, P.; Luckner, M.: Biosynthesis of pregnane derivatives in somatic embryos of Digitalis lanata. Phytochemistry, **46**, 507-513 (1997)

[2] Giacomini, M.; Wright, F.: The effects of progesterone and pregnanedione on the reductive metabolism of dihydrotestosterone in human skin. J. Steroid Biochem., **13**, 645-651 (1980)

[3] Ichihara, K.; Tanaka, C.: Progesterone metabolism in the gastric mucosa microsomes of guinea pig. J. Steroid Biochem., **32**, 835-840 (1989)

[4] Warneck, H.M.; Seitz, H.U.: 3β-Hydroxysteroid oxidoreductase in suspension cultures of Digitalis lanata EHRH. Z. Naturforsch. C, **45**, 963-972 (1990)

[5] Laine, M.; Ojanotko-Harri, A.: Progesterone metabolism by major salivary glands of rat. II. Parotid gland. J. Steroid Biochem. Mol. Biol., **37**, 605-611 (1990)

[6] Stuerenburg, H.J.; Fries, U.; Iglauer, F.; Kunze, K.: Effect of age on synthesis of the GABAergic steroids 5-α-pregnane-3,20-dione and 5-α-pregnane-3-α-ol-20-one in rat cortex in vitro. J. Neural Transm., **104**, 249-257 (1997)

[7] Wiebe, J.P.; Boushy, D.; Wolfe, M.: Synthesis, metabolism and levels of the neuroactive steroid, 3α-hydroxy-4-pregnen-20-one (3αHP), in rat pituitaries. Brain Res., **764**, 158-166 (1997)

[8] Petralia, S.M.; Jahagirdar, V.; Frye, C.A.: Inhibiting biosynthesis and/or metabolism of progestins in the ventral tegmental area attenuates lordosis of rats in behavioural oestrus. J. Neuroendocrinol., **17**, 545-552 (2005)

(R)-3-Hydroxyacid-ester dehydrogenase 1.1.1.279

1 Nomenclature

EC number
1.1.1.279

Systematic name
ethyl-(R)-3-hydroxyhexanoate:NADP$^+$ 3-oxidoreductase

Recommended name
(R)-3-hydroxyacid-ester dehydrogenase

Synonyms
EC 1.2.1.55
reductase, 3-oxo ester (R)-

CAS registry number
114705-02-1

2 Source Organism

<1> *Saccharomyces cerevisiae* (no sequence specified) [1]

3 Reaction and Specificity

Catalyzed reaction
ethyl (R)-3-hydroxyhexanoate + NADP$^+$ = ethyl 3-oxohexanoate + NADPH + H$^+$

Reaction type
oxidation
redox reaction
reduction

Substrates and products
S ethyl 3-oxobutanoate + NADPH <1> (Reversibility: ?) [1]
P ethyl (R)-3-hydroxybutanoate + NADP$^+$ <1> [1]
S ethyl 3-oxohexanoate + NADPH <1> (Reversibility: ?) [1]
P ethyl (R)-3-hydroxyhexanoate + NADP$^+$ <1> [1]
S Additional information <1> (<1> specific for 3-oxo-esters, no reaction with 3-oxo-acids [1]) (Reversibility: ?) [1]
P ?

Inhibitors
8-hydroxyquinoline 5-sulfonic acid <1> [1]
$HgCl_2$ <1> [1]
o-phenanthroline <1> [1]

Cofactors/prosthetic groups
NADPH <1> (<1> specific for [1]) [1]

Specific activity (U/mg)
1.958 <1> [1]

K_m-Value (mM)
2 <1> (ethyl 3-oxohexanoate) [1]
17 <1> (ethyl 3-oxobutanoate) [1]

pH-Optimum
6.1 <1> [1]

pH-Range
7.5 <1> (<1> rapid decrease of activity above [1]) [1]

4 Enzyme Structure

Molecular weight
780000-800000 <1> (<1> gel filtration [1]) [1]

Subunits
oligomer <1> (<1> x * 210000 + x * 200000, SDS-PAGE [1]) [1]

5 Isolation/Preparation/Mutation/Application

Localization
cytosol <1> [1]

Purification
<1> (using streptomycin sulfate treatment, Sephadex G-25 filtration, DEAE-Sepharose CL-6B chromatography, Sephadex G-150 filtration, Sepharose 6B filtration and hydroxyapatite chromatography) [1]

References

[1] Heidlas, J.; Engel, K.H.; Tressl, R.: Purification and characterization of two oxidoreductases involved in the enantioselective reduction of 3-oxo, 4-oxo and 5-oxo esters in bakers yeast. Eur. J. Biochem., **172**, 633-639 (1988)

(S)-3-Hydroxyacid-ester dehydrogenase 1.1.1.280

1 Nomenclature

EC number
1.1.1.280

Systematic name
ethyl-(S)-3-hydroxyhexanoate:NADP$^+$ 3-oxidoreductase

Recommended name
(S)-3-hydroxyacid-ester dehydrogenase

Synonyms
2-methyl-3-oxobutanoate reductase
3-oxo ester (S)-reductase
EC 1.2.1.56
reductase, 3-oxo ester (S)-

CAS registry number
114705-03-2

2 Source Organism

<1> *Saccharomyces cerevisiae* (no sequence specified) [1]
<2> *Klebsiella pneumoniae* (no sequence specified) [2]

3 Reaction and Specificity

Catalyzed reaction
ethyl (S)-3-hydroxyhexanoate + NADP$^+$ = ethyl 3-oxohexanoate + NADPH + H$^+$

Reaction type
oxidation
redox reaction
reduction

Substrates and products
S 4-oxopentanoate + NADPH <1> (Reversibility: ?) [1]
P (S)-4-hydroxypentanoate + NADP$^+$ <1> [1]
S 5-oxohexanoate + NADPH <1> (Reversibility: ?) [1]
P (S)-5-hydroxyhexanoate + NADP$^+$ <1> [1]

S benzyl 2-methyl-3-oxobutanoate + NADPH <2> (Reversibility: ?) [2]
P benzyl (2R,3S)-3-hydroxy-2-methylbutanoate + $NADP^+$ <2> [2]
S ethyl 2-methyl-3-oxobutanoate + NADPH <2> (Reversibility: ?) [2]
P ethyl (2R,3S)-3-hydroxy-2-methylbutanoate + $NADP^+$ <2> [2]
S ethyl 2-oxocyclohexanecarboxylate + NADPH <2> (Reversibility: ?) [2]
P ethyl (1R,2S)-2-hydroxycyclohexanecarboxylate + $NADP^+$ <2> [2]
S ethyl 2-oxocyclopentanecarboxylate + NADPH <2> (Reversibility: ?) [2]
P ethyl (1R,2S)-2-hydroxycyclopentanecarboxylate + $NADP^+$ <2> [2]
S ethyl 3-oxobutyrate + NADPH <1, 2> (Reversibility: ?) [1, 2]
P ethyl (S)-3-hydroxybutyrate + $NADP^+$ <1, 2> [1, 2]
S ethyl 3-oxohexanoate + NADPH <1> (Reversibility: ?) [1]
P ethyl (S)-3-hydroxyhexanoate + $NADP^+$ <1> [1]
S Additional information <2> (<2> ethyl (2R,3S)-3-hydroxy-2-methylbutanoate is a useful starting material for the synthesis of ferroelectric liquid crystal compounds and various other biologically active substances [2]) (Reversibility: ?) [2]
P ?

Inhibitors

8-hydroxyquinoline 5-sulfonic acid <1> [1]
Cu^{2+} <2> (<2> 1 mM completely inactivates the enzyme [2]) [2]
dipicolinate <1> [1]
Fe^{2+} <2> (<2> 1 mM completely inactivates the enzyme [2]) [2]
$HgCl_2$ <1> [1]
Mn^{2+} <2> (<2> 1 mM completely inactivates the enzyme [2]) [2]
Zn^{2+} <2> (<2> 1 mM completely inactivates the enzyme [2]) [2]
o-phenanthroline <1> [1]

Cofactors/prosthetic groups

NADPH <1, 2> (<1,2> specific for [1,2]) [1, 2]

Specific activity (U/mg)

0.411 <1> [1]

K_m-Value (mM)

0.9 <1> (ethyl 3-oxobutyrate) [1]
5.3 <1> (ethyl 3-oxohexanoate) [1]
5.6 <2> (2-methyl-3-oxobutanoate) [2]
12.5 <2> (benzyl 2-methyl-3-oxobutanoate) [2]
13.1 <1> (5-oxohexanoate) [1]
17.1 <1> (4-oxopentanoate) [1]

pH-Optimum

6.9 <1> [1]
7 <2> (<2> reduction reaction [2]) [2]
10 <2> (<2> oxidation reaction [2]) [2]

Temperature optimum (°C)

45 <2> [2]

4 Enzyme Structure

Molecular weight

30000 <2> (<2> gel filtration [2]) [2]
48000-50000 <1> (<1> gel filtration [1]) [1]

Subunits

dimer <1> (<1> 2 * 24000, SDS-PAGE [1]) [1]
monomer <2> (<2> 1 * 31000, SDS-PAGE [2]) [2]

5 Isolation/Preparation/Mutation/Application

Localization

cytosol <1> [1]

Purification

<1> (using streptomycin sulfate treatment, Sephadex G-25 filtration, DEAE-Sepharose CL-6B chromatography, Sephadex G-150 filtration, Sepharose 6B filtration and hydroxyapatite chromatography) [1]
<2> (using ammonium sulfate fractionation, ion exchange chromatography, affinity chromatography, and gel filtration) [2]

Application

synthesis <2> (<2> ethyl (2R,3S)-3-hydroxy-2-methylbutanoate is a useful starting material for the synthesis of ferroelectric liquid crystal compounds and various other biologically active substances [2]) [2]

6 Stability

pH-Stability

5-8 <2> [2]

Temperature stability

45 <2> (<2> stable up to 45°C, inactivated above 50°C, 10 min [2]) [2]

References

[1] Heidlas, J.; Engel, K.H.; Tressl, R.: Purification and characterization of two oxidoreductases involved in the enantioselective reduction of 3-oxo, 4-oxo and 5-oxo esters in bakers yeast. Eur. J. Biochem., **172**, 633-639 (1988)
[2] Miya, H.; Kawada, M.; Sugiyama, Y.: Purification and characterization of ethyl 2-methyl-3-oxobutanoate reductase from Klebsiella pneumoniae IFO3319. Biosci. Biotechnol. Biochem., **60**, 760-764 (1996)

GDP-4-dehydro-6-deoxy-D-mannose reductase 1.1.1.281

1 Nomenclature

EC number
1.1.1.281

Systematic name
GDP-6-deoxy-D-mannose:NAD(P)$^+$ 4-oxidoreductase (D-rhamnose-forming)

Recommended name
GDP-4-dehydro-6-deoxy-D-mannose reductase

Synonyms
GDP-4-keto-6-deoxy-D-mannose reductase (ambiguous) <3> [1]
GDP-6-deoxy-D-lyxo-4-hexulose reductase <3> [1]
RMD <3> [1]

CAS registry number
9075-56-3

2 Source Organism

<1> *Pseudomonas aeruginosa* (no sequence specified) [2]
<2> *Aneurinibacillus thermoaerophilus* (no sequence specified) [1]
<3> *Aneurinibacillus thermoaerophilus* (UNIPROT accession number: Q6T1X6) [1]

3 Reaction and Specificity

Catalyzed reaction
GDP-6-deoxy-D-mannose + NAD(P)$^+$ = GDP-4-dehydro-6-deoxy-D-mannose + NAD(P)H + H$^+$

Reaction type
oxidation
redox reaction
reduction

Natural substrates and products
S GDP-4-dehydro-6-deoxy-D-mannose + NADPH <1> (Reversibility: ?) [2]
P GDP-6-deoxy-D-mannose + NADP$^+$ <1> [2]

S GDP-6-deoxy-D-mannose + NAD(P)$^+$ <1, 2> (Reversibility: ?) [1, 2]
P GDP-4-dehydro-6-deoxy-D-mannose + NAD(P)H

Substrates and products

S GDP-4-dehydro-6-deoxy-D-mannose + NADPH <1> (Reversibility: ?) [2]
P GDP-6-deoxy-D-mannose + NADP$^+$ <1> [2]
S GDP-6-deoxy-D-lyxo-4-hexulose + NADH + H$^+$ <1, 3> (Reversibility: ir) [1, 2]
P GDP-α-D-rhamnose + NAD$^+$ <1, 3> [1, 2]
S GDP-6-deoxy-D-lyxo-4-hexulose + NADPH + H$^+$ <1, 3> (Reversibility: ir) [1, 2]
P GDP-α-D-rhamnose + NADP$^+$ <1, 3> [1, 2]
S GDP-6-deoxy-D-mannose + NAD(P)$^+$ <1, 2> (Reversibility: ?) [1, 2]
P GDP-4-dehydro-6-deoxy-D-mannose + NAD(P)H

Cofactors/prosthetic groups

NAD$^+$ <1, 2> [1,2]
NADH <1, 2> [1,2]
NADP$^+$ <1, 2> [1,2]
NADPH <1, 2> [1,2]

4 Enzyme Structure

Subunits

? <1, 3> (<1> x * 34000, recombinant FLAG-tagged enzyme [2]; <3> x * 38100, SDS-PAGE [1]; <1> x * 34000, SDS-PAGE of FLAG-tagged recombinant enzyme, x * 33900, calculated [2]) [1, 2]

5 Isolation/Preparation/Mutation/Application

Purification

<3> (recombinant enzyme) [1]

Cloning

<1> (expression in Saccharomyces cerevisiae) [2]
<3> (overexpression in Escherichia coli) [1]

Application

synthesis <1> (<1> generating GDP-rhamnose for in vitro rhamnosylation of glycoproteins and glycopeptides [2]; <1> the functional expression of the Pseudomonas aeruginosa enzyme in Saccharomyces cerevisiae will provide a tool for generating GDP-rhamnose for in vitro rhamnosylation of glycoprotein and glycopeptides [2]; <1> generation of GDP-rhamnose by coexpression of enzyme with GDP-mannose-4,6-dehydratase in Saccharomyces cerevisiae [2]) [2]

References

[1] Kneidinger, B.; Graninger, M.; Adam, G.; Puchberger, M.; Kosma, P.; Zayni, S.; Messner, P.: Identification of two GDP-6-deoxy-D-lyxo-4-hexulose reductases synthesizing GDP-D-rhamnose in Aneurinibacillus thermoaerophilus L420-91T. J. Biol. Chem., **276**, 5577-5583 (2001)

[2] Mäki, M.; Järvinen, N.; Räbinä, J.; Roos, C.; Maaheimo, H.; Mattila, P.; Renkonen, R.: Functional expression of Pseudomonas aeruginosa GDP-4-keto-6-deoxy-D-mannose reductase which synthesizes GDP-rhamnose. Eur. J. Biochem., **269**, 593-601 (2002)

Quinate/shikimate dehydrogenase 1.1.1.282

1 Nomenclature

EC number
1.1.1.282

Systematic name
L-quinate:NAD(P)$^+$ 3-oxidoreductase

Recommended name
quinate/shikimate dehydrogenase

Synonyms
YdiB <1, 2, 3, 8, 9, 10, 11, 13, 14> [6, 9]
dehydroquinate dehydratase-shikimate dehydrogenase <6> [8]
quinate/shikimate 5-dehydrogenase <3> [6]
quinate/shikimate dehydrogenase <1, 2, 3, 8, 9, 10, 11, 13, 14> [9]
Additional information <3> (<3> YdiB is a class A NAD(P)$^+$-dependent oxidoreductase [3]) [3]

CAS registry number
9026-87-3 (cf. EC 1.1.1.25)
9028-28-8 (cf. EC 1.1.1.24)

2 Source Organism

<1> *Salmonella typhimurium* (no sequence specified) [9]
<2> *Haemophilus influenzae* (no sequence specified) [9]
<3> *Escherichia coli* (no sequence specified) [2, 3, 6, 9]
<4> *Neurospora crassa* (no sequence specified) [1]
<5> *Pinus taeda* (no sequence specified) [7]
<6> *Arabidopsis thaliana* (no sequence specified) [8]
<7> *Rhodococcus rhodochrous* (no sequence specified) [4]
<8> *Lactobacillus plantarum* (no sequence specified) [9]
<9> *Streptococcus pyogenes* (no sequence specified) [9]
<10> *Enterococcus faecium* (no sequence specified) [9]
<11> *Enterococcus faecalis* (no sequence specified) [9]
<12> *Pinus sylvestris* (no sequence specified) [5]
<13> *Listeria monocytogenes* (no sequence specified) [9]
<14> *Shigella flexneri* (no sequence specified) [9]
<15> *Escherichia coli* (UNIPROT accession number: P15770) [2]
<16> *Larix sibirica* (no sequence specified) [5]

3 Reaction and Specificity

Catalyzed reaction

L-quinate + NAD(P)$^+$ = 3-dehydroquinate + NAD(P)H + H$^+$ (<15> reaction mechanism [2]; <3> reaction mechanism, in which an aspartate acts as the general acid/base catalyst during the hydride transfer [3]; <7> stereospecificity [4])
shikimate + NAD(P)$^+$ = 3-dehydroshikimate + NAD(P)H + H$^+$ (<15> reaction mechanism [2]; <3> reaction mechanism, in which an aspartate acts as the general acid/base catalyst during the hydride transfer [3]; <7> stereospecificity [4])

Reaction type

oxidation
redox reaction
reduction

Natural substrates and products

S 3-dehydroquinate + NADPH + H$^+$ <5> (<5> may be responsible for the synthesis of quinic acid from the intermediate compound of the shikimate pathway, dehydroquinic acid [7]) (Reversibility: r) [7]
P L-quinate + NADP$^+$
S 3-dehydroshikimate + NAD(P)H + H$^+$ <3> (<3> YdiB catalyzes the reduction of 3-dehydroshikimate to shikimate as part of the shikimate pathway [3]) (Reversibility: ?) [3]
P shikimate + NAD(P)$^+$
S Additional information <4, 7, 12, 15, 16> (<4> catalyzes the first reaction in the inducible quinic acid catabolic pathway [1]; <7> initial enzyme of the hydroaromatic pathway [4]; <12,16> role in quinic acid metabolism [5]; <15> YdiB may be involved in shikimate pathway or may be essential for growth of the organism with quinate as a sole carbon source [2]) (Reversibility: ?) [1, 2, 4, 5]
P ?

Substrates and products

S 3-dehydroquinate + NADPH + H$^+$ <3, 5> (<5> may be responsible for the synthesis of quinic acid from the intermediate compound of the shikimate pathway, dehydroquinic acid [7]) (Reversibility: r) [6, 7]
P L-quinate + NADP$^+$
S 3-dehydroshikimate + NAD(P)H + H$^+$ <3> (<3> YdiB catalyzes the reduction of 3-dehydroshikimate to shikimate as part of the shikimate pathway [3]) (Reversibility: ?) [3]
P shikimate + NAD(P)$^+$
S 3-dehydroshikimate + NADH + H$^+$ <3> (Reversibility: r) [6]
P shikimate + NAD$^+$
S L-quinate + 3-acetylpyridine adenine dinucleotide <7> (<7> 67% of the activity with NAD$^+$ [4]) (Reversibility: ?) [4]
P 3-dehydroquinate + ?

S L-quinate + NAD(P)$^+$ <3, 4, 12, 15, 16> (<4> bifunctional enzyme with a single binding site for both substrates quinate and shikimate, the velocity is approximately 3fold lower with quinate than with shikimate [1]; <3> detailed strucure of YdiB, specificity for binding NAD$^+$/NADH over NADP$^+$/NADPH [3]; <15> YdiB is a dual specific quinate/shikimate dehydrogenase that utilizes either NAD$^+$ or NADP$^+$ as cofactor, YdiB is equally active with shikimate or quinate, but has a tendency to be more efficient with NAD$^+$ than with NADP$^+$, detailed structure of YdiB, mechanism [2]) (Reversibility: r) [1, 2, 3, 5]

P 3-dehydroquinate + NAD(P)H + H$^+$

S L-quinate + NADP$^+$ <3, 5, 7> (<7> 0.3% of the activity with NAD$^+$ [4]; <5> both quinate and shikimate dehydrogenase activities are catalyzed by a single broad-specificity quinate (shikimate) dehydrogenase with a common substrate binding site, the velocity is 2fold greater with quinate than with shikimate [7]) (Reversibility: r) [4, 6, 7]

P 3-dehydroquinate + NADPH + H$^+$

S L-quinate + β-NAD$^+$ <7> (<7> highly stereospecific with regard to hydroaromatic substrates, oxidizes only the axial hydroxyl group at C-3 of the (-)-enantiomer, a single enzyme with both quinate and shikimate dehydrogenase activity [4]) (Reversibility: r) [4]

P 3-dehydroquinate + β-NADH + H$^+$ (<7> (-)-enantiomer, reverse reaction: lower activity than with t-4,c-5-dihydroxy-3-oxocyclohexane-c-1-carboxylate or 4-hydroxy-3-oxocyclohexane-c-1-carboxylate [4])

S L-quinate + nicotinamide 1,N^6-ethenoadenine dinucleotide <7> (<7> 69% of the activity with NAD$^+$ [4]) (Reversibility: ?) [4]

P 3-dehydroquinate + ?

S L-quinate + nicotinamide hypoxanthine dinucleotide <7> (<7> 1.3fold higher activity than with NAD$^+$ [4]) (Reversibility: ?) [4]

P 3-dehydroquinate + ?

S dihydroshikimate + NAD$^+$ <7> (<7> highly stereospecific with regard to hydroaromatic substrates, oxidizes only the axial hydroxyl group at C-3 of the (-)-enantiomer, 38% of the activity with (-)-quinate [4]) (Reversibility: r) [4]

P t-4,c-5-dihydroxy-3-oxocyclohexane-c-1-carboxylate + NADH + H$^+$ (<7> (-)-enantiomer, reverse reaction: 3.3fold higher activity than with (-)-3-dehydroquinate [4])

S shikimate + NAD(P)$^+$ <3, 4, 15> (<4> bifunctional enzyme with a single binding site for both substrates quinate and shikimate, the velocity is approximately 3fold higher with shikimate than with quinate [1]; <3> detailed strucure of YdiB, catalytic mechanism, specificity for binding NAD$^+$/NADH over NADP$^+$/NADPH [3]; <15> YdiB is a dual specific quinate/shikimate dehydrogenase that utilizes either NAD$^+$ or NADP$^+$ as cofactor, YdiB is equally active with shikimate or quinate, but has a tendency to be more efficient with NAD$^+$ than with NADP$^+$, detailed structure of YdiB, mechanism [2]) (Reversibility: r) [1, 2, 3]

P 3-dehydroshikimate + NAD(P)H + H$^+$ (<15> model for 3-dehydroshikimate recognition [2])

S shikimate + NAD^+ <1, 2, 3, 7, 8, 9, 10, 11, 13, 14> (<7> highly stereospecific with regard to hydroaromatic substrates, oxidizes only the axial hydroxyl group at C-3 of the (-)-enantiomer, a single enzyme with both quinate and shikimate dehydrogenase activity, 72% of the activity with (-)-quinate [4]) (Reversibility: r) [4, 6, 9]

P 3-dehydroshikimate + NADH + H^+ (<7> (-)-enantiomer, reverse reaction: 15% of the activity with (-)-3-dehydroquinate [4])

S shikimate + $NADP^+$ <5> (<5> both quinate and shikimate dehydrogenase activities are catalyzed by a single broad-specificity quinate (shikimate) dehydrogenase with a common substrate binding site, the velocity is 2fold lower with shikimate than with quinate [7]) (Reversibility: r) [7]

P 3-dehydroshikimate + NADPH + H^+

S t-3,t-4-dihydroxycyclohexane-c-1-carboxylate + NAD^+ <7> (<7> highly stereospecific with regard to hydroaromatic substrates, oxidizes only the axial hydroxyl group at C-3 of the (-)-enantiomer, 44% of the activity with (-)-quinate [4]) (Reversibility: r) [4]

P 4-hydroxy-3-oxocyclohexane-c-1-carboxylate + NADH + H^+ (<7> (-)-isomer, reverse reaction: 2.2fold higher activity than with (-)-3-dehydroquinate [4])

S t-3-hydroxy-4-oxocyclohexane-c-1-carboxylate + NAD^+ <7> (<7> 6% of the activity with (-)-quinate [4]) (Reversibility: ?) [4]

P ?

S Additional information <4, 7, 12, 15, 16> (<4> catalyzes the first reaction in the inducible quinic acid catabolic pathway [1]; <7> initial enzyme of the hydroaromatic pathway [4]; <12,16> role in quinic acid metabolism [5]; <15> YdiB may be involved in shikimate pathway or may be essential for growth of the organism with quinate as a sole carbon source [2]; <7> substrate specificity, enzyme is highly stereospecific with regard to hydroaromatic substrates, oxidizing only the axial hydroxyl group at C-3 of (-)-enantiomer of quinate, shikimate, dihydroshikimate and t-3,t-4-dihydroxycyclohexane-c-1-carboxylate, enzyme shows activity with several NAD^+ analogues, reverse reaction: not 4-hydroxy-3-oxocyclohex-4-ene-c-1-carboxylate, not: α-NAD^+, β-NMN or nicotinic acid dinucleotide [4]) (Reversibility: ?) [1, 2, 4, 5]

P ?

Inhibitors

L-quinate <5> (<5> competitive inhibitor with respect to shikimate [7]) [7]
Shikimate <5> (<5> competitive inhibitor with respect to L-quinate [7]) [7]

Cofactors/prosthetic groups

3-acetylpyridine adenine dinucleotide <7> (<7> 67% of the activity with NAD^+ [4]) [4]
NAD^+ <1, 2, 3, 6, 7, 8, 9, 10, 12, 13, 14, 15, 16> (<3> NAD^+ binding site, bound very tightly, NAD^+ is bound to the Rossmann domain in an elongated fashion with the nicotinamide ring in the pro-R conformation, specificity for binding NAD^+ over $NADP^+$ [3]; <15> utilizes either NAD^+ or $NADP^+$ as a

cofactor, tendency to be more efficient with NAD^+ than with $NADP^+$, mode of binding [2]) [2,3,4,5,6,8,9]
NADH <3, 7> (<3> specificity for binding NADH over NADPH [3]) [3,4,6]
$NADP^+$ <3, 5, 7, 12, 15, 16> (<7> 0.3% of the activity with NAD^+ [4]; <3> specificity for binding NAD^+ over $NADP^+$ [3]; <15> utilizes either NAD^+ or $NADP^+$ as a cofactor, tendency to be more efficient with NAD^+ than with $NADP^+$, mode of binding [2]) [2,3,4,5,6,7]
NADPH <3, 5> (<3> specificity for binding NADH over NADPH [3]) [3,6,7]
Nicotinamide hypoxanthine dinucleotide <7> (<7> 1.3fold higher activity than with NAD^+ [4]) [4]
nicotinamide 1,N^6-ethenoadenine dinucleotide <7> (<7> 69% of the activity with NAD^+ [4]) [4]
Additional information <7> (<7> absolute requirement for a nicotinamide nucleotide cofactor for the oxidation of quinate or shikimate, cofactor specificity, not: α-NAD^+, β-NMN or nicotinic acid dinucleotide [4]) [4]

Turnover number (min^{-1})

0.004 <3> (NAD^+, <3> mutant K71A, in the presence of L-quinate, 20°C, pH 9.0 [6]) [6]
0.011 <3> (shikimate, <3> mutant K71A, 20°C, pH 9.0 [6]) [6]
0.012 <3> (L-quinate, <3> mutant K71A, 20°C, pH 9.0 [6]) [6]
0.012 <3> (NAD^+, <3> mutant K71A, in the presence of shikimate, 20°C, pH 9.0 [6]) [6]
0.022 <3> (NAD^+, <3> mutant D107A, in the presence of shikimate, 20°C, pH 9.0 [6]; <3> mutant Q262A, in the presence of shikimate, 20°C, pH 9.0 [6]) [6]
0.023 <3> (NAD^+, <3> mutant S67A, in the presence of L-quinate, 20°C, pH 9.0 [6]) [6]
0.024 <3> (L-quinate, <3> mutant S67A, 20°C, pH 9.0 [6]) [6]
0.043 <3> (NAD^+, <3> mutant Q262A, in the presence of L-quinate, 20°C, pH 9.0 [6]) [6]
0.05 <3, 15> (L-quinate, <15> pH 9, 20°C, cosubstrate NAD^+ or $NADP^+$ [2]; <3> mutant Q262A, 20°C, pH 9.0 [6]) [2, 6]
0.05 <15> (NAD^+, <15> pH 9, 20°C, cosubstrate shikimate or L-quinate [2]) [2]
0.05 <15> ($NADP^+$, <15> pH 9, 20°C, cosubstrate L-quinate [2]) [2]
0.05 <3, 15> (shikimate, <15> pH 9, 20°C, cosubstrate NAD^+ [2]; <3> mutant Q262A, 20°C, pH 9.0 [6]) [2, 6]
0.054 <3> (shikimate, <3> mutant T106A, 20°C, pH 9.0 [6]) [6]
0.06 <3> (L-quinate, <3> mutant Y39F, 20°C, pH 9.0 [6]) [6]
0.068 <3> (L-quinate, <3> mutant S22A, 20°C, pH 9.0 [6]) [6]
0.082 <3> (NAD^+, <3> mutant S22A, in the presence of L-quinate, 20°C, pH 9.0 [6]) [6]
0.09 <3> (NAD^+, <3> mutant Y39F, in the presence of L-quinate, 20°C, pH 9.0 [6]) [6]
0.091 <3> (shikimate, <3> wild type enzyme, 20°C, pH 9.0 [6]) [6]
0.105 <3> (NAD^+, <3> wild type enzyme, in the presence of shikimate, 20°C, pH 9.0 [6]) [6]

0.108 <3> (NAD$^+$, <3> mutant T106A, in the presence of shihikimat, 20°C, pH 9.0 [6]) [6]
0.113 <3> (L-quinate, <3> wild type enzyme, 20°C, pH 9.0 [6]) [6]
0.117 <15> (NADP$^+$, <15> pH 9, 20°C, cosubstrate shikimate [2]) [2]
0.117 <15> (shikimate, <15> pH 9, 20°C, cosubstrate NADP$^+$ [2]) [2]
0.122 <3> (NAD$^+$, <3> mutant T106A, in the presence of L-quinate, 20°C, pH 9.0 [6]) [6]
0.132 <3> (shikimate, <3> mutant D107A, 20°C, pH 9.0 [6]) [6]
0.142 <3> (NAD$^+$, <3> wild type enzyme, in the presence of L-quinate, 20°C, pH 9.0 [6]) [6]
0.148 <3> (shikimate, <3> mutant S67A, 20°C, pH 9.0 [6]) [6]
0.153 <3> (NAD$^+$, <3> mutant S22A, in the presence of shikimate, 20°C, pH 9.0 [6]) [6]
0.178 <3> (NAD$^+$, <3> mutant S67A, in the presence of shikimate, 20°C, pH 9.0 [6]) [6]
0.187 <3> (shikimate, <3> mutant Y39F, 20°C, pH 9.0 [6]) [6]
0.233 <3> (L-quinate, <3> mutant T106A, 20°C, pH 9.0 [6]) [6]
0.24 <3> (NAD$^+$, <3> mutant Y39F, in the presence of shikimate, 20°C, pH 9.0 [6]) [6]

Specific activity (U/mg)

0.00383 <12> (<12> needles of 20 days old saplings, quinate and NADP$^+$ as substrates [5]) [5]
0.00391 <12> (<12> hypocotyls of 20 days old saplings, quinate and NADP$^+$ as substrates [5]) [5]
0.0664 <16> (<16> needles, in June, the period of maximal quinate dehydrogenase content, quinate and NADP$^+$ as substrates [5]) [5]
Additional information <5, 7, 15> (<15> YdiB has a low specific activity [2]) [2, 4, 7]

K_m-Value (mM)

0.0004 <3> (NAD$^+$, <3> mutant K71A, in the presence of shikimate, 20°C, pH 9.0 [6]) [6]
0.001 <5> (NADP$^+$, <5> pH 10, 20°C, cosubstrate shikimate, both forms of quinate (shikimate) dehydrogenase [7]) [7]
0.0017 <3> (shikimate, <3> mutant S67A, 20°C, pH 9.0 [6]) [6]
0.002 <3> (NAD$^+$, <3> mutant K71A, in the presence of L-quinate, 20°C, pH 9.0 [6]) [6]
0.0028 <3> (shikimate, <3> mutant Q262A, 20°C, pH 9.0 [6]) [6]
0.0029 <3> (shikimate, <3> wild type enzyme, 20°C, pH 9.0 [6]) [6]
0.0032 <3> (shikimate, <3> mutant S22A, 20°C, pH 9.0 [6]; <3> mutant Y39F, 20°C, pH 9.0 [6]) [6]
0.005-0.012 <5> (NADPH, <5> pH 10, 20°C, cosubstrate dehydroquinate, both forms of quinate (shikimate) dehydrogenase [7]) [7]
0.0054 <3> (L-quinate, <3> mutant Y39F, 20°C, pH 9.0 [6]) [6]
0.0057 <3> (L-quinate, <3> mutant S22A, 20°C, pH 9.0 [6]) [6]
0.0068 <3> (NAD$^+$, <3> mutant D107A, in the presence of shikimate, 20°C, pH 9.0 [6]) [6]

0.007 <5> ($NADP^+$, <5> pH 10, 20°C, cosubstrate L-quinate, both forms of quinate (shikimate) dehydrogenase [7]) [7]
0.0091 <3> (L-quinate, <3> wild type enzyme, 20°C, pH 9.0 [6]) [6]
0.0107 <3> (NAD^+, <3> mutant Q262A, in the presence of shikimate, 20°C, pH 9.0 [6]) [6]
0.0121 <3> (NAD^+, <3> mutant S67A, in the presence of L-quinate, 20°C, pH 9.0 [6]) [6]
0.0122 <3> (NAD^+, <3> wild type enzyme, in the presence of shikimate, 20°C, pH 9.0 [6]) [6]
0.0125 <3> (NAD^+, <3> mutant Y39F, in the presence of L-quinate, 20°C, pH 9.0 [6]) [6]
0.0126 <3> (NAD^+, <3> mutant S67A, in the presence of shikimate, 20°C, pH 9.0 [6]) [6]
0.0136 <3> (L-quinate, <3> mutant Q262A, 20°C, pH 9.0 [6]) [6]
0.0142 <3> (NAD^+, <3> mutant Q262A, in the presence of L-quinate, 20°C, pH 9.0 [6]) [6]
0.0157 <3> (L-quinate, <3> mutant S67A, 20°C, pH 9.0 [6]) [6]
0.0158 <3> (NAD^+, <3> mutant T106A, in the presence of shikimate, 20°C, pH 9.0 [6]; <3> mutant Y39F, in the presence of shikimate, 20°C, pH 9.0 [6]) [6]
0.0168 <3> (NAD^+, <3> mutant S22A, in the presence of L-quinate, 20°C, pH 9.0 [6]) [6]
0.018 <3> (NAD^+, <3> mutant S22A, in the presence of shikimate, 20°C, pH 9.0 [6]) [6]
0.0184 <3> (NAD^+, <3> wild type enzyme, in the presence of L-quinate, 20°C, pH 9.0 [6]) [6]
0.02 <3> (shikimate, <3> pH 9, 20°C, cosubstrate NAD^+ [2]) [2]
0.041 <3> (L-quinate, <3> pH 9, 20°C, cosubstrate NAD^+ [2]) [2]
0.0578 <3> (NAD^+, <3> mutant T106A, in the presence of L-quinate, 20°C, pH 9.0 [6]) [6]
0.087 <3> (NAD^+, <3> pH 9, 20°C, cosubstrate shikimate [2]) [2]
0.1 <3> ($NADP^+$, <3> pH 9, 20°C, cosubstrate shikimate [2]) [2]
0.116 <3> (NAD^+, <3> pH 9, 20°C, cosubstrate L-quinate [2]) [2]
0.12 <3> (shikimate, <3> pH 9, 20°C, cosubstrate $NADP^+$ [2]) [2]
0.15 <7> (β-NAD^+, <7> pH 10, cosubstrate (-)-quinate [4]) [4]
0.199 <3> (shikimate, <3> mutant T106A, 20°C, pH 9.0 [6]) [6]
0.37 <4> (L-quinate) [1]
0.48 <7> (Nicotinamide hypoxanthine dinucleotide, <7> pH 10, cosubstrate (-)-quinate [4]) [4]
0.5 <15> ($NADP^+$, <15> pH 9, 20°C, cosubstrate L-quinate [2]) [2]
0.51 <7> (nicotinamide 1,N^6-ethanoadenine dinucleotide, <7> pH 10, cosubstrate (-)-quinate [4]) [4]
0.513 <3> (shikimate, <3> mutant K71A, 20°C, pH 9.0 [6]) [6]
0.555 <3> (L-quinate, <3> pH 9, 20°C, cosubstrate $NADP^+$ [2]) [2]
0.65 <7> (3-acetylpyridine adenine dinucleotide, <7> pH 10, cosubstrate (-)-quinate [4]) [4]

0.7-0.8 <5> (shikimate, <5> pH 10, 20°C, cosubstrate $NADP^+$, both forms of quinate (shikimate) dehydrogenase [7]) [7]
1 <5> (dehydroquinate, <5> pH 10, 20°C, cosubstrate NADPH, form P1 of quinate (shikimate) dehydrogenase [7]) [7]
1.18 <4> (shikimate) [1]
2.47 <7> (t-3,t-4-dihydroxycyclohexane-c-1-carboxylate, <7> pH 10, (-)-enantiomer, cosubstrate NAD^+ [4]) [4]
2.56 <7> (dihydroshikimate, <7> pH 10, (-)-enantiomer, cosubstrate NAD^+ [4]) [4]
2.95 <7> (L-quinate, <7> pH 10, (-)-enantiomer, cosubstrate NAD^+ [4]) [4]
3.4-3.6 <5> (L-quinate, <5> pH 10, 20°C, cosubstrate $NADP^+$, both forms of quinate (shikimate) dehydrogenase [7]) [7]
5.25 <7> (shikimate, <7> pH 10, (-)-enantiomer, cosubstrate NAD^+ [4]) [4]
5.3 <5> (dehydroquinate, <5> pH 10, 20°C, cosubstrate NADPH, form P2 of quinate (shikimate) dehydrogenase [7]) [7]
20.52 <3> (L-quinate, <3> mutant T106A, 20°C, pH 9.0 [6]) [6]
32.68 <3> (L-quinate, <3> mutant K71A, 20°C, pH 9.0 [6]) [6]
40.07 <3> (shikimate, <3> mutant D107A, 20°C, pH 9.0 [6]) [6]
Additional information <12, 16> (<12,16> kinetic data [5]) [5]

pH-Optimum

7.5 <7> (<7> reduction reaction, assay at [4]) [4]
7.7 <5> (<5> back reaction, dehydroquinate or dehydroshikimate and NADPH as substrates [7]) [7]
9 <15> (<15> assay at [2]) [2]
10 <7> (<7> oxidation reaction, quinate or shikimate as substrates, 50 mM glycine-KOH [4]) [4]
10.3 <5> (<5> quinate or shikimate and $NADP^+$ as substrates [7]) [7]
10.6 <3> (<3> assay at [3]) [3]

pH-Range

7-8.5 <7> (<7> the rate of activity decreases more rapidly, maximal 25% at pH 7.5, at pH values from 8.5 to 7 when shikimate rather than quinate is used as substrate [4]) [4]

Temperature optimum (°C)

20 <5, 15> (<5,15> assay at [2,7]) [2, 7]
25 <3> (<3> assay at [3]) [3]

4 Enzyme Structure

Molecular weight

33000 <3> (<3> SDS-PAGE, mutant N92A [6]) [6]
35000 <5> (<5> two forms of the bifunctional quinate (shikimate) dehydrogenase, gel filtration [7]) [7]
44000 <7> (<7> gel filtration [4]) [4]

53000 <5> (<5> two forms of the bifunctional quinate (shikimate) dehydrogenase, gel filtration [7]) [7]
57000 <3> (<3> SDS-PAGE, wild type enzyme [6]) [6]
64000 <15> (<15> apoprotein, gel filtration [2]) [2]

Subunits

dimer <3, 15> (<3> 2 * 31350, wild type enzyme, gel filtration [6]) [2, 6]
homodimer <3> [3]
monomer <4, 7> (<7> 1 * 31500, SDS-PAGE [4]; <4> 1 * 41000 [1]) [1, 4]

5 Isolation/Preparation/Mutation/Application

Source/tissue

hypocotyl <12> (<12> of 20 days old saplings [5]) [5]
needle <5, 12, 16> (<5> from current-year shoots of 20 years old trees, the youngest basal parts of needles [7]; <16> of 12 years old larch auxiblasts [5]; <12> of 20 days old saplings [5]) [5, 7]

Purification

<3> [2, 3]
<4> [1]
<5> (3000fold, two forms of the bifunctional quinate (shikimate) dehydrogenase: P1 and P2) [7]
<7> (188fold) [4]

Crystallization

<15> (YdiB, with bound cofactors NAD^+ or $NADP^+$) [2]
<31> (YdiB complexed with NAD^+, hanging drop vapor diffusion method, X-ray analysis) [3]

Cloning

<3> (ydiB gene, expression in Escherichia coli BL21(DE3)) [3]
<15> (ydib gene, expression in Escherichia coli DL41, sequencing, ydiB is located between the genes b1691 and aroD) [2]

Engineering

D107A <3> (<3> site-directed mutagenesis [6]) [6]
K71G <3> (<3> site-directed mutagenesis [6]) [6]
N92A <3> (<3> site-directed mutagenesis [6]) [6]
Q262A <3> (<3> site-directed mutagenesis [6]) [6]
S22A <3> (<3> site-directed mutagenesis [6]) [6]
S67A <3> (<3> site-directed mutagenesis, increased activity compared to wild type enzyme [6]) [6]
T106A <3> (<3> site-directed mutagenesis [6]) [6]
Y39F <3> (<3> site-directed mutagenesis [6]) [6]

Application

agriculture <3> (<3> development of novel herbicides [6]) [6]
medicine <3> (<3> development of novel antimicrobial agents [6]) [6]

pharmacology <3> (<3> enzymes of the shikimate pathway has been promoted as a target for the development of antimicrobial agents [3]) [3]

6 Stability

Temperature stability

45 <7> (<7> $t_{1/2}$: 1.85 min [4]) [4]
Additional information <4> (<4> the thermal stability of enzyme is greatly enhanced by low concentrations of quinate, shikimate, NADH or high ionic strength [1]) [1]

References

[1] Barea, J.L.; Giles, N.H.: Purification and characterization of quinate (shikimate) dehydrogenase, an enzyme in the inducible quinic acid catabolic pathway of Neurospora crassa. Biochim. Biophys. Acta, **524**, 1-14 (1978)

[2] Michel, G.; Roszak, A.W.; Sauve, V.; Maclean, J.; Matte, A.; Coggins, J.R.; Cygler, M.; Lapthorn, A.J.: Structures of shikimate dehydrogenase AroE and its paralog YdiB. A common structural framework for different activitie. J. Biol. Chem., **278**, 19463-19472 (2003)

[3] Benach, J.; Lee, I.; Edstrom, W.; Kuzin, A.P.; Chiang, Y.; Acton, T.B.; Montelione, G.T.; Hunt, J.F.: The 2.3-Å crystal structure of the shikimate 5-dehydrogenase orthologue YdiB from Escherichia coli suggests a novel catalytic environment for an NAD-dependent dehydrogenase. J. Biol. Chem., **278**, 19176-19182 (2003)

[4] Bruce, N.C.; Cain, R.B.: Hydroaromatic metabolism in Rhodococcus rhodochrous: purification and characterization of its NAD-dependent quinate dehydrogenase. Arch. Microbiol., **154**, 179-186 (1990)

[5] Osipov, V.I.; Shein, I.V.: The role of quinate dehydrogenase in quinic acid metabolism in coniferous plants. Biokhimiya, **51**, 230-236 (1986)

[6] Lindner, H.A.; Nadeau, G.; Matte, A.; Michel, G.; Menard, R.; Cygler, M.: Site-directed mutagenesis of the active site region in the quinate/shikimate 5-dehydrogenase YdiB of Escherichia coli. J. Biol. Chem., **280**, 7162-7169 (2004)

[7] Ossipov, V.; Bonner, C.; Ossipova, S.; Jensen, R.: Broad-specificity quinate (shikimate) dehydrogenase from Pinus taeda needles. Plant Physiol. Biochem., **38**, 923-928 (2000)

[8] Singh, S.A.; Christendat, D.: Structure of Arabidopsis dehydroquinate dehydratase-shikimate dehydrogenase and implications for metabolic channeling in the shikimate pathway. Biochemistry, **45**, 10406 (2006)

[9] Singh, S.; Korolev, S.; Koroleva, O.; Zarembinski, T.; Collart, F.; Joachimiak, A.; Christendat, D.: Crystal structure of a novel shikimate dehydrogenase from Haemophilus influenzae. J. Biol. Chem., **280**, 17101-17108 (2005)

Methylglyoxal reductase (NADPH-dependent) 1.1.1.283

1 Nomenclature

EC number
1.1.1.283

Systematic name
lactaldehyde:NADP$^+$ oxidoreductase

Recommended name
methylglyoxal reductase (NADPH-dependent)

Synonyms
EC 1.1.1.78 <1> (<1> formerly [4]) [4]
AKR <3> [5]
SakR1 <3> [5]
aldo-keto reductase <3> [5]

CAS registry number
37250-16-1

2 Source Organism

<1> *Saccharomyces cerevisiae* (no sequence specified) [1, 2, 4]
<2> *Aspergillus niger* (no sequence specified) [3]
<3> *Synechococcus sp.* (no sequence specified) [5]

3 Reaction and Specificity

Catalyzed reaction
(R)-lactaldehyde + NAD$^+$ = methylglyoxal + NADH + H$^+$ (<1,2> similar enzyme with NADPH-requirement for methylglyoxal reduction, that is inactive with NAD$^+$, NADH and NADP$^+$ [1,2,3])
lactaldehyde + NADP$^+$ = methylglyoxal + NADPH + H$^+$

Natural substrates and products
S methylglyoxal + NADPH <1, 2> (<2> similar enzyme with NADPH requirement [3]) (Reversibility: ?) [1, 2, 3]
P lactaldehyde + NADP$^+$ <1, 2> [1, 2, 3]

Substrates and products
S 2-nitrobenzaldehyde + NADPH <3> (Reversibility: ?) [5]
P 2-nitrobenzyl alcohol + NADP$^+$

S 3-nitrobenzaldehyde+ NADPH <3> (Reversibility: ?) [5]
P 3-nitrobenzyl alcohol + $NADP^+$
S 4-nitrobenzaldehyde+ NADPH <3> (Reversibility: ?) [5]
P 4-nitrobenzyl alcohol + $NADP^+$
S D-arabinose + NADPH <3> (Reversibility: ?) [5]
P ? + $NADP^+$
S D-galactose + NADPH <3> (Reversibility: ?) [5]
P ? + $NADP^+$
S D-glucose + NADPH <3> (Reversibility: ?) [5]
P ? + $NADP^+$
S D-xylose + NADPH <3> (Reversibility: ?) [5]
P ? + $NADP^+$
S DL-glyceraldehyde + NADPH <3> (Reversibility: ?) [5]
P glycerol + $NADP^+$
S acetaldehyde + NADPH <2, 3> (<2> enzyme MGR II [3]) (Reversibility: ?) [3, 5]
P ethanol + $NADP^+$ <2> [3]
S benzaldehyde + NADPH <3> (Reversibility: ?) [5]
P benzyl alcohol + $NADP^+$
S crotonaldehyde + NADPH <3> (Reversibility: ?) [5]
P (2Z)-but-2-en-1-ol + $NADP^+$
S diacetyl + NADPH <3> (Reversibility: ?) [5]
P ? + $NADP^+$
S dihydroxyacetone + NADPH <3> (Reversibility: ?) [5]
P glycerol + $NADP^+$
S glyoxal + NADPH <2> (<2> enzymes MGR I and MGR II [3]) (Reversibility: ?) [3]
P glycolaldehyde + $NADP^+$ <2> [3]
S isatin + NADPH <3> (Reversibility: ?) [5]
P ? + $NADP^+$
S isopentaldehyde + NADPH <3> (Reversibility: ?) [5]
P isopentanol + $NADP^+$
S methylglyoxal + NADPH <1, 2, 3> (<2> similar enzyme with NADPH requirement [3]) (Reversibility: ?) [1, 2, 3, 5]
P lactaldehyde + $NADP^+$ <1, 2> [1, 2, 3]
S methylglyoxal + NADPH + H^+ <1> (Reversibility: ?) [4]
P (R)-lactataldehyde + $NADP^+$
S ninhydrin + NADPH <3> (Reversibility: ?) [5]
P ? + $NADP^+$
S phenylglyoxal + NADPH <2> (<2> enzymes MGR I and MGR II [3]) (Reversibility: ?) [3]
P hydroxyphenylacetaldehyde + NAD^+ <2> [3]
S propionaldehyde + NADPH <2> (<2> enzyme MGR II [3]) (Reversibility: ?) [3]
P propanol + $NADP^+$
S succinic semialdehyde + NADPH <3> (Reversibility: ?) [5]
P ? + $NADP^+$

S Additional information <2> (<2> enzyme MGR I: specific for 2-oxoaldehydes (glyoxal phenylglyoxal), enzyme MGR II: active towards 2-oxoaldehydes (glyoxal, methylglyoxal, phenylglyoxal), 4,5-dioxovalerate and some aldehydes (propionaldehyde and acetaldehyde) [3]) (Reversibility: ?) [3]
P ?

Inhibitors

2-mercaptoethanol <2> [3]
Ca^{2+} <2> (<2> enzymes MGR I (slightly), MGR II [3]) [3]
Co^{2+} <2> (<2> activates enzyme MGR I, slightly inhibits enzyme MGR II [3]) [3]
Cu^{2+} <2> (<2> enzymes MGR I and MGR II [3]) [3]
dithiothreitol <2> [3]
EDTA <2> [3]
glyoxal <2> (<2> in absence of NADPH [3]) [3]
iodoacetate <2> (<2> activation of enzyme MGR I, inhibition of enzyme MGR II [3]) [3]
methylglyoxal <2> (<2> in absence of NADPH [3]) [3]
Mg^{2+} <2> (<2> enzymes MGR I and MGR II [3]) [3]
Mn^{2+} <2> (<2> enzymes MGR I, MGR II [3]) [3]
N-ethylmaleimide <2> (<2> activation of enzyme MGR I, inhibition of enzyme MGR II [3]) [3]
$NADP^+$ <2> [3]
Ni^{2+} <2> (<2> enzyme MGR I [3]) [3]
phenylglyoxal <2> (<2> in absence of NADPH [3]) [3]
Zn^{2+} <2> (<2> enzyme MGR I [3]) [3]
p-chloromercuribenzoate <2> (<2> activation of enzyme MGR I, inhibition of enzyme MGR II [3]) [3]

Cofactors/prosthetic groups

NADPH <1, 2, 3> (<2> NADPH required [3]) [1,2,3,4,5]

Activating compounds

iodoacetate <2> (<2> activation of enzyme MGR I, inhibition of enzyme MGR II [3]) [3]
N-ethylmaleimide <2> (<2> activation of enzyme MGR I, inhibition of enzyme MGR II [3]) [3]
p-chloromercuribenzoate <2> (<2> activation of enzyme MGR I, inhibition of enzyme MGR II [3]) [3]

Metals, ions

Co^{2+} <2> (<2> activates enzyme MGR I, slightly inhibits enzyme MGR II [3]) [3]

Turnover number (min^{-1})

0.52 <3> (DL-glyceraldehyde, <3> 1 mM [5]) [5]
0.62 <3> (D-glucose, <3> 50 mM [5]) [5]
0.65 <3> (isopentaldehyde, <3> 1 mM [5]) [5]
0.68 <3> (D-galactose, <3> 50 mM [5]) [5]

0.85 <3> (methylglyoxal, <3> 1 mM [5]) [5]
1.05 <3> (benzaldehyde, <3> 1 mM [5]) [5]
1.12 <3> (3-nitrobenzaldehyde, <3> 1 mM [5]) [5]
1.22 <3> (D-xylose, <3> 50 mM [5]) [5]
1.24 <3> (4-nitrobenzaldehyde, <3> 1 mM [5]) [5]
1.27 <3> (L-arabinose, <3> 50 mM [5]) [5]
1.31 <3> (2-nitrobenzaldehyde, <3> 1 mM [5]) [5]

Specific activity (U/mg)
0.038 <3> (<3> 1 mM acetaldehyde as substrate [5]) [5]
0.047 <3> (<3> 1 mM crotonaldehyde as substrate [5]; <3> 1 mM isatin as substrate [5]) [5]
0.048 <3> (<3> 1 mM dihydroxyacetone as substrate [5]) [5]
0.109 <3> (<3> 50 mM D-glucose as substrate [5]) [5]
0.211 <3> (<3> 1 mM ninhydrin as substrate [5]) [5]
0.225 <3> (<3> 1 mM succinic semialdehyde as substrate [5]) [5]
0.24 <3> (<3> 50 mM D-galactose as substrate [5]) [5]
0.364 <3> (<3> 50 mM L-arabinose as substrate [5]) [5]
0.373 <3> (<3> 1 mM diacetyl as substrate [5]) [5]
0.507 <3> (<3> 1 mM DL-glyceraldehyde as substrate [5]) [5]
0.641 <3> (<3> 1 mM isopentaldehyde as substrate [5]) [5]
0.651 <3> (<3> 1 mM benzaldehyde as substrate [5]) [5]
0.66 <3> (<3> 50 mM D-xylose as substrate [5]) [5]
0.83 <2> (<2> enzyme MGR I [3]) [3]
1.072 <3> (<3> 1 mM 3-nitrobenzaldehyde as substrate [5]) [5]
1.091 <3> (<3> 1 mM 4-nitrobenzaldehyde as substrate [5]) [5]
1.206 <3> (<3> 1 mM 2-nitrobenzaldehyde as substrate [5]) [5]
1.278 <3> (<3> 1 mM methylglyoxal as substrate [5]) [5]
7.77 <2> (<2> enzyme MGR II [3]) [3]
8.35 <1> [4]

K_m-Value (mM)
0.0068 <2> (NADPH, <2> enzyme MGR II [3]) [3]
0.054 <2> (NADPH, <2> enzyme MGR I [3]) [3]
0.08 <3> (methylglyoxal, <3> 1 mM [5]) [5]
0.31 <3> (2-nitrobenzaldehyde, <3> 1 mM [5]) [5]
0.56 <3> (DL-glyceraldehyde, <3> 1 mM [5]) [5]
0.7 <2> (methylglyoxal) [3]
0.85 <3> (4-nitrobenzaldehyde, <3> 1 mM [5]) [5]
0.91 <3> (3-nitrobenzaldehyde, <3> 1 mM [5]) [5]
1.33 <3> (isopentaldehyde, <3> 1 mM [5]) [5]
1.43 <2> (methylglyoxal, <2> enzyme MGR I [3]) [3]
1.69 <3> (benzaldehyde, <3> 1 mM [5]) [5]
4.35 <2> (phenylglyoxal, <2> enzyme MGR II [3]) [3]
10 <2> (glyoxal, <2> enzyme MGR II [3]) [3]
15.4 <2> (methylglyoxal, <2> enzyme MGR I [3]) [3]
104 <3> (D-xylose, <3> 50 mM [5]) [5]
160 <3> (D-galactose, <3> 50 mM [5]) [5]

234 <3> (L-arabinose, <3> 50 mM [5]) [5]
445 <3> (D-glucose, <3> 50 mM [5]) [5]

pH-Optimum

6 <3> [5]
6.5 <2> (<2> enzyme MGR I [3]) [3]
9 <2> (<2> enzyme MGR II [3]) [3]

4 Enzyme Structure

Molecular weight

37000 <2> (<2> enzyme MGR I, similar enzyme with NADPH-requirement for methylglyoxal reduction, that is inactive with NAD^+, NADH and $NADP^+$, gel filtration [3]) [3]
38000 <2> (<2> enzyme MGR II, similar enzyme with NADPH-requirement for methylglyoxal reduction, that is inactive with NAD^+, NADH and $NADP^+$, gel filtration [3]) [3]

Subunits

monomer <2> (<2> 1 * 36000, enzyme MGR I, 1 * 38000, enzyme MGR II, similar enzyme with NADPH-requirement for methylglyoxal reduction, that is inactive with NAD^+, NADH and $NADP^+$, SDS-PAGE [3]) [3]

5 Isolation/Preparation/Mutation/Application

Purification

<1> (partial) [4]
<2> (two methylglyoxylate reductases: MGR I and MGR II, similar enzyme with NADPH-requirement for methylglyoxal reduction, that is inactive with NAD^+, NADH and $NADP^+$) [3]

Cloning

<1> (GRE2) [4]
<1> (similar enzyme with NADPH-requirement for methylglyoxal reduction, that is inactive with NAD^+, NADH and $NADP^+$) [1]

6 Stability

pH-Stability

5-7.5 <3> (<3> enzymic activity declined rapidly when pH values were higher than 7.5 or lower than 5 [5]) [5]

Storage stability

<2>, 4°C [3]

References

[1] Murata, K.; Fukuda, Y.; Shimosaka, M.; Watanabe, K.; Saikusa, T.; Kimura, A.: Phenotypic character of the methylglyoxal resistance gene in Saccharomyces cerevisiae: expression in Escherichia coli and application to breeding wild-type yeast strains. Appl. Environ. Microbiol., **50**, 1200-1207 (1985)
[2] Murata, K.; Inoue, Y.; Saikusa, T.; Watanabe, K.; Fukuda, Y.; Shimosaka, M.; Kimura, A.: Metabolism of α-ketoglutarate in yeasts: inducible formation of methylglyoxal reductase and its relation to growth arrest of Saccharomyces cerevisiae. J. Ferment. Technol., **64**, 1-4 (1986)
[3] Inoue, Y.; Rhee, H.; Watanabe, K.; Murata, K.; Kimura, A.: Metabolism of 2-oxoaldehyde in mold. Purification and characterization of two methylglyoxal reductases from Aspergillus niger. Eur. J. Biochem., **171**, 213-218 (1988)
[4] Chen, C.N.; Porubleva, L.; Shearer, G.; Svrakic, M.; Holden, L.G.; Dover, J.L.; Johnston, M.; Chitnis, P.R.; Kohl, D.H.: Associating protein activities with their genes: rapid identification of a gene encoding a methylglyoxal reductase in the yeast Saccharomyces cerevisiae. Yeast, **20**, 545-554 (2003)
[5] Xu, D.; Liu, X.; Guo, C.; Zhao, J.: Methylglyoxal detoxification by an aldo-keto reductase in the cyanobacterium Synechococcus sp. PCC 7002. Microbiology, **152**, 2013-2021 (2006)

S-(Hydroxymethyl)glutathione dehydrogenase 1.1.1.284

1 Nomenclature

EC number
1.1.1.284

Systematic name
S-(hydroxymethyl)glutathione:NAD^+ oxidoreductase

Recommended name
S-(hydroxymethyl)glutathione dehydrogenase

Synonyms
ADH3 <7, 36> [40, 50]
alcohol dehydrogenase SFA
EC 1.2.1.1 <7> (<7> formerly [40]) [40]
FALDH <15, 36> [46, 50, 51, 52]
FDH <7> (<7> incorrect [40]) [40, 49]
FLD <22> [45]
GD-FAlDH <7> [40]
GS-FDH <7> (<7> incorrect [40]) [40]
GSNO reductase <15> [51]
glutathione-dependent formaldehyde dehydrogenase <7, 15, 22, 36> (<7> incorrect [40]) [40, 41, 45, 46, 49, 50, 51]
NAD- and glutathione-dependent formaldehyde dehydrogenase <7> [40]
NAD-dependent formaldehyde dehydrogenase <7> [40]
PmFLD1 <44> [48]
S-nitrosoglutathione reductase <15> [43]
alcohol dehydrogenase class III <36> [50]
c-ADH <7> [40]
class III alcohol dehydrogenase <7, 15, 25> [40, 44, 52]
dehydrogenase, formaldehyde
formaldehyde dehydrogenase <7> (<7> incorrect [40]) [40]
formaldehyde dehydrogenase (glutathione) <7> (<7> incorrect [40]) [40]
formic dehydrogenase <7> (<7> incorrect [40]) [40]
glutathione (GSH)-dependent formaldehyde dehydrogenase <15> [52]

CAS registry number
9028-84-6

2 Source Organism

<1> *Gallus gallus* (no sequence specified) [25]
<2> *Drosophila melanogaster* (no sequence specified) [21, 36]
<3> *Paracoccus denitrificans* (no sequence specified) [28]
<4> *Haemophilus influenzae* (no sequence specified) [39]
<5> *Mus musculus* (no sequence specified) [5,29]
<6> *Escherichia coli* (no sequence specified) [18,21,39]
<7> *Homo sapiens* (no sequence specified) [6, 8, 10, 13, 14, 16, 21, 29, 34, 40, 41, 42, 49]
<8> *Rattus norvegicus* (no sequence specified) [10,21,29,38]
<9> *Saccharomyces cerevisiae* (no sequence specified) [9,21,35]
<10> *Bos taurus* (no sequence specified) [25, 27, 29, 31]
<11> *Oryctolagus cuniculus* (no sequence specified) [29, 37]
<12> *Ovis aries* (no sequence specified) [29]
<13> *Pisum sativum* (no sequence specified) [2, 21]
<14> *Zea mays* (no sequence specified) [32]
<15> *Arabidopsis thaliana* (no sequence specified) [21, 43, 46, 51, 52]
<16> *Pseudomonas aeruginosa* (no sequence specified) [18]
<17> *Pichia pastoris* (no sequence specified) [4, 7, 11]
<18> *Pseudomonas putida* (no sequence specified) [26]
<19> *Serratia marcescens* (no sequence specified) [18]
<20> *Canis familiaris* (no sequence specified) [32]
<21> *Corynebacterium sp.* (no sequence specified) [24]
<22> *Candida boidinii* (no sequence specified) [4, 7, 12, 17, 20, 28, 30, 45]
<23> *Hansenula polymorpha* (no sequence specified) [7, 15, 22, 28]
<24> *Klebsiella pneumoniae* (no sequence specified) [18]
<25> *Agaricus bisporus* (no sequence specified) [44]
<26> *Enterobacter cloacae* (no sequence specified) [18]
<27> *Citrobacter freundii* (no sequence specified) [18]
<28> *Rhodobacter sphaeroides* (no sequence specified) [33]
<29> *Pichia sp.* (no sequence specified) [7]
<30> *Methylobacillus glycogenes* (no sequence specified) [1]
<31> *Protaminobacter candidus* (no sequence specified) [1]
<32> *Methylophilus methanolovorus* (no sequence specified) [1]
<33> *Kloeckera sp.* (no sequence specified) [4]
<34> *Debaryomyces vanriji* (no sequence specified) [3, 26]
<35> *Trichosporon penicillatum* (no sequence specified) [3]
<36> *Branchiostoma floridae* (no sequence specified) [50]
<37> *Thiobacillus acidophilus* (no sequence specified) [23]
<38> *Methylobacterium organophilum* (no sequence specified) [19]
<39> *Methylomonas clara* (no sequence specified) [19]
<40> *Methylomonas albus* (no sequence specified) [19]
<41> *Methylobacillus sp.* (no sequence specified) [1]
<42> *Thiobacillus versutus* (no sequence specified) [28]
<43> *Hansenula polymorpha* (UNIPROT accession number: Q96V39) [47]
<44> *Pichia methanolica* (UNIPROT accession number: Q768R6) [48]

3 Reaction and Specificity

Catalyzed reaction

S-(hydroxymethyl)glutathione + NAD(P)$^+$ = S-formylglutathione + NAD(P)H + H$^+$ (<7> random sequential mechanism [16]; <22> ordered bi-bi mechanism [17]; <7> random bi-bi mechanism for oxidation of S-hydroxymethylglutathione and 12-hydroxydodecanoic acid [34]; <22> bi bi mechanism [7])

Reaction type

oxidation
redox reaction
reduction

Natural substrates and products

S S-(hydroxymethyl)glutathione + NAD$^+$ <7, 15> (<15> essential role in formaldehyde detoxifcation [46]; <7> the enzyme plays an important role in the formaldehyde detoxification and reduction of the nitric oxide metabolite [42]) (Reversibility: ?) [42, 46]

P S-formylglutathione + NADH

S S-nitrosoglutathione + NADH <15> (<15> the enzyme provides a defense mechanism against nitrosative stress, enzymatic pathway that modulates the bioactivity and toxicity of NO [43]) (Reversibility: ?) [43]

P S-amino-L-glutathione + NAD$^+$

S formaldehyde + glutathione + NAD$^+$ <3, 4, 6, 9, 13, 15, 16, 19, 21, 22, 23, 24, 26, 27, 34, 35, 38, 39, 40, 42> (<22> inducible enzyme [12]; <4,6> inducible enzyme, at least one function of the enzyme in Gram-negative bacteria is to detoxify exogenous formaldehyde encountered in their environment [39]; <22> enzyme is found only in methanol-grown cells and is absent in cells grown on ethanol or glucose as carbon source [30]; <9,13,15> main enzymatic system responsible for the formaldehyde elimination [21]; <6,16,19,24,26,27> enzymatic degradation of formaldehyde seems to play an important role in resistance against formaldehyde [18]; <3,22,23,42> principal enzyme for biological formaldehyde oxidation [28]; <21> enzyme is involved in methanol metabolism [24]; <22> key enzyme of the dissimilatory pathway of the methanol metabolism [12,17]; <22> the synthesis of the enzyme is induced by methanol and repressed by glucose [20]; <23> enzyme of methanol dissimilation. When cells are grown on glucose, the enzyme is not detected during the exponential growth, but is formed in the late stationary phase without addition of methanol. Enzyme is synthesized during growth on sorbitol, glycerol, ribose and xylose [22]; <22> key step of the methanol catabolism in yeast [7]; <34,35> resistance to formaldehyde is attributed to detoxification by oxidation [3]; <38,39,40> enzyme of the formaldehyde oxidation pathway via the linear sequence [19]; <9> enzyme is not essential but enhances the resistance against formaldehyde [35]) (Reversibility: r) [3, 7, 12, 17, 18, 19, 20, 21, 22, 24, 28, 30, 35, 39]

P S-formylglutathione + NADH

S Additional information <7, 14, 22, 43, 44> (<14> the enzyme may be involved in the detoxification of long-chain lipid peroxidation products [32]; <44> FLD1 is involved in the detoxification of formaldehyde in methanol metabolism, and Fld1p coordinates the formaldehyde level in methanol-grown cells according to the methanol concentration on growth. FLD activity is mainly induced by methanol, and this induction is not completely repressed by glucose [48]; <43> key enzyme required for the catabolism of methanol as a carbon source and certain primary amines, such as methylamine as nitrogen sources in methylotrophic yeasts. The expression of FLDH1 is strictly regulated and can be controlled at two expression levels by manipulation of the growth conditions. The gene is strongly induced under methylotrophic growth conditions. Moderate expression is obtained under conditions in which a primary amine, e.g. methylamine is used as nitrogen source [47]; <22> the enzyme is essential for growth on methanol [45]; <7> the enzyme plays an important role in the metabolism of glutathione adducts such as S-(hydroxymethyl)glutathione and S-nitrosoglutathione [41]) (Reversibility: ?) [32, 41, 45, 47, 48]

P ?

Substrates and products

S 10-hydroxydodecanoic acid + glutathione + NAD^+ <14> (Reversibility: ?) [32]

P ?

S 12-hydroxydodecanoic acid + NAD^+ <7> (Reversibility: ?) [49]

P ? + NADH

S 12-hydroxydodecanoic acid + glutathione + NAD^+ <7> (Reversibility: ?) [40]

P S-(11-carboxy)undecanyl-glutathione + NADH + H^+

S 12-hydroxydodecanoic acid + glutathione + NAD^+ <15> (Reversibility: ?) [46]

P S-(11-carboxy)undecanylglutathione + NADH

S 12-hydroxydodedecanoic acid + glutathione + NAD^+ <7, 14, 28> (Reversibility: ?) [32, 33, 34]

P 12-oxododecanoic acid + ? <7> [34]

S 3-ethoxy-2-hydroxybutyraldehyde + glutathione + NAD^+ <7> (<7> weak activity [13]) (Reversibility: ?) [13, 29]

P S-(3-ethoxy-2-hydroxybutyryl)-glutathione

S S-(hydroxymethyl)glutathione + NAD^+ <7, 15, 25> (<15> essential role in formaldehyde detoxifcation [46]; <7> the enzyme plays an important role in the formaldehyde detoxification and reduction of the nitric oxide metabolite [42]) (Reversibility: ?) [42, 43, 44, 46]

P S-formylglutathione + NADH

S S-formylglutathione + NADH <7, 8, 9, 17, 22> (Reversibility: r) [7, 9, 11, 12, 13, 14, 17, 29]

P formaldehyde + glutathione + NAD^+ <7, 8, 9, 17, 22> [7, 9, 11, 12, 13, 14, 17, 29]

S S-formylglutathione + NADPH <7> (Reversibility: r) [13, 29]

P formaldehyde + glutathione + NADP$^+$ <7> [13, 29]
S S-hydroxymethylglutathione + NAD$^+$ <7, 36> (Reversibility: ?) [49, 5]
P S-formylglutathione + NADH
S S-nitrosoglutathione + NAD$^+$ <7, 15, 36> (Reversibility: ?) [49, 50, 51]
P S-amino-L-glutathione + NADH
S S-nitrosoglutathione + NADH <15> (<15> the enzyme provides a defense mechanism against nitrosative stress, enzymatic pathway that modulates the bioactivity and toxicity of NO [43]) (Reversibility: ?) [43]
P S-amino-L-glutathione + NAD$^+$
S S-pyruvylglutathione + NADH <7> (Reversibility: ?) [13]
P methylglyoxal + glutathione + NAD$^+$ <7> [13]
S butanol + glutathione + NAD$^+$ <28> (Reversibility: ?) [33]
P ?
S cinnamyl alcohol + glutathione + NAD$^+$ <15> (Reversibility: ?) [46]
P S-cinnamylglutathione + NADH
S decanol + glutathione + NAD$^+$ <28> (Reversibility: ?) [33]
P ?
S farnesol + glutathione + NAD$^+$ <15> (Reversibility: ?) [46]
P S-farnesylglutathione + NADH
S formaldehyde + 6-mercaptohexanoate + NAD$^+$ <8> (<8> 30% of the activity with glutathione [38]) (Reversibility: ?) [38]
P S-formyl-6-mercaptohexenoate + NADH
S formaldehyde + 8-mercaptooctanoate + NAD$^+$ <8> (<8> 35% of the activity with glutathione [38]) (Reversibility: ?) [38]
P S-formyl-8-mercaptooctanoate + NADH
S formaldehyde + S-hydroxymethyl glutathione + NAD$^+$ <2, 6, 7, 8, 9, 10, 13> (Reversibility: ?) [21, 27, 31, 35]
P ?
S formaldehyde + captopril + NAD$^+$ <8> (<8> 8% of the activity with glutathione [38]) (Reversibility: ?) [38]
P S-formylcaptopril + NADH
S formaldehyde + glutathione + 3-acetylpyridine-adenine dinucleotide <7> (Reversibility: ?) [16]
P S-formylglutathione + ?
S formaldehyde + glutathione + NAD$^+$ <1, 2, 3, 4, 5, 6, 7, 8, 9, 10, 11, 12, 13, 14, 15, 16, 17, 18, 19, 20, 21, 22, 23, 24, 26, 27, 28, 29, 30, 31, 32, 33, 34, 35, 37, 38, 39, 40, 41, 42> (<7> at pH 8 the rate of the reverse reaction with S-formylglutathione and NADH is about the same as that of the forward reaction with formaldehyde, glutathione and NAD$^+$. At pH 5.7 the rate of the reverse reaction with S-formylglutathione and NADH is 3.9times and with S-formylglutathione and NADPH 2.0times, that of the forward reaction rate with NAD$^+$ at pH 8.0 [13]; <7, 8, 9, 10, 13, 14, 20, 22, 28> the true substrate is a hemimercaptal, S-hydroxymethylglutathione, spontaneously formed from formaldehyde and glutathione [7, 17, 21, 31, 32, 33, 34, 35, 38]; <22> inducible enzyme [12]; <4,6> inducible enzyme, at least one function of the enzyme in Gram-negative bacteria is to detoxify exogenous formaldehyde encountered in their environment [39]; <22> enzyme is

found only in methanol-grown cells and is absent in cells grown on ethanol or glucose as carbon source [30]; <9,13,15> main enzymatic system responsible for the formaldehyde elimination [21]; <6,16,19,24,26,27> enzymatic degradation of formaldehyde seems to play an important role in resistance against formaldehyde [18]; <3,22,23,42> principal enzyme for biological formaldehyde oxidation [28]; <21> enzyme is involved in methanol metabolism [24]; <22> key enzyme of the dissimilatory pathway of the methanol metabolism [12,17]; <22> the synthesis of the enzyme is induced by methanol and repressed by glucose [20]; <23> enzyme of methanol dissimilation. When cells are grown on glucose, the enzyme is not detected during the exponential growth, but is formed in the late stationary phase without addition of methanol. Enzyme is synthesized during growth on sorbitol, glycerol, ribose and xylose [22]; <22> key step of the methanol catabolism in yeast [7]; <34,35> resistance to formaldehyde is attributed to detoxification by oxidation [3]; <38,39,40> enzyme of the formaldehyde oxidation pathway via the linear sequence [19]; <9> enzyme is not essential but enhances the resistance against formaldehyde [35]) (Reversibility: r) [1, 2, 3, 4, 5, 6, 7, 8, 9, 10, 11, 12, 13, 14, 15, 16, 17, 18, 19, 20, 21, 22, 23, 24, 25, 26, 27, 28, 29, 30, 31, 32, 33, 34, 35, 36, 37, 38, 39, 52]

P S-formylglutathione + NADH <1, 2, 3, 4, 5, 6, 7, 8, 9, 10, 11, 12, 13, 14, 15, 16, 17, 18, 19, 20, 21, 22, 23, 24, 26, 27, 28, 29, 30, 31, 32, 33, 34, 35, 37, 38, 39, 40, 41, 42> [1, 2, 3, 4, 5, 6, 7, 8, 9, 10, 11, 12, 13, 14, 15, 16, 17, 18, 19, 20, 21, 22, 23, 24, 25, 26, 27, 28, 29, 30, 31, 32, 33, 34, 35, 36, 37, 38, 39]

S formaldehyde + glutathione + $NADP^+$ <7, 8> (Reversibility: r) [13, 29, 38]

P S-formylglutathione + NADPH <7, 8> [13, 29, 38]

S formaldehyde + glutathione + nicotinamide-hypoxanthine dinucleotide <7> (Reversibility: ?) [16]

P S-formylglutathione + ?

S formaldehyde + glutathione + thio-NAD^+ <7> (Reversibility: ?) [16]

P S-formylglutathione + thio-NADH

S formaldehyde + glutathione monomethyl ester + NAD^+ <8> (<8> 70% of the activity with glutathione [38]) (Reversibility: ?) [38]

P S-formylglutathione monomethyl ester + NADH

S geraniol + glutathione + NAD^+ <15> (Reversibility: ?) [46]

P S-geranylglutathione + NADH

S glyoxal + glutathione + NAD^+ <7, 9, 13> (<7> weak activity [13]; <13> 35% of the activity with formaldehyde [2]) (Reversibility: ?) [2, 9, 13, 29]

P S-oxoacetylglutathione + NADH

S hydroxypyruvaldehyde + glutathione + NAD^+ <7> (<7> weak activity [13]) (Reversibility: ?) [13, 29]

P S-hydroxypyruvylglutathione + NADH

S methylglyoxal + glutathione + NAD^+ <7, 9, 10, 13, 17, 18, 22, 34> (<13> 90% of the activity with formaldehyde [2]; <22> 89% of the activity with formaldehyde [7]; <7> 85% of the activity with formaldehyde [13]) (Reversibility: ?) [2, 7, 9, 11, 12, 13, 26, 29, 30, 31]

P S-pyruvylglutathione + NADH

S octanol + glutathione + NAD^+ <14, 28> (<14> octanol-1 [32]) (Reversibility: ?) [32, 33]
P ?
S pentanol + glutathione + NAD^+ <14> (Reversibility: ?) [32]
P ?
S Additional information <7, 14, 22, 43, 44> (<14> the enzyme may be involved in the detoxification of long-chain lipid peroxidation products [32]; <44> FLD1 is involved in the detoxification of formaldehyde in methanol metabolism, and Fld1p coordinates the formaldehyde level in methanol-grown cells according to the methanol concentration on growth. FLD activity is mainly induced by methanol, and this induction is not completely repressed by glucose [48]; <43> key enzyme required for the catabolism of methanol as a carbon source and certain primary amines, such as methylamine as nitrogen sources in methylotrophic yeasts. The expression of FLDH1 is strictly regulated and can be controlled at two expression levels by manipulation of the growth conditions. The gene is strongly induced under methylotrophic growth conditions. Moderate expression is obtained under conditions in which a primary amine, e.g. methylamine is used as nitrogen source [47]; <22> the enzyme is essential for growth on methanol [45]; <7> the enzyme plays an important role in the metabolism of glutathione adducts such as S-(hydroxymethyl)glutathione and S-nitrosoglutathione [41]) (Reversibility: ?) [32, 41, 45, 47, 48]
P ?

Inhibitors

1,10-Phenanthroline <8, 17> (<8> no effect [10]; <17> 1 mM, 30% inhibition [11]) [10, 11]
12-oxododecanoic acid <7, 8> (<7> substrate inhibition [34]; <8> linear competitive [38]) [34, 38]
2,2'-Dipyridyl <17> (<17> 1 mM, 20% inhibition [11]) [11]
ADP <22> (<22> competitive with respect to NAD^+, noncompetitive with respect to S-hydroxymethylglutathione [17]) [7, 12, 17]
ADP-ribose <7> (<7> competitive with respect to β-NAD^+ and non-competitive with glutathione [16]) [13, 16]
AMP <22> (<22> competitive with respect to NAD^+, noncompetitive with respect to S-hydroxymethylglutathione [17]) [7, 12, 17]
ATP <22> (<22> competitive with respect to NAD^+, noncompetitive with respect to S-hydroxymethylglutathione [17]) [7, 12, 17]
Ag^+ <22> [7]
$AgNO_3$ <7> (<7> NADH protects more effectively than NAD^+ [29]) [29]
α-NAD^+ <7> (<7> competitive with respect to β-NAD^+ and non-competitive with glutathione [16]) [13, 16]
CN^- <17, 22> (<22> 1 mM, 42% inhibition [30]; <17> 1 mM, 48% inhibition [4]) [4, 30]
Cd^{2+} <17, 22> (<17,22> 1 mM, complete inhibition [4,11,30]) [4, 7, 11, 12, 30]
Citrate <17> (<17> 1 mM, 24% inhibition [4]) [4]

Cu^{2+} <17, 22> (<22> 1 mM, complete inhibition [30]; <17> 1 mM, 76% inhibition [4]) [4, 7, 11, 12, 30]
dodecanoic acid <7> [40, 49]
EDTA <17> (<17> 1 mM, 12% inhibition [4]) [4]
folic acid <7, 8> (<7> no effect [10]) [10]
formaldehyde <7, 9> [9, 29]
Hg^{2+} <7, 17, 22> (<17,22> 1 mM, complete inhibition [11,30]; <17> 1 mM, 90% inhibition [4]; <7> NADH protects more effectively than NAD^+ [29]) [4, 7, 11, 12, 13, 29, 30]
iodoacetamide <17> (<17> 1 mM, 20% inhibition [11]) [11]
iodoacetate <7, 22> (<22> 1 mM, 15% inhibition [30]; <7> NADH protects more effectively than NAD^+ [29]) [29, 30]
methylglyoxal <7> [29]
NAD^+ <7> (<7> product inhibition [34]; <7> competitive, nonlinear inhibitor, when the concentration of NADH varies at constant S-formylglutathione concentration, noncompetitive inhibitor when the concentration of S-formylglutathione is varied at constant NADH concentration [13]; <7> inhibits reverse reaction with NAD^+ or $NADP^+$ [29]) [13, 29, 34]
NADH <7, 17, 22> (<7,17> product inhibition [4,34]; <22> competitive inhibition, product inhibition [7]; <7> competitive with NAD^+ [14]; <7> inhibits reaction with formaldehyde, glutathione and NAD^+ [29]) [4, 7, 14, 17, 29, 34]
NADPH <7> (<7> inhibits reaction with formaldehyde, glutathione and NAD^+ [29]) [29]
NEM <17> (<17> 1 mM, 35% inhibition [11]) [11]
NO_2^- <17> (<17> 1 mM, 14% inhibition [4]) [4]
PCMB <17, 22> (<17,22> 1 mM, complete inhibition [11,30]) [7, 11, 12, 30]
S-formylglutathione <7, 8, 22> (<7> inhibits reaction with formaldehyde, glutathione and NAD^+ [29]; <8> product inhibition with S-hydroxymethylglutathione [29,38]) [7, 17, 29, 38]
S-hydroxymethylglutathione <22> [17]
Zn^{2+} <17, 22> (<17> 1 mM, 40% inhibition [11]; <22> 1 mM, 50% inhibition [30]; <17> 1 mM, 28% inhibition [4]) [4, 11, 30]
imidazole <17> (<17> 1 mM, 32% inhibition [11]) [11]
p-hydroxymercuribenzoate <7> [13]
pyrazole <14, 20> [32]

Cofactors/prosthetic groups

3-acetylpyridine adenine dinucleotide <7> (<7> can substitute for β-NAD^+ [16]) [16]
NAD^+ <1, 2, 3, 4, 5, 6, 7, 8, 9, 10, 11, 12, 13, 14, 15, 16, 17, 18, 19, 20, 21, 22, 23, 24, 25, 26, 27, 28, 29, 30, 31, 32, 33, 34, 35, 36, 37, 38, 39, 40, 41, 42> (<7> His47 may function both as a catalytic base for proton transfer and in the binding of the adenosine phosphate of NAD(H) [6]) [1, 2, 3, 4, 5, 6, 7, 8, 9, 10, 11, 12, 13, 14, 15, 16, 17, 18, 19, 20, 21, 22, 23, 24, 25, 26, 27, 28, 29, 30, 31, 32, 33, 34, 35, 36, 37, 38, 39, 40, 42, 43, 44, 46, 49, 50, 51, 52]
NADH <7, 9, 17, 22> [7,9,11,12,13,14,17,29]

NADP$^+$ <7, 8, 9, 17, 22, 37> (<7, 8, 9, 17, 22, 37> no activity with NADP$^+$ [7, 9, 10, 11, 12, 23]; <7> at pH 8 the K_m-value is about 1000times greater than that for NAD$^+$. At pH 6 NADP$^+$ is used better but this reaction appears to be without physiological significance [13]) [7,9,10,11,12,13,23]
NADPH <7> [13,29]
nicotinamide hypoxanthine dinucleotide <7> (<7> can substitute for β-NAD$^+$ [16]) [16]
thio-NAD$^+$ <7> (<7> can substitute for β-NAD$^+$ [16]) [16]

Activating compounds

pyrazole <7, 8> (<8> stimulates [10]; <7> no effect [10]) [10]

Metals, ions

Zinc <7> (<7> contains a catalytic zinc ion [6]; <7> zinc-dependent enzyme [34]; <7> contains zinc at the active site [42]; <7> in the ternary complex with S-(hydroxymethyl)glutathione and the reduced coenzyme the active site zinc has a tetrahedral coordination environment with Cys44, His66 and Cys173 as the protein ligands in addition to S-(hydroxymethy)glutathione [41]; <7> mettaloenzyme [40]) [6, 34, 40, 41, 42]
Additional information <7, 8> (<7> contains metal [10]; <8> does not contain metal [10]) [10]

Turnover number (min^{-1})

0.1 <15> (farnesol, <15> pH 7.5, 25°C [46]) [46]
0.217 <7> (12-hydroxydodecanoic acid) [34]
0.217 <7> (NAD$^+$, <7> reaction with 12-hydroxydodecanoic acid [34]) [34]
0.233 <17> (glutathione) [4]
0.267 <17> (formaldehyde) [4]
0.92 <7> (NAD$^+$, <7> mutant E67L, pH 7.5 [49]) [49]
1.04 <7> (NAD$^+$) [34]
1.04 <7> (S-hydroxymethyl glutathione) [34]
1.05 <10> (S-hydroxymethyl glutathione, <10> enzyme form F6.0 [27]) [27]
1.13 <10> (S-hydroxymethyl glutathione, <10> enzyme form F6.2 [27]) [27]
1.4 <7> (S-hydroxymethylglutathione, <7> mutant E67L, pH 7.5, 30°C [49]) [49]
1.42 <10> (S-hydroxymethyl glutathione, <10> enzyme form F6.5 [27]) [27, 31]
2.1 <15> (farnesol, <15> pH 10, 25°C [46]) [46]
2.5 <7> (NAD$^+$, <7> wild type enzyme, pH 7.5 [49]) [49]
2.5 <7, 36> (S-hydroxymethylglutathione, <7> wild type enzyme, pH 7.5, 30°C [49]) [49, 50]
3.33 <7> (S-hydroxymethyl glutathione) [21]
3.6 <8> (S-hydroxymethyl glutathione) [21]
4.75 <9> (S-hydroxymethyl glutathione, <9> mutant D267E/T269I [35]) [35]
5.4 <15> (cinnamyl alcohol, <15> pH 7.5, 25°C [46]) [46]
5.6 <15> (12-hydroxydodecanoic acid, <15> pH 10, 25°C [46]) [46]
6.33 <13> (S-hydroxymethyl glutathione) [21]
15.3 <7> (S-nitrosoglutathione, <7> mutant R368L, pH 7.5, 25°C [49]) [49]

16 <2> (S-hydroxymethyl glutathione) [21]
16.7 <7> (NAD^+, <7> mutant R368L, pH 7.5 [49]) [49]
16.7 <36> (S-nitrosoglutathione) [50]
20 <15> (geraniol, <15> pH 10, 25°C [46]) [46]
20 <7> (S-hydroxymethylglutathione, <7> mutant R368L, pH 7.5, 30°C [49]) [49]
20.33 <15> (cinnamyl alcohol, <15> pH 10, 25°C [46]) [46]
22.5 <15> (S-(hydroxymethyl)glutathione, <15> pH 8, 25°C [46]) [46]
29.3 <7> (S-nitrosoglutathione, <7> wild type enzyme, pH 7.5, 25°C [49]) [49]
34.5 <7> (NADH, <7> wild type enzyme, pH 7.5 [49]) [49]
51.7 <9> (S-hydroxymethylglutathione, <9> wild-type enzyme [35]) [35]
88.3 <9> (S-hydroxymethyl glutathione) [21]
156 <6> (S-hydroxymethyl glutathione) [21]
230 <7> (NADH, <7> mutant R368L, pH 7.5 [49]) [49]
Additional information <14> [32]

Specific activity (U/mg)

0.092 <8> [10]
0.225 <7> [10]
0.64 <9> [9]
2.78 <13> [2]
3.17 <7> [13]
3.2 <7> [29]
4.68 <14> [32]
8.7 <22> [20]
10.2 <17> [4]
15 <15> [46]
38.3 <22> [7]
47.9 <22> [12, 30]
184 <34> [26]
185 <18> [26]
2090 <17> [11]
Additional information <7, 10> [8, 31]

K_m-Value (mM)

0.0002 <7> (NADH, <7> wild type enzyme, pH 7.5 [49]) [49]
0.0004 <7> (S-hydroxymethyl glutathione) [21]
0.00092 <8> (S-hydroxymethyl glutathione) [21]
0.0017 <7> (S-hydroxymethyl glutathione, <7> wild type enzyme, pH 7.5, 30°C [49]) [49]
0.002 <13> (S-hydroxymethyl glutathione) [21]
0.0025 <9> (S-hydroxymethyl glutathione, <9> mutant D267E/T269I [35]) [35]
0.003 <15> (farnesol, <15> pH 10, 25°C [46]) [46]
0.0033 <7> (NAD^+, <7> wild type enzyme [49]) [49]
0.004 <36> (S-hydroxymethyl glutathione, <36> in 0.1 M sodium diphosphate at pH 8.0 [50]) [50]

0.0042 <7> (NADH, <7> pH 5.7 [13,29]) [13, 29]
0.0048 <7> (S-nitrosoglutathione, <7> wild type enzyme, pH 7.5, 25°C [49]) [49]
0.0048 <7> (β-NAD$^+$) [16]
0.006 <8> (glutathione) [38]
0.006 <2> (S-hydroxymethyl glutathione) [21]
0.0065 <13> (NAD$^+$, <13> reaction with S-hydroxymethyl glutathione [21]) [21]
0.007 <7> (NAD$^+$) [10]
0.007 <15> (S-hydroxymethyl glutathione, <15> pH 8, 25°C [46]) [46]
0.0071 <8> (formaldehyde) [10]
0.0077 <15> (farnesol, <15> pH 7.5, 25°C [46]) [46]
0.008 <7> (glutathione, <7> pH 8 [13]) [13]
0.008 <36> (NAD$^+$, <36> in 0.1 M sodium diphosphate at pH 7.5 [50]) [50]
0.008 <10> (S-hydroxymethyl glutathione) [31]
0.0081 <7> (glutathione, <7> pH 8.0 [29]) [29]
0.0087 <7> (formaldehyde) [10]
0.009 <7> (formaldehyde, <7> pH 8 [13]) [13]
0.009 <7> (NAD$^+$, <7> pH 8.0 [29]) [29]
0.0093 <7> (NAD$^+$, <7> wild type enzyme, pH 10 [49]) [49]
0.01 <8> (NAD$^+$) [10]
0.011 <7> (NADH, <7> mutant R368L, pH 7.5 [49]) [49]
0.0136 <7> (NAD$^+$, <7> pH 5.7 [29]) [29]
0.0144 <7> (NAD$^+$, <7> mutant R368L, pH 7.5 [49]) [49]
0.016 <7> (NADH, <7> pH 8.0 [29]) [29]
0.019 <20> (formaldehyde) [32]
0.02 <14> (formaldehyde) [32]
0.021 <14> (formaldehyde) [32]
0.025 <22> (NAD$^+$) [7]
0.025 <22> (NADH) [7]
0.03 <9> (S-hydroxymethyl glutathione, <9> wild-type enzyme [35]) [35]
0.0301 <7> (thio-NAD$^+$) [16]
0.031 <14, 20> (NAD$^+$) [32]
0.0352 <7> (nicotinamide-hypoxanthine dinucleotide) [16]
0.037 <14> (glutathione) [32]
0.038 <7> (S-hydroxymethyl glutathione, <7> mutant R368L, pH 7.5, 30°C [49]) [49]
0.04 <9> (S-hydroxymethyl glutathione) [21]
0.041 <7> (12-hydroxydodecanoic acid, <7> wild type enzyme, pH 10 [49]) [49]
0.045 <9> (NAD$^+$, <9> wilde-type enzyme [35]) [35]
0.045 <36> (NADH, <36> in 0.1 M sodium diphosphate at pH 7.5 [50]) [50]
0.045 <7> (NADPH, <7> pH 5.7 [13,29]) [13, 29]
0.05 <22> (NAD$^+$) [20]
0.0509 <7> (3-acetylpyridine-adenine dinucleotide) [16]
0.054 <7> (NAD$^+$, <7> mutant E67L, pH 10 [49]) [49]

0.056 <36> (S-nitrosoglutathione, <36> in 0.1 M sodium diphosphate at pH 8.0 [50]) [50]
0.062 <20> (glutathione) [32]
0.068 <9> (NAD^+) [9]
0.09 <22> (NAD^+) [12, 30]
0.094 <6> (S-hydroxymethyl glutathione) [21]
0.116 <7> (NAD^+, <7> mutant E67L [49]) [49]
0.12 <17> (NAD^+) [11]
0.12 <22> (S-formylglutathione) [7]
0.125 <8> (glutathione monomethyl ester) [38]
0.13 <22> (glutathione) [12, 30]
0.14 <13> (NAD^+, <13> reaction with S-hydroxymethyl glutathione [21]) [21]
0.15 <18> (NAD^+) [26]
0.16 <7> (12-hydroxydodecanoic acid, <7> mutant R368L, pH 10 [49]) [49]
0.16 <7> (NAD^+, <7> mutant R368L, pH 10 [49]) [49]
0.16 <7> (S-nitrosoglutathione, <7> mutant R368L, pH 7.5, 25°C [49]) [49]
0.16 <17> (reduced glutathione) [11]
0.24 <17> (NAD^+) [4]
0.25 <22> (formaldehyde) [12, 30]
0.25 <17> (NADH) [11]
0.28 <7> (S-formylglutathione, <7> pH 5.7, with NADPH as cofactor [29]) [29]
0.29 <22> (formaldehyde) [7]
0.3 <7> (S-formylglutathione, <7> pH 5.7, with NADH as cofactor [13,29]) [13, 29]
0.31 <17> (formaldehyde) [11]
0.31 <34> (NAD^+) [26]
0.39 <18> (formaldehyde) [26]
0.43 <17> (formaldehyde) [4]
0.48 <17> (glutathione) [4]
0.57 <7> ($NADP^+$, <7> pH 5.7 [29]) [29]
0.6 <17> (S-formylglutathione) [11]
0.68 <7> (12-hydroxydodecanoic acid, <7> mutant E67L, pH 10 [49]) [49]
0.8 <15> (geraniol, <15> pH 10, 25°C [46]) [46]
1.14 <34> (formaldehyde) [26]
1.2 <22> (methylglyoxal) [12, 30]
2.65 <7> (S-hydroxymethyl glutathione, <7> mutant E67L, pH 7.5, 30°C [49]) [49]
2.8 <22> (methylglyoxal) [7]
3.5 <15> (cinnamyl alcohol, <15> pH 10, 25°C [46]) [46]
4 <7> (pyruvylglutathione, <7> pH 5.7, with NADH as cofactor [29]) [29]
4.7 <15> (12-hydroxydodecanoic acid, <15> pH 10, 25°C [46]) [46]
9.5 <7> ($NADP^+$, <7> pH 8.0 [29]) [29]
22 <15> (cinnamyl alcohol, <15> pH 7.5, 25°C [46]) [46]
Additional information <13> [2]

K_i-Value (mM)

0.148 <7> (dodecanoic acid, <7> wild type enzyme [49]) [49]
0.161 <7> (dodecanoic acid, <7> mutant R368L [49]) [49]
0.617 <7> (dodecanoic acid, <7> mutant E67L [49]) [49]

pH-Optimum

5.7 <7> (<7> reaction with S-formylglutathione and NADH [13,29]) [13, 29]
6 <7, 13> (<13> below, reaction with S-formylglutathione + NADH [2]; <7> reaction with $NADP^+$ as cofactor [29]) [2, 29]
7.5-8.4 <8> [10]
7.5-9 <22> [30]
7.6-8.8 <7> [10]
7.9 <17> [4]
7.9-8.2 <10> [25]
8 <3, 7, 9, 17, 22, 23, 33, 34, 42> (<7,17> oxidation of formaldehyde [11,13]; <7> reaction with NAD^+ as cofactor [29]) [4, 7, 9, 11, 13, 20, 26, 28, 29]
8-9 <13> (<13> reaction with formaldehyde + NAD^+ + glutathione [2]) [2]
8-9.5 <14, 20> [32]
8.5 <17, 22> [7, 12]
9 <18> [26]

pH-Range

6-8.5 <9> (<9> active from pH 6.0 to pH 8.5 [9]) [9]
6.4-9.3 <17> (<17> 50% of maximal activity at pH 6.4 and at pH 9.3 [4]) [4]
6.5-9.5 <7> (<7> pH 6.5: about 70% of maximal activity, pH 9.5: about 80% of maximal activity [10]) [10]
6.8-9.3 <8> (<8> pH 6.8: about 80% of maximal activity, pH 9.3: about 90% of maximal activity [10]) [10]
7-9 <17> (<17> pH 7.0: about 55% of maximal activity, pH 9.0: about 90% of maximal activity [11]) [11]

Temperature optimum (°C)

30 <33> [4]
34 <22> [12]
35 <22, 34> [4, 26]
45 <14, 18> [26, 32]
47 <17> [4]
55 <20> [32]

Temperature range (°C)

20-65 <17> (<17> 20°C: about 60% of maximal activity, 65°C: about 40% of maximal activity [4]) [4]

4 Enzyme Structure

Molecular weight

72000 <18, 34> (<18,34> gel filtration [26]) [26]
75000-78000 <10> (<10> gel filtration [31]) [31]

75000-80000 <28> (<28> non-denaturing PAGE [33]) [33]
80000 <9, 22> (<22> equilibrium sedimentation [12,30]; <9> non-denaturing PAGE [35]) [12, 30, 35]
81400 <7> (<7> gel filtration [29]) [29]
82000 <14, 22> (<14,22> gel filtration [7,32]) [7, 32]
82300 <13> (<13> gel filtration [2]) [2]
84000 <17> (<17> gel filtration [4,11]) [4, 11]
90000 <7> (<7> gel filtration [10]) [10]
110000 <8> (<8> gel filtration [10]) [10]

Subunits
? <10, 25> (<25> x * 40000, SDS-PAGE [44]; <10> x * 41000, enzyme form F6.0, F6.2 and F6.5, SDS-PAGE [27]) [27, 44]
dimer <10, 13, 14, 17, 18, 22, 34> (<17> 2 * 39000, SDS-PAGE [4]; <22,34> 2 * 40000, SDS-PAGE [12,26,30]; <10,17> 2 * 41000, SDS-PAGE [11,31]; <14,18> 2 * 42000, SDS-PAGE [26,32]; <22> 2 * 43000, SDS-PAGE [7]; <13> 2 * 42100, SDS-PAGE [2]) [2, 4, 7, 11, 12, 26, 30, 31, 32]
Additional information <7> (<7> the enzyme exists in a complex with other proteins or in a polymeric form until the ultimate steps in purification [10]) [10]

5 Isolation/Preparation/Mutation/Application

Source/tissue
BY-2 cell <15> [51]
adrenal gland <10> [29]
ascites tumor cell <5> [29]
brain <7, 8, 10, 11> [29]
cardiac muscle <8> [29]
cell suspension culture <14> [32]
epidermis <15> (<15> in root [52]) [52]
gut <36> [50]
kidney <8> [29]
leaf <15> [51, 52]
liver <1, 5, 7, 8, 10, 11, 12> [5, 10, 13, 14, 16, 25, 27, 29, 31, 37, 38]
lung <20> [32]
mesophyll <15> (<15> in leaf [52]) [52]
root <15> [51, 52]
seed <13> [2]
skeletal muscle <8> [29]
vascular system <15> [52]
Additional information <36> (<36> dorsal region of the club-shaped gland [50]) [50]

Localization
cytoplasm <28> [33]
cytosol <7> [29]
microtubule <15> [51]

Purification

<7> [10, 29, 34]
<7> (DE52 column chromatography, Q-Sepharose column chromatography and Sephacryl S-100HR column chromatography) [49]
<7> (wild-type recombinant enzyme) [8]
<8> [10]
<9> [9, 35]
<10> [31]
<13> [2]
<14> [32]
<15> [46]
<17> [4, 11]
<18> [26]
<20> [32]
<22> [7, 12, 30, 45]
<22> (partial) [20]
<25> [44]
<34> [26]
<36> (Talon metal affinity resin chromatography) [50]

Crystallization

<7> (crystal structure of the enzyme in a binary complex with $NAD^+(\gamma)$) [6]
<7> (crystals of the binary complex with NAD(H) are grown by a sitting drop vapor diffusion method, where the apoenzyme is equilibrated with a 2 mM NAD^+ containing crystallization buffer (0.1 M potassium phosphate, pH 7.1, 12-15% PEG 8000, 2 mM DTT, 0.05 mM $ZnSO_4$)) [42]
<7> (sitting-drop vapor diffusion at 4°C. A 15-20 mg/ml enzyme solution equilibrated with a mother liquor containing 0.1 M phosphate buffer, pH 7.1, 10 mM $ZnSO_4$, 1 mM DTT, and 12-16% PEG 8000. Ternary complex with S-(hydroxymethyl)glutathione and the reduced coenzyme to 2.6 A resolution) [41]
<7> (sitting-drop vapor-diffusion method at 4°C from a 15-20 mg/ml enzyme solution equilibrated with 0.1 M potassium phosphate buffer pH 6.9-7.1, 0.1 mM $ZnSO_4$, 1 mM dithiothreitol, 12-15% PEG 8000. Binary complex with substrate 12-hydroxydodecanoic acid and a ternary complex with NAD^+ and the inhibitor dodecanoic acid are determined at 2.0 and 2.3 A resolution by X-ray crystallography using the anomalous diffraction signal of zinc) [40]

Cloning

<2> (characterization of the DNA sequence of the gfd/odh genomic region containing the gene which encodes glutathione-dependent formaldehyde dehydrogenase) [36]
<7> [8]
<7> (expressed in Escherichia coli BL21 cells) [49]
<11> [37]
<15> (overexpression in Escherichia coli) [43]
<15> (overexpression in Saccharomyces cerevisiae) [46]
<22> (expression in Escherichia coli) [45]

<36> (expressed in Escherichia coli BL21 (DE3) cells) [50]
<43> [47]
<44> [48]

Engineering

D267E <9> (<9> kinetic properties identical to that of wild-type enzyme [35]) [35]
D267E/T269I <9> (<9> K_m-value for S-hydroxymethylglutathione is about 11fold lower than that of the wild-type enzyme, turnover-number for S-hydroxymethylglutathione is about 8fold lower than that of wild-type enzyme [35]) [35]
D269I <9> (<9> highly unstable enzyme, kinetic properties identical to that of wild-type enzyme [35]) [35]
D57L <7> (<7> considerable loss of formaldehyde dehydrogenase activity [8]) [8]
D57L/Y93F <7> (<7> fall in ratio k_{cat}/K_m for hydroxymethylglutathione by a factor of 1250, alcohol dehydrogenase activity of the mutant has gained a characteristic class I property, complete inhibition by 4-methylpyrazole at concentrations only partially reducing the activity of the wild-type class III enzyme [8]) [8]
E67L <7> (<7> mutant in which Glu67 is substituted with Leu [49]) [49]
R368L <7> (<7> mutant in which Arg368 is substituted with Leu [49]) [49]
T48A <7> (<7> enzyme has essentially no alcohol dehydrogenase activity but has some glutathione-dependent formaldehyde dehydrogenase activity [8]) [8]
Y93F <7> (<7> decreased turnover number for substrates in general, inhibition of alcohol dehydrogenase activity by 4-methylpyrazole, which is not found in the wild-type enzyme [8]) [8]

Application

analysis <23> (<23> direct enzymatic assay of formaldehyde dehydrogenase permits screening of yeast colonies [15]) [15]

6 Stability

pH-Stability

6-8.5 <18> (<18> 4°C, 3 d, stable [26]) [26]
6-10 <22> (<22> 20°C, 1 h, stable [7]) [7]
6.1 <17> (<17> 37°C, 10 min, 50% loss of activity [4]) [4]
6.5-8.5 <34> (<34> 4°C, 3 d, stable [26]) [26]
7 <17> (<17> 37°C, 10 min, about 20% loss of activity [4]) [4]
8.5 <17> (<17> 37°C, 10 min, about 30% loss of activity [4]) [4]
9.4 <17> (<17> 37°C, 10 min, about 50% loss of activity [4]) [4]

Temperature stability

30 <34> (<34> 10 min, stable [26]) [26]
35 <33> (<33> 10 min, 50% loss of activity [4]) [4]

37 <7> (<7> 15 min, without NADH, 40% loss of activity [29]) [29]
40 <18> (<18> 10 min, stable [26]) [26]
52 <17, 22> (<17,22> 10 min, 50% loss of activity [4]) [4]
64 <7> (<7> 15 min, in presence of NADH, 40% loss of activity [29]) [29]
68 <7> (<7> 15 min, in presence of NADH, 80% loss of activity [29]) [29]
72 <10> (<10> 5 min, complete inactivation [25]) [25]

General stability information

<7>, NAD^+ and NADH protect from denaturation by high temperatures, acid pH and storage at 0°C [13]
<7>, one freezing at -20°C results in full inactivation [29]
<7>, overnight dialysis to remove mercaptoethanol destroys 60% of its original activity [29]
<13>, enzyme is unstable in purified state [2]
<13>, freezing almost completely destroys activity [2]
<22>, inactivated by freezing and thawing [30]

Storage stability

<7>, -70°C, in presence of 2 mM mercaptoethanol, 50% loss of activity [29]
<7>, 0-5°C, when EDTA and dithiothreitol are used throughout the purification procedure, stable for at least 3 weeks [10]
<7>, 0°C, 50% loss of activity after 1 month [29]
<7>, 0°C, half-life in absence of a stabilizer is a few h, in presence of 1 mM NAD^+ or 40% v/v glycerol the half life is about 1 month, 0.5 mM NADH is even more effective [13]
<13>, 0°C, half-life is about 10 days [2]
<17>, 4°C, in presence of glycerol, 70-80% recovery of activity after 7 days [4]
<22>, -20°C, in presence of glycerol, stable for at least 6 months [7]
<22>, 50% glycerol, stable for a long time [30]
<18, 34>, 5°C, 10 mM potassium phosphate buffer, pH 7.5, 0.1% 2-mercaptoethanol, stable for at least 1 month [26]

References

[1] Trotsenko, Y.A.; Doronina, N.V.; Govorukhina, N.I.: Metabolism of non-motile obligatory methylotrophic bacteria. FEMS Microbiol. Lett., **33**, 293-297 (1986)

[2] Uotila, L.; Koivusalo, M.: Purification of formaldehyde and formate dehydrogenases from pea seeds by affinity chromatography and S-formylglutathione as the intermediate of formaldehyde metabolism. Arch. Biochem. Biophys., **196**, 33-45 (1979)

[3] Kato, N.; Miyawaki, N.; Sakazawa, C.: Oxidation of formaldehyde by the resistant yeasts Debaryomyces vanriji and Trichosporon penicillatum. Agric. Biol. Chem., **46**, 655-661 (1982)

[4] Allais, J.J.; Louktibi, A.; Baratti, J.: Oxidation of methanol by the yeast, Pichia pastoris. Purification and properties of the formaldehyde dehydrogenase. Agric. Biol. Chem., **47**, 1509-1516 (1983)
[5] Foglio, M.H.; Duester, G.: Characterization of the functional gene encoding mouse class III alcohol dehydrogenase (glutathione-dependent formaldehyde dehydrogenase) and an unexpressed processed pseudogene with an intact open reading frame. Eur. J. Biochem., **237**, 496-504 (1996)
[6] Yang, Z.N.; Bosron, W.F.; Hurley, T.D.: Structure of human $\chi\chi$ alcohol dehydrogenase: a glutathione-dependent formaldehyde dehydrogenase. J. Mol. Biol., **265**, 330-343. (1997)
[7] Kato, N.: Formaldehyde dehydrogenase from methylotrophic yeasts. Methods Enzymol., **188**, 455-462 (1990)
[8] Estonius, M.; Hoeoeg, J.O.; Danielsson, O.; Joernvall, H.: Residues specific for class III alcohol dehydrogenase. Site-directed mutagenesis of the human enzyme. Biochemistry, **33**, 15080-15085 (1994)
[9] Rose, Z.B.; Racker, E.: Formaldehyde dehydrogenase. Methods Enzymol., **9**, 357-360 (1966)
[10] Goodman, J.I.; Tephly, T.R.: A comparison of rat and human liver formaldehyde dehydrogenase. Biochim. Biophys. Acta, **252**, 489-505 (1971)
[11] Patel, R.N.; Hou, C.T.; Derelanko, P.: Microbial oxidation of methanol: purification and properties of formaldehyde dehydrogenase from a Pichia sp. NRRL-Y-11328. Arch. Biochem. Biophys., **221**, 135-142 (1983)
[12] Schuette, H.; Kula, M.R.; Sahm, H.: Formaldehyde dehydrogenase from Candida boidinii. Methods Enzymol., **89**, 527-531 (1982)
[13] Uotila, L.; Koivusalo, M.: Formaldehyde dehydrogenase. Methods Enzymol., **77**, 314-320 (1981)
[14] Uotila, L.; Mannervik, B.: Product inhibition studies of human liver formaldehyde dehydrogenase. Biochim. Biophys. Acta, **616**, 153-157 (1980)
[15] Eggeling, L.; Sahm, H.: Direct enzymatic assay for alcohol oxidase, alcohol dehydrogenase, and formaldehyde dehydrogenase in colonies of Hansenula polymorpha. Appl. Environ. Microbiol., **39**, 268-269 (1980)
[16] Uotila, L.; Mannervik, B.: A steady-state-kinetic model for formaldehyde dehydrogenase from human liver. A mechanism involving NAD^+ and the hemimercaptal adduct of glutathione and formaldehyde as substrates and free glutathione as an allosteric activator of the enzyme. Biochem. J., **177**, 869-878 (1979)
[17] Kato, N.; Sahm, H.; Wagner, F.: Steady-state kinetics of formaldehyde dehydrogenase and formate dehydrogenase from a methanol-utilizing yeast, Candida boidinii. Biochim. Biophys. Acta, **566**, 12-20 (1979)
[18] Kaulfers, P.M.; Marquardt, A.: Demonstration of formaldehyde dehydrogenase activity in formaldehyde-resistant Enterobacteriacea. FEMS Microbiol. Lett., **79**, 335-338 (1991)
[19] Roitsch, T.; Stolp, H.: Distribution of dissimilatory enzymes in methane and methanol oxidizing bacteria. Arch. Microbiol., **143**, 233-236 (1985)
[20] Sahm, H.; Wagner, F.: Microbial assimilation of methanol. Properties of formaldehyde dehydrogenase and formate dehydrogenase from Candida boidinii. Arch. Mikrobiol., **90**, 263-268 (1973)

[21] Fernandez, M.R.; Biosca, J.A.; Martinez, M.C.; Achkor, H.; Farres, J.; Pares, X.: Formaldehyde dehydrogenase from yeast and plant: implications for the general functional and structural significance of class III alcohol dehydrogenase. Adv. Exp. Med. Biol., **414**, 373-381 (1997)
[22] Eggeling, L.; Sahm, H.: Derepression and partial insensitivity to carbon catabolite repression of the methanol dissimilating enzymes in Hansenula polymorpha. Eur. J. Appl. Microbiol. Biotechnol., **5**, 197-202 (1978)
[23] Pronk, J.T.; de Bruijn, P.; van Dijken, J.P.; Bos, D.P.; Kuenen, J.G.: Energetics of mixotrophic and autotrophic C1-metabolism by Thiobacillus acidophilus. Arch. Microbiol., **154**, 576-583 (1990)
[24] Bastide, A.; Laget, M.; Patte, J.C.; Dumenil, G.: Methanol metabolism in Corynebacterium sp. XG, a facultatively methylotrophic strain. J. Gen. Microbiol., **135**, 2869-2874 (1989)
[25] Strittmatter, P.; Ball, E.G.: Formaldehyde dehydrogenase, a glutathione-dependent enzyme system. J. Biol. Chem., **213**, 445-461 (1955)
[26] Kato, N.; Miyawaki, N.; Sakazawa, C.: Formaldehyde dehydrogenase from formaldehyde-resistant Debaryomyces vanriji FT-1 and Pseudomonas putida F61. Agric. Biol. Chem., **47**, 415-416 (1983)
[27] Pourmotabbed, T.; Shih, M.J.; Creighton, D.J.: Bovine liver formaldehyde dehydrogenase. Kinetic and molecular properties. J. Biol. Chem., **264**, 17384-17388 (1989)
[28] Van Ophem, P.W.; Duine, J.A.: NAD- and co-substrate (GSH or factor)-dependent formaldehyde dehydrogenases from methylotrophic microorganisms act as a class III alcohol dehydrogenase. FEMS Microbiol. Lett., **116**, 87-93 (1994)
[29] Uotila, L.; Koivusalo, M.: Formaldehyde dehydrogenase from human liver. Purification, properties, and evidence for the formation of glutathione thiol esters by the enzyme. J. Biol. Chem., **249**, 7653-7663 (1974)
[30] Schuette, H.; Flossdorf, J.; Sahm, H.; Kula, M.R.: Purification and properties of formaldehyde dehydrogenase and formate dehydrogenase from Candida boidinii. Eur. J. Biochem., **62**, 151-160 (1976)
[31] Pourmotabbed, T.; Creighton, D.J.: Substrate specificity of bovine liver formaldehyde dehydrogenase. J. Biol. Chem., **261**, 14240-14244 (1986)
[32] Wippermann, U.; Fliegmann, J.; Bauw, G.; Langebartels, C.; Maier, K.; Sandermann, H., Jr.: Maize glutathione-dependent formaldehyde dehydrogenase. Protein sequence and catalytic properties. Planta, **208**, 12-18 (1999)
[33] Barber, R.D.; Rott, M.A.; Donohue, T.J.: Characterization of a glutathione-dependent formaldehyde dehydrogenase from Rhodobacter sphaeroides. J. Bacteriol., **178**, 1386-1393 (1996)
[34] Sanghani, P.C.; Stone, C.L.; Ray, B.D.; Pindel, E.V.; Hurley, T.D.; Bosron, W.F.: Kinetic mechanism of human glutathione-dependent formaldehyde dehydrogenase. Biochemistry, **39**, 10720-10729 (2000)
[35] Fernandez, M.R.; Biosca, J.A.; Torres, D.; Crosas, B.; Pares, X.: A double residue substitution in the coenzyme-binding site accounts for the different kinetic properties between yeast and human formaldehyde dehydrogenases. J. Biol. Chem., **274**, 37869-37875 (1999)

[36] Luque, T.; Atrian, S.; Danielsson, O.; Jornvall, H.; Gonzalez-Duarte, R.: Structure of the Drosophila melanogaster glutathione-dependent formaldehyde dehydrogenase/octanol dehydrogenase gene (class III alcohol dehydrogenase). Evolutionary pathway of the alcohol dehydrogenase genes. Eur. J. Biochem., **225**, 985-993 (1994)
[37] Svensson, S.; Hedberg, J.J.; Hoog, J.O.: Structural and functional divergence of class II alcohol dehydrogenase. Cloning and characterization of rabbit liver isoforms of the enzyme. Eur. J. Biochem., **251**, 236-243 (1998)
[38] Koivusalo, M.; Uotila, L.: Glutathione-dependent formaldehyde dehydrogenase/class III alcohol dehydrogenase: Further characterization of the rat liver enzyme. Adv. Exp. Med. Biol., **328**, 465-474 (1993)
[39] Gutheil, W.G.; Kasimoglu, E.; Nicholson, P.C.: Induction of glutathione-dependent formaldehyde dehydrogenase activity in Escherichia coli and Hemophilus influenza. Biochem. Biophys. Res. Commun., **238**, 693-696 (1997)
[40] Sanghani, P.C.; Robinson, H.; Bosron, W.F.; Hurley, T.D.: Human glutathione-dependent formaldehyde dehydrogenase. Structures of apo, binary, and inhibitory ternary complexes. Biochemistry, **41**, 10778-10786 (2002)
[41] Sanghani, P.C.; Bosron, W.F.; Hurley, T.D.: Human glutathione-dependent formaldehyde dehydrogenase. Structural changes associated with ternary complex formation. Biochemistry, **41**, 15189-15194 (2002)
[42] Sanghani, P.C.; Robinson, H.; Bennett-Lovsey, R.; Hurley, T.D.; Bosron, W.F.: Structure-function relationships in human class III alcohol dehydrogenase (formaldehyde dehydrogenase). Chem. Biol. Interact., **143-144**, 195-200 (2003)
[43] Sakamoto, A.; Ueda, M.; Morikawa, H.: Arabidopsis glutathione-dependent formaldehyde dehydrogenase is an S-nitrosoglutathione reductase. FEBS Lett., **515**, 20-24 (2002)
[44] Norin, A.; Shafqat, J.; El-Ahmad, M.; Alvelius, G.; Cederlund, E.; Hjelmqvist, L.; Jornvall, H.: Class III alcohol dehydrogenase: consistent pattern complemented with the mushroom enzyme. FEBS Lett., **559**, 27-32 (2004)
[45] Lee, B.; Yurimoto, H.; Sakai, Y.; Kato, N.: Physiological role of the glutathione-dependent formaldehyde dehydrogenase in the methylotrophic yeast Candida boidinii. Microbiology, **148**, 2697-2704 (2002)
[46] Achkor, H.; Diaz, M.; Fernandez, M.R.; Biosca, J.A.; Pares, X.; Martinez, M.C.: Enhanced formaldehyde detoxification by overexpression of glutathione-dependent formaldehyde dehydrogenase from Arabidopsis. Plant Physiol., **132**, 2248-2255 (2003)
[47] Baerends, R.J.; Sulter, G.J.; Jeffries, T.W.; Cregg, J.M.; Veenhuis, M.: Molecular characterization of the Hansenula polymorpha FLD1 gene encoding formaldehyde dehydrogenase. Yeast, **19**, 37-42 (2002)
[48] Nakagawa, T.; Ito, T.; Fujimura, S.; Chikui, M.; Mizumura, T.; Miyaji, T.; Yurimoto, H.; Kato, N.; Sakai, Y.; Tomizuka, N.: Molecular characterization of the glutathione-dependent formaldehyde dehydrogenase gene FLD1 from the methylotrophic yeast Pichia methanolica. Yeast, **21**, 445-453 (2004)
[49] Sanghani, P.C.; Davis, W.I.; Zhai, L.; Robinson, H.: Structure-function relationships in human glutathione-dependent formaldehyde dehydrogenase.

Role of Glu-67 and Arg-368 in the catalytic mechanism. Biochemistry, **45**, 4819-4830 (2006)

[50] Godoy, L.; Gonzalez-Duarte, R.; Albalat, R.: S-Nitrosogluthathione reductase activity of amphioxus ADH3: insights into the nitric oxide metabolism. Int. J. Biol. Sci., **2**, 117-124 (2006)

[51] Martinez, M.C.; Diaz, M.; Achkor, H.; Espunya, M.C.: Glutathione-dependent formaldehyde dehydrogenase/GSNO reductase from Arabidopsis. Expression pattern and functional implications in phytoremediation and pathogenesis. NATO Sci. Ser. Ser. I, **371**, 253-259 (2006)

[52] Espunya, M.C.; Diaz, M.; Moreno-Romero, J.; Martinez, M.C.: Modification of intracellular levels of glutathione-dependent formaldehyde dehydrogenase alters glutathione homeostasis and root development. Plant Cell Environ., **29**, 1002-1011 (2006)

3″-Deamino-3″-oxonicotianamine reductase 1.1.1.285

1 Nomenclature

EC number
1.1.1.285

Systematic name
2'-deoxymugineic acid:NAD(P)$^+$ 3"-oxidoreductase

Recommended name
3"-deamino-3"-oxonicotianamine reductase

2 Source Organism

<1> *Hordeum vulgare* (no sequence specified) [1]

3 Reaction and Specificity

Catalyzed reaction
2'-deoxymugineic acid + NAD(P)$^+$ = 3"-deamino-3"-oxonicotianamine + NAD(P)H + H$^+$

Natural substrates and products
- **S** 3"-deamino-3"-oxonicotianamine + NAD(P)H + <1> (<1> biosynthesis of mugineic acid-family phytosiderophores [1]) (Reversibility: ?) [1]
- **P** 2'-deoxymugineic acid + NAD(P)$^+$

Substrates and products
- **S** 3"-deamino-3"-oxonicotianamine + NAD(P)H + <1> (<1> biosynthesis of mugineic acid-family phytosiderophores [1]) (Reversibility: ?) [1]
- **P** 2'-deoxymugineic acid + NAD(P)$^+$

5 Isolation/Preparation/Mutation/Application

Source/tissue
root <1> (<1> tips of primary root [1]) [1]

References

[1] Shojima, S.; Nishizawa, N.; Fushiya, S.; Nozoe, S.; Irifune, T.; Mori, S.: Biosynthesis of phytosiderophores. In vitro biosynthesis of 2'-deoxymugineic acid from L-methionine and nicotianamine. Plant Physiol., **93**, 1497-1503 (1990)

Isocitrate-homoisocitrate dehydrogenase 1.1.1.286

1 Nomenclature

EC number
1.1.1.286

Systematic name
isocitrate(homoisocitrate):NAD^+ oxidoreductase (decarboxylating)

Recommended name
isocitrate-homoisocitrate dehydrogenase

Synonyms
HICDH <2> [1]

2 Source Organism

<1> *Deinococcus radiodurans* (no sequence specified) [3]
<2> *Thermus thermophilus* (UNIPROT accession number: Q8RQU4) [1]
<3> *Pyrocococcus horikoshii* (no sequence specified) (<3> BCA_2 [2]) [2]

3 Reaction and Specificity

Catalyzed reaction
(1R,2S)-1-hydroxybutane-1,2,4-tricarboxylate + NAD^+ = 2-oxoadipate + CO_2 + NADH + H^+
isocitrate + NAD^+ = 2-oxoglutarate + CO_2 + NADH

Substrates and products
S 3-isopropylmalate + NAD^+ <1> (<1> at 0.1% of the rate with homoisocitrate [3]) (Reversibility: ?) [3]
P ?
S homoisocitrate + NAD^+ <1, 2, 3> (<1> 1.5fold preferred over isocitrate [3]) (Reversibility: ?) [1, 2, 3]
P ? + NADH
S isocitrate + NAD^+ <1, 2, 3> (<2> about 20times more efficient than homoisocitrate [1]; <3> about 20times more efficient than homoisocitrate, NAD^+ is preferred over $NADP^+$ [2]) (Reversibility: ?) [1, 2, 3]
P 2-oxoglutarate + CO_2 + NADH
S isocitrate + $NADP^+$ <1, 3> (<1,3> NAD^+ is preferred over $NADP^+$ [2,3]) (Reversibility: ?) [2, 3]

P 2-oxoglutarate + CO_2 + NADPH
S Additional information <1, 2, 3> (<2,3> no substrate: 3-isopropylmalate [1,2]; <1> no substrate: $NADP^+$ [3]) (Reversibility: ?) [1, 2, 3]
P ?

Inhibitors

K^+ <1> (<1> absolute requirement, inhibitory above 0.4 M [3]) [3]
Mn^{2+} <1> (<1> inhibitory above 1 mM [3]) [3]
NH_4^+ <1> (<1> may substitute for K^+, inhibitory above 0.2 M [3]) [3]

Cofactors/prosthetic groups

NAD^+ <1> [3]

Metals, ions

K^+ <1> (<1> absolute requirement, inhibitory above 0.4 M [3]) [3]
Mg^{2+} <1> (<1> divalent cation required, Mg^{2+} is preferred over Mn^{2+} [3]) [3]
Mn^{2+} <1> (<1> divalent cation required, Mg^{2+} is preferred over Mn^{2+}, inhibitory above 1 mM [3]) [3]

Turnover number (min^{-1})

0.371 <1> (3-isopropylmalate, <1> 30°C, pH 7.8, wild-type [3]) [3]
5.46 <1> (homoisocitrate, <1> 30°C, pH 7.8, A80 deletion mutant [3]) [3]
10.9 <1> (isocitrate, <1> 30°C, pH 7.8, A80 deletion mutant [3]) [3]
13.7 <3> (homoisocitrate, <3> 70°C, pH 7.8 [2]) [2]
14.8 <3> (isocitrate, <3> 70°C, pH 7.8 [2]) [2]
20.8 <1> (homoisocitrate, <1> 30°C, pH 7.8, mutant R87V [3]) [3]
21.8 <3> (NAD^+, <3> 70°C, pH 7.8 [2]) [2]
27.8 <1> (homoisocitrate, <1> 30°C, pH 7.8, mutant R87T [3]) [3]
30.3 <1> (NAD^+, <1> 30°C, pH 7.8 [3]) [3]
42.8 <1> (isocitrate, <1> 30°C, pH 7.8, wild-type [3]) [3]
46.2 <1> (homoisocitrate, <1> 30°C, pH 7.8, wild-type [3]) [3]
171 <2> (isocitrate, <2> 60°C, pH 8.0 [1]) [1]
171 <2> (homoisocitrate, <2> 60°C, pH 8.0 [1]) [1]

K_m-Value (mM)

0.0164 <3> (isocitrate, <3> 70°C, pH 7.8 [2]) [2]
0.0183 <3> (homoisocitrate, <3> 70°C, pH 7.8 [2]) [2]
0.0771 <3> (NAD^+, <3> 70°C, pH 7.8 [2]) [2]
0.193 <1> (NAD^+, <1> 30°C, pH 7.8 [3]) [3]
0.211 <1> (homoisocitrate, <1> 30°C, pH 7.8, wild-type [3]) [3]
0.291 <1> (isocitrate, <1> 30°C, pH 7.8, wild-type [3]) [3]
0.405 <2> (isocitrate, <2> 60°C, pH 8.0 [1]) [1]
0.465 <1> (isocitrate, <1> 30°C, pH 7.8, A80 deletion mutant [3]) [3]
0.819 <1> (homoisocitrate, <1> 30°C, pH 7.8, A80 deletion mutant [3]) [3]
1 <1> (homoisocitrate, <1> 30°C, pH 7.8, mutant R87V [3]) [3]
1.33 <1> (3-isopropylmalate, <1> 30°C, pH 7.8, wild-type [3]) [3]
1.5 <1> (homoisocitrate, <1> 30°C, pH 7.8, mutant R87T [3]) [3]
7.486 <2> (homoisocitrate, <2> 60°C, pH 8.0 [1]) [1]

4 Enzyme Structure

Molecular weight

154000 <1> (<1> gel filtration [3]) [3]

Subunits

tetramer <1, 2> (<1> 4 * 35751, mass spectrometry, 4 * 35727, calculated [3]; <2> sedimentation equilibrium analysis, SDS-PAGE [1]) [1, 3]

5 Isolation/Preparation/Mutation/Application

Purification

<3> (recombinant enzyme expressed in Escherichia coli, purification of inclusion bodies requires N-laurylsarcosine) [2]

Cloning

<2> (expression in Escherichia coli) [1]
<3> [2]

Engineering

R85V <1, 2> (<2> complete loss of activity with citrate, retains activity with isocitrate [1]; <1> increased ability to use 3-isopropylmalate [3]) [1, 3]
R87T <1> (<1> uses substrate homoisocitrate, but not substrates isocitrate or 3-isopropylmalate [3]) [3]
R87V <1> (<1> uses substrate homoisocitrate, but not substrates isocitrate or 3-isopropylmalate [3]) [3]
Additional information <1> (<1> deletion of A80, inability of mutant to use 3-isopropylmalate, isocitrate is 4fold preferred over homoisocitrate [3]) [3]

6 Stability

Temperature stability

90 <2> (<2> half-life 16.7 h [1]) [1]

References

[1] Miyazaki, J.; Kobashi, N.; Nishiyama, M.; Yamane, H.: Characterization of homoisocitrate dehydrogenase involved in lysine biosynthesis of an extremely thermophilic bacterium, Thermus thermophilus HB27, and evolutionary implication of β-decarboxylating dehydrogenase. J. Biol. Chem., **278**, 1864-1871 (2003)

[2] Miyazaki, K.: Bifunctional isocitrate-homoisocitrate dehydrogenase: a missing link in the evolution of β-decarboxylating dehydrogenase. Biochem. Biophys. Res. Commun., **331**, 341-346 (2005)

[3] Miyazaki, K.: Identification of a novel trifunctional homoisocitrate dehydrogenase and modulation of the broad substrate specificity through site-directed mutagenesis. Biochem. Biophys. Res. Commun., **336**, 596-602 (2005)

D-Arabinitol dehydrogenase ($NADP^+$) 1.1.1.287

1 Nomenclature

EC number
1.1.1.287

Systematic name
D-arabinitol:$NADP^+$ oxidoreductase

Recommended name
D-arabinitol dehydrogenase ($NADP^+$)

Synonyms
ARD1p <1> [1]
ArDH <2> [2]
D-arabitol dehydrogenase <2> [2]
D-arabitol dehydrogenase 1 <1> [1]
NADP(+)-dependent D-arabitol dehydrogenase <1> [1]
Additional information <2> (<2> the enzyme belongs to the short-chain dehydrogenase family [2]) [2]

2 Source Organism

<1> *Uromyces fabae* (UNIPROT accession number: Q4R0J7) [1]
<2> *Gluconobacter oxydans* (UNIPROT accession number: Q58LW6) [2]

3 Reaction and Specificity

Catalyzed reaction
D-arabinitol + $NADP^+$ = D-ribulose + NADPH + H^+
D-arabinitol + $NADP^+$ = D-xylulose + NADPH + H^+

Natural substrates and products
S D-arabitinol + $NADP^+$ <2> (Reversibility: r) [2]
P D-xylulose + NADPH
S Additional information <1> (<1> the enzyme activity is highly increased during infection processes and pathogenesis, development after infection, overview [1]) (Reversibility: ?) [1]
P ?

Substrates and products
S D-arabitinol + $NADP^+$ <1> (Reversibility: r) [1]
P D-ribulose + NADPH

S D-arabitinol + $NADP^+$ <1, 2> (Reversibility: r) [1, 2]
P D-xylulose + NADPH
S D-arabitol + $NADP^+$ <2> (<2> preferred substrate for the forward and reverse reaction, respectively [2]) (Reversibility: r) [2]
P D-xylulose + NADPH
S D-mannitol + $NADP^+$ <1, 2> (<2> preferred substrate for the forward and reverse reaction, respectively [2]) (Reversibility: r) [1, 2]
P D-fructose + NADPH
S D-sorbitol + $NADP^+$ <2> (Reversibility: r) [2]
P sorbose + NADPH
S glycerol + $NADP^+$ <2> (Reversibility: r) [2]
P ? + NADPH
S meso-erythritol + $NADP^+$ <2> (Reversibility: r) [2]
P L-erythrose + NADPH
S Additional information <1, 2> (<1> the enzyme activity is highly increased during infection processes and pathogenesis, development after infection, overview [1]; <2> substrate specificity for different sugars, overview, no activity with xylitol, L-arabitol, and L-sorbitol [2]; <1> the enzyme shows high substrate specificity, the reverse reaction is preferred [1]) (Reversibility: ?) [1, 2]
P ?

Cofactors/prosthetic groups

$NADP^+$ <1, 2> (<2> dependent on [2]; <1> dependent on, absolutely specific for, forward reaction [1]) [1,2]
NADPH <1> (<1> dependent on, absolutely specific for, reverse reaction [1]) [1]

Turnover number (min^{-1})

1.28 <2> (meso-erythritol, <2> pH 8.5, 25°C [2]) [2]
1.33 <2> (glycerol, <2> pH 8.5, 25°C [2]) [2]
2.57 <2> (D-sorbitol, <2> pH 8.5, 25°C [2]) [2]
196 <2> (D-xylulose, <2> pH 6.0, 25°C [2]) [2]
224 <2> (D-fructose, <2> pH 6.0, 25°C [2]) [2]
243 <2> (D-mannitol, <2> pH 8.5, 25°C [2]) [2]
258 <2> (D-arabitol, <2> pH 8.5, 25°C [2]) [2]

Specific activity (U/mg)

2.8 <2> (<2> purified recombinant enzyme, substrate glycerol [2]) [2]
3.2 <2> (<2> purified recombinant enzyme, substrate meso-erythritol [2]) [2]
6.4 <2> (<2> purified recombinant enzyme, substrate D-sorbitol [2]) [2]
7.9 <2> (<2> purified recombinant enzyme, substrate D-fructose [2]) [2]
11.4 <2> (<2> purified recombinant enzyme, substrate D-xylulose [2]) [2]
23.5 <2> (<2> purified recombinant enzyme, substrate D-mannitol [2]) [2]
27.6 <2> (<2> purified recombinant enzyme, substrate D-arabitol [2]) [2]

K_m-Value (mM)

0.07 <1> (NADP$^+$, <1> pH 9.0, with D-mannitol [1]) [1]
0.08 <1> (NADP$^+$, <1> pH 9.0, with D-arabitinol [1]) [1]
0.16 <1> (NADPH, <1> pH 6.0, with D-fructose [1]) [1]
0.2 <1> (NADPH, <1> pH 6.0, with D-ribulose [1]) [1]
0.26 <1> (NADPH, <1> pH 6.0, with D-xylulose [1]) [1]
7.8 <1> (D-xylulose, <1> pH 6.0 [1]) [1]
12.5 <2> (D-arabitol, <2> pH 8.5, 25°C [2]) [2]
18.5 <2> (D-mannitol, <2> pH 8.5, 25°C [2]) [2]
25 <2> (D-xylulose, <2> pH 6.0, 25°C [2]) [2]
47 <2> (D-fructose, <2> pH 6.0, 25°C [2]) [2]
53.9 <1> (D-ribulose, <1> pH 6.0 [1]) [1]
125 <2> (D-sorbitol, <2> pH 8.5, 25°C [2]) [2]
160 <1> (D-fructose, <1> pH 6.0 [1]) [1]
266 <2> (meso-erythritol, <2> pH 8.5, 25°C [2]) [2]
384 <2> (glycerol, <2> pH 8.5, 25°C [2]) [2]
425 <1> (D-mannitol, <1> pH 9.0 [1]) [1]
834 <1> (D-arabitinol, <1> pH 9.0 [1]) [1]
Additional information <1> (<1> kinetics [1]) [1]

pH-Optimum

6 <1> (<1> assay at, reverse reaction [1]) [1]
6.5 <2> (<2> reverse reaction [2]) [2]
8.5 <2> (<2> forward reaction [2]) [2]
9 <1> (<1> assay at, forward reaction [1]) [1]

Temperature optimum (°C)

25 <2> (<2> assay at [2]) [2]

4 Enzyme Structure

Subunits

? <2> (<2> x * 29000, recombinant enzyme, SDS-PAGE [2]) [2]

5 Isolation/Preparation/Mutation/Application

Source/tissue

culture condition:arabitol-grown cell <2> [2]
haustorium <1> (<1> exclusively, in the lumen [1]) [1]
Additional information <1> (<1> cultivated on artificial surfaces in in vitro infection structures [1]) [1]

Localization

membrane <2> (<2> bound [2]) [2]
soluble <1> [1]

Purification

<1> (recombinant His-tagged enzyme from Escherichia coli by nickel affinity chromatography) [1]

<2> (recombinant His6-tagged enzyme from Escherichia coli) [2]

Cloning

<1> (gene ARD1, DNA and amino acid sequence determination and analysis, expression in Saccharomyces cerevisiae strain 23344c, overexpression in Escherichia coli strain BL21(DE3) as N-terminally His6-tagged protein) [1]

<2> (gene ArDH, DNA and amino acid sequence determination and analysis, gene ArDH is part of an operon with several components of a sugar transporter system, subcloning in Escherichia coli strain TG-1, functional expression in Escherichia coli strain BL21(DE3) as His6-tagged enzyme) [2]

References

[1] Link, T.; Lohaus, G.; Heiser, I.; Mendgen, K.; Hahn, M.; Voegele, R.T.: Characterization of a novel NADP(+)-dependent D-arabitol dehydrogenase from the plant pathogen Uromyces fabae. Biochem. J., **389**, 289-295 (2005)

[2] Cheng, H.; Jiang, N.; Shen, A.; Feng, Y.: Molecular cloning and functional expression of D-arabitol dehydrogenase gene from Gluconobacter oxydans in Escherichia coli. FEMS Microbiol. Lett., **252**, 35-42 (2005)

Xanthoxin dehydrogenase 1.1.1.288

1 Nomenclature

EC number
1.1.1.288

Systematic name
xanthoxin:NAD^+ oxidoreductase

Recommended name
xanthoxin dehydrogenase

Synonyms
ABA2 <4> [4]
SDR1 <4> [3]
xanthoxin oxidase <4> [4]

CAS registry number
129204-37-1

2 Source Organism

<1> *Phaseolus vulgaris* (no sequence specified) [5, 6]
<2> *Pisum sativum* (no sequence specified) [5]
<3> *Zea mays* (no sequence specified) [5]
<4> *Arabidopsis thaliana* (no sequence specified) [1, 2, 3, 4]
<5> *Vigna radiata* (no sequence specified) [5]
<6> *Lycopersicon esculentum* (no sequence specified) [6,7]
<7> *Cucurbita maxima* (no sequence specified) [5]

3 Reaction and Specificity

Catalyzed reaction
xanthoxin + NAD^+ = abscisic aldehyde + NADH + H^+

Natural substrates and products
S xanthoxin + NAD^+ <1, 4, 6> (<4> enzyme is involved abscisic acid biosynthesis [3]; <4> enzyme is involved in abscisic acid synthesis [1]; <1,6> enzyme is involved in biosynthesis of abscisic acid [6]; <4> last step in abscisic acid biosynthetic pathway, constitutively expressed, not upregulated in response to osmotic stress [2]; <1> neither water stress nor cy-

cloheximide significantly affects the level of enzyme activity [5]) (Reversibility: ?) [1, 2, 3, 5, 6, 7]

P abscisic aldehyde + NADH + H^+

Substrates and products

S xanthoxin + NAD^+ <1, 4, 6> (<4> enzyme is involved abscisic acid biosynthesis [3]; <4> enzyme is involved in abscisic acid synthesis [1]; <1,6> enzyme is involved in biosynthesis of abscisic acid [6]; <4> last step in abscisic acid biosynthetic pathway, constitutively expressed, not upregulated in response to osmotic stress [2]; <1> neither water stress nor cycloheximide significantly affects the level of enzyme activity [5]) (Reversibility: ?) [1, 2, 3, 5, 6, 7]

P abscisic aldehyde + NADH + H^+

Cofactors/prosthetic groups

NAD^+ <4> (<4> absolute requirement for [3]) [3,4]

K_m-Value (mM)

0.0202 <4> (xanthoxin) [3]

5 Isolation/Preparation/Mutation/Application

Source/tissue

leaf <1, 2, 3, 5, 6, 7> [5, 6]
root <1> [5]

Localization

cytosol <1, 4> [3, 5]
Additional information <1> (<1> no significant activity is observed in chloroplast [5]) [5]

Purification

<4> (recombinant enzyme) [3]

Cloning

<4> [2]
<4> (expression in Pichia pastoris) [3]

References

[1] Taylor, I.B.; Burbidge, A.; Thompson, A.J.: Control of abscisic acid synthesis. J. Exp. Bot., **51**, 1563-1574. (2000)

[2] Gonzalez-Guzman, M.; Apostolova, N.; Belles, J.M.; Barrero, J.M.; Piqueras, P.; Ponce, M.R.; Micol, J.L.; Serrano, R.; Rodriguez, P.L.: The short-chain alcohol dehydrogenase ABA_2 catalyzes the conversion of xanthoxin to abscisic aldehyde. Plant Cell, **14**, 1833-1846 (2002)

[3] Cheng, W.H.; Endo, A.; Zhou, L.; Penney, J.; Chen, H.C.; Arroyo, A.; Leon, P.; Nambara, E.; Asami, T.; Seo, M.; Koshiba, T.; Sheen, J.: A unique short-chain

dehydrogenase/reductase in Arabidopsis glucose signaling and abscisic acid biosynthesis and functions. Plant Cell, **14**, 2723-2743 (2002)
[4] Schwartz, S.H.; Leon-Kloosterziel, K.M.; Koornneef, M.; Zeevaart, J.A.: Biochemical characterization of the aba2 and aba3 mutants in Arabidopsis thaliana. Plant Physiol., **114**, 161-166 (1997)
[5] Sindhu, R.K.; Walton, D.C.: Conversion of xanthoxin to abscisic acid by cell-free preparations from bean leaves. Plant Physiol., **85**, 916-921 (1987)
[6] Sindhu, R.K.; Walton, D.C.: Xanthoxin metabolism in cell-free preparations from wild-type and wilty mutants of tomato. Plant Physiol., **88**, 178-182 (1988)
[7] Parry, A.D.; Neill, S.J.; Horgan, R.: Xanthoxin levels and metabolism in the wild-type and wilty mutants of tomato. Planta, **173**, 397-404 (1988)

Sorbose reductase 1.1.1.289

1 Nomenclature

EC number
1.1.1.289

Systematic name
D-glucitol:NADP$^+$ oxidoreductase

Recommended name
sorbose reductase

Synonyms
L-sorbose reductase <2> [2]

CAS registry number
138440-90-1

2 Source Organism

<1> *Candida albicans* (no sequence specified) [4]
<2> *Gluconobacter melanogenus* (no sequence specified) [1, 2]
<3> *Gluconobacter suboxydans* (no sequence specified) [3]

3 Reaction and Specificity

Catalyzed reaction
D-glucitol + NADP$^+$ = L-sorbose + NADPH + H$^+$

Substrates and products

S 5-keto-D-fructose + NADPH <2> (<2> 86.8% of the rate with L-sorbose [1]) (Reversibility: r) [1]
P ? + NADP$^+$
S D-fructose + NADPH <1> (<1> less effective than reaction with L-sorbose [4]) (Reversibility: ?) [4]
P D-mannitol + NADP$^+$
S D-mannitol + NADP$^+$ <2> (<2> 134% of the rate with L-sorbose [1]) (Reversibility: r) [1]
P L-fructose + NADPH
S D-sorbitol + NADP$^+$ <1, 2> (<2> 134% of the rate with L-sorbose [1]; <1> low efficiency [4]) (Reversibility: r) [1, 2, 4]
P L-sorbose + NADPH

S L-fructose + NADPH <2> (<2> 142% of the rate with L-sorbose [1]) (Reversibility: r) [1]
P D-mannitol + $NADP^+$
S L-sorbose + NADPH <2> (Reversibility: r) [1, 2]
P D-sorbitol + $NADP^+$
S L-sorbose + NADPH <1> (Reversibility: ?) [4]
P L-sorbitol + $NADP^+$
S erythritol + $NADP^+$ <2> (<2> 65.3% of the rate with L-sorbose [1]) (Reversibility: r) [1]
P ? + NADPH
S ribulose + NADPH <2> (<2> 2.4% of the rate with L-sorbose [1]) (Reversibility: r) [1]
P ? + $NADP^+$
S tagatose + NADPH <2> (<2> 0.8% of the rate with L-sorbose [1]) (Reversibility: r) [1]
P ? + $NADP^+$
S xylulose + NADPH <2> (<2> 25.7% of the rate with L-sorbose [1]) (Reversibility: r) [1]
P ? + $NADP^+$
S Additional information <2> (<2> highly specific for D-sorbitol and L-sorbose. Reaction rate in L-sorbose reduction highly predominates over L-sorbitol oxidation over a wide pH range [2]) (Reversibility: ?) [2]
P ?

Inhibitors

N-ethylmaleimide <2> (<2> 0.97 mM, 81% residual activity [1]) [1]
Na_2HAsO_4 <2> (<2> 0.97 mM, 90% residual activity [1]) [1]
quinine <2> (<2> 0.49 mM, complete loss of activity [1]) [1]
sodium azide <2> (<2> 0.97 mM, 90% residual activity [1]) [1]
p-chloromercuribenzoate <2> (<2> 0.19 mM, 43% residual activity [1]) [1]
Additional information <2> (<2> not inhibitory: sodium fluoroacetate, sodium fluoride, KCN, monoiodoacetate [1]) [1]

Cofactors/prosthetic groups

$NADP^+$ <2> [1,2]
NADPH <1, 2, 3> [1,2,3,4]
Additional information <2> (<2> no cofactor: NAD^+, NADH [2]) [2]

Turnover number (min^{-1})

154 <1> (D-fructose, <1> pH 7.5, 25°C, positive cooperativity [4]) [4]
782 <1> (L-sorbose, <1> pH 7.5, 25°C, Michaelis-Menten kinetics [4]) [4]

Specific activity (U/mg)

2.93 <1> (<1> substrate D-sorbitol, pH 6.2, 25°C [4]) [4]
33.1 <1> (<1> substrate D-fructose, pH 6.2, 25°C [4]) [4]
77 <2> (<2> pH 6.0, 25°C, substrate L-sorbose [2]) [2]
145 <1> (<1> substrate L-sorbose, pH 6.2, 25°C [4]) [4]

K_m-Value (mM)

0.032 <2> (NADPH, <2> pH 6.0, 25°C [2]) [2]
0.111 <2> (NADPH, <2> pH 7.0, 25°C [1]) [1]
7.3 <2> (D-mannitol, <2> pH 10.0, 25°C [1]) [1]
35 <2> (L-sorbose, <2> pH 6.0, 25°C [2]) [2]
160 <2> (D-fructose, <2> pH 7.0, 25°C [1]) [1]
167 <2> (D-sorbitol, <2> pH 10.0, 25°C [1]) [1]
328 <2> (L-sorbose, <2> pH 7.0, 25°C [1]) [1]
1873 <1> (D-fructose, <1> $K_{0.5}$-value, positive cooperativity [4]) [4]
3953 <1> (L-sorbose, <1> pH 7.5, 25°C [4]) [4]

pH-Optimum

6 <2> (<2> reduction of L-sorbose [2]) [2]
6.2 <1> [4]
7 <2> (<2> reduction of L-sorbose [1]) [1]
8 <2> (<2> oxidation of D-sorbitol [2]) [2]
10-10.5 <2> (<2> oxidation of D-sorbitol [1]) [1]

pH-Range

5-7 <2> (<2> reduction of L-sorbose [2]) [2]
7-9 <2> (<2> oxidation of D-sorbitol [2]) [2]

Temperature optimum (°C)

30 <2> [1]

4 Enzyme Structure

Molecular weight

58000 <2> (<2> gel filtration [1]) [1]
60000 <2> (<2> gel filtration [2]) [2]
124000 <1> (<1> MALDI-TOF [4]) [4]

Subunits

dimer <2> (<2> 2 * 30000, SDS-PAGE [2]) [2]
monomer <2> (<2> 1 * 60000, SDS-PAGE [1]) [1]
tetramer <1> (<1> 4 * 31000, SDS-PAGE and calculated [4]) [4]

5 Isolation/Preparation/Mutation/Application

Localization

cytosol <2> [1, 2]

Purification

<1> (recombinant enzyme with FLAG-tag) [4]
<2> [1, 2]

Crystallization

<2> [2]

Engineering

Additional information <3> (<3> gene disruption mutant, no enzymic activity, no assimilation of once-produced L-sorbose [3]) [3]

6 Stability

pH-Stability

6.5-7.5 <2> (<2> 30°C, stable for 30 min [1]) [1]

Temperature stability

30 <2> (<2> pH 6.5-7.5, 30 min, stable [1]) [1]
55 <2> (<2> stable up to, for 10 min [1]) [1]
65 <1> (<1> 2 min, more than 90% inactivation [4]) [4]

Storage stability

<2>, -20°C, 0.01 M potassium buffer, pH 7.0, stable for at least one month [1]

References

[1] Sugisawa, T.; Hoshino, T.; Fujiwara, A.: Purification and properties of NADPH-linked L-sorbose reductase from Gluconobacter melanogenus N44-1. Agric. Biol. Chem., **55**, 2043-2049 (1991)

[2] Adachi, O.; Ano, Y.; Moonmangmee, D.; Shinagawa, E.; Toyama, H.; Theeragool, G.; Lotong, N.; Matsushita, K.: Crystallization and properties of NADPH-dependent L-sorbose reductase from Gluconobacter melanogenus IFO 3294. Biosci. Biotechnol. Biochem., **63**, 2137-2143 (1999)

[3] Shinjoh, M.; Tazoe, M.; Hoshino, T.: NADPH-dependent L-sorbose reductase is responsible for L-sorbose assimilation in Gluconobacter suboxydans IFO 3291. J. Bacteriol., **184**, 861-863 (2002)

[4] Greenberg, J.R.; Price, N.P.; Oliver, R.P.; Sherman, F.; Rustchenko, E.: Candida albicans SOU1 encodes a sorbose reductase required for L-sorbose utilization. Yeast, **22**, 957-969 (2005)

4-Phosphoerythronate dehydogenase 1.1.1.290

1 Nomenclature

EC number
1.1.1.290

Systematic name
4-phospho-D-erythronate:NAD$^+$ 2-oxidoreductase

Recommended name
4-phosphoerythronate dehydogenase

Synonyms
4-O-phosphoerythronate dehydrogenase <3> [4]
PDXB <1> [1]
PdxB 4PE dehydrogenase <3> [4]
RdxB <3> [4]
erythronate-4-phosphate dehydrogenase <2, 3> [2, 5]
pdx gene product <1, 3> [1, 2]

CAS registry number
125858-75-5

2 Source Organism

<1> *Escherichia coli* (no sequence specified) [1]
<2> *Pseudomonas aeruginosa* (no sequence specified) [5]
<3> *Escherichia coli* (UNIPROT accession number: P05459) (<3> isozyme SMO2-2 [2,4]) [2, 4]
<4> *Escherichia coli* (UNIPROT accession number: P05450) (<4> large subunit [3]) [3]

3 Reaction and Specificity

Catalyzed reaction
4-phospho-D-erythronate + NAD$^+$ = (3R)-3-hydroxy-2-oxo-4-phosphonooxybutanoate + NADH + H$^+$

Reaction type
oxidation
redox reaction
reduction

Natural substrates and products

S erythronate-4-phosphate + NAD^+ <4> (<4> reaction in pathway leading from erythrose-4-phosphate and glutamate to nitrogen 1 and carbon 5 and 6 of the pyridoxine ring [3]) (Reversibility: ?) [3]

P (3R)-3-hydroxy-2-oxo-4-phosphonooxybutanoate + NADH

S Additional information <3> (<3> pdxB is the first gene in the pdxB-hisT operon [2]; <3> the enzyme mediates a step in the biosynthesis of the coenzyme pyridoxal 5'-phosphate. Transcription of pdxB gene is positively growth rate regulated. PdxB-specific transcript remains unchanged during amino acid starvation in wild-type and relA mutant strains [4]) (Reversibility: ?) [2, 4]

P ?

Substrates and products

S erythronate-4-phosphate + NAD^+ <2> (Reversibility: ?) [5]

P 3-hydroxy-4-phospho-hydroxy-α-ketobutyrate + NADH

S erythronate-4-phosphate + NAD^+ <4> (<4> reaction in pathway leading from erythrose-4-phosphate and glutamate to nitrogen 1 and carbon 5,5, and 6 of the pyridoxine ring [3]) (Reversibility: ?) [3]

P (3R)-3-hydroxy-2-oxo-4-phosphonooxybutanoate + NADH

S Additional information <3> (<3> pdxB is the first gene in the pdxB-hisT operon [2]; <3> the enzyme mediates a step in the biosynthesis of the coenzyme pyridoxal 5-phosphate. Transcription of pdxB gene is positively growth rate regulated. PdxB-specific transcript remains unchanged during amino acid starvation in wild-type and relA mutant strains [4]) (Reversibility: ?) [2, 4]

P ?

Cofactors/prosthetic groups

NAD^+ <2> [5]

4 Enzyme Structure

Molecular weight

83000 <2> (<2> estimated by dynamic light-scattering analysis [5]) [5]

Subunits

homodimer <2> (<2> 2 * 42 067, enzyme including a C-terminal tag, enzyme consists of two identical 380-residue subunits [5]) [5]

5 Isolation/Preparation/Mutation/Application

Purification

<2> (metal-chelate chromatography on Ni-NTA resin and gel filtration) [5]

Crystallization

<2> (hanging-drop vapour diffusion at 24°C using 0.7 M ammonium dihydrogen phosphate, 0.4 M ammonium tartrate, 0.1 M sodium citrate pH 5.6 and 10 mM cupric chloride) [5]

Cloning

<2> (expressed in Escherichia coli) [5]

References

[1] Grant, G.A.: A new family of 2-hydroxyacid dehydrogenases. Biochem. Biophys. Res. Commun., **165**, 1371-1374 (1989)

[2] Schoenlein, P.V.; Roa, B.B.; Winkler, M.E.: Divergent transcription of pdxB and homology between the pdxB and serA gene products in Escherichia coli K-12. J. Bacteriol., **171**, 6084-6092 (1989)

[3] Lam, H.M.; Winkler, M.E.: Metabolic relationships between pyridoxine (vitamin B6) and serine biosynthesis in Escherichia coli K-12. J. Bacteriol., **172**, 6518-6528 (1990)

[4] Pease, A.J.; Roa, B.R.; Luo, W.; Winkler, M.E.: Positive growth rate-dependent regulation of the pdxA, ksgA, and pdxB genes of Escherichia coli K-12. J. Bacteriol., **184**, 1359-1369 (2002)

[5] Ha, J.Y.; Lee, J.H.; Kim, K.H.; Kim, D.J.; Lee, H.H.; Kim, H.K.; Yoon, H.J.; Suh, S.W.: Overexpression, crystallization and preliminary X-ray crystallographic analysis of erythronate-4-phosphate dehydrogenase from Pseudomonas aeruginosa. Acta Crystallogr. Sect. F, **62**, 139-141 (2006)

2-Hydroxymethylglutarate dehydrogenase 1.1.1.291

1 Nomenclature

EC number
1.1.1.291

Systematic name
(S)-2-hydroxymethylglutarate:NAD^+ oxidoreductase

Recommended name
2-hydroxymethylglutarate dehydrogenase

Synonyms
HgD <1> [1]

2 Source Organism

<1> *Eubacterium barkeri* (no sequence specified) [1]

3 Reaction and Specificity

Catalyzed reaction
(S)-2-hydroxymethylglutarate + NAD^+ = 2-formylglutarate + NADH + H^+

Natural substrates and products

- **S** 2-formylglutarate + NADH + H^+ <1> (<1> the enzyme forms part of the nicotinate-fermentation catabolism pathway in Eubacterium barkeri [1]) (Reversibility: ?) [1]
- **P** 2-hydroxymethylglutarate + NAD^+

Substrates and products

- **S** 2-formylglutarate + NADH + H^+ <1> (<1> the enzyme forms part of the nicotinate-fermentation catabolism pathway in Eubacterium barkeri [1]) (Reversibility: ?) [1]
- **P** 2-hydroxymethylglutarate + NAD^+
- **S** 2-hydroxymethylglutarate + NAD^+ <1> (Reversibility: r) [1]
- **P** 2-formylglutarate + NADH + H^+

Cofactors/prosthetic groups
NAD^+ <1> (<1> no activity with $NADP^+$ [1]) [1]
NADH <1> [1]

K_m-Value (mM)

0.1 <1> (NAD^+) [1]
1.1 <1> (2-hydroxymethylglutarate) [1]

4 Enzyme Structure

Subunits

tetramer <1> (<1> 4 * 32000, SDS-PAGE [1]) [1]

5 Isolation/Preparation/Mutation/Application

Purification

<1> [1]

Cloning

<1> (expression in Escherichia coli) [1]

6 Stability

Oxidation stability

<1>, instability under anaerobic conditions [1]

References

[1] Alhapel, A.; Darley, D.J.; Wagener, N.; Eckel, E.; Elsner, N.; Pierik, A.J.: Molecular and functional analysis of nicotinate catabolism in Eubacterium barkeri. Proc. Natl. Acad. Sci. USA, **103**, 12341-12346 (2006)

1,5-Anhydro-D-fructose reductase (1,5-anhydro-D-mannitol-forming) 1.1.1.292

1 Nomenclature

EC number
1.1.1.292

Systematic name
1,5-anhydro-D-mannitol:$NADP^+$ oxidoreductase

Recommended name
1,5-anhydro-D-fructose reductase (1,5-anhydro-D-mannitol-forming)

Synonyms
1,5-anhydro-D-fructose reductase <1> [2]
1,5-anhydro-D-fructose reductase (ambiguous) <2> [1]
AFR <2> [1]

CAS registry number
206138-19-4

2 Source Organism

<1> *Sinorhizobium morelense* (no sequence specified) [2]
<2> *Sinorhizobium morelense* (UNIPROT accession number: Q2I8V6) [1]

3 Reaction and Specificity

Catalyzed reaction
1,5-anhydro-D-mannitol + $NADP^+$ = 1,5-anhydro-D-fructose + NADPH + H^+

Substrates and products
S 1,5-anhydro-D-fructose + NADPH + H^+ <1, 2> (Reversibility: r) [1, 2]
P 1,5-anhydro-D-mannitol + $NADP^+$
S 1,5-anhydro-D-glucitol + $NADP^+$ <2> (<2> 23% of the activity with 1,5-anhydro-D-mannitol [1]) (Reversibility: ?) [1]
P ?
S 6-deoxy-D-glucosone + NADPH + H^+ <2> (<2> 22% of the activity with 1,5-anhydro-D-fructose [1]) (Reversibility: ?) [1]
P 6-deoxy-D-mannose + $NADP^+$
S D-allosone + NADPH + H^+ <2> (<2> 10% of the activity with 1,5-anhydro-D-fructose [1]) (Reversibility: ?) [1]
P D-altrose + $NADP^+$

S D-glucosone + NADPH + H^+ <2> (<2> 22% of the activity with 1,5-anhydro-D-fructose [1]) (Reversibility: ?) [1]

P D-mannose + $NADP^+$

S D-xylosone + NADPH + H^+ <2> (<2> 17% of the activity with 1,5-anhydro-D-fructose [1]) (Reversibility: ?) [1]

P D-lyxose + $NADP^+$

Cofactors/prosthetic groups

NADH <1> (<1> wild-type enzyme and mutant enzymes S10G, S33D, K94G, D176A, H180A and G206I show no activity with NADH. Mutant enzymes A13G, S10G/A13G and S33D/A13G are active with NADH [2]) [2]

$NADP^+$ <2> [1]

NADPH <1, 2> (<2> inactive towards NADH [1]; <1> the N-terminal domain displays a Rossman fold and contains the cofactor binding site. The intact crystals contain the oxidized cofactor $NADP^+$ [2]) [1,2]

Turnover number (min^{-1})

1.3 <1> (1,5-anhydro-D-fructose, <1> pH 7.5, mutant enzyme D176A, cofactor: NADPH [2]) [2]

3.7 <1> (1,5-anhydro-D-fructose, <1> pH 8.0, mutant enzyme H180A, cofactor: NADPH [2]) [2]

4.2 <1> (1,5-anhydro-D-fructose, <1> pH 6.5, mutant enzyme K94G, cofactor: NADPH [2]) [2]

5.5 <1> (1,5-anhydro-D-fructose, <1> pH 6.5, mutant enzyme S10G/A13G, cofactor: NADH [2]) [2]

6.3 <1> (1,5-anhydro-D-fructose, <1> pH 6.5, mutant enzyme S33D/A13G, cofactor: NADPH [2]) [2]

12.4 <1> (1,5-anhydro-D-fructose, <1> pH 6.5, mutant enzyme A13G, cofactor: NADH [2]) [2]

13.5 <1> (1,5-anhydro-D-fructose, <1> pH 6.5, mutant enzyme S33D/A13G, cofactor: NADH [2]) [2]

63.2 <2> (D-glucosone, <2> pH 6.5, 30°C [1]) [1]

119 <1> (1,5-anhydro-D-fructose, <1> pH 6.5, mutant enzyme S10G, cofactor: NADPH [2]) [2]

145 <1> (1,5-anhydro-D-fructose, <1> pH 6.5 recombinant wild-type enzyme, cofactor: NADPH [2]) [2]

156 <1> (1,5-anhydro-D-fructose, <1> pH 6.5, mutant enzyme G206I, cofactor: NADPH [2]) [2]

216 <1> (1,5-anhydro-D-fructose, <1> pH 6.5, native wild-type enzyme, cofactor: NADPH [2]) [2]

369 <1> (1,5-anhydro-D-fructose, <1> pH 6.5, mutant enzyme S10G/A13G, cofactor: NADPH [2]) [2]

405 <1> (1,5-anhydro-D-fructose, <1> pH 6.5, mutant enzyme A13G, cofactor: NADPH [2]) [2]

1300 <2> (1,5-anhydro-D-fructose, <2> pH 6.5, 30°C [1]) [1]

Specific activity (U/mg)

288.9 <2> (<2> native enzyme [1]) [1]
484 <2> (<2> recombinant enzyme expressed in Escherichia coli [1]) [1]

K_m-Value (mM)

0.02 <1> (NADPH, <1> pH 6.5, mutant enzyme A13G [2]) [2]
0.06 <1> (NADPH, <1> pH 6.5 recombinant wild-type enzyme [2]; <1> pH 6.5, mutant enzyme G206I [2]) [2]
0.1 <1> (NADPH, <1> pH 6.5, native wild-type enzyme [2]) [2]
0.2 <1, 2> (NADPH, <2> pH 6.5, 30°C [1]; <1> pH 6.5, mutant enzyme K94G [2]) [1, 2]
0.27 <1> (NADPH, <1> pH 6.5, mutant enzyme S10G [2]) [2]
0.38 <1> (NADPH, <1> pH 6.5, mutant enzyme S10G/A13G [2]) [2]
1 <1> (NADPH, <1> pH 6.5, mutant enzyme S33D/A13G [2]) [2]
1.1 <1> (NADH, <1> pH 6.5, mutant enzyme A13G [2]; <1> pH 6.5, mutant enzyme S33D/A13G [2]) [2]
1.2 <1> (NADH, <1> pH 6.5, mutant enzyme S10G/A13G [2]) [2]
3.2 <1> (1,5-anhydro-D-fructose, <1> pH 6.5, mutant enzyme S33D/A13G, cofactor: NADH [2]) [2]
3.5 <1> (1,5-anhydro-D-fructose, <1> pH 6.5, mutant enzyme S10G, cofactor: NADPH [2]) [2]
6.4 <1> (1,5-anhydro-D-fructose, <1> pH 6.5 recombinant wild-type enzyme, cofactor: NADPH [2]) [2]
7.1 <1> (1,5-anhydro-D-fructose, <1> pH 6.5, mutant enzyme S10G/A13G, cofactor: NADPH [2]) [2]
8.3 <1> (1,5-anhydro-D-fructose, <1> pH 6.5, mutant enzyme G206I, cofactor: NADPH [2]; <1> pH 6.5, native wild-type enzyme, cofactor: NADPH [2]) [2]
8.4 <2> (1,5-anhydro-D-fructose, <2> pH 6.5, 30°C [1]) [1]
8.5 <1> (1,5-anhydro-D-fructose, <1> pH 6.5, mutant enzyme A13G, cofactor: NADPH [2]) [2]
8.9 <1> (1,5-anhydro-D-fructose, <1> pH 8.0, mutant enzyme H180A, cofactor: NADPH [2]) [2]
11 <2> (D-glucosone, <2> pH 6.5, 30°C [1]) [1]
11.1 <1> (1,5-anhydro-D-fructose, <1> pH 6.5, mutant enzyme A13G, cofactor: NADH [2]) [2]
20.2 <1> (1,5-anhydro-D-fructose, <1> pH 6.5, mutant enzyme S33D/A13G, cofactor: NADPH [2]) [2]
22.5 <1> (1,5-anhydro-D-fructose, <1> pH 6.5, mutant enzyme K94G, cofactor: NADPH [2]) [2]
39 <1> (1,5-anhydro-D-fructose, <1> pH 6.5, mutant enzyme S10G/A13G, cofactor: NADH [2]) [2]
49 <1> (1,5-anhydro-D-fructose, <1> pH 7.5, mutant enzyme D176A, cofactor: NADPH [2]) [2]

pH-Optimum

6.5 <2> [1]

pH-Range

5.2-8.8 <2> (<2> about 50% of maximal activity at pH 5.5 and at pH 8.8 [1]) [1]

4 Enzyme Structure

Molecular weight

38200 <2> (<2> gel filtration [1]) [1]

Subunits

monomer <2> (<2> 1 * 40000, SDS-PAGE [1]; <2> 1 * 35100, MALDI-TOF-MS [1]) [1]

5 Isolation/Preparation/Mutation/Application

Purification

<2> (native and recombinant enzyme) [1]

Crystallization

<1> (hanging drop vapor diffusion method, enzyme crystallized in complex with the cofactor NADP(H) and its structure is determined to 2.2 A resolution using selenomethionine single-wavelength anomalous dispersion) [2]

Cloning

<1> (expression in Escherichia coli) [2]
<2> (overexpression in Escherichia coli) [1]

Engineering

A13G <1> (<1> k_{cat}/K_m for reaction with 1,5-anhydro-D-fructose and NADPH is 1.8fold higher than wild type value. Mutant enzyme shows activity with NADH as cofactor [2]) [2]
G206I <1> (<1> k_{cat}/K_m for reaction with 1,5-anhydro-D-fructose and NADPH is 1.4fold lower than wild type value [2]) [2]
H180A <1> (<1> k_{cat}/K_m for reaction with 1,5-anhydro-D-fructose and NADPH is 61.9fold lower than wild type value [2]) [2]
K94G <1> (<1> k_{cat}/K_m for reaction with 1,5-anhydro-D-fructose and NADPH is 137fold lower than wild type value [2]) [2]
S10G <1> (<1> k_{cat}/K_m for reaction with 1,5-anhydro-D-fructose and NADPH is 1.3fold higher than wild type value [2]) [2]
S10G/A13G <1> (<1> k_{cat}/K_m for reaction with 1,5-anhydro-D-fructose and NADPH is 2fold higher than wild type value. Mutant enzyme shows activity with NADH as cofactor [2]) [2]
S176A <1> (<1> k_{cat}/K_m for reaction with 1,5-anhydro-D-fructose and NADPH is 1001fold lower than wild type value [2]) [2]
S33D <1> (<1> no activity [2]) [2]

S33D/A13G <1> (<1> k_{cat}/K_m for reaction with 1,5-anhydro-D-fructose and NADPH is 84fold lower than wild type value. Mutant enzyme shows activity with NADH as cofactor [2]) [2]

6 Stability

Storage stability

<2>, -20°C or 0°C, 50 days, 50% loss of activity [1]

References

[1] Kuehn, A.; Yu, S.; Giffhorn, F.: Catabolism of 1,5-anhydro-D-fructose in Sinorhizobium morelense S-30.7.5: discovery, characterization, and overexpression of a new 1,5-anhydro-D-fructose reductase and its application in sugar analysis and rare sugar synthesis. Appl. Environ. Microbiol., **72**, 1248-1257 (2006)

[2] Dambe, T.R.; Kuehn, A.M.; Brossette, T.; Giffhorn, F.; Scheidig, A.J.: Crystal structure of NADP(H)-dependent 1,5-anhydro-D-fructose reductase from Sinorhizobium morelense at 2.2 A resolution: construction of a NADH-accepting mutant and its application in rare sugar synthesis. Biochemistry, **45**, 10030-10042 (2006)

Chlorophyll(ide) b reductase 1.1.1.294

1 Nomenclature

EC number
1.1.1.294

Systematic name
71-hydroxychlorophyllide-a:NAD(P)$^+$ oxidoreductase

Recommended name
chlorophyll(ide) b reductase

Synonyms
CBR <1> [4]
Chl b reductase <1> [3, 4]
chlorophyll b reductase <1> [3]
chlorophyll(ide) b reductase <1> [1, 3]

CAS registry number
185915-08-6

2 Source Organism

<1> *Hordeum vulgare* (no sequence specified) [1, 3, 4]
<2> *Helianthus annuus* (no sequence specified) [2]

3 Reaction and Specificity

Catalyzed reaction
71-hydroxychlorophyllide a + NAD(P)$^+$ = chlorophyllide b + NAD(P)H + H$^+$

Natural substrates and products
S Additional information <1> (<1> chlorophyll b reduction is considered to be an early and obligatory step of chlorophyll b breakdown [1]) (Reversibility: ?) [1]
P ?

Substrates and products
S Zn-pheophorbide b + NADPH + H$^+$ <1> (Reversibility: ?) [1]
P ?

S zinc (132R)-pheophorbide b + NADPH + H^+ <1> (<1> zinc (132S)-pheophorbide b is a better substrate than zinc (132R)-pheophorbide b [3]) (Reversibility: ?) [3]
P ?
S zinc (132S)-pheophorbide b + NADPH + H^+ <1> (<1> zinc (132S)-pheophorbide b is a better substrate than zinc (132R)-pheophorbide b [3]) (Reversibility: ?) [3]
P ?
S Additional information <1> (<1> chlorophyll b reduction is considered to be an early and obligatory step of chlorophyll b breakdown [1]) (Reversibility: ?) [1]
P ?

Cofactors/prosthetic groups

NADH <1> (<1> reaction requires NADPH or NADH. NADH is less effective [3]) [3]
NADPH <1> (<1> reaction requires NADPH or NADH. NADH is less effective [3]) [3]

5 Isolation/Preparation/Mutation/Application

Source/tissue

cotyledon <2> (<2> of wild-type [2]) [2]
gerontoplast <1> [1]
leaf <1, 2> (<1> a pronounced maximum of Chl b reductase activity at day 2 of senescence [1]; <2> of a light-grown chlorophyll-deficient mutant of Helianthus annuus [2]) [1, 2]
seedling <1> [1]

Localization

etioplast <1> [3]
plastid <1> (<1> the bulk of activity (83%) is found in the thylakoids and only traces (5%) in the envelope fraction [1]) [1]

References

[1] Scheumann, V.; Schoch, S.; Rudiger, W.: Chlorophyll b reduction during senescence of barley seedlings. Planta, **209**, 364-370 (1999)
[2] Vezitskii, A.Y.; Lezhneva, L.A.; Scherbakov, R.A.; Rassadina, V.V.; Averina, N.G.: Activity of chlorophyll synthetase and chlorophyll b reductase in the chlorophyll-deficient plastome mutant of sunflower. Russ. J. Plant Physiol., **46**, 502-506 (1999)

[3] Scheumann, V.; Ito, H.; Tanaka, A.; Schoch, S.; Rudiger, W.: Substrate specificity of chlorophyll(ide) b reductase in etioplasts of barley (Hordeum vulgare L.). Eur. J. Biochem., **242**, 163-170 (1996)
[4] Ruediger, W.: Biosynthesis of chlorophyll b and the chlorophyll cycle. Photosynth. Res., **74**, 187-193 (2002)

Quinoprotein glucose dehydrogenase 1.1.5.2

1 Nomenclature

EC number
1.1.5.2

Systematic name
D-glucose:ubiquinone oxidoreductase

Recommended name
quinoprotein glucose dehydrogenase

Synonyms
Asd <1> [53]
D-glucose:(pyrroloquinoline-quinone) 1-oxidoreductase
EC 1.1.99.17
GDH <14, 21> [39, 40]
GDH-B <7> [35]
PQQ glucose dehydrogenase <7, 23> [35, 42, 43, 45, 51]
PQQ-GDH <7, 14, 19> [41, 52]
PQQ-dependent GDH <14> [36]
PQQ-dependent glucose dehydrogenase <14> [36, 40]
PQQ-dependent soluble glucose dehydrogenase <1, 7, 10> [50]
PQQ-glucose dehydrogenase <7, 14> [41]
PQQGDH <1, 7> [43, 44]
PQQGDH-B <1, 7, 23> [34, 42, 45, 46, 48, 49, 51]
aldose sugar dehydrogenase <1> [53]
dehydrogenase, glucose (pyrroloquinoline-quinone)
glucose dehydrogenase <1, 7, 23> [38, 48, 49]
glucose dehydrogenase (PQQ dependent)
glucose dehydrogenase (pyrroloquinoline-quinone)
m-GDH <14> [41]
mGDH <1> [38]
membrane glucose dehydrogenase <21> [39]
pyrrolo-quinoline-quinone-linked glucose dehydrogenase <19> [52]
pyrroloquinoline quinone glucose dehydrogenase <7> [46]
quinoprotein D-glucose dehydrogenase
quinoprotein glucose DH
quinoprotein glucose dehydrogenase <1, 7, 10, 19, 23> [50, 51, 52, 53]
s-GDH <7> [41]
sGDH <1, 7, 10, 22, 23> [37, 38, 47, 50]
soluble glucose dehydrogenase <7, 22> [37, 47]

water-soluble PQQ glucose dehydrogenase <1> [34]
water-soluble pyrroquinoline quinone glucose dehydrogenase <1> [44]

CAS registry number
81669-60-5

2 Source Organism

<1> *Escherichia coli* (no sequence specified) [4, 5, 13, 18, 20, 21, 23, 24, 25, 26, 28, 34, 38, 44, 49, 50, 53]
<2> *Acetobacter aceti* (no sequence specified) [13]
<3> *Pseudomonas fluorescens* (no sequence specified) [14, 15]
<4> *Pseudomonas aeruginosa* (no sequence specified) [2, 13]
<5> *Klebsiella aerogenes* (no sequence specified) [2,4]
<6> *Pseudomonas sp.* (no sequence specified) [4,8]
<7> *Acinetobacter calcoaceticus* (no sequence specified) (<7> BCA2 [29]) [1, 3, 4, 8, 10, 11, 12, 13, 17, 18, 22, 27, 29, 30, 31, 32, 33, 35, 37, 41, 42, 43, 45, 46, 48, 50]
<8> *Klebsiella pneumoniae* (no sequence specified) [6,13]
<9> *Azotobacter vinelandii* (no sequence specified) [7]
<10> *Streptomyces coelicolor* (no sequence specified) [50]
<11> *Agrobacterium tumefaciens* (no sequence specified) [7]
<12> *Rhizobium leguminosarum* (no sequence specified) [7]
<13> *Gluconobacter oxydans* (no sequence specified) [4]
<14> *Erwinia sp.* (no sequence specified) [36, 40, 41]
<15> *Agrobacterium radiobacter* (no sequence specified) [7]
<16> *Gluconobacter oxydans subsp. suboxydans* (no sequence specified) [9, 13, 16]
<17> *Sinorhizobium meliloti* (no sequence specified) [19]
<18> *Rhizobium tropici* (no sequence specified) [19]
<19> *Gluconacetobacter diazotrophicus* (no sequence specified) [52]
<20> *Acinetobacter Iwoffii* (no sequence specified) [7]
<21> *Escherichia coli* (UNIPROT accession number: P15877) [39]
<22> *Acinetobacter calcoalceticus* (no sequence specified) [47]
<23> *Acinetobacter calcoaceticus* (UNIPROT accession number: P13650) [38, 51]

3 Reaction and Specificity

Catalyzed reaction
D-glucose + ubiquinone = D-glucono-1,5-lactone + ubiquinol (<7> ping-pong mechanism [4]; <14> catalytic mechanism [36]; <21,23> active site structure [38,39]; <7> hexa uni ping-pong mechanism [3]; <1> active site structure contains a loop 6BC region which does not directly interact with the substrates [34]; <7> active site structure from crystal structure, and de-

tailed catalytic mechanism [37]; <1> active site structure, putative catalytic mechanism and cycle involving Asp466 and Lys493 [38]; <22> catalytic mechanism, His144, Arg228, and Asn229 are involved [47])

Reaction type

oxidation
redox reaction
reduction

Natural substrates and products

S D-glucose + ? <23> (<23> physiological electron acceptor in not known [38]) (Reversibility: ?) [38]
P D-glucono-1,5-lactone + ?
S D-glucose + pyrroloquinoline quinone <7, 16> (<16> enzyme donates electrons directly to ubiquinone in the respiratory chain [9]; <7> membrane-bound enzyme functions by linking to the respiratory chain via ubiquinone [12]; <16> linked to the respiratory chain of a wide variety of bacteria [9]) (Reversibility: ?) [9, 12]
P D-glucono-1,5-lactone + pyrroloquinoline quinol
S D-glucose + ubiquinone <1, 7, 10, 19, 21, 23> (<21> transfers electrons to the cytochrome oxidase through ubiquinone in the electron transport chain [39]) (Reversibility: ?) [35, 39, 45, 50, 51, 52, 53]
P D-glucono-1,5-lactone + ubiquinol
S D-glucose + ubiquinone-7 <1> (<1> a pivotal PQQ-containing quinoprotein coupled to the respiratory chain in the periplasmic oxidation of alcohols and sugars [38]) (Reversibility: ?) [38]
P D-glucono-1,5-lactone + ubiquinol
S β-D-glucose + ubiquinone <7> (Reversibility: ?) [37]
P β-D-glucono-1,5-lactone + ubiquinol
S Additional information <1, 8, 23> (<1> the membrane-bound enzyme donates electrons directly to ubiquinone during the oxidation of D-glucose, and these electrons are subsequently transferred to ubiquinol oxidase in the respiratory chain [20]; <8> activity of the enzyme is regulated by both the glucose dehydrogenase apo-enzyme synthesis and the synthesis of the cofactor pyrroloquinoline quinone [6]; <1> the enzyme has a ubiquinone reacting site close to the periplasmic side of the membrane and thus its electron transfer to ubiquinone appears to be incapable of forming a proton electrochemical gradient across the inner membrane [24]; <1,23> evolutionary analysis of PQQ-containing proteins, overview [38]) (Reversibility: ?) [6, 20, 24, 38]
P ?

Substrates and products

S 2-amino-D-glucose + pyrroloquinoline quinone <1> (Reversibility: ?) [26]
P 2-amino-D-glucono-1,5-lactone + pyrroloquinoline quinol
S 2-deoxy-D-glucose + 2,6-dichlorophenolindolphenol <7> (Reversibility: ?) [42]
P 2-deoxy-D-glucono-1,5-lactone + ?

S 2-deoxy-D-glucose + N-methylphenazonium methyl sulfate <7> (Reversibility: ?) [33]
P 2-deoxy-D-glucono-1,5-lactone + ?
S 2-deoxy-D-glucose + pyrroloquinoline quinone <1, 7> (Reversibility: ?) [8, 26, 29]
P 2-deoxy-D-glucono-1,5-lactone + pyrroloquinoline quinol
S 2-deoxy-D-glucose + ubiquinone <1, 7, 21> (<1> low activity [44]; <7> low activity, recombinant isozyme PQQGDH-B [46]) (Reversibility: ?) [39, 44, 46]
P 2-deoxy-D-glucono-1,5-lactone + ubiquinol
S 3-O-methyl-D-glucose + 2,6-dichlorophenolindolphenol <7> (Reversibility: ?) [42, 43, 48]
P 3-O-methyl-D-glucono-1,5-lactone + ?
S 3-O-methyl-D-glucose + pyrroloquinoline quinone <1, 7> (Reversibility: ?) [26, 29]
P 3-O-methyl-D-glucono-1,5-lactone + pyrroloquinoline quinol
S 3-O-methyl-D-glucose + ubiquinone <1, 7, 21> (<7> recombinant isozyme PQQGDH-B [46]) (Reversibility: ?) [34, 39, 44, 46]
P 3-O-methyl-D-glucono-1,5-lactone + ubiquinol
S 3-deoxy-D-glucose + pyrroloquinoline quinone <1> (Reversibility: ?) [26]
P 3-deoxy-D-glucono-1,5-lactone + pyrroloquinoline quinol
S 6-deoxy-D-glucose + pyrroloquinoline quinone <1> (Reversibility: ?) [26]
P 6-deoxy-D-glucono-1,5-lactone + pyrroloquinoline quinol
S D-allose + pyrroloquinoline quinone <1, 7> (Reversibility: ?) [8, 26, 29]
P D-allono-1,5-lactone + pyrroloquinoline quinol
S D-fucose + N-ethylphenazonium ethyl sulfate <7> (Reversibility: ?) [33]
P 6-deoxy-D-galactono-1,5-lactone + ?
S D-fucose + N-methylphenazonium methyl sulfate <7> (Reversibility: ?) [33]
P 6-deoxy-D-galactono-1,5-lactone + ?
S D-fucose + pyrrolquinoline quinone <1> (Reversibility: ?) [26]
P 6-deoxy-D-galactono-1,5-lactone + pyrroloquinoline quinol
S D-galactose + 2,6-dichlorophenolindolphenol <7> (Reversibility: ?) [42, 43, 48]
P D-galactono-1,5-lactone + ?
S D-galactose + N-methylphenazonium methyl sulfate <7> (Reversibility: ?) [33]
P D-galactono-1,5-lactone + ?
S D-galactose + pyrroloquinoline quinone <1, 3, 7> (<3> 6.5% of the activity with D-glucose [14]; <7> 30% of the activity with D-glucose [4]) (Reversibility: ?) [4, 8, 14, 26, 29]
P D-galactono-1,5-lactone + pyrroloquinoline quinol
S D-galactose + ubiquinone <23> (Reversibility: ?) [51]
P ?
S D-galactose + ubiquinone <1, 7, 14, 21> (<1> low activity, recombinant wild-type and mutant isozymes PQQGDH-B [34]; <7> recombinant isozyme PQQGDH-B [46]) (Reversibility: ?) [34, 39, 41, 44, 46]

P D-galactono-1,5-lactone + ubiquinol
S D-glucose + 1,4-naphthoquinone <14> (Reversibility: ?) [40]
P D-glucono-1,5-lactone + ?
S D-glucose + 2,3-dichloro-1,4-naphthoquinone <14> (<14> 2,3-dichloro-1,4-naphthoquinone shows low effectivity as redox mediator [40]) (Reversibility: ?) [40]
P D-glucono-1,5-lactone + ?
S D-glucose + 2,3-dimethoxy-5-methyl-1,4-benzoquinone <1> (Reversibility: ?) [28]
P D-glucono-1,5-lactone + ?
S D-glucose + 2,6-dichlorophenol-indophenol <1, 3, 7, 16> (Reversibility: ?) [4, 9, 10, 15, 16, 20, 24, 27, 28, 29, 33]
P D-glucono-1,5-lactone + ?
S D-glucose + 2,6-dichlorophenolindolphenol <7> (<7> best substrate [42]) (Reversibility: ?) [35, 42, 43, 45, 48]
P D-glucono-1,5-lactone + ?
S D-glucose + 2-methyl-6-methoxy-1,4-benzoquinone <14> (Reversibility: ?) [40]
P D-glucono-1,5-lactone + ?
S D-glucose + 4-(4-ferrocenylimino-methyl)phenol <7> (Reversibility: ?) [41]
P D-glucono-1,5-lactone + ?
S D-glucose + 4-ferrocenylnitrophenol <7> (Reversibility: ?) [41]
P D-glucono-1,5-lactone + ?
S D-glucose + 4-ferrocenylphenol <7> (Reversibility: ?) [41]
P D-glucono-1,5-lactone + ?
S D-glucose + 9,10-phenanthrenequinone <14> (<14> 9,10-phenanthrenequinone shows low effectivity as redox mediator [40]) (Reversibility: ?) [40]
P D-glucono-1,5-lactone + ?
S D-glucose + ? <23> (<23> physiological electron acceptor in not known [38]) (Reversibility: ?) [38]
P D-glucono-1,5-lactone + ?
S D-glucose + N-ethylphenazonium ethyl sulfate <7> (Reversibility: ?) [33]
P D-glucono-1,5-lactone + ?
S D-glucose + N-methylphenazonium methyl sulfate <7> (Reversibility: ?) [33]
P D-glucono-1,5-lactone + ?
S D-glucose + ferricyanide <1, 3, 16> (Reversibility: ?) [14, 16, 28]
P D-glucono-1,5-lactone + ferrocyanide
S D-glucose + phenazine methosulfate <1, 3, 7, 16> (Reversibility: ?) [9, 10, 14, 16, 21, 28, 29]
P D-glucono-1,5-lactone + ?
S D-glucose + pyrroloquinoline quinone <1, 7, 16> (<16> enzyme donates electrons directly to ubiquinone in the respiratory chain [9]; <7> membrane-bound enzyme functions by linking to the respiratory chain via

ubiquinone [12]; <16> linked to the respiratory chain of a wide variety of bacteria [9]) (Reversibility: ?) [3, 4, 8, 9, 11, 12, 16, 21, 23, 26, 32]
P D-glucono-1,5-lactone + pyrroloquinoline quinol <7> [3, 4]
S D-glucose + trimethyl-1,4-benzoquinone <14> (Reversibility: ?) [40]
P D-glucono-1,5-lactone + ?
S D-glucose + ubiquinone <1, 7, 10, 14, 19, 21, 22, 23> (<7,14> best substrate [41,42]; <21> transfers electrons to the cytochrome oxidase through ubiquinone in the electron transport chain [39]; <1> best substrate, recombinant wild-type and mutant isozymes PQQGDH-B [34]; <7> recombinant isozyme PQQGDH-B [46]) (Reversibility: ?) [34, 35, 36, 38, 39, 40, 41, 42, 43, 44, 45, 46, 47, 48, 50, 51, 52, 53]
P D-glucono-1,5-lactone + ubiquinol
S D-glucose + ubiquinone Q_1 <3, 7> (Reversibility: ?) [12, 14]
P D-glucono-1,5-lactone + ?
S D-glucose + ubiquinone Q_2 <1, 3, 7> (Reversibility: ?) [12, 14, 15, 20]
P D-glucono-1,5-lactone + ?
S D-glucose + ubiquinone Q_4 <3> (Reversibility: ?) [14]
P D-glucono-1,5-lactone + ?
S D-glucose + ubiquinone Q_6 <3, 7> (Reversibility: ?) [12, 14, 15]
P D-glucono-1,5-lactone + ?
S D-glucose + ubiquinone Q_9 <3, 7> (Reversibility: ?) [12, 15]
P D-glucono-1,5-lactone +
S D-glucose + ubiquinone-7 <1> (<1> a pivotal PQQ-containing quinoprotein coupled to the respiratory chain in the periplasmic oxidation of alcohols and sugars [38]; <1> with ubiquinone-8 as minor compound, PQQ acts in electron transfer between enzyme and ubiquinone [38]) (Reversibility: ?) [38]
P D-glucono-1,5-lactone + ubiquinol
S D-mannose + 2,6-dichlorophenolindolphenol <7> (Reversibility: ?) [42]
P ?
S D-mannose + pyrroloquinoline quinone <1, 3, 7, 16> (<16> no activity [16]; <3> 8.6% of the activity with D-glucose [14]) (Reversibility: ?) [8, 14, 16, 26, 29]
P D-mannono-1,5-lactone + pyrroloquinoline quinol
S D-mannose + ubiquinone <7, 21> (<7> recombinant isozyme PQQGDH-B [46]) (Reversibility: ?) [39, 46]
P ? + ubiquinol
S D-melibiose + pyrroloquinoline quinone <1, 7> (<7> 10% of the activity with D-glucose [4]) (Reversibility: ?) [4, 26]
P ?
S D-ribose + pyrroloquinoline quinone <1> (Reversibility: ?) [26]
P D-ribono-1,5-lactone + pyrroloquinoline quinol
S D-ribose + ubiquinone <21> (Reversibility: ?) [39]
P ? + ubiquinol
S D-xylose + 2,6-dichlorophenolindolphenol <7> (Reversibility: ?) [42]
P D-xylono-1,5-lactone + ?
S D-xylose + N-ethylphenazonium ethyl sulfate <7> (Reversibility: ?) [33]

P D-xylono-1,5-lactone + ?

S D-xylose + N-methylphenazonium methyl sulfate <7> (Reversibility: ?) [33]

P D-xylono-1,5-lactone + ?

S D-xylose + pyrroloquinoline quinone <1, 3, 7, 16> (<16> no activity [16]; <3> 13% of the activity with D-glucose [14]; <7> 20% of the activity with D-glucose [4]) (Reversibility: ?) [4, 8, 14, 16, 26, 29]

P D-xylono-1,5-lactone + pyrroloquinoline quinol

S D-xylose + ubiquinone <1, 7, 21> (<7> recombinant isozyme PQQGDH-B [46]) (Reversibility: ?) [39, 44, 46]

P D-xylono-1,5-lactone + ubiquinol

S L-arabinose + N-ethylphenazonium ethyl sulfate <7> (Reversibility: ?) [33]

P L-arabino-1,5-lactone + ?

S L-arabinose + N-methylphenazonium methyl sulfate <7> (Reversibility: ?) [33]

P L-arabino-1,5-lactone + ?

S L-arabinose + pyrroloquinoline quinone <1, 3, 7> (<3> 2.8% of the activity with D-glucose [14]; <7> 35% of the activity with D-glucose [4]) (Reversibility: ?) [4, 8, 14, 26]

P L-arabino-1,5-lactone + pyrroloquinoline quinol

S L-arabinose + ubiquinone <7, 14, 21> (Reversibility: ?) [39, 41]

P L-arabino-1,5-lactone + ubiquinol

S L-lyxose + pyrroloquinoline quinone <1> (Reversibility: ?) [26]

P ?

S L-rhamnose + pyrroloquinoline quinone <7, 16> (<16> no activity [16]; <7> 7.5% of the activity with D-glucose [4]) (Reversibility: ?) [4, 16]

P L-rhamnono-1,5-lactone + pyrroloquinoline quinol

S allose + 2,6-dichlorophenolindolphenol <7> (Reversibility: ?) [42, 43, 48]

P ?

S allose + ubiquinone <1, 7> (<7> recombinant isozyme PQQGDH-B [46]) (Reversibility: ?) [34, 44, 46]

P ? + ubiquinol

S β-D-glucose + ubiquinone <7> (<7> redox-related structural changes, overview [37]) (Reversibility: ?) [37]

P β-D-glucono-1,5-lactone + ubiquinol

S cellobiose + 2,6-dichlorophenolindolphenol <7> (Reversibility: ?) [42]

P ?

S cellobiose + pyrroloquinoline quinone <7> (<7> 70% of the activity with D-glucose [4]) (Reversibility: ?) [4]

P ?

S cellobiose + ubiquinone <7> (<7> recombinant isozyme PQQGDH-B [46]) (Reversibility: ?) [41, 46]

P ? + ubiquinol

S lactose + 2,6-dichlorophenol-indophenol <7> (Reversibility: ?) [27]

P ?

S lactose + 2,6-dichlorophenolindolphenol <7> (Reversibility: ?) [35, 42, 43, 48]
P ?
S lactose + pyrroloquinoline quinone <7, 16> (<16> no activity [16]; <7> 65% of the activity with D-glucose [4]) (Reversibility: ?) [4, 16, 29]
P ?
S lactose + ubiquinone <23> (Reversibility: ?) [51]
P ?
S lactose + ubiquinone <1, 7> (<7> recombinant isozyme PQQGDH-B [46]) (Reversibility: ?) [34, 41, 44, 46]
P ? + ubiquinol
S maltose + 2,6-dichlorophenolindolphenol <7> (Reversibility: ?) [35, 42, 43, 48]
P ?
S maltose + pyrroloquinoline quinone <3, 7, 16> (<3> 3.2% of the activity with D-glucose [14]; <16> 5% of the activity with D-glucose [16]; <7> 90% of the activity with D-glucose [4]) (Reversibility: ?) [4, 14, 15, 16, 29]
P ?
S maltose + ubiquinone <23> (Reversibility: ?) [51]
P ?
S maltose + ubiquinone <1, 7> (<7> recombinant isozyme PQQGDH-B [46]) (Reversibility: ?) [34, 41, 44, 46]
P ? + ubiquinol
S melibiose + 2,6-dichlorophenol-indophenol <7> (Reversibility: ?) [27]
P ?
S Additional information <1, 7, 8, 14, 21, 23> (<7> the oxidation of dissacharides by the enzyme can be considered as an in vitro artefact caused by the removal of the enzyme from its natural environment [8]; <7> absolute specificity with respect to the C1 position, only sugars are oxidized which have the same configuration of the H/OH substituents at this site as the β-anomer of glucose. Absolute specificity with respect to the overall conformation of the sugar molecule, sugars with a 4C1 chair conformation are substrates, those with a 1C4 one are not. The nature and configuration of the substituents at the 3-position are hardly relevant for activity, and an equatorial pyranose group at the 4-position exhibits only a specific hindering of the binding of the aldose moiety of a disaccharide [22]; <1> the membrane-bound enzyme donates electrons directly to ubiquinone during the oxidation of D-glucose, and these electrons are subsequently transferred to ubiquinol oxidase in the respiratory chain [20]; <8> activity of the enzyme is regulated by both the glucose dehydrogenase apo-enzyme synthesis and the synthesis of the cofactor pyrroloquinoline quinone [6]; <1> the enzyme has a ubiquinone reacting site close to the periplasmic side of the membrane and thus its electron transfer to ubiquinone appears to be incapable of forming a proton electrochemical gradient across the inner membrane [24]; <1,23> evolutionary analysis of PQQ-containing proteins, overview [38]; <14> 2-hydroxy-1,4-naphthoquinone, tetramethyl-1,4-benzoquinone, and 2-methyl-1,4-naphthoqui-

none are no redox mediators for the enzyme [40]; <7> substrate specificities of recombinant wild-type and mutant enzymes, overview [42]; <1> substrate specificities of wild-type and mutant isozyme PQQGDH-B, overview [34]; <7> substrate specificity of native and immobilized enzyme [41]; <14> substrate specificity of native and immobilized enzyme, maltose is a poor substrate [41]; <21> substrate specificity of recombinant wild-type and mutant enzymes [39]; <1> substrate specificity of the enzyme in presence or absence of different recombinant peptide ligands, overview [44]; <7> substrate specificity of the recombinant enzyme produced in Pichia pastoris [46]) (Reversibility: ?) [6, 8, 20, 22, 24, 34, 38, 39, 40, 41, 42, 44, 46]

P ?

Inhibitors

Ba^{2+} <21> (<21> competitive to Mg^{2+}, wild-type enzyme [39]) [39]
Ca^{2+} <21> (<21> competitive to Mg^{2+}, wild-type enzyme [39]) [39]
D-glucose <7> (<7> substrate inhibition at high concentrations [3]) [3]
EDTA <3> (<3> 3.3 mM, complete inhibition [14]) [14]
Mg^{2+} <21> (<21> competitive to Ca^{2+}, Sr^{2+}, or Ba^{2+}, mutants D354N, N355D, and D354N/N355D [39]) [39]
pyrroloquinoline quinone <7> (<7> substrate inhibition at high concentrations [3]) [3]
Sr^{2+} <21> (<21> competitive to Mg^{2+}, wild-type enzyme [39]) [39]
methylhydrazine <7> (<7> competitive [32]) [32]
p-benzoquinone <3> (<3> 1.7 mM, complete inhibition [14]) [14]
Additional information <3, 7> (<3> purified enzyme is inactivated upon removal of detergent by acetone treatment. The detergent-depleted enzyme is partially activated by Triton X-100 [15]; <7> the protein region responsible for complete EDTA tolerance is located between 32% and 59% from the N-terminus, A27 region [18]; <7> poor inhibition of isozyme PQQGDH-B by 5 mM EDTA [46]) [15, 18, 46]

Cofactors/prosthetic groups

nicotinamide adenine dinucleotide <19> (<19> intracellular oxidative route [52]) [52]
pyrroloquinoline quinone <1, 2, 3, 4, 5, 6, 7, 8, 9, 10, 11, 12, 13, 14, 15, 16, 17, 18, 19, 20, 21, 22, 23> (<1, 3, 4, 5, 6, 7, 8, 9, 11, 12, 13, 15, 16, 17, 18, 20> prosthetic group [1, 2, 3, 4, 5, 6, 7, 9, 10, 14, 15, 16, 19, 21, 23, 24, 25, 26, 27, 28, 30, 34, 46]; <1> during the processing of pyrroloquinoline quinone into the apoenzyme to give active enzyme, its affinity is markedly dependent on the pH, four groups with pK values between pH 7 and pH 8 are involved [26]; <7> each subunit of the dimer contains one molecule of pyrroloquinoline quinone [3]; <7> reconstitution of the apoenzyme to full activity is achieved with a stoichiometric amount of pyrroloquinoline quinone. Mg^{2+} anchors pyrroloquinoline quinone cofactor to the enzyme protein and activates the bound cofactor [33]; <7> cofactor required [32]; <1,5,6> type I enzyme: pyrroloquinoline quinone can be removed by dialysis against EDTA-containg buffers [4]; <7,13> type II enzyme: pyrroloquinoline quinone can not be re-

moved by dialysis against EDTA-containg buffers [4]; <17> Sinorhizobium meliloti is unable to synthesize pyrroloquinoline quinone, synthesis of the holoenzyme in alfalfa nodules [19]; <2,4,7,16> organisms with an active holoenzyme [13]; <1,8> organisms with an inactive apoenzyme [13]; <1> the apoenzyme is converted to the holoenzyme with exogenous pyrroloquinoline quinone and Mg^{2+}. The holoenzyme gradually returns to the apoenzyme in absence of pyrroloquinoline quinone and/or Mg^{2+} [28]; <7> i.e. PQQ [43]; <7,14,23> i.e. PQQ, dependent on [35,36,38,40,41,42,48]; <1> i.e. PQQ, dependent on exogenous PQQ, since Escherichia coli does not synthesize PQQ itself [38]; <21> i.e. PQQ, dependent on, Escherichia coli needs to be reconstituted with PQQ for activity [39]; <7> i.e. PQQ, dependent on, functions as a redox mediator by transfer of hydride ions and electrons, redox-related structural changes, overview [37]; <22> i.e. PQQ, or 2,7,9-tricarboxy-1H-pyrrolo [2,3-f]-quinoline-4,5-dione, residues Gln231, Gln246, Ala350, and Leu376 are involved in binding [47]; <19> periplasmatic oxidative route [52]; <1,7,10> PQQ [50,53]) [1, 2, 3, 4, 5, 6, 7, 9, 10, 13, 14, 15, 16, 19, 21, 23, 24, 25, 26, 27, 28, 30, 32, 33, 34, 35, 36, 37, 38, 39, 40, 41, 42, 43, 44, 45, 46, 47, 48, 50, 51, 52, 53]
Additional information <14> (<14> ruthenium(III) bispyridine compounds and ruthenium(III) 4-methyl-bispyridine compounds can act as artificial electron transfer mediator system [36]) [36]

Metals, ions

Ba^{2+} <21> (<21> mutants D354N, N355D, and D354N/N355D [39]) [39]
Ca^{2+} <1, 7, 9, 14, 20, 21, 22, 23> (<1, 7, 14, 22> required [34,41,42,43,44,47]; <7> required for dimerization of the subunits as well as for functionalization of the bound pyrroloquinoline quinone. Binding of Ca^{2+} is much stronger in the holoenzyme than in the apoenzyme [30]; <9,20> in vivo reconstitution of apoenzyme with the prosthetic group is dependent on the presence of Ca^{2+} or Mg^{2+} [7]; <1> catalytic activity of the membrane-bound enzyme with Ca^{2+} is very similar to that with Mg^{2+} [28]; <7> pyrroloquinoline quinone is bound at the active site via a Ca^{2+} bridge, enzyme contains 1.95 mol of Ca^{2+} per mol of subunit [11]; <7> Ca^{2+} is required for reactivation after thermal inactivation [11]; <21> mutants D354N, N355D, and D354N/N355D [39]; <7> required for structure stabilization of the holoform [35]; <1,23> required, interacts with PQQ cofactor [38]) [7, 11, 28, 30, 34, 35, 38, 39, 41, 42, 43, 44, 45, 47, 48, 51, 53]
Cd^{2+} <7> (<7> can replace Ca^{2+} in reactivation after thermal inactivation [11]) [11]
Mg^{2+} <1, 7, 9, 20, 21> (<9,20> in vivo reconstitution of apoenzyme with the prosthetic group is dependent on the presence of Ca^{2+} or Mg^{2+} [7]; <1> the apoenzyme is converted to the holoenzyme with exogenous pyrroloquinoline quinone and Mg^{2+}. The holoenzyme gradually returns to the apoenzyme in absence of pyrroloquinoline quinone and/or Mg^{2+}. Mg^{2+} allows the cofactor to take a more appropriate position in the active site [28]; <1> K_m: 0.022 mM for the wild-type enzyme [23]; <7> Mg^{2+} anchors pyrroloquinoline quinone cofactor to the enzyme protein and activates the bound cofactor [33]; <21>

required, bound at the active site, cannot be substituted by Ca^{2+}, Sr^{2+}, or Ba^{2+} [39]) [7, 23, 28, 33, 39]
Mn^{2+} <7> (<7> can replace Ca^{2+} in reactivation after thermal inactivation [11]) [11]
Sr^{2+} <21> (<21> mutants D354N, N355D, and D354N/N355D [39]) [39]

Turnover number (min^{-1})

0.8 <7> (D-glucose, <7> recombinant mutant H168Q, pH 7.0 [42]) [42]
2.5 <7> (D-glucose, <7> recombinant mutant H168C, pH 7.0 [42]) [42]
65 <7> (maltose, <7> recombinant mutant D167E/N452T, pH 7.0 [42]) [42]
69 <1> (D-galactose, <1> N462H mutant isozyme PQQGDH-B, pH 7.0, 25°C [34]) [34]
72 <1, 7> (D-galactose, <1> N452T mutant isozyme PQQGDH-B, pH 7.0, 25°C [34]; <7> recombinant mutant N452T, pH 7.0 [42]) [34, 42]
73 <7> (allose, <7> recombinant mutant D167E/N452T, pH 7.0 [42]) [42]
89 <7> (D-galactose, <7> recombinant mutant D167E/N452T, pH 7.0 [42]) [42]
121 <7> (D-galactose, <7> pH 7.0, wild-type enzyme [29]) [29]
167 <7> (lactose, <7> recombinant mutant D167E/N452T, pH 7.0 [42]) [42]
201 <7> (D-xylose, <7> pH 7.0, wild-type enzyme [29]) [29]
215 <7> (3-O-methyl-D-glucose, <7> recombinant mutant D167E/N452T, pH 7.0 [42]) [42]
226 <7> (cellobiose, <7> recombinant mutant D167E/N452T, pH 7.0 [42]) [42]
232 <1, 7> (D-galactose, <7> recombinant wild-type isozyme PQQGDH-B, pH 7.0 [42]; <1> wild-type isozyme PQQGDH-B, pH 7.0, 25°C [34]) [34, 42]
267 <7> (D-mannose, <7> pH 7.0, wild-type enzyme [29]) [29]
331 <7> (2-deoxy-D-glucose, <7> pH 7.0, wild-type enzyme [29]) [29]
436 <7> (maltose, <7> recombinant mutant D167E, pH 7.0 [42]) [42]
478 <7> (lactose, <7> recombinant mutant D167E, pH 7.0 [42]) [42]
541 <7> (3-O-methyl-D-glucose, <7> recombinant mutant D167E, pH 7.0 [42]) [42]
558 <7> (allose, <7> recombinant mutant D167E, pH 7.0 [42]) [42]
574 <1> (lactose, <1> N462H mutant isozyme PQQGDH-B, pH 7.0, 25°C [34]) [34]
588 <1> (maltose, <1> N462H mutant isozyme PQQGDH-B, pH 7.0, 25°C [34]) [34]
630 <7> (D-galactose, <7> pH 7.0, mutant enzyme E277K [29]) [29]
669 <7> (lactose, <7> pH 7.0, wild-type enzyme [29]) [29]
678 <7> (D-xylose, <7> pH 7.0, mutant enzyme E277K [29]) [29]
785 <7> (maltose, <7> pH 7.0, wild-type enzyme [29]) [29]
861 <7> (D-mannose, <7> pH 7.0, mutant enzyme E277K [29]) [29]
949 <1, 7> (allose, <1> N452T mutant isozyme PQQGDH-B, pH 7.0, 25°C [34]; <7> recombinant mutant N452T, pH 7.0 [42]) [34, 42]
1002 <1, 7> (maltose, <1> N452T mutant isozyme PQQGDH-B, pH 7.0, 25°C [34]; <7> recombinant mutant N452T, pH 7.0 [42]) [34, 42]
1020 <7> (maltose, <7> pH 7.0, mutant enzyme E277K [29]) [29]

1035 <1> (allose, <1> N462H mutant isozyme PQQGDH-B, pH 7.0, 25°C [34]) [34]
1038 <1, 7> (lactose, <1> N452T mutant isozyme PQQGDH-B, pH 7.0, 25°C [34]; <7> recombinant mutant N452T, pH 7.0 [42]) [34, 42]
1060 <7> (2-deoxy-D-glucose, <7> pH 7.0, mutant enzyme E277K [29]) [29]
1060 <7> (cellobiose, <7> recombinant mutant D167E, pH 7.0 [42]) [42]
1064 <1> (3-O-methyl-D-glucose, <1> N462H mutant isozyme PQQGDH-B, pH 7.0, 25°C [34]) [34]
1073 <7> (cellobiose, <7> recombinant mutant N452T, pH 7.0 [42]) [42]
1193 <7> (D-glucose, <7> recombinant mutant D167E/N452T, pH 7.0 [42]) [42]
1253 <1, 7> (3-O-methyl-D-glucose, <1> N452T mutant isozyme PQQGDH-B, pH 7.0, 25°C [34]; <7> recombinant mutant N452T, pH 7.0 [42]) [34, 42]
1355 <7> (cellobiose, <7> recombinant wild-type isozyme PQQGDH-B, pH 7.0 [42]) [42]
1399 <1> (D-glucose, <1> N462H mutant isozyme PQQGDH-B, pH 7.0, 25°C [34]) [34]
1440 <7> (D-allose, <7> pH 7.0, wild-type enzyme [29]) [29]
1450 <7> (3-O-methyl-D-glucose, <7> pH 7.0, wild-type enzyme [29]) [29]
1659 <1, 7> (lactose, <7> recombinant wild-type isozyme PQQGDH-B, pH 7.0 [42]; <1> wild-type isozyme PQQGDH-B, pH 7.0, 25°C [34]) [34, 42]
1724 <7> (D-glucose, <7> recombinant mutant D167E, pH 7.0 [42]) [42]
1791 <1, 7> (D-glucose, <1> N452T mutant isozyme PQQGDH-B, pH 7.0, 25°C [34]; <7> recombinant mutant N452T, pH 7.0 [42]) [34, 42]
1800 <7> (lactose, <7> pH 7.0, mutant enzyme E277K [29]) [29]
1930 <1, 7> (maltose, <7> recombinant wild-type isozyme PQQGDH-B, pH 7.0 [42]; <1> wild-type isozyme PQQGDH-B, pH 7.0, 25°C [34]) [34, 42]
2270 <1> (maltotriose) [53]
2509 <1, 7> (allose, <7> recombinant wild-type isozyme PQQGDH-B, pH 7.0 [42]; <1> wild-type isozyme PQQGDH-B, pH 7.0, 25°C [34]) [34, 42]
2870 <1> (maltose) [53]
3011 <1, 7> (3-O-methyl-D-glucose, <7> recombinant wild-type isozyme PQQGDH-B, pH 7.0 [42]; <1> wild-type isozyme PQQGDH-B, pH 7.0, 25°C [34]) [34, 42]
3070 <7> (D-glucose, <7> pH 7.0, mutant enzyme E277K [29]) [29]
3178 <7> (D-glucose, <7> recombinant cytochrome c-fusion protein, pH 7.0 [35]) [35]
3200 <7> (3-O-methyl-D-glucose, <7> pH 7.0, mutant enzyme E277K [29]) [29]
3360 <1> (D-glucose) [53]
3440 <7> (D-glucose, <7> pH 7.0, wild-type enzyme [29]) [29]
3778 <7> (D-glucose, <7> recombinant Arg-tagged enzyme, pH 7.0 [48]) [48]
3860 <1, 7> (D-glucose, <7> recombinant wild-type isozyme PQQGDH-B, pH 7.0 [42]; <1> wild-type isozyme PQQGDH-B, pH 7.0, 25°C [34]) [34, 42]
4424 <7> (D-glucose, <7> recombinant chimeric mutant enzyme, pH 7.0 [48]) [48]

4560 <7> (D-allose, <7> pH 7.0, mutant enzyme E277K [29]) [29]
Additional information <7> [22]

Specific activity (U/mg)

0.01 <19> (<19> NAD^+, hexokinase activity, PQQ-GDH-negative mutant, lag phase [52]; <19> NAD^+, hexokinase activity, wild type, lag phase [52]) [52]
0.014 <19> (<19> $NADP^+$, hexokinase activity, cell free extract, wild type, culture medium 10gl-1 gluconate, C-limited [52]) [52]
0.015 <19> (<19> NAD^+, hexokinase activity, PQQ-GDH-negative mutant, lag phase, N_2 as N-source [52]; <19> NAD^+, hexokinase activity, wild type, lag phase, N_2 as N-source [52]; <19> NAD^+, NAD-GDH, wild type, lag phase, N_2 as N-source [52]) [52]
0.018 <19> (<19> NAD^+, hexokinase activity, wild type, exponential growth, N_2 as N-source [52]; <19> NAD^+, NAD-GDH, wild type, exponential growth [52]) [52]
0.02 <19> (<19> NAD^+, hexokinase activity, wild type, exponential growth [52]; <19> NAD^+, NAD-GDH, wild type, exponential growth, N_2 as N-source [52]) [52]
0.023 <19> (<19> $NADP^+$, hexokinase activity, cell free extract, wild type, culture medium 10 g/l-1 glucose, C-limited [52]) [52]
0.025 <19> (<19> NAD^+, NAD-GDH, wild type, lag phase [52]) [52]
0.028 <19> (<19> $NADP^+$, hexokinase activity, cell free extract, wild type, culture medium 10 g/l-1 glucose, C-limited, N_2 as N-source [52]) [52]
0.03 <19> (<19> $NADP^+$, hexokinase activity, cell free extract, PQQ-GDH-negative mutant, culture medium 20 g/l-1 glucose, C-excess, N_2 as N-source [52]) [52]
0.04 <19> (<19> $NADP^+$, hexokinase activity, cell free extract, PQQ-GDH-negative mutant, culture medium 10 g/l-1 glucose, C-limited, N_2 as N-source [52]) [52]
0.165 <19> (<19> 2,6-dichlorophenol-indophenol, glucose concentration in the culture medium 20 g/l-1, C-excess [52]) [52]
0.18 <19> (<19> NAD^+, cell free extract, wild type, culture medium 10 g/l-1 gluconate, C-limited [52]) [52]
0.2 <19> (<19> 2,6-dichlorophenol-indophenol, glucose concentration in the culture medium 30 g/l-1, C-excess [52]) [52]
0.216 <19> (<19> gluconate, glucose concentration in the culture medium 20 g/l-1, C-excess [52]) [52]
0.234 <19> (<19> 2,6-dichlorophenol-indophenol, glucose concentration in the culture medium 50 g/l-1, C-excess [52]) [52]
0.238 <19> (<19> 2,6-dichlorophenol-indophenol, glucose concentration in the culture medium 40 g/l-1, C-excess [52]) [52]
0.255 <19> (<19> 2,6-dichlorophenol-indophenol, PQQ-GDH, wild type, lag phase [52]) [52]
0.284 <19> (<19> NAD^+, cell free extract, wild type, culture medium 10 g/l-1 gluconate, C-limited, N_2 as N-source [52]) [52]
0.29 <19> (<19> NAD^+, NAD-GDH, PQQ-GDH-negative mutant, lag phase, N_2 as N-source [52]) [52]

0.33 <19> (<19> NAD^+, NAD-GDH, PQQ-GDH-negative mutant, lag phase [52]) [52]
0.335 <19> (<19> 2,6-dichlorophenol-indophenol, PQQ-GDH, wild type, exponential growth, N_2 as N-source [52]) [52]
0.34 <19> (<19> NAD^+, cell free extract, wild type, culture medium 10 g/l-1 glucose, C-limited [52]) [52]
0.352 <19> (<19> gluconate, glucose concentration in the culture medium 30 g/l-1, C-excess [52]) [52]
0.4 <19> (<19> NAD^+, cell free extract, wild type, culture medium 10 g/l-1 glucose, C-limited, N_2 as N-source [52]) [52]
0.406 <19> (<19> NAD^+, cell free extract, PQQ-GDH-negative mutant, culture medium 10 g/l-1 glucose, C-limited, N_2 as N-source [52]) [52]
0.414 <19> (<19> NAD^+, cell free extract, PQQ-GDH-negative mutant, culture medium 10 g/l-1 glucose, C-limited [52]) [52]
0.464 <19> (<19> gluconate, glucose concentration in the culture medium 40 g/l-1, C-excess [52]) [52]
0.47 <19> (<19> 2,6-dichlorophenol-indophenol, PQQ-GDH, wild type, lag phase, N_2 as N-source [52]) [52]
0.544 <19> (<19> gluconate, glucose concentration in the culture medium 50 g/l-1, C-excess [52]) [52]
0.553 <19> (<19> 2,6-dichlorophenol-indophenol, glucose concentration in the culture medium 10 g/l-1, C-limitation [52]) [52]
0.67 <19> (<19> NAD^+, cell free extract, wild type, culture medium 50 g/l-1 glucose, C-excess, N_2 as N-source [52]) [52]
0.71 <19> (<19> gluconate, glucose concentration in the culture medium 10 g/l-1, C-limitation [52]) [52]
0.72 <19> (<19> NAD^+, cell free extract, wild type, culture medium 20 g/l-1 glucose, C-excess, N_2 as N-source [52]) [52]
0.76 <19> (<19> 2,6-dichlorophenol-indophenol, PQQ-GDH, wild type, exponential growth [52]) [52]
0.853 <19> (<19> NAD^+, cell free extract, PQQ-GDH-negative mutant, culture medium 20 g/l-1 glucose, C-excess, N_2 as N-source [52]) [52]
1.1 <19> (<19> NAD^+, cell free extract, wild type, culture medium 30 g/l-1 glucose, C-excess, N_2 as N-source [52]) [52]
12 <14> (<14> purified enzyme [36]; <14> purified native *m*-GDH [41]) [36, 41]
35 <16> [16]
150 <1> [20]
211 <1> [24]
250 <7> (<7> purified native s-GDH [41]) [41]
386 <3> [14]
570 <7> (<7> membrane bound enzyme form [12]) [12]
612 <7> [33]
616 <16> [9]
640 <7> [4]
2209 <7> (<7> soluble enzyme form [12]) [12]
2500 <7> (<7> mutant N340F/Y418I [45]) [45]

2600 <7> [3]
2642 <7> (<7> purified recombinant mutant chimeric isozyme PQQGDH-B [48]) [48]
2800 <7> (<7> mutant T416V/T417V [45]) [45]
3030 <7> (<7> wild-type isozyme PQQGDH-B [45]) [45]
3100 <7> (<7> mutant N340F/Y418F [45]) [45]
4512 <7> (<7> purified recombinant Arg-tagged isozyme PQQGDH-B [48]) [48]
4610 <7> (<7> purified recombinant wild-type isozyme PQQGDH-B [48]) [48]
5080 <7> (<7> purified recombinant isozyme PQQGDH-B [46]) [46]
7400 <7> (<7> recombinant enzyme [30]) [30]
Additional information <1, 7> [10, 34, 42, 44]

K_m-Value (mM)

5e-005 <1> (pyrroloquinoline quinone, <1> pH 7.0, 25°C, mutant K493A [21]) [21]
9e-005 <1> (pyrroloquinoline quinone, <1> 25°C, wild-type enzyme [23]) [23]
0.00011 <1> (pyrroliquinoline quinone, <1> pH 7.0, 25°C, wild-type enzyme [20,21]) [20, 21]
0.00012 <1> (pyrroloquinoline quinone, <1> C-terminal periplasmic domain of glucose dehydrogenase [20]; <1> 25°C, mutant enzyme D730N [23]) [20, 23]
0.00014 <1> (pyrroloquinoline quinone, <1> pH 7.0, 21°C, mutant D466E and D466N [21]) [21]
0.00022 <1> (pyrroloquinoline quinone, <1> pH 7.0, 25°C, mutant W607A [21]) [21]
0.0005 <1> (pyrroloquinoline quinone, <1> 25°C, mutant enzyme S357L [23]) [23]
0.00064 <1> (pyrroloquinoline quinone, <1> 25°C, mutant enzyme G689D [23]) [23]
0.00088 <1> (pyrroloquinoline quinone, <1> pH 7.0, 25°C, mutant H262A [21]) [21]
0.0034 <7> (ubiquinone Q_6, <7> membrane-bound enzyme form [12]) [12]
0.0044 <7> (ubiquinone Q_9, <7> membrane-bound enzyme form [12]) [12]
0.006 <1> (pyrroloquinoline quinone, <1> pH 7.0, 25°C, mutant W404A [21]) [21]
0.0097 <3> (ubiquinone Q_6, <3> pH 8.8, 25°C [14]) [14]
0.012 <7> (ubiquinone Q_2, <7> membrane-bound enzyme form [12]) [12]
0.016 <1> (pyrroloquinoline quinone, <1> pH 7.0, 25°C, mutant K493R [21]) [21]
0.0177 <7> (ubiquinone Q_2, <7> soluble enzyme form [12]) [12]
0.019 <3> (ubiquinone Q_6, <3> pH 6.5, 25°C [15]) [15]
0.02 <1> (pyrroloquinoline quinone, <1> pH 7.0, 25°C, mutant W404F [21]) [21]

0.021 <1> (pyrroloquinoline quinone, <1> 25°C, mutant enzyme H775R [23]) [23]
0.025 <1> (ubiquinone Q_2, <1> wild-type enzyme [20]) [20]
0.028 <3> (ubiquinone Q_4, <3> pH 6.5, 25°C [15]) [15]
0.03 <1> (ubiquinone Q_2, <1> C-terminal periplasmic domain of glucose dehydrogenase [20]) [20]
0.031 <3> (ubiquinone Q_4, <3> pH 8.8, 25°C [14]) [14]
0.04 <3> (ubiquinone Q_2, <3> pH 6.5, 25°C [15]) [15]
0.06 <3> (ubiquinone Q_1, <3> pH 8.8, 25°C [14]) [14]
0.061 <3> (ubiquinone Q_2, <3> pH 8.8, 25°C [14]) [14]
0.064 <7> (N-methylphenazonium methyl sulfate, <7> pH 8.5, reaction with D-galactose [33]) [33]
0.074 <7> (phenazine methosulfate, <7> membrane-bound enzyme form [10,12]) [10, 12]
0.118 <7> (L-arabinose, <7> pH 8.5, reaction with D-glucose [33]) [33]
0.12 <7> (pyrroloquinoline quinone) [3]
0.13 <3> (phenazine methosulfate, <3> pH 8.8, 25°C [14]) [14]
0.148 <7> (ubiquinone Q_1, <7> membrane-bound enzyme form [12]) [12]
0.156 <7> (N-methylphenazonium methyl sulfate, <7> pH 8.5, reaction with D-xylose [33]) [33]
0.178 <7> (N-methylphenazonium methyl sulfate, <7> pH 8.5, reaction with D-glucose [33]) [33]
0.178 <7> (ubiquinone Q_1, <7> soluble enzyme form [12]) [12]
0.22 <7> (N-methylphenazonium methyl sulfate, <7> pH 8.5, reaction with D-fucose [33]) [33]
0.23 <7> (D-glucose, <7> pH 8.5, reaction with 2,6-dichlorophenolindophenol as electron acceptor [33]) [33]
0.268 <7> (N-ethylphenazonium ethyl sulfate, <7> pH 8.5, reaction with D-xylose [33]) [33]
0.278 <7> (N-ethylphenazonium ethyl sulfate, <7> pH 8.5, reaction with D-glucose [33]) [33]
0.291 <7> (N-ethylphenazonium ethyl sulfate, <7> pH 8.5, reaction with D-fucose [33]) [33]
0.292 <7> (N-ethylphenazonium ethyl sulfate, <7> pH 8.5, reaction with L-arabinose [33]) [33]
0.3 <7> (D-glucose, <7> pH 7.0, mutant enzyme E277G [29]) [27, 29]
0.362 <7> (N-methylphenazonium methyl sulfate, <7> pH 8.5, reaction with 2-deoxy-D-glucose [33]) [33]
0.47 <3> (D-glucose, <3> pH 6.0, 25°C, reaction with 2,6-dichlorophenolindophenol [14]) [14]
0.56 <3> (N,N,N,N-tetramethyl-*o*-phenylenediamine, <3> pH 8.8, 25°C [14]) [14]
0.66 <7> (lactose) [27]
0.69 <3> (ferricyanide, <3> pH 8.8, 25°C [14]) [14]
0.73 <7> (4-(4-ferrocenylimino-methyl)phenol, <7> s-GDH [41]) [41]
0.78 <7> (pyrroloquinoline quinone) [4]

0.8 <1> (D-glucose, <1> pH 7.0, 25°C, mutant enzyme N607A [21]; <1> 25°C, mutant enzyme D730 N [23]) [21, 23]
0.83 <14> (4-ferrocenylphenol, <14> *m*-GDH [41]) [41]
0.89 <7> (4-ferrocenylnitrophenol, <7> s-GDH [41]) [41]
0.9 <1> (D-glucose, <1> pH 7.0, 25°C, wild-type enzyme [21]) [21]
0.91 <1> (D-glucose, <1> 25°C, wild-type enzyme [23]) [23]
0.95 <1> (D-glucose, <1> wild-type enzyme [20]) [20]
0.98 <1> (D-glucose, <1> C-terminal periplasmic domain of glucose dehydrogenase [20]) [20]
1 <7> (4-ferrocenylphenol, <7> s-GDH [41]) [41]
1 <1> (D-glucose, <1> 25°C, mutant enzyme S357L [23]) [23]
1.2 <1, 7> (D-glucose, <1> pH 7.0, 25°C, mutant W404A [21]; <7> mutant E277N [29]) [21, 29]
1.3 <1> (6-deoxy-D-glucose, <1> wild-type enzyme [26]) [26]
1.3 <1> (D-glucose, <1> 25°C, mutant enzyme H775R [23]) [23]
1.4 <1> (D-glucose, <1> 25°C, mutant enzyme G689D [23]; <1> pH 7.0, 25°C, mutant K493A [21]) [21, 23]
1.5 <7> (D-allose) [8]
1.5 <1, 7> (D-glucose, <1> pH 7.0, 25°C, mutant K493R [21]; <7> mutant enzyme E277A [29]) [21, 29]
1.6 <3> (2,6-dichlorophenolindophenol, <3> pH 8.8, 25°C [14]) [14]
1.6 <1> (2-deoxy-D-glucose, <1> wild-type enzyme [26]) [26]
1.7 <7> (D-glucose) [8]
1.9 <7> (phenazine methosulfate, <7> soluble enzyme form [10,12]) [10, 12]
2 <7> (D-galactose, <7> recombinant wild-type enzyme, pH 7.0, 25°C [43]) [43]
2 <1> (D-glucose, <1> pH 7.0, 21°C, mutant W404F [21]) [21]
2-3 <7> (3-O-methyl-D-glucose, <7> recombinant Arg-tagged enzyme, pH 7.0 [48]) [48]
2-3 <7> (D-glucose, <7> recombinant Arg-tagged enzyme, pH 7.0 [48]; <7> recombinant cytochrome c-fusion protein, pH 7.0 [35]) [35, 48]
2.1 <1> (D-glucose, <1> wild-type enzyme [26]) [26]
2.43 <14> (4-(4-ferrocenylimino-methyl)phenol, <14> *m*-GDH [41]) [41]
2.5 <1> (D-allose, <1> wild-type enzyme [26]) [26]
2.5 <7> (D-glucose, <7> pH 7.0, mutant enzyme E277V [29]) [29]
2.7 <1> (D-galactose, <1> N462H mutant isozyme PQQGDH-B, pH 7.0, 25°C [34]) [34]
2.73 <14> (4-ferrocenylnitrophenol, <14> *m*-GDH [41]) [41]
2.8 <1> (D-glucose, <1> recombinant wild-type enzyme with bound Mg^{2+}, pH 6.5, 25°C [39]) [39]
3 <1> (D-glucose, <1> pH 7.0, 25°C, mutant D466E [21]) [21]
3.3 <7> (D-glucose) [3]
3.5 <21> (2-deoxy-D-glucose, <21> recombinant wild-type enzyme with bound Mg^{2+}, pH 6.5, 25°C [39]) [39]
3.5 <7> (D-galactose) [8]
3.7 <1, 7> (D-galactose, <1> N452T mutant isozyme PQQGDH-B, pH 7.0, 25°C [34]; <7> recombinant mutant N452T, pH 7.0 [42]) [34, 42]

4 <7> (D-galactose, <7> recombinant tagged enzyme, pH 7.0, 25°C [43]) [43]
4 <7> (D-glucose, <7> pH 8.5, reaction with N-methylphenazonium methyl sulfate or N-ethylphenazonium ethal sulfate as electron acceptor [33]) [33]
4.2 <7> (D-glucose, <7> membrane-bound enzyme form [12]) [10, 12]
4.3 <7> (D-glucose, <7> pH 7.0, mutant enzyme E277Q [29]) [29]
5 <7> (D-fucose, <7> pH 8.5, reaction with N-ethylphenazonium ethyl sulfate as electron acceptor [33]) [33]
5 <7> (D-galactose, <7> recombinant triple mutant enzyme, pH 7.0, 25°C [43]) [43]
5.3 <1, 7> (D-galactose, <7> recombinant wild-type isozyme PQQGDH-B, pH 7.0 [42]; <1> wild-type isozyme PQQGDH-B, pH 7.0, 25°C [34]) [34, 42]
5.5 <7> (D-xylose) [8]
5.9 <7> (D-glucose, <7> pH 8.5, reaction with Wurster' Blue as electron acceptor [33]) [33]
6.8 <7> (D-galactose, <7> pH 7.0, wild-type enzyme and mutant enzyme E277K [29]) [29]
7 <7> (D-glucose, <7> pH 7.0, mutant enzyme I278F [29]) [29]
7 <7> (D-xylose, <7> pH 8.5, reaction with N-ethylphenazonium ethyl sulfate as electron acceptor [33]) [33]
7.4 <7> (D-glucose, <7> pH 7.0, mutant enzyme E277D [29]) [29]
7.5 <7> (lactose, <7> pH 7.0, mutant enzyme E277K [29]) [29]
7.7 <7> (D-glucose, <7> pH 7.0, mutant enzyme E277H [29]) [29]
8 <7> (D-galactose, <7> recombinant Arg-tagged enzyme, pH 7.0 [48]) [48]
8.3 <1> (D-fucose, <1> wild-type enzyme [26]) [26]
8.8 <7> (D-glucose, <7> pH 7.0, mutant enzyme E277K [29]) [29]
8.9 <7> (D-glucose, <7> pH 7.0, mutant enzyme E277K [29]) [29]
9 <7> (D-galactose, <7> recombinant chimeric mutant enzyme, pH 7.0 [48]) [48]
9.5 <1> (2-amino-D-glucose, <1> wild-type enzyme [26]) [26]
10 <1> (D-glucose, <1> pH 7.0, 25°C, mutant H262A [21]) [21]
10 <7> (maltose, <7> recombinant wild-type enzyme, pH 7.0, 25°C [43]) [43]
10.8 <1> (3-deoxy-D-glucose, <1> wild-type enzyme [26]) [26]
11 <7> (maltose, <7> recombinant tagged enzyme, pH 7.0, 25°C [43]) [43]
12 <7> (D-fucose, <7> pH 8.5, reaction with N-methylphenazonium methyl sulfate as electron acceptor [33]) [33]
12 <1> (D-glucose, <1> pH 7.0, 25°C, mutant D466N [21]) [21]
12 <7> (D-xylose, <7> pH 8.5, reaction with N-methylphenazonium methyl sulfate as electron acceptor [33]) [33]
12 <1> (lactose, <1> pH 7.0, 25°C, linked dimeric enzyme [25]) [25]
12.3 <1> (D-glucose, <1> N462H mutant isozyme PQQGDH-B, pH 7.0, 25°C [34]) [34]
12.5 <1, 7> (D-glucose, <1> N452T mutant isozyme PQQGDH-B, pH 7.0, 25°C [34]; <7> recombinant mutant N452T, pH 7.0 [42]) [34, 42]
13 <7> (maltose, <7> recombinant chimeric mutant enzyme, pH 7.0 [48]; <7> recombinant triple mutant enzyme, pH 7.0, 25°C [43]) [43, 48]
13.6 <7> (2-deoxy-D-glucose) [8]

14 <7> (cellobiose, <7> recombinant wild-type isozyme PQQGDH-B and mutant N452T, pH 7.0 [42]) [42]
14 <7> (maltose, <7> recombinant Arg-tagged enzyme, pH 7.0 [48]) [48]
14.3 <7> (D-xylose, <7> pH 7.0, wild-type enzyme [29]) [29]
14.3 <7> (lactose, <7> pH 7.0, wild-type enzyme [29]) [29]
14.3 <7> (maltose, <7> pH 7.0, mutant enzyme E277K [29]) [29]
15 <7> (maltose, <7> recombinant cytochrome c-fusion protein, pH 7.0 [35]) [35]
15.7 <7> (D-glucose, <7> pH 7.0, mutant enzyme N279H [29]) [29]
16 <7> (cellobiose, <7> recombinant mutant D167E/N452T, pH 7.0 [42]) [42]
16 <7> (D-glucose, <7> recombinant mutant T416V/T417V, pH 7.0, 25°C [45]) [45]
16 <1, 7> (maltose, <1> N462H mutant isozyme PQQGDH-B, pH 7.0, 25°C [34]; <7> recombinant mutant D167E/N452T, pH 7.0 [42]) [34, 42]
17 <7> (cellobiose, <7> recombinant mutant D167E, pH 7.0 [42]) [42]
17 <21> (D-xylose, <21> recombinant wild-type enzyme with bound Mg^{2+}, pH 6.5, 25°C [39]) [39]
17.5 <21> (D-galactose, <21> recombinant wild-type enzyme with bound Mg^{2+}, pH 6.5, 25°C [39]) [39]
17.7 <1> (D-melibiose, <1> wild-type enzyme [26]) [26]
18 <7> (L-arabinose, <7> pH 8.5, reaction with N-methylphenazonium methyl sulfate as electron acceptor [33]) [33]
18 <1> (lactose, <1> N462H mutant isozyme PQQGDH-B, pH 7.0, 25°C [34]) [34]
18.9 <1, 7> (lactose, <7> recombinant wild-type isozyme PQQGDH-B, pH 7.0 [42]; <1> wild-type isozyme PQQGDH-B, pH 7.0, 25°C [34]) [34, 42]
19 <7> (D-galactose, <7> pH 8.5, reaction with N-methylphenazonium methyl sulfate as electron acceptor [33]) [33]
19 <7> (D-mannose) [8]
19 <7> (L-arabinose, <7> pH 8.5, reaction with N-ethylphenazonium ethyl sulfate as electron acceptor [33]) [33]
19 <7> (lactose, <7> recombinant cytochrome c-fusion protein, pH 7.0 [35]) [35]
20 <1, 7> (D-glucose, <1> pH 7.0, 25°C, linked dimeric enzyme [25]; <7> recombinant wild-type enzyme and mutants N340F/Y418F and N340F/Y418I, pH 7.0, 25°C [45]) [25, 45]
20 <7> (lactose, <7> recombinant Arg-tagged enzyme, and recombinant chimeric mutant enzyme, pH 7.0 [48]) [48]
21 <7> (allose, <7> pH 7.0, mutant enzyme E277K [29]) [29]
22 <7> (2-deoxy-D-glucose, <7> pH 8.5, reaction with N-methylphenazonium methyl sulfate as electron acceptor [33]) [33]
22 <7> (3-O-methyl-D-glucose, <7> recombinant chimeric mutant enzyme, pH 7.0 [48]) [48]
22 <7> (D-glucose, <7> recombinant chimeric mutant enzyme, pH 7.0 [48]) [4, 48]

22 <7> (D-mannose, <7> pH 7.0, wild-type enzyme and mutant enzyme E277K [29]) [29]
22 <1> (D-xylose, <1> wild-type enzyme [26]) [26]
24 <7> (D-glucose, <7> pH 7.0, mutant enzyme D275E and D276E [29]) [29]
24.5 <7> (D-glucose, <7> soluble enzyme [10,12]) [10, 12]
25 <1, 7> (D-glucose, <7> recombinant triple mutant enzyme, pH 7.0, 25°C [43]; <7> recombinant wild-type isozyme PQQGDH-B, pH 7.0 [42]; <1> wild-type isozyme PQQGDH-B, pH 7.0, 25°C [34]) [34, 42, 43]
25 <7> (lactose, <7> recombinant wild-type enzyme, pH 7.0, 25°C [43]) [43]
26 <7> (D-glucose, <7> pH 7.0, wild-type enzyme [29]) [29]
26 <7> (lactose, <7> recombinant tagged enzyme, pH 7.0, 25°C [43]) [43]
26 <1, 7> (maltose, <7> recombinant wild-type isozyme PQQGDH-B, pH 7.0 [42]; <1> wild-type isozyme PQQGDH-B, pH 7.0, 25°C [34]) [34, 42]
26.7 <7> (lactose, <7> soluble enzyme [10,12]) [10, 12]
27 <7> (3-O-methyl-D-glucose, <7> pH 7.0, mutant enzyme E277K [29]) [29]
27 <7> (D-glucose, <7> recombinant wild-type enzyme and tagged enzyme, pH 7.0, 25°C [43]) [43]
27 <7> (melibiose) [27]
27.6 <1, 7> (3-O-methyl-D-glucose, <1> N452T mutant isozyme PQQGDH-B, pH 7.0, 25°C [34]; <7> recombinant mutant N452T, pH 7.0 [42]) [34, 42]
28.7 <1, 7> (3-O-methyl-D-glucose, <7> recombinant wild-type isozyme PQQGDH-B, pH 7.0 [42]; <1> wild-type isozyme PQQGDH-B, pH 7.0, 25°C [34]) [34, 42]
28.8 <1> (3-O-methyl-D-glucose, <1> N462H mutant isozyme PQQGDH-B, pH 7.0, 25°C [34]) [34]
29 <7> (D-allose, <7> pH 7.0, wild-type enzyme [29]) [29]
30.9 <7> (maltose, <7> pH 7.0, wild-type enzyme [29]) [29]
31 <21> (L-arabinose, <21> recombinant wild-type enzyme with bound Mg^{2+}, pH 6.5, 25°C [39]) [39]
32 <1> (2-deoxy-D-glucose, <1> mutant enzyme H262 [26]) [26]
32.5 <1> (allose, <1> N462H mutant isozyme PQQGDH-B, pH 7.0, 25°C [34]) [34]
33.6 <1, 7> (lactose, <1> N452T mutant isozyme PQQGDH-B, pH 7.0, 25°C [34]; <7> recombinant mutant N452T, pH 7.0 [42]) [34, 42]
34 <7> (D-xylose, <7> pH 7.0, mutant enzyme E277K [29]) [29]
34 <7> (allose, <7> recombinant Arg-tagged enzyme, pH 7.0 [48]) [48]
35.5 <1, 7> (allose, <7> recombinant wild-type isozyme PQQGDH-B, pH 7.0 [42]; <1> wild-type isozyme PQQGDH-B, pH 7.0, 25°C [34]) [34, 42]
36 <7> (lactose, <7> recombinant triple mutant enzyme, pH 7.0, 25°C [43]) [43]
36 <7> (allose, <7> recombinant chimeric mutant enzyme, pH 7.0 [48]) [48]
38.7 <1, 7> (allose, <1> N452T mutant isozyme PQQGDH-B, pH 7.0, 25°C [34]; <7> recombinant mutant N452T, pH 7.0 [42]) [34, 42]
39 <1> (D-galactose, <1> wild-type enzyme [26]) [26]
40 <7> (D-ribose) [8]
41 <7> (3-O-methyl-D-glucose, <7> recombinant wild-type enzyme, pH 7.0, 25°C [43]) [43]

46 <7> (3-O-methyl-D-glucose, <7> pH 7.0, wild-type enzyme [29]) [29]
46 <1> (L-arabinose, <1> wild-type enzyme [26]) [26]
46.5 <1, 7> (maltose, <1> N452T mutant isozyme PQQGDH-B, pH 7.0, 25°C [34]; <7> recombinant mutant N452T, pH 7.0 [42]) [34, 42]
48 <7> (D-glucose, <7> recombinant mutant D167E/N452T, pH 7.0 [42]) [42]
53 <7> (3-O-methyl-D-glucose, <7> recombinant tagged enzyme, pH 7.0, 25°C [43]) [43]
55 <7> (D-glucose, <7> recombinant mutant D167E, pH 7.0 [42]) [42]
55 <7> (lactose, <7> recombinant mutant D167E/N452T, pH 7.0 [42]) [42]
56 <7> (3-O-methyl-D-glucose, <7> recombinant triple mutant enzyme, pH 7.0, 25°C [43]) [43]
63 <7> (allose, <7> recombinant wild-type enzyme, pH 7.0, 25°C [43]) [43]
75 <7> (allose, <7> recombinant tagged enzyme, pH 7.0, 25°C [43]) [43]
76 <7> (allose, <7> recombinant triple mutant enzyme, pH 7.0, 25°C [43]) [43]
77 <7> (lactose, <7> recombinant mutant D167E, pH 7.0 [42]) [42]
78 <1> (D-mannose, <1> wild-type enzyme [26]) [26]
79 <1, 21> (3-O-methyl-D-glucose, <1> wild-type enzyme [26]; <21> recombinant wild-type enzyme with bound Mg^{2+}, pH 6.5, 25°C [39]) [26, 39]
88 <7> (2-deoxy-D-glucose, <7> pH 7.0, mutant enzyme E277K [29]) [29]
90 <7> (2-deoxy-D-glucose, <7> pH 7.0, wild-type enzyme [29]) [29]
99 <7> (3-O-methyl-D-glucose, <7> recombinant mutant D167E, pH 7.0 [42]) [42]
100 <1> (L-lyxose, <1> wild-type enzyme [26]) [26]
110 <1> (D-ribose, <1> wild-type enzyme [26]) [26]
116 <21> (D-mannose, <21> recombinant wild-type enzyme with bound Mg^{2+}, pH 6.5, 25°C [39]) [39]
145 <7> (D-galactose, <7> recombinant mutant D167E/N452T, pH 7.0 [42]) [42]
150 <1> (maltotriose) [53]
154 <7> (D-glucose, <7> recombinant mutant H168Q, pH 7.0 [42]) [42]
156 <7> (maltose, <7> recombinant mutant D167E, pH 7.0 [42]) [42]
166 <21> (D-ribose, <21> recombinant wild-type enzyme with bound Mg^{2+}, pH 6.5, 25°C [39]) [39]
170 <1> (maltose) [53]
182 <7> (allose, <7> recombinant mutant D167E/N452T, pH 7.0 [42]) [42]
193 <7> (D-glucose, <7> recombinant mutant H168C, pH 7.0 [42]) [42]
198 <7> (3-O-methyl-D-glucose, <7> recombinant mutant D167E/N452T, pH 7.0 [42]) [42]
199 <7> (allose, <7> recombinant mutant D167E, pH 7.0 [42]) [42]
400 <1> (D-glucose) [53]
460 <1> (D-glucose, <1> mutant enzyme H262Y [26]) [26]
810 <1> (D-allose, <1> mutant enzyme H262Y [26]) [26]
Additional information <7, 14, 21> (<7> kinetics [43]; <14> electrochemical data and kinetics for quinone derivatives as redox mediators [40]; <14> kinetic study, enzyme follows Michaelis-Menten kinetics using artificial electron transfer mediator system based on ruthenium(III) compounds for activ-

ity assays [36]; <7> kinetics, cooperativity in the recombinant chimeric mutant enzyme which possesses only 1 active subunit derived from the wild-type enzyme [48]; <21> metal binding kinetics, wild-type and mutant enzymes, K_m for the different substrates of mutant enzymes with different metal ions bound, overview [39]; <7> predicted binding energy of wild-type and mutant enzymes [45]) [36, 39, 40, 43, 45, 48]

K_i-Value (mM)

Additional information <21> (<21> inhibitory effects of the different metal ions on recombinant wild-type and mutant enzymes [39]) [39]

pH-Optimum

3 <16> (<16> reaction with potassium ferricyanide [16]) [16]
5.5 <14> (<14> native *m*-GDH [41]) [41]
6 <7, 14, 16> (<7> reaction with 2,6-dichlorophenol-indophenol [4]; <16> reaction with phenazine methosulfate, 2,6-dichlorophenol indophenol and pyrroloquinoline quinone [16]; <14> immobilized m-GDH [41]) [4, 16, 41]
6-9 <7> (<7> immobilized s-GDH, broad maximum [41]) [41]
6.5 <7, 21> (<7> soluble enzyme [10,12]; <21> assay at [39]) [10, 12, 39]
7 <1, 7, 23> (<1,7> assay at [34,35,42,43,44,45,46,48]; <7> D-glucose oxidation [22]; <7> native s-GDH [41]; <1,23> activity assay [51,53]) [22, 34, 35, 41, 42, 43, 44, 45, 46, 48, 51, 53]
8.5 <7> (<7> membrane-bound enzyme [10,12]) [10, 12]
9 <7> (<7> reaction with pyrroloquinoline quinone [4]) [4]
Additional information <7, 14> (<7,14> pH-dependence of native and immobilized enzyme [41]) [41]

pH-Range

4-8 <16> (<16> pH 4.0: about 50% of maximal activity, pH 8.0: about 35% of maximal activity [16]) [16]
7.5-9.5 <7> (<7> pH 7.5: about 40% of maximal activity, pH 9.5: about 85% of maximal activity, membrane-bound enzyme, reaction with phenazine methosulfate, 2,6-dichlorophenol indophenol and pyrroloquinoline quinone [10,12]) [10, 12]

Temperature optimum (°C)

25 <1, 7, 21> (<1,7,21> assay at [34,39,43,45]; <1> activity assay [53]) [34, 39, 43, 45, 53]

4 Enzyme Structure

Molecular weight

40000 <1> (<1> determined by SDS-PAGE [53]) [53]
50000 <23> (<23> monomer [51]) [51]
87000 <16> (<16> sucrose density gradient centrifugation [16]) [16]
93000 <3> (<3> sucrose density gradient centrifugation [14]) [14]
94000 <7> (<7> gel filtration [4]) [4]
110000 <7> (<7> equilibrium sedimentation [3]) [3]

Subunits

? <7> (<7> x * 55000, soluble enzyme SDS-PAGE [10,12]; <7> x * 83000, soluble enzyme, SDS-PAGE [10]; <7> x * 65000, recombinant cytochrome c-fusion protein, SDS-PAGE [35]) [10, 12, 35]
dimer <1, 7, 22, 23> (<7,22> 2 * 50000, SDS-PAGE [47,48]; <7> 2 * 48000, SDS-PAGE [4]; <7> 2 * 54000, SDS-PAGE [3]; <23> α_2, functional subunit composition, domain and overall structure analysis [38]; <7> 2 * 50000, deglycosylated recombinant isozyme PQQGDH-B, SDS-PAGE [46]; <7> structure analysis, 6-blade β-propeller protein with each blade consisting of a 4-stranded anti-parallel β-sheet [45]) [3, 4, 10, 38, 45, 46, 47, 48, 49]
homodimer <23> (<23> 2 * 50000 Da [51]) [51]
monomer <1, 3, 7> (<7> 1 * 83000, membrane-bound enzyme form, mainly monomeric in detergent solution [12]; <3> 1 * 87000, urea-SDS-PAGE [14]; <1> functional subunit composition, domain and overall structure analysis [38]) [12, 14, 38]
Additional information <1, 7> (<1> 3D-model prediction for wild-type and mutant isozyme PQQGDH-B [34]; <7> redox-related structural changes, overview [37]; <7> the monomeric enzyme is not active, 3D models of wild-type and mutant enzymes [45]) [34, 37, 45]

Posttranslational modification

Glycoprotein <7> (<7> secreted recombinant isozyme PQQGDH-B expressed in Pichia pastoris [46]) [46]

5 Isolation/Preparation/Mutation/Application

Source/tissue

bacteroid <17> [19]

Localization

cytoplasm <7, 19, 22, 23> (<7,22,23> soluble enzyme sGDH [37,38,47]) [37, 38, 47, 52]
cytoplasmic membrane <1, 3> (<1,3> outer surface of the cytoplasmic membrane [13,14,15]) [13, 14, 15]
membrane <1, 2, 3, 4, 7, 8, 14, 16, 21, 22> (<1,3> bound to [14,15,20,28]; <1> outer surface of cytoplasmic membrane [13]; <1> inner membrane, the enzyme has a ubiquinone reacting site close to the periplasmic side of the membrane, and thus its electron transfer to ubiquinone appears to be incapable of forming a proton electrochemical gradient across the inner membrane [24]; <1> it is likely that the C-terminal periplasmic domain of glucose dehydrogenase possesses a ubiquinone-reacting site and transfers electrons directly to ubiquinone [20]; <7> enzyme exists as a soluble form and a membrane-bound form [12]; <14> bound enzyme *m*-GDH [41]; <1> bound, enzyme mGDH [38]; <22> enzyme mGDH [47]; <7> isozyme PQQGDH-A [46]) [9, 10, 12, 13, 14, 15, 20, 24, 26, 28, 38, 39, 41, 46, 47]
periplasm <7, 17, 18, 19, 21> [19, 30, 39, 46, 52]

soluble <1, 7> (<7> enzyme s-GDH [41]; <1> isoyzme PQQGDH-B [34]; <7> isozyme PQQGDH-B [42,45,46,48]; <1> isozyme PQQGDH-B, water-soluble quinoprotein [49]; <1> water-soluble enzyme PQQGDH [44]) [12, 27, 34, 35, 41, 42, 44, 45, 46, 48, 49]

Purification

<1> [20, 50]
<1> (linked-dimeric enzyme) [25]
<1> (mutant enzymes) [23]
<1> (using a Ni-NTA column, the hexahistidine tag is removed by cleavage with tobacco edge virus protease, cleaved protein is purified by gel filtration) [53]
<1> (wild-type and mutant enzyme H262Y) [26]
<3> [14, 15]
<7> [1, 3, 4, 12, 30, 50]
<7> (optimization of the purification process involving cation exchange chromatography in presence of Zn^{2+}) [43]
<7> (partial) [8]
<7> (recombinant cytochrome c-fusion protein from Escherichia coli) [35]
<7> (recombinant isozyme PQQGDH-B from Pichia pastoris culture medium, optimization of down-stream processing) [46]
<7> (soluble enzyme and membrane-bound enzyme) [10]
<10> [50]
<16> [9, 16]
<19> (disrupted cell suspension is centrifuged and the supernatant is used as source for the enzyme, cell-free extract) [52]
<23> (using a His Microspin purification module) [51]
<273972> (recombinant wild-type and mutant enzymes from strain PP2418 by solubilization with Triton X-100 and ion exchange chromatography) [39]

Renaturation

<7> (protein dissociation and redimerization of recombinant wild-type and Arg-tagged enzyme, and the mutant H168Q, overview) [48]

Crystallization

<1> (apo- and holoenzyme at a resolution of 1.5 A) [53]
<1> (crystallization by microseeding, data set is collected at 2.0 A resolution) [50]
<7> [3]
<7> (apo form of the soluble glucose dehydrogenase) [17]
<7> (crystal structure analysis of enzyme in ternary complex with β-D-glucose and PQQ) [37]
<7> (crystallization by microseeding) [50]
<7> (ternary complex of the enzyme with pyrroloquinoline quinone and methylhydrazine) [32]
<10> (crystallization by microseeding, data set is collected at 1.8 A resolution) [50]

Cloning

<1> [5, 23, 49]
<1> (expression of wild-type and mutant isozymes PQQGDH-B) [34]
<1> (into the pET M-11 vector for transformation of DH5α and BL21DE3 cells) [53]
<1> (random peptide ligands M13-phage library display is used to expresss the enzyme in presence of peptide ligands, overview) [44]
<7> (expression in the periplasm of Escherichia coli strain PP2418, with and without a C-terminal tag of 3 Arg residues) [43]
<7> (expression of isozyme PQQGDH-B with and without an polyarginine tail in Escherichia coli strain PP2418, and of mutant H168Q) [48]
<7> (expression of isozyme PQQGDH-B, in Pichia pastoris using the α-factor signal sequence of Saccharomyces cerevisiae, recombinant enzyme is secreted to the medium, optimization of enzyme production) [46]
<7> (expression of the apoenzyme in Escherichia coli) [30]
<7> (expression of wild-type and mutant isozyme PQQGDH-B) [42]
<7> (expression of wild-type isozyme PQQGDH-B and mutant enzymes in Escherichia coli) [45]
<7> (gene gdhB, expression of the enzyme GDH-B as fusion protein C-terminally fused to cytochrome c domain from Comamonas testosteroni in Escherichia coli DH5α) [35]
<23> (into a His-tagged vector for expression in Escherichia coli JM109) [51]
<273972> (expression of wild-type and mutant enzymes in strain PP2418) [39]

Engineering

A71P/N454S <23> (<23> mutant, relative activity vs wild type, substrate glucose 0.95, substrate maltose 0.75 [51]) [51]
A98G/K126R/L445I/N454S <23> (<23> mutant, relative activity vs wild type, substrate glucose 1.00, substrate maltose 0.78 [51]) [51]
D167A <7> (<7> site-directed mutagenesis, substrate binding residue mutation, reduced activity compared to the wild-type enzyme [42]) [42]
D167C <7> (<7> site-directed mutagenesis, substrate binding residue mutation, reduced activity compared to the wild-type enzyme [42]) [42]
D167E <7> (<7> site-directed mutagenesis, substrate binding residue mutation, slightly reduced activity compared to the wild-type enzyme [42]) [42]
D167E/N452T <7> (<7> site-directed mutagenesis, reduced activity compared to the wild-type enzyme [42]) [42]
D167G <7> (<7> site-directed mutagenesis, substrate binding residue mutation, reduced activity compared to the wild-type enzyme [42]) [42]
D167H <7> (<7> site-directed mutagenesis, substrate binding residue mutation, reduced activity compared to the wild-type enzyme [42]) [42]
D167K <7> (<7> site-directed mutagenesis, substrate binding residue mutation, reduced activity compared to the wild-type enzyme [42]) [42]
D167N <7> (<7> site-directed mutagenesis, substrate binding residue mutation, reduced activity compared to the wild-type enzyme [42]) [42]

D167Q <7> (<7> site-directed mutagenesis, substrate binding residue mutation, reduced activity compared to the wild-type enzyme [42]) [42]
D167R <7> (<7> site-directed mutagenesis, substrate binding residue mutation, reduced activity compared to the wild-type enzyme [42]) [42]
D167S <7> (<7> site-directed mutagenesis, substrate binding residue mutation, reduced activity compared to the wild-type enzyme [42]) [42]
D167V <7> (<7> site-directed mutagenesis, substrate binding residue mutation, reduced activity compared to the wild-type enzyme [42]) [42]
D167W <7> (<7> site-directed mutagenesis, substrate binding residue mutation, reduced activity compared to the wild-type enzyme [42]) [42]
D167Y <7> (<7> site-directed mutagenesis, substrate binding residue mutation, reduced activity compared to the wild-type enzyme [42]) [42]
D276E <7> (<7> drastic decrease in EDTA tolerance [29]) [29]
D354N <21> (<21> site-directed mutagenesis, 9% of wild-type activity, mutant enzyme can be reconstituted with PQQ and Ca^{2+}, Sr^{2+}, or Ba^{2+}, but not with Mg^{2+}, which functions as a competitive inhibitor, in contrary to the wild-type enzyme [39]) [39]
D354N/N355D <21> (<21> site-directed mutagenesis, 10% of wild-type activity, mutant enzyme can be reconstituted with PQQ and Ca^{2+}, Sr^{2+}, or Ba^{2+}, but not with Mg^{2+}, which functions as a competitive inhibitor, in contrary to the wild-type enzyme [39]) [39]
D448N <1> (<1> site-directed mutagenesis, mutation in the active site loop 6BC region, mutant shows altered substrate specificity, but unaltered catalytic efficiency with D-glucose, compared to the wild-type isozyme PQQGDH-B [34]) [34]
D456N <1> (<1> site-directed mutagenesis, mutation in the active site loop 6BC region, mutant shows altered substrate specificity, but unaltered catalytic efficiency with D-glucose, compared to the wild-type isozyme PQQGDH-B [34]) [34]
D457N <1> (<1> site-directed mutagenesis, mutation in the active site loop 6BC region, mutant shows altered substrate specificity, but unaltered catalytic efficiency with D-glucose, compared to the wild-type isozyme PQQGDH-B [34]) [34]
D466E <1> (<1> very low glucose oxidase activity without influence on the affinity for pyrroloquinoline quinone, very low activity with ubiquinone Q_2 compared with the wild-type enzyme [21]) [21]
D466N <1> (<1> very low glucose oxidase activity without influence on the affinity for pyrroloquinoline quinone [21]) [21]
D730A <1> (<1> low glucose oxidase activity without influence on the affinity for pyrroloquinoline quinone, Mg^{2+} or substrate [23]) [23]
D730N <1> (<1> low glucose oxidase activity without influence on the affinity for pyrroloquinoline quinone, Mg^{2+} or substrate [23]) [23]
D730R <1> (<1> reduced affnity for pyrroloquinoline quinone [23]) [23]
E277A <7> (<7> decreased K_m value for glucose and altered substrate specificity, thermal stability is less than 20% of that of the wild-type enzyme [29]) [29]

E277D <7> (<7> decreased K_m value for glucose and altered substrate specificity, thermal stability is less than 20% of that of the wild-type enzyme [29]) [29]
E277G <7> (<7> drastic decrease in EDTA tolerance [29]) [29]
E277H <7> (<7> decreased K_m values for glucose and altered substrate specificity, thermal stability is less than 20% of that of the wild-type enzyme [29]) [29]
E277K <7> (<7> decreased K_m value for glucose and altered substrate specificity, significantly increased catalytic efficiency compared with the wild-type enzyme [29]) [29]
E277N <7> (<7> decreased K_m values for glucose and altered substrate specificity, thermal stability is less than 20% of that of the wild-type enzyme [29]) [29]
E277Q <7> (<7> decreased K_m values for glucose and altered substrate specificity, thermal stability is less than 20% of that of the wild-type enzyme [29]) [29]
E277V <7> (<7> decreased K_m values for glucose and altered substrate specificity, thermal stability is less than 20% of that of the wild-type enzyme [29]) [29]
E742G/P757L <1> (<1> slightly higher K_m value for Mg^{2+} [23]) [23]
G100R <23> (<23> mutant, relative activity vs wild type, substrate glucose 0.35, substrate maltose 0.26 [51]) [51]
G100W/G320E/M367P/A376T <23> (<23> mutant, relative activity vs wild type, substrate glucose 0.55, substrate maltose 0.20 [51]) [51]
G320D/M367P/A376T <23> (<23> mutant, relative activity vs wild type, substrate glucose 0.51, substrate maltose 0.16 [51]; <23> mutant, relative activity vs wild type, substrate glucose 0.69, substrate maltose 0.24 [51]) [51]
G320E <23> (<23> mutant, relative activity vs wild type, substrate glucose 0.92, substrate maltose 0.70 [51]) [51]
G320E/M367P/A376T <23> (<23> mutant, relative activity vs wild type, substrate glucose 0.69, substrate maltose 0.25 [51]) [51]
G320F/M367P/A376T <23> (<23> mutant, relative activity vs wild type, substrate glucose 0.48, substrate maltose 0.17 [51]) [51]
G320Y/M367P/A376T <23> (<23> mutant, relative activity vs wild type, substrate glucose 0.49, substrate maltose 0.16 [51]) [51]
G689D <1> (<1> significantly increased K_m for pyrroloquinoline quinone, slightly higher K_m value for Mg^{2+} [23]) [23]
H168C <7> (<7> site-directed mutagenesis, catalytic residue mutation, highly reduced activity compared to the wild-type enzyme [42]) [42]
H168Q <7> (<7> site-directed mutagenesis, catalytic residue mutation, nearly inactive mutant [42]; <7> site-directed mutagenesis, inactive mutant, a heterodimeric chimeric enzyme consisiting of 1 wild-type subunit and 1 mutant subunit shows decreased activity and a substrate specificity similar to the wild-type enzyme [48]) [42, 48]
H262A <1> (<1> reduced affinity both for glucose, 11fold, and pyrroloquinoline quinone, 8fold, without significant effect on glucose oxidase activity [21]) [21]

H262Y <1> (<1> greatly diminished catalytic efficiency for all substrates, rate of electron transfer to oxygen is unaffected, 230fold increased K_m value for glucose [26]) [26]
H775A <1> (<1> pronounced reduction of affinity for the prosthetic group pyrroloquinoline quinone [23]) [23]
H775R <1> (<1> pronounced reduction of affinity for the prosthetic group pyrroloquinoline quinone, 230fold higher K_m than wild-type enzyme [23]) [23]
K166E <7> (<7> site-directed mutagenesis, substrate binding residue mutation, altered substrate specificty compared to the wild-type enzyme [42]) [42]
K166G <7> (<7> site-directed mutagenesis, substrate binding residue mutation, altered substrate specificty compared to the wild-type enzyme [42]) [42]
K166I <7> (<7> site-directed mutagenesis, substrate binding residue mutation, altered substrate specificty compared to the wild-type enzyme [42]) [42]
K3E/E278G/G392C <23> (<23> mutant, relative activity vs wild type, substrate glucose 0.92, substrate maltose 0.53 [51]) [51]
K493A <1> (<1> very low glucose oxidase activity, without influence on the affinity for pyrroloquinoline quinone, very low activity with ubiquinone Q2 compared with the wild-type enzyme, very low activity of both phenazine methosulfate reductase and glucose oxidase in the membrane fractions compared with the wild type [21]) [21]
K493R <1> (<1> pronounced reduction of affinity for pyrroloquinoline quinone, very low activity of both phenazine methosulfate reductase and glucose oxidase in the membrane fractions compared with the wild type [21]) [21]
L194F/A376T <23> (<23> mutant, relative activity vs wild type, substrate glucose 0.39, substrate maltose 0.075 [51]) [51]
L194F/G320E/M367P <23> (<23> mutant, relative activity vs wild type, substrate glucose 0.38, substrate maltose 0.14 [51]) [51]
L194F/G320E/M367P/A376T <23> (<23> mutant, relative activity vs wild type, substrate glucose 0.36, substrate maltose 0.051 [51]) [51]
L194F/G320F <23> (<23> mutant, relative activity vs wild type, substrate glucose 0.38, substrate maltose 0.060 [51]) [51]
L194Q <23> (<23> mutant, relative activity vs wild type, substrate glucose 0.22, substrate maltose 0.15 [51]) [51]
M367P/A376T <23> (<23> mutant, relative activity vs wild type, substrate glucose 0.65, substrate maltose 0.24 [51]) [51]
N275E <7> (<7> drastic decrease in EDTA tolerance [29]) [29]
N340F/Y418F <7> (<7> site-directed mutagenesis, mutation of residues at the dimer interface, 2fold increased thermal stability at 55°C and unaltered catalytic efficiency compared to the wild-type enzyme [45]) [45]
N340F/Y418I <7> (<7> site-directed mutagenesis, mutation of residues at the dimer interface, 2fold increased thermal stability at 55°C and unaltered catalytic efficiency compared to the wild-type enzyme [45]) [45]
N355D <21> (<21> site-directed mutagenesis, 25% of wild-type activity, mutant enzyme can be reconstituted with PQQ and Ca^{2+}, Sr^{2+}, or Ba^{2+}, but not with Mg^{2+}, which functions as a competitive inhibitor, in contrary to the wild-type enzyme [39]) [39]

N452D <1> (<1> site-directed mutagenesis, mutation in the active site loop 6BC region, mutant shows altered substrate specificity, but unaltered catalytic efficiency with D-glucose, compared to the wild-type isozyme PQQGDH-B [34]) [34]
N452H <1> (<1> site-directed mutagenesis, mutation in the active site loop 6BC region, mutant shows altered substrate specificity, but unaltered catalytic efficiency with D-glucose, compared to the wild-type isozyme PQQGDH-B [34]) [34]
N452I <1> (<1> site-directed mutagenesis, mutation in the active site loop 6BC region, mutant shows altered substrate specificity, but unaltered catalytic efficiency with D-glucose, compared to the wild-type isozyme PQQGDH-B [34]) [34]
N452K <1> (<1> site-directed mutagenesis, mutation in the active site loop 6BC region, mutant shows altered substrate specificity, but unaltered catalytic efficiency with D-glucose, compared to the wild-type isozyme PQQGDH-B [34]) [34]
N452T <1, 7> (<7> site-directed mutagenesis, reduced activity compared to the wild-type enzyme [42]; <1> site-directed mutagenesis, mutation in the active site loop 6BC region, mutant shows narrowed substrate specificity, but unaltered catalytic efficiency, thermal stability, and EDTA tolerance compared to the wild-type isozyme PQQGDH-B [34]) [34, 42]
N454S <23> (<23> mutant, relative activity vs wild type, substrate glucose 0.87, substrate maltose 0.69 [51]) [51]
N462D <1> (<1> site-directed mutagenesis, mutation in the active site loop 6BC region, mutant shows altered substrate specificity, but unaltered catalytic efficiency with D-glucose, compared to the wild-type isozyme PQQGDH-B [34]) [34]
N462H <1> (<1> site-directed mutagenesis, mutation in the active site loop 6BC region, mutant shows altered substrate specificity, but unaltered catalytic efficiency with D-glucose, compared to the wild-type isozyme PQQGDH-B [34]) [34]
N462K <1> (<1> site-directed mutagenesis, mutation in the active site loop 6BC region, mutant shows altered substrate specificity, but unaltered catalytic efficiency with D-glucose, compared to the wild-type isozyme PQQGDH-B [34]) [34]
N462Y <1> (<1> site-directed mutagenesis, mutation in the active site loop 6BC region, mutant shows altered substrate specificity, but unaltered catalytic efficiency with D-glucose, compared to the wild-type isozyme PQQGDH-B [34]) [34]
Q169E <7> (<7> site-directed mutagenesis, substrate binding residue mutation, altered substrate specificty compared to the wild-type enzyme [42]) [42]
Q169K <7> (<7> site-directed mutagenesis, substrate binding residue mutation, altered substrate specificty compared to the wild-type enzyme [42]) [42]
Q193H <23> (<23> mutant, relative activity vs wild type, substrate glucose 0.41, substrate maltose 0.23 [51]) [51]
Q193S/G320E <23> (<23> mutant, relative activity vs wild type, substrate glucose 0.56, substrate maltose 0.19 [51]) [51]

Q209R/N240R/T389R <7> (<7> site-directed mutagenesis, increased thermal stability compared to the wild-type enzyme [43]) [43]
S145C <1> (<1> site-directed mutagenesis, introduction of a Cys residue in each monomer of the enzyme leads to formation of an intersubunit disulfide bridge at the dimer interface resulting in 30fold increased thermal stability at 55 °C compared to the wild-type enzyme [49]) [49]
S231C <7> (<7> increase in thermal stability [31]) [31]
S231D <7> (<7> increase in thermal stability [31]) [31]
S231H <7> (<7> increase in thermal stability [31]) [31]
S231K <7> (<7> more than 8fold increase in its half-life during the thermal inactivation at 55°C compared with the wild-type enzyme, retains catalytic activity similar to the wild-type enzyme [31]) [31]
S231L <7> (<7> increase in thermal stability [31]) [31]
S231M <7> (<7> increase in thermal stability [31]) [31]
S231N <7> (<7> increase in thermal stability [31]) [31]
S357L <1> (<1> significantly increased K_m for pyrroloquinoline quinone, slightly higher K_m value for Mg^{2+} [23]) [23]
T348G <7> (<7> mutant crystallized by microseeding, data set is collected at 2.36 A resolution [50]) [50]
T348G/N428P <7> (<7> mutant crystallized by microseeding, data set is collected at 2.15 A resolution [50]) [50]
T416V/T417V <7> (<7> site-directed mutagenesis, mutation of resides of the hydrophobic region, 2fold increased thermal stability at 55°C and unaltered catalytic efficiency compared to the wild-type enzyme [45]) [45]
V157I/M367V/T463S <23> (<23> mutant, relative activity vs wild type, substrate glucose 1.00, substrate maltose 0.78 [51]) [51]
V91A/W372R <23> (<23> mutant, relative activity vs wild type, substrate glucose 0.44, substrate maltose 0.22 [51]) [51]
W404A <1> (<1> pronounced reduction of affinity for pyrroloquinoline quinone, very low glucose oxidase activity and phenazine methosulfate reductase activity compared with wild-type enzyme [21]) [21]
W404F <1> (<1> pronounced reduction of affinity for pyrroloquinoline quinone, very weak activity of phenazine methosulfate reductase but still retains glucose oxidase activity equivalent to that of the wild-type [21]) [21]
Y171G/E245D/M341V/T348G/N428P <7> (<7> mutant crystallized by microseeding, data set is collected at 2.20 A resolution [50]) [50]
Y248F/N342D/A376T/A418V <23> (<23> mutant, relative activity vs wild type, substrate glucose 0.74, substrate maltose 0.43 [51]) [51]
Y302H <23> (<23> mutant, relative activity vs wild type, substrate glucose 0.47, substrate maltose 0.35 [51]) [51]
Additional information <1, 7> (<1,7> improved EDTA tolerance, thermal stability and substrate specificity of chimeric proteins [18]; <1> construction of a gene consisting of two identical subunits linked together by a DNA segment coding linker peptide region and production of a linked-dimeric enzyme, the linked-dimeric enzyme shows higher thermal stability than native dimeric enzyme [25]; <1> co-expression of peptide ligands in a random phage diplay modifies the substrate specificity of the enzyme towards mono- and disac-

charides, overview [44]; <7> engineering PQQ glucose dehydrogenase with improved substrate specificity [42]; <7> the recombinant cytochrome c-fusion protein shows a highly increased sensitivity when immobilized to the electrode as D-glucose sensor compared to the wild-type enzyme, overview [35]) [18, 25, 35, 42, 44]

Application

analysis <4, 5, 7> (<7> application of mutant enzyme S231K as a glucose sensor constituent. The mutant has more than 8fold increase in its half-life during the thermal inactivation at 55 °C compared with the wild-type enzyme and retains catalytic activity similar to the wild-type enzyme [31]; <4,5> apoenzyme is used as a biological test system for the detection of very low amounts of pyrroloquinoline quinone [2]; <7> enzyme is used as coupling enzyme for monitoring carbohydrate-transport reactions, the method is particularly suited for determining transport reactions that are not coupled to any form of metabolic energy such as uniport reactions, or for characterizing mutant proteins with a defective energy-coupling mechanism or system with high-affinity constants for sugars [27]) [2, 27, 31]

biotechnology <1, 7> (<1> bioengineering of water-soluble isozyme PQQGDH-B production at industrial level [34]; <7> engineering of the soluble enzyme GDH-B to enable the electron transfer to the electrode in absence of artificial electron mediator by mimicking the domain structure of the quinohemoprotein ethanol dehydrogenase from Comamonas testosteroni, which is composed of a PQQ-containing catalytic domain and a cytochrome c domain [35]; <7> engineering PQQ glucose dehydrogenase with improved substrate specificity [42]; <7> enzyme has a great potential for application as glucose sensor constituent [45]; <7> optimization of an expression system using Pichia pastoris for use in industrial level production [46]; <7> surface charge engineering of the enzyme for optimization of downstream processing in large scale enzyme production [43]) [34, 35, 42, 43, 45, 46]

diagnostics <7, 14> (<7> enzyme is industrially used as glucose sensor with high catalytic activity and insensitivity to oxygen [35]; <14> purified enzyme is immobilized on carbon electrodes modified with 4-ferrocenylphenol, 4-(4-ferrocenylimino-methyl)-phenol, or 4-ferrocenylnitrophenol, for use as glucose biosensors, kinetic behaviour during immobilization [41]; <7> purified enzyme is immobilized on carbon electrodes modified with 4-ferrocenylphenol, 4-(4-ferrocenylimino-methyl)phenol, or 4-ferrocenylnitrophenol, for use as glucose biosensors, kinetic behaviour during immobilization [41]) [35, 41]

6 Stability

pH-Stability

6 <16> (<16> activity with phenazine methosulfate, 2,6-dichlorophenol indophenol and pyrroloquinoline quinone [16]) [16]

Temperature stability

35 <7> (<7> pH 6, 10 min, in absence of Ca^{2+}, stable below, reversible inactivation above [11]) [11]
40.5 <21> (<21> 10 min, 50% inactivation of recombinant mutant D354N apoenzyme [39]) [39]
40.8 <21> (<21> 10 min, 50% inactivation of recombinant wild-type apoenzyme [39]) [39]
41 <21> (<21> 10 min, 50% inactivation of recombinant mutant D354N/N355D apoenzyme [39]) [39]
43.2 <21> (<21> 10 min, 50% inactivation of recombinant mutant N355D apoenzyme [39]) [39]
45 <1> (<1> 20 min, 20% loss of activity of native enzyme, about 50% loss of activity of linked-dimeric enzyme [25]) [25]
48.6 <21> (<21> 10 min, 50% inactivation of recombinant mutant D354N/N355D holoenzyme [39]) [39]
49 <21> (<21> 10 min, 50% inactivation of recombinant mutant D354N holoenzyme [39]) [39]
50 <7> (<7> pH 6, 10 min, in presence of Ca^{2+}, inactivation above [11]) [11]
51 <21> (<21> 10 min, 50% inactivation of recombinant wild-type holoenzyme [39]) [39]
55 <7, 21> (<7> $t_{1/2}$: 10 min for wild-type enzyme and mutant enzyme E277K, 4 min for mutant enzymes E277Q and N279H, 25 min for mutant enzyme I278F, less than 2 min for mutant enzymes E277G, E277A, E277D, E277H, E277N, E277V, D275E and D276E [29]; <21> 10 min, 50% inactivation of recombinant mutant N355D holoenzyme [39]; <7> isozyme PQQGDH-B, 50% residual activity after 10 min [46]; <7> recombinant tagged wild-type enzyme and mutant enzyme show about 90% remaining activity after 10 min, the nontagged wild-type enzyme shows about 40% remaining activity after 10 min [43]; <7> thermal stability of wild-type and mutant isozymes PQQGDH-B, overview [42]; <7> wild-type isozyme PQQGDH-B: half-life 9.5 min [45]) [29, 39, 42, 43, 45, 46]
70 <1> (<1> 10 min, mutant S145C retaines 90% of wild-type activity [49]) [49]
Additional information <1> (<1> the C-terminal 3% region, A3 region, plays an important role in the increase of thermal stability [18]) [18]

General stability information

<7>, the soluble enzyme is less stable than the membrane-bound form [41]
<14>, the membrane-bound enzyme is more stable than the soluble form [41]

References

[1] Duine, J.A.; Frank, J.; van Zeeland, J.K.: Glucose dehydrogenase from Acinetobacter calcoaceticus: a quinoprotein. FEBS Lett., **108**, 443-446 (1979)

[2] Duine, J.A.; Frank, J.; Jongejan, J.A.: Detection and determination of pyrroloquinoline quinone, the coenzyme of quinoproteins. Anal. Biochem., **133**, 239-243 (1983)
[3] Geiger, O.; Goerisch, H.: Crystalline quinoprotein glucose dehydrogenase from Acinetobacter calcoaceticus. Biochemistry, **25**, 6043-6048 (1986)
[4] Dokter, P.; Frank, J.; Duine, J.A.: Purification and characterization of quinoprotein glucose dehydrogenase from Acinetobacter calcoaceticus L.M.D. 79.41. Biochem. J., **239**, 163-167 (1986)
[5] Cleton-Jansen, A.M.; Goosen, N.; Fayet, O.; van de Putte, P.: Cloning, mapping, and sequencing of the gene encoding Escherichia coli quinoprotein glucose dehydrogenase. J. Bacteriol., **172**, 6308-6315 (1990)
[6] Hommes, R.W.J.; Herman, P.T.D.; Postma, P.W.; Tempest, D.W.; Neijssel, O.M.: The separate roles of PQQ and apo-enzyme syntheses in the regulation of glucose dehydrogenase activity in Klebsiella pneumoniae NCTC 418. Arch. Microbiol., **151**, 257-260 (1989)
[7] Van Schie, B.J.; de Mooy, O.H.; Linton, J.D.; van Dijken, J.P.; Kuenen, J.G.: PQQ-dependent production of gluconic acid by Acinetobacter, Agrobacterium and Rhizobium species. J. Gen. Microbiol., **133**, 867-875 (1987)
[8] Dokter, P.; Pronk, J.T.; van Schie, B.J.; van Dijken, J.P.; Duine, J.A.: The in vivo and in vitro substrate specificity of quinoprotein glucose dehydrogenase of Acinetobacter calcoaceticus LMD79.41. FEMS Microbiol. Lett., **43**, 195-200 (1987)
[9] Matsushita, K.; Shinagawa, E.; Adachi, O.; Amiyama, M.: Reactivity with ubiquinone of quinoprotein D-glucose dehydrogenase from Gluconobacter suboxydans. J. Biochem., **105**, 633-637 (1989)
[10] Matsushita, K.; Shinagawa, E.; Adachi, O.; Ameyama, M.: Quinoprotein D-glucose dehydrogenase of the Acinetobacter calcoaceticus respiratory chain: membrane-bound and soluble forms are different molecular species. Biochemistry, **28**, 6276-6280 (1989)
[11] Geiger, O.; Goerisch, H.: Reversible thermal inactivation of the quinoprotein glucose dehydrogenase from Acinetobacter calcoaceticus. Biochem. J., **261**, 415-421 (1989)
[12] Matsushita, K.; Shinagawa, E.; Adachi, O.; Ameyama, M.: Quinoprotein D-glucose dehydrogenase in Acinetobacetr calcoaceticus LMD 79.41: purification and characterization of the membrane-bound enzyme distinct from the soluble enzyme. Antonie Leeuwenhoek, **56**, 63-72 (1989)
[13] Matsushita, K.; Shinagawa, E.; Inoue, T.; Adachi, O.; Ameyama, M.: Immunological evidence for two types of PQQ-dependent D-glucose dehydrogenase in bacterial membranes and the location of the enzyme in Escherichia coli. FEMS Microbiol. Lett., **37**, 141-144 (1986)
[14] Matsushita, K.; Ameyama, M.: D-Glucose dehydrogenase from Pseudomonas fluorescens, membrane-bound. Methods Enzymol., **89**, 149-154 (1982)
[15] Matsushita, K.; Ohno, Y.; Shinagawa, E.; Adachi, O.; Ameyama, M.: Membrane-bound, electron transport-linked, D-glucose dehydrogenase of Pseudomonas fluorescens. Interaction of the purified enzyme with ubiquinone or phospholipid. Agric. Biol. Chem., **46**, 1007-1011 (1982)

[16] Ameyama, M.; Shinagawa, E.; Matsushita, K.; Adachi, O.: L-Glucose dehydrogenase of Gluconobacter suboxydans: solubilization, purification and characterization. Agric. Biol. Chem., **45**, 851-861 (1981)
[17] Oubrie, A.; Rozeboom, H.J.; Kalk, K.H.; Duine, J.A.; Dijkstra, B.W.: The 1.7 A crystal structure of the apo form of the soluble quinoprotein glucose dehydrogenase from Acinetobacter calcoaceticus reveals a novel internal conserved sequence repeat. J. Mol. Biol., **289**, 319-333 (1999)
[18] Yoshida, H.; Kojima, K.; Witarto, A.B.; Sode, K.: Engineering a chimeric pyrroloquinoline quinone glucose dehydrogenase: improvement of EDTA tolerance, thermal stability and substrate specificity. Protein Eng., **12**, 63-70 (1999)
[19] Bernardelli, C.E.; Luna, M.F.; Galar, M.L.; Boiardi, J.L.: Periplasmic PQQ-dependent glucose oxidation in free-living and symbiotic rhizobia. Curr. Microbiol., **42**, 310-315 (2001)
[20] Elias, M.; Tanaka, M.; Sakai, M.; Toyama, H.; Matsushita, K.; Adachi, O.; Yamada, M.: C-terminal periplasmic domain of Escherichia coli quinoprotein glucose dehydrogenase transfers electrons to ubiquinone. J. Biol. Chem., **276**, 48356-48361 (2001)
[21] Elias, M.D.; Tanaka, M.; Izu, H.; Matsushita, K.; Adachi, O.; Yamada, M.: Functions of amino acid residues in the active site of Escherichia coli pyrroloquinoline quinone-containing quinoprotein glucose dehydrogenase. J. Biol. Chem., **275**, 7321-7326 (2000)
[22] Olsthoorn, A.J.; Duine, J.A.: On the mechanism and specificity of soluble, quinoprotein glucose dehydrogenase in the oxidation of aldose sugars. Biochemistry, **37**, 13854-13861 (1998)
[23] Yamada, M.; Inbe, H.; Tanaka, M.; Sumi, K.; Matsushita, K.; Adachi, O.: Mutant isolation of the Escherichia coli quinoprotein glucose dehydrogenase and analysis of crucial residues Asp-730 and His-775 for its function. J. Biol. Chem., **273**, 22021-22027 (1998)
[24] Yamada, M.; Sumi, K.; Matsushita, K.; Adachi, O.; Yamada, Y.: Topological analysis of quinoprotein glucose dehydrogenase in Escherichia coli and its ubiquinone-binding site. J. Biol. Chem., **268**, 12812-12817 (1993)
[25] Sode, K.; Shirahane, M.; Yoshida, H.: Construction and characterization of a linked-dimeric pyrroloquinoline quinone glucose dehydrogenase. Biotechnol. Lett., **21**, 707-710 (1999)
[26] Cozier, G.E.; Salleh, R.A.; Anthony, C.: Characterization of the membrane quinoprotein glucose dehydrogenase from Escherichia coli and characterization of a site-directed mutant in which histidine-262 has been changed to tyrosine. Biochem. J., **340**, 639-647 (1999)
[27] Heuberger, E.H.M.L.; Poolman, B.: A spectroscopic assay for the analysis of carbohydrate transport reactions. Eur. J. Biochem., **267**, 228-234 (2000)
[28] Iswantini, D.; Kano, K.; Ikeda, T.: Kinetics and thermodynamics of activation of quinoprotein glucose dehydrogenase apoenzyme in vivo and catalytic activity of the activated enzyme in Escherichia coli cells. Biochem. J., **350**, 917-923 (2000)
[29] Igarashi, S.; Ohtera, T.; Yoshida, H.; Witarto, A.B.; Sode, K.: Construction and characterization of mutant water-soluble PQQ glucose dehydrogenases

with altered K(m) values–site-directed mutagenesis studies on the putative active site. Biochem. Biophys. Res. Commun., **264**, 820-824 (1999)
[30] Olsthoorn, A.J.J.; Duine, J.A.: Production, characterization, and reconstitution of recombinant quinoprotein glucose dehydrogenase (soluble type; EC 1.1.99.17) apoenzyme of Acinetobacter calcoaceticus. Arch. Biochem. Biophys., **336**, 42-48 (1996)
[31] Sode, K.; Ootera, T.; Shirahane, M.; Witarto, A.B.; Igarashi, S.; Yoshida, H.: Increasing the thermal stability of the water-soluble pyrroloquinoline quinone glucose dehydrogenase by single amino acid replacement. Enzyme Microb. Technol., **26**, 491-496 (2000)
[32] Dewanti, A.R.; Duine, J.A.: Reconstitution of membrane.integrated quinoprotein glucose dehydrogenase apoenzyme with PQQ and the holoenzyme's mechanism of action. Biochemistry, **37**, 6810-6818 (1998)
[33] Oubrie, A.; Rozeboom, H.J.; Dijkstra, B.W.: Active-site structure of the soluble quinoprotein glucose dehydrogenase complexed with methylhydrazine: A covalent cofactor-inhibitor complex. Proc. Natl. Acad. Sci. USA, **96**, 11787-11791 (1999)
[34] Sode, K.; Igarashi, S.; Morimoto, A.; Yoshida, H.: Construction of engineered water-soluble PQQ glucose dehydrogenase with improved substrate specificity. Biocatal. Biotransform., **20**, 405-412 (2002)
[35] Okuda, J.; Sode, K.: PQQ glucose dehydrogenase with novel electron transfer ability. Biochem. Biophys. Res. Commun., **314**, 793-797 (2004)
[36] Ivanova, E.V.; Ershov, A.Y.; Laurinavicius, V.; Meskus, R.; Ryabov, A.D.: Comparative kinetic study of D-glucose oxidation by ruthenium(III) compounds catalyzed by FAD-dependent glucose oxidase and PQQ-dependent glucose dehydrogenase. Biochemistry, **68**, 407-415 (2003)
[37] Oubrie, A.: Structure and mechanism of soluble glucose dehydrogenase and other PQQ-dependent enzymes. Biochim. Biophys. Acta, **1647**, 143-151 (2003)
[38] Yamada, M.; Elias, M.D.; Matsushita, K.; Migita, C.T.; Adachi, O.: Escherichia coli PQQ-containing quinoprotein glucose dehydrogenase: its structure comparison with other quinoproteins. Biochim. Biophys. Acta, **1647**, 185-192 (2003)
[39] James, P.L.; Anthony, C.: The metal ion in the active site of the membrane glucose dehydrogenase of Escherichia coli. Biochim. Biophys. Acta, **1647**, 200-205 (2003)
[40] Lapenaite, I.; Kurtinaitiene, B.; Anusevicius, Z.; Sarlauskas, J.; Bachmatova, I.; Marcinkeviciene, L.; Laurinavicius, V.; Ramanavicius, A.: Some quinone derivatives as redox mediators for PQQ-dependent glucose dehydrogenase. Biologia (Bratisl.), **1**, 20-22 (2004)
[41] Laurinavicius, V.; Razumiene, J.; Kurtinaitiene, B.; Gureviciene, V.; Marcinkeviciene, L.; Bachmatova, I.: Comparative characterization of soluble and membrane-bound PQQ-glucose dehydrogenases. Biologia (Bratisl.), **2**, 31-34 (2003)
[42] Igarashi, S.; Hirokawa, T.; Sode, K.: Engineering PQQ glucose dehydrogenase with improved substrate specificity. Site-directed mutagenesis studies

on the active center of PQQ glucose dehydrogenase. Biomol. Eng., **21**, 81-89 (2004)

[43] Koh, H.; Igarashi, S.; Sode, K.: Surface charge engineering of PQQ glucose dehydrogenase for downstream processing. Biotechnol. Lett., **25**, 1695-1701 (2003)

[44] Yoshida, H.; Yagi, Y.; Ikebukuro, K.; Sode, K.: Improved substrate specificity of water-soluble pyrroloquinoline quinone glucose dehydrogenase by a peptide ligand. Biotechnol. Lett., **25**, 301-305 (2003)

[45] Tanaka, S.; Igarashi, S.; Ferri, S.; Sode, K.: Increasing stability of water-soluble PQQ glucose dehydrogenase by increasing hydrophobic interaction at dimeric interface. BMC Biochem., **6**, 1 (2005)

[46] Yoshida, H.; Araki, N.; Tomisaka, A.; Sode, K.: Secretion of water soluble pyrroloquinoline quinone glucose dehydrogenase by recombinant Pichia pastoris. Enzyme Microb. Technol., **30**, 312-318 (2002)

[47] Reddy, S.Y.; Bruice, T.C.: Mechanism of glucose oxidation by quinoprotein soluble glucose dehydrogenase: insights from molecular dynamics studies. J. Am. Chem. Soc., **126**, 2431-2438 (2004)

[48] Igarashi, S.; Sode, K.: Construction and characterization of heterodimeric soluble quinoprotein glucose dehydrogenase. J. Biochem. Biophys. Methods, **61**, 331-338 (2004)

[49] Igarashi, S.; Sode, K.: Stabilization of quaternary structure of water-soluble quinoprotein glucose dehydrogenase. Mol. Biotechnol., **24**, 97-104 (2003)

[50] Sanchez-Weatherby, J.; Southall, S.; Oubrie, A.: Crystallization of quinoprotein glucose dehydrogenase variants and homologues by microseeding. Acta Crystallogr.Sect.F, **62**, 518-521 (2006)

[51] Hamamatsu, N.; Suzumura, A.; Nomiya, Y.; Sato, M.; Aita, T.; Nakajima, M.; Husimi, Y.; Shibanaka, Y.: Modified substrate specificity of pyrroloquinoline quinone glucose dehydrogenase by biased mutation assembling with optimized amino acid substitution. Appl. Microbiol. Biotechnol., **73**, 607-617 (2006)

[52] Luna, M.F.; Bernardelli, C.E.; Galar, M.L.; Boiardi, J.L.: Glucose metabolism in batch and continuous cultures of Gluconacetobacter diazotrophicus PAL 3. Curr.Microbiol., **52**, 163-168 (2006)

[53] Southall, S.M.; Doel, J.J.; Richardson, D.J.; Oubrie, A.: Soluble aldose sugar dehydrogenase from Escherichia coli: a highly exposed active site conferring broad substrate specificity. J. Biol. Chem., **281**, 30650-30659 (2006)

Pyranose dehydrogenase (acceptor) 1.1.99.29

1 Nomenclature

EC number
1.1.99.29

Systematic name
pyranose:acceptor oxidoreductase

Recommended name
pyranose dehydrogenase (acceptor)

Synonyms
PDH
dehydrogenase, pyranose 2/3- <1> [1]
pyranose 2-dehydrogenase <1> [1]
pyranose 2/3-dehydrogenase <1> [1]
pyranose 3-dehydrogenase <1> [1]
pyranose dehydrogenase <3> [6]
pyranose-quinone oxidoreductase
pyranose:quinone acceptor 2-oxidoreductase <1> [1]
quinone-dependent pyranose dehydrogenase

CAS registry number
190606-21-4

2 Source Organism

<1> *Agaricus bisporus* (no sequence specified) [1, 2, 3]
<2> *Macrolepiota rhacodes* (no sequence specified) [4]
<3> *Agaricus meleagris* (no sequence specified) [5, 6]

3 Reaction and Specificity

Catalyzed reaction
a pyranoside + acceptor = a 3-dehydropyranoside (or 3,4-didehydropyranoside) + reduced acceptor (<1> requires FAD, a number of aldoses and ketoses in pyranose form, as well as glycosides, gluco-oligosaccharides, sucrose and lactose can act as donor, 1,4-benzoquinone or ferricenium ion (ferrocene oxidized by removal of one electron) can serve as acceptor, unlike EC 1.1.3.10, pyranose oxidase, this fungal enzyme does not interact with O_2 and exhibits

extremely broad substrate tolerance with variable regioselectivity (C-3, C-2 or C_3 + C-2 or C-3 + C-4) for (di)oxidation of different sugars, D-glucose is exclusively or preferentially oxidized at C-3 (depending on the enzyme source), but can also be oxidized at C-2 + C-3, the enzyme also acts on 1,4-α- and 1,4-β-gluco-oligosaccharides, non-reducing gluco-oligosaccharides and L-arabinose, which are not substrates of EC 1.1.3.10, sugars are oxidized in their pyranose but not in their furanose form [1])
pyranose + acceptor = 2-dehydropyranose (or 3-dehydropyranose or 2,3-didehydropyranose) + reduced acceptor

Reaction type

oxidation
redox reaction
reduction

Natural substrates and products

S Additional information <1> (<1> oxidative carbohydrate metabolism, pyranose 2-dehydrogenase may play an important role in fungal lignocellulose decomposition by interconnecting ligninolysis with degradation of cell wall polysaccharide components [1]) (Reversibility: ?) [1]
P ?

Substrates and products

S 2-deoxy-D-glucose + 1,4-benzoquinone <2> (<2> 49% of the activity with D-glucose [4]) (Reversibility: ?) [4]
P ?
S 2-deoxy-D-glucose + ferricenium ion <2> (<2> 21% of the activity with D-glucose [4]) (Reversibility: ?) [4]
P ?
S 3-dehydro-D-glucose + 1,4-benzoquinone <2> (Reversibility: ?) [4]
P 2,3-didehydro-D-glucose + hydroquinone
S D-allose + 1,4-benzoquinone <1> (<1> 30% of the activity with D-glucose [1]) (Reversibility: ?) [1]
P 2-dehydro-D-allose + hydroquinone
S D-arabino-2-hexosulose + 1,4-benzoquinone <1> (<1> 59% of the activity with D-glucose [1]) (Reversibility: ?) [1]
P ?
S D-arabinose + 1,4-benzoquinone <1> (<1> 18% of the activity with D-glucose [1]) (Reversibility: ?) [1]
P 2-dehydro-D-arabinose + hydroquinone
S D-galactose + 1,4-benzoquinone <1> (<1> 88% of the activity with D-glucose [1]; <1> PDH oxidizes D-galactose exclusively at C-2 to produce D-lyxo-hexos-2-ulose [2]; <1> quinone-dependent PDH oxidizes D-galactose exclusively at C-2 [3]) (Reversibility: ?) [1, 2, 3]
P 2-dehydro-D-galactose + hydroquinone (<1> 2-keto-D-galactose [2])
S D-galactose + 1,4-benzoquinone <2> (<2> 66% of the activity with D-glucose [4]) (Reversibility: ?) [4]
P ?

S D-galactose + ferricenium ion <2> (<2> 60% of the activity with D-glucose [4]) (Reversibility: ?) [4]
P ?
S D-glucono-1,5-lactone + 1,4-benzoquinone <1, 2> (<2> 20% of the activity with D-glucose [4]; <1> 118% of the activity with D-glucose [1]) (Reversibility: ?) [1, 4]
P ?
S D-glucono-1,5-lactone + ferricenium ion <2> (<2> 8% of the activity with D-glucose [4]) (Reversibility: ?) [4]
P ?
S D-glucose + 1,4-benzoquinone <1> (Reversibility: ?) [1]
P 2-dehydro-D-glucose + hydroquinone (<1> D-arabino-2-hexosulose, 2-ketoglucose [1])
S D-glucose + 1,4-benzoquinone <1, 2, 3> (<2> oxidizes D-glucose exclusively at C-3, D-glucose is the preferred electron donor susbtrate [4]; <1> PDH catalyzes the C-2 and C-3 oxidation of D-glucose, quinone-dependent [2]; <1> quinone-dependent PDH catalyzes C-2 or, preferentially, C-3 dehydrogenation of D-glucose [3]; <3> simultaneous C-2 and C-3 oxidation of free D-glucose [5]) (Reversibility: ?) [2, 3, 4, 5]
P 3-dehydro-D-glucose + hydroquinone (<3> or 2-dehydro-D-glucose [5]; <1> or 2-dehydro-D-glucose or 2,3-didehydro-D-glucose [3]; <1> or 2-dehydro-D-glucose or 2,3-didehydro-D-glucose, products are 2-keto-D-glucose and, preferentially, 3-keto-D-glucose, both are subsequently oxidized by PDH to 2,3-diketo-D-glucose [2])
S D-glucose + ferricenium ion <2> (<2> oxidizes D-glucose exclusively at C-3, D-glucose is the preferred electron donor susbtrate [4]) (Reversibility: ?) [4]
P 3-dehydro-D-glucose + ferrocene
S D-lyxose + 1,4-benzoquinone <2> (<2> 5% of the activity with D-glucose [4]) (Reversibility: ?) [4]
P ?
S D-lyxose + ferricenium ion <2> (<2> 5% of the activity with D-glucose [4]) (Reversibility: ?) [4]
P ?
S D-mannose + 1,4-benzoquinone <2> (<2> 32% of the activity with D-glucose [4]) (Reversibility: ?) [4]
P ?
S D-mannose + ferricenium ion <2> (<2> 26% of the activity with D-glucose [4]) (Reversibility: ?) [4]
P ?
S D-ribose + 1,4-benzoquinone <1> (<1> 34% of the activity with D-glucose [1]) (Reversibility: ?) [1]
P 2-dehydro-D-ribose + hydroquinone
S D-xylose + 1,4-benzoquinone <1> (<1> 116% of the activity with D-glucose [1]; <1> quinone-dependent PDH oxidizes D-xylose at C-2 to 2-keto-D-xylose and successively at C-3 to 2,3-diketo-D-xylose [3]) (Reversibility: ?) [1, 3]

P 2-dehydro-D-xylose + hydroquinone (<1> or 2,3-didehydro-D-xylose [3])
S D-xylose + 1,4-benzoquinone <2> (<2> 72% of the activity with D-glucose [4]) (Reversibility: ?) [4]
P ?
S D-xylose + ferricenium ion <2> (<2> 81% of the activity with D-glucose [4]) (Reversibility: ?) [4]
P ?
S L-arabinose + 1,4-benzoquinone <1> (<1> 113% of the activity with D-glucose [1]) (Reversibility: ?) [1]
P 2-dehydro-L-arabinose + hydroquinone
S L-arabinose + 1,4-benzoquinone <2> (<2> 54% of the activity with D-glucose [4]) (Reversibility: ?) [4]
P ?
S L-arabinose + ferricenium ion <2> (<2> 72% of the activity with D-glucose [4]) (Reversibility: ?) [4]
P ?
S L-glucose + 1,4-benzoquinone <1> (<1> 61% of the activity with D-glucose [1]) (Reversibility: ?) [1]
P 2-dehydro-L-glucose + hydroquinone
S L-glucose + ferricenium ion <2> (<2> 11% of the activity with D-glucose [4]) (Reversibility: ?) [4]
P ?
S L-sorbose + 1,4-benzoquinone <1> (<1> 10% of the activity with D-glucose [1]) (Reversibility: ?) [1]
P 2-dehydro-L-sorbose + hydroquinone
S a pyranoside + acceptor <1, 2> (Reversibility: ?) [1, 2, 3, 4]
P a 3-dehydropyranoside (or 3,4-didehydropyranoside) + reduced acceptor
S aldopyranose + 1,4-benzoquinone <1> (<1> more active as electron acceptor than 2,6-dichlorophenolindophenol or ferricyanide, less active than 3,5-di-tert-butyl-1,2-benzoquinone, several oligosaccharides and glycopyranosides are converted to the corresponding 2-aldoketoses/aldosuloses [1]) (Reversibility: ?) [1]
P 2-aldoketose + hydroquinone
S aldopyranose + 2,6-dichlorophenolindophenol <1> (<1> 65.9% of the activity with 1,4-benzoquinone as electron acceptor, several oligosaccharides and glycopyranosides are converted to the corresponding 2-aldoketoses/aldosuloses [1]) (Reversibility: ?) [1]
P 2-aldoketose + reduced 2,6-dichlorophenolindophenol
S aldopyranose + 3,5-di-tert-butyl-1,2-benzoquinone <1> (<1> 4.3fold higher activity than with 1,4-benzoquinone as electron acceptor, several oligosaccharides and glycopyranosides are converted to the corresponding 2-aldoketoses/aldosuloses [1]) (Reversibility: ?) [1]
P 2-aldoketose + reduced 3,5-di-tert-butyl-1,2-benzoquinone
S aldopyranose + ferricyanide <1> (<1> 10.4% of the activity with 1,4-benzoquinone as electron acceptor, several oligosaccharides and glycopyranosides are converted to the corresponding 2-aldoketoses/aldosuloses [1]) (Reversibility: ?) [1]

P 2-aldoketose + reduced ferricyanide

S aldopyranose + quinone <1> (<1> C-2 specific sugar oxidoreductase, highly nonspecific, several oligosaccharides and glycopyranosides are converted to the corresponding 2-aldoketoses/aldosuloses [1]) (Reversibility: ?) [1]

P 2-aldoketose + hydroquinone

S α,α-trehalose + 1,4-benzoquinone <3> (<3> quinone-dependent PDH oxidizes α,α-trehalose at C-3 and C-3', 5% of the activity with D-glucose [5]) (Reversibility: ?) [5]

P 3,3'-didehydro-α,α-trehalose + hydroquinone (<3> with 3-dehydro-α,α-trehalose as reaction intermediate [5])

S cellobiose + 1,4-benzoquinone <1, 2> (<2> 13% of the activity with D-glucose [4]; <1> 94% of the activity with D-glucose [1]) (Reversibility: ?) [1, 4]

P ?

S cellobiose + ferricenium ion <2> (<2> 7% of the activity with D-glucose [4]) (Reversibility: ?) [4]

P ?

S cellotriose + 1,4-benzoquinone <1> (<1> 56% of the activity with D-glucose [1]) (Reversibility: ?) [1]

P ?

S erlose + 1,4-benzoquinone <3> (<3> quinone-dependent PDH oxidizes erlose at exclusively C-3 of its terminal glucopyranosyl moiety, 9% of the activity with D-glucose [5]) (Reversibility: ?) [5]

P 3-dehydroerlose + hydroquinone

S lactose + 1,4-benzoquinone <1> (<1> 21% of the activity with D-glucose [1]) (Reversibility: ?) [1]

P ?

S maltose + 1,4-benzoquinone <1, 2> (<2> 64% of the activity with D-glucose [4]; <1> 93% of the activity with D-glucose [1]) (Reversibility: ?) [1, 4]

P ?

S maltose + ferricenium ion <2> (<2> 45% of the activity with D-glucose [4]) (Reversibility: ?) [4]

P ?

S maltotriose + 1,4-benzoquinone <1, 2> (<1> 70% of the activity with D-glucose [1]; <2> 21% of the activity with D-glucose [4]) (Reversibility: ?) [1, 4]

P ?

S maltotriose + ferricenium ion <2> (<2> 14% of the activity with D-glucose [4]) (Reversibility: ?) [4]

P ?

S melizitose + 1,4-benzoquinone <3> (<3> quinone-dependent PDH oxidizes melezitose exclusively at C-3 of its terminal glucopyranosyl moiety, 7% of the activity with D-glucose [5]) (Reversibility: ?) [5]

P 3-dehydromelizitose + hydroquinone

S methyl-α-D-galactopyranoside + 1,4-benzoquinone <1, 3> (<1> 23% of the activity with D-glucose [1]; <3> formation of the C-3 carbonyl derivative, 7% of the activity with D-glucose [5]) (Reversibility: ?) [1, 5]
P methyl-α-3-dehydro-D-galactopyranoside + hydroquinone
S methyl-α-D-galactopyranoside + 1,4-benzoquinone <2> (<2> 11% of the activity with D-glucose [4]) (Reversibility: ?) [4]
P ?
S methyl-α-D-galactopyranoside + ferricenium ion <2> (<2> 8% of the activity with D-glucose [4]) (Reversibility: ?) [4]
P ?
S methyl-α-D-glucopyranoside + 1,4-benzoquinone <1, 3> (<1> 117% of the activity with D-glucose [1]; <3> formation of the C-3 carbonyl derivative, 126% of the activity with D-glucose [5]) (Reversibility: ?) [1, 5]
P methyl-α-3-dehydro-D-glucopyranoside + hydroquinone
S methyl-α-D-glucopyranoside + 1,4-benzoquinone <2> (<2> 141% of the activity with D-glucose [4]) (Reversibility: ?) [4]
P ?
S methyl-α-D-glucopyranoside + ferricenium ion <2> (<2> 143% of the activity with D-glucose [4]) (Reversibility: ?) [4]
P ?
S methyl-β-D-glucopyranoside + 1,4-benzoquinone <1, 3> (<1> 69% of the activity with D-glucose [1]; <3> formation of the C-3 carbonyl derivative, 20% of the activity with D-glucose [5]) (Reversibility: ?) [1, 5]
P methyl-β-3-dehydro-D-glucopyranoside + hydroquinone
S methyl-β-D-glucopyranoside + 1,4-benzoquinone <2> (<2> 30% of the activity with D-glucose [4]) (Reversibility: ?) [4]
P ?
S methyl-β-D-glucopyranoside + ferricenium ion <2> (<2> 17% of the activity with D-glucose [4]) (Reversibility: ?) [4]
P ?
S pyranose + acceptor <1, 2> (Reversibility: ?) [1, 2, 3, 4]
P 2-dehydropyranose (or 3-dehydropyranose or 2,3-didehydropyranose) + reduced acceptor
S sucrose + 1,4-benzoquinone <1, 2> (<2> 9% of the activity with D-glucose [4]; <1> 63% of the activity with D-glucose [1]) (Reversibility: ?) [1, 4]
P ?
S sucrose + 1,4-benzoquinone <3> (<3> quinone-dependent PDH oxidizes sucrose exclusively at C-3 of its terminal glucopyranosyl moiety, 11% of the activity with D-glucose [5]) (Reversibility: ?) [5]
P 3-dehydrosucrose + hydroquinone (<3> oxidation of sucrose at C-3 of its terminal glucopyranosyl moiety [5])
S sucrose + ferricenium ion <2> (<2> 6% of the activity with D-glucose [4]) (Reversibility: ?) [4]
P ?

S trehalose + 1,4-benzoquinone <1, 2> (<2> 7% of the activity with D-glucose [4]; <1> 54% of the activity with D-glucose [1]) (Reversibility: ?) [1, 4]

P ?

S trehalose + ferricenium ion <2> (<2> 8% of the activity with D-glucose [4]) (Reversibility: ?) [4]

P ?

S Additional information <1, 2, 3> (<1> oxidative carbohydrate metabolism, pyranose 2-dehydrogenase may play an important role in fungal lignocellulose decomposition by interconnecting ligninolysis with degradation of cell wall polysaccharide components [1]; <2> broad selectivity for sugar substrates, O_2 does not act as oxidant, not: D-allose, D-altrose, D-arabinose, D-ribose, D-fructose, L-sorbose, lactose, D-arabinitol, dulcitol, D-mannitol [4]; <1> enzyme is incapable of reducing O_2 to H_2O_2, not: cytochrome c, $NAD(P)^+$, nitroblue tetrazolium, low or no activity with D-altrose, D-fructose, D-lyxose, D-mannose, D-arabinitol, D-glucitol, D-mannitol or erythritol [1]; <3> extremely broad substrate tolerance [5]; <1> with 1,4-benzoquinone as electron acceptor, PDH catalyzes the oxidation of numerous monosaccharides and oligosaccharides [2]) (Reversibility: ?) [1, 2, 4, 5]

P ?

Inhibitors

Ag^+ <1> (<1> 10 mM, 100% inhibition [1]) [1]

CN^- <1> (<1> 10 mM, 83% inhibition [1]) [1]

Cu^{2+} <1> (<1> 10 mM, 90% inhibition [1]) [1]

$HgCl_2$ <1> (<1> 10 mM, 100% inhibition [1]) [1]

Additional information <1> (<1> not inhibited by Ca^{2+}, Co^{2+}, Fe^{2+}, Fe^{3+}, Mg^{2+}, Mn^{2+}, NH_4^+, Zn^{2+}, F^-, EDTA, NaN_3, dithiothreitol, phenanthroline and H_2O_2, each at 10 mM [1]) [1]

Cofactors/prosthetic groups

FAD <2> (<2> functions as the cofactor [4]) [4]

Flavin <1, 2> (<1> flavoprotein [2]; <1> contains a flavin prosthetic group [1]; <2> flavoprotein, FAD as cofactor [4]) [1,2,4]

Activating compounds

Acetate <1> (<1> 50 mM: activates, 100 mM: 100% increase in activity at pH 4 [1]) [1]

Specific activity (U/mg)

14.7 <3> (<3> pH 4.5, 25°C, D-glucose and 1,4-benzoquinone as substrates [5]) [5]

16.4 <1> (<1> pH 6.5, 25°C [1]) [1]

67 <2> (<2> 25°C, D-glucose and ferricenium ion as substrates [4]) [4]

K_m-Value (mM)

1.82 <2> (D-glucose, <2> 25°C, ferricenium ion as electron acceptor [4]) [4]

24 <2> (L-arabinose, <2> 25°C, ferricenium ion as electron acceptor [4]) [4]

29.2 <2> (D-xylose, <2> 25°C, ferricenium ion as electron acceptor [4]) [4]
29.8 <2> (D-galactose, <2> 25°C, ferricenium ion as electron acceptor [4]) [4]

pH-Optimum
3.5 <2> (<2> 50 mM sodium citrate buffer, 1,4-benzoquinone as electron acceptor [4]) [4]
4 <1> (<1> acetate buffer, two pH-optima [1]) [1]
4.5 <3> (<3> assay at [5]) [5]
5.5 <1> (<1> assay at [2]) [2]
6.7 <1> (<1> assay at [3]) [3]
9 <1> (<1> phosphate buffer, two pH-optima [1]) [1]
9.5 <2> (<2> 50 mM glycine buffer, ferricenium ion as electron acceptor [4]) [4]

pH-Range
3-9.5 <1> (<1> activity in the range is higher than 50% of the rate at the acidic optimum at pH 4 [1]) [1]

Temperature optimum (°C)
25 <1, 3> (<1,3> assay at [1,5]) [1, 5]
30 <1> (<1> assay at [2,3]) [2, 3]
65 <2> (<2> ferricenium ion as electron acceptor [4]) [4]
75 <2> (<2> 1,4-benzoquinone as electron acceptor [4]) [4]

4 Enzyme Structure

Molecular weight
76000 <2> (<2> gel filtration [4]) [4]
79000 <1> (<1> gel filtration [1]) [1]

Subunits
monomer <1, 2> (<1> 1 * 75000, SDS-PAGE [1]; <2> 1 * 78000, SDS-PAGE [4]) [1, 4]

Posttranslational modification
Glycoprotein <1, 2> (<2> heavily glycosylated, the sugar moiety of PDH comprises 17.3% of its total molecular mass [4]) [1, 2, 4]

5 Isolation/Preparation/Mutation/Application

Source/tissue
mycelium <1, 2, 3> (<1> 17 days old [1]) [1, 2, 3, 4, 5]

Localization
extracellular <1, 2> (<2> active extracellular secretion of PDH [4]; <1> enzyme is actively secreted into the extracellular fluid [1]) [1, 4]
intracellular <1> [3]

Purification

<1> [2, 3]
<1> (68.3fold) [1]
<2> (46fold) [4]
<3> (65fold) [5]

Application

synthesis <1, 3> (<3> PDH catalysis with 1,4-benzoquinone as an oxidant provides biocatalytic sugar chemistry with a new convenient tool for high yield production of 3-keto-oligosaccharides and 3-keto-glycosides [5]; <1> possible use in the preparation of rare di- and tricarbonyl sugar derivatives [3]) [3, 5]

6 Stability

General stability information

<1>, freezing in 50 mM sodium phosphate, 2.4 mg/ml, results in 90% loss of activity [1]
<2>, repeated freezing and thawing, two cycles, in the absence of glycerol has no effect on PDH activity [4]

Storage stability

<1>, 4°C, purified enzyme, 100 mM sodium phosphate, pH 6, 7, 8 or 9, 100 days, stable [1]
<2>, 4°C, purified PDH, 20 mM sodium phosphate buffer, pH 7, 0.02% ProClin 300, 2 months, no loss of activity [4]
<3>, 20°C, at least 14 days, stable [5]

References

[1] Volc, J.; Kubatova, E.; Wood, D.; Daniel, G.: Pyranose 2-dehydrogenase, a novel sugar oxidoreductase from the basidiomycete fungus Agaricus bisporus. Arch. Microbiol., **167**, 119-125 (1997)

[2] Volc, J.; Sedmera, P.; Halada, P.; Prikyrlova, V.; Daniel, G.: C-2 and C-3 oxidation of D-Glc, and C-2 oxidation of D-Gal by pyranose dehydrogenase from Agaricus bisporus. Carbohydr. Res., **310**, 151-156 (1998)

[3] Volc, J.; Sedmera, P.; Halada, P.; Prikyrlova¡, V.; Haltrich, D.: Double oxidation of D-xylose to D-glycero-pentos-2,3-diulose (2,3-diketo-D-xylose) by pyranose dehydrogenase from the mushroom Agaricus bisporus. Carbohydr. Res., **329**, 219-225 (2000)

[4] Volc, J.; Kubatova, E.; Daniel, G.; Sedmera, P.; Haltrich, D.: Screening of basidiomycete fungi for the quinone-dependent sugar C-2/C-3 oxidoreductase, pyranose dehydrogenase, and properties of the enzyme from Macrolepiota rhacodes. Arch. Microbiol., **176**, 178-186 (2001)

[5] Volc, J.; Sedmera, P.; Halada, P.; Daniel, G.; Prikyrlova, V.; Haltrich, D.: C-3 oxidation of non-reducing sugars by a fungal pyranose dehydrogenase: spectral characterization. J. Mol. Catal. B, **17**, 91-100 (2002)
[6] Sedmera, P.; Halada, P.; Kubatova, E.; Haltrich, D.; Prikrylova, V.; Volc, J.: New biotransformations of some reducing sugars to the corresponding (di)-dehydro(glycosyl) aldoses or aldonic acids using fungal pyranose dehydrogenase. J. Mol. Catal. B, **41**, 32-42 (2006)

2-Oxo-acid reductase 1.1.99.30

1 Nomenclature

EC number
1.1.99.30

Systematic name
(2R)-hydroxy-carboxylate:acceptor oxidoreductase

Recommended name
2-oxo-acid reductase

Synonyms
(2R)-hydroxycarboxlate viologen oxidoreductase <1> [4]
(2R)-hydroxycarboxlate-viologen-oxidoreductase <2> [2, 7]
(2R)-hydroxycarboxylate-viologen-oxidoreductase
2-oxoacid reductase <2> [7]
D-2-hydroxy acid dehydrogenase <2> [7]
HVOR <2> [1, 5, 6, 7]
dehydrogenase, D-2-hydroxy acid <2> [7]
hydroxy carboxylate viologen oxidoreductase <2> [6]
hydroxycarboxlate-oxidoreductase <2> [5]
hydroxycarboxlate-viologen-oxidoreductase <2> [7]

CAS registry number
9028-83-5

2 Source Organism

<1> *Proteus mirabilis* (no sequence specified) [3, 4]
<2> *Proteus vulgaris* (no sequence specified) [1, 2, 3, 5, 6, 7]
<3> *Proteus sp.* (no sequence specified) [3]

3 Reaction and Specificity

Catalyzed reaction
a (2R)-hydroxy-carboxylate + acceptor = a 2-oxo-carboxylate + reduced acceptor (<2> mechanism [7]; <1> aldonates and aldarates with an (R)-configured α-carbon atom are converted to the corresponding 2-oxocarboxylates [4])

Reaction type

oxidation
redox reaction
reduction

Substrates and products

S (2R)-hydroxy-4-methylpentanoate + 2-oxo-butyrate <2> (Reversibility: r) [5]
P 2-oxo-4-methylpentanoate + (2R)-hydroxybutyrate
S (2S,3R,4S)-2-hydroxy-3-halogenbutyrolactones + reduced benzyl viologen <2> (<2> highly stereospecific reaction [7]) (Reversibility: r) [7]
P ?
S (2S,3R,4S)-2-hydroxy-3-halogenbutyrolactones + reduced methyl viologen <2> (<2> highly stereospecific reaction [7]) (Reversibility: r) [7]
P ?
S (R)-lactate + oxidized antraquinone-2,6-disulfonate <2> (Reversibility: ?) [2]
P pyruvate + reduced antraquinone-2,6-disulfonate
S (R,S)-2-oxo-3-methylpentanoate + reduced benzyl viologen <2> (Reversibility: ?) [3]
P 2-hydroxy-3-methylpentanoate + oxidized benzyl viologen
S (R,S)-2-oxo-3-methylpentanoate + reduced methyl viologen <2> (Reversibility: ?) [3]
P 2-hydroxy-3-methylpentanoate + oxidized methyl viologen
S (S)-2-oxo-3-methylpentanoate + reduced benzyl viologen <1, 2> (Reversibility: ?) [3]
P 2-hydroxy-3-methylpentanoate + oxidized benzyl viologen
S (S)-2-oxo-3-methylpentanoate + reduced methyl viologen <1, 2> (Reversibility: ?) [3]
P 2-hydroxy-3-methylpentanoate + oxidized methyl viologen
S 2-hydroxyglutarate + oxidized acceptor <2> (Reversibility: r) [5]
P 2-oxoglutarate + reduced acceptor
S 2-oxo-3,3-dimethyl-4-hydroxybutanoate + reduced benzyl viologen <1, 2> (Reversibility: ?) [3]
P 3,3-dimethyl-2,4-dihydroxybutanoate + oxidized benzyl viologen
S 2-oxo-3,3-dimethyl-4-hydroxybutanoate + reduced methyl viologen <1, 2> (Reversibility: ?) [3]
P 3,3-dimethyl-2,4-dihydroxybutanoate + oxidized methyl viologen
S 2-oxo-3,5-dienoates + H_2 <2> (Reversibility: r) [5]
P ?
S 2-oxo-3-(3,4-dihydroxyphenyl)-propionate + H_2 <2> (Reversibility: r) [5]
P rosmarinic acid
S 2-oxo-3-enoates + H_2 <2> (Reversibility: r) [5]
P ?
S 2-oxo-4-(hydroxy-methyl-phosphinyl)-butanoate + reduced benzyl viologen <1> (Reversibility: ?) [3]

P 2-hydroxy-4-(hydroxymethylphosphinyl)-butanoate + oxidized benzyl viologen
S 2-oxo-4-(hydroxy-methyl-phosphinyl)-butanoate + reduced methyl viologen <1> (Reversibility: ?) [3]
P 2-hydroxy-4-(hydroxymethylphosphinyl)-butanoate + oxidized methyl viologen
S 2-oxo-4-methylpentanoate + reduced benzyl viologen <1> (Reversibility: ?) [3]
P 2-hydroxy-4-methylpentanoate + oxidized benzyl viologen
S 2-oxo-4-methylpentanoate + reduced methyl viologen <1> (Reversibility: ?) [3]
P 2-hydroxy-4-methylpentanoate + oxidized methyl viologen
S 2-oxo-carboxylate + reduced benzyl viologen <2> (<2> stereospecific, very wide substrate specificity, reduces 2-oxo-carboxylates with diverse types of residues at position 3, such as unbranched and branched alkyl, hydroxy, one or two conjugated carbon carbon double bonds, additional CO, and -OOC(CH_2)$_n$, overview [7]) (Reversibility: r) [7]
P (2R)-hydroxy-carboxylate + oxidized benzyl viologen
S 2-oxo-carboxylate + reduced methyl viologen <2> (<2> stereospecific, very wide substrate specificity, reduces 2-oxo-carboxylates with diverse types of residues at position 3, such as unbranched and branched alkyl, hydroxy, one or two conjugated carbon carbon double bonds, additional CO, and -OOC(CH_2)$_n$, overview [7]) (Reversibility: r) [7]
P (2R)-hydroxy-carboxylate + oxidized methyl viologen
S 2-oxoadipate + reduced benzyl viologen <1> (Reversibility: ?) [3]
P 2-hydroxyadipate + oxidized benzyl viologen
S 2-oxoadipate + reduced methyl viologen <1> (Reversibility: ?) [3]
P 2-hydroxyadipate + oxidized methyl viologen
S 2-oxoglutarate + reduced benzyl viologen <1, 2> (Reversibility: ?) [3]
P 2-hydroxyglutarate + oxidized benzyl viologen
S 2-oxoglutarate + reduced methyl viologen <1, 2> (Reversibility: ?) [3]
P 2-hydroxyglutarate + oxidized methyl viologen
S 2-oxononanoate + reduced benzyl viologen <1> (Reversibility: ?) [3]
P 2-hydroxynonanoate + oxidized benzyl viologen
S 2-oxononanoate + reduced methyl viologen <1> (Reversibility: ?) [3]
P 2-hydroxynonanoate + oxidized methyl viologen
S 3-fluoropyruvate + reduced benzyl viologen <1, 2> (Reversibility: ?) [3]
P 3-fluoro-2-hydroxypropionate + oxidized benzyl viologen
S 3-fluoropyruvate + reduced methyl viologen <1, 2> (Reversibility: ?) [3]
P 3-fluoro-2-hydroxypropionate + oxidized methyl viologen
S 5-benzyloxyindolylpyruvate + reduced benzyl viologen <1> (Reversibility: ?) [3]
P 3-(5-benzyloxyindolyl)-2-hydroxypropionate + oxidized benzyl viologen
S 5-benzyloxyindolylpyruvate + reduced methyl viologen <1> (Reversibility: ?) [3]
P 3-(5-benzyloxyindolyl)-2-hydroxypropionate + oxidized methyl viologen

S 6-phospho-D-gluconate + oxidized antraquinone-2,6-disulfonate <1> (Reversibility: r) [4]
P 6-phospho-D-arabino-hex-2-ulosonate + reduced antraquinone-2,6-disulfonate
S 6-phospho-D-gluconate + oxidized benzyl viologen <1> (Reversibility: r) [4]
P 6-phospho-D-arabino-hex-2-ulosonate + reduced benzyl viologen
S 6-phospho-D-gluconate + oxidized carboxamido-methyl viologen <1> (Reversibility: r) [4]
P 6-phospho-D-arabino-hex-2-ulosonate + reduced carboxamido-methyl viologen
S D-galactonate + oxidized anthraquinone-2,6-disulfonate <1> (Reversibility: r) [4]
P D-lyxo-hex-2-ulosonate + reduced anthraquinone-2,6-disulfonate
S D-galactonate + oxidized benzyl viologen <1> (Reversibility: r) [4]
P D-lyxo-hex-2-ulosonate + reduced benzyl viologen
S D-galactonate + oxidized carboxamido-methyl viologen <1> (Reversibility: r) [4]
P D-lyxo-hex-2-ulosonate + reduced carboxamido-methyl viologen
S D-glucarate + oxidized anthraquinone-2,6-disulfonate <1> (Reversibility: r) [4]
P 2-oxo-D-glucarate + reduced anthraquinone-2,6-disulfonate
S D-glucarate + oxidized benzyl viologen <1> (Reversibility: r) [4]
P 2-oxo-D-glucarate + reduced benzyl viologen
S D-glucarate + oxidized carboxamido-methyl viologen <1> (Reversibility: r) [4]
P 2-oxo-D-glucarate + reduced carboxamido-methyl viologen
S D-gluconate + oxidized 1,1'-carbamoyl methyl viologen <2> (<2> anaerobic conditions [1]) (Reversibility: r) [1]
P ? + reduced 1,1'-carbamoyl methyl viologen
S D-gluconate + oxidized anthraquinone-2,6-disulfonate <1> (Reversibility: r) [4]
P D-arabino-hex-2-ulosonate + reduced anthraquinone-2,6-disulfonate
S D-gluconate + oxidized benzyl viologen <1> (Reversibility: r) [4]
P D-arabino-hex-2-ulosonate + reduced benzyl viologen
S D-gluconate + oxidized carboxamido-methyl viologen <1> (Reversibility: r) [4]
P D-arabino-hex-2-ulosonate + reduced carboxamido-methyl viologen
S D-glucose 6-phosphate + oxidized 1,1'-carbamoyl methyl viologen <2> (<2> anaerobic conditions [1]) (Reversibility: r) [1]
P ? + reduced 1,1'-carbamoyl methyl viologen
S D-gulonate + oxidized anthraquinone-2,6-disulfonate <1> (Reversibility: r) [4]
P D-arabino-hex-2-ulosonate + reduced anthraquinone-2,6-disulfonate
S D-gulonate + oxidized benzyl viologen <1> (Reversibility: r) [4]
P D-arabino-hex-2-ulosonate + reduced benzyl viologen

S D-gulonate + oxidized carboxamido-methyl viologen <1> (Reversibility: r) [4]
P D-arabino-hex-2-ulosonate + reduced carboxamido-methyl viologen
S D-lactate + oxidized 1,1'-carbamoyl methyl viologen <2> (<2> anaerobic conditions [1]) (Reversibility: r) [1]
P pyruvate + reduced 1,1'-carbamoyl methyl viologen
S D-lactate + oxidized anthraquinone-2,6-disulfonate <1> (Reversibility: r) [4]
P pyruvate + reduced anthraquinone-2,6-disulfonate
S D-lactate + oxidized benzyl viologen <1> (Reversibility: r) [4]
P pyruvate + reduced benzyl viologen
S D-lactate + oxidized carboxamido-methyl viologen <1> (Reversibility: r) [4]
P pyruvate + reduced carboxamido-methyl viologen
S D-ribonate + oxidized anthraquinone-2,6-disulfonate <1> (Reversibility: r) [4]
P D-erythro-pent-2-ulosonate + reduced anthraquinone-2,6-disulfonate
S D-ribonate + oxidized benzyl viologen <1> (Reversibility: r) [4]
P D-erythro-pent-2-ulosonate + reduced benzyl viologen
S D-ribonate + oxidized carboxamido-methyl viologen <1> (Reversibility: r) [4]
P D-erythro-pent-2-ulosonate + reduced carboxamido-methyl viologen
S D-xylonate + oxidized anthraquinone-2,6-disulfonate <1> (Reversibility: r) [4]
P D-threo-pent-2-ulosonate + reduced anthraquinone-2,6-disulfonate
S D-xylonate + oxidized benzyl viologen <1> (Reversibility: r) [4]
P D-threo-pent-2-ulosonate + reduced benzyl viologen
S D-xylonate + oxidized carboxamido-methyl viologen <1> (Reversibility: r) [4]
P D-threo-pent-2-ulosonate + reduced carboxamido-methyl viologen
S L-arabinonate + oxidized anthraquinone-2,6-disulfonate <1> (Reversibility: r) [4]
P L-erythro-pent-2-ulosonate + reduced anthraquinone-2,6-disulfonate
S L-arabinonate + oxidized benzyl viologen <1> (Reversibility: r) [4]
P L-erythro-pent-2-ulosonate + reduced benzyl viologen
S L-arabinonate + oxidized carboxamido-methyl viologen <1> (Reversibility: r) [4]
P L-erythro-pent-2-ulosonate + reduced carboxamido-methyl viologen
S L-arabinuronate + oxidized anthraquinone-2,6-disulfonate <1> (Reversibility: r) [4]
P ? + reduced anthraquinone-2,6-disulfonate
S L-arabinuronate + oxidized benzyl viologen <1> (Reversibility: r) [4]
P ? + reduced benzyl viologen
S L-arabinuronate + oxidized carboxamido-methyl viologen <1> (Reversibility: r) [4]
P ? + reduced carboxamido-methyl viologen

S L-mannonate + oxidized anthraquinone-2,6-disulfonate <1> (Reversibility: r) [4]
P L-arabino-hex-2-ulosonate + reduced anthraquinone-2,6-disulfonate
S L-mannonate + oxidized benzyl viologen <1> (Reversibility: r) [4]
P L-arabino-hex-2-ulosonate + reduced benzyl viologen
S L-mannonate + oxidized carboxamido-methyl viologen <1> (Reversibility: r) [4]
P L-arabino-hex-2-ulosonate + reduced carboxamido-methyl viologen
S N-acetylneuraminate + oxidized 1,1'-carbamoyl methyl viologen <2> (<2> anaerobic conditions [1]) (Reversibility: r) [1]
P ? + reduced 1,1'-carbamoyl methyl viologen
S a (2R)-hydroxy-carboxylate + acceptor <2, 3> (Reversibility: ?) [1, 3]
P a 2-oxo-carboxylate + reduced acceptor
S chloropyruvic acid + reduced 1,1'-carbamoyl methyl viologen <2> (<2> stereospecific reaction [6]) (Reversibility: ?) [6]
P D-(S)-chlorolactate + oxidized 1,1'-carbamoyl methyl viologen
S galactarate + oxidized anthraquinone-2,6-disulfonate <1> (Reversibility: r) [4]
P 2-oxogalactarate + reduced anthraquinone-2,6-disulfonate
S galactarate + oxidized benzyl viologen <1> (Reversibility: r) [4]
P 2-oxogalactarate + reduced benzyl viologen
S galactarate + oxidized carboxamido-methyl viologen <1> (Reversibility: r) [4]
P 2-oxogalactarate + reduced carboxamido-methyl viologen
S indolylpyruvate + reduced benzyl viologen <1, 2> (Reversibility: ?) [3]
P 2-hydroxy-3-indolylpropionate + oxidized benzyl viologen
S indolylpyruvate + reduced methyl viologen <1, 2> (Reversibility: ?) [3]
P 2-hydroxy-3-indolylpropionat + oxidized methyl viologen
S lactate + oxidized acceptor <2> (Reversibility: r) [5]
P pyruvate + reduced acceptor
S lactobionate + oxidized anthraquinone-2,6-disulfonate <1> (Reversibility: r) [4]
P 4-O-(β-D-glactopyranosido)-D-arabino-hex-2-ulosonate + reduced anthraquinone-2,6-disulfonate
S lactobionate + oxidized benzyl viologen <1> (Reversibility: r) [4]
P 4-O-(β-D-glactopyranosido)-D-arabino-hex-2-ulosonate + reduced benzyl viologen
S lactobionate + oxidized carboxamido-methyl viologen <1> (Reversibility: r) [4]
P 4-O-(β-D-glactopyranosido)-D-arabino-hex-2-ulosonate + reduced carboxamido-methyl viologen
S mandelate + oxidized acceptor <2> (Reversibility: r) [5]
P oxo(phenyl)acetic acid + reduced acceptor
S pantolactone + oxidized acceptor <2> (Reversibility: r) [5]
P dehydropantolactone + reduced acceptor
S phenylglyoxylate + reduced benzyl viologen <1, 2> (Reversibility: ?) [3]
P (R)-mandelic acid

S phenylglyoxylate + reduced methyl viologen <1, 2> (Reversibility: ?) [3]
P (R)-mandelic acid
S phenylpyruvate + H_2 <2> (Reversibility: r) [5]
P (2R)-phenyllactate
S phenylpyruvate + reduced 1,1'-carbamoyl methyl viologen <2> (<2> anaerobic conditions [1]) (Reversibility: r) [1]
P phenyllactate + oxidized 1,1'-carbamoyl methyl viologen
S phenylpyruvate + reduced benzyl viologen <1, 2> (Reversibility: ?) [3]
P 2-hydroxy-3-phenylpropionate + oxidized benzyl viologen
S phenylpyruvate + reduced benzyl viologen <2> (Reversibility: r) [5]
P (2R)-phenyllactate + oxidized benzyl viologen
S phenylpyruvate + reduced benzyl viologen <2> (<2> anaerobic conditions [1]) (Reversibility: r) [1]
P phenyllactate + oxizized benzyl viologen
S phenylpyruvate + reduced carbamoylmethyl viologen <2> (Reversibility: r) [5]
P (2R)-phenyllactate + oxidized carbamoylmethyl viologen
S phenylpyruvate + reduced cobalt sepulchrate <2> (Reversibility: r) [5]
P (2R)-phenyllactate + oxidized cobalt sepulchrate
S phenylpyruvate + reduced indigocarmine <2> (Reversibility: r) [5]
P (2R)-phenyllactate + oxidized indigocarmine
S phenylpyruvate + reduced kresyl violet <2> (Reversibility: r) [5]
P (2R)-phenyllactate + oxidized kresyl violet
S phenylpyruvate + reduced meldolablue <2> (Reversibility: r) [5]
P (2R)-phenyllactate + oxidized meldolablue
S phenylpyruvate + reduced methyl viologen <1, 2> (Reversibility: ?) [3]
P 2-hydroxy-3-phenylpropionate + oxidized methyl viologen
S phenylpyruvate + reduced methylene blue <2> (Reversibility: r) [5]
P (2R)-phenyllactate + oxidized methylene blue
S phenylpyruvate + reduced phenosafranine <2> (Reversibility: r) [5]
P (2R)-phenyllactate + oxidized phenosafranine
S phenylpyruvate + reduced safranine TH <2> (Reversibility: r) [5]
P (2R)-phenyllactate + oxidized safranine TH
S pyruvate + reduced benzyl viologen <1, 2> (Reversibility: ?) [3]
P lactate + oxidized benzyl viologen
S pyruvate + reduced methyl viologen <1, 2> (Reversibility: ?) [3]
P lactate + oxidized methyl viologen
S Additional information <1, 2> (<2> 2-oxo acids with diverse substituents, e.g. carboxy, methyl, ethyl, halogen, methoxy, thiomethyl, NHCHO, unbranched and branched alkyl, OH^-containing residues in 3-position, double bonds containing residues, substituents with additional CO groups, $-OOC(CH_2)_n$ with n being at least 3, overview [7]; <1,2> broad substrate specificity, stereochemic specificity, no activity with 3-oxoglutarate and hydroxyacetone [3]; <1> product determination, stereochemistry and molecule conformations, products appear in pyranose, furanose or lactone form, overview [4]) (Reversibility: ?) [3, 4, 7]
P ?

Inhibitors

CN^- <1, 2> (<1,2> 30% inhibition at 10 mM [3]) [1, 3]
O_2 <2> (<2> oxygen-sensitive [1]) [1]
tungstate <2> (<2> complete inhibition in vivo [5]) [5]
Additional information <1, 2> (<1,2> no inhibition by CO [3]) [3]

Cofactors/prosthetic groups

molybdenum cofactor <2> (<2> i.e. MoCo [7]; <2> mononucleotide, 1 enzyme molecule contains 1 molybdenum atom [1]) [1,7]
Additional information <1, 2> (<2> enzyme contains iron sulfur clusters, with four iron and 4 sulfur atoms per enzyme molecule, enzyme belongs to the group of molybdoenzymes which possess no further prosthetic group than the iron sulfur clusters [1]; <2> neither NADH nor NADPH are cosubstrates for the enzyme, no flavin or heme cofactors [7]; <1,2> non-pyridine nucleotide-dependent [3]) [1,3,7]

Activating compounds

Additional information <2> (<2> highest activity in potassium phosphate buffer compared to Tris-buffer, MOPS, Pipes or diphosphate buffer [6]) [6]

Metals, ions

iron-sulfur cluster <2> [5]
molybdate <2> (<2> strong induction [5]) [5]
molybdenum <1, 2> (<2> required [7]; <1> dependent on [3]; <2> 4-12 atoms per enzyme molecule [5]) [3, 5, 7]

Specific activity (U/mg)

600 <1> (<1> purified enzyme [3]) [3]
810 <2> (<2> crude cell extract [5]) [5]
1000 <2> (<2> purified enzyme [7]) [7]
1800 <2> (<2> purified holoenzyme [1]) [1]
Additional information <2> (<2> production rates of (S)-chloroacetic acid by enzymatic and chemical or eletrochemical methods [6]) [6]

K_m-Value (mM)

0.025 <1> (carboxamido-methyl viologen, <1> pH 8.5, 38°C [4]) [4]
0.042 <1> (benzyl viologen, <1> pH 8.5, 38°C [4]) [4]
0.19 <1> (oxidized antraquinone-2,6-disulfonate, <1> pH 8.5, 38°C [4]) [4]
0.3 <1> (D-ribonate, <1> pH 8.5, 38°C [4]) [4]
0.42 <1> (galactarate, <1> pH 8.5, 38°C [4]) [4]
0.48 <1> (D-galactonate, <1> pH 8.5, 38°C [4]) [4]
0.64 <1> (6-phospho-D-gluconate, <1> pH 8.5, 38°C [4]) [4]
0.64 <1> (D-gulonate, <1> pH 8.5, 38°C [4]) [4]
0.66 <1> (lactobionate, <1> pH 8.5, 38°C [4]) [4]
0.68 <1> (D-lactate, <1> pH 8.5, 38°C [4]) [4]
0.92 <1> (L-mannonate, <1> pH 8.5, 38°C [4]) [4]
0.98 <1> (D-gluconate, <1> pH 8.5, 38°C [4]) [4]
1.27 <1> (D-xylonate, <1> pH 8.5, 38°C [4]) [4]
1.46 <1> (L-arabinuronate, <1> pH 8.5, 38°C [4]) [4]

1.68 <1> (D-glucarate, <1> pH 8.5, 38°C [4]) [4]
2.03 <1> (L-Arabinonate, <1> pH 8.5, 38°C [4]) [4]

pH-Optimum
7 <2> (<2> assay at [5,6]) [5, 6]
8.5 <1, 2> (<1,2> assay at [1,4]) [1, 4, 7]

Temperature optimum (°C)
25 <1, 2> (<1> assay at [4]; <2> dehydrogenase assay at [1]) [1, 4]
30 <2> (<2> assay at [6]) [6]
35 <2> (<2> assay at [5]) [5]
37 <1, 2> (<1> assay at [3]; <2> reductase assay at [1]) [1, 3]

4 Enzyme Structure

Molecular weight
600000 <2> (<2> gel filtration [1]; <2> above [7]) [1, 7]
750000 <2> [5]

Subunits
octamer <2> (<2> 8 * 80000, SDS-PAGE [1]) [1]
oligomer <2> (<2> x * 80000 + x * 64000, α,β-structure [7]) [7]
Additional information <2> (<2> subunit structure [1]) [1]

5 Isolation/Preparation/Mutation/Application

Source/tissue
Additional information <1, 2> (<1> aerobic growth conditions [3]; <2> anaerobic growth conditions [1,3]) [1, 3]

Localization
membrane <1, 2> (<1,2> bound [1,3,7]) [1, 3, 7]

Purification
<1> (167fold) [3]
<2> [3]
<2> (60fold) [5]
<2> (anaerobic conditions) [1]

Application
biotechnology <2> (<2> preparative scale production of pyruvate from (R)-lactate in an enzyme-membrane reactor with coupled electrochemical regeneration of the artificial mediator anthraquinone-2,6-disulfonate, process modeling and calculation [2]) [2]
synthesis <1, 2> (<1,2> enzyme can be used as a stereospecific biocatalyst for production of a wide range of 2-hydroxy-carboxylate compounds at preparative scale, with cofactor methyl or benzyl viologen being more stable than NAD(P)H [3]; <2> enzyme can be used as a stereospecific biocatalyst

for production of a wide range of compounds, e.g. 3-enoic-, 3,5-dienoic-, and 4-oxo-3R,S-hydroxyacids [7]; <2> large scale microbial production of D-(S)-chloroacetic acid [6]; <2> production of pyruvate from (R)-lactate in an enzyme-membrane reactor with coupled electrochemical regeneration of the artificial mediator anthraquinone-2,6-disulfonate [2]) [2, 3, 6, 7]

6 Stability

Temperature stability

22 <1, 2> (<1,2> room temperature, 30 h, stable [3]) [3]

Storage stability

<2>, -15°C, loss of 10% activity within 1-2 years [7]

<2>, -16°C, N_2-atmosphere, wet packed cells, 450 days, more than 90% remaining activity [5]

<2>, 30°C, 170 h, completely stable [6]

<2>, room temperature, freeze dried cells, 200 days, more than 80% remaining activity [5]

<2>, room temperature, freeze dried cells, loss of 10% activity within 1-2 years [7]

References

[1] Trautwein, T.; Krauss, F.; Lottspeich, F.; Simon, H.: The (2R)-hydroxycarboxylate-viologen-oxidoreductase from Proteus vulgaris is a molybdenum-containing iron-sulphur protein. Eur. J. Biochem., **222**, 1025-1032 (1994)

[2] Hekmat, D.; Danninger, J.; Simon, H.; Vortmeyer, D.: Production of pyruvate from (R)-lactate in an enzyme-membrane reactor with coupled electrochemical regeneration of the artificial mediator anthraquinone-2,6-disulfonate. Enzyme Microb. Technol., **24**, 471-479 (1999)

[3] Neumann, S.; Simon, H.: On a non-pyridine nucleotide-dependent 2-oxoacid reductase of broad specificity from two Proteus species. FEBS Lett., **167**, 29-32 (1985)

[4] Schinschel, C.; Simon, H.: Proteus mirabilis dehydrogenates aldonates and aldarates with an (R)-configured α-carbon atom to the corresponding 2-oxocarboxylates. Bioorg. Med. Chem., **2**, 483-491 (1994)

[5] Krauss, F.; Guenther, H.; Skopan, H.; Simon, H.: Additional experiences with Proteus vulgaris and its hydroxycarboxylate-oxidoreductase for the preparation of (2R)- or (2S)-hydroxycarboxylates. DECHEMA Biotechnol. Conf., **1**, 313-318 (1988)

[6] Andersson, M.; Holmberg, H.; Adlercreutz, P.: Microbial production of D-(S)-chlorolactic acid by Proteus vulgaris cells. Enzyme Microb. Technol., **22**, 170-178 (1998)

[7] Simon, H.: Properties and mechanistic aspects of newly found redox enzymes from anaerobes suitable for bioconversions on preparatory scale. Indian J. Chem. B, **32**, 170-175 (1993)

(S)-Mandelate dehydrogenase 1.1.99.31

1 Nomenclature

EC number
1.1.99.31

Systematic name
(S)-2-hydroxy-2-phenylacetate:acceptor 2-oxidoreductase

Recommended name
(S)-mandelate dehydrogenase

Synonyms
L-MDH <3> [1, 3]
L-mandelate dehydrogenase <2, 3> [1, 13]
MDH <1, 2> [4, 5, 13]

2 Source Organism

<1> *Pseudomonas putida* (no sequence specified) [4, 5, 6, 7, 8, 9, 10, 11, 12]
<2> *Acinetobacter calcoaceticus* (no sequence specified) [13]
<3> *Rhodotorula graminis* (no sequence specified) [1, 2, 3]

3 Reaction and Specificity

Catalyzed reaction
(S)-2-hydroxy-2-phenylacetate + acceptor = 2-oxo-2-phenylacetate + reduced acceptor

Reaction type
oxidation
redox reaction
reduction

Substrates and products
S (R,S)-*p*-chloromandelate + acceptor <1> (Reversibility: ?) [6]
P ?
S (S)-2-hydroxy-2-phenylacetate + O_2 <1> (<1> in absence of any other electron acceptor, oxygen is used by the wild-type enzyme for reoxidation of the reduced flavin, at a rather slow rate. Ping-pong kinetics [11]) (Reversibility: ?) [11]
P 2-oxo-2-phenylacetate + H_2O_2

S (S)-3-phenyllactate + acceptor <1> (Reversibility: ?) [7]
P ?
S (S)-mandelate + 2,6-dichloroindophenol <3> (Reversibility: ?) [1]
P 2-oxo-2-phenylacetate + ?
S (S)-mandelate + 2,6-dichlorophenolindophenol <1> (Reversibility: ?) [4]
P 2-oxo-2-phenylacetate + reduced 2,6-dichlorophenolindophenol
S (S)-mandelate + acceptor <1, 2> (<1> acceptor: 2,6-dichloroindophenol plus phenazine methosulfate. The MDH reaction has two rate-limiting step of similar activation energies: the formation and breakdown of a distinct intermediate, with the latter step being slightly more rate limiting. MDH is capable of catalyzing the reverse reaction, the reoxidation of reduced MDH by the product ketoacid, benzoylformate. The transient intermediate is observed during the reverse reaction as well [10]; <1> phenazine methosulfate and 2,6-dichloroindophenol as acceptor. In absence of any other electron acceptor, oxygen is used by the wild-type enzyme for reoxidation of the reduced flavin, at a rather slow rate [11]; <1,2> reduction of 2,6-dichloroindophenol in presence of N-methylphenazonium methosulfate [9,13]) (Reversibility: r) [5, 6, 7, 9, 10, 11, 12, 13]
P 2-oxo-2-phenylacetate + reduced acceptor
S (S)-mandelate + cytochrome c <3> (Reversibility: ?) [1, 3]
P 2-oxo-2-phenylacetate + ?
S (S)-mandelate + ferricyanide <3> (Reversibility: ?) [1]
P 2-oxo-2-phenylacetate + ferrocyanide + H_2O
S (S)-mandelate + ferricyanide <3> (Reversibility: ?) [3]
P 2-oxo-2-phenylacetate + ferrocyanide
S (S)-phenyllactate + 2,6-dichlorophenolindophenol <1> (Reversibility: ?) [4]
P ?
S 2-hydroxy-3-butynoate + acceptor <1> (Reversibility: ?) [5, 11]
P 2-oxo-3-butynoate + reduced acceptor
S 2-hydroxybutyrate + acceptor <1> (Reversibility: ?) [5, 11]
P 2-oxobutyrate + reduced acceptor
S 2-hydroxyhexanoate + acceptor <1> (Reversibility: ?) [5, 11]
P 2-oxohexanoate + reduced acceptor
S 2-hydroxyisocaproate + acceptor <1> (Reversibility: ?) [5, 11]
P 2-oxoisocaproate + reduced acceptor
S 2-hydroxyoctanoate + acceptor <1> (Reversibility: ?) [5, 11]
P 2-oxooctanoate + reduced acceptor
S 2-hydroxyvalerate + acceptor <1> (Reversibility: ?) [5, 11]
P 2-oxovalerate + reduced acceptor
S 3-hydroxy-DL-mandelate + ferricyanide <3> (Reversibility: ?) [2]
P 2-oxo-2-(3-hydroxyphenyl)acetate + ferrocyanide + H_2O
S 3-indolelactate + acceptor <1> (Reversibility: ?) [5, 11]
P 3-(1H-indol-2-yl)-2-oxopropenoate + reduced acceptor
S 3-methoxy-DL-mandelate + ferricyanide <3> (Reversibility: ?) [2]
P 2-oxo-2-(3-methoxyphenyl)acetate + ferrocyanide + H_2O
S 3-phenyllactate + acceptor <1> (Reversibility: ?) [5, 11]

P 2-oxo-3-phenylpropenoate + reduced acceptor
S 4-bromo-DL-mandelate + ferricyanide <3> (Reversibility: ?) [2]
P 2-oxo-2-(4-bromophenyl)acetate + ferrocyanide + H_2O
S 4-chloro-DL-mandelate + ferricyanide <3> (Reversibility: ?) [2]
P 2-oxo-2-(4-chlorophenyl)acetate + ferrocyanide + H_2O
S 4-chloromandelate + acceptor <1> (Reversibility: ?) [5, 11]
P 2-oxo-2-(4-chlorophenyl)acetate + reduced acceptor
S 4-fluoro-D,L-mandelate + ferricyanide <3> (Reversibility: ?) [2]
P 2-oxo-2-(4-fluorophenyl)acetate + ferrocyanide + H_2O
S 4-hydroxy-DL-mandelate + ferricyanide <3> (Reversibility: ?) [2]
P 2-oxo-2-(4-hydroxyphenyl)acetate + ferrocyanide + H_2O
S 4-methoxy-DL-mandelate + ferricyanide <3> (Reversibility: ?) [2]
P 2-oxo-2-(4-methoxyphenyl)acetate + ferrocyanide + H_2O
S 4-methyl-DL-mandelate + ferricyanide <3> (Reversibility: ?) [2]
P 2-oxo-2-(4-methylphenyl)acetate + ferrocyanide + H_2O
S DL-mandelate + ferricyanide <3> (Reversibility: ?) [2]
P 2-oxo-2-phenylacetate + ferrocyanide + H_2O
S L-mandelate + ferricyanide <3> (Reversibility: ?) [2]
P 2-oxo-2-phenylacetate + ferrocyanide
S indoleglycolate + acceptor <1> (Reversibility: ?) [5, 11]
P indol-2-yl(oxo)acetate + reduced acceptor
S mandelate + acceptor <1> (Reversibility: ?) [5, 11]
P 2-oxo-2-phenylacetate + reduced acceptor
S Additional information <3> (<3> enzyme-catalyzed reaction proceeds via the same transition state for each substrate and indicates that this transition state is relatively non-polar but has an electron-rich centre at the α-carbon position [2]) (Reversibility: ?) [2]
P ?

Inhibitors

(R)-mandelate <1> [9]
1-phenylacetate <1> [9]
hexanoate <3> [1]
L-lactate <3> [1]
L-phenyllactate <3> [1]
oxalate <3> [1]
sulfite <1> [11]
p-chloromercuribenzoate <2> (<2> 10-20% inhibition at 0.01 mM [13]) [13]
phenylethanediol <1> (<1> poor competitive inhibitor of wild-type enzyme [9]) [9]

Cofactors/prosthetic groups

FMN <1> (<1> dependent on. In the reductive half-reaction (S)-mandelate is oxidized to 2-oxo-2-phenylacetate, resulting in the reduction of FMN [11]) [4,9,10,11]
flavocytochrome b2 <3> (<3> the enzyme is a flavocytochrome b2 [1]) [1]

Turnover number (min^{-1})

0.03 <1> ((R,S)-*p*-chloromandelate, <1> mutant enzyme H274G, in presence of 20 mM imidazole [6]) [6]

0.03 <1> (2-hydroxyisovalerate, <1> pH 7.5, 20°C, MDH-GOX2, a chimeric mutant of (S)-mandelate dehydrogenase with residues 177-215 replaced by residues 176-195 of glycolate oxidase from spinach [5]) [5]

0.03 <1> (O_2, <1> pH 7.5, 20°C, mutant enzyme MDH-GOX2, a chimeric mutant of (S)-mandelate dehydrogenase with 205 residues 177-215 replaced by residues 176-195 of glycolate oxidase from spinach [11]) [11]

0.04 <1> ((S)-mandelate, <1> pH 7.5, 20°C, mutant enzyme R165G/R277K [9]) [9]

0.05 <1> ((S)-mandelate, <1> pH 7.5, 20°C, mutant enzyme G81V [11]) [11]

0.05 <1> (O_2, <1> pH 7.5, 20°C, mutant enzyme G81D [11]) [11]

0.07 <1> (O_2, <1> pH 7.5, 20°C, mutant enzyme G81V [11]) [11]

0.09 <1> ((S)-mandelate, <1> mutant enzyme H274G, in presence of 20 mM imidazole [6]) [6]

0.1 <1> (2-hydroxybutyrate, <1> pH 7.5, 20°C, MDH-GOX2, a chimeric mutant of (S)-mandelate dehydrogenase with residues 177-215 replaced by residues 176-195 of glycolate oxidase from spinach [5]; <1> pH 7.5, 20°C, wild-type enzyme [11]) [5, 11]

0.11 <1> ((S)-mandelate, <1> pH 7.5, 20°C, mutant enzyme G81D [11]) [11]

0.15 <1> (3-phenyllactate, <1> pH 7.5, 20°C, mutant enzyme G81A [11]) [11]

0.19 <1> (2-hydroxyhexanoate, <1> pH 7.5, 20°C, mutant enzyme G81A [11]) [11]

0.22 <1> (O_2, <1> pH 7.5, 20°C, mutant enzyme, G81A/MDH-GOX2, a chimeric mutant of (S)-mandelate dehydrogenase with residues 177-215 replaced by residues 176-195 of glycolate oxidase from spinach and a G81A mutation in MDH [11]) [11]

0.23 <1> ((S)-mandelate, <1> pH 7.5, 20°C, mutant enzyme R277G [9]) [9]

0.26 <1> (2-hydroxyoctanoate, <1> pH 7.5, 20°C, mutant enzyme G81A [11]) [11]

0.27 <1> ((S)-mandelate, <1> pH 7.5, 20°C, mutant enzyme R277L [7]) [7]

0.27 <1> (2-hydroxyvalerate, <1> pH 7.5, 20°C, mutant enzyme G81A [11]) [11]

0.27 <1> (3-indolelactate, <1> pH 7.5, 20°C, mutant enzyme G81A [11]) [11]

0.34 <1> (2-hydroxyhexanoate, <1> pH 7.5, 20°C, wild-type enzyme [11]) [11]

0.36 <1> (2-hydroxyvalerate, <1> pH 7.5, 20°C, wild-type enzyme [11]) [11]

0.37 <1> (2-hydroxybutyrate, <1> pH 7.5, 20°C, mutant enzyme G81A [11]) [11]

0.48 <1> (2-hydroxyhexanoate, <1> pH 7.5, 20°C, MDH-GOX2, a chimeric mutant of (S)-mandelate dehydrogenase with residues 177-215 replaced by residues 176-195 of glycolate oxidase from spinach [5]) [5]

0.48 <1> (2-hydroxyisocaproate, <1> pH 7.5, 20°C, mutant enzyme G81A [11]) [11]

0.5 <1> (2-hydroxy-3-butenoate, <1> pH 7.5, 20°C, MDH-GOX2, a chimeric mutant of (S)-mandelate dehydrogenase with residues 177-215 replaced by residues 176-195 of glycolate oxidase from spinach [5]) [5]
0.5 <1> (2-hydroxyoctanoate, <1> pH 7.5, 20°C, wild-type enzyme [11]) [11]
0.5 <1> (2-hydroxyvalerate, <1> pH 7.5, 20°C, MDH-GOX2, a chimeric mutant of (S)-mandelate dehydrogenase with residues 177-215 replaced by residues 176-195 of glycolate oxidase from spinach [5]) [5]
0.52 <1> (3-phenyllactate, <1> pH 7.5, 20°C, wild-type enzyme [11]) [11]
0.61 <1> (O_2, <1> pH 7.5, 20°C, mutant enzyme G81S [11]) [11]
0.64 <1> ((S)-mandelate, <1> pH 7.5, 20°C, mutant enzyme R277G [7]) [7]
0.8 <1> (2-hydroxyisocaproate, <1> pH 7.5, 20°C, wild-type enzyme [11]) [11]
0.87 <1> (3-phenyllactate, <1> pH 7.5, 20°C, MDH-GOX2, a chimeric mutant of (S)-mandelate dehydrogenase with residues 177-215 replaced by residues 176-195 of glycolate oxidase from spinach [5]) [5]
0.96 <1> (2-hydroxyoctanoate, <1> pH 7.5, 20°C, MDH-GOX2, a chimeric mutant of (S)-mandelate dehydrogenase with residues 177-215 replaced by residues 176-195 of glycolate oxidase from spinach [5]) [5]
1 <1> (3-indolelactate, <1> pH 7.5, 20°C, wild-type enzyme [11]) [11]
1.1 <1> ((S)-mandelate, <1> pH 7.5, 20°C, mutant enzyme R165K/R277K [9]) [9]
1.2 <1> (O_2, <1> pH 7.5, 20°C, wild-type enzyme [11]) [11]
1.2 <1> (indoleglycolate, <1> pH 7.5, 20°C, mutant enzyme G81A [11]) [11]
1.28 <1> (2-hydroxyisocaproate, <1> pH 7.5, 20°C, MDH-GOX2, a chimeric mutant of (S)-mandelate dehydrogenase with residues 177-215 replaced by residues 176-195 of glycolate oxidase from spinach [5]) [5]
1.52 <1> ((S)-mandelate, <1> pH 7.5, 20°C, mutant enzyme R277H [7]) [7]
1.6 <1> ((S)-mandelate, <1> pH 7.5, chimeric mutant of (S)-mandelate dehydrogenase with membrane anchoring loop replaced by a portion of glycolate oxidase from spinach [4]) [4]
2.3 <1> ((S)-mandelate, <1> pH 7.5, 20°C, mutant enzyme, G81A/MDH-GOX2, a chimeric mutant of (S)-mandelate dehydrogenase with residues 177-215 replaced by residues 176-195 of glycolate oxidase from spinach and a G81A mutation in MDH [11]) [11]
2.8 <1> ((S)-mandelate, <1> pH 7.5, 20°C, mutant enzyme G81S [11]) [11]
3-6 <3> ((S)-mandelate, <3> pH 7.5, 25°C, electron acceptor: 2,6-dichloroindophenol [1]) [1]
3.13 <1> (3-indolelactate, <1> pH 7.5, 20°C, MDH-GOX2, a chimeric mutant of (S)-mandelate dehydrogenase with residues 177-215 replaced by residues 176-195 of glycolate oxidase from spinach [5]) [5]
3.4 <1> (O_2, <1> pH 7.5, 20°C, mutant enzyme G81A [11]) [11]
3.9 <1> (2-hydroxy-3-butynoate, <1> pH 7.5, 20°C, wild-type enzyme [11]) [11]
4.4 <1> ((S)-mandelate, <1> pH 7.5, 20°C, mutant enzyme R165E [9]) [9]
6.6 <1> (2-hydroxy-3-butynoate, <1> pH 7.5, 20°C, MDH-GOX2, a chimeric mutant of (S)-mandelate dehydrogenase with residues 177-215 replaced by residues 176-195 of glycolate oxidase from spinach [5]) [5]

9 <1> ((S)-mandelate, <1> pH 7.5, (S)-mandelate dehydrogenase with residues 2-4 deleted [4]) [4]
9.3 <1> (4-chloromandelate, <1> pH 7.5, 20°C, mutant enzyme G81A [11]) [11]
13.2 <1> ((S)-mandelate, <1> pH 7.5, 20°C, mutant enzyme R165G [9]) [9]
14.8 <1> (2-hydroxy-3-butynoate, <1> pH 7.5, 20°C, mutant enzyme G81A [11]) [11]
15.2 <1> (mandelate, <1> pH 7.5, 20°C, mutant enzyme G81A [11]) [11]
18.4 <1> ((S)-mandelate, <1> pH 7.5, 20°C, mutant enzyme R165M [9]) [9]
19.2 <1> ((S)-mandelate, <1> pH 7.5, 20°C, mutant enzyme G81A [11]) [11]
24 <1> ((S)-mandelate, <1> pH 7.5, (S)-mandelate dehydrogenase with 17 residues deleted at the carboxy terminus [4]) [4]
38.5 <1> (indoleglycolate, <1> pH 7.5, 20°C, MDH-GOX2, a chimeric mutant of (S)-mandelate dehydrogenase with residues 177-215 replaced by residues 176-195 of glycolate oxidase from spinach [5]) [5]
49 <3> ((S)-mandelate, <3> pH 7.5, 25°C, electron acceptor: cytochrome c [1]) [1]
66 <1> ((S)-mandelate, <1> pH 7.5, 20°C, mutant enzyme R277K [9]) [9]
68 <3> (3-methoxy-DL-mandelate, <3> pH 7.5, 25°C, cosubstrate: ferricyanide [2]) [2]
68 <3> (4-chloro-DL-mandelate, <3> pH 7.5, 25°C, cosubstrate: ferricyanide [2]) [2]
73 <1> ((S)-mandelate, <1> pH 7.5, 20°C, mutant enzyme R277K [7]) [7]
80 <3> (4-methyl-DL-mandelate, <3> pH 7.5, 25°C, cosubstrate: ferricyanide [2]) [2]
88 <3> (DL-[2-2H]-mandelate, <3> pH 7.5, 25°C, cosubstrate ferricyanide, LMDH holoenzyme [3]) [3]
92 <3> (DL-[2-2H]-mandelate, <3> pH 7.5, 25°C, cosubstrate cytochrome, LMDH holoenzyme [3]) [3]
93 <3> (DL-[1-2H]mandelate, <3> pH 7.5, 25°C, cosubstrate: ferricyanide [2]) [2]
94 <3> (D,L-mandelate, <3> pH 7.5, 25°C, cosubstrate: ferricyanide [2]) [2]
98 <3> (4-fluoro-DL-mandelate, <3> pH 7.5, 25°C, cosubstrate: ferricyanide [2]) [2]
99 <3> (DL-[2-2H]-mandelate, <3> pH 7.5, 25°C, cosubstrate ferricyanide, L-MDH flavin domain [3]) [3]
106 <3> (4-methoxy-DL-mandelate, <3> pH 7.5, 25°C, cosubstrate: ferricyanide [2]) [2]
108 <3> (3-hydroxy-DL-mandelate, <3> pH 7.5, 25°C, cosubstrate: ferricyanide [2]) [2]
108 <3> (4-bromo-DL-mandelate, <3> pH 7.5, 25°C, cosubstrate: ferricyanide [2]) [2]
109 <3> ((S)-mandelate, <3> pH 7.5, 25°C, electron acceptor: ferricyanide [1]) [1]
114 <3> (L-mandelate, <3> pH 7.5, 25°C, cosubstrate: ferricyanide [2]) [2]
116 <3> (4-chloro-DL-mandelate, <3> pH 7.5, 25°C, cosubstrate: ferricyanide [2]) [2]

120 <1> ((S)-mandelate, <1> pH 7.5, 20°C, mutant enzyme R165K [9]) [9]
122 <1> (indoleglycolate, <1> pH 7.5, 20°C, wild-type enzyme [11]) [11]
134 <1> ((S)-mandelate, <1> pH 7.5, 4°C [10]) [10]
146 <3> (4-hydroxy-DL-mandelate, <3> pH 7.5, 25°C, cosubstrate: ferricyanide [2]) [2]
155 <3> (DL-[2-1H]-mandelate, <3> pH 7.5, 25°C, cosubstrate cytochrome c, LMDH holoenzyme [3]) [3]
174 <1> ((S)-mandelate, <1> pH 7.5, wild-type enzyme [4]) [4]
195 <1> (mandelate, <1> pH 7.5, 20°C, MDH-GOX2, a chimeric mutant of (S)-mandelate dehydrogenase with residues 177-215 replaced by residues 176-195 of glycolate oxidase from spinach [5]) [5]
201 <1> ((S)-3-phenyllactate, <1> pH 7.5, 20°C, mutant enzyme R277G [7]) [7]
205 <1> ((S)-mandelate, <1> pH 7.5, 20°C, cosubstrate: 2,6-dichloroindophenol, MDH-GOX2, a chimeric mutant of (S)-mandelate dehydrogenase with residues 177-215 replaced by residues 176-195 of glycolate oxidase from spinach [5]; <1> pH 7.5, 20°C, mutant enzyme MDH-GOX2, a chimeric mutant of (S)-mandelate dehydrogenase with 205 residues 177-215 replaced by residues 176-195 of glycolate oxidase from spinach [11]) [5, 11]
225 <3> (cytochrome c, <3> pH 7.5, 25°C [3]) [3]
270 <1> ((S)-3-phenyllactate, <1> pH 7.5, 20°C, wild-type enzyme [7]) [7]
270 <1> ((S)-mandelate, <1> pH 7.5, 20°C, wild-type enzyme [7]) [7]
290 <1> ((S)-mandelate, <1> pH 7.5, 20°C, cosubstrate: 2,6-dichloroindophenol, wild-type enzyme [5]) [5]
300 <3> (DL-[2-1H]-mandelate, <3> pH 7.5, 25°C, cosubstrate ferricyanide, LMDH holoenzyme [3]) [3]
313 <1> (2-oxo-2-phenylacetate) [12]
316 <3> (DL-[2-1H]-mandelate, <3> pH 7.5, 25°C, cosubstrate ferricyanide, L-MDH flavin domain [3]) [3]
334 <1> (4-chloromandelate, <1> pH 7.5, 20°C, wild-type enzyme [11]) [11]
350 <1> (mandelate, <1> pH 7.5, 20°C, wild-type enzyme [11]) [11]
360 <1> ((S)-mandelate, <1> pH 7.5, 20°C [10]; <1> pH 7.5, 20°C, wild-type enzyme [9,11]) [9, 10, 11]
402 <1> ((S)-mandelate) [12]
550 <3> (ferricyanide, <3> pH 7.5, 25°C [3]) [3]

K_m-Value (mM)

0.04 <1> ((S)-mandelate, <1> pH 7.5, 20°C, mutant enzyme, G81A/MDH-GOX2, a chimeric mutant of (S)-mandelate dehydrogenase with residues 177-215 replaced by residues 176-195 of glycolate oxidase from spinach and a G81A mutation in MDH [11]) [11]
0.04 <1> (O_2, <1> pH 7.5, 20°C, mutant enzyme G81V [11]) [11]
0.074 <1> ((S)-3-phenyllactate, <1> pH 7.5, 20°C, mutant enzyme R277G [7]) [7]
0.08 <3> (3-hydroxy-D,L-mandelate, <3> pH 7.5, 25°C, cosubstrate: ferricyanide [2]) [2]

0.085 <1> ((S)-mandelate, <1> pH 7.5, 20°C, cosubstrate: 2,6-dichloroindophenol, MDH-GOX2, a chimeric mutant of (S)-mandelate dehydrogenase with residues 177-215 replaced by residues 176-195 of glycolate oxidase from spinach [5]) [5]
0.09 <1> ((S)-mandelate, <1> pH 7.5, 20°C, mutant enzyme MDH-GOX2, a chimeric mutant of (S)-mandelate dehydrogenase with residues 177-215 replaced by residues 176-195 of glycolate oxidase from spinach [11]) [11]
0.11 <1> ((S)-mandelate, <1> pH 7.5, 4°C [10]) [10]
0.12 <1> ((S)-mandelate, <1> pH 7.5, 20°C [10]; <1> pH 7.5, 20°C, wild-type enzyme [7,9]) [7, 9, 10]
0.12 <3> (3-methoxy-D,L-mandelate, <3> pH 7.5, 25°C, cosubstrate: ferricyanide [2]) [2]
0.12 <3> (4-methoxy-D,L-mandelate, <3> pH 7.5, 25°C, cosubstrate: ferricyanide [2]) [2]
0.13 <1> ((S)-mandelate, <1> pH 7.5, 20°C, wild-type enzyme [11]) [11]
0.14 <1> (2-hydroxyoctanoate, <1> pH 7.5, 20°C, mutant enzyme G81A [11]) [11]
0.14 <1> (mandelate, <1> pH 7.5, 20°C, MDH-GOX2, a chimeric mutant of (S)-mandelate dehydrogenase with residues 177-215 replaced by residues 176-195 of glycolate oxidase from spinach [5]) [5]
0.15 <1> ((S)-mandelate, <1> pH 7.5, 20°C, mutant enzyme G81A [11]) [11]
0.15 <3> (4-chloro-D,L-mandelate, <3> pH 7.5, 25°C, cosubstrate: ferricyanide [2]) [2]
0.158 <1> ((S)-mandelate, <1> pH 7.5, (S)-mandelate dehydrogenase with 17 residues deleted at the carboxy terminus [4]) [4]
0.16 <3> (4-fluoro-D,L-mandelate, <3> pH 7.5, 25°C, cosubstrate: ferricyanide [2]) [2]
0.17 <1> ((S)-mandelate, <1> pH 7.5, 20°C, mutant enzyme G81V [11]) [11]
0.17 <1> (3-indolelactate, <1> pH 7.5, 20°C, mutant enzyme G81A [11]) [11]
0.17 <3> (4-methyl-D,L-mandelate, <3> pH 7.5, 25°C, cosubstrate: ferricyanide [2]) [2]
0.18 <1, 3> ((S)-mandelate, <1> pH 7.5, 20°C, cosubstrate: 2,6-dichloroindophenol, wild-type enzyme [5]; <3> pH 7.5, 25°C, electron acceptor: 2,6-dichloroindophenol [1]) [1, 5]
0.206 <1> ((S)-mandelate, <1> pH 7.5, wild-type enzyme [4]) [4]
0.22 <1> (O_2, <1> pH 7.5, 20°C, mutant enzyme G81D [11]) [11]
0.225 <1> ((S)-mandelate, <1> pH 7.5, (S)-mandelate dehydrogenase with residues 2-4 deleted [4]) [4]
0.229 <1> ((S)-mandelate, <1> pH 7.5, chimeric mutant of (S)-mandelate dehydrogenase with membrane anchoring loop replaced by a portion of glycolate oxidase from spinach [4]) [4]
0.24 <3> (L-mandelate, <3> pH 7.5, 25°C, cosubstrate: ferricyanide [2]) [2]
0.24 <1> (mandelate, <1> pH 7.5, 20°C, mutant enzyme G81A [11]; <1> pH 7.5, 20°C, wild-type enzyme [11]) [11]
0.24 <1> (indoleglycolate, <1> pH 7.5, 20°C, MDH-GOX2, a chimeric mutant of (S)-mandelate dehydrogenase with residues 177-215 replaced by residues 176-195 of glycolate oxidase from spinach [5]) [5]

0.26 <3> (4-bromo-D,L-mandelate, <3> pH 7.5, 25°C, cosubstrate: ferricyanide [2]) [2]
0.27 <3> ((S)-mandelate, <3> pH 7.5, 25°C, electron acceptor: ferricyanide [1]) [1]
0.27 <1> (4-chloromandelate, <1> pH 7.5, 20°C, wild-type enzyme [11]) [11]
0.29 <1> (2-hydroxyoctanoate, <1> pH 7.5, 20°C, MDH-GOX2, a chimeric mutant of (S)-mandelate dehydrogenase with residues 177-215 replaced by residues 176-195 of glycolate oxidase from spinach [5]) [5]
0.35 <3> (D,L-mandelate, <3> pH 7.5, 25°C, cosubstrate: ferricyanid [2]) [2]
0.37 <1> (3-phenyllactate, <1> pH 7.5, 20°C, mutant enzyme G81A [11]) [11]
0.38 <3> (4-chloro-D,L-mandelate, <3> pH 7.5, 25°C, cosubstrate: ferricyanide [2]) [2]
0.4 <1> (indoleglycolate, <1> pH 7.5, 20°C, wild-type enzyme [11]) [11]
0.43 <1> (4-chloromandelate, <1> pH 7.5, 20°C, mutant enzyme G81A [11]) [11]
0.47 <3> (DL-[2-1H]-mandelate, <3> pH 7.5, 25°C, cosubstrate ferricyanide, LMDH holoenzyme [3]) [3]
0.49 <3> ((S)-mandelate, <3> pH 7.5, 25°C, electron acceptor: cytochrome c [1]) [1]
0.49 <1> (2-hydroxyisocaproate, <1> pH 7.5, 20°C, MDH-GOX2, a chimeric mutant of (S)-mandelate dehydrogenase with residues 177-215 replaced by residues 176-195 of glycolate oxidase from spinach [5]) [5]
0.49 <1> (O_2, <1> pH 7.5, 20°C, mutant enzyme, G81A/MDH-GOX2, a chimeric mutant of (S)-mandelate dehydrogenase with residues 177-215 replaced by residues 176-195 of glycolate oxidase from spinach and a G81A mutation in MDH [11]) [11]
0.53 <3> (DL-[2-2H]-mandelate, <3> pH 7.5, 25°C, cosubstrate ferricyanide, LMDH holoenzyme [3]) [3]
0.63 <1> (indoleglycolate, <1> pH 7.5, 20°C, mutant enzyme G81A [11]) [11]
0.64 <1> (3-indolelactate, <1> pH 7.5, 20°C, MDH-GOX2, a chimeric mutant of (S)-mandelate dehydrogenase with residues 177-215 replaced by residues 176-195 of glycolate oxidase from spinach [5]) [5]
0.68 <1> (O_2, <1> pH 7.5, 20°C, mutant enzyme MDH-GOX2, a chimeric mutant of (S)-mandelate dehydrogenase with 205 residues 177-215 replaced by residues 176-195 of glycolate oxidase from spinach [11]) [11]
0.74 <3> (DL-[1-2H]mandelate, <3> pH 7.5, 25°C, cosubstrate: ferricyanide [2]) [2]
0.74 <3> (DL-[2-1H]-mandelate, <3> pH 7.5, 25°C, cosubstrate ferricyanide, L-MDH flavin domain [3]) [3]
0.78 <1> ((S)-phenyllactate, <1> pH 7.5, chimeric mutant of (S)-mandelate dehydrogenase with membrane anchoring loop replaced by a portion of glycolate oxidase from spinach [4]) [4]
0.78 <3> (DL-[2-1H]-mandelate, <3> pH 7.5, 25°C, cosubstrate cytochrome c, LMDH holoenzyme [3]) [3]
0.79 <3> (DL-[2-2H]-mandelate, <3> pH 7.5, 25°C, cosubstrate ferricyanide, L-MDH flavin domain [3]) [3]
0.8 <1> (2-hydroxyoctanoate, <1> pH 7.5, 20°C, wild-type enzyme [11]) [11]

0.89 <1> (2-hydroxyhexanoate, <1> pH 7.5, 20°C, MDH-GOX2, a chimeric mutant of (S)-mandelate dehydrogenase with residues 177-215 replaced by residues 176-195 of glycolate oxidase from spinach [5]) [5]
0.9 <1> (3-indolelactate, <1> pH 7.5, 20°C, wild-type enzyme [11]) [11]
1.1 <1> (O_2, <1> pH 7.5, 20°C, mutant enzyme G81A [11]) [11]
1.3 <1> (3-phenyllactate, <1> pH 7.5, 20°C, MDH-GOX2, a chimeric mutant of (S)-mandelate dehydrogenase with residues 177-215 replaced by residues 176-195 of glycolate oxidase from spinach [5]) [5]
1.33 <3> (DL-[2-2H]-mandelate, <3> pH 7.5, 25°C, cosubstrate cytochrome, LMDH holoenzyme [3]) [3]
1.4 <1> (2-hydroxyhexanoate, <1> pH 7.5, 20°C, mutant enzyme G81A [11]) [11]
1.4 <1> (O_2, <1> pH 7.5, 20°C, mutant enzyme G81S [11]) [11]
1.5 <1> ((S)-mandelate, <1> pH 7.5, 20°C, mutant enzyme R165K [9]) [9]
1.6 <1> (2-hydroxyisocaproate, <1> pH 7.5, 20°C, mutant enzyme G81A [11]) [11]
2 <1> ((S)-mandelate, <1> pH 7.5, 20°C, mutant enzyme G81D [11]) [11]
2 <1> (3-phenyllactate, <1> pH 7.5, 20°C, wild-type enzyme [11]) [11]
2.3 <1> ((S)-mandelate, <1> pH 7.5, 20°C, mutant enzyme G81S [11]) [11]
2.6 <1> ((S)-phenyllactate, <1> pH 7.5, wild-type enzyme [4]) [4]
2.6 <1> (2-hydroxyisovalerate, <1> pH 7.5, 20°C, MDH-GOX2, a chimeric mutant of (S)-mandelate dehydrogenase with residues 177-215 replaced by residues 176-195 of glycolate oxidase from spinach [5]) [5]
3.2 <1> (2-hydroxyvalerate, <1> pH 7.5, 20°C, MDH-GOX2, a chimeric mutant of (S)-mandelate dehydrogenase with residues 177-215 replaced by residues 176-195 of glycolate oxidase from spinach [5]) [5]
3.2 <1> (O_2, <1> pH 7.5, 20°C, wild-type enzyme [11]) [11]
4.3 <1> ((S)-3-phenyllactate, <1> pH 7.5, 20°C, wild-type enzyme [7]) [7]
4.3 <1> (2-hydroxy-3-butynoate, <1> pH 7.5, 20°C, mutant enzyme G81A [11]) [11]
4.3 <1> (2-hydroxyisocaproate, <1> pH 7.5, 20°C, wild-type enzyme [11]) [11]
4.4 <1> ((S)-mandelate, <1> pH 7.5, 20°C, mutant enzyme R165M [9]; <1> pH 7.5, 20°C, mutant enzyme R277K [7]) [7, 9]
4.4 <1> (2-hydroxybutyrate, <1> pH 7.5, 20°C, mutant enzyme G81A [11]) [11]
4.9 <1> (2-hydroxyhexanoate, <1> pH 7.5, 20°C, wild-type enzyme [11]) [11]
5.6 <1> ((S)-mandelate, <1> pH 7.5, 20°C, mutant enzyme R277K [9]) [9]
5.8 <1> ((S)-mandelate, <1> pH 7.5, 20°C, mutant enzyme R165G [9]) [9]
6.4 <1> (2-hydroxyvalerate, <1> pH 7.5, 20°C, mutant enzyme G81A [11]) [11]
10.3 <1> (2-hydroxy-3-butenoate, <1> pH 7.5, 20°C, MDH-GOX2, a chimeric mutant of (S)-mandelate dehydrogenase with residues 177-215 replaced by residues 176-195 of glycolate oxidase from spinach [5]) [5]
12 <1> ((S)-mandelate, <1> pH 7.5, 20°C, mutant enzyme R165E [9]) [9]
13 <1> ((S)-mandelate, <1> mutant enzyme H274G, in presence of 20 mM imidazole [6]) [6]

13.2 <1> (2-hydroxybutyrate, <1> pH 7.5, 20°C, MDH-GOX2, a chimeric mutant of (S)-mandelate dehydrogenase with residues 177-215 replaced by residues 176-195 of glycolate oxidase from spinach [5]) [5]
15.2 <1> ((S)-mandelate, <1> pH 7.5, 20°C, mutant enzyme R165K/R277K [9]) [9]
15.3 <1> (2-hydroxyvalerate, <1> pH 7.5, 20°C, wild-type enzyme [11]) [11]
17 <1> ((S)-mandelate, <1> pH 7.5, 20°C, mutant enzyme R277G [9]) [9]
17.2 <1> (2-hydroxy-3-butynoate, <1> pH 7.5, 20°C, MDH-GOX2, a chimeric mutant of (S)-mandelate dehydrogenase with residues 177-215 replaced by residues 176-195 of glycolate oxidase from spinach [5]) [5]
19.2 <1> ((S)-mandelate, <1> pH 7.5, 20°C, mutant enzyme R277H [7]) [7]
22 <1> (2-hydroxy-3-butynoate, <1> pH 7.5, 20°C, wild-type enzyme [11]) [11]
31.4 <1> ((S)-mandelate, <1> pH 7.5, 20°C, mutant enzyme R165G/R277K [9]) [9]
32 <1> (2-hydroxybutyrate, <1> pH 7.5, 20°C, wild-type enzyme [11]) [11]
47 <1> ((S)-mandelate, <1> pH 7.5, 20°C, mutant enzyme R277G [7]) [7]
73 <1> ((S)-mandelate, <1> pH 7.5, 20°C, mutant enzyme R277L [7]) [7]

K_i-Value (mM)

0.078 <3> (oxalate, <3> pH 7.5, 25°C [1]) [1]
0.37 <3> (hexanoate, <3> pH 7.5, 25°C [1]) [1]
0.4 <3> (L-lactate, <3> pH 7.5, 25°C [1]) [1]
1.9 <3> (L-phenyllactate, <3> pH 7.5, 25°C [1]) [1]
2.9 <1> ((R)-mandelate, <1> pH 7.5, 20°C, wild-type enzyme [9]) [9]
3.1 <1> ((R)-mandelate, <1> pH 7.5, 20°C, mutant enzyme R165K [9]) [9]
7.2 <1> ((R)-mandelate, <1> pH 7.5, 20°C, mutant enzyme R277K [9]) [9]
11 <1> (1-phenylacetate, <1> pH 7.5, 20°C, mutant enzyme R165K [9]; <1> pH 7.5, 20°C, mutant enzyme R165K/R277K [9]) [9]
12 <1> (1-phenylacetate, <1> pH 7.5, 20°C, mutant enzyme R277K [9]; <1> pH 7.5, 20°C, wild-type enzyme [9]) [9]
19.5 <1> ((R)-mandelate, <1> pH 7.5, 20°C, mutant enzyme R277G [9]) [9]
20.3 <1> (1-phenylacetate, <1> pH 7.5, 20°C, mutant enzyme R165G/R277K [9]) [9]
22.5 <1> (1-phenylacetate, <1> pH 7.5, 20°C, mutant enzyme R277G [9]) [9]
67 <1> (1-phenylacetate, <1> pH 7.5, 20°C, mutant enzyme R165G [9]) [9]
73 <1> ((R)-mandelate, <1> pH 7.5, 20°C, mutant enzyme R165G [9]) [9]
700 <1> ((R)-mandelate, <1> pH 7.5, 20°C, mutant enzyme R165G/R277K [9]) [9]

5 Isolation/Preparation/Mutation/Application

Localization

membrane <1> (<1> associated [5]; <1> associated, wild-type enzyme [4]) [4, 5]

Purification

<1> [9, 10]
<1> (mutant enzymes R277K, R277G, R277H and R277L) [7]
<1> (mutant enzymes h274G, H274A and H274N) [6]
<1> (wild-type and mutant enzymes) [4, 11]
<3> (recombinant enzyme) [3]

Crystallization

<1> (1.35 A resolution structure of oxidized form and of the substrate-reduced form of MDH-GOX2, a chimeric mutant of (S)-mandelate dehydrogenase with residues 177-215 replaced by residues 176-195 of glycolate oxidase from spinach) [12]
<1> (hanging-drop vapor-diffusion method. Crystal structure of MDH-GOX2, a chimeric mutant of MDH with residues 177-215 replaced by residues 176-195 of glycolate oxidase from spinach) [8]

Cloning

<1> (expression in Escherichia coli) [4]
<3> (expression in Escherichia coli) [3]

Engineering

G81A <1> (<1> higher specificity for small substrates, compared to that of wild-type enzyme, the affinity for (S)-mandelate is relatively unchanged. The rate of the first half-reaction is slower than the wild-type rate. 2fold increase in oxidative half-reaction. Affinity for oxygen increases 10-15fold [11]) [11]
G81D <1> (<1> extreme low activity, reduction in activity is due to the decrease in electrophilicity of the FMN [11]) [11]
G81S <1> (<1> the rate of the first half-reaction is slower than the wild-type rate. The rate of the first half-reaction is slower than the wild-type rate. Affinity for O_2 increases 10-15fold [11]) [11]
G81V <1> (<1> extreme low activity, reduction in activity is due to the decrease in electrophilicity of the FMN [11]) [11]
H274A <1> (<1> inactive mutant, activity can partially be restored by the addition of exogenous imidazole [6]) [6]
H274D <1> (<1> inactive mutant [6]) [6]
H274G <1> (<1> inactive mutant, activity can be restored by the addition of exogenous imidazole [6]) [6]
R165E <1> (<1> mutant is less stable than wild-type enzyme, mutant enzyme is only partially reduced by (S)-mandelate. 81.8fold decrease in k_{cat} for (S)-mandelate, 199fold increase in K_M-value for (S)-mandelate compared to wild-type enzyme [9]) [9]
R165G <1> (<1> mutant is less stable than wild-type enzyme, mutant enzyme is only partially reduced by (S)-mandelate. 27.3fold decrease in k_{cat} for (S)-mandelate, 48.3fold increase in K_M-value for (S)-mandelate compared to wild-type enzyme [9]) [9]
R165G/R277G <1> (<1> no activity. Double mutant is less stable than single Arg165 mutant in term of long-term storage [9]) [9]

R165G/R277K <1> (<1> double mutant is less stable than single Arg165 mutant in term of long-term storage. 9000fold decrease in k_{cat} for (S)-mandelate, 261.7fold increase in K_M-value for (S)-mandelate compared to wild-type enzyme [9]) [9]
R165K <1> (<1> mutant is less stable than wild-type enzyme, mutant enzyme is fully reduced on addition of (S)-mandelate. 3fold decrease in k_{cat} for (S)-mandelate, 12.5fold increase in K_M-value for (S)-mandelate compared to wild-type enzyme [9]) [9]
R165K/R277G <1> (<1> no activity. Double mutant is less stable than single Arg165 mutant in term of long-term storage [9]) [9]
R165K/R277K <1> (<1> double mutant is less stable than single Arg165 mutant in term of long-term storage. 327fold decrease in k_{cat} for (S)-mandelate, 126.7fold increase in K_M-value for (S)-mandelate compared to wild-type enzyme [9]) [9]
R165M <1> (<1> mutant is less stable than wild-type enzyme, mutant enzyme is fully reduced on addition of (S)-mandelate. 19.6fold decrease in k_{cat} for (S)-mandelate, 36.7fold increase in K_M-value for (S)-mandelate compared to wild-type enzyme [9]) [9]
R277G <1> (<1> 1565fold decrease in k_{cat} for (S)-mandelate, 141fold increase in K_M-value for (S)-mandelate compared to wild-type enzyme [9]; <1> k_{cat} for (S)-mandelate is 178fold lower than wild-type value, K_M-value for (S)-mandelate is 160fold higher than wild-type value, K_M-value for (S)-3-phenyllactate is 1.6fold lower than wild-type value [7]) [7, 9]
R277H <1> (<1> k_{cat} for (S)-mandelate is 1000fold lower than wild-type value, K_M-value for (S)-mandelate is 608fold higher than wild-type value [7]) [7]
R277K <1> (<1> 5.5 fold decrease in k_{cat} for (S)-mandelate, 46.7fold increase in K_M-value for (S)-mandelate compared to wild-type enzyme [9]; <1> k_{cat} for (S)-mandelate is 3.7fold lower than wild-type value, K_M-value for (S)-mandelate is 36.7fold higher than wild-type value [7]) [7, 9]
R277L <1> (<1> k_{cat} for (S)-mandelate is 421fold lower fold than wild-type value, K_M-value for (S)-mandelate is 392fold higher than wild-type value [7]) [7]
Additional information <1> (<1> chimeric mutant of (S)-mandelate dehydrogenase with membrane anchoring loop replaced by a portion of glycolate oxidase from spinach is soluble and retains partial catalytic activity (about 1%) using (S)-mandelate as substrate. The activities of the soluble mutant enzymes (S)-mandelate dehydrogenase with residues 2-4 deleted and (S)-mandelate dehydrogenase with 17 residues deleted at the carboxy terminus are nearly the same when (S)-phenyllactate is used as substrate [4]; <1> MDH-GOX2, a chimeric mutant of (S)-mandelate dehydrogenase with residues 177-215 replaced by residues 176-195 of glycolate oxidase from spinach. G81A/MDH-GOX2, a chimeric mutant of (S)-mandelate dehydrogenase with residues 177-215 replaced by residues 176-195 of glycolate oxidase from spinach and a G81A mutation in MDH [11]; <1> MDH-GOX2, a chimeric mutant of (S)-mandelate dehydrogenase with residues 177-215 replaced by residues 176-195 of glycolate oxidase from spinach. This mutant is very similar to

the wild-type membrane-bound enzyme in its spectroscopic properties, substrate specificity, catalytic activity, kinetic mechanism, and lack of reactivity toeards oxygen. It should prove to be a highly useful model for structural studies of MDH [5]) [4, 5, 11]

6 Stability

Temperature stability

30 <2> (<2> 4 h, little change of activity of extract [13]) [13]

General stability information

<1>, repeated freeze-thaw cycles cause significant loss of activity [4]

Storage stability

<1>, -70°C, wild-type enzyme can be stored frozen in 20% ethanediol for weeks without loss of activity [4]

<1>, -70°C, wild-type enzyme retains activity for more than 1 year. The single Arg165 mutant lose activity after 2-3 months, double mutants R165K/R277K and R165G/R277K are inactivated after 2-3 weeks in storage [9]

References

[1] Smekal, O.; Yasin, M.; Fewson, C.A.; Reid, G.A.; Chapman, S.K.: L-mandelate dehydrogenase from Rhodotorula graminis: comparisons with the L-lactate dehydrogenase (flavocytochrome b2) from Saccharomyces cerevisiae. Biochem. J., **290**, 103-107 (1993)

[2] Smekal, O.; Reid, G.A.; Chapman, S.K.: Substrate analogues as probes of the catalytic mechanism of L-mandelate dehydrogenase from Rhodotorula graminis. Biochem. J., **297**, 647-652 (1994)

[3] Illias, R.M.; Sinclair, R.; Robertson, D.; Neu, A.; Chapman, S.K.; Reid, G.A.: L-Mandelate dehydrogenase from Rhodotorula graminis: cloning, sequencing and kinetic characterization of the recombinant enzyme and its independently expressed flavin domain. Biochem. J., **333**, 107-115 (1998)

[4] Mitra, B.; Gerlt, J.A.; Babbitt, P.C.; Koo, C.W.; Kenyon, G.L.; Joseph, D.; Petsko, G.A.: A novel structural basis for membrane association of a protein: construction of a chimeric soluble mutant of (S)-mandelate dehydrogenase from Pseudomonas putida. Biochemistry, **32**, 12959-12967 (1993)

[5] Xu, Y.; Mitra, B.: A highly active, soluble mutant of the membrane-associated (S)-mandelate dehydrogenase from Pseudomonas putida. Biochemistry, **38**, 12367-12376 (1999)

[6] Lehoux, I.E.; Mitra, B.: (S)-Mandelate dehydrogenase from Pseudomonas putida: mutations of the catalytic base histidine-274 and chemical rescue of activity. Biochemistry, **38**, 9948-9955 (1999)

[7] Lehoux, I.E.; Mitra, B.: Role of arginine 277 in (S)-mandelate dehydrogenase from Pseudomonas putida in substrate binding and transition state stabilization. Biochemistry, **39**, 10055-10065 (2000)

[8] Sukumar, N.; Xu, Y.; Gatti, D.L.; Mitra, B.; Mathews, F.S.: Structure of an active soluble mutant of the membrane-associated (S)-mandelate dehydrogenase. Biochemistry, **40**, 9870-9878 (2001)
[9] Xu, Y.; Dewanti, A.R.; Mitra, B.: Arginine 165/arginine 277 pair in (S)-mandelate dehydrogenase from Pseudomonas putida: role in catalysis and substrate binding. Biochemistry, **41**, 12313-12319 (2002)
[10] Dewanti, A.R.; Mitra, B.: A transient intermediate in the reaction catalyzed by (S)-mandelate dehydrogenase from Pseudomonas putida. Biochemistry, **42**, 12893-12901 (2003)
[11] Dewanti, A.R.; Xu, Y.; Mitra, B.: Role of glycine 81 in (S)-mandelate dehydrogenase from Pseudomonas putida in substrate specificity and oxidase activity. Biochemistry, **43**, 10692-10700 (2004)
[12] Sukumar, N.; Dewanti, A.R.; Mitra, B.; Mathews, F.S.: High resolution structures of an oxidized and reduced flavoprotein. The water switch in a soluble form of (S)-mandelate dehydrogenase. J. Biol. Chem., **279**, 3749-3757 (2004)
[13] Fewson, C.A.; Allison, N.; Hamilton, I.D.; Jardine, J.; Scott, A.J.: Comparison of mandelate dehydrogenases from various strains of Acinetobacter calcoaceticus: similarity of natural and 'evolved' forms. J. Gen. Microbiol., **134**, 967-974 (1988)

L-Sorbose 1-dehydrogenase 1.1.99.32

1 Nomenclature

EC number
1.1.99.32

Systematic name
L-sorbose:acceptor 1-oxidoreductase

Recommended name
L-sorbose 1-dehydrogenase

CAS registry number
133249-51-1

3 Reaction and Specificity

Catalyzed reaction
L-sorbose + acceptor = 1-dehydro-L-sorbose + reduced acceptor

Glutamyl-tRNA reductase 1.2.1.70

1 Nomenclature

EC number
1.2.1.70

Systematic name
L-glutamate-semialdehyde:NADP$^+$ oxidoreductase (L-glutamyl-tRNA^{Glu}-forming)

Recommended name
glutamyl-tRNA reductase

Synonyms
GTR <1, 11> [18, 20, 21, 22]
GluTR <2, 3, 5, 8, 9, 12> [9, 11, 16, 17, 19, 23]
glutamate tRNA reductase <2> [16]
glutamate-specific tRNA reductase <2> [16]
glutamyl transfer RNA reductase <2> [16]
glutamyl-tRNA reductase <1, 2, 3, 5, 9, 11, 12> [16, 17, 18, 19, 20, 21, 22, 23]
reductase, glutamyl-transfer ribonucleate <2> [16]

CAS registry number
119940-26-0

2 Source Organism

<1> *Chlamydomonas reinhardtii* (no sequence specified) [5, 6, 20, 21, 22]
<2> *Escherichia coli* (no sequence specified) [9, 10, 16]
<3> *Hordeum vulgare* (no sequence specified) [1, 4, 12, 15, 23]
<4> *Arabidopsis thaliana* (no sequence specified) [2]
<5> *Streptomyces coelicolor* (no sequence specified) [19]
<6> *Cucumis sativus* (no sequence specified) [13,14]
<7> *Synechocystis sp.* (no sequence specified) (<7> gene luxB [7]) [7]
<8> *Methanopyrus kandleri* (no sequence specified) [3,11]
<9> *Acidithiobacillus ferrooxidans* (no sequence specified) [17]
<10> *Methanopyrus kandleri* (UNIPROT accession number: Q9UXR8) [8]
<11> *Chlorobium vibrioforme* (UNIPROT accession number: P28462) [18]
<12> *Streptomyces nodosus* (no sequence specified) [19]

3 Reaction and Specificity

Catalyzed reaction

L-glutamate 1-semialdehyde + NADP$^+$ + tRNA^{Glu} = L-glutamyl-tRNA^{Glu} + NADPH + H$^+$

Natural substrates and products

S L-glutamyl-tRNA^{Glu} + NADPH + H$^+$ <1, 2, 3, 4, 6, 8> (<1> complex formation between glutamyl-tRNA synthetase and glutamyl-tRNA reductase during the tRNA-dependent synthesis of 5-aminolevulinic acid [5]; <6> glutamyl-tRNA reductase is the solely light-regulated enzyme of 5-aminolevulinic acid-synthesis system, and the elevation of glutamate by light may contribute to the stimulation of 5-aminolevulinic acid synthesis [14]; <6> stimulation of the synthesis of 5-aminolevulinic acid by benzyladenine is caused by increased levels of glutamyl-tRNA reductase and that the reductase is the regulatory and rate-determining enzyme in the 5-aminolevulinic-synthesis system except in untreated etioplasts, in which the level of glutamyl-tRNA may be rate-determining factor [13]; <2,3,8> the enzyme catalyzes the initial step of tetrapyrrole biosynthesis [10,11,15]; <2,4> the enzyme catalyzes the initial step of tetrapyrrole biosynthesis in plants and prokaryotes [2,16]; <3> the enzyme directs glutamate to chlorophyll biosynthesis [4]; <1> the enzyme is involved in δ-aminolevulinic acid formation during chlorophyll biosynthesis [6]; <8> the enzyme is involved in the C5 pathway [3]) (Reversibility: ?) [2, 3, 4, 5, 6, 10, 11, 12, 13, 14, 15, 16]

P L-glutamate 1-semialdehyde + NADP$^+$ + tRNA^{Glu}

Substrates and products

S L-glutamyl-tRNA^{Glu} + NADH + H$^+$ <1, 7, 9, 11> (<7> much higher activity occurs with NADPH than with NADH [7]; <9> Acidithiobacillus ferrooxidans contains three tRNA^{Glu} where only 2 are a substrate for glutamyl-tRNA reductase [17]; <1> the enzyme also exhibits an esterase activity [21]) (Reversibility: ?) [7, 17, 18, 20, 21]

P L-glutamate 1-semialdehyde + NAD$^+$ + tRNA^{Glu}

S L-glutamyl-tRNA^{Glu} + NADPH + H$^+$ <1, 2, 3, 4, 6, 7, 8, 10> (<1> complex formation between glutamyl-tRNA synthetase and glutamyl-tRNA reductase during the tRNA-dependent synthesis of 5-aminolevulinic acid [5]; <6> glutamyl-tRNA reductase is the solely light-regulated enzyme of 5-aminolevulinic acid-synthesis system, and the elevation of glutamate by light may contribute to the stimulation of 5-aminolevulinic acid synthesis [14]; <6> stimulation of the synthesis of 5-aminolevulinic acid by benzyladenine is caused by increased levels of glutamyl-tRNA reductase and that the reductase is the regulatory and rate-determining enzyme in the 5-aminolevulinic-synthesis system except in untreated etioplasts, in which the level of glutamyl-tRNA may be rate-determining factor [13]; <2,3,8> the enzyme catalyzes the initial step of tetrapyrrole biosynthesis [10,11,15]; <2,4> the enzyme catalyzes the initial step of tetrapyrrole biosynthesis in plants and prokaryotes [2,16]; <3> the enzyme directs glutamate to chlorophyll biosynthesis [4]; <1> the enzyme is involved in δ-aminolevulinic acid formation during chlorophyll biosynthesis [6]; <8> the enzyme is involved in the C5 pathway [3]; <3> A7-U66, U29-A41, A53-U61 and U72 are expected to be required for recognition by the barley chloroplast glutamyl-tRNA(Glu) reductase [1]; <10> in absence of NADPH, an esterase activity of GluTR hydrolyzes the highly reactive thioester of $tRNA^{Glu}$ to release glutamate [8]; <2> in presence of NADPH, the end product, glutamate-1-semialdehyde is formed. In the absence of NADPH, Escherichia coli GluTR exhibits substrate esterase activity [9]; <2> the enzyme interacts directly with the amino-acylated acceptor stem and the D-stem, whereas the anticodon domain serves as a major recognition element of aminoacyl tRNA synthetases [10]; <3> the fusion protein with glutathione S-transferase uses $tRNA^{Glu}$ from Hordeum vulgare chloroplast preferentially to Escherichia coli $tRNA^{Glu}$ [15]) (Reversibility: ?) [1, 2, 3, 4, 5, 6, 7, 8, 9, 10, 11, 12, 13, 14, 15, 16]

P L-glutamate 1-semialdehyde + $NADP^+$ + $tRNA^{Glu}$

Inhibitors

1,10-phenanthroline <2> (<2> 5 mM, 25% inhibition [9]) [9]
2,2'-dipyridyl <2> (<2> 5 mM, 20% inhibition [9]) [9]
2,4-diphenyl-6-styryl-1-*p*-tolyl-pyridinium boron tetrafluoride <4> (<4> IC50: 0.01 mM [2]) [2]
4-[4-(3,4-dihydroxyphenyl)-2,3-dimethylbutyl]benzene-1,2-diol <4> (<4> IC50: 0.055 mM [2]) [2]
5,5'-dithiobis(2-nitrobenzoic acid) <2, 10> (<2> 1 mM, 80% inhibition [9]; <10> 1.0 mM, 90% inhibition [8]) [8, 9]
5,5'-dithiobis-(2-nitrobenzoic acid) <8> [3]
5,6-dichloro-1,3-benzodioxol-2-one <4> (<4> IC50: 0.055 mM [2]) [2]
Cd^{2+} <3> (<3> 1 mM Cd(Ac)2, 92% inhibition [4]) [4]
Co^{2+} <2, 3> (<3> 10 mM $CoCl_2$, 78% inhibition [4]) [4, 9]
Cu^{2+} <3> (<3> 1 mM $CuCl_2$, 84% inhibition [4]) [4]
EDTA <2> (<2> 10 mM, 55% inhibition [9]) [9]
EGTA <2> (<2> 10 mM, 35% inhibition [9]) [9]
Fe^{2+} <3> (<3> 5 mM $FeSO_4$, 66% inhibition [4]) [4]
Fe^{3+} <3> (<3> 10 mM $FeCl_3$, 80% inhibition [4]) [4]
heme <1, 3, 7, 10> (<7> 0.05 mM, 50% inhibition [7]; <3> 0.004 mM, 50% inhibition [4]; <3> 0.002 mM, 63% inhibition of non-truncated enzyme [12]; <10> 0.007 mM, 70% inhibition [8]; <1> 80% inhibition at 0.005 mM [21]) [4, 7, 8, 12, 21]
hemin <3> (<3> 50% inhibition at 0.0015 mM [15]) [15]
iodoacetamide <2, 10> (<2> 0.1 mM, complete inhibition [9]; <10> 0.01 mM, 30% inhibition, 0.1 mM, complete inhibition [8]) [8, 9]
N-tosyl-L-phenylalaninechloromethyl ketone <10> (<10> 0.1 mM, 90% inhibition, 1.0 mM, complete inhibition [8]) [8]
Ni^{2+} <2> [9]

$PbCl_2$ <10> (<10> 0.1 mM, 60% inhibition, 1.0 mM, complete inhibition [8]) [8]
PtCl4 <2, 10> (<2> 1 mM, 90% inhibition [9]; <10> 0.1 mM, 55% inhibition, 1.0 mM, 90% inhibition [8]) [8, 9]
Zn^{2+} <3> (<3> 2.5 mM, $ZnSO_4$, 94% inhibition [4]) [4]
$ZnCl_2$ <10> (<10> 0.2 mM, 45% inhibition, 5.0 mM, 90% inhibition [8]) [8]
glutamate 1-semialdehyde <3> (<3> 0.2 mM, 50% inhibition [4]) [4]
glutamate-1-semialdehyde <10> (<10> 1.0 mM, 50% inhibition [8]) [8]
glutamycin <2, 8, 10> (<8> competitive [3]; <10> 2.5 mM, 75% inhibition [8]; <2> 3 mM, 90% inhibition [9]) [3, 8, 9]

Cofactors/prosthetic groups

heme <3, 11> (<3> the fusion protein with glutathione S-transferase contains heme, which can be reduced by NADPH and oxidized by air [15]; <11> one molecule per glutamyl-tRNA reductase molecule [18]) [15,18]
NADH <1, 7> (<7> much higher activity occurs with NADPH than with NADH [7]) [7,21]
NADPH <1, 2, 3, 4, 6, 7, 8, 10> (<7> half-maximal rate at 0.005 mM, saturation is not reached even at 10 mM NADH [7]) [1, 2, 3, 5, 6, 7, 8, 9, 10, 11, 12, 13, 14, 15, 16]
Additional information <10> (<10> the enzyme does not possess a chromophoric prosthetic group [8]) [8]

Activating compounds

glutamate-1-semialdehyde aminotransferase <1, 11> (<1,11> 2.5fold activity in presence of glutamate-1-semialdehyde aminotransferase [18,20]) [18, 20]

Metals, ions

Ca^{2+} <2> (<2> restores activity after treatment with chelating agents [9]) [9]
Mg^{2+} <2> (<2> stimulates, restores activity after treatment with chelating agents [9]) [9]
Mn^{2+} <2> (<2> restores activity after treatment with chelating agents [9]) [9]
Additional information <8> (<8> no significant stimulation by high salt concentrations [3]) [3]

Turnover number (min^{-1})

0.13 <2> (L-glutamyl-$tRNA^{Glu}$, <2> pH 8.1, 37°C [9]) [9]
0.15 <2> (NADPH, <2> pH 8.1, 37°C [9]) [9]
Additional information <7> [7]

Specific activity (U/mg)

0.12 <3> (<3> fusion protein with glutathione S-transferase [15]) [15]
0.47 <2> [16]
Additional information <1, 3> [4, 6]

K_m-Value (mM)

0.024 <2> (L-glutamyl-$tRNA^{Glu}$, <2> pH 8.1, 37°C [9]) [9]
0.039 <2> (NADPH, <2> pH 8.1, 37°C [9]) [9]

pH-Optimum

8.1 <2, 8> [8, 16]

Temperature optimum (°C)

37 <2> [16]
90 <8, 10> [3, 8]

4 Enzyme Structure

Molecular weight

49000 <11> (<11> SDS-PAGE, recombinant protein [18]) [18]
52500 <1> (<1> deduced from sequence [21]) [21]
101000 <1> (<1> sucrose density gradient sedimentation, dimer [20]) [20]
130000 <1> (<1> glycerol density gradient sedimentation, gel filtration [6]) [6]
180000 <2> (<2> gel filtration [9]) [9]
190000 <10> (<10> gel filtration [8]) [8]
250000 <3> (<3> gel filtration [15]) [15]
270000 <3> (<3> gel filtration [4]) [4]
350000 <7> (<7> glycerol density gradient centrifugation [7]) [7]

Subunits

? <3, 7, 10> (<3> x * 54000, SDS-PAGE [4]; <7> x * 39000, SDS-PAGE [7]; <10> x * 45436, electrospray ionization mass spectrometry [8]) [4, 7, 8]
dimer <1, 2, 11> (<2> 2 * 48000, homodimer with an extended structure, SDS-PAGE [9]; <2> 2 * 48448, homodimer with an extended structure, electrospray ionization mass spectrometry [9]; <11> 2 * 49000, gel-filtration in presence of 5% glycerol [18]; <1> 2 * 52500, sucrose gradient sedimentation compared to calculation from sequence [20]) [9, 18, 20]
heterotetramer <1> (<1> 2 * 52500 + 2 * 46000, sucrose gradient sedimentation of glutamyl-tRNA reductase preincubated with glutamate-1-semialdehyde aminotransferase. Both enzyme also co-precipitate in immunoprecipitation experiments [20]) [20]
homodimer <1> (<1> 2 * 52500, gel filtration [21]) [21]
monomer <1, 11> (<1> 1 * 130000, SDS-PAGE [6]; <11> 1 * 49000, gel-filtration without glycerol [18]) [6, 18]
tetramer <3> (<3> 4 * 60000, SDS-PAGE [15]) [15]

5 Isolation/Preparation/Mutation/Application

Source/tissue

cell culture <1, 5, 9, 11, 12> [17, 18, 19, 20, 21, 22]
cotyledon <6> (<6> greening [14]) [13, 14]
leaf <3> (<3> kinetin but not cytokinin stimulates expression of glutamyl-tRNA reductase expression in etiolated plants [23]) [23]
seed <6> [13]

seedling <3> (<3> kinetin stimulates expression of glutamyl-tRNA reductase in etiolated seedlings [23]) [23]

Localization

chloroplast <1, 3> (<3> stroma of greening chloroplast [4]; <1> glutamyl-tRNA reductase protein is 2fold more abundant in light grown cells compared to dark grown cells [22]) [1, 4, 22]
inclusion body <2> [16]

Purification

<1> [6]
<1> (Ni-affinity chromatography) [20, 21]
<2> [9, 16]
<3> [4]
<3> (fusion protein with glutathione S-transferase) [15]
<7> [7]
<8> [8]
<9> (Ni-affinity chromatography) [17]
<11> (Ni-affinity chromatography) [18]

Crystallization

<8> (crystallized in presence of glutamycin, the structure is solved by the multiple isomorphous replacement method using three heavy atom derivatives. The structure is subsequently refined at a resolution of 1.95 A) [11]

Cloning

<1> (expression as His-tag fusion protein in Escherichia coli) [21]
<1> (expression in Escherichia coli) [20]
<2> (expression plasmid pBKCwt overexpression of a 6*His-tagged enzyme) [16]
<2> (overexpression in Escherichia coli) [9]
<3> (expression in Escherichia coli as a fusion protein with glutathione S-transferase) [15]
<8> (overexpression in Escherichia coli) [3, 8]
<9> (expression in Escherichia coli BL21) [17]
<11> (expression in Escherichia coli) [18]

Engineering

C170S <2> (<2> mutant enzyme with esterase activity 95% of the wild-type activity and reductase activity with 90% of the wild-type activity [9]) [9]
C393S <10> (<10> 95% of the GluTR reductase activity compared to wild-type enzyme, 100% of the GluTR esterase activity compared to wild-type enzyme [8]) [8]
C42S <10> (<10> no GluTR reductase and GluTR esterase activity [8]) [8]
C48A <1> (<1> no activity [20]) [20]
C48S <8, 10> (<8> complete loss of activity [3]; <10> 90% of the GluTR reductase activity compared to wild-type enzyme, 95% of the GluTR esterase activity compared to wild-type enzyme [8]) [3, 8]

C50S <2> (<2> mutant enzyme with no esterase and reductase activity [9]) [9]
C6S <10> (<10> 130% of the GluTR reductase activity compared to wild-type enzyme, 120% of the GluTR esterase activity compared to wild-type enzyme [8]) [8]
C74S <2> (<2> mutant enzyme with esterase activity 110% of the wild-type activity and reductase activity with 120% of the wild-type activity [9]) [9]
C90S <10> (<10> 85% of the GluTR reductase activity compared to wild-type enzyme, 105% of the GluTR esterase activity compared to wild-type enzyme [8]) [8]
E114K <2> (<2> mutant enzyme with no esterase and reductase activity [9]) [9]
G106N <2> (<2> mutant enzyme with no esterase and reductase activity [9]) [9]
G191D <2> (<2> mutant enzyme reveals esterase, 105% of the wild-type activity, but no reductase activity [9]) [9]
G44C/S105N/A326T <2> (<2> mutant enzyme with no esterase and reductase activity [9]) [9]
G7D <2> (<2> mutant enzyme with no esterase and reductase activity [9]) [9]
H84A <10> (<10> no GluTR reductase activity, 5% of the GluTR esterase activity compared to wild-type enzyme [8]) [8]
H84N <10> (<10> 30% of the GluTR reductase activity compared to wild-type enzyme, 15% of the GluTR esterase activity compared to wild-type enzyme [8]) [8]
I318L/R322G/N454D <3> (<3> mutant enzyme with greatly reduced activity [12]) [12]
I464P <3> (<3> mutant enzyme with greatly reduced activity [12]) [12]
L387H/L302S <3> (<3> mutant enzyme with greatly reduced activity [12]) [12]
M122K/K154N/F371L/E400K <3> (<3> mutant enzyme with greatly reduced activity [12]) [12]
R3^{14}C <2> (<2> mutant enzyme with no esterase and reductase activity [9]) [9]
S145F <2> (<2> mutant enzyme with no esterase and reductase activity [9]) [9]
S22L/S164F <2> (<2> mutant enzyme with no esterase and reductase activity [9]) [9]
Additional information <3, 5, 12> (<3> a 30 amino acid N-terminal deletion has no detrimental effect on the catalytic activity of the enzyme [12]; <5,12> the knockout mutant is absolutely dependent on supplementation with 5-aminolevulinic acid [19]) [12, 19]

6 Stability

Storage stability

<2>, -80°C, 6*His-tagged enzyme is stable for at least 6 months [16]

References

[1] Willows, R.D.; Kannangara, C.G.; Pontoppidan, B.: Nucleotides of tRNA(-Glu) involved in recognition by barley chloroplast glutamyl-tRNA synthetase and glutamyl-tRNA reductase. Biochim. Biophys. Acta, **1263**, 228-234 (1995)

[2] Loida, P.J.; Thompson, R.L.; Walker, D.M.; CaJacob, C.A.: Novel inhibitors of glutamyl-tRNA(Glu) reductase identified through cell-based screening of the heme/chlorophyll biosynthetic pathway. Arch. Biochem. Biophys., **372**, 230-237 (1999)

[3] Moser, J.; Schubert, W.D.; Heinz, D.W.; Jahn, D.: Structure and function of glutamyl-tRNA reductase involved in 5-aminolaevulinic acid formation. Biochem. Soc. Trans., **30**, 579-584 (2002)

[4] Pontoppida, B.; Kannangara, C.G.: Purification and partial characterization of barley glutamyl-$tRNA^{Glu}$ reductase,the enzyme that directs glutamate to chlorophyll biosynthesis. Eur. J. Biochem., **225**, 529-537 (1994)

[5] Jahn, D.: Complex formation between glutamyl-tRNA synthetase and glutamyl-tRNA reductase during the tRNA-dependent synthesis of 5-aminolevulinic acid in Chlamydomonas reinhardtii. FEBS Lett., **314**, 77-80 (1992)

[6] Chen, M.W.; Jahn, D.; O'Neill, G.P.; Soll, D.: Purification of the glutamyl-tRNA reductase from Chlamydomonas reinhardtii involved in δ-aminolevulinic acid formation during chlorophyll biosynthesis. J. Biol. Chem., **265**, 4058-4063 (1990)

[7] Rieble, S.; Beale, S.I.: Purification of glutamyl-tRNA reductase from Synechocystis sp. PCC 6803. J. Biol. Chem., **266**, 9740-9745 (1991)

[8] Moser, J.; Lorenz, S.; Hubschwerlen, C.; Rompf, A.; Jahn, D.: Methanopyrus kandleri glutamyl-tRNA reductase. J. Biol. Chem., **274**, 30679-30685 (1999)

[9] Schauer, S.; Chaturvedi, S.; Randau, L.; Moser, J.; Kitabatake, M.; Lorenz, S.; Verkamp, E.; Schubert, W.D.; Nakayashiki, T.; Murai, M.; Wall, K.; Thomann, H.U.; Heinz, D.W.; Inokuchi, H.; Soll, D.; Jahn, D.: Escherichia coli glutamyl-tRNA reductase. Trapping the thioester intermediate. J. Biol. Chem., **277**, 48657-48663 (2002)

[10] Randau, L.; Schauer, S.; Ambrogelly, A.; Salazar, J.C.; Moser, J.; Sekine, S.; Yokoyama, S.; Soll, D.; Jahn, D.: tRNA recognition by glutamyl-tRNA reductase. J. Biol. Chem., **279**, 34931-34937 (2004)

[11] Schubert, W.-D.; Moser, J.; Schauer, S.; Heinz, D.W.; Jahn, D.: Structure and function of glutamyl-tRNA reductase, the first enzyme of tetrapyrrole biosynthesis in plants and prokaryotes. Photosynth. Res., **74**, 205-215 (c) (2002)

[12] Vothknecht, U.C.; Kannangara, C.G.; von Wettstein, D.: Barley glutamyl $tRNA^{Glu}$ reductase: mutations affecting haem inhibition and enzyme activity. Phytochemistry, **47**, 513-519 (1998)
[13] Masuda, T.; Ohta, H.; Shioi, Y.; Tsuji, H.; Takamiya, K.-i.: Stimulation of glutamyl-tRNA reductase activity by benzyladenine in greening cucumber cotyledons. Plant Cell Physiol., **36**, 1237-1243 (1995)
[14] Masuda, T.; Ohta, H.; Shioi, Y.; Takamiya, K.-i.: Light regulation of 5-aminolevulinic acid-synthesis system in Cucumis sativus: light stimulates activity of glutamyl-tRNA reductase during greening. Plant Physiol. Biochem., **34**, 11-16 (1996)
[15] Vothknecht, U.C.; Kannangara, C.G.; von Wettstein, D.: Expression of catalytically active barley glutamyl $tRNA^{Glu}$ reductase in Escherichia coli as a fusion protein with glutathione S-transferase. Proc. Natl. Acad. Sci. USA, **93**, 9287-9291 (1996)
[16] Schauer, S.; Luer, C.; Moser, J.: Large scale production of biologically active Escherichia coli glutamyl-tRNA reductase from inclusion bodies. Protein Expr. Purif., **31**, 271-275 (2003)
[17] Levican, G.; Katz, A.; Valenzuela, P.; Soell, D.; Orellana, O.: A tRNA(Glu) that uncouples protein and tetrapyrrole biosynthesis. FEBS Lett., **579**, 6383-6387 (2005)
[18] Srivastava, A.; Beale, S.I.: Glutamyl-tRNA reductase of Chlorobium vibrioforme is a dissociable homodimer that contains one tightly bound heme per subunit. J. Bacteriol., **187**, 4444-4450 (2005)
[19] Petricek, M.; Petrickova, K.; Havlicek, L.; Felsberg, J.: Occurrence of two 5-aminolevulinate biosynthetic pathways in Streptomyces nodosus subsp. asukaensis is linked with the production of asukamycin. J. Bacteriol., **188**, 5113-5123 (2006)
[20] Nogaj, L.A.; Beale, S.I.: Physical and kinetic interactions between glutamyl-tRNA reductase and glutamate-1-semialdehyde aminotransferase of Chlamydomonas reinhardtii. J. Biol. Chem., **280**, 24301-24307 (2005)
[21] Srivastava, A.; Lake, V.; Nogaj, L.A.; Mayer, S.M.; Willows, R.D.; Beale, S.I.: The Chlamydomonas reinhardtii gtr gene encoding the tetrapyrrole biosynthetic enzyme glutamyl-trna reductase: structure of the gene and properties of the expressed enzyme. Plant Mol. Biol., **58**, 643-658 (2005)
[22] Nogaj, L.A.; Srivastava, A.; van Lis, R.; Beale, S.I.: Cellular levels of glutamyl-tRNA reductase and glutamate-1-semialdehyde aminotransferase do not control chlorophyll synthesis in Chlamydomonas reinhardtii. Plant Physiol., **139**, 389-396 (2005)
[23] Yaronskaya, E.; Vershilovskaya, I.; Poers, Y.; Alawady, A.E.; Averina, N.; Grimm, B.: Cytokinin effects on tetrapyrrole biosynthesis and photosynthetic activity in barley seedlings. Planta, **224**, 700-709 (2006)

Succinylglutamate-semialdehyde dehydrogenase 1.2.1.71

1 Nomenclature

EC number
1.2.1.71

Systematic name
N-succinyl-L-glutamate 5-semialdehyde:NAD^+ oxidoreductase

Recommended name
succinylglutamate-semialdehyde dehydrogenase

Synonyms
AstD <1> [3]
N-succinylglutamate 5-semialdehyde dehydrogenase <1> [3]
SGSD <1> [3]
aruD <1, 2> [1, 3]
succinic semialdehyde dehydrogenase <1> [3]
succinylglutamic semialdehyde dehydrogenase <1> [3]

CAS registry number
201426-13-3

2 Source Organism

<1> *Escherichia coli* (no sequence specified) [3]
<2> *Pseudomonas aeruginosa* (no sequence specified) [1, 2]
<3> *Klebsiella aerogenes* (no sequence specified) [5]
<4> *Pseudomonas cepacia* (no sequence specified) [4, 5]

3 Reaction and Specificity

Catalyzed reaction
N-succinyl-L-glutamate 5-semialdehyde + NAD^+ + H_2O = N-succinyl-L-glutamate + NADH + 2 H^+

Reaction type
oxidation
redox reaction
reduction

Natural substrates and products

- **S** N-succinyl-L-glutamate 5-semialdehyde + NAD^+ + H_2O <1, 4> (<1,4> fourth enzyme of arginine succinyltransferase pathway [3,4,5]) (Reversibility: ?) [3, 4, 5]
- **P** N-succinyl-L-glutamate + NADH + H^+
- **S** N-succinyl-L-glutamate 5-semialdehyde + NAD^+ + H_2O <3> (<3> fourth enzyme of arginine succinyltransferase pathway [5]) (Reversibility: ?) [5]
- **P** N-succinyl-L-glutamate + NADH + 2 H^+
- **S** Additional information <2> (<2> fourth enzyme of the arginine succinyltransferase pathway [1]) (Reversibility: ?) [1]
- **P** ?

Substrates and products

- **S** N-succinyl-L-glutamate 5-semialdehyde + NAD^+ + H_2O <1, 4> (<1,4> fourth enzyme of arginine succinyltransferase pathway [3,4,5]) (Reversibility: ?) [3, 4, 5]
- **P** N-succinyl-L-glutamate + NADH + H^+
- **S** N-succinyl-L-glutamate 5-semialdehyde + NAD^+ + H_2O <3> (<3> fourth enzyme of arginine succinyltransferase pathway [5]) (Reversibility: ?) [5]
- **P** N-succinyl-L-glutamate + NADH + 2 H^+
- **S** Additional information <2> (<2> fourth enzyme of the arginine succinyltransferase pathway [1]) (Reversibility: ?) [1]
- **P** ?

Cofactors/prosthetic groups

NAD^+ <1> [3]

References

[1] Vander Wauven, C.; Jann, A.; Haas, D.; Leisinger, T.; Stalon, V.: N^2-Succinylornithine in ornithine catabolism of Pseudomonas aeruginosa. Arch. Microbiol., **150**, 400-404 (1988)

[2] Tricot, C.; Vander Wauven, C.; Wattiez, R.; Falmagne, P.; Stalon, V.: Purification and properties of a succinyltransferase from Pseudomonas aeruginosa specific for both arginine and ornithine. Eur. J. Biochem., **224**, 853-861 (1994)

[3] Schneider, B.L.; Kiupakis, A.K.; Reitzer, L.J.: Arginine catabolism and the arginine succinyltransferase pathway in Escherichia coli. J. Bacteriol., **180**, 4278-4286 (1998)

[4] Vander Wauven, C.; Stalon, V.: Occurrence of succinyl derivatives in the catabolism of arginine in Pseudomonas cepacia. J. Bacteriol., **164**, 882-886 (1985)

[5] Cunin, R.; Glansdorff, N.; Pierard, A.; Stalon, V.: Biosynthesis and metabolism of arginine in bacteria. Microbiol. Rev., **50**, 314-352 (1986)

Erythrose-4-phosphate dehydrogenase 1.2.1.72

1 Nomenclature

EC number
1.2.1.72

Systematic name
D-erythrose 4-phosphate:NAD$^+$ oxidoreductase

Recommended name
erythrose-4-phosphate dehydrogenase

Synonyms
E4P dehydrogenase <1> [6]
E4PDH <1> [1, 6]
Epd dehydrogenase <1> [4, 6]
GapB <1> (<1> renamed as epd [1]) [1, 3, 6]
GapB-encoded protein <1> [5]
epd <1> [1]
erythrose 4-phosphate dehydrogenase <1> [6]
gap2 <1> (<1> renamed as epd [1]) [1]
gapB-encoded dehydrogenase <1> (<1> renamed as epd [1]) [1]

CAS registry number
131554-04-6

2 Source Organism

<1> *Escherichia coli* (no sequence specified) (<1> isozyme SMO2-2 [1,4]) [1, 3, 4, 5, 6]
<2> *Vibrio cholerae* (no sequence specified) [2]

3 Reaction and Specificity

Catalyzed reaction
D-erythrose 4-phosphate + NAD$^+$ + H_2O = 4-phosphoerythronate + NADH + 2 H$^+$

Reaction type
oxidation
redox reaction
reduction

Natural substrates and products

S D-erythrose 4-phosphate + NAD^+ <1> (<1> the enzyme possible plays a role in the de novo biosynthesis of pyridoxal 5-phosphate [1]) (Reversibility: ?) [1]

P 4-phosphoerythronate + NADH

S Additional information <1, 2> (<1> enzyme is involved in pyridoxal 5'-phosphate coenzyme biosynthesis in Escherichia coli K-12 [4]; <1> epd expression is very low in Escherichia coli. In presence of glucose, the 3 epd, pgk and fba ORFs are efficiently cotranscribed from promoter epd P0. Translational limitation of the epd expression in Escherichia coli [6]; <2> expression of epd in Vibrio cholera is not regulated by iron, nor is it required for virulence in an infant model [2]; <1> gapB is dispensable for glycolysis and pyridoxal biosynthesis pathway [3]) (Reversibility: ?) [2, 3, 4, 6]

P ?

Substrates and products

S D-erythrose 4-phosphate + 3-acetylpyridine adenine dinucleotide <1> (Reversibility: ?) [1]

P 4-phosphoerythronate + ?

S D-erythrose 4-phosphate + NAD^+ <1> (<1> the enzyme possible plays a role in the de novo biosynthesis of pyridoxal 5'-phosphate [1]; <1> the chemical mechanism of erythrose 4-phosphate oxidation by gap-encoded protein proceeds through a two-step mechanism involving covalent intermediates with Cys149, with rates associated to the acylation and deacylation process of 280 per s and 20 per s, respectively. No isotopic solvent effect is observed, suggesting that the rate-limiting step is not hydrolysis [5]) (Reversibility: ?) [1, 5]

P 4-phosphoerythronate + NADH

S Additional information <1, 2> (<1> enzyme is involved in pyridoxal 5'-phosphate coenzyme biosynthesis in Escherichia coli K-12 [4]; <1> epd expression is very low in Escherichia coli. In presence of glucose, the 3 epd, pgk and fba ORFs are efficiently cotranscribed from promoter epd P0. Translational limitation of the epd expression in Escherichia coli [6]; <2> expression of epd in Vibrio cholera is not regulated by iron, nor is it required for virulence in an infant model [2]; <1> gapB is dispensable for glycolysis and pyridoxal biosynthesis pathway [3]; <1> the enzyme also shows low phosphorylating glyceraldehyde-3-phosphate dehydrogenase activity [5]) (Reversibility: ?) [2, 3, 4, 5, 6]

P ?

Cofactors/prosthetic groups

3-acetylpyridine adenine dinucleotide <1> (<1> less efficient than NAD^+ [1]) [1]

NAD^+ <1> [1]

Activating compounds

Additional information <1> (<1> the enzyme requires the presence of a reducing agent, such as DTT or 2-mercaptoethanol, to maintain activity [1]) [1]

Turnover number (min^{-1})

0.21 <1> (D-erythrose 4-phosphate, <1> 25°C, mutant enzyme H176N [5]) [5]
0.21 <1> (NAD^+, <1> 25°C, mutant enzyme H176N [5]) [5]
0.6 <1> (D-erythrose 4-phosphate, <1> 25°C, mutant enzyme C311Y [5]) [5]
0.6 <1> (NAD^+, <1> 25°C, mutant enzyme C311Y [5]) [5]
3.5 <1> (D-erythrose 4-phosphate, <1> 25°C, mutant enzyme C311A [5]) [5]
3.5 <1> (NAD^+, <1> 25°C, mutant enzyme C311A [5]) [5]
4 <1> (D-erythrose 4-phosphate, <1> 25°C, mutant enzyme M179T [5]) [5]
4 <1> (NAD^+, <1> 25°C, mutant enzyme M179T [5]) [5]
7.4 <1> (D-erythrose 4-phosphate, <1> 25°C, mutant enzyme E32D [5]) [5]
7.4 <1> (NAD^+, <1> 25°C, mutant enzyme E32D [5]) [5]
20 <1> (D-erythrose 4-phosphate, <1> 25°C, wild-type enzyme [5]) [5]
20 <1> (NAD^+, <1> 25°C, wild-type enzyme [5]) [5]
169 <1> (NAD^+) [1]
200 <1> (D-erythrose 4-phosphate) [1]

Specific activity (U/mg)

14.7 <1> [5]

K_m-Value (mM)

0.074 <1> (NAD^+) [1]
0.24 <1> (D-erythrose 4-phosphate, <1> 25°C, mutant enzyme H176N [5]) [5]
0.5 <1> (NAD^+, <1> 25°C, mutant enzyme H176N [5]) [5]
0.51 <1> (D-erythrose 4-phosphate, <1> 25°C, wild-type enzyme [5]) [5]
0.7 <1> (D-erythrose 4-phosphate, <1> 25°C, mutant enzyme C311Y [5]; <1> 25°C, mutant enzyme M179T [5]) [5]
0.8 <1> (D-erythrose 4-phosphate, <1> 25°C, mutant enzyme C311A [5]) [5]
0.8 <1> (NAD^+, <1> 25°C, wild-type enzyme [5]) [5]
0.96 <1> (D-erythrose 4-phosphate) [1]
1.9 <1> (D-erythrose 4-phosphate, <1> 25°C, mutant enzyme E32D [5]) [5]
2.3 <1> (NAD^+, <1> 25°C, mutant enzyme E32D [5]) [5]
3.5 <1> (NAD^+, <1> 25°C, mutant enzyme M179T [5]) [5]
4 <1> (NAD^+, <1> 25°C, mutant enzyme C311A [5]) [5]
5 <1> (NAD^+, <1> 25°C, mutant enzyme C311Y [5]) [5]

pH-Optimum

7-10 <1> (<1> the pH profile rises steeply between pH 7 and 9 and then drops slightly at pH 10 [1]) [1]

Temperature optimum (°C)

50 <1> (<1> at pH 8.6 [1]) [1]

4 Enzyme Structure

Molecular weight

132000 <1> (<1> gel filtration [1]) [1]

Subunits

? <1> (<1> x * 37170, mass spectrometry [5]) [5]
tetramer <1> (<1> 4 * 37200, SDS-PAGE [1]) [1]

5 Isolation/Preparation/Mutation/Application

Purification

<1> [1]
<1> (recombinant) [5]

Cloning

<1> [1]

Engineering

C149A <1> (<1> no significant phosphorylating glyceraldehyde-3-phosphate dehydrogenase activity [5]) [5]
C149G <1> (<1> no significant phosphorylating glyceraldehyde-3-phosphate dehydrogenase activity [5]) [5]
C149V <1> (<1> no significant phosphorylating glyceraldehyde-3-phosphate dehydrogenase activity [5]) [5]
C311A <1> (<1> mutation does not drastically change the phosphorylating glyceraldehyde-3-phosphate dehydrogenase activity. 5.7fold decrease in turnover number for the activity with D-erythrose 4-phosphate and NAD^+ [5]) [5]
C311Y <1> (<1> mutation does not drastically change the phosphorylating glyceraldehyde-3-phosphate dehydrogenase activity. 33.3fold decrease in turnover number for the activity with D-erythrose 4-phosphate and NAD^+ [5]) [5]
E32D <1> (<1> mutation does not drastically change the phosphorylating glyceraldehyde-3-phosphate dehydrogenase activity. 2.7fold decrease in turnover number for the activity with D-erythrose 4-phosphate and NAD^+ [5]) [5]
H176N <1> (<1> turnover number of phosphorylating glyceraldehyde-3-phosphate dehydrogenase activity is decreased by the factor 40. 95.2fold decrease in turnover number for the activity with D-erythrose 4-phosphate and NAD^+ [5]) [5]
M179T <1> (<1> mutation does not drastically change the phosphorylating glyceraldehyde-3-phosphate dehydrogenase activity. 5fold decrease in turnover number for the activity with D-erythrose 4-phosphate and NAD^+ [5]) [5]

6 Stability

Temperature stability

50 <1> (<1> stable below [1]) [1]

Storage stability

<1>, 4°C, stable for weeks with negligible loss of activity [1]

References

[1] Zhao, G.; Pease, A.J.; Bharani, N.; Winkler, M.E.: Biochemical characterization of gapB-encoded erythrose 4-phosphate dehydrogenase of Escherichia coli K-12 and its possible role in pyridoxal 5'-phosphate biosynthesis. J. Bacteriol., **177**, 2804-2812 (1995)

[2] Carroll, P.A.; Zhao, G.; Boyko, S.A.; Winkler, M.E.; Calderwood, S.B.: Identification, sequencing, and enzymatic activity of the erythrose-4-phosphate dehydrogenase gene of Vibrio cholerae. J. Bacteriol., **179**, 293-296 (1997)

[3] Seta, F.D.; Boschi-Muller, S.; Vignais, M.L.; Branlant, G.: Characterization of Escherichia coli strains with gapA and gapB genes deleted. J. Bacteriol., **179**, 5218-5221 (1997)

[4] Yang, Y.; Zhao, G.; Man, T.K.; Winkler, M.E.: Involvement of the gapA- and epd (gapB)-encoded dehydrogenases in pyridoxal 5'-phosphate coenzyme biosynthesis in Escherichia coli K-12. J. Bacteriol., **180**, 4294-4299 (1998)

[5] Boschi-Muller, S.; Azza, S.; Pollastro, D.; Corbier, C.; Branlant, G.: Comparative enzymatic properties of GapB-encoded erythrose-4-phosphate dehydrogenase of Escherichia coli and phosphorylating glyceraldehyde-3-phosphate dehydrogenase. J. Biol. Chem., **272**, 15106-15112 (1997)

[6] Bardey, V.; Vallet, C.; Robas, N.; Charpentier, B.; Thouvenot, B.; Mougin, A.; Hajnsdorf, E.; Regnier, P.; Springer, M.; Branlant, C.: Characterization of the molecular mechanisms involved in the differential production of erythrose-4-phosphate dehydrogenase, 3-phosphoglycerate kinase and class II fructose-1,6-bisphosphate aldolase in Escherichia coli. Mol. Microbiol., **57**, 1265-1287 (2005)

Abscisic-aldehyde oxidase 1.2.3.14

1 Nomenclature

EC number
1.2.3.14

Systematic name
abscisic-aldehyde:oxygen oxidoreductase

Recommended name
abscisic-aldehyde oxidase

Synonyms
AAO3 <2, 5> [4, 6]
ABA aldehyde oxidase <3, 4> [2, 8]
AO-3 <1> [3]
AOd <2> [4]
AOδ <2> [1]
Arabidopsis aldehyde oxidase 3 <5> [6]
abscisic aldehyde oxidase <2> [4]
abscisic aldehyde oxidase 3 <2> [7]

CAS registry number
129204-36-0
9029-07-6

2 Source Organism

<1> *Pisum sativum* (no sequence specified) [3]
<2> *Arabidopsis sp.* (no sequence specified) [1, 4, 7]
<3> *Lycopersicon esculentum* (no sequence specified) [5, 8]
<4> *Nicotiana plumbaginifolia* (no sequence specified) [2]
<5> *Arabidopsis thaliana* (UNIPROT accession number: Q7G9P4) [6]

3 Reaction and Specificity

Catalyzed reaction
abscisic aldehyde + H_2O + O_2 = abscisate + H_2O_2

Reaction type
oxidation
redox reaction
reduction

Natural substrates and products

S abscisic aldehyde + H_2O + O_2 <2, 5> (<2> the AAO3 gene product plays a major role in abscisic acid biosynthesis in seed [7]; <2> the enzyme catalyzes the final step of abscisic acid biosynthesis, specifically in rosette leaves [1]; <2> the enzyme catalyzes the final step of abscisic acid biosynthesis. AAO3 is the AAO that plays a major role in abscisic acid biosynthesis in seeds as well as in leaves [4]; <5> the enzyme catalyzes the final step of abscisic acid biosynthesis. NO detectable activity in guard cells of nonstressed rosette or wet-control leaves [6]) (Reversibility: ?) [1, 4, 6, 7]

P abscisate + H_2O_2

Substrates and products

S abscisic aldehyde + H_2O + O_2 <2, 5> (<2> the AAO3 gene product plays a major role in abscisic acid biosynthesis in seed [7]; <2> the enzyme catalyzes the final step of abscisic acid biosynthesis, specifically in rosette leaves [1]; <2> the enzyme catalyzes the final step of abscisic acid biosynthesis. AAO3 is the AAO that plays a major role in abscisic acid biosynthesis in seeds as well as in leaves [4]; <5> the enzyme catalyzes the final step of abscisic acid biosynthesis. NO detectable activity in guard cells of nonstressed rosette or wet-control leaves [6]) (Reversibility: ?) [1, 4, 6, 7]

P abscisate + H_2O_2

Metals, ions

Co <3> (<3> MoCo-containing enzyme [5]) [5]
Mo <3> (<3> molybdoenzyme [8]; <3> MoCo-containing enzyme [5]) [5, 8]

5 Isolation/Preparation/Mutation/Application

Source/tissue

guard cell <5> (<5> AAO3 mRNA expression in guard cells of dehydrated rosette leaves [6]) [6]
leaf <1, 2, 5> (<5> AAO3 mRNA expression in guard cells of dehydrated rosette leaves [6]; <2> rosette [1]) [1, 3, 4, 6]
root <3> [5]
seed <2> [4]
shoot <3> [5]

References

[1] Seo, M.; Peeters, A.J.M.; Koiwai, H.; Oritani, T.; Marion-Poll, A.; Zeevaart, J.A.D.; Koornneef, M.; Kamiya, Y.; Koshiba, T.: The Arabidopsis aldehyde oxidase 3 (AAO3) gene product catalyzes the final step in abscisic acid biosynthesis in leaves. Proc. Natl. Acad. Sci. USA, **97**, 12908-12913 (2000)

[2] Leydecker, M.T.; Moureaux, T.; Kraepiel, Y.; Schnorr, K.; Caboche, M.: Molybdenum cofactor mutants, specifically impaired in xanthine dehydrogenase

activity and abscisic acid biosynthesis, simultaneously overexpress nitrate reductase. Plant Physiol., **107**, 1427-1431 (1995)

[3] Zdunek, E.; Lips, S.H.: Transport and accumulation rates of abscisic acid and aldehyde oxidase activity in Pisum sativum L. in response to suboptimal growth conditions. J. Exp. Bot., **52**, 1269-1276 (2001)

[4] Seo, M.; Aoki, H.; Koiwai, H.; Kamiya, Y.; Nambara, E.; Koshiba, T.: Comparative studies on the Arabidopsis aldehyde oxidase (AAO) gene family revealed a major role of AAO3 in ABA biosynthesis in seeds. Plant Cell Physiol., **45**, 1694-1703 (2004)

[5] Sagi, M.; Fluhr, R.; Lips, S.H.: Aldehyde oxidase and xanthine dehydrogenase in a flacca tomato mutant with deficient abscisic acid and wilty phenotype. Plant Physiol., **120**, 571-577 (1999)

[6] Koiwai, H.; Nakaminami, K.; Seo, M.; Mitsuhashi, W.; Toyomasu, T.; Koshiba, T.: Tissue-specific localization of an abscisic acid biosynthetic enzyme, AAO3, in Arabidopsis. Plant Physiol., **134**, 1697-1707 (2004)

[7] Gonzalez-Guzman, M.; Abia, D.; Salinas, J.; Serrano, R.; Rodriguez, P.L.: Two new alleles of the abscisic aldehyde oxidase 3 gene reveal its role in abscisic acid biosynthesis in seeds. Plant Physiol., **135**, 325-333 (2004)

[8] Marin, E.; Marion-Poll, A.: Tomato flacca mutant is impaired in ABA aldehyde oxidase and xanthine dehydrogenase activities. Plant Physiol. Biochem., **35**, 369-372 (1997)

Carbon-monoxide dehydrogenase (ferredoxin) 1.2.7.4

1 Nomenclature

EC number
1.2.7.4

Systematic name
carbon-monoxide,water:ferredoxin oxidoreductase

Recommended name
carbon-monoxide dehydrogenase (ferredoxin)

Synonyms
ACS/CODH <7> [18]
ACS/CODH Mt <7> [18]
CH_3-CO dehydrogenase <7> [10]
CO oxidation/H_2 evolution system <4> [12]
CO:MB oxidoreductase <5> [7]
CODH <4, 7> [10, 12, 13, 14, 15]
CODH/ACS <7> [6, 16, 17]
$\alpha_2\beta_2$acetyl-coenzyme A synthase/carbon monoxide dehydrogenase <7> [18]
carbon monoxide dehydrogenase/acetyl-CoA synthase <7> [6, 16, 17]
carbon monoxide:methylene blue oxidoreductase <5> [7]

CAS registry number
64972-88-9

2 Source Organism

<1> *Methanosarcina barkeri* (no sequence specified) [1]
<2> *Methanothrix soehngenii* (no sequence specified) [11]
<3> *Streptomyces sp.* (no sequence specified) [9]
<4> *Rhodospirillum rubrum* (no sequence specified) [12]
<5> *Pseudomonas carboxydovorans* (no sequence specified) [7]
<6> *Methanosarcina thermophila* (no sequence specified) [5]
<7> *Moorella thermoacetica* (no sequence specified) [2, 3, 4, 6, 8, 10, 13, 14, 15, 16, 17, 18]

3 Reaction and Specificity

Catalyzed reaction

$CO + H_2O$ + oxidized ferredoxin = CO_2 + reduced ferredoxin

Natural substrates and products

S $CO + H_2O$ + oxidized cytochrome b <7> (<7> membrane-bound b-type, native electron carrier [3]) (Reversibility: ?) [3]
P CO_2 + reduced cytochrome b
S $CO + H_2O$ + oxidized ferredoxin <1, 3, 4, 5, 6, 7> (<7> key enzyme of the acetyl-CoA pathway of autotrophic growth [10]; <7> also used in degradation of acetyl-CoA to form methane and CO_2 [17]; <7> acetate biosynthesis pathway, catalyzes an exchange reaction between CO and the carbonyl group of acetyl-CoA [4]; <7> involved inacetyl-CoA synthesis complexed with corrinoid/Fe-S protein [15]; <1> may participate in methanogenesis by cleavage of acetate, reverse of the reaction in acetate biosynthesis [1]; <7> key enzyme in the acetyl-CoA pathway, catalyzes not only the oxidation of CO to CO_2, but also the final step, synthesis of acetyl-CoA from a methyl group, CO, and CoA [14,17,18]; <7> initial step of CO metabolism in acetogenic bacteria [2]; <7> function of the CO dehydrogenase is to reduce the cobalt of the corrinoid enzyme to Co^{2+}, which is required for it to act as a methyl acceptor, essential for acetyl-CoA synthesis with CO as the substrate [8]) (Reversibility: ?) [1, 2, 3, 4, 5, 6, 7, 8, 9, 10, 12, 13, 14, 15, 16, 17, 18]
P CO_2 + reduced ferredoxin
S $CO + H_2O$ + oxidized ubiquinone Q_{10} <5> (<5> physiological electron acceptor [7]) (Reversibility: ?) [7]
P CO_2 + reduced ubiquinone Q_{10}

Substrates and products

S $CO + H_2O$ + FMN <1, 7> (Reversibility: ?) [1, 2, 3]
P CO_2 + FAD
S $CO + H_2O$ + oxidized 2,3,5-triphenyltetrazolium chloride <1> (Reversibility: ?) [1]
P CO_2 + reduced 2,3,5-triphenyltetrazolium chloride
S $CO + H_2O$ + oxidized benzyl viologen <3> (Reversibility: ?) [9]
P CO_2 + reduced benzyl viologen
S $CO + H_2O$ + oxidized cytochrome b <7> (<7> membrane-bound b-type, native electron carrier [3]) (Reversibility: ?) [3]
P CO_2 + reduced cytochrome b
S $CO + H_2O$ + oxidized cytochrome c <3> (Reversibility: ?) [9]
P CO_2 + reduced cytochrome c
S $CO + H_2O$ + oxidized cytochrome c_3 <7> (<7> Desulfovibrio vulgaris cytochrome c_3 [3]) (Reversibility: ?) [3]
P CO_2 + reduced cytochrome c_3
S $CO + H_2O$ + oxidized dichlorophenolindophenol <5> (Reversibility: ?) [7]
P CO_2 + reduced dichlorophenolindophenol

S CO + H_2O + oxidized ferredoxin <1, 3, 4, 5, 6, 7> (<7> ferredoxin I and II [2]; <7> ferredoxin from Clostridium pasteurianum is also a substrate [3,14]; <7> key enzyme of the acetyl-CoA pathway of autotrophic growth [10]; <7> also used in degradation of acetyl-CoA to form methane and CO_2 [17]; <7> acetate biosynthesis pathway, catalyzes an exchange reaction between CO and the carbonyl group of acetyl-CoA [4]; <7> involved inacetyl-CoA synthesis complexed with corrinoid/Fe-S protein [15]; <1> may participate in methanogenesis by cleavage of acetate, reverse of the reaction in acetate biosynthesis [1]; <7> key enzyme in the acetyl-CoA pathway, catalyzes not only the oxidation of CO to CO_2, but also the final step, synthesis of acetyl-CoA from a methyl group, CO, and CoA [14,17,18]; <7> initial step of CO metabolism in acetogenic bacteria [2]; <7> function of the CO dehydrogenase is to reduce the cobalt of the corrinoid enzyme to Co^{2+}, which is required for it to act as a methyl acceptor, essential for acetyl-CoA synthesis with CO as the substrate [8]) (Reversibility: ?) [1, 2, 3, 4, 5, 6, 7, 8, 9, 10, 12, 13, 14, 15, 16, 17, 18]

P CO_2 + reduced ferredoxin

S CO + H_2O + oxidized flavodoxin <1, 4, 7> (<4> Acetobacter vinelandii flavodoxin [12]) (Reversibility: ?) [1, 2, 3, 12]

P CO_2 + reduced flavodoxin

S CO + H_2O + oxidized methyl viologen <1, 4, 6, 7> (Reversibility: ?) [1, 2, 3, 5, 12, 13, 14]

P CO_2 + reduced methyl viologen

S CO + H_2O + oxidized methylene blue <3, 5, 7> (Reversibility: ?) [2, 7, 9]

P CO_2 + reduced methylene blue

S CO + H_2O + oxidized phenazine methosulfate <1> (Reversibility: ?) [1]

P CO_2 + reduced phenazine methosulfate

S CO + H_2O + oxidized pyocyanine <5> (<5> perchlorate [7]) (Reversibility: ?) [7]

P CO_2 + reduced pyocyanine

S CO + H_2O + oxidized rubredoxin <7> (<7> most efficient electron acceptor [2]) (Reversibility: ?) [2]

P CO_2 + reduced rubredoxin

S CO + H_2O + oxidized thionine <5> (<5> Lauths violet [7]) (Reversibility: ?) [7]

P CO_2 + reduced thionine

S CO + H_2O + oxidized toluylene blue <5> (Reversibility: ?) [7]

P CO_2 + reduced toluylene blue

S CO + H_2O + oxidized ubiquinone Q_{10} <5> (<5> physiological electron acceptor [7]) (Reversibility: ?) [7]

P CO_2 + reduced ubiquinone Q_{10}

S CO + H_2O + oxidized viologen <7> (Reversibility: ?) [2]

P CO_2 + reduced viologen

S Additional information <1, 3, 5, 6, 7> (<7> pure enzyme has no hydrogenase or formate dehydrogenase activity [2]; <5> does not catalyze reduction of pyridine or flavin nucleotides, incapable of oxidizing methane, methanol, or formaldehyde in presence of methylene blue as electron acceptor, methyl viologen, benzyl viologen, $NADP^+$, NAD^+, neutral red, FAD^+, FMN and ferricyanide are no electron acceptors [7]; <1> enzyme fails to reduce NAD^+, $NADP^+$ or the 8-hydroxy-5-deazaflavin factor F_{420} [1]; <7> no complex formation between CODH and ferredoxins from Spinacia oleracea and Spirulina platensis [14]; <7> CODH contains binding sites for the methyl, carbonyl, and CoA moieties of acetyl-CoA and catalyzes the assembly of acetyl-CoA from these enzyme-bound groups, a more appropriate name for CODH is "acetyl-CoA synthase" [13,17]; <7> spinach ferredoxin, FAD, NAD^+ and $NADP^+$ are not reduced [3]; <3> NAD^+, $NADP^+$ and clostridial ferredoxin are not reduced [9]; <6> coenzyme F420 does not substitute for ferredoxin, artificial electron acceptor metronidazole is not reduced by CO dehydrogenase complex alone, but ferredoxin coupled CO oxidation by CO dehydrogenase to the reduction of metronidazole, which provided a quantitative and convenient assay [5]) (Reversibility: ?) [1, 2, 3, 5, 7, 9, 13, 14, 17]

P ?

Inhibitors

2,3-butanedione <1> [1]
5,5'-dithiobis (2-nitrobenzoate) <7> [4]
glyoxaldehyde <1> (<1> only with CO and methyl viologen as substrates [1]) [1]
KCN <1, 5, 7> (<7> CO reverses cyanide inhibition, but promotes reaction with methyl iodide [2,3]) [1, 2, 3, 7]
mersalyl acid <7> [4]
methyl iodide <7> (<7> irreversible inhibition [2]) [2, 4]
NaN_3 <5> [7]
O_2 <1> [1]
Additional information <1, 5, 7> (<7> not affected by propyl iodide, methyl iodide, carbon tetrachloride,or metal chelators [3]; <7> pyruvate and ATP have no effect [2]; <7> phenylglyoxal, methylglyoxal, butanedione, mersalyl acid, methyl iodide, 5,5-dithiobis (2-nitrobenzoate) and sodium dithionite inhibits the exchange reaction between CO and acetyl-CoA [4]; <5> activity is not or only scarely effected by hypophosphite, chlorate and fluoride, formate, bicarbonate, $MgCl_2$ and molecular hydrogen have no effect on activity [7]; <1> dimethylglyoxime is no inhibitor, formaldehyde and acetaldehyde does not inactivate the enzyme [1]; <7> inhibition of the methylation of CO dehydrogenase with CH_3I or CH_3-corrinoid enzyme by *p*-chloromercuribenzoate, $Na_2S_2O_4$, dithioerythritol, oxygen and 5,5-dimethyl-1-pyrroline-1-oxide [10]) [1, 2, 3, 4, 7, 10]

Cofactors/prosthetic groups

FAD <3> [9]

Metals, ions

Co^{2+} <6> [5]
Cu^{2+} <7> (<7> Ni-Fe-Cu center as metallocofactor [17]) [17]

Fe^{2+} <1, 3, 4, 6, 7> (<1> tightly bound by the enzyme [1]; <4,6,7> Ni-and FeS-containg center, known as cluster C [5,6,12,16]) [1, 2, 5, 6, 8, 9, 12, 15, 16, 17]
Mo3+ <3> (<3> molybdenum hydroxylase [9]) [9]
Ni^{2+} <1, 4, 6, 7> (<1> tightly bound by the enzyme [1]; <4,6,7> Ni-and FeS-containg center, known as cluster C [5,6,12]) [1, 2, 3, 5, 6, 8, 12, 17, 18]
Zn^{2+} <6, 7> (<7> α subunit contains 1 Zn^{2+} ion and 1 Ni^{2+} ion [18]) [5, 8, 18]
Additional information <1> (<1> no stimulation by $MgSO_4$, $CoCl_2$, $NiCl_2$ and $ZnSO_4$ [1]) [1]

Turnover number (min^{-1})

Additional information <7> (<7> pH 5.3, 50°C, activity of CO oxidation to CO_2, ferredoxin II as electron carrier, turnover number is approximately 780 mol/mol of CO dehydrogenase [4]) [4]

Specific activity (U/mg)

0.82 <7> (<7> specific activity of acetyl-CoA synthesis [13]) [13]
1.94 <5> [7]
25.7 <3> [9]
133.6 <1> [1]

K_m-Value (mM)

0.053 <5> (CO, <5> pH 7.0, 30°C [7]) [7]
0.5 <1> (2,3,5-triphenyltetrazolium chloride, <1> pH 7.0, 25°C [1]) [1]
3.03 <7> (methyl viologen, <7> pH 8.4, 50°C [2]) [2]
3.4 <4> (methyl viologen, <4> pH 7.5, 25°C [12]) [12]
7.1 <1> (methyl viologen, <1> pH 7.0, 25°C [1]) [1]

pH-Optimum

5.3 <7> [4]
6 <7> (<7> complex formation between CODH and ferredoxin [14]) [14]
6.8-7 <5> (<5> methylene blue reduction [7]) [7]
7 <5> [7]
7-9 <1> [1]
8.2-8.4 <7> [2]
10.6 <7> (<7> metyl viologen as substrate [3]) [3]

pH-Range

5-11 <1> (<1> very little activity outside this range [1]) [1]

Temperature optimum (°C)

63 <5> [7]
70 <3, 7> [3, 9]

Temperature range (°C)

20-80 <5> [7]
30-74 <3> [9]

4 Enzyme Structure

Molecular weight

155000 <7> (<7> sedimentation equilibrium centrifugation [2]) [2]
161000 <1, 7> (<1> gel filtration [1]; <7> pore limit gel electrophoresis [2]) [1, 2]
190000 <2> [11]
205000 <1> (<1> gradient gel electrophoresis [1]) [1]
214000 <5> (<5> calculated from the number of subunits [7]) [7]
230000 <5> (<5> sucrose density gradient centrifugation [7]) [7]
234000 <5> (<5> sedimentation equilibrium [7]) [7]
268000 <3> (<3> gel filtration [9]) [9]
282000 <3> (<3> gel filtration [9]) [9]
310000 <7> (<7> $\alpha_2\beta_2$ [17]) [17]
410000 <7> (<7> gel filtration [3]) [3]
436000 <7> (<7> gel filtration [2]) [2]
440000 <7> (<7> $\alpha_3\beta_3$, gel filtration [14]) [14]

Subunits

dimer <4, 5> (<5> 2 * 107000, SDS-PAGE [7]; <4> 1 * 62000 + 1 * 22000, $\alpha\beta$, SDS-PAGE [12]) [7, 12]
heterotetramer <1, 2, 3, 7> (<7> $\alpha_2\beta_2$ [17,18]; <2> 2 * 79400 + 2 * 19400, SDS-PAGE [11]; <1> 2 * 84500 + 2 * 19700, $\alpha_2\beta_2$, SDS-PAGE [1]; <3> 2 * 11000 + 2 * 33000, $\alpha_2\beta_2$, SDS-PAGE [9]; <2> 2 * 79000 + 2 * 19000, recombinant enzyme, expressed in Escherichia coli or Desulfovibrio vulgaris, antibody reaction [11]) [1, 9, 11, 17, 18]
hexamer <7> (<7> 3 * 78000 + 3 * 71000, $\alpha_3\beta_3$, SDS-PAGE [2,15]; <7> 3 * 77000 + 3 * 71000, $\alpha_3\beta_3$, SDS-PAGE [13]) [2, 13, 15]

5 Isolation/Preparation/Mutation/Application

Source/tissue

mycelium <3> [9]

Localization

membrane <4> [12]

Purification

<1> [1]
<2> [11]
<3> [9]
<4> [12]
<5> [7]
<6> (purified as corrinoid complex) [5]
<7> [2, 3, 4, 8, 13, 14, 15]

Crystallization

<7> (2.2 A resolution crystal structure, coordinates deposited in the Brookhaven Protein Data Bank, accession code 1MJG) [17]
<7> (crystal structure of CO-treated ACS/CODH Mt solved at 1.9 A resolution, monoclinic space group C_2 crystal form with a = 245 A, b = 82 A, c = 167 A, hanging drops in an anaerobic glove box) [18]

Cloning

<2> (genes cdhA and cdhB coding for the large and small subunits of the enzyme isolated from a genomic library of Methanothrix soehngenii DNA in Escherichia coli, introduction and expression of the genes in Escherichia coli DH5α, TG1, Q359, S17-1 or Desulfovibrio vulgaris NCIMB8303 yields immunoreactive proteins, but no enzyme activity in heterologous hosts) [11]
<7> (gene encoding CODH cloned into Escherichia coli) [13]

6 Stability

Temperature stability

25-70 <7> (<7> purification can be done at 25°C without loss of activity, heating for 30 min at 67°C leads to apparent loss, heating for only a few min at 70°C activates the enzyme [2]) [2]
30-74 <3> (<3> rapidly inactivated at temperatures above [9]) [9]
80-104 <1> (<1> most of the activity remains after heating at 80°C, nearly complete inactivation after heating at 104°C, greater loss of activity at all temperatures when heated in presence of CO [1]) [1]

Oxidation stability

<1>, strong inhibition by oxygen, extremely sensitive, exposed to air activity is lost in less than 1 min [1]
<7>, addition of the enzyme to an aerobic buffer, 50 mM Tris/HCl, pH 7.6 results in a 98% loss of activity within 15 min, presence of CO has no apparent effect on the stability of the enzyme [2]
<7>, enzymatic activity is destroyed by air, a low amount of activity still is present after 2 h exposure to air [3]

Storage stability

<1>, -70°C, samples thawed under anaerobic conditions exhibits full activity even after 6 months of storage [1]
<3>, -20°C, purified enzyme stored in 50% ethanediol, little loss of activity is observed over several weeks [9]
<3>, 4°C, purified preparation loses most of its activity on overnight storage [9]
<5>, -20°C stored in 50 mM phosphate buffer, pH 7.0 under air, stable for 6 days, within 18 days 50% of the initial activity remains, addition of 50% glycerol is without effect [7]
<7>, -20°C, stored frozen in an oxygen-free atmosphere in buffer containing glycerol and 2 mM dithionite, not stable for more than 1 month [2]

<7>, 10°C, Tris/HCl, pH 7.6, enzyme is active for at least 2 days [2]
<7>, 5°C, stored in an oxygen-free atmosphere in buffer containing glycerol and 2 mM dithionite, not stable for more than 1 month [2]

References

[1] Grahame, D.A.; Stadtman, T.C.: Carbon monoxide dehydrogenase from Methanosarcina barkeri. Disaggregation, purification, and physicochemical properties of the enzyme. J. Biol. Chem., **262**, 3706-3712 (1987)
[2] Ragsdale, S.W.; Clark, J.E.; Ljungdahl, L.G.; Lundie, L.L.; Drake, H.L.: Properties of purified carbon monoxide dehydrogenase from Clostridium thermoaceticum, a nickel, iron-sulfur protein. J. Biol. Chem., **258**, 2364-2369 (1983)
[3] Drake, H.L.; Hu. S.I.; Wood, H.G.: Purification of carbon monoxide dehydrogenase, a nickel enzyme from Clostridium thermoaceticum. J. Biol. Chem., **255**, 7174-7180 (1980)
[4] Ragsdale, S.W.; Wood, H.G.: Acetate biosynthesis by acetogenic bacteria. Evidence that carbon monoxide dehydrogenase is the condensing enzyme that catalyzes the final steps of the synthesis. J. Biol. Chem., **260**, 3970-3977 (1985)
[5] Terlesky, K.C.; Fery, J.G.: Ferredoxin requirement for electron transport from the carbon monoxide dehydrogenase complex to a membrane-bound hydrogenase in acetate-grown Methanosarcina thermophila. J. Biol. Chem., **263**, 4075-4079 (1988)
[6] Seravalli, J.; Kumar, M.; Lu, W.P.; Ragsdale, S.W.: Mechanism of carbon monoxide oxidation by the carbon monoxide dehydrogenase/acetyl-CoA synthase from Clostridium thermoaceticum: Kinetic characterization of the intermediates. Biochemistry, **36**, 11241-11251 (1997)
[7] Meyer, O.; Schlegel, H.-G.: Carbon monoxide:methylene blue oxidoreductase from Pseudomonas carboxydovorans. J. Bacteriol., **141**, 74-80 (1980)
[8] Pezacka, E.; Wood, H.G.: Role of carbon monoxide dehydrogenase in the autotrophic pathway used by acetogenic bacteria. Proc. Natl. Acad. Sci. USA, **81**, 6261-6265 (1984)
[9] Bell, J.M.; Colby, J.; Williams, E.: Carbon monoxide oxidoreductase from Streptomyces strain G26 is a molybdenum hydroxylase. Biochem. J., **250**, 605-612 (1988)
[10] Pezacka, E.; Wood, H.G.: Acetyl-CoA pathway of autotrophic growth. Identification of the methyl-binding site of the carbon monoxide dehydrogenase. J. Biol. Chem., **263**, 16000-16006 (1988)
[11] Eggen, R.I.L.; Geerling, A.C.M.; Jetten, M.S.M.; De Vos, W.M.: Cloning, expression, and sequence analysis of the genes for carbon monoxide dehydrogenase of Methanothrix soehngenii. J. Biol. Chem., **266**, 6883-6887 (1991)
[12] Ensign, S.A.; Ludden, P.W.: Characterization of the CO oxidation/H_2 evolution system of Rhodospirillum rubrum. Role of a 22-kDa iron-sulfur protein in mediating electron transfer between carbon monoxide dehydrogenase and hydrogenase. J. Biol. Chem., **266**, 18395-18404 (1991)

[13] Roberts, J.R.; Lu, W.-P.; Ragsdale, S.W.: Acetyl-coenzyme A synthesis from methyltetrahydrofolate, CO, and coenzyme A by enzymes purified from Clostridium thermoaceticum: Attainment of in vivo rates and identification of rate-limiting steps. J. Bacteriol., **174**, 4667-4676 (1992)

[14] Shanmugasundaram, T.; Wood, H.G.: Interaction of ferredoxin with carbon monoxide dehydrogenase from Clostridium thermoaceticum. J. Biol. Chem., **267**, 897-900 (1992)

[15] Shanmugasundaram, T.; Sundaresh, C.S.; Kumar, G.K.: Identification of a cysteine involved in the interaction between carbon monoxide dehydrogenase and corrinoid/Fe-S protein from Clostridium thermoaceticum. FEBS Lett., **326**, 281-284 (1993)

[16] Menon, S.; Ragsdale, S.W.: Role of the [4Fe-4S] cluster in reductive activation of the cobalt center of the corrinoid iron-sulfur protein from Clostridium thermoaceticum during acetate biosynthesis. Biochemistry, **37**, 5689-5698 (1998)

[17] Doukov, T.I.; Iverson, T.M.; Seravalli, J.; Ragsdale, S.W.; Drennan, C.L.: A Ni-Fe-Cu center in a bifunctional carbon monoxide dehydrogenase/acetyl-CoA synthase. Science, **298**, 567-572 (2002)

[18] Darnault, C.; Volbeda, A.; Kim, E.J.; Legrand, P.; Vernede, X.; Lindahl, P.A.; Fontecilla-Camps, J.C.: Ni-Zn-[Fe4-S4] and Ni-Ni-[Fe4-S4] clusters in closed and open a subunits of acetyl-CoA synthase/carbon monoxide dehydrogenase. Nat. Struct. Biol., **10**, 271-279 (2003)

Aldehyde ferredoxin oxidoreductase 1.2.7.5

1 Nomenclature

EC number
1.2.7.5

Systematic name
aldehyde:ferredoxin oxidoreductase

Recommended name
aldehyde ferredoxin oxidoreductase

Synonyms
AOR <1, 3, 9> [13, 14, 16]
aldehyde ferredoxin oxidoreductase <1, 8> [12, 14]
aldehyde oxidoreductase <9> [16]
FOR <1, 2, 6> [3, 5, 9, 14, 15]
WOR5 <1> [14]
aldehyde oxidase (ferredoxin)
formaldehyde ferredoxin oxidoreductase <1, 2, 6> [3, 4, 5, 14, 15]
formaldehyde oxidoreductase <1> [15]
oxidase, aldehyde (ferredoxin)
Additional information <1> (<1> formaldehyde ferredoxin oxidoreductase and aldehyde ferredoxin oxidoreductase are 2 separate enzymes [4, 9, 10, 11]) [4, 9, 10, 11]

CAS registry number
138066-90-7

2 Source Organism

<1> *Pyrococcus furiosus* (no sequence specified) [2, 3, 4, 6, 7, 8, 9, 10, 11, 14, 15]
<2> *Thermococcus litoralis* (no sequence specified) [3, 4, 11]
<3> *Pyrobaculum aerophilum* (no sequence specified) [13]
<4> *Pyrococcus sp. ES-4* (no sequence specified) [4, 7, 11]
<5> *Pyrococcus furiosus* (UNIPROT accession number: Q51739) [5,11]
<6> *Thermococcus litoralis* (UNIPROT accession number: Q56303) [5]
<7> *Thermococcus sp. ES-1* (no sequence specified) [1,11]
<8> *Magnetospirillum magneticum* (no sequence specified) [12]
<9> *Desulfovibrio aminophilus* (no sequence specified) [16]

3 Reaction and Specificity

Catalyzed reaction

an aldehyde + H_2O + 2 oxidized ferredoxin = an acid + 2 H^+ + 2 reduced ferredoxin (<1,4> active site structure [7,10]; <1> intermolecular interaction analysis [10]; <1> substrate binding, catalytic site and reaction mechanism including cofactor action [6,10])

Reaction type

oxidation
redox reaction
reduction

Natural substrates and products

S an aldehyde + H_2O + oxidized ferredoxin <1, 2, 4, 7> (<7> enzyme oxidizes aldehydes generated during amino acid catabolism [1]; <1,2> formaldehyde ferredoxin oxidoreductase, enzyme plays a role in peptide fermentation in hyperthermophilic archaea [3,11]) (Reversibility: ir) [1, 2, 3, 4, 6, 9, 11]
P an acid + H^+ + reduced ferredoxin <1, 2, 4, 7> [2, 3, 4, 6, 9, 11]
S glyceraldehyde + H_2O + oxidized ferredoxin <1> (<1> involved in glycolysis [2]) (Reversibility: ir) [2]
P glycerate + H^+ + reduced ferredoxin <1> [2]
S Additional information <8> (<8> the enzyme is expressed under microaerobic conditions. The non-magnetic mutant, NMA21 is defective in aldehyde ferredoxin oxidoreductase gene. The enzyme may contribute to ferric iron reduction during synthesis of bacterial magnetic particles under microaerobic respiration [12]) (Reversibility: ?) [12]
P ?

Substrates and products

S 2-ethylhexanal + H_2O + oxidized methyl viologen <1> (<1> WOR5 [14]) (Reversibility: ?) [14]
P 2-ethylhexanoate + H^+ + reduced methyl viologen
S 2-methoxybenzaldehyde + H_2O + oxidized methyl viologen <1> (<1> WOR5 [14]) (Reversibility: ?) [14]
P 2-methoxybenzoate + H^+ + reduced methyl viologen
S 2-methylbutyraldehyde + H_2O + oxidized methyl viologen <1> (<1> WOR5 [14]) (Reversibility: ?) [14]
P 2-methylbutyrate + H^+ + reduced methyl viologen
S 2-methylvaleraldehyde + H_2O + oxidized methyl viologen <1> (<1> WOR5 [14]) (Reversibility: ?) [14]
P 2-methylpentanoate + H^+ + reduced methyl viologen
S 2-naphthaldehyde + H_2O + oxidized methyl viologen <1> (<1> WOR5 [14]) (Reversibility: ?) [14]
P 2-naphthoate + H^+ + reduced methyl viologen
S 2-phenylpropanal + H_2O + oxidized methyl viologen <1> (<1> WOR5 [14]) (Reversibility: ?) [14]

P 2-phenylpropanoate + H^+ + reduced methyl viologen

S 3-phenylbutyraldehyde + H_2O + oxidized methyl viologen <1> (<1> WOR5 [14]) (Reversibility: ?) [14]

P 3-phenylbutyrate + H^+ + reduced methyl viologen

S acetaldehyde + H_2O + oxidized benzyl viologen <1, 2, 7> (<1,2> formaldehyde ferredoxin oxidoreductase [9,11]) (Reversibility: ir) [1, 9, 11]

P acetate + H^+ + reduced benzyl viologen <1, 2, 7> [1, 9, 11]

S acetaldehyde + H_2O + oxidized benzyl viologen <3> (<3> at 5% of the activity with crotonaldehyde [13]) (Reversibility: ?) [13]

P acetate + reduced benzyl viologen

S acetaldehyde + H_2O + oxidized ferredoxin <1, 2, 7, 9> (<7> best substrate [1]; <1,2> formaldehyde ferredoxin oxidoreductase [3,9]) (Reversibility: ir) [1, 2, 3, 9, 16]

P acetate + H^+ + reduced ferredoxin <1, 2, 7> [1, 2, 3, 9]

S acetaldehyde + H_2O + oxidized methyl viologen <1> (<1> WOR5 [14]) (Reversibility: ?) [14]

P acetate + H^+ + reduced methyl viologen

S acetate + H^+ + reduced methyl viologen <7> (<7> very low activity [1]; <7> below pH 6.0 [1]) (Reversibility: ir) [1]

P acetaldehyde + H_2O + oxidized methyl viologen <7> [1]

S an aldehyde + H_2O + oxidized ferredoxin <1, 2, 4, 5, 6, 7> (<2> no activity with aldehyde phosphates [3]; <1,2> formaldehyde ferredoxin oxidoreductase, C_1-C_3 aldehydes [3,9,10]; <7> enzyme oxidizes aldehydes generated during amino acid catabolism [1]; <1,2> formaldehyde ferredoxin oxidoreductase, enzyme plays a role in peptide fermentation in hyperthermophilic archaea [3,11]) (Reversibility: ir) [1, 2, 3, 4, 5, 6, 7, 8, 9, 10, 11]

P an acid + H^+ + reduced ferredoxin <1, 2, 4, 5, 6, 7> [1, 2, 3, 4, 5, 6, 7, 8, 9, 10, 11]

S benzaldehyde + H_2O + oxidized benzyl viologen <2, 7> (<2> no activity [11]) (Reversibility: ir) [1, 11]

P benzoic acid + H^+ + reduced benzyl viologen

S benzaldehyde + H_2O + oxidized benzyl viologen <3> (<3> at 149% of the activity with crotonaldehyde [13]) (Reversibility: ?) [13]

P benzoate + reduced benzyl viologen

S benzaldehyde + H_2O + oxidized ferredoxin <1, 2, 9> (<2> very poor substrate [3]) (Reversibility: ir) [3, 16]

P benzoic acid + H^+ + reduced ferredoxin

S butyraldehyde + H_2O + oxidized ferredoxin <1, 2> (<1> low activity [9]; <1,2> formaldehyde ferredoxin oxidoreductase [3,9]) (Reversibility: ir) [2, 3, 9]

P butyrate + H^+ + reduced ferredoxin <1, 2> [2, 3, 9]

S cinnamaldehyde + H_2O + oxidized benzyl viologen <3> (<3> at 139% of the activity with crotonaldehyde [13]) (Reversibility: ?) [13]

P cinnamate + reduced benzyl viologen

S cinnamaldehyde + H_2O + oxidized methyl viologen <1> (<1> WOR5 [14]) (Reversibility: ?) [14]

P cinnamate + H^+ + reduced methyl viologen

S crotonaldehyde + H_2O + oxidized benzyl viologen <3> (Reversibility: ?) [13]
P crotonate + reduced benzyl viologen
S crotonaldehyde + H_2O + oxidized benzyl viologen <1, 2, 7> (<2> no activity [11]; <1> low activity, formaldehyde ferredoxin oxidoreductase [9]) (Reversibility: ir) [1, 9, 11]
P pyruvate + H^+ + reduced benzyl viologen
S crotonaldehyde + H_2O + oxidized ferredoxin <1> (<1> best substrate [2]) (Reversibility: ir) [2, 3]
P pyruvate + H^+ + reduced ferredoxin
S crotonaldehyde + H_2O + oxidized ferredoxin <3> (<3> utilizes a 7Fe ferredoxin as the putative physiological redox partner, instead of a 4Fe ferredoxin as in Pyrococcus furiosus, 17% of the activity with benzyl viologen [13]) (Reversibility: ?) [13]
P crotonate + reduced ferredoxin
S crotonaldehyde + H_2O + oxidized methyl viologen <1> (<1> WOR5 [14]) (Reversibility: ?) [14]
P crotonate + H^+ + reduced methyl viologen
S crotonaldehyde + H_2O + oxidized methyl viologen <1, 2> (<2> very poor substrate [3]; <1> artifical electron acceptor [2,3]) (Reversibility: ir) [2, 3]
P pyruvate + H^+ + reduced methyl viologen
S formaldehyde + H_2O + oxidized benzyl viologen <3> (Reversibility: ?) [13]
P formate + reduced benzyl viologen
S formaldehyde + H_2O + oxidized benzyl viologen <1, 2, 7> (<1,2> formaldehyde ferredoxin oxidoreductase [9,11]) (Reversibility: ir) [1, 9, 11]
P formate + H^+ + reduced benzyl viologen <1, 2, 7> [1, 9, 11]
S formaldehyde + H_2O + oxidized ferredoxin <1, 2> (<1> formaldehyde and aldehyde ferredoxin oxidoreductases [9]) (Reversibility: ir) [2, 3, 9, 15]
P formate + H^+ + reduced ferredoxin <1, 2> [2, 3, 9]
S formaldehyde + H_2O + oxidized methyl viologen <1> (<1> WOR5 [14]) (Reversibility: ?) [14]
P formate + H^+ + reduced methyl viologen
S glutaraldehyde + H_2O + oxidized methyl viologen <1> (<1> WOR5 [14]) (Reversibility: ?) [14]
P glutarate + H^+ + reduced methyl viologen
S glutaric dialdehyde + H_2O + oxidized benzyl viologen <2> (<2> formaldehyde ferredoxin oxidoreductase [11]) (Reversibility: ir) [11]
P ?
S glyceraldehyde + H_2O + oxidized ferredoxin <1, 2> (<1> involved in glycolysis [2]) (Reversibility: ir) [2, 3]
P glycerate + H^+ + reduced ferredoxin <1, 2> [2, 3]
S glyceraldehyde + H_2O + oxidized methyl viologen <1, 7> (<7> very poor substrate [1]) (Reversibility: ir) [1, 2]
P glycerate + H^+ + reduced methyl viologen <1> [2]

S hexanal + H_2O + oxidized ferredoxin <1> (<1> WOR5, AOR and FOR [14]) (Reversibility: ?) [14]
P hexanoate + H^+ + reduced ferredoxin
S hexanal + H_2O + oxidized methyl viologen <1> (<1> WOR5, AOR and FOR [14]) (Reversibility: ?) [14]
P hexanoate + H^+ + reduced methyl viologen
S indoleacetaldehyde + H_2O + oxidized benzyl viologen <2, 7> (<2> formaldehyde ferredoxin oxidoreductase [11]) (Reversibility: ir) [1, 11]
P indoleacetate + H^+ + reduced benzyl viologen
S isobutyraldehyde + H_2O + oxidized methyl viologen <1> (<1> WOR5 [14]) (Reversibility: ?) [14]
P isobutyrate + H^+ + reduced methyl viologen
S isovalerylaldehyde + H_2O + oxidized benzyl viologen <1, 2, 7> (<1,2> formaldehyde ferredoxin oxidoreductase, no activity [9,11]) (Reversibility: ir) [1, 9, 11]
P isovalerate + H^+ + reduced benzyl viologen
S phenylacetaldehyde + H_2O + oxidized benzyl viologen <2, 7> (<2> no activity [11]) (Reversibility: ir) [1, 11]
P phenylacetate + H^+ + reduced benzyl viologen
S phenylacetaldehyde + H_2O + oxidized ferredoxin <1, 2> (<2> very poor substrate [3]) (Reversibility: ir) [3]
P phenylacetate + H^+ + reduced ferredoxin
S phenylpropionaldehyde + H_2O + oxidized benzyl viologen <2> (<2> formaldehyde ferredoxin oxidoreductase [11]) (Reversibility: ir) [11]
P phenylpropanoate + H^+ + reduced benzyl viologen
S propionaldehyde + H_2O + oxidized benzyl viologen <1, 2, 7> (<1,2> formaldehyde ferredoxin oxidoreductase [9,11]) (Reversibility: ir) [1, 9, 11]
P propionate + H^+ + reduced benzyl viologen <1, 2, 7> [1, 9, 11]
S propionaldehyde + H_2O + oxidized ferredoxin <1, 2, 9> (<1,2> formaldehyde ferredoxin oxidoreductase [3,9]) (Reversibility: ir) [3, 9, 16]
P propionate + H^+ + reduced ferredoxin <1, 2> [3, 9]
S salicylaldehyde + H_2O + oxidized benzyl viologen <7> (Reversibility: ir) [1]
P salicylic acid + H^+ + reduced benzyl viologen
S succinic semialdehyde + H_2O + oxidized benzyl viologen <2> (<2> formaldehyde ferredoxin oxidoreductase [11]) (Reversibility: ir) [11]
P succinic acid + H^+ + reduced benzyl viologen
S Additional information <1, 2, 7, 8, 9> (<1> enzyme shows an active and an inactive form [2,8]; <7> acid reduction is strongly dependent on the overall reduction potential, therfore reverse reaction is only possible with reduced methyl viologen at a low rate [1]; <1> no oxidizing activity with glyceraldehyde-3-phosphate, glyoxylate, glucose, glucose 6-phosphate, CO, H_2, formate, pyruvate, 2-oxoglutarate [2]; <1,2> NAD(P) is no electron acceptor [2,3]; <1> no H_2 evolution from reduced methyl viologen [2]; <1,2> no activity with CoA [2,3]; <8> the enzyme is expressed under microaerobic conditions. The non-magnetic mutant, NMA21 is defective in aldehyde ferredoxin oxidoreductase gene. The enzyme may contribute

to ferric iron reduction during synthesis of bacterial magnetic particles under microaerobic respiration [12]; <1> no activity of WOR5 with glyceraldehyde-3-phosphate [14]; <9> no activity towards xanthine, allopurinol, and isoquinoline, inactive towards N-heterolytic compounds [16]) (Reversibility: ?) [1, 2, 3, 8, 12, 14, 16]

P ?

Inhibitors

2,2'-bipyridyl <7> (<7> competitive to methyl viologen [1]) [1]
acetaldehyde <7> (<7> 50% inhibition at 5 mM [1]) [1]
aldehydes <7> (<7> substrate inhibition at high concentrations, overview [1]) [1]
arsenite <1, 2, 7> (<2> weak inhibition [3]; <7> 50% inhibition at 0.5 mM [1]; <1> 50% inhibition at 1 mM [2]) [1, 2, 3]
crotonaldehyde <1, 7> (<1> substrate inhibition above 0.2 mM [2]; <7> 50% inhibition at 20 mM [1]) [1, 2]
cyanide <1, 2> (<1> 50% inhibition at 8 mM [2]) [2, 3]
iodoacetate <1, 2, 7> (<2> weak inhibition [3]; <7> 50% inhibition at 0.5 mM [1]; <1> 50% inhibition at 0.2 mM [2]) [1, 2, 3]
phenylacetaldehyde <7> (<7> 50% inhibition at 7 mM [1]) [1]
propionaldehyde <7> (<7> 50% inhibition at 0.6 mM [1]) [1]
Zn^{2+} <7> (<7> 50% inhibition at 0.25 mM [1]) [1]
isovalerylaldehyde <7> (<7> 50% inhibition at 4 mM [1]) [1]
p-chloromercuribenzoate <7> [1]
Additional information <1> (<1> no effect of CoA [2]) [2]

Cofactors/prosthetic groups

ferredoxin <1, 2, 4, 5, 6, 7> (<1> formaldehyde ferredoxin oxidoreductase, binding mechanism [10]) [1,2,3,4,5,6,7,8,9,10,11]
molybdopterin <1, 2, 4, 7> (<1,2,4> pterin ring system [4]) [1,4]
camMPT <1, 2, 4> (<1,2,4> pterin ring system [4]) [4]
form A cofactor <1, 2, 4, 7> (<1> formaldehyde ferredoxin oxidoreductase [9]; <1,2,4> dephospho [4]; <7> in presence of iodine [1]) [1,4,9]
form B cofactor <7> (<7> in absence of iodine [1]) [1]
tungsten cofactor <1, 2, 4, 5, 7> (<1,2> formaldehyde ferredoxin oxidoreductase [10,11]; <1,2,4,5,7> pterin ring system [4,11]) [4,10,11]
tungsten-molybdopterin <1, 2, 4, 5, 7> (<1,4> electron and redox properties [7]; <1> tungstopterin center [2,6,10]; <1> structural coordination of the tungsten by 4 sulfur ligands in the cofactor [6]; <1> interaction with of DXXGL(D,C)X motif and Gly341, Asp343, Asp338, Gly492, Cys494, and Asp489 [6]; <1,2,4,5,7> in each subunit tungsten is coordinated by 4 dithiolene sulfur atoms from 2 pterin molecules, together with a single [4Fe-4S]-cluster coordinated by 4 sulfur atoms from 4 cysteine residues [11]; <1> modeling of the tungsten sites of inactive and active enzyme forms [8]) [2, 6, 7, 8, 10, 11]

Activating compounds

sulfide <1> (<1> activation of formaldehyde ferredoxin oxidoreductase [9,11]; <1> 4-5fold increase in activity at room temperature and pH 8.0 at 20 mM, in presence of 20 mM dithionite [11]) [9, 11]
tungsten <1, 2> (<1,2> stimulates cell growth [3,11]) [3, 11]
Additional information <1> (<1> no effect of CoA [2]) [2]

Metals, ions

Ca^{2+} <1, 7> (<1> binding site, formaldehyde ferredoxin oxidoreductase [10]; <7> 0.7 gatoms per subunit [1]; <1> 0.4 mol per mol of subunit, formaldehyde ferredoxin oxidoreductase [9]; <1> 1 Ca^{2+} per subunit, formaldehyde ferredoxin oxidoreductase [11]) [1, 9, 10, 11, 14]
Fe^{2+} <1, 2, 4, 5, 6, 7, 9> (<1, 2, 4, 6, 7> contains 1 iron-sulfur cluster [4Fe-4S] per subunit [1, 3, 5, 6, 7, 10, 11]; <1,4> electron and redox properties of the [4Fe-4S]-centers [7]; <1> 6.6 gatoms per 85000g of protein [2]; <1, 2, 4, 5, 7> in each subunit tungsten is coordinated by 4 dithiolene sulfur atoms from 2 pterin molecules, together with a single [4Fe-4S]-cluster coordinated by 4 sulfur atoms from 4 cysteine residues [11]; <1> contains 1 additional Fe^{2+} besides the [4Fe-4S]-cluster [11]; <1> 3.8 gatoms per subunit, formaldehyde ferredoxin oxidoreductase [9]; <1> the [4Fe-4S]-cluster is coordinated by Cys287 of the formaldehyde ferredoxin oxidoreductase and Asp14 of ferredoxin, structure [10]; <9> two(2Fe-2S) clusters per monomer [16]; <1> WOR5 contains one (4Fe-4S) cluster per subunit [14]) [1, 2, 3, 5, 6, 7, 9, 10, 11, 14, 16]
iron <3> (<3> contains 3.3 mol Fe per mol of monomer, the enzyme contains a single [4Fe-4S]2+/1+ cluster per monomer [13]) [13]
Mg^{2+} <1, 7> (<1> 1.5 mol per mol of subunit, formaldehyde ferredoxin oxidoreductase [9]; <1> both enzyme forms [11]; <1,7> contains 1 Mg-atom per subunit [1,11]) [1, 9, 11, 14]
molybdenum <1, 2, 4, 7, 9> (<1,4> part of a tungsten-molybdopterin cofactor [2,6,7,8]; <9> molybdenum centre [16]) [1, 2, 4, 6, 7, 8, 16]
tungsten <1, 2, 3, 4, 5, 6, 7> (<1,2> part of a tungsten-molybdopterin cofactor [2, 3, 6, 7, 8]; <1, 2, 4, 5, 7> in each subunit tungsten is coordinated by 4 dithiolene sulfur atoms from 2 pterin molecules, together with a single [4Fe-4S]-cluster coordinated by 4 sulfur atoms from 4 cysteine residues [11]; <1> 1.0 gatoms per 85000g of protein [2]; <1,4> electron and redox properties of the 2 tungsten forms with different potential [7]; <1,2,7> 1 tungsten per subunit [1,3,11]; <1> 0.9 gatom per subunit, formaldehyde ferredoxin oxidoreductase [9]; <3> contains 0.7 mol tungsten per mol of monomer, the enzyme contains a single tungstopterin cofactor [13]; <1> WOR5 contains one tungstobispertin center per subunit [14]) [1, 2, 3, 4, 5, 6, 7, 8, 9, 10, 11, 13, 14]
Zn^{2+} <7> (<7> 0.2 gatoms per subunit [1]) [1]
Additional information <3> (<3> contains no detectable molybdenum [13]) [13]

Turnover number (min^{-1})

1.6 <7> (acetate, <7> pH 8.4, 85°C, benzyl viologen as electron acceptor [1]) [1]
2.8 <7> (glyceraldehyde, <7> pH 8.4, 45°C, benzyl viologen as electron acceptor [1]) [1]
13 <7> (salicylaldehyde, <7> pH 8.4, 85°C, benzyl viologen as electron acceptor [1]) [1]
22 <7> (crotonaldehyde, <7> pH 8.4, 45°C, benzyl viologen as electron acceptor [1]) [1]
55 <7> (indoleacetaldehyde, <7> pH 8.4, 85°C, benzyl viologen as electron acceptor [1]) [1]
62 <7> (methyl viologen, <7> pH 8.4, 85°C, with crotonaldehyde [1]) [1]
250 <7> (benzyl viologen, <7> pH 8.4, 85°C, with crotonaldehyde [1]) [1]
269 <7> (crotonaldehyde, <7> pH 8.4, 85°C, benzyl viologen as electron acceptor [1]) [1]
270 <7> (ferredoxin, <7> ES-1 ferredoxin, pH 8.4, 85°C, with crotonaldehyde [1]) [1]
272 <7> (isovalerylaldehyde, <7> pH 8.4, 85°C, benzyl viologen as electron acceptor [1]) [1]
343 <7> (acetaldehyde, <7> pH 8.4, 85°C, benzyl viologen as electron acceptor [1]) [1]
720 <7> (benzaldehyde, <7> pH 8.4, 85°C, benzyl viologen as electron acceptor [1]) [1]
950 <7> (formaldehyde, <7> pH 8.4, 85°C, benzyl viologen as electron acceptor [1]) [1]
960 <7> (phenylacetaldehyde, <7> pH 8.4, 85°C, benzyl viologen as electron acceptor [1]) [1]
1100 <7> (propionaldehyde, <7> pH 8.4, 85°C, benzyl viologen as electron acceptor [1]) [1]

Specific activity (U/mg)

0.34 <1> (<1> with acetaldehyde as substrate, at 60°C [14]) [14]
0.38 <1> (<1> WOR5, with 0.2 mM crotonaldehyde as substrate, at 60°C [14]) [14]
0.48 <1> (<1> FOR, with 5 mM hexanal as substrate, at 80°C [14]) [14]
0.93 <1> (<1> WOR5, with 0.2 mM crotonaldehyde as substrate, at 80°C [14]) [14]
1.1 <1> (<1> with crotonaldehyde as substrate, at 60°C [14]) [14]
1.4 <1> (<1> with glutaraldehyde as substrate, at 60°C [14]) [14]
3 <1> (<1> WOR5, with 5 mM hexanal as substrate, at 80°C [14]) [14]
3.1 <1> (<1> AOR, with 5 mM hexanal as substrate, at 80°C [14]) [14]
4.7 <1> (<1> WOR5, with 50 mM formaldehyde as substrate, at 60°C [14]) [14]
6.5 <3> [13]
6.8 <2> (<2> purified enzyme [3]) [3]
7.4 <1> (<1> with cinnamaldehyde as substrate, at 60°C [14]) [14]

7.7 <1> (<1> with 2-methylbutyraldehyde as substrate, at 60°C [14]; <1> with 2-naphthaldehyde as substrate, at 60°C [14]) [14]
8 <1> (<1> with 3-phenylbutyraldehyde as substrate, at 60°C [14]) [14]
8.3 <1> (<1> with 2-ethylhexanal as substrate, at 60°C [14]) [14]
8.5 <1> (<1> with formaldehyde as substrate, at 60°C [14]) [14]
9.3 <1> (<1> with 2-phenylpropanal as substrate, at 60°C [14]) [14]
11.4 <1> (<1> WOR5, with 50 mM formaldehyde as substrate, at 80°C [14]) [14]
11.8 <1> (<1> with isobutyraldehyde as substrate, at 60°C [14]) [14]
12.7 <1> (<1> with 2-methylvaleraldehyde as substrate, at 60°C [14]) [14]
15.1 <1> (<1> with 2-methoxybenzaldehyde as substrate, at 60°C [14]) [14]
15.6 <1> (<1> WOR5, with 5 mM hexanal as substrate, at 60°C [14]) [14]
42 <1> (<1> purified formaldehyde ferredoxin oxidoreductase [9]) [9]
53.6 <1> (<1> purified enzyme, substrate crotonaldehyde with methyl viologen as electron acceptor [2]) [2]
59 <1> (<1> purified enzyme, substrate crotonaldehyde with methyl viologen as electron acceptor [3]) [3]
238 <7> (<7> purified enzyme [1]) [1]
Additional information <1> (<1> formaldehyde ferredoxin oxidoreductase, substrate specificity [9]) [9]

K_m-Value (mM)

0.01 <7> (ferredoxin, <7> ES-1 ferredoxin [1]; <7> pH 8.4, 85°C, with crotonaldehyde [1]) [1]
0.016 <7> (acetaldehyde, <7> pH 8.4, 85°C, benzyl viologen as electron acceptor [1,11]) [1, 11]
0.028 <7> (isovalerylaldehyde, <7> pH 8.4, 85°C, benzyl viologen as electron acceptor [1,11]) [1, 11]
0.04 <1> (crotonaldehyde, <1> pH 8.4, 80°C, with methyl viologen as electron acceptor [2]) [2]
0.05 <7> (indoleacetaldehyde, <7> pH 8.4, 85°C, benzyl viologen as electron acceptor [1,11]) [1, 11]
0.057 <7> (benzaldehyde, <7> pH 8.4, 85°C, benzyl viologen as electron acceptor [1,11]) [1, 11]
0.065 <7> (salicylaldehyde, <7> pH 8.4, 85°C, benzyl viologen as electron acceptor [1]) [1]
0.076 <7> (phenylacetaldehyde, <7> pH 8.4, 85°C, benzyl viologen as electron acceptor [1,11]) [1, 11]
0.12 <1> (2-phenylpropanal, <1> at 60°C under anaerobic conditions [14]) [14]
0.13 <7> (crotonaldehyde, <7> pH 8.4, 45°C, benzyl viologen as electron acceptor [1]) [1]
0.14 <7> (crotonaldehyde, <7> pH 8.4, 85°C, benzyl viologen as electron acceptor [1,11]) [1, 11]
0.15 <7> (propionaldehyde, <7> pH 8.4, 85°C, benzyl viologen as electron acceptor [1,11]) [1, 11]
0.17 <1> (2-ethylhexanal, <1> at 60°C under anaerobic conditions [14]) [14]

0.17 <7> (benzyl viologen, <7> pH 8.4, 85°C, with crotonaldehyde [1]) [1]
0.18 <1> (hexanal, <1> at 60°C under anaerobic conditions [14]) [14]
0.2 <7> (glyceraldehyde, <7> pH 8.4, 45°C, benzyl viologen as electron acceptor [1]) [1]
0.27 <1> (2-methylvaleraldehyde, <1> at 60°C under anaerobic conditions [14]) [14]
0.3 <3> (crotonaldehyde, <3> pH 8.4, 60°C [13]) [13]
0.42 <1> (3-phenylbutyraldehyde, <1> at 60°C under anaerobic conditions [14]) [14]
0.43 <1> (2-methylbutyraldehyde, <1> at 60°C under anaerobic conditions [14]) [14]
0.79 <1> (isobutyraldehyde, <1> at 60°C under anaerobic conditions [14]) [14]
0.8 <1> (glutaric dialdehyde, <1> pH 8.4, 80°C, formaldehyde ferredoxin oxidoreductase, benzyl viologen as electron acceptor [11]) [11]
1 <1> (glyceraldehyde, <1> pH 8.4, 65°C, with methyl viologen as electron acceptor [2]) [2]
1.3 <1> (2-naphthaldehyde (β), <1> at 60°C under anaerobic conditions [14]) [14]
1.42 <7> (formaldehyde, <7> pH 8.4, 85°C, benzyl viologen as electron acceptor [1,11]) [1, 11]
1.5 <1> (acetaldehyde, <1> at 60°C under anaerobic conditions [14]) [14]
1.6 <1> (cinnamaldehyde, <1> at 60°C under anaerobic conditions [14]) [14]
1.8 <7> (acetate, <7> pH 8.4, 85°C, benzyl viologen as electron acceptor [1]) [1]
2.9 <7> (methyl viologen, <7> pH 8.4, 85°C, with crotonaldehyde [1]) [1]
4.8 <1> (2-methoxybenzaldehyde, <1> at 60°C under anaerobic conditions [14]) [14]
8 <1> (succinic semialdehyde, <1> pH 8.4, 80°C, formaldehyde ferredoxin oxidoreductase, benzyl viologen as electron acceptor [11]) [11]
9.4 <1> (glutaraldehyde, <1> at 60°C under anaerobic conditions [14]) [14]
15 <1> (phenylpropionaldehyde, <1> pH 8.4, 80°C, formaldehyde ferredoxin oxidoreductase, benzyl viologen as electron acceptor [11]) [11]
25 <1> (formaldehyde, <1> formaldehyde ferredoxin oxidoreductase, pH 8.4, 80°C, with benzyl viologen as electron acceptor [9,11]) [9, 11]
25 <1> (indoleacetaldehyde, <1> pH 8.4, 80°C, formaldehyde ferredoxin oxidoreductase, benzyl viologen as electron acceptor [11]) [11]
45 <1> (formaldehyde, <1> at 60°C under anaerobic conditions [14]) [14]
46 <1> (crotonaldehyde, <1> at 60°C under anaerobic conditions [14]) [14]
60 <1> (acetaldehyde, <1> pH 8.4, 85°C, benzyl viologen as electron acceptor [11]) [11]
60 <1> (propionaldehyde, <1> pH 8.4, 80°C, formaldehyde ferredoxin oxidoreductase, benzyl viologen as electron acceptor [11]) [11]
62 <2> (formaldehyde, <2> pH 8.4, 80°C [3]) [3]
Additional information <1> (<1> formaldehyde ferredoxin oxidoreductase, K_m for several substrates at different concentrations with purified and sulfide-activated enzyme, overview [9]) [9]

pH-Optimum

8.4 <1, 2, 4, 7> (<1,2,4,7> assay at [1,3,9,11]; <1> sharp maximum [2]) [1, 2, 3, 9, 11]
10 <2> (<2> above, at 80°C [3]) [3]

pH-Range

5.5-10 <1> (<1> formaldehyde ferredoxin oxidoreductase, activity increases linearly from pH 5.5 to pH 10.0 at 80°C [9]) [9]
6-10 <7> (<7> oxidation of acetaldehyde with benzyl viologen as electron acceptor [1]) [1]
7-8.4 <1> (<1> activity increases with increasing pH at increasing temperature, nearly no activity below pH 7.0 [2]) [2]

Temperature optimum (°C)

80 <1, 2, 4, 7> (<1,2,4,7> assay at [9,11]) [9, 11, 14]
85 <7> (<7> assay at [1]) [1]
90 <1, 2> (<1,2> above [2,3]; <2> at pH 8.4 [3]) [2, 3]
95 <1> (<1> above [6]) [6]
100 <3> [13]

Temperature range (°C)

30-100 <1> [14]
60-90 <1> (<1> formaldehyde ferredoxin oxidoreductase, activity increases 4.5fold from 60°C to 90°C at pH 8.4 [9]) [9]

4 Enzyme Structure

Molecular weight

90000 <1> (<1> gel filtration [2]) [2]
105000 <7> (<7> gel filtration [1]) [1]
130000 <3> (<3> gel filtration [13]) [13]
134000 <1> (<1> crystal structure [6]) [6]
135000 <1> (<1> WOR5, native PAGE [14]) [14]
136100 <5> (<5> DNA sequence determination, including cofactors [5]) [5, 11]
200000 <9> (<9> gel filtration [16]) [16]
275000 <1> (<1> formaldehyde ferredoxin oxidoreductase, gel filtration [9]) [9]
280000 <1, 2> (<2> formaldehyde ferredoxin oxidoreductase, gel filtration [3]) [3, 11]
281300 <6> (<6> formaldehyde ferredoxin oxidoreductase, DNA sequence determination [5]) [5]

Subunits

dimer <1, 3, 5, 7> (<7> 2 * 67000, SDS-PAGE [1,11]; <3> 2 * 68000, SDS-PAGE [13]; <1> 2 * 66000, crystal structure [6]; <5> 2 * 66630-85000, DNA sequence determination and SDS-PAGE [5]; <1> 2 * 67000 [11]) [1, 5, 6, 11, 13]

homodimer <1, 9> (<9> 2 * 100000, SDS-PAGE [16]; <1> WOR5, 2 * 67000, SDS-PAGE [14]) [14, 16]
monomer <1> (<1> 1 * 80000, SDS-PAGE [2]) [2]
tetramer <1, 2, 6> (<1> 4 * 68000, formaldehyde ferredoxin oxidoreductase, SDS-PAGE [9]; <2> 4 * 70000, formaldehyde ferredoxin oxidoreductase, SDS-PAGE [3]; <6> 4 * 68941-69000, formaldehyde ferredoxin oxidoreductase, DNA sequence determination and SDS-PAGE [5]) [3, 5, 9]

5 Isolation/Preparation/Mutation/Application

Source/tissue

Additional information <8> (<8> the enzyme is expressed under microaerobic conditions [12]) [12]

Localization

cytoplasm <8> [12]

Purification

<1> [4]
<1> (WOR5, by gel filtration) [14]
<1> (formaldehyde and aldehyde ferredoxin oxidoreductase) [11]
<1> (formaldehyde ferredoxin oxidoreductase, 17fold) [9]
<1> (formaldehyde ferredoxin oxidoreductase, highly reducing and strict anaerobic conditions are required) [11]
<1> (to homogeneity under strict anaerobic conditions) [2]
<2> (formaldehyde ferredoxin oxidoreductase, 17fold to homogeneity) [3]
<2> (formaldehyde ferredoxin oxidoreductase, highly reducing and strict anaerobic conditions are required) [11]
<3> [13]
<7> (about 120fold, to homogeneity) [1]
<9> (gel filtration) [16]

Crystallization

<1> [11]
<1> (X-ray crystal structure determination and analysis) [6]
<1> (crystallization of the 3 different cofactor model complexes: 1. $[Et_4N]_2[W^{VI}O(1,2\text{-dicyanoethylenedithiolate})_2]$, 2. $[Et_4N]_2[W^{IV}O(1,2\text{-dicyanoethylenedithiolate})_2]$, and 3. $[Et_4N]_2[W^{VI}O(S2)(1,2\text{-dicyanoethylenedithiolate})_2]$, X-ray structure determination and analysis) [8]
<1> (formaldehyde ferredoxin oxidoreductase, purified enzyme, modified melting-point capillary method, room temperature under argon atmosphere, protein solution: 55-65 mg/ml, 50 mM Tris-HCl, pH 8.0, 2 mM dithionite, 2 mM DTT, 0.2 M KCl, precipitant solution: 30% v/v glycerol, 20% w/v PEG 4000, 0.1 M sodium citrate, pH 5.6, 0.2 M NaCl, several weeks, X-ray structure determination and analysis) [10]

Cloning

<5> (gene aor, DNA sequence determination and analysis) [5]
<6> (gene for, DNA sequence determination and analysis) [5]

Application

Additional information <1> (<1> identification of WOR5 as a fifth aldehyde oxidorecutase with very broad substrate specifity [14]) [14]

6 Stability

Temperature stability

23 <1, 7> (<7> pure enzyme, 2 mg/ml, 50 mM Tris-HCl, pH 8.0, 2 mM sodium dithionite, 2 mM DTT, 10% v/v glycerol, exposure to air, irreversible loss of 50% activity after 1 min [1]; <1> pure enzyme, 5.2 mg/ml, 50 mM Tris-HCl, pH 8.0, 2 mM sodium dithionite, 2 mM DTT, 10% v/v glycerol, exposure to air, less than 20% remaining activity after 5 min, after 10 additional min 10% residual activity is reached, which stays constant for more than 30 min [2]) [1, 2]
80 <1> (<1> formaldehyde ferredoxin oxidoreductase, $t_{1/2}$: 6 h [11]; <1> aldehyde ferredoxin oxidoreductase, $t_{1/2}$: 15 min [11]; <1> formaldehyde ferredoxin oxidoreductase, $t_{1/2}$: 8 h, in presence of 2 mM dithionite and 2 mM DTT, pH 8.4 [9]) [9, 11]
Additional information <1, 2, 4, 5, 6, 7> (<1,2,4,5,6,7> the enzyme is thermostable [1,2,3,4,5,6,9]) [1, 2, 3, 4, 5, 6, 9]

Oxidation stability

<1>, aldehyde ferredoxin oxidoreductase, $t_{1/2}$: 30 min at 23 °C [11]
<1>, formaldehyde ferredoxin oxidoreductase, $t_{1/2}$: 9 h at 23°C [11]
<1>, sensitive against O_2, dithionite is added to avoid trace contamination with O_2 during purification and enzyme assay [2]
<7>, very sensitive to O_2, after 1 h at 60°C 3% activity remains, after 1 h at 4°C 39% activity remains [1]

General stability information

<1>, 10% glycerol and 2 mM DTT stabilize during purification [2, 9]

Storage stability

<1>, -196°C, pure enzyme pellet is thawed anaerobically after storage in liquid N_2, no loss of activity for several months [2]
<1>, 4°C or 23°C, pure enzyme, 50 mM Tris-HCl, pH 8.0, 2 mM sodium dithionite, 2 mM DTT, 10% v/v glycerol, loss of 25% activity after 6 h under strict anaerobic conditions [2]

References

[1] Heider, J.; Ma, K.; Adams, M.W.W.: Purification, characterization, and metabolic function of tungsten-containing aldehyde ferredoxin oxidoreductase

from the hyperthermophilic and proteolytic archaeon Thermococcus strain ES-1. J. Bacteriol., **177**, 4757-4764 (1995)

[2] Mukund, S.; Adams, M.W.W.: The novel tungsten-iron-sulfur protein of the hyperthermophilic archaebacterium, Pyrococcus furiosus, is an aldehyde ferredoxin oxidoreductase. Evidence for its participation in a unique glycolytic pathway. J. Biol. Chem., **266**, 14208-14216 (1991)

[3] Mukund, S.; Adams, M.W.: Characterization of a novel tungsten-containing formaldehyde ferredoxin oxidoreductase from the hyperthermophilic archaeon, Thermococcus litoralis. A role for tungsten in peptide catabolism. J. Biol. Chem., **268**, 13592-13600 (1993)

[4] Johnson, J.L.; Rajagopalan, K.V.; Mukund, S.; Adams, M.W.W.: Identification of molybdopterin as the organic component of the tungsten cofactor in four enzymes from hyperthermophilic Archaea. J. Biol. Chem., **268**, 4848-4852 (1993)

[5] Kletzin, A.; Mukund, S.; Kelley-Crouse, T.L.; Chan, M.K.; Rees, D.C.; Adams, M.W.: Molecular characterization of the genes encoding the tungsten-containing aldehyde ferredoxin oxidoreductase from Pyrococcus furiosus and formaldehyde ferredoxin oxidoreductase from Thermococcus litoralis. J. Bacteriol., **177**, 4817-4819 (1995)

[6] Chan, M.K.; Mukund, S.; Kletzin, A.; Adams, M.W.; Rees, D.C.: Structure of a hyperthermophilic tungstopterin enzyme, aldehyde ferredoxin oxidoreductase. Science, **267**, 1463-1469 (1995)

[7] Koehler, B.P.; Mukund, S.; Conover, R.C.; Dhawan, I.K.; Roy, R.; Adams, M.W.W.; Johnson, M.K.: Spectroscopic characterization of the tungsten and iron centers in aldehyde ferredoxin oxidoreductases from two hyperthermophilic archaea. J. Am. Chem. Soc., **118**, 12391-12405 (1996)

[8] Das, S.K.; Biswas, D.; Maiti, R.; Sarkar, S.: Modeling the tungsten sites of inactive and active forms of hyperthermophilic Pyrococcus furiosus aldehyde ferredoxin oxidoreductase. J. Am. Chem. Soc., **118**, 1387-1397 (1996)

[9] Roy, R.; Mukund, S.; Schut, G.J.; Dunn, D.M.; Weiss, R.; Adams, M.W.: Purification and molecular characterization of the tungsten-containing formaldehyde ferredoxin oxidoreductase from the hyperthermophilic archaeon Pyrococcus furiosus: the third of a putative five-member tungstoenzyme family. J. Bacteriol., **181**, 1171-1180 (1999)

[10] Hu, Y.; Faham, S.; Roy, R.; Adams, M.W.; Rees, D.C.: Formaldehyde ferredoxin oxidoreductase from Pyrococcus furiosus: the 1.85 A resolution crystal structure and its mechanistic implications. J. Mol. Biol., **286**, 899-914 (1999)

[11] Roy, R.; Menon, A.L.; Adams, M.W.W.: Aldehyde oxidoreductases from Pyrococcus furiosus. Methods Enzymol., **331**, 132-144 (2001)

[12] Wahyudi, A.T.; Takeyama, H.; Okamura, Y.; Fukuda, Y.; Matsunaga, T.: Characterization of aldehyde ferredoxin oxidoreductase gene defective mutant in Magnetospirillum magneticum AMB-1. Biochem. Biophys. Res. Commun., **303**, 223-229 (2003)

[13] Hagedoorn, P.L.; Chen, T.; Schroder, I.; Piersma, S.R.; Vries, S.D.; Hagen, W.R.: Purification and characterization of the tungsten enzyme aldehyde:-

ferredoxin oxidoreductase from the hyperthermophilic denitrifier Pyrobaculum aerophilum. J. Biol. Inorg. Chem., **10**, 259-269 (2005)

[14] Bevers, L.E.; Bol, E.; Hagedoorn, P.L.; Hagen, W.R.: WOR5, a novel tungsten-containing aldehyde oxidoreductase from Pyrococcus furiosus with a broad substrate specificity. J. Bacteriol., **187**, 7056-7061 (2005)

[15] Bol, E.; Bevers, L.E.; Hagedoorn, P.L.; Hagen, W.R.: Redox chemistry of tungsten and iron-sulfur prosthetic groups in Pyrococcus furiosus formaldehyde ferredoxin oxidoreductase. J. Biol. Inorg. Chem., **11**, 999-1006 (2006)

[16] Thapper, A.; Rivas, M.G.; Brondino, C.D.; Ollivier, B.; Fauque, G.; Moura, I.; Moura, J.J.: Biochemical and spectroscopic characterization of an aldehyde oxidoreductase isolated from Desulfovibrio aminophilus. J. Inorg. Biochem., **100**, 44-50 (2006)

Glyceraldehyde-3-phosphate dehydrogenase (ferredoxin) 1.2.7.6

1 Nomenclature

EC number
1.2.7.6

Systematic name
D-glyceraldehyde-3-phosphate:ferredoxin oxidoreductase

Recommended name
glyceraldehyde-3-phosphate dehydrogenase (ferredoxin)

Synonyms
GAPOR
dehydrogenase, glyceraldehyde phosphate (ferredoxin)
glyceraldehyde 3-phosphate oxidoreductase <1> [3]
glyceraldehyde phosphate dehydrogenase (ferredoxin)
glyceraldehyde-3-phosphate Fd oxidoreductase
glyceraldehyde-3-phosphate ferredoxin oxidoreductase <1, 2> [1, 2, 4]
glyceraldehyde-3-phosphate ferredoxin reductase

CAS registry number
162995-20-2

2 Source Organism

<1> *Pyrococcus furiosus* (no sequence specified) [1, 2, 3]
<2> *Pyrococcus furiosus* (UNIPROT accession number: O93720) [4]

3 Reaction and Specificity

Catalyzed reaction
D-glyceraldehyde-3-phosphate + H_2O + 2 oxidized ferredoxin = 3-phospho-D-glycerate + 2 H^+ + 2 reduced ferredoxin (<1,2> this enzyme is thought to function in place of glyceraldehyde-3-phosphate dehydrogenase and possibly phosphoglycerate kinase in the novel Embden-Meyerhoff-type glycolytic pathway found in Pyrococcus furiosus. It is specific for glyceraldehyde-3-phosphate [2,4]; <1> electron transfer, mechanism and reduction potential [3])

Reaction type
oxidation
redox reaction
reduction

Natural substrates and products

S D-glyceraldehyde-3-phosphate + H_2O + oxidized ferredoxin <1, 2> (<2> enzyme is a site for glycolytic regulation [4]; <1> might have a glycolytic role and functions in place of glyceraldehyde-3-phosphate dehydrogenase and possibly phosphoglycerate kinase [2]) (Reversibility: ?) [1, 2, 4]

P 3-phospho-D-glycerate + H^+ + reduced ferredoxin <1, 2> [1, 2, 4]

Substrates and products

S D-glyceraldehyde-3-phosphate + H_2O + oxidized benzyl viologen <1, 2> (<1,2> artificial electron acceptor [1,2,4]) (Reversibility: ?) [1, 2, 4]

P 3-phospho-D-glycerate + H^+ + reduced benzyl viologen <1, 2> [1, 2, 4]

S D-glyceraldehyde-3-phosphate + H_2O + oxidized ferredoxin <1, 2> (<2> enzyme is a site for glycolytic regulation [4]; <1> might have a glycolytic role and functions in place of glyceraldehyde-3-phosphate dehydrogenase and possibly phosphoglycerate kinase [2]) (Reversibility: ?) [1, 2, 3, 4]

P 3-phospho-D-glycerate + H^+ + reduced ferredoxin <1, 2> [1, 2, 3, 4]

S Additional information <1> (<1> no activity with $NAD(P)^+$ as electron acceptor [1,2]; <1> no activity with formaldehyde, propionaldehyde, phenylacetaldehyde, butyraldehyde, crotonaldehyde, acetaldehyde, dihydroxyacetone phosphate, glyceraldehyde, benzaldehyde, glucose, glucose 6-phosphate, or glyoxylate [1,2]) (Reversibility: ?) [1, 2]

P ?

Inhibitors

crotonaldehyde <1> (<1> 25% inhibition at 10 mM [2]) [2]
dithionite <1> (<1> reversible by 70% [2]; <1> 50% inhibition at 3 mM, pH 8.0 [2]; <1> O_2-scavenger [2]) [2, 3]
formaldehyde <1> (<1> 25% inhibition at 10 mM [2]) [2]
glyceraldehyde-3-phosphate <1> (<1> substrate inhibition above 0.5 mM [2]) [2]

Cofactors/prosthetic groups

ferredoxin <1, 2> (<2> dependent on [4]) [1,2,3,4]
cubane <1> [3]
tungsten-molybdopterin <1> (<1> tungstopterin center [1,2]; <1> structure [1]) [1,2]

Activating compounds

2,3-bisphosphoglycerate <1> (<1> stimulates [2]) [2]
3-phosphoglycerate <1> (<1> stimulates [2]) [2]
acetyl phosphate <1> (<1> stimulates [2]) [2]
ionic strength <1> (<1> particularly by polyanions [2]) [2]
potassium phosphate <1> (<1> 3.5fold activation at 200 mM [2]) [2]
sodium arsenate <1> (<1> 3.5fold activation at 200 mM [2]) [2]
sodium citrate <1> (<1> 3.5fold activation at 200 mM [2]) [2]
sodium sulfate <1> (<1> 3.5fold activation at 200 mM [2]) [2]
tungsten <1> (<1> stimulates cell growth when added to the medium [1]) [1]

potassium chloride <1> (<1> 3.5fold activation at 200 mM [2]) [2]
Additional information <2> (<2> enzyme is induced during glycolysis [4]) [4]

Metals, ions

Fe^{2+} <1, 2> (<1> determination of the redox potential [3]; <1,2> enzyme contains iron-sulfur cluster [2,4]; <1> 1 [4Fe-4S]2+ cluster per enzyme molecule [1,3]; <1> 5-6 iron g-atoms per mol of enzyme [1,2]) [1, 2, 3, 4]
Mg^{2+} <1> (<1> contains 1 molecule per enzyme [1]) [1]
Tungsten <1, 2> (<1> 1 tungsten center per molecule of enzyme [1,3]; <1> determination of the redox potential [3]; <1> 0.85 tungsten g-atoms per mol of enzyme, part of a tungsten-molybdopterin cofactor [2]; <2> single tungsten environment is afforded for the reaction [4]) [1, 2, 3, 4]
Zn^{2+} <1> (<1> contains 2 molecules per enzyme [1]) [1]

Specific activity (U/mg)

25 <1, 2> (<1,2> purified enzyme [1,4]) [1, 4]
30 <1> (<1> purified enzyme [3]) [3]
140 <1> (<1> purified enzyme [2]) [2]
Additional information <1> (<1> strictly anaerobic assay conditions [1]) [1]

K_m-Value (mM)

0.006 <1> (oxidized ferredoxin, <1> pH 8.4, 70°C [2]) [2]
0.028 <1> (D-glyceraldehyde-3-phosphate, <1> with ferredoxin as electron acceptor, pH 8.4, 70°C [1,2]) [1, 2]
0.043 <1> (D-glyceraldehyde-3-phosphate, <1> with benzyl viologen as electron acceptor, pH 8.4, 70°C [2]) [2]

pH-Optimum

8 <2> (<2> assay at [4]) [4]
8.4 <1> (<1> assay at [1,2]) [1, 2]

Temperature optimum (°C)

50 <2> (<2> assay at [4]) [4]
70 <1> (<1> assay at [1,2]) [1, 2]

4 Enzyme Structure

Molecular weight

56000 <2> (<2> gel filtration [4]) [4]
60000 <1> (<1> gel filtration [2]) [2]
73000 <1> (<1> gel filtration [1]) [1]
74000 <2> (<2> sequence determination [4]) [4]

Subunits

monomer <1, 2> (<1> 1 * 63000, SDS-PAGE [2]; <2> 1 * 64000, SDS-PAGE [4]; <1> 1 * 73000, SDS-PAGE [1]) [1, 2, 4]

5 Isolation/Preparation/Mutation/Application

Localization

cytosol <1> [2]

Purification

<1> [3]
<1> (30fold, under strictly anaerobic conditions) [2]
<1> (to homogeneity) [1]
<2> (35fold, about 95% purity) [4]

Cloning

<2> (DNA sequence determination and analysis) [4]

6 Stability

Temperature stability

80 <1> (<1> $t_{1/2}$: 15 min [1]) [1]

Oxidation stability

<1>, O_2-sensitive, $t_{1/2}$ at 23°C: 6 h [1]
<1>, aerobic conditions: loss of about 50% activity in cell extract after 12 h at 23°C [2]
<1>, sensitive against O_2 [2]

Storage stability

<1>, 23°C, under argon, 12 h, no loss of activity in cell extract [2]

References

[1] Roy, R.; Menon, A.L.; Adams, M.W.W.: Aldehyde oxidoreductases from Pyrococcus furiosus. Methods Enzymol., **331**, 132-144 (2001)
[2] Mukund, S.; Adams, M.W.: Glyceraldehyde-3-phosphate ferredoxin oxidoreductase, a novel tungsten-containing enzyme with a potential glycolytic role in the hyperthermophilic archaeon Pyrococcus furiosus. J. Biol. Chem., **270**, 8389-8392 (1995)
[3] Hagedoorn, P.L.; Freije, J.R.; Hagen, W.R.: Pyrococcus furiosus glyceraldehyde 3-phosphate oxidoreductase has comparable W6+/5+ and W5+/4+ reduction potentials and unusual [4Fe-4S] EPR properties. FEBS Lett., **462**, 66-70 (1999)
[4] van der Oost, J.; Schut, G.; Kengen, S.W.; Hagen, W.R.; Thomm, M.; de Vos, W.M.: The ferredoxin-dependent conversion of glyceraldehyde-3-phosphate in the hyperthermophilic archaeon Pyrococcus furiosus represents a novel site of glycolytic regulation. J. Biol. Chem., **273**, 28149-28154 (1998)

3-Methyl-2-oxobutanoate dehydrogenase (ferredoxin) 1.2.7.7

1 Nomenclature

EC number
1.2.7.7

Systematic name
3-methyl-2-oxobutanoate:ferredoxin oxidoreductase (decarboxylating; CoA-2-methylpropanoylating)

Recommended name
3-methyl-2-oxobutanoate dehydrogenase (ferredoxin)

Synonyms
2-ketoisovalerate ferredoxin oxidoreductase
2-oxoisovalerate ferredoxin reductase
2-oxoisovalerate oxidoreductase
3-methyl-2-oxobutanoate synthase (ferredoxin)
VOR
branched-chain ketoacid ferredoxin reductase
branched-chain oxo acid ferredoxin reductase
keto-valine-ferredoxin oxidoreductase
ketoisovalerate ferredoxin reductase

CAS registry number
60320-97-0

2 Source Organism

<1> *Pyrococcus sp.* (no sequence specified) [3]
<2> *Methanobacterium thermoautotrophicum* (no sequence specified) [4]
<3> *Pyrococcus furiosus* (no sequence specified) [1, 2, 3, 4]
<4> *Thermococcus sp.* (no sequence specified) [3]
<5> *Thermococcus litoralis* (no sequence specified) [2,3,4]

3 Reaction and Specificity

Catalyzed reaction
3-methyl-2-oxobutanoate + CoA + oxidized ferredoxin = S-(2-methylpropanoyl)-CoA + CO_2 + reduced ferredoxin (<3,5> pathway [2]; <1,2,3,4,5> this enzyme is one of four 2-oxoacid oxidoreductases: indolepyruvate oxidoreduc-

tase, EC 1.2.7.8, pyruvate oxidoreductase, EC 1.2.7.1., 2-oxogluterate oxidoreductase, EC 1.2.7.3 [1,2,3,4])

Reaction type

oxidative decarboxylation

Natural substrates and products

S Additional information <2, 3, 4, 5> (<3,5> enzyme catalyses the CoA-dependent oxidation of branched-chain 2-ketoacids [1,2,3]; <2> enzyme is involved in the biosynthesis of amino acids from fatty acids take up from the grown medium [4]; <5> enzyme functions in the biosynthesis of branched-chain amino acids from volatile fatty acids that are excreted under nutrient-sufficient conditions [3]) (Reversibility: ?) [1, 2, 3, 4]
P ?

Substrates and products

S 2-keto-3-methylvalerate + CoA + oxidized ferredoxin <1, 3, 4, 5> (Reversibility: ?) [3]
P S-(3-methylpropanoyl)-CoA + CO_2 + reduced ferredoxin
S 2-ketobutyrate + CoA + oxidized ferredoxin <1, 2, 3, 4, 5> (<2> benzyl viologen as electron acceptor [4]) (Reversibility: r) [2, 3, 4]
P propanoyl-CoA + CO_2 + reduced ferredoxin
S 2-ketocaproate + CoA + oxidized ferredoxin <1, 3, 4, 5> (Reversibility: ?) [3]
P pentanoyl-CoA + CO_2 + reduced ferredoxin
S 2-ketoisocaproate + CoA + oxidized ferredoxin <1, 3, 4, 5> (Reversibility: r) [2, 3]
P S-(3-methylbutanoyl)-CoA + CO_2 + reduced ferredoxin
S 2-ketoisohexanoate + CoA + benzyl viologen <2> (Reversibility: ?) [4]
P isopentanoyl-CoA + CO_2 + reduced benzyl viologen
S 2-ketoisovalerate + CoA + benzyl viologen <2> (Reversibility: ?) [4]
P S-(2-methylpropanoyl)-CoA + CO_2 + reduced benzyl viologen
S 2-ketoisovalerate + CoA + oxidized ferredoxin <1, 3, 4, 5> (<5> or benzyl viologen or methyl viologen as electron acceptor [3]) (Reversibility: r) [2, 3]
P S-(2-methylpropanoyl)-CoA + CO_2 + reduced ferredoxin
S 2-ketomethylthiobutyrate + CoA + oxidized ferredoxin <1, 3, 4, 5> (Reversibility: r) [2, 3]
P 2-ketomethylthiopropanoyl-CoA + CO_2 + reduced ferredoxin
S 2-ketovalerate + CoA + benzyl viologen <2> (Reversibility: ?) [4]
P butanoyl-CoA + CO_2 + reduced benzyl viologen
S glyoxylate + CoA + oxidized ferredoxin <1, 3, 4, 5> (Reversibility: ?) [3]
P formyl-CoA + CO_2 + reduced ferredoxin
S hydroxyphenylpyruvate + CoA + oxidized ferredoxin <1, 3, 4, 5> (Reversibility: r) [2, 3]
P 4-hydroxyphenylacetyl-CoA + CO_2 + reduced ferredoxin
S indolpyruvate + CoA + oxidized ferredoxin <1, 3, 4, 5> (Reversibility: r) [2, 3]
P indolacetyl-CoA + CO_2 + reduced ferredoxin

S isobutyryl-CoA + CO_2 + reduced benzyl viologen <1, 3, 4, 5> (Reversibility: ?) [3]
P 2-oxoisopentanoate + CoA + benzyl viologen
S phenylglyoxylate + CoA + oxidized ferredoxin <1, 3, 4, 5> (Reversibility: r) [2, 3]
P benzoyl-CoA + CO_2 + reduced ferredoxin
S phenylpyruvate + CoA + oxidized ferredoxin <1, 2, 3, 4, 5> (<2> benzyl viologen as electron acceptor [4]) (Reversibility: r) [2, 3, 4]
P phenylacetyl-CoA + CO_2 + reduced ferredoxin
S pyruvate + CoA + oxidized ferredoxin <1, 3, 4, 5> (Reversibility: r) [2, 3]
P acetyl-CoA + CO_2 + reduced ferredoxin
S Additional information <1, 2, 3, 4, 5> (<2> no substrate are 2-ketoglutarate and pyruvate [4]; <1,3,4,5> enzyme catalyses the CoA-dependent oxidation of branched-chain 2-ketoacids [1,2,3]; <1,3,4,5> no substrate are 2-ketoglutarate, 2-ketomalonate, imidazolpyruvate [3]; <3,5> enzyme catalyses also the nonoxidative decarboxylation of pyruvate to acetaldehyde [2]; <3,5> no substrate is 2-ketoglutarate [2]; <2> enzyme is involved in the biosynthesis of amino acids from fatty acids take up from the grown medium [4]; <5> enzyme functions in the biosynthesis of branched-chain amino acids from volatile fatty acids that are excreted under nutrient-sufficient conditions [3]) (Reversibility: ?) [1, 2, 3, 4]
P ?

Inhibitors

EDTA <5> (<5> 80°C, complete loss of activity [3]) [3]
iodoacetate <5> (<5> 23°C, in a concentration-dependent manner [3]) [3]

Activating compounds

thiamine diphosphate <1, 3, 5> (<5> enzyme contains 0.17 mol per 230000 g of protein [3]; <1> K_m 0.076 mM [3]; <5> K_m 0.3 mM [3]; <5> thiamine diphosphate can be replaced by thiamine or thiamine monophosphate [3]; <1,3,5> requires for maximal activity of enzyme [1,2,3]) [1, 2, 3]

Metals, ions

Fe <3, 5> (<3,5> each of the tetramer contains three [4Fe-4S] cluster [1,2]; <5> 11 iron atoms per heterotetramer and two types of FeS are evident [3]) [1, 2, 3]
Mg^{2+} <1, 5> (<1,5> required for maximal activity of enzyme [3]; <5> K_m 3.00 mM [3]; <1> K_m 0.89 mM [3]) [3]

Turnover number (min^{-1})

0.0133 <5> (glyoxylate, <5> pH 6.9, 85°C, under argon [3]) [3]
0.267 <5> (pyruvate, <5> pH 6.9, 85°C, under argon [3]) [3]
0.433 <5> (indolpyruvate, <5> pH 6.9, 85°C, under argon [3]) [3]
0.467 <1> (phenylpyruvate, <1> pH 6.9, 85°C, under argon [3]) [3]
0.6 <5> (phenylpyruvate, <5> pH 6.9, 85°C, under argon [3]) [3]
0.667 <5> (2-ketomethylthiobutyrate, <5> pH 6.9, 85°C, under argon [3]) [3]
0.833 <5> (hydroxyphenylpyruvate, <5> pH 6.9, 85°C, under argon [3]) [3]
0.833 <5> (phenylglyoxalate, <5> pH 6.9, 85°C, under argon [3]) [3]

0.933 <1> (2-ketocaproate, <1> pH 6.9, 85°C, under argon [3]) [3]
1 <5> (2-ketocaproate, <5> pH 6.9, 85°C, under argon [3]) [3]
1.07 <1> (2-ketomethylisobutyrate, <1> pH 6.9, 85°C, under argon [3]) [3]
1.27 <5> (2-keto-3-methylvalerate, <5> pH 6.9, 85°C, under argon [3]) [3]
1.33 <1> (phenylglyoxylate, <1> pH 6.9, 85°C, under argon [3]) [3]
1.47 <1> (2-ketobutyrate, <1> pH 6.9, 85°C, under argon [3]) [3]
1.53 <5> (gerredoxin, <5> pH 6.9, 85°C, under argon [3]) [3]
1.73 <1> (pyruvate, <1> pH 6.9, 85°C, under argon [3]) [3]
2 <5> (2-ketoisocaproate, <5> pH 6.9, 85°C, under argon [3]) [3]
2.07 <1> (2-keto-3-methylvalerate, <1> pH 6.9, 85°C, under argon [3]) [3]
2.47 <1> (2-ketoisocaproate, <1> pH 6.9, 85°C, under argon [3]) [3]
3 <5> (2-ketoisovalerate, <5> pH 6.9, 85°C, under argon [3]) [3]
3.67 <1> (2-ketoisovalerate, <1> pH 6.9, 85°C, under argon [3]) [3]
5 <5> (CoA, <5> pH 6.9, 85°C, under argon [3]) [3]
5.6 <1> (CoA, <1> pH 6.9, 85°C, under argon [3]) [3]
22.3 <5> (2-ketobutyrate, <5> pH 6.9, 85°C, under argon [3]) [3]

Specific activity (U/mg)

1 <5> [3]
13.8 <2> [4]
46 <5> [3]

K_m-Value (mM)

0.004 <2> (CoA, <2> pH 8.0, 65°C, anaerobic [4]) [4]
0.0096 <1> (CoA, <1> pH 6.9, 85°C, under argon [3]) [3]
0.012 <5> (2-keto-3-methylvalerate, <5> pH 6.9, 85°C, under argon [3]) [3]
0.017 <5> (ferredoxin, <5> pH 6.9, 85°C, under argon [3]) [3]
0.026 <5> (2-ketoisocaproate, <5> pH 6.9, 85°C, under argon [3]) [3]
0.048 <5> (CO_2, <5> pH 7.0, 85°C, under argon [3]) [3]
0.049 <5> (2-ketocaproate, <5> pH 6.9, 85°C, under argon [3]) [3]
0.05 <5> (CoA, <5> pH 6.9, 85°C, under argon [3]) [3]
0.058 <5> (2-ketomethylthiobutyrate, <5> pH 6.9, 85°C, under argon [3]) [3]
0.094 <5> (2-ketoisovalerate, <5> pH 6.9, 85°C, under argon [3]) [3]
0.11 <1> (2-ketoisovalerate, <1> pH 6.9, 85°C, under argon [3]) [3]
0.15 <1> (2-keto-3-methylvalerate, <1> pH 6.9, 85°C, under argon [3]) [3]
0.17 <1> (2-ketobutyrate, <1> pH 6.9, 85°C, under argon [3]) [3]
0.176 <1> (2-ketoisocaproate, <1> pH 6.9, 85°C, under argon [3]) [3]
0.22 <5> (phenylpyruvate, <5> pH 6.9, 85°C, under argon [3]) [3]
0.25 <5> (isobutyryl-CoA, <5> pH 7.0, 85°C, under argon [3]) [3]
0.276 <1> (2-ketomethylisobutyrate, <1> pH 6.9, 85°C, under argon [3]) [3]
0.311 <1> (2-ketocaproate, <1> pH 6.9, 85°C, under argon [3]) [3]
0.38 <5> (2-ketobutyrate, <5> pH 6.9, 85°C, under argon [3]) [3]
0.466 <1> (phenylglyoxylate, <1> pH 6.9, 85°C, under argon [3]) [3]
0.6 <5> (phenylglyoxalate, <5> pH 6.9, 85°C, under argon [3]) [3]
0.75 <5> (pyruvate, <5> pH 6.9, 85°C, under argon [3]) [3]
1.3 <2> (2-ketoisovalerate, <2> pH 8.0, 65°C, anaerobic [4]) [4]
2.8 <1> (pyruvate, <1> pH 6.9, 85°C, under argon [3]) [3]
Additional information <3, 5> [2]

pH-Optimum
7 <5> [3]
9.7 <2> (<2> at 65°C [4]) [4]

Temperature optimum (°C)
75 <2> (<2> at pH 8.0 [4]) [4]
90 <5> (<5> ketoisovalerate oxidation [3]) [3]
98 <1> (<1> ketoisovalerate oxidation [3]) [3]

Temperature range (°C)
20 <1, 5> (<1,5> enzyme is inactive at 20°C [3]) [3]

4 Enzyme Structure

Molecular weight
230000 <5> (<5> calculated from amino acid sequence [3]) [3]

Subunits
? <2, 3, 5> (<2> x * 55000 + x * 37000 + x * 15000, SDS-PAGE [4]; <2,3,5> comparison of N-terminal amino acid sequences [4]; <3> $\alpha,\beta,\gamma,\delta$, x * 43960 + x * 34766 + x * 20033 + x * 11851, SDS PAGE, amino acid sequences [1]) [1, 4]
octamer <1, 5> (<1,5> $\alpha,\beta,\gamma,\delta$, 2 * 47000 + 2 * 34000 + 2 * 23000 + 2 * 13000, SDS PAGE [2,3]) [2, 3]

5 Isolation/Preparation/Mutation/Application

Purification
<1> (under anaerobic conditions) [3]
<2> (under anaerobic conditions) [4]
<3> (under anaerobic conditions) [2, 3]
<4> (under anaerobic conditions) [3]
<5> (under anaerobic conditions) [3]

Cloning
<3> [1, 2]

6 Stability

pH-Stability
6-8 <5> (<5> pH 6.0: less than 50% loss of activity, pH 8.0: 25% loss of activity [3]) [3]

Temperature stability
85 <5> (<5> pH 8.0, Tris-HCl, half-life 3 h [3]) [3]
95 <5> (<5> pH 8.0, Tris-HCl, half-life 50 min [3]) [3]

Oxidation stability

<3>, enzyme is irreversibly inactivated by oxygen [2]
<5>, about 50% of the enzyme activity in the cell extract is lost after exposure to air for 2 min, no activity loss if the extract is under argon [3]

References

[1] Kletzin, A.; Adams, M.W.W.: Molecular and phylogenetic chararcterization of pyruvate and 2-ketoisovalerate ferredoxin oxidoreductases from Pyrococcus furiosus and pyruvate ferredoxin oxidoreductase from Thermotoga maritima. J. Bacteriol., **178**, 248-257 (1996)
[2] Schut, G.J.; Menon, A.L.; Adams, M.W.W.: 2-Keto acid oxidoreductases from Pyrococcus furiosus and Thermococcus litoralis. Methods Enzymol., **331**, 144-158 (2001)
[3] Heider, J.; Mai, X.; Adams, M.W.: Characterization of 2-ketoisovalerate ferredoxin oxidoreductase, a new and reversible coenzyme A-dependent enzyme involved in peptide fermentation by hyperthermophilic archaea. J. Bacteriol., **178**, 780-787 (1996)
[4] Tersteegen, A.; Linder, D.; Thauer, R.K.; Hedderich, R.: Structures and functions of four anabolic 2-oxoacid oxidoreductases in Methanobacterium thermoautotrophicum. Eur. J. Biochem., **244**, 862-868 (1997)

Indolepyruvate ferredoxin oxidoreductase 1.2.7.8

1 Nomenclature

EC number
1.2.7.8

Systematic name
3-(indol-3-yl)pyruvate:ferredoxin oxidoreductase (decarboxylating, CoA-indole-acetylating)

Recommended name
indolepyruvate ferredoxin oxidoreductase

Synonyms
3-(indol-3-yl)pyruvate synthase (ferredoxin)
IOR
arylpyruvate-ferredoxin oxidoreductase
hydroxyphenylpyruvate oxidase
indolepyruvate oxidase
phenylpyruvate oxidase

CAS registry number
158886-06-7

2 Source Organism

<1> *Methanobacterium thermoautotrophicum* (no sequence specified) [1, 2]
<2> *Pyrococcus furiosus* (no sequence specified) [1, 3, 5]
<3> *Pyrococcus kodakaraensis* (no sequence specified) [4, 5]
<4> *Pyrococcus kodakaraensis* (UNIPROT accession number: D86221) [4]

3 Reaction and Specificity

Catalyzed reaction
(indol-3-yl)pyruvate + CoA + oxidized ferredoxin = S-2-(indol-3-yl)acetyl-CoA + CO_2 + reduced ferredoxin

Reaction type
oxidative decarboxylation

Natural substrates and products

S (indol-3-yl)pyruvate + CoA + oxidized ferredoxin <1, 2, 3> (<1> biosynthesis of amino acids from fatty acids [2]; <2> involved in peptide fermentation and metabolism of aromatic amino acids [3]) (Reversibility: r) [1, 2, 3, 5]

P S-2-(indol-3-yl)acetyl-CoA + CO_2 + reduced ferredoxin

Substrates and products

S (indol-3-yl)pyruvate + CoA + oxidized ferredoxin <1, 2, 3> (<1> biosynthesis of amino acids from fatty acids [2]; <2> involved in peptide fermentation and metabolism of aromatic amino acids [3]) (Reversibility: r) [1, 2, 3, 5]

P S-2-(indol-3-yl)acetyl-CoA + CO_2 + reduced ferredoxin

S 2-keto-4-methylthiobutyrate + CoA + oxidized ferredoxin <2> (Reversibility: r) [1, 3]

P CO_2 + reduced ferredoxin + 3-methylthiopropionyl-CoA

S 2-ketoisocaproate + CoA + oxidized ferredoxin <2> (Reversibility: r) [1, 3]

P CO_2 + reduced ferredoxin + isovaleryl-CoA

S indolepyruvate + CoA + methylviologen <2, 3> (Reversibility: ?) [1, 5]

P indoleacetyl-CoA + CO_2 + reduced methylviologen

S *p*-hydroxyphenylpyruvate + CoA + oxidized ferredoxin <2, 3> (Reversibility: r) [1, 3, 5]

P *p*-hydroxyphenylacetyl-CoA + CO_2 + reduced ferredoxin

S phenylpyruvate + CoA + oxidized ferredoxin <1, 2, 3> (Reversibility: r) [1, 2, 3, 5]

P phenylacetyl-CoA + CO_2 + reduced ferredoxin

S Additional information <2> (<2> pyruvate and transaminated forms of the amino acids aspartate, valine, glutamate and alanine, 2-ketobutyrate, oxalacetate, 2-ketoisovalerate, 2-ketoglutarate, 2-ketomalonate and phenylglyoxylate are not oxidized [3]; <2> pyruvate, 2-ketoisovalerate, 2-ketoglutarate, 2 ketobutyrate, and phenylglyoxylate are no substrates [1]) (Reversibility: ?) [1, 3]

P ?

Inhibitors

Cu^{2+} <2> [3]
cyanide <2> [3]
KCN <2> [3]
Zn^{2+} <2> [3]
Additional information <2> (<2> not inhibited by CO, unaffected by sodium nitrite, not inhibited by sodium fluoride or sodium azide, sodium benzoate, phenylacetate, pyruvate, 2-ketoglutarate and oxamate have no effect on activity [3]) [3]

Activating compounds

thiamine diphosphate <2, 4> (<2> activity increases about 8fold, K_m for thiamine diphosphate 0.004 mM [3]) [1, 3, 4]

Specific activity (U/mg)

31.15 <1> [2]
38 <2> [3]

K_m-Value (mM)

0.017 <2> (CoA, <2> pH 8.4, 80°C, substrate indolepyruvate [3]) [1, 3]
0.02 <1> (CoA, <1> pH 8.0, 65°C, substrate phenylpyruvate, benzyl viologen as electron acceptor [2]) [2]
0.021 <2> (CoA, <2> pH 8.4, 80°C, substrate phenylpyruvate [3]) [3]
0.048 <2> (ferredoxin, <2> pH 8.4, 80°C [3]) [1, 3]
0.095 <2> (phenylpyruvate, <2> pH 8.4, 80°C [3]) [3]
0.11 <2> (*p*-hydroxyphenylpyruvate, <2> pH 8.4, 80°C [3]) [3]
0.25 <2> (indolepyruvate, <2> pH 8.4, 80°C [3]) [3]
1.4 <1> (phenylpyruvate, <1> pH 8.0, 65°C [2]) [2]

K_i-Value (mM)

7.5 <2> (KCN, <2> pH 8.4, 80°C [3]) [3]

pH-Optimum

8.5-10.5 <2> (<2> at 80°C [3]) [3]
10 <1> [2]

Temperature optimum (°C)

70 <3, 4> [4, 5]
80 <1> [2]
90 <2> (<2> exhibits optimal activity above at pH 8.0 [3]) [3]

Temperature range (°C)

45-80 <3> [5]
70-90 <2> (<2> shows dramatic increase in activity above 70°C [3]) [3]

4 Enzyme Structure

Molecular weight

180000 <2> (<2> gel filtration [3]) [3]
192000 <3> (<3> gel filtration [5]) [5]

Subunits

tetramer <1, 2, 3, 4> (<2> 2 * 66000 + 2 * 23000, α_2,β_2, SDS-PAGE [1,3]; <2> 2 * 71000 + 2 * 24000, α_2,β_2, SDS-PAGE [1,5]; <1> 2 * 67000 + 2 * 23000, α_2,β_2, SDS-PAGE [2]) [1, 2, 3, 4, 5]

5 Isolation/Preparation/Mutation/Application

Localization

cytoplasm <2> [3]

Purification

<1> [2]
<2> [1, 3]
<3> [5]

Cloning

<2> [1]
<3> (gene cloning and sequence analysis of the IOR gene, 2 genes iorA and iorB, encoding α and β subunits are tandemly arranged, KOD1-IOR overproduced in anaerobically incubated Escherichia coli BL21(DE3)) [4, 5]
<4> (gene cloning and sequence analysis of the IOR gene, 2 genes iorA and iorB, encoding α and β subunits are tandemly arranged, KOD1-IOR overproduced in anaerobically incubated Escherichia coli BL21(DE3)) [4]

6 Stability

pH-Stability

8-10 <2> (<2> not active at pH 5.2 and pH 11.0 [3]) [3]

Temperature stability

80 <2> (<2> virtually inactive at 25°C, quite thermostable, with a half-life of activity at 80°C of about 80 min under anaerobic conditions [3]) [3]

Oxidation stability

<2>, enzyme is sensitive to inactivation by O_2, losing 50% of its activity after exposure to air for 20 min at 23°C [3]
<2>, irreversibly inactivated by oxygen [1]
<4>, half-life in presence of air is 15 min at 25°C [4]

References

[1] Schut, G.J.; Menon, A.L.; Adams, M.W.W.: 2-Keto acid oxidoreductases from Pyrococcus furiosus and Thermococcus litoralis. Methods Enzymol., **331**, 144-158 (2001)
[2] Tersteegen, A.; Linder, D.; Thauer, R.K.; Hedderich, R.: Structures and functions of four anabolic 2-oxoacid oxidoreductases in Methanobacterium thermoautotrophicum. Eur. J. Biochem., **244**, 862-868 (1997)
[3] Mai, X.; Adams, M.W.: Indolepyruvate ferredoxin oxidoreductase from the hyperthermophilic archaeon Pyrococcus furiosus. A new enzyme involved in peptide fermentation. J. Biol. Chem., **269**, 16726-16732 (1994)
[4] Siddiqui, M.A.; Fujiwara, S.; Imanaka, T.: Indolepyruvate ferredoxin oxidoreductase from Pyrococcus sp. KOD1 possesses a mosaic structure showing features of various oxidoreductases. Mol. Gen. Genet., **254**, 433-439 (1997)
[5] Siddiqui, M.A.; Fujiwara, S.; Takagi, M.; Imanaka, T.: In vitro heat effect on heterooligomeric subunit assembly of thermostable indolepyruvate ferredoxin oxidoreductase. FEBS Lett., **434**, 372-376 (1998)

2-Oxoglutarate ferredoxin oxidoreductase 1.2.7.9

1 Nomenclature

EC number
1.2.7.9 (deleted 2005. This enzyme is identical to EC 1.2.7.3, 2-oxoglutarate synthase)

Recommended name
2-oxoglutarate ferredoxin oxidoreductase

2 Source Organism

<1> *Halobacterium halobium* (no sequence specified) [1, 5]
<2> *Hydrogenobacter thermophilus* (no sequence specified) [4]
<3> *Hydrogenobacter thermophilus* (UNIPROT accession number: Q9AJM0, Q9AJL9) [2]
<4> *Hydrogenobacter thermophilus* (UNIPROT accession number: Q93RA0, Q93RA1, Q93RA2, Q93RA4) [3]

3 Reaction and Specificity

Catalyzed reaction
2-oxoglutarate + CoA + 2 oxidized ferredoxin = succinyl-CoA + CO_2 + 2 reduced ferredoxin

References

[1] Kerscher, L.; Oesterhelt, D.: Purification and properties of two 2-oxoacid:ferredoxin oxidoreductases from Halobacterium halobium. Eur. J. Biochem., **116**, 587-594 (1981)
[2] Yun, N.-R.; Arai, H.; Ishii, M.; Igarashi, Y.: The genes for anabolic 2-oxoglutarate:ferredoxin oxidoreductase from Hydrogenobacter thermophilus TK-6. Biochem. Biophys. Res. Commun., **282**, 589-594 (2001)
[3] Yun, N.-R.; Yamamoto, M.; Arai, H.; Ishii, M.; Igarashi, Y.: A novel five-subunit-type 2-oxoglutalate:ferredoxin oxidoreductases from Hydrogenobacter thermophilus TK-6. Biochem. Biophys. Res. Commun., **292**, 280-286 (2002)

[4] Yamamoto, M.; Arai, H.; Ishii, M.; Igarashi, Y.: Characterization of two different 2-oxoglutarate:ferredoxin oxidoreductases from Hydrogenobacter thermophilus TK-6. Biochem. Biophys. Res. Commun., **312**, 1297-1302 (2003)
[5] Cammack, R.; Kerscher, L.; Oesterhelt, D.: A stable free radical intermediate in the reaction of 2-oxoacid:ferredoxin oxidoreductases of Halobacterium halobium. FEBS Lett., **118**, 271-273 (1980)

Aldehyde dehydrogenase (FAD-independent) 1.2.99.7

1 Nomenclature

EC number
1.2.99.7

Systematic name
aldehyde:acceptor oxidoreductase (FAD-independent)

Recommended name
aldehyde dehydrogenase (FAD-independent)

Synonyms
AOR <3> [4]
AORDd
aldehyde oxidoreductase
BV-AIDH <2> [1]
DgAOR <2> [10]
MOP <2> [11]
MOP molybdenum-containing protein <2> [9]
aldehyde oxidase

CAS registry number
9029-07-6 (cf. EC 1.2.3.1)

2 Source Organism

<1> *Pseudomonas sp.* (no sequence specified) [3]
<2> *Desulfovibrio gigas* (no sequence specified) [1, 2, 5, 6, 7, 8, 9, 10, 11, 12]
<3> *Desulfovibrio alaskensis* (no sequence specified) [4]

3 Reaction and Specificity

Catalyzed reaction
an aldehyde + H_2O + acceptor = a carboxylate + reduced acceptor

Reaction type
oxidation
redox reaction
reduction

Substrates and products

S 3-pyridinecarboxaldehyde + H_2O + 2,6-dichlorophenol-indophenol <2> (<2> 159% of activity with acetaldehyde [6]) (Reversibility: ?) [6]
P 3-pyridinecarboxylate + reduced 2,6-dichlorophenol-indophenol
S acetaldehyde + H_2O + 2,6-dichlorophenolindophenol <2> (Reversibility: ?) [2, 6, 12]
P acetate + reduced 2,6-dichlorophenolindophenol
S acetaldehyde + H_2O + acceptor <2> (Reversibility: ?) [11]
P acetate + reduced acceptor
S acetaldehyde + H_2O + acceptor <2> (<2> aldehyde oxidoreductase does not react with O_2 [5]) (Reversibility: ?) [5]
P actetate + reduced acceptor
S acetaldehyde + H_2O + benzylviologen <2> (<2> natural electron aceptor not yet known, the artificial electron acceptors benzylviologen, methylviologen, 2,3-bis-(2-methoxy-4-nitro-5-sulfophenyl) and 2,6-dichlorphenol-indophenol have relative activities of 100%, 19%, 5%, and 1.3%, respectively [1]) (Reversibility: ?) [1]
P acetate + reduced benzylviologen
S acetaldehyde + H_2O + electron acceptor <2> (Reversibility: ?) [7, 9]
P acetate + reduced acceptor
S acetaldehyde + H_2O + electron acceptor <2> (<2> DgAOR probably reduces aldehydes to alcohols [10]) (Reversibility: ?) [8, 1]
P acetate + reduced electron acceptor
S aldehyde + H_2O + acceptor <1, 3> (Reversibility: ?) [3, 4]
P carboxylate + reduced acceptor
S benzaldehyde + H_2O + 2,6-dichlorophenol-indophenol <2, 3> (Reversibility: ?) [4, 6]
P benzoate + reduced 2,6-dichlorophenol-indophenol
S benzaldehyde + H_2O + benzylviologen <2> (<2> 16% of activity with acetaldehyde [1]; <2> 93% of activity with acetaldehyde [1]) (Reversibility: ?) [1]
P benzoate + reduced benzylviologen
S benzaldehyde + H_2O + cytochrome c <2> (<2> from horse heart, 25% of activity with 2,6-dichlorophenol-indophenol [6]) (Reversibility: ?) [6]
P benzoate + reduced cytochrome c
S benzaldehyde + H_2O + cytochrome c_3 <2> (<2> from, Desulfovibrio gigas, 20% of activity with 2,6-dichlorophenol-indophenol [6]) (Reversibility: ?) [6]
P benzoate + reduced cytochrome c_3
S benzaldehyde + H_2O + electron acceptor <2> (Reversibility: ?) [10]
P benzoate + reduced electron acceptor
S benzaldehyde + H_2O + ferrocyanide <2> (<2> 156% of activity with 2,6-dichlorophenol-indophenol [6]) (Reversibility: ?) [6]
P benzoate + ferrocyanide
S benzaldehyde + H_2O + methylene blue <2> (<2> 156% of activity with 2,6-dichlorophenol-indophenol [6]) (Reversibility: ?) [6]
P benzoate + reduced methylene blue

S butyraldehyde + H_2O + benzylviologen <2> (<2> 93% of activity with acetaldehyde [1]) (Reversibility: ?) [1]
P butanoate + reduced benzylviologen
S formaldehyde + H_2O + benzylviologen <2> (<2> 16% of activity with acetaldehyde [1]) (Reversibility: ?) [1]
P formate + reduced benzylviologen
S furfural + H_2O + benzylviologen <2> (<2> 77% of activity with acetaldehyde [1]) (Reversibility: ?) [1]
P furfurate + reduced benzylviologen
S furfuraldehyde + H_2O + 2,6-dichlorophenol-indophenol <2> (<2> 27% of activity with acetaldehyde [6]) (Reversibility: ?) [6]
P furfurate + reduced 2,6-dichlorophenol-indophenol
S glutaraldehyde + H_2O + benzylviologen <2> (<2> 20% of activity with acetaldehyde [1]) (Reversibility: ?) [1]
P glutarate + reduced benzylviologen
S glycoaldehyde + H_2O + benzylviologen <2> (<2> 23% of activity with acetaldehyde [1]) (Reversibility: ?) [1]
P glycolate + reduced benzylviologen
S hexanal + H_2O + benzylviologen <2> (<2> 6% of activity with acetaldehyde [1]) (Reversibility: ?) [1]
P hexanoate + reduced benzylviologen
S octanal + H_2O + benzylviologen <2> (<2> 93% of activity with acetaldehyde [1]) (Reversibility: ?) [1]
P octanoate + reduced benzylviologen
S *p*-anisaldehyde + H_2O + benzylviologen <2> (<2> 5% of activity with acetaldehyde [1]) (Reversibility: ?) [1]
P *p*-anisidic acid + reduced benzylviologen
S pentanal + H_2O + benzylviologen <2> (<2> 9% of activity with acetaldehyde [1]) (Reversibility: ?) [1]
P pentanoate + reduced benzylviologen
S phenylacetaldehyde + H_2O + benzylviologen <2> (<2> 31% of activity with acetaldehyde [1]) (Reversibility: ?) [1]
P phenylacetate + reduced benzylviologen
S propionaldehyde + H_2O + 2,6-dichlorophenol-indophenol <2> (Reversibility: ?) [6]
P propionate + reduced 2,6-dichlorophenol-indophenol
S propionaldehyde + H_2O + 2,6-dichlorophenolindophenol <2> (Reversibility: ?) [2]
P propionate + reduced 2,6-dichlorophenolindophenol
S propionaldehyde + H_2O + benzylviologen <2> (<2> 117% of activity with acetaldehyde [1]) (Reversibility: ?) [1]
P propionate + reduced benzylviologen
S salicylaldehyde + H_2O + 2,6-dichlorophenol-indophenol <2> (Reversibility: ?) [6]
P salicylate + reduced 2,6-dichlorophenol-indophenol

S salicylaldehyde + H_2O + 2,6-dichlorophenolindophenol <2> (Reversibility: ?) [2]
P salicylate + reduced 2,6-dichlorophenolindophenol
S salicylaldehyde + H_2O + benzylviologen <2> (<2> 14% of activity with acetaldehyde [1]) (Reversibility: ?) [1]
P salicylate + reduced benzylviologen

Inhibitors

arsenite <2> (<2> 5 mM, complete inhibition [1]) [1]
iodoacetate <2> (<2> 5 mM, complete inhibition [1]) [1]
KCN <2> (<2> 5 mM, 64% inhibition [1]) [1]
methanol <2> (<2> 1 M, 62% inhibition [1]) [1]

Metals, ions

iron <2, 3> (<2> molybdenum iron-sulfur protein [2]; <2> 4.8 iron per subunit, probably a [Fe-4S]1+ cluster [1]; <3> contains Fe-S cluster, 3.6 iron atoms per monomer [4]; <2> contains two [2Fe-2S] cluster [12]; <2> contains two [2Fe-2S] clusters [11]; <2> contains [2Fe-2S] cluster [6]; <2> DgAOR contains two spectroscopically distinguishable [2Fe-2S] centers [10]; <2> the cluster ligated by a ferredoxin-type motif is close to the protein surface, whereas that ligated by an unusual cysteine motif is in contact with the molybtopterin [7]; <2> two iron-sulfur centers of the [2Fe-2S] type [5]; <2> two [2Fe-2S] center [9]; <2> [Fe-2S] cluster [8]) [1, 2, 4, 5, 6, 7, 8, 9, 10, 11, 12]
K^+ <2> (<2> 10 mM, 50% activation [1]) [1]
molybdenum <2, 3> (<2> contains molybdopterin [12]; <2> molybdenum iron-sulfur protein [2,9]; <2> bound to a pterin cofactor and [2Fe-2S] cluster [8]; <2> contains a molybdopterin cytosine dinucleotide [11]; <3> contains molybdopterin cytosine dinucleotide cluster, 0.6 molybdenum atoms per monomer [4]; <2> molybdenum reductase [6]; <2> redox center formed by a fivefold-co-ordinated Mo atom bound to two oxygen ligands, one sulfur and one molybdopterin cytosin dinucleotide [10]; <2> the cluster ligated by a ferredoxin-type motif is close to the protein surface, whereas that ligated by an unusual cysteine motif is in contact with the molybtopterin [7]) [2, 4, 5, 6, 7, 8, 9, 10, 11, 12]
tungsten <2> (<2> 0.68 tungsten per subunit [1]) [1]

Turnover number (min^{-1})

0.007 <2> (propionaldehyde, <2> 25°C, pH 8.2 [2]) [2]
0.12 <2> (acetaldehyde, <2> 25°C, pH 8.2 [2]) [2]
0.15 <2> (salicylaldehyde, <2> 25°C, pH 7.6 [6]) [6]
0.25 <2> (salicylaldehyde, <2> 25°C, pH 8.2 [2]) [2]
0.44 <2> (propionaldehyde, <2> 25°C, pH 7.6 [6]) [6]
0.96 <3> (benzaldehyde) [4]
1.14 <2> (acetaldehyde, <2> 25°C, pH 7.6 [6]) [6]
1.43 <2> (benzaldehyde, <2> 25°C, pH 7.6 [6]) [6]

Specific activity (U/mg)

0.58 <3> [4]
1.17 <2> [6]
38.5 <2> [1]

K_m-Value (mM)

0.00025 <2> (salicylaldehyde, <2> 25°C, pH 7.6 [6]) [6]
0.0067 <3> (benzaldehyde) [4]
0.013 <2> (acetaldehyde, <2> 25°C, pH 7.6 [6]) [6]
0.015 <2> (propionaldehyde, <2> 25°C, pH 7.6 [6]) [6]
0.052 <2> (benzaldehyde, <2> 25°C, pH 7.6 [6]) [6]
0.55 <2> (benzylviologen, <2> 30°C, pH 7.5 [1]) [1]
10.8 <2> (propionaldehyde, <2> 30°C, pH 7.5 [1]) [1]
12.5 <2> (acetaldehyde, <2> 30°C, pH 7.5 [1]) [1]
20 <2> (benzaldehyde, <2> 30°C, pH 7.5 [1]) [1]
30.8 <2> (furfural, <2> 30°C, pH 7.5 [1]) [1]

pH-Optimum

7.8 <2> [6]
8 <2> (<2> broad optimum [1]) [1]

Temperature range (°C)

30-65 <2> [1]

4 Enzyme Structure

Molecular weight

120000 <2> (<2> gelfiltration, Sephacryl S-300 [1]) [1]
126000 <2> (<2> gel filtration, Superose 12 HR [1]) [1]
200000 <2> (<2> gel filtration [8]) [8]
206000 <3> (<3> gel filtration [4]) [4]

Subunits

dimer <2, 3> (<2> 2 * 62000, SDS-PAGE [1]; <2> 2 * 100000, SDS-PAGE [8]; <3> 2 * 96000, SDS-PAGE [4]) [1, 4, 8]

5 Isolation/Preparation/Mutation/Application

Purification

<2> [2, 5, 7, 8, 10, 12]
<2> (DEAE-52, DEAE-Biogel, hydroxylapatite, HPLC) [6]
<2> (Q-Sepharose, Butyl-Sepharose, DEAE-memsep 1010) [1]
<2> (aerobic purification) [11]
<3> (DEAE-cellulose, Source-15, Superdex 200) [4]

Crystallization

<2> [7]

<2> (crystal structure at 2.25 A resolution) [12]

<2> (vapour diffusion on sitting drops using a mixture of purified protein in 10 mM Tris-HCl, pH 7.5, and a crystallization solution of 30% isopropanol as precipitant and 200 mM $MgCl_2$ as additive in 200 mM HEPES, pH 7.6, growth of crystals at 4°C takes approx. 3 weeks, crystals diffract to 1.28 A) [11]

<2> (vapour-diffusion using sitting drops and a reservoir containing 100 mM Hepes, pH 7.5, 200 mM $MgCl_2$ and 30% isopropanol, droplets are prepared by mixing 0.004 ml of a 13 mg/ml protein solution in 10 mM Tris, pH 7.6 with 0.002 ml reservoir solution, single crystals are obtained within 3-6 weeks at 4°C, crystals diffract to 3.0 A) [8]

Cloning

<2> [9]

6 Stability

Oxidation stability

<2>, exposure to air leads to inactivation [1]

Storage stability

<2>, -30°C, phosphate buffer, no loss of activity after thawing [6]

References

[1] Hensgens, C.M.H.; Hagen, W.R.; Hansen, T.A.: Purification and characterization of a benzylviologen-linked tungsten-containing aldehyde oxidoreductase from Desulfovibrio gigas. J. Bacteriol., **177**, 6195-6200 (1995)

[2] Turner, N.; Barata, B.; Bray, R.C.; Deistung, J.; Le Gall, J.; Moura, J.J.G.: The molybdenum iron-sulphur protein from Desulfovibrio gigas as a form of aldehyde oxidase. Biochem. J., **243**, 755-761 (1987)

[3] Uchida, H.; Kondo, D.; Yamashita, A.; Nagaosa, Y.; Sakurai, T.; Fujii, Y.; Fujishiro, K.; Aisaka, K.; Uwajima, T.: Purification and characterization of an aldehyde oxidase from Pseudomonas sp. KY 4690. FEMS Microbiol. Lett., **229**, 31-36 (2003)

[4] Andrade, S.L.; Brondino, C.D.; Feio, M.J.; Moura, I.; Moura, J.J.: Aldehyde oxidoreductase activity in Desulfovibrio alaskensis NCIMB 13491. EPR assignment of the proximal [2Fe-2S] cluster to the Mo site. Eur. J. Biochem., **267**, 2054-2061 (2000)

[5] Bray, R.C.; Turner, N.A.; Le Gall, J.; Barata, B.A.S.; Moura, J.J.G.: Information from EPR spectroscopy on the iron-sulfur centers of the iron-molybdenum protein (aldehyde oxidoreductase) of Desulfovibrio gigas. Biochem. J., **280**, 817-820 (1991)

[6] Barata, B.A.; LeGall, J.; Moura, J.J.: Aldehyde oxidoreductase activity in Desulfovibrio gigas: in vitro reconstitution of an electron-transfer chain from

aldehydes to the production of molecular hydrogen. Biochemistry, **32**, 11559-11568 (1993)
[7] Caldeira, J.; Belle, V.; Asso, M.; Guigliarelli, B.; Moura, I.; Moura, J.J.; Bertrand, P.: Analysis of the electron paramagnetic resonance properties of the $[2Fe\text{-}2S]^+$ centers in molybdenum enzymes of the xanthine oxidase family: assignment of signals I and II. Biochemistry, **39**, 2700-2707 (2000)
[8] Romao, M.J.; Barata, B.A.; Archer, M.; Lobeck, K.; Moura, I.; Carrondo, M.A.; LeGall, J.; Lottspeich, F.; Huber, R.; Moura, J.J.: Subunit composition, crystallization and preliminary crystallographic studies of the Desulfovibrio gigas aldehyde oxidoreductase containing molybdenum and [2Fe-2S] centers. Eur. J. Biochem., **215**, 729-732 (1993)
[9] Thoenes, U.; Flores, O.L.; Neves, a.; Devreese, B.; Van Beeumen, J.J.; Huber, R. ; Romao, M.J.; LeGall, J.; Moura, J.J.G.; Roudrigues-Pousada, C.: Molecular cloning and sequence analysis of the gene of the molybdenum-containing aldehyde oxido-reductase of Desulfovibrio gigas, the deduced amino acid sequence shows similarity to xanthine dehydrogenase. Eur. J. Biochem., **220**, 901-910 (1994)
[10] Correia dos Santos, M.M.; Sousa, P.M.; Goncalves, M.L.; Romao, M.J.; Moura, I.; Moura, J.J.: Direct electrochemistry of the Desulfovibrio gigas aldehyde oxidoreductase. Eur. J. Biochem., **271**, 1329-1338 (2004)
[11] Rebelo, J.M.; Dias, J.M.; Huber, R.; Moura, J.J.; Romao, M.J.: Structure refinement of the aldehyde oxidoreductase from Desulfovibrio gigas (MOP) at 1.28 A. J. Biol. Inorg. Chem., **6**, 791-800 (2001)
[12] Huber, R.; Hof, P.; Duarte, R.O.; Moura, J.J.; Moura, I.; Liu, M.Y.; LeGall, J.; Hille, R.; Archer, M.; Romao, M.J.: A structure-based catalytic mechanism for the xanthine oxidase family of molybdenum enzymes. Proc. Natl. Acad. Sci. USA, **93**, 8846-8851 (1996)

Precorrin-2 dehydrogenase 1.3.1.76

1 Nomenclature

EC number
1.3.1.76

Systematic name
precorrin-2:NAD^+ oxidoreductase

Recommended name
precorrin-2 dehydrogenase

Synonyms
CysG <1, 2, 4, 6> [4, 6, 7, 8]
CysG enzyme <1> [6]
EC 2.1.1.107 <4> (<4> multifunctional protein, first of three steps leading to formation of siroheme from uroporphyrinogen III [8]) [8]
Met8p <3> [1, 3]
S-adenosyl-L-methionine(SAM)-dependent bismethyltransferase, dehydrogenase and ferrochelatase <4> [8]
SirC <5> [5]
cobaltochelatase <3> [1]
cysG gene product <6> [7]
precorrin-2 dehydrogenase <5> [5]
siroheme synthetase <6> (<6> multifunctional protein catalyzing all 3 final steps leading to siroheme from uroporphyrinogen III [7]) [7]

CAS registry number
227184-47-6

2 Source Organism

<1> *Salmonella typhimurium* (no sequence specified) [6]
<2> *Escherichia coli* (no sequence specified) [4]
<3> *Saccharomyces cerevisiae* (no sequence specified) [1, 2, 3]
<4> *Salmonella enterica* (no sequence specified) [8]
<5> *Bacillus megaterium* (UNIPROT accession number: P61818) [5]
<6> *Rhizobium etli* (UNIPROT accession number: O50477) [7]

3 Reaction and Specificity

Catalyzed reaction

precorrin-2 + NAD^+ = sirohydrochlorin + NADH + H^+

Reaction type

oxidation
redox reaction
reduction

Natural substrates and products

S precorrin-2 + NAD^+ <3> (Reversibility: ?) [2, 3]
P sirohydrochlorin + NADH
S precorrin-2 + NAD^+ <1, 2, 3, 4, 5, 6> (<3> bifunctional dehydrogenase and ferrochelatase, second of three steps leading to formation of siroheme from uroporphyrinogen III [3]; <2> multifunctional protein involved in S-adenosyl-L-methionine-dependent methylation, pyridine dinucleotide dependent dehydrogenation, and ferrochelation, second of three steps leading to formation of siroheme from uroporphyrinogen III [4]; <1> precorrin-2 is the precursor of both siroheme and B_{12}, first reaction specific to B_{12} synthesis, second of three steps leading to formation of siroheme from uroporphyrinogen III [6]; <6> second of three steps leading to formation of siroheme from uroporphyrinogen III [7]; <3,5> siroheme and cobalamin biosynthesis, second of three steps leading to formation of siroheme from uroporphyrinogen III [1,5]; <4> tetrapyrrole and cobalamin biosynthesis, second of three steps leading to formation of siroheme from uroporphyrinogen III [8]) (Reversibility: ?) [1, 3, 4, 5, 6, 7, 8]
P sirohydrochlorin + NADH + H^+ <1, 2, 3, 4, 5, 6> [1, 3, 4, 5, 6, 7, 8]

Substrates and products

S precorrin-2 + NAD^+ <3> (Reversibility: ?) [2, 3]
P sirohydrochlorin + NADH
S precorrin-2 + NAD^+ <1, 2, 3, 4, 5, 6> (<5> unlike Met8p and CysG, SirC has no chelatase actvity [5]; <3> bifunctional dehydrogenase and ferrochelatase, second of three steps leading to formation of siroheme from uroporphyrinogen III [3]; <2> multifunctional protein involved in S-adenosyl-L-methionine-dependent methylation, pyridine dinucleotide dependent dehydrogenation, and ferrochelation, second of three steps leading to formation of siroheme from uroporphyrinogen III [4]; <1> precorrin-2 is the precursor of both siroheme and B_{12}, first reaction specific to B_{12} synthesis, second of three steps leading to formation of siroheme from uroporphyrinogen III [6]; <6> second of three steps leading to formation of siroheme from uroporphyrinogen III [7]; <3,5> siroheme and cobalamin biosynthesis, second of three steps leading to formation of siroheme from uroporphyrinogen III [1,5]; <4> tetrapyrrole and cobalamin biosynthesis, second of three steps leading to formation of siroheme from uroporphyrinogen III [8]) (Reversibility: ?) [1, 3, 4, 5, 6, 7, 8]
P sirohydrochlorin + NADH + H^+ <1, 2, 3, 4, 5, 6> [1, 3, 4, 5, 6, 7, 8]

S Additional information <4> (<4> large multifunctional protein that catalyzes four diverse reactions, 2 S-adensyl-L-methionine-dependent methylations, NAD^+-dependent tetrapyrrole dehydrogenation and metal chelation [8]) (Reversibility: ?) [8]
P ? <4> [8]

Cofactors/prosthetic groups
NAD^+ <2, 3> (<3> β-NAD^+-dependent dehydrogenation of precorrin-2 [3]) [3,4]

Metals, ions
Co^{2+} <1> (<1> enzyme catalyzes cobalt insertion leading to B_{12} synthesis [6]) [6]
Fe^{2+} <1> (<1> enzyme catalyzes iron insertion leading to siroheme synthesis [6]) [6]

Specific activity (U/mg)
0.00088 <4> (<4> dehydrogenase activity, mutant S128D [8]) [8]
0.00993 <4> (<4> dehydrogenase activity, wild-type [8]) [8]
0.0112 <3> (<3> wild-type [3]) [3]
0.0133 <3> (<3> mutant H237A [3]) [3]
0.0332 <4> (<4> dehydrogenase activity, mutant S128A [8]) [8]
0.06 <5> [5]

4 Enzyme Structure

Molecular weight
27990 <2> (<2> predicted gene derived molecular mass [4]) [4]
29000 <2> (<2> SDS-PAGE [4]) [4]
45000 <5> (<5> gel filtration [5]) [5]
60000 <2> (<2> gel filtration [4]) [4]

Subunits
dimer <5> (<5> 2 * 22000, homodimer, SDS-PAGE [5]) [5]
homodimer <2, 3, 4> (<2> 2 * 29000, SDS-PAGE [4]; <4> 2 * 50000 [8]) [3, 4, 8]

Posttranslational modification
phosphoprotein <4> [8]

5 Isolation/Preparation/Mutation/Application

Purification
<2> [4]
<3> [1, 3]
<5> [5]

Crystallization

<3> (space group C2, a : 156.1 A,b : 81.2 A, c : 104.2 A) [3]

<4> (orthorhombic space group P2(1)2(1)2(1), unit cell dimensions a = 60.73 A, b = 121.49 A, c = 130.79 A) [8]

Cloning

<2> (expressed as recombinant protein) [4]

<3> [3]

<3> (MET8 mutants complemented by Salmonella typhimurium cysG, cloned and expressed in Escherichia coli) [1]

<4> (recombinantly overproduced in Escherichia coli) [8]

<5> (sequenced and overproduced, sirABC gene cluster complements Escherichia coli cysG mutants) [5]

<6> [7]

Engineering

D141A <3> (<3> site-directed mutagenesis [3]) [3]

G22D <3> (<3> site-directed mutagenesis [3]) [3]

H237A <3> (<3> site-directed mutagenesis [3]) [3]

K270I <4> (<4> site-directed mutagenesis [8]) [8]

L250A <4> (<4> site-directed mutagenesis [8]) [8]

N385A <4> (<4> site-directed mutagenesis [8]) [8]

S128A <4> (<4> site-directed mutagenesis [8]) [8]

S128D <4> (<4> site-directed mutagenesis [8]) [8]

References

[1] Raux, E.; McVeigh, T.; Peters, S.E.; Leustek, T.; Warren, M.J.: The role of Saccharomyces cerevisiae MET1p and MET8p in sirohaem and cobalamin biosynthesis. Biochem. J., **338**, 701-708 (1999)

[2] Warren, M.J.; Raux, E.; Schubert, H.L.; Escalante-Semerena, J.C.: The biosynthesis of adenosylcobalamin (vitamin B_{12}). Nat. Prod. Rep., **19**, 390-412 (2002)

[3] Schubert, H.L.; Raux, E.; Brindley, A.A.; Leech, H.K.; Wilson, K.S.; Hill, C.P.; Warren, M.J.: The structure of Saccharomyces cerevisiae Met8p, a bifunctional dehydrogenase and ferrochelatase. EMBO J., **21**, 2068-2075 (2002)

[4] Warren, M.J.; Bolt, E.L.; Roessner, C.A.; Scott, A.I.; Spencer, J.B.; Woodcock, S.C.: Gene dissection demonstrates that the Escherichia coli cysG gene encodes a multifunctional protein. Biochem. J., **302**, 837-844 (1994)

[5] Raux, E.; Leech, H.K.; Beck, R.; Schubert, H.L.; Santander, P.J.; Roessner, C.A.; Scott, A.I.; Martens, J.H.; Jahn, D.; Thermes, C.; Rambach, A.; Warren, M.J.: Identification and functional analysis of enzymes required for precorrin-2 dehydrogenation and metal ion insertion in the biosynthesis of sirohaem and cobalamin in Bacillus megaterium. Biochem. J., **370**, 505-516 (2003)

[6] Fazzio, T.G.; Roth, J.R.: Evidence that the CysG protein catalyzes the first reaction specific to B_{12} synthesis in Salmonella typhimurium, insertion of cobalt. J. Bacteriol., **178**, 6952-6959 (1996)
[7] Tate, R.; Riccio, A.; Iaccarino, M.; Patriarca, E.J.: A cysG mutant strain of Rhizobium etli pleiotropically defective in sulfate and nitrate assimilation. J. Bacteriol., **179**, 7343-7350 (1997)
[8] Stroupe, M.E.; Leech, H.K.; Daniels, D.S.; Warren, M.J.; Getzoff, E.D.: CysG structure reveals tetrapyrrole-binding features and novel regulation of siroheme biosynthesis. Nat. Struct. Biol., **10**, 1064-1073 (2003)

Anthocyanidin reductase 1.3.1.77

1 Nomenclature

EC number
1.3.1.77

Systematic name
flavan-3-ol:NAD(P)$^+$ oxidoreductase

Recommended name
anthocyanidin reductase

Synonyms
ANR <1, 2, 8> [2, 3]
AtANR <1, 9> [1, 2]
MtANR <8, 9> [1, 2, 5]
VvANR <12> [6]

CAS registry number
93389-48-1

2 Source Organism

<1> *Arabidopsis thaliana* (no sequence specified) [2, 4]
<2> *Camellia sinensis* (no sequence specified) [3]
<3> *Persea americana* (no sequence specified) [7]
<4> *Malus domestica* (no sequence specified) [7]
<5> *Pyrus communis* (no sequence specified) [7]
<6> *Taxus baccata* (no sequence specified) [7]
<7> *Coffea arabica* (no sequence specified) [7]
<8> *Medicago truncatula* (no sequence specified) [2,5]
<9> *Medicago truncatula* (UNIPROT accession number: Q84XT1) [1]
<10> *Rosa hybrida* (no sequence specified) [7]
<11> *Hypericum perforatum* (no sequence specified) [7]
<12> *Vitis vinifera* (UNIPROT accession number: Q7PCC4) [6]
<13> *Prunus cerasus* (no sequence specified) [7]
<14> *Poinsettia pulcherrima* (no sequence specified) [7]

3 Reaction and Specificity

Catalyzed reaction

a flavan-3-ol + 2 $NAD(P)^+$ = an anthocyanidin + 2 NAD(P)H + H^+

Reaction type

oxidation
redox reaction
reduction

Natural substrates and products

S 2,3-cis-flavan-3-ol + $NAD(P)^+$ <9> (<9> the enzyme is involved in formation of condensed tannins. The enzyme competes with anthocyanidin synthase, for the pool of flavan-3,4-diol [1]) (Reversibility: ?) [1]
P anthocyanidin + NAD(P)H + H^+
S anthocyanidin + NAD(P)H <1, 8> (<1,8> enzyme of flavonoid pathway involved in the biosynthesis of condensed tannins [2]) (Reversibility: ?) [2]
P 2,3-cis-flavan-3-ol + $NAD(P)^+$

Substrates and products

S 2,3-cis-flavan-3-ol + $NAD(P)^+$ <9> (<9> the enzyme is involved in formation of condensed tannins. The enzyme competes with anthocyanidin synthase, for the pool of flavan-3,4-diol [1]) (Reversibility: ?) [1]
P anthocyanidin + NAD(P)H + H^+
S NADPH + H^+ + cyanidin <4> (Reversibility: ?) [7]
P $NADP^+$ + (-)-epicatechin
S NADPH + H^+ + delphinidin <4> (Reversibility: ?) [7]
P $NADP^+$ + (-)-epigallocatechin
S NADPH + H^+ + pelargonidin <4> (Reversibility: ?) [7]
P $NADP^+$ + (-)-epiafzelechin
S NADPH + H^+ + petunidin <4> (Reversibility: ?) [7]
P $NADP^+$ + ?
S anthocyanidin + NAD(P)H <1, 8> (<1,8> enzyme of flavonoid pathway involved in the biosynthesis of condensed tannins [2]) (Reversibility: ?) [2]
P 2,3-cis-flavan-3-ol + $NAD(P)^+$
S cyanidin + NADPH <2> (Reversibility: ?) [3]
P epicatechin + $NADP^+$
S cyanidin + NADPH <1, 8> (<8> preference of anthocyanidin substrates in decreasing order: cyanidin, pelargonidin and delphinidin [2]) (Reversibility: ?) [2]
P (-)-epicatechin + $NAD(P)^+$
S delphinidin + NADPH <1, 8> (<8> preference of anthocyanidin substrates in decreasing order: cyanidin, pelargonidin and delphinidin [2]) (Reversibility: ?) [2]
P (-)-epigallocatechin + $NAD(P)^+$

S pelargonidin + NADPH <1, 8> (<8> preference of anthocyanidin substrates in decreasing order: cyanidin, pelargonidin and delphinidin [2]) (Reversibility: ?) [2]
P (-)-epiafzelechin + $NAD(P)^+$
S Additional information <4> (<4> no activity with paeonidin and malvidin [7]) (Reversibility: ?) [7]
P ?

Inhibitors
(+)-catechin <1> (<1> 0.5 mM, 50% inhibition [2]) [2]
(+/-)-dihydroquercetin <1> (<1> 0.025 mM [2]) [2]
Na^+ <8> (<8> above 200 mM [2]) [2]
quercetin <1, 8> (<1,8> strong inhibition [2]) [2]
Additional information <1, 8> (<8> no inhibition by (+)-catechin and (+/-)-dihydroquercetin [2]; <1> no inhibition by Na^+ up to 400 mM [2]) [2]

Cofactors/prosthetic groups
NAD^+ <9> [1]
NADH <1, 8> (<1,8> slight preference for NADPH over NADH [2]) [2]
NADPH <1, 2, 8> (<1,8> slight preference for NADPH over NADH [2]) [2,3]

Turnover number (min^{-1})
0.058 <1> (delphinidin, <1> pH 8.0 [2]) [2]
0.18 <8> (cyanidin, <8> pH 6.0 [2]) [2]
1.045 <1> (cyanidin, <1> pH 8.0 [2]) [2]
1.34 <8> (delphinidin, <8> pH 6.0 [2]) [2]

K_m-Value (mM)
0.0129 <8> (cyanidin, <8> pH 6.0 [2]) [2]
0.0145 <8> (pelargonidin, <8> pH 6.0 [2]) [2]
0.0178 <1> (delphinidin, <1> pH 8.0 [2]) [2]
0.0498 <8> (delphinidin, <8> pH 6.0 [2]) [2]
0.0528 <1> (pelargonidin, <1> pH 8.0 [2]) [2]
0.0739 <1> (cyanidin, <1> pH 8.0 [2]) [2]
0.45 <8> (NADPH, <8> pH 6.0 [2]) [2]
0.94 <8> (NADH, <8> pH 6.0 [2]) [2]

pH-Optimum
5 <4> [7]
5.5 <8> (<8> in 50 mM citrate/phosphate buffer [2]) [2]
6 <8> (<8> in 50 mM MES buffer [2]) [2]
8 <1> [2]

pH-Range
7-8.9 <1> (<1> pH 7.0: about 40% of maximal activity, pH 8.9: about 90% of maximal activity [2]) [2]

Temperature optimum (°C)
45 <4, 8> [2, 7]
55 <1> [2]

5 Isolation/Preparation/Mutation/Application

Source/tissue

berry <12> (<12> VvANR is expressed throughout early flower and berry development, with expression increasing after fertilization. Expression in berry skin and seeds until the onset of the ripening [6]) [6]
bract <14> [7]
flower <9, 12> (<12> VvANR is expressed throughout early flower and berry development, with expression increasing after fertilization [6]) [1, 6]
flower bud <9> [1]
fruit <3> [7]
leaf <2, 4, 5, 7, 8, 9, 10, 11, 12, 13> (<9> very weak [1]; <12> expression in expanding leaves [6]; <8> leaves of transgenic Medicago truncatula constitutively expressing MtANR contain up to three times more proanthocyanidins than those of wild-type plants at the same stage of development [5]) [1, 3, 5, 6, 7]
needle <6> [7]
seed <6, 9> (<9> strong expression [1]) [1, 7]
Additional information <8> (<8> in Medicago truncatula expression of MtANR driven by the 35S promoter results in a decrease of approximately 50% in anthocyanin pigmentation in the red spot compared with that of wild-type plants at the same stage of development [5]) [5]

Cloning

<1> (expression in Escherichia coli as a fusion protein with maltose-binding protein) [2]
<1> (genetic transformation of Arabidopsis thaliana with the Arabidopsis TT2 MYB transcription factor results in ectopic expression of the BANYULS gene, encoding anthocyanidin reductase, AHA10 encoding a P-type proton-pump and TT12 encoding a transporter involved in proanthocyanidin biosynthesis. When coupled with constitutive expression of PAP1, a positive regulator of anthocyanin biosynthesis, TT2 expression in Arabidopsis leads to the accumulation of proanthocyanidins, but only in a subset of cells in which the BANYULS promoter is naturally expressed. Ectopic expression of the maize Lc MYC transcription factor weakly induces AHA10 but does not induce BANYULS, TT12 or accumulation of proanthocyanidins) [4]
<8> (constitutive expression of the enzyme under control of the cauliflower mosaic virus 35S promoter in Nicotiana tabacum and Arabidopsis. Tobacco lines expressing the enzyme from Medicago trunculata lose the pink flower pigmentation characteristics of wild-type and empty vector control plants) [1]
<8> (expression in Escherichia coli as a fusion protein with maltose-binding protein) [2]
<12> (the full-length VvANR cDNA is cloned into the vector pART7 and transformed into tobacco via the pART27 vector under the control of the CaMV 35S promoter) [6]

References

[1] Xie, D.-Y.; Sharma, S.B.; Paiva, N.L.; Ferreira, D.; Dixon, R.A.: Role of anthocyanidin reductase, encoded by BANYULS in plant flavonoid biosynthesis. Science, **299**, 396-399 (2003)

[2] Xie, D.Y.; Sharma, S.B.; Dixon, R.A.: Anthocyanidin reductases from Medicago truncatula and Arabidopsis thaliana. Arch. Biochem. Biophys., **422**, 91-102 (2004)

[3] Punyasiri, P.A.; Abeysinghe, I.S.; Kumar, V.; Treutter, D.; Duy, D.; Gosch, C.; Martens, S.; Forkmann, G.; Fischer, T.C.: Flavonoid biosynthesis in the tea plant Camellia sinensis: properties of enzymes of the prominent epicatechin and catechin pathways. Arch. Biochem. Biophys., **431**, 22-30 (2004)

[4] Sharma, S.B.; Dixon, R.A.: Metabolic engineering of proanthocyanidins by ectopic expression of transcription factors in Arabidopsis thaliana. Plant J., **44**, 62-75 (2005)

[5] Xie, D.Y.; Sharma, S.B.; Wright, E.; Wang, Z.Y.; Dixon, R.A.: Metabolic engineering of proanthocyanidins through co-expression of anthocyanidin reductase and the PAP1 MYB transcription factor. Plant J., **45**, 895-907 (2006)

[6] Bogs, J.; Downey, M.O.; Harvey, J.S.; Ashton, A.R.; Tanner, G.J.; Robinson, S.P.: Proanthocyanidin synthesis and expression of genes encoding leucoanthocyanidin reductase and anthocyanidin reductase in developing grape berries and grapevine leaves. Plant Physiol., **139**, 652-663 (2005)

[7] Pfeiffer, J.; Kuehnel, C.; Brandt, J.; Duy, D.; Punyasiri, P.A.; Forkmann, G.; Fischer, T.C.: Biosynthesis of flavan 3-ols by leucoanthocyanidin 4-reductases and anthocyanidin reductases in leaves of grape (Vitis vinifera L.), apple (Malus x domestica Borkh.) and other crops. Plant Physiol. Biochem., **44**, 323-334 (2006)

Arogenate dehydrogenase ($NADP^+$) 1.3.1.78

1 Nomenclature

EC number
1.3.1.78

Systematic name
L-arogenate:$NADP^+$ oxidoreductase (decarboxylating)

Recommended name
arogenate dehydrogenase ($NADP^+$)

Synonyms
TyrAAT <3> [12]
TyrAAT1 <3> [12]
TyrAa <14, 15, 17, 20, 22, 23, 24, 25> [11]
TyrAsy <18> [13]
arogenate dehydrogenase isoform 1 <3> [12]
arogenate dehydrogenase isoform 2 <3> [12]
cyclohexadienyl dehydrogenase <3> [9]

CAS registry number
64295-75-6

2 Source Organism

<1> *Euglena gracilis* (no sequence specified) [4]
<2> *Zea mays* (no sequence specified) (<2> ST6GalNAc VI [4]) [4]
<3> *Arabidopsis thaliana* (no sequence specified) [9, 12]
<4> *Pseudomonas sp.* (no sequence specified) [6]
<5> *Acinetobacter calcoaceticus* (no sequence specified) [3]
<6> *Brevibacterium flavum* (no sequence specified) [1,4]
<7> *Corynebacterium glutamicum* (no sequence specified) [1,4]
<8> *Vigna radiata* (no sequence specified) [2,4]
<9> *Sorghum bicolor* (no sequence specified) [4,5]
<10> *Acinetobacter sp.* (no sequence specified) [3]
<11> *Streptomyces phaeochromogenes* (no sequence specified) [4]
<12> *Flavobacterium devorans* (no sequence specified) [8]
<13> *Nicotiana sylvestris* (no sequence specified) [4, 7, 10]
<14> *Chlorella sorokiniana* (no sequence specified) [10, 11]
<15> *Porphyridium cruentum* (no sequence specified) [11]
<16> *Brevibacterium ammoniagenes* (no sequence specified) [1]
<17> *Anabaena sp.* (no sequence specified) [11]

<18> *Synechocystis sp.* (no sequence specified) [13]
<19> *Acinetobacter lwoffii* (no sequence specified) [3]
<20> *Prochlorothrix hollandica* (no sequence specified) [11]
<21> *Phenylobacterium immobile* (no sequence specified) [4]
<22> *Synechocystis sp. PCC 6803* (UNIPROT accession number: P73906) [11]
<23> *Synechocystis sp. PCC 6902* (no sequence specified) [11]
<24> *Synechocystis sp. PCC 6301* (no sequence specified) [11]
<25> *Fisherella sp.* (no sequence specified) [11]

3 Reaction and Specificity

Catalyzed reaction

L-arogenate + NADP$^+$ = L-tyrosine + NADPH + CO_2 (<22> mechanism is probably steady-state random order [11])

Reaction type

oxidation
redox reaction
reduction

Natural substrates and products

S L-arogenate + NAD$^+$ <1, 2, 3, 4, 5, 6, 7, 8, 9, 10, 11, 12, 13, 14, 16, 19, 21> (<13> exclusive metabolic route to tyrosine [7]; <9> important enzyme in regulation of tyrosine biosynthesis [5]; <9,13> i.e. 3-(1-carboxy-4-hydroxycyclohexa-2,5-dien-1-yl)-L-alanine, final biosynthetic step to tyrosine [5, 7]) (Reversibility: ?) [1, 2, 3, 4, 5, 6, 7, 8, 9, 10]
P L-tyrosine + NADH + CO_2 <1, 2, 3, 4, 5, 6, 7, 8, 9, 10, 11, 12, 13, 14, 16, 19, 21> [1, 2, 3, 4, 5, 6, 7, 8, 9, 10]
S L-arogenate + NADP$^+$ <14, 15, 17, 20, 22, 23, 24, 25> (<14, 15, 17, 20, 22, 23, 24, 25> enzyme of the arogenate pathway [11]) (Reversibility: ?) [11]
P L-tyrosine + NADPH + CO_2

Substrates and products

S L-arogenate + NAD(P)$^+$ <1, 2, 3, 4, 5, 6, 7, 8, 9, 10, 11, 12, 13, 14, 16, 19, 21> (Reversibility: ?) [1, 2, 3, 4, 5, 6, 7, 8, 9, 10]
P L-tyrosine + NAD(P)H + CO_2 <1, 2, 3, 4, 5, 6, 7, 8, 9, 10, 11, 12, 13, 14, 16, 19, 21> [1, 2, 3, 4, 5, 6, 7, 8, 9, 10]
S L-arogenate + NAD$^+$ <1, 2, 3, 4, 5, 6, 7, 8, 9, 10, 11, 12, 13, 14, 16, 19, 21> (<13> exclusive metabolic route to tyrosine [7]; <9> important enzyme in regulation of tyrosine biosynthesis [5]; <9, 13> i.e. 3-(1-carboxy-4-hydroxycyclohexa-2,5-dien-1-yl)-L-alanine, final biosynthetic step to tyrosine [5, 7]) (Reversibility: ?) [1, 2, 3, 4, 5, 6, 7, 8, 9, 10]
P L-tyrosine + NADH + CO_2 <1, 2, 3, 4, 5, 6, 7, 8, 9, 10, 11, 12, 13, 14, 16, 19, 21> [1, 2, 3, 4, 5, 6, 7, 8, 9, 10]
S L-arogenate + NADP$^+$ <18> (Reversibility: ?) [13]
P L-tyrosine + NADH + CO_2
S L-arogenate + NADP$^+$ <3, 14, 15, 17, 20, 22, 23, 24, 25> (<14, 15, 17, 20, 22, 23, 24, 25> enzyme of the arogenate pathway [11]; <3> rapid equili-

brium random mechanism in which two dead-end complexes, enzyme-NADPH-arogenate and enzyme-$NADP^+$-tyrosine are formed [12]; <14, 15, 17, 20, 22, 23, 24, 25> specific for, no activity with prephenate and NAD^+ [11]) (Reversibility: ?) [11, 12]

P L-tyrosine + NADPH + CO_2

S prephenate + $NAD(P)^+$ <3, 8, 13> (<3> not prephenate [9]) (Reversibility: ?) [2, 4, 9]

P ?

S prephenate + $NADP^+$ <3> (Reversibility: ?) [12]

P 3-(4-hydroxyphenyl)-2-oxopropanoate + CO_2 + NADPH

S Additional information <3, 14, 15, 17, 20, 22, 23, 24, 25> (<14, 15, 17, 20, 22, 23, 24, 25> no activity with prephenate and NAD^+ [11]; <3> TyrAAT1 is four times more efficient in catalyzing arogenate dehydrogenase reaction than TyrAAT2. TyrAAT2 presents a weak prephenate dehydrogenase activity whereas Tyr-AAT1 does not [12]) (Reversibility: ?) [11, 12]

P ?

Inhibitors

2',5'-ADP <22> [11]
AMP <3> [12]
D-tyrosine <13> [4, 7]
L-phenylalanine <12> (<12> slight inhibition [8]) [8]
L-Tyr <22> [11]
L-tyrosine <1, 2, 3, 4, 5, 6, 7, 8, 9, 12, 13, 14, 16> (<6, 7, 8, 12, 16> no inhibition [1, 2, 8]; <5> competitive inhibition [3]) [1, 2, 3, 4, 5, 6, 7, 8, 9, 10, 12]
N-acetyl-DL-tyrosine <13> [4, 7]
NADPH <3> [12]
prephenate <3, 9> (<9> no inhibition [5]) [5, 12]
cis-aconitate <3> [12]
m-fluoro-DL-tyrosine <13> [4, 7]
p-hydroxymercuribenzoate <6, 7, 8, 16> [1, 4]
Additional information <18> (<18> almost completely insensitive to feedback inhibition by tyrosine. Phenylalanine and tryptophan show no effect at concentrations of up to 1 mM [13]) [13]

Cofactors/prosthetic groups

NAD^+ <2, 3, 4, 6, 7, 9, 11, 12, 13, 14, 16, 21> (<9, 13, 14> not [5, 7, 10]) [1, 4, 5, 6, 7, 8, 9, 10]
$NADP^+$ <1, 3, 4, 5, 6, 7, 8, 9, 12, 13, 14, 15, 16, 17, 18, 20, 22, 23, 24, 25> (<14, 15, 17, 18, 20, 22, 23, 24, 25> no activity with NAD^+ [11,13]) [1, 2, 3, 4, 5, 6, 7, 8, 9, 10, 11, 12, 13]
NADPH <3> [12]

Turnover number (min^{-1})

3.4 <3> (prephenate) [12]
37.3 <3> (L-arogenate, <3> pH 7.5, 25°C, TyrAAT2 [12]) [12]
84.2 <3> (arogenate, <3> pH 7.5, 25°C, TyrAAT1 [12]) [12]

Specific activity (U/mg)

0.0011 <13> [10]
0.0021 <10> [3]
0.0024 <14, 20> [11]
0.0039 <5> (<5> ATCC 23055 [3]) [3]
0.0043 <5> (<5> ATCC 14987 [3]) [3]
0.0063 <19> [3]
0.0078 <17> [11]
0.0097 <15> [11]
0.0109 <25> [11]
0.0115 <24> [11]
0.0145 <23> [11]
0.43 <9> [5]
73 <3> [12]
88.1 <18> [13]
142 <3> [12]
220 <22> [11]
Additional information <6, 7, 16> [1]

K_m-Value (mM)

0.0001 <13> (L-arogenate) [4]
0.0002 <1> (L-arogenate) [4]
0.003 <2> (L-arogenate) [4]
0.005 <5> ($NADP^+$, <5> ATCC 23055 [3]) [3]
0.0087 <13> ($NADP^+$) [7]
0.009 <18> ($NADP^+$) [13]
0.0102 <3> ($NADP^+$, <3> pH 7.5, 25°C, TyrAAT1 [12]) [12]
0.011 <9> ($NADP^+$) [5]
0.0147 <3> ($NADP^+$, <3> pH 7.5, 25°C, TyrAAT2 [12]) [12]
0.02 <3> ($NADP^+$, <3> recombinant from plasmid TyrA-AT2 [9]) [9]
0.038 <22> ($NADP^+$, <22> pH 8.6, 25°C [11]) [11]
0.04 <3> ($NADP^+$, <3> recombinant from plasmid TyrA-ATcM36 [9]) [9]
0.045 <3> (L-arogenate, <3> recombinant from plasmids TyrA-AT2 and TyrA-ATlM36 [9]) [9]
0.0526 <3> (L-arogenate, <3> pH 7.5, 25°C, TyrAAT1 [12]) [12]
0.07 <3> (L-arogenate, <3> recombinant from plasmid TyrA-ATcM36 [9]) [9]
0.075 <3> ($NADP^+$, <3> recombinant from plasmid TyrA-ATlM36 [9]) [9]
0.0842 <3> (L-arogenate, <3> pH 7.5, 25°C, TyrAAT2 [12]) [12]
0.093 <18> (L-arogenate) [13]
0.1 <13> (L-arogenate) [7]
0.313 <5> (L-arogenate, <5> ATCC 23055 [3]) [3]
0.331 <22> (L-arogenate, <22> pH 8.6, 25°C [11]) [11]
0.34 <9> (L-arogenate) [5]
17 <3> (prephenate) [12]
Additional information <1, 2, 6, 7, 8, 9, 13, 16> [1, 2, 4]

K_i-Value (mM)

0.0075 <3> (L-tyrosine, <3> pH 7.5, 25°C, with 0.05 mM $NADP^+$ as fixed substrate and arogenate as varied substrate, TyrAAT2 [12]) [12]
0.008 <3> (L-tyrosine, <3> recombinant from plasmid TyrA-ATlM36 [9]) [9]
0.0081 <3> (L-tyrosine, <3> pH 7.5, 25°C, with 0.05 mM $NADP^+$ as fixed substrate and arogenate as varied substrate, TyrAAT1 [12]) [12]
0.009 <5> (L-tyrosine, <5> ATCC 23055, competitive inhibition [3]) [3]
0.012 <3> (L-tyrosine, <3> recombinant from plasmid TyrA-AT2 [9]) [9]
0.014 <3> (L-tyrosine, <3> recombinant from plasmid TyrA-ATcM36 [9]) [9]
0.0142 <3> (L-tyrosine, <3> pH 7.5, 25°C, with 0.07 mM arogenate as fixed substrate and $NADP^+$ as varied substrate, TyrAAT1 [12]) [12]
0.0166 <3> (L-tyrosine, <3> pH 7.5, 25°C, with 0.07 mM arogenate as fixed substrate and $NADP^+$ as varied substrate, TyrAAT2 [12]) [12]
0.0539 <3> (NADPH, <3> pH 7.5, 25°C, with 0.02 mM $NADP^+$ as fixed substrate and arogenate as varied substrate, TyrAAT1 [12]) [12]
0.0588 <3> (NADPH, <3> pH 7.5, 25°C, with 0.2 mM arogenate as fixed substrate and $NADP^+$ as varied substrate, TyrAAT1 [12]) [12]
0.06 <9> (L-tyrosine) [5]
0.078 <22> (L-Tyr, <22> pH 8.6, 25°C, versus $NADP^+$ [11]) [11]
0.089 <22> (L-Tyr, <22> pH 8.6, 25°C, versus L-arogenate [11]) [11]
2.4 <3> (prephenate, <3> pH 7.5, 25°C, with 0.1 mM $NADP^+$ as fixed substrate and arogenate as varied substrate, TyrAAT2 [12]) [12]
4.2 <3> (prephenate, <3> pH 7.5, 25°C, with 0.1 mM $NADP^+$ as fixed substrate and arogenate as varied substrate, TyrAAT1 [12]) [12]
5.8 <3> (cis-aconitate, <3> pH 7.5, 25°C, with 0.1 mM $NADP^+$ as fixed substrate and arogenate as varied substrate, TyrAAT2 [12]) [12]
17.8 <3> (AMP, <3> pH 7.5, 25°C, with 0.05 mM $NADP^+$ as fixed substrate and arogenate as varied substrate, TyrAAT2 [12]) [12]
20.2 <3> (prephenate, <3> pH 7.5, 25°C, with 0.07 mM arogenate as fixed substrate and $NADP^+$ as varied substrate, TyrAAT1 [12]) [12]
22.5 <3> (prephenate, <3> pH 7.5, 25°C, with 0.07 mM arogenate as fixed substrate and $NADP^+$ as varied substrate, TyrAAT2 [12]) [12]
25.3 <3> (cis-aconitate, <3> pH 7.5, 25°C, with 0.2 mM arogenate as fixed substrate and $NADP^+$ as varied substrate, TyrAAT2 [12]) [12]
25.9 <3> (AMP, <3> pH 7.5, 25°C, with 0.05 mM $NADP^+$ as fixed substrate and arogenate as varied substrate, TyrAAT1 [12]) [12]
27.7 <3> (cis-aconitate, <3> pH 7.5, 25°C, with 0.1 mM $NADP^+$ as fixed substrate and arogenate as varied substrate, TyrAAT1 [12]) [12]
42.9 <3> (cis-aconitate, <3> pH 7.5, 25°C, with 0.2 mM arogenate as fixed substrate and $NADP^+$ as varied substrate, TyrAAT1 [12]) [12]
73.9 <3> (AMP, <3> pH 7.5, 25°C, with 0.2 mM arogenate as fixed substrate and $NADP^+$ as varied substrate, TyrAAT2 [12]) [12]
115 <3> (AMP, <3> pH 7.5, 25°C, with 0.2 mM arogenate as fixed substrate and $NADP^+$ as varied substrate, TyrAAT1 [12]) [12]

pH-Optimum
7.5 <18> [13]
7.8 <6, 7, 16> [1]
8.3-8.8 <22> [11]

pH-Range
6.5-8.5 <18> (<18> enzyme activity drops below pH 6.5 and above pH 8.5 [13]) [13]

Temperature optimum (°C)
25 <3> (<3> assay at [9]) [9]
28-30 <22> [11]
30 <4, 12> (<4,12> assay at [6,8]) [6, 8]
33 <6, 7, 16> [1]

Temperature range (°C)
28-42 <22> (<22> 28-30°C: optimum, 42°C: 50% of maximal activity [11]) [11]

4 Enzyme Structure

Molecular weight
52000 <8> (<8> gel filtration [2]) [2]
67000 <3> (<3> recombinant TyrAAT1, gel filtration [12]) [12]
68000 <16> (<16> gel filtration [1]) [1]
70000 <18> (<18> gel filtration [13]) [13]
158000 <6, 7> (<6,7> gel filtration [1]) [1]
210000 <5> (<5> ATCC 23055, gel filtration [3]) [3]
600000 <3> (<3> above, recombinant TyrAAT2 [12]) [12]

Subunits
dimer <18, 22> (<18> 2 * 32000, SDS-PAGE [13]) [11, 13]
monomer <3> (<3> 1 * 66000, recombinant TyrAAT1, SDS-PAGE [12]) [12]
oligomer <3> (<3> recombinant TyrAAT2 [12]) [12]

5 Isolation/Preparation/Mutation/Application

Source/tissue
seedling <9> [5]

Purification
<3> (TyrAAT1) [12]
<3> (TyrAAT2) [12]
<5> (ATCC 23055, partial) [3]
<6> (partial) [1]
<7> (partial) [1]

<8> (partial, NADP-dependent prephenate and pretyrosine dehydrogenase copurified, probably multifunctional protein) [2]
<9> (partial) [5]
<13> [4]
<16> (partial) [1]
<18> (recombinant enzyme expressed in Escherichia coli) [13]
<22> [11]

Crystallization

<18> (sitting drop method, structure of TyrAsy in complex with $NADP^+$ is refined to 1.6 A. The asymmetric unit contains two TyrAsy homodimers, with each monomer consisting of a nucleotide binding N-terminal domain and a particularly unique α-helical C-terminal dimerization domain) [13]

Cloning

<3> (TyrAAT1 overproduced in Escherichia coli) [12]
<3> (TyrAAT2 overproduced in Escherichia coli) [12]
<3> (tyrA gene, overexpression in Escherichia coli, 3 different plasmids TyrA-ATcM36, TyrA-ATlM36, TyrA-AT2) [9]
<22> (expression in Escherichia coli) [11]

Engineering

Additional information <6> (<6> mutant pheA5 and double-mutant pheA5 shkA1 by mutagenesis and growth selection, reduced activity [1]) [1]

6 Stability

General stability information

<22>, full activity is retained in repeated freeze-thaw cycles, when concentrations of 0.01 mg of protein per ml or more are maintained [11]

Storage stability

<22>, purified enzyme is less stable during storage at 4°C than when maintained frozen at -20°C or at -70°C [11]

References

[1] Fazel, A.M.; Jensen, R.A.: Obligatory biosynthesis of L-tyrosine via the pretyrosine branchlet in coryneform bacteria. J. Bacteriol., **138**, 805-815 (1979)
[2] Rubin, J.L.; Jensen, R.A.: Enzymology of L-tyrosine biosynthesis in mung bean (Vigna radiata [L.] Wilczek). Plant Physiol., **64**, 727-734 (1979)
[3] Byng, G.S.; Berry, A.; Jensen, R.A.: Evolutionary implications of features of aromatic amino acid biosynthesis in the genus Acinetobacter. Arch. Microbiol., **143**, 122-129 (1985)
[4] Bonner, C.; Jensen, R.: Arogenate dehydrogenase. Methods Enzymol., **142**, 488-494 (1987)

[5] Connelly, J.A.; Conn, E.E.: Tyrosine biosynthesis in Sorghum bicolor: isolation and regulatory properties of arogenate dehydrogenase. Z. Naturforsch. C, **41**, 69-78 (1986)
[6] Keller, B.; Keller, E.; Klages, U.; Lingens, F.: Aromatic amino acid biosynthesis in a 4-chlorobenzoic acid degrading Pseudomonas species; phenylalanine and tyrosine synthesis via arogenate. Syst. Appl. Microbiol., **4**, 27-33 (1983)
[7] Gaines, C.G.; Byng, G.S.; Whitaker, R.J.; Jensen, R.A.: L-tyrosine regulation and biosynthesis via arogenate dehydrogenase in suspension-cultured cells of Nicotiana silvestris Speg. et Comes. Planta, **156**, 233-240 (1982)
[8] Keller, B.; Keller, E.; Sussmuth, R.; Lingens, F.: Arogenate (pretyrosine) as an obligatory precursor of L-tyrosine biosynthesis in Flavobacterium devorans. FEMS Microbiol. Lett., **13**, 3-4 (1982)
[9] Rippert, P.; Matringe, M.: Molecular and biochemical characterization of an Arabidopsis thaliana arogenate dehydrogenase with two highly similar and active protein domains. Plant Mol. Biol., **48**, 361-368 (2002)
[10] Bonner, C.A.; Fischer, R.S.; Schmidt, R.R.; Miller, P.W.; Jensen, R.A.: Distinctive enzymes of aromatic amino acid biosynthesis that are highly conserved in land plants are also present in the chlorophyte alga Chlorella sorokiniana. Plant Cell Physiol., **36**, 1013-1022 (1995)
[11] Bonner, C.A.; Jensen, R.A.; Gander, J.E.; Keyhani, N.O.: A core catalytic domain of the TyrA protein family: arogenate dehydrogenase from Synechocystis. Biochem. J., **382**, 279-291 (2004)
[12] Rippert, P.; Matringe, M.: Purification and kinetic analysis of the two recombinant arogenate dehydrogenase isoforms of Arabidopsis thaliana. Eur. J. Biochem., **269**, 4753-4761 (2002)
[13] Legrand, P.; Dumas, R.; Seux, M.; Rippert, P.; Ravelli, R.; Ferrer, J.L.; Matringe, M.: Biochemical characterization and crystal structure of Synechocystis arogenate dehydrogenase provide insights into catalytic reaction. Structure, **14**, 767-776 (2006)

Arogenate dehydrogenase [NAD(P)$^+$] 1.3.1.79

1 Nomenclature

EC number
1.3.1.79

Systematic name
L-arogenate:NAD(P)$^+$ oxidoreductase (decarboxylating)

Recommended name
arogenate dehydrogenase [NAD(P)$^+$]

CAS registry number
64295-75-6

2 Source Organism

<1> *Flavobacterium devorans* (no sequence specified) [1]
<2> *Flavobacterium aquatile* (no sequence specified) [1]
<3> *Pseudomonas aureofaciens* (no sequence specified) [2]
<4> *Flavobacterium capsulatum* (no sequence specified) [1]
<5> *Flavobacterium breve* (no sequence specified) [1]
<6> *Flavobacterium odoratum* (no sequence specified) [1]
<7> *Flavobacterium sp. CB 6* (no sequence specified) [1]
<8> *Flavobacterium sp. CB 60* (no sequence specified) [1]
<9> *Flavobacterium suaveolens* (no sequence specified) [1]
<10> *Flavobacterium lutescens* (no sequence specified) [1]
<11> *Flavobacterium paucimobilis* (no sequence specified) [1]
<12> *Flavobacterium dehydrogenans* (no sequence specified) [1]

3 Reaction and Specificity

Catalyzed reaction
L-arogenate + NAD(P)$^+$ = L-tyrosine + NAD(P)H + CO_2

Reaction type
oxidation
reduction

Natural substrates and products

S L-arogenate + NAD^+ <1, 2, 3, 4, 5, 6, 7, 8, 9, 10, 11, 12> (<1, 2, 4, 5, 6, 7, 8, 9, 10, 11, 12> i.e. 3-(1-carboxy-4-hydroxycyclohexa-2,5-dien-1-yl)-L-alanine, final biosynthetic step to tyrosine [1]) (Reversibility: ?) [1, 2]

P L-tyrosine + NADH + CO_2 <1, 2, 3, 4, 5, 6, 7, 8, 9, 10, 11, 12> [1, 2]

Substrates and products

S L-arogenate + $NAD(P)^+$ <1, 2, 3, 4, 5, 6, 7, 8, 9, 10, 11, 12> (Reversibility: ?) [1, 2]

P L-tyrosine + NAD(P)H + CO_2 <1, 2, 3, 4, 5, 6, 7, 8, 9, 10, 11, 12> [1, 2]

S L-arogenate + NAD^+ <1, 2, 3, 4, 5, 6, 7, 8, 9, 10, 11, 12> (<1, 2, 4, 5, 6, 7, 8, 9, 10, 11, 12> i.e. 3-(1-carboxy-4-hydroxycyclohexa-2,5-dien-1-yl)-L-alanine, final biosynthetic step to tyrosine [1]) (Reversibility: ?) [1, 2]

P L-tyrosine + NADH + CO_2 <1, 2, 3, 4, 5, 6, 7, 8, 9, 10, 11, 12> [1, 2]

Inhibitors

L-phenylalanine <1, 2, 4, 5, 10, 11, 12> [1]
L-tyrosine <1, 2, 3, 4, 5, 10, 11, 12> [1, 2]

Cofactors/prosthetic groups

NAD^+ <1, 2, 3, 4, 5, 6, 7, 8, 9, 10, 11, 12> (<2, 4, 5, 6, 7, 8, 9, 10, 11, 12> $NADP^+$ is a better electron acceptor than NAD^+ [1]; <3> NAD^+ and $NADP^+$ equally effective [2]) [1,2]
$NADP^+$ <1, 2, 3, 4, 5, 6, 7, 8, 9, 10, 11, 12> (<2, 4, 5, 6, 7, 8, 9, 10, 11, 12> $NADP^+$ is a better electron acceptor than NAD^+ [1]; <3> NAD^+ and $NADP^+$ equally effective [2]) [1,2]

Activating compounds

L-tyrosine <6, 7, 9> (<6,7,9> slightly enhanced activity [1]) [1]
phenylalanine <6, 7, 9> (<6,7,9> slightly enhanced activity [1]) [1]

Specific activity (U/mg)

Additional information <1, 2, 4, 5, 6, 7, 8, 9, 10, 11, 12> [1]

References

[1] Waldner-Sander, S.; Keller, B.; Keller, E.; Lingens, F.: Zur Biosynthese von Phenylalanin und Tyrosin bei Flavobakterien. Hoppe-Seyler's Z. Physiol. Chem., **364**, 1467-1473 (1983)

[2] Keller, B.; Keller, E.; Salcher, O.; Lingens, F.: Arogenate (pretyrosine) pathway of tyrosine and phenylalanine biosynthesis in Pseudomonas aureofaciens ATCC 15926. J. Gen. Microbiol., **128**, 1199-1202 (1982)

Red chlorophyll catabolite reductase 1.3.1.80

1 Nomenclature

EC number
1.3.1.80

Systematic name
primary fluorescent chlorophyll catabolite:NADP$^+$ oxidoreductase

Recommended name
red chlorophyll catabolite reductase

Synonyms
ACD2 protein <5> [8]
At-RCCR <3> [3]
AtRCCR <3> [4]
HvRCCR <1> [4]
RCC reductase <1, 2, 5, 6> [1]
RCCR <1, 2, 5, 6> [1]
red Chl catabolite reductase <1, 2, 3, 5, 6> [1, 7]

CAS registry number
199618-44-5

2 Source Organism

<1> *Hordeum vulgare* (no sequence specified) [1, 4, 6]
<2> *Spinacia oleracea* (no sequence specified) [1]
<3> *Arabidopsis thaliana* (no sequence specified) [3, 4, 7]
<4> *Brassica napus* (no sequence specified) [5]
<5> *Arabidopsis sp.* (no sequence specified) [1,8]
<6> *Lycopersicon esculentum* (no sequence specified) [1,2]
<7> *Selaginella sp.* (no sequence specified) [2]
<8> *Tropaeolum majus* (no sequence specified) [2]
<9> *Taxus sp.* (no sequence specified) [2]
<10> *Carex sp.* (no sequence specified) [2]
<11> *Equisetum sp.* (no sequence specified) [2]
<12> *Cycas sp.* (no sequence specified) [2]
<13> *Cleome graveolens* (no sequence specified) [2]

3 Reaction and Specificity

Catalyzed reaction

primary fluorescent chlorophyll catabolite + $NADP^+$ = red chlorophyll catabolite + NADPH + H^+

Natural substrates and products

S Additional information <1, 3, 4, 5, 6> (<1> barley RCCR produces the C_1 isomer pFCC-1 [1]; <5> cell death gene ACD2 encodes red chlorophyll catabolite reductase and suppresses the spread of disease symptoms [8]; <3> RCCR absence causes leaf cell death as a result of the accumulation of photodynamic RCC. RCCR (together with pheophorbide a oxygenase) is required for the detoxification of chlorophyll catabolites [3]; <4> the enzyme is involved in breakdown of chlorophyll [5]; <1> the enzyme is involved in chlorophyll breakdown [6]; <3> the enzyme is involved in chlorophyll breakdown in senescent Arabidopsis leaves [7]; <1> the major product of reduction of red chlorophyll catabolite is pFCC1, but small quantities of its C_1 epimer, pFCC-2, also accumulate. Red chlorophyll catabolite reductase and pheophorbide a oxygenase catalyse the key eaction of chlorophyll catabolism, porphin macrocycle cleavage of pheide a to a primary fluorescent catabolite [4]; <3> the major product of reduction of red chlorophyll catabolite is pFCC1, but small quantities of its C_1 epimer, pFCC-2, also accumulate. Red chlorophyll catabolite reductase and pheophorbide a oxygenase catalyse the key reaction of chlorophyll catabolism, porphin macrocycle cleavage of pheide a to a primary fluorescent catabolite [4]; <5> with Arabidopsis RCCR, the C_1 isomer pFCC-1 is formed. RCCR could be required to mediate an efficient interaction between red chlorophyll catabolite (still bound to pheophorbide a oxygenase) and ferredoxin, thereby enabling a fast, regio-, and stereoselective reduction to blue-fluorescing intermediate [1]; <6> with tomato RCCR, the C_1 isomer pFCC-2 is formed [1]) (Reversibility: ?) [1, 3, 4, 5, 6, 7, 8]

P ?

Substrates and products

S red chlorophyll catabolite + reduced acceptor <1> (<1> three different primary fluorescent chlorophyll catabolites are produced, two of which could be identified as the stereoisomeric pFCCs from canola (Brassica napus) (pFCC-1) and sweet pepper (Capsicum annuum) (pFCC-2), respectively [6]) (Reversibility: ?) [6]

P primary fluorescent chlorophyll catabolite + acceptor

S Additional information <1, 2, 3, 4, 5, 6> (<1> barley RCCR produces the C_1 isomer pFCC-1 [1]; <5> cell death gene ACD2 encodes red chlorophyll catabolite reductase and suppresses the spread of disease symptoms [8]; <3> RCCR absence causes leaf cell death as a result of the accumulation of photodynamic RCC. RCCR (together with pheophorbide a oxygenase) is required for the detoxification of chlorophyll catabolites [3]; <4> the enzyme is involved in breakdown of chlorophyll [5]; <1> the enzyme is involved in chlorophyll breakdown [6]; <3> the enzyme is involved in chlorophyll breakdown in senescent Arabidopsis leaves [7]; <1> the major product of reduction of red chlorophyll catabolite is pFCC1, but small quantities of its C_1 epimer, pFCC-2, also accumulate. Red chlorophyll catabolite reductase and pheophorbide a oxygenase catalyse the key eaction of chlorophyll catabolism, porphin macrocycle cleavage of pheide a to a primary fluorescent catabolite [4]; <3> the major product of reduction of red chlorophyll catabolite is pFCC1, but small quantities of its C_1 epimer, pFCC-2, also accumulate. Red chlorophyll catabolite reductase and pheophorbide a oxygenase catalyse the key reaction of chlorophyll catabolism, porphin macrocycle cleavage of pheide a to a primary fluorescent catabolite [4]; <5> with Arabidopsis RCCR, the C_1 isomer pFCC-1 is formed. RCCR could be required to mediate an efficient interaction between red chlorophyll catabolite (still bound to pheophorbide a oxygenase) and ferredoxin, thereby enabling a fast, regio-, and stereoselective reduction to blue-fluorescing intermediate [1]; <6> with tomato RCCR, the C_1 isomer pFCC-2 is formed [1]; <2> spinach RCCR produces the C_1 isomer pFCC-2 [1]) (Reversibility: ?) [1, 3, 4, 5, 6, 7, 8]

P ?

Inhibitors

O_2 <1, 3> (<1,3> reaction is sensitive to oxygen [4,6]) [4, 6]

Cofactors/prosthetic groups

ferredoxin <1, 3, 4, 5> (<4> required [5]; <1> reaction is dependent on reduced ferredoxin [6]; <5> the enzyme is ferredoxin dependent, but appears to lack a metal or flavin cofactor, indicating that electrons are directly transferred from ferredoxin to RCC [1]; <3> the reaction requires reduced ferredoxin [4]) [1,4,5,6]

NADPH <1> (<1> required for formation of primary fluorescent chlorophyll catabolite [6]) [6]

K_m-Value (mM)

0.6 <1> (red chlorophyll catabolite) [6]

4 Enzyme Structure

Molecular weight

58000 <1> (<1> gel filtration [4]) [4]

Subunits

dimer <1> (<1> 2 * 31000, SDS-PAGE [4]) [4]

Posttranslational modification

proteolytic modification <3> (<3> AtRCCR encodes a 35000 Da protein that is processed to a mature form of 31000 Da [4]) [4]

5 Isolation/Preparation/Mutation/Application

Source/tissue

cotyledon <4> [5]
gerontoplast <6, 7, 8, 9, 10, 11, 12, 13> (<6, 7, 8, 9, 10, 11, 12, 13> membrane [2]) [2]
leaf <1, 3, 4, 5, 6, 7, 8, 9, 10, 11, 12, 13> (<3> senescent [4,7]; <1> present in all stages of leaf development. The highest specific activity is found in senescent leaves [6]; <6,7,8,9,10,11,12,13> the enzyme is not only present in senescent leaves but also at other stages of leaf development [2]; <1> the gene is expressed at all stages of leaf development [4]) [2, 3, 4, 5, 6, 7, 8]
root <1> [4, 6]
seedling <5> [1]

Localization

chloroplast <1, 3, 4, 5> (<3> AtRCCR import is inhibited when the protein is missing 40 amino acids at the N-terminus [4]; <3> RCCR participates in chlorophyll breakdown inside the chloroplast [3]; <5> soluble protein of chloroplasts, in young Arabidopsis seedlings also associated with mitochondria [1]; <1,4> stroma protein [5,6]) [1, 3, 4, 5, 6, 8]
membrane <6, 7, 8, 9, 10, 11, 12, 13> [2]
mitochondrion <5> (<5> soluble protein of chloroplasts, in young Arabidopsis seedlings also associated with mitochondria [1]) [1, 8]

Purification

<1> [1, 4]
<1> (partial) [6]
<3> [4]
<5> [1]

Cloning

<1> [1]
<1> (a partial gene sequence) [4]
<3> (expression in Escherichia coli) [4]
<5> (chimeric RCCRs composed of portions of the Arabidopsis and the tomato proteins are expressed in Escherichia coli) [1]
<5> (overexpression of ACD2 protein makes the plants tolerant but not resistant to bacterial infection) [8]
<6> (chimeric RCCRs composed of portions of the Arabidopsis and the tomato proteins are expressed in Escherichia coli) [1]

Engineering

F218V <3> (<3> mutation switches At-RCCR stereospecificity from pFCC-1 to pFCC-2 production [3]) [3]
V218V <5> (<5> mutation changes the specificity of the protein from pFCC-1 (C_1 isomers of pFCC) to pFCC-2 (C_1 isomers of pFCC) production [1]) [1]
Additional information <3, 5, 6> (<3> chimeric RCCRs are produced in Escherichia coli by replacing parts of mature enzyme from Arabidopsis thali-

ana with the respective sequences from Lycopersicon esculentum. A reversal of specificity is found in protein M, in which a domain of 37 amino acids of At-RCCR have been replaced with the respective Le-RCCR domain [3]; <5,6> chimeric RCCRs composed of portions of the Arabidopsis and the tomato proteins are expressed in Escherichia coli [1]) [1, 3]

Application

agriculture <5> (<5> economically important plants overexpressing ACD2 might also show increased tolerance to pathogens and might be useful for increasing crop yields [8]) [8]

References

[1] Hörtensteiner, S.: Chlorophyll degradation during senescence. Annu. Rev. Plant Biol., **57**, 55-77 (2006)

[2] Hörtensteiner, S.; Rodoni, S.; Schellenberg, M.; Vicentini, F.; Nandi, O.I.; Qui, Y.L.; Matile, P.: Evolution of chlorophyll degradation: the significance of RCC reductase. Plant Biol., **2**, 63-67 (2000)

[3] Pruzinska, A.; Anders, I.; Aubry, S.; Schenk, N.; Tapernoux-Lüthi, E.; Müller, T.; Kräutler, B.; Hörtensteiner, S.: In vivo participation of red chlorophyll catabolite reductase in chlorophyll breakdown. Plant Cell, **19**, 369-387 (2007)

[4] Wüthrich, K.L.; Bovet, L.; Hunziker, P.E.; Donnison, I.S.; Hörtensteiner, S.: Molecular cloning, functional expression and characterisation of RCC reductase involved in chlorophyll catabolism. Plant J., **21**, 189-198 (2000)

[5] Rodoni, S.; Mühlecker, W.; Anderl, M.; Kräutler, B.; Moser, D.; Thomas, H.; Matile, P.; Hörtensteiner, S.: Chlorophyll breakdown in senescent chloroplasts (cleavage of pheophorbide a in two enzymic steps). Plant Physiol., **115**, 669-676 (1997)

[6] Rodoni, S.; Vicentini, F.; Schellenberg, M.; Matile, P.; Hörtensteiner, S.: Partial purification and characterization of red chlorophyll catabolite reductase, a stroma protein involved in chlorophyll breakdown. Plant Physiol., **115**, 677-682 (1997)

[7] Pruzinska, A.; Tanner, G.; Aubry, S.; Anders, I.; Moser, S.; Müller, T.; Ongania, K.H.; Kräutler, B.; Youn, J.Y.; Liljegren, S.J.; Hörtensteiner, S.: Chlorophyll breakdown in senescent Arabidopsis leaves. Characterization of chlorophyll catabolites and of chlorophyll catabolic enzymes involved in the degreening reaction. Plant Physiol., **139**, 52-63 (2005)

[8] Mach, J.M.; Castillo, A.R.; Hoogstraten, R.; Greenberg, J.T.: The Arabidopsis-accelerated cell death gene ACD2 encodes red chlorophyll catabolite reductase and suppresses the spread of disease symptoms. Proc. Natl. Acad. Sci. USA, **98**, 771-776 (2001)

Tryptophan α,β-oxidase 1.3.3.10

1 Nomenclature

EC number
1.3.3.10

Systematic name
L-tryptophan:oxygen α,β-oxidoreductase

Recommended name
tryptophan α,β-oxidase

Synonyms
EC 1.4.3.17
L-tryptophan 2',3'-oxidase
L-tryptophan α,β-dehydrogenase
tryptophan side chain oxidase II <1> [1, 2]

CAS registry number
156859-19-7

2 Source Organism

<1> *Pseudomonas sp.* (no sequence specified) [1, 2]
<2> *Chromobacterium violaceum* (no sequence specified) [3, 4]

3 Reaction and Specificity

Catalyzed reaction
L-tryptophan + O_2 = α,β-didehydrotryptophan + H_2O_2

Reaction type
oxidation
redox reaction
reduction

Substrates and products
S L-tryptophan + O_2 <2> (Reversibility: ?) [3]
P α,β-didehydro-L-tryptophan + H_2O_2 <2> [3]
S L-tryptophanamide + O_2 <2> (Reversibility: ?) [3]
P α,β-didehydro-L-tryptophanamide + H_2O_2 <2> [3]
S N-acetyl-L-tryptophan + O_2 <2> (Reversibility: ?) [3]
P N-acetyl-α,β-didehydro-L-tryptophan + H_2O_2 <2> [3]

S N-acetyl-L-tryptophanamide + O_2 <2> (Reversibility: ?) [3, 4]
P N-acetyl-α,β-didehydrotryptophanamide + H_2O_2 <2> [3, 4]
S adrenocorticotropic hormone + O_2 <2> (Reversibility: ?) [3]
P ? + H_2O_2 <2> [3]
S indole-3-propionate + O_2 <2> (Reversibility: ?) [3]
P ? + H_2O_2 <2> [3]
S luteinizing hormone-releasing hormone + O_2 <2> (Reversibility: ?) [3]
P ? + H_2O_2 <2> [3]
S pentagastrin + O_2 <2> (Reversibility: ?) [3]
P ? + H_2O_2 <2> [3]
S tryptophan No.14 of α-globin + O_2 <1> (Reversibility: ?) [1]
P α,β-dehydrotryptophan + H_2O_2 <1> [1]
S tryptophan No.15 of β-globin + O_2 <1> (Reversibility: ?) [1]
P α,β-dehydrotryptophan + H_2O_2 <1> [1]

Inhibitors

5-hydroxy-L-tryptophan <2> (<2> competitive inhibition, K_i: 0.037 mM [3]) [3, 4]
D-tryptophan <2> (<2> competitive inhibition, K_i: 0.0066 mM [3]) [3, 4]
Sodium dodecyl sulfate <1> (<1> no modification with 0.4% SDS [1]) [1]
tryptamine <2> (<2> competitive inhibition, K_i: 0.0054 mMl [3]) [3, 4]
urea <1> (<1> no modification with 8 M urea [1]) [1]
methyl-3-indole <2> (<2> competitive inhibition, K_i: 0.0013 mM [3]) [3, 4]
Additional information <2> (<2> no inhibition with L-histidine, N-acetyl-L-phenylalaninamide and N-acetyl-L-tyrosinamide [3,4]) [3, 4]

Cofactors/prosthetic groups

Additional information <2> (<2> NAD^+, $NADP^+$, FAD, or FMN show no effect on the activity [4]) [4]

Metals, ions

$MgCl_2$ <2> (<2> 10 mM, residual activity 103% [3]) [3]

Turnover number (min^{-1})

0.283 <2> (adrenocorticotropic hormone, <2> dehydrogenation of the tryptophanyl side chain [3]) [3]
0.333 <2> (luteinizing hormone-releasing hormone, <2> dehydrogenation of the tryptophanyl side chain [3]) [3]
0.533 <2> (pentagastrin, <2> dehydrogenation of the tryptophanyl side chain [3]) [3]
0.593 <2> (L-tryptophanamide) [3]
0.66 <2> (L-tryptophan) [3]
0.753 <2> (N-acetyl-L-tryptophanamide) [3, 4]
0.785 <2> (indole-3-propionate) [3]
0.79 <2> (N-acetyl-L-tryptophan) [3]

Specific activity (U/mg)

0.02 <1> (<1> substrate tryptophanamide [2]) [2]
0.07 <1> (<1> substrate tryptophan-leucine [2]) [2]

0.31 <1> (<1> substrate tryptophan [2]) [2]
0.74 <1> (<1> substrate leucin-tryptophan-leucin [2]) [2]
1.08 <1> (<1> substrate indolepropionic acid [2]) [2]
1.42 <1> (<1> substrate N-acetyltryptophanamide [2]) [2]
1.82 <1> (<1> substrate N-acetyltryptophan [2]) [2]
2.32 <1> (<1> substrate leucin-tryptophan [2]) [2]
30.3 <2> (<2> after purification [3]) [3]

K_m-Value (mM)
0.0058 <2> (L-tryptophan) [3]
0.0091 <2> (L-tryptophanamide) [3]
0.0195 <2> (N-acetyl-L-tryptophanamide) [3, 4]
0.02 <2> (indole-3-propionate) [3]
0.026 <2> (pentagastrin, <2> dehydrogenation of the tryptophanyl side chain [3]) [3]
0.071 <2> (luteinizing hormone-releasing hormone, <2> dehydrogenation of the tryptophanyl side chain [3]) [3]
0.0791 <2> (N-acetyl-L-tryptophan) [3]
0.93 <2> (adrenocorticotropic hormone, <2> dehydrogenation of the tryptophanyl side chain [3]) [3]

K_i-Value (mM)
0.0013 <2> (methyl-3-indole) [3, 4]
0.0054 <2> (tryptamine) [3, 4]
0.0066 <2> (D-tryptophan) [3, 4]
0.037 <2> (5-hydroxy-L-tryptophan) [3, 4]

pH-Optimum
5 <2> [3]

pH-Range
3-9.5 <2> [4]

Temperature optimum (°C)
80 <2> [4]

4 Enzyme Structure

Molecular weight
150000 <1> [2]
680000 <2> (<2> gel filtration [3,4]) [3, 4]

Subunits
octamer <2> (<2> α, 4 * 14000 + β 4 * 74000, organized in an α_4,β_4 manner, heterooligomeric structure, one heme molecule per α,β protomer, SDS-PAGE [3,4]) [3, 4]

5 Isolation/Preparation/Mutation/Application

Localization

cytoplasm <2> [4]

Purification

<1> (gel filtration, phosphocellulose column chromatography, SDS-PAGE) [2]
<2> (cell disruption with Eaton press, ultracentrifugation, ammonium sulfate precipitation, DEAE-column, concentration using centrifugal microcentrators, gel filtration, purified 108fold, yield 33.8%) [4]

6 Stability

pH-Stability

7 <2> (<2> after dilution in 0.1 M bis-Tris buffer [3]) [3]

Temperature stability

80 <2> (<2> active up to 80°C [3]) [3]

General stability information

<2>, active with 0.2 M dithiothreitol, 1% SDS, or 4.5 M urea [3]

Storage stability

<2>, -80°C, for several months and for 24-48 h at room temperature in the absence of any additives [4]

References

[1] Takai, K.; Sasai, Y.; Morimoto, H.; Yamazaki, H.; Yoshii, H.; Inoue, S.: Enzymatic dehydrogenation of tryptophan residues of human globins by tryptophan side chain oxidase II. J. Biol. Chem., **259**, 4452-4457 (1984)
[2] Ito, S.; Takai, K.; Tokuyama, T.; Hayaishi, O.: Enzymatic modification of tryptophan residues by tryptophan side chain oxidase I and II from Pseudomonas. J. Biol. Chem., **256**, 7834-7843 (1981)
[3] Genet, R.; Benetti, P.H.; Hammadi, A.; Menez, A.: L-tryptophan 2',3'-oxidase from Chromobacterium violaceum. J. Biol. Chem., **270**, 23540-23545 (1995)
[4] Genet, R.; Donoyelle, C.; Menez, A.: Purification and partial characterization of an amino acid α,β-dehydrogenase, L-tryptophan 2',3'-oxidase from Chromobacterium violaceum. J. Biol. Chem., **269**, 18177-18184 (1994)

Pyrroloquinoline-quinone synthase 1.3.3.11

1 Nomenclature

EC number
1.3.3.11

Systematic name
6-(2-amino-2-carboxyethyl)-7,8-dioxo-1,2,3,4,5,6,7,8-octahydroquinoline-2,4-dicarboxylate:oxygen oxidoreductase (cyclizing)

Recommended name
pyrroloquinoline-quinone synthase

Synonyms
PqqC <1, 2, 3, 4> [1, 2, 3, 4, 5, 6]

CAS registry number
353484-42-1

2 Source Organism

<1> *Methylobacterium extorquens* (no sequence specified) [1, 3]
<2> *Deinococcus radiodurans* (no sequence specified) [2]
<3> *Methylobacterium extorquens* (UNIPROT accession number: Q49150) [4]
<4> *Klebsiella pneumoniae* (UNIPROT accession number: P27505) [5, 6]

3 Reaction and Specificity

Catalyzed reaction
6-(2-amino-2-carboxyethyl)-7,8-dioxo-1,2,3,4,5,6,7,8-octahydroquinoline-2,4-dicarboxylate + 3 O_2 = 4,5-dioxo-3a,4,5,6,7,8,9,9b-octahydro-1H-pyrrolo[2,3-f]quinoline-2,7,9-tricarboxylate + 2 H_2O_2 + 2 H_2O

Reaction type
cyclization
oxidation
reduction

Natural substrates and products
S 3a-(2-amino-2-caboxyethyl)-4,5-dioxo-4,5,6,7,8,9-hexahydroquinoline-7,9-dicarboxylic acid + O_2 <1, 2, 3, 4> (<1,3,4> synthesis of the pyrroloquinoline quinone (PQQ), that serves as prosthetic group for many bac-

terial enzymes [1,3,4,5,6]; <2> synthesis of the pyrroloquinoline quinone (PQQ), that serves as prosthetic group for many bacterial enzymes, PQQ might also be involved in the protection against reactive oxygen species during growth under stress conditions [2]) (Reversibility: ?) [1, 2, 3, 4, 5, 6]

P 4,5-dioxo-1H-pyrrolo[2,3-f]quinoline-2,7,9-tricarboxylate + H_2O_2 + H_2O (<1,2,3,4> the product is pyrroloquinoline quinone (PQQ) [1,2,3,4,5,6])

Substrates and products

S 3a-(2-amino-2-caboxyethyl)-4,5-dioxo-4,5,6,7,8,9-hexahydroquinoline-7,9-dicarboxylic acid + O_2 <1, 2, 3, 4> (<1,3,4> synthesis of the pyrroloquinoline quinone (PQQ), that serves as prosthetic group for many bacterial enzymes [1,3,4,5,6]; <2> synthesis of the pyrroloquinoline quinone (PQQ), that serves as prosthetic group for many bacterial enzymes, PQQ might also be involved in the protection against reactive oxygen species during growth under stress coditions [2]) (Reversibility: ?) [1, 2, 3, 4, 5, 6]

P 4,5-dioxo-1H-pyrrolo[2,3-f]quinoline-2,7,9-tricarboxylate + H_2O_2 + H_2O (<1,2,3,4> the product is pyrroloquinoline quinone (PQQ) [1,2,3,4,5,6])

Activating compounds

NAD(P)H <3> (<3> activating in the presence of EDTA [4]) [4]

Additional information <1> (<1> no dependence on NADH or NADPH in purified enzyme preparation [1]) [1]

pH-Optimum

8 <3> [4]

4 Enzyme Structure

Molecular weight

113000 <1> (<1> gel filtration [1]) [1]

Subunits

? <2, 3> (<2> x * 42000, SDS-PAGE [2]; <3> x * 41577, calculated from the deduced amino acid sequence [4]; <3> x * 42000, SDS-PAGE, recombinant protein from Escherichia coli [4]) [2, 4]

dimer <4> (<4> 2 * 28910 [5]; <4> 2 * 28986, calculated from deduced amino acid sequence [6]) [5, 6]

trimer <1> (<1> 3 * 42000, SDS-PAGE [1]) [1]

5 Isolation/Preparation/Mutation/Application

Purification

<1> (partial purification of recombinant protein) [1]

<4> [6]

Crystallization

<4> (sitting drop vapor diffusion method) [5, 6]

Cloning

<1> (expressed in Escherichia coli DHα) [1]

<2> (expressed in Escherichia coli BL21(DE3)pLysS, complements the mineral phosphate solubilization phenotype) [2]

<3> (expressed in Escherichia coli DH α, complementation studies with PqqC deficient Methylobacterium extorquens and Methylobacterium organophilum) [4]

<4> (expressed in Escherichia coli) [5]

<4> (expressed in Escherichia coli BL21(DE3)) [6]

Engineering

Additional information <1> (<1> N60I mutation detected in PqqC deficiency mutant [1]) [1]

References

[1] Toyama, H.; Fukumoto, H.; Saeki, M.; Matsushita, K.; Adachi, O.; Lidstrom, M.E.: PqqC/D, which converts a biosynthetic intermediate to pyrroloquinoline quinone. Biochem. Biophys. Res. Commun., **299**, 268-272 (2002)

[2] Khairnar, N.P.; Misra, H.S.; Apte, S.K.: Pyrroloquinoline-quinone synthesized in Escherichia coli by pyrroloquinoline-quinone synthase of Deinococcus radiodurans plays a role beyond mineral phosphate solubilization. Biochem. Biophys. Res. Commun., **312**, 303-308 (2003)

[3] Magnusson, O.T.; Toyama, H.; Saeki, M.; Schwarzenbacher, R.; Klinman, J.P.: The structure of a biosynthetic intermediate of pyrroloquinoline quinone (PQQ) and elucidation of the final step of PQQ biosynthesis. J. Am. Chem. Soc., **126**, 5342-5343 (2004)

[4] Toyama, H.; Chistoserdova, L.; Lidstrom, M.E.: Sequence analysis of pqq genes required for biosynthesis of pyrroloquinoline quinone in Methylobacterium extorquens AM1 and the purification of a biosynthetic intermediate. Microbiology, **143**, 595-602 (1997)

[5] Magnusson, O.T.; Toyama, H.; Saeki, M.; Rojas, A.; Reed, J.C.; Liddington, R.C.; Klinman, J.P.; Schwarzenbacher, R.: Quinone biogenesis: Structure and mechanism of PqqC, the final catalyst in the production of pyrroloquinoline quinone. Proc. Natl. Acad. Sci. USA, **101**, 7913-7918 (2004)

[6] Schwarzenbacher, R.; Stenner-Liewen, F.; Liewen, H.; Reed, J.C.; Liddington, R.C.: Crystal structure of PqqC from Klebsiella pneumoniae at 2.1 ANG resolution. Proteins, **56**, 401-403 (2004)

L-Galactonolactone oxidase 1.3.3.12

1 Nomenclature

EC number
1.3.3.12

Systematic name
L-galactono-1,4-lactone:oxygen 3-oxidoreductase

Recommended name
L-galactonolactone oxidase

Synonyms
GL oxidase
L-galactono-1,4-lactone oxidase
L-galactono-γ-lactone oxidase
galactone-γ-lactone oxidase
Additional information

CAS registry number
69403-13-0

2 Source Organism

<1> *Saccharomyces cerevisiae* (no sequence specified) [1, 2, 3, 4, 5]
<2> *Avena sativa* (no sequence specified) [6]

3 Reaction and Specificity

Catalyzed reaction
L-galactono-1,4-lactone + O_2 = L-ascorbate + H_2O_2

Reaction type
oxidation
redox reaction
reduction

Natural substrates and products
S L-galactono-1,4-lactone + O_2 <1, 2> (<1,2> enzyme catalyzes the last step of L-ascorbic acid biosynthesis [4,6]) (Reversibility: ?) [1, 2, 3, 4, 6]
P L-ascorbate + H_2O_2

Substrates and products

S D-altrono-1,4-lactone + O_2 <1> (Reversibility: ?) [1]
P ?
S D-altronolactone + O_2 <1> (<1> 68% of the activity with L-galactono-1,4-lactone [4]) (Reversibility: ?) [4]
P ?
S D-arabinono-1,4-lactone + O_2 <1> (Reversibility: ?) [1]
P ?
S D-threono-1,4-lactone + O_2 <1> (Reversibility: ?) [1]
P ?
S L-fucono-1,4-lactone + O_2 <1> (Reversibility: ?) [1]
P ?
S L-galactono-1,4-lactone + O_2 <1, 2> (<1,2> enzyme catalyzes the last step of L-ascorbic acid biosynthesis [4,6]) (Reversibility: ?) [1, 2, 3, 4, 5, 6]
P L-ascorbate + H_2O_2 <1> [1, 2, 3, 4]
S L-gulonolactone + O_2 <1> (<1> 32% of the activity with L-galactono-1,4-lactone [4]) (Reversibility: ?) [4]
P ?

Inhibitors

2,2'-dipyridyl disulfide <1> [2]
4,4'-dipyridyl disulfide <1> [2]
5,5'-dithiobis(2-nitrobenzoate) <1> [2]
D-galactono-1,4-lactone <1> (<1> competitive inhibition [1]) [1]
$HgCl_2$ <1> [2]
iodoacetamide <1> [1, 2]
L-gulono-1,4-lactone <1> (<1> competitive inhibition [1]) [1]
N-ethylmaleimide <1> [1, 2]
PCMB <1> [2, 4]
sulfide <1> [1]
sulfite <1> [1]
p-chloromercuriphenyl sulfonate <1> [1]
Additional information <2> (<2> the synthesis of the enzyme is sensitive to both mitochondrial and cytoplasmic translation inhibitors [6]) [6]

Cofactors/prosthetic groups

FAD <1> (<1> enzyme contains a covalently bound flavin [1,2,3,4,5]; <1> the covalently bound flavin is 8α-[N(1)histidyl]FAD [3]) [1, 2, 3, 4, 5]

Metals, ions

iron <1> (<1> enzyme contains an iron-sulfur cluster [1]) [1]

Specific activity (U/mg)

0.76 <1> [4]
3.448 <1> [1]

K_m-Value (mM)

0.16 <1> (D-arabino-1,4-lactone) [1]
0.18 <1> (oxygen) [1]

0.3 <1> (L-galactono-1,4-lactone) [1]
0.36 <1> (L-fucono-1,4-lactone) [1]
2 <1> (D-altrono-1,4-lactone) [1]
15 <1> (D-threono-1,4-lactone) [1]

K_i-Value (mM)

3.3 <1> (D-galactono-1,4-lactone) [1]
6.52 <1> (L-gulono-1,4-lactone) [1]

pH-Optimum

8.9 <1> [1]

4 Enzyme Structure

Molecular weight

70000 <1> (<1> gel filtration in presence of deoxycholate [1]) [1]
74000 <1> (<1> non-denaturing gradient PAGE in presence of deoxycholate [1]) [1]
290000 <1> (<1> gel filtration [4]) [4]

Subunits

tetramer <1> (<1> 4 * 56000, SDS-PAGE [4]; <1> 4 * 18000, SDS-PAGE [1]) [1, 4]
Additional information <2> (<2> might be a multimeric enzyme, in which some polypeptide chains could be synthesized in the cytosol and others in the mitochondria [6]) [6]

5 Isolation/Preparation/Mutation/Application

Localization

mitochondrion <1> [1, 4]
Additional information <2> (<2> might be a multimeric enzyme, in which some polypeptide chains could be synthesized in the cytosol and others in the mitochondria [6]) [6]

Purification

<1> [1, 2, 3, 4]

Cloning

<1> (expressed as a fusion protein with glutathione S-transferase in Escherichia coli cells with the expression vector pGEX-5X-3) [5]

6 Stability

Storage stability

<1>, 5°C, 6 months, 50% loss of activity [1]

References

[1] Bleeg, H.S.; Christensen, F.: Biosynthesis of ascorbate in yeast. Purification of L-galactono-1,4-lactone oxidase with properties different from mammalian L-gulonolactone oxidase. Eur. J. Biochem., **127**, 391-396 (1982)
[2] Noguchi, E.; Nishikimi, M.; Yagi, K.: Studies on the sulfhydryl group of L-galactonolactone oxidase. J. Biochem., **90**, 33-38 (1981)
[3] Kenney, W.C.; Edmondson, D.E.; Singer, T.P.; Nishikimi, M.; Noguchi, E.; Yagi, K.: Identification of the covalently-bound flavin of L-galactonolactone oxidase from yeast. FEBS Lett., **97**, 40-42 (1979)
[4] Nishikimi, M.; Noguchi, E.; Yagi, K.: Occurrence in yeast of L-galactonolactone oxidase which is similar to a key enzyme for ascorbic acid biosynthesis in animals, L-gulonolactone oxidase. Arch. Biochem. Biophys., **191**, 479-486 (1978)
[5] Nishikimi, M.; Ohta, Y.; Ishikawa, T.: Identification by bacterial expression of the yeast genomic sequence encoding L-galactono-γ-lactone oxidase, the homolog of L-ascorbic acid-synthesizing enzyme of higher animals. Biochem. Mol. Biol. Int., **44**, 907-913 (1998)
[6] De Gara, L.; Tommasi, F.; Liso, R.; Arrigoni, O.: The biogenesis of galactone-γ-lactone oxidase in Avena sativa embryos. Phytochemistry, **31**, 755-756 (1992)

Coproporphyrinogen dehydrogenase 1.3.99.22

1 Nomenclature

EC number
1.3.99.22

Systematic name
coproporphyrinogen-III:S-adenosyl-L-methionine oxidoreductase (decarboxylating)

Recommended name
coproporphyrinogen dehydrogenase

Synonyms
CPO <1, 2, 3, 4, 5> [4]
coprogen oxidase <3> [3]
HemN <1, 2, 3, 4, 5> [3, 4, 5, 6]
coproporphyrinogen III oxidase <3> [3]
coproporphyrinogen oxidase <3> [3]
coproporphyrinogenase <3> [3]
oxidase, coproporphyrinogen <3> [3]
oxygen-independent CPO <3> [2]
oxygen-independent coproporphyrinogen III oxidase <1, 2, 3, 4, 5> [4, 5, 6]
oxygen-independent coproporphyrinogen-III oxidase <3> [3]
radical SAM enzyme <3> [3]
Additional information <3> (<3> belongs to the Radical SAM protein family [1]; <3> member of the Radical SAM protein family [3]; <3> Radical SAM enzyme [2]) [1, 2, 3]

CAS registry number
9076-84-0

2 Source Organism

<1> *Salmonella typhimurium* (no sequence specified) [4]
<2> *Bacillus subtilis* (no sequence specified) [4]
<3> *Escherichia coli* (no sequence specified) [1, 2, 3, 4, 5, 6]
<4> *Ralstonia eutropha* (no sequence specified) [4]
<5> *Rhodobacter sphaeroides* (no sequence specified) [4]

3 Reaction and Specificity

Catalyzed reaction

coproporphyrinogen III + 2 S-adenosyl-L-methionine = protoporphyrinogen IX + 2 CO_2 + 2 L-methionine + 2 5'-deoxyadenosine (<3> mechanism [1]; <3> catalytic, radical mechanism [2]; <3> this enzyme differs from EC 1.3.3.3, coproporphyrinogen oxidase, by using S-adenosyl-L-methionine, AdoMet, instead of oxygen as oxidant, it occurs mainly in bacteria, whereas eukaryotes use the oxygen-dependent oxidase, the reaction starts by using an electron from the reduced form of the enzyme's [4Fe-4S] cluster to split AdoMet into methionine and the radical 5'-deoxyadenosin-5'-yl, this radical initiates attack on the 2-carboxyethyl groups, leading to their conversion into vinyl groups. The conversion of α-CH-CH_2-COO- leading to α-CH=CH_2 + CO_2 + e- replaces the electron initially used, reaction mechanism [3])

Reaction type

decarboxylation
oxidation
redox reaction
reduction

Natural substrates and products

S coproporphyrinogen-III + S-adenosyl-L-methionine <3> (<3> HemN catalyzes the essential conversion of coproporphyrinogen-III to protoporphyrinogen IX during heme biosynthesis [1]; <3> HemN catalyzes the prepenultimate step in anaerobic heme biosynthesis [3]) (Reversibility: ?) [1, 3]

P protoporphyrinogen IX + CO_2 + L-methionine + 5'-deoxyadenosine

Substrates and products

S coproporphyrinogen-III + S-adenosyl-L-methionine <3, 5> (<3> HemN catalyzes the essential conversion of coproporphyrinogen-III to protoporphyrinogen IX during heme biosynthesis [1]; <3> HemN catalyzes the prepenultimate step in anaerobic heme biosynthesis [3]; <3> HemN catalyzes the oxygen-independent conversion of coproporphyrinogen-III to protoporphyrinogen IX, requires S-adenosyl-L-methionine, NAD(P)H and additional cytoplasmatic components for catalysis. Cys-62, Cys-66 and Cys-69 are part of the conserved CXXXCXXC motif and essential for iron-sulfur cluster formation and enzyme function. Gly-111 and Gly-113 are part of the potential GGGTP S-adenosyl-L-methionine binding motif and essential for enzymatic function, catalytic, radical mechanism [2]; <3> HemN requires the juxtaposition of the [4Fe-4S] cluster and the co-substrate S-adenosyl-L-methionine. The reaction involves the stereospecific hydrogen abstraction of the pro-S hydrogen from the propionate side chain β-C of coproporphyrinogen-III, involvement of a coproporphyrinogenyl III radical, which is then decarboxylated releasing CO_2 and forming the vinyl group, enzyme structure, two-domain enzyme consisting of the catalytic N- and an α-helical C-terminal domain, substrate binding mode

[3]; <3> mechanism, the S-adenosyl-L-methionine sulfonium sulfur is near both the Fe and neighboring sulfur of the cluster allowing single electron transfer from the 4Fe-4S cluster to the S-adenosyl-L-methionine sulfonium. S-adenosyl-L-methionine is cleaved yielding a highly oxidizing 5'-deoxyadenosyl radical, HemN binds a second S-adenosyl-L-methionine immediately adjacent to the first and may thus successively catalyze two propionate decarboxylations. Cofactor geometry required for Radical SAM catalysis, detailed enzyme structure, two distinct domains, domain structure, S-adenosyl-L-methionine binding mode [1]) (Reversibility: ?) [1, 2, 3, 4, 5, 6]

P protoporphyrinogen IX + CO_2 + L-methionine + 5'-deoxyadenosine

Inhibitors

EDTA <3> (<3> strong inhibition [2]) [2]
o-phenanthroline <3> (<3> strong inhibition [2]) [2]

Cofactors/prosthetic groups

4Fe-4S-center <3> (<3> HemN binds a 4Fe-4S cluster through three cysteine residues: Cys-62, Cys-66 and Cys-69, a juxtaposed S-adenosyl-L-methionine coordinates the fourth Fe ion through its amide nitrogen and carboxylate oxygen, detailed binding mode, cofactor geometry required for Radical SAM catalysis [1]; <3> requirement, oxygen-sensitive Fe-S cluster, Cys-62, Cys-66 and Cys-69 are part of the conserved CXXXCXXC motif and essential for Fe-S cluster formation and enzyme function, Tyr-56 and His-58 are important for the Fe-S cluster integrity, His-58 may provide the fourth ligand besides the three cysteine residues [2]; <3> structure, location and coordination of the cofactor, HemN requires the juxtaposition of the [4Fe-4S] cluster and the co-substrate S-adenosyl-L-methionine [3]) [1,2,3]
NADH <3> (<3> requires NAD(P)H, higher activity than with NADPH as cofactor [2]) [2,4]
NADPH <3> (<3> requires NAD(P)H, lower activity than with NADH as cofactor [2]) [2]
S-adenosyl-L-methionine <3> (<3> HemN contains two S-adenosyl-L-methionine molecules as cofactors, detailed binding mode [1]; <3> uses S-adenosyl-L-methionine as a cofactor [2]; <3> cosubstrate of coproporphyrinogen-III [5]; <3> HemN binds two SAM molecules [4]) [1,2,4,5]

Metals, ions

Fe^{2+} <3> (<3> HemN binds a (4Fe-4S) cluster [4]) [4]

Temperature optimum (°C)

37 <3> (<3> assay at [2]) [2]

4 Enzyme Structure

Molecular weight

52000 <3> (<3> gel filtration in combination with glycerol gradient centrifugation [2]) [2]

Subunits

monomer <3> (<3> 1* 52734, sequence calculation [2]; <3> monomeric enzyme consisting of two distinct domains [1]; <3> by crystallographic studies [4]) [1, 2, 3, 4]

5 Isolation/Preparation/Mutation/Application

Purification

<3> [6]
<3> (by a single chromatographic step under anaerobic conditions) [4]
<3> (by affinity chromatography) [5]
<3> (recombinant HemN) [1]
<3> (recombinant wild-type and mutant HemN) [2]

Crystallization

<3> (crystal structure) [3]
<3> (crystal structure, co-crystallized with S-adenosyl-L-methionine, hanging-drop vapor diffusion method, X-ray analysis) [1]
<3> (presence of an unusually coordinated iron-sulfur cluster and two molecules of S-adenosylmethionine, which is of functional importance) [5]
<3> (under strictly anaerobic conditions, comparison of the crystal structures of the three radical SAM enzymes HemN, BioB and MoaA show that the enzymes are structurally significantly different) [4]

Cloning

<3> [1]
<3> (hemeN gene, overexpression in Escherichia coli BL21(DE3)) [2]
<3> (mutated enzymes expressed in Escherichia coli BL21(DE3)) [5]

Engineering

C62S <3> (<3> inactive mutant, no Fe-S cluster formation [2]) [2]
C66S <3> (<3> inactive mutant, no Fe-S cluster formation [2]) [2]
C69S <3> (<3> inactive mutant, no Fe-S cluster formation [2]) [2]
C71S <3> (<3> inactive mutant, same Fe-S cluster formation as in wild-type HemN [2]) [2]
E145A <3> (<3> appears colorless, the [4Fe-4S] cluster content is slightly reduced, no detectable S-adenosylmethionine cleavage, no detectable CPO activity [5]) [5]
E145I <3> (<3> appears colorless, the [4Fe-4S] cluster content is slightly reduced, no detectable S-adenosylmethionine cleavage, no detectable CPO activity [5]) [5]
F310A <3> (<3> is slightly yellow, the [4Fe-4S] cluster content is slightly reduced, cleaves only one S-adenosylmethionine molecule per molecule protein, residual CPO activity [5]) [5]
F310L <3> (<3> is slightly yellow, the [4Fe-4S] cluster content is slightly reduced, cleaves only one S-adenosylmethionine molecule per molecule protein, no detectable CPO activity [5]) [5]

F68L <3> (<3> mutant with 89% of wild-type activity [2]) [2]
G111V/G113V <3> (<3> double mutation of Gly-111 and Gly-113, which are part of the potential GGGTP S-adenosyl-L-methionine binding motif, completely abolishes enzyme activity, reduced Fe-S cluster formation [2]) [2]
H58L <3> (<3> inactive mutant, no Fe-S cluster formation [2]) [2]
I329A <3> (<3> contains the same amount of iron-sulfur cluster as the wild-type HemN, but cleaves only one S-adenosylmethionine molecule per molecule protein, no detectable CPO activity [5]; <3> exhibits the same yellow-brown color as wild-type HemN, but the [4Fe-4S] cluster content is slightly reduced and cleaves only one S-adenosylmethionine molecule per molecule protein, no detectable CPO activity [5]) [5]
Q311A <3> (<3> contains the same amount of iron-sulfur cluster as the wild-type HemN, but cleaves only one S-adenosylmethionine molecule per molecule protein, no detectable CPO activity [5]; <3> exhibits the same yellow-brown color as wild-type HemN, but the [4Fe-4S] cluster content is slightly reduced and cleaves only one S-adenosylmethionine molecule per molecule protein, no detectable CPO activity [5]) [5]
Y56A <3> (<3> appears colorless, the [4Fe-4S] cluster content is slightly reduced, no detectable S-adenosylmethionine cleavage, no detectable CPO activity [5]) [5]
Y56F <3> (<3> mutant with 45% of wild-type activity and reduced Fe-S cluster formation [2]) [2]
Y56L <3> (<3> appears colorless, the [4Fe-4S] cluster content is slightly reduced, no detectable S-adenosylmethionine cleavage, no detectable CPO activity [5]) [5]

Application

medicine <3> (<3> EPR spectrum of a potential substrate radical for HemN [6]) [6]
pharmacology <3> (<3> the structure of HemN sets the stage for the development of inhibitors with antibacterial function due to the uniquely bacterial occurence of the enzyme [1]) [1]

References

[1] Layer, G.; Moser, J.; Heinz, D.W.; Jahn, D.; Schubert, W.D.: Crystal structure of coproporphyrinogen III oxidase reveals cofactor geometry of Radical SAM enzymes. EMBO J., **22**, 6214-6224 (2003)
[2] Layer, G.; Verfurth, K.; Mahlitz, E.; Jahn, D.: Oxygen-independent coproporphyrinogen-III oxidase HemN from Escherichia coli. J. Biol. Chem., **277**, 34136-34142 (2002)
[3] Layer, G.; Heinz, D.W.; Jahn, D.; Schubert, W.D.: Structure and function of radical SAM enzymes. Curr. Opin. Chem. Biol., **8**, 468-476 (2004)
[4] Layer, G.; Kervio, E.; Morlock, G.; Heinz, D.W.; Jahn, D.; Retey, J.; Schubert, W.D.: Structural and functional comparison of HemN to other radical SAM enzymes. Biol. Chem., **386**, 971-980 (2005)

[5] Layer, G.; Grage, K.; Teschner, T.; Schuenemann, V.; Breckau, D.; Masoumi, A.; Jahn, M.; Heathcote, P.; Trautwein, A.X.; Jahn, D.: Radical S-adenosylmethionine enzyme coproporphyrinogen III oxidase HemN: functional features of the [4Fe-4S] cluster and the two bound S-adenosyl-L-methionines. J. Biol. Chem., **280**, 29038-29046 (2005)
[6] Layer, G.; Pierik, A.J.; Trost, M.; Rigby, S.E.; Leech, H.K.; Grage, K.; Breckau, D.; Astner, I.; Jaensch, L.; Heathcote, P.; Warren, M.J.; Heinz, D.W.; Jahn, D.: The substrate radical of Escherichia coli oxygen-independent coproporphyrinogen III oxidase HemN. J. Biol. Chem., **281**, 15727-15734 (2006)

all-trans-Retinol 13,14-reductase 1.3.99.23

1 Nomenclature

EC number
1.3.99.23

Systematic name
all-trans-13,14-dihydroretinol:acceptor 13,14-oxidoreductase

Recommended name
all-trans-retinol 13,14-reductase

Synonyms
(13,14)-all-trans-retinol saturase <1> [2]
RetSat <1, 2> [1, 2, 3]
RetSat A <3> [3]
all-trans-retinol:all-trans-13,14-dihydroretinol saturase <1> [2]
retinol saturase <1, 3> [2, 3]

CAS registry number
149147-14-8

2 Source Organism

<1> *Mus musculus* (no sequence specified) [2, 3]
<2> *Mus musculus* (UNIPROT accession number: Q64FW2) [1]
<3> *Danio rerio* (UNIPROT accession number: Q6DBT4) [3]

3 Reaction and Specificity

Catalyzed reaction
all-trans-13,14-dihydroretinol + acceptor = all-trans-retinol + reduced acceptor

Reaction type
oxidation
redox reaction
reduction

Natural substrates and products
S all-trans-retinol + reduced acceptor <2> (Reversibility: ?) [1]
P all-trans-13,14-dihydroretinol + acceptor

Substrates and products

S all-trans-13,14-didehydroretinol + reduced acceptor <1, 3> (Reversibility: ?) [3]
P all-trans-13,14-dihydro-3,4-didehydroretinol + acceptor (<3> preferred product [3])
S all-trans-13,14-didehydroretinol + reduced acceptor <3> (Reversibility: ?) [3]
P all-trans-7,8-dihydro-3,4-didehydroretinol + acceptor
S all-trans-retinol + reduced acceptor <1, 2, 3> (Reversibility: ?) [1, 3]
P all-trans-13,14-dihydroretinol + acceptor
S all-trans-retinol + reduced acceptor <3> (Reversibility: ?) [3]
P all-trans-7,8-dihydroretinol + acceptor (<3> preferred product [3])

5 Isolation/Preparation/Mutation/Application

Source/tissue

adult <3> [3]
intestine <2, 3> [1, 3]
kidney <2> [1]
liver <1, 2, 3> [1, 2, 3]
Additional information <3> (<3> hatchling [3]) [3]

Localization

endoplasmic reticulum <3> [3]
membrane <2, 3> (<2> associated to [1]) [1, 3]

Cloning

<2> [1]
<3> (into the Xho I site of pCDNA4/TO and expressed in T-REx-293 cells, cloned into the vector pCRII-TOPO) [3]

Application

Additional information <1, 3> (<1,3> conserved function but altered specificity of RetSat in vertebrates [3]) [3]

References

[1] Moise, A.R.; Kuksa, V.; Imanishi, Y.; Palczewski, K.: Identification of all-trans-retinol:all-trans-13,14-dihydroretinol saturase. J. Biol. Chem., **279**, 50230-50242 (2004)

[2] Moise, A.R.; Kuksa, V.; Blaner, W.S.; Baehr, W.; Palczewski, K.: Metabolism and transactivation activity of 13,14-dihydroretinoic acid. J. Biol. Chem., **280**, 27815-27825 (2005)

[3] Moise, A.R.; Isken, A.; Dominguez, M.; Lera, A.R.; Lintig, J.; Palczewski, K.: Specificity of zebrafish retinol saturase: formation of all-trans-13,14-dihydroretinol and all-trans-7,8-dihydroretinol. Biochemistry, **46**, 1811-1820 (2007)

Aspartate dehydrogenase 1.4.1.21

1 Nomenclature

EC number
1.4.1.21

Systematic name
L-aspartate:NAD(P)$^+$ oxidoreductase (deaminating)

Recommended name
aspartate dehydrogenase

Synonyms
L-aspDH <3> [4]
L-aspartate dehydrogenase <3, 4> [3, 4]
NAD-dependent aspartate dehydrogenase <1, 2> [1, 2]
NADH$_2$-dependent aspartate dehydrogenase <1, 2> [1, 2]
NADP$^+$-dependent aspartate dehydrogenase <1> [1]

CAS registry number
37278-97-0

2 Source Organism

<1> *Klebsiella pneumoniae* (no sequence specified) [1]
<2> *Rhizobium lupini* (no sequence specified) [2]
<3> *Archaeoglobus fulgidus* (no sequence specified) [4]
<4> *Thermotoga maritima* (UNIPROT accession number: Q9X1X6) [3]

3 Reaction and Specificity

Catalyzed reaction
L-aspartate + H_2O + NAD(P)$^+$ = oxaloacetate + NH_3 + NAD(P)H + H^+

Reaction type
oxidation
redox reaction
reduction

Natural substrates and products
S L-aspartate + NAD(P)$^+$ <4> (Reversibility: r) [3]
P oxaloacetate + NH_4^+ + NAD(P)H

S L-aspartate + $NADP^+$ + H_2O <1> (Reversibility: ?) [1]
P oxaloacetate + NH_4^+ + NADPH <1> [1]
S oxaloacetate + NAD(P)H + NH_4^+ <4> (Reversibility: ?) [3]
P L-aspartate + $NAD(P)^+$
S oxaloacetate + NH_4^+ + NADH <2> (<2> biosynthesis of aspartate [2]) (Reversibility: ?) [2]
P L-aspartate + NAD^+ + H_2O <2> [2]

Substrates and products

S L-aspartate + $NAD(P)^+$ <4> (Reversibility: r) [3]
P oxaloacetate + NH_4^+ + NAD(P)H
S L-aspartate + $NAD(P)^+$ + H_2O <3> (<3> the enzyme shows pro-R (A-type) stereospecificity for hydrogen transfer from the C_4 position of the nicotinamide moiety of NADH [4]) (Reversibility: ?) [4]
P oxaloacetate + NH_4^+ + NAD(P)H
S L-aspartate + $NADP^+$ + H_2O <1> (Reversibility: ?) [1]
P oxaloacetate + NH_4^+ + NADPH <1> [1]
S oxaloacetate + NAD(P)H + NH_4^+ <4> (Reversibility: ?) [3]
P L-aspartate + $NAD(P)^+$
S oxaloacetate + NH_4^+ + NADH <2> (<2> strictly specific for oxaloacetate and NADH, not NADPH [2]; <2> biosynthesis of aspartate [2]) (Reversibility: ?) [2]
P L-aspartate + NAD^+ + H_2O <2> [2]
S Additional information <1, 3> (<1> not: D-aspartate, L-glutamate, L-glycine, L-alanine, L-threonine, L-serine, L-leucine, L-isoleucine, L-methionine, L-cysteine, L-proline, L-valine, L-phenylalanine, L-tyrosine, L-tryptophan, L-lysine, L-histidine, L-arginine [1]; <3> no activity with D-aspartate, L-glutamate, L-alanine, L-leucine, L-phenylalanine, L-proline, glycine, L-serine, L-lysine, L-norvaline, L-norleucine, L-homoserine and L-2-amino-n-butyrate [4]) (Reversibility: ?) [1, 4]
P ? <1> [1]

Inhibitors

EDTA <1> [1]
K^+ <1> (<1> potassium ion in phosphate buffer may be inhibitory [1]) [1]
L-malate <4> (<4> competitive [3]) [3]
NH_4^+ <4> (<4> competitive [3]) [3]

Cofactors/prosthetic groups

$NAD(P)^+$ <3> [4]
NAD^+ <4> [3]
NADH <2, 4> (<2> specific for NADH [2]) [2,3]
$NADP^+$ <1, 4> (<1> requires $NADP^+$ as coenzyme [1]) [1,3]
NADPH <4> [3]
Additional information <1, 2> (<1> not: NAD^+, FMN, FAD [1]; <2> not: NADPH [2]) [1,2]

Activating compounds

Additional information <3> (<3> unaffected by EDTA, $CaCl_2$, $NiCl_2$, $CoCl_2$, $CuSO_4$ or $ZnCl_2$ [4]) [4]

Turnover number (min^{-1})

0.78 <4> (L-Asp, <4> cofactor NAD^+ [3]) [3]
0.78 <4> (L-aspartate with NAD, <4> +/- 0.02, with NAD^+ [3]) [3]
1.2 <4> (NAD^+, <4> +/- 0.04, with L-aspartate [3]) [3]
4.9 <4> (L-Asp, <4> cofactor $NADP^+$ [3]) [3]
4.9 <4> (L-aspartate, <4> +/- 0.09, with $NADP^+$ [3]) [3]
7.2 <4> ($NADP^+$, <4> +/- 0.17, with L-aspartate [3]) [3]

Specific activity (U/mg)

0.045 <1> (<1> pH 7, 30°C, crude extract [1]) [1]
0.75 <2> (<2> bacteroid extract of Lupinus luteus nodules [2]) [2]
1.63 <4> (<4> +/- 0.15, L-aspartate with NAD^+ [3]) [3]
3.36 <4> (<4> +/- 0.25, NAD^+ with L-aspartate [3]) [3]
4.6 <3> (<3> purified enzyme, at 50 °C [4]) [4]
9.51 <4> (<4> +/- 0.17, L-aspartate with $NADP^+$ [3]) [3]
12.32 <4> (<4> +/- 0.88, $NADP^+$ with L-aspartate [3]) [3]

K_m-Value (mM)

0.014 <3> (NADH) [4]
0.067 <4> (L-Asp, <4> cofactor NAD^+ [3]) [3]
0.067 <4> (L-aspartate, <4> +/- 0.008, with NAD^+ [3]) [3]
0.11 <3> (NAD^+) [4]
0.19 <3> (L-aspartate, <3> with NAD^+ as the electron acceptor [4]) [4]
0.25 <4> (NAD^+, <4> +/- 0.02 [3]) [3]
0.32 <3> ($NADP^+$) [4]
0.72 <4> ($NADP^+$, <4> +/- 0.04 [3]) [3]
1.2 <4> (L-Asp, <4> cofactor: $NADP^+$ [3]) [3]
1.2 <4> (L-aspartate, <4> +/- 0.05, with $NADP^+$ [3]) [3]
1.2 <3> (oxaloacetate) [4]
4.3 <3> (L-aspartate, <3> with $NADP^+$ as the electron acceptor [4]) [4]
167 <3> (NH_4^+) [4]

K_i-Value (mM)

4.02 <4> (L-malate, <4> +/- 0.48 [3]) [3]
32.5 <4> (NH_4^+, <4> +/- 4.9 [3]) [3]

pH-Optimum

6 <2> [2]
7 <1> (<1> in Tris-HCl buffer, oxidative deamination of L-aspartate [1]) [1]
8 <1, 4> (<1> in potassium phosphate buffer, oxidative deamination of L-aspartate [1]; <4> oxaloacetate animation [3]; <4> oxaloacetate amination [3]) [1, 3]
9.8 <4> (<4> L-Asp oxidation [3]; <4> L-aspartate oxidation [3]) [3]
11.6 <3> [4]

pH-Range

8-10.5 <4> (<4> L-Asp oxidation [3]; <4> L-aspartate oxidation [3]) [3]

Temperature optimum (°C)

30 <1> (<1> assay at [1]) [1]
70 <4> [3]
80 <3> [4]

Temperature range (°C)

25-90 <4> [3]
37-100 <3> [4]

4 Enzyme Structure

Molecular weight

48000 <3> (<3> gel filtration [4]) [4]
124000 <1> (<1> gel filtration [1]) [1]

Subunits

dimer <4> (<4> crystal structure [3]; <4> from crystal structure [3]) [3]
homodimer <1, 3> (<1> 2 * 62000, SDS-PAGE [1]; <3> 2 * 26000, SDS-PAGE, 2 * 26208, sequence analysis [4]) [1, 4]

5 Isolation/Preparation/Mutation/Application

Purification

<1> (about 500fold) [1]
<2> [2]
<3> (to homogeneity by heat treatment and affinity chromatography) [4]

Cloning

<3> (ligated into the expression vector pET11a, expression in Escherichia coli strain BL21(DE3)) [4]
<4> [3]

Application

Additional information <3> (<3> first report of an archaeal L-aspartate dehydrogenase, within the archaeal domain, homologues in many methanogenic species, but not in Thermococcales or Sulfolobales species [4]) [4]

6 Stability

Temperature stability

20 <1> (<1> up to 20°C, 10 min, Tris-HCl buffer, pH 7.5, stable [1]) [1]
50 <1> (<1> 10 min, Tris-HCl buffer, pH 7.5, 70% loss of activity [1]) [1]
60 <1> (<1> 10 min, Tris-HCl buffer, pH 7.5, 100% loss of activity [1]) [1]

80 <3> (<3> stable for 1 h [4]) [4]
100 <1> (<1> boiling inactivates [1]) [1]

General stability information

<1>, enzyme activity in Tris-HCl buffer is about 7fold higher than in potassium phosphate buffer [1]
<4>, stable [3]

Storage stability

<4>, 4°C, 5% glycerol, 0.5 M NaCl, pH 7.5, no loss in activity after several months [3]

References

[1] Okamura, T.; Noda, H.; Fukuda, S.; Ohsugi, M.: Aspartate dehydrogenase in vitamin B_{12}-producing Klebsiella pneumoniae IFO 13541. J. Nutr. Sci. Vitaminol., **44**, 483-490 (1998)
[2] Kretovich, W.L.; Kariakina, T.I.; Weinova, M.K.; Sidelnikova, L.I.; Kazakova, O.W.: The synthesis of aspartic acid in Rhizobium lupini bacteroids. Plant Soil, **61**, 145-156 (1981)
[3] Yang, Z.; Savchenko, A.; Yakunin, A.; Zhang, R.; Edwards, A.; Arrowsmith, C.; Tong, L.: Aspartate dehydrogenase, a novel enzyme identified from structural and functional studies of TM1643. J. Biol. Chem., **278**, 8804-8808 (2003)
[4] Yoneda, K.; Kawakami, R.; Tagashira, Y.; Sakuraba, H.; Goda, S.; Ohshima, T.: The first archaeal L-aspartate dehydrogenase from the hyperthermophile Archaeoglobus fulgidus: Gene cloning and enzymological characterization. Biochim. Biophys. Acta, **1764**, 1087-1093 (2006)

L-Lysine 6-oxidase 1.4.3.20

1 Nomenclature

EC number
1.4.3.20

Systematic name
L-lysine:oxygen 6-oxidoreductase (deaminating)

Recommended name
L-lysine 6-oxidase

Synonyms
L-lysine-ε oxidase <2> [2]
Lod <2> [2]
LodA <2> [2]
marinocine <1> [1]
marinocine antimicrobial protein <3> [3]

CAS registry number
860791-20-4

2 Source Organism

<1> *Marinomonas mediterranea* (no sequence specified) [1]
<2> *Marinomonas mediteranea* (no sequence specified) [2]
<3> *Marinomonas mediterranea* (UNIPROT accession number: Q24K54) [3]

3 Reaction and Specificity

Catalyzed reaction
L-lysine + O_2 + H_2O = 2-aminoadipate 6-semialdehyde + H_2O_2 + NH_3

Natural substrates and products
S L-lysine + O_2 + H_2O <3> (<3> the enzyme shows antibacterial activity [3]) (Reversibility: ?) [3]
P 2-aminoadipate 6-semialdehyde + H_2O_2

Substrates and products
S L-lysine + O_2 + H_2O <2> (<2> high stereospecificity for L-lysine [2]) (Reversibility: ?) [2]
P 2-aminoadipate 6-semialdehyde + H_2O_2 + NH_3

S L-lysine + O_2 + H_2O <3> (<3> the enzyme shows antibacterial activity [3]) (Reversibility: ?) [3]
P 2-aminoadipate 6-semialdehyde + H_2O_2
S L-ornithine + O_2 + H_2O <2> (<2> 15.1% of activity with L-lysine [2]) (Reversibility: ?) [2]
P 5-oxo-L-norvaline + H_2O_2 + NH_3
S α-N-acetyl-L-lysine + H_2O <2> (<2> 91.9% of the activity with L-lysine [2]) (Reversibility: ?) [2]
P N-acetyl-6-oxo-L-norleucine + H_2O_2 + NH_3
S Additional information <1> (<1> the enzyme shows antibacterial activity [1]) (Reversibility: ?) [1]
P ?

Inhibitors
6-aminocaproic acid <2> (<2> 0.2 mM, 72% inhibition [2]) [2]
amiloride <2> [2]
aminoguanidine <2> [2]
cadaverine <2> (<2> 0.2 mM, 74% inhibition [2]) [2]
β-aminoproprionitrile <2> (<2> 0.2 mM, 91% inhibition [2]) [2]

pH-Range
Additional information <1> (<1> at pH 7, the incubation of the samples at temperatures below inactivation produces a characteristic and reliable increase in activity. This activation is already noticeable at 50°C for treatments longer than 1 h [1]) [1]

Temperature range (°C)
Additional information <1> (<1> at pH 7, the incubation of the samples at temperatures below inactivation produces a characteristic and reliable increase in activity. This activation is already noticeable at 50°C for treatments longer than 1 h [1]) [1]

4 Enzyme Structure

Molecular weight
160000 <1> (<1> gel filtration, non-denaturing [1]) [1]

Subunits
? <1> (<1> two major bands of 97000 Da and 185000 Da appear in denaturing SDS-PAGE [1]) [1]

5 Isolation/Preparation/Mutation/Application

Source/tissue
culture medium <1> (<1> at death phase of growth [1]) [1]
culture supernatant <2> [2]

Purification

<1> [1]
<2> [2]

Cloning

<3> [3]

6 Stability

Temperature stability

70 <1> (<1> pH 5 and 7, 1 h, antibacterial activity is stable up to 70 [1]) [1]
75 <1> (<1> heat exposure at temperatures up to 75°C and pH 7 cause a conformational change in the marinocine structure leading to this rare strong activation, but temperatures above 75°C cause the denaturation in marinocine clearly observed in the electrophoresis experiments, with appearance of the two bands of 97000 Da and 185000 Da [1]) [1]

General stability information

<1>, quite resistant to hydrolytic enzymes [1]

References

[1] Lucas-Elio, P.; Hernandez, P.; Sanchez-Amat, A.; Solano, F.: Purification and partial characterization of marinocine, a new broad-spectrum antibacterial protein produced by Marinomonas mediterranea. Biochim. Biophys. Acta, **1721**, 193-203 (2005)

[2] Gomez, D.; Lucas-Elio, P.; Sanchez-Amat, A.; Solano, F.: A novel type of lysine oxidase: L-lysine-ε-oxidase. Biochim. Biophys. Acta, **247; 1764**, 1577-1585 (2006)

[3] Lucas-Elio, P.; Gomez, D.; Solano, F.; Sanchez-Amat, A.: The antimicrobial activity of marinocine, synthesized by Marinomonas mediterranea, is due to hydrogen peroxide generated by its lysine oxidase activity. J. Bacteriol., **188**, 2493-2501 (2006)

1-Pyrroline dehydrogenase 1.5.1.35

1 Nomenclature

EC number
1.5.1.35 (deleted, identical to 1.2.1.19)

Recommended name
1-pyrroline dehydrogenase

Methylenetetrahydrofolate reductase (ferredoxin) 1.5.7.1

1 Nomenclature

EC number
1.5.7.1

Systematic name
5-methyltetrahydrofolate:ferredoxin oxidoreductase

Recommended name
methylenetetrahydrofolate reductase (ferredoxin)

Synonyms
5,10-methylenetetrahydrofolate reductase <1> [1]

CAS registry number
9028-69-7 (cf. EC 1.7.99.5)

2 Source Organism

<1> *Clostridium formicoaceticum* (no sequence specified) [1]

3 Reaction and Specificity

Catalyzed reaction
5-methyltetrahydrofolate + 2 oxidized ferredoxin = 5,10-methylenetetrahydrofolate + 2 reduced ferredoxin + 2 H^+

Substrates and products
S 5,10-methylenetetrahydrofolate + $FADH_2$ <1> (Reversibility: ?) [1]
P 5,10-methylenetetrahydrofolate + FAD
S 5,10-methylenetetrahydrofolate + reduced ferredoxin <1> (Reversibility: ?) [1]
P 5-methyltetrahydrofolate + ferredoxin
S 5-methyltetrahydrofolate + FAD <1> (<1> 11% of the activity with methylene blue [1]) (Reversibility: ?) [1]
P 5,10-methylenetetrahydrofolate + $FADH_2$
S 5-methyltetrahydrofolate + benzyl viologen <1> (<1> 11% of the activity with methylene blue [1]) (Reversibility: ?) [1]
P 5,10-methylenetetrahydrofolate + reduced benzyl viologen
S 5-methyltetrahydrofolate + menadione <1> (<1> 25% of the activity with methylene blue [1]) (Reversibility: ?) [1]

P 5,10-methylenetetrahydrofolate + reduced menadiol
S 5-methyltetrahydrofolate + rubredoxin <1> (<1> 11% of the activity with methylene blue [1]) (Reversibility: ?) [1]
P 5,10-methylenetetrahydrofolate + reduced rubredoxin
S Additional information <1> (<1> no activity with pyridine nucleotides [1]) (Reversibility: ?) [1]
P ?

Cofactors/prosthetic groups
FAD <1> (<1> contains 1.7 FAD per 237000 Da enzyme [1]) [1]

Metals, ions
Iron <1> (<1> iron-sulfur protein, contains 15.2 mol iron and 19.5 mol acid labile sulfide per 237000 Da enzyme [1]) [1]
Zinc <1> (<1> enzyme contains 2.3 mol zinc per 237000 Da enzyme [1]) [1]

Specific activity (U/mg)
139 <1> [1]

K_m-Value (mM)
0.01 <1> (5-Methyltetrahydrofolate) [1]
11.1 <1> (benzyl viologen) [1]

pH-Optimum
7.8-8 <1> [1]

4 Enzyme Structure

Molecular weight
237000 <1> (<1> gel filtration [1]) [1]

Subunits
octamer <1> (<1> $\alpha_4\beta_4$, 4 * 26000 + 4 * 35000, SDS-PAGE [1]) [1]

5 Isolation/Preparation/Mutation/Application

Purification
<1> [1]

6 Stability

pH-Stability
7.1-7.6 <1> (<1> the enzyme is more stable in Tris/HCl than in triethanolamine/HCl or phosphate buffer [1]) [1]

Oxidation stability

<1>, the enzyme loses 50% of the activity within 2 h in aerobic buffer (0.1 M Tris/HCl buffer, pH 7.4) with or without a reducing agent and in anaerobic buffer without a reducing agent [1]

Storage stability

<1>, 10°C, pH 7.4, 50 mM Tris/HCl containing 20% glycerol and 2 mM dithionite in an anaerobic chamber [1]

References

[1] Clark, J.E.; Ljungdahl, L.G.: Purification and properties of 5,10-methylenetetrahydrofolate reductase, an iron-sulfur flavoprotein from Clostridium formicoaceticum. J. Biol. Chem., **259**, 10845-10849 (1984)

PreQ1 synthase 1.7.1.13

1 Nomenclature

EC number
1.7.1.13

Systematic name
queuine:$NADP^+$ oxidoreductase

Recommended name
preQ1 synthase

Synonyms
7-aminomethyl-7-carbaguanine:$NADP^+$ oxidoreductase <4> [1]
7-cyano-7-deazaguanine reductase <4> [1]
QueF <2, 3, 4> [1, 5]
YkvM <4> [1]
$preQ_0$ reductase <4> [1]
$preQ_0$ oxidoreductase <4> [1]
queuine synthase

2 Source Organism

<1> *Salmonella typhimurium* (no sequence specified) [2]
<2> *Bacillus subtilis* (no sequence specified) [5]
<3> *Escherichia coli* (no sequence specified) [3, 4, 5]
<4> *Bacillus subtilis* (UNIPROT accession number: O31678) [1]

3 Reaction and Specificity

Catalyzed reaction
queuine + 2 $NADP^+$ = 7-cyano-7-carbaguanine + 2 NADPH + 2 H^+

Natural substrates and products
S 7-cyano-7-deazaguanine + NADPH <2, 3, 4> (<2,3,4> late step in biosynthesis of the modified tRNA nucleoside queuosine [1,5]) (Reversibility: ?) [1, 5]
P queuine + $NADP^+$ + H^+

Substrates and products

S 7-cyano-7-deazaguanine + NADPH <2, 3, 4> (<2,3,4> late step in biosynthesis of the modified tRNA nucleoside queuosine [1,5]) (Reversibility: ?) [1, 3, 4, 5]

P queuine + $NADP^+$ + H^+

Cofactors/prosthetic groups

NADPH <2, 3, 4> [1,5]

Turnover number (min^{-1})

0.01 <3> (NADPH) [5]

K_m-Value (mM)

0.036 <3> (NADPH) [5]

4 Enzyme Structure

Subunits

decamer <4> [1]

5 Isolation/Preparation/Mutation/Application

Purification

<2> [5]
<3> [5]

Crystallization

<4> (sitting-drop vapor-diffusion method, space group: P3(1)21. Unit-cell parameters: a = b = 93.52 A, c = 193.76 A) [1]

Cloning

<2> (overproduction in Escherichia coli) [5]
<3> (overproduction in Escherichia coli) [5]

References

[1] Swairjo, M.A.; Reddy, R.R.; Lee, B.; Van Lanen, S.G.; Brown, S.; de Crecy-Lagard, V.; Iwata-Reuyl, D.; Schimmel, P.: Crystallization and preliminary x-ray characterization of the nitrile reductase QueF: a queuosine-biosynthesis enzyme. Acta Crystallogr. Sect. F, **F61**, 945-948 (2005)

[2] Kuchino, Y.; Kasai, H.; Nihei, K.; Nishimura, S.: Biosynthesis of the modified nucleoside Q in transfer RNA. Nucleic Acids Res., **3**, 393-398 (1976)

[3] Okada, N.; Noguchi, S.; Nishimura, S.; Ohgi, T.; Goto, T.; Crain, P.F.; McCloskey, J.A.: Structure determination of a nucleoside Q precursor isolated from E. coli tRNA: 7-(aminomethyl)-7-deazaguanosine. Nucleic Acids Res., **5**, 2289-2296 (1978)

[4] Noguchi, S.; Yamaizumi, Z.; Ohgi, T.; Goto, T.; Nishimura, Y.; Hirota, Y.; Nishimura, S.: Isolation of Q nucleoside precursor present in tRNA of an E. coli mutant and its characterization as 7-(cyano)-7-deazaguanosine. Nucleic Acids Res., **5**, 4215-4223 (1978)
[5] van Lanen, S.G.; Reader, J.S.; Swairjo, M.A.; de Crecy-Lagard, V.; Lee, B.; Iwata-Reuyl, D.: From cyclohydrolase to oxidoreductase: discovery of nitrile reductase activity in a common fold. Proc. Natl. Acad. Sci. USA, **102**, 4264-4269 (2005)

Hydroxylamine oxidoreductase 1.7.99.8

1 Nomenclature

EC number
1.7.99.8

Systematic name
hydroxylamine:acceptor oxidoreductase

Recommended name
hydroxylamine oxidoreductase

Synonyms
HAO
HCR <5> [6]
hydroxylamine-cytochrome c reductase <5> [6]

CAS registry number
9075-43-8

2 Source Organism

<1> *Alcaligenes faecalis* (no sequence specified) [4]
<2> *Nitrosomonas europaea* (no sequence specified) [2, 3, 5]
<3> *Pseudomonas denitrificans* (no sequence specified) [1]
<4> *Cytophaga-Flexibacter phylum* (no sequence specified) [5]
<5> *Nitrosomonas communis* (no sequence specified) [6]

3 Reaction and Specificity

Catalyzed reaction
hydrazine + acceptor = N_2 + reduced acceptor
hydroxylamine + NH_3 = hydrazine + H_2O

Reaction type
redox reaction

Natural substrates and products
S hydroxylamine + NH_3 <2, 3> (<2> HAO exists in two forms, designated HAO-A and HAO-B. HAO-A binds only one cyanide, HAO-B forms mono and then dicyano complexes. Heme P460, the only other metal center in HAO, may be the moiety that reats with cyanide [3]; <2> HAO-B is con-

verted to HAO-A when fully reduced by dithionite and reoxidized during passage over a gel filtration column. HAO-B is sometimes converted to HAO-A by aging at 4°C or 25°C, especially under illumination by room light. Incubation of ferric HAO-A for 12 h with increasing amounts of hydrogen peroxide results in progressive conversion to HAO-B [3]; <2> 1 mol of HAO contains 26 mol heme c, type P-468 [2]; <3> the enzyme does not contain heme groups [1]) (Reversibility: ?) [1, 2, 3]

P hydrazine + H_2O <2, 3> [1, 2, 3]

Substrates and products

S NO + acceptor <2> (Reversibility: ?) [2]

P N_2O + H_2O + reduced acceptor <2> [2]

S hydrazine + 2,6-dichlorophenolindophenol <2> (Reversibility: ?) [2]

P ?

S hydrazine + cytochrome c <5> (Reversibility: ?) [6]

P N_2 + reduced cytochrome c

S hydrazine + horse heart cytochrome c <2> (Reversibility: ?) [2, 3]

P ?

S hydrazine + phenazine methosulfate/methylthiazolyltetrazolium bromide <2> (Reversibility: ?) [2]

P ?

S hydrazine + potassium ferricyanide <2> (Reversibility: ?) [2]

P ?

S hydroxylamine + 2,6-dichlorophenolindophenol <2, 3> (<3> 2,6-dichlorophenolindophenol does not serve as electron acceptor [1]) (Reversibility: ?) [1, 2]

P ?

S hydroxylamine + H_2O <2> (Reversibility: ?) [3]

P HNO_2 + e + H^+ <2> [3]

S hydroxylamine + NH_3 <2, 3> (<2> HAO exists in two forms, designated HAO-A and HAO-B. HAO-A binds only one cyanide, HAO-B forms mono and then dicyano complexes. Heme P460, the only other metal center in HAO, may be the moiety that reats with cyanide [3]; <2> HAO-B is converted to HAO-A when fully reduced by dithionite and reoxidized during passage over a gel filtration column. HAO-B is sometimes converted to HAO-A by aging at 4°C or 25°C, especially under illumination by room light. Incubation of ferric HAO-A for 12 h with increasing amounts of hydrogen peroxide results in progressive conversion to HAO-B [3]; <2> 1 mol of HAO contains 26 mol heme c, type P-468 [2]; <3> the enzyme does not contain heme groups [1]) (Reversibility: ?) [1, 2, 3]

P hydrazine + H_2O <2, 3> [1, 2, 3]

S hydroxylamine + horse heart cytochrome c <2, 3> (<3> 5% of the activity obtained with potassium ferricyanide [1]) (Reversibility: ?) [1, 2, 3]

P NO + N_2O + H_2O + reduced horse heart cytochrome c <2> [2, 3]

S hydroxylamine + phenazine methosulfate/methylthiazolyltetrazolium bromide <2, 3> (<3> phenazine methosulfate/methylthiazolyltetrazolium bromide does not serve as electron acceptor [1]) (Reversibility: ?) [1, 2]

P NO + N_2O + H_2O + reduced phenazine methosulfate/methylthiazolyltetrazolium bromide <2> [2]

S hydroxylamine + potassium ferricyanide <1, 2, 3> (Reversibility: ?) [1, 2, 4]

P NO + N_2O + NO_2^- + H_2O + potassium ferrocyanide <1, 2, 3> [1, 2, 4]

S Additional information <1, 2, 3> (<2> benzyl viologen and N,N,N,N-tetramethyl-O-phenylenediamine do not serve as electron acceptors [2]; <1,2,3> NAD^+ does not serve as electron acceptor [1,2,4]; <1> horse heart cytochrome c, cytochrome c, pseudoazurin from Pseudomonas denitrificans, MTT/PMS, 2,6-dichloroindophenol/PMS, tetramethyl-*p*-phenylenediamine and O_2 do not serve as electron acceptors [4]; <3> $NADP^+$ does not serve as electron acceptor [1]; <1,3> FAD^+ does not serve as electron acceptor [1,4]) (Reversibility: ?) [1, 2, 4]

P ?

Inhibitors

2,4-Dinitrophenol <2, 4> (<2,4> inhibition at 0-2.17 mM [5]) [5]
Acetylene <2, 4> (<2,4> inhibition at 6 mM [5]) [5]
Cyanide <2> (<2> addition of 0.035 mM, 0.085 mM and 0.35 mM results in 25%, 50% and 95% inhibition, respectively [2]) [2]
$HgCl_2$ <2, 4> (<2,4> inhibition at 1.1 mM [5]) [5]
Hydrazine <2> (<2> competitive with respect to hydroxylamine. Addition of 0.0015 mM, 0.0035 mM and 0.35 mM results in 15%, 50% and 85% inhibition, respectively [2]) [2]
Hydrogen peroxide <2> (<2> addition of 0.0004 mM, 0.004 mM and 0.04 mM results in 30%, 75% and 90% inhibition, respectively [2]) [2, 3]
Nitrite <2, 4> (<2,4> inhibition at concentrations higher than 20 mM [5]) [5]
O_2 <2, 4> (<2,4> inhibition at 0.2 mM, reversible [5]) [5]
Phosphate <2, 4> (<2,4> inhibition at more than 2 mM [5]) [5]
SDS <2> (<2> addition of 0.75%, 3% and 8% results in 20%, 50% and 95% inhibition, respectively [2]) [2]
Additional information <2, 4> (<2> sodium azide, methanol, sodium nitrite and EDTA do not inhibit HAO activity [2]; <2> F^-, Cl^-, Br^-, N_3^-, SCN^- or OCN^- do not inhibit HAO activity [3]; <2,4> pyruvate, methanol, ethanol, alanine and glucose also inhibit the enzyme [5]) [2, 3, 5]

Cofactors/prosthetic groups

Cytochrome c <5> [6]

Activating compounds

Hydrazine <2, 4> (<2,4> at 0-3 mM produces activation [5]) [5]
Additional information <2> (<2> F^-, Cl^-, Br^-, N_3^-, SCN^- or OCN^- do not affect HAO activity [3]) [3]

Specific activity (U/mg)

0.054 <1> (<1> activity recovered from the anion exchange column. Gel filtration produces 80% loss of activity [4]) [4]
0.1 <3> (<3> activity of cell extracts, with potassium ferricyanide as the electron acceptor [1]) [1]
0.47 <3> (<3> activity of the purified enzyme [1]) [1]

0.87 <2> (<2> with NO_2^- as substrate [2]) [2]
1.1 <2> (<2> with hydrazine as substrate and phenazine methosulfate/ methylthiazolyltetrazolium bromide as electron acceptor [2]) [2]
1.2 <2> (<2> with hydrazine as substrate and 2,6-dichlorophenolindophenol as electron acceptor [2]) [2]
2.1 <2> (<2> with hydrazine as substrate and horse heart cytochrome c as electron acceptor [2]) [2]
2.9 <2> (<2> at room temperature, with NO or N_2O as substrates [2]) [2]
6.2 <2> (<2> at 30°C, with hydrazine as substrate and potassium ferricyanide as the electron acceptor [2]) [2]
18 <2> (<2> with hydroxylamine as substrate and horse heart cytochrome c as electron acceptor [2]) [2]
20 <2> (<2> with hydroxylamine as substrate and 2,6-dichlorophenolindophenol as electron acceptor [2]) [2]
21 <2, 4> (<2> with hydroxylamine as substrate and phenazine methosulfate/ methylthiazolyltetrazolium bromide as electron acceptor [2]) [2, 5]
25.6 <5> [6]
40.3 <2> (<2> at 30°C, with hydroxylamine as substrate and potassium ferricyanide as the electron acceptor [2]) [2]
75 <2> [5]

K_m-Value (mM)
0.003 <2> (hydrazine, <2> pH 7.5, 25°C, with HAO-A [3]) [3]
0.005 <2> (hydrazine, <2> pH 7.5, 25°C, with HAO-B [3]) [3]
0.018 <2> (hydrazine, <2> pH 8.0, 35°C [2]) [2]
0.026 <2, 4> (hydroxylamine, <2> pH 8.0, 35°C [2]) [2, 5]
0.037 <3> (hydroxylamine, <3> pH 8.8, 25°C [1]) [1]
0.127 <2> (hydroxylamine, <2> pH 8.0, 65°C [2]) [2]

pH-Optimum
6 <2> (<2> for NO reduction [2]) [2]
8 <2> (<2> for hydroxylamine and hydrazine conversion [2]) [2, 5]
8-9 <1> (<1> with Tris/HCl as the optimal buffer [4]) [4]
9 <3> [1]

pH-Range
6.7-8.3 <2> [5]

Temperature range (°C)
20-43 <2, 4> [5]

4 Enzyme Structure

Molecular weight
125000-175000 <2> [5]
132000 <3> (<3> gel filtration [1]) [1]
150000 <2, 4> (<2> gel filtration [2]) [2, 5]
183000 <2> (<2> nondenaturing PAGE [2]) [2]

Subunits

dimer <3> (<3> α_2, 2 * 68000, SDS-PAGE [1]) [1]
multimer <2, 4> (<4> α_2-α3, x * 60000-95000 [5]; <2> α_2-α3, x * 63000 [5]) [5]
trimer <2> (<2> α_3, 3 * 58000, SDS-PAGE [2]) [2]

5 Isolation/Preparation/Mutation/Application

Source/tissue

Additional information <5> (<5> optimal growth at pH 8.5 and 30°C, ammonium sulfate is required as energy source [6]) [6]

Cloning

<5> (DNA sequence adetermination and analysis of the 16S rRNA gene for construction of a phylogenetic tree) [6]

Application

environmental protection <2, 4> (<2,4> removal of ammonium from sludge degester effluents [5]) [5]

6 Stability

Temperature stability

121 <2, 4> (<2,4> denaturation, loss of activity [5]) [5]

General stability information

<2, 4>, γ radiation inactivates the enzyme [5]

Storage stability

<2>, frozen in liquid nitrogen, 10 mM Tris-HCl and 200 mM KCl, pH 8.0, stable [2]
<3>, -70°C, frozen in liquid nitrogen, 10% glycerol [1]

References

[1] Jetten, M.S.M.; De Bruijn, P.; Kuenen, J.G.: Hydroxylamine metabolism in Pseudomonas PB16: involvement of a novel hydroxylamine oxidoreductase. Antonie Leeuwenhoek, **71**, 69-74 (1997)
[2] Schalk, J.; de Vries, S.; Kuenen, J.G.; Jetten, M.S.: Involvement of a novel hydroxylamine oxidoreductase in anaerobic ammonium oxidation. Biochemistry, **39**, 5405-5412 (2000)
[3] Logan, M.S.P.; Balny, C.; Hooper, A.B.: Reaction with cyanide of hydroxylamine oxidoreductase of Nitrosomonas europaea. Biochemistry, **34**, 9028-9037 (1995)

[4] Otte, S.; Schalk, J.; Kuenen, J.G.; Jetten, M.S.M.: Hydroxylamine oxidation and subsequent nitrous oxide production by the heterotrophic ammonia oxidizer Alcaligenes faecalis. Appl. Microbiol. Biotechnol., **51**, 255-261 (1999)
[5] Jetten, M.S.M.; Strous, M.; van de Pas-Schoonen, K.T.; Schalk, J.; van Dongen, U.G.J.M.; van de Graaf, A.A.; Logemann, S.; Muyzer, G.; van Loosdrecht, M.C.M.; Kuenen, J.G.: The anaerobic oxidation of ammonium. FEMS Microbiol. Rev., **22**, 421-437 (1999)
[6] Tokuyama, T.; Mine, A.; Kamiyama, K.; Yabe, R.; Satoh, K.; Matsumoto, H.; Takahashi, R.; Itonaga, K.: Nitrosomonas communis strain YNSRA, an ammonia-oxidizing bacterium, isolated from the reed rhizoplane in an aquaponics plant. J. Biosci. Bioeng., **98**, 309-312 (2004)

Peptide-methionine (S)-S-oxide reductase 1.8.4.11

1 Nomenclature

EC number

1.8.4.11

Systematic name

L-methionine:thioredoxin-disulfide S-oxidoreductase (S-form oxidizing)

Recommended name

peptide-methionine (S)-S-oxide reductase

Synonyms

FMsr <8> [5]
MSR <3, 8, 17, 18, 21, 29> [18, 31]
MetSO-L12 reductase <8> [3]
MsrA <1, 2, 4, 5, 6, 7, 8, 9, 10, 11, 12, 13, 14, 15, 16, 17, 18, 19, 20, 21, 22, 24, 25, 26, 27, 28, 30, 31, 34, 35, 36, 37, 41> [1, 2, 5, 6, 7, 8, 9, 10, 11, 12, 13, 14, 15, 17, 18, 19, 20, 21, 22, 23, 24, 25, 26, 27, 28, 30, 32, 36, 37, 38, 39, 40, 41, 42, 43, 44, 45, 46, 47, 48, 49, 51, 52, 53, 54]
MsrA/B <16> [18]
MsrB <31> [32]
PMSR <8, 15, 23> [33, 34, 35, 38]
PMSRA <38, 39, 40> [50]
peptide Met(O) reductase <37> [47]
PilA <16> [54]
PilB <16, 17, 31> [10, 18, 32]
ecdysone-induced protein 28/29 kDa <36> [41]
methionine S-oxide reductase (S-form oxidizing)
methionine sulfoxide reductase <1, 2, 3, 4, 5, 6, 7, 8, 9, 11, 12, 13, 14, 15, 16, 17, 18, 19, 21, 22, 24, 26, 29, 30, 31> [1, 2, 5, 10, 14, 16, 18, 19, 20, 21, 29, 30, 32, 36, 46]
methionine sulfoxide reductase A <2, 7, 8, 9, 11, 12, 13, 15, 17, 19, 25, 35, 36, 41> [26, 27, 37, 39, 41, 42, 43, 45, 51]
methionine sulfoxide-S-reductase <34> [13]
methionine sulphoxide reductase <21> [31]
methionine sulphoxide reductase A <10> [17]
methionine-S-sulfoxide reductase <7> [40]
peptide methionine S-sulfoxide reductase <9> [48]
peptide methionine sulfoxide reductase <1, 4, 8, 9, 10, 12, 13, 15, 16, 18, 19, 22, 23, 30, 36, 37> [4, 6, 7, 8, 9, 11, 12, 15, 23, 24, 28, 33, 34, 35, 38, 41, 44, 47, 49, 53]
peptide methionine sulfoxide reductase A <38, 39, 40> [50]

peptide methionine sulfoxide reductase type A <6, 8, 13, 16, 17, 19, 20, 28> [22]
peptide methionine sulphoxide reductase <27> [25]
peptide-methionine (S)-S-oxide reductase <37> [47]
peptide-methionine sulfoxide reductase <12> [52]
protein-methionine-S-oxide-reductase <37> [47]
sulindac reductase <13> [16]
Additional information <2, 7, 8, 9, 11, 12, 15, 38, 39, 40> (<2, 7, 8, 9, 11, 12, 15> the enzyme belongs to the Msr family of enzymes [42]; <38, 39, 40> the enzyme belongs to the peptide methionine sulfoxide reductase A, PMSRA, gene family [50]) [42, 50]

2 Source Organism

<1> *Drosophila melanogaster* (no sequence specified) [18, 23]
<2> *Staphylococcus aureus* (no sequence specified) [14, 19, 20, 29, 42]
<3> *Haemophilus influenzae* (no sequence specified) [18]
<4> *Mycoplasma genitalium* (no sequence specified) [8, 20]
<5> *Rhizobium meliloti* (no sequence specified) [20]
<6> *Bacillus subtilis* (no sequence specified) [20,21,22]
<7> *Mus musculus* (no sequence specified) [2,18,19,40,42]
<8> *Escherichia coli* (no sequence specified) [1, 3, 4, 5, 7, 12, 15, 16, 18, 19, 20, 21, 22, 23, 38, 42, 46]
<9> *Homo sapiens* (no sequence specified) [11, 18, 19, 23, 24, 37, 39, 41, 42, 44, 48]
<10> *Rattus norvegicus* (no sequence specified) [17, 23, 28]
<11> *Sus scrofa* (no sequence specified) [19, 42]
<12> *Saccharomyces cerevisiae* (no sequence specified) [5, 18, 19, 23, 30, 36, 42, 45, 52]
<13> *Bos taurus* (no sequence specified) (<13> enzyme precursor [16]) [6, 9, 16, 18, 22, 51]
<14> *Mycobacterium smegmatis* (no sequence specified) [20]
<15> *Arabidopsis thaliana* (no sequence specified) (<15> NarG, α subunit [34]) [19, 33, 34, 42]
<16> *Neisseria gonorrhoeae* (no sequence specified) [4, 18, 20, 22, 54]
<17> *Neisseria meningitidis* (no sequence specified) [10, 18, 20, 21, 22, 27]
<18> *Streptococcus pneumoniae* (no sequence specified) [4, 12, 18]
<19> *Mycobacterium tuberculosis* (no sequence specified) [7, 20, 22, 26]
<20> *Lycopersicon esculentum* (no sequence specified) [22]
<21> *Helicobacter pylori* (no sequence specified) [18, 20, 31]
<22> *Erwinia chrysanthemi* (no sequence specified) [18, 20, 46, 53]
<23> *Secale cereale* (no sequence specified) [35]
<24> *Xanthomonas campestris* (no sequence specified) [21]
<25> *Macaca mulatta* (no sequence specified) [43]
<26> *Vibrio cholerae* (no sequence specified) [20]
<27> *Ochrobactrum anthropi* (no sequence specified) [25]

<28> *Deinococcus radiodurans* (no sequence specified) [22]
<29> *Streptococcus gordonii* (no sequence specified) [18]
<30> *Actinobacillus actinomycetemcomitans* (no sequence specified) [20, 49]
<31> *Neisseria gonorrhoeae* (UNIPROT accession number: P14930) [32]
<32> no activity in *Aquifex aeolicus* [20]
<33> no activity in *Thermotoga maritima* [20]
<34> *Caenorhabditis elegans* (UNIPROT accession number: O02089) [13]
<35> *Homo sapiens* (UNIPROT accession number: Q66MI7) [41]
<36> *Drosophila melanogaster* (UNIPROT accession number: P08761) [41]
<37> *Xanthomonas campestris* (UNIPROT accession number: Q8VS50) [47]
<38> *Arabidopsis thaliana* (UNIPROT accession number: Q94A72) [50]
<39> *Arabidopsis thaliana* (UNIPROT accession number: P54150) [50]
<40> *Arabidopsis thaliana* (UNIPROT accession number: Q3E7T3) [50]
<41> *Mus musculus* (UNIPROT accession number: Q9D6Y7) [41]

3 Reaction and Specificity

Catalyzed reaction

L-methionine (S)-sulfoxide + thioredoxin = L-methionine + thioredoxin disulfide (<8,17> reaction mechanism [10,12]; <8,17> 3-step ping pong reaction mechanism involving catalytic and recycling cysteine residues, formation of a sulfenic acid reaction intermediate, overview [21]; <6,24> 3-step reaction mechanism involving catalytic and recycling cysteine residues, formation of a sulfenic acid reaction intermediate, overview [21]; <6, 8, 13, 16, 17, 19, 20, 28> catalytic mechanism and structural features, roles of cysteine residues, active site structure [22]; <2, 4, 5, 6, 8, 14, 16, 17, 19, 21, 22, 26, 30> catalytic mechanism involves the formation of a sulfenic acid intermediate [20]; <8> catalytic mechanism involving the formation of a sulfenic acid intermediate, Cys52 is involved [18]; <13> catalytic mechanism involving the formation of a sulfenic acid intermediate, Cys72, Cys218 and Cys228 are involved [18]; <19> catalytic mechanism of MsrA, active site structure, modeling of protein-bound methionine sulfoxide recognition and repair from the crystal structure [26]; <17> catalytic mechanism of MsrA, the rate limiting step occurs after formation of the sulfenic acid intermediate and is associated with either the Cys51/Cys198 disulfide bond formation or the thioredoxin reduction process [27]; <31> reaction mechanism, modeling of substrate binding at the active site, Cys72 is involved [32])

peptide-L-methionine + thioredoxin disulfide + H_2O = peptide-L-methionine (S)-S-oxide + thioredoxin (<13> catalytic mechanism, Cys72 is essential for activity forming disulfide bonds with either Cys218 or Cys227 [6]; <17> catalytic mechanism, rate-limiting reduction of the Cys51-Cys198 disulfide bond by thioredoxin and formation of the thiosulfenic acid intermediate on Cys51 [27])

Natural substrates and products

S Hsp21 L-methionine S-oxide + dithiothreitol <15> (<15> chloroplast-localized small heat shock protein, repair function for heat shock protein Hsp21 by restoring the structure, which is crucial for cellular resistance to oxidative stress, the enzyme can protect the chaperone-like activity of Hsp21 [33]) (Reversibility: ?) [33]

P Hsp21 L-methionine + dithiothreitol S-oxide

S L-methionine (R,S)-sulfoxide + thioredoxin <3, 16, 17, 18, 21, 29> (<16> the enzyme protect cells against oxidative damage and plays a role in age-related missfunctions [18]; <3, 17, 18, 21, 29> the enzyme protects cells against oxidative damage and plays a role in age-related misfunctions [18]) (Reversibility: ?) [18]

P L-methionine + thioredoxin disulfide

S L-methionine (S)-sulfoxide + thioredoxin <1, 2, 4, 5, 6, 7, 8, 9, 11, 12, 13, 14, 15, 16, 17, 18, 19, 21, 22, 24, 26, 30> (<8> enzyme is involved in repairing of oxidized methionine residues in proteins [1]; <8> FMsr is absolutely specific for the S-isomer of free methionine sulfoxide, no activity with protein bound methionine sulfoxide [5]; <21> important antioxidant enzyme and colonization factor in the gastric pathogen, a methionine repair enzyme responsible for stress resistance [31]; <8> membrane-bound enzyme form Mem-R,S-Msr, enzyme form MsrA is specific for the S-form, MsrA enzyme form variants with specificities for either free or protein-bound methionine [18]; <2, 4, 5, 6, 7, 8, 9, 11, 12, 14, 15, 16, 17, 19, 21, 22, 26, 30> MsrA is specific for the S-form [19,20]; <12> MsrA is specific for the S-form of the substrate [30]; <6, 8, 24> MsrA is specific for the S-form, active on free and protein-bound methionine, the latter is bound more efficiently [21]; <12,13> MsrA is specific for the S-form, enzyme variants with specificities for either free or protein-bound methionine [18]; <1,22> MsrA is specific for the S-form, free and protein-bound methionine [18]; <7,9> MsrA is specific for the S-form, there exist enzyme variants with specificities for either free or protein-bound methionine [18]; <12> MsrA is specific for the S-isomer [5]; <12> MsrA specifically reduces the S-form of methionine sulfoxide [36]; <2> MsrAs are specific for the (S)-form of the substrate [29]; <9> oxidation of protein-bound methionine results in loss of protein function, but can be reversed by the enzyme activity reducing methionine sulfoxide [37]; <8> substrates are several peptides and proteins, overview [12]) (Reversibility: ?) [1, 5, 12, 18, 19, 20, 21, 26, 29, 30, 31, 36, 37]

P L-methionine + thioredoxin disulfide

S L-methionine-(S)-S-oxide + thioredoxin <2, 7, 8, 9, 11, 12, 15> (<7> stereospecific reduction [2]; <9> substrates are HIV-2, which is inactivated by oxidation of its methionine residues M76 and M95, the potassium channel of the brain, the inhibitor IκB-α, or calmodulin, overview [42]) (Reversibility: ?) [2, 42]

P L-methionine + thioredoxin disulfide + H_2O

S L-methionine-(S)-sulfoxide + thioredoxin <1, 8, 9, 10, 12> (<1,8,9,10,12> MsrA is specific for the S-form [23]) (Reversibility: ?) [23]

P L-methionine + thioredoxin disulfide

S Tyr-Gly-Gly-Phe-L-methionine-(S)-S-oxide + thioredoxin <8> (<8> oxidized Met-enkephalin [3]) (Reversibility: ?) [3]

P Tyr-Gly-Gly-Phe-L-methionine + thioredoxin disulfide + H_2O

S calmodulin L-methionine-(S)-sulfoxide + thioredoxin <10> (<10> MsrA is specific for the S-form, enzyme provides protection against oxidative damage by reactive oxygen species and has a repair function for oxidized protein methionine residues, which restores the calmodulin binding to adenylate cyclase of the pathogen Bordetella pertussis, which is an essential step for the bacterium to enter host cells, overview [28]) (Reversibility: ?) [28]

P calmodulin L-methionine + thioredoxin disulfide

S calmodulin-L-methionine (S)-sulfoxide + thioredoxin <8> (Reversibility: ?) [12]

P calmodulin-L-methionine + thioredoxin disulfide

S peptide-L-methionine-(S)-S-oxide + thioredoxin <2, 4, 7, 8, 9, 11, 12, 13, 15, 16, 18, 19, 22, 30, 37, 38, 39, 40> (<7, 8, 13, 19, 30, 38, 39, 40> stereospecific reduction [2,7,49,50,51]; <9> MsrA is involved in regulation of protein function and in elimination of reactive oxygen species via reversible methionine formation besides protein repair in human skin [39]; <7> MsrA is involved in repair of oxidized proteins [40]; <8> MsrA is involved in the antioxidant defense [38]; <12> MsrA regulation, overview [45]; <37> physiological role, overview [47]; <22> stereospecific reduction of protein-bound methionine (S)-sulfoxide residues, the enzyme is involved in oxidized protein repair [53]; <4> stereospecific reduction of protein-bound methionine (S)-sulfoxide residues, the enzyme is involved in repair of oxidized proteins by reducing oxidized methionine residues, which is required for resistance to hydrogenperoxide and other reactive oxygen species, and for adherence to host cell surfaces [8]; <8,22> stereospecific reduction, MsrA is essential for protein repair and protection against oxidative damage [46]; <9> stereospecific reduction, the enzyme is involved in repair of oxidized proteins [11]; <15> substrate in vivo is e.g. the small heat shock protein Hsp-21 which loses its chaperone-like activity upon methionine oxidation [42]; <9> substrate is oxidized A-type potassium channel ShC/B whose activity strongly depends on the oxidative state of a methionine residue in the N-terminal part of the polypeptide [44]; <8> substrate is oxidized ribosomal L12 protein, stereospecific reduction [3]; <9> the enzyme protects the epidermis cells against irradiation and oxidative damages, overview [48]) (Reversibility: ?) [2, 3, 4, 7, 8, 11, 38, 39, 40, 42, 44, 45, 46, 47, 48, 49, 50, 51, 52, 53]

P peptide-L-methionine + thioredoxin disulfide + H_2O

S protein L-methionine (S)-sulfoxide + thioredoxin <27> (<27> enzyme provides protection against oxidative damage by reactive oxygen species and has a repair function for oxidized protein methionine residues [25]) (Reversibility: ?) [25]

P protein L-methionine + thioredoxin disulfide

S protein L-methionine-(S)-sulfoxide + thioredoxin <8, 10, 15> (<15> enzyme provides protection against oxidative damage by reactive oxygen species and has a repair function for oxidized protein methionine residues [34]; <8> MsrA and the soluble isozyme MsrA1 are specific for the S-form, the membrane-associated isozyme reduces both R- and S-stereoisomers of methionine sulfoxide, N-acetylmethionine sulfoxide, and D-Ala-Met-enkephalin [15]; <10> MsrA is specific for the S-form, enzyme provides protection against oxidative damage by reactive oxygen species and has a repair function for oxidized protein methionine residues [17]) (Reversibility: ?) [15, 17, 34]

P protein L-methionine + thioredoxin disulfide

S ribosomal protein L12-L-methionine (S)-sulfoxide + thioredoxin <8> (Reversibility: ?) [12]

P ribosomal protein L12-L-methionine + thioredoxin disulfide

S sulindac + thioredoxin <8, 13> (<13> activation of a methionine sulfoxide-containing prodrug, activity with membrane-bound enzyme form Mem-R,S-Msr [18]; <8> activation of a methionine sulfoxide-containing prodrug, activity with membrane-bound enzyme form Mem-R,S-Msr and MsrA [18]) (Reversibility: ?) [18]

P sulindac sulfide + thioredoxin disulfide (<8,13> activated drug which inhibits cyclooxygenase 1 and 2 and exhibiting anti-inflammatory activity [18])

S sulindac + thioredoxin disulfide <8, 13> (<8> activation of the antiinflammatory drug with anti-tumorigenic activity, which acts via inhibition of cyclooxygenases 1 and 2 [16]) (Reversibility: ?) [16]

P sulindac sulfide + thioredoxin

S Additional information <1, 2, 4, 5, 6, 7, 8, 9, 10, 11, 12, 13, 14, 15, 16, 17, 18, 19, 21, 22, 23, 26, 27, 30, 37> (<8> physiological role [1]; <27> detoxification enzyme [25]; <8> cellular system of balancing native proteins and oxidatively damaged proteins by use of protein biosynthesis, protein oxidative modification, protein elimination, and oxidized protein repair involving the enzyme, overview, enzyme protects against oxidative damage of proteins [23]; <12> cellular system of balancing native proteins and oxidatively damaged proteins by use of protein biosynthesis, protein oxidative modification, protein elimination, and oxidized protein repair involving the enzyme, overview, enzyme protects against oxidative damage of proteins, enzyme activity is not age-related [23]; <1,9,10> cellular system of balancing native proteins and oxidatively damaged proteins by use of protein biosynthesis, protein oxidative modification, protein elimination, and oxidized protein repair involving the enzyme, overview, enzyme protects against oxidative damage of proteins, loss of enzyme activity is age-related [23]; <9> downregulation of MsrA during replicative senescence of cells leads to accumulation of oxidized proteins and age-related increased oxidative damage [24]; <16> enzyme acts on free and protein-bound methionine [20]; <15> enzyme has regulatory function in the plant cell [33]; <4> enzyme repairs oxidatively damaged free and protein bound methionine and recycles it from methionine sulfoxide [20];

<19> enzyme repairs oxidatively damaged free and protein bound methionine and recycles it from methionine sulfoxide, e.g. the heat shock protein and chaperone Hsp16.3, role of the MsrA/MsrB repair pathway in cellular protein dynamics [20]; <5,6,14,21,22,26> enzyme repairs oxidatively damaged free and protein bound methionine and recycles it from methionine sulfoxide, role of the MsrA/MsrB repair pathway in cellular protein dynamics [20]; <2> enzyme repairs oxidatively damaged free and protein bound methionine and recycles it from methionine sulfoxide, role of the MsrA/MsrB repair pathway in cellular protein dynamics, MsrA is important for virulence in mice [20]; <8> enzyme repairs oxidatively damaged free and protein bound methionine and recycles it from methionine sulfoxide, role of the MsrA/MsrB repair pathway in cellular protein dynamics, the MsrA/MsrB repair pathway is involved in the signal recognition particle-dependent protein targeting pathway, regulation mechanism of gene expression, overview [20]; <17> enzymes acts on free and protein-bound methionine [20]; <23> potential role of the enzyme in cold-acclimation, enzyme may protect the cells from photodamage [35]; <12> protection of the cells against reactive oxidizing species, biological consequences of methionine oxidation, physiological role, overview [30]; <15> recycling of free methionine, enzyme reverses the oxidative damage at methionine protein residues oxidized to methionine sulfoxide being a major cause of aging and age-related diseases, Msr can regulate protein function, be involved in signal transduction, and prevent accumulation of faulty proteins [19]; <9,11> recycling of free methionine, enzyme reverses the oxidative damage at methionine protein residues oxidized to methionine sulfoxide being a major cause of aging and age-related diseases, Msr can regulate protein function, be involved in signal transduction, and prevent accumulation of faulty proteins, MsrA has several different physiological repair and regulatory functions, overview [19]; <7> recycling of free methionine, enzyme reverses the oxidative damage at methionine protein residues oxidized to methionine sulfoxide being a major cause of aging and age-related diseases, MsrA can regulate protein function, be involved in signal transduction, and prevent accumulation of faulty proteins, MsrB has several different physiological repair and regulatory functions, overview, oxidation of 2 essential methionine residues of HIV-2 particles can inactivate the virus and prevent infection of human cells [19]; <2,8,12> recycling of free methionine, enzyme reverses the oxidative damage at methionine protein residues oxidized to methionine sulfoxide being a major cause of aging, Msr can regulate protein function, be involved in signal transduction, and prevent accumulation of faulty proteins, MsrA has several different physiological repair and regulatory functions, overview [19]; <30> role of the MsrA/MsrB repair pathway in cellular protein dynamics, mutation of gene msrA has no effect on virulence, and on resistance to oxidative agents, and causes no defect in cell envelope, msrA is probably linked to biofilm formation, enzyme repairs oxidatively damaged free and protein bound methionine and recycles it from methionine sulfoxide [20]; <9> the enzyme is an essential regulator

of longevity and is important for lens cell viability and resistance to oxidative stress, methionine sulfoxide is the major oxidative stress product, up to 60%, in cataract while being essentially absent in clear lens [37]; <12> the enzyme is essential in protection of the cells against oxidative damage by reactive oxygen species, yeast cell life span analysis of wild-type and mutant cells, the latter either overexpress or lack enzyme activity, overview [36]; <8,18> the enzyme is important in protection of the cell against oxidative damage by oxidation of methionine residues in proteins, biological function [12]; <7> the enzyme protect cells against oxidative damage and plays a role in age-related diseases [18]; <22> the enzyme protect cells against oxidative damage and plays a role in age-related missfunctions [18]; <8> the enzyme protect cells against oxidative damage and plays a role in age-related missfunctions, the membrane-bound enzyme form Mem-R,S-Msr also utilizes the R-isomer of methionine sulfoxide as substrate [18]; <9> the enzyme protects cells against oxidative damage and plays a role in age-related and neurological diseases, like Parkinsons or Alzheimers disease [18]; <1,12,13> the enzyme protects cells against oxidative damage and plays a role in age-related diseases [18]; <2> the MsrA1/MsrB system is physiologically more significant in Staphylococcus aureus than MsrA2 [29]; <7> MsrA is a regulator of antioxidant defense and lifespan in mammals [2]; <4,22> MsrA is a virulence determinant for the plant pathogen required for full virulence [8,53]; <13> MsrA protects neuronal cells against brief hypoxia/reoxygenation, the enzyme is involved in oxidized protein repair and protects cells against reactive oxygen species and oxidative damage preventing apoptosis, overview [51]; <8,19> MsrA protects the bacterium against oxidative damage from reactive nitrogen intermediates [7]; <12> MsrA protects the cell against damage caused by oxidative stress through treatment with H_2O_2, paraquat, or 2,2-azobis-(2-amidinopropane) dihydrochloride [52]; <11> roles of methionine sulfoxide reductases in antioxidant defense, protein regulation via alternating it between active and inactive form, and survival, MsrA protects cells from the cytotoxic effects of reactive oxygen species, ROS, overview, enzyme involvement in protein repair and associated factors, protein regulation pathway, overview [42]; <12> roles of methionine sulfoxide reductases in antioxidant defense, protein regulation via alternating it between active and inactive form, and survival, MsrA protects cells from the cytotoxic effects of reactive oxygen species, ROS, overview, the enzyme is involved in age-related diseases such as Alzheimers or Parkinsons diseases as well as in diseases caused by prions, mechanism, overview, enzyme involvement in protein repair and associated factors, protein regulation pathway [42]; <2,7,8,9,15> roles of methionine sulfoxide reductases in antioxidant defense, protein regulation via alternating it between active and inactive form, and survival, MsrA protects cells from the cytotoxic effects of reactive oxygen species, ROS, overview, the enzyme is involved in age-related diseases such as Alzheimers or Parkinsons diseases as well as in diseases caused by prions, mechanism, overview, enzyme involvement in protein repair and associated factors, protein regulation pathway, overview [42]; <8,16,18> the enzyme contributes to the maintenance of adhesins in the pathogen, overview [4]; <30> the enzyme is not a major virulence determinant in the oral pathogen, MsrA is required for protein repair and protection against oxidative damage as well as for the proper expression or maintenance of functional adhesins [49]; <37> the enzyme plays an important role in the oxidative stress response [47]) (Reversibility: ?) [1, 2, 4, 7, 8, 12, 18, 19, 20, 23, 24, 25, 29, 30, 33, 35, 36, 37, 42, 47, 49, 51, 52, 53]

P ?

Substrates and products

S (S)-methyl 4-tolyl sulfoxide + thioredoxin <8, 12> (<8> FMsr is specific for the S-isomer [5]) (Reversibility: ?) [5]

P ?

S DL-methionine (S)-sulfoxide + thioredoxin <17> (<17> enzyme MsrA is specific for the S-form, active on free and protein-bound methionine, the latter is bound more efficiently [21]) (Reversibility: ?) [21]

P DL-methionine + thioredoxin disulfide

S Fmoc L-methionine-(S)-sulfoxide + dithiothreitol <15> (Reversibility: ?) [34]

P Fmoc L-methionine + dithiothreitol disulfide

S His6-Ala-Ala-Gln-MetO-Ile + DTT <9> (Reversibility: ?) [48]

P His6-Ala-Ala-Gln-Met-Ile + DTT disulfide + H_2O

S Hsp21 L-methionine S-oxide + dithiothreitol <15> (<15> chloroplast-localized small heat shock protein, repair function for heat shock protein Hsp21 by restoring the structure, which is crucial for cellular resistance to oxidative stress, the enzyme can protect the chaperone-like activity of Hsp21 [33]; <15> Hsp21 contains 6 methionine residues at positions 49, 52, 55, 59, 62, and 67, about half of the residues are reduced by the enzyme probably due to its stereospecificity [33]) (Reversibility: ?) [33]

P Hsp21 L-methionine + dithiothreitol S-oxide

S L-methionine (R,S)-sulfoxide + thioredoxin <3, 16, 17, 18, 21, 29> (<16> the enzyme protect cells against oxidative damage and plays a role in age-related missfunctions [18]; <3, 17, 18, 21, 29> the enzyme protects cells against oxidative damage and plays a role in age-related missfunctions [18]; <3, 16, 17, 18, 21, 29> enzyme MsrA/B shows both MsrA and MsrB activity, free and protein-bound methionine [18]) (Reversibility: ?) [18]

P L-methionine + thioredoxin disulfide

S L-methionine (S)-sulfoxide + dithiothreitol <8> (Reversibility: ?) [12]

P L-methionine + dithiothreitol disulfide

S L-methionine (S)-sulfoxide + thioredoxin <1, 2, 4, 5, 6, 7, 8, 9, 11, 12, 13, 14, 15, 16, 17, 18, 19, 21, 22, 24, 26, 30, 31> (<8> enzyme is involved in repairing of oxidized methionine residues in proteins [1]; <8> FMsr is absolutely specific for the S-isomer of free methionine sulfoxide, no activity with protein bound methionine sulfoxide [5]; <21> important antioxidant enzyme and colonization factor in the gastric pathogen, a methionine repair enzyme responsible for stress resistance [31]; <8> membrane-bound enzyme form Mem-R,S-Msr, enzyme form MsrA is specific for the S-form, MsrA enzyme form variants with specificities for either free or protein-bound methionine [18]; <8> membrane-bound enzyme form Mem-R,S-Msr, enzyme form MsrA is specific for the S-form, there exist MsrA enzyme form variants with specificities for either free or protein-bound methionine [18]; <2, 4, 5, 6, 7, 8, 9, 11, 12, 14, 15, 16, 17, 19, 21, 22, 26, 30> MsrA is specific for the S-form [19, 20]; <8,12> MsrA is specific for the S-form of the substrate [1,30]; <6, 8, 24> MsrA is specific for the S-form, active on free and protein-bound methionine, the latter is bound more efficiently [21]; <12,13> MsrA is specific for the S-form, enzyme variants with specificities for either free or protein-bound methionine [18]; <1,22> MsrA is specific for the S-form, free and protein-bound methionine [18]; <7,9> MsrA is specific for the S-form, there exist enzyme variants with specificities for either free or protein-bound methionine [18]; <12> MsrA is specific for the S-isomer [5]; <12> MsrA specifically reduces the S-form of methionine sulfoxide [36]; <2> MsrAs are specific for the (S)-form of the substrate [29]; <9> oxidation of protein-bound methionine results in loss of protein function, but can be reversed by the enzyme activity reducing methionine sulfoxide [37]; <8> substrates are several peptides and proteins, overview [12]; <19> absolute specificity for the S-form [26]; <17> enzyme MsrA shows absolute specificity for the S-form of free methionine sulfoxide, no activity with the R-form, enzyme MsrA is oxidized at Cys51/Cys198 forming a disulfide [27]; <19> enzyme MsrA, absolute specificity for the S-form [26]; <17> MsrA activity of the tandem domains of PilB, the MsrA domain alone does not utilize the R-isomer [10]; <31> MsrA activity of the tandem domains of PilB, the MsrA domain alone does very poorly utilize the R-isomer [32]; <2> the 2 MsrA enzymes are absolutely specific for the S-form of the substrate [14]) (Reversibility: ?) [1, 5, 10, 12, 14, 18, 19, 20, 21, 26, 27, 29, 30, 31, 32, 36, 37]

P L-methionine + thioredoxin disulfide

S L-methionine sulfoxide enkephalin + thioredoxin <8> (<8> membrane-bound enzyme form Mem-R,S-Msr [18]) (Reversibility: ?) [18]

P L-methionine enkephalin

S L-methionine-(R)-sulfoxide + thioredoxin <8> (<8> the membrane-associated isozyme reduces both R- and S-stereoisomers of methionine sulfoxide in proteins [15]) (Reversibility: ?) [15]

P L-methionine + thioredoxin disulfide

S L-methionine-(S)-S-oxide + DTT <39> (<39> stereospecific reduction, 9-fluorenylmethyl chloroformate-labeled substrate [50]) (Reversibility: ?) [50]

P L-methionine + DTT disulfide + H_2O

S L-methionine-(S)-S-oxide + thioredoxin <2, 7, 8, 9, 11, 12, 15, 17, 35, 41> (<2, 7, 8, 9, 11, 12, 15, 17, 35, 41> stereospecific reduction [2,27,41,42]; <9> substrates are HIV-2, which is inactivated by oxidation of its methionine residues M76 and M95, the potassium channel of the brain, the inhibitor IκB-α, or calmodulin, overview [42]; <7> stereospecific reduction, free methionine-(S)-S-oxide [2]) (Reversibility: ?) [2, 27, 41, 42, 52]

P L-methionine + thioredoxin disulfide + H_2O

S L-methionine-(S)-sulfoxide + thioredoxin <1, 8, 9, 10, 12> (<1, 8, 9, 10, 12> MsrA is specific for the S-form [23]; <8> MsrA and soluble isozyme MsrA1 are specific for the S-form, the membrane-associated isozyme reduces both R- and S-stereoisomers of methionine sulfoxide in proteins [15]) (Reversibility: ?) [15, 23]

P L-methionine + thioredoxin disulfide

S N-acetyl-L-methionine (R,S)-sulfoxide + thioredoxin <8, 16> (<16> enzyme MsrA/B shows both MsrA and MsrB activity, free and protein-bound methionine [18]; <8> membrane-bound enzyme form Mem-R,S-Msr [18]) (Reversibility: ?) [18]

P N-acetyl-L-methionine + thioredoxin disulfide

S N-acetyl-L-methionine (S)-sulfoxide methyl ester + thioredoxin <17> (<17> enzyme MsrA [21]) (Reversibility: ?) [21]

P N-acetyl-L-methionine methyl ester + thioredoxin disulfide

S N-acetyl-L-methionine-(R)-sulfoxide + thioredoxin <8> (<8> the membrane-associated isozyme reduces both R- and S-stereoisomer of methionine sulfoxide in proteins [15]) (Reversibility: ?) [15]

P N-acetyl-L-methionine + thioredoxin disulfide

S N-acetyl-L-methionine-(S)-S-oxide + DTT <8, 9, 13, 19> (<8, 9, 13, 19> stereospecific reduction [6,7,44]) (Reversibility: ?) [6, 7, 44]

P N-acetyl-L-methionine + DTT disulfide + H_2O

S N-acetyl-L-methionine-(S)-S-oxide + thioredoxin <8, 9, 13, 19, 25> (<8, 9, 13, 19, 25> stereospecific reduction [6,7,43,44]) (Reversibility: ?) [6, 7, 43, 44]

P N-acetyl-L-methionine + thioredoxin disulfide + H_2O

S N-acetyl-L-methionine-(S)-sulfoxide + thioredoxin <8, 12> (<12> MsrA is specific for the S-form [23]; <8> MsrA and soluble isozyme MsrA1 are specific for the S-form, the membrane-associated isozyme reduces both R- and S-stereoisomers [15]) (Reversibility: ?) [15, 23]

P N-acetyl-L-methionine + thioredoxin disulfide

S Tyr-Gly-Gly-Phe-L-methionine-(S)-S-oxide + DTT <8> (<8> oxidized Met-enkephalin [3]) (Reversibility: ?) [3]

P Tyr-Gly-Gly-Phe-L-methionine + DTT disulfide + H_2O

S Tyr-Gly-Gly-Phe-L-methionine-(S)-S-oxide + thioredoxin <8> (<8> oxidized Met-enkephalin [3]) (Reversibility: ?) [3]

P Tyr-Gly-Gly-Phe-L-methionine + thioredoxin disulfide + H_2O

S calmodulin L-methionine-(S)-sulfoxide + thioredoxin <10> (<10> MsrA is specific for the S-form, enzyme provides protection against oxidative damage by reactive oxygen species and has a repair function for oxidized protein methionine residues, which restores the calmodulin binding to adenylate cyclase of the pathogen Bordetella pertussis, which is an essential step for the bacterium to enter host cells, overview [28]; <10> MsrA is specific for the S-form, recombinant human calmodulin, recombinant rat enzyme, artificial system, determination of oxidized methionine residues being reduced by the enzyme, overview [28]) (Reversibility: ?) [28]

P calmodulin L-methionine + thioredoxin disulfide

S calmodulin-L-methionine (S)-sulfoxide + thioredoxin <8> (Reversibility: ?) [12]

P calmodulin-L-methionine + thioredoxin disulfide

S dabsyl L-methionine-(S)-sulfoxide + NADPH <34> (<34> synthetic substrate, MsrA is absolutely specific for the S-form, 7fold lower activity with NADPH compared to DTT [13]) (Reversibility: ?) [13]

P dabsyl L-methionine + $NADP^+$

S dabsyl L-methionine-(S)-sulfoxide + dithiothreitol <34> (<34> synthetic substrate, MsrA is absolutely specific for the S-form [13]) (Reversibility: ?) [13]

P dabsyl L-methionine + dithiothreitol disulfide

S dabsyl L-methionine-(S)-sulfoxide + thioredoxin <10, 34> (<10> synthetic substrate [17]; <34> synthetic substrate, MsrA is absolutely specific for the S-form [13]) (Reversibility: ?) [13, 17]

P dabsyl L-methionine + thioredoxin disulfide

S dabsyl-L-methionine (S)-sulfoxide + thioredoxin <8, 12> (<12> MsrA is specific for the S-isomer [5]; <12> MsrA specifically reduces the S-form of methionine sulfoxide [36]; <8> FMsr is specific for the S-isomer [5]) (Reversibility: ?) [5, 36]

P dabsyl-L-methionine + thioredoxin disulfide

S dabsyl-L-methionine-(S)-S-oxide + DTT <7, 12, 30> (<7,12,30> stereospecific reduction [40,45,49]) (Reversibility: ?) [40, 45, 49]

P dabsyl-L-methionine + DTT disulfide + H_2O

S dabsyl-L-methionine-(S)-S-oxide + thioredoxin <7, 12> (<7> stereospecific reduction [2]) (Reversibility: ?) [2, 52]

P dabsyl-L-methionine + thioredoxin disulfide + H_2O

S oxidized calmodulin + thioredoxin <8> (<8> enzyme reduces L-methionine (S)-sulfoxide of the protein substrate [1]) (Reversibility: ?) [1]

P partially reduced calmodulin + thioredoxin disulfide

S peptide-L-methionine-(S)-S-oxide + DTT <8, 9> (<9> stereospecific reduction [11]; <8> protein-bound substrate, stereospecific reduction, substrate is oxidized ribosomal L12 protein [3]) (Reversibility: ?) [3, 11, 38]

P peptide-L-methionine + DTT disulfide + H_2O

S peptide-L-methionine-(S)-S-oxide + thioredoxin <2, 4, 7, 8, 9, 11, 12, 13, 15, 16, 18, 19, 22, 30, 36, 37, 38, 39, 40> (<7, 8, 9, 12, 13, 15, 16, 18, 19, 22, 30, 36, 37, 38, 39, 40> stereospecific reduction [2, 4, 7, 11, 39, 40, 41, 42, 45, 46, 47, 48, 49, 50, 51]; <9> MsrA is involved in regulation of protein function and in elimination of reactive oxygen species via reversible methionine formation besides protein repair in human skin [39]; <7> MsrA is involved in repair of oxidized proteins [40]; <8> MsrA is involved in the antioxidant defense [38]; <12> MsrA regulation, overview [45]; <37> physiological role, overview [47]; <22> stereospecific reduction of protein-bound methionine (S)-sulfoxide residues, the enzyme is involved in oxidized protein repair [53]; <4> stereospecific reduction of protein-bound methionine (S)-sulfoxide residues, the enzyme is involved in repair of oxidized proteins by reducing oxidized methionine residues, which is required for resistance to hydrogenperoxide and other reactive oxygen species, and for adherence to host cell surfaces [8]; <8,22> stereospecific reduction, MsrA is essential for protein repair and protection against oxidative damage [46]; <9> stereospecific reduction, the enzyme is involved in repair of oxidized proteins [11]; <15> substrate in vivo is e.g. the small heat shock protein Hsp-21 which loses its chaperone-like activity upon methionine oxidation [42]; <9> substrate is oxidized A-type potassium channel ShC/B whose activity strongly depends on the oxidative state of a methionine residue in the N-terminal part of the polypeptide [44]; <8> substrate is oxidized ribosomal L12 protein, stereospecific reduction [3]; <9> the enzyme protects the epidermis cells against irradiation and oxidative damages, overview [48]; <8> protein-bound substrate, stereospecific reduction, substrates are oxidized ribosomal L12 protein or oxidized Met-enkephalin [3]; <4,22> stereospecific reduction of protein-bound methionine (S)-sulfoxide residues [8,53]) (Reversibility: ?) [2, 3, 4, 7, 8, 11, 38, 39, 40, 41, 42, 44, 45, 46, 47, 48, 49, 50, 51, 52, 53]

P peptide-L-methionine + thioredoxin disulfide + H_2O

S protein L-methionine (S)-sulfoxide + thioredoxin <27> (<27> enzyme provides protection against oxidative damage by reactive oxygen species and has a repair function for oxidized protein methionine residues [25]) (Reversibility: ?) [25]

P protein L-methionine + thioredoxin disulfide

S protein L-methionine-(S)-sulfoxide + thioredoxin <8, 10, 15> (<15> enzyme provides protection against oxidative damage by reactive oxygen species and has a repair function for oxidized protein methionine residues [34]; <8> MsrA and the soluble isozyme MsrA1 are specific for the S-form, the membrane-associated isozyme reduces both R- and S-stereoisomers of methionine sulfoxide, N-acetylmethionine sulfoxide, and D-Ala-Met-enkephalin [15]; <10> MsrA is specific for the S-form [17]; <10> MsrA is specific for the S-form, enzyme provides protection against oxidative damage by reactive oxygen species and has a repair function for oxidized protein methionine residues [17]) (Reversibility: ?) [15, 17, 34]

P protein L-methionine + thioredoxin disulfide

S ribosomal protein L12-L-methionine (S)-sulfoxide + thioredoxin <8> (Reversibility: ?) [12]

P ribosomal protein L12-L-methionine + thioredoxin disulfide

S sulindac + thioredoxin <8, 13> (<13> activation of a methionine sulfoxide-containing prodrug, activity with membrane-bound enzyme form Mem-R,S-Msr [18]; <8> activation of a methionine sulfoxide-containing prodrug, activity with membrane-bound enzyme form Mem-R,S-Msr and MsrA [18]; <8,13> activity with membrane-bound enzyme form Mem-R,S-Msr and MsrA [18]) (Reversibility: ?) [18]

P sulindac sulfide + thioredoxin disulfide (<8,13> activated drug which inhibits cyclooxygenase 1 and 2 and exhibiting anti-inflammatory activity [18])

S sulindac + thioredoxin disulfide <8, 13> (<8> activation of the antiinflammatory drug with anti-tumorigenic activity, which acts via inhibition of cyclooxygenases 1 and 2 [16]; <8> highest activity by enzyme MsrA, low activity by enzyme MsrA1 [16]) (Reversibility: ?) [16]

P sulindac sulfide + thioredoxin

S Additional information <1, 2, 4, 5, 6, 7, 8, 9, 10, 11, 12, 13, 14, 15, 16, 17, 18, 19, 21, 22, 23, 24, 26, 27, 30, 31, 34, 37> (<8,12> substrate specificity [5]; <8> physiological role [1]; <27> detoxification enzyme [25]; <8> cellular system of balancing native proteins and oxidatively damaged proteins by use of protein biosynthesis, protein oxidative modification, protein elimination, and oxidized protein repair involving the enzyme, overview, enzyme protects against oxidative damage of proteins [23]; <12> cellular system of balancing native proteins and oxidatively damaged proteins by use of protein biosynthesis, protein oxidative modification, protein elimination, and oxidized protein repair involving the enzyme, overview, enzyme protects against oxidative damage of proteins, enzyme activity is not age-related [23]; <1,9,10> cellular system of balancing native proteins and oxidatively damaged proteins by use of protein biosynthesis, protein oxidative modification, protein elimination, and oxidized protein repair involving the enzyme, overview, enzyme protects against oxidative damage of proteins, loss of enzyme activity is age-related [23]; <9> downregulation of MsrA during replicative senescence of cells leads to accumulation of oxidized proteins and age-related increased oxidative damage [24]; <2,4,5,6,8,14,16,17,19,21,22,26,30> enzyme acts on free and protein-bound methionine [20]; <15> enzyme has regulatory function in the plant cell [33]; <4> enzyme repairs oxidatively damaged free and protein bound methionine and recycles it from methionine sulfoxide [20]; <19> enzyme repairs oxidatively damaged free and protein bound methionine and recycles it from methionine sulfoxide, e.g. the heat shock protein and chaperone Hsp16.3, role of the MsrA/MsrB repair pathway in cellular protein dynamics [20]; <5,6,14,21,22,26> enzyme repairs oxidatively damaged free and protein bound methionine and recycles it from methionine sulfoxide, role of the MsrA/MsrB repair pathway in cellular protein dynamics [20]; <2> enzyme repairs oxidatively damaged free and protein bound methionine and recycles it from methionine sulfoxide, role of the MsrA/MsrB repair pathway in cellular protein dynamics, MsrA is important for virulence in mice [20]; <8> enzyme repairs oxidatively damaged free and protein bound methionine and recycles it from methionine sulfoxide, role of the MsrA/MsrB repair pathway in cellular protein dynamics, the MsrA/MsrB repair pathway is involved in the signal recognition particle-dependent protein targeting pathway, regulation mechanism of gene expression, overview [20]; <17> enzymes acts on free and protein-bound methionine [20]; <8> MsrA is specific for the S-form of the substrate [12]; <23> potential role of the enzyme in cold-acclimation, enzyme may protect the cells from photodamage [35]; <12> protection of the cells against reactive oxidizing species, biological consequences of methionine oxidation, physiological role, overview [30]; <15> recycling of free methionine, enzyme reverses the oxidative damage at methionine protein residues oxidized to methionine sulfoxide being a major cause of aging and age-related diseases, Msr can regulate protein function, be involved in signal transduction, and prevent accumulation of faulty proteins [19]; <9,11> recycling of free methionine, enzyme reverses the oxidative damage at methionine protein residues oxidized to methionine sulfoxide being a major cause of aging and age-related diseases, Msr can regulate protein function, be involved in signal transduction, and prevent accumulation of faulty proteins, MsrA has several different physiological repair and regulatory functions, overview [19]; <7> recycling of free methionine, enzyme reverses the oxidative damage at methionine protein residues oxidized to methionine sulfoxide being a major cause of aging and age-related diseases, MsrA can regulate protein function, be involved in signal transduction, and prevent accumulation of faulty proteins, MsrB has several different physiological repair and regulatory functions, overview, oxidation of 2 essential methionine residues of HIV-2 particles can inactivate the virus and prevent infection of human cells [19]; <2,8,12> recycling of free methionine, enzyme reverses the oxidative damage at methionine protein residues oxidized to methionine sulfoxide being a major cause of aging, Msr can regulate protein function, be involved in signal transduction, and prevent accumulation of faulty proteins, MsrA has several different physiological repair and regulatory functions, overview [19]; <30> role of the MsrA/MsrB repair pathway in cellular protein dynamics, mutation of gene msrA has no effect on virulence, and on resistance to oxidative agents, and causes no defect in cell envelope, msrA is probably linked to biofilm formation, enzyme repairs oxidatively damaged free and protein bound methionine and recycles it from methionine sulfoxide [20]; <9> the enzyme is an essential regulator of longevity and is important for lens cell viability and resistance to oxidative stress, methionine sulfoxide is the major oxidative stress product, up to 60%, in cataract while being essentially absent in clear lens [37]; <12> the enzyme is essential in protection of the cells against oxidative damage by reactive oxygen species, yeast cell life span analysis of wild-type and mutant cells, the latter either overexpress or lack enzyme activity, overview [36]; <8,18> the enzyme is important in protection of the cell against oxidative damage by oxidation of methionine residues in proteins, biological function [12]; <7> the enzyme protect cells against oxidative damage and plays a role in age-related diseases [18]; <22> the enzyme protect cells against oxidative damage and plays a role in age-related missfunctions [18]; <8> the enzyme protect cells against oxidative damage and plays a role in age-related missfunctions, the membrane-bound enzyme form Mem-R,S-Msr also utilizes the R-isomer of methionine sulfoxide as substrate [18]; <9> the enzyme protects cells against oxidative damage and plays a role in age-related and neurological diseases, like Parkinsons or Alzheimers disease [18]; <1,12,13> the enzyme protects cells against oxidative damage and plays a role in age-related diseases [18]; <2> the MsrA1/MsrB system is physiologically more significant in Staphylococcus aureus than MsrA2 [29];

<34> enzyme converts free and protein-bound methionine [13]; <9> enzyme reduces oxidized methionine residues of the α-1-proteinase inhibitor, calmodulin, and thrombomodulin, which become reversibly inactivated upon oxidation [18]; <1> enzyme reduces oxidized methionine residues of the shaker potassium channel, which becomes reversibly inactivated upon oxidation [18]; <17> substrate specificities of enzymes, the reduction step is rate-determining [21]; <16> substrate specificity and activity of MsrB/PilB in comparison to MsrA, overview [18]; <8> substrate specificity of enzyme forms with S-form of free and protein-bound methionine sulfoxide, overview [16]; <17> substrate specificity of MsrA activity, diverse substrates, overview [10]; <8> substrate specificity of the different enzyme forms, overview, the membrane-bound enzyme form Mem-R,S-Msr also utilizes the R-isomer of methionine sulfoxide as substrate, enzyme reduces oxidized methionine residues of the ribosomal protein L12, which becomes reversibly inactivated and forms monomers instead of dimers upon oxidation [18]; <21> the enzyme also exhibits MsrB activity utilizing L-methionine (R)-sulfoxide as substrate [31]; <8> the enzymes utilize free and protein-bound L-methionine and N-acetyl-L-methionine as substrates [15]; <6,8,24> the reduction step is rate-determining [21]; <31> the tandem domains of PilB also posesses MsrB activity utilizing L-methionine (R)-sulfoxide as substrate, the MsrB domain alone does not utilize the S-isomer [32]; <7> MsrA is a regulator of antioxidant defense and lifespan in mammals [2]; <4,22> MsrA is a virulence determinant for the plant pathogen required for full virulence [8,53]; <13> MsrA protects neuronal cells against brief hypoxia/reoxygenation, the enzyme is involved in oxidized protein repair and protects cells against reactive oxygen species and oxidative damage preventing apoptosis, overview [51]; <8,19> MsrA protects the bacterium against oxidative damage from reactive nitrogen intermediates [7]; <12> MsrA protects the cell against damage caused by oxidative stress through treatment with H_2O_2, paraquat, or 2,2-azobis-(2-amidinopropane) dihydrochloride [52]; <11> roles of methionine sulfoxide reductases in antioxidant defense, protein regulation via alternating it between active and inactive form, and survival, MsrA protects cells from the cytotoxic effects of reactive oxygen species, ROS, overview, enzyme involvement in protein repair and associated factors, protein regulation pathway, overview [42]; <12> roles of methionine sulfoxide reductases in antioxidant defense, protein regulation via alternating it between active and inactive form, and survival, MsrA protects cells from the cytotoxic effects of reactive oxygen species, ROS, overview, the enzyme is involved in age-related diseases such as Alzheimers or Parkinsons diseases as well as in diseases caused by prions, mechanism, overview, enzyme involvement in protein repair and associated factors, protein regulation pathway [42]; <2,7,8,9,15> roles of methionine sulfoxide reductases in antioxidant defense, protein regulation via alternating it between active and inactive form, and survival, MsrA protects cells from the cytotoxic effects of reactive oxygen species, ROS, overview, the enzyme is involved in age-related diseases such as Alzhei-

mers or Parkinsons diseases as well as in diseases caused by prions, mechanism, overview, enzyme involvement in protein repair and associated factors, protein regulation pathway, overview [42]; <8,16,18> the enzyme contributes to the maintenance of adhesins in the pathogen, overview [4]; <30> the enzyme is not a major virulence determinant in the oral pathogen, MsrA is required for protein repair and protection against oxidative damage as well as for the proper expression or maintenance of functional adhesins [49]; <37> the enzyme plays an important role in the oxidative stress response [47]; <7> role of subcellular localization in structure-function relationship of the isozymes, overview [40]; <16> the bifunctional enzyme catalyzes both reactions of MsrB or PilB, EC 1.8.4.12, and of MsrA or PilA, EC 1.8.4.11, the catalytic sites for the two different activities are localized separately on the enzyme molecule, overview [4,54]) (Reversibility: ?) [1, 2, 4, 5, 7, 8, 10, 12, 13, 15, 16, 18, 19, 20, 21, 23, 24, 25, 29, 30, 31, 32, 33, 35, 36, 37, 40, 42, 47, 49, 51, 52, 53, 54]

P ?

Inhibitors

3-carboxy 4-nitrobenzenethiol <8, 17> (<8,17> binds specifically to the sulfenic acid reaction intermediate [21]) [21]

dimedone <8, 17> (<8,17> binds specifically to the sulfenic acid reaction intermediate [21]) [21]

Additional information <15, 23, 40> (<15> effects of high light, ozone, and paraquat on enzyme activity and photosynthetic activity in wild-type and transgenic plants, overview [34]; <23> enzyme expression decreases during dehardening from 4°C to 22°C of cold-acclimated plants, effects of light and temperature on enzyme expression and activity, overview [35]; <40> isozyme PMSRA5 expression is slightly suppressed by ozone [50]) [34, 35, 50]

Cofactors/prosthetic groups

dithiothreitol <7, 8, 9, 13, 15, 17, 19, 30, 34, 39> (<15> absolutely dependent on in vitro and in vivo with substrate Hsp21 [33]; <8> MsrA can also utilize DTT as reductant, the membrane-isozyme shows only poor activity, while MsrA1 is not active with DTT [15]; <34> preferred cofactor in vitro [13]; <8> utilized in vitro [12]) [3,6,7,10,11,12,13,15,33,34,38,40,44,48,49,50]

NADPH <8, 34> (<8> membrane-bound enzyme form Mem-R,S-Msr [18]) [13,18]

thioredoxin <1, 2, 3, 4, 5, 6, 7, 8, 9, 10, 11, 12, 13, 14, 15, 16, 17, 18, 19, 21, 22, 24, 25, 26, 29, 30, 31, 34, 36, 37, 38, 39, 40> (<12> dependent on [30]; <8> preferred cofactor [18]; <6,8,17,24> physiological cofactor [21]; <8> physiologic cofactor [12]; <12> cytosolic thioredoxins 1 and 2 are involved in regulation of MsrA exression [45]; <17> rate-limiting reduction of the Cys51-Cys198 disulfide bond by thioredoxin in catalysis [27]) [1, 2, 3, 4, 5, 6, 7, 8, 10, 11, 12, 13, 14, 15, 16, 17, 18, 19, 20, 21, 23, 26, 27, 28, 29, 30, 31, 32, 34, 36, 37, 38, 39, 40, 41, 42, 43, 44, 45, 47, 48, 50, 51, 52, 53]

Additional information <6, 8, 17, 24> (<8> DTT can partially substitute for thioredoxin in vitro, low activity [21]; <6,17,24> DTT can substitute for thioredoxin in vitro [21]; <8> no activity with DTT as cofactor by membrane-bound enzyme form Mem-R,S-Msr [18]) [18,21]

Activating compounds

KCl <12> (<12> enzyme prefers high ionic strength, activation [5]) [5]
Na_2SO_4 <12> (<12> enzyme prefers high ionic strength, activation [5]) [5]
NaCl <12> (<12> enzyme prefers high ionic strength, activation [5]) [5]
NaF <12> (<12> enzyme prefers high ionic strength, activation [5]) [5]
Additional information <2, 8, 9, 11, 12, 15, 19, 23, 27, 37, 38, 39> (<23> 48 h exposure to high light at 22°C induces enzyme expression, effects of light and temperature on enzyme expression and activity, overview [35]; <15> effects of high light, ozone, and paraquat on enzyme activity and photosynthetic activity in wild-type and transgenic plants, overview [34]; <27> enzyme expression is induced, together with the glutathione S-transferase, by toxic concentrations of aromatic substrates such as phenol and 4-chlorophenol which cause production of reactive oxygen species [25]; <19> enzyme is induced under oxidative stress and heat shock [20]; <8> gene msrA is induced during biofilm formation [20]; <8> MsrA expression and enzyme formation is induced in the stationary growth phase or on starvation for amino acids, glucose, or nitrogen [19]; <12> MsrA expression is induced by oxidizing diamide treatment and γ-irradiation, the factor calcium-phosphate-binding protein CPBP, a homologue of elongation factor 1-γ participates in transcription of gene msrA as part of a transcription regulation complex [19]; <2> oxacillin induces MsrA expression and enzyme formation [19]; <9> oxidative stress, e.g. caused by H_2O_2 up to 0.4 mM, induces enzyme expresssion [24]; <2> the msrA1-msrB operon is induced by antibiotics, expression of msrA1 is increased in the stationary phase but not affected by H_2O_2 [20]; <11> chronic dietary intake of soybean protein induces the expression of MsrA [42]; <12> H_2O_2 induces MsrA expression in a thioredoxin-dependent manner, msrA promoter and calcium phospholipid binding protein, CPBP, form a complex and enhance msrA expression, cytosolic thioredoxins 1 and 2 are involved in the regulation [45]; <38> isozyme PMSRA2 is slightly induced by paraquat, ozone, and high light conditions [50]; <39> isozyme PMSRA3 expression is slightly induced by ozone [50]; <39> isozyme PMSRA4 is highly induced by high light conditions, and slightly by ozone, paraquat, and cercosporin [50]; <9> MsrA is upregulated by UVA radiation in keratinocytes [48]; <8> starvation induces the expression of MsrA [42]; <2> the antibiotic oxacillin induces MsrA expression [42]; <12> the calcium phospholipid-binding protein, CPBP, is an analogue of the elongation factor 1-γ and regulates MsrA expression by binding to the MsrA promoter, starvation induces MsrA expression, also diamide treatment and γ-irradiation induce the enzyme [42]; <37> the enzyme shows an oxidant-inducible expression pattern independent of gene oxyR [47]) [19, 20, 24, 25, 34, 35, 42, 45, 47, 48, 50]

Metals, ions

KCl <8> (<8> enzyme prefers high ionic strength, activation [5]) [5]
Mg^{2+} <8> (<8> activates [3]) [3]
Na_2SO_4 <8> (<8> enzyme prefers high ionic strength, activation [5]) [5]

NaCl <8> (<8> enzyme prefers high ionic strength, activation [5]) [5]
NaF <8> (<8> enzyme prefers high ionic strength, activation [5]) [5]
Additional information <13, 17> (<17> content of free cysteinyl residues in wild-type and mutant enzymes, MsrA and MsrB domains, overview [10]; <13> the enzyme does not require meal ions for activity, free sulfhydryl content and disulfide bond numbers in wild-type and mutant enzymes, overview [6]) [6, 10]

Turnover number (min^{-1})

0.06 <34> (dabsyl L-methionine-(S)-sulfoxide, <34> pH 7.4, 37°C, recombinant enzyme [13]) [13]
0.28 <7> (dabsyl-L-methionine-(S)-S-oxide, <7> recombinant wild-type MsrA, with DTT, pH 7.5, 37°C [40]) [40]
0.6 <17> (L-methionine (S)-sulfoxide, <17> pH 5.5, 25°C, MsrA, steady-state conditions [27]) [27]
0.78 <7> (dabsyl-L-methionine-(S)-S-oxide, <7> recombinant truncated MsrA Δ(1-46), with DTT, pH 7.5, 37°C [40]) [40]
0.91 <39> (9-fluorenylmethyl chloroformate-labeled L-methionine-(S)-S-oxide, <39> recombinant isozyme PMSRA3, with DTT, pH 7.5, 22°C [50]) [50]
1.59 <39> (9-fluorenylmethyl chloroformate-labeled L-methionine-(S)-S-oxide, <39> recombinant isozyme PMSRA4, with DTT, pH 8.0, 22°C [50]) [50]
3.7 <8> (L-methionine (S)-sulfoxide, <8> pH 8.0, 25°C, cofactor thioredoxin [21]) [21]
3.7 <17> (L-methionine (S,R)-sulfoxide, <17> MsrB activity of PILB, pH 8.0, 25°C [10]) [10]
7 <17> (L-methionine (S)-sulfoxide, <17> pH 8.0, 25°C, MsrA, steady-state conditions [27]) [27]
Additional information <17> [27]

Specific activity (U/mg)

0.000015 <12> (<12> wild-type Saccharomyces cerevisiae strain, substrate L-methionine-(S)-S-oxide [52]) [52]
0.00006 <2> (<2> msrA1/msrA2 double knockout strain RN450 [29]; <2> msrA1/msrA$_2$ double knockout strain RN450, substrate L-methionine (S)-sulfoxide [29]) [29]
0.000068 <12> (<12> wild-type Saccharomyces cerevisiae strain, substrate dabsyl-L-methionine-(S)-S-oxide [52]) [52]
0.00008 <2> (<2> msrA1 knockout strain RN450 [29]; <2> msrA1 knockout strain RN450, substrate L-methionine (S)-sulfoxide [29]) [29]
0.00009 <8> (<8> enzyme form Mem-R,S-Msr, substrate sulindac [18]) [18]
0.00019 <8> (<8> enzyme form MsrA [16]; <8> enzyme form MsrA, substrate sulindac [18]) [16, 18]
0.0002 <10> (<10> cytosol, overall Msr activity, substrate is a mixture of S- and R-form of dabsyl L-methionine sulfoxide [17]) [17]
0.00021 <10> (<10> mitochondria, overall Msr activity, substrate is a mixture of S- and R-form of dabsyl L-methionine sulfoxide [17]) [17]

0.00024 <2> (<2> msrA2 knockout strain RN450 [29]; <2> msrA2 knockout strain RN450, substrate L-methionine (S)-sulfoxide [29]) [29]
0.00026 <2> (<2> wild-type strain RN450 [29]; <2> wild-type strain RN450, substrate L-methionine (S)-sulfoxide [29]) [29]
0.0003 <12> (<12> recombinant Saccharomyces cerevisiae strain overexpressing MsrA, substrate L-methionine-(S)-S-oxide [52]) [52]
0.0004 <8> (<8> membrane vesicles, substrate N-acetyl-L-methionine-(R)-sulfoxide [15]) [15]
0.00044 <8> (<8> MsrA, substrate N-acetyl-L-methionine-(S)-sulfoxide [15]) [15]
0.00047 <8> (<8> membrane vesicles, substrate N-acetyl-L-methionine-(S)-sulfoxide [15]) [15]
0.0017 <12> (<12> recombinant Saccharomyces cerevisiae strain overexpressing MsrA, substrate dabsyl-L-methionine-(S)-S-oxide [52]) [52]
0.0018 <8> (<8> wild-type strain, substrate L-methionine (S)-sulfoxide [18]) [18]
0.063 <2> (<2> purified recombinant MsrA1, substrate L-methionine (S)-sulfoxide [14]) [14]
0.083 <31> (<31> purified recombinant MsrA/MsrB tandem domain, substrate L-methionine (S)-sulfoxide [32]) [32]
0.099 <7> (<7> purified recombinant truncated MsrA Δ(1-46), substrates DTT and dabsyl-L-methionine-(S)-S-oxide [40]) [40]
0.15 <31> (<31> purified recombinant MsrA domain alone, substrate L-methionine (S)-sulfoxide [32]) [32]
0.238 <7> (<7> purified recombinant wild-type MsrA, substrates DTT and dabsyl-L-methionine-(S)-S-oxide [40]) [40]
0.24 <2> (<2> purified recombinant MsrA2, substrate L-methionine (S)-sulfoxide [14]) [14]
0.33 <8> (<8> in vitro, substrate free L-methionine (R)-sulfoxide [1]) [1]
0.96 <34> (<34> purified enzyme, substrate dabsyl L-methionine-(S)-sulfoxide with dithiothreitol [13]) [13]
3 <17> (<17> recombinant wild-type MsrA domain, cosubstrate dithiothreitol [10]) [10]
4.2 <12, 17> (<12> purified recombinant wild-type enzyme, substrate L-methionine (S)-sulfoxide [5]; <17> recombinant wild-type MsrA/MsrB, cosubstrate dithiothreitol [10]) [5, 10]
43 <12> (<12> purified recombinant wild-type enzyme, substrate dabsyl-L-methionine (S)-sulfoxide [5]) [5]
170 <17> (<17> recombinant wild-type MsrA/MsrB, cosubstrate thioredoxin [10]) [10]
220 <17> (<17> recombinant wild-type MsrA domain, cosubstrate thioredoxin [10]) [10]
Additional information <2, 7, 8, 9, 12, 13, 15, 18, 19, 22, 25> (<2> 69% of wild-type strain RB450 MsrA activity belongs to MsrA1, 8% to MsrA2, and 23% to MsrA3 [29]; <8> activity in mutant strains [18]; <12> recombinant activity in overexpressing yeast cells [36]; <15> relative activity in chloroplasts of wild-type and transgenic plants [34]; <13> subcellular sulindac re-

ducing activity in calf liver [18]; <7> activities in wild-type and mutant mice in different organs during 2 days, overview [2]; <8> activity in wild-type and recombinant strain [7]; <19> activity in wild-type strain and recombinant Escherichia coli strain [7]; <12> cell survival rates after treatement with H_2O_2, paraquat, or 2,2-azobis-(2-amidinopropane) dihydrochloride [52]; <25> subcellular distribution of MsrA activity in eye tissue, overview [43]; <22> virulence on chicory leaves of wild-type and recombinant strains, overview [53]) [2, 4, 5, 7, 18, 24, 29, 34, 36, 43, 44, 52, 53]

K_m-Value (mM)

0.01 <8> (thioredoxin, <8> pH 8.0, 25°C, substrate L-methionine (S)-sulfoxide [21]) [21]
0.044 <12> (L-methionine (S)-sulfoxide, <12> recombinant wild-type enzyme, pH 7.5, 37°C [5]) [5]
0.075 <17> (thioredoxin, <17> MsrA activity of PILB, pH 8.0, 25°C [10]) [10]
0.12 <8> (L-methionine (S)-sulfoxide, <8> recombinant wild-type enzyme, pH 7.5, 37°C [5]) [5]
0.34 <7> (dabsyl-L-methionine-(S)-S-oxide, <7> recombinant wild-type MsrA, with DTT, pH 7.5, 37°C [40]) [40]
1.1 <39> (9-fluorenylmethyl chloroformate-labeled L-methionine-(S)-S-oxide, <39> recombinant isozyme PMSRA4, with DTT, pH 8.0, 22°C [50]) [50]
1.18 <34> (dabsyl L-methionine-(S)-sulfoxide, <34> pH 7.4, 37°C, recombinant enzyme [13]) [13]
1.9 <8> (L-methionine (S)-sulfoxide, <8> pH 8.0, 25°C, cofactor thioredoxin [21]) [21]
2.12 <39> (9-fluorenylmethyl chloroformate-labeled L-methionine-(S)-S-oxide, <39> recombinant isozyme PMSRA3, with DTT, pH 7.5, 22°C [50]) [50]
4.3 <7> (dabsyl-L-methionine-(S)-S-oxide, <7> recombinant truncated MsrA Δ(1-46), with DTT, pH 7.5, 37°C [40]) [40]
9 <17> (L-methionine (R,S)-sulfoxide, <17> MsrB activity of PILB, pH 8.0, 25°C [10]) [10]
Additional information <6, 8, 17, 24> (<17> kinetics [10]; <6,8,24> kinetic mechanism [21]; <17> kinetics of disulfide formation in MsrA at Cys52/Cys198 at pH 5.5 and pH 8.0, 25°C, single turnover experiments, steady-state kinetics [27]; <17> detailed kinetics, wild-type and mutants MsrAs [27]) [10, 21, 27]

pH-Optimum

5.5 <17> (<17> assay at [27]) [27]
6.9 <31> (<31> assay at [32]) [32]
7 <12> (<12> assay at [45]) [45]
7-7.5 <34> [13]
7.4 <2, 8, 9, 13, 21, 22, 30> (<2, 8, 9, 13, 21, 22, 30> assay at [3, 6, 11, 14, 15, 16, 31, 44, 49, 51, 53]) [3, 6, 11, 14, 15, 16, 31, 44, 49, 51, 53]
7.5 <2, 7, 8, 9, 12, 39> (<2, 7, 8, 9, 12, 39> assay at [5, 29, 36, 40, 48, 50, 52]) [5, 29, 36, 40, 48, 50, 52]
7.5-8 <38> (<38> assay at [50]) [50]

7.8 <15> (<15> assay at [33]) [33]
8 <8, 15, 17, 39> (<8,15,17,39> assay at [21,34,50]; <17> L-methionine formation [27]) [21, 27, 34, 50]

pH-Range
5.5-8 <17> [27]
6.5-8.5 <34> (<34> sharp decrease below pH 6.5 and above pH 8.5 [13]) [13]

Temperature optimum (°C)
22 <39> (<39> assay at [50]) [50]
25 <8, 15, 17, 22> (<8,15,17,22> assay at [21,27,33,53]) [21, 27, 33, 53]
37 <2, 7, 8, 9, 12, 13, 30, 31, 34> (<2, 7, 8, 9, 12, 13, 30, 31> assay at [3, 5, 6, 11, 14, 15, 16, 29, 32, 36, 40, 44, 45, 48, 49, 51, 52]; <34> about [13]) [3, 5, 6, 11, 13, 14, 15, 16, 29, 32, 36, 40, 44, 45, 48, 49, 51, 52]

Temperature range (°C)
17-57 <34> (<34> activity rapidly decreases below 17°C and above 57°C [13]) [13]

4 Enzyme Structure

Subunits
? <1, 3, 7, 8, 9, 12, 13, 16, 17, 18, 19, 21, 22, 29, 34> (<22> x * 27000, SDS-PAGE [46]; <17> x * 21898, about, recombinant MsrA domain, mass spectrometry [10]; <1,3,7,8,9,12,13,17,18,21,22,29> x * 25000, MsrA [18]; <34> x * 29700, recombinant MsrA, SDS-PAGE [13]; <21> x * 43000, Msr, SDS-PAGE [31]; <16> x * 57000, MsrA/B [18]; <19> x * 21000-27000, about, recombinant MsrA, SDS-PAGE [7]) [7, 10, 13, 18, 31, 46]
monomer <6, 8, 17, 24> [21, 22]
Additional information <6, 8, 13, 16, 17, 19, 20, 21, 22, 28, 31> (<17,20,28> amino acid sequence comparison [22]; <8> amino acid sequence comparison, analysis of three-dimensional structures, N-terminal coil, structure comparison to enzymes from various species [22]; <13,16,19> amino acid sequence comparison, analysis of three-dimensional structures, structure comparison to enzymes from various species [22]; <6> amino acid sequence comparison, structure analysis, active enzyme possesses no N-terminal coil before the core [22]; <31> enzyme domain and active site structure, MsrA domain comprises residues 181-362, overview [32]; <17> MsrA activity is located on the central domain of PILB, the fused domains are folded entities [10]; <17> MsrA and enzyme MsrB, methionine S-oxide reductase (R-form oxidizing), form domains of a single polypeptide together with a third thioredoxin-like domain [20]; <21> MsrA and MsrB, EC 1.8.4.B1, are fused together [20]; <16> the 2 enzyme activities, MsrA and MsrB, form domains of a single polypeptide together with a third thioredoxin-like domain [20]; <8,13> the Cys residue within the conserved sequence motif GCFWG at the N-terminus is essential for catalytic activity [18]; <17> the enzyme forms are produced as individual folded entities, but in vivo the enzyme is part of a three-domain protein

named PILB, with the central domain exhibiting MsrA activity, and the C-terminal domain showing MsrB activity [21]; <8,22> residual dipolar coupling determination and analysis, comparison with the crystal structure, protein fold and tertiary structure determination [46]) [10, 18, 20, 21, 22, 32, 46]

Posttranslational modification

proteolytic modification <10> (<10> enzyme precursor contains a cleavable N-terminal signal sequence for targeting to the mitochondria [17]) [17]

5 Isolation/Preparation/Mutation/Application

Source/tissue

HLE cell <9> (<9> human lens epithelial cell line [37]) [37]
HaCaT cell <9> (<9> primary keratinocytes, expression level at normal light and UV-light conditions [48]) [48]
WI-38 cell <9> (<9> embryonic fibroblast, activity during development: downregulation during replicative senescence [24]; <9> fibroblast, young and old cells, enzyme expression pattern during cell development [23]) [23, 24]
bone marrow <9> (<9> high MsrA expression level [44]) [44]
brain <7, 9, 10, 13, 41> (<13> calf, sulindac reducing activity [18]; <10> MsrA [23]) [2, 16, 18, 19, 23, 41, 42]
cauline leaf <38, 39, 40> (<39> highest expression level of isozyme PMSRA4 [50]) [50]
cerebellum <9> (<9> high MsrA expression level [44]) [44]
culture condition:4-chlorophenol-grown cell <27> [25]
epidermis <9> [39, 48]
epithelium <9> (<9> lens [37]) [37]
eye <25> [43]
fibroblast <9> (<9> cell line WI-38, young and old cells, enzyme expression pattern during cell development, senescent cells show decreased enzyme expression and activity [23]; <9> WI-38 [24]) [23, 24]
flower <38, 39, 40> (<38,39,40> mature, after anthesis [50]) [50]
flower bud <38, 39, 40> [50]
heart <9> (<9> ventricle, high MsrA expression level [44]) [44]
hippocampus <9> (<9> high MsrA expression level [44]) [44]
kidney <7, 9, 10, 13, 41> (<13> calf, sulindac reducing activity [18]; <13> highest sulindac reductase activity [16]; <10> MsrA [23]; <9> high MsrA expression level [44]) [2, 16, 18, 19, 23, 41, 44]
leaf <23, 38, 39, 40> (<38,39,40> young and mature [50]; <23> high expression level in cold-hardened plants at 4°C [35]) [35, 50]
lens <9> [37]
lens fiber <9> (<9> cortical and nuclear components [37]) [37]
liver <7, 9, 10, 13, 41> (<13> calf, sulindac reducing activity [18]; <10> MsrA [23]; <9> high MsrA expression level, especially in fetal liver [44]) [2, 16, 17, 18, 19, 23, 41, 44]

lung <7> [2, 19]
macula <25> [43]
melanocyte <9> (<9> epidermal [39]) [39]
retina <25> (<25> peripheral [43]) [43]
root <38, 39, 40> (<39> highest expression level of isozyme PMSRA3 [50]) [50]
seedling <38, 39, 40> (<38,39,40> germinated [50]; <39> germinated, highest expression level of isozyme PMSRA4 [50]) [50]
skin <9> (<9> accumulation in the upper dermis [48]) [39, 48]
stem <38, 39, 40> (<38> highest expression level of isozyme PMSRA2 [50]; <40> highest expression level of isozyme PMSRA5 [50]) [50]
Additional information <9, 10, 15, 23, 38, 39, 40> (<23> expression pattern analysis [35]; <9> enzyme expression level and methionine sulfoxide content in fibroblasts during development, overview [24]; <15> high expression level of the plastidic isozyme pPMSR in photosynthetic active tissue [33]; <10> organ-specific expression patterns [23]; <9> expression analysis of MsrA [39]; <38,39,40> quantitative expression profile analysis [50]; <9> tissue expression analysis of MsrA, wide tissue distribution, no expression in leukemia and lymphoma cell lines [44]) [23, 24, 33, 35, 39, 44, 50]

Localization

cell envelope <27> (<27> associated [25]) [25]
chloroplast <15, 39> (<15> plastidial isozyme PMSR4, soluble fraction [34]; <15> plastidic isozyme pPMSR [33]; <39> isozyme PMSRA3 [50]; <39> isozyme PMSRA4 [50]) [33, 34, 50]
cytoplasmic membrane <27> (<27> inner or outer side [25]) [25]
cytosol <7, 9, 10, 23, 25, 30, 35, 36, 38, 39, 41> (<10> cytosolic isozyme of MsrA [17]; <39> cytosolic isozyme PMsrA1 [50]; <35> isozyme MsrA3 [41]; <38> isozyme PMSRA2 [50]; <41> splicing variant MsrA(S) [41]; <30> subcellular localization analysis, overview [49]) [17, 23, 35, 39, 40, 41, 43, 49, 50]
extracellular <21> (<21> secretion of MsrA and MsrB, methionine S-oxide reductase (R-form oxidizing), fused together [20]) [20]
membrane <8, 13, 16, 21> (<16> outer cell membrane [20]; <8> enzyme form Mem-R,S-Msr [18]; <8> membrane-associated isozyme, MsrA [15]) [15, 16, 18, 20, 31]
microsome <8, 13> (<13> calf, sulindac reducing activity [18]) [16, 18]
mitochondrion <7, 8, 9, 10, 13, 25, 41> (<10> matrix [23]; <10> matrix, mitochondrial isozyme of MsrA [17]; <41> mitochondrial splicing variant MsrA [41]; <9> MsrA contains a mitochondrial targeting sequence which is not necessary for catalytic activity [11]) [11, 16, 17, 23, 40, 41, 43]
nucleus <9, 41> (<41> splicing variant MsrA(S) [41]) [39, 41]
outer membrane <16> [54]
periplasmic space <27> [25]
soluble <8> (<8> soluble isozyme MsrA1 [15]) [15]
Additional information <7, 9, 13, 21, 25, 27, 35, 36, 38, 39, 40, 41> (<21> no activity in the cytoplasm [31]; <27> subcellular distribution of the enzyme in cells grown on 4-chlorophenol [25]; <13> subcellular sulindac reducing ac-

tivity distribution in calf liver [18]; <7> post-translationally occuring subcellular targeting and distribution of isozymes, overview, several forms are generated from a single translation form [40]; <25> subcellular distribution of isozymes MsrA1-3 expression in eye tissue, overview, no localization in endosomes, lysosomes, endoplasmic reticulum, and Golgi apparatus [43]; <39> subcellular lcalization of isozymes, overview [50]; <9> subcellular localization analysis, the subcellular localization is not tissue-dependent [11]; <9,35,36,41> subcellular localization is regulated by alternative splicing, overview [41]; <38,40> subcellular localization of isozymes, overview [50]) [11, 18, 25, 31, 40, 41, 43, 50]

Purification

<2> (recombinant His-tagged MsrA1, and MsrA2 from Escherichia coli by nickel affinity chromatography) [14]
<4> (recombinant His10-tagged MsrA from Escherichia coli strain BL21(DE3) by nickel affinity chromatography) [8]
<8> (native enzyme by ammonium sulfate fractionation, dialysis, anion exchange chromatography, and gel filtration) [3]
<8> (partially) [5]
<8> (recombinant His-tagged enzyme from Escherichia coli strain BL21(DE3)) [1]
<8> (recombinant MsrA) [16]
<8> (recombinant MsrA from strain B834(DE3)) [38]
<9> (partially by subcellular fractionation) [39]
<9> (recombinant His-tagged wild-type and Δ(1-22) deletion mutant from Escherichia coli strain M15 by nickel affinity chromatography) [11]
<10> (MsrA from cytosolic and mitochondrial fractions) [17]
<10> (recombinant MsrA from Escherichia coli strain BL21(DE3)) [28]
<12> (recombinant His-tagged wild-type and mutant enzymes from Escherichia coli) [5]
<13> (partially, cell fragmentation) [16]
<13> (recombinant N-terminally His-tagged wild-type and mutant MsrAs from Escherichia coli by nickel affinity chromatography and dialysis) [6]
<15> (purification of chloroplasts from wild-type and transgenic plants) [34]
<15> (recombinant plastidic isozyme pPMSR from Escherichia coli strain BL21(DE3)) [33]
<17> (recombinant wild-type and mutant MsrAs from Escherichia coli strain BE002) [27]
<19> (recombinant N-terminally 10His-tagged enzyme MsrA from Escherichia coli strain Bl21(DE3) by nickel affinity chromatography) [26]
<22> (recombinant His6-tagged MsrA from Escherichia coli strain BL21 by nickel affinity chromatography to homogeneity) [53]
<23> (recombinant enzyme from Escherichia coli) [35]
<25> (native enzyme partially from retina by subcellular fractionation) [43]
<31> (recombinant His-tagged full length tandem enzyme, MsrA, and MsrB domains from Escherichia coli, tags are removed by thrombin digestion) [32]

<34> (recombinant His-tagged MsrA from Escherichia coli strain BL21(DE3) by nickel affinity and ion exchange chromatography, 425fold) [13]
<39> (recombinant His-tagged isozyme PMSRA3 from Escherichia coli by nickel affinity chromatography) [50]
<39> (recombinant His-tagged isozyme PMSRA4 from Escherichia coli by nickel affinity chromatography) [50]

Crystallization

<8> (15-50 mg/ml purified recombinant MsrA in 50 mM Tris-HCl, pH 8.0, 2 mM EDTA, and 10 mM DTT, hanging drop vapour diffusion method, droplet size is 0.004-0.008 ml, equal volumes of protein and precipitant solution, X-ray diffraction structure determination and analysis at 1.9 A resolution) [38]
<19> (single crystals of recombinant N-terminally 10His-tagged enzyme MsrA complexed with protein-bound methionine, hanging drop method, 30 mg/ml protein in 25 mM Tris-HCl, pH 8.0, 1 mM EDTA, 1 mM tris(carboxyethyl)phosphine hydrochloride, precipitant solution contains 2.0 M sodium formate, 0.1 M sodium citrate, pH 6.0, 4°C, 1 week, prior to data collection, crystals are soaked in 6.3 M sodium formate, 0.1 M sodium citrate, pH 6.0, for 2 min, and are flash-cooled, X-ray diffraction structure determination and analysis at 1.5 A resolution, polycrystalline clusters are obtained by sitting drop vapor diffusion method) [26]

Cloning

<2> (3 genes msrA and 1 gene msrB form an operon, one of the 3 msrA genes is fused to the msrB gene, genetic organization and regulation, overview) [20]
<2> (complementation of the msrA1/msrA2 double mutant with the msr1 construct, promotor exchange between msrA1 and msrA2) [29]
<2> (mrsA1, and msrA2 are transcribed as polycistronic transcript, overexpression of MsrA1, and MsrA2 as His-tagged proteins in Escherichia coliBL21(DE3)) [14]
<4> (overexpression of His10-tagged MsrA in Escherichia coli strain BL21(DE3)) [8]
<5> (the chromosome contains 2 copies of gene msrA, a plasmid harbors 1 copy of gene msrA) [20]
<6> (genes msrA and msrB, EC 1.8.4.12, form an operon) [20]
<7> (msrA, DNA and amino acid sequence determination and analysis, expression of C-terminally His-tagged or GFP-tagged wild-type or truncated MsrA in Escherichia coli strain BL21(DE3), expression of MsrA isozymes in Saccharomyces cerevisiae, subcellular localization of the recombinant enzymes in cytosol and mitochondria, overview) [40]
<7> (overexpression of MsrA leads to increased resistance to reactive oxygen species) [18]
<8> (gene msrA) [12]
<8> (gene msrA, DNA and amino acid sequence determination and analysis, recombinant expression, functional overexpression of MsrB from gene msrB or yeaA) [18]

<8> (gene msrA, expression in Escherichia coli strain BL21(DE3) as His-tagged enzyme) [1]
<8> (gene msrA, expression in an msrA-deficient Escherichia coli mutant strain Tn903::msrA conferring resistance against oxidative damage from reactive nitrogen intermediates) [7]
<8> (gene msrA, expression in strain B834(DE3)) [38]
<8> (gene msrA, located in the chromosome at 95.69 min, respectively, recombinant expression of msrA, regulation mechanism of gene expression, overview) [20]
<9> (expression analysis in suncellular fractions of melanocytes, overview) [39]
<9> (expression of GFP-tagged wild-type MsrA and several GFP-tagged deletion mutants in several mammalian cell lines, overview, expression of His-tagged wild-type and Δ(1-22) deletion mutant in Escherichia coli strain M15) [11]
<9> (gene msrA, DNA and amino acid sequence determination and analysis, expression analysis, co-expression of A-type potassium channel ShC/B and MsrA in Xenopus laevis oocytes significantly accelerating the inactivation of the channel protein, functional expression of MsrA as GST-fusion protein in Escherichia coli strain BL21) [44]
<9> (genomic structure, alternative splicing variants, overview) [41]
<9> (overexpression of MsrA in T-lymphocytes and PC12 cells leads to increased resistance of the cells to reactive oxygen species and apoptotic death) [18]
<9> (overexpression of msrA gene in HLE cells protects against oxidative stress, while silencing of the gene by short interfering RNA-targeted gene silencing method renders the lens epithelial cells more sensitive to oxidative stress damage) [37]
<10> (expression of MsrA in Escherichia coli strain BL21(DE3)) [28]
<10> (mitochondrial and cytosolic isozymes are encoded by a single gene) [17]
<10> (mitochondrial and cytosolic isozymes are encoded on a single gene with 2 initiations sites, delivering an N-terminal signal peptide to the mitochondrial enzyme form) [23]
<12> (expression of His-tagged wild-type and mutant enzymes in Escherichia coli) [5]
<12> (gene msrA, msrA promoter and calcium phopsholipid binding protein, CPBP, form a complex and enhance msrA expression in absence of presence of H_2O_2, cytosolic thioredoxins 1 and 2 are involved in the regulation, overview) [45]
<12> (gene msrA, subcloning in Escherichia coli, stable functional overexpression of MsrA in Saccharomyces cerevisiae and human T cells) [52]
<12> (overexpression of MsrA in a yeast strain, expression of MsrA as N-terminally 6His-tagged protein in Escherichia coli strain BL-21) [36]
<13> (functional overexpression of EGFP-tagged MSRA in PC-12 cells using adenovirus-mediated gene transfer protecting the cells against apoptosis caused by hyperoxia, overview) [51]

<13> (gene bmsrA, expression of GFP-tagged enzyme in syncytial blastoderm-stage embryos of Drosophila melanogaster) [9]
<13> (gene msrA, DNA and amino acid sequence determination and analysis, recombinant expression, overexpression of MsrA leads to increased resistance to reactive oxygen species) [18]
<13> (overexpression of N-terminally His-tagged wild-type and mutant MsrAs in Escherichia coli) [6]
<15> (expression of the plastidic isozyme pPMSR in Escherichia coli strain BL21(DE3) without the chloroplast signal sequence) [33]
<16> (gene msr or pilB, DNA sequence determination and analysis) [18]
<16> (gene pilB, expression in Escherichia coli) [4]
<16> (genes msrA and msrB are translationally fused) [20]
<17> (expression of wild-type and mutant MsrAs in Escherichia coli strain BE002) [27]
<17> (genes msrA and msrB, methionine S-oxide reductase (R-form oxidizing), are translationally fused) [20]
<17> (overexpression of wild-type and mutant enzymes in Escherichia coli) [10]
<18> (gene msrA, expression in Escherichia coli as GST-fusion protein) [4]
<19> (gene msrA, DNA and amino acid sequence determination and analysis, expression in an msrA-deficient Escherichia coli mutant strain Tn903::msrA conferring resistance against oxidative damage from reactive nitrogen intermediates) [7]
<19> (overexpression of N-terminally 10His-tagged enzyme MsrA in Escherichia coli strain Bl21(DE3)) [26]
<21> (msr gene, DNA sequence determination and analysis, subcloning in Escherichia coli strain DH5-α, functional complementation of the enzyme-deficient mutant with the wild-type gene) [31]
<22> (gene msrA, DNA and amino acid sequence determination and analysis, expression of His6-tagged MsrA in Escherichia coli strain BL21) [53]
<23> (DNA and amino acid sequence determination and analysis, expression in Escherichia coli) [35]
<25> (gene msrA, genetic structure, DNA and amino acid sequence determination and analysis, expression of isozymes MsrA1-3 in ARPE cells in the cytosol and in mitochondria) [43]
<26> (chromosome 1 contains 1 gene msrA, chromosome 2 contains 1 gene msrA) [20]
<30> (gene msrA, DNA and amino acid sequence determination and analysis) [49]
<31> (overexpression of the full length tandem enzyme, the MsrA, and the MsrB domains, all His-tagged, in Escherichia coli) [32]
<34> (gene msrA, expression in Escherichia coli strain BL21(DE3) as N-terminally His-tagged enzyme, the HIs-tag does not influence enzyme activity) [13]
<35> (genomic structure, alternative splicing variants, overview) [41]
<36> (genomic structure, alternative splicing variants, overview) [41]

<37> (gene msrA, DNA and amino acid sequence determination and analysis, determination of expression pattern) [47]
<38> (locus At5g07460, isozyme PMSRA2, DNA and amino acid sequence determination and analysis, phylogenetic tree, prediction of cis-elements in the promoter, overview) [50]
<39> (locus At4g25130, isozyme PMSRA4, DNA and amino acid sequence determination and analysis, phylogenetic tree, expression of His-tagged isozyme PMSRA4 in Escherichia coli, prediction of cis-elements in the promoter, overview) [50]
<39> (locus At5g07470, isozyme PMSRA3, DNA and amino acid sequence determination and analysis, phylogenetic tree, expression of His-tagged isozyme PMSRA3 in Escherichia coli, prediction of cis-elements in the promoter, overview) [50]
<39> (locus At5g61640, isozyme PMSRA1, DNA and amino acid sequence determination and analysis, phylogenetic tree, prediction of cis-elements in the promoter, overview) [50]
<40> (locus At2g18030, isozyme PMSRA5, DNA and amino acid sequence determination and analysis, phylogenetic tree, expression in Escherichia coli strain BL21(DE3), prediction of cis-elements in the promoter, overview) [50]
<41> (genomic structure, alternative splicing variants, overview, expression of GFP-tagged and/or His-tagged mitochondrial and cytosolic MsrAs in CV-1 cells) [41]

Engineering

C107S <13> (<13> site-directed mutagenesis, the mutant shows 14% increased activity with DTT and 4% with thioredoxin compared to the wild-type enzyme [6]) [6]
C107S/C218S <13> (<13> site-directed mutagenesis, the mutant shows 78% reduced activity with DTT and 94% with thioredoxin compared to the wild-type enzyme [6]) [6]
C107S/C218S/C$_{22}$7S <13> (<13> site-directed mutagenesis, the mutant shows 61% reduced activity with DTT and 92% with thioredoxin compared to the wild-type enzyme [6]) [6]
C107S/C227S <13> (<13> site-directed mutagenesis, the mutant shows 4% reduced activity with DTT and 86% with thioredoxin compared to the wild-type enzyme [6]) [6]
C151A <15> (<15> site-directed mutagenesis, activity with Hsp21 is similar to the wild-type enzyme [33]) [33]
C198S <8, 17> (<8> MsrA mutant, mutation of one recycling Cys to Ser results in an enzyme forming methionine but without recycling activity, probably due to formation of a nonproductive complex between sulfenic intermediate and thioredoxin [21]; <17> site-directed mutagenesis, altered kinetics compared to the wild-type enzyme [27]; <17> site-directed mutagenesis, altered kinetics and disulfide bond formation compared to the wild-type enzyme [27]) [21, 27]
C206S <17> (<17> site-directed mutagenesis, MsrA domain of PILB, inactive mutant [10]) [10]

C218S <13> (<13> site-directed mutagenesis, the mutant shows 65% reduced activity with DTT and 78% with thioredoxin compared to the wild-type enzyme [6]) [6]
C218S/C227S <13> (<13> site-directed mutagenesis, the mutant shows 58% reduced activity with DTT and 96% with thioredoxin compared to the wild-type enzyme [6]) [6]
C227S <13> (<13> site-directed mutagenesis, the mutant shows 11% reduced activity with DTT and 81% with thioredoxin compared to the wild-type enzyme [6]) [6]
C25S <12> (<12> site-directed mutagenesis, inactive mutant [5]) [5]
C348S <17> (<17> site-directed mutagenesis, MsrA domain of PILB, mutant is inactive with thioredoxin, but about 10fold more active than the wild-type enzyme MsrA domain [10]) [10]
C52S <8> (<8> site-directed mutagenesis, inactive mutant, no protection of the cell against reactive oxygen species [18]; <8> site-directed mutagenesis, reduced activity compared to the wild-type, no complementation of a msrA knockout mutant [12]) [12, 18]
C72S <13> (<13> site-directed mutagenesis, inactive mutant, no disulfide bond in the mutant enzyme [6]) [6]
C72S/C107S/C227S <13> (<13> site-directed mutagenesis, inactuve mutant [6]) [6]
C72S/C218S <13> (<13> site-directed mutagenesis, inactive mutant [6]) [6]
F26A <12> (<12> site-directed mutagenesis, inactive mutant [5]) [5]
F26H <12> (<12> site-directed mutagenesis, inactive mutant [5]) [5]
G24A <12> (<12> site-directed mutagenesis, 60% reduced activity compared to the wild-type enzyme [5]) [5]
G28A <12> (<12> site-directed mutagenesis, 81% reduced activity compared to the wild-type enzyme [5]) [5]
W27A <12> (<12> site-directed mutagenesis, inactive mutant [5]) [5]
W35F <17> (<17> site-directed mutagenesis, altered kinetics and disulfide bond formation compared to the wild-type enzyme [27]) [27]
W53F <17> (<17> site-directed mutagenesis, altered kinetics compared to the wild-type enzyme [27]) [27]
Additional information <1, 2, 3, 4, 7, 8, 12, 13, 14, 15, 16, 17, 18, 19, 21, 22, 29, 30> (<12> a knockout MsrA mutant strain is sensitive to reactive oxygen species [18]; <18> a knockout MsrA mutant strain is sensitive to reactive oxygen species and is 60% reduced binding to host lung cells [18]; <3, 16, 17, 21, 29> a knockout MsrA mutant strain is sensitive to reactive oxygen species and shows decreased adherence to host cells [18]; <8> a knockout MsrA mutant strain is sensitive to reactive oxygen species and shows decreased adherence to host cells, construction of a msrA/msrB double mutant for detection of additional enzyme form activities [18]; <22> a knockout MsrA mutant strain is sensitive to reactive oxygen species and shows defective interaction with plant host cells [18]; <18> a knockout mutant shows reduced ability to attach to host lung and vein epithelial cells [12]; <4> a msrA mutant exhibited reduced adherence to erythrocytes as well as increased sensitivity to H_2O_2 and tert-butyl hydroperoxide killing, the mutant

is not able to survive in hamster lungs [20]; <14> construction of a msrA mutant which shows drastically reduced ability to survive in macrophages compared to the wild-type strain, the mutant shows increased sensitivity to cumene hydroperoxide and tert-butyl hydroperoxide, but not to H_2O_2, paraquat, or sodium nitrite [20]; <7> construction of a MsrA null mutant which exhibits a neurological disorder in the form of ataxia, is more sensitive to oxidative stress, and has by about 40% shorter life span than the wild-type mice at normal oxygen conditions and 10% at hyperoxic conditions [19]; <8> construction of a MsrA/MsrB double mutant [15]; <2> construction of a msrA2 knockout mutant, a msrA1 knockout mutant, and a msrA1/msrA2 double knockout mutant, the $msrA_2$ mutant strain and the msrA1/msrA2 double mutant strain show H_2O_2 tolerance like the wild-type parent strain, but the double mutant strain complemented by $msrA_2$ is H_2O_2 susceptible, effects of H_2O_2 or oxidative stress are related to the promotors, overview [29]; <21> construction of an enzyme-deficient mutant strain which shows diminished growth in presence of chemical oxidants with rapid loss of viability compared to the wild-type strain, activity can be recovered by complementation with the wild-type gene, study of oxidative stress resistance and colonization activity [31]; <8> construction of knockout mutants which show higher sensitivity to hydrogen peroxide compared to wild-type cells which can be compensated by complementation with the wild-type msrA gene from either Escherichia coli or Mycobacterium tuberculosis, but a mutant C52S msrA gene cannot restore activity in the knockout mutant strain, mutants show reduced type 1 fimbriae-mediated mannose-dependent agglutination of erythrocytes [12]; <22> construction of several mutants with reduced virulence via transposon mutagenesis involving mutation of msrA, mutation of msrA leads o increased sensitivity to oxidative agents, non-motility, and reduced spreading-out and life-span of the pathogen in plants [20]; <15> construction of several transgenic plant lines with altered expression level of the plastidic isozyme PMSR4 of 40-600% compared to wild-type expression level, which results in no phenotype under optimal growing conditions, but at oxidative stress conditions differences in the photosynthesis rate, and the rate of oxidized methionine residues in the chloroplast occur, overexpressing plants are more resistant to oxidative stress, while antisense plants show increased sensitivity, expression analysis [34]; <2,8,12> H_2O_2 shortens the life span of cells in constructed null mutants [19]; <7> knockout mutants show shortened life span and have neurological lesions [18]; <19> msrA gene disruption leads to loss of pathogenicity of the bacterium due to loss of ability to colonize host cells in humans [26]; <16> mutant strain produce a truncated version of fused MsrA/MsrB with increased sensitivity to H_2O_2 and siperoxide anions [20]; <30> mutation of gene msrA has no effect on virulence, and on resistance to oxidative agents, and causes no defect in cell envelope, msrA is probably linked to biofilm formation [20]; <8> mutation of msrA results in reduced development of mature biofilm [20]; <8> mutation of the 2 recycling Cys to Ser results in an enzyme forming methionine but without recycling activity, while exchange of the catalytic Cys for Ser causes complete loss of activity [21]; <2> mutation of the msr genes impair virulence, overview, mutation of the msrA1 operon leads to increased susceptibility to H_2O_2 [20]; <17> mutation of the recycling Cys to Ser results in an enzyme forming methionine but without recycling activity, while exchange of the catalytic Cys for Ser causes complete loss of activity [21]; <1> overexpression of MsrA leads to extended life span of the flies up to 70%, the resistance against paraquat-induced oxidative stress is increased [23]; <1> transgenic flies overexpressing MsrA show increased extended life span, with extended time of physical and sexual activity, and increased resistance to paraquat [18]; <2,8> bacterial cells lacking MsrA show increased sensitivity to oxidative damage, a shortened lifespan under hyperoxic conditions, and methionine-(R)-S-oxide accumulation [42]; <7> construction of the truncated MsrA Δ(1-46) mutant [40]; <13> expression of GFP-tagged MsrA in syncytial blastoderm-stage embryos of Drosophila melanogaster leads to a phenotype with extended lifespan of the mutant fruit flies, resistance to paraquat-induced oxidative stress, and to deleyed senescence-induced decline in general activity and reproductive capacity, overview [9]; <13> free sulfhydryl content and disulfide bond numbers in wild-type and mutant enzymes, overview [6]; <7> generation of MsrA-deficient mutant mice, the mutant mice show tip-toe-walking, reduced lifespan under hyperoxic conditions and increased sensitivity to oxidative stress compared to wild-type mice, phenotype overview [2]; <4> msrA disruption mutants, constructed by insertion mutagenesis, show highly increased sensitivity to H_2O_2 and reduced capability to adhere to host cell surfaces for infection, overview [8]; <7> msrA gene disruption leads to a shortened life span both under normoxic and hyperoxic conditions, MsrA null mutant mice shows greater sensitivity to hyperoxic conditions compared to wild-type mice, construction of MsrA overexpressing strains, phenotypes, overview [42]; <30> msrA gene inactivation by chromosomal insertion via allele replacement mutagenesis does not lead to increased sensitivity against oxidative stress but to complete loss of enzymatic activity with synthetic substrates [49]; <8> mutants defective in the pilA-pilB locus show affected ligand binding and impaired hemagglutination, the mutants ability to bind eukaryotic receptors is altered, overview [4]; <18> mutation of msrA affects the adherence of pneumococci to epithelial cells, overview [4]; <16> mutation of pilB affects the adherence of gonococci to epithelial cells, overview [4]; <12> overexpression of peptide-methionine sulfoxide reductase in Saccharomyces cerevisiae, resulting in a strain with 25fold increased activity, and MOLT-4 human lymphocyte cells provides them with high resistance to oxidative stress by H_2O_2, paraquat, or 2,2-azobis-(2-amidinopropane) dihydrochloride, the recombinant yeast strain grows better and and shows a higher survival rate compared to the wild-type parent strain [52]; <12> thioredoxin-deficient yeast strains show increased sensitivity to H_2O_2 [45]; <12> yeast cells lacking MsrA show increased sensitivity to oxidative damage, a shortened lifespan under hyperoxic conditions, and methionine-(S)-S-oxide accumulation [42]) [2, 4, 6, 8, 9, 12, 15, 18, 19, 20, 21, 23, 26, 29, 31, 34, 40, 42, 45, 49, 52]

Application

biotechnology <15> (<15> enzyme is a target for modification of redox-dependent regulation [33]) [33]

synthesis <8, 13> (<8,13> enzyme can be useful in the development and action of anti-cancer and anti-inflammation drugs [18]) [18]

References

[1] Grimaud, R.; Ezraty, B.; Mitchell, J.K.; Lafitte, D.; Briand, C.; Derrick, P.J.; Barras, F.: Repair of oxidized proteins: identification of a new methionine sulfoxide reductase. J. Biol. Chem., **276**, 48915-48920 (2001)

[2] Moskovitz, J.; Bar-Noy, S.; Williams, W.M.; Requena, J.; Berlett, B.S.; Stadtman, E.R.: Methionine sulfoxide reductase (MsrA) is a regulator of antioxidant defense and lifespan in mammals. Proc. Natl. Acad. Sci. USA, **98**, 12920-12925 (2001)

[3] Brot, N.; Weissbach, L.; Werth, J.; Weissbach, H.: Enzymatic reduction of protein-bound methionine sulfoxide. Proc. Natl. Acad. Sci. USA, **78**, 2155-2158 (1981)

[4] Wizemann, T.M.; Moskovitz, J.; Pearce, B.J.; Cundell, D.; Arvidson, C.G.; So, M.; Weissbach, H.; Brot, N.; Masure, H.R.: Peptide methionine-sulfoxide reductase contributes to the maintenance of adhesions in three major pathogens. Proc. Natl. Acad. Sci. USA, **93**, 7985-7990 (1996)

[5] Moskovitz, J.; Poston, J.M.; Berlett, B.S.; Nosworthy, N.J.; Szczepanowski, R.; Stadtman, E.R.: Identification and characterization of a putative active site for peptide methionine sulfoxide reductase (MsrA) and its substrate stereospecificity. J. Biol. Chem., **275**, 14167-14172 (2000)

[6] Lowther, W.T.; Brot, N.; Weissbach, H.; Honek, J.F.; Matthews, B.W.: Thiol-disulfide exchange is involved in the catalytic mechanism of peptide methionine sulfoxide reductase. Proc. Natl. Acad. Sci. USA, **97**, 6463-6468 (2000)

[7] St. John, G.; Brot, N.; Ruan, J.; Erdjument-Bromage, H.; Tempst, P.; Weissbach, H.; Nathan, C.: Peptide methionine sulfoxide reductase from Escherichia coli and Mycobacterium tuberculosis protects bacteria against oxidative damage from reactive nitrogen intermediates. Proc. Natl. Acad. Sci. USA, **98**, 9901-9906 (2001)

[8] Dhandayuthapani, S.; Blaylock, M.W.; Bebear, C.M.; Rasmussen, W.G.; Baseman, J.B.: Peptide methionine sulfoxide reductase (MsrA) is a virulence determinant in Mycoplasma genitalium. J. Bacteriol., **183**, 5645-5650 (2001)

[9] Ruan, H.; Tang, X.D.; Chen, M.L.; Joiner, M.A.; Sun, G.; Brot, N.; Weissbach, H.; Heinemann, S.H.; Iverson, L.; Wu, C.F.; Hoshi, T.: High-quality life extension by the enzyme peptide methionine sulfoxide reductase. Proc. Natl. Acad. Sci. USA, **99**, 2748-2753 (2002)

[10] Olry, A.; Boschi-Muller, S.; Marraud, M.; Sanglier-Cianferani, S.; Van Dorsselear, A.; Branlant, G.: Characterization of the methionine sulfoxide reduc-

tase activities of PILB, a probable virulence factor from Neisseria meningitidis. J. Biol. Chem., **277**, 12016-12022 (2002)
[11] Hansel, A.; Kuschel, L.; Hehl, S.; Lemke, C.; Agricola, H.J.; Hoshi, T.; Heinemann, S.H.: Mitochondrial targeting of the human peptide methionine sulfoxide reductase (MSRA), an enzyme involved in the repair of oxidized proteins. FASEB J., **16**, 911-913 (2002)
[12] Weissbach, H.; Etienne, F.; Hoshi, T.; Heinemann, S.H.; Lowther, W.T.; Matthews, B.; St John, G.; Nathan, C.; Brot, N.: Peptide methionine sulfoxide reductase: structure, mechanism of action, and biological function. Arch. Biochem. Biophys., **397**, 172-178 (2002)
[13] Lee, B.C.; Lee, Y.K.; Lee, H.J.; Stadtman, E.R.; Lee, K.H.; Chung, N.: Cloning and characterization of antioxidant enzyme methionine sulfoxide-S-reductase from Caenorhabditis elegans. Arch. Biochem. Biophys., **434**, 275-281 (2005)
[14] Moskovitz, J.; Singh, V.K.; Requena, J.; Wilkinson, B.J.; Jayaswal, R.K.; Stadtman, E.R.: Purification and characterization of methionine sulfoxide reductases from mouse and Staphylococcus aureus and their substrate stereospecificity. Biochem. Biophys. Res. Commun., **290**, 62-65 (2002)
[15] Spector, D.; Etienne, F.; Brot, N.; Weissbach, H.: New membrane-associated and soluble peptide methionine sulfoxide reductases in Escherichia coli. Biochem. Biophys. Res. Commun., **302**, 284-289 (2003)
[16] Etienne, F.; Resnick, L.; Sagher, D.; Brot, N.; Weissbach, H.: Reduction of Sulindac to its active metabolite, sulindac sulfide: assay and role of the methionine sulfoxide reductase system. Biochem. Biophys. Res. Commun., **312**, 1005-1010 (2003)
[17] Vougier, S.; Mary, J.; Friguet, B.: Subcellular localization of methionine sulphoxide reductase A (MsrA): evidence for mitochondrial and cytosolic isoforms in rat liver cells. Biochem. J., **373**, 531-537 (2003)
[18] Weissbach, H.; Resnick, L.; Brot, N.: Methionine sulfoxide reductases: history and cellular role in protecting against oxidative damage. Biochim. Biophys. Acta, **1703**, 203-212 (2005)
[19] Moskovitz, J.: Methionine sulfoxide reductases: ubiquitous enzymes involved in antioxidant defense, protein regulation, and prevention of aging-associated diseases. Biochim. Biophys. Acta, **1703**, 213-219 (2005)
[20] Ezraty, B.; Aussel, L.; Barras, F.: Methionine sulfoxide reductases in prokaryotes. Biochim. Biophys. Acta, **1703**, 221-229 (2005)
[21] Boschi-Muller, S.; Olry, A.; Antoine, M.; Branlant, G.: The enzymology and biochemistry of methionine sulfoxide reductases. Biochim. Biophys. Acta, **1703**, 231-238 (2005)
[22] Kauffmann, B.; Aubry, A.; Favier, F.: The three-dimensional structures of peptide methionine sulfoxide reductases: current knowledge and open questions. Biochim. Biophys. Acta, **1703**, 249-260 (2005)
[23] Petropoulos, I.; Friguet, B.: Protein maintenance in aging and replicative senescence: a role for the peptide methionine sulfoxide reductases. Biochim. Biophys. Acta, **1703**, 261-266 (2005)
[24] Picot, C.R.; Perichon, M.; Cintrat, J.C.; Friguet, B.; Petropoulos, I.: The peptide methionine sulfoxide reductases, MsrA and MsrB (hCBS-1), are down-

regulated during replicative senescence of human WI¯38 fibroblasts. FEBS Lett., **558**, 74-78 (2004)
[25] Tamburro, A.; Robuffo, I.; Heipieper, H.J.; Allocati, N.; Rotilio, D.; Di Ilio, C.; Favaloro, B.: Expression of glutathione S-transferase and peptide methionine sulphoxide reductase in Ochrobactrum anthropi is correlated to the production of reactive oxygen species caused by aromatic substrates. FEMS Microbiol. Lett., **241**, 151-156 (2004)
[26] Taylor, A.B.; Benglis, D.M., Jr.; Dhandayuthapani, S.; Hart, P.J.: Structure of Mycobacterium tuberculosis methionine sulfoxide reductase A in complex with protein-bound methionine. J. Bacteriol., **185**, 4119-4126 (2003)
[27] Antoine, M.; Boschi-Muller, S.; Branlant, G.: Kinetic characterization of the chemical steps involved in the catalytic mechanism of methionine sulfoxide reductase A from Neisseria meningitidis. J. Biol. Chem., **278**, 45352-45357 (2003)
[28] Vougier, S.; Mary, J.; Dautin, N.; Vinh, J.; Friguet, B., Ladant, D.: Essential role of methionine residues in calmodulin binding to Bordetelle pertussis adenylate cyclase, as probed by selective oxidation and repair by the peptide methionine sulfoxide reductases. J. Biol. Chem., **279**, 30210-30218 (2004)
[29] Singh, V.K.; Moskovitz, J.: Multiple methionine sulfoxide reductase genes in Staphylococcus aureus: expression of activity and roles in tolerance of oxidative stress. Microbiology, **149**, 2739-2747 (2003)
[30] Stadtman, E.R.; Moskovitz, J.; Berlett, B.S.; Levine, R.L.: Cyclic axidation and reduction of protein methionine residues is an important antioxidant mechanism. Mol. Cell. Biochem., **2347235**, 3-9 (2002)
[31] Alamuri, P.; Maier, R.J.: Methionine sulphoxide reductase is an important antioxidant enzyme in the gastric pathogen Helicobacter pylori. Mol. Microbiol., **53**, 1397-1406 (2004)
[32] Lowther, W.T.; Weissbach, H.; Etienne, F.; Brot, N.; Matthews, B.W.: The mirrored methionine sulfoxide reductases of Neisseria gonorrhoeae pilB. Nat. Struct. Biol., **9**, 348-352 (2002)
[33] Gustavsson, N.; Kokke, B.P.; Harndahl, U.; Silow, M.; Bechtold, U.; Poghosyan, Z.; Murphy, D.; Boelens, W.C.; Sundby, C.: A peptide methionine sulfoxide reductase highly expressed in photosynthetic tissue in Arabidopsis thaliana can protect the chaperone-like activity of a chloroplast-localized small heat shock protein. Plant J., **29**, 545-553 (2002)
[34] Romero, H.M.; Berlett, B.S.; Jensen, P.J.; Pell, E.J.; Tien, M.: Investigations into the role of the plastidial peptide methionine sulfoxide reductase in response to oxidative stress in Arabidopsis. Plant Physiol., **136**, 3784-3794 (2004)
[35] In, O.; Berberich, T.; Romdhane, S.; Feierabend, J.: Changes in gene expression during dehardening of cold-hardened winter rye (Secale cereale L.) leaves and potential role of a peptide methionine sulfoxide reductase in cold-acclimation. Planta, **220**, 941-950 (2005)
[36] Koc, A.; Gasch, A.P.; Rutherford, J.C.; Kim, H.Y.; Gladyshev, V.N.: Methionine sulfoxide reductase regulation of yeast lifespan reveals reactive oxygen

species-dependent and -independent components of aging. Proc. Natl. Acad. Sci. USA, **101**, 7999-8004 (2004)

[37] Kantorow, M.; Hawse, J.R.; Cowell, T.L.; Benhamed, S.; Pizarro, G.O.; Reddy, V.N.; Hejtmancik, J.F.: Methionine sulfoxide reductase A is important for lens cell viability and resistance to oxidative stress. Proc. Natl. Acad. Sci. USA, **101**, 9654-9659 (2004)

[38] Tete-Favier, F.; Cobessi, D.; Leonard, G.A.; Azza, S.; Talfournier, F.; Boschi-Muller, S.; Branlant, G.; Aubry, A.: Crystallization and preliminary X-ray diffraction studies of the peptide methionine sulfoxide reductase from Escherichia coli. Acta Crystallogr. Sect. D, **56**, 1194-1197 (2000)

[39] Schallreuter, K.U.; Rubsam, K.; Chavan, B.; Zothner, C.; Gillbro, J.M.; Spencer, J.D.; Wood, J.M.: Functioning methionine sulfoxide reductases A and B are present in human epidermal melanocytes in the cytosol and in the nucleus. Biochem. Biophys. Res. Commun., **342**, 145-152 (2006)

[40] Kim, H.Y.; Gladyshev, V.N.: Role of structural and functional elements of mouse methionine-S-sulfoxide reductase in its subcellular distribution. Biochemistry, **44**, 8059-8067 (2005)

[41] Kim, H.-Y.; Gladyshev, V.N.: Alternative first exon splicing regulates subcellular distribution of methionine sulfoxide reductases. BMC Mol. Biol., **7**, 11 (2006)

[42] Moskovitz, J.: Roles of methionine suldfoxide reductases in antioxidant defense, protein regulation and survival. Curr. Pharm. Des., **11**, 1451-1457 (2005)

[43] Lee, J.W.; Gordiyenko, N.V.; Marchetti, M.; Tserentsoodol, N.; Sagher, D.; Alam, S.; Weissbach, H.; Kantorow, M.; Rodriguez, I.R.: Gene structure, localization and role in oxidative stress of methionine sulfoxide reductase A (MSRA) in the monkey retina. Exp. Eye Res., **82**, 816-827 (2006)

[44] Kuschel, L.; Hansel, A.; Schonherr, R.; Weissbach, H.; Brot, N.; Hoshi, T.; Heinemann, S.H.: Molecular cloning and functional expression of a human peptide methionine sulfoxide reductase (hMsrA). FEBS Lett., **456**, 17-21 (1999)

[45] Hanbauer, I.; Moskovitz, J.: The yeast cytosolic thioredoxins are involved in the regulation of methionine sulfoxide reductase A. Free Radic. Biol. Med., **40**, 1391-1396 (2006)

[46] Beraud, S.; Bersch, B.; Brutscher, B.; Gans, P.; Barras, F.; Blackledge, M.: Direct structure determination using residual dipolar couplings: reaction-site conformation of methionine sulfoxide reductase in solution. J. Am. Chem. Soc., **124**, 13709-13715 (2002)

[47] Vattanaviboon, P.; Seeanukun, C.; Whangsuk, W.; Utamapongchai, S.; Mongkolsuk, S.: Important role for methionine sulfoxide reductase in the oxidative stress response of Xanthomonas campestris pv. phaseoli. J. Bacteriol., **187**, 5831-5836 (2005)

[48] Ogawa, F.; Sander, C.S.; Hansel, A.; Oehrl, W.; Kasperczyk, H.; Elsner, P.; Shimizu, K.; Heinemann, S.H.; Thiele, J.J.: The repair enzyme peptide methionine-S-sulfoxide reductase is expressed in human epidermis and upregulated by UVA radiation. J. Invest. Dermatol., **126**, 1128-1134 (2006)

[49] Mintz, K.P.; Moskovitz, J.; Wu, H.; Fives-Taylor, P.M.: Peptide methionine sulfoxide reductase (MsrA) is not a major virulence determinant for the oral pathogen Actinobacillus actinomycetemcomitans. Microbiology, **148**, 3695-3703 (2002)
[50] Romero, H.M.; Pell, E.J.; Tien, M.: Expression profile analysis and biochemical properties of the peptide methionine sulfoxide reductase A (PMSRA) gene family in Arabidopsis. Plant Sci., **170**, 705-714 (2006)
[51] Yermolaieva, O.; Xu, R.; Schinstock, C.; Brot, N.; Weissbach, H.; Heinemann, S.H.; Hoshi, T.: Methionine sulfoxide reductase A protects neuronal cells against brief hypoxia/reoxygenation. Proc. Natl. Acad. Sci. USA, **101**, 1159-1164 (2004)
[52] Moskovitz, J.; Flescher, E.; Berlett, B.S.; Azare, J.; Poston, J.M.; Stadtman, E.R.: Overexpression of peptide-methionine sulfoxide reductase in Saccharomyces cerevisiae and human T cells provides them with high resistance to oxidative stress. Proc. Natl. Acad. Sci. USA, **95**, 14071-14075 (1998)
[53] El Hassouni, M.; Chambost, J.P.; Expert, D.; Van Gijsegem, F.; Barras, F.: The minimal gene set member msrA, encoding peptide methionine sulfoxide reductase, is a virulence determinant of the plant pathogen Erwinia chrysanthemi. Proc. Natl. Acad. Sci. USA, **96**, 887-892 (1999)
[54] Skaar, E.P.; Tobiason, D.M.; Quick, J.; Judd, R.C.; Weissbach, H.; Etienne, F.; Brot, N.; Seifert, H.S.: The outer membrane localization of the Neisseria gonorrhoeae MsrA/B is involved in survival against reactive oxygen species. Proc. Natl. Acad. Sci. USA, **99**, 10108-10113 (2002)

Peptide-methionine (R)-S-oxide reductase 1.8.4.12

1 Nomenclature

EC number

1.8.4.12

Systematic name

L-methionine:thioredoxin-disulfide S-oxidoreductase (R-form oxidizing)

Recommended name

peptide-methionine (R)-S-oxide reductase

Synonyms

CBS-1 <8, 33> [20, 25, 32]
CBS1 <8> [19]
MSR <3, 7, 17, 18, 20, 27> [5, 13, 24]
MsrA/B <16> [13]
MsrB <1, 2, 4, 5, 6, 7, 8, 9, 10, 11, 12, 13, 14, 15, 16, 17, 19, 20, 21, 23, 24, 25, 26, 28, 29, 35> [1, 4, 6, 7, 8, 9, 10, 12, 13, 14, 15, 16, 17, 18, 19, 20, 21, 23, 25, 28, 29, 30, 31, 32, 33, 34, 35, 36, 37, 38, 39]
PMSR <14, 22> [26, 27]
PilB <16, 17, 29> [4, 5, 13, 16, 25, 33, 38, 39]
Sel-X <6, 8, 11, 12> [13]
SelR <6, 7, 32> [1, 2]
YeaA <7, 34> [7, 25]
cysteine-containing methionine-R-sulfoxide reductase <6, 8> [35]
methionine sulfoxide reductase <2, 3, 4, 5, 6, 7, 8, 10, 11, 12, 13, 14, 16, 17, 18, 19, 20, 21, 23, 25, 27, 28, 29, 33, 34> [1, 3, 4, 8, 11, 13, 14, 15, 16, 22, 23, 25, 28]
methionine sulfoxide reductase B <6, 8, 11, 14, 17, 26, 29> [6, 9, 12, 25, 30, 31, 32, 34, 35]
methionine sulphoxide reductase <20> [24]
methionine-R-sulfoxide reductase <6, 32> [2, 29]
methionine-R-sulfoxide reductase B <7, 17> [39]
peptide methionine sulfoxide reductase <1, 7, 8, 9, 11, 14, 17, 22> [5, 7, 10, 18, 20, 21, 26, 27]
peptide methionine sulfoxide reductase type B <1, 7, 8, 15, 16, 17, 24, 35> [17, 19]
selenocysteine-containing methionine-R-sulfoxide reductase <6, 8> [35]
selenoprotein R <6, 8, 32> [2, 32]
sulindac reductase <12> [11]
Additional information <6, 8, 11, 14> (<6,8,11,14> the enzyme belongs to the Msr family of enzymes [31]) [31]

2 Source Organism

<1> *Drosophila melanogaster* (no sequence specified) [17, 18]
<2> *Staphylococcus aureus* (no sequence specified) [8, 14, 15, 22]
<3> *Haemophilus influenzae* (no sequence specified) [13]
<4> *Rhizobium meliloti* (no sequence specified) [15]
<5> *Bacillus subtilis* (no sequence specified) [15,16]
<6> *Mus musculus* (no sequence specified) [2, 8, 9, 13, 14, 23, 29, 31, 35]
<7> *Escherichia coli* (no sequence specified) [1, 7, 10, 11, 13, 14, 15, 16, 17, 18, 36, 39]
<8> *Homo sapiens* (no sequence specified) [3, 13, 14, 18, 19, 20, 21, 30, 31, 32, 35, 37]
<9> *Rattus norvegicus* (no sequence specified) [18]
<10> *Sus scrofa* (no sequence specified) [14]
<11> *Saccharomyces cerevisiae* (no sequence specified) [13, 14, 18, 28, 31]
<12> *Bos taurus* (no sequence specified) (<12> enzyme precursor [11]) [11, 13]
<13> *Mycobacterium smegmatis* (no sequence specified) [15]
<14> *Arabidopsis thaliana* (no sequence specified) [14, 26, 31, 34]
<15> *Pseudomonas aeruginosa* (no sequence specified) [17]
<16> *Neisseria gonorrhoeae* (no sequence specified) [13, 15, 17, 33, 38]
<17> *Neisseria meningitidis* (no sequence specified) [4, 5, 12, 13, 15, 16, 17, 39]
<18> *Streptococcus pneumoniae* (no sequence specified) [13]
<19> *Mycobacterium tuberculosis* (no sequence specified) [15]
<20> *Helicobacter pylori* (no sequence specified) [13, 15, 24]
<21> *Erwinia chrysanthemi* (no sequence specified) [15]
<22> *Secale cereale* (no sequence specified) [27]
<23> *Xanthomonas campestris* (no sequence specified) [16]
<24> *Caulobacter crescentus* (no sequence specified) [17]
<25> *Vibrio cholerae* (no sequence specified) [15]
<26> *Lactobacillus reuteri* (no sequence specified) [6]
<27> *Streptococcus gordonii* (no sequence specified) [13]
<28> *Actinobacillus actinomycetemcomitans* (no sequence specified) [15]
<29> *Neisseria gonorrhoeae* (UNIPROT accession number: P14930) [25]
<30> no activity in *Aquifex aeolicus* [15]
<31> no activity in *Thermotoga maritima* [15]
<32> *Drosophila melanogaster* (UNIPROT accession number: Q81NK9) [2]
<33> *Homo sapiens* (UNIPROT accession number: Q9Y3D2) [25]
<34> *Escherichia coli* (UNIPROT accession number: P39903) [25]
<35> *Synechococcus elongatus* (no sequence specified) [17]

3 Reaction and Specificity

Catalyzed reaction

L-methionine (R)-sulfoxide + thioredoxin = L-methionine + thioredoxin disulfide (<7, 17> 3-step ping pong reaction mechanism involving catalytic and recycling cysteine residues, formation of a sulfenic acid reaction intermediate, overview [16]; <5, 23> 3-step reaction mechanism involving catalytic and recycling cysteine residues, formation of a sulfenic acid reaction intermediate, overview [16]; <2, 4, 5, 7, 13, 16, 17, 19, 20, 21, 25, 28> catalytic mechanism involves the formation of a sulfenic acid intermediate [15]; <7, 12> catalytic mechanism involving the formation of a sulfenic acid intermediate [13]; <1, 7, 16, 17> mechanism, active site structure, conserved catalytic Cys residues are essential for activity, Cys residue recycling, overview [17]; <15, 24, 35> mechanism, active site structure, conserved catalytic Cys residues, situated in the C-terminal end, are essential for activity, Cys residue recycling, overview [17]; <17> reaction mechanism, Cys494 and Cys439 are involved [4]; <29> reaction mechanism, modeling of substrate binding at the active site, Cys444 and Cys495 are involved [25]; <6> selenomethionine is essential for MsrB activity [9]; <8> the active site selenocysteine SeC169 is essential for enzyme activity [19]; <17> three-step catalytic mechanism, influence of pH on reaction mechanism, overview [12])

peptide-L-methionine + thioredoxin disulfide + H_2O = peptide-L-methionine (R)-S-oxide + thioredoxin (<7,8> catalytic mechanism and the role of cofactor recycling in vivo [36,37]; <6,8> catalytic mechanism involving residues at positions 95, 41, 97, 77, and 80, molecular modeling, role of selenocysteine- and cysteine residues in catalysis, overview [35]; <7,17> catalytic mechanism, overview [39])

Natural substrates and products

S Hsp21 L-methionine S-oxide + dithiothreitol <14> (<14> chloroplast-localized small heat shock protein, repair function for heat shock protein Hsp21 by restoring the structure, which is crucial for cellular resistance to oxidative stress, the enzyme can protect the chaperone-like activity of Hsp21 [26]) (Reversibility: ?) [26]

P Hsp21 L-methionine + dithiothreitol S-oxide

S L-methionine (R)-sulfoxide + thioredoxin <2, 4, 5, 6, 7, 8, 10, 11, 12, 13, 14, 16, 17, 19, 20, 21, 23, 25, 26, 28, 32> (<12> enzyme form MsrB is specific for the R-form, enzyme form variants with specificities for either free or protein-bound methionine [13]; <7> enzyme is involved in repairing of oxidized methionine residues in proteins [1]; <20> important antioxidant enzyme and colonization factor in the gastric pathogen, a methionine repair enzyme responsible for stress resistance [24]; <7> membrane-bound enzyme form Mem-R,S-Msr, enzyme form MsrB is specific for the R-form, MsrB enzyme form variants with specificities for either free or protein-bound methionine, Mem-R,S-Msr also posesses MsrA activity utilizing L-methionine (S)-sulfoxide as substrate [13]; <8> Msr is specific for the R-isomer [3]; <6,32> MsrB is absolute specific for the R-form, no activity with the S-form, pathway overview [2]; <2> MsrB is specific for the (R)-form of the substrate [22]; <2, 4, 5, 6, 7, 8, 10, 11, 13, 14, 16, 17, 19, 20, 21, 25, 28> MsrB is specific for the R-form [14,15]; <6> MsrB is specific for the R-form of the substrate [23]; <5,7,23> MsrB is specific for the R-form, active on free and protein-bound methionine, the latter is bound more efficiently [16]; <6,8,11> MsrB is specific for the R-form, enzyme variants with specificities for either free or protein-bound methionine [13]; <11> MsrB specifically reduces the R-form of methionine sulfoxide [28]; <26> protein-bound methionine residues [6]; <7> substrates are peptides and proteins [7]; <6> together with the enzyme MsrA, EC 1.8.4.11, which is absolutely specific for the S-form substrate, the enzyme can repair methionine-damaged proteins and salvage free methionine under oxidative stress int the living cell [9]) (Reversibility: ir) [1, 2, 3, 6, 7, 9, 13, 14, 15, 16, 22, 23, 24, 28]

P L-methionine + thioredoxin disulfide

S L-methionine (R,S)-sulfoxide + thioredoxin <3, 16, 17, 18, 20, 27> (<16> the enzyme protect cells against oxidative damage and plays a role in age-related misfunctions [13]; <3, 17, 18, 20, 27> the enzyme protects cells against oxidative damage and plays a role in age-related misfunctions [13]) (Reversibility: ?) [13]

P L-methionine + thioredoxin disulfide

S L-methionine-(R)-S-oxide + thioredoxin <6, 7, 8, 11, 14, 17> (<7,17> stereospecific reduction [39]; <14> absolute stereospecific reduction [34]; <8> MsrB is involved in regulation of protein function and in elimination of reactive oxygen species via reversible methionine formation besides protein repair in human skin [30]; <8> stereospecific reduction, the isozymes of MsrB are involved in lens cell viability and oxidative stress protection [32]; <14> substrate in vivo is e.g. the small heat shock protein Hsp-21 which loses its chaperone-like activity upon methionine oxidation [31]; <8> substrates are HIV-2, which is inactivated by oxidation of its methionine residues M76 and M95, the potassium channel of the brain, the inhibitor IκB-α, or calmodulin, overview [31]; <7,8> the cofactor thioredoxin can be recycled in vivo by thionein due to its high content of cysteines, overview [36,37]) (Reversibility: r) [30, 31, 32, 34, 35, 36, 37, 39]

P L-methionine + thioredoxin disulfide + H_2O

S L-methionine-(R)-sulfoxide + thioredoxin <1, 7, 8, 9, 11> (<1, 7, 8, 9, 11> MsrB is specific for the R-form [18]) (Reversibility: ?) [18]

P L-methionine + thioredoxin disulfide

S calmodulin L-methionine-(R)-sulfoxide + thioredoxin <8> (<8> MsrB is specific for the R-form, enzyme provides protection against oxidative damage by reactive oxygen species and has a repair function for oxidized protein methionine residues, which restores the calmodulin binding to adenylate cyclase of the pathogen Bordetella pertussis, which is an essential step for the bacterium to enter host cells, overview [21]) (Reversibility: ?) [21]

P calmodulin L-methionine + thioredoxin disulfide

S calmodulin-L-methionine (R)-sulfoxide + thioredoxin <7> (Reversibility: ?) [7]

P calmodulin-L-methionine + thioredoxin disulfide

S protein L-methionine (R)-sulfoxide + dithiothreitol <8> (<8> type B enzyme CBS1 is stereospecific for the R-stereomer of methionine residues of peptides and proteins [19]) (Reversibility: ?) [19]

P protein L-methionine + dithiothreitol disulfide

S protein L-methionine (R)-sulfoxide + thioredoxin <8> (<8> type B enzyme CBS1 is stereospecific for the R-stereomer of methionine residues of peptides and proteins [19]) (Reversibility: ?) [19]

P protein L-methionine + thioredoxin disulfide

S protein L-methionine-(R)-sulfoxide + thioredoxin <7> (<7> MsrB is specific for the R-form, the membrane-associated isozyme reduces both R- and S-stereoisomers of methionine sulfoxide, N-acetylmethionine sulfoxide, and D-Ala-Met-enkephalin [10]) (Reversibility: ?) [10]

P protein L-methionine + thioredoxin disulfide

S sulindac + thioredoxin <7, 12> (<12> activation of a methionine sulfoxide-containing prodrug, activity with membrane-bound enzyme form Mem-R,S-Msr [13]; <7> activation of a methionine sulfoxide-containing prodrug, activity with membrane-bound enzyme form Mem-R,S-Msr and MsrA [13]) (Reversibility: ?) [13]

P sulindac sulfide + thioredoxin disulfide (<7,12> activated drug which inhibits cyclooxygenase 1 and 2 and exhibiting anti-inflammatory activity [13])

S sulindac + thioredoxin disulfide <7, 12> (<7> activation of the antiinflammatory drug with anti-tumorigenic activity, which acts via inhibition of cyclooxygenases 1 and 2 [11]) (Reversibility: ?) [11]

P sulindac sulfide + thioredoxin

S Additional information <1, 2, 4, 5, 6, 7, 8, 9, 10, 11, 12, 13, 14, 16, 17, 19, 20, 21, 22, 25, 26, 28, 32> (<7> cellular system of balancing native proteins and oxidatively damaged proteins by use of protein biosynthesis, protein oxidative modification, protein elimination, and oxidized protein repair involving the enzyme, overview, enzyme protects against oxidative damage of proteins [18]; <11> cellular system of balancing native proteins and oxidatively damaged proteins by use of protein biosynthesis, protein oxidative modification, protein elimination, and oxidized protein repair involving the enzyme, overview, enzyme protects against oxidative damage of proteins, enzyme activity is not age-related [18]; <1,8,9> cellular system of balancing native proteins and oxidatively damaged proteins by use of protein biosynthesis, protein oxidative modification, protein elimination, and oxidized protein repair involving the enzyme, overview, enzyme protects against oxidative damage of proteins, loss of enzyme activity is age-related [18]; <8> downregulation of CBS-1 during replicative senescence of cells leads to accumulation of oxidized proteins and age-related increased oxidative damage [20]; <16,17> enzyme acts on free and protein-bound methionine [15]; <7> enzyme contributes to resistance against cadmium, physiological role [1]; <14> enzyme has regulatory function in the plant cell [26]; <6,32> enzyme provides protection for the cell against oxidative stress [2]; <19> enzyme repairs oxidatively damaged free and protein bound methionine and recycles it from methionine sulfoxide, e.g. the heat shock protein and chaperone Hsp16.3, role of the MsrA/MsrB repair pathway in cellular protein dynamics [15]; <2, 4, 5, 13, 20, 21, 25> enzyme repairs oxidatively damaged free and protein bound methionine and recycles it from methionine sulfoxide, role of the MsrA/MsrB repair pathway in cellular protein dynamics [15]; <7> enzyme repairs oxidatively damaged free and protein bound methionine and recycles it from methionine sulfoxide, role of the MsrA/MsrB repair pathway in cellular protein dynamics, the MsrA/MsrB repair pathway is involved in the signal recognition particle-dependent protein targeting pathway, regulation mechanism of gene expression, overview [15]; <22> potential role of the enzyme in cold-acclimation, enzyme may protect the cells from photodamage [27]; <6> protection of the cells against reactive oxidizing species, biological consequences of methionine oxidation, physiological role, overview [23]; <14> recycling of free methionine, enzyme reverses the oxidative damage at methionine protein residues oxidized to methionine sulfoxide being a major cause of aging and age-related diseases, Msr can regulate protein function, be involved in signal transduction, and prevent accumulation of faulty proteins [14]; <8,10> recycling of free methionine, enzyme reverses the oxidative damage at methionine protein residues oxidized to methionine sulfoxide being a major cause of aging and age-related diseases, Msr can regulate protein function, be involved in signal transduction, and prevent accumulation of faulty proteins, MsrB has several different physiological repair and regulatory functions, overview [14]; <6> recycling of free methionine, enzyme reverses the oxidative damage at methionine protein residues oxidized to methionine sulfoxide being a major cause of aging and age-related diseases, Msr can regulate protein function, be involved in signal transduction, and prevent accumulation of faulty proteins, MsrB has several different physiological repair and regulatory functions, overview, oxidation of 2 essential methionine residues of HIV-2 particles can inactivate the virus and prevent infection of human cells [14]; <2,7,11> recycling of free methionine, enzyme reverses the oxidative damage at methionine protein residues oxidized to methionine sulfoxide being a major cause of aging, Msr can regulate protein function, be involved in signal transduction, and prevent accumulation of faulty proteins, MsrB has several different physiological repair and regulatory functions, overview [14]; <28> role of the MsrA/MsrB repair pathway in cellular protein dynamics, enzyme repairs oxidatively damaged free and protein bound methionine and recycles it from methionine sulfoxide [15]; <26> the enzyme contributes to the ecological performance of Lactobacillus reuteri in gastrointestinal ecosystems together with the high-molecular-mass surface protein Lsp, enzyme expression is induced in vivo [6]; <11> the enzyme is essential in protection of the cells against oxidative damage by reactive oxygen species, yeast cell life span analysis of wild-type and mutant cells, the latter either overexpress or lack enzyme activity, overview [28]; <6> the enzyme protect cells against oxidative damage and plays a role in age-related diseases [13]; <7> the enzyme protect cells against oxidative damage and plays a role in age-related missfunctions [13]; <8> the enzyme protects cells against oxidative damage and plays a role in age-related and neurological diseases, like Parkinsons or Alzheimers disease [13]; <11,12> the enzyme protects cells against oxidative damage and plays a role in age-related diseases [13]; <2> the MsrA1/MsrB system is physiologically more significant in Staphylococcus aureus than MsrA2 [22]; <16> PilB affects the survival of the organism to reactive oxygen species, PilB is not involved in piliation, pilin production, or adherence [38]; <14> roles of methionine suldfoxide reductases in antioxidant defense, protein regulation via alternating it between active and inactive form, and survival, Msr B protects cells from the cytotoxic effects of reactive oxygen species, ROS, overview [31]; <6> roles of methionine sulfoxide reductases in antioxidant defense, protein regulation via alternating it between active and inactive form, and survival, MsrB protects cells from the cytotoxic effects of reactive oxygen species, ROS, overview, enzyme involvement in protein repair and associated factors, protein regulation pathway, overview [31]; <11> roles of methionine sulfoxide reductases in antioxidant defense, protein regulation via alternating it between active and inactive form, and survival, MsrB protects cells from the cytotoxic effects of reactive oxygen species, ROS, overview, regulation of MsrB expression, overview [31]; <8> roles of methionine sulfoxide reductases in antioxidant defense, protein regulation via alternating it between active and inactive form, and survival, MsrB protects cells from the cytotoxic effects of reactive oxygen species, ROS, overview, the enzyme is involved in age-related diseases such as Alzheimers or Parkinsons diseases as well as in diseases caused by prions, mechanism, overview, enzyme involvement in protein repair and associated factors, protein regulation pathway, overview [31]; <16> The thioredoxin domain of PilB can use electrons from DsbD to reduce downstream methionine sulfoxide reductases, overview [33]) (Reversibility: ?) [1, 2, 6, 13, 14, 15, 18, 20, 22, 23, 26, 27, 28, 31, 33, 38]

P ?

Substrates and products

S (R)-methyl 4-tolyl sulfoxide + thioredoxin <8> (Reversibility: ?) [3]

P ?

S DL-methionine (R)-sulfoxide + thioredoxin <17> (<17> enzyme MsrB is specific for the R-form, active on free and protein-bound methionine, the latter is bound more efficiently [16]) (Reversibility: ?) [16]

P DL-methionine + thioredoxin disulfide

S Hsp21 L-methionine S-oxide + dithiothreitol <14> (<14> chloroplast-localized small heat shock protein, repair function for heat shock protein Hsp21 by restoring the structure, which is crucial for cellular resistance to oxidative stress, the enzyme can protect the chaperone-like activity of Hsp21 [26]; <14> Hsp21 contains 6 methionine residues at positions 49,

52, 55, 59, 62, and 67, about half of the residues are reduced by the enzyme probably due to its stereospecificity [26]) (Reversibility: ?) [26]

P Hsp21 L-methionine + dithiothreitol S-oxide

S L-methionine (R)-sulfoxide + dithiothreitol <7> (Reversibility: ?) [7]

P L-methionine + dithiothreitol disulfide

S L-methionine (R)-sulfoxide + thioredoxin <2, 4, 5, 6, 7, 8, 10, 11, 12, 13, 14, 16, 17, 19, 20, 21, 23, 25, 26, 28, 29, 32, 33, 34> (<12> enzyme form MsrB is specific for the R-form, enzyme form variants with specificities for either free or protein-bound methionine [13]; <7> enzyme is involved in repairing of oxidized methionine residues in proteins [1]; <20> important antioxidant enzyme and colonization factor in the gastric pathogen, a methionine repair enzyme responsible for stress resistance [24]; <7> membrane-bound enzyme form Mem-R,S-Msr, enzyme form MsrB is specific for the R-form, MsrB enzyme form variants with specificities for either free or protein-bound methionine, Mem-R,S-Msr also posesses MsrA activity utilizing L-methionine (S)-sulfoxide as substrate [13]; <8> Msr is specific for the R-isomer [3]; <6, 32> MsrB is absolute specific for the R-form, no activity with the S-form, pathway overview [2]; <2> MsrB is specific for the (R)-form of the substrate [22]; <2, 4, 5, 6, 7, 8, 10, 11, 13, 14, 16, 17, 19, 20, 21, 25, 28> MsrB is specific for the R-form [14, 15]; <6, 7> MsrB is specific for the R-form of the substrate [1, 23]; <5, 7, 23> MsrB is specific for the R-form, active on free and protein-bound methionine, the latter is bound more efficiently [16]; <6,8,11,12> MsrB is specific for the R-form, enzyme variants with specificities for either free or protein-bound methionine [13]; <11,17> MsrB specifically reduces the R-form of methionine sulfoxide [12,28]; <26> protein-bound methionine residues [6]; <7> substrates are peptides and proteins [7]; <6> together with the enzyme MsrA, EC 1.8.4.11, which is absolutely specific for the S-form substrate, the enzyme can repair methionine-damaged proteins and salvage free methionine under oxidative stress int the living cell [9]; <7> membrane-bound enzyme form Mem-R,S-Msr [13]; <17,29> MsrB activity of the tandem domains of PilB, the MsrB domain alone does not utilize the S-isomer [4,25]; <6,32> MsrB is absolute specific for the R-form, no activity with the S-form [2]; <33,34> MsrB is specific for the R-isomer, no activity with the S-isomer [25]; <2> PilB shows absolute specificity for the R-form of free and protein-bound methionine sulfoxide [8]; <6> the native MsrB as well as the recombinant modified MsrB show absolute specificity for the R-form of free and protein-bound methionine sulfoxide, no activity with the S-form [8,9]) (Reversibility: ir) [1, 2, 3, 4, 6, 7, 8, 9, 12, 13, 14, 15, 16, 22, 23, 24, 25, 28]

P L-methionine + thioredoxin disulfide

S L-methionine (R,S)-sulfoxide + thioredoxin <3, 16, 17, 18, 20, 27> (<16> the enzyme protect cells against oxidative damage and plays a role in age-related misfunctions [13]; <3,17,18,20,27> the enzyme protects cells against oxidative damage and plays a role in age-related misfunctions [13]; <3,16,17,18,20,27> enzyme MsrA/B shows both MsrA and MsrB activity, free and protein-bound methionine [13]) (Reversibility: ?) [13]

P L-methionine + thioredoxin disulfide

S L-methionine sulfoxide enkephalin + thioredoxin <7> (<7> membrane-bound enzyme form Mem-R,S-Msr [13]) (Reversibility: ?) [13]

P L-methionine enkephalin

S L-methionine-(R)-S-oxide + DTT <7, 8, 17> (<17> stereospecific reduction [39]) (Reversibility: ?) [36, 37, 39]

P L-methionine + DTT disulfide + H_2O

S L-methionine-(R)-S-oxide + DTT <6> (<6> isozymes MsrB1, MsrB2, and MsrB3 [35]) (Reversibility: ?) [35]

P L-methionine + thioredoxin disulfide + H_2O

S L-methionine-(R)-S-oxide + dithioerythritol <14> (<14> absolute stereospecific reduction, MsrB1 and MsrB2 [34]) (Reversibility: ?) [34]

P L-methionine + dithioerythritol disulfide + H_2O

S L-methionine-(R)-S-oxide + thioredoxin <6, 7, 8, 11, 14, 17> (<6, 7, 8, 11, 14, 17> stereospecific reduction [29,31,32,39]; <14> absolute stereospecific reduction [34]; <8> MsrB is involved in regulation of protein function and in elimination of reactive oxygen species via reversible methionine formation besides protein repair in human skin [30]; <8> stereospecific reduction, the isozymes of MsrB are involved in lens cell viability and oxidative stress protection [32]; <14> substrate in vivo is e.g. the small heat shock protein Hsp-21 which loses its chaperone-like activity upon methionine oxidation [31]; <8> substrates are HIV-2, which is inactivated by oxidation of its methionine residues M76 and M95, the potassium channel of the brain, the inhibitor IκB-α, or calmodulin, overview [31]; <7,8> the cofactor thioredoxin can be recycled in vivo by thionein due to its high content of cysteines, overview [36,37]; <14> absolute stereospecific reduction, isozyme MsrB2, no activity with isozyme MsrB1 [34]; <6,8> isozymes MsrB1, MsrB2, and MsrB3 [35]; <8> stereospecific reduction, MsrB accepts free and protein-bound substrates [30]) (Reversibility: r) [29, 30, 31, 32, 34, 35, 36, 37, 39]

P L-methionine + thioredoxin disulfide + H_2O

S L-methionine-(R)-sulfoxide + thioredoxin <1, 7, 8, 9, 11> (<1, 7, 8, 9, 11> MsrB is specific for the R-form [18]; <7> MsrB is specific for the R-form, the membrane-associated isozyme reduces both R- and S-stereoisomers of methionine sulfoxide in proteins [10]) (Reversibility: ?) [10, 18]

P L-methionine + thioredoxin disulfide

S N-acetyl-L-methionine (R)-sulfoxide methyl ester + thioredoxin <17> (<17> enzyme MsrB [16]) (Reversibility: ?) [16]

P N-acetyl-L-methionine methyl ester + thioredoxin disulfide

S N-acetyl-L-methionine (R,S)-sulfoxide + thioredoxin <7, 16> (<16> enzyme MsrA/B shows both MsrA and MsrB activity, free and protein-bound methionine [13]; <7> membrane-bound enzyme form Mem-R,S-Msr [13]) (Reversibility: ?) [13]

P N-acetyl-L-methionine + thioredoxin disulfide

S N-acetyl-L-methionine-(R)-S-oxide + thioredoxin <14, 16> (<14> protein-bound substrate, preferred substrate of isozyme MsrB2 [34]) (Reversibility: ?) [34, 38]

P N-acetyl-L-methionine + thioredoxin disulfide + H_2O

S N-acetyl-L-methionine-(R)-sulfoxide + thioredoxin <7, 11> (<11> MsrB is specific for the R-form [18]; <7> MsrB is specific for the R-form, the membrane-associated isozyme reduces both R- and S-stereoisomers [10]) (Reversibility: ?) [10, 18]

P N-acetyl-L-methionine + thioredoxin disulfide

S acetyl-L-methionine (R)-sulfoxide methyl ester + thioredoxin <17> (<17> the affinity of MsrB to acetyl-L-methionine (R)-sulfoxide methyl ester is higher than to L-methionine (R)-sulfoxide [12]) (Reversibility: ?) [12]

P L-methionine methyl ester + thioredoxin disulfide

S calmodulin L-methionine-(R)-sulfoxide + thioredoxin <8> (<8> MsrB is specific for the R-form, enzyme provides protection against oxidative damage by reactive oxygen species and has a repair function for oxidized protein methionine residues, which restores the calmodulin binding to adenylate cyclase of the pathogen Bordetella pertussis, which is an essential step for the bacterium to enter host cells, overview [21]; <8> MsrB is specific for the R-form, recombinant human calmodulin, recombinant human enzyme, artificial system, determination of oxidized methionine residues being reduced by the enzyme, overview [21]) (Reversibility: ?) [21]

P calmodulin L-methionine + thioredoxin disulfide

S calmodulin-L-methionine (R)-sulfoxide + thioredoxin <7> (Reversibility: ?) [7]

P calmodulin-L-methionine + thioredoxin disulfide

S dabsyl-L-methionine (R)-sulfoxide + thioredoxin <6, 8, 11> (<11> MsrB specifically reduces the R-form of methionine sulfoxide [28]; <8> FMsr is specific for the R-isomer [3]; <6> the native MsrB as well as the recombinant modified MsrB show absolute specificity for the R-form of free and protein-bound methionine sulfoxide, no activity with the S-form [8]) (Reversibility: ?) [3, 8, 28]

P dabsyl-L-methionine + thioredoxin disulfide

S oxidized calmodulin + thioredoxin <7> (<7> enzyme reduces L-methionine (R)-sulfoxide of the protein substrate [1]) (Reversibility: ?) [1]

P partially reduced calmodulin + thioredoxin disulfide

S protein L-methionine (R)-sulfoxide + dithiothreitol <8> (<8> type B enzyme CBS1 is stereospecific for the R-stereomer of methionine residues of peptides and proteins [19]) (Reversibility: ?) [19]

P protein L-methionine + dithiothreitol disulfide

S protein L-methionine (R)-sulfoxide + thioredoxin <8> (<8> type B enzyme CBS1 is stereospecific for the R-stereomer of methionine residues of peptides and proteins [19]) (Reversibility: ?) [19]

P protein L-methionine + thioredoxin disulfide

S protein L-methionine-(R)-sulfoxide + thioredoxin <7> (<7> MsrB is specific for the R-form, the membrane-associated isozyme reduces both R- and S-stereoisomers of methionine sulfoxide, N-acetylmethionine sulfoxide, and D-Ala-Met-enkephalin [10]) (Reversibility: ?) [10]

P protein L-methionine + thioredoxin disulfide

S sulindac + thioredoxin <7, 12> (<12> activation of a methionine sulfoxide-containing prodrug, activity with membrane-bound enzyme form Mem-R,S-Msr [13]; <7> activation of a methionine sulfoxide-containing prodrug, activity with membrane-bound enzyme form Mem-R,S-Msr and MsrA [13]; <12> activity with membrane-bound enzyme form Mem-R,S-Msr [13]; <7> activity with membrane-bound enzyme form Mem-R,S-Msr and MsrA [13]) (Reversibility: ?) [13]

P sulindac sulfide + thioredoxin disulfide (<7,12> activated drug which inhibits cyclooxygenase 1 and 2 and exhibiting anti-inflammatory activity [13])

S sulindac + thioredoxin disulfide <7, 12> (<7> activation of the antiinflammatory drug with anti-tumorigenic activity, which acts via inhibition of cyclooxygenases 1 and 2 [11]; <7> highest activity by a membrane bound enzyme form Mem-R,S-Msr, which preferentially reduces the R-substrate form, no activity by enzyme forms fRMsr, fSMsr, low activity by enzyme forms MsrB [11]) (Reversibility: ?) [11]

P sulindac sulfide + thioredoxin

S Additional information <1, 2, 4, 5, 6, 7, 8, 9, 10, 11, 12, 13, 14, 16, 17, 19, 20, 21, 22, 23, 25, 26, 28, 29, 32> (<6, 8> substrate specificity [3, 9]; <7> cellular system of balancing native proteins and oxidatively damaged proteins by use of protein biosynthesis, protein oxidative modification, protein elimination, and oxidized protein repair involving the enzyme, overview, enzyme protects against oxidative damage of proteins [18]; <11> cellular system of balancing native proteins and oxidatively damaged proteins by use of protein biosynthesis, protein oxidative modification, protein elimination, and oxidized protein repair involving the enzyme, overview, enzyme protects against oxidative damage of proteins, enzyme activity is not age-related [18]; <1,8,9> cellular system of balancing native proteins and oxidatively damaged proteins by use of protein biosynthesis, protein oxidative modification, protein elimination, and oxidized protein repair involving the enzyme, overview, enzyme protects against oxidative damage of proteins, loss of enzyme activity is age-related [18]; <8> down-regulation of CBS-1 during replicative senescence of cells leads to accumulation of oxidized proteins and age-related increased oxidative damage [20]; <2, 4, 5, 7, 13, 16, 17, 19, 20, 21, 25, 28> enzyme acts on free and protein-bound methionine [15]; <7> enzyme contributes to resistance against cadmium, physiological role [1]; <14> enzyme has regulatory function in the plant cell [26]; <6,32> enzyme provides protection for the cell against oxidative stress [2]; <19> enzyme repairs oxidatively damaged free and protein bound methionine and recycles it from methionine sulfoxide, e.g. the heat shock protein and chaperone Hsp16.3, role of the MsrA/MsrB repair pathway in cellular protein dynamics [15]; <2, 4, 5, 13, 20, 21, 25> enzyme repairs oxidatively damaged free and protein bound methionine and recycles it from methionine sulfoxide, role of the MsrA/MsrB repair pathway in cellular protein dynamics [15]; <7> enzyme repairs oxidatively damaged free and protein bound methionine and recycles it from methionine sulfoxide, role of the MsrA/MsrB repair pathway in cellular protein dynamics, the MsrA/MsrB repair pathway is involved in the signal recognition particle-dependent protein targeting pathway, regulation mechanism of gene expression, overview [15]; <7> MsrB is specific for the R-form of the substrate [7]; <22> potential role of the enzyme in cold-acclimation, enzyme may protect the cells from photodamage [27]; <6> protection of the cells against reactive oxidizing species, biological consequences of methionine oxidation, physiological role, overview [23]; <14> recycling of free methionine, enzyme reverses the oxidative damage at methionine protein residues oxidized to methionine sulfoxide being a major cause of aging and age-related diseases, Msr can regulate protein function, be involved in signal transduction, and prevent accumulation of faulty proteins [14]; <8,10> recycling of free methionine, enzyme reverses the oxidative damage at methionine protein residues oxidized to methionine sulfoxide being a major cause of aging and age-related diseases, Msr can regulate protein function, be involved in signal transduction, and prevent accumulation of faulty proteins, MsrB has several different physiological repair and regulatory functions, overview [14]; <6> recycling of free methionine, enzyme reverses the oxidative damage at methionine protein residues oxidized to methionine sulfoxide being a major cause of aging and age-related diseases, Msr can regulate protein function, be involved in signal transduction, and prevent accumulation of faulty proteins, MsrB has several different physiological repair and regulatory functions, overview, oxidation of 2 essential methionine residues of HIV-2 particles can inactivate the virus and prevent infection of human cells [14]; <2,7,11> recycling of free methionine, enzyme reverses the oxidative damage at methionine protein residues oxidized to methionine sulfoxide being a major cause of aging, Msr can regulate protein function, be involved in signal transduction, and prevent accumulation of faulty proteins, MsrB has several different physiological repair and regulatory functions, overview [14]; <28> role of the MsrA/MsrB repair pathway in cellular protein dynamics, enzyme repairs oxidatively damaged free and protein bound methionine and recycles it from methionine sulfoxide [15]; <26> the enzyme contributes to the ecological performance of Lactobacillus reuteri in gastrointestinal ecosystems together with the high-molecular-mass surface protein Lsp, enzyme expression is induced in vivo [6]; <11> the enzyme is essential in protection of the cells against oxidative damage by reactive oxygen species, yeast cell life span analysis of wild-type and mutant cells, the latter either overexpress or lack enzyme activity, overview [28]; <6> the enzyme protect cells against oxidative damage and plays a role in age-related diseases [13]; <7> the enzyme protect cells against oxidative damage and plays a role in age-related missfunctions [13]; <8> the enzyme protects cells against oxidative damage and plays a role in age-related and neurological diseases, like Parkinsons or Alzheimers disease [13]; <11,12> the enzyme protects cells against oxidative damage and plays a role in age-related diseases [13]; <2> the MsrA1/MsrB system is physiologically more significant in Staphylococcus aureus than MsrA2 [22]; <8> enzyme reduces oxidized methionine residues of the α-1-proteinase inhibitor, calmodulin, and thrombomodulin, which become reversibly inactivated upon oxidation [13]; <17> substrate specificities of enzymes, the reduction step is rate-determining [16]; <16> substrate specificity and activity of MsrB/PilB in comparison to MsrA, overview [13]; <7> substrate specificity of enzyme forms with R-form of free and protein-bound methionine sulfoxide, overview [11]; <17> substrate specificity of MsrB activity, diverse substrates, overview [4]; <7> substrate specificity of the different enzyme forms, overview, enzyme reduces oxidized methionine residues of the ribosomal protein L12, which becomes reversibly inactivated and forms monomers instead of dimers upon oxidation, Mem-R,S-Msr also posesses MsrA activity utilizing L-methionine (S)-sulfoxide as substrate [13]; <20> the enzyme also exhibits MsrA activity utilizing L-methionine (S)-sulfoxide as substrate [24]; <7> the enzymes utilize free and protein-bound L-methionine and N-acetyl-L-methionine as substrates, the membrane-associated isozyme also shows MsrA activity utilizing L-methionine (S)-sulfoxide and N-acetyl-L-methionine (S)-sulfoxide as substrates [10]; <5,7,23> the reduction step is rate-determining [16]; <29> the tandem domains of PilB also possess MsrA activity utilizing L-methionine (S)-sulfoxide as substrate, the MsrA domain alone does very poorly utilize the R-isomer [25]; <16> PilB affects the survival of the organism to reactive oxygen species, PilB is not involved in piliation, pilin production, or adherence [38]; <14> roles of methionine sulfoxide reductases in antioxidant defense, protein regulation via alternating it between active and inactive form, and survival, MsrB protects cells from the cytotoxic effects of reactive oxygen species, ROS, overview [31]; <6> roles of methionine sulfoxide reductases in antioxidant defense, protein regulation via alternating it between active and inactive form, and survival, MsrB protects cells from the cytotoxic effects of reactive oxygen species, ROS, overview, enzyme involvement in protein repair and associated factors, protein regulation pathway, overview [31]; <11> roles of methionine sulfoxide reductases in antioxidant defense, protein regulation via alternating it between active and inactive form, and survival, MsrB protects cells from the cytotoxic effects of reactive oxygen species, ROS, overview, regulation of MsrB expression, overview [31]; <8> roles of methionine sulfoxide reductases in antioxidant defense, protein regulation via alternating it between active and inactive form, and survival, MsrB protects cells from the cytotoxic effects of reactive oxygen species, ROS, overview, the enzyme is involved in age-related diseases such as Alzheimers or Parkinsons diseases as well as in diseases caused by prions, mechanism, overview, enzyme involvement in protein repair and associated factors, protein regulation pathway, overview [31]; <16> The thioredoxin domain of PilB can use electrons from DsbD to reduce downstream methionine sulfoxide reductases, overview [33]; <14> substrate specificity, overview, isozyme MsrB1 is not able to reduce free L-methionine-(R)-S-oxide or N-acetyl-L-methionine-(R)-S-oxide, while isozyme MsrB2 prefers protein-bound substrates such as N-acetyl-L-methionine (R)-S-oxide, overview [34]; <16> the bifunctional enzyme catalyzes both reactions of MsrB or PilB, EC 1.8.4.12, and of MsrA or PilA, EC 1.8.4.11, the catalytic sites for the two different activities are localized separately on the enzyme molecule, overview [33]; <16,17> the bifunctional enzyme catalyzes both reactions of MsrB or PilB, EC 1.8.4.12, and of MsrA or PilA, EC 1.8.4.11, the catalytic sites for the two different activities are localized separatly on the enzyme molecule, overview [38,39]; <6,8,11,14> the enzyme utilizes free and protein-bound methionine-(R)-S-oxide as substrate, but prefers the latter, methionine oxidation inactivates the proteins showing equal distribution of S-MetO and R-MetO [31]; <6,8> the thioredoxin dependence is different for selenocysteine- and cysteine-containing enzyme, overview [35]) (Reversibility: ?) [1, 2, 3, 4, 6, 7, 9, 10, 11, 13, 14, 15, 16, 18, 20, 22, 23, 24, 25, 26, 27, 28, 31, 33, 34, 35, 38, 39]

P ?

Inhibitors

3-carboxy 4-nitrobenzenethiol <7, 17> (<7,17> binds specifically to the sulfenic acid reaction intermediate [16]) [16]
dimedone <7, 17> (<7,17> binds specifically to the sulfenic acid reaction intermediate [16]) [16]
Additional information <6, 7, 10, 22> (<22> enzyme expression decreases during dehardening from 4°C to 22°C of cold-acclimated plants, effects of light and temperature on enzyme expression and activity, overview [27]; <6,10> selenium-adequate diet retains MsrB mRNA and protein expression at basal levels [14]; <7> no inhibition by EDTA, 1,10-phenanthroline, and pyridine 2,6-dicarboxylic acid [39]) [14, 27, 39]

Cofactors/prosthetic groups

dithiothreitol <6, 7, 8, 14, 17> (<14> absolutely dependent on in vitro and in vivo with substrate Hsp21 [26]; <7> MsrB can also utilize DTT as reductant, the membrane-isozyme shows only poor activity [10]; <7> utilized in vitro [7]; <8> can act as cofactor only in vitro showing a much higher activity than thioredoxin [37]; <17> can substitute for thioredoxin [39]; <7> can substitute for thioredoxin in vitro [36]; <6> results in higher activity compared to cofactor thioredoxin [35]) [4,7,10,19,26,35,36,37,39]
NADPH <7> (<7> membrane-bound enzyme form Mem-R,S-Msr [13]) [13]
dithioerythritol <14> (<14> DTE [34]) [34]
thioredoxin <1, 2, 3, 4, 5, 6, 7, 8, 9, 10, 11, 12, 13, 14, 16, 17, 18, 19, 20, 21, 23, 25, 26, 27, 28, 29, 32, 33, 34> (<6,17> dependent on [12,23]; <7> preferred cofactor [13]; <5,7,17,23> physiological cofactor [16]; <7> physiologic cofactor [7]; <14> isozyme MsrB2, no activity with isozyme MsrB1 [34]; <8> natural cofactor, the cofactor can be recycled in vivo by reduction via zinc-containing metallothionein, Zn-MT, after removal of the zinc ion due to its high content of cysteines, mechanism, overview [37]; <16> PilB contains an N-terminal thioredoxin-like domain, the NT domain, fused to the MsrA and MSrB domains, structure overview [33]; <7> preferred and natural cofactor, the cofactor can be recycled in vivo by reduction via zinc-containing metallothionein, Zn-MT, after removal of the zinc ion due to its high content of

cysteines, mechanism, overview [36]) [1, 2, 3, 4, 6, 7, 8, 9, 10, 11, 12, 13, 14, 15, 16, 18, 19, 21, 22, 23, 24, 25, 28, 30, 31, 32, 33, 34, 35, 36, 37, 38, 39]
Additional information <5, 7, 17, 23> (<7> DTT can partially substitute for thioredoxin in vitro, low activity [16]; <5,17,23> DTT can substitute for thioredoxin in vitro [16]; <7> no activity with DTT as cofactor by membrane-bound enzyme form Mem-R,S-Msr [13]) [13,16]

Activating compounds

Additional information <2, 8, 11, 14, 19, 22> (<22> 48 h exposure to high light at 22°C induces enzyme expression, effects of light and temperature on enzyme expression and activity, overview [27]; <19> enzyme is induced under oxidative stress and heat shock [15]; <14> expression of MsrB is induced by dehydration and H_2O_2 [14]; <11> MsrB expression is induced by heat shock and alkylating methyl-methanesulfonate treatment [14]; <8> oxidative stress, e.g. caused by H_2O_2 up to 0.4 mM, induces enzyme expresssion [20]; <2> the msrA1-msrB operon is induced by antibiotics [15]; <14> heat shock, dehydration, and reactive oxygen species like H_2O_2 induce the expression of MsrB [31]; <11> starvation induces MsrB expression, also heat treatment and methylmethanesulfonate induce the enzyme [31]) [14, 15, 20, 27, 31]

Metals, ions

Fe^{2+} <7> (<7> with Zn^{2+} in a ratio of 1 mol per mole of enzyme, tight metal binding [39]) [39]
Se^{2+} <6, 8, 11, 12, 32> (<6, 8, 11, 12, 32> selenoprotein [2, 13, 19]) [2, 13, 19]
Zn^{2+} <5, 6, 7, 15, 17, 23, 24, 32, 35> (<7, 17> about 50% of MsrB binds a zinc atom in opposite direction of the active site, enzyme contains the CXXC motif, binding of Zn^{2+} modulates the catalytic efficiency via structural changes [16]; <5, 23> about 50% of MsrBs binds a zinc atom in opposite direction of the active site [16]; <7> binding by residues at positions 45, 48, 94, and 97, 2 CXXC-motifs, role in catalysis [17]; <15, 24, 35> enzyme belongs to the metal-containing MsrB group I, metal binding by 2 CXXC-motifs, role in catalysis [17]; <6,32> metalloenzyme, content determination [2]; <7> with Fe^{2+} in a ratio of 1 mol per mole of enzyme, tight metal binding, the metal binding site is composed of two CXXC motifs located at the opposite side of the active site, role in catalysis and structural stability, overview [39]) [2, 16, 17, 39]
Additional information <6, 7, 8, 11, 14, 16, 17> (<17> content of free cysteinyl residues in wild-type and mutant enzymes, MsrA and MsrB domains, overview [4]; <7> enzyme belongs to the metal-containing MsrB group I [17]; <6> isozyme MsrB1 contains selenocysteine, while isozymes MsrB2 and MsrB3 contain cysteine residues, the thioredoxin dependence is different for selenocysteine- and cysteine-containing enzyme, overview [35]; <8> native isozyme MsrB1 contains selenocysteine, while native isozymes MsrB2 and MsrB3 contain cysteine residues, the thioredoxin dependence is different for selenocysteine- and cysteine-containing enzyme, overview [35]; <16> PilB is a selenocysteine-containing enzyme [33]; <6,8,11,14> the major isozyme of MsrB, MsrB1, is a selenoprotein, selenium affects the expression of MsrB [31]; <7> the zinc:iron ratio is 8:2 to 6:4, metal content of wild-type and mutant enzymes, overview [39]; <17> wild-type MsrB of Neisseria me-

ningitidis is no metal-binding enzyme, but contains a preformed metal binding site, metal binding to MsrB results in inhibition of binary complex formation between oxidized MsrB and reduced thioredoxin but not between reduced MsrB and substrate, metal content of wild-type and mutant enzymes, overview [39]) [4, 17, 31, 33, 35, 39]

Turnover number (min^{-1})

0.002 <6> (L-methionine-(R)-S-oxide, <6> recombinant wild-type cysteine-containing isozyme MsrB1 expressed in NIH 3T3 cells [35]) [35]
0.02 <14> (L-methionine-(R)-S-oxide, <14> recombinant isozyme MsrB2, with dithioerythritol [34]) [34]
0.07 <14> (L-methionine-(R)-S-oxide, <14> recombinant isozyme MsrB1, with dithioerythritol [34]) [34]
0.22 <6> (L-methionine-(R)-S-oxide, <6> recombinant wild-type selenocysteine-containing isozyme MsrB1 expressed in NIH 3T3 cells [35]) [35]
0.46 <6> (L-methionine-(R)-S-oxide, <6> recombinant wild-type isozyme MsrB2 expressed in Escherichia coli [35]) [35]
0.83 <8> (L-methionine-(R)-S-oxide, <8> recombinant wild-type isozyme MsrB3 expressed in Escherichia coli [35]) [35]
2.8 <17> (L-methionine (R,S)-sulfoxide, <17> MsrB activity of PILB, pH 8.0, 25°C [4]) [4]
Additional information <17> [12]

Specific activity (U/mg)

0.000029 <2> (<2> wild-type strain RN450, substrate L-methionine (R)-sulfoxide [22]) [22]
0.00003 <2> (<2> $msrA_2$ knockout strain RN450, substrate L-methionine (R)-sulfoxide [22]) [22]
0.000046 <6> (<6> purified recombinant wild-type cysteine-containing isozyme MsrB1 expressed in NIH 3T3 cells, substrates are L-methionine-(R)-S-oxide and thioredoxin [35]) [35]
0.000085 <7> (<7> enzyme form Mem-R,S-Msr [11]) [11]
0.00009 <7> (<7> enzyme form Mem-R,S-Msr, substrate sulindac [13]) [13]
0.00022 <7> (<7> MsrB, substrate N-acetyl-L-methionine-(R)-sulfoxide [10]) [10]
0.0004 <7> (<7> membrane vesicles, substrate N-acetyl-L-methionine-(R)-sulfoxide [10]) [10]
0.0007 <6> (<6> kidney, substrate L-methionine (R)-sulfoxide [8]) [8]
0.0012 <2> (<2> msrA1 knockout strain RN450, substrate L-methionine (R)-sulfoxide [22]) [22]
0.0015 <2> (<2> msrA1/msrA2 double knockout strain RN450, substrate L-methionine (R)-sulfoxide [22]) [22]
0.0019 <6> (<6> liver, substrate L-methionine (R)-sulfoxide [8]) [8]
0.002 <6> (<6> purified recombinant wild-type cysteine-containing isozyme MsrB1 expressed in NIH 3T3 cells, substrates are L-methionine-(R)-S-oxide and DTT [35]) [35]

0.0048 <7> (<7> wild-type strain, substrate L-methionine (R)-sulfoxide [13]) [13]
0.01 <6> (<6> purified recombinant wild-type isozyme MsrB2 expressed in Escherichia coli, substrates are L-methionine-(R)-S-oxide and thioredoxin [35]) [35]
0.023 <8> (<8> purified recombinant wild-type isozyme MsrB3 expressed in Escherichia coli, substrates are L-methionine-(R)-S-oxide and thioredoxin [35]) [35]
0.045 <6> (<6> purified recombinant wild-type selenocysteine-containing isozyme MsrB1 expressed in NIH 3T3 cells, substrates are L-methionine-(R)-S-oxide and thioredoxin [35]) [35]
0.075 <29> (<29> purified recombinant MsrA/MsrB tandem domain, substrate L-methionine (R)-sulfoxide [25]) [25]
0.085 <34> (<34> purified recombinant CBS-1, substrate L-methionine (R)-sulfoxide [25]) [25]
0.092 <33> (<33> purified recombinant CBS-1, substrate L-methionine (R)-sulfoxide [25]) [25]
0.11 <29> (<29> purified recombinant MsrB domain alone, substrate L-methionine (R)-sulfoxide [25]) [25]
0.17 <6> (<6> purified recombinant wild-type selenocysteine-containing isozyme MsrB1 expressed in NIH 3T3 cells, substrates are L-methionine-(R)-S-oxide and DTT [35]) [35]
0.386 <6> (<6> purified recombinant wild-type isozyme MsrB2 expressed in Escherichia coli, substrates are L-methionine-(R)-S-oxide and DTT [35]) [35]
0.452 <8> (<8> purified recombinant wild-type isozyme MsrB3 expressed in Escherichia coli, substrates are L-methionine-(R)-S-oxide and DTT [35]) [35]
1.78 <2> (<2> purified recombinant PilB, substrate L-methionine (R)-sulfoxide [8]) [8]
3 <17> (<17> recombinant wild-type MsrB domain, cosubstrate dithiothreitol [4]) [4]
4.2 <17> (<17> recombinant wild-type MsrA/MsrB, cosubstrate dithiothreitol [4]) [4]
12 <17> (<17> recombinant wild-type MsrB domain, cosubstrate thioredoxin [4]) [4]
170 <17> (<17> recombinant wild-type MsrA/MsrB, cosubstrate thioredoxin [4]) [4]
Additional information <6, 7, 8, 11, 12, 14> (<7> in vivo MsrB activity is below detection limit [7]; <11> recombinant activity in overexpressing yeast cells [28]; <12> subcellular sulindac reducing activity in calf liver [13]; <6> tissue specific activity, activity in tissues of MsrA-deficient mutant mice [8]; <6,8> thioredoxin- and DTT-dependent activities of mutant enzymes, overview [35]) [7, 8, 13, 20, 28, 34, 35]

K_m-Value (mM)

0.026 <17> (thioredoxin, <17> pH 5.5, 25°C [12]) [12]
0.034 <17> (thioredoxin, <17> MsrB activity of PILB, pH 8.0, 25°C [4]) [4]

0.054 <14> (L-methionine-(R)-S-oxide, <14> recombinant isozyme MsrB2, with dithioerythritol [34]) [34]
0.058 <17> (thioredoxin, <17> MsrB, pH 8.0, 25°C, substrate acetyl-L-methionine (R)-sulfoxide N-methyl ester [16]) [16]
0.079 <14> (L-methionine-(R)-S-oxide, <14> recombinant isozyme MsrB1, with dithioerythritol [34]) [34]
0.14 <17> (acetyl-L-methionine (R)-S-oxide methyl ester, <17> pH 5.5, 25°C [12]) [12]
0.31 <6> (L-methionine-(R)-S-oxide, <6> recombinant wild-type isozyme MsrB2 expressed in Escherichia coli [35]) [35]
0.5 <6> (L-methionine-(R)-S-oxide, <6> recombinant wild-type selenocysteine-containing isozyme MsrB1 expressed in NIH 3T3 cells [35]) [35]
0.8 <8> (L-methionine-(R)-S-oxide, <8> recombinant wild-type isozyme MsrB3 expressed in Escherichia coli [35]) [35]
1.3 <6> (L-methionine-(R)-S-oxide, <6> recombinant wild-type cysteine-containing isozyme MsrB1 expressed in NIH 3T3 cells [35]) [35]
2.2 <17> (acetyl-L-methionine (R)-S-oxide methyl ester, <17> pH 8.0, 25°C [12]) [12]
7 <17> (thioredoxin, <17> pH 8.0, 25°C [12]) [12]
56 <17> (L-methionine (R,S)-sulfoxide, <17> MsrB activity of PILB, pH 8.0, 25°C [4]) [4]
Additional information <5, 6, 7, 8, 17, 23> (<17> kinetics [4]; <5,17,23> kinetic mechanism [16]; <17> stopped flow kinetics, steady-state kinetics [12]; <7,17> thermodynamics and kinetics of wild-type and mutant MsrB [39]; <6,8> thioredoxin- and DTT-dependent kinetics of wild-type and mutant enzymes, overview [35]) [4, 12, 16, 35, 39]

pH-Optimum

6.9 <29, 33, 34> (<29,33,34> assay at [25]) [25]
7.4 <2, 6, 7, 8, 12, 16, 20> (<2,6,7,8,12,16,20> assay at [8,9,10,11,19,24,32,33]) [8, 9, 10, 11, 19, 24, 32, 33]
7.5 <2, 7, 8, 11> (<2,7,8,11> assay at [1,3,22,28]) [1, 3, 22, 28]
7.8 <14> (<14> assay at [26]) [26]
8 <7, 17> (<7,17> assay at [16]) [16]

Temperature optimum (°C)

25 <7, 14, 17> (<7,14,17> assay at [16,26,39]) [16, 26, 39]
37 <2, 6, 7, 8, 11, 12, 16, 29, 33, 34> (<2,6,7,8,11,12,16,29,33,34> assay at [1,3,8,9,10,11,19,22,25,28,33]) [1, 3, 8, 9, 10, 11, 19, 22, 25, 28, 33]

4 Enzyme Structure

Subunits

? <6, 16, 17, 20> (<6> x * 13000, native wild-type MsrB, SDS-PAGE [9]; <17> x * 16372.0, recombinant MsrB, mass spectrometry, x * 16467.2, recombinant selenomethionine-MsrB, mass spectrometry [5]; <17> x * 16374, about, recombinant wild-type MsrB domain, mass spectrometry [4]; <20> x * 43000,

Msr, SDS-PAGE [24]; <16> x * 57000, MsrA/B [13]; <16> x * 15000-35000, recombinant His-tagged PilB forms, SDS-PAGE [38]) [4, 5, 9, 13, 24, 38]
monomer <5, 17, 23> [12, 16]
Additional information <1, 7, 16, 17, 20, 29> (<1,7,16,17> analysis of three-dimensional structure and active site structure [17]; <29> enzyme domain and active site structure, MsrB domain comprises residues 375-522, overview [25]; <20> MsrA, EC 1.8.4.11, and MsrB are fused together [15]; <17> MsrB activity is located on the C-terminal domain of PILB, the fused domains are folded entities [4]; <17> MsrB and enzyme MsrA, EC 1.8.4.11, form domains of a single polypeptide together with a third thioredoxin-like domain [15]; <7> MsrB has a highly conserved sequence among organisms [1]; <16> the 2 enzyme activities, MsrA and MsrB, form domains of a single polypeptide together with a third thioredoxin-like domain [15]; <17> the enzyme forms are produced as individual folded entities, but in vivo the enzyme is part of a three-domain protein named PILB, with the central domain exhibiting MsrA activity, and the C-terminal domain showing MsrB activity [16]) [1, 4, 15, 16, 17, 25]

5 Isolation/Preparation/Mutation/Application

Source/tissue

SRA01/04 cell <8> (<8> i.e. HLE cell, transformed human lens epithelial cells [32]) [32]
WI^{-}38 cell <8> (<8> embryonic fibroblast, activity during development: downregulation during replicative senescence [20]; <8> fibroblast, young and old cells, enzyme expression pattern during cell development [18]) [18, 20]
bone marrow <8> (<8> isozyme MsrB1 [32]) [32]
brain <6, 8, 12> (<12> calf, sulindac reducing activity [13]) [8, 11, 13, 14, 31]
cerebellum <6, 8> (<8> isozyme MsrB2 [32]) [8, 32]
colon <8> (<8> isozymes MsrB3 and MsrB2 [32]) [32]
epidermis <8> [30]
fibroblast <8> (<8> cell line WI-38, young and old cells, enzyme expression pattern during cell development, senescent cells show decreased enzyme expression and activity [18]; <8> WI^{-}38 [20]) [18, 20]
heart <8> (<8> most abundant in ventricles, interventricular septum, and apex [19]; <8> isozymes MsrB3 and MsrB2 [32]) [19, 32]
heart ventricle <8> [19]
interventricular septum <8> [19]
kidney <6, 8, 12> (<12> calf, sulindac reducing activity [13]; <12> highest sulindac reductase activity [11]; <8> isozyme MsrB1 and MsrB2 [32]) [8, 11, 13, 14, 19, 32]
leaf <14, 22> (<22> high expression level in cold-hardened plants at 4°C [27]) [27, 34]

lens <8> (<8> epithelia and fibers, isozymes MsrB1 or selenoprotein R, MsrB2 or CBS-1, and MsrB3, differential expression patterns of isozymes, overview [32]) [32]
leukocyte <8> (<8> polymorphonuclear [3]) [3]
liver <6, 8, 12> (<6> highest activity [8]; <12> calf, sulindac reducing activity [13]) [8, 9, 11, 13, 14, 19]
lung <6> [14]
melanocyte <8> (<8> epidermal [30]) [30]
skeletal muscle <8> (<8> isozymes MsrB1 and MsrB2 [32]) [19, 32]
skin <8> [30]
small intestine <8> (<8> isozyme MsrB3 [32]) [32]
spleen <8> (<8> isozyme MsrB1 [32]) [32]
Additional information <8, 14, 22> (<22> expression pattern analysis [27]; <8> enzyme expression level and methionine sulfoxide content in fibroblasts during development, overview [20]; <14> high expression level of the plastidic isozyme pPMSR in photosynthetic active tissue [26]; <8> tissue distribution, main expression of CBS1 in muscle tissues, overview [19]; <8> expression analysis of MsrB isozymes [30]; <8> tissue-specific expression of MsrB isozymes, overview [32]) [19, 20, 26, 27, 30, 32]

Localization

chloroplast <14> (<14> plastidic isozyme pPMSR [26]; <14> plastidic isozymes MsrB1 and MsrB2, analysis of subcellular localization [34]) [26, 34]
cytosol <6, 8, 9, 11, 12, 22> (<8,11> Sel-X, a MsrB enzyme form variant [13]; <12> Sel-X, a MsrB enzyme form variant, sulindac reducing activity in calf [13]; <6> Sel-X, a MsrB enzyme variant [13]; <6,8> isozyme MsrB1 [31]; <8> isozyme MsrB1 or selenoprotein R [32]) [13, 18, 27, 30, 31, 32]
endoplasmic reticulum <6, 8> (<6,8> isozyme MsrB3 [31,32]; <6> specific isozyme MsrB3 [29]) [29, 31, 32]
extracellular <20> (<20> secretion of MsrA, EC 1.8.4.11, and MsrB fused together [15]) [15]
intracellular <11, 14> [31]
membrane <7, 12, 16, 20> (<16> outer cell membrane [15]; <7> enzyme form Mem-R,S-Msr [13]; <7> membrane-associated enzyme MsrB [11]) [10, 11, 13, 15, 24]
microsome <7, 12> (<12> calf, sulindac reducing activity [13]) [11, 13]
mitochondrion <6, 7, 8, 9, 11, 12> (<9> matrix [18]; <8,11> Sel-X, a MsrB enzyme form variant [13]; <12> Sel-X, a MsrB enzyme form variant, sulindac reducing activity in calf [13]; <6> Sel-X, a MsrB enzyme variant [13]; <8> isozyme MsrB2 or CBS-1, and isozyme MsrB3 [32]; <6,8> isozymes MsrB2 and MsrB3 [31]) [11, 13, 18, 31, 32]
nucleus <6, 8> (<6,8> isozyme MsrB1 [31]; <8> isozyme MsrB1 or selenoprotein R [32]) [30, 31, 32]
outer membrane <16> [38]
periplasm <16> [33]
Additional information <6, 8, 12, 20> (<20> no activity in the cytoplasm [24]; <12> subcellular sulindac reducing activity distribution in calf liver

[13]; <6> subcellular localization analysis, no localization of MsrB3 in the mitochondria [29]; <6,8> subcellular targeting is determined by alternative splicing [31]) [13, 24, 29, 31]

Purification

<2> (recombinant His-tagged PilB from Escherichia coli by nickel affinity chromatography) [8]
<6> (partial purification of native MsrB from liver, recombinant modified MsrB from Escherichia coli) [8]
<6> (recombinant C-terminally His-tagged wild-type and mutant MsrB to homogeneity) [9]
<6> (recombinant wild-type and mutant enzymes from Escherichia coli strain BL21(DE3) by heparin affinity chromatography) [2]
<6> (recombinant wild-type and mutant isozymes MsrB1 and MsrB2 from Escherichia coli strain BL21(DE3) and NIH-3T3 cells) [35]
<7> (recombinant MsrB) [11]
<7> (recombinant enzyme from Escherichia coli strain DH5α) [1]
<8> (partially by subcellular fractionation) [30]
<8> (recombinant His-tagged wild-type and mutant enzymes from Escherichia coli by nickel affinity chromatography) [19]
<8> (recombinant MsrB from Escherichia coli strain BL21(DE3)) [21]
<8> (recombinant wild-type and mutant isozyme MsrB3 from Escherichia coli strain BL21(DE3)) [35]
<12> (partially, cell fragmentation) [11]
<14> (recombinant plastidic isozyme pPMSR from Escherichia coli strain BL21(DE3)) [26]
<14> (recombinant plastidic isozymes MsrB1 and MsrB2 from Escherichia coli) [34]
<16> (recombinant His-tagged PilB forms in Escherichia coli strain Xl-1 blue by nickel affinity chromatography) [38]
<16> (recombinant N-terminally His-tagged full-length PilB and MsrB domain variants from Escherichia coli by nickel affinity chromatography followed by cleavage of the His-Tag through thrombin, followed by gel filtration and ion exchange chomatography) [33]
<17> (recombinant MsrB domain from Escherichia coli) [5]
<22> (recombinant enzyme from Escherichia coli) [27]
<29> (recombinant His-tagged full length tandem enzyme, MsrA, and MsrB domains from Escherichia coli, tags are removed by thrombin digestion) [25]
<32> (recombinant His-tagged enzyme from Escherichia coli strain BL21(DE3)) [2]
<33> (recombinant His-tagged enzyme from Escherichia coli) [25]
<34> (recombinant His-tagged enzyme from Escherichia coli) [25]

Crystallization

<16> (purified recombinant PilB mutant L38M/L41M, vapour diffusion method, 30 mg/ml protein in 20 mM HEPES, pH 7.5, and 100 mM NaCl, is mixed with well solution containing 0.1 M MES, pH 6.5, 0.2 M ammonium sulfate, 26% PEG 2000 monomethylester, and 25% glycerol, X-ray diffraction

structure determination and analysis at 1.6 A resolution, multiwavelength anomalous dispersion at -170°C) [33]
<17> (selenomethionine-substituted peptide methionine sulfoxide reductase B domain, hanging drop vapour diffusion method in multiwell tissue-culture plates, 0.004 ml protein solution containing 75 mg/ml protein in 50 mM Tris-HCl, pH 8.0, mixed with 0.004 ml precipitant solution at 20°C, 3 days, X-ray diffraction structure determination and analysis at 1.8 A resolution) [5]
<29> (purified recombinant detagged MsrB domain containing SeMet39, hanging drop vapour-diffusion method, 15 mg/ml protein in 20 mM Tris, pH 8.5, 10% v/v glycerol, against a well solution containing 0.1 M sodium cacodylate, pH 6.5, 30% w/v PEG 4000, room temperature, X-ray diffraction structure determination and analysis at 1.8 A resolution, modeling) [25]

Cloning

<2> (3 genes msrA and 1 gene msrB form an operon, one of the 3 msrA genes is fused to the msrB gene, genetic organization and regulation, overview) [15]
<2> (pilB is transcribed as polycistronic transcript, overexpression of PilB as His-tagged protein in Escherichia coliBL21(DE3)) [8]
<4> (the chromosome contains 2 copies of gene msrB, a plasmid harbors 1 copy of gene msrB) [15]
<5> (genes msrA, EC 1.8.4.11, and msrB form an operon) [15]
<6> (endoplasmic reticulum isozyme MsrB3, DNA and amino acid sequence determination and analysis, expression of different constructs of GFP-tagged MsrB3 in monkey kidney CV-1 cells or in mouse fibroblast NIH 3T3 cells) [29]
<6> (expression of wild-type and mutant isozymes MsrB1 and MsrB2 in Escherichia coli strain BL21(DE3) and in NIH-3T3 cells) [35]
<6> (gene msrB with the codon for selenomehinonine is exchanged for methionine, overexpression in Escherichia coli) [8]
<6> (overexpression of wild-type and mutant enzymes in Escherichia coli strain BL21(DE3)) [2]
<6> (overexpression of wild-type and mutants in Escherichia coli, expression as N- or C-terminally 6His-tagged protein lowers the recombinant expression level to 3% of total enzyme expressed, labeling of expressed wild-type with 75SeMet) [9]
<7> (gene msrB, located in the chromosome at 40.09 min, respectively, regulation mechanism of gene expression, overview) [15]
<7> (gene msrB, overexpression of wild-type enzyme and mutant enzymes in Escherichia coli strain DH5α) [1]
<8> (expression analysis in subcellular fractions of melanocytes, overview) [30]
<8> (expression of MsrB in Escherichia coli strain BL21(DE3)) [21]
<8> (expression of wild-type and mutant isozyme MsrB3 in Escherichia coli strain BL21(DE3)) [35]

<8> (overexpression of truncated wild-type and mutant enzymes as His-tagged proteins in Escherichia coli, functional coexpression of CBS_1 in oocytes with Drosophila melanogaster ShC/B potassium channel) [19]
<9> (mitochondrial and cytosolic isozymes are encoded on a single gene with 2 initiations sites, delivering an N-terminal signal peptide to the mitochondrial enzyme form) [18]
<11> (overexpression of MsrB in a yeast strain, expression of MsrB as N-terminally 6His-tagged protein in Escherichia coli strain BL-21) [28]
<14> (DNA and amino acid sequence determination and analysis of plastidic isozymes MsrB1 and MsrB2, functional expression in Escherichia coli) [34]
<14> (expression of the plastidic isozyme pPMSR in Escherichia coli strain BL21(DE3) without the chloroplast signal sequence) [26]
<16> (gene msr or pilB, DNA sequence determination and analysis) [13]
<16> (gene pilB, expression of N-terminally His-tagged full-length wild-type and mutant PilB and MsrB domain variants in Escherichia coli) [33]
<16> (gene pilB, transposon insertion, truncated PilB enzyme forms of the enzyme lacking the MsrA domain from strain MS11, variant VD300, overexpression of the His-tagged PilB forms in Escherichia coli strain Xl-1 blue) [38]
<16> (genes msrA and msrB are translationally fused) [15]
<17> (expression of wild-type and mutant enzymes in Escherichia coli) [12]
<17> (genes msrB and msrA, EC 1.8.4.11, are translationally fused) [15]
<17> (overexpression of the MsrB domain in Escherichia coli, strain B834(DE3) produces a selenomethionine-substituted MsrB) [5]
<17> (overexpression of wild-type and mutant enzymes in Escherichia coli) [4]
<20> (msr gene, DNA sequence determination and analysis, subcloning in Escherichia coli strain DH5-α, functional complementation of the enzyme-deficient mutant with the wild-type gene) [24]
<22> (DNA and amino acid sequence determination and analysis, expression in Escherichia coli) [27]
<25> (chromosome 1 contains 2 genes msrB, chromosome 2 contains 1 gene msrB) [15]
<26> (subcloning in Escherichia coli) [6]
<29> (overexpression of the full length tandem enzyme, the MsrA, and the MsrB domains, all His-tagged, in Escherichia coli) [25]
<32> (gene selR, DNA and amino acid sequence determination and analysis, expression as His-tagged enzyme in Escherichia coli strain BL21(DE3)) [2]
<33> (CBS-1, overexpression of the His-tagged enzyme in Escherichia coli) [25]
<34> (yeaA, overexpression of the His-tagged enzyme in Escherichia coli) [25]

Engineering

C105S <8> (<8> site-directed mutagenesis, unaltered activity compared to the wild-type enzyme [19]) [19]

C151A <14> (<14> site-directed mutagenesis, activity with Hsp21 is similar to the wild-type enzyme [26]) [26]
C169S <8> (<8> site-directed mutagenesis, active site mutant, completely inactive mutant [19]) [19]
C439S <17> (<17> site-directed mutagenesis, MsrA domain of PILB, mutant is inactive with thioredoxin, but about 10fold more active than the wild-type enzyme MsrA domain [4]) [4]
C45D/C48S/C94S/C97S <7> (<7> site-directed mutagenesis, the mutant MsrB loses binding ability for Zn^{2+} and Fe^{2+}, and shows no catalytic activity in presence of thioredoxin or DTT, substitution of the two cysteine residues of MsrB results in complete loss of the enzymes metal binding and reductase activity [39]) [39]
C494S <17> (<17> site-directed mutagenesis, MsrA domain of PILB, inactive mutant [4]) [4]
C63S <17> (<17> site-directed mutagenesis, the mutant accumulates the sulfenic acid intermediate, while the wild-type accumulates the disulfide intermediate [12]) [12]
D45C/S48C/S94C/A97C <17> (<17> site-directed mutagenesis, the mutant MsrB shows increased binding of Zn^{2+} and Fe^{2+} compared to the wild-type enzyme, overview, introduction of two cysteine residues into Neisseria meningitidis MsrB analogously to the Escherichia coli enzyme results in increased tight binding of zinc to and strongly increased thermal stability with wild-type reductase actvity but no thioredoxin recycling activity [39]) [39]
E81V <6> (<6> site-directed mutagenesis, mutation in the selenocysteine-containing or the cysteine-containing isozyme MsrB1, both mutants show reduced activity with either cofactor thioredoxin and DTT compared to wild-type MsrB1s [35]) [35]
F97N <6> (<6> site-directed mutagenesis, mutation in the selenocysteine-containing or the cysteine-containing isozyme MsrB1, both mutants show altered activity and kinetics compared to wild-type MsrB1s [35]) [35]
G77H <6> (<6> site-directed mutagenesis, mutation in the selenocysteine-containing or the cysteine-containing isozyme MsrB1, the selenocysteine-containing mutant shows reduced activity with either cofactor thioredoxin and DTT compared to wild-type selenocysteine MsrB1, while the cysteine-containing mutant shows activity and kinetics similar to the wild-type cysteine MsrB1 [35]) [35]
G77H/E81V/F97N <6> (<6> site-directed mutagenesis, mutation in the selenocysteine-containing or the cysteine-containing isozyme MsrB1, both mutants show altered activity and kinetics compared to wild-type MsrB1s [35]) [35]
G77H/F97N <6> (<6> site-directed mutagenesis, mutation in the selenocysteine-containing or the cysteine-containing isozyme MsrB1, both mutants show altered activity and kinetics compared to wild-type MsrB1s [35]) [35]
H77G <6, 8> (<6> site-directed mutagenesis, mutation of isozyme MsrB2 leads to highly reduced activity with either cofactor thioredoxin and DTT compared to wild-type MsrB2 [35]; <8> site-directed mutagenesis, mutation

of isozyme MsrB3 leads to highly reduced activity with cofactor thioredoxin or DTT compared to wild-type MsrB3 [35]) [35]

H77G/I81E/N97F <8> (<8> site-directed mutagenesis, mutation of isozyme MsrB3, inactive mutant [35]) [35]

H77G/N97F <6, 8> (<6> site-directed mutagenesis, mutation of isozyme MsrB2, inactive mutant [35]; <8> site-directed mutagenesis, mutation of isozyme MsrB3, inactive mutant [35]) [35]

H77G/V81E/N97F <6> (<6> site-directed mutagenesis, mutation of isozyme MsrB2, inactive mutant [35]) [35]

I81E <8> (<8> site-directed mutagenesis, mutation of isozyme MsrB3 leads to slightly increased activity with cofactor thioredoxin and reduced activcity with DTT compared to wild-type MsrB3 [35]) [35]

L38M/L41M <16> (<16> site-directed mutagenesis, mutation of the NT domain of PilB, thioredoxin binding structure, crystal structure analysis, overview [33]) [33]

N97F <6, 8> (<6> site-directed mutagenesis, mutation of isozyme MsrB2, the mutant is inactive with cofactor thioredoxin and shows highly reduced activity with cofactor DTT compared to wild-type MsrB2 [35]; <8> site-directed mutagenesis, mutation of isozyme MsrB3, the mutant is inactive with cofactor thioredoxin and shows highly reduced activity with cofactor DTT compared to wild-type MsrB3 [35]) [35]

N97Y <6, 8> (<6> site-directed mutagenesis, mutation of isozyme MsrB2, the mutant is inactive with cofactor thioredoxin and shows highly reduced activity with cofactor DTT compared to wild-type MsrB2 [35]; <8> site-directed mutagenesis, mutation of isozyme MsrB3, the mutant shows highly reduced activity with cofactor DTT or thioredoxin compared to wild-type MsrB3 [35]) [35]

V81E <6> (<6> site-directed mutagenesis, mutation of isozyme MsrB2 leads to reduced activity with either cofactor thioredoxin and DTT compared to wild-type MsrB2 [35]) [35]

W110A <8> (<8> site-directed mutagenesis, reduced activity compared to the wild-type enzyme [19]) [19]

W65F <17> (<17> site-directed mutagenesis, structural change of substrate binding and active site structure compared to the wild-type enzyme [12]) [12]

Additional information <2, 6, 7, 8, 11, 14, 16, 17, 20, 26> (<7> construction of a MsrA/MsrB double mutant [10]; <7> construction of a msrA/msrB double mutant for detection of additional enzyme form activities [13]; <6> construction of a non-selenomethionine mutant of MsrB by site-directed mutagenesis, exchange of the selenomethionine by Cys, Ala, or Ser, the Cys-enzyme shows reduced activity, the Ser- and Ala-enzymes are inactive, substrate specificity, overview [9]; <20> construction of an enzyme-deficient mutant strain which shows diminished growth in presence of chemical oxidants with rapid loss of viability compared to the wild-type strain, activity can be recovered by complementation with the wild-type gene, study of oxidative stress resistance and colonization activity [24]; <26> enzyme inactivation by insertional mutagenesis in strain 100-23, reduction of ecological performance of the mutant strain in the gut in vivo, mutant phenotype analysis [6]; <2,7> H_2O_2 shortens the life span of cells in constructed null mutants [14]; <11> H_2O_2 shortens the life span of cells in constructed null mutants, the mutants show decreased MsrB activity with age compared to the wild-type enzyme [14]; <16> mutant strains produce a truncated version of fused MsrA/MsrB with increased sensitivity to H_2O_2 and superoxide anions [15]; <2> mutation of the msr genes impair virulence, overview, mutation of the msrA1 operon leads to increased susceptibility to H_2O_2 [15]; <17> mutation of the recycling Cys to Ser results in an enzyme forming methionine but without recycling activity, while exchange of the catalytic Cys for Ser causes complete loss of activity [16]; <14> cells lacking MsrB show increased sensitivity to oxidative damage, and methionine-(R)-S-oxide accumulation [31]; <6> construction of MsrB null mutant and of overexpressing strains, phenotypes, overview [31]; <16> construction of truncated PilB domain forms, which show decreased survival of the organism to reactive oxygen species, the mutations do not affect piliation, pilin production, or adherence, overview [38]; <6,8> substitution of Cys residues abolish the enzymes activity with thioredoxin and increase the DTT-dependent activity, overview [35]; <11> yeast cells lacking MsrB show increased sensitivity to oxidative damage, and methionine-(R)-S-oxide accumulation [31]) [6, 9, 10, 13, 14, 15, 16, 24, 31, 35, 38]

Application

biotechnology <14> (<14> enzyme is a target for modification of redox-dependent regulation [26]) [26]

synthesis <7> (<7> enzyme can be useful in the development and action of anti-cancer and anti-inflammation drugs [13]) [13]

References

[1] Grimaud, R.; Ezraty, B.; Mitchell, J.K.; Lafitte, D.; Briand, C.; Derrick, P.J.; Barras, F.: Repair of oxidized proteins: identification of a new methionine sulfoxide reductase. J. Biol. Chem., **276**, 48915-48920 (2001)

[2] Kryukov, G.V.; Kumar, R.A.; Koc, A.; Sun, Z.; Gladyshev, V.N.: Selenoprotein R is a zinc-containing stereo-specific methionine sulfoxide reductase. Proc. Natl. Acad. Sci. USA, **99**, 4245-4250 (2002)

[3] Moskovitz, J.; Poston, J.M.; Berlett, B.S.; Nosworthy, N.J.; Szczepanowski, R.; Stadtman, E.R.: Identification and characterization of a putative active site for peptide methionine sulfoxide reductase (MsrA) and its substrate stereospecificity. J. Biol. Chem., **275**, 14167-14172 (2000)

[4] Olry, A.; Boschi-Muller, S.; Marraud, M.; Sanglier-Cianferani, S.; Van Dorsselear, A.; Branlant, G.: Characterization of the methionine sulfoxide reductase activities of PILB, a probable virulence factor from Neisseria meningitidis. J. Biol. Chem., **277**, 12016-12022 (2002)

[5] Kauffmann, B.; Favier, F.; Olry, A.; Boschi-Muller, S.; Carpentier, P.; Branlant, G.; Aubry, A.: Crystallization and preliminary X-ray diffraction studies

of the peptide methionine sulfoxide reductase B domain of Neisseria meningitidis PILB. Acta Crystallogr. Sect. D, **58**, 1467-1469 (2002)

[6] Walter, J.; Chagnaud, P.; Tannock, G.W.; Loach, D.M.; Dal Bello, F.; Jenkinson, H.F.; Hammes, W.P.; Hertel, C.: A high-molecular-mass surface protein (Lsp) and methionine sulfoxide reductase B (MsrB) contribute to the ecological performance of Lactobacillus reuteri in the murine gut. Appl. Environ. Microbiol., **71**, 979-986 (2005)

[7] Weissbach, H.; Etienne, F.; Hoshi, T.; Heinemann, S.H.; Lowther, W.T.; Matthews, B.; St John, G.; Nathan, C.; Brot, N.: Peptide methionine sulfoxide reductase: structure, mechanism of action, and biological function. Arch. Biochem. Biophys., **397**, 172-178 (2002)

[8] Moskovitz, J.; Singh, V.K.; Requena, J.; Wilkinson, B.J.; Jayaswal, R.K.; Stadtman, E.R.: Purification and characterization of methionine sulfoxide reductases from mouse and Staphylococcus aureus and their substrate stereospecificity. Biochem. Biophys. Res. Commun., **290**, 62-65 (2002)

[9] Bar-Noy, S.; Moskovitz, J.: Mouse methionine sulfoxide reductase B: effect of selenocysteine incorporation on its activity and expression of the seleno-containing enzyme in bacterial and mammalian cells. Biochem. Biophys. Res. Commun., **297**, 956-961 (2002)

[10] Spector, D.; Etienne, F.; Brot, N.; Weissbach, H.: New membrane-associated and soluble peptide methionine sulfoxide reductases in Escherichia coli. Biochem. Biophys. Res. Commun., **302**, 284-289 (2003)

[11] Etienne, F.; Resnick, L.; Sagher, D.; Brot, N.; Weissbach, H.: Reduction of Sulindac to its active metabolite, sulindac sulfide: assay and role of the methionine sulfoxide reductase system. Biochem. Biophys. Res. Commun., **312**, 1005-1010 (2003)

[12] Olry, A.; Boschi-Muller, S.; Branlant, G.: Kinetic characterization of the catalytic mechanism of methionine sulfoxide reductase B from Neisseria meningitidis. Biochemistry, **43**, 11616-11622 (2004)

[13] Weissbach, H.; Resnick, L.; Brot, N.: Methionine sulfoxide reductases: history and cellular role in protecting against oxidative damage. Biochim. Biophys. Acta, **1703**, 203-212 (2005)

[14] Moskovitz, J.: Methionine sulfoxide reductases: ubiquitous enzymes involved in antioxidant defense, protein regulation, and prevention of aging-associated diseases. Biochim. Biophys. Acta, **1703**, 213-219 (2005)

[15] Ezraty, B.; Aussel, L.; Barras, F.: Methionine sulfoxide reductases in prokaryotes. Biochim. Biophys. Acta, **1703**, 221-229 (2005)

[16] Boschi-Muller, S.; Olry, A.; Antoine, M.; Branlant, G.: The enzymology and biochemistry of methionine sulfoxide reductases. Biochim. Biophys. Acta, **1703**, 231-238 (2005)

[17] Kauffmann, B.; Aubry, A.; Favier, F.: The three-dimensional structures of peptide methionine sulfoxide reductases: current knowledge and open questions. Biochim. Biophys. Acta, **1703**, 249-260 (2005)

[18] Petropoulos, I.; Friguet, B.: Protein maintenance in aging and replicative senescence: a role for the peptide methionine sulfoxide reductases. Biochim. Biophys. Acta, **1703**, 261-266 (2005)

[19] Jung, S.; Hansel, A.; Kasperczyk, H.; Hoshi, T.; Heinemann, S.H.: Activity, tissue distribution and site-directed mutagenesis of a human peptide methionine sulfoxide reductase of type B: hCBS1. FEBS Lett., **527**, 91-94 (2002)
[20] Picot, C.R.; Perichon, M.; Cintrat, J.C.; Friguet, B.; Petropoulos, I.: The peptide methionine sulfoxide reductases, MsrA and MsrB (hCBS-1), are downregulated during replicative senescence of human WI-38 fibroblasts. FEBS Lett., **558**, 74-78 (2004)
[21] Vougier, S.; Mary, J.; Dautin, N.; Vinh, J.; Friguet, B., Ladant, D.: Essential role of methionine residues in calmodulin binding to Bordetelle pertussis adenylate cyclase, as probed by selective oxidation and repair by the peptide methionine sulfoxide reductases. J. Biol. Chem., **279**, 30210-30218 (2004)
[22] Singh, V.K.; Moskovitz, J.: Multiple methionine sulfoxide reductase genes in Staphylococcus aureus: expression of activity and roles in tolerance of oxidative stress. Microbiology, **149**, 2739-2747 (2003)
[23] Stadtman, E.R.; Moskovitz, J.; Berlett, B.S.; Levine, R.L.: Cyclic axidation and reduction of protein methionine residues is an important antioxidant mechanism. Mol. Cell. Biochem., **2347235**, 3-9 (2002)
[24] Alamuri, P.; Maier, R.J.: Methionine sulphoxide reductase is an important antioxidant enzyme in the gastric pathogen Helicobacter pylori. Mol. Microbiol., **53**, 1397-1406 (2004)
[25] Lowther, W.T.; Weissbach, H.; Etienne, F.; Brot, N.; Matthews, B.W.: The mirrored methionine sulfoxide reductases of Neisseria gonorrhoeae pilB. Nat. Struct. Biol., **9**, 348-352 (2002)
[26] Gustavsson, N.; Kokke, B.P.; Harndahl, U.; Silow, M.; Bechtold, U.; Poghosyan, Z.; Murphy, D.; Boelens, W.C.; Sundby, C.: A peptide methionine sulfoxide reductase highly expressed in photosynthetic tissue in Arabidopsis thaliana can protect the chaperone-like activity of a chloroplast-localized small heat shock protein. Plant J., **29**, 545-553 (2002)
[27] In, O.; Berberich, T.; Romdhane, S.; Feierabend, J.: Changes in gene expression during dehardening of cold-hardened winter rye (Secale cereale L.) leaves and potential role of a peptide methionine sulfoxide reductase in cold-acclimation. Planta, **220**, 941-950 (2005)
[28] Koc, A.; Gasch, A.P.; Rutherford, J.C.; Kim, H.Y.; Gladyshev, V.N.: Methionine sulfoxide reductase regulation of yeast lifespan reveals reactive oxygen species-dependent and -independent components of aging. Proc. Natl. Acad. Sci. USA, **101**, 7999-8004 (2004)
[29] Kim, H.Y.; Gladyshev, V.N.: Characterization of mouse endoplasmic reticulum methionine-R-sulfoxide reductase. Biochem. Biophys. Res. Commun., **320**, 1277-1283 (2004)
[30] Schallreuter, K.U.; Rubsam, K.; Chavan, B.; Zothner, C.; Gillbro, J.M.; Spencer, J.D.; Wood, J.M.: Functioning methionine sulfoxide reductases A and B are present in human epidermal melanocytes in the cytosol and in the nucleus. Biochem. Biophys. Res. Commun., **342**, 145-152 (2006)

[31] Moskovitz, J.: Roles of methionine suldfoxide reductases in antioxidant defense, protein regulation and survival. Curr. Pharm. Des., **11**, 1451-1457 (2005)

[32] Marchetti Maria, A.; Pizarro Gresin, O.; Sagher, D.; Deamicis, C.; Brot, N.; Hejtmancik, J.F.; Weissbach, H.; Kantorow, M.: Methionine sulfoxide reductases B1, B2, and B3 are present in the human lens and confer oxidative stress resistance to lens cells. Invest. Ophthalmol. Vis. Sci., **46**, 2107-2112 (2005)

[33] Brot, N.; Collet, J.F.; Johnson, L.C.; Jonsson, T.J.; Weissbach, H.; Lowther, W.T.: The thioredoxin domain of Neisseria gonorrhoeae PilB can use electrons from DsbD to reduce downstream methionine sulfoxide reductases. J. Biol. Chem., **281**, 32668-32675 (2006)

[34] Vieira Dos Santos, C.; Cuine, S.; Rouhier, N.; Rey, P.: The Arabidopsis plastidic methionine sulfoxide reductase B proteins. Sequence and activity characteristics, comparison of the expression with plastidic methionine sulfoxide reductase A, and induction by photooxidative stress. Plant Physiol., **138**, 909-922 (2005)

[35] Kim, H.-Y.; Gladyshev, V.N.: Different catalytic mechanisms in mammalian selenocysteine- and cysteine-containing methionine-R-sulfoxide reductases. PLoS Biol., **3**, 2080-2089 (2005)

[36] Sagher, D.; Brunell, D.; Hejtmancik, J.F.; Kantorow, M.; Brot, N.; Weissbach, H.: Thionein can serve as a reducing agent for the methionine sulfoxide reductases. Proc. Natl. Acad. Sci. USA, **103**, 8656-8661 (2006)

[37] Sagher, D.; Brunell, D.; Hejtmancik, J.F.; Kantorow, M.; Brot, N.; Weissbach, H.: Thionein can serve as a reducing agent for the methionine sulfoxide reductases. Proc. Natl. Acad. Sci. USA, **103**, 8656-8661. (2006)

[38] Skaar, E.P.; Tobiason, D.M.; Quick, J.; Judd, R.C.; Weissbach, H.; Etienne, F.; Brot, N.; Seifert, H.S.: The outer membrane localization of the Neisseria gonorrhoeae MsrA/B is involved in survival against reactive oxygen species. Proc. Natl. Acad. Sci. USA, **99**, 10108-10113 (2002)

[39] Olry, A.; Boschi-Muller, S.; Yu, H.; Burnel, D.; Branlant, G.: Insights into the role of the metal binding site in methionine-R-sulfoxide reductases B. Protein Sci., **14**, 2828-2837 (2005)

L-Methionine (S)-S-oxide reductase 1.8.4.13

1 Nomenclature

EC number
1.8.4.13

Systematic name
L-methionine:thioredoxin-disulfide S-oxidoreductase

Recommended name
L-methionine (S)-S-oxide reductase

Synonyms
methionine sulfoxide reductase <1, 4> [1]

2 Source Organism

<1> *Escherichia coli* (no sequence specified) [1, 4]
<2> *Saccharomyces cerevisiae* (no sequence specified) [2]
<3> *Escherichia coli K-12* (no sequence specified) (<3> HpdA protein [3]) [3]
<4> *Escherichia coli B* (no sequence specified) [1]

3 Reaction and Specificity

Catalyzed reaction
L-methionine + thioredoxin disulfide + H_2O = L-methionine (S)-S-oxide + thioredoxin

Natural substrates and products

S L-(-)-methionine S-oxide + NADPH <2> (<2> the enzyme is specific for the L-(-)-stereoisomer [2]) (Reversibility: ir) [2]
P L-methionine + $NADP^+$ + H_2O
S L-methionine (S)-sulfoxide + NADPH <1> (<1> membrane-bound enzyme form Mem-R,S-Msr [4]) (Reversibility: ?) [4]
P L-methionine + $NADP^+$ + H_2O
S sulindac + NADPH <1> (<1> activation of a methionine sulfoxide-containing prodrug, activity with membrane-bound enzyme form Mem-R,S-Msr [4]) (Reversibility: ?) [4]
P sulindac sulfide + $NADP^+$ + H_2O (<1> activated drug which inhibits cyclooxygenase 1 and 2 and exhibits anti-inflammatory activity [4])

S Additional information <1, 2, 3, 4> (<1> the enzyme protects cells against oxidative damage and plays a role in age-related missfunctions [4]; <4> methionine sulfoxide reduction pathway, overview [1]; <2> possible pathway, overview [2]; <3> the enzyme might act in vivo on the free methionine sulfoxide substrate or on protein-bound methionine sulfoxide residues [3]) (Reversibility: ?) [1, 2, 3, 4]

P ?

Substrates and products

S L-(-)-methionine S-oxide + NADPH <2> (<2> the enzyme is specific for the L-(-)-stereoisomer [2]; <2> the enzyme is absolutely specific for the L-(-)-stereoisomer [2]) (Reversibility: ir) [2]

P L-methionine + $NADP^+$ + H_2O

S L-methionine (S)-S-oxide + NADPH <3, 4> (<3> substrate is L-Met-DL-sulfoxide or L-Met-L-sulfoxide [3]) (Reversibility: ?) [1, 3]

P L-methionine + $NADP^+$ + H_2O

S L-methionine (S)-S-oxide + reduced DTT <4> (<4> highest activity at 10 mM [1]) (Reversibility: ?) [1]

P L-methionine + oxidized DTT + H_2O

S L-methionine (S)-sulfoxide + NADPH <1> (<1> membrane-bound enzyme form Mem-R,S-Msr [4]) (Reversibility: ?) [4]

P L-methionine + $NADP^+$ + H_2O

S L-methionine sulfoxide enkephalin + NADPH <1> (<1> membrane-bound enzyme form Mem-R,S-Msr [4]) (Reversibility: ?) [4]

P L-methionine enkephalin + $NADP^+$ + H_2O

S N-acetyl-L-methionine (S)-sulfoxide + thioredoxin <1> (<1> membrane-bound enzyme form Mem-R,S-Msr [4]) (Reversibility: ?) [4]

P N-acetyl-L-methionine + NADPH

S sulindac + NADPH <1> (<1> activation of a methionine sulfoxide-containing prodrug, activity with membrane-bound enzyme form Mem-R,S-Msr [4]; <1> activity with membrane-bound enzyme form Mem-R,S-Msr [4]) (Reversibility: ?) [4]

P sulindac sulfide + $NADP^+$ + H_2O (<1> activated drug which inhibits cyclooxygenase 1 and 2 and exhibits anti-inflammatory activity [4])

S Additional information <1, 2, 3, 4> (<1> the enzyme protects cells against oxidative damage and plays a role in age-related missfunctions [4]; <4> methionine sulfoxide reduction pathway, overview [1]; <2> possible pathway, overview [2]; <3> the enzyme might act in vivo on the free methionine sulfoxide substrate or on protein-bound methionine sulfoxide residues [3]; <1> substrate specificity, overview, Mem-R,S-Msr can also catalyze the reduction of L-methionine (R)-sulfoxide, EC 1.8.4.14, enzyme reduces oxidized methionine residues of the ribosomal protein L12, which becomes reversibly inactivated and forms monomers instead of dimers upon oxidation [4]; <2> the purified enzyme may contain thioredoxin [2]) (Reversibility: ?) [1, 2, 3, 4]

P ?

Inhibitors

4-Chloromercuribenzoate <2> (<2> complete inhibition at 1 mM [2]) [2]
Arsenite <2> (<2> strong inhibition [2]) [2]
Bovine serum albumin <2> [2]
DL-methionine sulfoximine <2> (<2> 50% inhibition at 1 mM [2]) [2]
Iodoacetic acid <2> (<2> complete inhibition at 10 mM [2]) [2]
L-(-)-methionine S-oxide <2> (<2> competitively inhibits the activity with a disulfide about 2.5fold [2]) [2]
hydroxyethyl disulfide <2> (<2> competitively inhibits the activity with L-(-)-methionine S-oxide about 2.5fold [2]) [2]

Cofactors/prosthetic groups

NADPH <1, 2, 3, 4> (<3> required [3]; <2> specific for [2]; <1> membrane-bound enzyme form Mem-R,S-Msr [4]; <4> required, can be substituted by DTT, but not by GSH or 2-mercaptoethanol [1]) [1,2,3,4]
Additional information <1> (<1> no activity with DTT as cofactor by membrane-bound enzyme form Mem-R,S-Msr [4]) [4]

Metals, ions

Mg^{2+} <4> [1]

Specific activity (U/mg)

0.00009 <1> (<1> enzyme form Mem-R,S-Msr, substrate sulindac [4]) [4]
0.00452 <4> (<4> purified enzyme [1]) [1]

K_m-Value (mM)

0.0003 <4> (L-methionine (S)-S-oxide, <4> pH 7.5, 37°C [1]) [1]
0.2 <2> (L-(-)-methionine S-oxide, <2> pH 7.0, 22°C [2]) [2]

pH-Optimum

6.9 <3> (<3> assay at [3]) [3]
7 <2> [2]
7.5 <4> (<4> assay at [1]) [1]

Temperature optimum (°C)

22 <2> (<2> assay at room temperature [2]) [2]
37 <3, 4> (<3,4> assay at [1,3]) [1, 3]

4 Enzyme Structure

Subunits

? <4> (<4> x * 21000, SDS-PAGE [1]) [1]

5 Isolation/Preparation/Mutation/Application

Localization

membrane <1> (<1> enzyme form Mem-R,S-Msr [4]) [4]
particle-bound <3> [3]
soluble <3> [3]

Purification

<2> (two native enzyme forms by ammonium sulfate fractionation, anion exchange chromatography, and a calcium phosphate resin chromatography) [2]
<3> (native enzyme partially by subcellular fractionation) [3]
<4> (native enzyme about 1100fold by ammonium sulfate fractionation, anion exchange chromatography, gel filtration, again anion exchange chromatography, heat treatment, and hydroxyapatite chromatography to near homogeneity) [1]

Application

synthesis <1> (<1> enzyme can be useful in the development and action of anti-cancer and anti-inflammation drugs [4]) [4]

References

[1] Ejiri, S.; Weissbach, H.; Brot, N.: The purification of methionine sulfoxide reductase from Escherichia coli. Anal. Biochem., **102**, 393-398 (1980)
[2] Black, S.; Harte, E.M.; Hudson, B.; Wartolofsky, L.: A specific enzymatic reduction of L(-)methionine sulfoxide and a related nonspecific reduction of disulfides. J. Biol. Chem., **235**, 2910-2916 (1960)
[3] Ejiri, S.; Weissbach, H.; Brot, N.: Reduction of methionine sulfoxide to methionine by Escherichia coli. J. Bacteriol., **139**, 161-164 (1979)
[4] Weissbach, H.; Resnick, L.; Brot, N.: Methionine sulfoxide reductases: history and cellular role in protecting against oxidative damage. Biochim. Biophys. Acta, **1703**, 203-212 (2005)

L-Methionine (R)-S-oxide reductase 1.8.4.14

1 Nomenclature

EC number
1.8.4.14

Systematic name
L-methionine:thioredoxin-disulfide S-oxidoreductase [L-methionine (R)-S-oxide-forming]

Recommended name
L-methionine (R)-S-oxide reductase

Synonyms
fRMsr <1> [1]
methionine sulfoxide reductase <1> [1]

2 Source Organism

<1> *Escherichia coli* (no sequence specified) [1]

3 Reaction and Specificity

Catalyzed reaction
L-methionine + thioredoxin disulfide + H_2O = L-methionine (R)-S-oxide + thioredoxin

Substrates and products
S L-methionine + thioredoxin disulfide + H_2O <1> (<1> enzyme specific for free L-methionine-(R)-sulfoxide and not met-R-(o) in peptide linkage [1]) (Reversibility: ?) [1]
P L-methionine (R)-S-oxide + thioredoxin

Cofactors/prosthetic groups
Dithiothreitol <1> (<1> DTT can replace NADPH, but is much less effective [1]) [1]
NADPH <1> (<1> absolute requirement [1]) [1]
thioredoxin <1> (<1> omission results in a 70% drop in activity [1]) [1]

5 Isolation/Preparation/Mutation/Application

Purification

<1> (partially purified by $(NH_4)_2SO_4$ precipitation (30-60% saturation) from MsrA/MsrB double mutants) [1]

References

[1] Etienne, F.; Spector, D.; Brot, N.; Weissbach, H.: A methionine sulfoxide reductase in Escherichia coli that reduces the R enantiomer of methionine sulfoxide. Biochem. Biophys. Res. Commun., **300**, 378-382 (2003)

Thiosulfate dehydrogenase (quinone) 1.8.5.2

1 Nomenclature

EC number
1.8.5.2

Systematic name
thiosulfate:6-decylubiquinone oxidoreductase

Recommended name
thiosulfate dehydrogenase (quinone)

Synonyms
TQO <1> [1]
thiosulfate oxidoreductase <1> [1]
thiosulfate oxidoreductase tetrathionate-forming <1> [1]
thiosulfate:quinone oxidoreductase <1> [1]
thiosulphate:quinone oxidoreductase <1> [1]

CAS registry number
9076-88-4

2 Source Organism

<1> *Acidianus ambivalens* (no sequence specified) [1]
<2> *Acidithiobacillus ferrooxidans* (no sequence specified) [2]

3 Reaction and Specificity

Catalyzed reaction
2 thiosulfate + 2 6-decylubiquinone = tetrathionate + 2 6-decylubiquinol

Natural substrates and products
S thiosulfate + 6-decylubiquinone <1> (<1> the enzyme couples sulfur compound oxidation with quinone reduction [1]) (Reversibility: r) [1]
P tetrathionate + 6-decylubiquinol
S thiosulfate + caldariella quinone <1> (<1> the physiological electron acceptor is most probably a caldariella quinone type quinone [1]) (Reversibility: r) [1]
P tetrathionate + caldariella quinole

Substrates and products

- **S** tetrathionate + reduced methylene blue <1> (Reversibility: r) [1]
- **P** thiosulfate + oxidized methylene blue
- **S** thiosulfate + 6-decylubiquinone <1> (<1> the enzyme couples sulfur compound oxidation with quinone reduction [1]) (Reversibility: r) [1]
- **P** tetrathionate + 6-decylubiquinol
- **S** thiosulfate + caldariella quinone <1> (<1> the physiological electron acceptor is most probably a caldariella quinone type quinone [1]) (Reversibility: r) [1]
- **P** tetrathionate + caldariella quinole
- **S** thiosulfate + ferricyanide <1> (Reversibility: r) [1]
- **P** tetrathionate + ferrocyanide
- **S** Additional information <1> (<1> horse heart cytochrome c is not reduced [1]) (Reversibility: ?) [1]
- **P** ?

Inhibitors

dithionite <1> (<1> 1 mM, complete inhibition [1]) [1]
metabisulfite <1> (<1> 1 mM, complete inhibition [1]) [1]
N-ethylmaleimide <1> (<1> 1 mM, 54% inhibition [1]) [1]
sulfite <1> (<1> 0.005 mM, 48% inhibition, 0.05 mM, complete inhibition [1]) [1]
Triton X-100 <1> (<1> 1%, complete inhibition [1]) [1]
Zn^{2+} <1> (<1> 1 mM, 37% inhibition, 5 mM, 82% inhibition [1]) [1]
reduced titanium citrate <1> (<1> 1 mM, 65% inhibition [1]) [1]
Additional information <1> (<1> no inhibition by sulfate or tetrathionate [1]) [1]

Cofactors/prosthetic groups

Additional information <1> (<1> a mixture of caldariella quinone, Sulfolobus quinone and menaquinone is non-covalently bound to the protein [1]) [1]

Turnover number (min^{-1})

167 <1> (thiosulfate, <1> 80°C, pH 6.0 [1]) [1]

Specific activity (U/mg)

0.397 <1> (<1> activity with decyl ubiquinone at pH 6, 92°C [1]) [1]
49.9 <1> (<1> with ferricyanide as electron acceptor [1]) [1]
73.4 <1> (<1> activity with ferricyanide at pH 6, 92°C [1]) [1]
397 <1> (<1> with 6-decylubiquinone as electron acceptor [1]) [1]

K_m-Value (mM)

0.00587 <1> (6-decylubiquinone) [1]
0.0059 <1> (6-decylubiquinone, <1> 80°C, pH 6.0 [1]) [1]
2.6 <1> (thiosulfate, <1> 80°C, pH 6.0 [1]) [1]
3.4 <1> (ferricyanide, <1> 80°C, pH 6.0 [1]) [1]

K_i-Value (mM)

0.005 <1> (sulfite, <1> pH 6.0, 80°C [1]) [1]

pH-Optimum

4.5-5 <1> [1]
5 <1> [1]

pH-Range

3.5-6.8 <1> (<1> approx. 55% of maximal activity at pH 3.5, approx. 22% of maximal activity at pH 6.5 [1]) [1]
4.5-5.5 <1> (<1> pH 4.5: about 70% of maximal activity, pH 5.5: about 35% of maximal activity [1]) [1]

Temperature optimum (°C)

20-92 <1> (<1> increasing activity is observed in the range between 20 and 92°C, a maximum is not observed because of technical reasons [1]) [1]

Temperature range (°C)

20-92 <1> (<1> increasing activity is observed in the range between 20 and 92°C, a maximum is not observed because of technical reasons [1]) [1]

4 Enzyme Structure

Molecular weight

102000 <1> (<1> gel filtration [1]) [1]

Subunits

tetramer <1> (<1> α_2,β_2, 2 * 16000 + 2 * 28000, SDS-PAGE [1]; <1> α_2,β_2, 2 * 20400 + 2 * 18700, deduced from nucleotide sequence [1]; <1> $\alpha_2\beta_2$, the 28000 Da subunit and the 16000 Da subunit are identical to to DoxA and DoxD from Acidianus ambivalens quinol:oxygen oxidoreductase, 2 * 16000 + 2 * 28000, SDS-PAGE [1]) [1]

Posttranslational modification

glycoprotein <1> (<1> the larger subunit (28000 Da) appears to be glycosylated [1]) [1]
phosphoprotein <1> (<1> glycosylation of subunit DoxA, i.e. β subunit of TQO [1]) [1]

5 Isolation/Preparation/Mutation/Application

Localization

membrane <1> [1]

Purification

<1> [1]
<1> (solubilization, Q-Sepharose, hydroxylapatite, DEAE-Sepharose) [1]
<2> (affinity matrix HR 5/5 column chromatography, matrices based on cytochrome c immobilized on crosslinked triazine (2,4,6-tris(aminoethylamine)-1,3,5-triazine), cytochrome c immobilized on Silasorb-amine with carbodiimide activation and cytochrome c immobilized on Sepharose CL-4B) [2]

References

[1] Muller, F.H.; Bandeiras, T.M.; Urich, T.; Teixeira, M.; Gomes, C.M.; Kletzin, A.: Coupling of the pathway of sulphur oxidation to dioxygen reduction: characterization of a novel membrane-bound thiosulphate:quinone oxidoreductase. Mol. Microbiol., **53**, 1147-1160 (2004)

[2] Janiczek, O.; Pokorna, B.; Zemanova, J.; Mandl, M.: Use of immobilized cytochrome c as a ligand for affinity chromatography of thiosulfate dehydrogenase from Acidithiobacillus ferrooxidans. J. Biotechnol., **117**, 293-298 (2005)

CoB-CoM Heterodisulfide reductase 1.8.98.1

1 Nomenclature

EC number
1.8.98.1

Systematic name
coenzyme B:coenzyme M:methanophenazine oxidoreductase

Recommended name
CoB-CoM heterodisulfide reductase

Synonyms
CO:heterodisulfide oxidoreductase system <5> [18]
Co:CoM-S-S-CoB oxidoreductase system <5> [18]
CoM-S-S-HTP reductase <2> [2]
DsrK <3> (<3> homologue to the HdrD subunit of heterodisulfide reductase [22]) [22]
$F_{420}H_2$-dependent CoM-S-S-HTP reductase <12, 13> [14]
$F_{420}H_2$-dependent heterodisulfide reductase <12, 13> [14]
H_2:heterodisulfide oxidoreductase <1, 9> [5, 7, 19]
HDR <1, 3, 4, 5, 11> [3, 4, 11, 16, 21, 22, 23]
HdrB <11> (<11> catalytic subunit of HDR [23]) [23]
HdrC <11> (<11> catalytic subunit of HDR [23]) [23]
HdrD <1> (<1> catalytic subunit of HDR [23]) [23]
coenzyme F_{420}:heterodisulfide oxidoreductase <7> [20]
heterodisulfide reductase <1, 3, 4, 5, 11> [4, 21, 22, 23]
soluble heterodisulfide reductase
Additional information <1, 2> (<2> H_2:heterodisulfide oxidoreductase complex consists of F_{420}-non-reducing hydrogenase and heterodisulfide reductase [8]; <1> the H_2:heterodisulfide oxidoreductase multienzyme complex is composed of nine polypeptides with MW 46000 Da, 39000 Da, 28000 Da, 25000 Da, 23000 Da, 21000 Da, 20000 Da, 16000 Da, 15000 Da [7]) [7, 8]

CAS registry number
116515-35-6
128172-71-4

2 Source Organism

<1> *Methanosarcina barkeri* (no sequence specified) [2, 7, 9, 11, 16, 23]
<2> *Methanobacterium thermoautotrophicum* (no sequence specified) (<2> β-subunit [2,8,13]) [2, 8, 13]

<3> *Desulfovibrio desulfuricans* (no sequence specified) [22]
<4> *Desulfovibrio vulgaris* (no sequence specified) [21]
<5> *Methanosarcina thermophila* (no sequence specified) [3,4,18,23]
<6> *Methanococcus voltae* (no sequence specified) [6]
<7> *methanogenic bacterium strain Gö1* (no sequence specified) [20]
<8> *Methanothermobacter thermautotrophicus* (no sequence specified) [12]
<9> *Methanosarcina mazei* (no sequence specified) [1,5,19]
<10> *Methanosarcina barkeri* (UNIPROT accession number: P96796) [10]
<11> *Methanothermobacter marburgensis* (no sequence specified) [11, 15, 16, 17, 23]
<12> *Methanolobus tindarius* (no sequence specified) [14]
<13> *bacterium strain Gö1* (no sequence specified) [14]

3 Reaction and Specificity

Catalyzed reaction

coenzyme B + coenzyme M + methanophenazine = N-{7-[(2-sulfoethyl)-dithio]heptanoyl}-3-O-phospho-L-threonine + dihydromethanophenazine

Reaction type

oxidation

reduction (<1,5,11> the enzyme reduces the heterodisulfide CoM-S-S-CoB of the methanogenic thiol-coenzymes, coenzyme M and coenzyme B [23])

Natural substrates and products

S N-(7-[(2-sulfoethyl)dithio]heptanoyl)-3-O-phospho-L-threonine + H_2 <1, 2, 9> (<2> energy-conserving step [8]; <1> energy-conserving step in methanohenic archaea [7]; <9> reaction is coupled with transfer of protons across the cytoplasmic membrane. The electrochemical proton gradient thereby generated drives ATP synthesis from ADP plus phosphate. Two different proton-translocating segments are present in the H_2:hetero-disulfide oxidoreductase system. The first involves the 2-hydroxyphenazine-dependent hydrogenase, and the second involves the heterodisulfide reductase [19]) (Reversibility: ?) [7, 8, 9, 19]

P coenzyme B + coenzyme M

S N-(7-[(2-sulfoethyl)dithio]heptanoyl)-3-O-phospho-L-threonine + coenzyme $F_{420}H_2$ <6, 7> (<7> reaction is coupled with the transfer of protons across the membrane of washed vesicles, which drives ATP formation from ADP [20]) (Reversibility: ?) [6, 2]

P coenzyme B + coenzyme M + coenzyme F_{420}

S N-(7-[(2-sulfoethyl)dithio]heptanoyl)-3-O-phospho-L-threonine + dihydromethanophenazine <5> (Reversibility: ?) [4]

P coenzyme B + coenzyme M + methanophenazine

S N-(7-[(2-sulfoethyl)dithio]heptanoyl)-3-O-phospho-L-threonine + reduced cytochrome b <5> (<5> electron transfer by a CO dehydrogenase in complex with the heterodisulfide oxidoreductase, no H_2 uptake in vivo, reaction is part of the overall reduction of acetate to methane and H_2O, also providing a proton-translocation redox system to drive ATP formation [18]) (Reversibility: ?) [18]

P coenzyme B + coenzyme M + oxidized cytochrome b

S coenzyme B + coenzyme M + methanophenazine <2, 5, 9> (Reversibility: ?) [1, 2, 3, 4]

P N-(7-[(2-sulfoethyl)dithio]heptanoyl)-3-O-phospho-L-threonine + dihydromethanophenazine

Substrates and products

S CoM-S-S-CoB + ? <1, 5, 11> (Reversibility: ?) [23]

P CoM-SH + CoB-SH + ?

S N-(7-[(2-sulfoethyl)dithio]heptanoyl)-3-O-phospho-L-threonine + H_2 <8> (Reversibility: ?) [12]

P coenzyme B + coenzyme

S N-(7-[(2-sulfoethyl)dithio]heptanoyl)-3-O-phospho-L-threonine + H_2 <1, 2, 9> (<2> energy-conserving step [8]; <1> energy-conserving step in methanohenic archaea [7]; <9> reaction is coupled with transfer of protons across the cytoplasmic membrane. The electrochemical proton gradient thereby generated drives ATP synthesis from ADP plus phosphate. Two different proton-translocating segments are present in the H_2:heterodisulfide oxidoreductase system. The first involves the 2-hydroxyphenazine-dependent hydrogenase, and the second involves the heterodisulfide reductase [19]; <9> reaction is independent of coenzyme F_{420}. Reaction is coupled to proton translocation across the cytoplasmic membrane into the lumen of the everted vesicle. The transmembrane electrochemical gradient thereby generated unduces ATP synthesis from ADP + phosphate [5]; <2> the D-enantiomer of the heterodisulfide is reduced at approximately 35% of the rate observed for the L-form [2]) (Reversibility: ?) [2, 5, 7, 8, 9, 13, 19]

P coenzyme B + coenzyme M

S N-(7-[(2-sulfoethyl)dithio]heptanoyl)-3-O-phospho-L-threonine + coenzyme $F_{420}H_2$ <6, 7, 12, 13> (<7> reaction is coupled with the transfer of protons across the membrane of washed vesicles, which drives ATP formation from ADP [20]) (Reversibility: ?) [6, 14, 20]

P coenzyme B + coenzyme M + coenzyme F420

S N-(7-[(2-sulfoethyl)dithio]heptanoyl)-3-O-phospho-L-threonine + dihydro-2-hydroxyphenazine <9> (Reversibility: ?) [19]

P coenzyme B + coenzyme M

S N-(7-[(2-sulfoethyl)dithio]heptanoyl)-3-O-phospho-L-threonine + dihydromethanophenazine <5> (<5> ping-pong mechanism [4]) (Reversibility: ?) [4]

P coenzyme B + coenzyme M + methanophenazine

S N-(7-[(2-sulfoethyl)dithio]heptanoyl)-3-O-phospho-L-threonine + reduced 2-bromophenazine <9> (Reversibility: ?) [1]

P coenzyme B + coenzyme M + 2-bromophenazine

S N-(7-[(2-sulfoethyl)dithio]heptanoyl)-3-O-phospho-L-threonine + reduced 2-hydroxyphenazine <9> (Reversibility: ?) [1]

P coenzyme B + coenzyme M + 2-hydroxyphenazine

S N-(7-[(2-sulfoethyl)dithio]heptanoyl)-3-O-phospho-L-threonine + reduced benzyl viologen <1, 11> (<11> reaction intermediate is a [4Fe-4S]3+ cluster that is coordinated by one of the cysteines of nearby active-site disulfide or by the sulfur of coenzyme M [11]) (Reversibility: ?) [7, 11]

P coenzyme B + coenzyme M + benzyl viologen

S N-(7-[(2-sulfoethyl)dithio]heptanoyl)-3-O-phospho-L-threonine + reduced benzylviologen <2> (Reversibility: r) [2]

P coenzyme B + coenzyme M + benzylviologen

S N-(7-[(2-sulfoethyl)dithio]heptanoyl)-3-O-phospho-L-threonine + reduced benzylviologen <1> (<1> reaction intermediate is a [4Fe-4S]3+ cluster that is coordinated by one of the cysteines of nearby active-site disulfide or by the sulfur of coenzyme M [11]) (Reversibility: ?) [11]

P coenzyme B + coenzyme M + benzy viologen

S N-(7-[(2-sulfoethyl)dithio]heptanoyl)-3-O-phospho-L-threonine + reduced cytochrome b <5> (<5> electron transfer by a CO dehydrogenase in complex with the heterodisulfide oxidoreductase, no H_2 uptake in vivo, reaction is part of the overall reduction of acetate to methane and H_2O, also providing a proton-translocation redox system to drive ATP formation [18]) (Reversibility: ?) [18]

P coenzyme B + coenzyme M + oxidized cytochrome b

S N-(7-[(2-sulfoethyl)dithio]heptanoyl)-3-O-phospho-L-threonine + reduced methyl viologen <5> (<5> ping-pong mechanism [4]) (Reversibility: ?) [4]

P coenzyme B + coenzyme M + methyl viologen

S N-(7-[(2-sulfoethyl)dithio]heptanoyl)-3-O-phospho-L-threonine + reduced methylene blue <1, 2> (Reversibility: r) [2]

P coenzyme B + coenzyme M + methylene blue

S N-(7-[(2-sulfoethyl)dithio]heptanoyl)-3-O-phospho-L-threonine + reduced methylviologen <1, 9> (Reversibility: ?) [1, 9]

P coenzyme B + coenzyme M + methyl viologen

S N-(7-[(2-sulfoethyl)dithio]heptanoyl)-3-O-phospho-L-threonine + reduced methylviologen <5> (Reversibility: ?) [18]

P coenzyme B + coenzyme M + oxidized methylviologen

S N-(7-[(2-sulfoethyl)dithio]heptanoyl)-3-O-phospho-L-threonine + reduced methylviologen <5, 6> (Reversibility: ?) [3, 6]

P coenzyme B + coenzyme M + methylviologen

S N-(7-[(2-sulfoethyl)dithio]heptanoyl)-3-O-phospho-L-threonine + reduced phenazine-1-carboxylic acid <9> (Reversibility: ?) [1]

P coenzyme B + coenzyme M + phenazine-1-carboxylic acid

S coenzyme B + 6-mercaptohexanoyl-L-threonine phosphate + methylene blue <2> (Reversibility: ir) [2]

P ? + methylene blue

S coenzyme B + coenzyme M + methanophenazine <1, 2, 5, 9, 11> (Reversibility: ?) [1, 2, 3, 4, 23]

P N-(7-[(2-sulfoethyl)dithio]heptanoyl)-3-O-phospho-L-threonine + dihydromethanophenazine

S coenzyme B + coenzyme M + methylene blue <1> (Reversibility: ?) [9]

P N-(7-[(2-sulfoethyl)dithio]heptanoyl)-3-O-phospho-L-threonine + reduced methylene blue

S Additional information <2, 12, 13> (<2> homodisulfides of coenzyme M and coenzyme B are not reduced [2]; <12,13> no activity with the homodisulfides of coenzyme M and coenzyme B [14]; <2> the enzyme does not catalyze the reduction of N-(7-[(2-sulfoethyl)dithio]heptanoyl)-3-O-phospho-L-threonine with NADH, NADPH or reduced coenzyme F_{420} [2]) (Reversibility: ?) [2, 14]

P ?

Inhibitors

(2R,3R)-3-hydroxy-2-([6-[(2-sulfoethyl)dithio]hexanoyl]amino)butanoic acid <2> (<2> K_i above 2 mM [2]) [2]
1,3-propanesulfone <1> (<1> 0.3 mM, 50% inhibition [11]) [11]
iodoacetamide <1> (<1> 0.1 mM, 50% inhibition [11]) [11]
mersalyl acid <5> [4]
NEM <1> (<1> 0.4 mM, 50% inhibition [11]) [11]
Additional information <1, 5, 11> (<11> no inhibition by NEM, iodoacetamide or 1,3-propanesulfone up to 2 mM [11]; <1,5,11> HDR is not inhibited by cysteine alkylating reagents at concentrations up to 2 mM [23]) [11, 23]

Cofactors/prosthetic groups

cytochrome b <1, 10> (<10> the 23000 da subunit is cytochrome b [10]; <1> the 23000 Da subunit is the cytochrome b, cytochrome b serves as electron donor for the heterodisulfide reduction [9]; <1> the H2:heterodisulfide oxidoreductase multienzyme complex contains 3.2 nmol cytochrome b per mg protein [7]) [7, 9, 10]
FAD <1, 2, 5, 11> (<2> 1 mol H2: heterodisulfide oxidoreductase complex with MW 250000 Da contains approximately 0.9 mol FAD [8]; <2> enzyme contains 4 mol FAD per mol of native enzyme [2]; <1> the enzyme contains 3 mol of FAD per mg of protein [9]; <1> the H_2:heterodisulfide oxidoreductase multienzyme complex contiams 0.6 nmol per mg of protein [7]) [2, 7, 8, 9, 23]
heme <5> (<5> contains two b-hemes. Only the low spin heme and the high potential FeS cluster are involved in reduction of N-(7-[(2-sulfoethyl)dithio]-heptanoyl)-3-O-phospho-L-threonine [4]; <5> the enzyme contains two discrete b-type hemes in the small subunit. One heme is high-spin and has a high midpoint potential, -23 mV, whereas the other heme is low-spin and exhibits a relatively low midpoint potential, -180 mV [3]) [3, 4]
coenzyme F_{420} <6> [6]
Additional information <1, 2, 5, 10> (<5> coenzyme F_{420} is not required [18]; <2> no reaction with conenzyme F_{420} [8]; <10> not a flavoenzyme, both subunits lack an FAD-binding motif [10]; <1> the H_2:heterodisukfide oxidoreductase complex does not mediate the reduction of coenzyme F_{420} with H_2 nor the oxidation of reduced coenzyme F_{420} with the heterodisulfide [7]) [7, 8, 10, 18]

Activating compounds

Aquacobalamin <2> (<2> stimulates the activity of the purified H_2: heterodisulfide oxidoreductase complex [8]) [8]

Metals, ions

iron <1, 2, 5, 11> (<2> 1 mol H2: heterodisulfide oxidoreductase complex with MW 250000 Da contains approximately 26 mol non-heme iron [8]; <11> active-site [4Fe-4S] cluster that is directly involved in mediating heterodisulfide reduction [15]; <1,5,11> coenzyme M binds to a [4Fe-4S] cluster in the active site [16,23]; <11> contains an active-site [4Fe-4S] cluster that is directly involved in mediating heterodisulfide reduction [17]; <5> contains two [Fe4S4] clusters. Only the low spin heme and the high potential FeS cluster are involved in reduction of N-(7-[(2-sulfoethyl)dithio]heptanoyl)-3-O-phospho-L-threonine. A 4 Fe cluster is the initial electron acceptor [4]; <1> the enzyme contains 280 nmol of non-heme iron per mg of protein [9]; <5> the enzyme contains two distinct [Fe4S4]2+/1+ clusters in the large subunit [3]; <1> the H2:heterodisulfide oxidoreductase multienzyme complex contains 70-80 nmol non-heme iron and acid labile sulfur per mg of protein [7]) [3, 4, 7, 8, 9, 15, 16, 17, 23]

Ni <1, 2> (<2> 1 mol H_2: heterodisulfide oxidoreductase complex with MW 250000 Da contains approximately 0.6 mol nickel [8]; <1> the H_2:heterodisulfide oxidoreductase multienzyme complex contains 0.6 nmol Ni per mg of protein [7]) [7, 8]

Turnover number (min^{-1})

74 <5> (dihydromethanophenazine, <5> 25°C [4]) [4]

Additional information <2> [2]

Specific activity (U/mg)

0.075 <12, 13> [14]

0.198 <9> [5]

0.2 <2> [13]

0.6 <1> (<1> formation of the heterodisulfide, pH 7.6, 40°C [2]) [2]

0.62 <1> (<1> reduction of the heterodisulfide, pH 7.6, 40°C [2]) [2]

3 <1> [9]

6 <1> (<1> reduction of the heterodisulfide with H_2 [7]) [7]

10 <2> [2]

24 <1> (<1> reduction of the heterodisulfide with benzylviologen [7]) [7]

29 <2> [2]

52 <5> [4]

200 <5> [3]

Additional information <6> [6]

K_m-Value (mM)

0.03 <2> (H2, <2> pH 7, 65°C, H2: heterodisulfide oxidoreductase complex [8]) [8]

0.092 <5> (dihydromethanophenazine, <5> 25°C [4]) [4]

0.1 <2> (N-(7-[(2-sulfoethyl)dithio]heptanoyl)-3-O-phospho-L-threonine, <2> pH 7.6, 54°C [2]) [2]

0.13 <2> (N-(7-[(2-sulfoethyl)dithio]heptanoyl)-3-O-phospho-L-threonine, <2> pH 7.6, 65°C [2]) [2]
0.144 <5> (N-(7-[(2-sulfoethyl)dithio]heptanoyl)-3-O-phospho-L-threonine, <5> 25°C [4]) [4]
0.2 <2> (6-mercaptohexanoyl-L-threonine phosphate, <2> pH 7.6, 54°C [2]) [2]
0.2 <2> (coenzyme M, <2> pH 7.6, 65°C [2]) [2]
0.2 <2> (N-(7-[(2-sulfoethyl)dithio]heptanoyl)-3-O-phospho-L-threonine, <2> pH 7, 65°C, H_2: heterodisulfide oxidoreductase complex [8]) [8]
0.25 <1> (coenzyme B, <1> pH 7.0, 37°C [9]) [9]
0.4 <1> (N-(7-[(2-sulfoethyl)dithio]heptanoyl)-3-O-phospho-L-threonine, <1> pH 7.0, 37°C, reaction with reduced benzyl viologen [9]) [9]
0.8 <1> (coenzyme M, <1> pH 7.0, 37°C [9]) [9]
3 <2> (N-(7-[(2-sulfoethyl)dithio]heptanoyl)-3-O-phospho-D-threonine, <2> pH 7.6, 65°C [2]) [2]
Additional information <2> (<2> K_m-value for N-(7-[(2-sulfoethyl)dithio]-heptanoyl)-3-O-phospho-L-threonine is below 0.1 mM [13]; <2> the K_M-value for coenzyme B is below 0.05 mM at pH 7.6, 65°C [2]; <2> the K_M-value for reduced benzyl viologen is below 0.05 mM [2]) [2, 13]

pH-Optimum

6.3 <2> (<2> Pipes/KOH buffer [13]) [13]
7 <2> (<2> H_2:heterodisulfide oxidoreductase activity [8]) [8]
7.6 <2> (<2> 20°C, Tris-HCl buffer [13]) [13]

pH-Range

5-7.5 <1> (<1> activity with reduced benzyl viologen and N-(7-[(2-sulfoethyl)dithio]heptanoyl)-3-O-phospho-L-threonine increases continously with decreasing pH in the range between pH 7.5 and pH 5.0 [9]) [9]

Temperature optimum (°C)

50 <1> [9]
75-80 <2> [2]

4 Enzyme Structure

Molecular weight

61000 <3> (<3> SDS-PAGE [22]) [22]
155000 <5> (<5> gel filtration [3]) [3]
206000 <5> (<5> equilibrium sedimentation [3]) [3]
250000 <2> (<2> hexameric complex, H_2:heterodisulfide oxidoreductase complex consisting of F_{420}-non-reducing hydrogenase and heterodisulfide reductase, 20% is eluted as a dodecameric complex with MW 250000 Da, gel filtration [8]) [8]
350000-400000 <2> (<2> gel filtration [2]) [2]
550000 <2> (<2> gel filtration [2]; <2> dodecameric complex, H_2:heterodisulfide oxidoreductase complex consisting of F_{420}-non-reducing hydrogenase

and heterodisulfide reductase, 20% is eluted as a hexameric complex with MW 250000 Da, gel filtration [8]) [2, 8]
Additional information <1> (<1> MW of H2:heterodisulfide oxidoreductase multienzyme complex is 800000-1300000, gel filtration [7]) [7]

Subunits

? <1, 2, 5, 10> (<2> $\alpha_4\beta_4\gamma 4$, x * 80000 + x * 36000 + x * 21000, the native enzyme is probably a tetramer of trimers [2]; <1> the H_2:heterodisulfide oxidoreductase multienzyme complex is composed of nine polypepetides with MW 46000 Da, 39000 Da, 28000 Da, 25000 Da, 23000 Da, 21000 Da, 20000 Da, 16000 Da, 15000 Da, SDS-PAGE [7]; <10> x * 23000 + x * 46000, the 230000 Da protein is cytochrome b, the 46000 Da protein is an iron-sulfur protein [10]; <1> x * 46000 + x * 23000, the 23000 Da subunit is cytochrome b, SDS-PAGE [9]; <5> x * 53000 + x * 27000, SDS-PAGE [3]) [2, 3, 7, 9, 10]
Additional information <2> (<2> four major protein bands of 40000 Da, 50000 Da, 55000 Da and 66000 Da are detected by SDS-PAGE [2]; <2> H_2:heterodisulfide oxidoreductase complex consisting of F_{420}-non-reducing hydrogenase and heterodisulfide reductase is composed of six different subunits of MW 80000 Da, 51000 Da, 41000 Da, 36000 Da, 21000 Da and 17000 Da. The complex partially dissociates into two subcomplexes. The first subcomplex is composed of the 51000 Da, 41000 Da and 17000 Da subunits. The other subcomplex is composed of the 80000 Da, 36000 Da and 21000 Da subunits [8]) [2, 8]

Posttranslational modification

Lipoprotein <5> [3]

5 Isolation/Preparation/Mutation/Application

Source/tissue

cell suspension <8> [12]
culture condition:acetate-grown cell <5> [3, 18]
culture condition:methanol-grown cell <1, 10> [9, 10]

Localization

cytoplasmic membrane <2> (<2> loosely associated with [8]) [8]
membrane <1, 5, 6, 9, 10, 12, 13> (<1,6,9,10,12,13> bound to [1,6,7,10,14,19]) [1, 3, 6, 7, 9, 10, 14, 19]
soluble <2, 5> [13, 18]

Purification

<1> [7, 9, 10, 11, 16, 23]
<2> [2, 13]
<2> (H_2:heterodisulfide oxidoreductase complex) [8]
<2> (partial) [2]
<3> (DEAE column chromatography, Q-Sepharose column chromatography and Superdex S200 column gel filtration) [22]
<5> [3, 23]

<6> [6]
<11> [11, 16, 23]

Cloning
<1> [10]
<11> (overexpression of HdrB subunit in Bacillus subtilis) [11]

6 Stability

Temperature stability
70 <2> (<2> 30 min, 50% loss of activity [2]) [2]
80 <2> (<2> rapid inactivation above [13]) [13]

Oxidation stability
<2>, partially purified enzyme is insensitive towards O_2 [2]
<6>, incubation of anaerobically prepared membranes in the presence of O_2 results in a complete loss of activity within a few min [6]

General stability information
<1>, more stable in presence of CHAPS than in the presence of dodecyl β-D-maltoside [9]

Storage stability
<2>, 4°C, more than 80% inactivation within 12 h under aerobic conditions [2]
<2>, 4°C, under N_2, partially purified enzyme, 50 mM Tris-HCl, pH 7.6, 0.5 M NaCl, loss of 20% activity within 24 h [13]

References

[1] Abken, H.J.; Tietze, M.; Brodersen, J.; Baumer, S.; Beifuss, U.; Deppenmeier, U.: Isolation and characterization of methanophenazine and function of phenazines in membrane-bound electron transport of Methanosarcina mazei Go1. J. Bacteriol., **180**, 2027-2032 (1998)
[2] Hedderich, R.; Berkessel, A.; Thauer, R.K.: Purification and properties of heterodisulfide reductase from Methanobacterium thermoautotrophicum (strain Marburg). Eur. J. Biochem., **193**, 255-261 (1990)
[3] Simianu, M.; Murakami, E.; Brewer, J.M.; Ragsdale, S.W.: Purification and properties of the heme- and iron-sulfur-containing heterodisulfide reductase from Methanosarcina thermophila. Biochemistry, **37**, 10027-10039 (1998)
[4] Murakami, E.; Deppenmeier, U.; Ragsdale, S.W.: Characterization of the intramolecular electron transfer pathway from 2-hydroxyphenazine to the heterodisulfide reductase from Methanosarcina thermophilo. J. Biol. Chem., **276**, 2432-2439 (2001)

[5] Deppenmeier, U.; Blaut, M.; Gottschalk, G.: H2:heterodisulfide oxidoreductase, a second energy-conserving system in the methanogenic strain Goe1. Arch. Microbiol., **155**, 272-277 (1991)

[6] Brodersen, J.; Gottschalk, G.; Deppenmeier, U.: Membrane-bound $F_{420}H_2$-dependent heterodisulfide reduction in Methanococcus voltae. Arch. Microbiol., **171**, 115-121 (1999)

[7] Heiden, S.; Hedderich, R.; Setzke, E.; Thauer, R.K.: Purification of a cytochrome b containing H_2:heterodisulfide oxidoreductase complex from membranes of Methanosarcina barkeri. Eur. J. Biochem., **213**, 529-535 (1993)

[8] Setzke, E.; Hedderich, R.; Heiden, S.; Thauer, R.K.: H_2:heterodisulfide oxidoreductase complex from Methanobacterium thermoautotrophicum. Composition and properties. Eur. J. Biochem., **220**, 139-148 (1994)

[9] Heiden, S.; Hedderich, R.; Setzke, E.; Thauer, R.K.: Purification of a two-subunit cytochrome-b-containing heterodisulfide reductase from methanol-grown Methanosarcina barkeri. Eur. J. Biochem., **221**, 855-861 (1994)

[10] Kunkel, A.; Vaupel, M.; Heim, S.; Thauer, R.K.; Hedderich, R.: Heterodisulfide reductase from methanol-grown cells of Methanosarcina barkeri is not a flavoenzyme. Eur. J. Biochem., **244**, 226-234 (1997)

[11] Madadi-Kahkesh, S.; Duin, E.C.; Heim, S.; Albracht, S.P.; Johnson, M.K.; Hedderich, R.: A paramagnetic species with unique EPR characteristics in the active site of heterodisulfide reductase from methanogenic archaea. Eur. J. Biochem., **268**, 2566-2577 (2001)

[12] de Poorter, L.M.; Geerts, W.G.; Theuvenet, A.P.; Keltjens, J.T.: Bioenergetics of the formyl-methanofuran dehydrogenase and heterodisulfide reductase reactions in Methanothermobacter thermautotrophicus. Eur. J. Biochem., **270**, 66-75 (2003)

[13] Hedderich, R.; Thauer, R.K.: Methanobacterium thermoautotrophicum contains a soluble enzyme system that specifically catalyzes the reduction of the heterodisulfide of coenzyme M and 7-mercaptoheptanoylthreonine phosphate with H_2. FEBS Lett., **234**, 223-227 (1988)

[14] Deppenmeier, U.; Blaut, M.; Mahlmann, A.; Gottschalk, G.: Membrane-bound $F_{420}H_2$-dependent heterodisulfide reductase in methanogenic bacterium strain Go1 and Methanolobus tindarius. FEBS Lett., **261**, 199-203 (1990)

[15] Duin, E.C.; Madadi-Kahkesh, S.; Hedderich, R.; Clay, M.D.; Johnson, M.K.: Heterodisulfide reductase from Methanothermobacter marburgensis contains an active-site [4Fe-4S] cluster that is directly involved in mediating heterodisulfide reduction. FEBS Lett., **512**, 263-268 (2002)

[16] Duin, E.C.; Bauer, C.; Jaun, B.; Hedderich, R.: Coenzyme M binds to a [4Fe-4S] cluster in the active site of heterodisulfide reductase as deduced from EPR studies with the [^{33}S]coenzyme M-treated enzyme. FEBS Lett., **538**, 81-84 (2003)

[17] Shokes, J.E.; Duin, E.C.; Bauer, C.; Jaun, B.; Hedderich, R.; Koch, J.; Scott, R.A.: Direct interaction of coenzyme M with the active-site Fe-S cluster of heterodisulfide reductase. FEBS Lett., **579**, 1741-1744 (2005)

[18] Peer, C.W.; Painter, M.H.; Rasche, M.E.; Ferry, J.G.: Characterization of a CO:heterodisulfide oxidoreductase system from acetate-grown Methanosarcina thermophila. J. Bacteriol., **176**, 6974-6979 (1994)
[19] Ide, T.; Baumer, S.; Deppenmeier, U.: Energy conservation by the H_2:heterodisulfide oxidoreductase from Methanosarcina mazei Gö1: identification of two proton-translocating segments. J. Bacteriol., **181**, 4076-4080 (1999)
[20] Deppenmeier, U.; Blaut, M.; Mahlmann, A.; Gottschalk, G.: Reduced coenzyme F420: heterodisulfide oxidoreductase, a proton-translocating redox system in methanogenic bacteria. Proc. Natl. Acad. Sci. USA, **87**, 9449-9453 (1990)
[21] Zhang, W.; Culley, D.E.; Scholten, J.C.; Hogan, M.; Vitiritti, L.; Brockman, F.J.: Global transcriptomic analysis of Desulfovibrio vulgaris on different electron donors. Antonie Leeuwenhoek, **89**, 221-237 (2006)
[22] Pires, R.H.; Venceslau, S.S.; Morais, F.; Teixeira, M.; Xavier, A.V.; Pereira, I.A.: Characterization of the Desulfovibrio desulfuricans ATCC 27774 DsrMKJOP complex-a membrane-bound redox complex involved in the sulfate respiratory pathway. Biochemistry, **45**, 249-262 (2006)
[23] Hedderich, R.; Hamann, N.; Bennati, M.: Heterodisulfide reductase from methanogenic archaea: a new catalytic role for an iron-sulfur cluster. Biol. Chem., **386**, 961-970 (2005)

Sulfiredoxin **1.8.98.2**

1 Nomenclature

EC number
1.8.98.2

Systematic name
peroxiredoxin-(S-hydroxy-S-oxocysteine):thiol oxidoreductase [ATP-hydrolysing; peroxiredoxin-(S-hydroxycysteine)-forming]

Recommended name
sulfiredoxin

Synonyms
Srx <1, 2, 3, 5, 6, 7> [1, 3, 6, 8, 9, 11]
Srx1 <2, 4> [7, 10]
cysteine-sulfinic acid reductase <5> [11]
peroxiredoxin-(S-hydroxy-S-oxocysteine) reductase <2> [1]
sulfiredoxin 1 <2> [7]
sulphiredoxin <2, 4> [1, 5]

CAS registry number
710319-61-2

2 Source Organism

<1> *Mus musculus* (no sequence specified) [3]
<2> *Homo sapiens* (no sequence specified) [1, 2, 3, 6, 7, 9, 10]
<3> *Rattus norvegicus* (no sequence specified) [3, 4]
<4> *Saccharomyces cerevisiae* (no sequence specified) [5, 10]
<5> *Arabidopsis thaliana* (no sequence specified) (<5> NarG, α subunit [11]) [11]
<6> *Oryza sativa* (no sequence specified) [8]
<7> *Arabidopsis thaliana* (UNIPROT accession number: Q8GY89) [8]

3 Reaction and Specificity

Catalyzed reaction
peroxiredoxin-(S-hydroxy-S-oxocysteine) + ATP + 2 R-SH = peroxiredoxin-(S-hydroxycysteine) + ADP + phosphate + R-S-S-R

Reaction type

reduction (<2> ATP-dependent reduction of cysteinesulfinic acid of hyperoxidized peroxiredoxin [6]; <2,4> reduction of hyperoxidized 2-Cys peroxiredoxin [10])

Natural substrates and products

S peroxiredoxin-(S-hydroxy-S-oxocysteine) + ATP + 2 R-SH <2, 3, 4> (<2> antioxidant protein with a role in signaling through catalytic reduction of oxidative modifications. Srx also has a role in the reduction of glutathionylation a post-translational, oxidative modification that occurs on numerous proteins and has been implicated in a wide variety of pathologies, including Parkinson's disease. Unlike the reduction of peroxiredoxin overoxidation, Srx-dependent deglutathionylation appears to be nonspecific [2]; <3> reduction of Cys-SO_2H by Srx is specific to 2-Cys peroxiredoxin isoforms. For proteins such as Prx VI and GAPDH, sulfinic acid formation might be an irreversible process that causes protein damage [4]; <2> Srx is largely responsible for reduction of the Cys-SO_2H of peroxiredoxin in A549 human cells [3]; <4> sulphiredoxin is important for the antioxidant function of peroxiredoxins, and is likely to be involved in the repair of proteins containing cysteine sulphinic acid modifications, and in signalling pathways involving protein oxidation [5]) (Reversibility: r) [2, 3, 4, 5]

P peroxiredoxin-(S-hydroxycysteine) + ADP + phosphate + R-S-S-R

S peroxiredoxin-(S-hydroxy-S-oxocysteine) + ATP + 2 R-SH <2> (<2> repairs the inactivated forms of typical two-Cys peroxiredoxins implicated in hydrogen peroxide-mediated cell signaling [1]) (Reversibility: ?) [1]

P peroxiredoxin-(S-hydroxycysteine) + ADP + phosphate + S-S-S-R

Substrates and products

S peroxiredoxin-(S-hydroxy-S-oxocysteine) + ATP + 2 R-SH <1, 2, 3, 4, 5, 6, 7> (<2> antioxidant protein with a role in signaling through catalytic reduction of oxidative modifications. Srx also has a role in the reduction of glutathionylation a post-translational, oxidative modification that occurs on numerous proteins and has been implicated in a wide variety of pathologies, including Parkinson's disease. Unlike the reduction of peroxiredoxin overoxidation, Srx-dependent deglutathionylation appears to be nonspecific [2]; <3> reduction of Cys-SO_2H by Srx is specific to 2-Cys peroxiredoxin isoforms. For proteins such as Prx VI and GAPDH, sulfinic acid formation might be an irreversible process that causes protein damage [4]; <2> Srx is largely responsible for reduction of the Cys-SO_2H of peroxiredoxin in A549 human cells [3]; <4> sulphiredoxin is important for the antioxidant function of peroxiredoxins, and is likely to be involved in the repair of proteins containing cysteine-sulphinic acid modifications, and in signalling pathways involving protein oxidation [5]; <2> both glutathione and thioredoxin are potential physiological electron donors [3]; <2> the ATP molecule is cleaved between the β- and γ-phosphate groups [1]) (Reversibility: r) [1, 2, 3, 4, 5, 6, 7, 8, 9, 10, 11]

P peroxiredoxin-(S-hydroxycysteine) + ADP + phosphate + R-S-S-R

S peroxiredoxin-(S-hydroxy-S-oxocysteine) + ATP + 2 R-SH <2> (<2> repairs the inactivated forms of typical two-Cys peroxiredoxins implicated in hydrogen peroxide-mediated cell signaling [1]) (Reversibility: ?) [1]
P peroxiredoxin-(S-hydroxycysteine) + ADP + phosphate + S-S-S-R
S peroxiredoxin-(S-hydroxy-S-oxocysteine) + GTP + 2 R-SH <2> (<2> both glutathione and thioredoxin are potential physiological electron donors [3]) (Reversibility: ?) [3]
P peroxiredoxin-(S-hydroxycysteine) + GDP + phosphate + R-S-S-R
S peroxiredoxin-(S-hydroxy-S-oxocysteine) + dATP + 2 R-SH <2> (<2> both glutathione and thioredoxin are potential physiological electron donors [3]) (Reversibility: ?) [3]
P peroxiredoxin-(S-hydroxycysteine) + dADP + phosphate + R-S-S-R
S peroxiredoxin-(S-hydroxy-S-oxocysteine) + dGTP + 2 R-SH <2> (<2> both glutathione and thioredoxin are potential physiological electron donors [3]) (Reversibility: ?) [3]
P peroxiredoxin-(S-hydroxycysteine) + GDP + phosphate + R-S-S-R
S Additional information <2> (<2> no activity with CTP, UTP, dCTP, or dTTP [3]) (Reversibility: ?) [3]
P ?

Cofactors/prosthetic groups
ATP <2, 4> [10]

Activating compounds
ATP <2> [9]

Metals, ions
Mg^{2+} <2, 4> (<4> required [5]) [5, 9, 10]

Turnover number (min^{-1})
0.003 <2> (peroxiredoxin-(S-hydroxy-S-oxocysteine), <2> at 30°C [9]) [9]

K_m-Value (mM)
0.0012 <2> (thioredoxin 1) [3]
0.03 <2> (ATP) [3]
1.8 <2> (GSH) [3]

4 Enzyme Structure

Molecular weight
12000 <2> (<2> SDS-PAGE [9]) [9]
13910 <5> [11]
14000 <2> [6]

5 Isolation/Preparation/Mutation/Application

Source/tissue

A-549 cell <2> [3]
HeLa cell <2> [3]
kidney <2> [3]
leaf <6, 7> [8]
lung <2> [3]
spleen <2> [3]
thymus <2> [3]

Localization

chloroplast <6, 7> [8]
cytosol <2> [3]
plastid <5> [11]

Purification

<2> [1, 3]
<2> (Ni^{2+}-NTA column chromatography) [7]
<4> (recombinant) [5]

Crystallization

<2> (crystals of the wild-type and SeMet forms of ET-hSrx were obtained by the vapor diffusion method. 1.65 A crystal structure of human Srx) [1]

Cloning

<1> (expression in Escherichia coli) [3]
<2> (expressed in HEK293 cells and Escherichia coli) [7]
<2> (expression in Escherichia coli) [3]
<3> (expression in Escherichia coli) [3]
<6> (expressed in Saccharomyces cerevisiae) [8]
<7> (expressed in Saccharomyces cerevisiae) [8]

Engineering

C99A <2> (<2> Cys99 replaced by Ala [9]) [9]
C99S <2> (<2> Cys99 replaced by Ser [9]; <2> mutant created to show the importance of the cytosine residue in the deglutathionylation function of the protein [7]) [7, 9]
D57N <2> (<2> Asp57 replaced by Asn [9]) [9]
D79N <2> (<2> Asp60 replaced by Asn [9]) [9]
H99N <2> (<2> His99 replaced by Asn [9]) [9]
K60R <2> (<2> Lys60 replaced by Arg [9]) [9]
R100M <2> (<2> Arg100 replaced by Met [9]) [9]
R50M <2> (<2> Arg50 replaced by Met [9]) [9]

References

[1] Jönsson, T.J.; Murray, M.S.; Johnson, L.C.; Poole, L.B.; Lowther, W.T.: Structural basis for the retroreduction of inactivated peroxiredoxins by human sulfiredoxin. Biochemistry, **44**, 8634-8642 (2005)

[2] Findlay, V.J.; Tapiero, H.; Townsend, D.M.: Sulfiredoxin: a potential therapeutic agent?. Biomed. Pharmacother., **59**, 374-379 (2005)

[3] Chang, T.S.; Jeong, W.; Woo, H.A.; Lee, S.M.; Park, S.; Rhee, S.G.: Characterization of mammalian sulfiredoxin and its reactivation of hyperoxidized peroxiredoxin through reduction of cysteine sulfinic acid in the active site to cysteine. J. Biol. Chem., **279**, 50994-51001 (2004)

[4] Woo, H.A.; Jeong, W.; Chang, T.S.; Park, K.J.; Park, S.J.; Yang, J.S.; Rhee, S.G.: Reduction of cysteine sulfinic acid by sulfiredoxin is specific to 2-Cys peroxiredoxins. J. Biol. Chem., **280**, 3125-3128 (2005)

[5] Biteau, B.; Labarre, J.; Toledano, M.B.: ATP-dependent reduction of cysteine-sulphinic acid by S. cerevisiae sulphiredoxin. Nature, **425**, 980-984 (2003)

[6] Lee, D.Y.; Park, S.J.; Jeong, W.; Sung, H.J.; Oho, T.; Wu, X.; Rhee, S.G.; Gruschus, J.M.: Mutagenesis and modeling of the peroxiredoxin (Prx) complex with the NMR structure of ATP-bound human sulfiredoxin implicate aspartate 187 of Prx I as the catalytic residue in ATP hydrolysis. Biochemistry, **45**, 15301-15309 (2006)

[7] Findlay, V.J.; Townsend, D.M.; Morris, T.E.; Fraser, J.P.; He, L.; Tew, K.D.: A novel role for human sulfiredoxin in the reversal of glutathionylation. Cancer Res., **66**, 6800-6806 (2006)

[8] Liu, X.P.; Liu, X.Y.; Zhang, J.; Xia, Z.L.; Liu, X.; Qin, H.J.; Wang, D.W.: Molecular and functional characterization of sulfiredoxin homologs from higher plants. Cell Res., **16**, 287-296 (2006)

[9] Jeong, W.; Park, S.J.; Chang, T.S.; Lee, D.Y.; Rhee, S.G.: Molecular mechanism of the reduction of cysteine sulfinic acid of peroxiredoxin to cysteine by mammalian sulfiredoxin. J. Biol. Chem., **281**, 14400-14407 (2006)

[10] Jang, H.H.; Chi, Y.H.; Park, S.K.; Lee, S.S.; Lee, J.R.; Park, J.H.; Moon, J.C.; Lee, Y.M.; Kim, S.Y.; Lee, K.O.; Lee, S.Y.: Structural and functional regulation of eukaryotic 2-Cys peroxiredoxins including the plant ones in cellular defense-signaling mechanisms against oxidative stress. Physiol. Plant., **126**, 549-559 (2006)

[11] Rey, P.; Becuwe, N.; Barrault, M.B.; Rumeau, D.; Havaux, M.; Biteau, B.; Toledano, M.B.: The Arabidopsis thaliana sulfiredoxin is a plastidic cysteine-sulfinic acid reductase involved in the photooxidative stress response. Plant J., **49**, 505-514 (2007)

Ribosyldihydronicotinamide dehydrogenase (quinone) 1.10.99.2

1 Nomenclature

EC number
1.10.99.2

Systematic name
1-(β-D-ribofuranosyl)-1,4-dihydronicotinamide:quinone oxidoreductase

Recommended name
ribosyldihydronicotinamide dehydrogenase (quinone)

Synonyms
N-ribosyldihydronicotinamide dehydrogenase (quinone) <4> [9]
NAD(P)H:quinone oxidoreductase-2 <2, 4> (<4> misleading [9]) [1, 9]
NAD(P)H:quinone oxidoreductase2 <4> (<4> misleading [9]) [9]
NQO2 <1, 2, 3, 4, 6> [1, 5, 7, 8, 9, 11, 14, 15, 16, 17, 18]
NRH-oxidizing enzyme <4> [9]
NRH:quinone oxidoreductase <2, 6> [8, 11]
NRH:quinone oxidoreductase 2 <1, 2, 3, 4> [7, 9, 15, 16, 17, 18]
NRH:quinone oxireductase 2 <2> [14]
NRH:quinone reductase 2 <1, 7> [2]
QR2 <1, 2, 4, 7> [2, 3, 4, 9, 10, 12, 13, 14]
bQR2 <4> [13]
dihydronicotinamide riboside:quinone oxidoreductase 2 <2> [18]
dihydronicotinamide riboside:quinone reductase 2 <1> [7]
melatonin-binding site MT3 <2, 5> [10]
quinone reductase 2 <1, 2, 4, 7> [2, 4, 6, 9, 10, 12, 14]
quinone reductase type 2 <2, 4> [13]

CAS registry number
667919-86-0

2 Source Organism

<1> *Mus musculus* (no sequence specified) [2, 7, 15, 17]
<2> *Homo sapiens* (no sequence specified) [1, 3, 4, 5, 6, 10, 11, 12, 13, 14, 16, 18]
<3> *Rattus norvegicus* (no sequence specified) [9]
<4> *Bos taurus* (no sequence specified) [9, 13]
<5> *Mesocricetus auratus* (no sequence specified) [10]
<6> *Mus musculus* (UNIPROT accession number: Q9JI75) [8]
<7> *Homo sapiens* (UNIPROT accession number: P16083) [2]

3 Reaction and Specificity

Catalyzed reaction

1-(β-D-ribofuranosyl)-1,4-dihydronicotinamide + a quinone = 1-(β-D-ribofuranosyl)nicotinamide + a hydroquinone

Reaction type

oxidation
redox reaction
reduction

Natural substrates and products

S Additional information <1, 2, 5, 7> (<1> by deleting QR2 it seems that mice become increasingly susceptible to polycyclic aromatic hydrocarbon-induced skin carcinogenesis [2]; <2,5> enzyme exhibits melatonin-binding activity [10]; <7> high QR2 activity might make individuals more susceptible to Parkinson's disease. By inhibiting QR2, it seems that anti-malarial compounds such as quinacrine favour the red blood cell oxidative stress leading to the death of the parasite [2]; <2> knockdown K562 cells exhibit increased antioxidant and detoxification enzyme expression and reduced proliferation rates [4]; <1> organs of mice deleted for NQO2 are depleted of MT3 binding sites. NQO2 is part of the melatonin receptor MT3 binding sites [7]; <2> the expression of the NQO2 gene is induced in response to 2,3,7,8-tetrachlorodibenzo-*p*-dioxin [5]) (Reversibility: ?) [2, 4, 5, 7, 10]
P ?

Substrates and products

S 2,6-dichloroindophenol + dihydronicotinamide riboside <1> (Reversibility: ?) [15]
P ? + nicotinamide riboside
S 5-(aziridin-1-yl)-2,4-dinitrobenzamide + dihydronicotinamide riboside <1> (Reversibility: ?) [15]
P ? + nicotinamide riboside
S 5-(aziridin-1-yl)-2,4-dinitrobenzamide + menadiol <2> (Reversibility: ?) [14]
P ? + reduced menadiol
S ? + dihydronicotinamide riboside <2> (Reversibility: ?) [16]
P ? + nicotinamide riboside
S N-benzyldihydronicotinamide + coenzyme Q0 <2> (Reversibility: ?) [6]
P ?
S N-benzyldihydronicotinamide + menadione <1, 2> (Reversibility: ?) [2, 6, 7]
P ?
S N-benzylnicotinamide + menadiol <2> (Reversibility: ?) [6]
P ?
S N-methyldihydronicotinamide + menadione <2> (Reversibility: ?) [12, 13]

P ?

S N-methyldihydronicotinamide + menadione <7> (Reversibility: ?) [2]

S N-ribosylnicotinamide + menadiol <2> (<2> N-ribosylnicotinamide is a poor substrate [6]) (Reversibility: ?) [6]

P ?

S benzo(a)pyrene-3,6-quinone + dihydronicotinamide riboside <1> (Reversibility: ?) [15]

P ? + nicotinamide riboside

S dihydrobenzylnicotinamide + menadione <2> (Reversibility: ?) [3]

P ?

S dihydronicotinamide riboside + 2,6-dichlorophenolindophenol <2> (Reversibility: ?) [1]

P nicotinamide riboside + reduced 2,6-dichlorophenolindophenol

S dihydronicotinamide riboside + 3-(4,5-dimethylthiazaol-2-yl)-2,5-diphenyltetrazolium <2> (Reversibility: ?) [1]

P nicotinamide riboside + ?

S dihydronicotinamide riboside + CB 1954 <2> (Reversibility: ?) [1]

P nicotinamide riboside + ?

S dihydronicotinamide riboside + menadione <4> (Reversibility: ?) [9]

P ribosyl nicotinamide + menadiol

S dihydronicotinamide riboside + menadione/3-(4,5-dimethylthiazaol-2-yl)-2,5-diphenyltetrazolium <2> (Reversibility: ?) [1]

P nicotinamide riboside + ?

S dihydronicotinamide riboside + methyl red <2> (Reversibility: ?) [1]

P nicotinamide riboside + ?

S menadione + 1-(2-hydroxyethyl)dihydronicotinamide <2> (Reversibility: ?) [4]

P ?

S menadione + dihydrobenzylnicotinamide <2, 5> (Reversibility: ?) [10]

P ?

S menadione + dihydronicotinamide riboside <1> (Reversibility: ?) [15]

P menadiol + nicotinamide riboside

S mitomycin C + dihydronicotinamide riboside <1> (Reversibility: ?) [15]

P ? + nicotinamide riboside

S reduced N-methyldihydronicotinamide + menadione <4> (Reversibility: ?) [13]

P ? + menadiol

S reduced N^{1}-(benzyl)-nicotinamide + menadione <4> (Reversibility: ?) [9]

P N^{1}-(benzyl)-nicotinamide + menadiol

S reduced N^{1}-(methyl)-nicotinamide + menadione <4> (Reversibility: ?) [9]

P N^{1}-(methyl)-nicotinamide + reduced menadiol

S reduced N^{1}-(n-propyl)-nicotinamide + menadione <4> (Reversibility: ?) [9]

P N^{1}-(n-propyl)-nicotinamide + menadiol

S tetrahydrofolate + menadiol <2> (Reversibility: ?) [6]

P ?

S Additional information <1, 2, 4, 5, 7> (<1> by deleting QR2 it seems that mice become increasingly susceptible to polycyclic aromatic hydrocarbon-induced skin carcinogenesis [2]; <2,5> enzyme exhibits melatonin-binding activity [10]; <7> high QR2 activity might make individuals more susceptible to Parkinson's disease. By inhibiting QR2, it seems that anti-malarial compounds such as quinacrine favour the red blood cell oxidative stress leading to the death of the parasite [2]; <2> knockdown K562 cells exhibit increased antioxidant and detoxification enzyme expression and reduced proliferation rates [4]; <1> organs of mice deleted for NQO2 are depleted of MT3 binding sites. NQO2 is part of the melatonin receptor MT3 binding sites [7]; <2> the expression of the NQO2 gene is induced in response to 2,3,7,8-tetrachlorodibenzo-*p*-dioxin [5]; <7> enzyme cannot use NADH or NADPH as reducing agent [2]; <4> no activity with NADH, NADPH, NMNH, reduced 3-acetylpyridine adenine dinucleotide, xanthine or hypoxanthine. No activity with 1,4-benzoquinone and potassium ferricyanide [9]; <2> no activity with one-electron acceptors such as potassium ferricyanide [1]) (Reversibility: ?) [1, 2, 4, 5, 7, 9, 10]

P ?

Inhibitors

1,4-dimethylphenanthrene <4> (<4> 0.00001 mM, 15% inhibition of the reaction with N^1-(n-propyl)-nicotinamide [9]) [9]
12-methylbenz[a]anthracene <4> (<4> 0.00001 mM, 51% inhibition of the reaction with N^1-(n-propyl)-nicotinamide [9]) [9]
2-(2-methoxy-6H-pyrido[2',3':4,5]pyrrolo[2,1-a]isoindol-11-yl)ethylamine <2> (<2> IC50: 0.00087 mM [3]) [3]
2-hydroxyestradiol <2> (<2> 0.01 mM, 17% inhibition [1]) [1]
2-iodo-5-methyoxycarbonylamino-N-acetyltryptamine <2> [3]
2-iodo-melatonin <2> (<2> IC50: 0.016 mM [3]) [3]
2-iodomelatonin <2, 5> [10]
4',5,7-trihydroxyflavone <2> (<2> 0.01 mM, 61% inhibition [1]) [1]
4-hydroxyestradiol <2> (<2> 0.01 mM, 18% inhibition [1]) [1]
4-hydroxyestrone <2> (<2> 0.01 mM, 22% inhibition [1]) [1]
5-hydroxyflavone <2> (<2> IC50: 340 nM [6]) [6]
5-methoxycarbonylamino-N-acetyltryptamine <2, 5> [10]
5-methyoxycarbonylamino-N-acetyltryptamine <2> (<2> IC50: 0.295 mM [3]) [3]
7,12-dimethylbenz[a]anthracene <4> (<4> 0.00001 mM, 51% inhibition of the reaction with N^1-(n-propyl)-nicotinamide [9]) [9]
7,8-dihydroxyflavone <2> (<2> 0.01 mM, 7% inhibition [1]) [1]
7-methylbenz[a]anthracene <4> (<4> 0.00001 mM, 66% inhibition of the reaction with N^1-(n-propyl)-nicotinamide [9]) [9]
9,10-dimethylanthracene <4> (<4> 0.00001 mM, 25% inhibition of the reaction with N^1-(n-propyl)-nicotinamide [9]) [9]
9-methylanthracene <4> (<4> 0.00001 mM, 26% inhibition of the reaction with N^1-(n-propyl)-nicotinamide [9]) [9]

apigenin <2> (<2> IC50: 430 nM [6]) [6]
chloroquine <2> [12]
dicoumarol <2, 5> (<2> 0.01 mM, 25% inhibition [1]; <2> 0.01 mM, marginal inhibition [13]; <2> IC50: 0.59 mM [3]) [1, 3, 10, 13]
estradiol <2> (<2> 0.01 mM, 13% inhibition [1]) [1]
flavone <2> (<2> 0.01 mM, 14% inhibition [1]) [1]
genistein <2> [6]
kaempferol <2> (<2> IC50: 380 nM [6]) [6]
luteolin <2> (<2> IC50: 780 nM [6]) [6]
melatonin <2, 5> (<2> IC50: 0.13 mM [3]) [3, 10]
morin <2> (<2> 0.01 mM, 95% inhibition [1]) [1]
N-acetylserotonin <2, 5> (<2> IC50: 0.099 mM [3]) [3, 10]
N-[2-(2-iodo-5-methoxy-1-methyl-4-nitroindol-3-yl)ethyl]acetamide <2> (<2> inhibits enzymatic mechanism of the enzyme through the MT3 binding site, IC50: 0.0003 mM [3]) [3]
N-[2-(2-methoxy-6H-dipyrido[2,3-a:3,2-e]pyrrolizin-11-yl)ethyl]-2-furamide <2> (<2> IC50: 14 nM [3]) [3]
N-[2-(2-methoxy-6H-pyrido[2',3':4,5]pyrrolo[2,1-a]isoindol-11-yl)ethyl]2-furamide <2> (<2> IC50: 0.0002 mM [3]) [3]
N-[2-(5-methoxy-4-nitro-1H-indol-3-yl)ethyl]acetamide <2> (<2> IC50: 0.0015 mM [3]) [3]
N-[2-(5-methoxy-7-nitro-1H-indol-3-yl)ethyl]acetamide <2> (<2> IC50: 0.044 mM [3]) [3]
N-[2-(7-methylaminosulfonyl-1-naphthyl)ethyl]acetamide <2> (<2> IC50: 0.038 mM [3]) [3]
N-[2-(8-methoxy-3,4-dihydro-2H-pyrido[2',3':4,5]pyrrolo[2,1-b][1,3]oxazin-10-yl)ethyl]-2-furamide <2> (<2> IC50: 0.005 mM [3]) [3]
N^1-[2-(2-methoxy-6H-pyrido[2',3':4,5]pyrrolo[2,1-a]isoindol-11-yl)ethyl]-acetamide <2> (<2> IC50: 0.0019 mM [3]) [3]
quercetin <1, 2> (<2> 0.01 mM, complete inhibition [1]; <2> IC50: 0.0014 mM [6]) [1, 6, 15, 16]
quinacrine <2> [12]
quinine <2> [12]
S26553 <2, 5> (<5> i.e. N-methyl-[1-[2-(acetylamino)ethyl]napthalen-7-yl]-carbamate [10]) [10]
S28128 <2> (<2> IC50: 0.00091 mM [3]) [3]
serotonin <2> [10]
α-naphthoflavone <2> (<2> 0.01 mM, 75% inhibition [1]) [1]
benz[a]anthracene <4> (<4> 0.00001 mM, 40% inhibition of the reaction with N^1-(n-propyl)-nicotinamide [9]) [9]
benzo(a)pyrene <1, 2> (<1> 20 mM, 90% inhibition [15]) [15, 16]
benzo[a]pyrene <2> (<2> 0.0001 mM, 75% inhibition [13]) [13]
β-naphthoflavone <2> (<2> 0.01 mM, 76% inhibition [1]) [1]
chrysin <2> (<2> 0.01 mM, complete inhibition [1]) [1]
chrysin-dimethylether <2> (<2> IC50: 0.0013 mM [6]) [6]
chrysoeriol <2> (<2> IC50: 300 nM [6]) [6]
galangin <2> (<2> 0.01 mM, complete inhibition [1]) [1]

isorhamnetin <2> (<2> IC50: 860 nM [6]) [6]
mefloquine <2> [12]
methyl-1-(2-acetamidoethyl)-7-naphthylcarbamate <2> (<2> IC50: 0.015 mM [3]) [3]
primaquine <2> [12]
quercetin-3-O-β-glucopyranosyl <2> (<2> IC50: 0.0015 mM [6]) [6]
resveratrol <2> (<2> the chemopreventive and cardioprotective properties of resveratrol are possibly the results of QR2 activity inhibition, which in turn, up-regulates the expression of cellular antioxidant enzymes and cellular resistance to oxidative stress. All three resveratrol hydroxyl groups form hydrogen bonds with amino acids from QR2, anchoring a flat resveratrol molecule in parallel with the isoalloxazine ring of FAD [4]) [4]
Additional information <1> (<1> not inhibited by dicoumarol, Cibacron blue, phenindone [15]) [15]

Cofactors/prosthetic groups

FAD <1, 2, 4, 7> (<7> contains 1 FAD per monomer firmly bound to the enzyme through multiple interactions [2]; <2> mediates hydride transfer, very tightly bound to the enzyme. Activity of the purified enzyme is not further increased by added FAD [13]; <2> one prosthetic group per subunit [1]; <4> the enzyme contains FAD as the sole bound flavin. The prosthetic group can be removed by treatment with acid in ammonium sulfate, and the resolved enzyme may be reactivated by FAD or by higher concentrations FMN [9]; <4> very tightly bound to the enzyme. Activity of the purified enzyme is not further increased by added FAD [13]) [1,2,4,6,9,13,14,15]
FMN <4> (<4> the enzyme contains FAD as the sole bound flavin. The prosthetic group can be removed by treatment with acid in ammonium sulfate, and the resolved enzyme may be reactivated by FAD or by higher concentrations FMN [9]) [9]
NADH <1, 2> (<2> very poor electron donor [1]) [1,17]
nicotinamide riboside <1, 2> (<1> reduced nicotinamide riboside [15]) [15, 17, 18]
Additional information <2, 7> (<2> completely inert to NADPH, NADH and NMNH [13]; <7> enzyme cannot use NADH or NADPH as reducing agent [2]) [2,13]

Metals, ions

Zn^{2+} <2> [14]

Turnover number (min^{-1})

2.6 <2> (NADH) [1]
3.83 <2> (methyl red) [1]
6 <2> (5-(aziridin-1-yl)-2,4-dinitrobenzamide, <2> at 25°C in pH 8.5 buffer (50 mM Tris, 140 mM NaCl, and 0.1% Tween 20) with 0.1 mM menadione as co-substrate [14]) [14]
6 <2> (CB 1954) [1]
11 <2> (3-(4,5-dimethylthiazaol-2-yl)-2,5-diphenyltetrazolium) [1]

38.33 <2> (menadione/3-(4,5-dimethylthiazaol-2-yl)-2,5-diphenyltetrazolium) [1]
43.33 <2> (dihydronicotinamide riboside) [1]
46.67 <2> (2,6-dichlorophenolindophenol) [1]
91.8 <2> (2,6-dichlorophenolindophenol, <2> chimeric enzyme of the human NQO2 with 43 amino acids from the carboxyl terminus of human DT-diaphorase [1]) [1]

Specific activity (U/mg)
242 <2> [13]
Additional information <4> [9]

K_m-Value (mM)
0.0016 <4> (menadione, <4> pH 7.6, with reduced N^1-(n-propyl)-nicotinamide as acceptor [9]) [9]
0.0023 <2> (menadione/3-(4,5-dimethylthiazaol-2-yl)-2,5-diphenyltetrazolium) [1]
0.0047 <2> (methyl red) [1]
0.0067 <2> (menadione, <2> pH 8.5, 25°C, fluorescence assay [3]) [3, 6]
0.007 <2> (2,6-dichlorophenolindophenol) [1]
0.0089 <2> (menadione, <2> pH 7.5 [10]) [10]
0.012 <2> (3-(4,5-dimethylthiazaol-2-yl)-2,5-diphenyltetrazolium) [1]
0.017 <2> (menadione, <2> pH 8.5, HPLC approach [3]) [3]
0.019 <2> (2,6-dichlorophenolindophenol, <2> chimeric enzyme of the human NQO2 with 43 amino acids from the carboxyl terminus of human DT-diaphorase [1]) [1]
0.02 <2> (N-benzyldihydronicotinamide) [6]
0.02 <2> (dihydrobenzylnicotinamide, <2> pH 8.5, 25°C, fluorescence assay [3]) [3]
0.023 <4> (reduced N^1-(n-propyl)-nicotinamide, <4> pH 7.6 [9]) [9]
0.0252 <5> (menadione, <5> pH 7.5 [10]) [10]
0.028 <2> (dihydronicotinamide riboside) [1]
0.04 <2> (dihydrobenzylnicotinamide, <2> pH 8.5, HPLC approach [3]) [3]
0.0576 <2> (dihydrobenzylnicotinamide, <2> pH 7.5 [10]) [10]
0.0608 <5> (dihydrobenzylnicotinamide, <5> pH 7.5 [10]) [10]
0.061 <2> (5-(aziridin-1-yl)-2,4-dinitrobenzamide, <2> at 25°C in pH 8.5 buffer (50 mM Tris, 140 mM NaCl, and 0.1% Tween 20) with 0.1 mM menadione as co-substrate [14]) [14]
0.252 <2> (NADH) [1]
0.26 <2> (CB 1954) [1]
2 <2> (tetrahydrofolate) [6]

K_i-Value (mM)
0.000021 <2> (quercetin) [1]
0.00003 <2> (N-[2-(2-methoxy-6H-dipyrido[2,3-a:3,2-e]pyrrolizin-11-yl)-ethyl]-2-furamide, <2> reduced form hQR2-$FADH_2$ [3]) [3]
0.00004 <2> (N-[2-(2-methoxy-6H-dipyrido[2,3-a:3,2-e]pyrrolizin-11-yl)-ethyl]-2-furamide, <2> oxidized form hQR2-FAD [3]) [3]

0.00005 <2> (resveratrol) [4]
0.00008 <2> (N-[2-(2-iodo-5-methoxy-1-methyl-4-nitroindol-3-yl)ethyl]acetamide, <2> oxidized form hQR2-FAD [3]) [3]
0.00018 <2> (S28128, <2> oxidized form hQR2-FAD [3]) [3]
0.00031 <2> (N-[2-(2-iodo-5-methoxy-1-methyl-4-nitroindol-3-yl)ethyl]acetamide, <2> reduced form hQR2-$FADH_2$ [3]) [3]
0.00051 <2> (quinacrine) [12]
0.00061 <2> (chloroquine) [12]
0.0009 <2> (S28128, <2> reduced form hQR2-$FADH_2$ [3]) [3]
0.00104 <2> (primaquine) [12]
0.0012 <5> (2-iodomelatonin, <5> pH 7.5 [10]) [10]
0.0028 <5> (S26553, <5> pH 7.5 [10]) [10]
0.004 <2> (2-iodo-melatonin, <2> reduced form hQR2-$FADH_2$ [3]) [3]
0.012 <2> (2-iodo-5-methyoxycarbonylamino-N-acetyltryptamine, <2> oxidized form hQR2-FAD [3]) [3]
0.017 <2> (mefloquine) [12]
0.029 <2> (2-iodomelatonin, <2> pH 7.5 [10]) [10]
0.033 <2> (2-iodo-5-methyoxycarbonylamino-N-acetyltryptamine, <2> reduced form hQR2-$FADH_2$ [3]) [3]
0.0414 <2> (S26553, <2> pH 7.5 [10]) [10]
0.043 <5> (melatonin, <5> pH 7.5 [10]) [10]
0.0435 <5> (5-methoxycarbonylamino-N-acetyltryptamine, <5> pH 7.5 [10]) [10]
0.05 <2> (2-iodo-melatonin, <2> oxidized form hQR2-FAD [3]) [3]
0.092 <5> (N-acetylserotonin, <5> pH 7.5 [10]) [10]
0.15 <2> (melatonin, <2> reduced form hQR2-$FADH_2$ [3]) [3]
0.204 <2> (N-acetylserotonin, <2> pH 7.5 [10]) [10]
0.246 <2> (melatonin, <2> pH 7.5 [10]) [10]
0.252 <2> (quinine) [12]
0.376 <2> (5-methoxycarbonylamino-N-acetyltryptamine, <2> pH 7.5 [10]) [10]
0.566 <5> (dicoumarol, <5> pH 7.5 [10]) [10]
0.62 <2> (dicoumarol, <2> pH 7.5 [10]) [10]
2.933 <2> (serotonin) [10]

4 Enzyme Structure

Molecular weight

40000 <4> (<4> gel filtration [13]) [13]
50000 <2> (<2> gel filtration [1]) [1]

Subunits

? <2> (<2> x * 24600, SDS-PAGE [13]) [13]
dimer <2, 4> (<2> 2 * 26000, SDS-PAGE [1]) [1, 13]

5 Isolation/Preparation/Mutation/Application

Source/tissue

bile <2> (<2> no significant difference in expression in normal and malignant biliary tissue. Activity of NQO2 is significatly lower than in hepatocellular tissue [11]) [11]
brain <2, 3, 7> (<2,7> minimal expression [2,5]) [2, 5, 9]
erythrocyte <3, 7> (<7> modest level [2]) [2, 9]
heart <2, 3, 7> [2, 5, 9]
kidney <2, 3, 4, 5, 7> [2, 5, 9, 10, 13]
liver <2, 3, 6, 7> (<7> high expression level [2]; <2> in normal hepatocellular tissue, the two NQO isoforms are differentially regulated with a higher expression of NQO2 than NQO1. In malignant hepatocellular tissue NQO1 is up-regulated and NQO2 is down-regulated [11]) [2, 5, 8, 9, 11, 13, 16]
lung <2> [5]
pancreas <2, 7> (<2,7> minimal expression [2,5]) [2, 5]
prostate gland ventral lobe <3> [9]
seminal vesicle <3> [9]
skeletal muscle <2, 3, 7> (<2> highest expression [5]; <7> high expression level [2]) [2, 5, 9]
testis <6> [8]
Additional information <1, 2, 3, 7> (<3> no activity in blood plasma [9]; <2> no activity in placenta [5]; <7> no expression in placenta [2]; <1> no expression in skeletal muscle [2]) [2, 5, 9]

Localization

cytosol <2> [6, 11]

Purification

<2> [1, 10]
<2> (DEAE Sepharose, Superdex 75 column chromatography and Mono-Q column chromatography) [14]
<4> [9, 13]
<5> [10]

Crystallization

<2> [14]
<2> (crystal structure of the enzyme in complex with resveratrol. Crystals of the native QR2 and QR2-resveratrol complex all belong to the $P2_12_12_1$ space group) [4]
<7> (2.1 A resolution) [2]

Cloning

<1> (expressed in Chinese hamster ovary cells and Escherichia coli) [15]
<2> [5, 11]
<2> (expressed in Escherichia coli) [14]
<2> (expressed in Hep-G2 cells) [16]
<2> (expression in CHO cells) [10]

<2> (overexpression in Escherichia coli) [13]
<6> (the mouse NQO2 cDNA is subcloned into the pMT2 eukaryotic expression vector which, upon transfection in monkey kidney COS1 cells, produce a significant increase in NQO2 activity. Deletion of 54 amino acids from the N-terminus of the mouse NQO2 protein results in the loss of NQO2 expression and activity in transfected COS1 cells) [8]

Engineering

N161H <2> (<2> results in the total loss of the enzymatic activity towards activation of 5-(aziridin-1-yl)-2,4-dinitrobenzamide, whereas the rates of reduction towards menadione are not altered [14]) [14]
Additional information <2> (<2> construction of a chimeric enzyme of the human NQO2 with 43 amino acids from the carboxyl terminus of human DT-diaphorase. HNQO2-hDT43 still uses dihydronicotinamide riboside as an electron donor. The chimeric enzyme is inhibited by quercetin but not dicoumarol [1]) [1]

Application

medicine <2> (<2> anti-cancer therapy [14]) [14]

References

[1] Wu, K.; Knox, R.; Sun, X.Z.; Joseph, P.; Jaiswal, A.K.; Zhang, D.; Deng, P.S.K.; Chen, S.: Catalytic properties of NAD(P)H:quinone oxidoreductase-2 (NQO2), a dihydronicotinamide riboside dependent oxidoreductase. Arch. Biochem. Biophys., **347**, 221-228 (1997)

[2] Vella, F.; Ferry, G.; Delagrange, P.; Boutin, J.A.: NRH:quinone reductase 2: an enzyme of surprises and mysteries. Biochem. Pharmacol., **71**, 1-12 (2005)

[3] Mailliet, F.; Ferry, G.; Vella, F.; Berger, S.; Coge, F.; et. al.: Characterization of the melatoninergic MT3 binding site on the NRH:quinone oxidoreductase 2 enzyme. Biochem. Pharmacol., **71**, 74-88 (2005)

[4] Buryanovskyy, L.; Fu, Y.; Boyd, M.; Ma, Y.; Hsieh, T.; Wu, J.M.; Zhang, Z.: Crystal structure of quinone reductase 2 in complex with resveratrol. Biochemistry, **43**, 11417-11426 (2004)

[5] Long, D.J., 2nd; Jaiswal, A.K.: NRH:quinone oxidoreductase2 (NQO2). Chem. Biol. Interact., **129**, 99-112 (2000)

[6] Boutin, J.A.; Chatelein-Egger, F.; Vella, F.; Delagrange, P.; Ferry, G.: Quinone reductase 2 substrate specificity and inhibitgion pharmacology. Chem. Biol. Interact., **151**, 213-228 (2005)

[7] Mailliet, F.; Ferry, G.; Vella, F.; Thiam, K.; Delagrange, P.; Boutin, J.A.: Organs from mice deleted for NRH:quinone oxidoreductase 2 are deprived of the melatonin binding site MT3. FEBS Lett., **578**, 116-120 (2004)

[8] Long, D.J.; Jaiswal, A.K.: Mouse NRH:quinone oxidoreductase (NQO2): cloning of cDNA and gene- and tissue-specific expression. Gene, **252**, 107-117 (2000)

[9] Liao, S.; Dulaney, J.T.; Williams-Ashman, H-.G.: Purification and properties of a flavoprotein catalyzing the oxidation of reduced ribosyl nicotinamide. J. Biol. Chem., **237**, 2981-2987 (1962)
[10] Nosjean, O.; Ferro, M.; Coge, F.; Beauverger, P.; et al.: Identification of the melatonin-binding site MT3 as the quinone reductase 2. J. Biol. Chem., **275**, 31311-31317 (2000)
[11] Strassburg, A.; Strassburg, C.P.; Manns, M.P.; Tukey, R.H.: Differential gene expression of NAD(P)H:quinone oxidoreductase and NRH:quinone oxidoreductase in human hepatocellular and biliary tissue. Mol. Pharmacol., **61**, 320-325 (2002)
[12] Graves, P.R.; Kwiek, J.J.; Fadden, P.; et al.: Discovery of novel targets of quinoline drugs in the human purine binding proteome. Mol. Pharmacol., **62**, 1364-1372 (2002)
[13] Zhao, Q.; Yang, X.L.; Holtzclaw, W.D.; Talalay, P.: Unexpected genetic and structural relationships of a long-forgotten flavoenzyme to NAD(P)H:quinone reductase (DT-diaphorase). Proc. Natl. Acad. Sci. USA, **94**, 1669-1674 (1997)
[14] Fu, Y.; Buryanovskyy, L.; Zhang, Z.: Crystal structure of quinone reductase 2 in complex with cancer prodrug CB1954. Biochem. Biophys. Res. Commun., **336**, 332-338 (2005)
[15] Celli, C.M.; Tran, N.; Knox, R.; Jaiswal, A.K.: NRH:quinone oxidoreductase 2 (NQO2) catalyzes metabolic activation of quinones and anti-tumor drugs. Biochem. Pharmacol., **72**, 366-376 (2006)
[16] Wang, W.; Jaiswal, A.K.: Nulear factor Nrf2 and antioxidant response element regulate NRH:quinone oxidoreductase 2 (NQO2) gene expression and antioxidant induction. Free Radic. Biol. Med., **40**, 1119-1130 (2006)
[17] Iskander, K.; Li, J.; Han, S.; Zheng, B.; Jaiswal, A.K.: NQO1 and NQO2 regulation of humoral immunity and autoimmunity. J. Biol. Chem., **281**, 30917-30924 (2006)
[18] Okada, S.; Farin, F.M.; Stapleton, P.; Viernes, H.; Quigley, S.D.; Powers, K.M.; Smith-Weller, T.; Franklin, G.M.; Longstreth, W.T.; Swanson, P.D.; Checkoway, H.: No associations between Parkinsons disease and polymorphisms of the quinone oxidoreductase (NQO1, NQO2) genes. Neurosci. Lett., **375**, 178-180 (2005)

Violaxanthin de-epoxidase 1.10.99.3

1 Nomenclature

EC number
1.10.99.3

Systematic name
violaxanthin:ascorbate oxidoreductase

Recommended name
violaxanthin de-epoxidase

Synonyms
VDE <1, 2, 4, 5, 6, 7, 8, 9, 10> [8, 10, 11, 12, 13, 14, 17, 18, 20, 21, 22, 23, 24]
Vio de-epoxidase <2> [15]

CAS registry number
57534-73-3

2 Source Organism

<1> *Triticum aestivum* (no sequence specified) [2, 3, 10, 20]
<2> *Spinacia oleracea* (no sequence specified) [4, 7, 8, 10, 11, 14, 15, 17, 19, 23]
<3> *Pisum sativum* (no sequence specified) [16]
<4> *Nicotiana tabacum* (no sequence specified) [13]
<5> *Arabidopsis thaliana* (no sequence specified) [18,24]
<6> *Lycopersicon esculentum* (no sequence specified) [6,21]
<7> *Lactuca sativa* (no sequence specified) [1,12]
<8> *Prunus armeniaca* (no sequence specified) [21]
<9> *Lemna trisulca* (no sequence specified) (<9> ST6GAL2, fragment [22]) [21,22]
<10> *Mantoniella squamata* (no sequence specified) [17, 19]
<11> *Arabidopsis thaliana* (UNIPROT accession number: Q39249) [5]
<12> *Nicotiana tabacum* (UNIPROT accession number: Q40593) [5]
<13> *Spinacia oleracea* (UNIPROT accession number: Q9SM43) [9]

3 Reaction and Specificity

Catalyzed reaction
antheraxanthin + ascorbate = zeaxanthin + dehydroascorbate + H_2O
violaxanthin + ascorbate = antheraxanthin + dehydroascorbate + H_2O

Natural substrates and products

- **S** antheraxanthin + ascorbate <1, 2, 6, 10, 11, 12> (<6> de-epoxidation reaction of the xanthophyll cycle plays an important role in the protection of the chloroplast against photooxidative damage. Violaxanthin is bound to the antenna proteins of both photosystems. In photosystem II, the formation of zeaxanthin is essential for the pH-dependent dissipation of excess light energy as heat. Violaxanthin bound to site V1 and N1 is easily accessible for de-epoxidation, whereas violaxanthin bound to L2 is only partially and/or with the slower kinetics converible to zeaxanthin [6]; <1> reaction of the xanthophyll cycle [3]; <10> the violaxanthin/antheraxanthin cycle in Mantionella is caused by the interaction of the slow second de-epoxidation step and the relatively fast epoxidation of antheraxanthin to violaxanthin [17]; <11,12> violaxanthin de-epoxidase and zeaxanthin epoxidase catalyze the addition and removal of epoxide groups in carotenoids of xanthophyll cycle in plants. The xanthophyll cycle is implicated in protecting the photosynthetoic apparatus from excessive light [5]) (Reversibility: ?) [3, 5, 6, 10, 14, 17]
- **P** zeaxanthin + dehydroascorbate + H_2O
- **S** violaxanthin + ascorbate <1, 2, 6, 10, 11, 12> (<6> de-epoxidation reaction of the xanthophyll cycle plays an important role in the protection of the chloroplast against photooxidative damage. Violaxanthin is bound to the antenna proteins of both photosystems. In photosystem II, the formation of zeaxanthin is essential for the pH-dependent dissipation of excess light energy as heat. Violaxanthin bound to site V1 and N1 is easily accessible for de-epoxidation, whereas violaxanthin bound to L2 is only partially and/or with the slower kinetics converible to zeaxanthin [6]; <1> reaction of the xanthophyll cycle [3]; <11,12> violaxanthin de-epoxidase and zeaxanthin epoxidase catalyze the addition and removal of epoxide groups in carotenoids of xanthophyll cycle in plants. The xanthophyll cycle is implicated in protecting the photosynthetic apparatus from excessive light [5]) (Reversibility: ?) [3, 5, 6, 10, 14, 17]
- **P** antheraxanthin + dehydroascorbate + H_2O
- **S** Additional information <4> (<4> the level of violaxanthin de-epoxidase changes in an inverse, nonlinear relationship with respect to the VAZ pool (violaxanthin + antheraxanthin + zeaxanthin), suggesting that enzyme levels can be indirectly regulated by the VAZ pool [13]) (Reversibility: ?) [13]
- **P** ?

Substrates and products

- **S** all-trans-neoxanthin + ascorbate <2, 1> (<10> 0.21% of the activity with violaxanthin [19]; <2> 2.5% of the activity with violaxanthin [19]) (Reversibility: ?) [19]
- **P** ? + dehydroascorbate + H_2O
- **S** antheraxanthin + ascorbate <1, 2, 5, 6, 7, 9, 10, 11, 12> (<6> de-epoxidation reaction of the xanthophyll cycle plays an important role in the protection of the chloroplast against photooxidative damage. Violaxanthin is bound to the antenna proteins of both photosystems. In photosystem II,

the formation of zeaxanthin is essential for the pH-dependent dissipation of excess light energy as heat. Violaxanthin bound to site V1 and N1 is easily accessible for de-epoxidation, whereas violaxanthin bound to L2 is only partially and/or with the slower kinetics converible to zeaxanthin [6]; <1> reaction of the xanthophyll cycle [3]; <10> the violaxanthin/antheraxanthin cycle in Mantionella is caused by the interaction of the slow second de-epoxidation step and the relatively fast epoxidation of antheraxanthin to violaxanthin [17]; <11,12> violaxanthin de-epoxidase and zeaxanthin epoxidase catalyze the addition and removal of epoxide groups in carotenoids of xanthophyll cycle in plants. The xanthophyll cycle is implicated in protecting the photosynthetoic apparatus from excessive light [5]; <10> 11% of the activity with violaxanthin [19]; <2> 146% of the activity with violaxanthin [19]; <7> the activity is 5.25fold higher than the activity of violaxanthin [1]) (Reversibility: ?) [1, 3, 5, 6, 7, 10, 12, 14, 17, 19, 20, 22, 23, 24]

P zeaxanthin + dehydroascorbate + H_2O

S cryptoxanthin epoxide + ascorbate <7> (<7> 92% of the activity with violaxanthin [1]) (Reversibility: ?) [1]

P ?

S cryptoxanthin-5,6,5',6'-di-epoxide + ascorbate <2, 1> (<10> 36.5% of the activity with violaxanthin [19]; <2> 54% of the activity with violaxanthin [19]) (Reversibility: ?) [19]

P ? + dehydroascorbate + H_2O

S cryptoxanthin-5,6-epoxide + ascorbate <2, 1> (<10> 1% of the activity with violaxanthin [19]; <2> 12.5% of the activity with violaxanthin [19]) (Reversibility: ?) [19]

P ? + dehydroascorbate + H_2O

S diadinoxanthin + ascorbate <2, 1> (<10> 11% of the activity with violaxanthin [19]; <2> 69% of the activity with violaxanthin [19]) (Reversibility: ?) [19]

P ? + dehydroascorbate + H_2O

S diadinoxanthin + ascorbate <7> (<7> the activity is 2.25fold higher than the activity of violaxanthin [1]) (Reversibility: ?) [1]

P ?

S lutein epoxide + ascorbate <7> (<7> the activity is 1.33fold higher than the activity of violaxanthin [1]) (Reversibility: ?) [1]

P ?

S lutein-5,6-epoxide + ascorbate <2, 1> (<2> 110% of the activity with violaxanthin [19]; <10> 20% of the activity with violaxanthin [19]) (Reversibility: ?) [19]

P ? + dehydroascorbate + H_2O

S neoxanthin + ascorbate <7> (<7> 10% of the activity with violaxanthin [1]) (Reversibility: ?) [1]

P ?

S violaxanthin + ascorbate <1, 2, 5, 6, 7, 8, 9, 10, 11, 12> (<6> de-epoxidation reaction of the xanthophyll cycle plays an important role in the protection of the chloroplast against photooxidative damage. Violaxanthin is bound to the antenna proteins of both photosystems. In photosystem II, the formation of zeaxanthin is essential for the pH-dependent dissipation of excess light energy as heat. Violaxanthin bound to site V1 and N1 is easily accessible for de-epoxidation, whereas violaxanthin bound to L2 is only partially and/or with the slower kinetics converible to zeaxanthin [6]; <1> reaction of the xanthophyll cycle [3]; <11,12> violaxanthin de-epoxidase and zeaxanthin epoxidase catalyze the addition and removal of epoxide groups in carotenoids of xanthophyll cycle in plants. The xanthophyll cycle is implicated in protecting the photosynthetic apparatus from excessive light [5]) (Reversibility: ?) [1, 3, 4, 5, 6, 7, 10, 12, 14, 17, 19, 20, 21, 22, 23, 24]

P antheraxanthin + dehydroascorbate + H_2O

S Additional information <2, 4, 7, 10> (<4> the level of violaxanthin de-epoxidase changes in an inverse, nonlinear relationship with respect to the VAZ pool (violaxanthin + antheraxanthin + zeaxanthin), suggesting that enzyme levels can be indirectly regulated by the VAZ pool [13]; <2,10> no activity with 9-cis neoxanthin and 9-cis violaxanthin [19]; <2> only the epoxy-oxygen at the 5,6(5,6) position of xanthophylls are cleaved by the VDE, whereas ring-spanning epoxides at position 3,6(3,6) are not accessible to the enzyme. The structure and chemical ligands of the second jonon ring are insignificant for the de-epoxidation of the 5,6-epoxy groups of the first ring. The epoxy-free second jonon ring is not involved in the binding of the xanthophyll to the catalytic center and does not affect the enzyme reaction. Due to steric hindrance, any tested cis-configuration in the polyene chain of the xanthophylls, as well as the 8-oxy group, in fucoxanthin, prevents the de-epoxidation [7]; <7> the enzyme is inactive against monoepoxy and diepoxy β-carotene, antheraxanthin-A and lutein epoxide [1]) (Reversibility: ?) [1, 7, 13, 19]

P ?

Inhibitors

Cys <2> (<2> 2.7 mM, 50% inhibition [14]) [14]

Dithiothreitol <2> (<2> 0.055 mM, 50% inhibition. The rate of conversion of violaxanthin to antheraxanthin is relatively unchanged whereas the conversion of antheraxanthin to zeaxanthin is 50-80% inhibited [14]) [14]

Pepstatin <2> (<2> 50% inhibition at 0.12 mM, reversible, protonation-induced structural change of the enzyme [4]) [4]

Pepstatin A <2> [11]

mercaptoethanol <2> (<2> 0.68 mM, 50% inhibition [14]) [14]

o-Phenanthroline <2> (<2> 0.025 mM, 50% inhibition. The rate of conversion of violaxanthin to antheraxanthin is relatively unchanged whereas the conversion of antheraxanthin to zeaxanthin is 50-80% inhibited [14]) [14]

zeaxanthin <2> (<2> product inhibition [14]) [14]

Additional information <6, 8, 9> (<9> no inhibition by glucose, glucosamine, galactose, galactosamine, mannose and mannosamine at a concentration of 0.5% [22]; <6,8,9> no inhibitory effect by 10 mM Zn^{2+} and 10 mM Cd^{2+} [21]) [21, 22]

Activating compounds

Lipid <1> (<1> the enzyme requires lipid inverted hexagonal structures for activity [3]) [3]
Monogalactosyldiacylglycerol <1, 7> (<7> required for an active enzyme. Optimal ratio of monogalactosyldiacylglycerol to the substrate violaxanthin is 28:1 [1]; <1> 0.0116 mM is needed to achieve saturation of the de-epoxidation reaction [20]) [1, 20]
Phosphatidylcholine <5> (<5> supports slow and incomplete de-epoxidation [24]) [24]
digalactosyldiacylglyceride <5> (<5> 0.0067 mM, supports slow but nevertheless complete to nearly complete de-epoxidation [24]) [24]
monogalactosyldiacylglyceride <5> (<5> 0.01 mM, supports rapid and complete de-epoxidation [24]) [24]

Metals, ions

Additional information <6, 8, 9> (<6,8,9> no stimulatory effect by 10 mM Zn^{2+} and 10 mM Cd^{2+} [21]) [21]

Specific activity (U/mg)

19 <2> [14]
916 <7> [12]

K_m-Value (mM)

0.000049 <7> (violaxanthin, <7> pH 5.1, 26°C, at the optimal ratio of monogalactosyldiglyceride to the substrate violaxanthin is 28:1 [1]) [1]
0.000352 <7> (violaxanthin) [12]
0.00145 <7> (violaxanthin, <7> pH 5.1, 26°C, at an ratio of monogalactosyldiacylglyceride to the substrate violaxanthin is 45:1 [1]) [1]
0.0016 <2> (antheraxanthin, <2> pH 5.2, 29°C [17]) [17]
0.0033 <2, 10> (violaxanthin, <2,10> pH 5.2, 29°C [17]) [17]
0.005 <2> (violaxanthin) [14]
0.0053 <2> (antheraxanthin) [14]
0.006 <10> (antheraxanthin, <10> pH 5.2, 29°C [17]) [17]
2.3 <2> (ascorbate, <2> at pH 5.0, with 0.0004 mM violaxanthin and 0.0116 mM monogalactosyldiacylglycerol in reaction medium containing 10 mM KCl, 5 mM $MgCl_2$ and 40 mM morpholinoethanesulfonic acid [23]) [23]
4.4 <7> (ascorbate) [12]
5.3 <2> (antheraxanthin) [7]
11.1 <2> (violaxanthin) [7]

pH-Optimum

5-5.2 <2> (<2> irrespective of the presence of high or low ascorbate concentrations [23]) [23]
5.2 <1, 2, 7> [12, 14, 20]

pH-Range

Additional information <3> (<3> dicyclohexylcarbodiimide alters the pH dependence of violaxanthin de-epoxidation [16]) [16]

4 Enzyme Structure

Molecular weight

43000 <2> [8]
45800 <2> (<2> gel filtration [14]) [14]
60000 <7> (<7> gel filtration [1]) [1]

Subunits

? <7> (<7> x * 43000, SDS-PAGE [12]) [12]
monomer <2> (<2> 1 * 43300, SDS-PAGE [14]) [14]

5 Isolation/Preparation/Mutation/Application

Source/tissue

leaf <1, 2, 3, 4, 6, 7, 8, 9> (<4> grown under high light [13]) [2, 3, 8, 12, 13, 14, 15, 16, 17, 20, 21, 23]

Localization

chloroplast <1, 2, 5, 7> [1, 12, 15, 20, 24]
chloroplast thylakoid <2> [23]
thylakoid <1, 2, 3, 5, 6> (<5> lumen [18]; <1,2> regulation of the violaxanthin de-epoxidase involves a conformational change at low lumenal pH, followed by binding of the enzyme to the thylakoid membrane. Protonation of His at low pH induces the conformational change of the enzyme, and hence indirectly regulates binding of the enzyme to the thylakoid membrane [10]; <2> the enzyme is mobile within the thylakoid lumen at neutral pH values which occur under in-vivo conditions in the dark. However, upon strong illumination, when the lumen pH drops below pH 6.5 due to formation of a proton gradient, the properties of the de-epoxidase are altered and the enzyme becomes tightly bound to the membrane thus gaining access to its substrate [15]) [4, 6, 7, 10, 14, 15, 16, 18]
thylakoid membrane <1, 2, 4, 10> (<1> violaxanthin de-epoxidation most probably takes place inside MGDG-rich domains of the thylakoid membrane [2]) [2, 11, 13, 19]

Purification

<2> [8, 14]
<2> (in presence of Tween 20. The enzyme has more than one disulfide bond and takes multiple forms depending on the extent of the reduction) [11]
<7> [12]

Cloning

<5> (a full-length violaxanthin de-epoxidase and deletion mutants of the N- and C-terminal regions are expressed in Escherichia coli and Nicotiana tabacum L. cv. Xanthi. High expression of the enzyme in Escherichia coli is achieved after adding the argU gene that encodes the Escherichia coli arginine AGA tRNA. The specific activity of the violaxanthin de-epoxidase ex-

pressed in Escherichia coli is low, possibly due to incorrect folding. The transformed tobacco exhibits a 13fold to 19fold increase in VDE specific activity, indicating correct protein folding) [18]
<5> (expressed in Nicotiana tabacum) [24]
<13> (expression in Escherichia coli) [9]

Engineering

Δ1-4 <5> (<5> removal of 4 amino acids from the N-terminal region abolishes all violaxanthin de-epoxidase activity [18]) [18]
Δ258-349 <5> (<5> 71 C-terminal amino acid can be removed without affecting activity [18]) [18]
H121A/H124A <2> (<2> considerably lower pH dependence for binding than wild-type, cooperativity value around 2 compared to wild-type value of 3.7. K_m-value for ascorbate is 3.2 mM compared to 1.9 mM for wild-type enzyme [10]) [10]
H121R/H124R <13> (<13> inactive mutant enzyme [9]) [9]
H124R <2> (<2> considerably lower pH dependence for binding than wild-type, cooperativity value around 2 compared to wild-type value of 3.7. K_m-value for ascorbate is 2.1 mM compared to 1.9 mM for wild-type enzyme [10]; <2> considerably lower pH dependence for binding than wild-type, cooperativity value around 2 compared to wild-type value of 3.7.K_m-value for ascorbate is 1.5 mM compared to 1.9 mM for wild-type enzyme [10]) [10]
H134A <2> (<2> considerably lower pH dependence for binding than wild-type, cooperativity value around 2 compared to wild-type value of 3.7 [10]) [10]
H167A/H173A <2> (<2> considerably lower pH dependence for binding than wild-type, cooperativity value of 1.6 compared to wild-type value of 3.7. K_m-value for ascorbate is 8.3 mM compared to 1.9 mM for wild-type enzyme [10]) [10]
H167R/H173R <2> (<2> considerably lower pH dependence for binding than wild-type, cooperativity value around 2 compared to wild-type value of 3.7. K_m-value for ascorbate is 6.3 mM compared to 1.9 mM for wild-type enzyme [10]) [10]

6 Stability

pH-Stability

6.5 <2> (<2> no activity above pH 6.5 [23]) [23]

References

[1] Yamamoto, H.Y.; Higashi, R.M.: Violaxanthin de-epoxidase. Lipid composition and substrate specificity. Arch. Biochem. Biophys., **190**, 514-522 (1978)

[2] Latowski, D.; Kostecka, A.; Strzalka, K.: Effect of monogalactosyldiacylglycerol and other thylakoid lipids on violaxanthin de-epoxidation in liposomes. Biochem. Soc. Trans., **28**, 810-812 (2000)
[3] Latowski, D.; Akerlund, H.E.; Strzalka, K.: Violaxanthin de-epoxidase, the xanthophyll cycle enzyme, requires lipid inverted hexagonal structures for its activity. Biochemistry, **43**, 4417-4420 (2004)
[4] Kawano, M.; Kuwabara, T.: pH-dependent reversible inhibition of violaxanthin de-epoxidase by pepstatin related to protonation-induced structural change of the enzyme. FEBS Lett., **481**, 101-104 (2000)
[5] Bugos, R.C.; Hieber, A.D.; Yamamoto, H.Y.: Xanthophyll cycle enzymes are members of the lipocalin family, the first identified from plants. J. Biol. Chem., **273**, 15321-15324 (1998)
[6] Wehner, A.; Storf, S.; Jahns, P.; Schmid, V.H.: De-epoxidation of violaxanthin in light-harvesting complex I proteins. J. Biol. Chem., **279**, 26823-26829 (2004)
[7] Grotz, B.; Molnar, P.; Stransky, H.; Hager, A.: Substrate specificity and functional aspects of violaxanthin-de-epoxidase, an enzyme of the xanthophyll cycle. J. Plant Physiol., **154**, 437-446 (1999)
[8] Arvidsson, P.-O.; Eva Bratt, C.; Carlsson, M.; Aakerlund, H.-E.: Purification and identification of the violaxanthin deepoxidase as a 43 kDa protein. Photosynth. Res., **49**, 119-129 (1996)
[9] Emanuelsson, A.; Eskling, M.; Akerlund, H.-E.: Chemical and mutational modification of histidines in violaxanthin de-epoxidase from Spinacia oleracea. Physiol. Plant., **119**, 97-104 (2003)
[10] Gisselsson, A.; Szilagyi, A.; Akerlund, H.-E.: Role of histidines in the binding of violaxanthin de-epoxidase to the thylakoid membrane as studied by site-directed mutagenesis. Physiol. Plant., **122**, 337-343 (2004)
[11] Kuwabara, T.; Hasegawa, M.; Kawano, M.; Takaichi, S.: Characterization of violaxanthin de-epoxidase purified in the presence of Tween 20: effects of dithiothreitol and pepstatin A. Plant Cell Physiol., **40**, 1119-1126 (1999)
[12] Rockholm, D.C.; Yamamoto, H.Y.: Violaxanthin de-epoxidase. Plant Physiol., **110**, 697-703 (1996)
[13] Bugos, R.C.; Chang, S.H.; Yamamoto, H.Y.: Developmental expression of violaxanthin de-epoxidase in leaves of tobacco growing under high and low light. Plant Physiol., **121**, 207-214 (1999)
[14] Havir, E.A.; Tausta, S.L.; Peterson, R.B.: Purification and properties of violaxanthin de-epoxidase from spinach. Plant Sci., **123**, 57-66 (1997)
[15] Hager, A.; Holocher, K.: Localization of the xanthophyll-cycle enzyme violaxanthin de-epoxidase within the thylakoid lumen and abolition of its mobility by a (light-dependent) pH decrease. Planta, **192**, 581-589 (1994)
[16] Jahns, P.; Heyde, S.: Dicyclohexylcarbodiimide alters the pH dependence of violaxanthin de-epoxidation. Planta, **207**, 393-400 (1999)
[17] Frommolt, R.; Goss, R.; Wilhelm, C.: The de-epoxidase and epoxidase reactions of Mantoniella squamata (Prasinophyceae) exhibit different substrate-specific reaction kinetics compared to spinach. Planta, **213**, 446-456 (2001)

[18] Hieber, A.D.; Bugos, R.C.; Verhoeven, A.S.; Yamamoto, H.Y.: Overexpression of violaxanthin de-epoxidase: properties of C-terminal deletions on activity and pH-dependent lipid binding. Planta, **214**, 476-483 (2002)
[19] Goss, R.: Substrate specificity of the violaxanthin de-epoxidase of the primitive green alga Mantoniella squamata (Prasinophyceae). Planta, **217**, 801-812 (2003)
[20] Goss, R.; Lohr, M.; Latowski, D.; Grzyb, J.; Vieler, A.; Wilhelm, C.; Strzalka, K.: Role of hexagonal structure-forming lipids in diadinoxanthin and violaxanthin solubilization and de-epoxidation. Biochemistry, **44**, 4028-4036 (2005)
[21] Latowski, D.; Kruk, J.; Strzaska, K.: Inhibition of zeaxanthin epoxidase activity by cadmium ions in higher plants. J. Inorg. Biochem., **99**, 2081-2087 (2005)
[22] Latowski, D.; Banas, A.K.; Strzaska, K.; Gabrys, H.: Amino sugars - new inhibitors of zeaxanthin epoxidase, a violaxanthin cycle enzyme. J. Plant Physiol., **164**, 231-237 (2007)
[23] Grouneva, I.; Jakob, T.; Wilhelm, C.; Goss, R.: Influence of ascorbate and pH on the activity of the diatom xanthophyll cycle-enzyme diadinoxanthin de-epoxidase. Physiol. Plant., **126**, 205-211 (2006)
[24] Yamamoto, H.Y.: Functional roles of the major chloroplast lipids in the violaxanthin cycle. Planta, **224**, 719-724 (2006)

Peroxiredoxin 1.11.1.15

1 Nomenclature

EC number

1.11.1.15

Systematic name

thiol-containing-reductant:hydroperoxide oxidoreductase

Recommended name

peroxiredoxin

Synonyms

1-Cys Prx <6, 19> [13, 46]
1-Cys peroxiredoxin <19, 21, 28, 46> [17, 31, 38, 39, 46, 49]
1Cys-peroxiredoxin <20> [4]
2-CP <17> [27]
2-Cys Prx <9, 10, 47> [18, 28, 32]
2-Cys peroxiredoxin <6, 8, 9, 10, 19, 22, 25, 28> [17, 18, 22, 28, 29, 36, 40, 43, 51]
2-Cys peroxiredoxin TPx-1 <22> [51]
2-cysteine peroxiredoxin <17> [27]
2Cys-peroxiredoxin <37> [26]
Ac-1-Cys Prx <46> [31]
AhpC <1, 29> [9, 34]
AsPrx <32> [7]
Bcp2 <21> [39]
EcTpx <5> (<5> an atypical 2Cys Tpx with an intramolecular disulfide bond formed between the peroxidatic Cys61 and the resolving Cys95 residue [16]) [16]
FhePrx <25> [40]
Gpx2 <8> [43]
LimTXNPx <24> [6]
MtTPx <18> [44]
PRDX5 <6> [19]
PRx IV <7> [8, 10]
PcPrx-1 <23> [37]
Pf1-Cys-Prx <19> [38]
PfAOP <19> [46]
PfTrx-Px1 <19> [22]
PfTrx-Px2 <19> [53]
Prdx6 <40, 41, 42, 43, 44> [41]
Prx <3, 8> [3, 15]

Prx I <4, 6, 7> (<6,7> 2-Cys-Prx [8]) [8, 12, 45]
Prx II <6, 7> (<6,7> 2-Cys-Prx [8]) [8, 12]
Prx III <6> [12]
Prx Q <11> [54]
Prx V <6, 7> (<6,7> 2-Cys-Prx [8]) [8]
Prx VI <6, 7> (<6,7> 1-Cys-Prx [8]) [8]
PrxII F <11> [42]
PrxV <6> [14]
TPx-1 <22> [51]
TPx-B <6> [29]
TPx1 <16, 19> [36, 55]
TgPrx2 <28> [49]
TgTrx-Px1 <28> [17]
TgTrx-Px2 <28> [17]
Ts2-CysPrx <38> [48]
cPrx I <8> [3]
cPrx II <8> [3]
peroxiredoxin <24> [6]
peroxiredoxin 5 <6> [19, 20]
peroxiredoxin 6 <40, 41, 42, 43, 44> [41]
peroxiredoxin I <4> [45]
peroxiredoxin II <4, 39> [2, 33]
peroxiredoxin IV <7> [10]
peroxiredoxin Q <11> [54]
peroxiredoxin VI <6> [1]
rDiPrx-1 <31> [24]
thioredoxin peroxidase <2, 47> [11, 32]
thioredoxin peroxidase B <6> [29]
thioredoxin-dependent alkyl hydroperoxide reductase <29> [9]
thioredoxin-dependent peroxidase <2> [11]
type II peroxiredoxin <10, 12> [25, 30]
type II peroxiredoxin F <11> [42]

CAS registry number

207137-51-7

2 Source Organism

<1> *Salmonella typhimurium* (no sequence specified) [34]
<2> *Haemophilus influenzae* (no sequence specified) [11]
<3> *Thermus aquaticus* (no sequence specified) [15]
<4> *Mus musculus* (no sequence specified) [2, 45]
<5> *Escherichia coli* (no sequence specified) [16]
<6> *Homo sapiens* (no sequence specified) [1,8,12,13,14,19,20,29]
<7> *Rattus norvegicus* (no sequence specified) [8,10]
<8> *Saccharomyces cerevisiae* (no sequence specified) [3,43]

<9> *Hordeum vulgare* (no sequence specified) [28]
<10> *Pisum sativum* (no sequence specified) [18, 30]
<11> *Arabidopsis thaliana* (no sequence specified) (<11> NarG, α subunit [54]) [42, 54]
<12> *Populus euramericana* (no sequence specified) [25]
<13> *Entamoeba histolytica* (no sequence specified) [23]
<14> *Trichomonas vaginalis* (no sequence specified) [50]
<15> *Neisseria meningitidis* (no sequence specified) [5]
<16> *Schizosaccharomyces pombe* (no sequence specified) [55]
<17> *Arabidopsis sp.* (no sequence specified) [27]
<18> *Mycobacterium tuberculosis* (no sequence specified) [44]
<19> *Plasmodium falciparum* (no sequence specified) [22, 36, 38, 46, 53]
<20> *Oryza sativa* (no sequence specified) [4]
<21> *Sulfolobus solfataricus* (no sequence specified) [39]
<22> *Plasmodium berghei* (no sequence specified) [51]
<23> *Phanerochaete chrysosporium* (no sequence specified) [37]
<24> *Leishmania infantum* (no sequence specified) [6]
<25> *Fasciola hepatica* (no sequence specified) [40]
<26> *Aeropyrum pernix* (no sequence specified) [47]
<27> *Populus trichocarpa* (no sequence specified) [35]
<28> *Toxoplasma gondii* (no sequence specified) [17, 49]
<29> *Helicobacter pylori* (UNIPROT accession number: P21762) [9]
<30> *Plasmodium falciparum* (UNIPROT accession number: Q9GSV3) [21]
<31> *Dirofilaria immitis* (UNIPROT accession number: Q9U5A1) [24]
<32> *Ascaris suum* (UNIPROT accession number: Q9NL98) [7]
<33> *Plasmodium falciparum* (UNIPROT accession number: Q9N699) [21]
<34> *Entamoeba moshkovskii* (UNIPROT accession number: Q5W9B7) [23]
<35> *Entamoeba moshkovskii* (UNIPROT accession number: Q5W9B5) [23]
<36> *Entamoeba moshkovskii* (UNIPROT accession number: Q5W9B6) [23]
<37> *Brassica rapa* (UNIPROT accession number: Q9SQJ4) [26]
<38> *Taenia solium* (no sequence specified) [48]
<39> *Populus sp.* (UNIPROT accession number: Q8S3L0) [33]
<40> *Homo sapiens* (UNIPROT accession number: P30041) [41]
<41> *Mus musculus* (UNIPROT accession number: O08709) [41]
<42> *Rattus norvegicus* (UNIPROT accession number: O35244) [41]
<43> *Bos taurus* (UNIPROT accession number: O77834) [41]
<44> *Sus scrofa* (UNIPROT accession number: Q9TSX9) [41]
<45> *Rhizobium etli* (UNIPROT accession number: O69777) [52]
<46> *Antrodia camphorata* (UNIPROT accession number: A1KZ88) [31]
<47> *Antrodia camphorata* (UNIPROT accession number: A1KXU3) [32]

3 Reaction and Specificity

Catalyzed reaction

2 R'-SH + ROOH = R'-S-S-R' + H_2O + ROH (<25> the enzyme exhibits a saturable, single-displacement-like reaction mechanism rather than non-sa-

turable double displacement (ping-pong) enzyme substitution mechanism [40])

Reaction type

oxidation
redox reaction
reduction

Natural substrates and products

S Additional information <23> (<23> PcPrx-1 may play a protective role against oxidative stress [37]) (Reversibility: ?) [37]

P ?

S Additional information <3, 4, 6, 7, 8, 9, 11, 16, 17, 19, 20, 21, 22, 28, 29, 30, 32, 35, 40, 41, 42, 43, 44, 45> (<30> 1-Cys peroxiredoxin protects DNA from degradation by reactive O_2 species in presence of low molecular mass thiols such as dithiothreitol and glutathione [21]; <19> after subjection to exogenous and endogenous oxidative stress, the Plasmodium falciparum blood stage form shows a marked elevation of PfTrx-Px1 mRNA and protein levels consistent with the structure of related proteins [22]; <20> antioxidant activity may be the primary function of the enzyme [4]; <28> peroxidases belonging to the class of 1-Cys and 2-Cys peroxiredoxins play crucial roles in maintaining redox balance. TgTrx-Px1 is an extremely potent antioxidant [17]; <3> peroxiredoxin and NADH:peroxiredoxin oxidoreductase together catalyze the anaerobic reduction of H_2O_2 [15]; <7> Prx IV is a secretable protein and may exert its protective function against oxidative damage by scavenging reactive oxygen species in the extracellular space [10]; <4> the enzyme is essential for sustaining life span of erythrocytes in mice by protecting them from oxidative stress [2]; <35> the enzyme is important for protection against endogenously generated H_2O_2 [23]; <32> the enzyme might function as a major antioxidant enzyme in Ascaris suum [7]; <29> the enzyme plays a critical role in defending the organism against oxygen toxicity [9]; <6> the enzyme promotes potassium efflux and down-regulates apoptosis and the recruitment of monocytes by endothelial tissue [29]; <9> the midpoint redox potential of -315 mV places 2-Cys Prx reduction after Calvin cycle activation and before switching the male valve for export of excess reduction equivalents to the cytosol. The activity of the enzyme is also linked to chloroplastic NAD(P)H metabolism. Saline-stress-induced oligomerization of the enzyme triggers membrane attachment and allows for detoxification of peroxides at the site of production in immediate vicinity of the thylakoid membrane [28]; <17> the photosynthetic machinery needs high levels of enzyme during leaf development to protect it from oxidative damage. The damage is reduced by the accumulation of 2-CP protein, by the de novo synthesis and replacement of damaged proteins, and by the induction of other antioxidant defenses in 2-CP mutants [27]; <6> together with glutathione peroxidase and catalase, Prx enzymes likely play an important role in eliminating peroxides generated during metabolism. In addition Prx I and II might participate in the signaling cascades of growth factors and tumor necrosis factor-α by regulating the intracellular concentration of H_2O_2 [12]; <21> Bcp2 plays an important role in the peroxide-scavaging system in Sulfolobus solfataricus. The enzyme protects plasmid DNA from nicking by the metal-catalysed oxidation system [39]; <45> constitutive expression of prxS confers enhanced survival and growth to Rhizobium etli in presence of H_2O_2. Defence of Rhizobium etli bacteroids against oxidative stress involves a complexly regulated atypical 2-Cys peroxiredoxin [52]; <8> Gpx2 is likely to play an important role in the protection of cells from oxidative stress in the presence of Ca^{2+} [43]; <19> Pf1-Cys-Prx protects the parasite against oxidative stress by binding to ferriprotoporphyrin [38]; <4> Prx I has at least two distinct roles: as an antioxidant enzyme and as a regulator of p38 MAPK [45]; <11> Prx Q attaches to photosystem II and has a specific function distinct from 2-Cys peroxiredoxin in protecting photosynthesis. Its absence causes metabolic changes that are sensed and trigger appropriate compensatory responses [54]; <40,41,42,43,44> the enzyme functions in antioxidant defense and lung phospholipid metabolism [41]; <11> the enzyme is essential for redox homeostasis and root growth of Arabidopsis thaliana under stress [42]; <19> the enzyme is involved in the detoxification of reactive oxygen species and of reactive nitrogen species [36]; <22> Tpx-1 is required for normal gametocyte development but does not affect the male/female gametocyte ratio or male gametogenesis [51]; <16> Tpx1 is an upstream activator of Pap1. At low H_2O_2 concentrations, this oxidant scavenger can transfer a redox signal to Pap1, whereas higher concentrations of the oxidant inhibit the Tpx1-Pap1 redox relay through the temporal inactivation of Tpx1 by oxidation of its catalytic cysteine to a sulfinic acid. This cysteine modification can be reversed by the sulfiredoxin Srx1, its expression in response to high doses of H_2O_2 strictly depending on active Sty1. Tpx1 oxidation to the cysteine-sulfinic acid and its reversion by Srx1 constitutes a redox switch in H_2O_2 signaling, restricting Pap1 activation within a narrow range of H_2O_2 concentrations [55]) (Reversibility: ?) [2, 4, 7, 9, 10, 12, 15, 17, 21, 22, 23, 27, 28, 29, 36, 38, 39, 41, 42, 43, 45, 51, 52, 54, 55]

P ?

Substrates and products

S 1-palmitoyl-2-arichidonoyl-sn-glycero-3-phosphocholine + GSH <40> (Reversibility: ?) [41]

P ?

S 1-palmitoyl-2-linolenoyl-sn-glycero-3-phosphocholine hydroperoxide + GSH <40> (Reversibility: ?) [41]

P ?

S H_2O_2 + GSH <7, 8, 40, 41, 42, 43, 44> (Reversibility: ?) [10, 41, 43]

P H_2O + GSSG

S H_2O_2 + NADPH <15> (<15> reaction is driven by glutathione which is maintained reduced via NADPH and glutathione reductase. Both the peroxiredoxin and glutaredoxin domains are biochemically active in the natural hybrid protein which contains both a peroxiredoxin and a glutaredoxin domain. When expressed separately, the glutaredoxin domain is catalytically active and the peroxiredoxin domain posseses a weak activity when supplemented with exogenous glutaredoxin [5]) (Reversibility: ?) [5]

P H_2O + $NADP^+$

S H_2O_2 + dithiothreitol <6> (Reversibility: ?) [13]

P ?

S H_2O_2 + dithiothreitol <1, 10, 11, 21, 35, 45, 47> (Reversibility: ?) [18, 23, 32, 34, 39, 42, 52]

P H_2O + oxidized dithiothreitol

S H_2O_2 + ferrithiocyanate <23> (Reversibility: ?) [37]

P ?

S H_2O_2 + reduced glutaredoxin <12> (Reversibility: ?) [25]

P H_2O + oxidized glutaredoxin

S H_2O_2 + reduced plasmoredoxin <19> (Reversibility: ?) [36]

P H_2O + oxidized plasmoredoxin

S H_2O_2 + reduced thioredoxin <2, 6, 7, 8, 9, 11, 19, 25, 28, 29, 38, 45> (<28> ping-pong mechanism [17]; <19> ping-pong kinetics [22]; <2> reaction is dependent on thioredoxin [11]; <9> thioredoxin from Escherichia coli and thioredoxin f and m from spinach [28]; <28> the enzyme works on the basis of a monothiol mechanism [49]; <19> the peroxiredoxin preferrs PfTrx2 to PfTrx1 as a reducing partner [53]) (Reversibility: ?) [9, 10, 11, 12, 14, 17, 22, 28, 36, 40, 43, 48, 49, 52, 53, 54]

P H_2O + oxidized thioredoxin

S H_2O_2 + thioredoxin <30> (<30> 1-Cys peroxiredoxin is less active than 2-Cys peroxiredoxin [21]) (Reversibility: ?) [21]

P ?

S H_2O_2 + tryparedoxin 2 <24> (Reversibility: ?) [6]

P H_2O + oxidized tryparedoxin 2

S NADPH + H_2O_2 <37> (Reversibility: ?) [26]

P $NADP^+$ + H_2O

S arachidonoyl hydroperoxide + GSH <40> (Reversibility: ?) [41]

P ?

S cumene hydroperoxide + GSH <40> (Reversibility: ?) [41]

P 2-phenylpropan-2-ol + GSSG

S cumene hydroperoxide + dithiothreitol <11> (<11> 40% of the activity with H_2O_2 [42]) (Reversibility: ?) [42]

P 2-phenylpropan-2-ol + oxidized dithiothreitol

S cumene hydroperoxide + reduced glutaredoxin <12> (Reversibility: ?) [25]

P 2-phenylpropan-2-ol + oxidized glutaredoxin

S cumene hydroperoxide + reduced thioredoxin <25, 38> (Reversibility: ?) [40, 48]

P ?

S cumene hydroperoxide + reduced thioredoxin <2, 11> (<2> about 10% of the activity with H_2O_2 [11]) (Reversibility: ?) [11, 54]
P 2-phenylpropan-2-ol + oxidized thioredoxin
S cumene hydroperoxide + tryparedoxin 2 <24> (<24> 57% of the activity with H_2O_2 [6]) (Reversibility: ?) [6]
P 2-phenylpropan-2-ol + oxidized tryparedoxin 2
S dithiothreitol + H_2O_2 <31> (Reversibility: ?) [24]
P oxidized dithiothreitol + H_2O
S linoleic acid hydroperoxide + tryparedoxin 2 <24> (<24> 7.8% of the activity with H_2O_2 [6]) (Reversibility: ?) [6]
P ?
S linolenoyl hydroperoxide + GSH <40> (Reversibility: ?) [41]
P ?
S linoleoyl hydroperoxide + reduced thioredoxin <11> (Reversibility: ?) [54]
P ?
S paraquat + dithiothreitol <21> (Reversibility: ?) [39]
P ?
S peroxinitrite + thioredoxin <18> (Reversibility: ?) [44]
P ?
S phosphatidyl choline hydroperoxide + tryparedoxin 2 <24> (<24> 3.8% of the activity with H_2O_2 [6]) (Reversibility: ?) [6]
P ?
S phosphatidylcholine hydroperoxide + reduced glutaredoxin <12> (Reversibility: ?) [25]
P ? + oxidized glutaredoxin
S tert-butyl hydroperoxide + GSH <4, 8, 39> (Reversibility: ?) [33, 41, 43]
P tert-butanol + GSSG
S tert-butyl hydroperoxide + dithiothreitol <10, 11, 21, 46> (<11> 50% of the activity with H_2O_2 [42]) (Reversibility: ?) [18, 31, 39, 42]
P tert-butanol + oxidized dithiothreitol
S tert-butyl hydroperoxide + reduced glutaredoxin <12> (Reversibility: ?) [25]
P tert-butanol + oxidized glutaredoxin
S tert-butyl hydroperoxide + reduced thioredoxin <8, 25> (Reversibility: ?) [40, 43]
P t-butanol + oxidized thioredoxin
S tert-butyl hydroperoxide + reduced thioredoxin <28> (Reversibility: ?) [49]
P ? + oxidized thioredoxin
S tert-butyl hydroperoxide + reduced thioredoxin <2, 11, 19, 24, 28> (<24> 9.5% of the activity with tert-butyl hydroperoxide and tryparedoxin 2 [6]; <2> about 60% of the activity with H_2O_2 [11]; <19> the peroxiredoxin preferrs PfTrx2 to PfTrx1 as a reducing partner [53]) (Reversibility: ?) [6, 11, 17, 53, 54]
P tert-butanol + oxidized thioredoxin

S tert-butyl hydroperoxide + tryparedoxin 2 <24> (<24> 95% of the activity with H_2O_2 [6]) (Reversibility: ?) [6]
P tert-butanol + oxidized tryparedoxin 2
S Additional information <11, 23> (<23> PcPrx-1 may play a protective role against oxidative stress [37]; <11> insignificant affinity towards complex phospholipid hydroperoxide [54]) (Reversibility: ?) [37, 54]
S Additional information <3, 4, 6, 7, 8, 9, 10, 11, 16, 17, 19, 20, 21, 22, 24, 28, 29, 30, 32, 35, 40, 41, 42, 43, 44, 45> (<30> 1-Cys peroxiredoxin protects DNA from degradation by reactive O_2 species in presence of low molecular mass thiols such as dithiothreitol and glutathione [21]; <19> after subjection to exogenous and endogenous oxidative stress, the Plasmodium falciparum blood stage form shows a marked elevation of PfTrx-Px1 mRNA and protein levels consistent with the structure of related proteins [22]; <20> antioxidant activity may be the primary function of the enzyme [4]; <28> peroxidases belonging to the class of 1-Cys and 2-Cys peroxiredoxins play crucial roles in maintaining redox balance. TgTrx-Px1 is an extremely potent antioxidant [17]; <3> peroxiredoxin and NADH:peroxiredoxin oxidoreductase together catalyze the anaerobic reduction of H_2O_2 [15]; <7> Prx IV is a secretable protein and may exert its protective function against oxidative damage by scavenging reactive oxygen species in the extracellular space [10]; <4> the enzyme is essential for sustaining life span of erythrocytes in mice by protecting them from oxidative stress [2]; <35> the enzyme is important for protection against endogenously generated H_2O_2 [23]; <32> the enzyme might function as a major antioxidant enzyme in Ascaris suum [7]; <29> the enzyme plays a critical role in defending the organism against oxygen toxicity [9]; <6> the enzyme promotes potassium efflux and down-regulates apoptosis and the recruitment of monocytes by endothelial tissue [29]; <9> the midpoint redox potential of -315 mV places 2-Cys Prx reduction after Calvin cycle activation and before switching the male valve for export of excess reduction equivalents to the cytosol. The activity of the enzyme is also linked to chloroplastic NAD(P)H metabolism. Saline-stress-induced oligomerization of the enzyme triggers membrane attachment and allows for detoxification of peroxides at the site of production in immediate vicinity of the thylakoid membrane [28]; <17> the photosynthetic machinery needs high levels of enzyme during leaf development to protect it from oxidative damage. The damage is reduced by the accumulation of 2-CP protein, by the de novo synthesis and replacement of damaged proteins, and by the induction of other antioxidant defenses in 2-CP mutants [27]; <6> together with glutathione peroxidase and catalase, Prx enzymes likely play an important role in eliminating peroxides generated during metabolism. In addition Prx I and II might participate in the signaling cascades of growth factors and tumor necrosis factor-α by regulating the intracellular concentration of H_2O_2 [12]; <8> cPrx I and cPrx II function both as peroxidases and as molecular chaperones. The peroxidase function predominates in the lower molecular weight forms, whereas the chaperone function predominates in the higher molecular weight complexes. Oxidative stress and heat shock

exposure of yeasts cause the protein structures of cPrxl and cPrx II to shift from low MW species to high molecular weight complexes. This triggers a peroxidase-to-chaperone functional switch [3]; <6> no activity with glutaredoxin or glutathione [14]; <6> the enzyme exhibits a low level of phospholipiase A_2 activity at acidic pH [13]; <10> thioredoxin m is more efficent than thioredoxin f in reducing the enzyme [18]; <24> tryparedoxin peroxidase activity. The enzyme does not follow a classic ping-pong mechanism [6]; <21> Bcp2 plays an important role in the peroxide-scavaging system in Sulfolobus solfataricus. The enzyme protects plasmid DNA from nicking by the metal-catalysed oxidation system [39]; <45> constitutive expression of prxS confers enhanced survival and growth to Rhizobium etli in presence of H_2O_2. Defence of Rhizobium etli bacteroids against oxidative stress involves a complexly regulated atypical 2-Cys peroxiredoxin [52]; <8> Gpx2 is likely to play an important role in the protection of cells from oxidative stress in the presence of Ca^{2+} [43]; <19> Pf1-Cys-Prx protects the parasite against oxidative stress by binding to ferriprotoporphyrin [38]; <4> Prx I has at least two distinct roles: as an antioxidant enzyme and as a regulator of p38 MAPK [45]; <11> Prx Q attaches to photosystem II and has a specific function distinct from 2-Cys peroxiredoxin in protecting photosynthesis. Its absence causes metabolic changes that are sensed and trigger appropriate compensatory responses [54]; <40,41,42,43,44> the enzyme functions in antioxidant defense and lung phospholipid metabolism [41]; <11> the enzyme is essential for redox homeostasis and root growth of Arabidopsis thaliana under stress [42]; <19> the enzyme is involved in the detoxification of reactive oxygen species and of reactive nitrogen species [36]; <22> Tpx-1 is required for normal gametocyte development but does not affect the male/female gametocyte ratio or male gametogenesis [51]; <16> Tpx1 is an upstream activator of Pap1. At low H_2O_2 concentrations, this oxidant scavenger can transfer a redox signal to Pap1, whereas higher concentrations of the oxidant inhibit the Tpx1-Pap1 redox relay through the temporal inactivation of Tpx1 by oxidation of its catalytic cysteine to a sulfinic acid. This cysteine modification can be reversed by the sulfiredoxin Srx1, its expression in response to high doses of H_2O_2 strictly depending on active Sty1. Tpx1 oxidation to the cysteine-sulfinic acid and its reversion by Srx1 constitutes a redox switch in H_2O_2 signaling, restricting Pap1 activation within a narrow range of H_2O_2 concentrations [55]; <19> cumene hydroperoxide is not accepted as a substrate [53]; <19> PfTPx1 also possesses peroxynitrite reductase activity [36]; <28> TgPrx2 has general antioxidant properties as indicated by its ability to protect glutamine synthetase against a dithiothreitol Fe^{3+}-catalyzed system. TgPrx2 does not reduce H_2O_2 or tert-butyl hydroperoxide at the expense of glutaredoxin, thioredoxin or glutathione [49]) (Reversibility: ?) [2, 3, 4, 6, 7, 9, 10, 12, 13, 14, 15, 17, 18, 21, 22, 23, 27, 28, 29, 36, 38, 39, 41, 42, 43, 45, 49, 51, 52, 53, 54, 55]

P ?

Inhibitors

NEM <31> [24]

SDS <46, 47> (<47> 1%, slight decrease in activity [32]; <46> activity is lost completely in the presence of 1% SDS or higher [31]) [31, 32]

imidazole <46, 47> (<47> 0.1 M, 50% inhibition [32]; <46> approximately 70% of the Prx activity is lost in the presence of 0.4 M imidazole or higher. The presence of 1.6 M imidazole seems to promote protein degradation [31]) [31, 32]

tert-butyl hydroperoxide <28> (<28> substrate inhibition [17]) [17]

Cofactors/prosthetic groups

thioredoxin <2, 10> (<2> reaction is dependent on [11]; <10> thioredoxin m is more efficent than thioredoxin f in reducing the enzyme [18]) [11,18]

Turnover number (min^{-1})

0.69 <10> (H_2O_2) [18]
0.83 <28> (H_2O_2) [17]
0.9 <12> (cumene hydroperoxide) [25]
1 <1> (H_2O_2, <1> pH 7.0, 25°C, wild-type enzyme [34]) [34]
1.4 <8> (H_2O_2, <8> pH 7.0, reaction with GSH [43]) [43]
1.5 <12> (tert-butyl hydroperoxide) [25]
1.67 <19> (H_2O_2, <19> pH 7.6, 30°C [22]) [22]
2.1 <12> (phosphatidylcholine hydroperoxide) [25]
2.6 <12> (H_2O_2) [25]
2.79 <28> (tert-butyl hydroperoxide) [17]
6.3 <25> (cumene hydroperoxide, <25> 22°C, pH 7.0 [40]) [40]
7.6 <25> (tert-butyl hydroperoxide, <25> 22°C, pH 7.0 [40]) [40]
9.65 <19> (thioredoxin, <19> pH 7.6, 30°C [22]) [22]
15.8 <25> (H_2O_2, <25> 22°C, pH 7.0 [40]) [40]
25 <1> (H_2O_2, <1> pH 7.0, 25°C, mutant enzyme T77I [34]) [34]
31.5 <1> (H_2O_2, <1> pH 7.0, 25°C, mutant enzyme T77D [34]) [34]
37 <24> (tryparedoxin 2, <24> pH 7.6, 25°C [6]) [6]
75.8 <1> (H_2O_2, <1> pH 7.0, 25°C, mutant enzyme T77V [34]) [34]
109 <8> (tert-butyl hydroperoxide, <8> pH 7.0, reaction with GSH [43]) [43]
368 <8> (tert-butyl hydroperoxide, <8> pH 7.0, reaction with thioredoxin [43]) [43]
957 <8> (H_2O_2, <8> pH 7.0, reaction with thioredoxin [43]) [43]

K_m-Value (mM)

0.00004 <28> (tert-butyl hydroperoxide) [17]
0.00038 <28> (H_2O_2) [17]
0.00078 <19> (H_2O_2, <19> pH 7.6, 30°C [22]) [22]
0.001 <6> (thioredoxin, <6> pH 7.0, 37°C [14]) [14]
0.0014 <1> (H_2O_2, <1> pH 7.0, 25°C, wild-type enzyme [34]) [34]
0.0016 <1> (H_2O_2, <1> pH 7.0, 25°C, mutant enzyme T77V [34]) [34]
0.0042 <28> (thioredoxin) [17]
0.0091 <24> (tert-butyl hydroperoxide, <24> pH 7.6, 25°C [6]) [6]
0.0127 <25> (cumene hydroperoxide, <25> 22°C, pH 7.0 [40]) [40]

0.0186 <25> (tert-butyl hydroperoxide, <25> 22°C, pH 7.0 [40]) [40]
0.0189 <19> (thioredoxin, <19> pH 7.6, 30°C [22]) [22]
0.02 <8> (H_2O_2, <8> pH 7.0, reaction with thioredoxin [43]) [43]
0.023 <12> (phosphatidylcholine hydroperoxide, <12> pH 8.0, 30°C [25]) [25]
0.0276 <10> (H_2O_2) [18]
0.03 <25> (H_2O_2, <25> 22°C, pH 7.0 [40]) [40]
0.0319 <24> (tryparedoxin 2, <24> pH 7.6, 25°C [6]) [6]
0.035 <13> (H_2O_2, <13> pH 7.0, 37°C [23]) [23]
0.036 <34> (H_2O_2, <34> pH 7.0, 37°C [23]) [23]
0.062 <1, 45> (H_2O_2, <1> pH 7.0, 25°C, mutant enzyme T77D [34]) [34, 52]
0.0625 <8> (tert-butyl hydroperoxide, <8> pH 7.0, reaction with thioredoxin [43]) [43]
0.0934 <1> (H_2O_2, <1> pH 7.0, 25°C, mutant enzyme T77I [34]) [34]
0.12 <40> (1-palmitoyl-2-linolenoyl-sn-glycero-3-phosphocholine hydroperoxide) [41]
0.12 <40> (cumene hydroperoxide) [41]
0.129 <40> (1-palmitoyl-2-arichidonoyl-sn-glycero-3-phosphocholine) [41]
0.135 <40> (arachidonoyl hydroperoxide) [41]
0.141 <40> (linolenoyl hydroperoxide) [41]
0.142 <40> (tert-butyl hydroperoxide) [41]
0.17 <8> (H_2O_2, <8> pH 7.0, reaction with GSH [43]) [43]
0.18 <40> (H_2O_2) [41]
0.313 <8> (tert-butyl hydroperoxide, <8> pH 7.0, reaction with GSH [43]) [43]
16.28 <31> (H_2O_2) [24]
Additional information <6, 19> (<6> the K_m-value for H_2O_2 is below 0.02 mM [14]) [14, 36, 53]

pH-Optimum

9 <46> [31]

pH-Range

5.4-9 <46> (<46> no activity at or below pH 5.4, optimal activity at pH 9.0 [31]) [31]

4 Enzyme Structure

Molecular weight

25000 <28> (<28> TgTrx-Px2 monomer, gel filtration [17]) [17]
34000 <6> (<6> gel filtration [14]) [14]
49000 <10> (<10> dimer, gel filtration [18]) [18]
52000 <28> (<28> TgTrx-Px2 dimer, gel filtration [17]) [17]
150000 <28> (<28> TgTrx-Px2 hexamer, gel filtration [17]) [17]
235000 <3> (<3> gel filtration [15]) [15]
250000 <28> (<28> TgTrx-Px1 decamer, gel filtration [17]) [17]

388000 <10> (<10> decamer, gel filtration [18]) [18]
400000 <19> (<19> gel filtration [22]) [22]

Subunits

? <6, 7, 8, 10, 14, 21, 23, 30, 31, 32, 45, 46> (<8> x * 18000, SDS-PAGE [43]; <6,7> x * 27000, SDS-PAGE [1,10]; <31> x * 26000, SDS-PAGE [24]; <46> x * 25000, SDS-PAGE [31]; <32> x * 21589, calculation from nucleotide sequence [7]; <30> x * 21800, 2-Cys peroxiredoxin, calculation from nucleotide sequence [21]; <10> x * 21913, calculation from nucleotide sequence [18]; <30> x * 24700, 1-Cys peroxiredoxin, calculation from nucleotide sequence [21]; <6> x * 25034, calculation from nucleotide sequence [1]; <14> x * 18000, calculated from sequence [50]; <45> x * 20080, calculated from sequence [52]; <23> x * 22100, calculated from sequence [37]; <21> x * 24745, calculated from sequence [39]; <46> x * 25081, calculated from sequence [31]) [1, 7, 10, 18, 21, 24, 31, 37, 39, 43, 50, 52]
decamer <1, 6, 19, 24, 28> (<28> 10 * 24600, TgTrx-Px1, SDS-PAGE [17]; <24> 10 * 27345, the enzyme exists as a homodecameric ring structure composed of five dimers, MALDI-TOF spectrometry [6]; <6> a stable decamer forms in vivo under conditions of oxidative stress [29]; <19> consisting of pentamers of homodimers that have a doughnut-like shape [22]) [6, 17, 22, 29, 34]
dimer <5, 6, 19, 27, 28, 37, 38, 47> (<47> 2 * 21000, SDS-PAGE [32]; <6> 2 * 17000, SDS-PAGE [14]; <28> 1 * 26500, TgTrx-Px2, SDS-PAGE [17]; <6> 2 * 17030, calculation from nucleotide sequence [14]; <37> 2 * 22000, mutant enzyme Δ66-273 migrates as a dimer under non-reducing SDS-PAGE [26]; <47> 2 * 20965, calculated from sequence [32]; <38> 2 * 22000, the homodimeric oxidized enzyme form is reduced to a monomeric form by thioredoxin and by dithiothreitol and is converted to a homodimeric oxidized form by H_2O_2, SDS-PAGE [48]; <28> 2 * 25650, the dimeric TgPrx2 is able to form tetramers and hexamers which are non-covalently associated, calculated from sequence [49]; <27> the crystal structure and solution NMR provide evidence that the reduced protein is a specific noncovalent homodimer both in the crystal and in solution [35]) [14, 16, 17, 26, 32, 35, 36, 46, 48, 49]
hexamer <28> (<28> 1 * 26500, TgTrx-Px2, SDS-PAGE [17]) [17]
monomer <28, 37> (<37> 1 * 22000, mutant enzyme Δ66-273 migrates as a monomer under reducing conditions [26]; <28> 1 * 26500, TgTrx-Px2, SDS-PAGE [17]) [17, 26]
multimer <3> (<3> x * 21000, SDS-PAGE [15]) [15]
Additional information <2, 6, 8, 10, 19, 26> (<2> MW of open reading frame HI0572 determined by SDS-PAGE is 20000 Da [11]; <19> no dimeric form detectable [22]; <8> recombinant cPrx I produces in Escherichia coli forms differently sized high molecular weight protein structures. cPrx I and cPrx II function both as peroxidases and as molecular chaperones. The peroxidase function predominates in the lower molecular weight forms, whereas the chaperone function predominates in the higher molecular weight complexes. Oxidative stress and heat shock exposure of yeasts cause the protein structures of cPrxl and cPrx II to shift from low MW species to high molecular

weight complexes. This triggers a peroxidase-to-chaperone functional switch [3]; <6> the crystal structure shows that peroxiredoxin 5 does not form a dimer [19]; <10> the transition dimer-decamer produced in vitro between pH 7.5 and 8.0 suggests that a great change in the enzyme quarternary structure of the enzyme may take place in the chloroplast during the dark-light transition. The dimer-decamer equilibrium depends on NaCl concentration and concentration of dithiothreitol [18]; <26> Prx is a toroid-shaped pentamer of homodimers, or an $(\alpha_2)_5$ decamer [47]; <19> the enzyme occurs in both dimeric and decameric forms when purified under non-reducing conditions [53]) [3, 11, 18, 19, 22, 47, 53]

5 Isolation/Preparation/Mutation/Application

Source/tissue

A-431 cell <6> (<6> 0.003 mg Prx I per mg of soluble protein Prx I, 0.0027 mg Prx II per mg of soluble protein, 0.001 mg Prx III per mg of soluble protein, 0.00003 mg Prx V per mg of soluble protein and 0.0002 mg Prx VI per mg of soluble protein [8]) [8]
HeLa cell <6> (<6> more than 0.004 mg Prx I per mg of soluble protein Prx I, 0.0033 mg Prx II per mg of soluble protein, less than 0.0003 mg Prx III per mg of soluble protein, 0.0005 mg Prx V per mg of soluble protein and above 0.003 mg Prx VI per mg of soluble protein [8]) [8]
Hep-G2 cell <6> (<6> 0.0033 mg Prx I per mg of soluble protein Prx I, 0.0013 mg Prx II per mg of soluble protein, 0.0005 mg Prx III per mg of soluble protein, 0.0005 mg Prx V per mg of soluble protein and 0.0017 mg Prx VI per mg of soluble protein [8]) [8]
J774A.1 cell <4> [45]
JURKAT cell <6> (<6> 0.0027 mg Prx I per mg of soluble protein Prx I, 0.001 mg Prx II per mg of soluble protein, less than 0.0003 mg Prx III per mg of soluble protein, 0.0005 mg Prx V per mg of soluble protein and 0.0001 mg Prx VI per mg of soluble protein [8]) [8]
K-562 cell <6> (<6> 0.0033 mg Prx I per mg of soluble protein Prx I, 0.002 mg Prx II per mg of soluble protein, less than 0.0003 mg Prx III per mg of soluble protein, 0.0003 mg Prx V per mg of soluble protein and more than 0.005 mg Prx VI per mg of soluble protein [8]) [8]
PC-12 <7> (<7> 0.002 mg Prx I per mg of soluble protein Prx I, 0.0033 mg Prx II per mg of soluble protein, 0.0007 mg Prx III per mg of soluble protein, 0.0002 mg Prx V per mg of soluble protein and 0.0002 mg Prx VI per mg of soluble protein [8]) [8]
U-937 cell <6> (<6> 0.0027 mg Prx I per mg of soluble protein Prx I, 0.0005 mg Prx II per mg of soluble protein, less than 0.0003 mg Prx III per mg of soluble protein, 0.0005 mg Prx V per mg of soluble protein and 0.0013 mg Prx VI per mg of soluble protein [8]) [8]
adrenal gland <7> (<7> less than 0.002 mg Prx I per mg of soluble protein Prx I, 0.002 mg Prx II per mg of soluble protein, less than 0.004 mg Prx III per

mg of soluble protein, 0.0003 mg Prx V per mg of soluble protein and 0.0003 mg Prx VI per mg of soluble protein [8]) [8]
bacteroid <45> (<45> the enzyme is strongly expressed under microaerobic conditions and during the sambiotic interaction with Phaseolus vulgaris [52]) [52]
bloodstream form <19> (<19> expressed at high levels during the haem-digesting stage [38]) [38]
brain <7> (<7> 0.0013 mg Prx I per mg of soluble protein Prx I, 0.0013 mg Prx II per mg of soluble protein, 0.0005 mg Prx III per mg of soluble protein, 0.001 mg Prx V per mg of soluble protein and 0.0017 mg Prx VI per mg of soluble protein [8]) [8]
cysticercus <38> (<38> tegumentary and muscle cells [48]) [48]
erythrocyte <4, 6> [2, 29]
flower <11> (<11> containing green sepals [54]) [54]
fruiting body <46, 47> [31, 32]
heart <7> (<7> 0.0013 mg Prx I per mg of soluble protein Prx I, 0.002 mg Prx II per mg of soluble protein, 0.0033 mg Prx III per mg of soluble protein, 0.0005 mg Prx V per mg of soluble protein and 0.0003 mg Prx VI per mg of soluble protein [8]) [8]
hypothalamus <7> (<7> 0.0013 mg Prx I per mg of soluble protein Prx I, 0.0013 mg Prx II per mg of soluble protein, 0.0005 mg Prx III per mg of soluble protein, 0.0007 mg Prx V per mg of soluble protein and 0.0007 mg Prx VI per mg of soluble protein [8]) [8]
kidney <7> (<7> 0.002 mg Prx I per mg of soluble protein Prx I, 0.0007 mg Prx II per mg of soluble protein, 0.0007 mg Prx III per mg of soluble protein, 0.0013 mg Prx V per mg of soluble protein and 0.0003 mg Prx VI per mg of soluble protein [8]) [8]
lateral hypodermal chord <31> (<31> lateral hypodermal chords of both male and female worms [24]) [24]
leaf <9, 11, 17, 37> (<37> the gene is predominantly expressed in leaf tissue of seedlings [26]; <11> about equal acttivity in young and mature leaves. The amount of Prx Q is decreased in senescent leaves [54]) [26, 27, 28, 54]
liver <7> (<7> 0.0007 mg Prx I per mg of soluble protein Prx I, 0.0005 mg Prx II per mg of soluble protein, 0.0007 mg Prx III per mg of soluble protein, 0.0007 mg Prx V per mg of soluble protein and 0.0003 mg Prx VI per mg of soluble protein [8]) [8, 10]
lung <7, 42> (<7> 0.001 mg Prx I per mg of soluble protein Prx I, 0.0013 mg Prx II per mg of soluble protein, 0.0003 mg Prx III per mg of soluble protein, 0.0003 mg Prx V per mg of soluble protein and 0.0017 mg Prx VI per mg of soluble protein [8]) [8, 10, 41]
muscle <31> (<31> afibrillar muscle cells in male worms [24]) [24]
muscle cell <38> (<38> of cysticercus [48]) [48]
pancreas <7> (<7> 0.0003 mg Prx III per mg of soluble protein, 0.0002 mg Prx V per mg of soluble protein and 0.00003 mg Prx VI per mg of soluble protein [8]) [8, 10]
placenta <7> (<7> 0.0002 mg Prx I per mg of soluble protein Prx I, 0.0007 mg Prx II per mg of soluble protein, less than 0.0003 mg Prx III per mg of

soluble protein, 0.0007 mg Prx V per mg of soluble protein and 0.0002 mg Prx VI per mg of soluble protein [8]) [8]
seed <20> (<20> transcript liver rapidly decreases after imbibition of seeds, but the protein is detected for 15 days after imbibition [4]) [4]
seedling <17> [27]
spleen <7> (<7> 0.0007 mg Prx I per mg of soluble protein Prx I, 0.002 mg Prx II per mg of soluble protein, 0.0003 mg Prx III per mg of soluble protein, 0.0003 mg Prx V per mg of soluble protein and 0.00003 mg Prx VI per mg of soluble protein [8]) [8, 10]
stem <11> [54]
tachyzoite <28> [17]
tegument <38> [48]
testis <7> (<7> 0.0007 mg Prx I per mg of soluble protein Prx I, 0.001 mg Prx II per mg of soluble protein, 0.0003 mg Prx III per mg of soluble protein, 0.0002 mg Prx V per mg of soluble protein and 0.001 mg Prx VI per mg of soluble protein [8]) [8, 10]
thymus <7> (<7> 0.0003 mg Prx I per mg of soluble protein Prx I, 0.0007 mg Prx II per mg of soluble protein, 0.0003 mg Prx III per mg of soluble protein, 0.0003 mg Prx V per mg of soluble protein and 0.00003 mg Prx VI per mg of soluble protein [8]) [8]
thyroid gland <7> (<7> 0.0007 mg Prx I per mg of soluble protein Prx I, 0.0007 mg Prx II per mg of soluble protein, 0.0005 mg Prx III per mg of soluble protein, 0.0003 mg Prx V per mg of soluble protein and 0.0002 mg Prx VI per mg of soluble protein [8]) [8]
trophozoite <13, 35> [23]
uterine <31> (<31> uterine wall in female worms [24]) [24]
Additional information <11, 37> (<11> no activity in root [54]; <37> the mRNA is generally expressed in most tissues of mature plant, except root [26]) [26, 54]

Localization

chloroplast <9, 11, 17> (<11> peroxiredoxin Q represents about 0.3% of chloroplast proteins. It attaches to the thylakoid membrane and is detected in preparations enriched in photosystem II complexes [54]) [27, 28, 54]
cytoplasm <35> [23]
cytosol <6, 8, 19, 28> (<6> Prx I and Prx II [12]) [3, 12, 13, 14, 17, 19, 38, 49]
endoplasmic reticulum <7> [10]
extracellular <7> (<7> Prx IV [8]) [8]
hydrogenosome <14> [50]
membrane <6> (<6> associated with erythrocyte membrane [29]) [29]
mitochondrion <6, 10, 11, 19, 24> (<11> matrix [42]; <6> Prx III [12]) [6, 12, 14, 19, 30, 42, 53]
nucleus <35> [23]
peroxisome <6> [14, 19]
soluble <15> (<15> when expressed in Escherichia coli [5]) [5]
thylakoid <9> (<9> reversible binding of the oligomeric form of the enzyme to the thylakoid membrane [28]) [28]

Purification

<2> (recombinant open reading frame HI0572) [11]
<3> [15]
<6> [1, 13, 14]
<8> [43]
<10> [18, 30]
<11> [54]
<15> [5]
<21> [39]
<23> [37]
<25> (recombinant enzyme) [40]
<28> [17, 49]
<29> [9]
<38> [48]
<39> (recombinant) [33]
<45> [52]
<46> [31]
<47> (recombinant) [32]

Crystallization

<1> (hanging drop method, wild-type enzyme and mutant enzymes T77V, T77I and T77D) [34]
<5> (hanging-drop vapor-diffusion method, structure determined at 2.2 A resolution) [16]
<6> (1.5 A resolution crystal structure of peroxiredoxin 5 in the reduced form) [19]
<6> (crystal form of human peroxiredoxin 5 is described at 2.0 A resolution) [20]
<6> (structure of decameric 2-Cys peroxiredoxin at 1.7 A resolution) [29]
<10> (hanging-drop vapour-diffusion technique, resolution of 2.8 A, the protein crystallizes in space group P1, with unit-cell parameters a = 61.88 A, b = 66.40 A, c = 77.23 A, $\alpha = 102.90$, $\beta = 104.40$, $\gamma = 99.07$) [30]
<19> (hanging-drop vapor-diffusion method, 1.8 A resolution crystal structure) [46]
<19> (sitting-drop vapour diffusion) [53]
<26> (2.3 A resolution, microbatch method at 18°C, using a crystallization robot, TERA) [47]
<27> [35]
<40> [41]

Cloning

<2> (expression of open reading frame HI0572 in Escherichia coli) [11]
<3> [15]
<6> (expression in Escherichia coli) [1, 14]
<6> (expression in Escherichia coli or in NIH 3T3 cells) [13]
<6> (expression of Prx I, Prx II and Prx III in Escherichia coli) [12]
<7> (expression in COS cells and in Sf21 cells) [10]
<8> (expression in Escherichia coli) [3, 43]

<10> (expression using vector pET3d) [18]
<10> (the 171-amino-acid mature protein (estimated molecular weight 18600 Da) is cloned into the pET3d vector and overexpressed in Escherichia coli) [30]
<11> [54]
<12> (no post-translational modification occurs in Escherichia coli) [25]
<15> (1. full-length protein with 345 amino acids, 2. a construction which stretches from the N-terminus to the end of the peroxiredoxin domain, contains 165 amino acids, 3. the grx module, that starts with MAQESVA and ends with the C-terminus of the fusion with 77 amino acids overall, expression in Escherichia coli) [5]
<17> [27]
<19> [53]
<19> (expression in Escherichia coli) [38]
<19> (expression in Eschericia coli) [46]
<20> (overexpression in Nicotiana tabacum cv. Xanthi-nc constitutively expressing the enzyme. Transgenic plants show a germination frequency similar to control plants. The transgenic lines exhibit higher resistance against oxidative stress) [4]
<21> (expression in Escherichia coli) [39]
<23> (the recombinant PcPrx-1 protein is expressed as a histidine fusion protein in Escherichia coli) [37]
<24> (expression in Escherichia coli as an N-terminally His-tagged protein) [6]
<25> [40]
<26> (expression in Escherichia coli) [47]
<28> (TgTrx-Px1 and TgTrx-Px2, expression in Escherichia coli) [17]
<28> (expression in Escherichia coli strain BL21) [49]
<29> (expression in Escherichia coli) [9]
<31> (expression in Escherichia coli as a polyhistidine fusion protein) [24]
<32> [7]
<34> (expression in Escherichia coli) [23]
<37> (expression of N-terminally truncated enzyme in Escherichia coli, Δ66-273) [26]
<38> (expression in Escherichia coli) [48]
<39> (expression in Escherichia coli) [33]
<45> (expression in Escherichia coli) [52]
<46> (expression in Escherichia coli) [31]
<47> (coding region is subcloned into pAVD10, transformed into Escherichia coli, and expressed as a His-tagged fusion protein) [32]

Engineering

C152S <6> (<6> mutant enzyme shows no detectable thioredoxin-dependent peroxidase activity [14]) [14]
C156S <45> (<45> no peroxidase activity [52]) [52]

C170A <19> (<19> k_{cat}/K_m of mutant enzyme is reduced for thioredoxin as a substrate approximately 50fold. In contrast to the wildtype enzyme, covalently linked dimers are not formed [36]) [36]
C185S <15> (<15> inactive mutant enzyme [5]) [5]
C47S <6> (<6> mutation abolishes peroxidase activity,mutation has no effect on lipase activity [13]) [13]
C48S <6> (<6> mutant enzyme shows no detectable thioredoxin-dependent peroxidase activity [14]) [14]
C56S <45> (<45> no peroxidase activity [52]) [52]
C60S <18> (<18> inactive mutant enzyme [44]) [44]
C73S <6> (<6> mutation has no effect on activity [14]) [14]
C80S <18> (<18> mutant is indistinguishable from the wild-type enzyme [44]) [44]
C81S <45> (<45> although the PrxS C81S mutant protein can be overexpressed and purified under denaturing conditions, it is not possible to obtain any active C81S PrxS. the C81S mutant is prone to inclusion body formation [52]) [52]
C91S <6> (<6> mutation has no effect on lipase activity [13]) [13]
C93S <18> (<18> the mutant is fully active as a thioredoxin-dependent peroxidase and remains active despite exposure to peroxynitrite, pronounced instability of the mutant enzyme under oxidizing conditions [44]) [44]
Δ66-273 <37> (<37> the mutant enzyme prevents the inactivation of glutamine synthetase and the DNA cleavage in the metal-catalyzed oxidation system. In the yeast thioredoxin system containing thioredoxin reductase, thioredoxin, and NADPH, the ΔC2C-Prx exhibits peroxidase activity on H_2O_2 [26]) [26]
R129Q <12> (<12> inactive with H_2O_2 or tert-butyl hydroperoxide, catalyzes degradation of cumene hydroperoxide with low efficiency [25]) [25]
S32A <6> (<6> mutation has no effect on lipase activity [13]) [13]
S32G <6> (<6> mutation has no effect on lipase activity [13]) [13]
T48V <12> (<12> inactive with H_2O_2 or tert-butyl hydroperoxide, catalyzes degradation of cumene hydroperoxide with low efficiency [25]) [25]
T77D <1> (<1> mutation disrupts the decamer stability, mutant enzyme exhibits ca. 100fold lower catalytic efficiency than wild-type enzyme [34]) [34]
T77I <1> (<1> mutation disrupts the decamer stability, mutant enzyme exhibits ca. 100fold lower catalytic efficiency than wild-type enzyme [34]) [34]
T77V <1> (<1> mutation enhances the decamer stability, mutant enzyme has slightly higher activity than wild-type enzyme [34]) [34]

6 Stability

pH-Stability

5-11 <47> (<47> 37°C, 30 min, stable [32]) [32]
7.8-10.2 <46> (<46> 37°C, 30 min, stable [31]) [31]

Temperature stability

60 <46, 47> (<47> 2 min, Prx retains 68% activity [32]; <46> half-life: 15.5 min [31]) [31, 32]
80 <21> (<21> 6 h, the 63% of the activity is retained [39]) [39]
90 <21> (<21> half-life: 3 h [39]) [39]
95 <21> (<21> 30 min, 70% loss of activity [39]) [39]

General stability information

<46>, 1.6 M imidazole seems to promote protein degradation [31]
<46>, sensitive to trypsin and chymotrypsin treatment. Very little or no activity detected after 40 min incubation with either protease [31]
<47>, susceptible to chymotrypsin treatment but resistant to digestion by trypsin. The enzyme retains approximately 65%, 50%, and 40% activity after treating with chymotrypsin for 1, 2, and 3 h, respectively [32]

Storage stability

<28>, 4°C, stable for several weeks [49]

References

[1] Merkulova, M.I.; Shuvaeva, T.M.; Radchenko, V.V.; Yanin, B.A.; Bondar, A.A.; Sofin, A.D.; Lipkin, V.M.: Recombinant human peroxiredoxin VI: preparation and protective properties in vitro. Biochemistry, **67**, 1235-1239 (2002)

[2] Lee, T.H.; Kim, S.U.; Yu, S.L.; Kim, S.H.; Park do, S.; Moon, H.B.; Dho, S.H.; Kwon, K.S.; Kwon, H.J.; Han, Y.H.; Jeong, S.; Kang, S.W.; Shin, H.S.; Lee, K.K.; Rhee, S.G.; Yu, D.Y.: Peroxiredoxin II is essential for sustaining life span of erythrocytes in mice. Blood, **101**, 5033-5038 (2003)

[3] Jang, H.H.; Lee, K.O.; Chi, Y.H.; Jung, B.G.; Park, S.K.; Park, J.H.; Lee, J.R.; Lee, S.S.; Moon, J.C.; Yun, J.W.; Choi, Y.O.; Kim, W.Y.; Kang, J.S.; Cheong, G.-W.; Yun, D.-J.; Rhee, S.G.; Cho, M.J.; Lee, S.Y.: Two enzymes in one: Two yeast peroxiredoxins display oxidative stress-dependent switching from a peroxidase to a molecular chaperone function. Cell, **117**, 625-635 (2004)

[4] Lee, K.O.; Jang, H.H.; Jung, B.G.; Chi, Y.H.; Lee, J.Y.; Choi, Y.O.; Lee, J.R.; Lim, C.O.; Cho, M.J.; Lee, S.Y.: Rice 1Cys-peroxiredoxin over-expressed in transgenic tobacco does not maintain dormancy but enhances antioxidant activity. FEBS Lett., **486**, 103-106 (2000)

[5] Rouhier, N.; Jacquot, J.P.: Molecular and catalytic properties of a peroxiredoxin-glutaredoxin hybrid from Neisseria meningitidis. FEBS Lett., **554**, 149-153 (2003)

[6] Castro, H.; Budde, H.; Flohe, L.; Hofmann, B.; Lunsdorf, H.; Wissing, J.; Tomas, A.M.: Specificity and kinetics of a mitochondrial peroxiredoxin of Leishmania infantum. Free Radic. Biol. Med., **33**, 1563-1574 (2002)

[7] Tsuji, N.; Kasuga-Aoki, H.; Isobe, T.; Yoshihara, S.: Cloning and characterisation of a peroxiredoxin from the swine roundworm Ascaris suum. Int. J. Parasitol., **30**, 125-128 (2000)

[8] Rhee, S.G.; Kang, S.W.; Chang, T.-S.; Jeong, W.; Kim, K.: Peroxiredoxin, a novel family of peroxidases. IUBMB Life, **52**, 35-41 (2001)
[9] Baker, L.M.; Raudonikiene, A.; Hoffman, P.S.; Poole, L.B.: Essential thioredoxin-dependent peroxiredoxin system from Helicobacter pylori: genetic and kinetic characterization. J. Bacteriol., **183**, 1961-1973 (2001)
[10] Okado-Matsumoto, A.; Matsumoto, A.; Fujii, J.; Taniguchi, N.: Peroxiredoxin IV is a secretable protein with heparin-binding properties under reduced conditions. J. Biochem., **127**, 493-501 (2000)
[11] Hwang, Y.S.; Chae, H.Z.; Kim, K.: Characterization of Haemophilus influenzae peroxiredoxins. J. Biochem. Mol. Biol., **33**, 514-518 (2000)
[12] Kang, S.W.; Chae, H.Z.; Seo, M.S.; Kim, K.; Baines, I.C.; Rhee, S.G.: Mammalian peroxiredoxin isoforms can reduce hydrogen peroxide generated in response to growth factors and tumor necrosis factor-α. J. Biol. Chem., **273**, 6297-6302 (1998)
[13] Kang, S.W.; Baines, I.C.; Rhee, S.G.: Characterization of a mammalian peroxiredoxin that contains one conserved cysteine. J. Biol. Chem., **273**, 6303-6311 (1998)
[14] Seo, M.S.; Kang, S.W.; Kim, K.; Baines, I.C.; Lee, T.H.; Rhee, S.G.: Identification of a new type of mammalian peroxiredoxin that forms an intramolecular disulfide as a reaction intermediate. J. Biol. Chem., **275**, 20346-20354 (2000)
[15] Logan, C.; Mayhew, S.G.: Cloning, overexpression, and characterization of peroxiredoxin and NADH peroxiredoxin reductase from Thermus aquaticus. J. Biol. Chem., **275**, 30019-30028 (2000)
[16] Choi, J.; Choi, S.; Choi, J.; Cha, M.-K.; Kim, I.-H.; Shin, W.: Crystal structure of Escherichia coli thiol peroxidase in the oxidized state: insights into intramolecular disulfide formation and substrate binding in atypical 2-Cys peroxiredoxins. J. Biol. Chem., **278**, 49478-49486 (2003)
[17] Akerman, S.E.; Müller, S.: Peroxiredoxin-linked detoxification of hydroperoxides in Toxoplasma gondii. J. Biol. Chem., **280**, 564-570 (2005)
[18] Bernier-Villamor, L.; Navarro, E.; Sevilla, F.; Lazaro, J.J.: Cloning and characterization of a 2-Cys peroxiredoxin from Pisum sativum. J. Exp. Bot., **55**, 2191-2199 (2004)
[19] Declercq, J.P.; Evrard, C.; Clippe, A.; Stricht, D.V.; Bernard, A.; Knoops, B.: Crystal structure of human peroxiredoxin 5, a novel type of mammalian peroxiredoxin at 1.5 A resolution. J. Mol. Biol., **311**, 751-759 (2001)
[20] Evrard, C.; Capron, A.; Marchand, C.; Clippe, A.; Wattiez, R.; Soumillion, P.; Knoops, B.; Declercq, J.P.: Crystal structure of a dimeric oxidized form of human peroxiredoxin 5. J. Mol. Biol., **337**, 1079-1090 (2004)
[21] Krnajski, Z.; Walter, R.D.; Muller, S.: Isolation and functional analysis of two thioredoxin peroxidases (peroxiredoxins) from Plasmodium falciparum. Mol. Biochem. Parasitol., **113**, 303-308 (2001)
[22] Akerman, S.E.; Müller, S.: 2-Cys peroxiredoxin PfTrx-Px1 is involved in the antioxidant defence of Plasmodium falciparum. Mol. Biochem. Parasitol., **130**, 75-81 (2003)

[23] Cheng, X.J.; Yoshihara, E.; Takeuchi, T.; Tachibana, H.: Molecular characterization of peroxiredoxin from Entamoeba moshkovskii and a comparison with Entamoeba histolytica. Mol. Biochem. Parasitol., **138**, 195-203 (2004)
[24] Chandrashekar, R.; Tsuji, N.; Morales, T.H.; Carmody, A.B.; Ozols, V.O.; Welton, J.; Tang, L.: Removal of hydrogen peroxide by a 1-cysteine peroxiredoxin enzyme of the filarial parasite Dirofilaria immitis. Parasitol. Res., **86**, 200-206 (2000)
[25] Rouhier, N.; Gelhaye, E.; Corbier, C.; Jacquot, J.P.: Active site mutagenesis and phospholipid hydroperoxide reductase activity of poplar type II peroxiredoxin. Physiol. Plant., **120**, 57-62 (2004)
[26] Cheong, N.E.; Choi, Y.O.; Lee, K.O.; Kim, W.Y.; Jung, B.G.; Chi, Y.H.; Jeong, J.S.; Kim, K.; Cho, M.J.; Lee, S.Y.: Molecular cloning, expression, and functional characterization of a 2Cys-peroxiredoxin in Chinese cabbage. Plant Mol. Biol., **40**, 825-834 (1999)
[27] Baier, M.; Dietz, K.J.: Protective function of chloroplast 2-cysteine peroxiredoxin in photosynthesis. Evidence from transgenic Arabidopsis. Plant Physiol., **119**, 1407-1414 (1999)
[28] Konig, J.; Baier, M.; Horling, F.; Kahmann, U.; Harris, G.; Schurmann, P.; Dietz, K.J.: The plant-specific function of 2-Cys peroxiredoxin-mediated detoxification of peroxides in the redox-hierarchy of photosynthetic electron flux. Proc. Natl. Acad. Sci. USA, **99**, 5738-5743 (2002)
[29] Schroder, E.; Littlechild, J.A.; Lebedev, A.A.; Errington, N.; Vagin, A.A.; Isupov, M.N.: Crystal structure of decameric 2-Cys peroxiredoxin from human erythrocytes at 1.7 A resolution. Structure Fold. Des., **8**, 605-615 (2000)
[30] Barranco-Medina, S.; Lopez-Jaramillo, F.J.; Bernier-Villamor, L.; Sevilla, F.; LaÂ´zaro, J.-J.: Cloning, overexpression, purification and preliminary crystallographic studies of a mitochondrial type II peroxiredoxin from Pisum sativum. Acta Crystallogr. Sect. F, **62**, 695-698 (2006)
[31] Wen, L.; Huang, H.M.; Juang, R.H.; Lin, C.T.: Biochemical characterization of 1-Cys peroxiredoxin from Antrodia camphorata. Appl. Microbiol. Biotechnol., **73**, 1314-1322 (2007)
[32] Huang, J.K.; Ken, C.F.; Huang, H.M.; Lin, C.T.: Biochemical characterization of a novel 2-Cys peroxiredoxin from Antrodia camphorata. Appl. Microbiol. Biotechnol., **74**, 84-92 (2007)
[33] Rouhier, N.; Gama, F.; Wingsle, G.; Gelhaye, E.; Gans, P.; Jacquot, J.P.: Engineering functional artificial hybrid proteins between poplar peroxiredoxin II and glutaredoxin or thioredoxin. Biochem. Biophys. Res. Commun., **341**, 1300-1308 (2006)
[34] Parsonage, D.; Youngblood, D.S.; Sarma, G.N.; Wood, Z.A.; Karplus, P.A.; Poole, L.B.: Analysis of the link between enzymatic activity and oligomeric state in AhpC, a bacterial peroxiredoxin. Biochemistry, **44**, 10583-10592 (2005)
[35] Echalier, A.; Trivelli, X.; Corbier, C.; Rouhier, N.; Walker, O.; Tsan, P.; Jacquot, J.P.; Aubry, A.; Krimm, I.; Lancelin, J.M.: Crystal structure and solution NMR dynamics of a D (type II) peroxiredoxin glutaredoxin and thioredoxin dependent: a new insight into the peroxiredoxin oligomerism. Biochemistry, **44**, 1755-1767 (2005)

[36] Nickel, C.; Trujillo, M.; Rahlfs, S.; Deponte, M.; Radi, R.; Becker, K.: Plasmodium falciparum 2-Cys peroxiredoxin reacts with plasmoredoxin and peroxynitrite. Biol. Chem., **386**, 1129-1136 (2005)
[37] Jiang, Q.; Yan, Y.H.; Hu, G.K.; Zhang, Y.Z.: Molecular cloning and characterization of a peroxiredoxin from Phanerochaete chrysosporium. Cell. Mol. Biol. Lett., **10**, 659-668 (2005)
[38] Kawazu, S.; Ikenoue, N.; Takemae, H.; Komaki-Yasuda, K.; Kano, S.: Roles of 1-Cys peroxiredoxin in haem detoxification in the human malaria parasite Plasmodium falciparum. FEBS J., **272**, 1784-1791 (2005)
[39] Limauro, D.; Pedone, E.; Pirone, L.; Bartolucci, S.: Identification and characterization of 1-Cys peroxiredoxin from Sulfolobus solfataricus and its involvement in the response to oxidative stress. FEBS J., **273**, 721-731 (2006)
[40] Sekiya, M.; Mulcahy, G.; Irwin, J.A.; Stack, C.M.; Donnelly, S.M.; Xu, W.; Collins, P.; Dalton, J.P.: Biochemical characterisation of the recombinant peroxiredoxin (FhePrx) of the liver fluke, Fasciola hepatica. FEBS Lett., **580**, 5016-5022 (2006)
[41] Manevich, Y.; Fisher, A.B.: Peroxiredoxin 6, a 1-Cys peroxiredoxin, functions in antioxidant defense and lung phospholipid metabolism. Free Radic. Biol. Med., **38**, 1422-1432 (2005)
[42] Finkemeier, I.; Goodman, M.; Lamkemeyer, P.; Kandlbinder, A.; Sweetlove, L.J.; Dietz, K.J.: The mitochondrial type II peroxiredoxin F is essential for redox homeostasis and root growth of Arabidopsis thaliana under stress. J. Biol. Chem., **280**, 12168-12180 (2005)
[43] Tanaka, T.; Izawa, S.; Inoue, Y.: GPX2, encoding a phospholipid hydroperoxide glutathione peroxidase homologue, codes for an atypical 2-Cys peroxiredoxin in Saccharomyces cerevisiae. J. Biol. Chem., **280**, 42078-42087 (2005)
[44] Trujillo, M.; Mauri, P.; Benazzi, L.; Comini, M.; De Palma, A.; Flohe, L.; Radi, R.; Stehr, M.; Singh, M.; Ursini, F.; Jaeger, T.: The mycobacterial thioredoxin peroxidase can act as a one-cysteine peroxiredoxin. J. Biol. Chem., **281**, 20555-20566 (2006)
[45] Conway, J.P.; Kinter, M.: Dual role of peroxiredoxin I in macrophage-derived foam cells. J. Biol. Chem., **281**, 27991-28001 (2006)
[46] Sarma, G.N.; Nickel, C.; Rahlfs, S.; Fischer, M.; Becker, K.; Karplus, P.A.: Crystal structure of a novel Plasmodium falciparum 1-Cys peroxiredoxin. J. Mol. Biol., **346**, 1021-1034 (2005)
[47] Mizohata, E.; Sakai, H.; Fusatomi, E.; Terada, T.; Murayama, K.; Shirouzu, M.; Yokoyama, S.: Crystal structure of an archaeal peroxiredoxin from the aerobic hyperthermophilic crenarchaeon Aeropyrum pernix K1. J. Mol. Biol., **354**, 317-| (2005)
[48] Molina-Lopez, J.; Jimenez, L.; Ochoa-Sanchez, A.; Landa, A.: Molecular cloning and characterization of a 2-Cys peroxiredoxin from Taenia solium. J. Parasitol., **92**, 796-802 (2006)
[49] Deponte, M.; Becker, K.: Biochemical characterization of Toxoplasma gondii 1-Cys peroxiredoxin 2 with mechanistic similarities to typical 2-Cys Prx. Mol. Biochem. Parasitol., **140**, 87-96 (2005)

[50] Puetz, S.; Gelius-Dietrich, G.; Piotrowski, M.; Henze, K.: Rubrerythrin and peroxiredoxin: two novel putative peroxidases in the hydrogenosomes of the microaerophilic protozoon Trichomonas vaginalis. Mol. Biochem. Parasitol., **142**, 212-223 (2005)
[51] Yano, K.; Komaki-Yasuda, K.; Tsuboi, T.; Torii, M.; Kano, S.; Kawazu, S.: 2-Cys Peroxiredoxin TPx-1 is involved in gametocyte development in Plasmodium berghei. Mol. Biochem. Parasitol., **148**, 44-51 (2006)
[52] Dombrecht, B.; Heusdens, C.; Beullens, S.; Verreth, C.; Mulkers, E.; Proost, P.; Vanderleyden, J.; Michiels, J.: Defence of Rhizobium etli bacteroids against oxidative stress involves a complexly regulated atypical 2-Cys peroxiredoxin. Mol. Microbiol., **55**, 1207-1221 (2005)
[53] Boucher, I.W.; McMillan, P.J.; Gabrielsen, M.; Akerman, S.E.; Brannigan, J.A.; Schnick, C.; Brzozowski, A.M.; Wilkinson, A.J.; Mueller, S.: Structural and biochemical characterization of a mitochondrial peroxiredoxin from Plasmodium falciparum. Mol. Microbiol., **61**, 948-959 (2006)
[54] Lamkemeyer, P.; Laxa, M.; Collin, V.; Li, W.; Finkemeier, I.; Schoettler, M.A.; Holtkamp, V.; Tognetti, V.B.; Issakidis-Bourguet, E.; Kandlbinder, A.; Weis, E.; Miginiac-Maslow, M.; Dietz, K.J.: Peroxiredoxin Q of Arabidopsis thaliana is attached to the thylakoids and functions in context of photosynthesis. Plant J., **45**, 968-981 (2006)
[55] Vivancos, A.P.; Castillo, E.A.; Biteau, B.; Nicot, C.; Ayte, J.; Toledano, M.B.; Hidalgo, E.: A cysteine-sulfinic acid in peroxiredoxin regulates H_2O_2-sensing by the antioxidant Pap1 pathway. Proc. Natl. Acad. Sci. USA, **102**, 8875-8880 (2005)

Versatile peroxidase 1.11.1.16

1 Nomenclature

EC number

1.11.1.16

Systematic name

reactive-black-5:hydrogen-peroxide oxidoreductase

Recommended name

versatile peroxidase

Synonyms

versatile peroxidase MnP2 <1> [17, 19]
versatile peroxidase VPL2 precursor <5> [18]

CAS registry number

114995-15-2
42613-30-9

2 Source Organism

<1> *Pleurotus ostreatus* (no sequence specified) [3, 4, 17, 19]
<2> *Bjerkandera adusta* (no sequence specified) [8, 9, 10]
<3> *Pleurotus eryngii* (no sequence specified) [2, 5, 7, 11, 13, 14, 15, 16]
<4> *Bjerkandera sp.* (no sequence specified) [1]
<5> *Pleurotus eryngii* (UNIPROT accession number: O94753) [6,18]
<6> *Pleurotus eryngii* (UNIPROT accession number: Q9UVP6) [12]

3 Reaction and Specificity

Catalyzed reaction

Reactive Black 5 + H_2O_2 = oxidized Reactive Black 5 + 2 H_2O
donor + H_2O_2 = oxidized donor + 2 H_2O (<3> NMR study of enzyme both in resting state and cyanide-inhibited form. Analysis of interaction with Mn^{2+} [15]; <2> study of reaction of H_2O_2 with the enzyme in the absence of substrate suggests an amino acid radical in moderate distance from ferryl heme. A tryptophan radical is formed during the catalytic mechanism [8]; <3> the radical intermediate in the reaction of enzyme with H_2O_2 is a tryptophan neutral radical at W164 [14])

Substrates and products

S 1,4-benzohydroquinone + H_2O_2 <3> (Reversibility: ?) [11]
P ? + H_2O
S 1,4-dimethoxybenzene + H_2O_2 <4> (<4> Mn^{2+}-independent activity [1]) (Reversibility: ?) [1]
P 1,4-benzoquinone + H_2O
S 1-methylanthracene + H_2O_2 <2> (<2> at 43% of the rate with 9-methylanthracene [10]) (Reversibility: ?) [10]
P 1-methylanthraquinone + H_2O
S 2,2'-azino-bis(3-ethylbenzothiazoline-6-sulfonic acid) + H_2O_2 <4> (<4> Mn^{2+}-independent activity [1]) (Reversibility: ?) [1]
P ? + H_2O
S 2,6-dimethoxybenzohydroquinone + H_2O_2 <3> (Reversibility: ?) [11]
P ? + H_2O
S 2,6-dimethoxyphenol + H_2O_2 <1> (Reversibility: ?) [4]
P ? + H_2O_2
S 2,6-dimethoxyphenol + H_2O_2 <3, 4> (<3> Mn^{2+}-dependent and independent activity [2]; <4> Mn^{2+}-independent activity [1]) (Reversibility: ?) [1, 2]
P ? + H_2O
S 2-chloro-1,4-dimethoxybenzene + H_2O_2 <4> (<4> Mn^{2+}-independent activity [1]) (Reversibility: ?) [1]
P 2-chloro-1,4-benzoquinone + H_2O
S 2-methoxy-1,4-benzohydroquinone + H_2O_2 <3> (Reversibility: ?) [11]
P ? + H_2O
S 2-methylanthracene + H_2O_2 <2> (<2> at 24% of the rate with 9-methylanthracene [10]) (Reversibility: ?) [10]
P 2-methylanthraquinone + H_2O
S 3-hydroxyanthranilic acid + H_2O_2 <4> (<4> Mn^{2+}-independent activity [1]) (Reversibility: ?) [1]
P ? + H_2O
S 9-methylanthracene + H_2O_2 <2> (Reversibility: ?) [10]
P 9-methylanthraquinone + H_2O
S Mn^{2+} + H_2O_2 <1, 3, 4, 5> (Reversibility: ?) [1, 2, 6, 13, 16, 18, 19]
P Mn^{3+} + H_2O
S NADH + + H_2O_2 <3> (Reversibility: ?) [2]
P NAD^+ + H_2O
S Poly R-478 <1> (<1> no redox mediators involved [3]) (Reversibility: ?) [3]
P oxidized Poly R-478
S RNase A + H_2O_2 <1> (Reversibility: ?) [19]
P ?
S Reactive Black 5 + H_2O_2 <3, 5> (<3> catalyzed by isoform PS1 [13]) (Reversibility: ?) [13, 16, 18]
P oxidized Reactive Black 5 + H_2O
S anthracene + H_2O_2 <2> (<2> at 4.8% of the rate with 9-methylanthracene [10]) (Reversibility: ?) [10]

P anthraquinone + H_2O
S bovine pancreatic RNase <1> (<1> no redox mediators involved [3]) (Reversibility: ?) [3]
P oxidized bovine pancreatic RNase
S carbazole + H_2O_2 <2> (<2> at 4.8% of the rate with 9-methylanthracene [10]) (Reversibility: ?) [10]
P ? + H_2O
S guaiacol + H_2O_2 <3, 4> (<4> Mn^{2+}-independent activity [1]) (Reversibility: ?) [1, 13]
P ? + H_2O
S methoxyhydroquinone + H_2O_2 <3, 5> (Reversibility: ?) [6, 13]
P ? + H_2O
S *o*-anisidine + H_2O_2 <4> (<4> Mn^{2+}-independent activity [1]) (Reversibility: ?) [1]
P ? + H_2O
S *p*-anisidine + H_2O_2 <4> (<4> Mn^{2+}-independent activity [1]) (Reversibility: ?) [1]
P ? + H_2O
S 1,4-dimethoxybenzene + H_2O_2 <3> (<3> catalyzed by isoforms PS3, PS1 [13]) (Reversibility: ?) [13]
P benzoquinone + H_2O
S phenol red + H_2O_2 <3> (<3> Mn^{2+}-dependent activity [2]) (Reversibility: ?) [2]
P oxidized phenol red + H_2O
S syringol + H_2O_2 <3> (Reversibility: ?) [13]
P ? + H_2O
S vanillylidenacetone + H_2O_2 <3> (<3> Mn^{2+}-dependent activity [2]) (Reversibility: ?) [2]
P ? + H_2O
S veratryl alcohol + H_2O_2 <1, 3, 5> (<3> catalyzed by isoforms PS2, PS_1 [13]; <3> Mn^{2+}-independent activity [2]) (Reversibility: ?) [2, 3, 13, 16, 18, 19]
P veratraldehyde + H_2O
S veratryl alcohol + H_2O_2 <4, 5> (<4> Mn^{2+}-independent activity [1]) (Reversibility: ?) [1, 6]
P veratraldehyde + H_2O
S Additional information <3> (<3> in the absence of Mn^{2+}, efficient hydroquinone oxidation is dependent on exogenous H_2O_2. In the presence of Mn^{2+}, exogenous H_2O_2 is not required for complete oxidation of hydroquinones [11]) (Reversibility: ?) [11]
P ?

Inhibitors

Mn^{2+} <4> (<4> above 0.1 mM, severe inhibition of oxidation of veratryl alcohol [1]) [1]
bovine pancreatic RNase <1> (<1> competitive to oxidation of veratryl alcohol, non-competitive to oxidation of Mn^{2+} [3]) [3]

Cofactors/prosthetic groups

Heme <3, 5> (<5> enzyme contains heme [18]) [13,18]

Metals, ions

Ca^{2+} <2, 5> (<5> enzyme contains Ca^{2+} [18]; <2> required, Ca^{2+}-depleted enzyme is able to form the active intermediate compound I but its long range electron transfer has been disrupted [9]) [9, 18]
Iron <2> (<2> at pH 4.5, Ca^{2+}-depleted enzyme has a high-spin Fe^{3+} [9]) [9]
Mn^{2+} <3, 6> (<3> isoenzyme Pl, putative Mn-interaction site near E35, E39, D175. Isoenzyme Ps1, putative Mn-interaction site near E36, E40, D181 [7]; <6> Mn-binding site involving E36, E40, and D181 [12]; <3> putative Mn^{2+}-binding site, isoenzyme PS_1 [13]) [7, 12, 13]

Turnover number (min^{-1})

1.3 <4> (2,2'-azino-bis(3-ethylbenzothiazoline-6-sulfonic acid), <4> pH 4.5 [1]) [1]
1.4 <4> (veratryl alcohol, <4> pH 4.5 [1]) [1]
2 <5> (Mn^{2+}, <5> 25°C, pH 5, mutant E36A/E40A/D175A [18]; <5> 25°C, pH 5, mutant E36A/E40A/D175A/P327ter [18]) [18]
2.3 <4> (2,6-dimethoxyphenol, <4> pH 4.5 [1]) [1]
2.4 <2> (2-methylanthracene, <2> pH 4.0 [10]) [10]
2.8 <4> (veratryl alcohol, <4> pH 3.0 [1]) [1]
3 <3> (syringol, <3> pH 3.0, isoenzyme PS3 [13]) [13]
4 <3> (veratryl alcohol, <3> pH 3.0, isoenzyme PS_1 [13]) [13]
4 <3> (methoxyhydroquinone, <3> pH 3.0, isoenzyme PS1 [13]) [13]
4 <3> (*p*-dimethoxybenzene, <3> pH 3.0, isoenzyme PS1 [13]) [13]
4.7 <3> (Reactive Black 5, <3> mutant H232F, pH 3.5 [16]) [16]
5 <5> (Mn^{2+}, <5> 25°C, pH 5, mutant E36A/E40A [18]) [18]
5 <3> (Reactive Black 5, <3> pH 3.0, isoenzyme PS_1 [13]; <3> wild-type, pH 3.5 [16]) [13, 16]
6 <3> (Reactive Black 5, <3> mutant P76H, pH 3.5 [16]) [16]
6 <3> (syringol, <3> pH 3.0, isoenzyme PS1 [13]) [13]
8 <3> (veratryl alcohol, <3> wild-tpye, pH 3.0 [16]) [16]
11 <3> (veratryl alcohol, <3> mutant P76H, pH 3.0 [16]) [16]
13 <5> (veratryl alcohol) [6]
14 <3> (veratryl alcohol, <3> mutant H232F, pH 3.0 [16]) [16]
15 <5> (Mn^{2+}, <5> 25°C, pH 5, mutant E40A [18]) [18]
17 <5> (methoxyhydroquinone) [6]
19 <3> (methoxyhydroquinone, <3> pH 3.0, isoenzyme PS3 [13]) [13]
32 <5> (Mn^{2+}, <5> 25°C, pH 5, mutant D175A [18]) [18]
59 <4> (Mn^{2+}, <4> pH 4.5 [1]) [1]
75 <4> (H_2O_2, <4> pH 4.5 [1]) [1]
78 <3> (Mn^{2+}, <3> pH 5.0, isoenzyme PS3 [13]) [13]
79 <3> (Mn^{2+}, <3> pH 5.0, isoenzyme PS1 [13]) [13]
85 <5> (Mn^{2+}, <5> 25°C, pH 5, mutant E36A [18]) [18]
103 <5> (Mn^{2+}, <5> 25°C, pH 5, mutant E36D [18]) [18]
110 <5> (Mn^{2+}) [6]

145 <5> (Mn^{2+}, <5> 25°C, pH 5, mutant E40D [18]) [18]
207 <3> (Mn^{2+}, <3> mutant W164S, pH 5.0 [16]) [16]
247 <3> (Mn^{2+}, <3> mutant W164S/P76H, pH 5.0 [16]) [16]
291 <3> (Mn^{2+}, <3> mutant P76H, pH 5.0 [16]) [16]
298 <3> (Mn^{2+}, <3> wild-tpye, pH 5.0 [16]) [16]
308 <3> (Mn^{2+}, <3> mutant H232F, pH 5.0 [16]) [16]
320 <3> (Mn^{2+}, <3> mutant W164H, pH 5.0 [16]) [16]
467 <5> (Mn^{2+}, <5> 25°C, pH 5, mutant A173R [18]) [18]

Specific activity (U/mg)

8.4 <3> (<3> substrate 1,4-benzohydroquinone, pH 5.0, absence of Mn^{2+} [11]) [11]
9.6 <3> (<3> substrate 2-methoxy-1,4-benzohydroquinone, pH 5.0, absence of Mn^{2+} [11]) [11]
10.8 <3> (<3> substrate 2,6-dimthoxy-1,4-benzohydroquinone, pH 5.0, absence of Mn^{2+} [11]) [11]
80 <4> (<4> pH 4.5 [1]) [1]
148.5 <5> [6]
334 <1> (<1> growth on rich peptone medium containing 0.5 mM Mn^{2+} [4]) [4]
559 <1> (<1> after growth on rich peptone medium [4]) [4]

K_m-Value (mM)

0.002 <3> (H_2O_2, <3> cosubstrate aromatic compound, pH 3.0, isoenzyme PS1 [13]) [13]
0.002 <3> (Reactive Black 5, <3> pH 3.0, isoenzyme PS1 [13]) [13]
0.0028 <3> (Reactive Black 5, <3> wild-tpye, pH 3.5 [16]) [16]
0.003 <3> (H_2O_2, <3> cosubstrate aromatic compound, pH 3.0, isoenzyme PS3 [13]) [13]
0.0031 <3> (Reactive Black 5, <3> mutant P76H, pH 3.5 [16]) [16]
0.0036 <3> (Reactive Black 5, <3> mutant H232F, pH 3.5 [16]) [16]
0.005 <3> (vanillylidenacetone, <3> presence of Mn^{2+}, isoenzyme MP-1, pH 5.0 [2]; <3> presence of Mn^{2+}, isoenzyme MP-2, pH 5.0 [2]) [2]
0.006 <3> (H_2O_2, <3> cosubstrate Mn^{2+}, isoenzyme MP-1, pH 5.0 [2]) [2]
0.009 <1, 3> (H_2O_2, <1> native enzyme [19]; <3> pH 5.0, cosubstrate Mn^{2+}, isoenzyme PS1 [13]) [13, 19]
0.01 <3> (2,6-dimethoxyphenol, <3> presence of Mn^{2+}, isoenzyme MP-1, pH 5.0 [2]; <3> presence of Mn^{2+}, isoenzyme MP-2, pH 5.0 [2]) [2]
0.01 <3> (H_2O_2, <3> cosubstrate Mn^{2+}, isoenzyme MP-2, pH 5.0 [2]; <3> pH 5.0, cosubstrate Mn^{2+}, isoenzyme PS3 [13]) [2, 13]
0.0103 <1> (H_2O_2, <1> recombinant enzyme [19]) [19]
0.012 <5> (Mn^{2+}) [6]
0.015 <3> (Mn^{2+}, <3> cosubstrate dimethoxyphenol, isoenzyme MP-1, pH 5.0 [2]; <3> cosubstrate dimethoxyphenol, isoenzyme MP-2, pH 5.0 [2]) [2]
0.017 <3> (methoxyhydroquinone, <3> pH 3.0, isoenzyme PS1 [13]) [13]
0.019 <5> (methoxyhydroquinone) [6]

0.02 <3> (Mn^{2+}, <3> cosubstrate H_2O_2, isoenzyme MP-1, pH 5.0 [2]; <3> cosubstrate H_2O_2, isoenzyme MP-2, pH 5.0 [2]) [2]
0.0223 <1> (Mn^{2+}, <1> recombinant enzyme [19]) [19]
0.0263 <1> (Mn^{2+}, <1> native enzyme [19]) [19]
0.031 <4> (H_2O_2, <4> pH 4.5 [1]) [1]
0.037 <4> (2,2'-azino-bis(3-ethylbenzothiazoline-6-sulfonic acid), <4> pH 4.5 [1]) [1]
0.041 <4> (2,6-dimethoxyphenol, <4> pH 4.5 [1]) [1]
0.048 <3> (Mn^{2+}, <3> pH 5.0, isoenzyme PS1 [13]) [13]
0.051 <4> (Mn^{2+}, <4> pH 4.5 [1]) [1]
0.11 <3> (Mn^{2+}, <3> mutant W164S, pH 5.0 [16]) [16]
0.116 <4> (veratryl alcohol, <4> pH 3.0 [1]) [1]
0.126 <3> (Mn^{2+}, <3> mutant W164H, pH 5.0 [16]) [16]
0.16 <3> (2,6-dimethoxyphenol, <3> absence of Mn^{2+}, isoenzyme MP-1, pH 3.0 [2]) [2]
0.189 <3> (Mn^{2+}, <3> wild-tpye, pH 5.0 [16]) [16]
0.2 <3> (Mn^{2+}, <3> pH 5.0, isoenzyme PS3 [13]) [13]
0.2 <3> (syringol, <3> pH 3.0, isoenzyme PS1 [13]) [13]
0.218 <3> (Mn^{2+}, <3> mutant H232F, pH 5.0 [16]) [16]
0.25 <3> (2,6-dimethoxyphenol, <3> absence of Mn^{2+}, isoenzyme MP-2, pH 3.0 [2]) [2]
0.262 <3> (Mn^{2+}, <3> mutant P76H, pH 5.0 [16]) [16]
0.351 <3> (Mn^{2+}, <3> mutant W164S/P76H, pH 5.0 [16]) [16]
0.417 <5> (Mn^{2+}, <5> 25°C, pH 5, mutant A173R [18]) [18]
0.534 <4> (veratryl alcohol, <4> pH 4.5 [1]) [1]
1 <3> (syringol, <3> pH 3.0, isoenzyme PS3 [13]) [13]
2.17 <5> (veratryl alcohol) [6]
2.4 <3> (*p*-dimethoxybenzene, <3> pH 3.0, isoenzyme PS1 [13]) [13]
2.41 <3> (veratryl alcohol, <3> mutant P76H, pH 3.0 [16]) [16]
2.5-3 <3> (methoxyhydroquinone, <3> pH 3.0, isoenzyme PS3 [13]) [13]
2.75 <3> (veratryl alcohol, <3> wild-tpye, pH 3.0 [16]) [16]
3 <3> (veratryl alcohol, <3> absence of Mn^{2+}, isoenzyme MP-2, pH 3.0 [2]) [2]
3.5 <3> (veratryl alcohol, <3> absence of Mn^{2+}, isoenzyme MP-1, pH 3.0 [2]; <3> pH 3.0, isoenzyme PS_1 [13]) [2, 13]
3.58 <3> (veratryl alcohol, <3> mutant H232F, pH 3.0 [16]) [16]
4.754 <1> (veratryl alcohol, <1> native enzyme [19]) [19]
4.91 <5> (Mn^{2+}, <5> 25°C, pH 5, mutant E40D [18]) [18]
4.98 <5> (Mn^{2+}, <5> 25°C, pH 5, mutant E36D [18]) [18]
5.08 <1> (veratryl alcohol, <1> recombinant enzyme [19]) [19]
11.26 <5> (Mn^{2+}, <5> 25°C, pH 5, mutant E40A [18]) [18]
13.84 <5> (Mn^{2+}, <5> 25°C, pH 5, mutant E36A [18]) [18]
16.1 <5> (Mn^{2+}, <5> 25°C, pH 5, mutant D175A [18]) [18]
46.86 <5> (Mn^{2+}, <5> 25°C, pH 5, mutant E36A/E40A [18]) [18]
69.97 <5> (Mn^{2+}, <5> 25°C, pH 5, mutant E36A/E40A/D175A/P327ter [18]) [18]
76.4 <5> (Mn^{2+}, <5> 25°C, pH 5, mutant E36A/E40A/D175A [18]) [18]

pH-Optimum

3 <2, 3> (<2> Mn^{2+}-dependent reaction [10]; <3> Mn^{2+}-independent oxidation of substituted phenols and aromatic molecules [13]; <3> oxidation of 2,6-dimethoxyphenol, absence of Mn^{2+} [2]) [2, 10, 13]
4 <2, 3> (<2> Mn^{2+}-independent reaction [10]; <3> oxidation of 2,6-dimethoxyphenol, presence of Mn^{2+} [2]) [2, 10]
4.5 <2> [9]
5 <3> (<3> oxidation of Mn^{2+} [2,13]) [2, 13]

4 Enzyme Structure

Subunits

? <3, 5> (<3> x * 43000, SDS-PAGE, both isoenzyme MP-1 and MP-2 [2]; <5> x * 43000, SDS-PAGE, x * 41000, SDS-PAGE of deglycosylated enzyme [6]; <3> x * 45000, isoenzymes PS1, PS2, x * 42000, isoenzyme PS3 [13]) [2, 6, 13]
Additional information <1> (<1> W170 exposed on enzyme surface is a substrate-binding site both for veratryl alcohol and for polymeric substrates [3]) [3]

Posttranslational modification

glycoprotein <3, 5> (<5> N-glycosylated [6]; <3> MP-1, up to 5% carbohydrate content, MP-2, up to 7% carbohydrate [2]) [2, 6]

5 Isolation/Preparation/Mutation/Application

Purification

<1> [19]
<3> [2, 13]
<4> [1]
<5> (recombinant) [18]

Crystallization

<5> (hanging drop vapor diffusion method, crystal structures of untreated versatile peroxidase (immediately after expression in Escherichia coli and in vitro reconstitution), native versatile peroxidase (treated with Mn^{2+}), D175A variant, and wild-type verstile peroxidase (from Pleurotus eryngii culture)) [18]

Cloning

<1> (a recombinant mnp2 construct under the control of Pleurotus ostreatus sdi1 expression signals is introduced into the wild-type Pleurotus ostreatus strain by cotransformation with a carboxin-resistant marker plasmid. Recombinant Pleurotus ostreatus strains with elevated manganese peroxidase (MnP) productivity are successfully isolated. The productivity of the recombinant

MnP2 in the present system is not high enough to meet the requirements for industrial applications) [17]
<1> (recombinant MnP2 is exclusively expressed and no endogenous MnP isozymes are secreted by the recombinant Pleurotus oseatus srain TM2-18) [19]
<3> [7]
<5> (expression in Escherichia coli) [18]

Engineering

A173R <5> (<5> k_{cat}/K_M for Mn^{2+} is 1.4fold lower than wild-type value, k_{cat}/K_m for veratryl alcohol is 1.4fold higher than wild-type value, k_{cat}/K_m for Reactive Black 5 is 1.3fold higher than wild-type value [18]) [18]
D175A <5> (<5> k_{cat}/K_M for Mn^{2+} is 842fold lower than wild-type value, k_{cat}/K_m for veratryl alcohol is3.2 fold higher than wild-type value, k_{cat}/K_m for Reactive Black 5 is 1.8fold higher than wild-type value [18]) [18]
E36A <5> (<5> k_{cat}/K_M for Mn^{2+} is 258fold lower than wild-type value, k_{cat}/K_m for veratryl alcohol is identical to wild-type value, k_{cat}/K_m for Reactive Black 5 is 1.2fold higher than wild-type value [18]) [18]
E36A/E40A <5> (<5> k_{cat}/K_M for Mn^{2+} is 16000fold lower than wild-type value, k_{cat}/K_m for veratryl alcohol is 1.3fold higher than wild-type value, k_{cat}/K_m for Reactive Black 5 is 1.1fold higher than wild-type value [18]) [18]
E36A/E40A/D175A <5> (<5> k_{cat} for Mn^{2+} is 149fold lower than wild-type value, k_{cat}/K_m for veratryl alcohol is nearly identical to wild-type value, k_{cat}/K_m for Reactive Black 5 is 2fold higher than wild-type value [18]) [18]
E36A/E40A/D175A/P327ter <5> (<5> k_{cat} for Mn^{2+} is 149fold lower than wild-type value, k_{cat}/K_m for veratryl alcohol is 1.6fold lower than wild-type value, k_{cat}/K_m for Reactive Black 5 is 2.4fold higher than wild-type value [18]) [18]
E36D <5> (<5> k_{cat}/K_M for Mn^{2+} is 77fold lower than wild-type value, k_{cat}/K_m for veratryl alcohol is 1.3fold higher than wild-type value, k_{cat}/K_m for Reactive Black 5 is 3.5fold higher than wild-type value [18]) [18]
E40A <5> (<5> k_{cat}/K_M for Mn^{2+} is 1231fold lower than wild-type value, k_{cat}/K_m for veratryl alcohol is 1.2fold lower than wild-type value, k_{cat}/K_m for Reactive Black 5 is nearly identical to wild-type value [18]) [18]
E40D <5> (<5> k_{cat}/K_M for Mn^{2+} is 54fold lower than wild-type value, k_{cat}/K_m for veratryl alcohol is 1.3fold lower than wild-type value, k_{cat}/K_m for Reactive Black 5 is 2.4fold higher than wild-type value [18]) [18]
H232F <3> (<3> not involved in long-range electron transfer [16]) [16]
P76H <3> (<3> not involved in long-range electron transfer [16]) [16]
W164H <3> (<3> no enzymic activity with veratryl alcohol or Reactive Black 5 [16]) [16]
W164S <3> (<3> no enzymic activity with veratryl alcohol or Reactive Black 5 [16]) [16]
W164S/P76H <3> (<3> no enzymic activity with veratryl alcohol or Reactive Black 5 [16]) [16]

6 Stability

Storage stability

<3>, -20°C, stable [2]
<3>, 4°C, stable for at least 72 h [2]

References

[1] Mester, T.; Field, J.A.: Characterization of a novel manganese peroxidase-lignin peroxidase hybrid isozyme produced by Bjerkandera species strain BOS55 in the absence of manganese. J. Biol. Chem., **273**, 15412-15417 (1998)

[2] Martinez, M.J.; Ruiz-Duenas, F.J.; Guillen, F.; Martinez, A.T.: Purification and catalytic properties of two manganese peroxidase isoenzymes from Pleurotus eryngii. Eur. J. Biochem., **237**, 424-432 (1996)

[3] Kamitsuji, H.; Watanabe, T.; Honda, Y.; Kuwahara, M.: Direct oxidation of polymeric substrates by multifunctional manganese peroxidase isoenzyme from Pleurotus ostreatus without redox mediators. Biochem. J., **386**, 387-393 (2005)

[4] Cohen, R.; Persky, L.; Hazan-Eitan, Z.; Yarden, O.; Hadar, Y.: Mn^{2+} alters peroxidase profiles and lignin degradation by the white-rot fungus Pleurotus ostreatus under different nutritional and growth conditions. Appl. Biochem. Biotechnol., **102-103**, 415-429 (2002)

[5] Ruiz-Duenas, F.J.; Guillen, F.; Camarero, S.; Perez-Boada, M.; Martinez, M.J.; Martinez, A.T.: Regulation of peroxidase transcript levels in liquid cultures of the ligninolytic fungus Pleurotus eryngii. Appl. Environ. Microbiol., **65**, 4458-4463 (1999)

[6] Ruiz-Duenas, F.J.; Martinez, M.J.; Martinez, A.T.: Heterologous expression of Pleurotus eryngii peroxidase confirms its ability to oxidize Mn(2+) and different aromatic substrates. Appl. Environ. Microbiol., **65**, 4705-4707 (1999)

[7] Ruiz-Duenas, F.J.; Camarero, S.; Perez-Boada, M.; Martinez, M.J.; Martinez, A.T.: A new versatile peroxidase from Pleurotus. Biochem. Soc. Trans., **29**, 116-122 (2001)

[8] Pogni, R.; Baratto, M.C.; Giansanti, S.; Teutloff, C.; Verdin, J.; Valderrama, B.; Lendzian, F.; Lubitz, W.; Vazquez-Duhalt, R.; Basosi, R.: Tryptophan-based radical in the catalytic mechanism of versatile peroxidase from Bjerkandera adusta. Biochemistry, **44**, 4267-4274 (2005)

[9] Verdin, J.; Pogni, R.; Baeza, A.; Baratto, M.C.; Basosi, R.; Vazquez-Duhalt, R.: Mechanism of versatile peroxidase inactivation by Ca(2+) depletion. Biophys. Chem., **121**, 163-170 (2006)

[10] Wang, Y.; Vazquez-Duhalt, R.; Pickard, M.A.: Manganese-lignin peroxidase hybrid from Bjerkandera adusta oxidizes polycyclic aromatic hydrocarbons more actively in the absence of manganese. Can. J. Microbiol., **49**, 675-682 (2003)

[11] Gomez-Toribio, V.; Martinez, A.T.; Martinez, M.J.; Guillen, F.: Oxidation of hydroquinones by the versatile ligninolytic peroxidase from Pleurotus eryngii. H_2O_2 generation and the influence of Mn^{2+}. Eur. J. Biochem., **268**, 4787-4793 (2001)
[12] Camarero, S.; Ruiz-Duenas, F.J.; Sarkar, S.; Martinez, M.J.; Martinez, A.T.: The cloning of a new peroxidase found in lignocellulose cultures of Pleurotus eryngii and sequence comparison with other fungal peroxidases. FEMS Microbiol. Lett., **191**, 37-43 (2000)
[13] Camarero, S.; Sarkar, S.; Ruiz-Duenas, F.J.; Martinez, M.J.; Martinez, A.T.: Description of a versatile peroxidase involved in the natural degradation of lignin that has both manganese peroxidase and lignin peroxidase substrate interaction sites. J. Biol. Chem., **274**, 10324-10330 (1999)
[14] Pogni, R.; Baratto, M.C.; Teutloff, C.; Giansanti, S.; Ruiz-Duenas, F.J.; Choinowski, T.; Piontek, K.; Martinez, A.T.; Lendzian, F.; Basosi, R.: A tryptophan neutral radical in the oxidized state of versatile peroxidase from Pleurotus eryngii: a combined multifrequency EPR and density functional theory study. J. Biol. Chem., **281**, 9517-9526 (2006)
[15] Banci, L.; Camarero, S.; Martinez, A.T.; Martinez, M.J.; Perez-Boada, M.; Pierattelli, R.; Ruiz-Duenas, F.J.: NMR study of manganese(II) binding by a new versatile peroxidase from the white-rot fungus Pleurotus eryngii. J. Biol. Inorg. Chem., **8**, 751-760 (2003)
[16] Perez-Boada, M.; Ruiz-Duenas, F.J.; Pogni, R.; Basosi, R.; Choinowski, T.; Martinez, M.J.; Piontek, K.; Martinez, A.T.: Versatile peroxidase oxidation of high redox potential aromatic compounds: site-directed mutagenesis, spectroscopic and crystallographic investigation of three long-range electron transfer pathways. J. Mol. Biol., **354**, 385-402 (2005)
[17] Tsukihara, T.; Honda, Y.; Watanabe, T.: Molecular breeding of white rot fungus Pleurotus ostreatus by homologous expression of its versatile peroxidase MnP2. Appl. Microbiol. Biotechnol., **71**, 114-120 (2006)
[18] Ruiz-Duenas, F.J.; Morales, M.; Perez-Boada, M.; Choinowski, T.; Martinez, M.J.; Piontek, K.; Martinez, A.T.: Manganese oxidation site in Pleurotus eryngii versatile peroxidase: a site-directed mutagenesis, kinetic, and crystallographic study. Biochemistry, **46**, 66-77 (2007)
[19] Tsukihara, T.; Honda, Y.; Sakai, R.; Watanabe, T.; Watanabe, T.: Exclusive overproduction of recombinant versatile peroxidase MnP2 by genetically modified white rot fungus, Pleurotus ostreatus. J. Biotechnol., **126**, 431-439 (2006)

9-*cis*-Epoxycarotenoid dioxygenase **1.13.11.51**

1 Nomenclature

EC number

1.13.11.51

Systematic name

9-cis-epoxycarotenoid 11,12-dioxygenase

Recommended name

9-cis-epoxycarotenoid dioxygenase

Synonyms

9-cis epoxycarotenoid dioxygenase <4, 12, 13, 14, 15, 16, 17, 18, 30, 31, 32, 33> [17]
9-cis-epoxicarotenoid dioxygenase <4> [11]
AtCCD1 <4> [4, 19]
AtNCED3 <2> [14]
CCD <24> [26]
CitNCED2 <19, 20, 21> [20]
CitNCED3 <26, 27, 28> [20]
CsNCED1 <7> [21]
CsNCED2 <7> [21]
LeNCED1 <6> [5]
NCED <2, 3, 4, 6, 9, 10, 11, 12, 13, 14, 15, 16, 17, 18, 22, 24, 25, 29, 30, 31, 32, 33> [2, 3, 7, 10, 14, 16, 17, 18, 22, 24, 25, 26, 27]
NCED1 <5, 23> [15, 23]
NCED2 <23> [23]
NSC_1 <9> [18]
NSC_2 <9> [18]
NSC_3 <9> [18]
PvNCED1 <1, 2> [13, 14]
STO1 <4> [11]
VP14 <1, 2> [1, 9, 12, 14]
VP14 epoxy-carotenoid dioxygenase <2> [6]
VP14 protein <2> [14]
VuNCED1 protein <8> [8]
VvNCED1 <10> (<10> two genes encoding NCED: VvNCED1 and VvNCED2 [3]) [3]
nine-cis-epoxy dioxygenase <9> [18]
nine-cis-epoxycarotenoid dioxygenase <2> [14]
putative carotenoid cleavage dioxygenase <24> [26]

CAS registry number
199877-10-6

2 Source Organism

<1> *Phaseolus vulgaris* (no sequence specified) [9, 13]
<2> *Zea mays* (no sequence specified) [1, 6, 12, 14]
<3> *Solanum tuberosum* (no sequence specified) [25]
<4> *Arabidopsis thaliana* (no sequence specified) [4, 11, 17, 19, 24]
<5> *Arachis hypogaea* (no sequence specified) [15]
<6> *Lycopersicon esculentum* (no sequence specified) [5,7]
<7> *Citrus sinensis* (no sequence specified) [21]
<8> *Vigna unguiculata* (no sequence specified) [8]
<9> *Nostoc sp.* (no sequence specified) [18]
<10> *Vitis vinifera* (no sequence specified) [3]
<11> *Galium aparine* (no sequence specified) [22]
<12> *Eucalyptus globulus* (no sequence specified) [17]
<13> *Eucalyptus camaldulensis* (no sequence specified) [17]
<14> *Vigna unguiculata* (UNIPROT accession number: Q9FS24) [2, 10, 17]
<15> *Lycopersicon esculentum* (UNIPROT accession number: O24023) [17]
<16> *Persea americana* (UNIPROT accession number: Q9AXZ3) [17]
<17> *Solanum tuberosum* (UNIPROT accession number: Q9M3Z9) [17]
<18> *Zea mays* (UNIPROT accession number: O24592) [17]
<19> *Citrus limon* (UNIPROT accession number: Q1XHJ4) [20]
<20> *Citrus sinensis* (UNIPROT accession number: Q1XHJ5) [20]
<21> *Citrus unshiu* (UNIPROT accession number: Q1XHJ6) [20]
<22> *Arachis hypogaea* (UNIPROT accession number: Q70KK0) [16]
<23> *Gentiana lutea* (no sequence specified) [23]
<24> *Citrus clementina* (no sequence specified) [26]
<25> *Oncidium sp.* (UNIPROT accession number: Q52QS7) [27]
<26> *Citrus limon* (UNIPROT accession number: Q1XHJ1) [20]
<27> *Citrus sinensis* (UNIPROT accession number: Q1XHJ2) [20]
<28> *Citrus unshiu* (UNIPROT accession number: Q1XHJ3) [20]
<29> *Oncidium sp.* (UNIPROT accession number: Q52QS5) [27]
<30> *Eucalyptus grandis* (no sequence specified) [17]
<31> *Eucalyptus saligna* (no sequence specified) [17]
<32> *Eucalyptus urophylla* (no sequence specified) [17]
<33> *Phaseolus vulgaris* (UNIPROT accession number: Q9M6E8) [17]

3 Reaction and Specificity

Catalyzed reaction
9'-cis-neoxanthin + O_2 = 2-cis,4-trans-xanthoxin + (3S,5R,6R)-5,6-dihydroxy-6,7-didehydro-5,6-dihydro-12'-apo-β-caroten-12'-al

9-cis-violaxanthin + O_2 = 2-cis,4-trans-xanthoxin + (3S,5R,6S)-5,6-epoxy-3-hydroxy-5,6-dihydro-12'-apo-β-caroten-12'-al
a 9-cis-epoxycarotenoid + O_2 = 2-cis,4-trans-xanthoxin + a 12'-apo-carotenal

Natural substrates and products

S 9'-cis-neoxanthin + O_2 <1> (<1> drought-induced biosynthesis of abscisic acid is regulated by the 9-cis-epoxy-carotenoid cleavage reaction [13]) (Reversibility: ?) [13]
P 2-cis,4-trans-xanthoxin + (3S,5R,6R)-5,6-dihydroxy-6,7-didehydro-5,6-dihydro-12'-apo-β-caroten-12'al
S 9-cis-violaxanthin + O_2 <1> (<1> drought-induced biosynthesis of abscisic acid is regulated by the 9-cis-epoxy-carotenoid cleavage reaction [13]) (Reversibility: ?) [13]
P 2-cis,4-trans-xanthoxin + (3S,5R,6S)-5,6-epoxy-3-hydroxy-5,6-dihydro-12'apo-β-caroten-12'-al
S Additional information <2, 4, 6, 8, 10> (<2> enzyme is involved in abscisic acid pathway [1]; <8> expression is strongly induced by drough stress in the 8-day old plant. The enzyme has a key role in the synthesis of abscisic acid under drough stress [8]; <4> key enzyme in the abscisic acid biosynthetic pathway [11]; <10> key enzyme involved in the biosynthetic pathway of abscisic acid [3]; <6> key regulatory enzyme in abscisic acid biosynthesis in leaves [5]; <2> the enzyme functions in root and leaf in abscisic acid synthesis [12]; <2> the oxidative cleavage of 9-cis epoxy-carotenoids is the first commited step and key regulatory step in the abscisic acid biosynthesis [6]) (Reversibility: ?) [1, 3, 5, 6, 8, 11, 12]
P ?

Substrates and products

S 9'-cis-neoxanthin + O_2 <2, 4, 5, 7, 11, 12, 13, 14, 22, 24, 25, 29, 30, 31, 32> (Reversibility: ?) [1, 2, 15, 16, 17, 21, 22, 24, 26, 27]
P 2-cis,4-trans-xanthoxin + (3S,5R,6R)-5,6-dihydroxy-6,7-didehydro-5,6-dihydro-12'-apo-β-caroten-12'-al
S 9'-cis-neoxanthin + O_2 <4> (Reversibility: ?) [4]
P 5,6-epoxy-3-hydroxy-12'-apo-β-caroten-12'-al + 3,5-dihydroxy-6,7-didehydro-12'-apo-β-caroten-12'-al + ?
S 9'-cis-neoxanthin + O_2 <1, 14> (<1> drought-induced biosynthesis of abscisic acid is regulated by the 9-cis-epoxy-carotenoid cleavage reaction [13]) (Reversibility: ?) [10, 13]
P 2-cis,4-trans-xanthoxin + (3S,5R,6R)-5,6-dihydroxy-6,7-didehydro-5,6-dihydro-12'-apo-β-caroten-12'al
S 9-cis-8'R-luteoxanthin + O_2 <2> (Reversibility: ?) [1]
P ?
S 9-cis-8'S-luteoxanthin + O_2 <2> (Reversibility: ?) [1]
P ?
S 9-cis-antheraxanthin + O_2 <2> (Reversibility: ?) [1]
P ?

S 9-cis-violaxanthin + O_2 <2, 4, 5, 7, 11, 12, 13, 22, 24, 30, 31, 32> (Reversibility: ?) [1, 15, 16, 17, 21, 22, 24, 26]
P 2-cis,4-trans-xanthoxin + (3S,5R,6S)-5,6-epoxy-3-hydroxy-5,6-dihydro-12'-apo-β-caroten-12'-al
S 9-cis-violaxanthin + O_2 <4> (Reversibility: ?) [4]
P 5,6-epoxy-3-hydroxy-12'-apo-β-caroten-12'-al + 5,6-epoxy-3-hydroxy-9-apo-β-caroten-9-one + ?
S 9-cis-violaxanthin + O_2 <1, 19, 20, 21, 26, 27, 28> (<1> drought-induced biosynthesis of abscisic acid is regulated by the 9-cis-epoxy-carotenoid cleavage reaction [13]; <21,28> CitCCD1 [20]) (Reversibility: ?) [13, 2]
P 2-cis,4-trans-xanthoxin + (3S,5R,6S)-5,6-epoxy-3-hydroxy-5,6-dihydro-12'apo-β-caroten-12'-al
S all-trans-violaxanthin + O_2 <4, 19, 20, 21, 26, 27, 28> (Reversibility: ?) [4, 2]
P 4,9-dimethyldodeca-2,4,6,8,10-pentaene-1,12-dial + ?
S β,β-carotene + O_2 <9> (Reversibility: ?) [18]
P all-trans retinal + ?
S β,β-carotene + O_2 <4> (Reversibility: ?) [4]
P 4,9-dimethyldodeca-2,4,6,8,10-pentaene-1,12-dial + 3-hydroxy-9-apo-β-caroten-9-one + ?
S β-apo-8'-carotenal + O_2 <9> (Reversibility: ?) [18]
P ?
S β-apo-8'-carotenal + O_2 <4> (Reversibility: ?) [19]
P β-ionone + C17-dialdehyde
S β-carotene + O_2 <4> (Reversibility: ?) [19]
P ?
S β-cryptoxanthin + O_2 <19, 20, 21, 26, 27, 28> (Reversibility: ?) [20]
P ?
S diapocarotenedial + O_2 <9> (Reversibility: ?) [18]
P ?
S lutein + O_2 <4> (Reversibility: ?) [4]
P 4,9-dimethyldodeca-2,4,6,8,10-pentaene-1,12-dial + ?
S lycopene + O_2 <9> (Reversibility: ?) [18]
P ?
S torulene + O_2 <9> (Reversibility: ?) [18]
P ?
S zeaxanthin + O_2 <4> (Reversibility: ?) [19]
P ?
S zeaxanthin + O_2 <4, 9, 19, 20, 21, 26, 27, 28> (Reversibility: ?) [4, 18, 20]
P 4,9-dimethyldodeca-2,4,6,8,10-pentaene-1,12-dial + ?
S Additional information <2, 4, 6, 8, 10> (<2> enzyme is involved in abscisic acid pathway [1]; <8> expression is strongly induced by drough stress in the 8-day old plant. The enzyme has a key role in the synthesis of abscisic acid under drough stress [8]; <4> key enzyme in the abscisic acid biosynthetic pathway [11]; <10> key enzyme involved in the biosynthetic pathway of abscisic acid [3]; <6> key regulatory enzyme in abscisic acid biosynthesis in leaves [5]; <2> the enzyme functions in root and leaf in

abscisic acid synthesis [12]; <2> the oxidative cleavage of 9-cis epoxy-carotenoids is the first commited step and key regulatory step in the abscisic acid biosynthesis [6]) (Reversibility: ?) [1, 3, 5, 6, 8, 11, 12]

P ?

Inhibitors

(3,4-dimethoxybenzyl)[2-(4-methoxy-phenyl)-ethyl]methylamine <14> (<14> IC50: 0.087 mM [2]) [2]

2-([[2-(4-hydroxyphenyl)ethyl]amino]methyl)phenol <14> (<14> IC50: 0.16 mM [2]) [2]

N-[(2E)-3-(3,4-dimethoxyphenyl)prop-2-en-1-yl]-4-methoxyaniline <14> (<14> IC50: 0.145 mM [2]) [2]

nordihydroguaiaretic acid <14> (<14> IC50: 0.183 mM [2]) [2, 10]

[[(E)-3-(3,4-dimethoxy-phenyl)-allyl]-(4-fluoro-benzyl)-amino]-acetic acid methyl ester <14> (<14> IC50: 0.055 mM. Inhibitor of abscisic acid biosynthesis inhibits abscisic acid accumulation and stomatal closing [2]) [2]

[[3-(3,4-dimethoxyphenyl)allyl]-(4-fluorobenzyl)amino]acetic acid methyl ester <14> (<14> competitive [10]) [10]

K_m-Value (mM)

0.008 <2> (9-cis-8'R-luteoxanthin) [1]

0.015 <2> (9-cis-8'S-luteoxanthin) [1]

0.02 <2> (9-cis-antheraxanthin) [1]

0.027 <2> (9'-cis-neoxanthin) [1]

0.049 <14> (9'-cis-neoxanthin, <14> pH 7.0, 20°C [10]) [10]

0.058 <2> (9-cis-violaxanthin) [1]

K_i-Value (mM)

0.0388 <14> ([[3-(3,4-dimethoxyphenyl)allyl]-(4-fluorobenzyl)amino]acetic acid methyl ester, <14> pH 7.0, 20°C [10]) [10]

pH-Optimum

7.2 <9> [18]

4 Enzyme Structure

Molecular weight

66860 <22> (<22> calculated from sequence of cDNA [16]) [16]

Posttranslational modification

proteolytic modification <2> (<2> VP14 is imported into chloroplast with cleavage of a short stroma-targeting domain. Mature VP14 exists in two forms, one which is soluble in stroma and the other bound to thylakoid [6]) [6]

5 Isolation/Preparation/Mutation/Application

Source/tissue

bark <12, 13, 30, 31, 32> [17]
callus <12, 13, 30, 31, 32> [17]
embryo <2, 4> [12, 24]
endosperm <4> [24]
flavedo <7> (<7> CsNCED1 and CsNCED2 [21]) [21]
flower bud <25, 29> [27]
fruit <19, 20, 21, 24, 26, 27, 28> [20, 26]
leaf <1, 2, 6, 7, 8, 10, 12, 13, 22, 24, 30, 31, 32> (<6> during drought stress, NCED mRNA increrases. NCED mRNA has a diurnal cycle coincident with the light/dark transition [7]; <1> PvNCED1 mRNA is increased in water stressed leaves, rehydration causes a rapid decrease in PvNCED1 mRNA [13]; <2> strongly induced by water stress [12]; <8> the VuNCED1 transcript is strongly induced in stems and leaves by drought treatment, but less in roots [8]; <7> only CsNCED1 [21]) [3, 7, 8, 12, 13, 16, 17, 21, 26]
ovary <24> (<24> in developing ovary [26]) [26]
petal <23, 25, 29> [23, 27]
plant <5> [15]
root <2, 6, 8, 11, 12, 13, 24, 30, 31, 32> (<6> during drought stress, NCED mRNA increrases [7]; <8> the VuNCED1 transcript is strongly induced in stems and leaves by drought treatment, but less in roots [8]) [7, 8, 12, 17, 22, 26]
seedling <12, 13, 24, 30, 31, 32> [17, 26]
shoot <11> [22]
stem <8, 12, 13, 22, 30, 31, 32> (<8> the VuNCED1 transcript is strongly induced in stems and leaves by drought treatment, but less in roots [8]) [8, 16, 17]
tuber cortex <3> [25]
tuber meristem <3> [25]
tuber periderm <3> [25]

Localization

chloroplast <2> (<2> VP14 is imported into chloroplast with cleavage of a short stroma-targeting domain. Mature VP14 exists in two forms, one which is soluble in stroma and the other bound to thylakoid [6]) [6]
thylakoid <1> [13]

Purification

<1> (recombinant) [13]
<2> [1]
<9> (Talon resin immobilized metal affinity chromatography) [18]

Cloning

<1> (enzyme produced in transgenic Nicotiana plumbaginifolia. Constitutive expression of PvNCED1 results in an increase in abscisic acid and its catabolite, phaseic acid. When the PcNCED1 gene is driven by dexamethasone-in-

ducible promoter, a transient induction of PvNCED1 message and accumulation of) [9]
<1> (expression in Escherichia coli. Import of PvNCED1 into pea chloroplasts) [13]
<4> (expressed in Escherichia coli JM109 cells) [19]
<4> (expression in Escherichia coli as a fusion to glutathione S-transferase) [4]
<5> (expressed in Arabidopsis thaliana) [15]
<6> (creation of transgenic tobacco plants in which expression of the LeNCED1 coding region is under tetracycline-inducible control. Transgenic tomato plants are also produced containing the leNCED1 coding region under the control of one of two strong constitutive promoters, either the doubly enhanced CaMV 35S promoter or the chimaeric/super-promoter) [5]
<8> (expression in Escherichia coli) [8]
<9> (expressed in Escherichia coli strain JM109 and strain BL21) [18]
<23> (expressed in Nicotiana tabacum) [23]

Engineering
H173A <9> (<9> inactive enzyme [18]) [18]
H221A <9> (<9> inactive enzyme [18]) [18]
H285A <9> (<9> inactive enzyme [18]) [18]
H68A <9> (<9> inactive enzyme [18]) [18]

Application
agriculture <1> (<1> it is possible to manipulate abscisic acid levels in plants by overexpressing the key regulatory gene in abscisic acid biosynthesis. Stress tolerance can be improved by increasing abscisic acid level [9]) [9]

References

[1] Schwartz, S.H.; Tan, B.C.; McCarty, D.R.; Welch, W.; Zeevaart, J.A.: Substrate specificity and kinetics for VP14, a carotenoid cleavage dioxygenase in the ABA biosynthetic pathway. Biochim. Biophys. Acta, **1619**, 9-14 (2003)
[2] Han, S.Y.; Kitahata, N.; Saito, T.; Kobayashi, M.; Shinozaki, K.; Yoshida, S.; Asami, T.: A new lead compound for abscisic acid biosynthesis inhibitors targeting 9-cis-epoxycarotenoid dioxygenase. Bioorg. Med. Chem. Lett., **14**, 3033-3036 (2004)
[3] Soar, C.J.; Speirs, J.; Maffei, S.M.; Loveys, B.R.: Gradients in stomatal conductance, xylem sap ABA and bulk leaf ABA along canes of Vitis vinifera cv. Shiraz: Molecular and physiological studies investigating their source. Funct. Plant Biol., **31**, 659-669 (2004)
[4] Schwartz, S.H.; Qin, X.; Zeevaart, J.A.D.: Characterization of a novel carotenoid cleavage dioxygenase from plants. J. Biol. Chem., **276**, 25208-25211 (2001)
[5] Thompson, A.J.; Jackson, A.C.; Symonds, R.C.; Mulholland, B.J.; Dadswell, A.R.; Blake, P.S.; Burbidge, A.; Taylor, I.B.: Ectopic expression of a tomato 9-

cis-epoxycarotenoid dioxygenase gene causes over-production of abscisic acid. Plant J., **23**, 363-374 (2000)
[6] Tan, B.C.; Cline, K.; McCarty, D.R.: Localization and targeting of the VP14 epoxy-carotenoid dioxygenase to chloroplast membranes. Plant J., **27**, 373-382 (2001)
[7] Thompson, A.J.; Jackson, A.C.; Parker, R.A.; Morpeth, D.R.; Burbidge, A.; Taylor, I.B.: Abscisic acid biosynthesis in tomato: regulation of zeaxanthin epoxidase and 9-cis-epoxycarotenoid dioxygenase mRNAs by light/dark cycles, water stress and abscisic acid. Plant Mol. Biol., **42**, 833-845 (2000)
[8] Iuchi, S.; Kobayashi, M.; Yamaguchi-Shinozaki, K.; Shinozaki, K.: A stress-inducible gene for 9-cis-epoxycarotenoid dioxygenase involved in abscisic acid biosynthesis under water stress in drought-tolerant cowpea. Plant Physiol., **123**, 553-562 (2000)
[9] Qin, X.; Zeevaart, J.A.: Overexpression of a 9-cis-epoxycarotenoid dioxygenase gene in Nicotiana plumbaginifolia increases abscisic acid and phaseic acid levels and enhances drought tolerance. Plant Physiol., **128**, 544-551 (2002)
[10] Han, S.Y.; Kitahata, N.; Sekimata, K.; Saito, T.; Kobayashi, M.; Nakashima, K.; Yamaguchi-Shinozaki, K.; Shinozaki, K.; Yoshida, S.; Asami, T.: A novel inhibitor of 9-cis-epoxycarotenoid dioxygenase in abscisic acid biosynthesis in higher plants. Plant Physiol., **135**, 1574-1582 (2004)
[11] Ruggiero, B.; Koiwa, H.; Manabe, Y.; Quist, T.M.; Inan, G.; Saccardo, F.; Joly, R.J.; Hasegawa, P.M.; Bressan, R.A.; Maggio, A.: Uncoupling the effects of abscisic acid on plant growth and water relations. Analysis of sto1/nced3, an abscisic acid-deficient but salt stress-tolerant mutant in Arabidopsis. Plant Physiol., **136**, 3134-3147 (2004)
[12] Tan, B.C.; Schwartz, S.H.; Zeevaart, J.A.; McCarty, D.R.: Genetic control of abscisic acid biosynthesis in maize. Proc. Natl. Acad. Sci. USA, **94**, 12235-12240 (1997)
[13] Qin, X.; Zeevaart, J.A.: The 9-cis-epoxycarotenoid cleavage reaction is the key regulatory step of abscisic acid biosynthesis in water-stressed bean. Proc. Natl. Acad. Sci. USA, **96**, 15354-15361 (1999)
[14] Schwartz, S.H.; Tan, B.C.; Gage, D.A.; Zeevaart, J.A.; McCarty, D.R.: Specific oxidative cleavage of carotenoids by VP14 of maize. Science, **276**, 1872-1874 (1997)
[15] Wan, X.R.; Li, L.: Regulation of ABA level and water-stress tolerance of Arabidopsis by ectopic expression of a peanut 9-cis-epoxycarotenoid dioxygenase gene. Biochem. Biophys. Res. Commun., **347**, 1030-1038 (2006)
[16] Wan, X.; Li, L.: Molecular cloning and characterization of a dehydration-inducible cDNA encoding a putative 9-cis-epoxycarotenoid dioxygenase in Arachis hygogaea L. DNA Seq., **16**, 217-223 (2005)
[17] Guerrini, I.A.; Trigueiro, R.M.; Leite, R.M.; Wilcken, C.F.; Velini, E.D.; Mori, E.S.; Furtado, E.L.; Marino, C.L.; Maia, I.G.: Eucalyptus ESTs involved in the production of 9-cis epoxycarotenoid dioxygenase, a regulatory enzyme of abscisic acid production. Genet. Mol. Biol., **28**, 640-643 (2005)

[18] Marasco, E.K.; Vay, K.; Schmidt-Dannert, C.: Identification of carotenoid cleavage dioxygenases from Nostoc sp. PCC 7120 with different cleavage activities. J. Biol. Chem., **281**, 31583-31593 (2006)
[19] Schmidt, H.; Kurtzer, R.; Eisenreich, W.; Schwab, W.: The carotenase AtCCD1 from Arabidopsis thaliana is a dioxygenase. J. Biol. Chem., **281**, 9845-9851 (2006)
[20] Kato, M.; Matsumoto, H.; Ikoma, Y.; Okuda, H.; Yano, M.: The role of carotenoid cleavage dioxygenases in the regulation of carotenoid profiles during maturation in citrus fruit. J. Exp. Bot., **57**, 2153-2164 (2006)
[21] Rodrigo, M.J.; Alquezar, B.; Zacarias, L.: Cloning and characterization of two 9-cis-epoxycarotenoid dioxygenase genes, differentially regulated during fruit maturation and under stress conditions, from orange (Citrus sinensis L. Osbeck). J. Exp. Bot., **57**, 633-643 (2006)
[22] Kraft, M.; Kuglitsch, R.; Kwiatkowski, J.; Frank, M.; Grossmann, K.: Indole-3-acetic acid and auxin herbicides up-regulate 9-cis-epoxycarotenoid dioxygenase gene expression and abscisic acid accumulation in cleavers (Galium aparine): interaction with ethylene. J. Exp. Bot., **58**, 1497-1503 (2007)
[23] Zhu, C.; Kauder, F.; Roemer, S.; Sandmann, G.: Cloning of two individual cDNAS encoding 9-cis-epoxycarotenoid dioxygenase from Gentiana lutea, their tissue-specific expression and physiological effect in transgenic tobacco. J. Plant Physiol., **164**, 195-204 (2007)
[24] Lefebvre, V.; North, H.; Frey, A.; Sotta, B.; Seo, M.; Okamoto, M.; Nambara, E.; Marion-Poll, A.: Functional analysis of Arabidopsis NCED6 and NCED9 genes indicates that ABA synthesized in the endosperm is involved in the induction of seed dormancy. Plant J., **45**, 309-319 (2006)
[25] Destefano-Beltran, L.; Knauber, D.; Huckle, L.; Suttle, J.C.: Effects of post-harvest storage and dormancy status on ABA content, metabolism, and expression of genes involved in ABA biosynthesis and metabolism in potato tuber tissues. Plant Mol. Biol., **61**, 687-697 (2006)
[26] Agusti, J.; Zapater, M.; Iglesias, D.J.; Cercos, M.; Tadeo, F.R.; Talon, M.: Differential expression of putative 9-cis-epoxycarotenoid dioxygenases and abscisic acid accumulation in water stressed vegetative and reproductive tissues of citrus. Plant Sci., **172**, 85-94 (2006)
[27] Hieber, A.D.; Mudalige-Jayawickrama, R.G.; Kuehnle, A.R.: Color genes in the orchid Oncidium Gower Ramsey: identification, expression, and potential genetic instability in an interspecific cross. Planta, **223**, 521-531 (2006)

Indoleamine 2,3-dioxygenase 1.13.11.52

1 Nomenclature

EC number
1.13.11.52

Systematic name
D-tryptophan:oxygen 2,3-oxidoreductase (decyclizing)

Recommended name
indoleamine 2,3-dioxygenase

Synonyms
IDO <1, 2, 3, 6, 11, 12> [3, 5, 7, 8, 9, 10, 11, 12, 17, 18, 20, 26, 27, 28, 29, 30, 33, 38, 39, 40, 41, 42]
indoleamine-pyrrole 2,3-dioxygenase <11> [38]
hIDO <2> [34, 35]

CAS registry number
9014-51-1

2 Source Organism

<1> *Mus musculus* (no sequence specified) [13, 14, 20, 23, 28, 30]
<2> *Homo sapiens* (no sequence specified) [3, 5, 6, 7, 8, 9, 10, 11, 14, 17, 19, 26, 27, 29, 30, 31, 32, 33, 34, 35, 36, 37, 40, 41, 42]
<3> *Rattus norvegicus* (no sequence specified) [4, 14, 25, 39]
<4> *Sus scrofa* (no sequence specified) [14]
<5> *Bos taurus* (no sequence specified) [14]
<6> *Oryctolagus cuniculus* (no sequence specified) [1, 2, 12, 14, 15, 16, 21, 22, 24, 30, 33]
<7> *Pseudomonas fluorescens* (no sequence specified) [14]
<8> *Capra hircus* (no sequence specified) [14]
<9> *Macaca fuscata fuscata* (no sequence specified) [14]
<10> *Meriones unguiculatus* (no sequence specified) [4]
<11> *Mus musculus* (UNIPROT accession number: P28776) [38]
<12> *Human Immunodeficiency Virus* (no sequence specified) [18]

3 Reaction and Specificity

Catalyzed reaction

D-tryptophan + O_2 = N-formyl-D-kynurenine
L-tryptophan + O_2 = N-formyl-L-kynurenine

Natural substrates and products

S 5-hydroxy-L-tryptophan + O_2 <6> (Reversibility: ?) [12]
P N-formyl-5-hydroxy-L-kynurenine
S D-tryptophan + O_2 <1, 2, 3, 6, 10> (Reversibility: ?) [1, 4, 12, 13, 15, 16, 17, 20, 21, 25, 26, 27, 28, 29]
P N-formyl-D-kynurenine
S L-tryptophan + O_2 <1, 2, 3, 6, 10, 12> (<6> higher affinity at alkaline pH than at acidic pH [1]; <3,10> no activity with oxygen concentrations below 0.1 mM, maximum activity at 1.15 mM oxygen [4]) (Reversibility: ?) [1, 2, 4, 5, 7, 11, 12, 13, 15, 16, 17, 18, 19, 20, 21, 25, 26, 27, 28, 29, 30]
P N-formyl-L-kynurenine
S L-tryptophan + O_2^- <3, 1> (Reversibility: ?) [4]
P N-formyl-L-kynurenine
S serotonin + O_2 <6> (Reversibility: ?) [12]
P ?
S tryptamine + O_2 <6> (Reversibility: ?) [12]
P ?
S Additional information <2> (<2> the enzyme plays an important physiological role in the defense mechanism against a variety of infectious pathogens, in the regulation of T-cell function by macrophages and a subset of dendritic cells, and in the synthesis of UV filters in human lenses. Serious problems arise from the unregulated over-expression of the enzyme, which often results in a deleterious systemic Trp depletion and/or the accumulation of neurotoxin, quinolinic acidn the brain. Enzyme expression in malignant tumors helps them to avoid the immune surveillance through a local Trp depletion. The kynurenilation of the lens protein with UV filters thus appears to be the major cause of age-related cataract [33]) (Reversibility: ?) [33]
P ?

Substrates and products

S 1-benzofuran-DL-tryptophan + O_2 <2> (<2> 1 mM, 22% activity relative to L-tryptophan, 50 mM potassium phosphate, pH 6.5, 10 mM ascorbic acid, 0.01 mM methylene blue, 0.1 mg catalase, 37°C, 10 min [26]) (Reversibility: ?) [26]
P ?
S 1-benzothiophene-DL-tryptophan + O_2 <2> (<2> 1 mM, 19% activity relative to L-tryptophan, 50 mM potassium phosphate, pH 6.5, 10 mM ascorbic acid, 0.01 mM methylene blue, 0.1 mg catalase, 37°C, 10 min [26]) (Reversibility: ?) [26]
P ?

S 1-methyl-DL-tryptophan + O_2 <2> (<2> 1 mM, 7% activity relative to L-tryptophan, 50 mM potassium phosphate, pH 6.5, 10 mM ascorbic acid, 0.01 mM methylene blue, 0.1 mg catalase, 37°C, 10 min [26]) (Reversibility: ?) [26]
P ?
S 2-bromo-L-tryptophan + O_2 <2> (<2> 1 mM, 21% activity relative to L-tryptophan, 50 mM potassium phosphate, pH 6.5, 10 mM ascorbic acid, 0.01 mM methylene blue, 0.1 mg catalase, 37°C, 10 min [26]) (Reversibility: ?) [26]
P ?
S 2-chloro-L-tryptophan + O_2 <2> (<2> 1 mM, 33% activity relative to L-tryptophan, 50 mM potassium phosphate, pH 6.5, 10 mM ascorbic acid, 0.01 mM methylene blue, 0.1 mg catalase, 37°C, 10 min [26]) (Reversibility: ?) [26]
P ?
S 2-hydroxy-L-tryptophan + O_2 <2> (<2> 1 mM, 4% activity relative to L-tryptophan, 50 mM potassium phosphate, pH 6.5, 10 mM ascorbic acid, 0.01 mM methylene blue, 0.1 mg catalase, 37°C, 10 min [26]) (Reversibility: ?) [26]
P ?
S 3-N-aminoethyl-tryptophan + O_2 <2> (<2> 1 mM, 32% activity relative to L-tryptophan, 50 mM potassium phosphate, pH 6.5, 10 mM ascorbic acid, 0.01 mM methylene blue, 0.1 mg catalase, 37°C, 10 min [26]) (Reversibility: ?) [26]
P ?
S 3-aceto-tryptophan + O_2 <2> (<2> 1 mM, 8% activity relative to L-tryptophan, 50 mM potassium phosphate, pH 6.5, 10 mM ascorbic acid, 0.01 mM methylene blue, 0.1 mg catalase, 37°C, 10 min [26]) (Reversibility: ?) [26]
P ?
S 3-indoleethanol + O_2^- <6> (Reversibility: ?) [24]
P ?
S 4-methyl-DL-tryptophan + O_2 <2> (<2> 1 mM, 33% activity relative to L-tryptophan, 50 mM potassium phosphate, pH 6.5, 10 mM ascorbic acid, 0.01 mM methylene blue, 0.1 mg catalase, 37°C, 10 min [26]) (Reversibility: ?) [26]
P ?
S 5-benzyloxy-DL-tryptophan + O_2 <2> (<2> 1 mM, 1% activity relative to L-tryptophan, 50 mM potassium phosphate, pH 6.5, 10 mM ascorbic acid, 0.01 mM methylene blue, 0.1 mg catalase, 37°C, 10 min [26]) (Reversibility: ?) [26]
P ?
S 5-bromo-DL-tryptophan + O_2 <2> (<2> 1 mM, 36% activity relative to L-tryptophan, 50 mM potassium phosphate, pH 6.5, 10 mM ascorbic acid, 0.01 mM methylene blue, 0.1 mg catalase, 37°C, 10 min [26]) (Reversibility: ?) [26]
P ?

S 5-fluoro-DL-tryptophan + O_2 <2> (<2> 1 mM, 46% activity relative to L-tryptophan, 50 mM potassium phosphate, pH 6.5, 10 mM ascorbic acid, 0.01 mM methylene blue, 0.1 mg catalase, 37°C, 10 min [26]) (Reversibility: ?) [26]
P ?
S 5-hydroxy-L-tryptophan + O_2 <2> (Reversibility: ?) [3]
P N-formyl-5-hydroxy-L-kynurenine
S 5-hydroxy-L-tryptophan + O_2 <2, 6> (Reversibility: ?) [8, 12]
P N-formyl-5-hydroxy-L-kynurenine
S 5-hydroxy-L-tryptophan + O_2 <2> (<2> 1 mM, 59% activity relative to L-tryptophan, 50 mM potassium phosphate, pH 6.5, 10 mM ascorbic acid, 0.01 mM methylene blue, 0.1 mg catalase, 37°C, 10 min [26]) (Reversibility: ?) [6, 26]
P ?
S 5-hydroxytryptamine + O_2 <2> (Reversibility: ?) [6]
P ?
S 5-hydroxytryptophan + O_2^- <6> (Reversibility: ?) [22]
P N-formyl-5-hydroxykynurenine
S 5-methoxy-DL-tryptophan + O_2 <2> (<2> 1 mM, 70% activity relative to L-tryptophan, 50 mM potassium phosphate, pH 6.5, 10 mM ascorbic acid, 0.01 mM methylene blue, 0.1 mg catalase, 37°C, 10 min [26]) (Reversibility: ?) [26]
P ?
S 5-methyl-DL-tryptophan + O_2 <2> (<2> 1 mM, 123% activity relative to L-tryptophan, 50 mM potassium phosphate, pH 6.5, 10 mM ascorbic acid, 0.01 mM methylene blue, 0.1 mg catalase, 37°C, 10 min [26]) (Reversibility: ?) [26]
P ?
S 6-fluoro-DL-tryptophan + O_2 <2> (<2> 1 mM, 38% activity relative to L-tryptophan, 50 mM potassium phosphate, pH 6.5, 10 mM ascorbic acid, 0.01 mM methylene blue, 0.1 mg catalase, 37°C, 10 min [26]) (Reversibility: ?) [26]
P ?
S 6-methyl-DL-tryptophan + O_2 <2> (<2> 1 mM, 72% activity relative to L-tryptophan, 50 mM potassium phosphate, pH 6.5, 10 mM ascorbic acid, 0.01 mM methylene blue, 0.1 mg catalase, 37°C, 10 min [26]) (Reversibility: ?) [26]
P ?
S 6-nitro-L-tryptophan + O_2 <2> (<2> 1 mM, 2% activity relative to L-tryptophan, 50 mM potassium phosphate, pH 6.5, 10 mM ascorbic acid, 0.01 mM methylene blue, 0.1 mg catalase, 37°C, 10 min [26]) (Reversibility: ?) [26]
P ?

S 7-methyl-DL-tryptophan + O_2 <2> (<2> 1 mM, 18% activity relative to L-tryptophan, 50 mM potassium phosphate, pH 6.5, 10 mM ascorbic acid, 0.01 mM methylene blue, 0.1 mg catalase, 37°C, 10 min [26]) (Reversibility: ?) [26]

P ?

S D-Trp + O_2 <2, 6> (Reversibility: ?) [33]

P D-formylkynurenine

S D-tryptophan + O_2 <1, 2, 3, 6, 10> (Reversibility: ?) [1, 3, 4, 12, 13, 15, 16, 17, 20, 21, 25, 26, 27, 28, 29]

P N-formyl-D-kynurenine

S D-tryptophan + O_2 <1> (Reversibility: ?) [14]

P N-formyl-D-kynurenine

S D-tryptophan + O_2 <2, 6> (Reversibility: ?) [2, 8]

P N-formyl-D-kynurenine

S D-tryptophan + O_2 <2> (Reversibility: ?) [6]

P D-kynurenine

S D-tryptophan + O_2^- <6> (<6> O_2^- binds first to the ferric enzyme and is followed by rapid binding of L-tryptophan [24]) (Reversibility: ?) [22, 24]

P N-formyl-D-kynurenine

S L-Trp + O_2 <2, 6> (<2> proposed reaction mechanism involves the proton abstraction by iron-bound dixoygen. The O-O bond is precisely controlled by the heme proximal and distal environment and is not cleaved before the incorporation of both oxygen atoms into the substrate [42]) (Reversibility: ?) [33, 41, 42]

P L-formylkynurenine

S L-tryptophan + O_2 <2, 3> (Reversibility: ?) [6, 25]

P L-kynurenine

S L-tryptophan + O_2 <1, 2, 3, 4, 5, 6, 7, 8, 9, 10, 11, 12> (<6> higher affinity at alkaline pH than at acidic pH [1]; <3,10> no activity with oxygen concentrations below 0.1 mM, maximum activity at 1.15 mM oxygen [4]) (Reversibility: ?) [1, 2, 3, 4, 5, 7, 8, 11, 12, 13, 14, 15, 16, 17, 18, 19, 20, 21, 23, 26, 27, 28, 29, 30, 34, 35, 38]

P N-formyl-L-kynurenine

S L-tryptophan + O_2- <3, 6, 10> (Reversibility: ?) [4, 22, 24]

P N-formyl-L-kynurenine

S L-tryptophan ethyl ester + O_2 <2> (<2> 1 mM, 14% activity relative to L-tryptophan, 50 mM potassium phosphate, pH 6.5, 10 mM ascorbic acid, 0.01 mM methylene blue, 0.1 mg catalase, 37°C, 10 min [26]) (Reversibility: ?) [26]

P ?

S L-tryptophan methyl ester + O_2 <2> (<2> 1 mM, 15% activity relative to L-tryptophan, 50 mM potassium phosphate, pH 6.5, 10 mM ascorbic acid, 0.01 mM methylene blue, 0.1 mg catalase, 37°C, 10 min [26]) (Reversibility: ?) [26]

P ?

S N-acetyl-L-tryptophan + O_2 <2> (<2> 1 mM, 3% activity relative to L-tryptophan, 50 mM potassium phosphate, pH 6.5, 10 mM ascorbic acid, 0.01 mM methylene blue, 0.1 mg catalase, 37°C, 10 min [26]) (Reversibility: ?) [26]
P ?
S Trp + O_2 <2> (Reversibility: ?) [36]
P formylkynurenine
S α-N-methyl-L-tryptophan + O_2 <2> (<2> 1 mM, 21% activity relative to L-tryptophan, 50 mM potassium phosphate, pH 6.5, 10 mM ascorbic acid, 0.01 mM methylene blue, 0.1 mg catalase, 37°C, 10 min [26]) (Reversibility: ?) [26]
P ?
S α-methyl-DL-tryptophan + O_2 <2> (<2> 1 mM, 35% activity relative to L-tryptophan, 50 mM potassium phosphate, pH 6.5, 10 mM ascorbic acid, 0.01 mM methylene blue, 0.1 mg catalase, 37°C, 10 min [26]) (Reversibility: ?) [26]
P ?
S α-methyl-DL-tryptophan + O_2^- <6> (Reversibility: ?) [24]
P ?
S β-methyl-DL-tryptophan + O_2 <2> (<2> 1 mM, 32% activity relative to L-tryptophan, 50 mM potassium phosphate, pH 6.5, 10 mM ascorbic acid, 0.01 mM methylene blue, 0.1 mg catalase, 37°C, 10 min [26]) (Reversibility: ?) [26]
P ?
S indole + O_2^- <6> (Reversibility: ?) [24]
P ?
S kynurenine <1> (Reversibility: ?) [23]
P ?
S serotonin + O_2 <2, 6> (Reversibility: ?) [12, 33]
P ?
S serotonin + O_2^- <6> (Reversibility: ?) [22]
P ?
S tryptamine + O_2 <2, 6> (Reversibility: ?) [6, 12, 33]
P ?
S tryptamine + O_2^- <6> (Reversibility: ?) [22]
P ?
S Additional information <2, 6> (<6> not active with indoleacetic acid and skatole [22]; <2> the enzyme plays an important physiological role in the defense mechanism against a variety of infectious pathogens, in the regulation of T-cell function by macrophages and a subset of dendritic cells, and in the synthesis of UV filters in human lenses. Serious problems arise from the unregulated over-expression of the enzyme, which often results in a deleterious systemic Trp depletion and/or the accumulation of neurotoxin, quinolinic acidn the brain. Enzyme expression in malignant tumors helps them to avoid the immune surveillance through a local Trp depletion. The kynurenilation of the lens protein with UV filters thus appears

to be the major cause of age-related cataract [33]) (Reversibility: ?) [22, 33]

P ?

Inhibitors

1-methyltryptophan <1> (<1> treatment of pregnant mice result in rapid T-cell-induced rejection of allogeneic concepti [30]) [30]
1-benzofuran-DL-tryptophan <2> (<2> 1 mM, 43% inhibition relative to L-tryptophan, 50 mM potassium phosphate, pH 6.5, 10 mM ascorbic acid, 0.01 mM methylene blue, 0.1 mg catalase, 37°C, 10 min [26]) [26]
1-benzothiophene-DL-tryptophan <2> (<2> 1 mM, 16% inhibition relative to L-tryptophan, 50 mM potassium phosphate, pH 6.5, 10 mM ascorbic acid, 0.01 mM methylene blue, 0.1 mg catalase, 37°C, 10 min [26]) [26]
1-methyl-DL-tryptophan <1, 2> (<1> competitive inhibitor [20]; <2> 1 mM, 26% inhibition relative to L-tryptophan, 50 mM potassium phosphate, pH 6.5, 10 mM ascorbic acid, 0.01 mM methylene blue, 0.1 mg catalase, 37°C, 10 min [26]; <2> competitive inhibition with L-tryptophan, 2 mM, 70% inhibition [6]) [6, 11, 20, 26]
2-bromo-L-tryptophan <2> (<2> 1 mM, 11% inhibition relative to L-tryptophan, 50 mM potassium phosphate, pH 6.5, 10 mM ascorbic acid, 0.01 mM methylene blue, 0.1 mg catalase, 37°C, 10 min [26]) [26]
2-chloro-L-tryptophan <2> (<2> 1 mM, 20% inhibition relative to L-tryptophan, 50 mM potassium phosphate, pH 6.5, 10 mM ascorbic acid, 0.01 mM methylene blue, 0.1 mg catalase, 37°C, 10 min [26]) [26]
2-hydroxy-L-tryptophan <2> (<2> 1 mM, 30% inhibition relative to L-tryptophan, 50 mM potassium phosphate, pH 6.5, 10 mM ascorbic acid, 0.01 mM methylene blue, 0.1 mg catalase, 37°C, 10 min [26]) [26]
2-hydroxygarveatin E <2> [41]
2-hydroxygarvin A <2> [41]
3-N-aminoethyl-tryptophan <2> (<2> 1 mM, 28% inhibition relative to L-tryptophan, 50 mM potassium phosphate, pH 6.5, 10 mM ascorbic acid, 0.01 mM methylene blue, 0.1 mg catalase, 37°C, 10 min [26]) [26]
3-indoleethanol <6> (<6> lowers K_m value for D-tryptophan by 25% at pH 7, enhances V_{max} by 40-60% [16]) [16]
4-methyl-DL-tryptophan <2> (<2> 1 mM, 26% inhibition relative to L-tryptophan, 50 mM potassium phosphate, pH 6.5, 10 mM ascorbic acid, 0.01 mM methylene blue, 0.1 mg catalase, 37°C, 10 min [26]) [26]
4-phenylimidazole <6> (<6> heme ligand, non competitive, 0.1 M potassium phosphate buffer (pH 5.5-8), 0.1 M Tris-HCl buffer (pH 7.5-8), 25°C [15]) [15]
4-phenylimidazole and cyanide <2> [42]
5-hydroxy-L-tryptophan <1, 2> (<1> 0.1 mM, 40% inhibition of cleavage of D-tryptophan, 53% inhibition of cleavage of L-tryptophan [14]; <2> 1 mM, 12% inhibition relative to L-tryptophan, 50 mM potassium phosphate, pH 6.5, 10 mM ascorbic acid, 0.01 mM methylene blue, 0.1 mg catalase, 37°C, 10 min [26]) [14, 26]

5-benzyloxy-DL-tryptophan <2> (<2> 1 mM, 2% inhibition relative to L-tryptophan, 50 mM potassium phosphate, pH 6.5, 10 mM ascorbic acid, 0.01 mM methylene blue, 0.1 mg catalase, 37°C, 10 min [26]) [26]
5-bromo-DL-tryptophan <2> (<2> 1 mM, 56% inhibition relative to L-tryptophan, 50 mM potassium phosphate, pH 6.5, 10 mM ascorbic acid, 0.01 mM methylene blue, 0.1 mg catalase, 37°C, 10 min [26]) [26]
5-fluoro-DL-tryptophan <2> (<2> 1 mM, 32% inhibition relative to L-tryptophan, 50 mM potassium phosphate, pH 6.5, 10 mM ascorbic acid, 0.01 mM methylene blue, 0.1 mg catalase, 37°C, 10 min [26]) [26]
5-hydroxy-D-tryptophan <1> (<1> 0.1 mM, 26% inhibition of cleavage of D-tryptophan, 29% inhibition of cleavage of L-tryptophan [14]) [14]
5-methoxy-DL-tryptophan <2> (<2> 1 mM, 35% inhibition relative to L-tryptophan, 50 mM potassium phosphate, pH 6.5, 10 mM ascorbic acid, 0.01 mM methylene blue, 0.1 mg catalase, 37°C, 10 min [26]) [26]
5-methyl-DL-tryptophan <2> (<2> 1 mM, 6% inhibition relative to L-tryptophan, 50 mM potassium phosphate, pH 6.5, 10 mM ascorbic acid, 0.01 mM methylene blue, 0.1 mg catalase, 37°C, 10 min [26]) [26]
6-fluoro-DL-tryptophan <2> (<2> 1 mM, 54% inhibition relative to L-tryptophan, 50 mM potassium phosphate, pH 6.5, 10 mM ascorbic acid, 0.01 mM methylene blue, 0.1 mg catalase, 37°C, 10 min [26]) [26]
6-methyl-DL-tryptophan <2> (<2> 1 mM, 20% inhibition relative to L-tryptophan, 50 mM potassium phosphate, pH 6.5, 10 mM ascorbic acid, 0.01 mM methylene blue, 0.1 mg catalase, 37°C, 10 min [26]) [26]
6-nitro-D-tryptophan <2> (<2> 1 mM, 7% inhibition relative to L-tryptophan, 50 mM potassium phosphate, pH 6.5, 10 mM ascorbic acid, 0.01 mM methylene blue, 0.1 mg catalase, 37°C, 10 min [26]) [26]
6-nitro-L-tryptophan <2> (<2> 1 mM, competitive inhibition, 52% inhibition relative to L-tryptophan, 50 mM potassium phosphate, pH 6.5, 10 mM ascorbic acid, 0.01 mM methylene blue, 0.1 mg catalase, 37°C, 10 min [26]) [26]
7-methyl-DL-tryptophan <2> (<2> 1 mM, 36% inhibition relative to L-tryptophan, 50 mM potassium phosphate, pH 6.5, 10 mM ascorbic acid, 0.01 mM methylene blue, 0.1 mg catalase, 37°C, 10 min [26]) [26]
azide <6> [16]
CO <6> (<6> negatively cooperative [1]; <6> reversed by illumination [21]) [1, 16, 21]
cyanide <6> (<6> heme ligand, competitive for the ferric enzyme [15,16]) [15, 16]
dexamethasone <1, 6> (<1> inhibits IDO activity in brain and lung after systemic pokeweed mitogen administration and brains of mice suffering from malaria [30]) [30]
H_2O_2 <1, 2> [30]
indole <6> (<6> lowers K_m value for D-tryptophan by 60% at pH 7, lowers V_{max} by 12-24%, at 1 mM lowers activity by 13% [16]) [16]
indole-3-acetic acid <1> (<1> 1 mM, 19% inhibition of cleavage of D-tryptophan, 27% inhibition of cleavage of L-tryptophan [14]) [14]
indole-3-propionic acid <2> [26]

iodoacetic acid <1> (<1> 1 mM, 12% inhibition of cleavage of D-tryptophan, 7% inhibition of cleavage of L-tryptophan [14]) [14]
KCN <1, 6> (<1> 0.01 mM, 39% inhibition of cleavage of D-tryptophan, 48% inhibition of cleavage of L-tryptophan [14]) [14, 21]
L-tryptophan <1, 6> (<6> substrate inhibition [21]; <1> substrate inhibition at concentrations above 0.04 mM [13]) [13, 21]
L-tryptophan ethyl ester <2> (<2> 1 mM, 7% inhibition relative to L-tryptophan, 50 mM potassium phosphate, pH 6.5, 10 mM ascorbic acid, 0.01 mM methylene blue, 0.1 mg catalase, 37°C, 10 min [26]) [26]
L-tryptophan methyl ester <2> (<2> 1 mM, 30% inhibition relative to L-tryptophan, 50 mM potassium phosphate, pH 6.5, 10 mM ascorbic acid, 0.01 mM methylene blue, 0.1 mg catalase, 37°C, 10 min [26]) [26]
L-tryptphan <2> (<2> substrate inhibition at 0.14 mM and greater [26]) [26]
N-acetyl-L-tryptophan <2> (<2> 1 mM, 7% inhibition relative to L-tryptophan, 50 mM potassium phosphate, pH 6.5, 10 mM ascorbic acid, 0.01 mM methylene blue, 0.1 mg catalase, 37°C, 10 min [26]) [26]
N-ethylmaleimide <1> (<1> 1 mM, 58% inhibition of cleavage of D-tryptophan, 45% inhibition of cleavage of L-tryptophan [14]) [14]
NO <1, 2, 6> (<1> inhibits the transcription of IDO in murine macrophages [30]; <2> can inhibit hIDO by directly binding to the heme iron [35]) [16, 20, 30, 35]
NSC 401366 <2> (<2> competitive with respect to tryptophan [36]) [36]
NaN_3 <1> (<1> 10 mM, 33% inhibition of cleavage of D-tryptophan, 32% inhibition of cleavage of L-tryptophan [14]) [14]
O_2 <6> [16]
serotonin <1> (<1> 1 mM, 36% inhibition of cleavage of D-tryptophan, 37% inhibition of cleavage of L-tryptophan [14]) [14]
sodium azide <6> (<6> 1 mM causes 20% inhibition [21]) [21]
superoxide dismutase <6> (<6> 80-98% inhibition [2]; <6> can be inhibited by diethyldithiocarbamate [12]) [2, 12]
tiron <1, 6> (<6> 1 mM causes 40% inhibition [21]; <1> 10 mM, 2% inhibition of cleavage of D-tryptophan, 6% inhibition of cleavage of L-tryptophan [14]) [14, 21]
tryptamine <1, 2> (<1> 0.1 mM, 16% inhibition of cleavage of D-tryptophan, 11% inhibition of cleavage of L-tryptophan [14]) [14, 26]
α-N-methyl-L-tryptophan <2> (<2> 1 mM, 33% inhibition relative to L-tryptophan, 50 mM potassium phosphate, pH 6.5, 10 mM ascorbic acid, 0.01 mM methylene blue, 0.1 mg catalase, 37°C, 10 min [26]) [26]
α-methyl-DL-tryptophan <2> (<2> 1 mM, 1% inhibition relative to L-tryptophan, 50 mM potassium phosphate, pH 6.5, 10 mM ascorbic acid, 0.01 mM methylene blue, 0.1 mg catalase, 37°C, 10 min [26]) [26]
annulin A <2> [41]
annulin B <2> [41]
β-methyl-DL-tryptophan <2> (<2> 1 mM, 7% inhibition relative to L-tryptophan, 50 mM potassium phosphate, pH 6.5, 10 mM ascorbic acid, 0.01 mM methylene blue, 0.1 mg catalase, 37°C, 10 min [26]) [26]

exiguamine A <2> (<2> inihibor isolated from the marine sponge Neopetrosia exigua [40]) [40]
garveatin C <2> [41]
garveatin E <2> [41]
garveatinA <2> [41]
hydroxylamine sulfate <6> (<6> 0.1 mM causes 79% inhibition, 1 mM causes 94% inhibition [21]) [21]
imidazole <6> (<6> negatively cooperative or competitive [1]) [1]
indole-3-acrylic acid <1> (<1> competitive inhibitor [14]) [14]
interleukin-4 <2> [30]
methyl-thiohydantoin-tryptophan <2> (<2> limited selectivity towards indoleamine 2,3-dioxygenase [36]) [36]
n-butyl isocyanide <6> (<6> heme ligand which binds tightly to the ferrous enzyme [2]) [2]
norharman <2, 6> (<6> can bind to the ferric and ferrous enzyme to from a quarternary complex [16]; <6> non competitive, competitive for the ferric enzyme, 0.1 M potassium phosphate buffer (pH 5.5-8), 0.1 M Tris-HCl buffer (pH 7.5-8), 25°C [15]; <2> uncompetitive inhibition, 2 mM, 98% inhibition [6]) [6, 15, 16]
transforming growth factor-β <2> (<2> inhibits expression in skin and synovial fibroblasts [30]) [30]
tyron <6> [12]
Additional information <1, 2, 6> (<6> 3-indoleethanol and indole affect the L-tryptophan affinity of the ferrous enzyme, indole enhances affinity of enzyme for cyanide and azide and norharman, L-tryptophan lowers norharman affinity for the ferrous dioxygenase [16]; <2> not inhibited by 6-nitro-D-tryptophan and indole-nitrogen substituted analogues of trytophan [26]; <6> not inhibited by catalase and D-tryptophan [21]; <2> not inhibited by indole-3-acetamide, β-indoylacetonitrile snd 3-β-indoleacrylic acid [6]; <1> not inhibited by superoxide dismutase [14]) [6, 14, 16, 21, 26]

Cofactors/prosthetic groups

$FMNH_2$ <1> (<1> 4-5-fold higher activity than acitivity with methylene blue as the electron donor [13]) [13]
heme <2, 6> (<2> none of the polar amino acid residues in the distal heme pocket are essential for activity. The O-O bond is precisely controlled by the heme proximal and distal environment and is not cleaved before the incorporation of both oxygen atoms into the substrate [42]) [33,34,35,42]
methylene blue <6> (<6> stimulates the formation of D-kynurenine [21]) [21]
tetrahydrobiopterin <1> [13]
Additional information <1, 2, 6> (<6> ascorbic acid in combination with methylene blue or toluidine blue [16]; <6> ascorbic acid, can be replaced by xanthine oxidase with hypoxanthine [21]; <6> not affected by hydrogen peroxide [21]; <1,2,6> O_2^- [30]; <6> superoxide anion in combination with methylene blue or toluidine blue [16]; <6> toluidine blue, can replace methylene blue [21]) [16,21,30]

Activating compounds

3-indoleethanol <6> (<6> 1 mM, enhances activity by 87% [16]) [16]
cortisone <3> (<3> subcutaneously [25]) [25]
D-tryptophan <3> (<3> 2g subcutaneously [25]) [25]
diethyldithiocarbamate <6> (<6> inhibits superoxide dismutase [12]) [12]
inosine <6> (<6> 0.1 mM, increases activity 5fold [12]) [12]
L-tryptophan <3> (<3> 2g subcutaneously [25]) [25]
O_2^- <1, 3, 6, 10> [2, 4, 20]
α-interferon <2> (<2> induces de novo synthesis of IDO, but 2 to 3 orders of magnitude less than γ-IFN [29]; <2> induces IDO expression in monocytes [30]) [29, 30]
bacterial lipopolysaccharide <1> [23]
β-interferon <2> (<2> induces IDO expression in monocytes [30]) [30]
γ-interferon <1, 2, 6, 12> (<2> activates expression of IDO [11,19]; <2> induces de novo synthesis of IDO, 30fold increase in IDO activity [29]; <1,2,6> induces IDO expression [28,30]; <2> induces IDO expression in a dose-dependent manner [17]) [8, 11, 17, 18, 19, 20, 28, 29, 30]
hematine <3> [25]
hyperbaric oxygen <3, 10> [4]
hyperoxia <6> (<6> induces IDO expression in the lung [30]) [30]
lipopolysaccharide <2, 6> (<2> induces IDO expression in monocytes [30]; <6> induces IDO expression in the lung [30]) [30]
methylene blue <6> (<6> acts as electron mediator from donors to the ferric dioxygenase [2]) [2]
paraquat <6> (<6> induces IDO expression in the lung [30]) [30]
tumor-necrosis-factor-α <2> (<2> costimulation with γ-interferon [19]) [19]
Additional information <6> (<6> IDO can bind following substances as ligand in effector binding site: alkyl isocyanides, CN^-, F^-, N_3^-, imidazoles, pyridines, formate, benzhydroxamate, phosphines [1]; <6> not induced by either tryptophan or glucocorticoid [12]; <6> requires molecular oxygen and superoxide anion [12]) [1, 12]

Metals, ions

CN^- <6> (<6> enhances the affinity for L-tryptophan for the ferric enzyme in reziprocal manner, positive cooperativity [1]) [1]
F^- <6> (<6> enhances the affinity for L-tryptophan for the ferric enzyme in reziprocal manner, positive cooperativity [1]) [1]
N_3^- <6> (<6> enhances the affinity for L-tryptophan for the ferric enzyme in reziprocal manner, positive cooperativity [1]) [1]

Turnover number (min^{-1})

0.0015 <1> (kynurenine, <1> LPS-treated mice [23]) [23]
0.002 <1> (kynurenine) [23]

Specific activity (U/mg)

0.000011 <2> (<2> pulmonary IDO activity in pneumothorax patients [29]) [29]

0.000014 <2> (<2> pulmonary IDO activity in benign tumor patient [29]) [29]
0.000015 <2> (<2> pulmonary IDO activity in tuberculosis patient [29]) [29]
0.000039 <2> (<2> pulmonary IDO activity in bronchiectasis patient [29]) [29]
0.00005 <2> (<2> pulmonary IDO activity in lung cancer patients [29]) [29]
0.000067 <2> (<2> pulmonary IDO activity in pulmonary abscess patient [29]) [29]
0.000192 <3> (<3> L-tryptophan, Tris-buffer, pH 8.0, 0.0001 mM methylene blue and 0.001 mM each of ATP and Mg^{2+}, 37°C, 90 min, no rat-pretreatment [25]) [25]
0.000208 <3> (<3> D-tryptophan, Tris-buffer, pH 8.0, 0.0001 mM methylene blue and 0.001 mM each of ATP and Mg^{2+}, 37°C, 90 min, no rat-pretreatment [25]) [25]
0.000224 <3> (<3> D-tryptophan, Tris-buffer, pH 8.0, 0.0001 mM methylene blue and 0.001 mM each of ATP and Mg^{2+}, 37°C, 90 min, rat-pretreatment with neomycin [25]) [25]
0.000245 <3> (<3> L-tryptophan, Tris-buffer, pH 8.0, 0.0001 mM methylene blue and 0.001 mM each of ATP and Mg^{2+}, 37°C, 90 min, rat-pretreatment with cortisone [25]) [25]
0.000256 <3> (<3> D-tryptophan, Tris-buffer, pH 8.0, 0.0001 mM methylene blue and 0.001 mM each of ATP and Mg^{2+}, 37°C, 90 min, rat-pretreatment with cortisone [25]) [25]
0.000272 <3> (<3> L-tryptophan, Tris-buffer, pH 8.0, 0.0001 mM methylene blue and 0.001 mM each of ATP and Mg^{2+}, 37°C, 90 min, rat-pretreatment with D-tryptophan [25]) [25]
0.000277 <3> (<3> D-tryptophan, Tris-buffer, pH 8.0, 0.0001 mM methylene blue and 0.001 mM each of ATP and Mg^{2+}, 37°C, 90 min, rat-pretreatment with D-tryptophan [25]) [25]
0.00033 <2> (<2> crude extract of human lung slices treated with γ-IFN [29]) [29]
0.000598 <2> (<2> supernatant, 0.8 mM L-tryptophan, 40 mM ascorbic acid, 0.02 mM methylene blue, 200 units/ml catalase, 100 mM potassium phosphate buffer, pH 6.5, 37°C, 30 min [6]) [6]
0.01 <6> (<6> ascending colon, 0.05 mM potassium phosphate buffer (pH 7.5), 0.005 mM methylene blue, 0.01 mM ascorbic acid, 0.003 mM tryptophan, 37°C [21]; <6> caecum, 0.05 mM potassium phosphate buffer (pH 7.5), 0.005 mM methylene blue, 0.01 mM ascorbic acid, 0.003 mM tryptophan, 37°C [21]) [21]
0.011 <10> (<10> brain, in air, 50 mM potassium phosphate buffer, pH 6.5, 0.025 mM methylene blue, 20 mM ascorbate, 0.05 mM catalase, 0.4 mM L-tryptophan, 37°C [4]) [4]
0.014 <10> (<10> brain, in 1.2 atm oxygen, 50 mM potassium phosphate buffer, pH 6.5, 0.025 mM methylene blue, 20 mM ascorbate, 0.05 mM catalase, 0.4 mM L-tryptophan, 37°C [4]) [4]
0.022 <2> (<2> sample 14 of 22 cervical mucus sample, determined by HPLC/UV detection, reaction mixture include 0.025 mM methylene blue,

20 mM ascorbic acid, 0.04 mM L-tryptophan solution, 0.05 ml catalase and 50 mM potassium phosphate buffer (pH 6.5), 30 min, 37°C [27]) [27]

0.024 <2> (<2> sample 17 of 22 cervical mucus sample, determined by HPLC/UV detection, reaction mixture includes 0.025 mM methylene blue, 20 mM ascorbic acid, 0.04 mM L-tryptophan solution, 0.05 ml catalase and 50 mM potassium phosphate buffer (pH 6.5), 30 min, 37°C [27]) [27]

0.04 <6> (<6> small intestine IV, 0.05 mM potassium phosphate buffer (pH 7.5), 0.005 mM methylene blue, 0.01 mM ascorbic acid, 0.003 mM tryptophan, 37°C [21]) [21]

0.05 <6> (<6> small intestine proximal portion, 0.05 mM potassium phosphate buffer (pH 7.5), 0.005 mM methylene blue, 0.01 mM ascorbic acid, 0.003 mM tryptophan, 37°C [21]) [21]

0.07 <1, 6, 10> (<1> esophagus [23]; <10> brain, in 5.2 atm oxygen, 50 mM potassium phosphate buffer, pH 6.5, 0.025 mM methylene blue, 20 mM ascorbate, 0.05 mM catalase, 0.4 mM L-tryptophan, 37°C [4]; <6> small intestine V, 0.05 mM potassium phosphate buffer (pH 7.5), 0.005 mM methylene blue, 0.01 mM ascorbic acid, 0.003 mM tryptophan, 37°C [21]) [4, 21, 23]

0.072 <2> (<2> assuming an average wet weight of the lens of 0.2 g, assay in 50 mM potassium phosphate buffer (pH 6.5), 20 mM ascorbic acid, 0.01 mM methylene blue, 0.1 mg/ml catalase, 0.4 mM L-tryptophan, 90 min, 37°C [5]) [5]

0.08 <6> (<6> appendix, 0.05 mM potassium phosphate buffer (pH 7.5), 0.005 mM methylene blue, 0.01 mM ascorbic acid, 0.003 mM tryptophan, 37°C [21]) [21]

0.122 <1> (<1> submandibular gland [23]) [23]

0.13 <1> (<1> testis [23]) [23]

0.145 <3> (<3> small intestine, in 5.2 atm oxygen, 50 mM potassium phosphate buffer, pH 6.5, 0.025 mM methylene blue, 20 mM ascorbate, 0.05 mM catalase, 0.4 mM L-tryptophan, 37°C [4]) [4]

0.15 <6> (<6> appendix, 0.05 mM potassium phosphate buffer (pH 7.5), 0.005 mM methylene blue, 0.01 mM ascorbic acid, 0.003 mM tryptophan, 37°C [21]) [21]

0.155 <1> (<1> kidney [23]) [23]

0.165 <1> (<1> brain [23]) [23]

0.183 <1> (<1> heart [23]) [23]

0.198 <10> (<10> lung, in air, 50 mM potassium phosphate buffer, pH 6.5, 0.025 mM methylene blue, 20 mM ascorbate, 0.05 mM catalase, 0.4 mM L-tryptophan, 37°C [4]) [4]

0.232 <1> (<1> brain, lipopolysaccharide-treated mice [23]) [23]

0.24 <1, 3> (<1> kidney, lipopolysaccharide-treated mice [23]; <3> small intestine, in 1.2 atm oxygen, 50 mM potassium phosphate buffer, pH 6.5, 0.025 mM methylene blue, 20 mM ascorbate, 0.05 mM catalase, 0.4 mM L-tryptophan, 37°C [4]) [4, 23]

0.278 <1> (<1> esophagus, lipopolysaccharide-treated mice [23]) [23]

0.342 <1> (<1> submandibular gland, lipopolysaccharide-treated mice [23]) [23]

0.375 <1> (<1> stomach [23]) [23]

0.38 <3> (<3> small intestine, in air, 50 mM potassium phosphate buffer, pH 6.5, 0.025 mM methylene blue, 20 mM ascorbate, 0.05 mM catalase, 0.4 mM L-tryptophan, 37°C [4]) [4]
0.425 <1> (<1> thymus [23]) [23]
0.45 <1> (<1> lung [23]) [23]
0.468 <1> (<1> cecum [23]) [23]
0.474 <2> (<2> sample 11 of 22 cervical mucus sample, determined by HPLC/UV detection, reaction mixture includes 0.025 mM methylene blue, 20 mM ascorbic acid, 0.04 mM L-tryptophan solution, 0.05 ml catalase and 50 mM potassium phosphate buffer (pH 6.5), 30 min, 37°C [27]) [27]
0.479 <2> (<2> sample 19 of 22 cervical mucus sample, determined by HPLC/UV detection, reaction mixture includes 0.025 mM methylene blue, 20 mM ascorbic acid, 0.04 mM L-tryptophan solution, 0.05 ml catalase and 50 mM potassium phosphate buffer (pH 6.5), 30 min, 37°C [27]) [27]
0.55 <1> (<1> 25°C, 50 mM HEPES/NaOH (pH 7.3), 5 mM D-tryptophan 0.01 mM methylene blue, 5 mM ascorbic acid, 0.001 mM catalase [14]) [14]
0.56 <1> (<1> testis, lipopolysaccharide-treated mice [23]) [23]
0.57 <1> (<1> heart, lipopolysaccharide-treated mice [23]) [23]
0.636 <2> (<2> sample 15 of 22 cervical mucus sample, determined by HPLC/UV detection, reaction mixture includes 0.025 mM methylene blue, 20 mM ascorbic acid, 0.04 mM L-tryptophan solution, 0.05 ml catalase and 50 mM potassium phosphate buffer (pH 6.5), 30 min, 37°C [27]) [27]
0.689 <2> (<2> sample 21 of 22 cervical mucus sample, determined by HPLC/UV detection, reaction mixture includes 0.025 mM methylene blue, 20 mM ascorbic acid, 0.04 mM L-tryptophan solution, 0.05 ml catalase and 50 mM potassium phosphate buffer (pH 6.5), 30 min, 37°C [27]) [27]
0.72 <1> (<1> seminal vesicle [23]) [23]
0.735 <1> (<1> trachea, lipopolysaccharide-treated mice [23]) [23]
0.755 <1> (<1> spleen, lipopolysaccharide-treated mice [23]) [23]
0.8 <1> (<1> thymus, lipopolysaccharide-treated mice [23]) [23]
0.873 <10> (<10> lung, in 5.2 atm oxygen, 50 mM potassium phosphate buffer, pH 6.5, 0.025 mM methylene blue, 20 mM ascorbate, 0.05 mM catalase, 0.4 mM L-tryptophan, 37°C [4]) [4]
0.903 <1> (<1> colon [23]) [23]
0.905 <10> (<10> lung, in 1.2 atm oxygen, 50 mM potassium phosphate buffer, pH 6.5, 0.025 mM methylene blue, 20 mM ascorbate, 0.05 mM catalase, 0.4 mM L-tryptophan, 37°C [4]) [4]
0.913 <2> (<2> sample 13 of 22 cervical mucus sample, determined by HPLC/UV detection, reaction mixture includes 0.025 mM methylene blue, 20 mM ascorbic acid, 0.04 mM L-tryptophan solution, 0.05 ml catalase and 50 mM potassium phosphate buffer (pH 6.5), 30 min, 37°C [27]) [27]
0.92 <1> (<1> spleen [23]) [23]
0.968 <1> (<1> trachea [23]) [23]
1.39 <1> (<1> urinary bladder, lipopolysaccharide-treated mice [23]) [23]
1.455 <1> (<1> urinary bladder [23]) [23]
1.902 <1> (<1> stomach, lipopolysaccharide-treated mice [23]) [23]
2.017 <1> (<1> intestine [23]) [23]

2.13 <1> (<1> 25°C, 50 mM HEPES/NaOH (pH 7.3), 2 mM L-tryptophan 0.01 mM methylene blue, 5 mM ascorbic acid, 0.001 mM catalase [14]) [14]
2.48 <2> (<2> kynurenine, 50 mM potassium phosphate buffer (pH 6.5), 20 mM ascorbic acid, 0.2 mg/ml catalase, 0.01 mM methylene blue, 0.4 mM tryptophan, 37°C, 10-60 min [3]) [3]
2.67 <2> (<2> kynurenine, 50 mM potassium phosphate buffer (pH 6.5), 20 mM ascorbic acid, 0.2 mg/ml catalase, 0.01 mM methylene blue, 0.4 mM tryptophan, 37°C, 60 min [10]) [10]
4.07 <1> (<1> intestine, lipopolysaccharide-treated mice [23]) [23]
5.19 <6> (<6> small intestine VI, 0.05 mM potassium phosphate buffer (pH 7.5), 0.005 mM methylene blue, 0.01 mM ascorbic acid, 0.003 mM tryptophan, 37°C [21]) [21]
5.34 <1> (<1> lung, lipopolysaccharide-treated mice [23]) [23]
6.97 <6> (<6> small intestine distal portion, 0.05 mM potassium phosphate buffer (pH 7.5), 0.005 mM methylene blue, 0.01 mM ascorbic acid, 0.003 mM tryptophan, 37°C [21]) [21]
8.33 <1> (<1> seminal vesicle, lipopolysaccharide-treated mice [23]) [23]
15.31 <1> (<1> cecum, lipopolysaccharide-treated mice [23]) [23]
26.68 <1> (<1> colon, lipopolysaccharide-treated mice [23]) [23]
32.22 <1> (<1> epididymis [23]) [23]
104.9 <1> (<1> epididymis, lipopolysaccharide-treated mice [23]) [23]
234 <6> (<6> enzyme after purification, 0.05 mM potassium phosphate buffer (pH 7.5), 0.005 mM methylene blue, 0.01 mM ascorbic acid, 0.003 mM tryptophan, 37°C [21]) [21]
Additional information <2> (<2> 0.0142 nmol/mg/lens, assay in 50 mM potassium phosphate buffer (pH 6.5), 20 mM ascorbic acid, 0.01 mM methylene blue, 0.1 mg/ml catalase, 0.4 mM L-tryptophan, 90 min, 37°C [5]) [5]

K_m-Value (mM)

0.0019 <2> (D-tryptophan, <2> H303A mutant, 50 mM potassium phosphate buffer (pH 6.5), 20 mM ascorbic acid, 0.3 mg/ml catalase, 0.01 mM methylene blue, 0.4 mM tryptophan, 37°C, 10-60 min [8]) [8]
0.0024 <2> (D-tryptophan, <2> H16A, 50 mM potassium phosphate buffer (pH 6.5), 20 mM ascorbic acid, 0.3 mg/ml catalase, 0.01 mM methylene blue, 0.4 mM tryptophan, 37°C, 10-60 min [8]) [8]
0.0044 <2> (D-tryptophan, <2> K352A mutant, 50 mM potassium phosphate buffer (pH 6.5), 20 mM ascorbic acid, 0.3 mg/ml catalase, 0.01 mM methylene blue, 0.4 mM tryptophan, 37°C, 10-60 min [8]) [8]
0.005 <2> (D-tryptophan, <2> wild type, 50 mM potassium phosphate buffer (pH 6.5), 20 mM ascorbic acid, 0.3 mg/ml catalase, 0.01 mM methylene blue, 0.4 mM tryptophan, 37°C, 10-60 min [8]) [8]
0.005 <6> (L-tryptophan, <6> pH 6.0 [16]) [16]
0.0053 <2> (D-tryptophan, <2> V109A mutant, 50 mM potassium phosphate buffer (pH 6.5), 20 mM ascorbic acid, 0.3 mg/ml catalase, 0.01 mM methylene blue, 0.4 mM tryptophan, 37°C, 10-60 min [8]) [8]

0.0056 <2> (L-tryptophan, <2> H16A mutant, 50 mM potassium phosphate buffer (pH 6.5), 20 mM ascorbic acid, 0.3 mg/ml catalase, 0.01 mM methylene blue, 0.4 mM tryptophan, 37°C, 10-60 min [8]) [8]
0.0065 <6> (L-tryptophan, <6> pH 8.0 [1]) [1]
0.0089 <2> (L-tryptophan, <2> V109A mutant, 50 mM potassium phosphate buffer (pH 6.5), 20 mM ascorbic acid, 0.3 mg/ml catalase, 0.01 mM methylene blue, 0.4 mM tryptophan, 37°C, 10-60 min [8]) [8]
0.009 <6> (L-tryptophan, <6> pH 7.5 [1]) [1]
0.0094 <2> (L-tryptophan, <2> H303A mutant, 50 mM potassium phosphate buffer (pH 6.5), 20 mM ascorbic acid, 0.3 mg/ml catalase, 0.01 mM methylene blue, 0.4 mM tryptophan, 37°C, 10-60 min [8]) [8]
0.01 <1> (L-tryptophan, <1> pH 7.5, 37 °C, 5 min incubation time [13]) [13]
0.013 <6> (L-tryptophan, <6> pH 7.0 [1]) [1]
0.018 <2> (L-tryptophan, <2> 50 mM potassium phosphate, pH 6.5, 10 mM ascorbic acid, 0.01 mM methylene blue, 0.1 mg catalase, 37°C, 10 min [26]) [26]
0.02 <2> (L-tryptophan, <2> wild type, 50 mM potassium phosphate buffer (pH 6.5), 20 mM ascorbic acid, 0.3 mg/ml catalase, 0.01 mM methylene blue, 0.4 mM tryptophan, 37°C, 10-60 min [8]) [3, 8]
0.0201 <2> (L-tryptophan, <2> K352A mutant, 50 mM potassium phosphate buffer (pH 6.5), 20 mM ascorbic acid, 0.3 mg/ml catalase, 0.01 mM methylene blue, 0.4 mM tryptophan, 37°C, 10-60 min [8]) [8]
0.0246 <2> (L-tryptophan) [6]
0.025 <6> (L-tryptophan, <6> pH 6.5 [1]) [1]
0.05 <6> (L-tryptophan, <6> pH 6.0 [1]) [1]
0.056 <2> (6-methyl-DL-tryptophan, <2> 50 mM potassium phosphate, pH 6.5, 10 mM ascorbic acid, 0.01 mM methylene blue, 0.1 mg catalase, 37°C, 10 min [26]) [26]
0.062 <1> (D-tryptophan, <1> pH 7.5, 37 °C, 40 min incubation time [13]) [13]
0.088 <2> (5-methyl-DL-tryptophan, <2> 50 mM potassium phosphate, pH 6.5, 10 mM ascorbic acid, 0.01 mM methylene blue, 0.1 mg catalase, 37°C, 10 min [26]) [26]
0.1 <2> (5-hydroxy-L-tryptophan, <2> H303A mutant, 50 mM potassium phosphate buffer (pH 6.5), 20 mM ascorbic acid, 0.3 mg/ml catalase, 0.01 mM methylene blue, 0.4 mM tryptophan, 37°C, 10-60 min [8]; <2> V109A mutant, 50 mM potassium phosphate buffer (pH 6.5), 20 mM ascorbic acid, 0.3 mg/ml catalase, 0.01 mM methylene blue, 0.4 mM tryptophan, 37°C, 10-60 min [8]) [8]
0.113 <2> (5-methoxy-DL-tryptophan, <2> 50 mM potassium phosphate, pH 6.5, 10 mM ascorbic acid, 0.01 mM methylene blue, 0.1 mg catalase, 37°C, 10 min [26]) [26]
0.14 <2> (5-hydroxy-L-tryptophan, <2> H16A, 50 mM potassium phosphate buffer (pH 6.5), 20 mM ascorbic acid, 0.3 mg/ml catalase, 0.01 mM methylene blue, 0.4 mM tryptophan, 37°C, 10-60 min [8]) [8]

0.17 <2> (5-hydroxy-L-tryptophan, <2> K352A mutant, 50 mM potassium phosphate buffer (pH 6.5), 20 mM ascorbic acid, 0.3 mg/ml catalase, 0.01 mM methylene blue, 0.4 mM tryptophan, 37°C, 10-60 min [8]) [8]
0.21 <1> (L-tryptophan) [14]
0.25 <6> (L-tryptophan, <6> pH 5.5 [1]) [1]
0.4 <2> (5-hydroxy-L-tryptophan, <2> wild type, 50 mM potassium phosphate buffer (pH 6.5), 20 mM ascorbic acid, 0.3 mg/ml catalase, 0.01 mM methylene blue, 0.4 mM tryptophan, 37°C, 10-60 min [8]) [8]
0.44 <2> (5-hydroxy-L-tryptophan) [3]
0.66 <6> (D-tryptophan, <6> pH 7.0, 1 mM indole [16]) [16]
0.83 <6> (D-tryptophan, <6> 0.1 M potassium phosphate buffer (pH 6-8), 0.025 mM methylene blue, 0.2 mg catalase, 10 mM ascorbic acid, 50 nM dioxygenase, 25°C [15]) [15]
0.96 <1> (D-tryptophan) [14]
1.3 <6> (D-tryptophan, <6> pH 7.0, 1 mM 3-indoleethanol [16]) [16]
1.7 <6> (D-tryptophan, <6> pH 7.0 [16]) [16]
2.8 <3, 10> (oxygen, <3,10> pH 6.5, 37°C [4]) [4]
3.795 <2> (D-tryptophan, <2> 50 mM potassium phosphate, pH 6.5, 10 mM ascorbic acid, 0.01 mM methylene blue, 0.1 mg catalase, 37°C, 10 min [26]) [26]
5 <2> (D-Tryptophan) [3]

K_i-Value (mM)

0.00012 <2> (annulin B, <2> 25°C, pH 6.5 [41]) [41]
0.00014 <2> (annulin A, <2> 25°C, pH 6.5 [41]) [41]
0.00021 <2> (exiguamine A) [40]
0.00069 <2> (annulin A, <2> 25°C, pH 6.5 [41]) [41]
0.0012 <2> (garveatin C, <2> 25°C, pH 6.5 [41]) [41]
0.0014 <2> (2-hydroxygarveatin E, <2> 25°C, pH 6.5 [41]) [41]
0.0015 <2> (imidodicarbonimidic diamide, N-methyl-N"-9-phenanthrenyl-, monohydrochloride) [36]
0.0023 <2> (2-hydroxygarvin A, <2> 25°C, pH 6.5 [41]) [41]
0.0031 <2> (garveatin E, <2> 25°C, pH 6.5 [41]) [41]
0.0032 <2> (garveatin A, <2> 25°C, pH 6.5 [41]) [41]
0.0045 <1> (indole-3-acrylic acid, <1> against L-tryptophan [14]) [14]
0.0068 <1> (indole-3-acrylic acid, <1> against D-tryptophan [14]) [14]
0.05 <6> (D-tryptophan, <6> 0.1 M potassium phosphate buffer (pH 6-8), 0.025 mM methylene blue, 0.2 mg catalase, 10 mM ascorbic acid, 50 nM dioxygenase, 25°C [15]) [15]
0.068 <2> (1-methyl-DL-tryptophan) [6]
0.176 <2> (norharman) [6]
0.18 <2> (6-nitro-L-tryptophan, <2> 1 mM, competitive inhibition, 52% inhibition relative to L-tryptophan, 50 mM potassium phosphate, pH 6.5, 10 mM ascorbic acid, 0.01 mM methylene blue, 0.1 mg catalase, 37°C, 10 min [26]) [26]

pH-Optimum

6.5 <2, 6> (<2,6> L-tryptophan [6,21]) [6, 21]
6.6-6.8 <2> (<2> in 50 mM phosphate buffer [26]) [26]
7 <6> [15]
7-8 <1> [13]
7.3 <1> (<1> 25°C [14]) [14]
7.5 <6> (<6> D-tryptophan [21]) [21]

pH-Range

5.5-8 <6> (<6> 5% activity loss at pH 5.5, 10% activity loss at pH 7, 20% activity loss at pH 8 [21]) [15, 21]

4 Enzyme Structure

Molecular weight

43000 <2> (<2> gel filtration, lung [29]) [29]
45000 <2> (<2> SDS-PAGE [5,8,27]) [5, 8, 27]
45440 <2> (<2> ESI mass spectrometry [3]) [3]
46980 <2> (<2> ESI mass spectrometry [3]) [3]
150000 <1> (<1> $\alpha_2\beta_2$ subunit [14]) [14]
320000 <2> (<2> gel filtration, liver [29]) [29]

5 Isolation/Preparation/Mutation/Application

Source/tissue

B-lymphocyte <2> [11]
HSVEC cell <2> (<2> i.e. human umbilical vein endothelial cell, express little indoleamine 2,3-dioxygenase, which is poorly upregulated upon activation (except by mycoplasma) [32]) [32]
HUVEC cell <2> (<2> in HUVEC cells the enzyme is upregulated by incubation with cytokines or in mycoplasma-infected cells. Inhibition of indoleamine 2,3-dioxygenase improves ability of HUVEC cells to stimulate T-cell proliferation [32]) [32]
bone marrow <2> (<2> marrow stromal cell or mesenchymal stem cells (MSC) [17]) [17]
brain <1, 2, 3, 6, 10> (<2> microvascular endothelial cell, HBMEC [19]; <3> in the group of rats aged 12 months, the highest specific activity is found in the small intestine [39]) [4, 19, 23, 28, 30, 33, 39]
cecum <1> [23]
cerebrospinal fluid <2> (<2> from humans suffering from inflammatory neurological disorders exhibit increased IDO activity [30]) [30]
cervical carcinoma cell <2> [7]
cervical epithelium <2> (<2> very irregular activity [27]) [27]
colon <1, 6> [23, 33]

corneal epithelium <11> (<11> consistently expresses the enzyme at low levels. Exposure to UV-light leads to a dose-responding upregulation. Overexpression of indoleamine-pyrrole 2,3-dioxygenase could reduce apoptosis significantly following UV-B irradiation [38]) [38]

decidua <2> (<2> in glandular epithelium in first trimester pregnancy [27]; <2> strong expression in stromal and glandular epithelial cells of the first trimester decidua [9]) [9, 27]

dendritic cell <2, 11> (<2> tumor-associated dendritic cell, indoleamine 2,3-dioxygenase is up-regulated after maturation of dendritic cells in presence of prostaglandin E_2 [37]) [37, 38]

endometrioid carcinoma cell <2> (<2> epithelium [7]) [7]

endothelial cell <2> (<2> arterially derived, express little indoleamine 2,3-dioxygenase, which is poorly upregulated upon activation (except by mycoplasma) [32]) [32]

epididymis <1> [13, 23, 30]

esophagus <1> [23]

fallopian tube <2> [27]

heart <1> [23]

hippocampus <1> (<1> neurons, level of activity is lower than in monocytic cells in lymph nodes [28]; <1> the enzyme is upregulated by interferon-γ in neurons of the hippocampus [28]) [28]

intestine <1, 2, 3, 6> (<3> intestinal mucosa of rat exhibits lower activity than in rabbit [25]) [12, 21, 22, 23, 25, 33]

kidney <1, 3, 6> (<3> low activity in rats aged 2-3 months or 12 months [39]) [23, 33, 39]

large intestine <6> [33]

lens cortex <2> (<2> anterior and bow region [5]) [5]

liver <1, 2, 3, 4, 5, 6, 8, 9> [14, 25, 29]

lung <1, 2, 3, 6, 10> (<2> cancer bearing lung exhibits 20fold increase in IDO activity [30]; <3> low activity, activity is not age-related [39]) [4, 23, 29, 30, 33, 39]

macrophage <1> [28]

mesenchymal epithelium <2> [27]

microglia <1> [28]

monocyte <2> (<2> induced by α-, β- and γ-interferon [30]) [30]

natural killer cell <2> [11]

neuron <1> (<1> the enzyme is upregulated by interferon-γ in neurons of the hippocampus. The enzyme could contribute to the vulnerability of neurons to inflammatory conditions [28]) [28]

peripheral blood mononuclear cell <2> (<2> K562 from human erythroleukemia, human hepatocellular carcinoma cell line HepG2 [11]) [11]

peritoneal macrophage <1> [20]

placenta <2> (<2> term placenta [33]; <2> brush border microvillous plasma membranes of placental syncytiotrophoblast [6]; <2> in the first and second trimester placenta the enzyme is localized to the syncytiotrophoblast, stroma and macrophages. In term placenta confined mainly to vascular en-

dothelial cells of villous blood vessels, and to macrophages within the fetal villus [9]) [6, 9, 27, 33]
reproductive system <2> [27]
seminal vesicle <1> [23]
skin fibroblast <2> [30]
small intestine <1, 3, 6> (<3> in newborn rats IDO activity is present only in small intestine. In the group of rats aged 2-3 months, the highest specific activity is found in the small intestine, decrease of enzyme activity in relation to age [39]) [1, 4, 15, 16, 24, 30, 33, 39]
spleen <1, 6> (<1> dentritic cells [20]) [20, 23, 33]
stomach <1, 6> [23, 33]
stromal cell <2> (<2> peritumoral infiltrate [7]) [7]
submandibular gland <1> [23]
syncytiotrophoblast <2> (<2> brush border microvillous plasma membranes of placental syncytiotrophoblast [6]) [6, 27]
synovial fibroblast <2> [30]
testis <1> [23]
thymus <1> [23]
trachea <1> [23]
urinary bladder <1> [23]
Additional information <1, 2> (<2> endometrial glandular and surface epithelial cells exhibit strong activity during the secretory phase [27]; <1> entorhinal cortex [28]; <2> T-lymphocyte [11]) [11, 27, 28]

Localization

cytoplasm <2> [6]
dendrite <1> [20]
Additional information <2> (<2> expressed in epithelial cells but not in mature fiber cells [5]) [5]

Purification

<1> (ammonium sulfate precipitation, hydroxyapatite column chromatography (two times), gel filtration, DEAE cellulose column chromatography, third hydroxyapatite column chromatography, Sephadex G-200 gel filtration, second DEAE cellulose column chromatography) [14]
<1> (homogenization) [23]
<1> (homogenization, ammonium sulfate precipitation, gel filtration with Sephacryl S200, Bio-gel HT column, stored at -75°C) [13]
<2> [26, 34]
<2> (0-5°C, homogenization, Protein G-Sepharose) [5]
<2> (Ni-NTA-agarose and gel filtration (Superdex 75)) [10]
<2> (cryosections deparaffinized, rehydrated, cooked in 0.01 M citrate buffer (pH 6.0), incubated with IDO antibody at room temperature for 30 min) [27]
<2> (homogenization, filtered through cotton gauze, centrifugation, collection, stored at -20°C) [6]
<2> (phosphocellulose and Ni-NTA-agarose affinity chromatography) [3]
<2> (recombinant) [31]
<3> (dialysis, lyophilization) [25]

<3> (tissues thawed at room temperature, homogenized in 0.14 M potassium chloride plus 20 mM potassium phosphate, pH 7.0) [4]
<6> [15, 16, 24]
<6> (homogenization, ammonium sulfate precipitation) [21]
<10> (tissues thawed at room temperature, homogenized in 0.14 M potassium chloride plus 20 mM potassium phosphate, pH 7.0) [4]

Crystallization
<2> (hanging-drop vapour-diffusion method, crystal structure of the enzyme complexed with the ligand inhibitor 4-phenylimidazole and cyanide at resolutions of 2.3 and 3.4 A respectively) [42]
<2> (sitting-drop vapour-diffusion method. Crystals belong to the orthorhombic space group P2(1)2(1)2, with unit-cell parameters a = 86.1, b = 98.0, c = 131.0 A. 2.3 A resolution) [31]

Cloning
<2> (expression in Escherichia coli) [31]
<2> (expression in Escherichia coli E538, expressed with His-Tag) [3]
<2> (expression in Escherichia coli E538, formation of inclusion bodies at 37°C with reduced formation, at 30°C, expressed with His-Tag) [10]
<2> (expression in Escherichia coli strain YA21) [26]
<2> (expression in Saccharomyces cerevisiae tryptophan auxotroph can restore growth in presence of low tryptophan concentrations) [36]
<11> [38]

Engineering
D274A <2> (<2> mutant without enzyme activity, may be distal ligand or essential in maintaining the conformation of the heme pocket [8]) [8]
F226A <2> (<2> drastically reduced dioxygenase activity [42]) [42]
F227A <2> (<2> drastically reduced dioxygenase activity [42]) [42]
H16A <2> (<2> does not act as proximal ligand [8]) [8]
H303A <2> (<2> does not act as proximal ligand [8]; <2> Fe^{3+}/Fe^{2+} reduction potential of H303A variant is about 70 mV lower than that of recombinant wild-type enzyme, leading to a destabilization of the ferrous-oxy complex [34]) [8, 34]
H346A <2> (<2> mutant without enzyme activity, His346 is essential for heme binding [8]) [8]
K352A <2> (<2> mutant with diminished heme binding ability [8]) [8]
R231A <2> (<2> drastically reduced dioxygenase activity [42]) [42]
S263A <2> (<2> activity is reduced to 15% [42]) [42]
V109A <2> (<2> mutant maintains heme binding ability [8]) [8]
medicine <2> (<2> IDO expression as a T-cell inhibitory effector pathway in professional antigen-presenting cells [17]) [17]

Application
medicine <2, 3, 10, 12> (<2> enzyme catalyzes the first and rate-limiting step in kynurenine pathway [8]; <2> first and probably rate-limiting enzyme in UV filter biosynthesis [5]; <2> first and rate-limiting enzyme in tryptophan metabolism, plays a role in pathogenesis of many diseases [10]; <2> first and

rate-limiting enzyme in tryptophan metabolism, plays a role in pathogenesis of many diseases [3]; <2> IDO induces suppression of antitumoral immune response, inhibits tumor cell proliferation by tryptophan depletion [7]; <2> inhibits T-cell activation and proliferation [11]; <2> plays a role in the antiparasitic defense in humans [19]; <2> plays a role in the antiparasitic defense in humans, inhibits the growth of tumor cell lines in vitro, inhibits T lymphocyte proliferation [30]; <3,10> rate-limiting enzyme in kynurenine pathway [4]; <12> reduced tryptophan and increased kynurenine concentrations, chronic immune activation which can be referred to increased IDO activation [18]) [3, 4, 5, 7, 8, 10, 11, 18, 19, 30]
pharmacology <2> (<2> first reaction in the tryptophan catabolic pathway in mammals [26]) [26]

6 Stability

Temperature stability

-20 <2> (<2> losing 70% activity after 24 h, addition of 5 mg/ml bovine serum albumin maintains 80% activity after 5 days [26]) [26]
4 <2> (<2> maintains 70% activity with addition of 2 mg/ml catalase for 4 days [26]) [26]

Oxidation stability

<6>, highly autoxidizable, requires superoxide anion or a reductant such as ascorbic acid and methylene blue as cofactors [16]
<6>, leuco-methylene blue reduces ferric enzyme [2]

Storage stability

<1>, -75°C [13]
<2>, -20°C, losing 70% activity after 24 h, addition of 5 mg/ml bovine serum albumin maintains 80% activity after 5 days [26]

References

[1] Sono, M.: Spectroscopic and equilibrium studies of ligand and organic substrate binding to indolamine 2,3-dioxygenase. Biochemistry, **29**, 1451-1460 (1990)
[2] Sono, M.: The roles of superoxide anion and methylene blue in the reductive activation of indoleamine 2,3-dioxygenase by ascorbic acid or by xanthine oxidase-hypoxanthine. J. Biol. Chem., **264**, 1616-1622 (1989)
[3] Littlejohn, T.K.; Takikawa, O.; Skylas, D.; Jamie, J.F.; Walker, M.J.; Truscott, R.J.W.: Expression and purification of recombinant human indoleamine 2,3-dioxygenase. Protein Expr. Purif., **19**, 22-29 (2000)
[4] Dang, Y.; Dale, W.E.; Brown, O.R.: Comparative effects of oxygen on indoleamine 2,3-dioxygenase and tryptophan 2,3-dioxygenase of the kynurenine pathway. Free Radic. Biol. Med., **28**, 615-624 (2000)

[5] Takikawa, O.; Littlejohn, T.K.; Truscott, R.J.W.: Indoleamine 2,3-dioxygenase in the human lens, the first enzyme in the synthesis of UV filters. Exp. Eye Res., **72**, 271-277 (2001)
[6] Kudo, Y.; Boyd, C.A.R.: Human placental indoleamine 2,3-dioxygenase: cellular localization and characterization of an enzyme preventing fetal rejection. Biochim. Biophys. Acta, **1500**, 119-124 (2000)
[7] Sedlmayr, P.; Semlitsch, M.; Gebru, G.; Karpf, E.; Reich, O.; Tang, T.; Wintersteiger, R.; Takikawa, O.; Dohr, G.: Expression of indoleamine 2,3-dioxygenase in carcinoma of human endometrium and uterine cervix. Adv. Exp. Med. Biol., **527**, 91-95 (2003)
[8] Littlejohn, T.K.; Takikawa, O.; Truscott, R.J.W.; Walker, M.J.: Asp274 and His346 are essential for heme binding and catalytic function of human indolamine 2,3-dioxygenase. J. Biol. Chem., **278**, 29525-29531 (2003)
[9] Ligam, P.; Manuelpillai, U.; Wallace, E.M.; Walker, D.: Localisation of indoleamine 2,3-dioxygenase and kynurenine hydroxylase in the human placenta and decidua: implications for role of the kynurenine pathway in pregnancy. Placenta, **26**, 498-504 (2005)
[10] Austin, C.J.; Mizdrak, J.; Matin, A.; Sirijovski, N.; Kosim-Satyaputra, P.; Willows, R.D.; Roberts, T.H.; Truscott, R.J.; Polekhina, G.; Parker, M.W.; Jamie, J.F.: Optimised expression and purification of recombinant human indoleamine 2,3-dioxygenase. Protein Expr. Purif., **37**, 392-398 (2004)
[11] Kai, S.; Goto, S.; Tahara, K.; Sasaki, A.; Tone, S.; Kitano, S.: Indolamine 2,3-dioxygenase is necessary for cytolytic activity of natural killer cells. Scand. J. Immunol., **59**, 177-182 (2004)
[12] Hayaishi, O.: Utilization of superoxide anion by indoleamine oxygenase-catalyzed tryptophan and indoleamine oxidation. Adv. Exp. Med. Biol., **398**, 285-289 (1996)
[13] Ozaki, Y.; Reinhard, J.F., Jr.; Nichol, C.A.: Cofactor activity of dihydroflavin mononucleotide and tetrahydrobiopterin for murine epididymal indoleamine 2,3-dioxygenase. Biochem. Biophys. Res. Commun., **137**, 1106-1111 (1986)
[14] Watanabe, Y.; Fujiwara, M.; Yoshida, R.; Hayaishi, O.: Stereospecificity of hepatic L-tryptophan 2,3-dioxygenase. Biochem. J., **189**, 393-405 (1980)
[15] Sono, M.; Cady, S.G.: Enzyme kinetic and spectroscopic studies of inhibitor and effector interactions with indoleamine 2,3-dioxygenase. 1. Norharman and 4-phenylimidazole binding to the enzyme as inhibitors and heme ligands. Biochemistry, **28**, 5392-5399 (1989)
[16] Sono, M.: Enzyme kinetic and spectroscopic studies of inhibitor and effector interactions with indoleamine 2,3-dioxygenase. 2. Evidence for the existence of another binding site in the enzyme for indole derivative effectors. Biochemistry, **28**, 5400-5407 (1989)
[17] Meisel, R.; Zibert, A.; Laryea, M.; Gobel, U.; Daubener, W.; Dilloo, D.: Human bone marrow stromal cells inhibit allogeneic T-cell responses by indoleamine 2,3-dioxygenase-mediated tryptophan degradation. Blood, **103**, 4619-4621 (2004)
[18] Fuchs, D.; Moller, A.A.; Reibnegger, G.; Werner, E.R.; Werner-Felmayer, G.; Dierich, M.P.; Wachter, H.: Increased endogenous interferon-γ and neopter-

in correlate with increased degradation of tryptophan in human immunodeficiency virus type 1 infection. Immunol. Lett., **28**, 207-211 (1991)

[19] Daubener, W.; Spors, B.; Hucke, C.; Adam, R.; Stins, M.; Kim, K.S.; Schroten, H.: Restriction of Toxoplasma gondii growth in human brain microvascular endothelial cells by activation of indoleamine 2,3-dioxygenase. Infect. Immun., **69**, 6527-6531 (2001)

[20] Fallarino, F.; Vacca, C.; Orabona, C.; Belladonna, M.L.; Bianchi, R.; Marshall, B.; Keskin, D.B.; Mellor, A.L.; Fioretti, M.C.; Grohmann, U.; Puccetti, P.: Functional expression of indoleamine 2,3-dioxygenase by murine CD8 $\alpha(+)$ dendritic cells. Int. Immunol., **14**, 65-68 (2002)

[21] Yamamoto, S.; Hayaishi, O.: Tryptophan pyrrolase of rabbit intestine. J. Biol. Chem., **242 (22)**, 5260-5266 (1967)

[22] Hirata, F.; Ohnishi, T.; Hayaishi: Indoleamine 2,3-dioxygenase. Characterization and properties of enzyme. O_2- complex. J. Biol. Chem., **252**, 4637-4642 (1977)

[23] Takikawa, O.; Yoshida, R.; Kido, R.; Hayaishi, O.: Tryptophan degradation in mice initiated by indoleamine 2,3-dioxygenase. J. Biol. Chem., **261**, 3648-3653 (1986)

[24] Kobayashi, K.; Hayashi, K.; Sono, M.: Effects of tryptophan and pH on the kinetics of superoxide radical binding to indoleamine 2,3-dioxygenase studied by pulse radiolysis. J. Biol. Chem., **264**, 15280-15283 (1989)

[25] Loh, H.H.; Berg, C.P.: D-Tryptophan pyrrolase activity in the liver and the intestine of the rat. J. Nutr., **102**, 1331-1339 (1972)

[26] Southan, M.D.; Truscott, R.J.W.; Jamie, J.F.; Pelosi, L.; Walker, M.J.; Maeda, H.; Iwamoto, Y.; Tone, S.: Structural requirements of the competitive binding site of recombinant human indoleamine 2,3-dioxygenase. Med. Chem. Res., **6**, 343-352 (1996)

[27] Sedlmayr, P.; Blaschitz, A.; Wintersteiger, R.; Semlitsch, M.; Hammer, A.; MacKenzie, C.R.; Walcher, W.; Reich, O.; Takikawa, O.; Dohr, G.: Localization of indoleamine 2,3-dioxygenase in human female reproductive organs and the placenta. Mol. Hum. Reprod., **8**, 385-391 (2002)

[28] Roy, E.J.; Takikawa, O.; Kranz, D.M.; Brown, A.R.; Thomas, D.L.: Neuronal localization of indoleamine 2,3-dioxygenase in mice. Neurosci. Lett., **387**, 95-99 (2005)

[29] Yasui, H.; Takai, K.; Yoshida, R.; Hayaishi, O.: Interferon enhances tryptophan metabolism by inducing pulmonary indoleamine 2,3-dioxygenase: its possible occurrence in cancer patients. Proc. Natl. Acad. Sci. USA, **83**, 6622-6626 (1986)

[30] Thomas, S.R.; Stocker, R.: Redox reactions related to indoleamine 2,3-dioxygenase and tryptophan metabolism along the kynurenine pathway. Redox Rep., **4**, 199-220 (1999)

[31] Oda, S.; Sugimoto, H.; Yoshida, T.; Shiro, Y.: Crystallization and preliminary crystallographic studies of human indoleamine 2,3-dioxygenase. Acta Crystallogr. Sect. F, **62**, 221-223 (2006)

[32] Beutelspacher, S.C.; Tan, P.H.; McClure, M.O.; Larkin, D.F.; Lechler, R.I.; George, A.J.: Expression of indoleamine 2,3-dioxygenase (IDO) by endothe-

lial cells: implications for the control of alloresponses. Am. J. Transplant., **6**, 1320-1330 (2006)
[33] Takikawa, O.: Biochemical and medical aspects of the indoleamine 2,3-dioxygenase-initiated L-tryptophan metabolism. Biochem. Biophys. Res. Commun., **338**, 12-19 (2005)
[34] Papadopoulou, N.D.; Mewies, M.; McLean, K.J.; Seward, H.E.; Svistunenko, D.A.; Munro, A.W.; Raven, E.L.: Redox and spectroscopic properties of human indoleamine 2,3-dioxygenase and a His303Ala variant: implications for catalysis. Biochemistry, **44**, 14318-14328 (2005)
[35] Samelson-Jones, B.J.; Yeh, S.R.: Interactions between nitric oxide and indoleamine 2,3-dioxygenase. Biochemistry, **45**, 8527-8538 (2006)
[36] Vottero, E.; Balgi, A.; Woods, K.; Tugendreich, S.; Melese, T.; Andersen, R.J.; Mauk, A.G.; Roberge, M.: Inhibitors of human indoleamine 2,3-dioxygenase identified with a target-based screen in yeast. Biotechnol. J., **1**, 282-288 (2006)
[37] von Bergwelt-Baildon, M.S.; Popov, A.; Saric, T.; Chemnitz, J.; Classen, S.; Stoffel, M.S.; Fiore, F.; Roth, U.; Beyer, M.; Debey, S.; Wickenhauser, C.; Hanisch, F.G.; Schultze, J.L.: CD25 and indoleamine 2,3-dioxygenase are upregulated by prostaglandin E_2 and expressed by tumor-associated dendritic cells in vivo: additional mechanisms of T-cell inhibition. Blood, **108**, 228-237 (2006)
[38] Serbecic, N.; Beutelspacher, S.C.: Indoleamine 2,3-dioxygenase protects corneal endothelial cells from UV mediated damage. Exp. Eye Res., **82**, 416-426 (2006)
[39] Comai, S.; Bertazzo, A.; Ragazzi, E.; Caparrotta, L.; Costa, C.V.; Allegri, G.: Influence of age on Cu/Zn-superoxide dismutase and indole 2,3-dioxygenase activities in rat tissues. Ital. J. Biochem., **54**, 232-239 (2006)
[40] Brastianos, H.C.; Vottero, E.; Patrick, B.O.; Van Soest, R.; Matainaho, T.; Mauk, A.G.; Andersen, R.J.: Exiguamine A, an indoleamine-2,3-dioxygenase (IDO) inhibitor isolated from the marine sponge Neopetrosia exigua. J. Am. Chem. Soc., **128**, 16046-16047 (2006)
[41] Pereira, A.; Vottero, E.; Roberge, M.; Mauk, A.G.; Andersen, R.J.: Indoleamine 2,3-dioxygenase inhibitors from the Northeastern Pacific Marine Hydroid Garveia annulata. J. Nat. Prod., **69**, 1496-1499 (2006)
[42] Sugimoto, H.; Oda, S.; Otsuki, T.; Hino, T.; Yoshida, T.; Shiro, Y.: Crystal structure of human indoleamine 2,3-dioxygenase: catalytic mechanism of O_2 incorporation by a heme-containing dioxygenase. Proc. Natl. Acad. Sci. USA, **103**, 2611-2616 (2006)

Acireductone dioxygenase (Ni^{2+}-requiring) 1.13.11.53

1 Nomenclature

EC number

1.13.11.53

Systematic name

1,2-dihydroxy-5-(methylthio)pent-1-en-3-one:oxygen oxidoreductase (formate- and CO-forming)

Recommended name

acireductone dioxygenase (Ni^{2+}-requiring)

Synonyms

2-hydroxy-3-keto-5-thiomethylpent-1-ene dioxygenase <4> (<4> ambigous [5]) [5]
ADI1 <2, 3> [8, 14]
ARD <4, 5> [1, 2, 3, 5, 10, 11]
ARD1 <7> [12]
Ni(II)-ARD <5> [15, 16]
aci-reductone dioxygenase <2> [13, 14]
acidoreductone dioxygenase <4> (<4> ambigouos [5]) [5]
membrane-type 1 matrix metalloproteinase cytoplasmic tail binding protein-1 <2> (<2> MTCBP-1 [13]) [13]

CAS registry number

221681-64-7

2 Source Organism

<1> *Mus musculus* (no sequence specified) [17]
<2> *Homo sapiens* (no sequence specified) [13, 14]
<3> *Saccharomyces cerevisiae* (no sequence specified) [8]
<4> *Klebsiella pneumoniae* (no sequence specified) [1, 2, 3, 4, 5, 6, 7, 9, 11]
<5> *Klebsiella sp.* (no sequence specified) [10,15,16]
<6> *Mus musculus* (UNIPROT accession number: Q99JT9) [18]
<7> *Oryza sativa* (UNIPROT accession number: A2Z7C4) [12]

3 Reaction and Specificity

Catalyzed reaction

1,2-dihydroxy-5-(methylthio)pent-1-en-3-one + O_2 = 3-(methylthio)propanoate + formate + CO

Reaction type

dioxygenation

oxidation (<2> ARD (Ni) converts aci-reductone to 3-(methylthio)propanoate, carbon monoxide, and formate, and the reaction is not part of the methionine salvage pathway cycle (off-pathway). [13]; <1> if Ni^{2+} is bound in the active site, the substrate acireductone reacts with O_2 to yield formate and 4-methylthio-2-ketobutyrate [17])

Natural substrates and products

S 1,2-dihydroxy-5-(methylthio)pent-1-en-3-one + O_2 <4> (<4> enzyme of the methionine salvage pathway [7]; <4> reaction is a shunt out of the methionine salvage pathway [3]; <4> the enzyme represents a branch point in the methionine salvage pathway leading from methylthioadenosine to methionine [11]) (Reversibility: ir) [1, 3, 7, 11]

P 3-(methylthio)propanoate + formate + CO

S Additional information <3> (<3> it is proposed that Rnt1p cleavage and/or degradation by exonucleases helps prevent the accumulation of ADI1 mRNA prior to heat shock conditions. The ribonucleolytic pathways provide a mechanism to eliminate 3-extended forms that arise from poor 3-end processing signals present at the end of the ADI1 gene [8]) (Reversibility: ?) [8]

P ?

Substrates and products

S 1,2-dihydroxy-3-oxo-1-hexene + O_2 <4> (<4> incorporation of O_2 into C1 and C3 of 1,2-dihydroxy-3-oxo-1-hexene [1]) (Reversibility: ?) [1]

P ?

S 1,2-dihydroxy-3-oxo-3-phenyl-1-propene + O_2 <5> (Reversibility: ?) [16]

P ?

S 1,2-dihydroxy-5-(methylthio)pent-1-en-3-one + O_2 <1> (Reversibility: ?) [17]

P 4-methylthio-2-oxobutyrate + formate + CO

S 1,2-dihydroxy-5-(methylthio)pent-1-en-3-one + O_2 <2, 4, 5, 6, 7> (<4> enzyme of the methionine salvage pathway [7]; <4> reaction is a shunt out of the methionine salvage pathway [3]; <4> the enzyme represents a branch point in the methionine salvage pathway leading from methylthioadenosine to methionine [11]; <4> Ni^{2+} or Co^{2+} bound to the enzyme protein. If Fe^{2+} is bound instead of Ni^{2+} reaction catalyzed by EC 1.13.11.54 occurs instead [7]; <4> ordered-sequential mechanism [1]) (Reversibility: ir) [1, 3, 4, 5, 6, 7, 11, 12, 13, 14, 15, 16, 18]

P 3-(methylthio)propanoate + formate + CO

S 1,2-dihydroxyhex-1-en-3-one + O_2 <4> (Reversibility: ?) [5]

P butyrate + formate + CO

S Additional information <3, 4> (<3> it is proposed that Rnt1p cleavage and/or degradation by exonucleases helps prevent the accumulation of ADI1 mRNA prior to heat shock conditions. The ribonucleolytic pathways provide a mechanism to eliminate 3-extended forms that arise from poor 3-end processing signals present at the end of the ADI1 gene [8]; <4> aliphatic carbon-carbon bond cleavage reactivity of a mononuclear Ni(II) cis-β-keto-enolate complex in the presence of base and O_2: a model reaction for acireductone dioxygenase [3]) (Reversibility: ?) [3, 8]

P ?

Inhibitors

cycloheximide <7> (<7> 0.0711 mM or higher [12]) [12]

Activating compounds

ethylene <7> [12]

Metals, ions

Co^{2+} <4> (<4> apoenzyme is catalytically inactive. Addition of Ni^{2+} or Co^{2+} yields activity. Production in intact Escherichia coli of E-2 depends on the availability of the Fe^{2+}. Enzyme contains 1.1 Ni^{2+} per enzyme molecule [7]) [7]

Mg^{2+} <4> (<4> required [5]) [5]

Ni^{2+} <1, 2, 4, 5, 6, 7> (<5> required for activity [16]; <4> apoenzyme is catalytically inactive. Addition of Ni^{2+} or Co^{2+} yields activity. Production in intact Escherichia coli of E-2 depends on the availability of the Fe^{2+}. Enzyme contains 1.1 Ni^{2+} per enzyme molecule [7]; <4> enzyme contains 1 atom of Ni [1]; <4> enzyme contains Ni^{2+} [9]; <5> model for the solution structure of the paramagnetic Ni^{2+}-containing enzyme [10]; <4> Ni^{2+}-containg enzyme [3]; <4> solution structure of the nickel-containing enzyme is determined using NMR methods. X-ray absorption spectroscopy, assignment of hyperfine shifted NMR resonance and conserved domain homology are used to model the metal-binding site because of the paramagnetism of the bound Ni^{2+} [11]; <4> structure of the Ni site in resting Ni-ARD as containing a six coordinate Ni site composed of O/N-donor ligands including 3-4 histidine residues. The substrate binds to the Ni center in a bidentate fashion by displacing two ligands, at least one of which is a histidine ligand [2]; <1> Ni^{2+} can be conservatively replaced by Mn2 +or Co^{2+}, giving rise to ARD activity (CO production) [17]) [1, 2, 3, 7, 9, 10, 11, 12, 13, 16, 17, 18]

Turnover number (min^{-1})

500 <4> (1,2-dihydroxy-3-oxo-1-hexene) [1]
500 <4> (O_2) [1]

Specific activity (U/mg)

18.2 <4> [5]

K_m-Value (mM)

0.044 <7> (1,2-dihydroxy-5-(methylthio)pent-1-en-3-one) [12]
0.05 <4> (1,2-dihydroxy-3-oxo-1-hexene) [1]

0.11 <4> (O_2) [1]
500 <4> (1,2-dihydroxy-5-(methylthio)pent-1-en-3-one) [7]
500 <4> (O_2) [7]

pH-Optimum
7.4 <7> [12]

4 Enzyme Structure

Molecular weight
20250 <4> (<4> mass spectrometry [1,7]) [1, 7]
21390 <6> [18]

Subunits
monomer <1, 4, 6> (<4> 1 * 20000, SDS-PAGE [1]; <1> X-ray crystallography [17]; <4> 1 * 18500, SDS-PAGE [5]; <6> 1 * 21391 [18]) [1, 5, 17, 18]
oligomer <7> [12]

5 Isolation/Preparation/Mutation/Application

Source/tissue
COS-7 cell <2> [14]
HT-1080 cell <2> [14]
internode <7> [12]
root <7> [12]

Localization
cytoplasm <2> [14]
nucleus <2> [14]

Purification
<1> [17]
<4> [1, 2, 5, 6]
<4> (recombinant) [7]
<6> (Superdex 200 column chromatography and nickel-chelating resin column chromatography) [18]
<7> (DEAE cellulose column chromatography and Phenyl-Sepharose column chromatography) [12]

Renaturation
<7> (unfolded using 8 mM urea and refolded in the presence of 20 mM Ni^{2+}) [12]

Crystallization
<6> (nanodroplet vapour diffusion method using 19.0% (w/v) polyethylene glycol 4000, 19.0% (w/v) isopropanol, 5.0% glycerol, and 0.095 M Na-citrate pH 4.2 (final pH 5.6)) [18]

Cloning

<1> (expressed in Escherichia coli strain BL21(DE3)pLysS) [17]
<4> (expression in Escherichia coli) [2, 7]
<6> (expressed in Escherichia coli) [18]
<7> (expressed in Escherichia coli BL21 cells) [12]

Engineering

E94A <2> (<2> no activity [13]) [13]
H98S <1> (<1> mutation results in the formation of a stable soluble protein that while structurally different from ARD-Ni^{2+} shows a high degree of similarity to the ARD(Fe) enzyme [17]) [17]

References

[1] Dai, Y.; Pochapsky, T.C.; Abeles, R.H.: Mechanistic studies of two dioxygenases in the methionine salvage pathway of Klebsiella pneumoniae. Biochemistry, **40**, 6379-6387 (2001)
[2] Al-Mjeni, F.; Ju, T.; Pochapsky, T.C.; Maroney, M.J.: XAS investigation of the structure and function of Ni in acireductone dioxygenase. Biochemistry, **41**, 6761-6769 (2002)
[3] Szajna, E.; Arif, A.M.; Berreau, L.M.: Aliphatic carbon-carbon bond cleavage reactivity of a mononuclear Ni(II) cis-β-keto-enolate complex in the presence of base and O_2: a model reaction for acireductone dioxygenase (ARD). J. Am. Chem. Soc., **127**, 17186-17187 (2005)
[4] Furfine, E.S.; Abeles, R.H.: Intermediates in the conversion of 5'-S-methylthioadenosine to methionine in Klebsiella pneumonia. J. Bacteriol., **263**, 9598-9606 (1988)
[5] Wray, J.W.; Abeles, R.H.: A bacterial enzyme that catalyzes formation of carbon monoxide. J. Biol. Chem., **268**, 21466-21469 (1993)
[6] Wray, J.W.; Abeles, R.B.: The methionine salvage pathway in Klebsiella pneumoniae and rat liver. J. Biol. Chem., **270**, 3147-3153 (1995)
[7] Dai, Y.; Wensink, P.C.; Abeles, R.H.: One protein, two enzymes. J. Biol. Chem., **274**, 1193-1195 (1999)
[8] Zer, C.; Chanfreau, G.: Regulation and surveillance of normal and 3'-extended forms of the yeast acireductone dioxygenase mRNA by RNase III cleavage and exonucleolytic degradation. J. Biol. Chem., **280**, 28997-29003 (2005)
[9] Mo, H.; Dai, Y.; Pochapsky, S.S.; Pochapsky, T.C.: ^{1}H, ^{13}C and ^{15}N NMR assignments for a carbon monoxide generating metalloenzyme from Klebsiella pneumoniae. J. Biomol. NMR, **14**, 287-288 (1999)
[10] Pochapsky, T.C.; Pochapsky, S.S.; Ju, T.; Hoefler, C.; Liang, J.: A refined model for the structure of acireductone dioxygenase from Klebsiella ATCC 8724 incorporating residual dipolar couplings. J. Biomol. NMR, **34**, 117-127 (2006)

[11] Pochapsky, T.C.; Pochapsky, S.S.; Ju, T.; Mo, H.; Al-Mjeni, F.; Maroney, M.J.: Modeling and experiment yields the structure of acireductone dioxygenase from Klebsiella pneumoniae. Nat. Struct. Biol., **9**, 966-972 (2002)
[12] Sauter, M.; Lorbiecke, R.; Ouyang, B.; Pochapsky, T.C.; Rzewuski, G.: The immediate-early ethylene response gene OsARD1 encodes an acireductone dioxygenase involved in recycling of the ethylene precursor S-adenosylmethionine. Plant J., **44**, 718-729 (2005)
[13] Hirano, W.; Gotoh, I.; Uekita, T.; Seiki, M.: Membrane-type 1 matrix metalloproteinase cytoplasmic tail binding protein-1 (MTCBP-1) acts as an eukaryotic aci-reductone dioxygenase (ARD) in the methionine salvage pathway. Genes Cells, **10**, 565-574 (2005)
[14] Gotoh, I.; Uekita, T.; Seiki, M.: Regulated nucleo-cytoplasmic shuttling of human aci-reductone dioxygenase (hADI1) and its potential role in mRNA processing. Genes Cells, **12**, 105-117 (2007)
[15] Szajna-Fuller, E.; Chambers, B.M.; Arif, A.M.; Berreau, L.M.: Carboxylate coordination chemistry of a mononuclear Ni(II) center in a hydrophobic or hydrogen bond donor secondary environment: Relevance to acireductone dioxygenase. Inorg. Chem., **46**, 5486-5498 (2007)
[16] Szajna-Fuller, E.; Rudzka, K.; Arif, A.M.; Berreau, L.M.: Acireductone dioxygenase-(ARD-) type reactivity of a nickel(II) complex having monoanionic coordination of a model substrate: Product identification and comparisons to unreactive analogues. Inorg. Chem., **46**, 5499-5507 (2007)
[17] Ju, T.; Goldsmith, R.B.; Chai, S.C.; Maroney, M.J.; Pochapsky, S.S.; Pochapsky, T.C.: One protein, two enzymes revisited: A structural entropy switch interconverts the two isoforms of acireductone dioxygenase. J. Mol. Biol., **363**, 823-834 (2006)
[18] Xu, Q.; Schwarzenbacher, R.; Krishna, S.S.; McMullan, D.; Agarwalla, S.; Quijano, K.; Abdubek, P.; Ambing, E.; Axelrod, H.; Biorac, T.; Canaves, J.M.; Chiu, H.; Elsliger, M.; Grittini, C.; Grzechnik, S.K.; DiDonato, M.; Hale, J.; Hampton, E.; Han, G.W.; H: Crystal structure of acireductone dioxygenase (ARD) from Mus musculus at 2.06 A resolution. Proteins, **64**, 808-813 (2006)

Acireductone dioxygenase [iron(II)-requiring] 1.13.11.54

1 Nomenclature

EC number
1.13.11.54

Systematic name
1,2-dihydroxy-5-(methylthio)pent-1-en-3-one:oxygen oxidoreductase (formate-forming)

Recommended name
acireductone dioxygenase [iron(II)-requiring]

Synonyms
ARD' <1, 4> [1, 5]
MTCBP-1 <2> [4]
OsARD1 <5> [3]
Ymr009p <3> [4]
aci-reductone dioxygenase <2, 3> [4]

CAS registry number
221681-63-6
221681-64-7

2 Source Organism

<1> *Mus musculus* (no sequence specified) [5]
<2> *Homo sapiens* (no sequence specified) [4]
<3> *Saccharomyces cerevisiae* (no sequence specified) [4]
<4> *Klebsiella pneumoniae* (no sequence specified) [1]
<5> *Oryza sativa* (no sequence specified) [3]
<6> *Klebsiella oxytoca* (UNIPROT accession number: Q9ZFE7) [2]

3 Reaction and Specificity

Catalyzed reaction
1,2-dihydroxy-5-(methylthio)pent-1-en-3-one + O_2 = 4-(methylthio)-2-oxobutanoate + formate

Reaction type

oxidation (<2,3> ARD (Fe), converts the substrate to formate and a keto-acid precursor of methionine that is then converted to methionine (on-pathway) [4]; <1> if Fe^{2+} is bound in the active site, the substrate acireductone reacts with O_2 to yield formate and 3-(methylthio)propanoate [5])

Natural substrates and products

S 1,2-dihydroxy-5-(methylthio)pent-1-en-3-one + O_2 <4> (Reversibility: ?) [1]
P 3-(methylthio)propanoate + formate + CO
S 1,2-dihydroxy-5-(methylthio)pent-1-en-3-one + O_2 <6> (<6> enzyme of the methionine salvage pathway [2]) (Reversibility: ?) [2]
P 4-(methylthio)oxobutanoate + formate
S Additional information <5> (<5> OsARD1 is a primary ethylene response gene. Enzyme catalyzes the penultimate step in the methionine cycle. [3]) (Reversibility: ?) [3]
P ?

Substrates and products

S 1,2-dihydroxy-3-oxopent-1-ene + O_2 <5> (Reversibility: ?) [3]
P 2-oxo-butanoate + formate
S 1,2-dihydroxy-5-(methylthio)pent-1-en-3-one + O_2 <4> (Reversibility: ?) [1]
P 3-(methylthio)propanoate + formate + CO
S 1,2-dihydroxy-5-(methylthio)pent-1-en-3-one + O_2 <2, 3> (Reversibility: ir) [4]
P 2-keto-4-methyl thiobutyrate + formate
S 1,2-dihydroxy-5-(methylthio)pent-1-en-3-one + O_2 <1> (Reversibility: ?) [5]
P 3-(methylthio)propanoate + formate
S 1,2-dihydroxy-5-(methylthio)pent-1-en-3-one + O_2 <4, 6> (<6> enzyme of the methionine salvage pathway [2]; <6> Fe^{2+} bound to the enzyme protein. If Ni^{2+} or Co^{2+} is bound instead of Fe+, the reaction catalzed by EC 1.13.11.53 occurrs instead [2]; <4> ordered-sequential mechanism [1]) (Reversibility: ?) [1, 2]
P 4-(methylthio)oxobutanoate + formate
S 1,2-dihydroxyhex-1-en-3-one + O_2 <4> (<4> incorporation of O_2 into C_1 and C_2 of 1,2-dihydroxy-3-keto-1-hexene [1]) (Reversibility: ?) [1]
P 2-oxopentanoate + formate
S Additional information <5> (<5> OsARD1 is a primary ethylene response gene. Enzyme catalyzes the penultimate step in the methionine cycle. [3]) (Reversibility: ?) [3]
P ?

Metals, ions

Fe <4> (<4> enzyme contains 1 atom of Fe [1]) [1]
Fe^{2+} <1, 5, 6> (<6> apoenzyme is catalytically inactive. Addition of Fe^{2+} yields activity. Production of the enzyme in intact Escherichia coli depends

on the availability of the Fe^{2+}. Enzyme contains 0.9 Fe^{2+} per enzyme molecule [2]; <5> bacterially expressed AsARD1 preferentially binds Fe^{2+} rather than Ni^{2+} [3]; <1> Fe^{2+} can be replaced by Mg^{2+}, albeit with lower activity [5]) [2, 3, 5]

Turnover number (min^{-1})

11.7 <5> (1,2-dihydroxy-3-oxopent-1-ene) [3]
11.8 <5> (O_2) [3]
260 <4> (1,2-dihydroxyhex-1-en-3-one) [1]
260 <4> (O_2) [1]

K_m-Value (mM)

0.047 <4> (O_2) [1]
0.052 <4> (1,2-dihydroxyhex-1-en-3-one) [1]
1.1 <5> (O_2) [3]
25.9 <5> (1,2-dihydroxy-3-oxopent-1-ene) [3]
210 <6> (1,2-dihydroxy-5-(methylthio)pent-1-en-3-one) [2]
210 <6> (O_2) [2]

4 Enzyme Structure

Molecular weight

20240 <4, 6> (<4,6> mass spectrometry [1,2]) [1, 2]

Subunits

monomer <1, 4> (<4> 1 * 20000, SDS-PAGE [1]; <1> X-ray crystallography [5]) [1, 5]

5 Isolation/Preparation/Mutation/Application

Source/tissue

internode <5> (<5> strongly induced in adventitious roots and in the youngest internode of partially submerged plants [3]) [3]
root <5> (<5> strongly induced in adventitious roots and in the youngest internode of partially submerged plants [3]) [3]

Purification

<1> [5]
<4> [1]
<5> (recombinant) [3]
<6> (recombinant) [2]

Cloning

<1> (expressed in Escherichia coli strain BL21(DE3)pLysS) [5]
<5> (expression in Escherichia coli) [3]
<6> (expression in Escherichia coli) [2]

Engineering

E91A <3> (<3> no activity [4]) [4]
E94A <2> (<2> no activity [4]) [4]
H98S <1> (<1> mutation results in the formation of a stable soluble protein that while structurally different from ARD shows a high degree of similarity to the ARD enzyme [5]) [5]

References

[1] Dai, Y.; Pochapsky, T.C.; Abeles, R.H.: Mechanistic studies of two dioxygenases in the methionine salvage pathway of Klebsiella pneumoniae. Biochemistry, **40**, 6379-6387 (2001)
[2] Dai, Y.; Wensink, P.C.; Abeles, R.H.: One protein, two enzymes. J. Biol. Chem., **274**, 1193-1195 (1999)
[3] Sauter, M.; Lorbiecke, R.; Ouyang, B.; Pochapsky, T.C.; Rzewuski, G.: The immediate-early ethylene response gene OsARD1 encodes an acireductone dioxygenase involved in recycling of the ethylene precursor S-adenosylmethionine. Plant J., **44**, 718-729 (2005)
[4] Hirano, W.; Gotoh, I.; Uekita, T.; Seiki, M.: Membrane-type 1 matrix metalloproteinase cytoplasmic tail binding protein-1 (MTCBP-1) acts as an eukaryotic aci-reductone dioxygenase (ARD) in the methionine salvage pathway. Genes Cells, **10**, 565-574 (2005)
[5] Ju, T.; Goldsmith, R.B.; Chai, S.C.; Maroney, M.J.; Pochapsky, S.S.; Pochapsky, T.C.: One protein, two enzymes revisited: A structural entropy switch interconverts the two isoforms of acireductone dioxygenase. J. Mol. Biol., **363**, 823-834 (2006)

Sulfur oxygenase/reductase 1.13.11.55

1 Nomenclature

EC number
1.13.11.55

Systematic name
sulfur:oxygen oxidoreductase (hydrogen-sulfide- and sulfite-forming)

Recommended name
sulfur oxygenase/reductase

Synonyms
SOR <1, 2, 4, 5> [1, 2, 4, 5, 7]
sulfur oxygenase reductase <1> [5]

CAS registry number
120598-92-7

2 Source Organism

<1> *Acidianus ambivalens* (no sequence specified) [4, 5, 6, 8]
<2> *Acidianus sp.* (no sequence specified) [2]
<3> *Acidianus tengchongensis* (no sequence specified) [3]
<4> *Acidianus ambivalens* (UNIPROT accession number: P29082) [5, 7]
<5> *Acidianus sp.* (UNIPROT accession number: Q977W3) [1]

3 Reaction and Specificity

Catalyzed reaction
4 sulfur + 4 H_2O + O_2 = 2 hydrogen sulfide + 2 bisulfite + 2 H^+

Natural substrates and products
S sulfur + H_2O + O_2 <1> (<1> initial enzyme in the sulfur oxidation pathway [5]) (Reversibility: ?) [5]
P ?

Substrates and products
S S + O_2 <1, 2, 5> (Reversibility: ?) [1, 2, 4]
P SO_3^{2-} + $S_2O_3^{2-}$ + H_2S
S S + O_2 + H_2O <1, 3> (Reversibility: r) [3, 4]
P HSO_3^- + H_2S + H^+

S S + OH^- + O_2 <4> (Reversibility: r) [7]
P HSO_3^- + $S_2O_3^{2-}$ + HS- + H^+
S sulfur + H_2O + O_2 <1> (<1> initial enzyme in the sulfur oxidation pathway [5]) (Reversibility: ?) [5]
P ?

Inhibitors

2-iodoacetic acid <4> [7]
Co^{2+} <3> (<3> 1 mM,89.2% inhibition [3]) [3]
$CoCl_2$ <3> (<3> 1 mM, 70°C, 10.8% activity [3]) [3]
Cu^{2+} <3> (<3> 1 mM, 99.3% inhibition [3]) [3]
$CuCl_2$ <3> (<3> 1 mM, 70°C, 0.7% activity [3]) [3]
FAD <4> [7]
Fe^{2+} <4> [7]
Fe^{3+} <4> [7]
Mg^{2+} <3> (<3> 1 mM, 22.8% inhibition [3]) [3]
$MgCl_2$ <3> (<3> 1 mM, 70°C, 77.2% activity [3]) [3]
Mn^{2+} <3> (<3> 1 mM, 65.5% inhibition [3]) [3]
$MnCl_2$ <3> (<3> 1 mM, 70°C, 34.5% activity [3]) [3]
N-ethylmaleimide <3, 4> (<3> 0.1-1 mM, 70°C, 0.02-0.03% activity [3]) [3, 7]
NEM <3> (<3> 0.1 mM, complete [3]) [3]
Ni^{2+} <3> (<3> 1 mM, 87.5% inhibition [3]) [3]
$NiCl_2$ <3> (<3> 1 mM, 70°C, 12.5% activity [3]) [3]
Zn^{2+} <3, 4> (<3> 1 mM, 73% inhibition [3]; <3> 1 mM, 70°C, 27% activity [3]; <4> concentrations above 100 micromol [7]) [3, 7]
p-chloromercuribenzoic acid <4> [7]
Additional information <4> (<4> not inhibited by CN^-, N_2 and reduced glutathione [7]) [7]

Activating compounds

dithiothreitol <3> (<3> 0.05 mM, 70°C, 109.3% activity [3]; <3> 0.05 mM, 9.3% increase of activity [3]) [3]
EDTA <1, 3> (<1> 1-100 mM, activating, 85°C, pH 7.2, EDTA washed sulfur [4]; <3> 10 mM, 70Â°C, 108.8% activity [3]; <3> 0.05 mM, 8.8% increase of activity [3]) [3, 4]
GSH <3> (<3> 0.05 mM, 10.8% increase of activity [3]) [3]
glutathione <3> (<3> 0.05 mM, 70°C, 110% activity [3]) [3]

Metals, ions

Fe <1> (<1> iron content: 0.45 mol per mol subunit for recombinant wild-type enzyme, below 0.1 mol per mol subunit for mutant enzyme H86A, below 0.02 mol per mol subunit for mutant enzyme H90A, below 0.01 mol per mol subunit for mutant enzyme E114A, 0.02 mol per mol subunit for mutant enzyme E114D, 0.47 mol per mol subunit for mutant enzyme C31A, 0.42 mol per mol subunit for mutant enzyme C31S, 0.22 mol per mol subunit for mutant enzyme C101A, below 0.03 mol per mol subunit for mutant enzyme C101S, 0.19 mol per mol subunit for mutant enzyme C104A, 0.3 mol per mol subunit for mutant enzyme C104S, 0.56 mol per mol subunit for mutant

enzyme C101A/C104A, 0.4 mol per mol subunit for mutant enzyme C101S/ C104S [5]) [5]
Fe^{2+} <1, 4> (<1> low potential mononuclear non-heme iron center [4]; <4> mononucler non-heme iron site [5]) [4, 5]
Zn^{2+} <4> (<4> 0.01 mM or 1-2 mM in crude extracts [7]) [7]

Specific activity (U/mg)

0.02 <4> (<4> mutant C101S, U/mg, reductase reaction [5]) [5]
0.03 <4> (<4> mutant E114D, U/mg, reductase reaction [5]) [5]
0.04 <3> (<3> mutant C104S, formation of sulfite plus thiosulfate, 70°C [3]) [3]
0.074 <3> (<3> wild type in cellular lysate, formation of sulfite plus thiosulfate, 70°C [3]) [3]
0.08 <3> (<3> mutant C101S, formation of sulfite plus thiosulfate, 70°C [3]) [3]
0.1078 <3> (<3> wild type in supernatant, formation of sulfite plus thiosulfate, 70°C [3]) [3]
0.11 <4> (<4> mutant E114D, U/mg, oxygenase reaction [5]) [5]
0.12 <4> (<4> mutant C101S, U/mg, oxygenase reaction [5]) [5]
0.23 <4> (<4> mutant C104A, U/mg, reductase reaction [5]) [5]
0.43 <4> (<4> mutant C101A, U/mg, reductase reaction [5]) [5]
0.476 <3> (<3> wild type in pellet, formation of sulfite plus thiosulfate, 70°C [3]) [3]
0.5 <4> (<4> cytoplasm, pH 7.4, 85°C, sulfur reduction [7]) [7]
0.6 <1, 4> (<4> mutant C104S, U/mg, reductase reaction [5]; <1> wild type, hydrogen sulfide for reductase reaction, aerobically, 85°C, 70 mM Tris/HCl, pH 7.2, 0.1% (v/v) Tween 20 [4]) [4, 5]
0.64 <4> (<4> mutant C101/104A, U/mg, reductase reaction [5]) [5]
0.66 <1> (<1> recombinant soluble SOR, hydrogen sulfide for reductase reaction, aerobically, 85°C, 70 mM Tris/HCl, pH 7.2, 0.1% (v/v) Tween 20 [4]) [4]
1.12 <4> (<4> mutant C101/104A, U/mg, oxygenase reaction [5]) [5]
1.17 <4> (<4> mutant C101/104S, U/mg, reductase reaction [5]) [5]
1.35 <1, 4> (<4> mutant C104A, U/mg, oxygenase reaction [5]; <1> recombinant SOR from inclusion bodies, hydrogen sulfide for reductase reaction, aerobically, 85°C, 70 mM Tris/HCl, pH 7.2, 0.1% (v/v) Tween 20 [4]) [4, 5]
1.47 <4> (<4> mutant C101, U/mg, oxygenase reaction [5]) [5]
1.89 <4> (<4> cytoplasm, pH 7.4, 85°C, sulfur oxidation [7]) [7]
2.28 <4> (<4> mutant C104S, U/mg, oxygenase reaction [5]) [5]
2.29 <4> (<4> mutant C101/104S, U/mg, oxygenase reaction [5]) [5]
2.82 <1> (<1> recombinant soluble SOR, sulfite plus thiosulfite for oxgenase reaction, aerobically, 85°C, 70 mM Tris/HCl, pH 7.2, 0.1% (v/v) Tween 20 [4]) [4]
2.96 <1> (<1> wild type, sulfite plus thiosulfite for oxgenase reaction, aerobically, 85°C, 70 mM Tris/HCl, pH 7.2, 0.1% (v/v) Tween 20 [4]) [4]
4.19 <4> (<4> wild type, U/mg, reductase reaction [5]) [5]
4.85 <3> (<3> wild type, formation of sulfite plus thiosulfate, 70°C [3]) [3]

5.98 <1> (<1> recombinant SOR from inclusion bodies, sulfite plus thiosulfite for oxgenase reaction, aerobically, 85°C, 70 mM Tris/HCl, pH 7.2, 0.1% (v/v) Tween 20 [4]) [4]
10.6 <4> (<4> pH 7.4, 85°C, formation of sulfite [7]) [7]
11.52 <4> (<4> wild type, U/mg, oxygenase reaction [5]) [5]
186.7 <2> (<2> formation of sulfite and thiosulfate [2]) [2]
3100 <5> (<5> wild type, cell lysate, 65°C, 20 mM Tris-HCl, pH 8.0, formation of H_2S [1]) [1]
3300 <5> (<5> wild type, cell lysate treated at 75°C for 15 min, 65°C, 20 mM Tris-HCl, pH 8.0, formation of H_2S [1]) [1]
6300 <5> (<5> Escherichia coli HB101, cell lysate, 65°C, 20 mM Tris-HCl, pH 8.0, formation of H_2S [1]) [1]
28600 <5> (<5> wild type, cell lysate, 65°C, 20 mM Tris-HCl, pH 8.0, formation of thiosulfate and sulfite [1]) [1]
29700 <5> (<5> wild type, cell lysate treated at 75°C for 15 min, 65°C, 20 mM Tris-HCl, pH 8.0, formation of thiosulfate and sulfite [1]) [1]
45200 <5> (<5> Escherichia coli HB101, cell lysate treated at 75°C for 15 min, 65°C, 20 mM Tris-HCl, pH 8.0, formation of H_2S [1]) [1]
75000 <5> (<5> Escherichia coli HB101, cell lysate, 65°C, 20 mM Tris-HCl, pH 8.0, formation of thiosulfate and sulfite [1]) [1]
753000 <5> (<5> Escherichia coli HB101, cell lysate treated at 75°C for 15 min, 65°C, 20 mM Tris-HCl, pH 8.0, formation of thiosulfate and sulfite [1]) [1]

K_m-Value (mM)
2-3 <1> (sulfur, <1> oxygenase, 65°C, 0.5% (v/v) CHAPS [4]) [4]
13 <1> (sulfur, <1> reductase, 65°C, 0.5% (v/v) CHAPS [4]) [4]

pH-Optimum
5 <2> [2]
7-7.4 <4> [7]

pH-Range
3.5-9 <2> [2]
4-8 <4> [7]

Temperature optimum (°C)
70 <2> [2]
85 <4> [7]

Temperature range (°C)
50-90 <2> [2]
50-108 <4> [7]

4 Enzyme Structure

Molecular weight

35000 <2, 5> (<2,5> SDS-PAGE [1,2]) [1, 2]
35320 <1, 4> (<1> calculated from sequencing the sor gene [6]; <4> calculated from sor gene code [7]) [6, 7]
72000 <1> (<1> SDS-PAGE, homodimer [4]) [4]
550000 <1> (<1> gel filtration, oligomer with ferritin-like assembly [4]) [4]
560000 <4> (<4> gel filtration [7]) [7]
732000 <1> (<1> Blue-native PAGE [4]) [4]
871000 <1> (<1> calculated from crystal unit cell volume, the non-crystallographic symmetry operators and electron micsroscopy studies [4]) [4, 8]

Subunits

hexadecamer <4> (<4> 16 * 35317, calculated from sor gene code [7]) [7]
homodimer <1> (<1> 2 * 37000, SDS-PAGE [4]) [4]
icosatetramer <1> [8]
tetradecamer <4> (<4> 14 * 40000, SDS-PAGE [7]) [7]
tetramer <1> (<1> icosatetramer [4]) [4]

5 Isolation/Preparation/Mutation/Application

Localization

cytoplasm <1, 3, 4> (<3> also associated with the cytoplasmic membrane [3]) [3, 4, 7]
cytoplasmic membrane <3> (<3> membrane-associated enzyme activity is colocalized with the activities of sulfite:acceptor oxidoreductase and thiosulfate:acceptor oxidoreductase to catalyze sulfur oxidation [3]) [3]

Purification

<1> [4, 6]
<1> (dialysation and DEAE Sepharose at recombinant SOR, dialysation, phenyl-Sepharose, DEAE sepharose and superose 6HR 10/30 gel-filtration at wild type) [4]
<2> (heat treatment, DEAE-52 anion exchange and Sephadex G-200 chromatography) [2]
<3> [3]
<3> (Superdex 200 gel-filtration) [3]
<4> (one-step procedure over Strep-Tactin columns) [5]
<4> (sucrose density gradient centrifugation and preparative, non-denaturing PAGE) [7]

Crystallization

<1> (single colorless polyhedral crystals by sitting drop vapour diffusion techniques at -75°C and 32°C in either 0.1 M sodium acetate pH 4.5 with 8% MPD and 50 mM NaCl or 0.1 M sodium citrte pH 5.5 with 0.3 M $MgSO_4$, 1.7 A resolution) [4]

<1> (sitting drop vapour diffusion technique of recombinant enzyme, 1.7 A resolution) [8]

Cloning

<1> (expression in Escherichia coli) [4, 6]
<1> (expression in Escherichia coli BL21) [4]
<2> (expressed in Escherichia coli HB101) [2]
<3> (expression in Escherichia coli HB101) [3]
<3> (overexpression of wild-type and mutant enzymes in Escherichia coli) [3]
<4> [7]
<4> (expression in Escherichia coli BL21) [5]
<5> (overexpressed by Escherichia coli HB 101 opon a temperature shift from 30-42°C) [1]

Engineering

C101/104A <4> (<4> enzyme with decreased specific activity [5]) [5]
C101/104S <4> (<4> enzyme with decreased specific activity [5]) [5]
C101A <1, 4> (<4> enzyme with decreased specific activity [5]; <1> iron content is 49% of that of the recombinant wild-type enzyme, oxygenase activity is 12.8% of the activity of recombinant wild-type enzyme, reductase activity is 10.3% of the activity of recombinant wild-type enzyme [5]) [5]
C101A/C104A <1> (<1> iron content is 124% of that of the recombinant wild-type enzyme, oxygenase activity is 9.7% of the activity of recombinant wild-type enzyme, reductase activity is 15.3% of the activity of recombinant wild-type enzyme [5]) [5]
C101S <1, 3, 4> (<4> enzyme with decreased specific activity and a proportional decrease in iron content [5]; <3> mutant with reduced activity [3]; <3> 98.4% loss of activity [3]; <1> mutant enzyme contains no iron, oxygenase activity is 1.04% of the activity of recombinant wild-type enzyme, reductase activity is 0.5% of the activity of recombinant wild-type enzyme [5]) [3, 5]
C101S/C104S <1> (<1> iron content is 89% of that of the recombinant wild-type enzyme, oxygenase activity is 19.8% of the activity of recombinant wild-type enzyme, reductase activity is 27.9% of the activity of recombinant wild-type enzyme [5]) [5]
C104A <1, 4> (<4> enzyme with decreased specific activity [5]; <1> iron content is 42% of that of the recombinant wild-type enzyme, oxygenase activity is 11.8% of the activity of recombinant wild-type enzyme, reductase activity is 5.5% of the activity of recombinant wild-type enzyme [5]) [5]
C104S <1, 3, 4> (<4> enzyme with decreased specific activity [5]; <3> mutant with reduced activity [3]; <3> 99.2% loss of activity [3]; <1> iron content is 67% of that of the recombinant wild-type enzyme, oxygenase activity is 19.8% of the activity of recombinant wild-type enzyme, reductase activity is 14.3% of the activity of recombinant wild-type enzyme [5]) [3, 5]
C31A <1, 3, 4> (<3> complete loss of activity [3]; <4> inactive enzyme [5]; <3> mutant with reduced activity [3]; <1> inactive mutant enzyme, iron content is similar to that of recombinant wild-type enzyme [5]) [3, 5]

C31S <1, 3, 4> (<3> complete loss of activity [3]; <4> inactive enzyme [5]; <3> mutant with reduced activity [3]; <1> inactive mutant enzyme, iron content is similar to that of recombinant wild-type enzyme [5]) [3, 5]
E114A <1, 4> (<4> inactive enzyme with no iron incorporated [5]; <1> mutation results in inactive enzyme with no measurable iron found [5]) [5]
E114D <1, 4> (<4> enzyme with 1% of wild type activity [5]; <1> iron content is 4.4% of wild-type value, sulfur-oxidizing and sulfur-reducing activity is about 1% of the activity of activity of the recombinant wild-type enzyme [5]) [5]
H86A <1, 4> (<4> inactive enzyme with no iron incorporated [5]; <1> mutation results in inactive enzyme with no measurable iron found [5]) [5]
H90A <1, 4> (<4> inactive enzyme with no iron incorporated [5]; <1> mutation results in inactive enzyme with no measurable iron found [5]) [5]

Application

degradation <1, 2, 4> (<2> enzyme in the sulfur-oxidation pathway [2]; <1,4> initial enzyme in the sulfur-oxidation pathway [4,5,7]) [2, 4, 5, 7]
pharmacology <1> (<1> sulfur metabolism [6]) [6]

6 Stability

Storage stability

<1, 3>, -20°C [3, 4]

References

[1] He, Z.; Li, Y.; Zhou, P.; Liu, S.: Cloning and heterologous expression of a sulfur oxygenase/reductase gene from the thermoacidophilic archaeon Acidianus sp. S5 in Escherichia coli. FEMS Microbiol. Lett., **193**, 217-221 (2000)

[2] Sun, C.W.; Chen, Z.W.; He, Z.G.; Zhou, P.J.; Liu, S.J.: Purification and properties of the sulfur oxygenase/reductase from the acidothermophilic archaeon, Acidianus strain S5. Extremophiles, **7**, 131-134 (2003)

[3] Chen, Z.W.; Jiang, C.Y.; She, Q.; Liu, S.J.; Zhou, P.J.: Key role of cysteine residues in catalysis and subcellular localization of sulfur oxygenase-reductase of Acidianus tengchongensis. Appl. Environ. Microbiol., **71**, 621-628 (2005)

[4] Urich, T.; Coelho, R.; Kletzin, A.; Frazao, C.: The sulfur oxygenase reductase from Acidianus ambivalens is an icosatetramer as shown by crystallization and Patterson analysis. Biochim. Biophys. Acta, **1747**, 267-270 (2005)

[5] Urich, T.; Kroke, A.; Bauer, C.; Seyfarth, K.; Reuff, M.; Kletzin, A.: Identification of core active site residues of the sulfur oxygenase reductase from Acidianus ambivalens by site-directed mutagenesis. FEMS Microbiol. Lett., **248**, 171-176 (2005)

[6] Kletzin, A.: Molecular characterization of the sor gene, which encodes the sulfur oxygenase/reductase of the thermoacidophilic Archaeum Desulfurolobus ambivalens. J. Bacteriol., **174**, 5854-5859 (1992)

[7] Kletzin, A.: Sulfur oxidation and reduction in Archaea: Sulfur oxygenase/-reductase and hydrogenases from the extremely thermophilic and facultatively anaerobic archaeon Desulfurolobus ambivalens. Syst. Appl. Microbiol., **16**, 534-543 (1994)
[8] Urich, T.; Coelho, R.; Kletzin, A.; Frazao, C.: The sulfur oxygenase reductase from Acidianus ambivalens is an icosatetramer as shown by crystallization and Patterson analysis. Biochim. Biophys. Acta, **1747**, 267-| (2005)

Oplophorus-luciferin 2-monooxygenase 1.13.12.13

1 Nomenclature

EC number
1.13.12.13

Systematic name
Oplophorus-luciferin:oxygen 2-oxidoreductase (decarboxylating)

Recommended name
Oplophorus-luciferin 2-monooxygenase

Synonyms
Oplophorus luciferase

2 Source Organism

<1> *Oplophorus gracilorostris* (no sequence specified) [1, 2]
<2> *Diaphus gigas* (no sequence specified) [3]

3 Reaction and Specificity

Catalyzed reaction
Oplophorus luciferin + O_2 = oxidized Oplophorus luciferin + CO_2 + hν

Substrates and products
S 3-hydroxy-2-methylimidazol[1,2-a]pyridine + O_2 <1> (Reversibility: ?) [1]
P oxidized 3-hydroxy-2-methylimidazol[1,2-a]pyridine + CO_2 + hν
S Oplophorus luciferin + O_2 <2> (Reversibility: ?) [3]
P oxidized Oplophorus luciferin + CO_2 + hν
S bisdeoxycoelenterazine + O_2 <1> (Reversibility: ?) [2]
P oxidized bisdeoxycoelenterazine + CO_2 + hν
S coelenterazine + O_2 <1> (Reversibility: ?) [2]
P oxidized coelenterazine + CO_2 + hν

5 Isolation/Preparation/Mutation/Application

Source/tissue
muscle <2> [3]

Purification

<1> [1]

References

[1] Yamaguchi, I.: Oplophorus oxyluciferin and a model luciferin compound biologically active with Oplophorus luciferase. Biochem. J., **151**, 9-15 (1975)
[2] Nakamura, H.; Wu, C.; Murai, A.; Inouye, S.; Shimomura, O.: Efficient bioluminescence of bisdeoxycoelenterazine with the luciferase of a deep-sea shrimp Oplophorus. Tetrahedron Lett., **38**, 6405-6406 (1997)
[3] Kakoi, H.; Okada, K.: Chemical studies of fish bioluminescence. ITE Lett. Batteries New Technol. Med., **6**, 38-45 (2005)

Chlorophyllide-a oxygenase 1.13.12.14

1 Nomenclature

EC number
1.13.12.14

Systematic name
chlorophyllide-a:oxygen 7-oxidoreductase

Recommended name
chlorophyllide-a oxygenase

Synonyms
CAO <1, 3, 4, 5, 6, 8, 9, 10> [1, 2, 3, 5, 6, 7, 8, 9, 10, 11, 12, 13, 14, 15, 18]
chlorophyll a oxygenase <5, 8, 9, 10> [11, 12, 15, 16]
chlorophyllide a oxygenase

CAS registry number
216503-73-0

2 Source Organism

<1> *Chlamydomonas reinhardtii* (no sequence specified) [2]
<2> *Zea mays* (no sequence specified) [4]
<3> *Nicotiana tabacum* (no sequence specified) [1]
<4> *Arabidopsis thaliana* (no sequence specified) (<4> NarG, α subunit [7]) [1, 2, 5, 6, 7, 8, 9, 10, 14]
<5> *Oryza sativa* (no sequence specified) [3,12]
<6> *Prochlorothrix hollandica* (no sequence specified) [5,13,18]
<7> *Prochlorococcus marinus* (no sequence specified) [17]
<8> *Arabidopsis thaliana* (UNIPROT accession number: Q9MBA1) [13,16]
<9> *Chlamydomonas reinhardtii* (UNIPROT accession number: Q9ZWM5) [15]
<10> *Dunaliella salina* (UNIPROT accession number: Q9XJ38) [11]

3 Reaction and Specificity

Catalyzed reaction
7-hydroxychlorophyllide a + O_2 + NADPH + H^+ = chlorophyllide b + 2 H_2O + $NADP^+$ (<9> reaction mechanism [15]; <2> reaction mechanism, mechanism of 7-formyl group formation [4]; <1> reaction mechanism, the catalytic

site harbors a stable radical formed by Tyr422 which is involved in the catalytic reaction [2]; <4> reaction mechanism, the catalytic site harbors a stable radical formed by Tyr518 which is involved in the catalytic reaction [2]; <8> reaction mechanism, the enzyme C-terminal C domain is sufficient for catalytic activity [13])
chlorophyllide a + O_2 + NADPH + H^+ = 7-hydroxychlorophyllide a + H_2O + $NADP^+$ (<2,9> reaction mechanism [4,15]; <1> reaction mechanism, the catalytic site harbors a stable radical formed by Tyr422 which is involved in the catalytic reaction [2]; <4> reaction mechanism, the catalytic site harbors a stable radical formed by Tyr518 which is involved in the catalytic reaction [2]; <8> reaction mechanism, the enzyme C-terminal C domain is sufficient for catalytic activity [13])

Natural substrates and products

S 7-hydroxychlorophyllide a + O_2 + NADPH <1, 2, 3, 4, 6, 8, 9> (Reversibility: ?) [1, 2, 4, 5, 7, 9, 10, 13, 15]
P chlorophyllide b + H_2O + $NADP^+$
S chlorophyllide a + O_2 + NADPH <1, 2, 3, 4, 6, 8, 9> (Reversibility: ?) [1, 2, 4, 5, 7, 9, 10, 13, 15]
P 7-hydroxychlorophyllide a + H_2O + $NADP^+$
S chlorophyllide a + O_2 + NADPH <5> (Reversibility: ?) [12]
P chlorophyllide b + H_2O + $NADP^+$
S Additional information <1, 2, 3, 4, 5, 6, 8, 9, 10> (<10> chlorophyll antenna size adjustments by irradiance involve coordinate regulation of CAO and Lhcb gene expression, cytosolic signal transduction pathway for rapid enzyme regulation/short term photoacclimation involves phospholipase C activity, mechanism, the redox state of the plastoquinone pool is also involved in enzyme regulation via the long term photoacclimation, mechanism of photoacclimation, overview [11]; <1,4> chlorophyll b is required for assembly of the light-harvesting complexes [2]; <9> chlorophyll b production is important for plants to adapt to varying light environments [15]; <4> chlorophyll b synthesis by the enzyme is a key regulatory step in the control of antenna size, changing the antenna size of photosystems allows plants to acclimate to variations in light intensity for efficient photosynthesis creating a protective strategy to minimize photodamage, the chlorophyll b feedback regulates enzyme accumulation via the N-terminal domain A [8]; <4> chlorophyll b synthesis is involved in regulation of photosystem II antenna size via chlorophyll b content in the light-harvesting complex II [10]; <4> enzyme mRNA and protein levels correlate with the chlorophyll a/b ratio, enzyme expression is regulated by irradiation, enzyme activity in required for enlargment of light harvesting complex II during photoacclimation [6]; <5> isozyme CAO1 is important in chlorophyll b biosynthesis [12]; <4> key enzyme for chlorophyll b biosynthesis [9]; <4> land plants change the compositions of light-harvesting complexes and chlorophyll a/b ratios in response to the variable light environments which they encounter, and the enzyme is part of the regulatory mechanism, overview [7]; <2> pathway validation [4]; <4> the en-

zyme is involved in antenna pigment synthesis required for light-harvesting complex II and photosynthetic growth, chlorophyll b has a regulatory role in pigment shuffling of antenna systems, overview [5]; <6> the enzyme is involved in antenna pigment synthesis required for light-harvesting complexe II and photosynthetic growth, chlorophyll b has a regulatory role in pigment shuffling of antenna systems, overview [5]; <4> the enzyme is involved in light-dependent regulation of chlorophyll b biosynthesis which is related to the cab gene expression, chlorophyllide b is bound to the light harvesting complex II in vivo [1]; <3> the enzyme is involved in light-dependent regulation of chlorophyll b biosynthesis, chlorophyllide b is bound to the light harvesting complex II in vivo [1]; <4> the enzyme is part of a separate translocon complex involved in the regulated import and stabilization of the chlorophyllide b-binding light-harvesting proteins Lhcb1, of LHCII, and Lhcb4, of CP29, in chlorplasts [14]; <8> the enzyme is responsible for synthesis of chlorophyllide b which is a photosynthetic antenna pigment required for regulation of antenna size, overview [13]; <6> a chloroplast Clp protease is involved in regulating chlorophyll b biosynthesis through the destabilization of chlorophyllide a oxygenase in response to the accumulation of chlorophyll b [18]) (Reversibility: ?) [1, 2, 4, 5, 6, 7, 8, 9, 10, 11, 12, 13, 14, 15, 18]

P ?

Substrates and products

S 7-hydroxychlorophyllide a + O_2 + NADPH <1, 2, 3, 4, 6, 8, 9> (Reversibility: ?) [1, 2, 4, 5, 7, 9, 10, 13, 14, 15]

P chlorophyllide b + H_2O + $NADP^+$ (<2,4> product identification [4,9])

S Zn-7-hydroxy-pheide a + O_2 + NADPH <4> (Reversibility: ?) [9]

P Znpheide b + H_2O + $NADP^+$ (<4> product identification [9])

S Znpheide a + O_2 + NADPH <4> (Reversibility: ?) [9]

P Zn-7-hydroxy-pheide a + H_2O + $NADP^+$ (<4> product identification [9])

S chlorophyllide a + O_2 + NADPH <1, 2, 3, 4, 6, 8, 9> (Reversibility: ?) [1, 2, 4, 5, 7, 9, 10, 13, 14, 15]

P 7-hydroxychlorophyllide a + H_2O + $NADP^+$ (<2,4> product identification [4,9])

S chlorophyllide a + O_2 + NADPH <5, 8> (Reversibility: ?) [12, 16]

P chlorophyllide b + H_2O + $NADP^+$

S protochlorophyllide a + O_2 + NADPH <4> (<4> low activity [14]) (Reversibility: ?) [14]

P 7-hydroxychlorophyllide a + H_2O + $NADP^+$

S Additional information <1, 2, 3, 4, 5, 6, 8, 9, 10> (<10> chlorophyll antenna size adjustments by irradiance involve coordinate regulation of CAO and Lhcb gene expression, cytosolic signal transduction pathway for rapid enzyme regulation/short term photoacclimation involves phospholipase C activity, mechanism, the redox state of the plastoquinone pool is also involved in enzyme regulation via the long term photoacclimation, mechanism of photoacclimation, overview [11]; <1,4> chlorophyll b is required for assembly of the light-harvesting complexes [2]; <9> chlorophyll b production is important for plants to adapt to varying light environments [15]; <4> chlorophyll b synthesis by the enzyme is a key regulatory step in the control of antenna size, changing the antenna size of photosystems allows plants to acclimate to variations in light intensity for efficient photosynthesis creating a protective strategy to minimize photodamage, the chlorophyll b feedback regulates enzyme accumulation via the N-terminal domain A [8]; <4> chlorophyll b synthesis is involved in regulation of photosystem II antenna size via chlorophyll b content in the light-harvesting complex II [10]; <4> enzyme mRNA and protein levels correlate with the chlorophyll a/b ratio, enzyme expression is regulated by irradiation, enzyme activity in required for enlargment of light harvesting complex II during photoacclimation [6]; <5> isozyme CAO1 is important in chlorophyll b biosynthesis [12]; <4> key enzyme for chlorophyll b biosynthesis [9]; <4> land plants change the compositions of light-harvesting complexes and chlorophyll a/b ratios in response to the variable light environments which they encounter, and the enzyme is part of the regulatory mechanism, overview [7]; <2> pathway validation [4]; <4> the enzyme is involved in antenna pigment synthesis required for light-harvesting complex II and photosynthetic growth, chlorophyll b has a regulatory role in pigment shuffling of antenna systems, overview [5]; <6> the enzyme is involved in antenna pigment synthesis required for light-harvesting complexe II and photosynthetic growth, chlorophyll b has a regulatory role in pigment shuffling of antenna systems, overview [5]; <4> the enzyme is involved in light-dependent regulation of chlorophyll b biosynthesis which is related to the cab gene expression, chlorophyllide b is bound to the light harvesting complex II in vivo [1]; <3> the enzyme is involved in light-dependent regulation of chlorophyll b biosynthesis, chlorophyllide b is bound to the light harvesting complex II in vivo [1]; <4> the enzyme is part of a separate translocon complex involved in the regulated import and stabilization of the chlorophyllide b-binding light-harvesting proteins Lhcb1, of LHCII, and Lhcb4, of CP29, in chlorplasts [14]; <8> the enzyme is responsible for synthesis of chlorophyllide b which is a photosynthetic antenna pigment required for regulation of antenna size, overview [13]; <4> expression analysis and chlorophyll a/b ratios under various light conditions [7]; <4> reaction mechanism of the two-step reaction, no activity with chlorophyll a or protochlorophyllide a [9]; <6> a chloroplast Clp protease is involved in regulating chlorophyll b biosynthesis through the destabilization of chlorophyllide a oxygenase in response to the accumulation of chlorophyll b [18]) (Reversibility: ?) [1, 2, 4, 5, 6, 7, 8, 9, 10, 11, 12, 13, 14, 15, 18]

P ?

Inhibitors

Additional information <4, 5, 8, 10> (<8> enzyme expression is reduced under dim-light conditions [16]; <4> enzyme expression is regulated by irradiation and is reduced with increasing light intensity [6]; <10> enzyme expression is regulated by irradiation, mechanism of long term and short term photoacclimation, determination of the redox state of the plastoquinone pool influencing the enzyme activity, phospholipase C antagonists inhibit CAO induction [11]; <5> gene CAO1 expression is reduced in the dark, while expression of gene CAO2 is induced [12]; <4> the chlorophyll b feedback regulates enzyme accumulation via the N-terminal domain A [8]; <4> the enzyme activity is light-regulated, increased activity at low light, reduced at high light [1]) [1, 6, 8, 11, 12, 16]

Cofactors/prosthetic groups

NADPH <1, 2, 3, 4, 5, 6, 8, 9> (<4> using a ferredoxin-NADPH reduction system in activity assays [14]) [1,2,4,5,7,9,10,12,13,14,15,16]

Activating compounds

Additional information <5, 8, 10> (<10> enzyme expression is regulated by irradiation, mechanism of long term and short term photoacclimation, determination of the redox state of the plastoquinone pool influencing the enzyme activity [11]; <5> gene CAO1 is induced by light, gene CAO_2 is down-regulated after exposure to light [12]; <8> the enzyme expression is light inducible [16]) [11, 12, 16]

Metals, ions

Fe^{2+} <1, 4, 5, 6, 8, 9> (<4> the enzyme contains a [2Fe-2S] Rieske cluster [14]; <8> the enzyme contains a [2Fe-2S] Rieske cluster and a mononuclear iron-binding site at the C-terminal C domain, required for activity [13]; <6> the enzyme contains a [2Fe-2S] Rieske cluster and a mononuclear iron-binding site, required for activity [13]; <1,4> the enzyme contains a [2Fe-2S] Rieske cluster and a mononuclear non-hem iron-binding site [2]; <5> the enzyme contains a [2Fe-2S] Rieske cluster and a non-hem mononuclear iron-binding site, required for activity [12]; <9> the enzyme might contain a [2Fe-2S] Rieske cluster [15]; <5> the enzyme might contain a [2Fe-2S] Rieske cluster and a mononuclear iron-binding site, required for activity [3]) [2, 3, 12, 13, 14, 15]

Specific activity (U/mg)

Additional information <4, 5> (<5> chlorophyll a nd chlorophyll b contents in wild-type and mutants plants [3]; <4> expression analysis and chlorophyll a/b ratios under various light conditions [7]; <4> the assay is performed in darkness [14]) [3, 7, 14]

K_m-Value (mM)

Additional information <10> (<10> kinetics of photoacclimation [11]) [11]

Temperature optimum (°C)

23 <4> (<4> assay at [14]) [14]

4 Enzyme Structure

Subunits

? <4, 6> (<6> x * 40830, amino acid sequence calculation [13]; <4> x * 54000, recombinant enzyme, SDS-PAGE [2]) [2, 13]
Additional information <4, 8> (<8> the enzyme consists of 3 domains, the N-terminal A domain, the B domain, and the C-terminal C domain [13]; <4> the enzyme consists of three domains A, B, and C, domain structure, the N-terminal domain A of the enzyme confers protein instability in response to chlorophyll B accumulation [8]; <4> the enzyme is part of a separate translocon complex in chlorplasts [14]) [8, 13, 14]

5 Isolation/Preparation/Mutation/Application

Source/tissue

glume <5> (<5> low enzyme expression [3]) [3]
leaf <2, 3, 4, 5, 8> (<4> primary [2]; <5> high enzyme expression [3]; <2> etiolated, greening [4]; <4> low activity in etiolated leaves, higher activity in green mature leaves [10]) [1, 2, 3, 4, 5, 6, 10, 13]
shoot <5> (<5> etiolated, high enzyme expression [3]) [3]
Additional information <4, 5> (<5> CAO1 is expressed in photosynthetic tissue [12]; <5> no enzyme expression in roots [3]; <4> wild-type enzyme expression begins in de-etiolation phase [10]) [3, 10, 12]

Localization

chloroplast <1, 4, 8> (<4> envelope inner membrane [2]; <1> envelope inner membrane, in light and dark grown cells [2]; <4> envelope inner membrane, the N-terminal 56 amino acid residues contain a transit peptide sequence [8]; <4> intrinsic in inner envelope and thylakoid membrane [14]) [2, 5, 6, 8, 10, 13, 14, 16]
thylakoid membrane <1, 3, 4, 8> (<1> only in light grown cells [2]; <8> the enzyme is localized in the light-harvesting complexes [16]) [1, 2, 5, 8, 10, 14, 16]
Additional information <4> (<4> the enzyme is part of a separate translocon complex in chlorplasts [14]) [14]

Purification

<4> (partially by purification of chloroplast envelope inner membrane and thylakoid membranes) [8]
<4> (refolded recombinant His6-tagged CAO from Escherichia coli by nickel affinity chromatography to about 85% purity) [14]
<6> (recombinant enzyme partially by purification of transgenic plant thylakoid membranes) [5]

Renaturation

<4> (solubilization of recombinant His6-tagged enzyme from inclusion bodies after expression in Escherichia coli by 8 M urea and refolding) [14]

Cloning

<1> (expression of GFP-tagged enzyme in Chlamydomonas reinhardtii cells, phylogenetic tree) [2]
<4> (expression in Escherichia coli in membranes, phylogenetic tree) [2]
<4> (expression of GFP-tagged wild-type and GFP-tagged domain mutant enzymes in wild-type and enzyme-deficient chlorin-1 mutant plants) [8]
<4> (expression of His6-tagged CAO in Escherichia coli in inclusion bodies, in vitro expression by wheat germ lysates) [14]
<4> (functional expression in Escherichia coli) [9]
<4> (functional overexpression in transgenic tobacco plants via Agrobacterium tumefaciens LBA4404 transfection, the CAO gene is expressed under control of the 35S CaMV promoter) [1]
<4> (overexpression of the enzyme in Arabidopsis thaliana plants) [10]
<5> (DNA and amino acid sequence determination and analysis of wild-type and mutants genes, gene structures, overview) [3]
<6> (Arabidopsis which overexpress a CAO-GFP fusion are mutagenized) [18]
<6> (expression of the enzyme under control of the 35S CaMV promoter, using the Agrobacterium tumefaciens infection system, in transgenic Arabidopsis thaliana plants renders the transgenic plants capable of normal rate photosynthetic growth also under low light conditions in contrast to the wild-type Arabidopsis thaliana plants, transgenic plants show altered chlorophyll a/b ratio, chloroplast structure, and thylakoid membrane composition) [5]
<6> (introductionof the prokaryotic chlorophyll b synthesis gene for chlorophyllide a oxygenase (CAO) into Arabidopsis. In the transgenic plants (Prochlirothrix hollandica CAO plants), about 40% of chlorophyll a of the core antenna complexes is replaced by chlorophyll b in both photosystems) [5]
<6> (single-copy CAO gene, DNA and amino acid sequence determination and analysis, amino aid sequence and N-terminal extensions comparison with higher plant sequences, e.g. of Arabidopsis thaliana, functional expression in Synechocystis sp. PCC6803) [13]
<7> (introduction of the Prochlorococcus gene into Synechocystis cells) [17]
<8> (gene CAO, DNA and amino acid sequence determination and analysis, genetic organization) [16]
<8> (single-copy CAO gene, amino aid sequence and N-terminal extensions comparison with other higher plant sequences and the sequence of the algae Prochlorothrix hollandica, functional expression of ABC, BC, and C domain constructs in Synechocystis sp. PCC6803 reveals that the C domain is sufficient for catalytic activity) [13]
<9> (construction of a genomic library, DNA and amino acid sequence determination) [15]
<10> (expression analysis with isolated nuclei) [11]

Engineering

Additional information <3, 4, 5, 8, 9> (<5> construction of CAO1 and CAO_2 knockout mutants tagged by T-DNA or Tos-17, the CAO1 deficient mutant shows pale green leaves, the CAO_2 deficient mutant show no altered leaf phenotype compared to the wild-type plants [12]; <4> construction of domain mutants by combinating the three domains A, B, and C for construction of transgenic plants expressing the wild-type enzyme or domain A, or B, or C or a combinantion thereof, domain structure, transgenic subcellular localization, regulatory effects on chlorophyll a/b ratio and protein accumulation, overview [8]; <9> construction of six chlorophyll-b less mutants by insertional mutagenesis, analysis of DNA rearrangements, overview [15]; <4> construction of transgenic plants expressing the enzyme of Prochlirothrix hollandica in thylakoid membranes under control of the 35S CaMV promoter using the Agrobacterium tumefaciens infection system, the transgenic plants are capable of normal rate photosynthetic growth also under low light conditions in contrast to the wild-type plants, transgenic plants show an altered chlorophyll a/b ratio in photosystems I and II, chloroplast structure, and thylakoid membrane composition, phenotype, overview [5]; <4> construction of transgenic plants functionally overexpressing the enzyme under control of the 35S CaMV promoter leading to 10-20% enlarged antenna size of photosystem II [10]; <3> construction of transgenic tobacco plants overexpressing the Arabidopsis thaliana enzyme, the transgenic plants show 72% increased CAO activity and light-dependent regulation of chlorophyll b biosynthesis [1]; <4> construction of transgenic tobacco plants overexpressing the enzyme, the transgenic plants show 72% increased CAO activity and light-dependent regulation of chlorophyll b biosynthesis [1]; <4> enzyme level and activity is reduced in the conditional chlorina 1 mutant cch1 [6]; <8> isolation of 3 mutant plants with increased enzyme mRNA levels but reduced chlorophyll b levels [16]; <5> isolation of pale green rice mutants Y-15, a frame-shift mutant, and G-52, with a mutation of the lysine residue in the mononuclear iron-binding site, both mutant plants lack chlorophyll b, the pseudogene CAO-11 mutant shows no pale green leaf phenotype but a defect in the Rieske cluster [3]; <4> transgenic enzyme-overexpressing plants show accumulation of light-harvesting complexes and reduced β-carotene levels compared to wild-type plants, while it is increased in enzyme-deficient chlorina-1 mutant plants, in overexpressing plants the chlorophyll a/b ratio remains low under high-light conditions in contrast to wild-type plants [7]) [1, 3, 5, 6, 7, 8, 10, 12, 15, 16]

6 Stability

General stability information

<4>, the N-terminal domain of the enzyme confers protein instability in response to chlorophyll B accumulation [8]

<6>, a chloroplast Clp protease is involved in regulating chlorophyll b biosynthesis through the destabilization of CAO protein in response to the accumulation of chlorophyll b [18]

References

[1] Pattanayak, G.K.; Biswal, A.K.; Reddy, V.S.; Tripathy, B.C.: Light-dependent regulation of chlorophyll b biosynthesis in chlorophyllide a oxygenase over-expressing tobacco plants. Biochem. Biophys. Res. Commun., **326**, 466-471 (2005)

[2] Eggink, L.L.; LoBrutto, R.; Brune, D.C.; Brusslan, J.; Yamasato, A.; Tanaka, A.; Hoober, J.K.: Synthesis of chlorophyll b: localization of chlorophyllide a oxygenase and discovery of a stable radical in the catalytic subunit. BMC Plant Biol., **4**, 1-16 (2004)

[3] Morita, R.; Kusaba, M.; Yamaguchi, H.; Amano, E.; Miyao, A.; Hirochika, H.; Nishimura, M.: Characterization of chlorophyllide a oxygenase (CAO) in rice. Breed. Sci., **55**, 361-364 (2005)

[4] Porra, R.J.; Schäfer, W.; Cmiel, E.; Katheder, I.; Scheer, H.: The derivation of the formylgroup oxygen of chlorophyll b in higher plants from molecular oxygen. Achievement of high enrichment of the 7-formyl group oxygen from $18O_2$ in greening maize leaves. Eur. J. Biochem., **219**, 671-679 (1994)

[5] Hirashima, M.; Satoh, S.; Tanaka, R.; Tanaka, A.: Pigment shuffling in antenna systems achieved by expressing prokaryotic chlorophyllide a oxygenase in Arabidopsis. J. Biol. Chem., **281**, 15385-15393 (2006)

[6] Harper, A.L.; von Gesjen, S.E.; Linford, A.S.; Peterson, M.P.; Faircloth, R.S.; Thissen, M.M.; Brusslan, J.A.: Chlorophyllide a oxygenase mRNA and protein levels correlate with the chlorophyll a/b ratio in Arabidopsis thaliana. Photosynth. Res., **79**, 149-159 (2004)

[7] Tanaka, R.; Tanaka, A.: Effects of chlorophyllide a oxygenase overexpression on light acclimation in Arabidopsis thaliana. Photosynth. Res., **85**, 327-340 (2005)

[8] Yamasato, A.; Nagata, N.; Tanaka, R.; Tanaka, A.: The N-terminal domain of chlorophyllide a oxygenase confers protein instability in response to chlorophyll B accumulation in Arabidopsis. Plant Cell, **17**, 1585-1597 (2005)

[9] Oster, U.; Tanaka, R.; Tanaka, A.; Rudiger, W.: Cloning and functional expression of the gene encoding the key enzyme for chlorophyll b biosynthesis (CAO) from Arabidopsis thaliana. Plant J., **21**, 305-310 (2000)

[10] Tanaka, R.; Koshino, Y.; Sawa, S.; Ishiguro, S.; Okada, K.; Tanaka, A.: Overexpression of chlorophyllide a oxygenase (CAO) enlarges the antenna size of photosystem II in Arabidopsis thaliana. Plant J., **26**, 365-373 (2001)

[11] Masuda, T.; Tanaka, A.; Melis, A.: Chlorophyll antenna size adjustments by irradiance in Dunaliella salina involve coordinate regulation of chlorophyll a oxygenase (CAO) and Lhcb gene expression. Plant Mol. Biol., **51**, 757-771 (2003)

[12] Lee, S.; Kim, J.H.; Yoo, E.S.; Lee, C.H.; Hirochika, H.; An, G.: Differential regulation of chlorophyll a oxygenase genes in rice. Plant Mol. Biol., **57**, 805-818 (2005)

[13] Nagata, N.; Satoh, S.; Tanaka, R.; Tanaka, A.: Domain structures of chlorophyllide a oxygenase of green plants and Prochlorothrix hollandica in relation to catalytic functions. Planta, **218**, 1019-1025 (2004)

[14] Reinbothe, C.; Bartsch, S.; Eggink, L.L.; Hoober, J.K.; Brusslan, J.; Andrade-Paz, R.; Monnet, J.; Reinbothe, S.: A role for chlorophyllide a oxygenase in the regulated import and stabilization of light-harvesting chlorophyll a/b proteins. Proc. Natl. Acad. Sci. USA, **103**, 4777-4782 (2006)
[15] Tanaka, A.; Ito, H.; Tanaka, R.; Tanaka, N.K.; Yoshida, K.; Okada, K.: Chlorophyll a oxygenase (CAO) is involved in chlorophyll b formation from chlorophyll a. Proc. Natl. Acad. Sci. USA, **95**, 12719-12723 (1998)
[16] Espineda, C.E.; Linford, A.S.; Devine, D.; Brusslan, J.A.: The AtCAO gene, encoding chlorophyll a oxygenase, is required for chlorophyll b synthesis in Arabidopsis thaliana. Proc. Natl. Acad. Sci. USA, **96**, 10507-10511 (1999)
[17] Satoh, S.; Tanaka, A.: Identification of chlorophyllide a oxygenase in the prochlorococcus genome by a comparative genomic approach. Plant Cell Physiol., **47**, 1622-1629 (2006)
[18] Nakagawara, E.; Sakuraba, Y.; Yamasato, A.; Tanaka, R.; Tanaka, A.: Clp protease controls chlorophyll b synthesis by regulating the level of chlorophyllide a oxygenase. Plant J., **49**, 800-809 (2007)

Flavone synthase 1.14.11.22

1 Nomenclature

EC number
1.14.11.22

Systematic name
flavanone,2-oxoglutarate:oxygen oxidoreductase (dehydrating)

Recommended name
flavone synthase

Synonyms
FNS <1, 2> [3, 5]
FNS I <1, 2> [1, 3, 8, 9]
FS I <1> [1]
FSI <1, 2> [6, 7]
flavone synthase I <1, 2> [1, 6, 7, 8, 9]

CAS registry number
138263-98-6

2 Source Organism

<1> *Petroselinum crispum* (no sequence specified) [1, 2, 4, 5, 7, 8, 9]
<2> *Petroselinum crispum* (UNIPROT accession number: Q7XZQ8) [3, 6]

3 Reaction and Specificity

Catalyzed reaction
a flavanone + 2-oxoglutarate + O_2 = a flavone + succinate + CO_2 + H_2O

Substrates and products
S (2S)-eriodictyol + 2-oxoglutarate + O_2 <1> (Reversibility: ?) [1]
P ? + CO_2 + H_2O
S (2S)-eriodictyol + 2-oxoglutarate + O_2 <1> (Reversibility: ir) [7]
P luteolin + succinate + CO_2 + H_2O
S (2S)-naringenin + 2-oxoglutarate + O_2 <1, 2> (Reversibility: ir) [2, 3, 4, 6, 7, 8, 9]
P apigenin + succinate + CO_2 + H_2O
S (2S)-pinocembrin + 2-oxoglutarate + O_2 <1> (Reversibility: ?) [8]
P chrysin + succinate + CO_2 + H_2O

S 2-hydroxynaringenin + 2-oxoglutarate + O_2 <1> (Reversibility: ?) [1]
P ? + CO_2 + H_2O
S cis-dihydrokaempferol + 2-oxoglutarate + O_2 <2> (Reversibility: ?) [3]
P kaempferol + succinate + CO_2 + H_2O
S eriodictyol + 2-oxoglutarate + O_2 <2> (Reversibility: ?) [6]
P luteolin + succinate + CO_2 + H_2O
S eriodictyol + 2-oxoglutarate + O_2 <1> (<1> 24% of the activity with naringenin [5]) (Reversibility: ?) [5]
P ?
S hesperetin + 2-oxoglutarate + O_2 <1> (<1> 26% of the activity with naringenin [5]) (Reversibility: ?) [5]
P 3,5-dihydroxy-2-(3-hydroxy-4-methoxyphenyl)-chromen-4-one + succinate + CO_2 + H_2O
S homoeriodictyol + 2-oxoglutarate + O_2 <1> (<1> 97% of the activity with naringenin [5]) (Reversibility: ?) [5]
P 5,7,4'-trihydroxy-3'-methoxyflavone + succinate + CO_2 + H_2O
S naringenin + 2-oxoglutarate + O_2 <1> (Reversibility: ?) [5]
P apigenin + succinate + CO_2 + H_2O
S pinocembin + 2-oxoglutarate + O_2 <1> (<1> 24% of the activity with naringenin [5]) (Reversibility: ?) [5]
P 5-acetoxy-7-hydroxyflavone + succinate + CO_2 + H_2O
S pinocembrin + 2-oxoglutarate + O_2 <2> (Reversibility: ?) [6]
P chrysin + succinate + CO_2 + H_2O
S Additional information <1> (<1> no activity with 5,7-dihydroxy-3,4,5-O-trimethyl-flavanone,7-O-methyl-pinocembrin, 5-O-methyl-pinocembrin, 5,7-O-dimethyl-pinocembrin [5]) (Reversibility: ?) [5]
P ?

Inhibitors

2,4-pyridinedicarboxylate <1> (<1> potent competitive [1]) [1]
Cu^{2+} <1> (<1> 0.02 mM, complete inhibition in presence of 0.01 mM Fe^{2+} [1]) [1]
Zn^{2+} <1> (<1> 0.02 mM, complete inhibition in presence of 0.01 mM Fe^{2+} [1]) [1]
prunin <1> (<1> competitive with respect to (2S)-naringenin [1]) [1]

Activating compounds

ascorbate <1> (<1> required for full activity, can not be replaced as a reductant by 2-mercaptoethanol or dithiothreitol [1]) [1]

Metals, ions

Fe^{2+} <1> (<1> required for full activity, cannot be replaced by Mn^{2+}, Co^{2+} or Ni^{2+} [1]; <1> Fe^{2+}-dependent enzyme [9]) [1, 9]

Specific activity (U/mg)

0.00402 <1> [2]
0.0057 <1> [1]

K_m-Value (mM)

0.005 <1> (2S-naringenin) [1]
0.008 <1> (2S-eriodictyol) [1]
0.016 <1> (2-oxoglutarate, <1> in presence of 0.02 mM 2S-naringenin [1]) [1]

K_i-Value (mM)

0.0018 <1> (pyridine 2,4-dicarboxylate) [1]
0.057 <1> (prunin) [1]

pH-Optimum

8.5-8.6 <1> [1]

4 Enzyme Structure

Molecular weight

48000 <1> (<1> gel filtration [1]) [1]

Subunits

dimer <1> (<1> 2 * 24000-25000, SDS-PAGE [1]) [1]

5 Isolation/Preparation/Mutation/Application

Source/tissue

cell <1> [1]
cell culture <1> (<1> induced by continous irradiation with ultraviolet/blue light for 20 h [2]) [2]
leaf <2> [3, 6]
seedling <1> (<1> leaflet [4]) [4]

Localization

soluble <1> [1]

Purification

<1> [1, 2]

Cloning

<1> (expressed in Escherichia coli BLR (DE3) cells) [8]
<1> (expressed in Escherichia coli strain BL21Star) [7]
<1> (expression in yeast) [3, 4]
<2> (expressed in Saccharomyces cerevisiae) [6]

References

[1] Britsch, L.: Purification and characterization of flavone synthase I, a 2-oxoglutarate-dependent desaturase. Arch. Biochem. Biophys., **282**, 152-160 (1990)

[2] Lukacin, R.; Matern, U.; Junghanns, K.T.; Heskamp, M.L.; Britsch, L.; Forkmann, G.; Martens, S.: Purification and antigenicity of flavone synthase I from irradiated parsley cells. Arch. Biochem. Biophys., **393**, 177-183 (2001)
[3] Martens, S.; Forkmann, G.; Britsch, L.; Wellmann, F.; Matern, U.; Lukacin, R.: Divergent evolution of flavonoid 2-oxoglutarate-dependent dioxygenases in parsley. FEBS Lett., **544**, 93-98 (2003)
[4] Martens, S.; Forkmann, G.; Matern, U.; Lukacin, R.: Cloning of parsley flavone synthase I. Phytochemistry, **58**, 43-46 (2001)
[5] Schroeder, G.; Wehinger, E.; Lukacin, R.; Wellmann, F.; Seefelder, W.; Schwab, W.; Schroeder, J.: Flavonoid methylation: a novel 4'-O-methyltransferase from Catharanthus roseus, and evidence that partially methylated flavanones are substrates of four different flavonoid dioxygenases. Phytochemistry, **65**, 1085-1094 (2004)
[6] Leonard, E.; Yan, Y.; Lim, K.H.; Koffas, M.A.: Investigation of two distinct flavone synthases for plant-specific flavone biosynthesis in Saccharomyces cerevisiae. Appl. Environ. Microbiol., **71**, 8241-8248 (2005)
[7] Leonard, E.; Chemler, J.; Lim, K.H.; Koffas, M.A.: Expression of a soluble flavone synthase allows the biosynthesis of phytoestrogen derivatives in Escherichia coli. Appl. Microbiol. Biotechnol., **70**, 85-91 (2006)
[8] Miyahisa, I.; Funa, N.; Ohnishi, Y.; Martens, S.; Moriguchi, T.; Horinouchi, S.: Combinatorial biosynthesis of flavones and flavonols in Escherichia coli. Appl. Microbiol. Biotechnol., **71**, 53-58 (2006)
[9] Martens, S.; Mithoefer, A.: Flavones and flavone synthases. Phytochemistry, **66**, 2399-2407 (2005)

Flavonol synthase 1.14.11.23

1 Nomenclature

EC number
1.14.11.23

Systematic name
dihydroflavonol,2-oxoglutarate:oxygen oxidoreductase

Recommended name
flavonol synthase

Synonyms
FLS <1, 2, 3, 4, 5, 6> [3, 4, 7, 8, 9, 10]
flavonoid 2-oxoglutarate-dependent dioxygenase <2> [7]

CAS registry number
146359-76-4

2 Source Organism

<1> *Arabidopsis thaliana* (no sequence specified) [3, 5, 9]
<2> *Dianthus caryophyllus* (no sequence specified) [7]
<3> *Citrus unshiu* (no sequence specified) [4, 6]
<4> *Malus domestica* (no sequence specified) [10]
<5> *Pyrus communis* (no sequence specified) [10]
<6> *Vitis vinifera* (no sequence specified) [8]
<7> *Citrus unshiu* (UNIPROT accession number: Q9ZWQ9) [1]
<8> *Petroselinum crispum* (UNIPROT accession number: Q7XZQ6) [2]

3 Reaction and Specificity

Catalyzed reaction
a dihydroflavonol + 2-oxoglutarate + O_2 = a flavonol + succinate + CO_2 + H_2O

Natural substrates and products
S Additional information <1> (<1> the enzyme is involved in the biosynthesis of flavonoids [3]) (Reversibility: ?) [3]
P ?

Substrates and products

- **S** (+)-trans-dihydrokaempferol + 2-oxoglutarate + O_2 <3> (Reversibility: ?) [6]
- **P** kaempferol + succinate + CO_2 + H_2O
- **S** (2R)-naringenin + 2-oxoglutarate + O_2 <1> (Reversibility: ?) [3, 5]
- **P** (2S,3S)-dihydrokaempferol + succinate + CO_2 + H_2O (<1> + low amounts of kaempferol [5])
- **S** (2R)-naringenin + 2-oxoglutarate + O_2 <3> (Reversibility: ?) [6]
- **P** (-)-trans-dihydrokaempferol + succinate + CO_2 + H_2O
- **S** (2R)-naringenin + 2-oxoglutarate + O_2 <1> (<1> transhydroxylation [3]) (Reversibility: ?) [3]
- **P** (2S,3R)-dihydrokaempferol + succinate + CO_2 + H_2O
- **S** (2R,3R)-trans-dihydroquercetin + 2-oxoglutarate + O_2 <1> (Reversibility: ?) [3]
- **P** quercetin + succinate + CO_2 + H_2O
- **S** (2R,3S,4R)-trans-leucocyanidin + 2-oxoglutarate + O_2 <1> (Reversibility: ?) [3]
- **P** ? + succinate + CO_2 + H_2O
- **S** (2S)-naringenin + 2-oxoglutarate + O_2 <1> (Reversibility: ?) [3]
- **P** ? + succinate + CO_2 + H_2O
- **S** (2S)-naringenin + 2-oxoglutarate + O_2 <1> (Reversibility: ?) [3, 9]
- **P** (2R,3S)-cis-dihydrokaempferol + (2R,3R)-trans-dihydrokaempferol + apigenin + kaempferol + succinate + CO_2 + H_2O (<1> (2R,3S)-dihydrokaempferol is the predominant two-electron oxidation product [3])
- **S** (2S)-naringenin + 2-oxoglutarate + O_2 <1> (Reversibility: ?) [5]
- **P** (2R,3S)-dihydrokaempferol + (2S,3S)-dihydrokaempferol + kaempferol + succinate + CO_2 + H_2O
- **S** (2S)-naringenin + 2-oxoglutarate + O_2 <3> (Reversibility: ?) [6]
- **P** kaempferol + succinate + CO_2 + H_2O
- **S** cis-dihydrokaempferol + 2-oxoglutarate + O_2 <8> (<8> weak reaction, only traces of kaempferol produced [2]) (Reversibility: ?) [2]
- **P** kaempferol + succinate + CO_2 + H_2O
- **S** dihydrokaempferol + 2-oxoglutarate + O_2 <1, 2, 4, 5, 6, 7> (Reversibility: ir) [1, 8, 9, 10]
- **P** kaempferol + succinate + CO_2 + H_2O
- **S** dihydromyricetin + 2-oxoglutarate + O_2 <1, 4, 5, 6> (Reversibility: ir) [8, 9, 10]
- **P** myricetin + succinate + CO_2 + H_2O
- **S** dihydroquercetin + 2-oxoglutarate + O_2 <1, 2, 4, 5, 6, 7> (Reversibility: ir) [1, 8, 9, 10]
- **P** quercetin + succinate + CO_2 + H_2O
- **S** trans-dihydrokaempferol + 2-oxoglutarate + O_2 <8> (Reversibility: ?) [2]
- **P** kaempferol + succinate + CO_2 + H_2O
- **S** Additional information <1> (<1> the enzyme is involved in the biosynthesis of flavonoids [3]) (Reversibility: ?) [3]
- **P** ?

Inhibitors

2,2'-dipyridyl <2> (<2> 2 mM, 90% inhibition [7]) [7]
3-hydroxy-5-oxo-4-butyryl-cyclohex-3-ene-1-carboxylic acid ethyl ester <4, 5> (<4,5> 1 mM 48% activity, 0.1 mM 88% activity [10]) [10]
3-hydroxy-5-oxo-4-cyclopropanecarbonyl-cyclohex-3-ene-1-carboxylic acid ethyl ester <4, 5> (<4,5> 1 mM 50% activity, 0.1 mM 77% activity [10]) [10]
3-hydroxy-5-oxo-4-propionyl-clohex-3-ene-1-(2-dimethylamino)-thiazole <4, 5> (<4,5> 1 mM 44% activity, 0.1 mM 73% activity [10]) [10]
3-hydroxy-5-oxo-4-propionyl-cyclohex-3-ene-1-carbothioic acid S-ethyl ester <4, 5> (<4,5> 1 mM 33% activity, 0.1 mM 90% activity [10]) [10]
3-hydroxy-5-oxo-4-propionyl-cyclohex-3-ene-1-pentanoic acid <4, 5> (<4,5> 1 mM 23% activity, 0.1 mM 64% activity [10]) [10]
3-hydroxy-5-oxo-4-propionyl-cylohex-3-ene-1-carbaldehyde <4, 5> (<4,5> 1 mM 56% activity, 0.1 mM 82% activity [10]) [10]
diethyldicarbonate <2> (<2> 2 mM, 16% inhibition [7]) [7]
diethyldithiocarbamate <2> (<2> 2 mM, 80% inhibition [7]) [7]
EDTA <2> (<2> 5 mM, 94% inhibition [7]) [7]
KCN <2> (<2> 5 mM, 63% inhibition [7]) [7]
benzene-1,2,4,5-tetracarboxylic acid <4, 5> (<4,5> 1 mM 95% activity [10]) [10]
calcium 3-hydroxy-5-oxo-4-propionyl-cyclohex-3-ene-1-carboxylate <4, 5> (<5> 1 mM 11% activity, 0.1 mM 29% activity [10]; <4> 1 mM 11% activity, 0.1 mM 29% relative activity [10]) [10]
p-hydroxymercuribenzoate <2> (<2> 0.1 mM, 89% inhibition [7]) [7]
pyrazole-3,5-dicarboxylic acid <4, 5> (<4,5> 1 mM 22% activity, 0.1 mM 79% activity [10]) [10]
pyridine-2,4-dicarboxylic acid ester <4, 5> (<4,5> 1 mM 3% activity, 0.1 mM 13% activity [10]) [10]
pyridine-2,5-dicarboxylic acid <4, 5> (<4,5> 1 mM 3% activity, 0.1 mM 11% activity [10]) [10]
pyridine-3,4-dicarboxylic acid <4, 5> (<4,5> 1 mM 86% activity, 0.1 mM 85% activity [10]) [10]
sodium 4,6-dioxo-2,2-dimethyl-5-(1-alloxyamino-butylidene)-cyclohexane-1-carboxylic acid methyl ester <4, 5> (<4,5> 1 mM 78% activity, 0.1 mM 84% activity [10]) [10]

Cofactors/prosthetic groups

ascorbate <4, 5> (<4,5> required [10]) [10]

Activating compounds

ascorbate <2> (<2> required [7]) [7]
sorbitol <4, 5> (<4,5> 0.6 mM [10]) [10]
piperazine-1,4-bis-(2-ethane sulfonic acid) <4, 5> (<4, 5> 1 mM, 115% activity [10]) [10]

pyridine-2,3-dicarboxylic acid <4, 5> (<4,5> 1 mM 171% activity, 0.1 mM 115% activity [10]) [10]
pyridine-2,6-dicarboxylic acid <4, 5> (<4,5> 1 mM 150% activity, 0.1 mM 144% activity [10]) [10]
pyridine-2,6-dicarboxylic acid chloride <4, 5> (<4,5> 1 mM, 120% activity [10]) [10]
pyridine-3,5-dicarboxylic acid <4, 5> (<4,5> 1 mM, 128% activity [10]) [10]
pyrrole-3,4-dicarboxylic acid diethyl ester <4, 5> (<4,5> 1 mM, 113% activity [10]) [10]

Metals, ions
Fe^{2+} <2, 4, 5, 7> (<2,4,5> required [7,10]; <7> required, K_m-value: 0.011 mM [1]) [1, 7, 10]

K_m-Value (mM)
0.001 <4, 5> (dihydrokaempferol, <5> in 0.1 M K_2HPO_4/KH_2PO_4 (0.6 M sorbitol, 0.4% Na-ascorbate, pH 7.25) with 0.095 nmol dihydrokaempferol, 50 nmol 2-oxoglutarate and 10 nmol $FeSO_4$ hydrate at 30°C [10]; <4> in 0.1 M K_2HPO_4/KH_2PO_4 (0.6 M sorbitol, 0.4% Na-ascorbate, pH 7.25), 50 nmol 2-oxoglutarate and 10 nmol $FeSO_4$ hydrate at 30°C [10]) [10]
0.0012 <4, 5> (dihydroquercetin, <4,5> in 0.1 M K_2HPO_4/KH_2PO_4 (0.6 M sorbitol, 0.4% Na-ascorbate, pH 7.25) with 0.095 nmol dihydrokaempferol, 50 nmol 2-oxoglutarate and 10 nmol $FeSO_4$ hydrate at 30°C [10]) [10]
0.036 <3> (2-oxoglutarate) [1]
0.045 <3> (dihydrokaempferol) [1]
0.272 <3> (dihydroquercetin) [1]

pH-Optimum
5 <7> (<7> reaction with dihydroquercetin [1]) [1]
7.25 <4, 5> [10]
7.4 <2> [7]

Temperature optimum (°C)
30 <4, 5> [10]
37 <7> (<7> reaction with dihydroquercetin [1]) [1]

4 Enzyme Structure

Subunits
? <7> (<7> x * 38000, SDS-PAGE [1]; <7> x * 37899, MALDI-TOF mass spectrometry [1]) [1]

5 Isolation/Preparation/Mutation/Application

Source/tissue

epidermis <3> (<3> the level of the CitFLS transcript is high at the early developmental stage and low at the mature stage in the juice sacs/segment epidermis (edible part) [4]) [4]
flower bud <2> [7]
fruit <3, 6> (<6> peel [8]; <3> the level of the CitFLS transcript increases in the peel during fruit maturation [4]) [4, 8]
leaf <3, 4, 5> (<3> the level of the CitFLS transcript is higher in the young leaves than in the old leaves [4]) [4, 10]
peel <3> (<3> the level of the CitFLS transcript increases in the peel during fruit maturation [4]) [4]

Purification

<1> [5]
<3> [6]

Cloning

<1> (expressed in Escherichia coli strain BL21Star) [9]
<1> (expression in Escherichia coli) [5]
<3> [4]
<3> (expression in Escherichia coli) [6]
<8> (expression in yeast) [2]
<261070> (expression in Escherichia coli) [1]

Engineering

G261A <7> (<7> mutant enzyme with 10% of wild-type activity [1]) [1]
G261P <7> (<7> no immunoreactive FLS polypeptide detected [1]) [1]
G68A <7> (<7> mutant enzyme with 6% of wild-type activity [1]) [1]
G68P <7> (<7> no immunoreactive FLS polypeptide detected [1]) [1]
P207G <7> (<7> no effect on activity [1]) [1]

References

[1] Wellmann, F.; Lukacin, R.; Moriguchi, T.; Britsch, L.; Schiltz, E.; Matern, U.: Functional expression and mutational analysis of flavonol synthase from Citrus unshiu. Eur. J. Biochem., **269**, 4134-4142 (2002)
[2] Martens, S.; Forkmann, G.; Britsch, L.; Wellmann, F.; Matern, U.; Lukacin, R.: Divergent evolution of flavonoid 2-oxoglutarate-dependent dioxygenases in parsley. FEBS Lett., **544**, 93-98 (2003)
[3] Turnbull, J.J.; Nakajima, J.; Welford, R.W.; Yamazaki, M.; Saito, K.; Schofield, C.J.: Mechanistic studies on three 2-oxoglutarate-dependent oxygenases of flavonoid biosynthesis: anthocyanidin synthase, flavonol synthase, and flavanone 3β-hydroxylase. J. Biol. Chem., **279**, 1206-1216 (2004)

[4] Moriguchi, T.; Kita, M.; Ogawa, K.; Tomono, Y.; Endo, T.; Omura, M.: Flavonol synthase gene expression during citrus fruit development. Physiol. Plant., **114**, 251-258 (2002)
[5] Prescott, A.G.; Stamford, N.P.; Wheeler, G.; Firmin, J.L.: In vitro properties of a recombinant flavonol synthase from Arabidopsis thaliana. Phytochemistry, **60**, 589-593 (2002)
[6] Lukacin, R.; Wellmann, F.; Britsch, L.; Martens, S.; Matern, U.: Flavonol synthase from Citrus unshiu is a bifunctional dioxygenase. Phytochemistry, **62**, 287-292 (2003)
[7] Stich, K.; Eidenberger, T.; Wurst, F.; Forkmann, G.: Flavonol synthase activity and the regulation of flavonol and anthocyanin biosynthesis during flower development in Dianthus caryophyllus L. (carnation). Z. Naturforsch. C, **47**, 553-560 (1992)
[8] Fujita, A.; Goto-Yamamoto, N.; Aramaki, I.; Hashizume, K.: Organ-specific transcription of putative flavonol synthase genes of grapevine and effects of plant hormones and shading on flavonol biosynthesis in grape berry skins. Biosci. Biotechnol. Biochem., **70**, 632-638 (2006)
[9] Leonard, E.; Yan, Y.; Koffas, M.A.: Functional expression of a P_{450} flavonoid hydroxylase for the biosynthesis of plant-specific hydroxylated flavonols in Escherichia coli. Metab. Eng., **8**, 172-181 (2006)
[10] Halbwirth, H.; Fischer, T.C.; Schlangen, K.; Rademacher, W.; Schleifer, K.; Forkmann, G.; Stich, K.: Screening for inhibitors of 2-oxoglutarate-dependent dioxygenases: Flavanone 3β-hydroxylase and flavonol synthase. Plant Sci., **171**, 194-205 (2006)

2'-Deoxymugineic-acid 2'-dioxygenase 1.14.11.24

1 Nomenclature

EC number
1.14.11.24

Systematic name
2'-deoxymugineic acid,2-oxoglutarate:oxygen oxidoreductase (2-hydroxylating)

Recommended name
2'-deoxymugineic-acid 2'-dioxygenase

Synonyms
IDS3 <2> [2]

CAS registry number
37292-90-3

2 Source Organism

<1> *Hordeum vulgare* (no sequence specified) [1, 3]
<2> *Hordeum vulgare* (UNIPROT accession number: Q98LU11) [2]

3 Reaction and Specificity

Catalyzed reaction
2'-deoxymugineic acid + 2-oxoglutarate + O_2 = mugineic acid + succinate + CO_2

Substrates and products
- **S** 2'-deoxymugineic acid + 2-oxoglutarate + O_2 <1, 2> (Reversibility: ?) [2, 3]
- **P** mugineic acid + succinate + CO_2
- **S** 3-epihydroxy-2'-deoxymugineic acid + 2-oxoglutarate + O_2 <2> (Reversibility: ?) [2]
- **P** ? + succinate + CO_2

4 Enzyme Structure

Subunits

? <2> (<2> x * 37700, calculation from nucleotide sequence [2]) [2]

5 Isolation/Preparation/Mutation/Application

Source/tissue

root <1> (<1> specifically expressed in roots of iron-deficient barley [1]) [1, 3]

Cloning

<1> (introduction of the Ids3 gene into Oryza sativa cultivar Nipponbare and Tsukinohikari which lack Ids3 homologues) [3]
<2> (expression in Escherichia coli) [2]

References

[1] Nakanishi, H.; Okumura, N.; Emehara, Y.; Nishizawa, N.-K.; Chino, M.; Mori, S.: Expression of a gene specific for iron deficiency (Ids3) in the root of Hordeum vulgare. Plant Cell Physiol., **34**, 401-410 (1993)

[2] Nakanishi, H.; Yamaguchi, H.; Sasakuma, T.; Nishizawa, N.K.; Mori, S.: Two dioxygenase genes, Ids3 and Ids2, from Hordeum vulgare are involved in the biosynthesis of mugineic acid family phytosiderophores. Plant Mol. Biol., **44**, 199-207 (2000)

[3] Kobayashi, T.; Nakanishi, H.; Takahashi, M.; Kawasaki, S.; Nishizawa, N.-K.; Mori, S.: In vivo evidence that Ids3 from Hordeum vulgare encodes a dioxygenase that converts 2'-deoxymugineic acid to mugineic acid in transgenic rice. Planta, **212**, 864-871 (2001)

Mugineic-acid 3-dioxygenase 1.14.11.25

1 Nomenclature

EC number
1.14.11.25

Systematic name
mugineic acid,2-oxoglutarate:oxygen oxidoreductase (3-hydroxylating)

Recommended name
mugineic-acid 3-dioxygenase

Synonyms
IDS2 <1> [2]

CAS registry number
158540-24-0

2 Source Organism

<1> *Hordeum vulgare* (no sequence specified) [1, 2]

3 Reaction and Specificity

Catalyzed reaction
2'-deoxymugineic acid + 2-oxoglutarate + O_2 = 2-epihydroxy-2'-deoxymugineic acid + succinate + CO_2
mugineic acid + 2-oxoglutarate + O_2 = 3-epihydroxymugineic acid + succinate + CO_2

Substrates and products
S 2'-deoxymugineic acid + 2-oxoglutarate + O_2 <1> (Reversibility: ?) [2]
P 2-epihydroxy-2'-deoxymugineic acid + succinate + CO_2
S mugineic acid + 2-oxoglutarate + O_2 <1> (Reversibility: ?) [2]
P 3-epihydroxymugineic acid + succinate + CO_2

5 Isolation/Preparation/Mutation/Application

Source/tissue
root <1> (<1> Ids2 gene is expressed under iron deficiency conditions [1]) [1]

References

[1] Okumura, N.; Nishizawa, N.K.; Umehara, Y.; Ohata, T.; Nakanishi, H.; Yamaguchi, T.; Chino, M.; Mori, S.: A dioxygenase gene (Ids2) expressed under iron deficiency conditions in the roots of Hordeum vulgare. Plant Mol. Biol., **25**, 705-719 (1994)

[2] Nakanishi, H.; Yamaguchi, H.; Sasakuma, T.; Nishizawa, N.K.; Mori, S.: Two dioxygenase genes, Ids3 and Ids2, from Hordeum vulgare are involved in the biosynthesis of mugineic acid family phytosiderophores. Plant Mol. Biol., **44**, 199-207 (2000)

Deacetoxycephalosporin-C hydroxylase 1.14.11.26

1 Nomenclature

EC number
1.14.11.26

Systematic name
deacetoxycephalosporin-C,2-oxoglutarate:oxygen oxidoreductase (3-hydroxylating)

Recommended name
deacetoxycephalosporin-C hydroxylase

Synonyms
DAOCS <3> [11, 12]
deacetoxycephalosporin C synthase <3> [11, 12]

CAS registry number
69772-89-0
85746-10-7

2 Source Organism

<1> *Cephalosporium acremonium* (no sequence specified) [1, 3, 5, 6, 9]
<2> *Acremonium chrysogenum* (no sequence specified) [7]
<3> *Streptomyces clavuligerus* (no sequence specified) (<3> isozyme 2S [8]) [2, 8, 10, 11, 12]
<4> *Nocardia lactamdurans* (no sequence specified) [4]

3 Reaction and Specificity

Catalyzed reaction
deacetoxycephalosporin C + 2-oxoglutarate + O_2 = deacetylcephalosporin C + succinate + CO_2

Reaction type
hydroxylation

Natural substrates and products
S penicillin N + 2-oxoglutarate + O_2 <3> (Reversibility: ?) [12]
P ? + succinate + CO_2

Substrates and products

S 3'-methylcephem + 2-oxoglutarate + O_2 <4> (Reversibility: ?) [4]
P 3'-hydroxymethylcephem + succinate + CO_2
S 3-exomethylenecephalosporin C + 2-oxoglutarate + O_2 <3> (Reversibility: ?) [2]
P deacetylcephalosporin C + succinate + CO_2 + H_2O
S 3-oxomethylenecephalosporin C + 2-oxoglutarate + O_2 <3> (<3> 37% of activty with deacetoxycephalosporin C [10]) (Reversibility: ?) [10]
P deacetoxycephalosporin C + succinate + CO_2
S 7-aminodeacetoxycephalosporanic acid + 2-oxoglutarate + O_2 <1, 2> (<2> poor substrate [7]) (Reversibility: ?) [3, 7]
P 7-aminodeacetylcephalosporin + succinate + CO_2
S ampicillin + 2-oxoglutarate + O_2 <3> (Reversibility: ?) [11]
P ? + succinate + CO_2
S cephalexin + 2-oxoglutarate + O_2 <1> (Reversibility: ?) [3]
P ? + succinate + CO_2
S deacetoxycephalosporin C + 2-oxoglutarate + O_2 <1, 3> (Reversibility: ?) [3, 9, 10, 11, 12]
P deacetylcephalosporin C + succinate + CO_2
S deacetoxycephalosporin C + 2-oxoglutarate + O_2 <1> (Reversibility: ?) [6]
P deacetylcephalosporin C + succinate + CO_2 + H_2O
S penicillin G + 2-oxoglutarate + O_2 <3> (Reversibility: ?) [11, 12]
P ? + succinate + CO_2
S penicillin N + 2-oxoglutarate + O_2 <3> (Reversibility: ?) [11, 12]
P succinate + CO_2 + ?
S phenylacetyl-7-aminodeacetoxycephalosporanic acid + 2-oxoglutarate + O_2 <1, 3> (Reversibility: ?) [3, 11, 12]
P phenylacetyl-7-aminodeacetylcephalosporin + succinate + CO_2

Inhibitors

(2S,4S,6R,7R)-1-aza-7-((5R)-5-carboxypentanamidol)-4-methyl-8-oxo-5-thia-tricyclo-(4,2,0,0)octane-2-carboxylate <1> (<1> reversible, 0.04 mM, 90% inhibition [6]) [6]
5,5'-dithiobis-2-nitrobenzoic acid <1, 3> (<3> reactivation by dithiothreitol [2]; <3> 1 mM, no residual activity [10]; <1> 1 mM, 100% inhibition of hydroxylation reaction, EC 1.14.11.26 [9]) [2, 9, 10]
ampicillin <3> [2]
bathophenanthroline <1> [6]
Ca^{2+} <1> (<1> weak [9]) [9]
Co^{2+} <3> (<3> inhibition in decreasing order, Zn^{2+}, Co^{2+}, Ni^{2+} [2]) [2]
EDTA <1, 3> (<3> 0.5 mM, no residual activity [10]; <1> 0.6 mM, 83% inhibition of hydroxylation reaction, EC 1.14.11.26 [9]) [2, 9, 10]
iodoacetic acid <1, 3> (<3> 1 mM, 19% residual activity [10]; <1> 1 mM, 2% inhibition of hydroxylation reaction, EC 1.14.11.26 [9]) [2, 9, 10]

N-ethylmaleimide <1, 3> (<3> 1 mM, no residual activity [10]; <1> 1 mM, 97% inhibition of hydroxylation reaction, EC 1.14.11.26 [9]) [2, 6, 9, 10]
Ni^{2+} <3> (<3> inhibition in decreasing order, Zn^{2+}, Co^{2+}, Ni^{2+} [2]) [2]
penicillin G <3> [2]
penicillin N <3> (<3> 1 mM, 62% inhibition, competitive with deacetoxycephalosporin C [10]) [10]
penicillin V <3> [2]
Zn^{2+} <1, 3> (<1> strong [9]; <3> inhibition in decreasing order, Zn^{2+}, Co^{2+}, Ni^{2+} [2]) [2, 9]
ammonium hydrogencarbonate <1> (<1> 100-500 mM, complete inactivation [6]) [6]
o-phenanthroline <1, 3> (<3> 0.5 mM, no residual activity [10]; <1> 0.6 mM, 100% inhibition of hydroxylation reaction, EC 1.14.11.26 [9]) [2, 9, 10]
p-hydroxymercuribenzoate <1, 3> (<3> 0.1 mM, no residual activity [10]; <1> 1 mM, 100% inhibition of hydroxylation reaction, EC 1.14.11.26 [9]) [2, 9, 10]

Activating compounds

ammonium sulfate <1> [6]
ascorbate <1, 3> (<1> required [6]) [2, 6]
dithiothreitol <1, 3> (<1> up to 30% activation in presence of dithiothreitol [6]) [2, 6, 10]
glutathione <3> (<3> reduced [10]) [10]
ammonium chloride <1> [6]

Metals, ions

Fe^{2+} <1, 3> (<1> required for both activities [9]; <3> required, Ka-value 0.008 mM [2]; <3> required, without Fe^{2+}, less than 2% of maximum activity [10]; <3> Fe^{2+}-dependent [12]) [2, 9, 10, 12]
Additional information <1, 3> (<1,3> Fe^{2+} cannot be replaced by Mg^{2+}, Mn^{2+}, Ca^{2+}, Co^{2+}, Cu^{2+}, Ni^{2+} or Zn^{2+} [9,10]) [9, 10]

Turnover number (min^{-1})

0.048 <3> (penicillin N, <3> mutant enzyme C155Y/Y184/V275I/C281Y [12]; <3> mutant enzyme Y184H [12]) [12]
0.0507 <3> (penicillin G, <3> mutant enzyme M188I [12]) [12]
0.084 <3> (penicillin N, <3> mutant enzyme M188V [12]) [12]
0.178 <3> (penicillin N, <3> mutant enzyme G79E [12]) [12]
0.235 <3> (penicillin N, <3> mutant enzyme C155Y [12]) [12]
0.239 <3> (penicillin N, <3> mutant enzyme M73T [12]) [12]
0.252 <3> (penicillin N, <3> mutant enzyme V275I [12]) [12]
0.26 <3> (penicillin N, <3> mutant enzyme H244Q [12]) [12]
0.273 <3> (penicillin N, <3> mutant enzyme C281Y [12]) [12]
0.277 <3> (penicillin N, <3> mutant enzyme A106T [12]; <3> mutant enzyme M188I [12]) [12]
0.284 <3> (penicillin N, <3> mutant enzyme I305L [12]) [12]
0.297 <3> (penicillin N, <3> mutant enzyme L277Q [12]) [12]
0.307 <3> (penicillin N, <3> wild type enzyme [12]) [12]

0.31 <3> (penicillin N, <3> mutant enzyme I305M [12]) [12]
0.316 <3> (penicillin N, <3> mutant enzyme V275I/I305M [12]) [12]
0.366 <3> (penicillin N, <3> mutant enzyme N304K [12]) [12]

Specific activity (U/mg)
0.127 <1> (<1> pH 7.5, 36°C [9]) [9]

K_m-Value (mM)
0.004 <3> (penicillin N, <3> mutant enzyme N304K [12]) [12]
0.006 <3> (penicillin N, <3> mutant enzyme C281Y [12]; <3> mutant enzyme I305L [12]; <3> mutant enzyme M73T [12]) [12]
0.009 <3> (penicillin N, <3> mutant enzyme G79E [12]; <3> mutant enzyme H244Q [12]; <3> mutant enzyme T91A [12]) [12]
0.011 <3> (penicillin N, <3> mutant enzyme L277Q [12]; <3> mutant enzyme M188I [12]) [12]
0.012 <3> (penicillin N, <3> mutant enzyme I305M [12]; <3> mutant enzyme V275I [12]) [12]
0.013 <3> (penicillin N, <3> mutant enzyme A106T [12]; <3> mutant enzyme V275I/I305M [12]) [12]
0.014 <3> (penicillin N, <3> wild type enzyme [12]) [12]
0.016 <3> (penicillin N, <3> mutant enzyme C155Y [12]) [12]
0.02 <1> (deacetoxycephalosporin C, <1> pH 7.5, 36°C [9]) [9]
0.022 <1> (2-oxoglutarate, <1> pH 7.5, 36°C [9]) [9]
0.047 <3> (penicillin N, <3> mutant enzyme M188V [12]) [12]
0.092 <3> (penicillin N, <3> mutant enzyme C155Y/Y184/V275I/C281Y [12]) [12]
0.19 <3> (penicillin G, <3> mutant enzyme C155Y/Y184/V275I/C281Y [12]) [12]
0.22 <3> (penicillin G, <3> mutant enzyme N304K [12]) [12]
0.25 <3> (penicillin G, <3> mutant enzyme V275I/I305M [12]) [12]
0.295 <3> (penicillin N, <3> mutant enzyme Y184H [12]) [12]
0.66 <3> (penicillin G, <3> mutant enzyme I305L [12]) [12]
0.68 <3> (penicillin G, <3> mutant enzyme C281Y [12]) [12]
0.74 <3> (penicillin G, <3> mutant enzyme M73T [12]) [12]
0.75 <3> (penicillin G, <3> mutant enzyme G79E [12]; <3> mutant enzyme I305M [12]) [12]
0.76 <3> (penicillin G, <3> mutant enzyme M188V [12]) [12]
0.95 <3> (penicillin G, <3> mutant enzyme Y184H [12]) [12]
1.02 <3> (penicillin G, <3> mutant enzyme L277Q [12]) [12]
1.36 <3> (penicillin G, <3> mutant enzyme T91A [12]) [12]
1.39 <3> (penicillin G, <3> mutant enzyme H244Q [12]) [12]
1.68 <3> (penicillin G, <3> mutant enzyme V275I [12]) [12]
1.76 <3> (penicillin G, <3> mutant enzyme C155Y [12]) [12]
1.77 <3> (penicillin G, <3> mutant enzyme A106T [12]) [12]
1.96 <3> (penicillin G, <3> mutant enzyme M188I [12]) [12]
2.58 <3> (penicillin G, <3> wild type enzyme [12]) [12]

K_i-Value (mM)
0.12 <3> (penicillin N, <3> pH 7.3, 29°C [10]) [10]

pH-Optimum
7 <3> [2]
7.3 <1> (<1> hydroxylation reaction, EC 1.14.11.26 [9]) [9]
7.5 <1> [6]

Temperature optimum (°C)
36-38 <1> [9]

Temperature range (°C)
36 <3> [2]

4 Enzyme Structure

Molecular weight
26200 <3> (<3> gel filtration [8]) [8]
35000 <3> (<3> gel filtration [2]) [2]
43000 <1> (<1> gel filtration [9]) [9]

Subunits
? <1, 4> (<1> x * 40000, SDS-PAGE [6]; <4> x * 34366, calculated [4]; <1> x * 40000, SDS-PAGE, x * 36642, calculated [5]) [4, 5, 6]
monomer <1> (<1> 1 * 41000, SDS-PAGE [9]) [9]
Additional information <1> (<1> structural model of enzyme [3]) [3]

5 Isolation/Preparation/Mutation/Application

Localization
soluble <3> [8]

Purification
<1> [5, 6]
<1> (expression in Escherichia coli, overexpression as an insoluble and inactive enzyme, elaboration of refolding scheme resulting in highly active and moderately stable enzyme) [1]
<3> (anion-exchange chromatography and gel filtration) [12]
<3> (native enzyme, partial, recombinant enzyme, to near homogenity) [2]
<3> (partial) [8]

Renaturation
<3> (enzyme inactivated by 5,5'-dithiobis-2-nitrobenzoic acid, complete reactivation by dithiothreitol) [2]

Cloning
<1> [5]
<3> (expressed in Escherichia coli Tuner(DE3) and BL21(DE3) cells) [12]

<3> (expression in Escherichia coli) [2]
<4> (expression in Streptomyces lividans) [4]

Engineering

A106T <3> (<3> 80% relative activity compared to the wild type enzyme using 1 mM penicillin G as substrate [12]) [12]
C155Y <3> (<3> 90% relative activity compared to the wild type enzyme using 1 mM penicillin G as substrate [12]) [12]
C155Y/Y184H/V275I/C281Y <3> (<3> 580% relative activity compared to the wild type enzyme using 1 mM penicillin G as substrate [12]) [12]
C281Y <3> (<3> 200% relative activity compared to the wild type enzyme using 1 mM penicillin G as substrate [12]) [12]
G300V <3> (<3> 410% relative activity compared to the wild type enzyme using 1 mM penicillin G as substrate [12]) [12]
G79E <3> (<3> 90% relative activity compared to the wild type enzyme using 1 mM penicillin G as substrate [12]) [12]
H244Q <3> (<3> 140% relative activity compared to the wild type enzyme using 1 mM penicillin G as substrate [12]) [12]
I305L <3> (<3> 230% relative activity compared to the wild type enzyme using 1 mM penicillin G as substrate [12]) [12]
I305M <3> (<3> 380% relative activity compared to the wild type enzyme using 1 mM penicillin G as substrate [12]) [12]
L277Q <3> (<3> 270% relative activity compared to the wild type enzyme using 1 mM penicillin G as substrate [12]) [12]
M188I <3> (<3> 90% relative activity compared to the wild type enzyme using 1 mM penicillin G as substrate [12]) [12]
M188V <3> (<3> 150% relative activity compared to the wild type enzyme using 1 mM penicillin G as substrate [12]) [12]
M306I <1> (<1> hydroxylation reaction of deacetoxycephalosporin C, EC 1.14.11.26, is abolished, 59% of wild-type ring expansion activity [3]) [3]
M73T <3> (<3> 180% relative activity compared to the wild type enzyme using 1 mM penicillin G as substrate [12]) [12]
N304K <3> (<3> 220% relative activity compared to the wild type enzyme using 1 mM penicillin G as substrate [12]) [12]
N305L <1, 2> (<1> 107% of wild-type ring expansion activity, 85% of wild-type hydroxylation activity [3]; <2> improved ability to convert penicillin analogs in ring expansion reaction of EC 1.14.20.1 [7]) [3, 7]
R308L <2> (<2> improved ability to convert penicillin analogs in ring expansion reaction of EC 1.14.20.1 [7]) [7]
T91A <3> (<3> 110% relative activity compared to the wild type enzyme using 1 mM penicillin G as substrate [12]) [12]
V275I <3> (<3> 270% relative activity compared to the wild type enzyme using 1 mM penicillin G as substrate [12]) [12]
V275I/I305M <3> (<3> 500% relative activity compared to the wild type enzyme using 1 mM penicillin G as substrate [12]) [12]

W82A <1> (<1> 5.5% of wild-type ring expansion activity, 71% of wild-type hydroxylation activity [3]; <1> ring expansion reaction of EC 1.14.20.1 is reduced [3]) [3]
W82S <1> (<1> 44% of wild-type ring expansion activity, 18% of wild-type hydroxylation activity [3]) [3]
Y184H <3> (<3> 200% relative activity compared to the wild type enzyme using 1 mM penicillin G as substrate [12]) [12]
Additional information <1> (<1> truncation of C-terminus to residue 310, 2fold enhancement of ring expansion reaction of penicillin G. Double mutant with truncation at residue 310 and M306I, selective catalyzation of ring expansion. Triple mutant with truncation at residue 310, M306I and N305L, selective catalization of ring expansion with improved kinetic parameters [3]) [3]

Application

synthesis <1> (<1> expression in Escherichia coli, overexpression as an insoluble and inactive enzyme, elaboration of refolding scheme resulting in highly active and moderately stable enzyme [1]) [1]

6 Stability

pH-Stability

6.5-9 <3> [2]

Temperature stability

40 <3> (<3> stable up to [2]) [2]

General stability information

<1>, sodium phosphate and Bis/Tris buffers are inhibitory [6]

Storage stability

<1>, -70°C, presence of dithiothreitol, stable for 4 weeks [6]
<1>, 4°C, presence of dithiothreitol, half-life is 48 h [6]

References

[1] Ghag, S.K.; Brems, D.N.; Hassell, T.C.; Yeh, W.K.: Refolding and purification of Cephalosporium acremonium deacetoxycephalosporin C synthetase/hydroxylase from granules of recombinant Escherichia coli. Biotechnol. Appl. Biochem., **24**, 109-119 (1996)

[2] Dotzlaf, J.E.; Yeh, W.K.: Purification and properties of deacetoxycephalosporin C synthase from recombinant Escherichia coli and its comparison with the native enzyme purified from Streptomyces clavuligerus. J. Biol. Chem., **264**, 10219-10227 (1989)

[3] Lloyd, M.D.; Lipscomb, S.J.; Hewitson, K.S.; Hensgens, C.M.H.; Baldwin, J.E.; Schofield, C.J.: Controlling the substrate selectivity of deacetoxycepha-

losporin/deacetylcephalosporin C synthase. J. Biol. Chem., **279**, 15420-15426 (2004)
[4] Coque, J.J.; Enguita, F.J.; Cardoza, R.E.; Martin, J.F.; Liras, P.: Characterization of the cefF gene of Nocardia lactamdurans encoding a 3'-methylcephem hydroxylase different from the 7-cephem hydroxylase. Appl. Microbiol. Biotechnol., **44**, 605-609 (1996)
[5] Samson, S.M.; Dotzlaf, J.E.; Slisz, M.L.; Becker, G.W.; Van Frank, R.M.; Veal, L.E.; Yeh, W.K.; Miller, J.R.; Queener, S.W.; Ingolia, T.D.: Cloning and expression of the fungal expandase/hydroxylase gene involved in cephalosporin biosynthesis. Bio/Technology, **5**, 1207-1214 (1987)
[6] Baldwin, J.E.; Adlington, R.M.; Coates, J.B.; Crabbe, M.J.; Crouch, N.P.; Keeping, J.W.; Knight, G.C.; Schofield, C.J.; Ting, H.H.; Vallejo, C.A.; et al.: Purification and initial characterization of an enzyme with deacetoxycephalosporin C synthetase and hydroxylase activities. Biochem. J., **245**, 831-841 (1987)
[7] Wu, X.B.; Fan, K.Q.; Wang, Q.H.; Yang, K.Q.: C-terminus mutations of Acremonium chrysogenum deacetoxy/deacetylcephalosporin C synthase with improved activity toward penicillin analogs. FEMS Microbiol. Lett., **246**, 103-110 (2005)
[8] Jensen, S.E.; Westlake, D.W.S.; Wolfe, S.: Deacetoxycephalosporin C synthetase and deacetoxycephalosporin C hydroxylase are two separate enzymes in Streptomyces clavuligerus. J. Antibiot., **38**, 263-265 (1985)
[9] Dotzlaf, J.E.; Yeh, W.K.: Copurification and characterization of deacetoxycephalosporin C synthetase/hydroxylase from Cephalosporium acremonium. J. Bacteriol., **169**, 1611-1618 (1987)
[10] Baker, B.J.; Dotzlaf, J.E.; Yeh, W.K.: Deacetoxycephalosporin C hydroxylase of Streptomyces clavuligerus. Purification, characterization, bifunctionality, and evolutionary implication. J. Biol. Chem., **266**, 5087-5093 (1991)
[11] Stok, J.E.; Baldwin, J.E.: Development of enzyme-linked immunosorbent assays for the detection of deacetoxycephalosporin C and isopenicillin N synthase activity. Anal. Chim. Acta, **577**, 153-162 (2006)
[12] Wei, C.; Yang, Y.; Deng, C.; Liu, W.; Hsu, J.; Lin, Y.; Liaw, S.; Tsai, Y.: Directed evolution of Streptomyces clavuligerus deacetoxycephalosporin C synthase for enhancement of penicillin G expansion. Appl. Environ. Microbiol., **71**, 8873-8880 (2005)

[histone-H3]-Lysine-36 demethylase 1.14.11.27

1 Nomenclature

EC number
1.14.11.27

Systematic name
protein-N^6,N^6-dimethyl-L-lysine,2-oxoglutarate:oxygen oxidoreductase

Recommended name
[histone-H3]-lysine-36 demethylase

Synonyms
H3-K36 demethylase <1, 2> [1]
H3-K36-specific demethylase <1> [1]
JHDM1A <1> [1]
JmjC domain-containing histone demethylase 1 <1> [1]
JmjC domain-containing histone demethylase 1A <1> [1]
histone-lysine(H3-K36) demethylase <1> [1]
scJHDM1 <2> [1]

CAS registry number
55071-98-2

2 Source Organism

<1> *Homo sapiens* (no sequence specified) [1]
<2> *Saccharomyces cerevisiae* (no sequence specified) [1]

3 Reaction and Specificity

Catalyzed reaction
protein N^6,N^6-dimethyl-L-lysine + 2-oxoglutarate + O_2 = protein N^6-methyl-L-lysine + succinate + formaldehyde + CO_2
protein N^6-methyl-L-lysine + 2-oxoglutarate + O_2 = protein L-lysine + succinate + formaldehyde + CO_2

Substrates and products
S protein N^6,N^6-dimethyl-L-lysine + 2-oxoglutarate + O_2 <2> (<2> specifically demethylates Lys36 of histone H3 [1]) (Reversibility: ?) [1]
P protein N^6-methyl-L-lysine + succinate + formaldehyde + CO_2

S protein N^6-methyl-L-lysine + 2-oxoglutarate + O_2 <2> (<2> specifically demethylates Lys36 of histone H3 [1]) (Reversibility: ?) [1]

P protein L-lysine + succinate + formaldehyde + CO_2

Activating compounds

ascorbate <1> (<1> required for optimal activity [1]) [1]

Metals, ions

Fe^{2+} <1> (<1> required [1]) [1]

5 Isolation/Preparation/Mutation/Application

Source/tissue

HeLa cell <1> [1]

Localization

nucleus <1> [1]

Purification

<1> (recombinant enzyme purified from baculovirus-infected Sf9 cells) [1]

Engineering

H212A <1> (<1> mutation completely abolishes the enzymatic activity [1]) [1]

H305A <2> (<2> mutation completely abolishes the enzymatic activity [1]) [1]

References

[1] Tsukada, Y.; Fang, J.; Erdjument-Bromage, H.; Warren, M.E.; Borchers, C.H.; Tempst, P.; Zhang, Y.: Histone demethylation by a family of JmjC domain-containing proteins. Nature, **439**, 811-816 (2006)

Proline 3-hydroxylase 1.14.11.28

1 Nomenclature

EC number
1.14.11.28

Systematic name
L-proline,2-oxoglutarate:oxygen oxidoreductase (3-hydroxylating)

Recommended name
proline 3-hydroxylase

Synonyms
P-3-H <3> [3]
P3H1 <1> [5]
proline 3-hydroxylase <2, 3, 4> [1, 3, 4]
proline 3-hydroxylase type I <3> [2, 6]
proline 3-hydroxylase type II <3> [2, 6]
prolyl 3-hydrolase <1> [5]

CAS registry number
162995-24-6

2 Source Organism

<1> *Gallus gallus* (no sequence specified) [5]
<2> *Bacillus sp.* (no sequence specified) [1]
<3> *Streptomyces sp.* (no sequence specified) [1, 2, 3, 4, 6]
<4> *Streptomyces canus* (no sequence specified) [1]

3 Reaction and Specificity

Catalyzed reaction
L-proline + 2-oxoglutarate + O_2 = cis-3-hydroxy-L-proline + succinate + CO_2

Natural substrates and products
S L-proline + 2-oxoglutarate + O_2 <3> (Reversibility: ?) [3]
P cis-3-hydroxy-L-proline + succinate + CO_2

Substrates and products
S 3,4-dehydro-L-proline + 2-oxoglutarate + O_2 <3> (Reversibility: ?) [2, 6]
P cis-3,4-epoxy-L-proline + succinate + CO_2

S L-2-azetidinecarboxylic acid + 2-oxoglutarate + O_2 <3> (Reversibility: ?) [2, 6]
P cis-3-hydroxyazetidine-L-2-carboxylic acid + succinate + CO_2
S L-pipecolic acid + 2-oxoglutarate + O_2 <3> (Reversibility: ?) [2, 6]
P cis-3-hydroxy-L-pipecolic acid + succinate + CO_2
S L-proline + 2-oxoglutarate + O_2 <2, 3, 4> (Reversibility: ?) [1, 2, 3, 4, 6]
P cis-3-hydroxy-L-proline + succinate + CO_2

Inhibitors

citric acid <3> (<3> 30% inhibition at 5 mM [4]) [4]
Co^{2+} <3> (<3> more than 50% inhibition at 1 mM [2]; <3> more than 90% inhibition at 1 mM [4]) [2, 4]
$CoCl_2$ <3> (<3> 82% inhibition at 0.1 mM, 98% inhibition at 1 mM, 5 mM 2-oxoglutarate, 1 mM ferrous sulfate, 5 mM L-ascorbic acid, 100 mM TES pH 7.5, 30°C [1]) [1]
Cu^{2+} <3> (<3> more than 50% inhibition at 1 mM [2]; <3> more than 90% inhibition at 1 mM [4]) [2, 4]
$CuSO_4$ <3> (<3> 59% inhibition at 0.1 mM, complete inhibition at 1 mM, 5 mM 2-oxoglutarate, 1 mM ferrous sulfate, 5 mM L-ascorbic acid, 100 mM TES pH 7.5, 30°C [1]) [1]
EDTA <3> (<3> 95% inhibition at 2 mM [4]; <3> more than 50% inhibition at 1 mM [2]; <3> 95% inhibition at 2 mM, 5 mM 2-oxoglutarate, 1 mM ferrous sulfate, 5 mM L-ascorbic acid, 100 mM TES pH 7.5, 30°C [1]) [1, 2, 4]
Ni^{2+} <3> (<3> more than 50% inhibition at 1 mM [2]) [2]
Zn^{2+} <3> (<3> more than 50% inhibition at 1 mM [2]; <3> more than 90% inhibition at 1 mM [4]) [2, 4]
$ZnSO_4$ <3> (<3> 95% inhibition at 0.1 mM, complete inhibition at 1 mM, 5 mM 2-oxoglutarate, 1 mM ferrous sulfate, 5 mM L-ascorbic acid, 100 mM TES pH 7.5, 30°C [1]) [1]

Cofactors/prosthetic groups

2-oxoglutarate <2, 3, 4> (<3> strictly required [2]; <2,3,4> strictly required, highest activity at 4 mM [1,4]) [1,2,4]

Activating compounds

L-ascorbate <3> (<3> 2 mM, no further stimulation above 2 mM [4]) [4]

Metals, ions

Fe^{2+} <2, 3, 4> (<2,3> 1 mM added to cell extracts, 60% inhibition without addition of Fe^{2+} [1]; <2> 1 mM added to cell extracts, 85% inhibition without addition of Fe^{2+} [1]; <4> 1 mM added to cell extracts, complete inhibition without addition of Fe^{2+} [1]; <3> strictly required [2]; <3> strictly required, optimal stimulation at 1 mM [4]) [1, 2, 4]

Turnover number (min^{-1})

3.2 <3> (L-proline, <3> 100 mM N-tris(hydroxymethyl)methyl-2-aminoethanesulfonic acid pH 7.0, 5 mM 2-oxoglutarate, 5 mM L-proline, 5 mM L-ascorbate, 1 mM $FeSO_4$, 35°C [4]) [4]

Specific activity (U/mg)

0.00029 <4> [1]
0.00119 <3> (<3> 5 mM 2-oxoglutarate, 1 mM ferrous sulfate, 5 mM L-ascorbic acid, 100 mM TES pH 7.5, 30°C [1]) [1]
0.00132 <2> [1]
0.00149 <2> [1]
7.2 <3> [2]
12.7 <3> [2]

K_m-Value (mM)

0.047 <3> (2-oxoglutarate) [2]
0.083 <3> (2-oxoglutarate) [2]
0.11 <3> (2-oxoglutarate, <3> 100 mM N-tris(hydroxymethyl)methyl-2-aminoethanesulfonic acid pH 7.0, 5 mM 2-oxoglutarate, 5 mM L-proline, 5 mM L-ascorbate, 1 mM $FeSO_4$, 35°C [4]) [4]
0.2 <3> (L-proline) [2, 6]
0.4 <3> (L-proline) [6]
0.43 <3> (L-proline) [2]
0.56 <3> (L-proline, <3> 100 mM N-tris(hydroxymethyl)methyl-2-aminoethanesulfonic acid pH 7.0, 5 mM 2-oxoglutarate, 5 mM L-proline, 5 mM L-ascorbate, 1 mM $FeSO_4$, 35°C [4]) [4]
2.1 <3> (L-2-azetidinecarboxylic acid) [6]
3.8 <3> (L-pipecolic acid) [6]
8.4 <3> (3,4-dehydro-L-proline) [6]
100 <3> (L-pipecolic acid) [6]

pH-Optimum

6 <3> [2]
7 <3> [4]

Temperature optimum (°C)

35 <3> [2, 4]
40 <3> [2]

4 Enzyme Structure

Molecular weight

33150 <3> (<3> calculated [2]) [2]
33570 <3> (<3> calculated [2]) [2]
35000 <3> (<3> SDS-PAGE [4]; <3> apparent molecular weight on SDS-PAGE [2]) [2, 4]

Subunits

dimer <3> (<3> X-ray structure [3]) [3]

5 Isolation/Preparation/Mutation/Application

Source/tissue
cell culture <2, 3, 4> [1, 2, 3, 4, 6]
embryo <1> [5]

Localization
endoplasmic reticulum <1> [5]

Purification
<1> [5]
<3> [3]
<3> (2 chromatographic steps to homogeneity) [2, 6]
<3> (5 chromatographic steps near to homogeneity) [4]

Crystallization
<3> [3]

Cloning
<3> (expression in Escherichia coli) [2, 6]

6 Stability

pH-Stability
5.5-8.5 <3> (<3> activity remained above 85% after 16 h at 4°C [2]) [2]
5.5-9 <3> (<3> activity remained above 85% after 16 h at 4°C [2]) [2]

Temperature stability
30 <3> (<3> below 30°C activity remained above 85% after 30 min at pH 6.0 [2]) [2]
40 <3> (<3> below 40°C activity remained above 85% after 30 min at pH 6.0 [2]) [2]

References

[1] Mori, H.; Shibasaki, T.; Uozaki, Y.; Ochiai, K.; Ozaki, A.: Detection of novel proline 3-hydroxylase activities in Streptomyces and Bacillus spp. by regio- and stereospecific hydroxylation of L-proline. Appl. Environ. Microbiol., **62**, 1903-1907 (1996)
[2] Shibasaki, T.; Mori, H.; Ozaki, A.: Cloning of an isozyme of proline 3-hydroxylase and its purification from recombinant Escherichia coli. Biotechnol. Lett., **22**, 1967-1973 (2000)
[3] Clifton, I.J.; Hsueh, L.C.; Baldwin, J.E.; Harlos, K.; Schofield, C.J.: Structure of proline 3-hydroxylase. Evolution of the family of 2-oxoglutarate dependent oxygenases. Eur. J. Biochem., **268**, 6625-6636 (2001)

[4] Mori, H.; Shibasaki, T.; Yano, K.; Ozaki, A.: Purification and cloning of a proline 3-hydroxylase, a novel enzyme which hydroxylates free L-proline to cis-3-hydroxy-L-proline. J. Bacteriol., **179**, 5677-5683 (1997)
[5] Vranka, J.A.; Sakai, L.Y.; Bachinger, H.P.: Prolyl 3-hydroxylase 1, enzyme characterization and identification of a novel family of enzymes. J. Biol. Chem., **279**, 23615-23621 (2004)
[6] Shibasaki, T.; Sakurai, W.; Hasegawa, A.; Uosaki, Y.; Mori, H.; Yoshida, M.; Ozaki, A.: Substrate selectivities of proline hydroxylases. Tetrahedron Lett., **40**, 5227-5230 (1999)

3-Phenylpropanoate dioxygenase 1.14.12.19

1 Nomenclature

EC number
1.14.12.19

Systematic name
3-phenylpropanoate,NADH:oxygen oxidoreductase (2,3-hydroxylating)

Recommended name
3-phenylpropanoate dioxygenase

Synonyms
3-phenylpropionate dioxygenase <1> [1, 2, 3, 4, 5]
HcA1A2CD <1> [4, 5]
Hca dioxygenase <1> [1, 2, 3, 4, 5]
HcaA1A2CD <1> [1, 2, 3]

CAS registry number
105503-66-0

2 Source Organism

<1> *Escherichia coli* (no sequence specified) (<1> isozyme SMO_2-2 [3]) [1, 2, 3, 4, 5]

3 Reaction and Specificity

Catalyzed reaction
3-phenylpropanoate + NADH + H^+ + O_2 = 3-(cis-5,6-dihydroxycyclohexa-1,3-dien-1-yl)propanoate + NAD^+

Natural substrates and products
S 3-phenylpropionate + NADH + O_2 <1> (<1> enables utilization of 3-phenylpropionate as carbon source [1]) (Reversibility: ?) [1]
P 3-(2,3-dihydroxyphenyl)propionate + NAD^+
S 3-phenylpropionate + NADH + O_2 <1> (<1> enables utilization of 3-phenylpropionate as carbon source [2]) (Reversibility: ?) [2]
P 3-(cis-5,6-dihydroxycyclohexa-1,3-dien-1-yl)propanoate + NAD^+
S 3-phenylpropionate + NADH + O_2 <1> (<1> enables utilization of 3-phenylpropionate as carbon source [4]; <1> enables utilization of 3-phenylpropionate as carbon source, important role in wine making, aging and

storage [3]; <1> involved in oxidative stress response [5]) (Reversibility: ?) [3, 4, 5]

P 3-[(5S,6R)-5,6-dihydroxycyclohexa-1,3-dien-1-yl]propanoic acid + NAD^+

Substrates and products

S 3-phenylpropionate + NADH + O_2 <1> (<1> enables utilization of 3-phenylpropionate as carbon source [1]) (Reversibility: ?) [1]

P 3-(2,3-dihydroxyphenyl)propionate + NAD^+

S 3-phenylpropionate + NADH + O_2 <1> (<1> enables utilization of 3-phenylpropionate as carbon source [2]) (Reversibility: ?) [2]

P 3-(cis-5,6-dihydroxycyclohexa-1,3-dien-1-yl)propanoate + NAD^+

S 3-phenylpropionate + NADH + O_2 <1> (<1> enables utilization of 3-phenylpropionate as carbon source [4]; <1> enables utilization of 3-phenylpropionate as carbon source, important role in wine making, aging and storage [3]; <1> involved in oxidative stress response [5]) (Reversibility: ?) [3, 4, 5]

P 3-[(5S,6R)-5,6-dihydroxycyclohexa-1,3-dien-1-yl]propanoic acid + NAD^+

S cinnamic acid + NADH + O_2 <1> (Reversibility: ?) [3]

P cinnamic acid-2,3-dihydrodiol

Cofactors/prosthetic groups

NADH <1> [1,2,3,4,5]

4 Enzyme Structure

Subunits

tetramer <1> (<1> gene products from hcaA1, hcaA2, hcaC and hcaD, 1 * 51109 + 1 * 20579 + 1 * 11328 + 1 * 43978, subunit masses calculated from deduced amino acid sequences [3]) [3]

5 Isolation/Preparation/Mutation/Application

Cloning

<1> (expressed in Salmonella typhimurium) [3, 4]

References

[1] Burlingame, R., Chapmen, P.J.: Catabolism of phenylpropionic acid and its 3-hydroxy-derivative by Escherichia coli. J. Bacteriol., **155**, 113-121 (1983)

[2] Burlingame, R.P.; Wyman, L.; Chapman, P.J.: Isolation and characterization of Escherichia coli mutants defective for phenylpropionate degradation. J. Bacteriol., **168**, 55-64 (1986)

[3] Diaz, E.; Ferrandez, A.; Garcia, J.L.: Characterization of the hca cluster encoding the dioxygenolytic pathway for initial catabolism of 3-phenylpropionic acid in Escherichia coli K-12. J. Bacteriol., **180**, 2915-2923 (1998)

[4] Turlin, E.; Perrotte-Piquemal, M.; Danchin, A.; Biville, F.: Regulation of the early steps of 3-phenylpropionate catabolism in Escherichia coli. J. Mol. Microbiol. Biotechnol., **3**, 127-133 (2001)
[5] Turlin, E.; Sismeiro, O.; Le Caer, J.P.; Labas, V.; Danchin, A.; Biville, F.: 3-phenylpropionate catabolism and the Escherichia coli oxidative stress response. Res. Microbiol., **156**, 312-321 (2005)

Pheophorbide a oxygenase 1.14.12.20

1 Nomenclature

EC number
1.14.12.20

Systematic name
pheophorbide-a,NADPH:oxygen oxidoreductase (biladiene-forming)

Recommended name
pheophorbide a oxygenase

Synonyms
AtPaO <1, 3> [1, 7]
Lls1 <3> [1]
PAO <3> [1]
accelerated cell death 1 <3> [1]
chloroplast pheophorbide a oxygenase PaO1 <4> [6]
chloroplast pheophorbide a oxygenase PaO_2 <5> [6]
lethal leaf-spot 1 homolog <3> [1]
pheide a monooxygenase <3> [1]
pheide a oxygenase <3> [1]

CAS registry number
211049-58-0

2 Source Organism

<1> *Arabidopsis thaliana* (no sequence specified) [3, 4, 7]
<2> *Brassica napus* (no sequence specified) [2, 5]
<3> *Arabidopsis thaliana* (UNIPROT accession number: Q9FYC2) [1]
<4> *Brassica napus* (UNIPROT accession number: Q0ZKW3) [6]
<5> *Brassica napus* (UNIPROT accession number: Q0ZKW4) [6]

3 Reaction and Specificity

Catalyzed reaction
pheophorbide a + NADPH + H^+ + O_2 = red chlorophyll catabolite + $NADP^+$

Natural substrates and products
S Additional information <1, 2> (<1> ACD1 is involved in PaO activity, and its inhibition led to photooxidative destruction of the cell [4]; <2> key

step in chlorophyll breakdown [2]; <1> PaO expression is correlated positively with senescence. The in vivo function of PaO is the degradation of pheide a during senescence [7]; <1> rred chlorophyll catabolite reductase together with pheophorbide a oxygenase is required for the detoxification of chlorophyll catabolites [3]) (Reversibility: ?) [2, 3, 4, 7]

P ?

Substrates and products

S pheophorbide a + NADPH + H^+ + O_2 <2> (Reversibility: ?) [2, 5]

P red chlorophyll catabolite + $NADP^+$

S Additional information <1, 2> (<1> ACD1 is involved in PaO activity, and its inhibition lead to photooxidative destruction of the cell [4]; <2> key step in chlorophyll breakdown [2]; <1> PaO expression is correlated positively with senescence. The in vivo function of PaO is the degradation of pheide a during senescence [7]; <1> red chlorophyll catabolite reductase together with pheophorbide a oxygenase is required for the detoxification of chlorophyll catabolites [3]) (Reversibility: ?) [2, 3, 4, 7]

P ?

Cofactors/prosthetic groups

ferredoxin <3> (<3> enzyme requires ferredoxin as the source of electrons [1]) [1]

Metals, ions

Fe <1, 3> (<3> enzyme contains a Rieske-type iron-sulfur cluster [1]; <1> Rieske-type iron-sulfur cluster-containing protein [7]) [1, 7]

4 Enzyme Structure

Posttranslational modification

phosphoprotein <4, 5> (<4,5> phosphorylation is a posttranslational control mechanism [6]) [6]

5 Isolation/Preparation/Mutation/Application

Source/tissue

flower <3> [1]

gerontoplast <3> (<3> envelope membrane [1]) [1]

leaf <1, 3, 4, 5> (<4> BnPaO1 is identified in senescing canola leaves [6]; <5> $BnPaO_2$ is identified in senescing canola leaves [6]; <3> PAO mRNA and PAO proteins are present at low levels in presenescent leaves, but are highly enriched during senescence. PAO is upregulated rapidly upon wounding [1]) [1, 4, 6]

seed <4, 5> (<4> BnPaO1 is identified during early seed development. Increase in PaO activity during seed maturation corresponds to a decrease in the phosphorylation of the PaO enzyme [6]; <5> $BnPaO_2$ is identified during

early seed development and in maturing, degreening seeds. Increase in PaO activity during seed maturation corresponds to a decrease in the phosphorylation of the PaO enzyme [6]) [6]
seedling <2> [2]
silique <3> [1]

Localization
chloroplast <3> [1]
membrane <2> [5]

Purification
<2> (partial) [2, 5]

Cloning
<1> (expression in Escherichia coli) [7]
<3> (expression in Escherichia coli) [1]
<4> [6]
<5> [6]

References

[1] Hörtensteiner, S.: Chlorophyll degradation during senescence. Annu. Rev. Plant Biol., **57**, 55-77 (2006)
[2] Hörtensteiner, S.; Wüthrich, K.L.; Matile, P.; Ongania, K.H.; Kräutler, B.: The key step in chlorophyll breakdown in higher plants. Cleavage of pheophorbide a macrocycle by a monooxygenase. J. Biol. Chem., **273**, 15335-15339 (1998)
[3] Pruzinska, A.; Anders, I.; Aubry, S.; Schenk, N.; Tapernoux-Lüthi, E.; Müller, T.; Kräutler, B.; Hörtensteiner, S.: In vivo participation of red chlorophyll catabolite reductase in chlorophyll breakdown. Plant Cell, **19**, 369-387 (2007)
[4] Tanaka, R.; Hirashima, M.; Satoh, S.; Tanaka, A.: The Arabidopsis-accelerated cell death gene ACD1 is involved in oxygenation of pheophorbide a: inhibition of the pheophorbide a oxygenase activity does not lead to the "stay-green" phenotype in Arabidopsis. Plant Cell Physiol., **44**, 1266-1274 (2003)
[5] Rodoni, S.; Mühlecker, W.; Anderl, M.; Kräutler, B.; Moser, D.; Thomas, H.; Matile, P.; Hörtensteiner, S.: Chlorophyll breakdown in senescent chloroplasts (cleavage of pheophorbide a in two enzymic steps). Plant Physiol., **115**, 669-676 (1997)
[6] Chung, D.W.; Pruzinska, A.; Hörtensteiner, S.; Ort, D.R.: The role of pheophorbide a oxygenase expression and activity in the canola green seed problem. Plant Physiol., **142**, 88-97 (2006)
[7] Pruzinska, A.; Tanner, G.; Anders, I.; Roca, M.; Hortensteiner, S.: Chlorophyll breakdown: pheophorbide a oxygenase is a Rieske-type iron-sulfur protein, encoded by the accelerated cell death 1 gene. Proc. Natl. Acad. Sci. USA, **100**, 15259-15264 (2003)

Vanillate monooxygenase 1.14.13.82

1 Nomenclature

EC number
1.14.13.82

Systematic name
vanillate:oxygen oxidoreductase (demethylating)

Recommended name
vanillate monooxygenase

Synonyms
4-hydroxy-3-methoxybenzoate demethylase
EC 1.2.3.12
IvaAB <9, 10> [10]
LigM <7> [9]
Mtv <6> [6]
VanA <8> [10]
tetrahydrofolate-dependent O-demethylase <7> [9]
vanAB <3, 5> [5, 7]
vanillate O-demethylase
vanillate demethylase <5> [7]
vanillate-O-demethylase <2> [8]
vanillate/3-O-methylgallate O-demethylase <7> [9]
vanillate/3MGA O-demethylase <7> [9]
vanillic acid O-demethylase
Additional information <8, 9, 10> (<8,9,10> the enzyme belongs to the phthalate family oxygenase [10]) [10]

CAS registry number
39307-11-4

2 Source Organism

<1> *Pseudomonas sp.* (no sequence specified) [1, 4]
<2> *Pseudomonas putida* (no sequence specified) [8]
<3> *Corynebacterium glutamicum* (no sequence specified) [5]
<4> *Acinetobacter sp.* (no sequence specified) [2, 3]
<5> *Streptomyces sp.* (no sequence specified) [7]
<6> *Moorella thermoacetica* (no sequence specified) [6]
<7> *Sphingomonas paucimobilis* (UNIPROT accession number: Q60FX1) [9]

<8> *Comamonas testosteroni* (UNIPROT accession number: Q2KQ82) [10]
<9> *Comamonas testosteroni* (UNIPROT accession number: Q2KQ78) [10]
<10> *Comamonas testosteroni* (UNIPROT accession number: Q2KQ79) [10]

3 Reaction and Specificity

Catalyzed reaction

vanillate + O_2 + NADH + H^+ = 3,4-dihydroxybenzoate + NAD^+ + H_2O + formaldehyde

Reaction type

oxidation
redox reaction
reduction

Natural substrates and products

S isovanillic acid + NAD(P)H + O_2 <9, 1> (<9,10> O-demethylation by IvaAB [10]) (Reversibility: ?) [10]
P procatechuic acid + $NAD(P)^+$ + H_2O + formaldehyde
S vanillate + O_2 + NADH <1, 3, 4> (<4> able to demethylate one methoxy group or to monohydroxylate one methyl group in the meta position [2]) (Reversibility: r) [1, 2, 3, 4, 5]
P 3,4-dihydroxybenzoate + NAD^+ + H_2O + formaldehyde <1, 4> [1, 2, 3, 4]
S vanillate + O_2 + electron donor + tetrahydrofolate <6> (Reversibility: ?) [6]
P 3,4-dihydroxybenzoate + oxidized electron donor + H_2O + methyltetrahydrofolate
S vanillic acid + NAD(P)H + O_2 <2, 8> (<2> functional coupling between vanillate-O-demethylase and formaldehyde detoxification pathway, formaldehyde is further converted to formate by glutathione-dependent formaldehyde dehydrogenase, encoded by gene frmA [8]; <8> O-demethylation by VanA [10]) (Reversibility: ?) [8, 1]
P protocatechuic acid + $NAD(P)^+$ + H_2O + formaldehyde (<2> protocatechuic acid is 3,4-dihydroxybenzoate [8])
S vanillic acid + NADH + O_2 <5, 7> (Reversibility: ?) [7, 9]
P protocatechuic acid + NAD^+ + H_2O + formaldehyde (<5> protocatechuic acid is 3,4-dihydroxybenzoate [7]; <7> protocatechuic acid is 3,4-dihydroxybenzoate, which is further degraded via the protocatechiuc acid 4,5-cleavage pathway [9])
S veratric acid + NAD(P)H + O_2 <8> (<8> O-demethylation by VanA [10]) (Reversibility: ?) [10]
P isovanillic acid + $NAD(P)^+$ + H_2O + formaldehyde
S veratric acid + NAD(P)H + O_2 <9, 1> (<9,10> O-demethylation by IvaAB [10]) (Reversibility: ?) [10]
P vanillic acid + $NAD(P)^+$ + H_2O + formaldehyde
S Additional information <5> (<5> the enzyme forms a vanillate demethylase complex of VanA, a terminal oxygenase subunit, with VanB, a ferre-

doxin-like subunit, performing demethylation of vanillic acid and veratric acid [7]) (Reversibility: ?) [7]

P ?

Substrates and products

S 3,4,5-trimethoxybenzoate + O_2 + NADH <4> (<4> 70% of the activity compared to vanillate [2]) (Reversibility: ?) [2]

P 3-hydroxy-4,5-dimethoxybenzoate + NAD^+ + H_2O + formaldehyde <4> [2]

S 3,4-dimethoxybenzoate + O_2 + NADH <4> (Reversibility: ?) [2]

P isovanillate + NAD^+ + H_2O + formaldehyde <4> [2]

S 3-(hydroxymethyl)-benzoate + NAD^+ + H_2O + formaldehyde <4> (Reversibility: ?) [2]

P *m*-toluate + O_2 + NADH <4> [2]

S 3-O-methylgallate + NADH + O_2 <7> (Reversibility: ?) [9]

P ?

S 3-hydroxymethyl-4-hydroxy-5-methylbenzoate + NADH + H_2O + formaldehyde <4> (<4> 85% of the activity compared to vanillate [2]) (Reversibility: ?) [2]

P 4-hydroxy-3,5-dimethylbenzoate + O_2 + NADH <4> [2]

S dicamba + O_2 + electron donor + tetrahydrofolate <6> (<6> titanium III citrate can act as electron donor, inducible three-component system consisting of MtvA, MtvB and MtvC, that catalyzes methyl transfer from vanillate to tetrahydrofolate [6]) (Reversibility: ?) [6]

P ? + oxidized electron donor + H_2O + methyltetrahydrofolate

S isovanillic acid + NAD(P)H + O_2 <9, 1> (<9,10> O-demethylation by IvaAB [10]; <9,10> i.e. 3-hydroxy-4-methoxybenzoate, O-demethylation by IvaAB [10]) (Reversibility: ?) [10]

P procatechuic acid + $NAD(P)^+$ + H_2O + formaldehyde (<9,10> procatechuic acid is 3,4-dihydroxybenzoate [10])

S *m*-anisate + O_2 + NADH <4> (Reversibility: ?) [2]

P *m*-hydroxybenzoate + NAD^+ + H_2O + formaldehyde <4> [2]

S vanillate + O_2 + NADH <1, 3, 4> (<4> able to demethylate one methoxy group or to monohydroxylate one methyl group in the meta position [2]) (Reversibility: r) [1, 2, 3, 4, 5]

P 3,4-dihydroxybenzoate + NAD^+ + H_2O + formaldehyde <1, 4> [1, 2, 3, 4]

S vanillate + O_2 + electron donor + tetrahydrofolate <6> (<6> titanium III citrate can act as electron donor, inducible three-component system consisting of MtvA, MtvB and MtvC, that catalyzes methyl transfer from vanillate to tetrahydrofolate [6]) (Reversibility: ?) [6]

P 3,4-dihydroxybenzoate + oxidized electron donor + H_2O + methyltetrahydrofolate

S vanillic acid + NAD(P)H + O_2 <2, 8> (<2> functional coupling between vanillate-O-demethylase and formaldehyde detoxification pathway, formaldehyde is further converted to formate by glutathione-dependent formaldehyde dehydrogenase, encoded by gene frmA [8]; <8> O-demethylation by VanA [10]; <8> i.e. 4-hydroxy-3-methoxybenzoate, O-demethylation by VanA [10]) (Reversibility: ?) [8, 1]

P procatechuic acid + $NAD(P)^+$ + H_2O + formaldehyde (<2,8> procatechuic acid is 3,4-dihydroxybenzoate [8,10])

S vanillic acid + NADH + O_2 <5, 7> (Reversibility: ?) [7, 9]

P procatechuic acid + NAD^+ + H_2O + formaldehyde (<5,7> procatechuic acid is 3,4-dihydroxybenzoate [7,9]; <7> procatechuic acid is 3,4-dihydroxybenzoate, which is further degraded via the procatechiuc acid 4,5-cleavage pathway [9])

S veratric acid + NAD(P)H + O_2 <8> (<8> O-demethylation by VanA [10]; <8> i.e. 3,4-dimethoxybenzoate, O-demethylation by VanA [10]) (Reversibility: ?) [10]

P isovanillic acid + $NAD(P)^+$ + H_2O + formaldehyde

S veratric acid + NAD(P)H + O_2 <9, 1> (<9,10> O-demethylation by IvaAB [10]; <9,10> i.e. 3,4-dimethoxybenzoate, O-demethylation by IvaAB [10]) (Reversibility: ?) [10]

P vanillic acid + $NAD(P)^+$ + H_2O + formaldehyde

S Additional information <5, 7> (<5> the enzyme forms a vanillate demethylase complex of VanA, a terminal oxygenase subunit, with VanB, a ferredoxin-like subunit, performing demethylation of vanillic acid and veratric acid [7]; <7> syringate is no substrate [9]) (Reversibility: ?) [7, 9]

P ?

Inhibitors

4-hydroxy-3,5-dimethylbenzoate <4> (<4> 20% competitive inhibition at 3 mM [2]) [2]

m-anisate <4> (<4> 15-77.5% competitive inhibition of vanillate demethylation between 1-3 mM [2]) [2]

m-toluate <4> (<4> 45% competitive inhibition at 3 mM [2]) [2]

Cofactors/prosthetic groups

NAD(P)H <2, 8, 9, 10> [8, 10]

NADH <3, 5, 7> (<5> binding sequence motif, overview [7]) [5, 7, 9]

Activating compounds

tetrahydrofolate <7> (<7> dependent on [9]) [9]

Additional information <8, 9, 10> (<8,9,10> the enzyme is inducible by vanillic acid and veratric acid [10]) [10]

Metals, ions

Fe^{2+} <5> (<5> binding sequence motif of iron-sulfur cluster, overview [7]) [7]

Specific activity (U/mg)

0.003 <2> (<2> recombinant whole cell activities with both cofactors NADPH or NADH [8]) [8]

K_m-Value (mM)

0.085 <6> (vanillate, <6> 55°C, pH 6.6 [6]) [6]

9 <6> (dicamba, <6> 55°C, pH 6.6 [6]) [6]

pH-Optimum

6.6 <6> [6]
8 <7> (<7> assay at [9]) [9]

Temperature optimum (°C)

30 <2, 7> (<2,7> assay at [8,9]) [8, 9]
55 <6> [6]

4 Enzyme Structure

Subunits

dimer <8, 9, 10> (<8,9,10> presence of both subunits is essential [10]) [10]

5 Isolation/Preparation/Mutation/Application

Source/tissue

cell culture <1, 4> [1, 2, 3]
Additional information <8, 9, 10> (<8> in succinate- and procatechuic acid-grown cells, there is negligible degradative activity towards vanillic acid, veratric acid, and isovanillic acid and little to no expression of genes van, ivaA, and ivaB, growth on vanillic acid or veratric acid results in production of active oxygenases and expression of vanA, ivaA, and ivaB [10]; <9,10> in succinate- and protpcatechuic acid-grown cells, there is negligible degradative activity towards vanillic acid, veratric acid, and isovanillic acid and little to no expression of genes vanA, ivaA, and ivaB, growth on vanillic acid or veratric acid results in production of active oxygenases and expression of vanA, ivaA, and ivaB [10]) [10]

Cloning

<2> (genes vanA and vanB, DNA and amino acid sequence determination and analysis, functional coexpression of genes vanA and vanB in Escherichia coli K12 strain WV1181) [8]
<4> (expression in Escherichia coli) [2]
<5> (genes vanA and vanB, DNA and amino acid sequence determination and analysis, genetic organization, sequence comparison, functional coexpression of genes vanA and vanB in Streptomyces lividans strain 1326 from the high-copy plasmid pIJ702, selection on veratric acid) [7]
<7> (gene ligM, DNA and amino acid sequence determination and analysis, functional expression in Escherichia coli strains JM109 and BL21(DE3)) [9]
<8> (gene vanA, DNA and amino acid sequence determination and analysis, genetic organization, physical map, expression analysis, recombinant expression in Escherichia coli) [10]
<9> (gene ivaB, DNA and amino acid sequence determination and analysis, genetic organization, physical map, expression analysis, recombinant expression in Escherichia coli) [10]

<10> (gene ivaA, DNA and amino acid sequence determination and analysis, genetic organization, physical map, expression analysis, recombinant expression in Escherichia coli) [10]

Engineering

H156R <4> (<4> substituting an iron-binding ligand [2]) [2]
W217R <4> (<4> functional at 22°C but not at 37°C [2]) [2]
Additional information <2, 4, 7> (<4> several amino acid substitutions give indications of the extent to which amino acid substitutions can be tolerated at specified positions [2]; <7> disruption of ligM leads to significant growth retardation on both vanillate and syringate, a ligM desA double mutant completely looses the ability to transform vanillate, phenotypes, overview [9]; <2> the reaction of the recombinant enzyme is one of the limiting steps of the prospective bioconversion from lignin-derived compounds to valuable products in recombinant Escherichia coli [8]) [2, 8, 9]

References

[1] Priefert, H.; Rabenhorst, J.; Steinbuchel, A.: Molecular characterization of genes of Pseudomonas sp. strain HR199 involved in bioconversion of vanillin to protocatechuate. J. Bacteriol., **179**, 2595-2607 (1997)
[2] Morawski, B.; Segura, A.; Ornston, L.N.: Substrate range and genetic analysis of Acinetobacter vanillate demethylase. J. Bacteriol., **182**, 1383-1389 (2000)
[3] Morawski, B.; Segura, A.; Ornston, L.N.: Repression of Acinetobacter vanillate demethylase synthesis by VanR, a member of the GntR family of transcriptional regulators. FEMS Microbiol. Lett., **187**, 65-68 (2000)
[4] Brunel, F.; Davison, J.: Cloning and sequencing of Pseudomonas genes encoding vanillate demethylase. J. Bacteriol., **170**, 4924-4930 (1988)
[5] Merkens, H.; Beckers, G.; Wirtz, A.; Burkovski, A.: Vanillate Metabolism in Corynebacterium glutamicum. Curr. Microbiol., **51**, 59-65 (2005)
[6] Naidu, D.; Ragsdale, S.W.: Characterization of a three-component vanillate O-demethylase from Moorella thermoacetica. J. Bacteriol., **183**, 3276-3281 (2001)
[7] Nishimura, M.; Ishiyama, D.; Davies, J.: Molecular cloning of Streptomyces genes encoding vanillate demethylase. Biosci. Biotechnol. Biochem., **70**, 2316-2319 (2006)
[8] Hibi, M.; Sonoki, T.; Mori, H.: Functional coupling between vanillate-O-demethylase and formaldehyde detoxification pathway. FEMS Microbiol. Lett., **253**, 237-242 (2005)
[9] Abe, T.; Masai, E.; Miyauchi, K.; Katayama, Y.; Fukuda, M.: A tetrahydrofolate-dependent O-demethylase, LigM, is crucial for catabolism of vanillate and syringate in Sphingomonas paucimobilis SYK-6. J. Bacteriol., **187**, 2030-2037 (2005)
[10] Providenti, M.A.; OBrien, J.M.; Ruff, J.; Cook, A.M.; Lambert, I.B.: Metabolism of isovanillate, vanillate, and veratrate by Comamonas testosteroni strain BR6020. J. Bacteriol., **188**, 3862-3869 (2006)

Precorrin-3B synthase 1.14.13.83

1 Nomenclature

EC number
1.14.13.83

Systematic name
precorrin-3A,NADH:oxygen oxidoreductase (20-hydroxylating)

Recommended name
precorrin-3B synthase

Synonyms
CobG <2> [1, 3, 4]
CobZ <1> [5]
precorrin-3X synthase <2> [1]

CAS registry number
152787-63-8

2 Source Organism

<1> *Rhodobacter capsulatus* (no sequence specified) [5]
<2> *Pseudomonas denitrificans* (no sequence specified) [1, 2, 3, 4]

3 Reaction and Specificity

Catalyzed reaction
precorrin-3A + NADH + H^+ + O_2 = precorrin-3B + NAD^+ + H_2O (<1> reaction mechanism of CobZ [5])

Natural substrates and products
S precorrin-3 + NADH + H^+ + O_2 <2> (<2> biosynthesis of vitamin B_{12}, part of the ring contractase system for oxidative ring contraction [2]) (Reversibility: ir) [2]
P precorrin-3x + NAD^+ + H_2O <2> [2]
S precorrin-3A + NADH + H^+ + O_2 <2> (<2> biosynthesis of vitamin B_{12} [4]) (Reversibility: ir) [4]
P precorrrin-3B + NAD^+ + H_2O <2> (<2> aerobic incubation with CobG alone, if CobJ and S-adenosyl-L-methionine are included the product is precorrin-4 [4]) [4]

S precorrin-3A + NADH + H^+ + O_2 <2> (<2> pathway to coenzyme B_{12}, biosynthesis of the corrin macrocycle, catalyzes a complex oxidative reaction involving C20 hydroxylation and γ-lactone formation from ring-A acetate to C1 [3]) (Reversibility: ir) [3]
P precorrin-3B + NAD^+ + H_2O <2> [3]
S precorrin-3A + NADH + O_2 <1, 2> (<1> the enzyme is involved in ring contraction, whereby an integral carbon atom of the tetrapyrrole-derived macrocycle is removed, and cobalt chelation during aerobic synthesis of vitamin B_{12}, the catalytic cycle of CobZ, overview [5]) (Reversibility: ?) [3, 5]
P precorrin-3B + NAD^+ + H_2O
S Additional information <2> (<2> biosynthesis of vitamin B_{12} [1]) (Reversibility: ?) [1]
P ? <2> [1]

Substrates and products

S precorrin-3 + NADH + H^+ + O_2 <2> (<2> biosynthesis of vitamin B_{12}, part of the ring contractase system for oxidative ring contraction [2]) (Reversibility: ir) [1, 2]
P precorrin-3x + NAD^+ + H_2O <2> [1, 2]
S precorrin-3A + NADH + H^+ + O_2 <2> (<2> biosynthesis of vitamin B_{12} [4]) (Reversibility: ir) [4]
P precorrrin-3B + NAD^+ + H_2O <2> (<2> aerobic incubation with CobG alone, if CobJ and S-adenosyl-L-methionine are included the product is precorrin-4 [4]) [4]
S precorrin-3A + NADH + H^+ + O_2 <2> (<2> pathway to coenzyme B_{12}, biosynthesis of the corrin macrocycle, catalyzes a complex oxidative reaction involving C20 hydroxylation and γ-lactone formation from ring-A acetate to C1 [3]) (Reversibility: ir) [3]
P precorrin-3B + NAD^+ + H_2O <2> [3]
S precorrin-3A + NADH + O_2 <1, 2> (<1> the enzyme is involved in ring contraction, whereby an integral carbon atom of the tetrapyrrole-derived macrocycle is removed, and cobalt chelation during aerobic synthesis of vitamin B_{12}, the catalytic cycle of CobZ, overview [5]; <1> the enzyme generates a hydroxy lactone intermediate [5]) (Reversibility: ?) [3, 5]
P precorrin-3B + NAD^+ + H_2O
S Additional information <2> (<2> biosynthesis of vitamin B_{12} [1]) (Reversibility: ?) [1]
P ? <2> [1]

Cofactors/prosthetic groups

flavin <1> (<1> enzyme-bound [5]) [5]
NADH <2> [1]

Metals, ions

Fe^{2+} <1, 2> (<2> iron-sulfur protein [3,4]; <2> contains Fe_4-S_4 [1]; <1> the enzyme contains both a non-haem iron and an Fe-S centre [5]) [1, 3, 4, 5]

Specific activity (U/mg)
0.048 <2> [3]

4 Enzyme Structure

Molecular weight
45000 <2> (<2> SDS-PAGE [2]) [2]
46000 <2> (<2> SDS-PAGE [4]) [4]
46690 <2> (<2> calculated from the gene sequence [1]) [1]
46990 <2> (<2> predicted from the DNA sequence [2]) [2]
48000 <2> (<2> SDS-PAGE [1]) [1]

5 Isolation/Preparation/Mutation/Application

Purification
<2> [1, 3, 4]

Cloning
<2> (expressed in Escherichia coli BL21DE3) [2]
<2> (gene cobG heterologously expressed in Escherichia coli) [4]
<2> (genetic and sequence analysis) [3]
<2> (overexpressed in recombinant strains of Escherichia coli) [1]

6 Stability

General stability information
<2>, upon handling or storing CobG preparations, the sulfur content, the A400, and the enzymatic activity drops simultaneously [3]

References

[1] Roessner, C.A.; Spencer, J.B.; Ozaki, S.; Min, C.; Atshaves, B.P.; Nayar, P.; Anousis, N.; Stolowich, N.J.; Holderman, M.T.; Scott, A.I.: Overexpression in Escherichia coli of 12 vitamin B_{12} biosynthetic enzymes. Protein Expr. Purif., **6**, 155-163 (1995)
[2] Scott, A.I.; Roessner, C.A.; Stolowich, N.J.; Spencer, J.B.; Min, C.; Ozaki, S.I.: Biosynthesis of vitamin B_{12}. Discovery of the enzymes for oxidative ring contraction and insertion of the fourth methyl group. FEBS Lett., **331**, 105-108 (1993)
[3] Debussche, L.; Thibaut, D.; Cameron, B.; Crouzet, J.; Blanche, F.: Biosynthesis of the corrin macrocycle of coenzyme B_{12} in Pseudomonas denitrificans. J. Bacteriol., **175**, 7430-7440 (1993)

[4] Stamford, N.P.; Duggan, S.; Li, Y.; Alanine, A.I.; Crouzet, J.; Battersby, A.R.: Biosynthesis of vitamin B_{12}: the multi-enzyme synthesis of precorrin-4 and factor IV. Chem. Biol., **4**, 445-451 (1997)
[5] Heldt, D.; Lawrence, A.D.; Lindenmeyer, M.; Deery, E.; Heathcote, P.; Rigby, S.E.; Warren, M.J.: Aerobic synthesis of vitamin B_{12}: ring contraction and cobalt chelation. Biochem. Soc. Trans., **33**, 815-819 (2005)

4-Hydroxyacetophenone monooxygenase 1.14.13.84

1 Nomenclature

EC number
1.14.13.84

Systematic name
(4-hydroxyphenyl)ethan-1-one,NADPH:oxygen oxidoreductase (ester-forming)

Recommended name
4-hydroxyacetophenone monooxygenase

Synonyms
HAPMO <1, 2, 3> [1, 2, 3, 4, 5, 6, 7]
arylketone monooxygenase <2> [4]
oxygenase, 4-hydroxyacetophenone mono- <2> [4]
Additional information <1, 2> (<2> Baeyer-Villiger-type monooxygenase [4]; <1> the enzyme is a Baeyer-Villiger monooxygenase [7]) [4, 7]

CAS registry number
156621-13-5

2 Source Organism

<1> *Pseudomonas fluorescens* (no sequence specified) [1, 2, 3, 5, 6, 7]
<2> *Pseudomonas putida* (no sequence specified) [4]
<3> *Pseudomonas fluorescens* (UNIPROT accession number: Q93TJ5) [1]

3 Reaction and Specificity

Catalyzed reaction
(4-hydroxyphenyl)ethan-1-one + NADPH + H^+ + O_2 = 4-hydroxyphenyl acetate + $NADP^+$ + H_2O (<1> reaction mechanism [6]; <1> enantioselectivity [2]; <2> contains FAD, the enzyme from Pseudomonas fluorescens ACB catalyses the conversion of a wide range of acetophenone derivatives, highest activity occurs with compounds bearing an electron-donating substituent at the para position of the aromatic ring, in the absence of substrate, the enzyme can act as an NADPH oxidase (EC 1.6.3.1) [4])

Reaction type
Bayer-Villiger reaction
oxidation

redox reaction
reduction

Natural substrates and products

S (4-hydroxyphenyl)ethan-1-one + NADPH + O_2 <1> (Reversibility: ?) [5, 6, 7]
P 4-hydroxyphenyl acetate + $NADP^+$ + H_2O
S 4-hydroxyacetophenone + NADPH + H^+ + O_2 <1, 2, 3> (<1> physiological substrate [2]; <3> catalyzes the first step in the degradation of 4-hydroxyacetophenone [1]; <2> involved in the Baeyer-Villiger oxygenation of 4-hydroxyacetophenone, catalyzes one of the steps in the catabolism of 4-ethylphenol, inducible enzyme [4]) (Reversibility: ?) [1, 2, 4]
P 4-hydroxyphenyl acetate + $NADP^+$ + H_2O <1, 2, 3> [1, 2, 4]

Substrates and products

S (4-hydroxyphenyl)ethan-1-one + NADPH + O_2 <1> (<1> Baeyer-Villiger oxidation, stereospecific reaction, regioselectivity [5]) (Reversibility: ?) [1, 5, 6, 7]
P 4-hydroxyphenyl acetate + $NADP^+$ + H_2O
S 2,4-pentanedione + NADPH + H^+ + O_2 <1> (<1> very poor substrate [2]) (Reversibility: ?) [2]
P ? <1> [2]
S 2-acetylpyridine + NADPH + H^+ + O_2 <1> (Reversibility: ?) [2]
P ? <1> [2]
S 2-acetylpyrrole + NADPH + H^+ + O_2 <1> (Reversibility: ?) [2]
P ? <1> [2]
S 2-chloro-thioanisole + NADPH + O_2 <1> (Reversibility: ?) [7]
P 2-chloro-methyl phenyl sulfoxide + $NADP^+$ + H_2O
S 2-hydroxyacetophenone + NADPH + H^+ + O_2 <1> (Reversibility: ?) [2]
P 2-hydroxyphenyl acetate + $NADP^+$ + H_2O <1> [2]
S 2-oxabicyclo[3.2.0]heptan-6-one + NADPH + O_2 <1> (Reversibility: ?) [5]
P ?
S 2-oxabicyclo[4.2.0]octan-7-one + NADPH + O_2 <1> (Reversibility: ?) [5]
P ?
S 2-pyrrole carboxaldehyde + NADPH + H^+ + O_2 <1> (Reversibility: ?) [2]
P ? <1> [2]
S 3-chloro-2-butanone + NADPH + H^+ + O_2 <1> (<1> very poor substrate [2]) (Reversibility: ?) [2]
P ? <1> [2]
S 3-chloro-thioanisole + NADPH + O_2 <1> (Reversibility: ?) [7]
P 3-chloro-methyl phenyl sulfoxide + $NADP^+$ + H_2O
S 3-hydroxyacetophenone + NADPH + H^+ + O_2 <1> (Reversibility: ?) [2]
P 3-hydroxyphenyl acetate + $NADP^+$ + H_2O <1> [2]
S 3-oxabicyclo[3.2.0]heptan-6-one + NADPH + O_2 <1> (Reversibility: ?) [5]
P ?
S 4-acetylpyridine + NADPH + H^+ + O_2 <1> (<1> very poor substrate [2]) (Reversibility: ?) [2]
P ? <1> [2]

S 4-amino-thioanisole + NADPH + O_2 <1> (Reversibility: ?) [7]
P 4-amino-methyl phenyl sulfoxide + $NADP^+$ + H_2O
S 4-aminoacetophenone + NADPH + H^+ + O_2 <1, 3> (<1> best substrate [2]; <3> strictly NADPH-dependent [1]) (Reversibility: ?) [1, 2]
P 4-aminophenyl acetate + $NADP^+$ + H_2O <1, 3> [1, 2]
S 4-chloro-thioanisole + NADPH + O_2 <1> (Reversibility: ?) [7]
P 4-chloro-methyl phenyl sulfoxide + $NADP^+$ + H_2O
S 4-cyano-thioanisole + NADPH + O_2 <1> (Reversibility: ?) [7]
P 4-cyano-methyl phenyl sulfoxide + $NADP^+$ + H_2O
S 4-fluoroacetophenone + NADPH + H^+ + O_2 <1, 3> (<3> strictly NADPH-dependent, poor substrate [1]) (Reversibility: ?) [1, 2]
P 4-fluorophenyl acetate + $NADP^+$ + H_2O <1, 3> [1, 2]
S 4-hydroxy-3-methylacetophenone + NADPH + H^+ + O_2 <3> (<3> strictly NADPH-dependent [1]) (Reversibility: ?) [1]
P 4-hydroxy-3-methylphenyl acetate + $NADP^+$ + H_2O <3> [1]
S 4-hydroxyacetophenone + NADH + H^+ + O_2 <1> (<1> 700fold preference for NADPH over NADH [3]) (Reversibility: ?) [3]
P 4-hydroxyphenyl acetate + NAD^+ + H_2O <1> [3]
S 4-hydroxyacetophenone + NADPH + H^+ + O_2 <1, 2, 3> (<1> 700fold preference for NADPH over NADH, Arg-440 plays an important role in catalysis [3]; <2> consumption of one molecule of oxygen and oxidation of one molecule of NADPH per substrate molecule, specific for NADPH as electron donor, requirement for O_2 [4]; <1> NADPH- and oxygen-dependent Baeyer-Villiger oxidation, good substrate [2]; <3> strictly NADPH-dependent, tight coupling between NADPH oxidation and substrate oxygenation [1]; <1> physiological substrate [2]; <3> catalyzes the first step in the degradation of 4-hydroxyacetophenone [1]; <2> involved in the Baeyer-Villiger oxygenation of 4-hydroxyacetophenone, catalyzes one of the steps in the catabolism of 4-ethylphenol, inducible enzyme [4]) (Reversibility: ?) [1, 2, 3, 4]
P 4-hydroxyphenyl acetate + $NADP^+$ + H_2O <1, 2, 3> (<3> product is instable [1]) [1, 2, 3, 4]
S 4-hydroxybenzaldehyde + NADPH + H^+ + O_2 <1> (Reversibility: ?) [2]
P 4-hydroxyphenyl formate + $NADP^+$ + H_2O <1> [2]
S 4-hydroxybenzaldehyde + NADPH + H^+ + O_2 <3> (<3> strictly NADPH-dependent [1]) (Reversibility: ?) [1]
P ? <3> [1]
S 4-hydroxypropiophenone + NADPH + H^+ + O_2 <1, 2, 3> (<1> good substrate [2]; <2> specific for NADPH as electron donor, requirement for O_2 [4]; <3> strictly NADPH-dependent [1]) (Reversibility: ?) [1, 2, 4]
P 4-hydroxyphenyl propionate + $NADP^+$ + H_2O <1, 2, 3> [1, 2, 4]
S 4-methoxy-thioanisole + NADPH + O_2 <1> (Reversibility: ?) [7]
P 4-methoxy-methyl phenyl sulfoxide + $NADP^+$ + H_2O
S 4-methoxyacetophenone + NADPH + H^+ + O_2 <1, 3> (<3> strictly NADPH-dependent [1]) (Reversibility: ?) [1, 2]
P 4-methoxyphenyl acetate + $NADP^+$ + H_2O <1, 3> [1, 2]
S 4-methyl-thioanisole + NADPH + O_2 <1> (Reversibility: ?) [7]

P 4-methyl-methyl phenyl sulfoxide + $NADP^+$ + H_2O
S 4-methylacetophenone + NADPH + H^+ + O_2 <1, 3> (<3> strictly NADPH-dependent [1]) (Reversibility: ?) [1, 2]
P 4-methylphenyl acetate + $NADP^+$ + H_2O <1, 3> [1, 2]
S 4-nitro-thioanisole + NADPH + O_2 <1> (Reversibility: ?) [7]
P 4-nitro-methyl phenyl sulfoxide + $NADP^+$ + H_2O
S acetophenone + NADPH + H^+ + O_2 <1, 2, 3> (<2> specific for NADPH as electron donor, requirement for O_2 [4]; <3> strictly NADPH-dependent [1]) (Reversibility: ?) [1, 2, 4]
P phenyl acetate + $NADP^+$ + H_2O <1, 2, 3> [1, 2, 4]
S acetylcyclohexane + NADPH + H^+ + O_2 <1> (<1> poor substrate [2]) (Reversibility: ?) [2]
P ? <1> [2]
S benzaldehyde + NADPH + H^+ + O_2 <1> (Reversibility: ?) [2]
P phenyl formate + $NADP^+$ + H_2O <1> [2]
S bicyclo[3.2.0]hept-2-en-6-one + NADPH + O_2 <1> (Reversibility: ?) [5]
P ?
S bicyclo[3.2.0]heptan-6-one + NADPH + O_2 <1> (Reversibility: ?) [5]
P ?
S bicyclo[4.2.0]octan-7-one + NADPH + O_2 <1> (Reversibility: ?) [5]
P ?
S bicyclohept-2-en-6-one + NADPH + H^+ + O_2 <1> (<1> enantioselectivity, preferably converts (1R,5S)-bicyclohept-2-en-6-one with an enantiomeric ratio (E) of 20 [2]) (Reversibility: ?) [2]
P ? <1> [2]
S butyrophenone + NADPH + H^+ + O_2 <1> (Reversibility: ?) [2]
P phenyl butyrate + $NADP^+$ + H_2O <1> [2]
S cyclohexane carboxaldehyde + NADPH + H^+ + O_2 <1> (Reversibility: ?) [2]
P ? <1> [2]
S hydroxyacetone + NADPH + H^+ + O_2 <1> (<1> very poor substrate [2]) (Reversibility: ?) [2]
P ? <1> [2]
S isobutyrophenone + NADPH + H^+ + O_2 <1> (Reversibility: ?) [2]
P phenyl isobutyrate + $NADP^+$ + H_2O <1> [2]
S methyl 4-tolyl sulfide + NADPH + H^+ + O_2 <1> (<1> enantioselectivity, HAPMO is efficient and highly selective in the asymmetric formation of the corresponding (S)-sulfoxide [2]) (Reversibility: ?) [2]
P ? <1> [2]
S methylphenyl sulfide + NADPH + H^+ + O_2 <1> (<1> enantioselectivity, HAPMO is efficient and highly selective in the asymmetric formation of the corresponding (S)-sulfoxide [2]) (Reversibility: ?) [2]
P ? <1> [2]
S propiophenone + NADPH + H^+ + O_2 <1> (Reversibility: ?) [2]
P phenyl propionate + $NADP^+$ + H_2O <1> [2]
S thioanisole + NADPH + O_2 <1> (Reversibility: ?) [7]
P methyl phenyl sulfoxide + $NADP^+$ + H_2O

S tricyclo[4.2.1.02,5]nonan-3-one + NADPH + O_2 <1> (Reversibility: ?) [5]
P ?
S Additional information <1, 2, 3> (<1> a broad range of carbonylic compounds that are structurally more or less similar to 4-hydroxyacetophenone are substrates, catalyzes Baeyer-Villiger reaction with aromatic ketones and aldehydes, enzyme is capable of enantioselective formation of lactones from ketones and is also able to catalyze stereoselective sulfoxidation reactions by using aromatic sulfides, enantioselectivity, not: 2-nitroacetophenone, 4-nitroacetophenone, benzophenone, benzoin, 2'-acetonaphthone, 3-acetylpyridine, benzoic acid, methyl 4-hydroxybenzoate, benzamide, N,N-dimethylaniline, phenylacetone, 1-indanone, 4-hydroxy-1-indanone, 1,3-indanone, 4-chromanone, cyclopentanone, cyclohexanone, progesterone, dihydrocarvone, acetone, butanone, 4-heptanone [2]; <1> catalyzes Baeyer-Villiger oxidation reactions on various ketones, oxidizes a variety of aromatic ketones and sulfides [3]; <3> catalyzes the Baeyer-Villiger oxidation of aromatic compounds, converts a wide range of acetophenones via a Baeyer-Villiger rearrangement reaction into the corresponding phenyl acetates, the highest catalytic efficiency is observed with compounds bearing an electron donating substituent at the para position of the aromatic ring, in the absence of substrate the enzyme can act as an NADPH oxidase forming hydrogen peroxide, not: cyclohexanone, cyclopentanone, NADH [1]; <2> not: benzophenone, cyclohexylacetone, NADH [4]; <1> substrate specificity, regioselectivity, the enzyme preferably and stereospecifically catalyzes Baeyer-Villiger oxidations of of ketones bearing a cyclobutanone structural motif, e.g. oxidation of several prochiral cyclobutanones to antipodal butyrolactones, overview [5]; <1> the enzyme catalyzes Baeyer-Villiger oxidations of a wide range of ketones, thereby generating esters or lactones, overview [6]; <1> the recombinant enzyme is highly enantioselective in the synthesis of chiral phenyl and benzyl sulfoxides, oxidation of aromatic sulfides, substrate specificity and enantioselectivity, overview [7]) (Reversibility: ?) [1, 2, 3, 4, 5, 6, 7]
P ? <1, 2, 3> [1, 2, 3, 4]

Inhibitors

3-aminopyridine adenine dinucleotide phosphate <1> (<1> cofactor analogue, tight binding to the wild-type enzyme and mutant R440A [6]) [6]
acetyl-$NADP^+$ <1> (<1> 0.5 mM, 35% inhibition [3]) [3]
amino-$NADP^+$ <1> (<1> very effective inhibitor, 0.005 mM, 80% inhibition [3]) [3]
Additional information <1> (<1> not inhibited by $NADP^+$ [3]) [3]

Cofactors/prosthetic groups

FAD <1, 2, 3> (<1> flavoenzyme [7]; <1> contains one FAD molecule per subunit [3]; <2> contains one molecule of FAD per enzyme molecule, FAD is reduced specifically by NADPH and not by NADH [4]; <3> FAD-dependent, each subunit contains a noncovalently bound FAD molecule, both molecules participate in the reduction reaction, molecular oxygen is able to reoxidize the flavin cofactor [1]; <1> required for activity, each subunit con-

tains a non-covalently, tightly bound FAD cofactor, binding of FAD is important for the octameric conformation [6]) [1,3,4,5,6,7]
flavin <1> (<1> flavin-containing enzyme [2]) [2]
NADH <1> (<1> 700fold preference for NADPH over NADH [3]) [3]
NADPH <1, 2, 3> (<1> dependent on [7]; <1> NADPH-dependent [2]; <1> specific for NADPH as coenzyme, Arg-339 and Lys-439 are involved in coenzyme recognition, Arg-440 not, Lys-439 plays a role in recognizing the 2'-phosphate of NADPH, 700fold preference for NADPH over NADH [3]; <3> strictly NADPH-dependent [1]; <2> uses NADPH as cofactor [4]) [1,2,3,4,5,6,7]
Additional information <1, 2, 3> (<2,3> not: NADH [1,4]; <1> complex formation between the cofactors NADPH or 3-aminopyridine adenine dinucleotide phosphate [6]) [1,4,6]

Turnover number (min^{-1})

0.1 <1> (NADPH, <1> above, pH 7.5, 25°C, cosubstrate 4-hydroxyacetophenone, R339A mutant [3]) [3]
0.19 <1> (NADH, <1> pH 7.5, 25°C, cosubstrate 4-hydroxyacetophenone, R339A mutant [3]) [3]
0.2 <1> (NADPH, <1> pH 7.5, 25°C, cosubstrate 4-hydroxyacetophenone, K439P mutant [3]) [3]
0.5 <1> (4-acetylpyridine, <1> pH 7.5, 30°C [2]) [2]
0.6 <1, 3> (4-fluoroacetophenone, <1,3> pH 8, 30°C [1,2]) [1, 2]
0.7 <1> (NADH, <1> pH 7.5, 25°C, cosubstrate 4-hydroxyacetophenone, K439P mutant [3]) [3]
1 <1> (NADPH, <1> pH 7.5, 25°C, cosubstrate 4-hydroxyacetophenone, K439F mutant [3]) [3]
1.2 <1> (butyrophenone, <1> pH 7.5, 30°C [2]) [2]
1.6 <1> (NADPH, <1> pH 7.5, 25°C, cosubstrate 4-hydroxyacetophenone, K439A mutant [3]) [3]
1.7 <1, 3> (4-methoxyacetophenone, <1,3> pH 8, 30°C [1,2]) [1, 2]
1.7 <1> (NADH, <1> pH 7.5, 25°C, cosubstrate 4-hydroxyacetophenone, wild-type HAPMO [3]) [3]
2.1 <1> (2,4-pentanedione, <1> pH 7.5, 30°C [2]) [2]
2.2 <1> (benzaldehyde, <1> pH 7.5, 30°C [2]) [2]
2.4 <1> (NADH, <1> pH 7.5, 25°C, cosubstrate 4-hydroxyacetophenone, K439A mutant [3]; <1> pH 7.5, 25°C, cosubstrate 4-hydroxyacetophenone, K439F mutant [3]) [3]
2.5 <1> (isobutyrophenone, <1> pH 7.5, 30°C [2]) [2]
3.5 <1> (NADPH, <1> pH 7.5, 25°C, cosubstrate 4-hydroxyacetophenone, K439N mutant [3]) [3]
3.9 <1> (acetylcyclohexane, <1> pH 7.5, 30°C [2]) [2]
4.1 <1> (hydroxyacetone, <1> pH 7.5, 30°C [2]) [2]
4.5 <3> (acetophenone, <3> pH 8, 30°C [1]) [1]
4.7 <1> (methylphenyl sulfide, <1> pH 7.5, 30°C [2]) [2]
4.8 <1> (3-hydroxyacetophenone, <1> pH 7.5, 30°C [2]) [2]
5 <1> (cyclohexane carboxaldehyde, <1> pH 7.5, 30°C [2]) [2]

5.1 <1> (NADH, <1> pH 7.5, 25°C, cosubstrate 4-hydroxyacetophenone, K439N mutant [3]) [3]
5.4 <3> (4-hydroxy-3-methylacetophenone, <3> pH 8, 30°C [1]) [1]
6.3 <1, 3> (4-methylacetophenone, <1,3> pH 8, 30°C [1,2]) [1, 2]
6.7 <1> (2-hydroxyacetophenone, <1> pH 7.5, 30°C [2]) [2]
6.7 <1> (bicyclohept-2-en-6-one, <1> pH 7.5, 30°C [2]) [2]
7.3 <1> (methyl 4-tolyl sulfide, <1> pH 7.5, 30°C [2]) [2]
7.4 <1> (3-chloro-2-butanone, <1> pH 7.5, 30°C [2]) [2]
7.6 <1, 3> (4-hydroxybenzaldehyde, <1,3> pH 8, 30°C [1,2]) [1, 2]
7.8 <1> (2-acetylpyridine, <1> pH 7.5, 30°C [2]) [2]
8.6 <1> (2-pyrrole carboxaldehyde, <1> pH 7.5, 30°C [2]) [2]
9.3 <3> (NADPH, <3> pH 8, 30°C [1]) [1]
9.4 <1> (2-acetylpyrrole, <1> pH 7.5, 30°C [2]) [2]
10.1 <3> (4-hydroxyacetophenone, <3> pH 8, 30°C [1]) [1]
10.1 <1> (NADPH, <1> pH 7.5, 25°C, cosubstrate 4-hydroxyacetophenone, wild-type HAPMO [3]) [3]
10.6 <3> (4-hydroxypropiophenone, <3> pH 8, 30°C [1]) [1]
11 <1> (propiophenone, <1> pH 7.5, 30°C [2]) [2]
11.9 <1> (4-hydroxypropiophenone, <1> pH 7.5, 30°C [2]) [2]
12.3 <3> (4-aminoacetophenone, <3> pH 8, 30°C [1]) [1]
12.6 <1> (4-hydroxyacetophenone, <1> pH 7.5, 30°C [2]) [2]
12.7 <1> (4-aminoacetophenone, <1> pH 7.5, 30°C [2]) [2]
13.2 <1> (acetophenone, <1> pH 7.5, 30°C [2]) [2]

Specific activity (U/mg)

5.5 <3> (<3> pH 8, 30°C [1]) [1]
8.1 <2> (<2> pH 8, 30°C [4]) [4]

K_m-Value (mM)

0.0024 <1> (4-hydroxypropiophenone, <1> pH 7.5, 30°C [2]) [2]
0.003 <1> (4-aminoacetophenone, <1> pH 8, 30°C [1]) [1]
0.0082 <1> (4-aminoacetophenone, <1> pH 7.5, 30°C [2]) [2]
0.0092 <1> (4-hydroxyacetophenone, <1> pH 7.5, 30°C, pH-dependent [2]) [2]
0.012 <1> (4-hydroxypropiophenone, <1> pH 8, 30°C [1]) [1]
0.012 <1> (NADPH, <1> pH 7.5, 25°C, cosubstrate 4-hydroxyacetophenone, wild-type HAPMO [3]) [3]
0.017 <1> (NADPH, <1> pH 7.5, 25°C, cosubstrate 4-hydroxyacetophenone, K439P mutant [3]) [3]
0.0175 <2> (NADPH, <2> pH 8, 30°C, 4-hydroxyacetophenone as substrate [4]) [4]
0.023 <2> (4-hydroxypropiophenone, <2> pH 8, 30°C [4]) [4]
0.039 <1> (4-hydroxyacetophenone, <1> pH 8, 30°C [1]) [1]
0.047 <2> (4-hydroxyacetophenone, <2> pH 8, 30°C [4]) [4]
0.064 <1> (NADPH, <1> pH 8, 30°C [1]) [1]
0.078 <1> (NADPH, <1> pH 7.5, 25°C, cosubstrate 4-hydroxyacetophenone, K439N mutant [3]) [3]
0.08 <1> (NADPH, <1> pH 7.5, 25°C, cosubstrate 4-hydroxyacetophenone, K439F mutant [3]) [3]

0.1 <1> (4-hydroxybenzaldehyde, <1> pH 8, 30°C [2]) [2]
0.101 <1> (4-hydroxybenzaldehyde, <1> pH 8, 30°C [1]) [1]
0.16 <1> (4-methylacetophenone, <1> pH 8, 30°C [2]) [2]
0.161 <1> (4-methylacetophenone, <1> pH 8, 30°C [1]) [1]
0.2 <1> (NADPH, <1> pH 7.5, 25°C, cosubstrate 4-hydroxyacetophenone, K439A mutant [3]) [3]
0.3 <1> (NADH, <1> pH 7.5, 25°C, cosubstrate 4-hydroxyacetophenone, K439F mutant [3]) [3]
0.33 <1> (2-acetylpyrrole, <1> pH 7.5, 30°C [2]) [2]
0.37 <1> (methyl 4-tolyl sulfide, <1> pH 7.5, 30°C [2]) [2]
0.38 <1> (4-hydroxy-3-methylacetophenone, <1> pH 8, 30°C [1]) [1]
0.384 <2> (acetophenone, <2> pH 8, 30°C [4]) [4]
0.41 <1> (2-pyrrole carboxaldehyde, <1> pH 7.5, 30°C [2]) [2]
0.52 <1> (NADH, <1> pH 7.5, 25°C, cosubstrate 4-hydroxyacetophenone, K439A mutant [3]) [3]
0.53 <1> (propiophenone, <1> pH 7.5, 30°C [2]) [2]
0.54 <1> (4-methoxyacetophenone, <1> pH 8, 30°C [2]) [2]
0.54 <1> (isobutyrophenone, <1> pH 7.5, 30°C [2]) [2]
0.541 <1> (4-methoxyacetophenone, <1> pH 8, 30°C [1]) [1]
0.61 <1> (2-hydroxyacetophenone, <1> pH 7.5, 30°C [2]) [2]
0.78 <1> (NADH, <1> pH 7.5, 25°C, cosubstrate 4-hydroxyacetophenone, K439N mutant [3]) [3]
1 <1> (4-fluoroacetophenone, <1> pH 8, 30°C [2]) [2]
1.04 <1> (4-fluoroacetophenone, <1> pH 8, 30°C [1]) [1]
1.2 <1> (2-acetylpyridine, <1> pH 7.5, 30°C [2]) [2]
1.28 <1> (NADH, <1> pH 7.5, 25°C, cosubstrate 4-hydroxyacetophenone, K439P mutant [3]) [3]
1.4 <1> (3-hydroxyacetophenone, <1> pH 7.5, 30°C [2]) [2]
1.4 <1> (methylphenyl sulfide, <1> pH 7.5, 30°C [2]) [2]
1.44 <1> (NADH, <1> pH 7.5, 25°C, cosubstrate 4-hydroxyacetophenone, wild-type HAPMO [3]) [3]
1.6 <1> (benzaldehyde, <1> pH 7.5, 30°C [2]) [2]
1.9 <1> (4-acetylpyridine, <1> pH 7.5, 30°C [2]) [2]
2 <1> (butyrophenone, <1> pH 7.5, 30°C [2]) [2]
2-3 <1> (3-chloro-2-butanone, <1> pH 7.5, 30°C [2]) [2]
2.27 <1> (acetophenone, <1> pH 8, 30°C [1]) [1]
2.3 <1> (acetophenone, <1> pH 7.5, 30°C [2]) [2]
3 <1> (NADH, <1> pH 7.5, 25°C, cosubstrate 4-hydroxyacetophenone, R339A mutant [3]) [3]
3 <1> (NADPH, <1> above, pH 7.5, 25°C, cosubstrate 4-hydroxyacetophenone, R339A mutant [3]) [3]
3 <1> (cyclohexane carboxaldehyde, <1> pH 7.5, 30°C [2]) [2]
3.3 <1> (bicyclohept-2-en-6-one, <1> pH 7.5, 30°C [2]) [2]
4.8 <1> (acetylcyclohexane, <1> pH 7.5, 30°C [2]) [2]
4.9 <1> (2,4-pentanedione, <1> pH 7.5, 30°C [2]) [2]
29 <1> (hydroxyacetone, <1> pH 7.5, 30°C [2]) [2]

pH-Optimum

7.5 <1, 3> (<1> assay at [2,3]) [1, 2, 3]
8 <2> (<2> Tris-HCl buffer [4]) [4]
9 <1> (<1> assay at [7]) [7]

Temperature optimum (°C)

25 <1> (<1> assay at [3,7]) [3, 7]
30 <1, 2, 3> (<1,2> assay at [2,4]) [1, 2, 4]

4 Enzyme Structure

Molecular weight

70000 <2> (<2> gel filtration [4]) [4]
140000 <3> (<3> gel filtration [1]) [1]
145000 <1> [3]
145100 <1> (<1> mutant R339A, dimeric structure, mass spectrometry [6]; <1> mutant R440A, dimeric structure, mass spectrometry [6]) [6]
145200 <1> (<1> wild-type enzyme, dimeric structure, mass spectrometry [6]) [6]

Subunits

dimer <1> [3, 6]
homodimer <3> (<3> 2 * 70000, SDS-PAGE, 2 * 71884, sequence calculation [1]) [1]
monomer <2> (<2> 1 * 71000, SDS-PAGE [4]) [4]
octamer <1> [6]
Additional information <1> (<1> quaternary structure of wild-type and mutant enzymes, overview [6]) [6]

5 Isolation/Preparation/Mutation/Application

Localization

soluble <3> (<3> recombinant HAPMO [1]) [1]

Purification

<1> (recombinant HAPMO) [2]
<1> (wild-type and mutant HAPMO) [3]
<1> (wild-type, 25fold, and recombinant HAPMO) [1]
<2> (98fold) [4]

Cloning

<1> (expressed in Escherichia coli TOP10) [5]
<1> (gene hapE, expression of wild-type and mutant enzymes in Escherichia coli) [6]
<1> (gene hapE, overexpression in Escherichia coli) [5]

<1> (hapE gene, expression in Escherichia coli TOP10) [2, 3]
<1> (hapE, expression in Escherichia coli BL21(DE3)pLysS, sequence analysis) [1]

Engineering

G490A <3> (<3> strictly conserved glycine, dramatic effect of mutation on binding and oxidation of NADPH [1]) [1]
H61T <1> (<1> the H61T mutant is purified as apo-enzyme and mainly exists as a dimeric species, the binding of FAD to the enzyme restores the octameric conformation [6]) [6]
K439A <1> (<1> mutant with 100fold decrease in catalytic efficiency with NADPH, mainly caused by increased K_m, 4fold increased efficiency with NADH [3]) [3]
K439F <1> (<1> mutant with higher activity with NADH compared to wild-type HAPMO [3]) [3]
K439N <1> (<1> mutant with higher activity with NADH compared to wild-type HAPMO [3]) [3]
K439P <1> (<1> mutant with higher activity with NADH compared to wild-type HAPMO [3]) [3]
R339A <1> (<1> mutant with largely decreased affinity for NADPH and decreased catalytic efficiency with NADH [3]; <1> site-directed mutagenesiss, the mutant shows decreased activity and affinity to NADPH compared to the wild-type enzyme, the mutant only weakly interacts with 3-aminopyridine adenine dinucleotide phosphate [6]) [3, 6]
R440A <1> (<1> totally inactive mutant, Arg-440 is crucial for completing the catalytic cycle [3]; <1> site-directed mutagenesiss, inactive mutant, the mutant shows highly increased affinity to NADPH compared to the wild-type enzyme [6]) [3, 6]

Application

synthesis <1> (<1> the enzyme can be useful in the highly enantioselective in the synthesis of chiral phenyl and benzyl sulfoxides [7]) [7]
Additional information <1, 3> (<3> as acylcatechols are valuable synthons for the fine chemical industry, HAPMO might develop as a useful biocatalytic tool [1]; <1> potential of HAPMO for biotechnological applications [2]) [1, 2]

6 Stability

General stability information

<1>, the association with the coenzyme NADPH is crucial for enzyme stability, 3-aminopyridine adenine dinucleotide phosphate highly stabilizes the inactive dimeric state of the enzyme [6]

References

[1] Kamerbeek, N.M.; Moonen, M.J.; van der Ven, J.G.; van Berkel, W.J.H.; Fraaije, M.W.; Janssen, D.B.: 4-Hydroxyacetophenone monooxygenase from Pseudomonas fluorescens ACB: a novel flavoprotein catalyzing Baeyer-Villiger oxidation of aromatic compounds. Eur. J. Biochem., **268**, 2547-2557 (2001)

[2] Kamerbeek, N.M.; Olsthoorn, A.J.J.; Fraaije, M.W.; Janssen, D.B.: Substrate specificity and enantioselectivity of 4-hydroxyacetophenone monooxygenase. Appl. Environ. Microbiol., **69**, 419-426 (2003)

[3] Kamerbeek, N.M.; Fraaije, M.W.; Janssen, D.B.: Identifying determinants of NADPH specificity in Baeyer-Villiger monooxygenases. Eur. J. Biochem., **271**, 2107-2116 (2004)

[4] Tanner, A.; Hopper, D.J.: Conversion of 4-hydroxyacetophenone into 4-phenyl acetate by a flavin adenine dinucleotide-containing Baeyer-Villiger-type monooxygenase. J. Bacteriol., **182**, 6565-6569 (2000)

[5] Mihovilovic, M.D.; Kapitan, P.; Rydz, J.; Rudroff, F.; Ogink, F.H.; Fraaije, M.W.: Biooxidation of ketones with a cyclobutanone structural motif by recombinant whole-cells expressing 4-hydroxyacetophenone monooxygenase. J. Mol. Catal. B, **32**, 135-140 (2005)

[6] van den Heuvel, R.H.; Tahallah, N.; Kamerbeek, N.M.; Fraaije, M.W.; van Berkel, W.J.; Janssen, D.B.; Heck, A.J.: Coenzyme binding during catalysis is beneficial for the stability of 4-hydroxyacetophenone monooxygenase. J. Biol. Chem., **280**, 32115-32121 (2005)

[7] de Gonzalo, G.; Torres Pazmino, D.E.; Ottolina, G.; Fraaije, M.W.; Carrea, G.: 4-Hydroxyacetophenone monooxygenase from Pseudomonas fluorescens ACB as an oxidative biocatalyst in the synthesis of optically active sulfoxides. Tetrahedron, **17**, 130-135 (2006)

Glyceollin synthase 1.14.13.85

1 Nomenclature

EC number
1.14.13.85

Systematic name
2-(or 4-)dimethylallyl-(6aS,11aS)-3,6a,9-trihydroxypterocarpan,NADPH:oxygen oxidoreductase (cyclizing)

Recommended name
glyceollin synthase

Synonyms
dimethylallyl-3,6a,9-trihydroxypterocarpan cyclase <1> [1]

CAS registry number
115110-88-8

2 Source Organism

<1> *Glycine max* (no sequence specified) [1]

3 Reaction and Specificity

Catalyzed reaction
2-(or 4-)dimethylallyl-(6aS,11aS)-3,6a,9-trihydroxypterocarpan + NADPH + H^+ + O_2 = glyceollin + $NADP^+$ + 2 H_2O (<1> a heme-thiolate protein, P_{450}, glyceollins II and III are formed from 2-dimethylallyl-(6aS,11aS)-3,6a,9-trihydroxypterocarpan whereas glyceollin I is formed from the 4-isomer [1])

Reaction type
cyclization
oxidation
reduction

Natural substrates and products
S Additional information <1> (<1> last committed step in the glyceollin biosynthesis [1]) (Reversibility: ?) [1]
P ?

Substrates and products

S 2-(or 4-)dimethylallyl-(6aS,11aS)-3,6a,9-trihydroxypterocarpan + NADPH + H^+ + O_2 <1> (<1> enzyme catalyzes an NADPH-dependent and oxygen-dependent cyclization of a mixture of 2- and 4-dimethylallylglycinols to the glyceollin isomers I-III [1]) (Reversibility: ?) [1]

P glyceollin + $NADP^+$ + 2 H_2O (<1> glyceollin isomers I-III, glyceollin isomer I is formed from 4-dimethylallylglycinol and glyceollin isomers II and III from 2-dimethylallylglycinol [1])

S Additional information <1> (<1> last committed step in the glyceollin biosynthesis [1]; <1> not: NADH, enzyme does not catalyze the conversion of rot-2-enoic acid to deguelin in the presence or absence of NADPH [1]) (Reversibility: ?) [1]

P ?

Inhibitors

ancymidol <1> (<1> 0.01 mM: 15% inhibition, 0.1 mM: 85% inhibition [1]) [1]

BAS 110 <1> (<1> 0.1 mM, 15% inhibition [1]) [1]

BAS 111 <1> (<1> 0.1 mM, 39% inhibition [1]) [1]

CO <1> (<1> inhibits in a white light-reversible manner in the presence of oxygen [1]) [1]

cytochrome c <1> (<1> 0.01 mM: 25% inhibition, 0.025 mM: 64% inhibition, 0.05 mM: 85% inhibition [1]) [1]

LAB 978 <1> (<1> 0.01 mM: 17% inhibition, 0.1 mM: 49% inhibition [1]) [1]

metyrapone <1> (<1> 0.01 mM: 15% inhibition, 0.1 mM: 37% inhibition [1]) [1]

$NADP^+$ <1> (<1> strong inhibition by $NADP^+$ at low NADPH concentrations [1]) [1]

SKF 525 <1> (<1> 0.1 mM: 11% inhibition, 1 mM: 100% inhibition [1]) [1]

glyceollin <1> (<1> product inhibition, 0.058 mM, 50% inhibition [1]) [1]

ketoconazol <1> (<1> 0.01 mM: 34% inhibition, 0.1 mM: 100% inhibition [1]) [1]

potassium isothiocyanate <1> (<1> 0.01 mM: 19% inhibition, 0.1 mM: 73% inhibition, 1 mM: 100% inhibition [1]) [1]

tetcyclasis <1> (<1> 0.01 mM: 9% inhibition, 0.1 mM: 67% inhibition [1]) [1]

Cofactors/prosthetic groups

cytochrome P_{450} <1> (<1> cytochrome P-450-dependent monooxygenase, average content of cytochrome P-450: 0.16 nmol/mg protein [1]) [1]

cytochrome b <1> (<1> average content of b-type cytochromes: 0.26 nmol/mg protein [1]) [1]

NADPH <1> (<1> NADPH-dependent, absolute requirement [1]) [1]

O_2 <1> (<1> oxygen-dependent [1]) [1]

Activating compounds

NADH <1> (<1> a synergistic effect of NADH, 94% increase in cyclase activity, is observed at low NADPH concentrations of about 0.01 mM [1]) [1]

Additional information <1> (<1> challenge with either a glucan elicitor from Phytophthora megasperma f.sp. glycinea or with yeast extract causes strong stimulation of cyclase activity with a maximum at about 24 h after elicitor addition [1]) [1]

Specific activity (U/mg)

Additional information <1> [1]

K_m-Value (mM)

0.002 <1> (2-(or 4-)dimethylallyl-(6aS,11aS)-3,6a,9-trihydroxypterocarpan, <1> pH 7.5, 20°C, mixture of the two glyceollidin isomers [1]) [1]
0.03 <1> (NADPH + H^+, <1> pH 7.5, 20°C [1]) [1]

pH-Optimum

7.5-9.5 <1> (<1> very broad pH-optimum [1]) [1]

pH-Range

7-9.5 <1> (<1> pH 7: 50% of maximum activity, no activity measurable above pH 9.5, probably because of the instability of the substrate [1]) [1]

Temperature optimum (°C)

20 <1> (<1> assay at [1]) [1]

5 Isolation/Preparation/Mutation/Application

Source/tissue

cell suspension culture <1> (<1> elicitor-challenged, unstimulated cell culture does not contain detectable cyclase activity [1]) [1]

Localization

endoplasmic reticulum <1> (<1> the cyclase is associated with [1]) [1]
microsome <1> [1]

References

[1] Welle, R.; Grisebach, H.: Induction of phytoalexin synthesis in soybean: enzymatic cyclization of prenylated pterocarpans to glyceollin isomers. Arch. Biochem. Biophys., **263**, 191-198 (1988)

2-Hydroxyisoflavanone synthase 1.14.13.86

1 Nomenclature

EC number
1.14.13.86

Systematic name
apigenin,NADPH:oxygen oxidoreductase (isoflavanone-forming)

Recommended name
2-hydroxyisoflavanone synthase

Synonyms
2-HIS <5> [1]
2-hydroxyisoflavanone synthase <6> [6]
CYP Ge-8 <6> [5]
CYP93C <6> [6]
CYP93C1v2 <5> [1]
CYP93C2 <2> [4]
CYP93C2 protein <6> [5]
F3H <6> [6]
IFS <1, 2, 3, 6, 7> [3, 4, 5, 6, 7, 9, 10]
IFS1 <7> [3]
P_{450} 2-hydroxyisoflavanone synthase <6> [6]
P_{450} isoflavonoid synthase <6> [6]
cyp93c1 protein <8> [3]
flavanone 3β-hydroxylase <6> [6]
isoflavone synthase <1, 3, 7, 8> [3, 7, 8, 9, 10]

CAS registry number
168680-18-0

2 Source Organism

<1> *Glycine max* (no sequence specified) [2, 8, 9, 10]
<2> *Glycyrrhiza echinata* (no sequence specified) [4]
<3> *Trifolium pratense* (no sequence specified) [7]
<4> no activity in *Oryza sativa* [8]
<5> *Glycine max* (UNIPROT accession number: Q9SWR5) [1]
<6> *Glycyrrhiza echinata* (UNIPROT accession number: Q9SXS3) [5,6]
<7> *Glycine max* (UNIPROT accession number: Q9M6D6) [3]
<8> *Glycine max* (UNIPROT accession number: O48926) [3]

3 Reaction and Specificity

Catalyzed reaction

apigenin + 2 NADPH + 2 H^+ + O_2 = 2-hydroxy-2,3-dihydrogenistein + 2 $NADP^+$ + H_2O (<2> mechanism [4]; <5> a heme-thiolate protein, P_{450}, EC 4.2.1.105, 2-hydroxyisoflavanone dehydratase, acts on 2-hydroxy-2,3-dihydrogenistein with loss of water and formation of genistein, this may occur spontaneously [1]; <6> stereoselectivity [5]; <6> the active-site residues Ser310 and Lys375 are critical for unusual aryl migration of the flavanone substrate of CYP93C2, reaction mechanism [6])

Reaction type

1,2-aryl migration
hydroxylation
oxidation
redox reaction
reduction

Natural substrates and products

S (2S)-flavanone + NADPH + H^+ + O_2 <2> (<2> isoflavonoid biosynthesis [4]) (Reversibility: ?) [4]
P 2-hydroxyisoflavanone + $NADP^+$ + H_2O
S (2S)-liquiritigenin + NADPH + O_2 <6> (<6> i.e. (2S)-7,4-dihydroxyflavanone, aryl migration with hydrogen abstraction, 2-step reaction via radical intermediate, overview [6]) (Reversibility: ?) [6]
P 2,7,4'-trihydroxyflavanone + $NADP^+$ + H_2O (<6> a 2-hydroxyisoflavanone [6])
S (2S)-liquiritigenin + NADPH + O_2 <6> (<6> i.e. (2S)-7,4-dihydroxyflavanone, aryl migration with hydrogen abstraction, 2-step reaction via radical intermediate, overview [6]) (Reversibility: ?) [6]
P 3β,7,4'-trihydroxyflavanone + $NADP^+$ + H_2O (<6> a 3-hydroxyflavanone [6])
S liquiritigenin + NADPH + O_2 <1> (Reversibility: ?) [8, 1]
P daidzein + $NADP^+$ + H_2O
S naringenin + NADPH + O_2 <1> (Reversibility: ?) [8, 1]
P genistein + $NADP^+$ + H_2O
S Additional information <1, 5, 6, 7, 8> (<6> involved in the biosynthesis of the isoflavonoid skeleton, key enzyme in the isoflavonoid biosynthesis [5]; <7,8> isoflavone synthase catalyzes the first committed step of isoflavone biosynthesis, a branch of the phenylpropanoid pathway [3]; <5> the first specific reaction in the biosynthesis of isoflavonoid compounds is the 2-hydroxylation, coupled to aryl migration, of a flavanone [1]; <6> flavanone pathways, overview [6]; <1> key enzyme for the formation of the isoflavones, RNA interference of isoflavone synthase genes leads to silencing in tissues distal to the transformation site and to enhanced susceptibility to Phytophthora sojae, overview [9]; <1> secretion of flavonoids is a first step in the legume-Rhizobium interactions, overview, key enzyme

that redirects phenylpropanoid pathway intermediates from flavonoids to isoflavonoids, isoflavonoids, secreted by the legumes, play a critical role in plant development and defence response, they also play an important role in promoting the formation of nitrogen-fixing nodules by symbiotic rhizobia, the enzyme enhances expression of nod genes, exppression analysis in different rhizobia, overview [8]; <1> the enzyme is important in the phenylpropanoid pathway, overview [10]) (Reversibility: ?) [1, 3, 5, 6, 8, 9, 10]

P ?

Substrates and products

S (2S)-flavanone + NADPH + H^+ + O_2 <2, 6> (<2> isoflavonoid biosynthesis [4]; <6> IFS catalyzes the 2-hydroxylation associated with 1,2-aryl migration, substrate stereoselectivity, $CYP93C_2$ acts only on natural substrates with (2S)-chirality [5]; <2> the reaction consists of hydroxylation of the flavanone molecule at C-2 and an intramolecular 1,2-aryl migration from C-2 to C-3 to yield 2-hydroxyisoflavanone, roles of Ser-310 and Lys-375 in the catalysis, active site structure [4]) (Reversibility: ?) [4, 5]

P 2-hydroxyisoflavanone + $NADP^+$ + H_2O (<2> product is highly unstable, 3-hydroxyflavanone is a minor by-product of the reaction [4])

S (2S)-liquiritigenin + NADPH + H^+ + O_2 <2, 6> (<6> recombinant CYP93C2, catalyzes the 2-hydroxylation associated with 1,2-aryl migration, stereoselectivity [5]; <2> roles of Ser-310 and Lys-375 in the catalysis, active site structure [4]) (Reversibility: ?) [4, 5]

P 2,7,4'-trihydroxyisoflavanone + $NADP^+$ + H_2O (<2> 2,7,4'-trihydroxyisoflavanone dehydrates to daidzein by acid treatment, 3,7,4'-trihydroxyflavanone is a minor by-product of the reaction [4]; <6> the product is rapidly converted to daidzein by acid treatment [5])

S (2S)-liquiritigenin + NADPH + O_2 <6> (<6> i.e. (2S)-7,4-dihydroxyflavanone, aryl migration with hydrogen abstraction, 2-step reaction via radical intermediate, overview [6]) (Reversibility: ?) [6]

P 2,7,4'-trihydroxyflavanone + $NADP^+$ + H_2O (<6> a 2-hydroxyisoflavanone [6])

S (2S)-liquiritigenin + NADPH + O_2 <6> (<6> i.e. (2S)-7,4-dihydroxyflavanone, aryl migration with hydrogen abstraction, 2-step reaction via radical intermediate, overview [6]) (Reversibility: ?) [6]

P 3β,7,4'-trihydroxyflavanone + $NADP^+$ + H_2O (<6> a 3-hydroxyflavanone [6])

S (2S)-naringenin + NADPH + H^+ + O_2 <1, 6> (<1> absolute requirement for oxygen and NADPH, NADPH cannot be replaced by NADH, FAD, FMN or ascorbate [2]; <6> recombinant CYP93C2, catalyzes the 2-hydroxylation associated with 1,2-aryl migration [5]) (Reversibility: ?) [2, 5]

P 2,5,7,4'-tetrahydroxyisoflavanone + $NADP^+$ + H_2O (<1> the product is converted in a second step to genistein, formally a dehydration without requirement of NADPH or oxygen [2]; <6> the product is rapidly converted to genistein by acid treatment [5])

S liquiritigenin + NADPH + H^+ + O_2 <1> (Reversibility: ?) [2]

P ?

S liquiritigenin + NADPH + H^+ + O_2 <5> (<5> recombinant 2-HIS converts liquiritigenin to daidzein, most likely via 2,4',7-trihydroxyisoflavanone [1]) (Reversibility: ?) [1]

P 2,4',7-trihydroxyisoflavanone + $NADP^+$ + H_2O (<5> 2,4',7-trihydroxyisoflavanone spontaneously dehydrates to daidzein [1])

S liquiritigenin + NADPH + H^+ + O_2 <7, 8> (<8> 2-hydroxyisoflavanone intermediate [3]; <7> 2-hydroxyisoflavanone intermediate, better substrate than naringenin [3]) (Reversibility: ?) [3]

P daidzein + $NADP^+$ + H_2O

S liquiritigenin + NADPH + O_2 <1> (<1> i.e. 7,4-dihydroxyflavanone [8,10]) (Reversibility: ?) [8, 1]

P daidzein + $NADP^+$ + H_2O (<1> 7,4-dihydroxyisoflavone [8,10])

S naringenin + NADPH + H^+ + O_2 <5, 7, 8> (<8> oxidative aryl migration, 2-hydroxyisoflavanone intermediate [3]; <7> oxidative aryl migration, 2-hydroxyisoflavanone intermediate, less efficient substrate than liquiritigenin [3]; <5> recombinant 2-HIS converts naringenin to genistein at a lower rate than liquiritigenin, converts naringenin via the 2-hydroxyisoflavanone to genistein [1]) (Reversibility: ?) [1, 3]

P genistein + $NADP^+$ + H_2O

S naringenin + NADPH + O_2 <1, 3> (<1> i.e. 5,7,4-trihydroxyflavanone [8,10]) (Reversibility: ?) [7, 8, 10]

P genistein + $NADP^+$ + H_2O (<1> i.e. 5,7,4-trihydroxyisoflavone [8,10])

S Additional information <1, 3, 5, 6, 7, 8> (<6> involved in the biosynthesis of the isoflavonoid skeleton, key enzyme in the isoflavonoid biosynthesis [5]; <7,8> isoflavone synthase catalyzes the first committed step of isoflavone biosynthesis, a branch of the phenylpropanoid pathway [3]; <5> the first specific reaction in the biosynthesis of isoflavonoid compounds is the 2-hydroxylation, coupled to aryl migration, of a flavanone [1]; <1> not: (2R)-naringenin, apigenin [2]; <6> flavanone pathways, overview [6]; <1> key enzyme for the formation of the isoflavones, RNA interference of isoflavone synthase genes leads to silencing in tissues distal to the transformation site and to enhanced susceptibility to Phytophthora sojae, overview [9]; <1> secretion of flavonoids is a first step in the legume-Rhizobium interactions, overview, key enzyme that redirects phenylpropanoid pathway intermediates from flavonoids to isoflavonoids, isoflavonoids, secreted by the legumes, play a critical role in plant development and defence response, they also play an important role in promoting the formation of nitrogen-fixing nodules by symbiotic rhizobia, the enzyme enhances expression of nod genes, exppression analysis in different rhizobia, overview [8]; <1> the enzyme is important in the phenylpropanoid pathway, overview [10]; <6> the enzyme also shows flavanone 3β-hydroxylase activity, overview [6]; <3> the enzyme interacts with the flavoprotein NADPH-cytochrome P_{450} reductase to transfer electrons from NADPH [7]) (Reversibility: ?) [1, 2, 3, 5, 6, 7, 8, 9, 10]

P ?

Inhibitors

2-mercaptoethanol <1> (<1> 14 mM: 66% inhibition [2]) [2]
ancymidol <1> (<1> 0.1 mM: 100% inhibition, 0.01 mM: 94% inhibition [2]) [2]
BAS 978 <1> (<1> a triazol plant growth regulator from BASF, 0.1 mM: 100% inhibition, 0.01 mM: 90% inhibition [2]) [2]
CO <1> (<1> inhibition in the presence of oxygen, reversible by white light [2]) [2]
cytochrome c <1> (<1> 0.01 mM: 51% inhibition, 0.05 mM: 75% inhibition [2]) [2]
dithiothreitol <1> (<1> 1.4 mM: 68% inhibition [2]) [2]
KSCN <1> (<1> 0.1 mM: 74% inhibition [2]) [2]
ketoconazole <1> (<1> 0.1 mM: 100% inhibition, 0.01 mM: 79% inhibition [2]) [2]
SKF 525 A <1> (<1> 1 mM: 40% inhibition [2]) [2]
tetcyclacis <1> (<1> 0.1 mM: 74% inhibition, 0.01 mM: 48% inhibition [2]) [2]
Additional information <1> (<1> metal chelators cause only weak inhibition [2]) [2]

Cofactors/prosthetic groups

cytochrome P_{450} <1, 2, 5, 6, 7, 8> [1-5]
heme <2> [4]
NADPH <1, 3, 5, 6, 7, 8> (<1> absolute requirement for NADPH, NADPH cannot be replaced by NADH, FAD, FMN or ascorbate [2]) [1,2,3,5,6,7,8,9,10]
O_2 <1, 7, 8> (<1> absolute requirement for oxygen [2]) [2,3]

Activating compounds

rice cytochrome P_{450} reductase <3> (<3> co-expression with the enzyme leads to a 4.3fold increase in IFS activity in recombinant yeast microsomes due to increased supply of NADPH to the enzyme, overview [7]) [7]
Additional information <6> (<6> CYP93C2 transcripts transiently accumulate upon elicitation [5]) [5]

Specific activity (U/mg)

Additional information <1> [2, 10]

K_m-Value (mM)

0.006 <2> ((2S)-liquiritigenin, <2> 30°C, S310T mutant $CYP93C_2$ [4]) [4]
0.008 <2> ((2S)-liquiritigenin, <2> 30°C, recombinant, wild-type $CYP93C_2$ [4]) [4]
0.011 <2> ((2S)-liquiritigenin, <2> 30°C, K375T mutant $CYP93C_2$ [4]) [4]
Additional information <1> [2]

pH-Optimum

7.5 <2, 6> (<6> assay at [6]) [4, 6]
8 <1, 7, 8> (<1,7,8> assay at [3,8,10]) [3, 8, 10]
8-8.6 <1> [2]

pH-Range

7.1-9 <1> (<1> half-maximal activities at pH 7.1 and 9 [2]) [2]
Additional information <2> (<2> the reaction proceeds in a wide pH-range [4]) [4]

Temperature optimum (°C)

10 <1> (<1> assay at [2]) [2]
15 <1> (<1> assay at [10]) [10]
16 <5> (<5> assay at [1]) [1]
22 <1> (<1> assay at room temperature [8]) [8]
30 <1, 2, 6> (<1,2,6> assay at [2,4,5,6]) [2, 4, 5, 6]
Additional information <7, 8> (<7,8> assay at room temperature [3]) [3]

4 Enzyme Structure

Subunits

? <5, 6> (<5> x * 58800, sequence calculation [1]; <6> x * 59000, SDS-PAGE, x * 59428, sequence calculation [5]) [1, 5]
Additional information <6> (<6> three-dimensional structure modelling [6]) [6]

5 Isolation/Preparation/Mutation/Application

Source/tissue

callus <6> (<6> callus cultures from young leaves and petioles [5]) [5]
cell culture <6> (<6> elicited cultured cells, callus cultures from young leaves and petioles [5]) [5]
cell suspension culture <1> (<1> elicitor-treated [2]) [2]
cotyledon <1> [9]
hypocotyl <1> (<1> from 5 days old seedlings, elicitor-treated [2]) [2]
leaf <1, 6> (<6> callus cultures from young leaves and petioles [5]) [5, 9]
nodule <1> [8]
petiole <6> (<6> callus cultures from young leaves and petioles [5]) [5]
root <1> [9, 10]
root hair <1> [10]
seedling <1> (<1> 5 days old, elicitor-treated hypocotyls [2]) [2]

Localization

endoplasmic reticulum <1> (<1> enzyme activity is associated with the endoplasmic reticulum [2]) [2]
membrane <1, 7> (<1> membrane fraction [2]; <7> microsomal, recombinant enzyme expressed in yeast [3]) [2, 3]
microsome <1, 2, 3, 5, 6, 7, 8> (<7> microsomal membranes, recombinant enzyme expressed in yeast [3]; <2> recombinant CYP93C2 expressed in Saccharomyces cerevisiae [4]; <6> recombinant CYP93C2, expressed in Saccharomyces cerevisiae BJ2168 [5]; <8> recombinant enzyme expressed in yeast [3]) [1, 2, 3, 4, 5, 7, 8, 10]

Purification

<1> (native enzyme partially by microsome preparation) [10]

Cloning

<1> (functional transgenic expression in Oryza sativa cv. Murasaki, strain R86, plants using the Agrobacterium tumefaciens, strain EHA105, transfection system) [8]

<1> (genes IFS-1 and IFS-2, cotyledon and root tissues are transformed with Agrobacterium rhizogenes strain K599 carrying an RNAi silencing construct designed to silence expression of both copies of IFS genes, all tissue show silenced accumulation of isoflavone, overview) [9]

<1> (transgenic antisense expression in soybean plants for enzyme downregulation using the Agrobacterium rhizogenes, strain K599, transformation system via cotyledons, overview) [10]

<2> (wild-type and mutant CYP93C2, expression in Saccharomyces cerevisiae) [4]

<3> (functional co-expression of IFS with Oryza sativa cytochrome P450 reductase in Saccharomyces cerevisiae) [7]

<5> (CYP93C1v2, expression in Sf9 cells) [1]

<6> (CYP93C2, expression in Saccharomyces cerevisiae BJ2168, sequencing) [5]

<6> (CYP93C2, phylogenetic analysis, expression of wild-type and mutant enzymes in Saccharomyces cerevisiae strain BJ2168 microsomes) [6]

<7> (cDNA, IFS1 expression in yeast and in Arabidopsis thaliana ecotype WS) [3]

<8> (cyp93c1 cDNA, expression in yeast) [3]

Engineering

K375T <2> (<2> mutant shows only formation of the by-product 3-hydroxyflavanone of the 2-hydroxyisoflavanone synthase reaction, no formation of the 2-hydroxyisoflavanone, kinetic data, reduced maximum velocity compared with wild-type enzyme, pH-dependence [4]) [4]

L371V <6> (<6> site-directed mutagenesis, inactive mutant [6]) [6]

L371V/K375T <6> (<6> site-directed mutagenesis, the mutant shows reduced activity and altered substrate specificity compared to the wild-type enzyme [6]) [6]

R104A <2> (<2> inactive mutant, mutant cannot yield an active conformation of P_{450} [4]) [4]

S310T <2> (<2> mutant shows increased formation of the by-product 3-hydroxyflavanone, kinetic data, reduced maximum velocity compared with wild-type enzyme, pH-dependence [4]) [4]

S310T/L371V <6> (<6> site-directed mutagenesis, the mutant shows reduced activity and altered substrate specificity compared to the wild-type enzyme [6]) [6]

S310T/L371V/K375T <6> (<6> site-directed mutagenesis, the triple mutant shows reduced activity, altered substrate specificity and higher thermal stability compared to the wild-type enzyme [6]) [6]

Additional information <1> (<1> down-regulation of the enzyme in root hairs leads to modificated phenolic metabolism, extraction and analysis of phenolic compounds, isoflavone concentrations are reduced by about 90%, overview [10]; <1> metabolic engineering of Oryza sativa with Glycine max isoflavone synthase for promoting nodulation gene expression in rhizobia, overview [8]; <1> RNA interference of isoflavone synthase genes leads to silencing in tissues distal to the transformation site and to enhanced susceptibility to Phytophthora sojae, overview [9]) [8, 9, 10]

6 Stability

Temperature stability

4 <1> (<1> half-life in the presence of 14 mM mercaptoethanol and 20% sucrose: 200 min [2]) [2]
10 <1> (<1> half-life in the presence of 14 mM mercaptoethanol and 20% sucrose: 200 min [2]) [2]
20 <1> (<1> half-life in the presence of 14 mM mercaptoethanol and 20% sucrose: 60 min [2]) [2]
30 <1> (<1> half-life in the presence of 14 mM mercaptoethanol and 20% sucrose: 13 min [2]) [2]
37 <6> (<6> half-life recombinant wild-type enzyme: 20.4 min, half-life recombinant mutant S310T/L371V/K375T: 83.6 min [6]) [6]

General stability information

<1>, unstable enzyme, 14 mM mercaptoethanol stabilizes the enzyme, but decreases its activity, 20% sucrose and 0.5 mM glutathione stabilize [2]

References

[1] Steele, C.L.; Gijzen, M.; Qutob, D.; Dixon, R.A.: Molecular characterization of the enzyme catalyzing the aryl migration reaction of isoflavonoid biosynthesis in soybean. Arch. Biochem. Biophys., **367**, 146-150 (1999)

[2] Kochs, G.; Grisebach, H.: Enzymic synthesis of isoflavones. Eur. J. Biochem., **115**, 311-318 (1986)

[3] Jung, W.; Yu, O.; Lau, S.-M.C.; O'Keefe, D.P.; Odell, J.; Fader, G.; McGonigle, B.: Identification and expression of isoflavone synthase, the key enzyme for biosynthesis of isoflavones in legumes. Nat. Biotechnol., **18**, 208-212 (2000)

[4] Sawada, Y.; Kinoshita, K.; Akashi, T.; Aoki, T.; Ayabe, S.: Key amino acid residues required for aryl migration catalysed by the cytochrome P_{450} 2-hydroxyisoflavanone synthase. Plant J., **31**, 555-564 (2002)

[5] Akashi, T.; Aoki, T.; Ayabe, S.: Cloning and functional expression of a cytochrome P450 cDNA encoding 2-hydroxyisoflavanone synthase involved in biosynthesis of the isoflavonoid skeleton in licorice. Plant Physiol., **121**, 821-828 (1999)

[6] Sawada, Y.; Ayabe, S.: Multiple mutagenesis of P_{450} isoflavonoid synthase reveals a key active-site residue. Biochem. Biophys. Res. Commun., **330**, 907-913 (2005)
[7] Kim, D.H.; Kim, B.G.; Lee, H.J.; Lim, Y.; Hur, H.G.; Ahn, J.H.: Enhancement of isoflavone synthase activity by co-expression of P_{450} reductase from rice. Biotechnol. Lett., **27**, 1291-1294 (2005)
[8] Sreevidya, V.S.; Srinivasa Rao, C.; Sullia, S.B.; Ladha, J.K.; Reddy, P.M.: Metabolic engineering of rice with soybean isoflavone synthase for promoting nodulation gene expression in rhizobia. J. Exp. Bot., **57**, 1957-1969 (2006)
[9] Subramanian, S.; Graham, M.Y.; Yu, O.; Graham, T.L.: RNA interference of soybean isoflavone synthase genes leads to silencing in tissues distal to the transformation site and to enhanced susceptibility to Phytophthora sojae. Plant Physiol., **137**, 1345-1353 (2005)
[10] Lozovaya, V.V.; Lygin, A.V.; Zernova, O.V.; Ulanov, A.V.; Li, S.; Hartman, G.L.; Widholm, J.M.: Modification of phenolic metabolism in soybean hairy roots through down regulation of chalcone synthase or isoflavone synthase. Planta, **225**, 665-679 (2007)

Licodione synthase 1.14.13.87

1 Nomenclature

EC number
1.14.13.87

Systematic name
liquiritigenin,NADPH:oxygen oxidoreductase (licodione-forming)

Recommended name
licodione synthase

Synonyms
(2S)-flavanone 2-hydroxylase <2> [1]
F2H <2> [1]
LS <1, 2> [1, 2]
Additional information <1, 2> (<2> CYP93B1 encodes (2S)-flavanone 2-hydroxylase, which has previously been designated to licodione synthase and flavone synthase II [1]; <1> LS is likely to be identical with flavone synthase II from soybean [2]) [1, 2]

CAS registry number
157972-05-9

2 Source Organism

<1> *Glycyrrhiza echinata* (no sequence specified) [2]
<2> *Glycyrrhiza echinata* (UNIPROT accession number: P93149) [1]

3 Reaction and Specificity

Catalyzed reaction
liquiritigenin + NADPH + H^+ + O_2 = licodione + $NADP^+$ + H_2O (<1> a heme-thiolate protein, P_{450}, it probably forms 2-hydroxyliquiritigenin which spontaneously forms licodione, NADH can act instead of NADPH, but more slowly, reaction mechanism [2]; <2> mechanism, stereoselectivity [1])

Reaction type
oxidation
redox reaction
reduction

Natural substrates and products

S Additional information <1, 2> (<2> flavone biosynthesis [1]; <1> involved in flavonoid biosynthesis, may play a role symbiotic plant-microbe interactions [2]) (Reversibility: ?) [1, 2]

P ?

Substrates and products

S (2S)-liquiritigenin + NADH + H^+ + O_2 <1> (<1> requires NAD(P)H and O_2 for activity, the reaction mechanism is likely to be 2-hydroxylation of the flavanone molecule to yield 2-hydroxyliquiritigenin and the subsequent hemiacetal opening to give licodione, introduces one atom of oxygen into the substrate, NADH serves as hydride donor less effectively than NADPH [2]) (Reversibility: ?) [2]

P licodione + NAD^+ + H_2O (<1> i.e. 1-(2,4-dihydroxyphenyl)-3-(4-hydroxyphenyl)-1,3-propanedione [2])

S (2S)-liquiritigenin + NADPH + H^+ + O_2 <1, 2> (<2> hydroxylation of liquiritigenin at C-2 and formation of licodione as a result of non-enzymatic hemiacetal opening, selectivity towards (2S)-flavanone [1]; <1> requires NAD(P)H and O_2 for activity, the reaction mechanism is likely to be 2-hydroxylation of the flavanone molecule to yield 2-hydroxyliquiritigenin and the subsequent hemiacetal opening to give licodione, introduces one atom of oxygen into the substrate, NADPH serves as hydride donor more effectively than NADH [2]) (Reversibility: ?) [1, 2]

P licodione + $NADP^+$ + H_2O (<1> i.e. 1-(2,4-dihydroxyphenyl)-3-(4-hydroxyphenyl)-1,3-propanedione [2])

S (2S)-naringenin + NADPH + H^+ + O_2 <1, 2> (<2> selectivity towards (2S)-flavanone [1]) (Reversibility: ?) [1, 2]

P 2-hydroxynaringenin + $NADP^+$ + H_2O (<1> the product is further converted into apigenin when treated with acid [2]; <2> the product is further converted into apigenin when treated with HCl [1])

S eriodictyol + NADPH + H^+ + O_2 <2> (Reversibility: ?) [1]

P luteolin + $NADP^+$ + H_2O (<2> product is formed after acid treatment [1])

S Additional information <1, 2> (<2> flavone biosynthesis [1]; <1> involved in flavonoid biosynthesis, may play a role symbiotic plant-microbe interactions [2]) (Reversibility: ?) [1, 2]

P ?

Inhibitors

ancymidol <1> (<1> 0.001 mM: 33% inhibition, 0.01 mM: 68% inhibition, 0.1 mM: 81% inhibition [2]) [2]

ketoconazole <1> (<1> 0.001 mM: 44% inhibition, 0.01 mM: 86% inhibition, 0.1 mM: 72% inhibition [2]) [2]

metyrapone <1> (<1> 0.001 mM: 23% inhibition, 0.01 mM: 57% inhibition, 0.1 mM: 74% inhibition [2]) [2]

Cofactors/prosthetic groups

cytochrome P_{450} <1, 2> (<1> cytochrome P_{450}-dependent, cytochrome P_{450} monooxygenase [2]) [1,2]
NAD(P)H <1> (<1> requirement [2]) [2]

Activating compounds

Additional information <1> (<1> yeast extract induces LS activity [2]) [2]

Temperature optimum (°C)

25 <1, 2> (<1,2> assay at [1,2]) [1, 2]

5 Isolation/Preparation/Mutation/Application

Source/tissue

cell suspension culture <1> (<1> yeast extract-induced [2]) [2]

Localization

microsome <1, 2> (<2> recombinant enzyme, expressed in Sf9 insect cells or yeast cells [1]) [1, 2]

Cloning

<1> (CYP Ge-5/CYP93B1, expression in Sf9 insect cells and in Saccharomyces cerevisiae BJ2168) [1]

References

[1] Akashi, T.; Aoki, T.; Ayabe, S.: Identification of a cytochrome P450 cDNA encoding (2S)-flavanone 2-hydroxylase of licorice (Glycyrrhiza echinata L.; Fabaceae) which represents licodione synthase and flavone synthase II. FEBS Lett., **431**, 287-290 (1998)

[2] Otani, K.; Takahashi, T.; Furuya, T.; Ayabe, S.: Licodione synthase, a cytochrome P_{450} monooxygenase catalyzing 2-hydroxylation of 5-deoxyflavanone, in cultured Glycyrrhiza echinata L. cells. Plant Physiol., **105**, 1427-1432 (1994)

Flavonoid 3',5'-hydroxylase 1.14.13.88

1 Nomenclature

EC number
1.14.13.88

Recommended name
flavonoid 3',5'-hydroxylase

Synonyms
F3',5'H <2, 3, 4, 8, 11, 12, 13, 14, 15, 16, 17, 18, 19, 20, 22, 23, 24, 28, 29, 30> [4, 9, 10, 11, 12, 13]
F3'5'H <5, 6, 9, 10, 21, 25, 26, 27> [2, 5, 6, 8]
flavonoid 3',5'-hydroxylase <24> [13]
plant P_{450} flavonoid 3',5'-hydroxylase <4> [9]
Additional information <8, 11, 12, 13, 14, 15, 16, 17, 18, 19, 20, 22, 29, 30> (<8,11,12,13,14,15,16,17,18,19,20,22,29,30> the enzyme belongs to the vast and versatile cytochrome P_{450} protein family of enzymes [11]) [11]

CAS registry number
94047-23-1

2 Source Organism

<1> *Solanum tuberosum* (no sequence specified) [7]
<2> *Petunia hybrida* (no sequence specified) [4]
<3> *Gentiana triflora* (no sequence specified) [10]
<4> *Catharanthus roseus* (no sequence specified) [9]
<5> *Petunia sp.* (UNIPROT accession number: Q9ZSP7) [5]
<6> *Petunia sp.* (UNIPROT accession number: P48418) [1,2]
<7> *Eustoma grandiflorum* (UNIPROT accession number: O04790) [1]
<8> *Vinca major* (UNIPROT accession number: Q76LL4) [3,11]
<9> *Eustoma grandiflorum* (no sequence specified) [6]
<10> *Vitis vinifera* (UNIPROT accession number: Q2PWV0) [8]
<11> *Pericallis cruenta* (UNIPROT accession number: Q304Q5) [11]
<12> *Delphinium grandiflorum* (UNIPROT accession number: Q52YL8) [11]
<13> *Glycine max* (UNIPROT accession number: Q6YLS3) [11]
<14> *Gossypium hirsutum* (UNIPROT accession number: Q84NG3) [11]
<15> *Nierembergia sp.* (UNIPROT accession number: Q8LP20) [11]
<16> *Petunia hybrida* (UNIPROT accession number: P48418) [11]
<17> *Lycianthes rantonnei* (UNIPROT accession number: Q9FPN3) [11]

<18> *Solanum melongena* (UNIPROT accession number: P37120) [11]
<19> *Solanum tuberosum* (UNIPROT accession number: Q5EWY2) [11]
<20> *Torenia hybrida* (UNIPROT accession number: Q9FS35) [11]
<21> *Vitis vinifera* (UNIPROT accession number: Q2PWV1) [8]
<22> *Verbena hybrida* (UNIPROT accession number: Q6J210) [11]
<23> *Vitis vinifera* (UNIPROT accession number: Q2UYU7) [12]
<24> *Vitis vinifera* (UNIPROT accession number: Q3C210) [13]
<25> *Vitis vinifera* (UNIPROT accession number: Q2PWV2) [8]
<26> *Vitis vinifera* (UNIPROT accession number: Q2PWV3) [8]
<27> *Vitis vinifera* (UNIPROT accession number: Q2PWV4) [8]
<28> *Gentiana scabra* (UNIPROT accession number: Q25C80) [10]
<29> *Callistephus chinensis* (UNIPROT accession number: Q9FPN4) [11]
<30> *Osteospermum hybrida* (UNIPROT accession number: Q304Q4) [11]

3 Reaction and Specificity

Catalyzed reaction

a 3'-hydroxyflavanone + NADPH + H^+ + O_2 = a 3',5'-dihydroxyflavanone + $NADP^+$ + H_2O
a flavanone + NADPH + H^+ + O_2 = a 3'-hydroxyflavanone + $NADP^+$ + H_2O

Natural substrates and products

S dihydrokaempferol + NADPH + O_2 <23, 24> (Reversibility: ?) [12, 13]
P dihydromyricetin + $NADP^+$ + H_2O
S dihydrokaempferol + NADPH + O_2 <10, 21, 25, 26, 27> (Reversibility: ?) [8]
P 3,5-dihydroxy-dihydrokaempferol + $NADP^+$ + H_2O
S dihydroquercetin + NADPH + O_2 <10, 21, 25, 26, 27> (Reversibility: ?) [8]
P 3,5-dihydroxy-dihydroquercetin + $NADP^+$ + H_2O
S naringenin + NADPH + O_2 <23, 24> (Reversibility: ?) [12, 13]
P 5,7,3',4',5'-pentahydroxyflavanone + $NADP^+$ + H_2O
S Additional information <1, 3, 5, 6, 7, 8, 10, 11, 12, 13, 14, 15, 16, 17, 18, 19, 20, 21, 22, 23, 24, 25, 26, 27, 28, 29, 30> (<6,7> key enzyme in the synthesis of 3,5-hydroxylated anthocyanins, which are generally required for blue or purple flowers [1]; <6> key enzyme in the synthesis of 3,5-hydroxylated anthocyanins, which are generally required for the expression of blue or purple flower color. Pigment composition analysis of transgenic plants suggests that the F35H transgene not only creates or inhibits the biosynthetic pathway to 3,5-hydroxylated anthocyanins but switches the pathway to 3,5-hydroxylated or 3-hydroxylated anthocyanins [2]; <5> the enzyme catalyzes the 3,5-hydroxylation of dihydroflavonols, the precursors of purple anthocyanins [5]; <1> enzyme is involved in anthocyanin biosynthesis. When expressed as a transgene in the red-skinned cultivar Desiree changes tuber skin color from red to purple [7]; <24> the enzyme is involved in the biosynthetic pathway of delphinidin-based anthocyanins, as well as of the flavonols quercetin and myricetin and procyanidin

and prodelphinidin, correlation of enzyme expression pattern and flavonoid composition, flavonoid composition in organs of Vitis vinifera, flavonoid biosynthetic pathways, overview [13]; <10,21,25,26,27> the enzyme plays a role in the phenyl-propanoid pathway and the biosynthesis of cyanidin- and delphinidin-based anthocyanin pigments in the socalled red cultivars of grapevine, metabolic profiling and colour variations, overview, the enzyme is responsible for flavonoid accumulation in tissues [8]; <3,28> the enzyme the key enzyme for delphinidin biosynthesis in the flavonoid biosynthetic pathway, biosynthesis of colour pigments, overview [10]; <8, 11, 12, 13, 14, 15, 16, 17, 18, 19, 20, 22, 29, 30> the hydroxylation pattern of the B-ring of flavonoids is determined by the activity of the flavonoid 3-hydroxylase and flavonoid 3,5-hydroxylase, phylogenetic analysis of sequences of both enzymes indicate that F3,5H is recruited from F3H before the divergence of angiosperms and gymnosperms, flavonoid biosynthesis pathways, overview [11]; <23> transcriptional regulation of the enzyme in fruits, the enzyme is expressed after flowering, when proanthocyanidins are synthesized, accumulation of hydroxylated anthocyaniins in grape berries, flavonoid biosynthetic pathways, overview [12]) (Reversibility: ?) [1, 2, 5, 7, 8, 10, 11, 12, 13]

P ?

Substrates and products

S (2S)-naringenin + NADPH + O_2 <8, 11, 12, 13, 14, 15, 16, 17, 18, 19, 20, 22, 29, 30> (Reversibility: ?) [11]

P 5,7,3',4',5'-pentahydroxyflavanone + $NADP^+$ + H_2O

S 5,7,3',4'-tetrahydroxyflavanone + NADPH + H^+ + O_2 <2> (Reversibility: ?) [4]

P 5,7,3',4',5'-pentahydroxyflavanone + $NADP^+$ + H_2O

S 5,7,4'-trihydroxyflavanone + NADPH + H^+ + O_2 <2> (Reversibility: ?) [4]

P 5,7,3',4'-tetrahydroxyflavanone + $NADP^+$ + H_2O

S apigenin + NADPH + O_2 <8, 11, 12, 13, 14, 15, 16, 17, 18, 19, 20, 22, 29, 30> (Reversibility: ?) [11]

P 5,7,3',4',5'-pentahydroxyflavone + $NADP^+$ + H_2O

S dihydrokaempferol + NADPH + O_2 <8, 11, 12, 13, 14, 15, 16, 17, 18, 19, 20, 22, 23, 24, 29, 30> (Reversibility: ?) [11, 12, 13]

P dihydromyricetin + $NADP^+$ + H_2O

S dihydrokaempferol + NADPH + O_2 <10, 21, 25, 26, 27> (Reversibility: ?) [8]

P 3,5-dihydroxy-dihydrokaempferol + $NADP^+$ + H_2O

S dihydroquercetin + NADPH + H^+ + O_2 <2> (Reversibility: ?) [4]

P 5'-hydroxyquercetin + $NADP^+$ + H_2O

S dihydroquercetin + NADPH + O_2 <10, 21, 25, 26, 27> (Reversibility: ?) [8]

P 3,5-dihydroxy-dihydroquercetin + $NADP^+$ + H_2O

S kaempferol + NADPH + O_2 <8, 11, 12, 13, 14, 15, 16, 17, 18, 19, 20, 22, 29, 30> (Reversibility: ?) [11]

P 5,7,3',4',5'-pentahydroxyflavone + $NADP^+$ + H_2O

S naringenin + NADPH + H^+ + O_2 <2> (Reversibility: ?) [4]

P 5,7,3',4',5'-pentahydroxyflavanone + $NADP^+$ + H_2O

S naringenin + NADPH + O_2 <23, 24> (Reversibility: ?) [12, 13]

P 5,7,3',4',5'-pentahydroxyflavanone + $NADP^+$ + H_2O

S Additional information <1, 3, 5, 6, 7, 8, 10, 11, 12, 13, 14, 15, 16, 17, 18, 19, 20, 21, 22, 23, 24, 25, 26, 27, 28, 29, 30> (<6,7> key enzyme in the synthesis of 3,5-hydroxylated anthocyanins, which are generally required for blue or purple flowers [1]; <6> key enzyme in the synthesis of 3,5-hydroxylated anthocyanins, which are generally required for the expression of blue or purple flower color. Pigment composition analysis of transgenic plants suggests that the F35H transgene not only creates or inhibits the biosynthetic pathway to 3,5-hydroxylated anthocyanins but switches the pathway to 3,5-hydroxylated or 3-hydroxylated anthocyanins [2]; <5> the enzyme catalyzes the 3,5-hydroxylation of dihydroflavonols, the precursors of purple anthocyanins [5]; <1> enzyme is involved in anthocyanin biosynthesis. When expressed as a transgene in the red-skinned cultivar Desiree changes tuber skin color from red to purple [7]; <24> the enzyme is involved in the biosynthetic pathway of delphinidin-based anthocyanins, as well as of the flavonols quercetin and myricetin and procyanidin and prodelphinidin, correlation of enzyme expression pattern and flavonoid composition, flavonoid composition in organs of Vitis vinifera, flavonoid biosynthetic pathways, overview [13]; <10,21,25,26,27> the enzyme plays a role in the phenyl-propanoid pathway and the biosynthesis of cyanidin- and delphinidin-based anthocyanin pigments in the socalled red cultivars of grapevine, metabolic profiling and colour variations, overview, the enzyme is responsible for flavonoid accumulation in tissues [8]; <3, 28> the enzyme is the key enzyme for delphinidin biosynthesis in the flavonoid biosynthetic pathway, biosynthesis of colour pigments, overview [10]; <8, 11, 12, 13, 14, 15, 16, 17, 18, 19, 20, 22, 29, 30> the hydroxylation pattern of the B-ring of flavonoids is determined by the activity of the flavonoid 3-hydroxylase and flavonoid 3,5-hydroxylase, phylogenetic analysis of sequences of both enzymes indicate that F3,5H is recruited from F3H before the divergence of angiosperms and gymnosperms, flavonoid biosynthesis pathways, overview [11]; <23> transcriptional regulation of the enzyme in fruits, the enzyme is expressed after flowering, when proanthocyanidins are synthesized, accumulation of hydroxylated anthocyaniins in grape berries, flavonoid biosynthetic pathways, overview [12]; <11, 12, 13, 14, 15, 16, 17, 18, 19, 20, 22, 29, 30> the enzyme hydroxylates a broad range of flavonoid substrates in vitro, 3,5-hydroxylation, 3,4-hydroxylation and 3,4,5-hydroxylation occurs, overview [11]; <8,13> the enzyme hydroxylates a broad range of flavonoid substrates in vitro, overview [11]) (Reversibility: ?) [1, 2, 5, 7, 8, 10, 11, 12, 13]

P ?

Inhibitors

1-aminobenzotriazole <2> [4]
CO <2> [4]
cytochrome c <2> (<2> 0.05 mM, complete inhibition [4]) [4]
diethyldithiocarbamate <2> [4]
digitonin <2> (<2> dose-dependent inhibition [4]) [4]
dodecyl maltoside <2> (<2> dose-dependent inhibition [4]) [4]
Emulgen 911 <2> (<2> dose-dependent inhibition [4]) [4]
Emulgen 913 <2> (<2> dose-dependent inhibition [4]) [4]
Lubrol Px <2> (<2> dose-dependent inhibition [4]) [4]
Mega-10 <2> (<2> dose-dependent inhibition [4]) [4]
Mega-8 <2> (<2> dose-dependent inhibition [4]) [4]
Mega-9 <2> (<2> dose-dependent inhibition [4]) [4]
NEM <2> [4]
Nonidet P40 <2> (<2> dose-dependent inhibition [4]) [4]
octyl-β-D-glucopyranoside <2> (<2> dose-dependent inhibition [4]) [4]
sodium cholate <2> (<2> dose-dependent inhibition [4]) [4]
sodium deoxycholate <2> (<2> dose-dependent inhibition [4]) [4]
Tetcyclacis <2> [4]
Triton X-100 <2> (<2> dose-dependent inhibition [4]) [4]
Triton X-114 <2> (<2> dose-dependent inhibition [4]) [4]
Tween 20 <2> (<2> dose-dependent inhibition [4]) [4]
Tween 80 <2> (<2> dose-dependent inhibition [4]) [4]
Zwittergent <2> (<2> dose-dependent inhibition [4]) [4]
Zwittergent 3-08 <2> (<2> dose-dependent inhibition [4]) [4]
Zwittergent 3-10 <2> (<2> dose-dependent inhibition [4]) [4]
Zwittergent 3-12 <2> (<2> dose-dependent inhibition [4]) [4]
Zwittergent 3-14 <2> (<2> dose-dependent inhibition [4]) [4]
Zwittergent 3-16 <2> (<2> dose-dependent inhibition [4]) [4]
decyl-β-D-glucopyranoside <2> (<2> dose-dependent inhibition [4]) [4]
heptyl-β-D-glucopyranoside <2> (<2> dose-dependent inhibition [4]) [4]
hexyl-β-D-glucopyranoside <2> (<2> dose-dependent inhibition [4]) [4]
nonyl-β-D-glucopyranoside <2> (<2> dose-dependent inhibition [4]) [4]
Additional information <2> (<2> a polyclonal antibody that inhibits higher plant NADPH-cytochrome P_{450} reductase inhibits the flavonoid 3,5-hydroxylase. No effect by diethylpyrocarbonate or phenylmethylsulfonyl fluoride [4]) [4]

Cofactors/prosthetic groups

cytochrome P_{450} <2, 5, 6, 7> (<5> enzyme is dependent on [5]; <2> inhibition by a polyclonal antibody that inhibits higher plant NADPH-cytochrome P_{450} reductase is consistent with the suggestion that flavonoid 3,5-hydroxylase is a monooxygenase consisting of cytochrome P_{450} and a NADPH-cytochrome P_{450} reductase [4]; <6,7> the enzyme is a member of the cytochrome P_{450} family [1]) [1,4,5]
NADH <2> (<2> slight activity [4]) [4]

NADPH <2, 8, 10, 11, 12, 13, 14, 15, 16, 17, 18, 19, 20, 21, 22, 23, 24, 25, 26, 27, 29, 30> (<2> dependent on [4]) [4,8,11,12,13]
cytochrome b_5 <5, 23> (<5> required for full activity of flavonoid 3,5-hydroxylase [5]; <23> expression of cytchrome b_5 in grape berries is correlated to enzyme expression during fruit development phases, overview [12]) [5,12]

Activating compounds

2-mercaptoethanol <2> (<2> maximal activation at 20 mM [4]) [4]
Additional information <10, 21, 25, 26, 27> (<10,21,25,26,27> transcriptional induction of the enzyme is temporally coordinated with the beginning of anthocyanin biosynthesis, the expression being 50fold higher in red berries versus green berries [8]) [8]

pH-Optimum

7.5 <8, 11, 12, 13, 14, 15, 16, 17, 18, 19, 20, 22, 29, 30> (<8, 11, 12, 13, 14, 15, 16, 17, 18, 19, 20, 22, 29, 30> assay at [11]) [11]
8 <2> [4]

Temperature optimum (°C)

30 <8, 11, 12, 13, 14, 15, 16, 17, 18, 19, 20, 22, 29, 30> (<8, 11, 12, 13, 14, 15, 16, 17, 18, 19, 20, 22, 29, 30> assay at [11]) [11]

5 Isolation/Preparation/Mutation/Application

Source/tissue

flower <5, 8, 10, 11, 12, 13, 14, 15, 16, 17, 18, 19, 20, 21, 22, 24, 25, 26, 27, 29, 30> (<10, 21, 25, 26, 27> immature [8]; <5> corolla sectors [5]) [5, 8, 11, 13]
flower bud <2, 7> (<2> activity varies with the development of flowers, peaking immediately prior to and during anthesis, but is absent in mature flowers [4]) [1, 4]
fruit <10, 21, 23, 24, 25, 26, 27> (<10, 21, 25, 26, 27> berry skin and flesh [8]; <24> berry skin of young and small, at the pre-veraison stage, and ripe fruits [13]; <23> berry skin, transcriptional regulation of the enzyme, onset of expression after flowering, low expression at the onset of ripening, high expression after veraison concomitant with the accumulation of 3- and 3,5-hydroxylated anthocyanins, overview [12]) [8, 12, 13]
leaf <8, 10, 21, 24, 25, 26, 27> (<10, 21, 25, 26, 27> apex, apical leaflet, reddish pigmented shoot tips [8]; <24> young, small, and middle size leaves [13]) [3, 8, 13]
petal <8, 11, 12, 13, 14, 15, 16, 17, 18, 19, 20, 22, 29, 30> [11]
seed <10, 21, 25, 26, 27> [8]
stem <24> (<24> of young shoots [13]) [13]
tendril <24> (<24> young [13]) [13]
tuber <1> (<1> expressed in tuber skin only in presence of the anthocyanin regulatory locus I [7]) [7]
Additional information <23, 24> (<24> tissue-specific expression analysis [13]; <23> no expression in seeds [12]) [12, 13]

Localization

microsome <2, 5, 8, 11, 12, 13, 14, 15, 16, 17, 18, 19, 20, 22, 23, 29, 30> (<2,5> microsomal membrane [4,5]) [4, 5, 11, 12]

Cloning

<3> (DNA and amino acid sequence determination and analysis, sequence comparison, genetic organization, and phylogenetic analysis) [10]
<4> (functional expression of the enzyme as fusion enzyme with a P_{450} reductase leading to biosynthesis of plant-specific di- and trihydroxylated flavonols in Escherichia coli strain BL21(DE3), feeding experiments and determination of the flavonoid spectra in different recombinant bacterial lines in vivo, overview) [9]
<5> (subcloning of AK14, encoding flavonoid-3',5'-hydroxylase, into a plant expression vector and transforming it to pink tobacco (Nicotiana tabacum cv. Petit Havana SR1) and pink petunia (var. Falcon), both of which originally lack the enzyme) [1]
<5> (when sense constructs are introduced into pink flower varieties that are deficient in the enzyme, transgenic plants show flower color changes from pink to magenta along with changes ihn anthocyanin composition. Some transgenic plants show novel pigmentation patterns, e.g. a star-shaped pattern. When sense constructs are introduced into blue flower petunia varieties, the flower color of the transgenic plants changes from deep blue to pale blue or even pale pink. Pigment composition analysis of transgenic plants suggests that the F3'5'H transgene not only creates or inhibits the biosynthetic pathway to 3',5'-hydroxylated anthocyanins but switches the pathway to 3',5'-hydroxylated or 3'-hydroxylated anthocyanins) [2]
<7> (subcloning of TG1, encoding flavonoid-3',5'-hydroxylase, into a plant expression vector and transforming it to pink tobacco (Nicotiana tabacum cv. Petit Havana SR1) which originally lacks the enzyme) [1]
<8> (DNA and amino acid sequence determination and analysis, phylogenetic analysis of sequences of both enzymes indicate that F3',5'H is recruited from F3'H before the divergence of angiosperms and gymnosperms, overview) [11]
<8> (when VmFH1, encoding the flavonoid 3',5'-hydroxylase is expressed in transgenic petunia hybrida under the control of the cauliflower mosaic virus 35S promoter, some transgenic plants show drastic flower color alteration from red to deep red with deep purple sectors) [3]
<9> (introduction of a flavonoid 3'5'-hydroxylase sequence into Lotus root cultures. Expression of the transgene is associated with increased levels of condensed tannins, no alteration in polymer hydroxylation) [6]
<10> (DNA and amino acid sequence determination and analysis, genetic mapping, phylogenetic analysis, expression analysis and metabolic profiling, phenotypes) [8]
<11> (DNA and amino acid sequence determination and analysis, phylogenetic analysis of sequences of both enzymes indicate that F3',5'H is recruited from F3'H before the divergence of angiosperms and gymnosperms, overview) [11]

<12> (DNA and amino acid sequence determination and analysis, phylogenetic analysis of sequences of both enzymes indicate that F3',5'H is recruited from F3'H before the divergence of angiosperms and gymnosperms, overview) [11]
<13> (DNA and amino acid sequence determination and analysis, phylogenetic analysis of sequences of both enzymes indicate that F3',5'H is recruited from F3'H before the divergence of angiosperms and gymnosperms, overview) [11]
<14> (DNA and amino acid sequence determination and analysis, phylogenetic analysis of sequences of both enzymes indicate that F3',5'H is recruited from F3'H before the divergence of angiosperms and gymnosperms, overview) [11]
<15> (DNA and amino acid sequence determination and analysis, phylogenetic analysis of sequences of both enzymes indicate that F3',5'H is recruited from F3'H before the divergence of angiosperms and gymnosperms, overview) [11]
<16> (DNA and amino acid sequence determination and analysis, phylogenetic analysis of sequences of both enzymes indicate that F3',5'H is recruited from F3'H before the divergence of angiosperms and gymnosperms, overview) [11]
<17> (DNA and amino acid sequence determination and analysis, phylogenetic analysis of sequences of both enzymes indicate that F3',5'H is recruited from F3'H before the divergence of angiosperms and gymnosperms, overview) [11]
<18> (DNA and amino acid sequence determination and analysis, phylogenetic analysis of sequences of both enzymes indicate that F3',5'H is recruited from F3'H before the divergence of angiosperms and gymnosperms, overview) [11]
<19> (DNA and amino acid sequence determination and analysis, phylogenetic analysis of sequences of both enzymes indicate that F3',5'H is recruited from F3'H before the divergence of angiosperms and gymnosperms, overview) [11]
<20> (DNA and amino acid sequence determination and analysis, phylogenetic analysis of sequences of both enzymes indicate that F3',5'H is recruited from F3'H before the divergence of angiosperms and gymnosperms, overview) [11]
<21> (DNA and amino acid sequence determination and analysis, genetic mapping, phylogenetic analysis, expression analysis and metabolic profiling, phenotypes) [8]
<22> (DNA and amino acid sequence determination and analysis, phylogenetic analysis of sequences of both enzymes indicate that F3',5'H is recruited from F3'H before the divergence of angiosperms and gymnosperms, overview) [11]
<23> (gene F3',5'H1, DNA and amino acid sequence determination and analysis, phylogenetic tree, developmental expression analysis, comparison of red and white cultivar enzyme expression levels, functional expression in Petunia hybrida altering the hosts' flower color and flavonoid composition) [12]

<24> (gene F3',5'h, DNA and amino acid sequence determination and analysis, genomic structure, expression pattern analysis) [13]
<25> (DNA and amino acid sequence determination and analysis, genetic mapping, phylogenetic analysis, expression analysis and metabolic profiling, phenotypes) [8]
<26> (DNA and amino acid sequence determination and analysis, genetic mapping, phylogenetic analysis, expression analysis and metabolic profiling, phenotypes) [8]
<27> (DNA and amino acid sequence determination and analysis, genetic mapping, phylogenetic analysis, expression analysis and metabolic profiling, phenotypes) [8]
<28> (DNA and amino acid sequence determination and analysis, sequence comparison, genetic organization, and phylogenetic analysis) [10]
<29> (DNA and amino acid sequence determination and analysis, phylogenetic analysis of sequences of both enzymes indicate that F3',5'H is recruited from F3'H before the divergence of angiosperms and gymnosperms, overview) [11]
<30> (DNA and amino acid sequence determination and analysis, phylogenetic analysis of sequences of both enzymes indicate that F3',5'H is recruited from F3'H before the divergence of angiosperms and gymnosperms, overview) [11]

Engineering

Additional information <3, 23, 28> (<3,28> pink-flowered Gentiana scabra plants are bred from spontaneous mutations of Gentiana triflora blue-flowered gentian plants, two different transposable elements, terminal-repeat retrotransposon in miniature, inserted in flavonoid 3,5-hydroxylase gene contribute to pink flower coloration, overview [10]; <23> transgenic Petunia hybrida lines, overexpressing the enzyme from Vitis vinifera, produce high levels of the 3,4,5-hydroxylated anthocyanin malvidin and very little 3,4-hydroxylated anthocyanins, they also show a shift from kaempferol to quercetin when compared to the nontransgenic control, flower colours and phenotypes, white and red grape varieties exhibit different temporal expression of F3,5H1 and cytochrome b_5, overview [12]) [10, 12]

Application

Additional information <23> (<23> understanding the regulation of flavonoid hydroxylases could be used to modify flavonoid composition of fruits [12]) [12]

6 Stability

General stability information

<2>, activity declines when microsomal suspensions are subjected to freezing and thawing [4]

Storage stability

<2>, 4°C, stable overnight in presence of 20% glycerol or sucrose at pH 7.5 [4]

<2>, 75°C, storage in presence of 20% sucrose for periods of greater than 1 month results in significant loss of enzyme activity [4]

References

[1] Shimada, Y.; Nakano-Shimada, R.; Ohbayashi, M.; Okinaka, Y.; Kiyokawa, S.; Kikuchi, Y.: Expression of chimeric P450 genes encoding flavonoid-3', 5'-hydroxylase in transgenic tobacco and petunia plants(1). FEBS Lett., **461**, 241-245 (1999)

[2] Shimada, Y.; Ohbayashi, M.; Nakano-Shimada, R.; Okinaka, Y.; Kiyokawa, S.; Kikuchi, Y.: Genetic engineering of the anthocyanin biosynthetic pathway with flavonoid-3',5'-hydroxylase: specific switching of the pathway in petunia. Plant Cell Rep., **20**, 456-462 (2001)

[3] Mori, S.; Kobayashi, H.; Hoshi, Y.; Kondo, M.; Nakano, M.: Heterologous expression of the flavonoid 3',5'-hydroxylase gene of Vinca major alters flower color in transgenic Petunia hybrida. Plant Cell Rep., **22**, 415-421 (2004)

[4] Menting, J.G.T.; Scopes, R.K.; Stevenson, T.W.: Characterization of flavonoid 3',5'-hydroxylase in microsomal membrane fraction of Petunia hybrida flowers. Plant Physiol., **106**, 633-642 (1994)

[5] de Vetten, N.; ter Horst, J.; van Schaik, H.P.; de Boer, A.; Mol, J.; Koes, R.: A cytochrome b_5 is required for full activity of flavonoid 3', 5'-hydroxylase, a cytochrome P_{450} involved in the formation of blue flower colors. Proc. Natl. Acad. Sci. USA, **96**, 778-783 (1999)

[6] Robbins, M.P.; Bavage, A.D.; Allison, G.; Davies, T.; Hauck, B.; Morris, P.: A comparison of two strategies to modify the hydroxylation of condensed tannin polymers in Lotus corniculatus L. Phytochemistry, **66**, 991-999 (2005)

[7] Jung, C.S.; Griffiths, H.M.; De Jong, D.M.; Cheng, S.; Bodis, M.; De Jong, W.S.: The potato P locus codes for flavonoid 3',5'-hydroxylase. Theor. Appl. Genet., **110**, 269-275 (2005)

[8] Castellarin, S.D.; Di Gaspero, G.; Marconi, R.; Nonis, A.; Peterlunger, E.; Paillard, S.; Adam-Blondon, A.F.; Testolin, R.: Colour variation in red grapevines (Vitis vinifera L.): genomic organisation, expression of flavonoid 3-hydroxylase, flavonoid 3,5-hydroxylase genes and related metabolite profiling of red cyanidin-/blue delphinidin-based anthocyanins in berry skin. BMC Genomics, **7**, 12 (2006)

[9] Leonard, E.; Yan, Y.; Koffas, M.A.: Functional expression of a P_{450} flavonoid hydroxylase for the biosynthesis of plant-specific hydroxylated flavonols in Escherichia coli. Metab. Eng., **8**, 172-181 (2006)

[10] Nakatsuka, T.; Nishihara, M.; Mishiba, K.; Hirano, H.; Yamamura, S.: Two different transposable elements inserted in flavonoid 3,5-hydroxylase gene

contribute to pink flower coloration in Gentiana scabra. Mol. Genet. Genomics, **275**, 231-241 (2006)

[11] Seitz, C.; Eder, C.; Deiml, B.; Kellner, S.; Martens, S.; Forkmann, G.: Cloning, functional identification and sequence analysis of flavonoid 3'-hydroxylase and flavonoid 3',5'-hydroxylase cDNAs reveals independent evolution of flavonoid 3,5-hydroxylase in the Asteraceae family. Plant Mol. Biol., **61**, 365-381 (2006)

[12] Bogs, J.; Ebadi, A.; McDavid, D.; Robinson, S.P.: Identification of the flavonoid hydroxylases from grapevine and their regulation during fruit development. Plant Physiol., **140**, 279-291 (2006)

[13] Jeong, S.T.; Goto-Yamamoto, N.; Hashizume, K.; Esaka, M.: Expression of the flavonoid 3'-hydroxylase and flavonoid 3',5'-hydroxylase genes and flavonoid composition in grape (Vitis vinifera). Plant Sci., **170**, 61-69 (2006)

Isoflavone 2'-hydroxylase 1.14.13.89

1 Nomenclature

EC number
1.14.13.89

Systematic name
isoflavone,NADPH:oxygen oxidoreductase (2'-hydroxylating)

Recommended name
isoflavone 2'-hydroxylase

Synonyms
CYP Ge-3 <2> [1]
CYP81E1 <2> [1]
I2'H <1, 3> [2, 4]
LjCY-2 protein <1> [2]
MtCYP81E7 <3> [4]
isoflavone 2'-monooxygenase <2> [1]

CAS registry number
110183-49-8

2 Source Organism

<1> *Lotus japonicus* (no sequence specified) [2]
<2> *Glycyrrhiza echinata* (no sequence specified) [1, 3]
<3> *Medicago truncatula* (UNIPROT accession number: Q6WNR0) [4]

3 Reaction and Specificity

Catalyzed reaction
an isoflavone + NADPH + H^+ + O_2 = a 2'-hydroxyisoflavone + $NADP^+$ + H_2O

Natural substrates and products
S Additional information <1, 2> (<1> the enzyme is essential for isoflavane pterocarpan phytoalexin biosynthesis [2]; <2> the enzyme is involved in biosynthesis of flavonoids [3]; <2> the enzyme is involved in the biosynthesis of isoflavonoid-derived antimicrobial compounds of legumes [1]) (Reversibility: ?) [1, 2, 3]
P ?

Substrates and products

S 2'-hydroxyformononetin + NADPH + O_2 <3> (<3> 9.1% of the activity with biochanin A [4]) (Reversibility: ?) [4]

P ?

S 3'-hydroxyformononetin + NADPH + O_2 <3> (<3> 4.7% of the activity with biochanin A [4]) (Reversibility: ?) [4]

P ?

S biochanin A + NADPH + O_2 <3> (Reversibility: ?) [4]

P 2'-hydroxybichanin A + $NADP^+$ + H_2O

S daidzein + NADPH + O_2 <2, 3> (<3> 18.6% of the activity with biochanin A [4]) (Reversibility: ?) [1, 4]

P 2'-hydroxydaidzein + $NADP^+$ + H_2O (<3> + low concentrations of 3-hydroxydaidzein [4])

S formononetin + NADPH + O_2 <2, 3> (<3> 37.2% of the activity with biochanin A [4]) (Reversibility: ?) [1, 4]

P 2'-hydroxyformononetin + $NADP^+$ + H_2O

S genistein + NADPH + O_2 <2, 3> (<3> 10.2% of the activity with biochanin A [4]) (Reversibility: ?) [1, 4]

P 2'-hydroxygenistein + $NADP^+$ + H_2O (<3> + low concentrations of 3-hydroxygenistaien [4])

S pseudobaptigenin + NADPH + O_2 <3> (<3> 35% of the activity with biochanin A [4]) (Reversibility: ?) [4]

P ?

S Additional information <1, 2, 3> (<1> the enzyme is essential for isoflavan pterocarpan phytoalexin biosynthesis [2]; <2> the enzyme is involved in biosynthesis of flavonoids [3]; <2> the enzyme is involved in the biosynthesis of isoflavonoid-derived antimicrobial compounds of legumes [1]; <3> no activity with isoformononetin, prunetin, 6,7,4-trihydroxyisoflavone, 3,4,7-trihydroxyisoflavone [4]; <2> no activity with naringenin (5,7,4-trihydroxyflavanone), liquiritigenin (7,4-dihydroxyflavanone), trans-cinnamic acid and 4-coumaric acid [1]) (Reversibility: ?) [1, 2, 3, 4]

P ?

Cofactors/prosthetic groups

cytochrome P_{450} <2, 3> [1,4]
NADPH <2, 3> [1,4]

Turnover number (min^{-1})

0.015 <3> (formononetin, <3> pH 8.0 [4]) [4]
0.033 <3> (biochanin A, <3> pH 8.0 [4]) [4]

K_m-Value (mM)

0.00017 <3> (NADPH, <3> pH 8.0 [4]) [4]
0.051 <3> (biochanin A, <3> pH 8.0 [4]) [4]
0.067 <3> (formononetin, <3> pH 8.0 [4]) [4]

pH-Optimum

8 <3> [4]

5 Isolation/Preparation/Mutation/Application

Source/tissue

cell suspension culture <2> (<2> suspension, induction with yeast extract [3]) [3]
root <3> [4]
seedling <1> (<1> activity increases 10 h after elicitation with reduced glutathione. Only a low level of transcription is observed in untreated seedlings [2]) [2]

Localization

endoplasmic reticulum <3> [4]

Cloning

<1> (expression in yeast) [2]
<2> [3]
<2> (expression in Saccharomyces cerevisiae) [1]
<3> (expression in yeast strain WAT11) [4]

References

[1] Akashi, T.; Aoki, T.; Ayabe, S.I.: CYP81E1, a cytochrome P_{450} cDNA of licorice (Glycyrrhiza echinata L.), encodes isoflavone 2'-hydroxylase. Biochem. Biophys. Res. Commun., **251**, 67-70 (1998)
[2] Shimada, N.; Akashi, T.; Aoki, T.; Ayabe, S.i.: Induction of isoflavonoid pathway in the model legume Lotus japonicus: molecular characterization of enzymes involved in phytoalexin biosynthesis. Plant Sci., **160**, 37-47 (2000)
[3] Nakamura, K.; Akashi, T.; Aoki, T.; Kawaguchi, K.; Ayabe, S.: Induction of isoflavonoid and retrochalcone branches of the flavonoid pathway in cultured Glycyrrhiza echinata cells treated with yeast extract. Biosci. Biotechnol. Biochem., **63**, 1618-1620 (1999)
[4] Liu, C.J.; Huhman, D.; Sumner, L.W.; Dixon, R.A.: Regiospecific hydroxylation of isoflavones by cytochrome P_{450} 81E enzymes from Medicago truncatula. Plant J., **36**, 471-484 (2003)

Zeaxanthin epoxidase 1.14.13.90

1 Nomenclature

EC number
1.14.13.90

Systematic name
zeaxanthin,NAD(P)H:oxygen oxidoreductase

Recommended name
zeaxanthin epoxidase

Synonyms
ABA2 protein <42> [3]
CHL <44> [14]
LHCII <6> [1]
TIL <45> [14]
ZE <12, 15, 16, 17, 18, 19, 20, 21, 22, 23, 24, 25, 26, 27, 28, 29, 30, 31, 32, 33, 34, 35, 36, 37, 38, 39, 40, 41> [9]
ZEP <4, 10, 44, 45, 48> [7, 13, 14, 15]
Zea-epoxidase <14> [6]
chloroplastic lipocalin <44> [14]
lipocalin-like protein <8, 11, 44> [14]
temperature stress-induced lipocalin <46> [14]
temperature-induced lipocalin <45> [14]
violaxanthin cycle enzyme <9> [12]
zeaxanthin-epoxidase <1> [4]
Additional information <8, 11, 44, 45, 46> (<8,11,44,45,46> the enzyme belongs to the lipocalin family [14]) [14]

CAS registry number
149718-34-3

2 Source Organism

<1> *Spinacia oleracea* (no sequence specified) [4]
<2> *Arabidopsis thaliana* (no sequence specified) [5]
<3> *Arabidopsis sp.* (no sequence specified) [2]
<4> *Lycopersicon esculentum* (no sequence specified) [2, 7, 11]
<5> *Capsicum annuum* (no sequence specified) [2]
<6> *Secale cereale* (no sequence specified) [1]
<7> *Prunus armeniaca* (no sequence specified) [11]

<8> *Debaryomyces hansenii* (no sequence specified) [14]
<9> *Lemna trisulca* (no sequence specified) [11,12]
<10> *Nicotiana plumbaginifolia* (no sequence specified) [15]
<11> *Porphyra yezoensis* (no sequence specified) [14]
<12> *Viscum album* (no sequence specified) [9]
<13> *Mantoniella squamata* (no sequence specified) [8]
<14> *Lycopersicon esculentum* (UNIPROT accession number: Z83835) [6]
<15> *Amyema miquelii* (no sequence specified) [9]
<16> *Amyema cambagei* (no sequence specified) [9]
<17> *Amyema pendulum* (no sequence specified) [9]
<18> *Muellerina eucalyptoides* (no sequence specified) [9]
<19> *Nuytsia floribunda* (no sequence specified) [9]
<20> *Phthirusa ovata* (no sequence specified) [9]
<21> *Psittachanthus dichrous* (no sequence specified) [9]
<22> *Psittachanthus robustus* (no sequence specified) [9]
<23> *Struthanthus flexicaulis* (no sequence specified) [9]
<24> *Struthanthus marginatus* (no sequence specified) [9]
<25> *Viscum laxum* (no sequence specified) [9]
<26> *Arceuthobium oxycedri* (no sequence specified) [9]
<27> *Arceuthobium divaricatum* (no sequence specified) [9]
<28> *Phoradendron californicum* (no sequence specified) [9]
<29> *Phoradendron juniperinum* (no sequence specified) [9]
<30> *Phoradendron dipterum* (no sequence specified) [9]
<31> *Phoradendron emarginatum* (no sequence specified) [9]
<32> *Phoradendron piperoides* (no sequence specified) [9]
<33> *Phoradendron tunaeforme* (no sequence specified) [9]
<34> *Phoradendron hexastichon* (no sequence specified) [9]
<35> *Phoradendron perrottetii* (no sequence specified) [9]
<36> *Phoradendron piauhyanum* (no sequence specified) [9]
<37> *Phoradendron semivenosum* (no sequence specified) [9]
<38> *Phoradendron undulatum* (no sequence specified) [9]
<39> *Cassytha filiformis* (no sequence specified) [9]
<40> *Exocarpos cupressiformis* (no sequence specified) [9]
<41> *Melampyrum pratense* (no sequence specified) [9]
<42> *Nicotiana plumbaginifolia* (UNIPROT accession number: Q40412) [3]
<43> *Prunus americana* (no sequence specified) [2]
<44> *Triticum aestivum* (UNIPROT accession number: Q38JB3) [14]
<45> *Triticum aestivum* (UNIPROT accession number: Q38JE5) [14]
<46> *Triticum aestivum* (UNIPROT accession number: Q8S9H0) [14]
<47> *Arabidopsis thaliana* (UNIPROT accession number: Q9FGC7) [10]
<48> *Solanum tuberosum* (UNIPROT accession number: Q^{3}HNF5) [13]

3 Reaction and Specificity

Catalyzed reaction

antheraxanthin + NAD(P)H + H^+ + O_2 = violaxanthin + $NAD(P)^+$ + H_2O (<9> creation of an unstable carotenoid carbocation in the molecular mechanism of epoxidation [12])
zeaxanthin + NAD(P)H + H^+ + O_2 = antheraxanthin + $NAD(P)^+$ + H_2O (<9> creation of an unstable carotenoid carbocation in the molecular mechanism of epoxidation [12])

Reaction type

oxidation
redox reaction
reduction

Natural substrates and products

S antheraxanthin + NAD(P)H + O_2 <2, 4, 7, 9, 10, 12, 13, 14, 15, 16, 17, 18, 19, 20, 21, 22, 23, 24, 25, 26, 27, 28, 29, 30, 31, 32, 33, 34, 35, 36, 37, 38, 39, 40, 41, 42, 47, 48> (<14, 42> early step in abscisic acid biosynthetic pathway [3, 6]; <12,15,16,17, 18, 19, 20, 21, 22, 23, 24, 25, 26, 27, 28, 29, 30, 31, 32, 33, 34, 35, 36, 37, 38, 39, 40, 41> reaction in violaxanthin cycle and in pigment/abscisic acid synthesis [9]; <2> the enzyme catalyzes the removal of epoxide groups in carotenoids of xanthophyll cycle in plants [5]; <13> the violaxanthin/antheraxanthin cycle in Mantionella is caused by the interaction of the slow second de-epoxidation step and the relatively fast epoxidation of antheraxanthin to violaxanthin [8]) (Reversibility: ?) [3, 5, 6, 8, 9, 10, 11, 13, 15]
P violaxanthin + $NAD(P)^+$ + H_2O
S zeaxanthin + NAD(P)H + H^+ + O_2 <2, 12, 15, 16, 17, 18, 19, 20, 21, 22, 23, 24, 25, 26, 27, 28, 29, 30, 31, 32, 33, 34, 35, 36, 37, 38, 39, 40, 41> (<12,15,16, 17, 18, 19, 20, 21, 22, 23, 24, 25, 26, 27, 28, 29, 30, 31, 32, 33, 34, 35, 36, 37, 38, 39, 40, 41> reaction in violaxanthin cycle and in pigment/abscisic acid synthesis [9]; <2> the enzyme catalyzes the removal of epoxide groups in caroteids of xanthophyll cycle in plants [5]) (Reversibility: ?) [5, 9]
P antheraxanthin + $NAD(P)^+$ + H_2O
S zeaxanthin + NAD(P)H + O_2 <4, 7, 9, 10, 13, 14, 42, 47, 48> (<14,42> early step in abscisic acid biosynthetic pathway [3,6]) (Reversibility: ?) [3, 6, 8, 10, 11, 13, 15]
P antheraxanthin + $NAD(P)^+$ + H_2O
S Additional information <4, 9, 10, 44, 45, 46, 47, 48> (<4> drought stress causes an increase of zeaxanthin epoxidase mRNA in root but not in leafs. Strong diurnal expression pattern for zeaxanthin epoxidase in leafs, oscillation with a phase very similar to light-harvesting complex II mRNA, oscillation continues in a 48 h dark period [7]; <47> the enzyme catalyses the epoxidation of zeaxanthin to antheraxanthin and violaxanthin, generating the epoxycarotenoid precursor of the ABA biosynthetic pathway,

overview [10]; <10> the enzyme catalyses the epoxidation of zeaxanthin to antheraxanthin and violaxanthin, generating the epoxycarotenoid precursor of the abscisic acid biosynthetic pathway, but has no regulating function on abscisic acid synthesis, overview [15]; <45,46> the enzyme has a protective function of the photosynthetic system against temperature stress, the expression of lipocalins and lipocalin-like proteins is associated with abiotic stress response and is correlated with the plant's capacity to develop freezing tolerance, overview [14]; <44> the enzyme has a protective function of the photosynthetic system against temperature stress, the expression of lipocalins and lipocalin-like proteins is associated with abiotic stress response and is correlated with the plant's capacity to develop freezing tolerance, overview, the enzyme is under regulation during the diurnal cycle [14]; <9> the enzyme is important in photoprotective mechanisms in higher plants [12]; <48> the enzyme is involved in the biosynthesis of the plant hormone ascisic acid, overview [13]) (Reversibility: ?) [7, 10, 12, 13, 14, 15]

P ?

Substrates and products

S antheraxanthin + NAD(P)H + O_2 <1, 2, 3, 4, 5, 7, 9, 10, 12, 13, 14, 15, 16, 17, 18, 19, 20, 21, 22, 23, 24, 25, 26, 27, 28, 29, 30, 31, 32, 33, 34, 35, 36, 37, 38, 39, 40, 41, 42, 43, 47, 48> (<14, 42> early step in abscisic acid biosynthetic pathway [3, 6]; <12,15,16,17, 18, 19, 20, 21, 22, 23, 24, 25, 26, 27, 28, 29, 30, 31, 32, 33, 34, 35, 36, 37, 38, 39, 40, 41> reaction in violaxanthin cycle and in pigment/abscisic acid synthesis [9]; <2> the enzyme catalyzes the removal of epoxide groups in caroteids of xanthophyll cycle in plants [5]; <13> the violaxanthin/antheraxanthin cycle in Mantionella is caused by the interaction of the slow second de-epoxidation step and the relatively fast epoxidation of antheraxanthin to violaxanthin [8]) (Reversibility: ?) [2, 3, 4, 5, 6, 8, 9, 10, 11, 13, 15]

P violaxanthin + $NAD(P)^+$ + H_2O

S zeaxanthin + NAD(P)H + H^+ + O_2 <2, 12, 15, 16, 17, 18, 19, 20, 21, 22, 23, 24, 25, 26, 27, 28, 29, 30, 31, 32, 33, 34, 35, 36, 37, 38, 39, 40, 41> (<12, 15, 16, 17, 18, 19, 20, 21, 22, 23, 24, 25, 26, 27, 28, 29, 30, 31, 32, 33, 34, 35, 36, 37, 38, 39, 40, 41> reaction in violaxanthin cycle and in pigment/abscisic acid synthesis [9]; <2> the enzyme catalyzes the removal of epoxide groups in caroteids of xanthophyll cycle in plants [5]) (Reversibility: ?) [5, 9]

P antheraxanthin + $NAD(P)^+$ + H_2O

S zeaxanthin + NAD(P)H + O_2 <1, 3, 4, 5, 7, 9, 10, 13, 14, 42, 43, 47, 48> (<14, 42> early step in abscisic acid biosynthetic pathway [3,6]) (Reversibility: ?) [2, 3, 4, 6, 8, 10, 11, 13, 15]

P antheraxanthin + $NAD(P)^+$ + H_2O

S Additional information <4, 9, 10, 44, 45, 46, 47, 48> (<4> drought stress causes an increase of zeaxanthin epoxidase mRNA in root but not in leafs. Strong diurnal expression pattern for zeaxanthin epoxidase in leafs, oscillation with a phase very similar to light-harvesting complex II mRNA, oscillation continues in a 48 h dark period [7]; <47> the enzyme catalyses the epoxidation of zeaxanthin to antheraxanthin and violaxanthin, generating the epoxycarotenoid precursor of the ABA biosynthetic pathway, overview [10]; <10> the enzyme catalyses the epoxidation of zeaxanthin to antheraxanthin and violaxanthin, generating the epoxycarotenoid precursor of the abscisic acid biosynthetic pathway, but has no regulating function on abscisic acid synthesis, overview [15]; <45,46> the enzyme has a protective function of the photosynthetic system against temperature stress, the expression of lipocalins and lipocalin-like proteins is associated with abiotic stress response and is correlated with the plant's capacity to develop freezing tolerance, overview [14]; <44> the enzyme has a protective function of the photosynthetic system against temperature stress, the expression of lipocalins and lipocalin-like proteins is associated with abiotic stress response and is correlated with the plant's capacity to develop freezing tolerance, overview, the enzyme is under regulation during the diurnal cycle [14]; <9> the enzyme is important in photoprotective mechanisms in higher plants [12]; <48> the enzyme is involved in the biosynthesis of the plant hormone ascisic acid, overview [13]; <4,9> a conserved cysteine residue is involved in the catalytic reaction [11]) (Reversibility: ?) [7, 10, 11, 12, 13, 14, 15]

P ?

Inhibitors

Cd^{2+} <4, 9> (<9> inhibition by interaction with a cysteine residue of the protein that is important for the enzyme activity [11]; <4> inhibition by interaction with a cysteine residue of the protein that is important for the enzyme activity, inhibition by cadmium ions is reversible by Zn^{2+} ions [11]) [11]

diphenyleneiodoniumchloride <1> (<1> IC50: 0.0023 mM [4]) [4]

Additional information <7, 9> (<9> no effect of sugars and amino sugars on violaxanthin de-epoxidase is observed, but epoxidation of zeaxanthin to violaxanthin is inhibited, incubation with amino sugars under a 6 d photoperiod enhances the inhibitory effect, zeaxanthin epoxidation is completely inhibited under such conditions, overview [12]; <7> no inhibition by Cd^{2+} [11]) [11, 12]

Cofactors/prosthetic groups

FAD <1, 42> (<1> essential cofactor [4]; <42> enzyme has a FAD-binding domain [3]) [3,4]

NAD(P)H <1, 2, 3, 4, 5, 7, 9, 10, 12, 14, 15, 16, 17, 18, 19, 20, 21, 22, 23, 24, 25, 26, 27, 28, 29, 30, 31, 32, 33, 34, 35, 36, 37, 38, 39, 40, 41, 42, 43, 47, 48> [2, 3, 4, 5, 6, 9, 10, 11, 13, 15]

Additional information <44> (<44> induction by abiotic stress [14]) [14]

Activating compounds

Additional information <45, 46> (<45,46> induction by abiotic stress [14]) [14]

Metals, ions

Additional information <4, 7, 9> (<9> no effect on the enzyme by Zn^{2+} [11]; <4> no effect on the enzyme by Zn^{2+} alone [11]; <7> no effect on the enzyme by Zn^{2+} and Cd^{2+} [11]) [11]

Specific activity (U/mg)

Additional information <9> [12]

pH-Optimum

5.1 <4, 7, 9> (<4,7,9> assay at [11]) [11]

Temperature optimum (°C)

25 <4, 7, 9> (<4,7,9> assay at [11]) [11]

4 Enzyme Structure

Subunits

? <48> (<48> x * 73100, about, amino acid sequence calculation [13]) [13]
Additional information <2> (<2> sequence analysis establishes the enzyme as a member of the lipocalin family of proteins, a diverse group of proteins that bind small hydrophobic molecules and share a conserved tertiary structure of eight β-strands forming a barrel configuration [5]) [5]

5 Isolation/Preparation/Mutation/Application

Source/tissue

leaf <1, 4, 7, 9, 44, 45, 46, 47> (<44,45,46> green [14]; <4> drought stress does not cause an increase of zeaxanthin epoxidase mRNA. Strong diurnal expression pattern for zeaxanthin epoxidase, oscillation with a phase very similar to light-harvesting complex II mRNA, oscillation continues in a 48 h dark period [7]) [4, 7, 10, 11, 12, 14]
root <4> (<4> drought stress causes an increase of zeaxanthin epoxidase mRNA [7]) [7]
seed <10> (<10> maximal accumulation of xanthophylls occurrs at mid-development [15]) [15]
tuber <48> (<48> meristem, periderm, and cortex, completely dormant, partly dormant, and completely non-dormant [13]) [13]
Additional information <48> (<48> quantitative expression analysis in tuber tissues during dormancy, overview [13]) [13]

Localization

chloroplast <44> (<44> the enzyme contains a transit peptide that targets it to the chloroplast [14]) [14]
plasma membrane <45, 46> (<45, 46> temperature-induced enzyme [14]) [14]
thylakoid <1> (<1> stromal side [4]) [4]
thylakoid membrane <6> (<6> model of a spatial organization of the xanthophyll cycle in the thylakoid membrane [1]) [1]

Cloning

<14> (structure and expression of cDNA) [6]
<42> (expression in Escherichia coli) [3]
<44> (DNA and amino acid sequence comparison and analysis, expression analysis, phylogenetic analysis and evolution of lipocalins) [14]
<45> (DNA and amino acid sequence comparison and analysis, expression analysis, phylogenetic analysis and evolution of lipocalins) [14]
<46> (DNA and amino acid sequence comparison and analysis, expression analysis, phylogenetic analysis and evolution of lipocalins) [14]
<47> (genetic mapping of aba1 mutants, complementation of a mutant strain) [10]
<48> (DNA and amino acid sequence determination and analysis, expression analysis in tuber tissues during dormancy) [13]

Engineering

Additional information <10, 47> (<47> screening and analysis of gene ABA1 mutations, phenotypes, overview [10]; <10> seed dormancy is impaired in mutants of conversion of zeaxanthin into violaxanthin by zeaxanthin epoxidase, seed dormancy is restored by the introduction of the Arabidopsis thaliana ZEP gene under the control of promoters inducing expression during later stages of seed development compared to wild type Nicotiana plumbaginifolia ZEP, and in dry and imbibed seeds, alterations in the timing and level of ZEP expression do not highly affect the temporal regulation of ABA accumulation in transgenic seeds, despite notable perturbations in xanthophyll accumulation, overview [15]) [10, 15]

References

[1] Gruszecki, W.I.; Krupa, Z.: LHCII, the major light-harvesting pigment-protein complex is a zeaxanthin epoxidase. Biochim. Biophys. Acta, **1144**, 97-101 (1993)

[2] Hieber, A.D.; Bugos, R.C.; Yamamoto, H.Y.: Plant lipocalins: violaxanthin de-epoxidase and zeaxanthin epoxidase. Biochim. Biophys. Acta, **1482**, 84-91 (2000)

[3] Marin, E.; Nussaume, L.; Quesada, A.; Gonneau, M.; Sotta, B.; Hugueney, P.; Frey, A.; Marion-Poll, A.: Molecular identification of zeaxanthin epoxidase of Nicotiana plumbaginifolia, a gene involved in abscisic acid biosynthesis and corresponding to the ABA locus of Arabidopsis thaliana. EMBO J., **15**, 2331-2342 (1996)

[4] Büch, K.; Stransky, H.; Hager, A.: FAD is a further essential cofactor of the NAD(P)H and O_2-dependent zeaxanthin-epoxidase. FEBS Lett., **376**, 45-48 (1995)

[5] Bugos, R.C.; Hieber, A.D.; Yamamoto, H.Y.: Xanthophyll cycle enzymes are members of the lipocalin family, the first identified from plants. J. Biol. Chem., **273**, 15321-15324 (1998)

[6] Burbidge, A.; Greieve, T.; Terry, C.; Corlett, J.; Thompson, A.; Taylor, I.: Structure and expression of a cDNA encoding zeaxanthin epoxidase, isolated from a wilt-related tomato (Lycopersicon esculentum Mill.) library. J. Exp. Bot., **48**, 1749-1750 (1997)

[7] Thompson, A.J.; Jackson, A.C.; Parker, R.A.; Morpeth, D.R.; Burbidge, A.; Taylor, I.B.: Abscisic acid biosynthesis in tomato: regulation of zeaxanthin epoxidase and 9-cis-epoxycarotenoid dioxygenase mRNAs by light/dark cycles, water stress and abscisic acid. Plant Mol. Biol., **42**, 833-845 (2000)

[8] Frommolt, R.; Goss, R.; Wilhelm, C.: The de-epoxidase and epoxidase reactions of Mantoniella squamata (Prasinophyceae) exhibit different substrate-specific reaction kinetics compared to spinach. Planta, **213**, 446-456 (2001)

[9] Matsubara, S.; Morosinotto, T.; Bassi, R.; Christian, A.L.; Fischer-Schliebs, E.; Luttge, U.; Orthen, B.; Franco, A.C.; Scarano, F.R.; Forster, B.; Pogson, B.J.; Osmond, C.B.: Occurrence of the lutein-epoxide cycle in mistletoes of the Loranthaceae and Viscaceae. Planta, **217**, 868-879 (2003)

[10] Barrero, J.M.; Piqueras, P.; Gonzalez-Guzman, M.; Serrano, R.; Rodriguez, P.L.; Ponce, M.R.; Micol, J.L.: A mutational analysis of the ABA1 gene of Arabidopsis thaliana highlights the involvement of ABA in vegetative development. J. Exp. Bot., **56**, 2071-2083 (2005)

[11] Latowski, D.; Kruk, J.; Strzaska, K.: Inhibition of zeaxanthin epoxidase activity by cadmium ions in higher plants. J. Inorg. Biochem., **99**, 2081-2087 (2005)

[12] Latowski, D.; Banas, A.K.; Strzaska, K.; Gabrys, H.: Amino sugars - new inhibitors of zeaxanthin epoxidase, a violaxanthin cycle enzyme. J. Plant Physiol., **164**, 231-237 (2007)

[13] Destefano-Beltran, L.; Knauber, D.; Huckle, L.; Suttle, J.C.: Effects of post-harvest storage and dormancy status on ABA content, metabolism, and expression of genes involved in ABA biosynthesis and metabolism in potato tuber tissues. Plant Mol. Biol., **61**, 687-697 (2006)

[14] Charron, J.F.; Ouellet, F.; Pelletier, M.; Danyluk, J.; Chauve, C.; Sarhan, F.: Identification, expression, and evolutionary analyses of plant lipocalins. Plant Physiol., **139**, 2017-2028 (2005)

[15] Frey, A.; Boutin, J.P.; Sotta, B.; Mercier, R.; Marion-Poll, A.: Regulation of carotenoid and ABA accumulation during the development and germination of Nicotiana plumbaginifolia seeds. Planta, **224**, 622-632 (2006)

Deoxysarpagine hydroxylase 1.14.13.91

1 Nomenclature

EC number
1.14.13.91

Systematic name
10-deoxysarpagine,NADPH:oxygen oxidoreductase (10-hydroxylating)

Recommended name
deoxysarpagine hydroxylase

2 Source Organism

<1> *Rauvolfia serpentina* (no sequence specified) [1]

3 Reaction and Specificity

Catalyzed reaction
10-deoxysarpagine + NADPH + H^+ + O_2 = sarpagine + $NADP^+$ + H_2O

Natural substrates and products
S 10-deoxysarpagine + NADPH + O_2 <1> (<1> most probably the natural substrate [1]) (Reversibility: ?) [1]
P sarpagine + $NADP^+$ + H_2O

Substrates and products
S 10-deoxysarpagine + NADPH + O_2 <1> (<1> most probably the natural substrate [1]) (Reversibility: ?) [1]
P sarpagine + $NADP^+$ + H_2O

Inhibitors
ancymidole <1> (<1> 1.0 mM, 49% inhibition [1]) [1]
BAS 110 W <1> (<1> 1 mM, complete inhibition [1]) [1]
BAS 111 W <1> (<1> 1 mM, 84% inhibition [1]) [1]
CO <1> (<1> ratio CO:O_2 of 9:1, 70% inhibition in the light, 80% inhibition under red light, 77% inhibition under blue light [1]) [1]
cytochrome c <1> (<1> 0.01 mM, 46% inhibition [1]) [1]
ketoconazole <1> (<1> 1 mM, 69% inhibition [1]) [1]
LAB 1500978 <1> (<1> 1 mM, 74% inhibition [1]) [1]
metyrapone <1> (<1> 1 mM, 83% inhibition [1]) [1]
$NADP^+$ <1> (<1> competitive inhibition [1]) [1]
tetcyclacis <1> (<1> 1 mM, 45% inhibition [1]) [1]

Cofactors/prosthetic groups
NADPH <1> (<1> can not be replaced by NADH [1]) [1]

K_m-Value (mM)
0.0074 <1> (10-deoxysarpagine, <1> 35°C, PH 8.0 [1]) [1]
0.025 <1> (NADPH, <1> 35°C, PH 8.0 [1]) [1]

pH-Optimum
8 <1> [1]

Temperature optimum (°C)
35 <1> [1]

5 Isolation/Preparation/Mutation/Application

Localization
membrane <1> (<1> associated [1]) [1]

Purification
<1> [1]

6 Stability

Storage stability
<1>, -25°C, 20% sucrose, more than 3 months, no loss of activity [1]

References

[1] Yu, B.; Ruppert, M.; Stockigt, J.: Deoxysarpagine hydroxylase–a novel enzyme closing a short side pathway of alkaloid biosynthesis in Rauvolfia. Bioorg. Med. Chem., **10**, 2479-2483 (2002)

Phenylacetone monooxygenase 1.14.13.92

1 Nomenclature

EC number
1.14.13.92

Systematic name
phenylacetone,NADPH:oxygen oxidoreductase

Recommended name
phenylacetone monooxygenase

Synonyms
BVMO <2> [1]
Baeyer-Villiger monooxygenase <2> [1]
EtaA <1> [2]
PAMO <2> [1, 3, 4, 5, 6, 7]
Additional information <2> (<2> the enzyme belongs to the Baeyer-Villiger monooxygenases [4,7]; <2> the enzyme belongs to the Baeyer-Villiger monooxygenases, BVMOs [6]) [4, 6, 7]

2 Source Organism

<1> *Mycobacterium tuberculosis* (no sequence specified) [2]
<2> *Thermobifida fusca* (no sequence specified) [1, 3, 4, 5, 6, 7]

3 Reaction and Specificity

Catalyzed reaction
phenylacetone + NADPH + H^+ + O_2 = benzyl acetate + $NADP^+$ + H_2O (<2> reaction mechanism [4])

Natural substrates and products
S phenylacetone + NADPH + O_2 <2> (Reversibility: ?) [6, 7]
P benzyl acetate + $NADP^+$ + H_2O

Substrates and products
S (2-methylphenyl)acetone + NADPH + O_2 <2> (Reversibility: ?) [1]
P ? + $NADP^+$ + O_2
S (R)-1-acetoxy-phenylacetone + NADPH + O_2 <2> (Reversibility: ?) [7]
P (R)-1-hydroxy-1-phenylacetone + $NADP^+$ + H_2O
S (S)-nicotine + NADPH + O_2 <2> (Reversibility: ?) [7]

P ?
S 2-decanone + NADPH + O_2 <1> (Reversibility: ?) [2]
P methyl nonanoate + octyl acetate + $NADP^+$ + H_2O
S 2-dodecanone + NADPH + O_2 <2> (Reversibility: ?) [1]
P nonanol ethyl ester + $NADP^+$ + O_2
S 2-dodecanone + NADPH + O_2 <1> (Reversibility: ?) [2]
P nonyl acetate + methyl decanoate + $NADP^+$ + H_2O
S 2-heptanone + NADPH + O_2 <1> (Reversibility: ?) [2]
P pentyl acetate + $NADP^+$ + H_2O
S 2-hexanone + NADPH + O_2 <1> (Reversibility: ?) [2]
P butyl acetate + $NADP^+$ + H_2O
S 2-methylphenylcyclohexanone + NADPH + O_2 <2> (<2> mutant P3 prefers the R-isomer [4]) (Reversibility: ?) [4]
P 2-methylphenyl-hexano-6-lactone + NADP + H_2O
S 2-octanone + NADPH + O_2 <1> (Reversibility: ?) [2]
P heptyl acetate + $NADP^+$ + H_2O
S 2-phenylcyclohexanone + NADPH + O_2 <2> (<2> molecular modeling of the Criegee intermediate, the wild-type enzyme prefers the S-isomer, while mutants P1-P3 all prefer the R-isomer [4]) (Reversibility: ?) [4]
P 2-phenyl-hexano-6-lactone + NADP + H_2O
S 3-octanone + NADPH + O_2 <1> (Reversibility: ?) [2]
P ethyl hexanoate + pentyl propanoate + $NADP^+$ + H_2O
S 3-phenylpenta-2,4-dione + NADPH + O_2 <2> (Reversibility: ?) [7]
P (R)-phenylacetylcarbinol + $NADP^+$ + H_2O (<2> the product is a well-known precursor in the synthesis of ephedrine and pseudoephedrine [7])
S 4-heptanone + NADPH + O_2 <1> (Reversibility: ?) [2]
P propanyl butanoate + $NADP^+$ + H_2O
S 4-hydroxyacetophenone + NADPH + O_2 <2> (Reversibility: ?) [1]
P acetic acid 4-hydroxyphenyl ester + $NADP^+$ + O_2
S 4-phenylcyclohexanone + NADPH + O_2 <2> (Reversibility: ?) [4]
P 4-phenyl-hexano-6-lactone + NADP + H_2O
S N,N-dimethylbenzylamine + NADPH + O_2 <2> (Reversibility: ?) [7]
P N,N-dimethylbenzylamine N-oxide + $NADP^+$ + H_2O
S benzylacetone + NADPH + O_2 <1, 2> (Reversibility: ?) [1, 2]
P ? + $NADP^+$ + H_2O
S benzylacetone + NADPH + O_2 <2> (<2> low activity [7]) (Reversibility: ?) [7]
P ?
S bicyclohept-2-en-6-one + NADPH + O_2 <2> (Reversibility: ?) [1]
P ? + $NADP^+$ + O_2
S bicyclohept-2-en-6-one + NADPH + O_2 <1> (Reversibility: ?) [2]
P 3-oxabicyclo[3.3.0]oct-6-en-2-one + $NADP^+$ + H_2O
S diketone + NADPH + O_2 <2> (Reversibility: ?) [7]
P (R)-1-acetoxy-phenylacetone + $NADP^+$ + H_2O
S ethionamide + NADPH + O_2 <2> (Reversibility: ?) [1]
P ? + $NADP^+$ + O_2
S ethionamide + NADPH + O_2 <1> (Reversibility: ?) [2]

P ? + $NADP^+$ + H_2O
S methyl 4-tolylsulfide + NADPH + O_2 <2> (Reversibility: ?) [1]
P ? + $NADP^+$ + O_2
S methyl-*p*-tolylsulfide + NADPH + O_2 <1> (Reversibility: ?) [2]
P ? + $NADP^+$ + H_2O
S phenylacetone + NADPH + O_2 <1, 2> (Reversibility: ?) [1, 2, 3, 4, 6, 7]
P benzyl acetate + $NADP^+$ + H_2O
S phenylboronic acid + NADPH + O_2 <2> (<2> formation of phenol [7]) (Reversibility: ?) [7]
P ?
S rac-bicyclo [3.2.0]hept-2-en-6-one + NADPH + O_2 <2> (<2> activity and stereoselectivity of wild-type and mutant enzymes, overview [5]) (Reversibility: ?) [5]
P ?
S Additional information <2> (<2> enzyme activity in a variety of different aqueous-organic media using organic sulfides as substrates, enantioselectivity, overview [6]; <2> substrate selectivity and stereospecificity of wild-type and mutant enzymes, overview [4]; <2> the recombinant His-tagged enzyme is active with a large range of sulfides and ketones, as well as with several sulfoxides, an amine and an organoboron compound, high enantioselectivity dependeing on the substrate, substrate specificity, overview [7]) (Reversibility: ?) [4, 6, 7]
P ?

Inhibitors

1,4-dioxane <2> (<2> 33% inhibition at 10% concentration as co-solvent, 53% inhibition at 30% [6]) [6]
3-acetyl-$NADP^+$ <1> (<1> 0.5 mM, 50% inhibition [2]) [2]
3-amino-$NADP^+$ <1> (<1> 0.05 mM, 74% inhibition [2]) [2]
acetone <2> (<2> 45% inhibition at 10% concentration as co-solvent inpresence of substrate, 93% in absence of substrate [6]) [6]
acetonitrile <2> (<2> 88.5% inhibition at 30% concentration as co-solvent [6]) [6]
ethanol <2> (<2> 45% inhibition at 10% concentration as co-solvent [6]) [6]
isopropanol <2> (<2> 52% inhibition at 30% concentration as co-solvent [6]) [6]
methanol <2> (<2> 40% inhibition at 30% concentration as co-solvent [6]) [6]
n-propanol <2> (<2> 80% inhibition at 30% concentration as co-solvent [6]) [6]

Cofactors/prosthetic groups

FAD <1, 2> (<2> binding structure involving Arg337, overview [4]) [1,2,4]
NADPH <1, 2> (<2> can not be replaced by NADH [1]) [1,2,4,5,6,7]

Activating compounds

Bovine serum albumin <1> (<1> 2 mg7ml, 930% activation [2]) [2]

Turnover number (min^{-1})

0.017 <1> (phenylacetone, <1> 25°C, pH 8.5 [2]) [2]
0.021 <1> (benzylacetone, <1> 25°C, pH 8.5 [2]) [2]
0.023 <1> (2-dodecanone, <1> 25°C, pH 8.5 [2]) [2]
0.027 <1> (ethionamide, <1> 25°C, pH 8.5 [2]) [2]
0.22 <2> (phenylacetone, <2> mutant P3 [4]) [4]
0.23 <2> (2-dodecanone, <2> 30°C, pH 7.5, 0.1 mM NADPH [1]) [1]
0.25 <2> (2-phenylcyclohexanone, <2> mutant P3 [4]) [4]
0.25 <2> (phenylacetone, <2> mutant P1 [4]) [4]
0.26 <1> (phenylacetone, <1> 25°C, pH 8.5, in the presence of 0.006 mM bovine serum albumin [2]) [2]
0.31 <2> (2-phenylcyclohexanone, <2> mutant P1 [4]) [4]
0.34 <2> (4-hydroxyacetophenone, <2> 30°C, pH 7.5, 0.1 mM NADPH [1]) [1]
0.4 <2> (phenylacetone, <2> mutant P2 [4]) [4]
0.5 <2> (2-phenylcyclohexanone, <2> mutant P2 [4]) [4]
1.1 <2> (bicyclohept-2-en-6-one, <2> 30°C, pH 7.5, 0.1 mM NADPH [1]) [1]
1.8 <2> (benzylacetone, <2> 30°C, pH 7.5, 0.1 mM NADPH [1]) [1]
1.9 <2> (phenylacetone, <2> wild-type enzyme [4]; <2> 30°C, pH 7.5, 0.1 mM NADPH [1]) [1, 4]
2 <2> ((2-methylphenyl)acetone, <2> 30°C, pH 7.5, 0.1 mM NADPH [1]) [1]
2.1 <2> (methyl 4-tolylsulfide, <2> 30°C, pH 7.5, 0.1 mM NADPH [1]) [1]
Additional information <2> (<2> all substrates show a turnover between 1.2 s-1 and 3.6 s-1 [7]) [7]

Specific activity (U/mg)

Additional information <2> (<2> catalytic efficiency of wild-type and mutant PAMO in in vitro catalysis with two-liquid phases, overview [5]; <2> substrate specificity, diverse substrates, overview [7]) [4, 5, 7]

K_m-Value (mM)

0.059 <2> (phenylacetone, <2> wild-type enzyme [4]; <2> 30°C, pH 7.5, 0.1 mM NADPH [1]) [1, 4]
0.061 <1> (phenylacetone, <1> 25°C, pH 8.5 [2]) [2]
0.07 <2> (2-phenylcyclohexanone, <2> mutant P3 [4]) [4]
0.2 <1> (2-dodecanone, <1> 25°C, pH 8.5 [2]) [2]
0.26 <2> (2-dodecanone, <2> 30°C, pH 7.5, 0.1 mM NADPH [1]) [1]
0.28 <1> (phenylacetone, <1> 25°C, pH 8.5, in the presence of 0.006 mM bovine serum albumin [2]) [2]
0.34 <1> (ethionamide, <1> 25°C, pH 8.5 [2]) [2]
0.36 <2> (benzylacetone, <2> 30°C, pH 7.5, 0.1 mM NADPH [1]) [1]
0.5 <2> (2-phenylcyclohexanone, <2> mutant P2 [4]) [4]
0.52 <1> (benzylacetone, <1> 25°C, pH 8.5 [2]) [2]
0.83 <2> ((2-methylphenyl)acetone, <2> 30°C, pH 7.5, 0.1 mM NADPH [1]) [1]
0.86 <2> (methyl 4-tolylsulfide, <2> 30°C, pH 7.5, 0.1 mM NADPH [1]) [1]
2.2 <2> (4-hydroxyacetophenone, <2> 30°C, pH 7.5, 0.1 mM NADPH [1]) [1]
2.3 <2> (2-phenylcyclohexanone, <2> mutant P1 [4]) [4]

2.5 <2> (phenylacetone, <2> mutant P3 [4]) [4]
3 <2> (phenylacetone, <2> mutant P1 [4]) [4]
4 <2> (phenylacetone, <2> mutant P2 [4]) [4]
15 <2> (bicyclohept-2-en-6-one, <2> 30°C, pH 7.5, 0.1 mM NADPH [1]) [1]
Additional information <2> (<2> steady-state kinetics [7]) [7]

pH-Optimum
8 <2> (<2> assay at [5]) [1, 5]
9 <2> (<2> assay at [6]) [6]

pH-Range
7-9 <2> (<2> more than 80% of maximal activity [1]) [1]

Temperature optimum (°C)
25 <2> (<2> assay at [4,6,7]) [4, 6, 7]
40 <2> (<2> assay at [5]) [5]
70 <2> (<2> 4fold higher activity when compared with the activity at 25°C [1]) [1]

4 Enzyme Structure

Subunits
monomer <2> (<2> 1 * 65000, SDS-PAGE, small amount of approx. 7% is present as a dimer [1]) [1]
Additional information <2> (<2> structure analysis, the enzyme exhibits a two-domain architecture, and a bulge loop region [4]) [4]

5 Isolation/Preparation/Mutation/Application

Localization
membrane <1> (<1> associated [2]) [2]

Purification
<1> (Q-Sepharose, hydroxyapatite, Superdex 200) [2]
<2> (recombinant PAMO) [1, 3]

Crystallization
<2> (PAMO crystal structure analysis, comparison to cyclohexanone monooxygenase, EC 1.14.13.22) [4]
<2> (crystals are grown at 20°C by vapor diffusion, hanging drops are formed by mixing equal volumes of 18 mg/ml protein in 5 mM FAD and 50 mM sodium phosphate, pH 7.0, and of a well solution consisting of 1.5 M ammonium sulfate and 500 mM lithium chloride, crystals diffract to 1.7 A resolution) [3]

Cloning
<1> (expression in Escherichia coli) [2]
<2> (expression in Escherichia coli) [1, 3]

Engineering

Additional information <2> (<2> construction of three mutants P1-P3 by elimination of a bulge loop region, involving residues Ser441, Ala442, and Leu443, leading to enhanced substrate enantioselectivity of Baeyer-Villiger reactions while maintaining high thermal stability, overview [4]; <2> engineering of three highly stereoselective mutants of the thermally stable phenylacetone monooxygenase as practical catalysts for enantioselective Baeyer-Villiger oxidations of several ketones on a preparative scale under in vitro conditions, optimization of the method including a coupled cofactor-regeneration system, reaction mechanism, overview [5]) [4, 5]

Application

synthesis <2> (<2> the enzyme has catalytic potential as a biocatalyst [7]; <2> the thermally stable phenylacetone monooxygenase and engineered mutants can be used as a practical catalysts for enantioselective Baeyer-Villiger oxidations of several ketones on a preparative scale under in vitro conditions, overview [5]) [5, 7]

6 Stability

Temperature stability

52 <2> (<2> 50% loss of activity after 24 h and 48 h in the absence and presence of FAD, respectively [1]) [1]

Organic solvent stability

2-pentanol <2> (<2> as a sacrificial substrate [5]) [5]
cyclohexane <2> (<2> stabilizes the enzyme, best organic solvent, inactivation of the enzyme within 5 min [5]) [5]
isopropanol <2> (<2> the most effective stoichiometric sacrificial electron donor, optimal at 5% v/v [5]) [5]
methyl tert-butyl ether <2> (<2> stabilizes the enzyme, inactivation of the enzyme within below 5 min [5]) [5]

General stability information

<2>, PAMO that has been activated at 50°C can be stored for several days at room temperature without any loss in activity [1]
<2>, detergents stabilize the enzyme, best by Tween-20 at 0.1% v/v [5]

References

[1] Fraaije, M.W.; Wu, J.; Heuts, D.P.; van Hellemond, E.W.; Spelberg, J.H.; Janssen, D.B.: Discovery of a thermostable Baeyer-Villiger monooxygenase by genome mining. Appl. Microbiol. Biotechnol., **66**, 393-400 (2005)
[2] Fraaije, M.W.; Kamerbeek, N.M.; Heidekamp, A.J.; Fortin, R.; Janssen, D.B.: The prodrug activator EtaA from Mycobacterium tuberculosis is a Baeyer-Villiger monooxygenase. J. Biol. Chem., **279**, 3354-3360 (2004)

[3] Malito, E.; Alfieri, A.; Fraaije, M.W.; Mattevi, A.: Crystal structure of a Baeyer-Villiger monooxygenase. Proc. Natl. Acad. Sci. USA, **101**, 13157-13162 (2004)
[4] Bocola, M.; Schulz, F.; Leca, F.; Vogel, A.; Fraaije, M.W.; Reetz, M.T.: Converting phenylacetone monooxygenase into phenylcyclohexanone monooxygenase by rational design: towards practical Baeyer-Villiger monooxygenases. Adv. Synth. Catal., **347**, 979-986 (2005)
[5] Schulz, F.; Leca, F.; Hollmann, F.; Reetz, M.T.: Towards practical biocatalytic Baeyer-Villiger reactions: applying a thermostable enzyme in the gram-scale synthesis of optically active lactones in a two-liquid-phase system. Beilstein J. Org. Chem., **1**, 10 (2005)
[6] De Gonzalo, G.; Ottolina, G.; Zambianchi, F.; Fraaije, M.W.; Carrea, G.: Biocatalytic properties of Baeyer-Villiger monooxygenases in aqueous-organic media. J. Mol. Catal. B, **39**, 91-97 (2006)
[7] de Gonzalo, G.; Torres Pazmino, D.E.; Ottolina, G.; Fraaije, M.W.; Carrea, G.: Oxidations catalyzed by phenylacetone monooxygenase from Thermobifida fusca. Tetrahedron, **16**, 3077-3083 (2005)

(+)-Abscisic acid 8'-hydroxylase 1.14.13.93

1 Nomenclature

EC number
1.14.13.93

Systematic name
abscisate,NADPH:oxygen oxidoreductase (8'-hydroxylating)

Recommended name
(+)-abscisic acid 8'-hydroxylase

Synonyms
ABAQ 8'-hydroxylase <3> [4]
CYP707A <3> [4]

CAS registry number
153190-37-5

2 Source Organism

<1> *Zea mays* (no sequence specified) [1, 3, 5, 7]
<2> *Solanum tuberosum* (no sequence specified) [6]
<3> *Arabidopsis thaliana* (no sequence specified) [2, 4, 6, 8]

3 Reaction and Specificity

Catalyzed reaction
(+)-abscisate + NADPH + H^+ + O_2 = 8'-hydroxyabscisate + $NADP^+$ + H_2O

Substrates and products
- **S** (+)-S-abscisate + NADPH + H^+ + O_2 <3> (Reversibility: ?) [2, 4]
- **P** 8'-hydroxyabscisate + $NADP^+$ + H_2O
- **S** (+)-S-abscisate + NADPH + H^+ + O_2 <1, 3> (<3> isoform CYP707A3 is specific for (+)-isomer [8]) (Reversibility: ?) [1, 5, 8]
- **P** 8'-hydroxyabscisate + $NADP^+$ + H_2O (<3> isoform CYP707A3, no hydroxylation at 7' position [8])
- **S** (+-)-3'-methyl-abscisate + NADPH + H^+ + O_2 <3> (<3> 32% of the activity with (+)-S-abscisate [2]) (Reversibility: ?) [2]
- **P** ? + $NADP^+$ + H_2O
- **S** (+-)-abscisic acid-3'-thio-n-butyl thiol + NADPH + H^+ + O_2 <3> (<3> 2% of the activity with (+)-S-abscisate [2]) (Reversibility: ?) [2]
- **P** ? + $NADP^+$ + H_2O

S 1'-deoxy-(+)-S-abscisate + NADPH + H^+ + O_2 <3> (<3> 99% of the activity with (+)-S-abscisate [2]) (Reversibility: ?) [2]

P ? + $NADP^+$ + H_2O

S 1'-deoxy-1'-fluoro-(+)-S-abscisate + NADPH + H^+ + O_2 <3> (<3> 94% of the activity with (+)-S-abscisate [2]) (Reversibility: ?) [2]

P ? + $NADP^+$ + H_2O

S 2'α,3'α-dihydro-2'α,3'α-epoxy-(+)-S-abscisate + NADPH + H^+ + O_2 <3> (<3> 19% of the activity with (+)-S-abscisate [2]) (Reversibility: ?) [2]

P ? + $NADP^+$ + H_2O

S 3'-bromo-(+)-S-abscisate + NADPH + H^+ + O_2 <3> (<3> 5% of the activity with (+)-S-abscisate [2]) (Reversibility: ?) [2]

P ? + $NADP^+$ + H_2O

S 3'-chloro-(+)-S-abscisate + NADPH + H^+ + O_2 <3> (<3> 19% of the activity with (+)-S-abscisate [2]) (Reversibility: ?) [2]

P ? + $NADP^+$ + H_2O

S 3'-fluoro-(+)-S-abscisate + NADPH + H^+ + O_2 <3> (<3> 68% of the activity with (+)-S-abscisate [2]) (Reversibility: ?) [2]

P ? + $NADP^+$ + H_2O

S 6-nor-(+)-S-abscisate + NADPH + H^+ + O_2 <3> (<3> 60% of the activity with (+)-S-abscisate [2]) (Reversibility: ?) [2]

P ? + $NADP^+$ + H_2O

S 7'-methyl-(+)-S-abscisate + NADPH + H^+ + O_2 <3> (<3> 15% of the activity with (+)-S-abscisate [2]) (Reversibility: ?) [2]

P ? + $NADP^+$ + H_2O

S 7'-nor-(+)-S-abscisate + NADPH + H^+ + O_2 <3> (<3> 15% of the activity with (+)-S-abscisate [2]) (Reversibility: ?) [2]

P ? + $NADP^+$ + H_2O

S 8'-fluoro-(+)-S-abscisate + NADPH + H^+ + O_2 <3> (<3> 11% of the activity with (+)-S-abscisate [2]) (Reversibility: ?) [2]

P ? + $NADP^+$ + H_2O

S 8'-methylene-(+)-S-abscisate + NADPH + H^+ + O_2 <3> (<3> 4% of the activity with (+)-S-abscisate [2]) (Reversibility: ?) [2]

P ? + $NADP^+$ + H_2O

S 9',9'-difluoro-(+)-S-abscisate + NADPH + H^+ + O_2 <3> (<3> 3% of the activity with (+)-S-abscisate [2]) (Reversibility: ?) [2]

P ? + $NADP^+$ + H_2O

S 9'-fluoro-(+)-S-abscisate + NADPH + H^+ + O_2 <3> (<3> 33% of the activity with (+)-S-abscisate [2]) (Reversibility: ?) [2]

P ? + $NADP^+$ + H_2O

S 9'-methyl-(+)-S-abscisate + NADPH + H^+ + O_2 <3> (<3> 3% of the activity with (+)-S-abscisate [2]) (Reversibility: ?) [2]

P ? + $NADP^+$ + H_2O

S Additional information <1, 3> (<3> isomerisation of 8'-hydroxy-abscisic acid to phaseic acid is not catalyzed by enzyme [8]; <1> specific for (+)-isomer of abscisate [5]; <3> substrate recognition strictly requires the 6-methyl groups [2]) (Reversibility: ?) [2, 5, 8]

P ?

Inhibitors

(+)-1'-methoxy-abscisate <1> (<1> competitive [1]) [1]
(+)-8',8'-difluoroabscisate <3> (<3> competitive [2]) [2]
(+)-8'-acetylene-abscisate <1> (<1> suicide inhibitor [1]; <1> about 60% inactivation [3]) [1, 3]
(+)-8'-cyano-abscisate <1> (<1> competitive [1]) [1]
(+)-8'-ethyl-abscisate <1> (<1> competitive [1]) [1]
(+)-8'-methylacetylene-abscisate <1> (<1> competitive [1]) [1]
(+)-8'-methylene-abscisate <1> (<1> competitive [1]) [1]
(+)-8'-propargyl-abscisate <1> (<1> competitive [1]) [1]
(+)-9'-allyl-abscisate <1> (<1> suicide inhibitor [1]) [1]
(+)-9'-propargyl-abscisate <1> (<1> suicide inhibitor [1]) [1]
(+-)-3'-methyl-abscisate <3> (<3> competitive, 64% inhibition at 0.05 mM [2]) [2]
(+-)-abscisic acid-3'-thio-n-butyl thiol <3> (<3> competitive, 51% inhibition at 0.05 mM [2]) [2]
(-)-8'-propargyl-abscisate <1> (<1> competitive [1]) [1]
(-)-9'-propargyl-abscisate <1> (<1> suicide inhibitor [1]) [1]
(1'S*,2'S*,6'S*)-(+-)-6-nor-2',3'-dihydro-4'-deoxo-8',8'-difluoro-abscisate <3> (<3> 50% inhibition at 0.00063 mM [2]) [2]
(1'S*,2'S*,6'S*)-(+-)-6-nor-2',3'-dihydro-4'-deoxo-abscisate <3> (<3> 50% inhibition at 0.00091 mM [2]) [2]
1'-deoxy-(+)-S-abscisate <3> (<3> competitive, 73% inhibition at 0.05 mM [2]) [2]
1'-deoxy-1'-fluoro-(+)-S-abscisate <3> (<3> competitive, 100% inhibition at 0.05 mM [2]) [2]
2'α,3'α-dihydro-2'α,3'α-epoxy-(+)-S-abscisate <3> (<3> competitive, 56% inhibition at 0.05 mM [2]) [2]
3'-azido-(+)-S-abscisate <3> (<3> competitive, 38% inhibition at 0.05 mM [2]) [2]
3'-bromo-(+)-S-abscisate <3> (<3> competitive, 65% inhibition at 0.05 mM [2]) [2]
3'-chloro-(+)-S-abscisate <3> (<3> competitive, 70% inhibition at 0.05 mM [2]) [2]
3'-fluoro-(+)-S-abscisate <3> (<3> competitive, 84% inhibition at 0.05 mM [2]) [2]
3'-iodo-(+)-S-abscisate <3> (<3> competitive, 54% inhibition at 0.05 mM [2]) [2]
5'α,8'-cyclo-(+)-S-abscisate <3> (<3> competitive, 28% inhibition at 0.05 mM [2]) [2]
6-nor-(+)-S-abscisate <3> (<3> competitive, 88% inhibition at 0.05 mM [2]) [2]
7'-methyl-(+)-S-abscisate <3> (<3> competitive, 83% inhibition at 0.05 mM [2]) [2]
7'-nor-(+)-S-abscisate <3> (<3> competitive, 28% inhibition at 0.05 mM [2]) [2]

8',8',8'-trifluoro-(+)-S-abscisate <3> (<3> competitive, 38% inhibition at 0.05 mM [2]) [2]
8',8'-difluoro-(+)-S-abscisate <3> (<3> competitive, 83% inhibition at 0.05 mM [2]) [2]
8'-fluoro-(+)-S-abscisate <3> (<3> competitive, 83% inhibition at 0.05 mM [2]) [2]
8'-methyl-(+)-S-abscisate <3> (<3> competitive, 35% inhibition at 0.05 mM [2]) [2]
8'-methylene-(+)-S-abscisate <3> (<3> competitive, 33% inhibition at 0.05 mM [2]) [2]
9',9',9'-trifluoro-(+)-S-abscisate <3> (<3> competitive, 55% inhibition at 0.05 mM [2]) [2]
9',9'-difluoro-(+)-S-abscisate <3> (<3> competitive, 76% inhibition at 0.05 mM [2]) [2]
9'-fluoro-(+)-S-abscisate <3> (<3> competitive, 83% inhibition at 0.05 mM [2]) [2]
9'-methyl-(+)-S-abscisate <3> (<3> competitive, 26% inhibition at 0.05 mM [2]) [2]
CO <1> (<1> inhibition is reversible by blue and amber light [7]) [7]
cytochrome c <1, 3> (<3> 0.1 mM, complete inhibition [8]; <1> oxidized form [7]) [7, 8]
tetcyclacis <1> (<1> 50% inhibition at 0.001 mM [7]) [7]
abscisic aldehyde <3> (<3> competitive, 31% inhibition at 0.05 mM [2]) [2]
Additional information <1> (<1> inhibition of reaction at O_2 concentrations less than 10% v/v [7]; <1> not inactivating: (+)-8'-methylene-abscisate, (+)-8'-methylacetylene-abscisate [3]) [3, 7]

Cofactors/prosthetic groups

cytochrome P_{450} <1> [7]
NADPH <1, 3> [4,7,8]
Additional information <1, 3> (<1,3> NADH may not substitute for NADPH [7,8]) [7,8]

Activating compounds

NADH <1> [7]

K_m-Value (mM)

0.0013 <3> ((+)-S-abscisate, <3> pH 7.25, 30°C, isoform CYP707A3 [8]) [8]
0.016 <1> ((+)-S-abscisate, <1> pH 7.6, 30°C [1]) [1]

K_i-Value (mM)

0.00016 <3> (6-nor-(+)-S-abscisate, <3> pH 7.25, 30°C [2]) [2]
0.00017 <3> ((+)-8',8'-difluoroabscisate, <3> pH 7.25, 30°C [2]) [2]
0.00017 <3> (8',8'-difluoro-(+)-S-abscisate, <3> pH 7.25, 30°C [2]) [2]
0.00025 <3> (9',9'-difluoro-(+)-S-abscisate, <3> pH 7.25, 30°C [2]) [2]
0.00027 <1> ((+)-9'-propargyl-abscisate, <1> pH 7.6, 30°C [1]) [1]
0.00027 <3> (8'-fluoro-(+)-S-abscisate, <3> pH 7.25, 30°C [2]) [2]
0.0004 <3> ((1'S*,2'S*,6'S*)-(+-)-6-nor-2',3'-dihydro-4'-deoxo-abscisate, <3> pH 7.25, 30°C [2]) [2]

0.00041 <3> ((1'S*,2'S*,6'S*)-(+-)-6-nor-2',3'-dihydro-4'-deoxo-8',8'-difluoro-abscisate, <3> pH 7.25, 30°C [2]) [2]
0.00071 <3> (8',8',8'-trifluoro-(+)-S-abscisate, <3> pH 7.25, 30°C [2]) [2]
0.00078 <3> (9'-fluoro-(+)-S-abscisate, <3> pH 7.25, 30°C [2]) [2]
0.00094 <3> (8'-methyl-(+)-S-abscisate, <3> pH 7.25, 30°C [2]) [2]
0.00106 <3> (9',9',9'-trifluoro-(+)-S-abscisate, <3> pH 7.25, 30°C [2]) [2]
0.0011 <1> ((+)-8'-propargyl-abscisate, <1> pH 7.6, 30°C [1]) [1]
0.00429 <3> (9'-methyl-(+)-S-abscisate, <3> pH 7.25, 30°C [2]) [2]
0.00543 <3> (8'-methylene-(+)-S-abscisate, <3> pH 7.25, 30°C [2]) [2]
0.0055 <1> ((+)-9'-allyl-abscisate, <1> pH 7.6, 30°C [1]) [1]
0.0111 <1> ((+)-1'-methoxy-abscisate, <1> pH 7.6, 30°C [1]) [1]
0.0129 <1> ((+)-8'-ethyl-abscisate, <1> pH 7.6, 30°C [1]) [1]
0.0135 <1> ((-)-9'-propargyl-abscisate, <1> pH 7.6, 30°C [1]) [1]
0.019 <1> ((+)-8'-acetylene-abscisate, <1> pH 7.6, 30°C [1]) [1]
0.056 <1> ((-)-8'-propargyl-abscisate, <1> pH 7.6, 30°C [1]) [1]
0.122 <1> ((+)-8'-methylene-abscisate, <1> pH 7.6, 30°C [1]) [1]
0.187 <1> ((+)-8'-cyano-abscisate, <1> pH 7.6, 30°C [1]) [1]
0.284 <1> ((+)-8'-methylacetylene-abscisate, <1> pH 7.6, 30°C [1]) [1]

pH-Optimum

7.4-7.8 <1> [7]

4 Enzyme Structure

Subunits

? <3> (<3> x * 53000, isoform CYC707A1, x * 52000, isoform CYP707A3, SDS-PAGE [8]) [8]

5 Isolation/Preparation/Mutation/Application

Source/tissue

cell suspension culture <1, 2, 3> [1, 3, 5, 6, 7]

Localization

microsome <1, 3> (<1> integral membrane protein [7]) [7, 8]

Renaturation

<1> (inhibition by CO is reversible by blue and amber light) [7]

Cloning

<3> (expression in baculovirus system, isoforms CYP707A1, CYP707A3, CYP707A4) [8]

Engineering

Additional information <3> (<3> CYP707A2 knockout mutant, hyperdormancy in seeds, accumulates 6fold higher levels of abscisic acid than wild-type [4]) [4]

References

[1] Cutler, A.J.; Rose, P.A.; Squires, T.M.; Loewen, M.K.; Shaw, A.C.; Quail, J.W.; Krochko, J.E.; Abrams, S.R.: Inhibitors of abscisic acid 8'-hydroxylase. Biochemistry, **39**, 13614-13624 (2000)

[2] Ueno, K.; Yoneyama, H.; Saito, S.; Mizutani, M.; Sakata, K.; Hirai, N.; Todoroki, Y.: A lead compound for the development of ABA 8'-hydroxylase inhibitors. Bioorg. Med. Chem. Lett., **15**, 5226-5229 (2005)

[3] Rose, P.A.; Cutler, A.J.; Irvine, N.M.; Shaw, A.C.; Squires, T.M.; Loewen, M.K.; Abrams, S.R.: 8'-Acetylene ABA: an irreversible inhibitor of ABA 8'-hydroxylase. Bioorg. Med. Chem. Lett., **7**, 2543-2546 (1997)

[4] Kushiro, T.; Okamoto, M.; Nakabayashi, K.; Yamagishi, K.; Kitamura, S.; Asami, T.; Hirai, N.; Koshiba, T.; Kamiya, Y.; Nambara, E.: The Arabidopsis cytochrome P450 CYP707A encodes ABA 8'-hydroxylases: key enzymes in ABA catabolism. EMBO J., **23**, 1647-1656 (2004)

[5] Cutler, A.J.; Squires, T.M.; Loewen, M.K.; Balsevich, J.J.: Induction of (+)-abscisic acid 8' hydroxylase by (+)-abscisic acid in cultured maize cells. J. Exp. Bot., **48**, 1787-1795 (1997)

[6] Windsor, M.L.; Zeevaart, J.A.: Induction of ABA 8'-hydroxylase by (+)-S-, (-)-R- and 8'-8'-8'-trifluoro-S-abscisic acid in suspension cultures of potato and Arabidopsis. Phytochemistry, **45**, 931-934 (1997)

[7] Krochko, J.E.; Abrams, G.D.; Loewen, M.K.; Abrams, S.R.; Cutler, A.J.: (+)-Abscisic acid 8'-hydroxylase is a cytochrome P_{450} monooxygenase. Plant Physiol., **118**, 849-860 (1998)

[8] Saito, S.; Hirai, N.; Matsumoto, C.; Ohigashi, H.; Ohta, D.; Sakata, K.; Mizutani, M.: Arabidopsis CYP707As encode (+)-abscisic acid 8'-hydroxylase, a key enzyme in the oxidative catabolism of abscisic acid. Plant Physiol., **134**, 1439-1449 (2004)

Lithocholate 6β-hydroxylase 1.14.13.94

1 Nomenclature

EC number
1.14.13.94

Systematic name
lithocholate,NADPH:oxygen oxidoreductase (6β-hydroxylating)

Recommended name
lithocholate 6β-hydroxylase

Synonyms
3A10/lithocholic acid 6β-hydroxylase <2> [6]
6β-hydroxylase <2> [2]
CYP3A10 <2> [2, 4]
cytochrome P_{450} 3A10/lithocholic acid <2> [2]
cytochrome P_{450} 3A10/lithocholic acid 6β-hydroxylase <2> [5]
lithocholate 6β-monooxygenase <2> [2]
lithocholic acid 6β-hydroxylase <1, 2> [1, 4]

CAS registry number
9075-83-6

2 Source Organism

<1> *Rattus norvegicus* (no sequence specified) [1, 3]
<2> *Mesocricetus auratus* (no sequence specified) [2, 4, 5, 6]

3 Reaction and Specificity

Catalyzed reaction
lithocholate + NADPH + H^+ + O_2 = 6β-hydroxylithocholate + $NADP^+$ + H_2O

Reaction type
oxidation
redox reaction
reduction

Natural substrates and products
S lithocholate + NADPH + H^+ + O_2 <2> (<2> a STAT factor mediates the sexually dimorphic regulation of hepatic cytochrome P_{450} 3A10/litho-

cholic acid 6β-hydroxylase gene expression by growth hormone [5]; <2> in male hamsters, 6β-hydroxylation is the major pathway for detoxification of lithocholate. Likely CYP3A10 is responsible for this activity [4]; <2> the male-specific P_{450} catalyzes the reaction. The pattern of growth hormone secretion is directly responsible for male-specific expression of this gene. STAT 5a and STAT 5b mediates GH-dependent regulation of CYP3A10/6β-hydroxylase promoter activity [6]) (Reversibility: ?) [4, 5, 6]

P 6β-hydroxylithocholate + $NADP^+$ + H_2O

Substrates and products

S lithocholate + NADPH + H^+ + O_2 <1, 2> (<2> a STAT factor mediates the sexually dimorphic regulation of hepatic cytochrome P_{450} 3A10/lithocholic acid 6β-hydroxylase gene expression by growth hormone [5]; <2> in male hamsters, 6β-hydroxylation is the major pathway for detoxification of lithocholate. Likely CYP3A10 is responsible for this activity [4]; <2> the male-specific P_{450} catalyzes the reaction. The pattern of growth hormone secretion is directly responsible for male-specific expression of this gene. STAT 5a and STAT 5b mediates GH-dependent regulation of CYP3A10/6β-hydroxylase promoter activity [6]) (Reversibility: ?) [1, 3, 4, 5, 6]

P 6β-hydroxylithocholate + $NADP^+$ + H_2O

S Additional information <1> (<1> hydroxylations of bile acids by reconstituted systems from rat liver microsomes [3]) (Reversibility: ?) [3]

P ?

Cofactors/prosthetic groups

cytochrome P_{450} <1, 2> [1,2,3,5,6]
NADPH <1, 2> [1,3,4,5,6]

5 Isolation/Preparation/Mutation/Application

Source/tissue

liver <1, 2> (<2> expression of CYP 3A10 protein is male-specific in hamster liver microsome [2]; <2> male-specific enzyme [4]; <2> male-specific enzyme. The pattern of growth hormone secretion is directly responsible for male-specific expression of this gene [6]) [1, 2, 3, 4, 5, 6]

Localization

microsome <1, 2> (<2> expression of CYP 3A10 protein is male-specific in hamster liver microsome [2]) [2, 3, 4]

Cloning

<2> (expression in COS cells) [2]

References

[1] Zimniak, P.; Holsztynska, E.J.; Radominska, A.; Iscan, M.; Lester, R.; Waxman, D.J.: Distinct forms of cytochrome P-450 are responsible for 6β-hydroxylation of bile acids and of neutral steroids. Biochem. J., **275** **(Pt 1)**, 105-111. (1991)

[2] Chang, T.K.; Teixeira, J.; Gil, G.; Waxman, D.J.: The lithocholic acid 6β-hydroxylase cytochrome P_{450}, CYP3A10, is an active catalyst of steroid-hormone 6β-hydroxylation. Biochem. J., **291** **(Pt 2)**, 429-433. (1993)

[3] Björkhem, I.; Danielsson, H.; Wikvall, K.: Hydroxylations of bile acids by reconstituted systems from rat liver microsomes. J. Biol. Chem., **249**, 6439-6445 (1974)

[4] Teixeira, J.; Gil, G.: Cloning, expression, and regulation of lithocholic acid 6β-hydroxylase. J. Biol. Chem., **266**, 21030-21036 (1991)

[5] Subramanian, A.; Teixeira, J.; Wang, J.; Gil, G.: A STAT factor mediates the sexually dimorphic regulation of hepatic cytochrome P450 3A10/lithocholic acid 6β-hydroxylase gene expression by growth hormone. Mol. Cell. Biol., **15**, 4672-4682 (1995)

[6] Subramanian, A.; Wang, J.; Gil, G.: STAT 5 and NF-Y are involved in expression and growth hormone-mediated sexually dimorphic regulation of cytochrome P_{450} 3A10/lithocholic acid 6β-hydroxylase. Nucleic Acids Res., **26**, 2173-2178 (1998)

7α-Hydroxycholest-4-en-3-one 12α-hydroxylase 1.14.13.95

1 Nomenclature

EC number
1.14.13.95

Systematic name
7α-hydroxycholest-4-en-3-one,NADPH:oxygen oxidoreductase (12α-hydroxylating)

Recommended name
7α-hydroxycholest-4-en-3-one 12α-hydroxylase

Synonyms
7α-hydroxy-4-cholest-3-on3 12α-monooxygenase <3> [5]
7α-hydroxy-4-cholesten-3-one 12α-monooxygenase <3> [3]
CYP12 <3> [5]
HCO 12α-hydroxylase <2, 3> [5, 6]
sterol 12α-hydroxylase <3> [5]

CAS registry number
39369-22-7
55963-45-6

2 Source Organism

<1> *Mus musculus* (no sequence specified) [5]
<2> *Rattus norvegicus* (no sequence specified) [2, 5, 6]
<3> *Oryctolagus cuniculus* (no sequence specified) [1, 3, 4, 5]

3 Reaction and Specificity

Catalyzed reaction
7α-hydroxycholest-4-en-3-one + NADPH + H^+ + O_2 = 7α,12α-dihydroxycholest-4-en-3-one + $NADP^+$ + H_2O

Reaction type
oxidation
redox reaction
reduction

Natural substrates and products

- **S** 7α-hydroxycholest-4-en-3-one + NADPH + H^+ + O_2 <2, 3> (<2> key enzyme for the formation of cholic acid [2]; <3> microsomal enzyme activity was markedly elevated by starvation or streptozotocin administration to the animals [4]) (Reversibility: ?) [2, 4]
- **P** 7α,12α-dihydroxycholest-4-en-3-one + $NADP^+$ + H_2O

Substrates and products

- **S** 7α-hydroxycholest-4-en-3-one + NADH + H^+ + O_2 <2> (<2> reaction with NADH shows 15% of the activity with NADPH [6]) (Reversibility: ?) [6]
- **P** 7α,12α-dihydroxycholest-4-en-3-one + NAD^+ + H_2O
- **S** 7α-hydroxycholest-4-en-3-one + NADPH + H^+ + O_2 <1, 2, 3> (<2> key enzyme for the formation of cholic acid [2]; <3> microsomal enzyme activity was markedly elevated by starvation or streptozotocin administration to the animals [4]; <2> NADPH is the preferred electron source [6]) (Reversibility: ?) [1, 2, 3, 4, 5, 6]
- **P** 7α,12α-dihydroxycholest-4-en-3-one + $NADP^+$ + H_2O

Inhibitors

5β-cholestane-3α,7α,12α-triol <3> (<3> 0.066 mM, 62% inhibition [3]) [3]
5β-cholestane-3α,7α-diol <3> (<3> 0.066 mM, 30% inhibition [3]) [3]
Emulgen 913 <3> (<3> 0.1% w/v, complete inhibition [3]) [3]
$MgCl_2$ <3> (<3> 1 mM, 10% inhibition [3]) [3]

Cofactors/prosthetic groups

cytochrome P_{450} <3> [3,5]
cytochrome P_{450} <3> (<3> omission results in complete loss of activity [4]) [4]
NADH <2> (<2> 15% of the activity with NADPH [6]) [6]
NADPH <1, 2, 3> (<2> preferred electron source [6]) [1,2,3,4,5,6]
cytochrome b_5 <3> (<3> omission of cytochrome b_5 resulted in 40% loss of activity [4]) [4]

Activating compounds

Additional information <3> (<3> NADPH-cytochrome P_{450} reductase, EC 1.6.2.4, is required for maximal activity [3,4]) [3, 4]

Specific activity (U/mg)

0.000885 <2> (<2> liver microsome [6]) [6]
0.484 <3> [4]

K_m-Value (mM)

0.027 <3> (7α-hydroxycholest-4-en-3-one) [3]

pH-Optimum

6.7-7 <3> [3]

pH-Range

6-7.5 <3> (<3> pH 6.0: about 40% of maximal activity, pH 7.5: about 65% of maximal activity [3]) [3]

4 Enzyme Structure

Molecular weight

56000 <3> (<3> non-denaturing PAGE [3]) [3]

Subunits

? <3> (<3> x * 50000, SDS-PAGE [4]) [4]

Posttranslational modification

Additional information <3> (<3> immunoblotting experiment show no correlation between the enzyme activity and the amount of protein, suggesting that post-translational modification may occur [4]) [4]

5 Isolation/Preparation/Mutation/Application

Source/tissue

liver <2, 3> (<2> activity in diabetic group is about half of the activity in the normal group [2]; <3> exclusively expressed in [5]) [1, 2, 3, 5, 6]

Localization

microsome <2, 3> [1, 2, 3, 6]

Purification

<3> [3, 4]
<3> (partially) [1]

Cloning

<3> (expression in COS cells) [1]
<3> (transfection of COS-M6 cells with the coding part of the cDNA) [5]

6 Stability

General stability information

<3>, stabilized at least 4 weeks under high buffer concentrations such as 300 mM phosphate buffer [3]

References

[1] Andersson, U.; Eggertsen, G.; Bjorkhem, I.: Rabbit liver contains one major sterol 12α-hydroxylase with broad substrate specificity. Biochim. Biophys. Acta, **1389**, 150-154 (1998)
[2] Kimura, K.; Ogura, M.: Effect of streptozotocin-induced diabetes on the activity of 7α-hydroxy-4-cholesten-3-one-specific 12α-hydroxylase in rats. Biol. Chem. Hoppe-Seyler, **369**, 1117-1120 (1988)

[3] Murakami, K.; Okada, Y.; Okuda, K.: Purification and characterization of 7α-hydroxy-4-cholesten-3-one 12α-monooxygenase. J. Biol. Chem., **257**, 8030-8035 (1982)
[4] Ishida, H.; Noshiro, M.; Okuda, K.; Coon, M.J.: Purification and characterization of 7α-hydroxy-4-cholesten-3-one 12α-hydroxylase. J. Biol. Chem., **267**, 21319-21323 (1992)
[5] Eggertsen, G.; Olin, M.; Andersson, U.; Ishida, H.; Kubota, S.; Hellman, U.; Okuda, K.I.; Bjorkhem, I.: Molecular cloning and expression of rabbit sterol 12α-hydroxylase. J. Biol. Chem., **271**, 32269-32275 (1996)
[6] Noshiro, M.; Ishida, H.; Hayashi, S.; Okuda, K.: Assays for cholesterol 7α-hydroxylase and 12α-hydroxylase using high performance liquid chromatography. Steroids, **45**, 539-550 (1985)

5β-Cholestane-$3\alpha,7\alpha$-diol 12α-hydroxylase 1.14.13.96

1 Nomenclature

EC number
1.14.13.96

Systematic name
5β-cholestane-$3\alpha,7\alpha$-diol,NADPH:oxygen oxidoreductase (12α-hydroxylating)

Recommended name
5β-cholestane-$3\alpha,7\alpha$-diol 12α-hydroxylase

Synonyms
12α-hydroxylase CYP8B1 <3> [3]
12α-hydroxylase gene <3> [4]
5β-cholestane-$3\alpha,7\alpha$-diol 12α-monooxygenase <2> [10]
CYP8B1 <1, 2, 3, 4> [1, 4, 5, 6, 7, 10, 12, 13, 14, 15, 16, 17]
cytochrome P-450 LM4 <5> [8]
cytochrome P_{450} 8B1 <2> [10]
sterol 12α-hydroxylase <1, 2, 3, 4> (<2> ambiguous [10]) [5, 6, 7, 10, 13, 14, 15, 16, 17]
sterol 12α-hydroxylase CYP8B1 <1, 3> [2]
sterol 12α-hydroxylase/CYP8b1 <3> [9]

CAS registry number
39369-22-7
440354-83-6

2 Source Organism

<1> *Mus musculus* (no sequence specified) [2, 12, 13, 14, 15, 16]
<2> *Homo sapiens* (no sequence specified) [1, 7, 10, 11, 17, 18]
<3> *Rattus norvegicus* (no sequence specified) [2, 3, 4, 6, 9, 16]
<4> *Sus scrofa* (no sequence specified) [5]
<5> *Oryctolagus cuniculus* (no sequence specified) [8]

3 Reaction and Specificity

Catalyzed reaction

5β-cholestane-3α,7α-diol + NADPH + H^+ + O_2 = 5β-cholestane-3α,7α,12α-triol + $NADP^+$ + H_2O

Reaction type

oxidation
redox reaction
reduction

Natural substrates and products

S cholesterol + NADPH + O_2 <1, 2, 3> (<1> dietary cholsterol and cholic acid have an influence of enzyme regulation [13]; <1> the enzyme channels bile acid precursors into cholic acid, but is not affected on mRNA and protein levels by increased cholic acid or reduced β-muricholic acid levels, overview [15]; <1> the enzyme is necessary for the synthesis of cholic acid [14]; <2> thyroid hormone triiodothyronine has a regulatory function on the enzyme expressiona and bile acid metabolism, overview [18]) (Reversibility: ?) [13, 14, 15, 16, 17, 18]

P 12-hydroxycholesterol + $NADP^+$ + H_2O

S Additional information <1, 2, 3, 4> (<2> bile acids repress human CYP8B1 transcription by reducing the transactivation activity of HNF4α through interaction of HNF4α with SHP and reduction of HNF4α expression in the liver [10]; <1> HNF4α directly binds to the promoter region of the mouse cyp8b1 and positively regulates its expression. CYP8B1 enzyme activity is absent in HNF4$\alpha\Delta$L mice [12]; <2> IL-1β inhibits CYP8B1 gene transcription via a mitogen-activated protein kinase/c-Jun NH_2-terminal kinase pathway that inhibits HNF4α gene expression and its DNA-binding ability. This mechanism may play an important role in the adaptive response to inflammatory cytokines and the protection of the liver during cholestasis [1]; <4> key enzyme in cholic acid biosynthesis [5]; <3> key enzyme of the bile acid biosynthetic pathway. It regulates the composition of bile acids in bile, i.e. ratio between cholic acid and chenodeoxycholic acid. Bile acids inhibit CYP8B1 gene transcription by inducing α-fetoprotein transcription factor and inhibiting HNF4α expression [4]; <3> specific enzyme required for cholic acid synthesis. The levels of this enzyme determine the ratio of cholic acid to chenodeoxycholic acid and thus the hydrophobicity of the circulating bile acid pool [9]; <1,3> sterol regulatory element binding proteins 1a and 1c enhance transcription of the CYP8B1 gene through binding to sterol regulatory element. Cholesterol loading reduces sterol regulatory element binding proteins 1 mRNA expression in addition to reducing functional sterol regulatory element binding proteins 1 protein, and results in decreasing CYP8B1 gene transcription [2]; <3> the enzyme is involved in bile acid synthesis, critical importance for the composition of bile acids formed in the liver, regulated by thyroid hormone at the mRNA level [3]; <3> the enzyme is required for biosynthesis of cholic acid. CYP8b1 is transcriptionally down-regulated by hydrophobic but not hydrophilic bile acids. Cholesterol feeding and a single thyroid hormone injection repress CYP8b1 in face of cholesterol 7α-hydroxylase specific activities [6]; <2> variation in the ratio between cholic acid and chenodeoxycholic acid in human bile are not due to polymorphisms in gene CYP8B1, overview [17]) (Reversibility: ?) [1, 2, 3, 4, 5, 6, 9, 10, 12, 17]

P ?

Substrates and products

S 5β-cholestane-3α,7α-diol + NADPH + O_2 <5> (Reversibility: ?) [8]

P 5β-cholestane-3α,7α,12α-triol + $NADP^+$ + H_2O

S cholesterol + NADPH + O_2 <1, 2, 3> (<1> dietary cholsterol and cholic acid have an influence of enzyme regulation [13]; <1> the enzyme channels bile acid precursors into cholic acid, but is not affected on mRNA and protein levels by increased cholic acid or reduced β-muricholic acid levels, overview [15]; <1> the enzyme is necessary for the synthesis of cholic acid [14]; <2> thyroid hormone triiodothyronine has a regulatory function on the enzyme expressiona and bile acid metabolism, overview [18]) (Reversibility: ?) [13, 14, 15, 16, 17, 18]

P 12-hydroxycholesterol + $NADP^+$ + H_2O

S Additional information <1, 2, 3, 4> (<2> bile acids repress human CYP8B1 transcription by reducing the transactivation activity of HNF4α through interaction of HNF4α with SHP and reduction of HNF4α expression in the liver [10]; <1> HNF4α directly binds to the promoter region of the mouse cyp8b1 and positively regulates its expression. CYP8B1 enzyme activity is absent in HNF4$\alpha\Delta$L mice [12]; <2> IL-1β inhibits CYP8B1 gene transcription via a mitogen-activated protein kinase/c-Jun NH_2-terminal kinase pathway that inhibits HNF4α gene expression and its DNA-binding ability. This mechanism may play an important role in the adaptive response to inflammatory cytokines and the protection of the liver during cholestasis [1]; <4> key enzyme in cholic acid biosynthesis [5]; <3> key enzyme of the bile acid biosynthetic pathway. It regulates the composition of bile acids in bile, i.e. ratio between cholic acid and chenodeoxycholic acid. Bile acids inhibit CYP8B1 gene transcription by inducing α-fetoprotein transcription factor and inhibiting HNF4α expression [4]; <3> specific enzyme required for cholic acid synthesis. The levels of this enzyme determine the ratio of cholic acid to chenodeoxycholic acid and thus the hydrophobicity of the circulating bile acid pool [9]; <1,3> sterol regulatory element binding proteins 1a and 1c enhance transcription of the CYP8B1 gene through binding to sterol regulatory element. Cholesterol loading reduces sterol regulatory element binding proteins 1 mRNA expression in addition to reducing functional sterol regulatory element binding proteins 1 protein, and results in decreasing CYP8B1 gene transcription [2]; <3> the enzyme is involved in bile acid synthesis, critical importance for the composition of bile acids formed in the liver, regulated by thyroid hormone at the mRNA level [3]; <3> the enzyme is required for biosynthesis of cholic acid. CYP8b1 is transcriptionally down-regulated by hydrophobic but not hydrophilic bile acids. Cholesterol feeding and a single thyroid hormone injection repress CYP8b1 in face of cholesterol 7α-hydroxylase specific activities [6]; <2> variation in the ratio between cholic acid and chenodeoxycholic acid in human bile are not due to polymorphisms in gene CYP8B1, overview [17]) (Reversibility: ?) [1, 2, 3, 4, 5, 6, 9, 10, 12, 17]

P ?

Inhibitors

Additional information <1, 2, 3> (<1> cholate supresses enzyme expression [14]; <1> cholic acid suppresses expression of CYP8B1 [13]; <1> feeding of cholic acid suppresses expression of CYP8B1, the effect is partly reversed by taurine, feeding of taurine alone has no effect on enzyme expression [16]; <3> feeding of cholic acid, taurine, or cholesterol does not affect expression of CYP8B1 [16]; <2> thyroid hormone triiodothyronine, i.e. T3, supresses enzyme expression in vivo [18]) [13, 14, 16, 18]

Cofactors/prosthetic groups

cytochrome P_{450} <3, 5> [3,8]
NADPH <1, 2, 3, 5> [8,13,14,15,16,17,18]

Activating compounds

Additional information <1, 3> (<3> feeding of cholic acid, taurine, or cholesterol does not affect expression of CYP8B1 [16]; <1> feeding of taurine alone has no effect on enzyme expression [16]) [16]

5 Isolation/Preparation/Mutation/Application

Source/tissue

hepatocyte <2> (<2> primary [18]; <2> mainly expressed in [7]) [7, 18]
liver <1, 2, 3, 4, 5> (<2> fetal [11]; <4> enzyme is expressed in a development-dependent fashion, expression appears restricted to the early fetal stages [5]; <1> enzyme expression pattern within the liver architecture, zonal distribution pattern, overview [14]) [1, 3, 4, 5, 7, 8, 11, 13, 14, 15, 16, 17, 18]

Localization

cytoplasm <2> [7]
microsome <1, 2, 3, 5> [3, 8, 13, 14, 15, 16, 18]
mitochondrion <2> [11]

Cloning

<1> (gene CYP8B1, expression analysis in liver tissues of wild-type, regulation of gene expression, overview) [14]
<1> (gene CYP8B1, genotyping of different strains of mice, e.g. C57BL/6J and CASA/Rk mice, and a hybrid of both) [15]
<2> [10]
<3> [9]

<3> (expression in COS-7 cells) [3]
<4> [5]

Engineering

Additional information <1, 2> (<1> construction of CYP8B1 null mice which show increased expression of cholesterol 7α-hydroxylase, CYP7A1, EC 1.14.13.17 [14]; <1> construction of male CYP8B1 knockout mice, the mutant mice show increased cholesterol synthesis with normal diet, and decreased cholesterol absorption under cholesterol diet and feeding of cholic acid, they accumulate less cholesteryl esters compared to wild-type mice, phenotype, overview [13]; <2> screening for genetic polymorphisms of gene CYP8B1, overview [17]) [13, 14, 17]

References

[1] Jahan, A.; Chiang, J.Y.: Cytokine regulation of human sterol 12α-hydroxylase (CYP8B1) gene. Am. J. Physiol., **288**, G685-695 (2005)

[2] Yang, Y.; Eggertsen, G.; Gafvels, M.; Andersson, U.; Einarsson, C.; Bjorkhem, I.; Chiang, J.Y.: Mechanisms of cholesterol and sterol regulatory element binding protein regulation of the sterol 12α-hydroxylase gene (CYP8B1). Biochem. Biophys. Res. Commun., **320**, 1204-1210 (2004)

[3] Andersson, U.; Yang, Y.Z.; Bjorkhem, I.; Einarsson, C.; Eggertsen, G.; Gafvels, M.: Thyroid hormone suppresses hepatic sterol 12α-hydroxylase (CYP8B1) activity and messenger ribonucleic acid in rat liver: failure to define known thyroid hormone response elements in the gene. Biochim. Biophys. Acta, **1438**, 167-174 (1999)

[4] Yang, Y.; Zhang, M.; Eggertsen, G.; Chiang, J.Y.: On the mechanism of bile acid inhibition of rat sterol 12α-hydroxylase gene (CYP8B1) transcription: roles of α-fetoprotein transcription factor and hepatocyte nuclear factor 4α. Biochim. Biophys. Acta, **1583**, 63-73 (2002)

[5] Lundell, K.; Wikvall, K.: Gene structure of pig sterol 12α-hydroxylase (CYP8B1) and expression in fetal liver: comparison with expression of taurochenodeoxycholic acid 6α-hydroxylase (CYP4A21). Biochim. Biophys. Acta, **1634**, 86-96 (2003)

[6] Vlahcevic, Z.R.; Eggertsen, G.; Bjorkhem, I.; Hylemon, P.B.; Redford, K.; Pandak, W.M.: Regulation of sterol 12α-hydroxylase and cholic acid biosynthesis in the rat. Gastroenterology, **118**, 599-607 (2000)

[7] Wang, J.; Greene, S.; Eriksson, L.C.; Rozell, B.; Reihner, E.; Einarsson, C.; Eggertsen, G.; Gafvels, M.: Human sterol 12α-hydroxylase (CYP8B1) is mainly expressed in hepatocytes in a homogenous pattern. Histochem. Cell Biol., **123**, 441-446 (2005)

[8] Hansson, R.; Wikvall, K.: Hydroxylations in biosynthesis and metabolism of bile acids. Catalytic properties of different cytochrome P-450. J. Biol. Chem., **255**, 1643-1649 (1980)

[9] del Castillo-Olivares, A.; Gil, G.: α_1-Fetoprotein transcription factor is required for the expression of sterol 12α-hydroxylase, the specific enzyme for

cholic acid synthesis. Potential role in the bile acid-mediated regulation of gene transcription. J. Biol. Chem., **275**, 17793-17799 (2000)
[10] Zhang, M.; Chiang, J.Y.: Transcriptional regulation of the human sterol 12α-hydroxylase gene (CYP8B1): roles of heaptocyte nuclear factor 4α in mediating bile acid repression. J. Biol. Chem., **276**, 41690-41699 (2001)
[11] Gustafsson, J.: Bile acid synthesis during development. Mitochondrial 12α-hydroxylation in human fetal liver. J. Clin. Invest., **75**, 604-607 (1985)
[12] Inoue, Y.; Yu, A.M.; Yim, S.H.; Ma, X.; Krausz, K.W.; Inoue, J.; Xiang, C.C.; Brownstein, M.J.; Eggertsen, G.; Bjorkhem, I.; Gonzalez, F.J.: Regulation of bile acid biosynthesis by hepatocyte nuclear factor 4α. J. Lipid Res., **47**, 215-227 (2006)
[13] Murphy, C.; Parini, P.; Wang, J.; Bjoerkhem, I.; Eggertsen, G.; Gafvels, M.: Cholic acid as key regulator of cholesterol synthesis, intestinal absorption and hepatic storage in mice. Biochim. Biophys. Acta, **1735**, 167-175 (2005)
[14] Wang, J.; Olin, M.; Rozell, B.; Bjoerkhem, I.; Einarsson, C.; Eggertsen, G.; Gafvels, M.: Differential hepatocellular zonation pattern of cholesterol 7α-hydroxylase (Cyp7a1) and sterol 12α-hydroxylase (Cyp8b1) in the mouse. Histochem. Cell Biol., **127**, 253-261 (2007)
[15] Sehayek, E.; Hagey, L.R.; Fung, Y.Y.; Duncan, E.M.; Yu, H.J.; Eggertsen, G.; Bjoerkhem, I.; Hofmann, A.F.; Breslow, J.L.: Two loci on chromosome 9 control bile acid composition: evidence that a strong candidate gene, Cyp8b1, is not the culprit. J. Lipid Res., **47**, 2020-2027 (2006)
[16] Chen, W.; Suruga, K.; Nishimura, N.; Gouda, T.; Lam, V.N.; Yokogoshi, H.: Comparative regulation of major enzymes in the bile acid biosynthesis pathway by cholesterol, cholate and taurine in mice and rats. Life Sci., **77**, 746-757 (2005)
[17] Abrahamsson, A.; Gafvels, M.; Reihner, E.; Bjoerkhem, I.; Einarsson, C.; Eggertsen, G.: Polymorphism in the coding part of the sterol 12α-hydroxylase gene does not explain the marked differences in the ratio of cholic acid and chenodeoxycholic acid in human bile. Scand. J. Clin. Lab. Invest., **65**, 595-600 (2005)
[18] Ellis, E.C.: Suppression of bile acid synthesis by thyroid hormone in primary human hepatocytes. World J. Gastroenterol., **12**, 4640-4645 (2006)

Taurochenodeoxycholate 6α-hydroxylase 1.14.13.97

1 Nomenclature

EC number
1.14.13.97

Systematic name
taurochenodeoxycholate,NADPH:oxygen oxidoreductase (6α-hydroxylating)

Recommended name
taurochenodeoxycholate 6α-hydroxylase

Synonyms
CYP3A4 <1, 2> [1, 3]
CYP4A21 <2, 4> [2, 3, 5]
taurochenodeoxycholate 6α-monooxygenase <2> [3]
taurochenodeoxycholic acid 6α-hydroxylase <2> [3, 5]

CAS registry number
105669-85-0

2 Source Organism

<1> *Homo sapiens* (no sequence specified) [1]
<2> *Sus scrofa* (no sequence specified) [3, 5]
<3> *Oryctolagus cuniculus* (no sequence specified) [4]
<4> *Sus scrofa* (UNIPROT accession number: Q70BZ7) [2]

3 Reaction and Specificity

Catalyzed reaction
lithocholate + NADPH + H^+ + O_2 = hyocholate + $NADP^+$ + H_2O
lithocholate + NADPH + H^+ + O_2 = hyodeoxycholate + $NADP^+$ + H_2O
taurochenodeoxycholate + NADPH + H^+ + O_2 = taurohyocholate + $NADP^+$ + H_2O

Substrates and products
S lithocholate + NADPH + H^+ + O_2 <2> (<2> activity is low compared to activity with taurochenodeoxycholate [3]) (Reversibility: ?) [3]
P hyocholate + $NADP^+$ + H_2O
S taurochenodeoxycholate + NADPH + H^+ + O_2 <2> (Reversibility: ?) [3, 5]
P taurohyocholate + $NADP^+$ + H_2O

Cofactors/prosthetic groups

cytochrome P_{450} <2, 4> [2,3]

K_m-Value (mM)

0.0629 <2> (taurochenodeoxycholate) [3]

4 Enzyme Structure

Subunits

? <2> (<2> x * 53000, SDS-PAGE [3]) [3]

5 Isolation/Preparation/Mutation/Application

Source/tissue

liver <2, 4> [2, 3, 5]

Localization

microsome <2> [3]

Purification

<2> [3]

Cloning

<2> (transfection into COS cells) [5]
<4> (6α-hydroxylase (CYP4A21) gene: evolution by gene duplication and gene conversion) [2]

References

[1] Russell, D.W.: The enzymes, regulation, and genetics of bile acid synthesis. Annu. Rev. Biochem., **72**, 137-174 (2003)

[2] Lundell, K.: The porcine taurochenodeoxycholic acid 6α-hydroxylase (CYP4A21) gene: evolution by gene duplication and gene conversion. Biochem. J., **378**, 1053-1058 (2004)

[3] Araya, Z.; Hellman, U.; Hansson, R.: Characterisation of taurochenodeoxycholic acid 6α-hydroxylase from pig liver microsomes. Eur. J. Biochem., **231**, 855-861 (1995)

[4] Kramer, W.; Sauber, K.; Baringhaus, K.-H.; Kurz, M.; Stengelin, S.; Lange, G.; Corsiero, D.; Girbig F.; König, W.; Weyland, C.: Identification of the bile acid-binding site of the ileal lipid-binding protein by photoaffinity labeling, matrix-assisted laser desorption ionization-mass spectrometry, and NMR structure. J. Biol. Chem., **276**, 7291-7301 (2001)

[5] Lundell, K.; Hansson, R.; Wikvall, K.: Cloning and expression of a pig liver taurochenodeoxycholic acid 6α-hydroxylase (CYP4A21): a novel member of the CYP4A subfamily. J. Biol. Chem., **276**, 9606-9612 (2001)

Cholesterol 24-hydroxylase 1.14.13.98

1 Nomenclature

EC number

1.14.13.98

Systematic name

cholesterol,NADPH:oxygen oxidoreductase (24-hydroxylating)

Recommended name

cholesterol 24-hydroxylase

Synonyms

CH24H <2> [8]
CYP46 <2, 5> [1, 3, 7]
CYP46A1 <1, 2, 4> [1, 5, 6, 10, 11]
cholesterol 24-monooxygenase <2> [1]
cholesterol 24S-hydroxylase <2> [1]
cytochrome P-450 46A1 <2> [1]
cytochrome P_{450} 46A1 <2> [2]

CAS registry number

213327-78-7
50812-30-1

2 Source Organism

<1> *Mus musculus* (no sequence specified) [4, 10, 13]
<2> *Homo sapiens* (no sequence specified) [1, 2, 5, 6, 7, 8, 10, 11]
<3> *Rattus norvegicus* (no sequence specified) [12]
<4> *Sus scrofa* (no sequence specified) [10]
<5> *Homo sapiens* (UNIPROT accession number: Q9Y6A2) [3,9]
<6> *Mus musculus* (UNIPROT accession number: Q9WVK8) [3,9]

3 Reaction and Specificity

Catalyzed reaction

cholesterol + NADPH + H^+ + O_2 = (24S)-24-hydroxycholesterol + $NADP^+$ + H_2O (<2> structure-function relationship [11])

Reaction type

oxidation
redox reaction
reduction

Natural substrates and products

S cholesterol + NADPH + H^+ + O_2 <1, 2, 5, 6> (<5,6> 24-hydroxylation represents an important pathway by which cholesterol is secreted from the brain. The enzyme contributes little to overall bile acid synthesis but is important in the turnover of cholesterol in the brain [3]; <1> cholesterol 24-hydroxylase constitutes a major tissue-specific pathway for cholesterol turnover in the brain [4]; <2> initiates cholesterol degradation in the brain. In addition to the involvement in cholesterol homeostasis in the brain, this enzyme may participate in metabolism of neurosteroids and drugs that can cross the blood-brain barrier and are targeted to the central nervous system [2]) (Reversibility: ?) [2, 3, 4]

P (24S)-24-hydroxycholesterol + $NADP^+$ + H_2O

S cholesterol + NADPH + O_2 <1, 3> (<1> enzyme deficiency leads to a suppression of the mevalonate pathway and a defect in learning [13]) (Reversibility: ?) [12, 13]

P 24-hydroxycholesterol + $NADP^+$ + H_2O

S cholesterol + NADPH + O_2 <1, 2, 4> (<1> the enzyme is important in cholesterol metabolism in the brain, cholesterol dynamics in the fetal and neonatal brain, overview [10]; <2,4> the enzyme is important in cholesterol metabolism in the brain, overview [10]; <2> the enzyme is responsible for cholesterol elimination from brain, (24S)-hydroxycholesterol can cross the blood-brain barrier, enter the circulation, and then be delivered to the liver for further degradation, cholesterol metabolism, overview, CYP46A1 may participate in metabolism of neurosteroids and drugs that are targeted to the central nervous system [11]) (Reversibility: ?) [10, 11]

P (24S)-hydroxycholesterol + $NADP^+$ + H_2O (<1,2,4> the product is able to travers the blood-brain barrier, while the substrate is not, transport, and excretion of (24S)-hydroxycholesterol, oxycholesterol transport mechanisms, overview [10])

S Additional information <2, 5, 6> (<2> CYP46A1 affects the pathophysiology of AD and provides insight into how polymorphisms in the CYP46A1 gene might influence the pathophysiology of this prevalent disease [5]; <2> regulation of cholesterol synthesis is more important for maintenance of cholesterol homeostasis in brain than is the corresponding regulation of cholesterol removal [6]; <5,6> the enzyme is a mediator of cholesterol homeostasis in the brain [9]; <2> (24S)-hydroxycholesterol is a potent activator of the LXR receptor, CYP46A1-deficiency plays a role in the pathogenesis of this neurological disorder, such as Alzheimers disease [11]; <2> cholesterol metabolism is important in Alzheimers disease pathogenesis, the enzyme is a risk factor for Alzheimers disease, along with cholesterol 25-hydroxylase and ATP-binding cassette transporter A1, overview [8]) (Reversibility: ?) [5, 6, 8, 9, 11]

P ?

Substrates and products

S (24S)-24-hydroxycholesterol + NADPH + O_2 <2> (Reversibility: ?) [1, 2]

P 24,25-dihydroxycholesterol + $NADP^+$ + H_2O

S (24S)-24-hydroxycholesterol + NADPH + O_2 <2> (Reversibility: ?) [1, 2]

P 24,27-dihydroxycholesterol + $NADP^+$ + H_2O

S (24S)-hydroxycholesterol + NADPH + O_2 <2> (<2> preferred substrate in vitro [11]) (Reversibility: ?) [11]

P 24,25-dihydroxycholesterol + 24,27-dihydroxycholesterol + $NADP^+$ + H_2O

S cholesterol + NADPH + H^+ + O_2 <1, 2, 5, 6> (<5,6> 24-hydroxylation represents an important pathway by which cholesterol is secreted from the brain. The enzyme contributes little to overall bile acid synthesis but is important in the turnover of cholesterol in the brain [3]; <1> cholesterol 24-hydroxylase constitutes a major tissue-specific pathway for cholesterol turnover in the brain [4]; <2> initiates cholesterol degradation in the brain. In addition to the involvement in cholesterol homeostasis in the brain, this enzyme may participate in metabolism of neurosteroids and drugs that can cross the blood-brain barrier and are targeted to the central nervous system [2]) (Reversibility: ?) [1, 2, 3, 4]

P (24S)-24-hydroxycholesterol + $NADP^+$ + H_2O

S cholesterol + NADPH + O_2 <1, 3> (<1> enzyme deficiency leads to a suppression of the mevalonate pathway and a defect in learning [13]) (Reversibility: ?) [12, 13]

P 24-hydroxycholesterol + $NADP^+$ + H_2O

S cholesterol + NADPH + O_2 <1, 2, 4> (<1> the enzyme is important in cholesterol metabolism in the brain, cholesterol dynamics in the fetal and neonatal brain, overview [10]; <2,4> the enzyme is important in cholesterol metabolism in the brain, overview [10]; <2> the enzyme is responsible for cholesterol elimination from brain, (24S)-hydroxycholesterol can cross the blood-brain barrier, enter the circulation, and then be delivered to the liver for further degradation, cholesterol metabolism, overview, CYP46A1 may participate in metabolism of neurosteroids and drugs that are targeted to the central nervous system [11]) (Reversibility: ?) [10, 11]

P (24S)-hydroxycholesterol + $NADP^+$ + H_2O (<1,2,4> the product is able to travers the blood-brain barrier, while the substrate is not, transport, and excretion of (24S)-hydroxycholesterol, oxycholesterol transport mechanisms, overview [10])

S dextromethorphan + NADPH + O_2 <2> (<2> O- and N-demethylation [11]) (Reversibility: ?) [11]

P ?

S diclofenac + NADPH + O_2 <2> (<2> 4-hydroxylation [11]) (Reversibility: ?) [11]

P ?

S phenacetin + NADPH + O_2 <2> (<2> O-deethylation [11]) (Reversibility: ?) [11]

P ?

S Additional information <2, 5, 6> (<2> CYP46A1 affects the pathophysiology of Alzheimer's disease and provides insight into how polymorphisms in the CYP46A1 gene might influence the pathophysiology of this prevalent disease [5]; <2> regulation of cholesterol synthesis is more important for maintenance of cholesterol homeostasis in brain than is the corresponding regulation of cholesterol removal [6]; <5,6> the enzyme is a mediator of cholesterol homeostasis in the brain [9]; <2> the enzyme is able to carry out side chain hydroxylations of two endogenous C27-steroids with and without a double bond between C5-C6 (7R-hydroxycholesterol and cholestanol, respectively) and introduce a hydroxyl group on the steroid nucleus of the C21-steroid hormones with the C4-C5 double bond (progesterone and testosterone). P450 46A1 metabolizes xenobiotics carrying out dextromethorphan O-demethylation and N-demethylations, diclofenac 4-hydroxylation, and phenacetin O-deethylation [2]; <2> (24S)-hydroxycholesterol is a potent activator of the LXR receptor, CYP46A1-deficiency plays a role in the pathogenesis of this neurological disorder, such as Alzheimers disease [11]; <2> cholesterol metabolism is important in Alzheimers disease pathogenesis, the enzyme is a risk factor for Alzheimers disease, along with cholesterol 25-hydroxylase and ATP-binding cassette transporter A1, overview [8]; <2> CYP46A1 shows broad substrate specificity [11]) (Reversibility: ?) [2, 5, 6, 8, 9, 11]

P ?

Cofactors/prosthetic groups

Cytochrome P-450 <2, 5, 6> [1,3,9]
NADPH <1, 2, 3, 4, 5, 6> [1,2,3,4,10,11,12,13]

Activating compounds

lovastatin <3> (<3> induces and activates the enzyme, acts neuroprotectively in hippocampus after kainate injury, overview [12]) [12]

Turnover number (min^{-1})

0.00117 <2> ((24S)-24-hydroxycholesterol, <2> full length enzyme [1]) [1]
0.11 <2> (Cholesterol) [11]
Additional information <2> (<2> turnover-numbers of truncated enzyme forms [1]) [1]

K_m-Value (mM)

0.0022 <2> ((24S)-24-hydroxycholesterol, <2> full length enzyme [1]) [1]
0.0053 <2> (Cholesterol) [11]
Additional information <2> (<2> K_m-value of truncated enzyme forms [1]) [1]

4 Enzyme Structure

Subunits

? <5, 6> (<5,6> x * 56821, calculation from nucleotide sequence [3]; <5> x * 56842 [3]) [3]
monomer <2> (<2> 1 * 49000, truncated enzyme form, SDS-PAGE [1]) [1]
Additional information <2> (<2> structure-function relationship [11]) [11]

5 Isolation/Preparation/Mutation/Application

Source/tissue

astrocyte <2> (<2> in normal brain [5]) [5]
brain <1, 2, 4, 5, 6> (<4> low activity [10]; <6> predominantly expressed in [9]; <2> specific [11]; <2> abnormal induction of the enzyme in patients with Alzheimer's disease [7]; <2> in neurons and some astrocytes in the normal brain, in Alzheimers disease the enzyme shows prominent expression in astrocytes and around amyloid plaques [5]; <5> predominantly expressed in. The 24-hydroxylase protein is first detected in the brain of mice at birth and continues to accumulate with age [9]; <2> cholesterol 24-hydroxylase is expressed predominantly in the brain [10]) [2, 3, 4, 5, 7, 9, 10, 11, 13]
glial cell <2> (<2> abnormal induction of the enzyme in patients with Alzheimer's disease [7]) [7]
hippocampus <1, 3> (<3> after kainate lesions [12]) [12, 13]
liver <1, 2, 4, 5> [3, 10]
neuron <1, 2, 3, 4> (<2> in normal brain [5]; <3> hippocampal [12]) [5, 10, 12]

Localization

cytosol <2> (<2> truncated enzymes are distributed at about a 1:1 ratio between the membrane fraction and the cytosol in low ionic strength buffer, when expressed in Escherichia coli [1]) [1]
endoplasmic reticulum <1, 2, 5, 6> [3, 9, 10, 11]
membrane <2> (<2> truncated enzymes are distributed at about a 1:1 ratio between the membrane fraction and the cytosol in low ionic strength buffer, when expressed in Escherichia coli [1]) [1]
microsome <4> (<4> of brain [10]) [10]
mitochondrion <4> (<4> of liver [10]) [10]

Purification

<2> (partial) [2]
<2> (truncated enzyme forms) [1]

Cloning

<1> (DNA sequence determination and analysis) [10]
<2> (5'-upstream region of human CYP46 gene) [6]

<2> (DNA and amino acid sequence determnination and analysis, determination of single nucleotide polymorphisms in North American Caucasians and Caribbean Hispanic Alzheimer patients, overview) [8]
<2> (DNA sequence determination and analysis, located on chromosome14q32.1) [10]
<2> (expression Escherichia coli, HEK293 cells transfected with CYP46A1) [2]
<2> (expression in Escherichia coli) [11]
<2> (expression in Escherichia coli, wild-type and trucation mutatants. All four mutants lack the N-terminal transmembrane region (residues 3-27), and, in addition, Δ46A1 has a 4 His-tag fused to the C-terminus, HΔ46A1 has the N-terminal 4 His-tag, HΔ46A1Δ has a 4 His-tag at the N-terminus and does not contain a proline-rich region at the C-terminus (residues 494-499), and Δ46A1Δ lacks the C-terminal proline-rich region) [1]
<5> [3, 9]
<6> [9]

Engineering

Additional information <1, 2> (<2> four truncation mutants: all lack the N-terminal transmembrane region (residues 3 -27), and, in addition, Δ46A1 has a 4 His-tag fused to the C-terminus, HΔ46A1 has the N-terminal 4 His-tag, HΔ46A1Δ has a 4 His-tag at the N-terminus and does not contain a proline-rich region at the C-terminus (residues 494-499), and Δ46A1Δ lacks the C-terminal proline-rich region. Truncated enzymes have moderately decreased catalytic efficiencies for either cholesterol or 24S-hydroxycholesterol or both, whereas their substrate-binding constants are either unchanged or decreased 2fold.The two forms, Δ64A1Δ and HΔ46A1Δ both lacking the C-terminal proline-rich region are good candidates for future crystallographic studies because they contain only 0.3 -0.8% of high molecular weight aggregates and their catalytic efficiencies are decreased no more than 2.3fold [1]; <1> brains from mice lacking 24-hydroxylase excrete cholesterol more slowly, and the tissue compensates by suppressing the mevalonate pathway, this suppression causes a defect in learning, 24-hydroxylase knockout mice exhibit severe deficiencies in spatial, associative, and motor learning, and in hippocampal long-term potentiation, the effects of genetic elimination of 24-hydroxylase on long-term potentiation are reversed by a 20-min treatment with geranylgeraniol but not by cholesterol, phenotype, overview [13]; <1> construction of knockout mice, concentration and the pool of cholesterol in the CNS is unchanged compared with control animals, but synthesis is suppressed by about 25% in the CYP46a1-deficient mice, but not in those lacking 7α- or 27-hydroxylase activity, the concentrations of 24S-hydroxycholesterol in the brain and plasma decline to very low levels, the CNS in the mouse apparently responds appropriately to loss of this excretory pathway by suppressing endogenous synthesis by an amount exactly equal to the mass of cholesterol that would normally be excreted as 24S-hydroxycholesterol [10]; <2> determination of single nucleotide polymorphisms in North American Caucasians and Caribbean Hispanic Alzheimer patients, overview [8]) [1, 8, 10, 13]

Application

medicine <2> (<2> CYP46A1 affects the pathophysiology of AD and provides insight into how polymorphisms in the CYP46A1 gene might influence the pathophysiology of this prevalent disease [5]) [5]

References

[1] Mast, N.; Andersson, U.; Nakayama, K.; Bjorkhem, I.; Pikuleva, I.A.: Expression of human cytochrome P450 46A1 in Escherichia coli: effects of N- and C-terminal modifications. Arch. Biochem. Biophys., **428**, 99-108 (2004)

[2] Mast, N.; Norcross, R.; Andersson, U.; Shou, M.; Nakayama, K.; Bjorkhem, I.; Pikuleva, I.A.: Broad substrate specificity of human cytochrome P450 46A1 which initiates cholesterol degradation in the brain. Biochemistry, **42**, 14284-14292 (2003)

[3] Russell, D.W.: Oxysterol biosynthetic enzymes. Biochim. Biophys. Acta, **1529**, 126-135 (2000)

[4] Lund, E.G.; Xie, C.; Kotti, T.; Turley, S.D.; Dietschy, J.M.; Russell, D.W.: Knockout of the cholesterol 24-hydroxylase gene in mice reveals a brain-specific mechanism of cholesterol turnover. J. Biol. Chem., **278**, 22980-22988 (2003)

[5] Brown, J., 3rd; Theisler, C.; Silberman, S.; Magnuson, D.; Gottardi-Littell, N.; Lee, J.M.; Yager, D.; Crowley, J.; Sambamurti, K.; Rahman, M.M.; Reiss, A.B.; Eckman, C.B.; Wolozin, B.: Differential expression of cholesterol hydroxylases in Alzheimer's disease. J. Biol. Chem., **279**, 34674-34681 (2004)

[6] Ohyama, Y.; Meaney, S.; Heverin, M.; Ekstrom, L.; Brafman, A.; Shafir, M.; Andersson, U.; Olin, M.; Eggertsen, G.; Diczfalusy, U.; Feinstein, E.; Bjorkhem, I.: Studies on the transcriptional regulation of cholesterol 24-hydroxylase (CYP46A1): Marked insensitivity towards different regulatory axes. J. Biol. Chem., **281**, 3810-3820 (2006)

[7] Bogdanovic, N.; Bretillon, L.; Lund, E.G.; Diczfalusy, U.; Lannfelt, L.; Winblad, B.; Russell, D.W.; Bjorkhem, I.: On the turnover of brain cholesterol in patients with Alzheimer's disease. Abnormal induction of the cholesterol-catabolic enzyme CYP46 in glial cells. Neurosci. Lett., **314**, 45-48 (2001)

[8] Shibata, N.; Kawarai, T.; Lee, J.H.; Lee, H.S.; Shibata, E.; Sato, C.; Liang, Y.; Duara, R.; Mayeux, R.P.; St George-Hyslop, P.H.; Rogaeva, E.: Association studies of cholesterol metabolism genes (CH25H, ABCA1 and CH24H) in Alzheimer's disease. Neurosci. Lett., **391**, 142-146 (2006)

[9] Lund, E.G.; Guileyardo, J.M.; Russell, D.W.: cDNA cloning of cholesterol 24-hydroxylase, a mediator of cholesterol homeostasis in the brain. Proc. Natl. Acad. Sci. USA, **96**, 7238-7243 (1999)

[10] Luetjohann, D.: Cholesterol metabolism in the brain: importance of 24S-hydroxylation. Acta Neurol. Scand., **114 (Suppl. 185)**, 33-42 (2006)

[11] Pikuleva, I.A.: Cholesterol-metabolizing cytochromes P_{450}. Drug Metab. Dispos., **34**, 513-520 (2006)

[12] He, X.; Jenner, A.M.; Ong, W.Y.; Farooqui, A.A.; Patel, S.C.: Lovastatin modulates increased cholesterol and oxysterol levels and has a neuroprotective

effect on rat hippocampal neurons after kainate injury. J. Neuropathol. Exp. Neurol., **65**, 652-663 (2006)

[13] Kotti, T.J.; Ramirez, D.M.; Pfeiffer, B.E.; Huber, K.M.; Russell, D.W.: Brain cholesterol turnover required for geranylgeraniol production and learning in mice. Proc. Natl. Acad. Sci. USA, **103**, 3869-3874 (2006)

24-Hydroxycholesterol 7α-hydroxylase 1.14.13.99

1 Nomenclature

EC number
1.14.13.99

Systematic name
(24R)-cholest-5-ene-3β,24-diol,NADPH:oxygen oxidoreductase (7α-hydroxylating)

Recommended name
24-hydroxycholesterol 7α-hydroxylase

Synonyms
24-hydroxycholesterol 7α-monooxygenase <1, 3> [1]
CYP39A1 oxysterol 7α-hydroxylase <1, 3> [1]
Cyp39a1 <1, 2, 3> [1, 2]
oxysterol 7α-hydroxylase CYP39A1 <2> [2]

CAS registry number
149316-80-3
288309-90-0

2 Source Organism

<1> *Homo sapiens* (no sequence specified) [1]
<2> *Bos taurus* (no sequence specified) [2]
<3> *Mus musculus* (UNIPROT accession number: Q9JKJ9) [1]

3 Reaction and Specificity

Catalyzed reaction
(24R)-cholest-5-ene-3β,24-diol + NADPH + H^+ + O_2 = (24R)-cholest-5-ene-3β,7α,24-triol + $NADP^+$ + H_2O

Reaction type
oxidation
redox reaction
reduction

Substrates and products

S (24R)-cholest-5-ene-3β,24-diol + NADPH + H^+ + O_2 <1> (<1> activity is largely restricted to 24-hydroxycholesterol [1]) (Reversibility: ?) [1]

P (24R)-cholest-5-ene-3β,7α,24-triol + $NADP^+$ + H_2O

S 24-cholest-5-ene-3β,24-diol + NADPH + H^+ + O_2 <2, 3> (<2,3> activity is largely restricted to 24-hydroxycholesterol [1,2]) (Reversibility: ?) [1, 2]

P (24R)-cholest-5-ene-3β,7α,24-triol + $NADP^+$ + H_2O

Cofactors/prosthetic groups

NADPH <2> [2]

4 Enzyme Structure

Subunits

? <2> (<2> x * 44000, SDS-PAGE [2]) [2]

5 Isolation/Preparation/Mutation/Application

Source/tissue

epithelium <2> (<2> ciliary nonpigmented of eye [2]) [2]
eye <2> (<2> ciliary nonpigmented epithelium [2]) [2]
liver <1, 3> (<3> sexually dimorphic expression pattern, higher expression in female than in male [1]) [1]

Cloning

<1> [1]
<3> [1]

References

[1] Li-Hawkins, J.; Lund, E.G.; Bronson, A.D.; Russell, D.W.: Expression cloning of an oxysterol 7α-hydroxylase selective for 24-hydroxycholesterol. J. Biol. Chem., **275**, 16543-16549 (2000)

[2] Ikeda, H.; Ueda, M.; Ikeda, M.; Kobayashi, H.; Honda, Y.: Oxysterol 7α-hydroxylase (CYP39A1) in the ciliary nonpigmented epithelium of bovine eye. Lab. Invest., **83**, 349-355 (2003)

25-Hydroxycholesterol 7α-hydroxylase 1.14.13.100

1 Nomenclature

EC number
1.14.13.100

Systematic name
cholest-5-ene-3β,25-diol,NADPH:oxygen oxidoreductase (7α-hydroxylating)

Recommended name
25-hydroxycholesterol 7α-hydroxylase

Synonyms
25-hydroxycholesterol 7α-monooxygenase <4> [4]
CYP7B1 <1, 2, 3, 4> [1, 3, 4, 5, 7, 8, 9]
CYP7B1 oxysterol 7α-hydroxylase <4> [4]
CYPB1 <3> [10]

CAS registry number
149316-80-3
440356-82-1

2 Source Organism

<1> *Mus musculus* (no sequence specified) [1, 7, 8]
<2> *Homo sapiens* (no sequence specified) [4, 5, 8, 9]
<3> *Rattus norvegicus* (no sequence specified) [3, 4, 6, 10]
<4> *Sus scrofa* (no sequence specified) [4]
<5> *Homo sapiens* (UNIPROT accession number: O75881) [2]
<6> *Mus musculus* (UNIPROT accession number: Q60991) [2]

3 Reaction and Specificity

Catalyzed reaction
cholest-5-ene-3β,25-diol + NADPH + H^+ + O_2 = cholest-5-ene-3β,7α,25-triol + $NADP^+$ + H_2O
cholest-5-ene-3β,27-diol + NADPH + H^+ + O_2 = cholest-5-ene-3β,7α,27-triol + $NADP^+$ + H_2O

Reaction type
oxidation
redox reaction
reduction

Natural substrates and products

S cholest-5-ene-3β,25-diol + NADPH + H^+ + O_2 <1, 6> (<6> CYP7B1 oxysterol 7α-hydroxylase pathway synthesizes 25% to 30% of all bile acids [2]; <1> dietary cholesterol or colestipol do not affect oxysterol 7α-hydroxylase enzyme activity, mRNA, or protein levels [1]; <1> loss of the enzyme in liver is compensated for by increases in the synthesis of bile acids by other pathways [7]) (Reversibility: ?) [1, 2, 7]
P cholest-5-ene-3β,7α,25-triol + $NADP^+$ + H_2O
S cholest-5-ene-3β,27-diol + NADPH + H^+ + O_2 <1, 6> (<6> CYP7B1 oxysterol 7α-hydroxylase pathway synthesizes 25% to 30% of all bile acids [2]; <1> loss of the enzyme in liver is compensated for by increases in the synthesis of bile acids by other pathways [7]) (Reversibility: ?) [2, 7]
P cholest-5-ene-3β,7α,27-triol + $NADP^+$ + H_2O
S Additional information <2, 3> (<3> both phorbol myristate acetate and cAMP decreases CYP7B1 activity by 60% and 34% respectively, in a time-dependent fashion. Changes in CYP7B1 messenger RNA levels correlate with changes in specific activities. CYP7B1 specific activity is highly regulated but does not seem to be rate limiting for bile acid synthesis [6]; <2> regulation of CYP7B1 transcription by Sp1 may play a pivotal role in regulating oxysterol levels, which regulate cholesterol metabolism [5]; <3> the enzyme is regulated by bile acids and cholesterol [10]) (Reversibility: ?) [5, 6, 10]
P ?

Substrates and products

S 17β-estradiol + NADPH + H^+ + O_2 <3> (Reversibility: ?) [3]
P 1,3,5(10)-estratrien-3,7α-17β-triol + $NADP^+$ + H_2O
S 22-hydroxycholesterol + NADPH + H^+ + O_2 <1> (Reversibility: ?) [8]
P 7α-22-dihydroxycholesterol + $NADP^+$ + H_2O
S 24,25-epoxycholesterol + NADPH + H^+ + O_2 <1> (Reversibility: ?) [8]
P 7α-hydroxy-24,25-epoxycholesterol + $NADP^+$ + H_2O
S 24-hydroxycholesterol + NADPH + H^+ + O_2 <1> (Reversibility: ?) [8]
P 7α-24-dihydroxycholesterol + NADPH + H_2O
S 4-androstene-3,17-dione + NADPH + H^+ + O_2 <3> (Reversibility: ?) [3]
P 7α-hydroxy-4-androstene-3,17-dione + $NADP^+$ + H_2O
S 5α-androstane-3β,17β-diol + NADPH + H^+ + O_2 <3> (Reversibility: ?) [3]
P 5α-androstane-3β,7α,17β-triol + $NADP^+$ + H_2O
S cholest-5-ene-3β,25-diol + NADPH + H^+ + O_2 <1, 2, 3, 4, 5, 6> (<6> CYP7B1 oxysterol 7α-hydroxylase pathway synthesizes 25% to 30% of all bile acids [2]; <1> dietary cholesterol or colestipol do not affect oxysterol 7α-hydroxylase enzyme activity, mRNA, or protein levels [1]; <1> loss of the enzyme in liver is compensated for by increases in the synthesis of bile acids by other pathways [7]; <1> substrates in descending order of activity: cholest-5-ene-3β,25-diol, cholest-5-ene-3β,27-diol, 24,25-epoxycholesterol, 22-hydroxycholesterol, 24-hydroxycholesterol [8]) (Reversibility: ?) [1, 2, 3, 4, 7, 8]
P cholest-5-ene-3β,7α,25-triol + $NADP^+$ + H_2O

S cholest-5-ene-3β,27-diol + NADPH + H^+ + O_2 <1, 5, 6> (<6> CYP7B1 oxysterol 7α-hydroxylase pathway synthesizes 25% to 30% of all bile acids [2]; <1> loss of the enzyme in liver is compensated for by increases in the synthesis of bile acids by other pathways [7]) (Reversibility: ?) [2, 7, 8]
P cholest-5-ene-3β,7α,27-triol + $NADP^+$ + H_2O
S dehydroepiandrosterone + NADPH + H^+ + O_2 <3> (Reversibility: ?) [3]
P 3β,7α-dihydroxy-5-androsten-17-one + $NADP^+$ + H_2O
S Additional information <2, 3> (<3> both phorbol myristate acetate and cAMP decreases CYP7B1 activity by 60% and 34% respectively, in a time-dependent fashion. Changes in CYP7B1 messenger RNA levels correlate with changes in specific activities. CYP7B1 specific activity is highly regulated but does not seem to be rate limiting for bile acid synthesis [6]; <2> regulation of CYP7B1 transcription by Sp1 may play a pivotal role in regulating oxysterol levels, which regulate cholesterol metabolism [5]; <3> the enzyme is regulated by bile acids and cholesterol [10]) (Reversibility: ?) [5, 6, 10]
P ?

Cofactors/prosthetic groups

cytochrome P_{450} <1, 3, 5, 6> [1,2,3]
NADPH <1, 2, 3, 4, 5, 6> [1,2,3,4,7]

K_m-Value (mM)

0.001 <3> (cholest-5-ene-3β,25-diol, <3> pH 7.4, 37°C [3]) [3]
0.004 <3> (4-androstene-3,17-dione, <3> pH 7.4, 37°C [3]) [3]
0.011 <3> (17β-estradiol, <3> pH 7.4, 37°C [3]) [3]
0.014 <3> (dehydroepiandrosterone, <3> pH 7.4, 37°C [3]) [3]
0.022 <3> (5α-androstane-3β,17β-diol, <3> pH 7.4, 37°C [3]) [3]

4 Enzyme Structure

Subunits

? <5, 6> (<5,6> x * 58255, calculation from nucleotide sequence [2]) [2]

5 Isolation/Preparation/Mutation/Application

Source/tissue

brain <2> [9]
colon <2> [9]
epithelial cell <3> (<3> of prostate [3]) [3]
hepatocyte <3> [6]
kidney <2> [9]
liver <1, 2, 3, 4, 6> (<3> CYP7B1 shows diurnal variation [10]; <1> enzyme is induced in the third week of life [1]; <6> hepatic expression is induced during the third week of life and thereafter exhibits a sexually dimorphic ex-

pression pattern in this and other tissues in which expression is higher in the male [2]; <1,2> higher activity in male than in female [8]) [1, 2, 4, 7, 8, 9, 10]
ovary <2> [9]
prostate <2, 3> (<3> epithelial cell [3]) [3, 9]
small intestine <2> [9]
testis <2> [9]

Localization
endoplasmic reticulum <5, 6> [2]
microsome <2, 3, 4> [4]

Cloning
<1> [8]
<2> [8]
<2> (14 deletion mutants of the regulatory sequences of CYP7B1/Luc reporter gene are constructed for transfection assays. The constructs are transiently transfected into HepG2, NT1088, and HEK 293 cells) [5]
<2> (expression in 293/T cells) [9]

References

[1] Schwarz, M.; Lund, E.G.; Lathe, R.; Björkhem, I.; Russell, D.W.: Identification and characterization of a mouse oxysterol 7α-hydroxylase cDNA. J. Biol. Chem., **272**, 23995-24001 (1997)

[2] Russell, D.W.: The enzymes, regulation, and genetics of bile acid synthesis. Annu. Rev. Biochem., **72**, 137-174 (2003)

[3] Martin, C.; Bean, R.; Rose, K.; Habib, F.; Seckl, J.: cyp7b1 catalyses the 7α-hydroxylation of dehydroepiandrosterone and 25-hydroxycholesterol in rat prostate. Biochem. J., **355**, 509-515 (2001)

[4] Toll, A.; Wikvall, K.; Sudjana-Sugiaman, E.; Kondo, K.-H.; Björhem, I.: 7α-Hydroxylation of 25-hydroxycholesterol in liver microsomes. Evidence that the enzyme involved is different from cholesterol 7α-hydoxylase. Eur. J. Biochem., **224**, 309-316 (1994)

[5] Wu, Z.; Chiang, J.Y.: Transcriptional regulation of human oxysterol 7α-hydroxylase gene (CYP7B1) by Sp1. Gene, **272**, 191-197 (2001)

[6] Pandak William, M.; Hylemon Phillip, B.; Ren, S.; Marques, D.; Gil, G.; Redford, K.; Mallonee, D.; Vlahcevic, Z.R.: Regulation of oxysterol 7α-hydroxylase (CYP7B1) in primary cultures of rat hepatocytes. Hepatology, **35**, 1400-1408 (2002)

[7] Li-Hawkins, J.; Lund, E.G.; Turley, S.D.; Russell, D.W.: Disruption of the oxysterol 7α-hydroxylase gene in mice. J. Biol. Chem., **275**, 16536-16542 (2000)

[8] Li-Hawkins, J.; Lund, E.G.; Bronson, A.D.; Russell, D.W.: Expression cloning of an oxysterol 7α-hydroxylase selective for 24-hydroxycholesterol. J. Biol. Chem., **275**, 16543-16549 (2000)

[9] Wu, Z.; Martin, K.O.; Javitt, N.B.; Chiang, J.Y.: Structure and functions of human oxysterol 7α-hydroxylase cDNAs and gene CYP7B1. J. Lipid Res., **40**, 2195-2203 (1999)
[10] Ren, S.; Marques, D.; Redford, K.; Hylemon, P.B.; Gil, G.; Vlahcevic, Z.R.; Pandak, W.M.: Regulation of oxysterol 7α-hydroxylase (CYP7B1) in the rat. Metabolism, **52**, 636-642 (2003)

Senecionine N-oxygenase 1.14.13.101

1 Nomenclature

EC number
1.14.13.101

Systematic name
senecionine,NADPH:oxygen oxidoreductase (N-oxide-forming)

Recommended name
senecionine N-oxygenase

Synonyms
SNO <1> [2]
oxygenase, senecionine N- <1> [2]
senecionine N-oxygenase <1> [2]
senecionine monooxygenase (N-oxide-forming)

CAS registry number
220591-70-4

2 Source Organism

<1> *Tyria jacobaeae* (UNIPROT accession number: Q8MP06) [2]
<2> no activity in *Spodoptera littoralis* [1]
<3> *Tyria jacobaeae* (no sequence specified) [1]
<4> *Creatonotos transiens* (no sequence specified) [1]
<5> *Arctia caja* (no sequence specified) [1]
<6> *Callimorpha dominula* (no sequence specified) [1]
<7> *Arctia intercalaris* (no sequence specified) [1]
<8> *Amerila phaedra* (no sequence specified) [1]
<9> *Idea leuconoe* (no sequence specified) [1]
<10> *Zonocerus variegatus* (no sequence specified) [1]

3 Reaction and Specificity

Catalyzed reaction
senecionine + NADPH + H^+ + O_2 = senecionine N-oxide + $NADP^+$ + H_2O

Reaction type
N-oxidation
oxidation

redox reaction
reduction

Natural substrates and products

S Additional information <1> (<1> the enzyme allows the larvae to feed on pyrrolizidine alkaloid-containing plants and to accumulate predation-deterrent pyrrolizidine alkaloids in the hemolymph [2]) (Reversibility: ?) [2]

P ? <1> [2]

Substrates and products

S axillaridine + NADPH + O_2 <3, 4, 5> (<3> 76% of the activity with senecionine [1]; <5> 79% of the activity with senecionine [1]; <4> 93% of the activity with senecionine [1]) (Reversibility: ?) [1]

P axillaridine N-oxide + $NADP^+$ + H_2O <3, 4, 5> [1]

S axillarine + NADPH + O_2 <1, 3, 4, 5> (<4> 38% of the activity with senecionine [1]; <5> 55% of the activity with senecionine [1]; <3> 74% of the activity with senecionine [1]; <1> 74% of the activity with senecionine, native enzyme. 83% of the activity with senecionine, recombinant enzyme [2]) (Reversibility: ?) [1, 2]

P axillarine N-oxide + $NADP^+$ + H_2O <1, 3, 4, 5> [1, 2]

S heliotrine + NADPH + O_2 <1, 3, 4, 5> (<4> 115% of the activity with senecionine [1]; <3> 25% of the activity with senecionine [1]; <1> 25% of the activity with senecionine, native enzyme. 49% of the activity with senecionine, recombinant enzyme [2]; <5> 86% of the activity with senecionine [1]) (Reversibility: ?) [1, 2]

P heliotrine N-oxide + $NADP^+$ + H_2O <1, 3, 4, 5> [1, 2]

S indicine + NADPH + O_2 <3, 4, 5> (<3> 35% of the activity with senecionine [1]; <4> 78% of the activity with senecionine [1]; <5> 90% of the activity with senecionine [1]) (Reversibility: ?) [1]

P indicine N-oxide + $NADP^+$ + H_2O <3, 4, 5> [1]

S lycopsamine + NADPH + O_2 <3, 4, 5> (<3> 20% of the activity with senecionine [1]; <4> 75% of the activity with senecionine [1]; <5> 93% of the activity with senecionine [1]) (Reversibility: ?) [1]

P lycopsamine N-oxide + $NADP^+$ + H_2O <3, 4, 5> [1]

S monocrotaline + NADPH + O_2 <1, 3, 4, 5> (<5> 103% of the activity with senecionine [1]; <4> 110% of the activity with senecionine [1]; <3> 92% of the activity with senecionine [1]; <1> 92% of the activity with senecionine, native enzyme. 94% of the activity with senecionine, recombinant enzyme [2]) (Reversibility: ?) [1, 2]

P monocrotaline N-oxide + $NADP^+$ + H_2O <1, 3, 4, 5> [1, 2]

S retrorsine + NADPH + O_2 <3, 4, 5> (<4> 93% of the activity with senecionine [1]; <3> 97% of the activity with senecionine [1]; <5> as active as senecionine [1]) (Reversibility: ?) [1]

P retrorsine N-oxide + $NADP^+$ + H_2O <3, 4, 5> [1]

S rinderine + NADPH + O_2 <3, 4, 5> (<3> 23% of the activity with senecionine [1]; <5> 79% of the activity with senecionine [1]; <4> 81% of the activity with senecionine [1]) (Reversibility: ?) [1]

P rinderine N-oxide + $NADP^+$ + H_2O <3, 4, 5> [1]

S senecionine + NADPH + O_2 <1, 3, 4, 5> (Reversibility: ?) [1, 2]
P senecionine N-oxide + $NADP^+$ + H_2O <1, 3, 4, 5> [1, 2]
S seneciphylline + NADPH + O_2 <1, 3, 4, 5> (<5> 117% of the activity with senecionine [1]; <4> 119% of the activity with senecionine [1]; <3> 95% of the activity with senecionine [1]; <1> 95% of the activity with senecionine, native enzyme. 93% of the activity with senecionine, recombinant enzyme [2]) (Reversibility: ?) [1, 2]
P seneciphylline N-oxide + $NADP^+$ + H_2O <1, 3, 4, 5> [1, 2]
S senecivernine + NADPH + O_2 <3, 4, 5> (<5> 107% of the activity with senecionine [1]; <4> 113% of the activity with senecionine [1]; <3> 81% of the activity with senecionine [1]) (Reversibility: ?) [1]
P senecivernine N-oxide + $NADP^+$ + H_2O <3, 4, 5> [1]
S triangularine + NADPH + O_2 <3> (<3> 60% of the activity with senecionine [1]) (Reversibility: ?) [1]
P triangularine N-oxide + $NADP^+$ + H_2O <3> [1]
S Additional information <1, 3, 4, 5> (<1> the enzyme is highly specific for toxic pyrrolizidine alkaloids. No activity with: senkirkine, dimethylaniline, L-Pro, caffeine, atropine, supinidine, retronecine, phalaenopsine [2]; <3> the enzyme N-oxidizes only alkaloids with structural elements which are essential for hepatotoxic and genotoxic pyrrolizidine alkaloids, i.e. 1,2-double bond, esterification of the allylic hydroxyl group, presence of a second free or esterified hydroxyl group at carbon 7. No activity with: senkirkine, sarracine, supinine, phalaenopsine, retronecine, heliotridine, supinidine, isoretronecanol, nicotine, caffeine, tropane alkaloids, quinolizidine alkaloids, indole alkaloids, isoquinolines and synthetic heterocyclic tertiary amines [1]; <4,5> the enzyme N-oxidizes only alkaloids with structural elements which are essential for hepatotoxic and genotoxic pyrrolizidine alkaloids, i.e. 1,2-double bond, esterification of the allylic hydroxyl group, presence of a second free or esterified hydroxyl group at carbon 7. No activity with: triangularine, senkirkine, sarracine, supinine, phalaenopsine, retronecine, heliotridine, supinidine, isoretronecanol, nicotine, caffeine, tropane alkaloids, quinolizidine alkaloids, indole alkaloids, isoquinolines and synthetic heterocyclic tertiary amines [1]; <1> the enzyme allows the larvae to feed on pyrrolizidine alkaloid-containing plants and to accumulate predation-deterrent pyrrolizidine alkaloids in the hemolymph [2]) (Reversibility: ?) [1, 2]
P ? <1, 3, 4, 5> [1, 2]

Cofactors/prosthetic groups

FAD <3> (<3> molar apoenzyme/FAD ratio of 0.4 [1]) [1]
NADPH <1, 3, 4, 5> (<1> dependent on [2]; <4,5> strictly dependent on [1]; <3> strictly dependent on, no reaction with NADH [1]) [1,2]

Specific activity (U/mg)

4.67 <1> [1]

K_m-Value (mM)

0.0013 <1> (NADPH, <1> pH 7.0, pH 37°C [1]) [1]
0.0014 <1> (senecionine, <1> pH 7.0, 37°C [1]) [1]
0.0029 <4> (monocrotaline, <4> pH 7.0, 37°C [1]) [1]
0.0031 <4> (senecionine, <4> pH 7.0, 37°C [1]) [1]
0.0048 <5> (senecionine, <5> pH 7.0, 37°C [1]) [1]
0.0051 <5> (monocrotaline, <5> pH 7.0, 37°C [1]) [1]
0.0125 <1> (monocrotaline, <1> pH 7.0, 37°C [1]) [1]
0.0186 <5> (heliotrine, <5> pH 7.0, 37°C [1]) [1]
0.0229 <4> (heliotrine, <4> pH 7.0, 37°C [1]) [1]
0.284 <1> (heliotrine, <1> pH 7.0, 37°C [1]) [1]

pH-Optimum

7 <3> [1]

Temperature optimum (°C)

40-45 <3> [1]

4 Enzyme Structure

Molecular weight

200000 <3> (<3> gel filtration [1]) [1]

Subunits

tetramer <3> (<3> 4 * 51000, SDS-PAGE [1]) [1]

5 Isolation/Preparation/Mutation/Application

Source/tissue

adult <8> [1]
fat body <10> [1]
hemolymph <1, 3> (<3> from late instar larvae [1]) [1, 2]
larva <4, 5, 6, 7> [1]
pupa <9> [1]

Localization

soluble <3> [1]

Purification

<1> [1]

Cloning

<1> (expression in active form in Escherichia coli BL21) [2]

References

[1] Lindigkeit, R.; Biller, A.; Buch, M.; Schiebel, H..M.; Boppre, M.; Hartmann, T.: The two faces of pyrrolizidine alkaloids: the role of the tertiary amine and its N-oxide in chemical defense of insects with aquired alkaloids. Eur. J. Biochem., **245**, 626-636 (1997)
[2] Naumann, C.; Hartmann, T.; Ober, D.: Evolutionary recruitment of a flavin-dependent monooxygenase for the detoxification of host plant-defended arctiid moth Tyria jacobaeae. Proc. natl. Acad. Sci. USA, **99**, 6085-6090 (2002)

Psoralen synthase 1.14.13.102

1 Nomenclature

EC number
1.14.13.102

Systematic name
(+)-marmesin,NADPH:oxygen oxidoreductase

Recommended name
psoralen synthase

Synonyms
CYP71AJ1 <3> [3]

CAS registry number
106527-99-5

2 Source Organism

<1> *Petroselinum crispum* (no sequence specified) [1]
<2> *Ammi majus* (no sequence specified) [2]
<3> *Ammi majus* (UNIPROT accession number: Q6QNI4) [3]

3 Reaction and Specificity

Catalyzed reaction
(+)-marmesin + NADPH + H^+ + O_2 = psoralen + $NADP^+$ + acetone + 2 H_2O (<1> reaction mechanism [1]; <3> reaction mechanism, analogy modeling and docking analysis of substrate binding, alignmant of recognition sites, modeling of the catalytic site, overview [3])

Natural substrates and products
S (+)-marmesin + NADPH + H^+ + O_2 <1, 2, 3> (<3> the enzyme is the first committed monooxygenase of linear furanocoumarin biosynthesis, pathway overview [3]; <2> the substrate is synthesized from demethylsuberosin by the marmesin synthase, biosynthesis of furanocoumarin phytoalexin which are then excreted from the endoplasmic reticulum lumen into the cell culture medium [2]) (Reversibility: ?) [1, 2, 3]
P psoralen + $NADP^+$ + acetone + H_2O

S Additional information <2> (<2> umbelliferone is a better precursor of psoralens than either demethylsuberosin or marmesin [2]) (Reversibility: ?) [2]

P ?

Substrates and products

S (+)-marmesin + NADPH + H^+ + O_2 <1, 2, 3> (<3> the enzyme is the first committed monooxygenase of linear furanocoumarin biosynthesis, pathway overview [3]; <2> the substrate is synthesized from demethylsuberosin by the marmesin synthase, biosynthesis of furanocoumarin phytoalexin which are then excreted from the endoplasmic reticulum lumen into the cell culture medium [2]; <3> release of acetone by syn-elimination [3]; <1> very poor activity with (-)-marmesin, i.e. nodakenetin [1]) (Reversibility: ?) [1, 2, 3]

P psoralen + $NADP^+$ + acetone + H_2O

S 5-hydroxymarmesin + NADPH + H^+ + O_2 <3> (<3> low activity, psoralen 5-monoxygenase activity [3]) (Reversibility: ?) [3]

P bergaptol + $NADP^+$ + H_2O

S Additional information <1, 2, 3> (<2> umbelliferone is a better precursor of psoralens than either demethylsuberosin or marmesin [2]; <3> no activity with (+)-columbianetin, the psoralen synthase is a cytochrome-P_{450}-dependent monooxygenase [3]; <1,2> the psoralen synthase is a cytochrome-P_{450}-dependent monooxygenase [1,2]) (Reversibility: ?) [1, 2, 3]

P ?

Inhibitors

(+)-columbianetin <3> (<3> i.e. (+)-3,4,2,3-D4-columbianetin, competitive inhibition, binds to the active site, analogy modeling and docking analysis, mechanism-based inactivation [3]) [3]

ancymidole <1, 2> (<2> 13% inhibition of the psoralen synthase at 0.05 mM [2]; <1> 48% inhibition of the psoralen synthase at 0.05 mM, 29% at 0.005 mM [1]) [1, 2]

BAS 110 <1, 2> (<1> 34% inhibition of the psoralen synthase at 0.001 mM, 30% at 0.0002 mM [1]; <2> 72% inhibition of the psoralen synthase at 0.05 mM, 38% at 0.005 mM [2]) [1, 2]

BAS 111 <1, 2> (<1> 52% inhibition of the psoralen synthase at 0.001 mM, 33% at 0.0002 mM [1]; <2> complete inhibition of the psoralen synthase at 0.001 mM, 86% at 0.0002 mM [2]) [1, 2]

BAS 978 <1, 2> (<1> 10% inhibition of the psoralen synthase at 0.001 mM [1]; <2> complete inhibition of the psoralen synthase at 0.05 mM, 68% at 0.005 mM [2]) [1, 2]

cytochrome c <1> (<1> 72% inhibition at 0.5 nM [1]) [1]

ketoconazole <1, 2> (<1> 38% inhibition of the psoralen synthase at 0.050 mM, 20% at 0.005 mM [1]; <2> complete inhibition of the psoralen synthase at 0.001 mM, 74% at 0.0002 mM [2]) [1, 2]

$NADP^+$ <1> (<1> 43% product inhibition at 1 mM [1]) [1]

tetcyclacis <1, 2> (<1> 47% inhibition of the psoralen synthase at 0.05 mM, 51% at 0.005 mM [1]; <2> 85% inhibition of the psoralen synthase at 0.001 mM, 25% at 0.0002 mM [2]) [1, 2]
carbon monoxide <1> [1]
Additional information <1, 3> (<1> no effect on enzyme activity by 2-mercaptoethanol and DTT [1]; <3> no inhibition by 5- and 8-methoxypsoralen, and angelicin [3]) [1, 3]

Cofactors/prosthetic groups
NADPH <1, 2, 3> (<1> dependent on, NADH acts synergisticallyy, but cannot substitute for NADPH [1]) [1,2,3]
Additional information <1, 2, 3> (<1,2,3> the psoralen synthase is a cytochrome-P_{450}-dependent monooxygenase [1,2,3]) [1,2,3]

Activating compounds
NADH <1> (<1> NADH acts synergistically with NADPH, but cannot substitute for NADPH [1]) [1]
Additional information <1, 2, 3> (<2> elicitor preparations from either Alternaria carthami Chowdhury or Phytophthora megasperma f.sp. glycinea induce the production of linear furanocoumarins in Ammi majus suspension culture [2]; <1> no effect on enzyme activity by 2-mercaptoethanol and DTT, elicitor fractions isolated from cell walls of Alternaria carthami Chowdhury and of Phythophthora megasperma Drechsler f. sp. glycinea induce the enzyme, the endoplasmic reticulum of cultured parsley cells is the primary target in the differential induction by elicitors from these two nonpathogenic strains of fungi. [1]; <3> psoralen synthase is induced rapidly from negligible background levels upon elicitation, e.g. by crude Phytophthora sojae cell wall elicitor fraction, of cell cultures with transient maxima at 9-10 h, furanocoumarin composition in induced cells, overview [3]) [1, 2, 3]

Metals, ions
iron <3> (<3> the psoralen synthase is a cytochrome-P_{450}-dependent monooxygenase [3]) [3]
Additional information <1> (<1> Mg^{2+} does not affect the enzyme activity [1]) [1]

Turnover number (min^{-1})
340 <3> ((+)-marmesin, <3> pH 7.0, 27°C, recombinant enzyme in microsomes [3]) [3]

K_m-Value (mM)
0.0015 <3> ((+)-marmesin, <3> pH 7.0, 27°C, recombinant enzyme in microsomes [3]) [3]
0.052 <1> (NADPH, <1> pH 7.0, 20°C, microsomes [1]) [1]

pH-Optimum
7 <2, 3> (<2,3> assay at [2,3]) [2, 3]
7-7.25 <1> [1]

Temperature optimum (°C)

20 <1, 2> (<1,2> assay at [1,2]) [1, 2]
27 <3> (<3> assay at [3]) [3]

4 Enzyme Structure

Subunits

? <3> (<3> x * 55891, sequence calculation [3]) [3]
Additional information <3> (<3> three-dimensional active site structure [3]) [3]

5 Isolation/Preparation/Mutation/Application

Source/tissue

cell suspension culture <1, 2, 3> [1, 2, 3]
leaf <3> [3]

Localization

endoplasmic reticulum <1> [1]
endoplasmic reticulum lumen <2> [2]
microsome <1, 2, 3> [1, 2, 3]

Purification

<1> (native enzyme partially by preparation of microsomes) [1]
<2> (native enzyme partially by preparation of microsomes) [2]
<3> (native enzyme partially by preparation of microsomes) [3]

Cloning

<3> (gene CYP71AJ1, DNA and amino acid sequence determination and analysis, phylogenetic analysis, functional expression of the wild-type enzyme in Escherichia coli and Saccharomyces cerevisiae strain WAT11 requiring swapping the N-terminal membrane anchor domain with that of CYP73A1, expression of the enzyme mutant in yeast cells) [3]

Engineering

M120V <3> (<3> site-directed mutagenesis [3]) [3]

6 Stability

Temperature stability

95 <1> (<1> short exposure leads to complete inactivation [1]) [1]

References

[1] Wendorff, H.; Matern, U.: Differential response of cultured parsley cells to elicitors from two non-pathogenic strains of fungi. Microsomal conversion of (+)marmesin into psoralen. Eur. J. Biochem., **161**, 391-398 (1986)
[2] Hamerski, D.; Matern, U.: Elicitor-induced biosynthesis of psoralens in Ammi majus L. suspension cultures. Microsomal conversion of demethylsuberosin into (+)marmesin and psoralen. Eur. J. Biochem., **171**, 369-375 (1988)
[3] Larbat, R.; Kellner, S.; Specker, S.; Hehn, A.; Gontier, E.; Hans, J.; Bourgaud, F.; Matern, U.: Molecular cloning and functional characterization of psoralen synthase, the first committed monooxygenase of furanocoumarin biosynthesis. J. Biol. Chem., **282**, 542-554 (2007)

8-Dimethylallylnaringenin 2'-hydroxylase 1.14.13.103

1 Nomenclature

EC number
1.14.13.103

Systematic name
sophoraflavanone-B,NADPH:oxygen oxidoreductase (2'-hydroxylating)

Recommended name
8-dimethylallylnaringenin 2'-hydroxylase

Synonyms
8-DMAN 2'-hydroxylase <1> [1]
8-dimethylallylnaringenin 2'-hydroxylase <1> [2]

CAS registry number
400886-31-9

2 Source Organism

<1> *Sophora flavescens* (no sequence specified) [1, 2]

3 Reaction and Specificity

Catalyzed reaction
sophoraflavanone B + NADPH + H^+ + O_2 = leachianone G + $NADP^+$ + H_2O

Natural substrates and products
S 8-dimethylallylnaringenin + NADPH + H^+ + O_2 <1> (<1> the enzyme is crucial for lavandulylated flavanone formation, the enzyme catalyzes the 2-hydroxylation of 8-DMAN, which is then formed to sophoraflavanone G, pathway overview [1,2]) (Reversibility: ?) [1, 2]
P leachianone G + $NADP^+$ + H_2O

Substrates and products
S (-)-(2S)-8-dimethylallylnaringenin + NADPH + H^+ + O_2 <1> (<1> i.e. sophoraflavanone B, the enzyme catalyzes the 2-hydroxylation of 8-DMAN and is O_2-dependent [1]) (Reversibility: ?) [1]
P leachianone G + $NADP^+$ + H_2O
S 8-dimethylallylnaringenin + NADPH + H^+ + O_2 <1> (<1> the enzyme is crucial for lavandulylated flavanone formation, the enzyme catalyzes the

2-hydroxylation of 8-DMAN, which is then formed to sophoraflavanone G, pathway overview [1,2]) (Reversibility: ?) [1, 2]
P leachianone G + $NADP^+$ + H_2O
S Additional information <1> (<1> the enzyme is a cytochrome P_{450} monooxygenase [1]) (Reversibility: ?) [1]
P ?

Inhibitors
cytochrome c <1> (<1> 70% inhibition at 10 mM [1]) [1]
metyrapone <1> (<1> 34% inhibition at 1 mM [1]) [1]
miconazole <1> (<1> 36% inhibition at 1 mM [1]) [1]
carbon monoxide <1> [1]
tropolon <1> (<1> 30% inhibition at 1 mM [1]) [1]
Additional information <1> (<1> no or poor inhibition by ancymidol and KCN [1]) [1]

Cofactors/prosthetic groups
NADPH <1> (<1> required, cannot be substituted by NADH, FAD, and FMN [1]) [1,2]
Additional information <1> (<1> the enzyme is a cytochrome P_{450} monooxygenase [1,2]) [1,2]

K_m-Value (mM)
0.034 <1> (NADPH) [1]
0.055 <1> (8-dimethylallylnaringenin) [1]

pH-Optimum
8.5 <1> [1]

Temperature optimum (°C)
30 <1> (<1> assay at [1]) [1]

5 Isolation/Preparation/Mutation/Application

Source/tissue
cell culture <1> [1]
cell suspension culture <1> [2]

Localization
endoplasmic reticulum <1> [2]
membrane <1> (<1> the enzyme is tightly bound to the membrane [1]) [1]
Additional information <1> (<1> subcellular localization of the enzymes involved in the biosynthetic route from naringenin to sophoraflavanone G, overview [2]) [2]

Purification
<1> (native enzyme partially by microsome preparation) [2]

References

[1] Yamamoto, H.; Yatou, A.; Inoue, K.: 8-dimethylallylnaringenin 2'-hydroxylase, the crucial cytochrome P450 mono-oxygenase for lavandulylated flavanone formation in Sophora flavescens cultured cells. Phytochemistry, **58**, 671-676 (2001)

[2] Zhao, P.; Inoue, K.; Kouno, I.; Yamamoto, H.: Characterization of leachianone G 2''-dimethylallyltransferase, a novel prenyl side-chain elongation enzyme for the formation of the lavandulyl group of sophoraflavanone G in Sophora flavescens Ait. cell suspension cultures. Plant Physiol., **133**, 1306-1313 (2003)

CMP-N-Acetylneuraminate monooxygenase 1.14.18.2

1 Nomenclature

EC number
1.14.18.2

Systematic name
CCMP-N-acetylneuraminate,ferrocytochrome-b_5:oxygen oxidoreductase (N-acetyl-hydroxylating)

Recommended name
CMP-N-acetylneuraminate monooxygenase

Synonyms
CMAH
CMP-N-acetylneuraminate,NAD(P)H:oxygen oxidoreductase (hydroxylating)
CMP-N-acetylneuraminic acid
CMP-N-acetylneuraminic acid hydroxylase
CMP-Neu5Ac hydroxylase
EC 1.14.13.45
oxygenase, cytidine monophosphoacetylneuraminate mono-

CAS registry number
116036-67-0

2 Source Organism

<1> *Cricetulus griseus* (no sequence specified) [25]
<2> *Mus musculus* (no sequence specified) [1, 3, 4, 5, 8, 10, 11, 13, 15, 16, 17, 23]
<3> *Homo sapiens* (no sequence specified) [26]
<4> *Rattus norvegicus* (no sequence specified) [1, 5]
<5> *Sus scrofa* (no sequence specified) [2,5,7,8,18,19,22,24]
<6> *Bos taurus* (no sequence specified) [5,8]
<7> *Asterias rubens* (no sequence specified) [6,9,14,20,21]
<8> no activity in *Homo sapiens* [12,13,27]
<9> *Ctenodiscus crispatus* (no sequence specified) [20]
<10> *Strongylocentrotus pallidus* (no sequence specified) [20]
<11> *Muelleria sp.* (no sequence specified) [20]

3 Reaction and Specificity

Catalyzed reaction

CMP-N-acetylneuraminate + 2 ferrocytochrome b_5 + O_2 + 2 H^+ = CMP-N-glycoloylneuraminate + 2 ferricytochrome b_5 + H_2O

Reaction type

oxidation
redox reaction
reduction
Additional information (<3> enzyme is used for the detection of human cancer like colon cancer, melanoma, retinoblastoma cell lines and tissues, germ cell tumours and breast cancer by detection of N-glycolylneuraminic acid-containing glycolipids [26])

Natural substrates and products

S CMP-N-acetylneuraminate + NADH + O_2 <2, 5, 6, 7> (<2> binding of CMP-N-acetylneuraminate to CMP-N-acetylneuraminate hydroxylase changes conformation of the enzyme so as to construct a recognition site for cytochrome b_5, followed by the formation of a ternary complex through this domain. Then the transport of electrons from NAD(P)H to the enzyme through cytochrome b_5 takes place, CMP-N-acetylneuraminate is converted to CMP-N-glycoloylneuraminic acid and finally the ternary complex dissociates into its components to release CMP-N-glycoloylneuraminic acid [15]; <2> the enzyme is the key for regulation of the overall velocity of CMP-NeuAc hydroxylation and consequently for the expression of N-glycoloylneuraminic acid glycoconjugates [10]; <2> a regulation of CMP-N-acetylneuraminate hydroxylation and thus the ratio of glycoconjugate-bound N-acetylneuraminate and N-acetylglycoloylneuraminate might occur by varying the amount of hydroxylase protein within the cell, possibly by controlling the expression of the hydroxylase gene [16]; <2> the enzyme plays a decisive role in governing the relative amounts of N-acetylneuraminate and N-acetylglycolylneuraminate occuring in the glycoconjugates of a tissue [1]; <5> the biosynthesis of the sialic acid N-glycolylneuraminic acid occurs by the action of cytidine monophosphate-N-acetylneuraminate hydroxylase. Incorporation of N-glycoloylneuraminic acid into glycoconjugates is generally controlled by the amount of hydroxylase protein expressed in a tissue [18]; <2> key enzyme for the expression of N-glycoloylneuraminic acid [17]) (Reversibility: ?) [1, 2, 3, 8, 9, 10, 15, 16, 17, 18]

P CMP-N-glycoloylneuraminate + NAD^+ + H_2O

Substrates and products

S CMP-N-acetylneuraminate + 6,7-dimethyl-5,6,7,8-tetrahydrobiopterin + O_2 <5> (<5> less effectiove than NADH or NADPH [2]) (Reversibility: ?) [2]

P CMP-N-glycoloylneuraminate + ?

S CMP-N-acetylneuraminate + NADH + O_2 <2, 5, 6, 7> (<2> binding of CMP-N-acetylneuraminate to CMP-N-acetylneuraminate hydroxylase changes conformation of the enzyme so as to construct a recognition site for cytochrome b_5, followed by the formation of a ternary complex through this domain. Then the transport of electrons from NAD(P)H to the enzyme through cytochrome b_5 takes place, CMP-N-acetylneuraminate is converted to CMP-N-glycoloylneuraminic acid and finally the ternary complex dissociates into its components to release CMP-N-glycoloylneuraminic acid [15]; <5> NADPH and NADH are by far the most effective cofactors [2]; <2> the enzyme is the key for regulation of the overall velocity of CMP-NeuAc hydroxylation and consequently for the expression of N-glycoloylneuraminic acid glycoconjugates [10]; <2> a regulation of CMP-N-acetylneuraminate hydroxylation and thus the ratio of glycoconjugate-bound N-acetylneuraminate and N-acetylglycoloylneuraminate might occur by varying the amount of hydroxylase protein within the cell, possibly by controlling the expression of the hydroxylase gene [16]; <2> the enzyme plays a decisive role in governing the relative amounts of N-acetylneuraminate and N-acetylglycolylneuraminate occuring in the glycoconjugates of a tissue [1]; <5> the biosynthesis of the sialic acid N-glycolylneuraminic acid occurs by the action of cytidine monophosphate-N-acetylneuraminate hydroxylase. Incorporation of N-glycoloylneuraminic acid into glycoconjugates is generally controlled by the amount of hydroxylase protein expressed in a tissue [18]; <2> key enzyme for the expression of N-glycoloylneuraminic acid [17]) (Reversibility: ?) [1, 2, 3, 4, 8, 9, 10, 14, 15, 16, 17, 18, 20, 21, 22, 23]

P CMP-N-glycoloylneuraminate + NAD^+ + H_2O

S CMP-N-acetylneuraminate + NADPH + O_2 <2, 5> (<2> binding of CMP-N-acetylneuraminate to CMP-N-acetylneuraminate hydroxylase changes conformation of the enzyme so as to construct a recognition site for cytochrome b_5, followed by the formation of a ternary complex through this domain. Then the transport of electrons from NAD(P)H to the enzyme through cytochrome b_5 takes place, CMP-N-acetylneuraminate is converted to CMP-N-glycoloylneuraminic acid and finally the ternary complex dissociates into its components to release CMP-N-glycoloylneuraminic acid [15]; <5> NADPH and NADH are by far the most effective cofactors [2]) (Reversibility: ?) [2, 3, 4, 15]

P CMP-N-glycoloylneuraminate + $NADP^+$ + H_2O

S CMP-N-acetylneuraminate + ascorbic acid + O_2 <2, 5> (<2> ascorbate is ineffective [3]; <5> less effective than NADH or NADPH [2]) (Reversibility: ?) [2, 3]

P CMP-N-glycoloylneuraminate + dehydroascorbate + H_2O

S CMP-N-acetylneuraminate + ferrocytochrome b_5 + O_2 <2, 5, 7> (<2> binding of CMP-N-acetylneuraminate to CMP-N-acetylneuraminate hydroxylase changes conformation of the enzyme so as to construct a recognition site for cytochrome b_5, followed by the formation of a ternary complex through this domain. Then the transport of electrons from NAD(P)H to the enzyme through cytochrome b_5 takes place, CMP-N-acetylneuraminate is converted to CMP-N-glycoloylneuraminic acid and finally the ternary complex dissociates into its components to release CMP-N-glycoloylneuraminic acid [15]; <7> evidence for the formation of a ternary complex of hydroxylase, CMP-Neu5Ac and cytochrome b_5 [21]; <2,5> pure hydrophilic cytochrome b_5 interacts more effectively with the hydroxylase than isolated amphiphytic cytochrome b_5 [8]; <2> 2 electrons are donated by either NADH or NADPH are transported via cytochrome b_5 in the CMP-NeuAc hydroxylation system of mouse liver cytosol [4]; <2> the electron carrier cytochrome b_5 is essential for activity [23]) (Reversibility: ?) [4, 8, 15, 16, 20, 21, 23]

P CMP-N-glycoloylneuraminate + ferricytrochrome b_5 + H_2O

S Additional information <2> (<2> no activity towards free or α-glycosidically bound N-acetylneuraminic acid [5]) (Reversibility: ?) [5]

P ?

Inhibitors

1,10-phenanthroline <2, 5, 7> (<2> 5 mM, complete inhibition [23]; <7> 2 mM, 88% inhibition [20]) [14, 20, 21, 22, 23]
2,2'-dipyridyl <7> (<7> 2 mM, 27% inhibition [20]) [20, 21]
3'-AMP <5> (<5> 5 mM, complete inhibition [5]) [5]
3'-CMP <5> (<5> 5 mM, about 20% inhibition [5]) [5]
3'-UMP <5> (<5> 5 mM, about 90% inhibition [5]) [5]
5'-AMP <5> (<5> 5 mM, about 78% inhibition [5]) [5]
5'-UMP <5> (<5> 5 mM, about 15% inhibition [5]) [5]
azide <7> [14]
CHAPS <7> (<7> 15 mM, complete inhibition [14]) [14]
CMP-N-glycoloylneuraminate <2> [23]
Ca^{2+} <7> (<7> 0.5 mM, 12% inhibition [14]) [14]
cardiolipin <2> (<2> 3 mM, 15% inhibition [23]) [23]
cholic acid <7> (<7> 10 mM, 89% inhibition [14]) [14]
Co^{2+} <2> [3]
Cu^{2+} <2> [3]
EDTA <2> [3, 13]
ferrozine <2, 5, 7> (<7> 2 mM, 83% inhibition [20]; <2> 0.7 mM, 82% inhibition [23]) [14, 20, 21, 22, 23]
Hg^{2+} <7> (<7> slight inhibition [14]) [14]
KCN <2, 5, 7> (<2> 5 mM, 49% inhibition [23]; <7> 2 mM, 80% inhibition [20]) [14, 20, 22, 23]
Mg^{2+} <2> [3]
Mn^{2+} <2, 7> (<7> 0.5 mM, slight inhibition [14]) [3, 14]
Na_2HPO_4 <2> (<2> 2 mM, 86% inhibition [23]) [23]
$Na_4P_2O_7$ <2> (<2> 2 mM, 54% loss of activity [23]) [23]
NaCl <7> (<7> above 100 mM [20]) [20]
Ni^{2+} <2> [3]
phosphatidic acid <2> (<2> 3 mM, 20% inhibition [23]) [23]
phosphatidylinositol <2> (<2> 3 mM, 68% inhibition [23]) [23]
Tiron <2, 5, 7> (<2> 5 mM, 82% inhibition [23]) [14, 22, 23]

Zn^{2+} <2> [3]
Zwittergent 3-12 <7> (<7> 5 mM, 40% inhibition [14]) [14]
anti-(rat cytochrome b_5) antiserum <2, 5> [8, 23]
decylglucopyranoside <7> (<7> 5 mM, 22% inhibition [14]) [14]
octylglucopyranoside <7> (<7> 30 mM, 48% inhibition [14]) [14]
Additional information <7> (<7> not inhibited by increased ionic strength, no inhibition by 1 M NaCl [14]) [14]

Cofactors/prosthetic groups

NADH <2, 5, 6, 7> (<7> most effective cofactor, optimal activity at 0.4 mM NADH, higher concentrations are slightly inhibitory [14]; <5> NADPH and NADH are by far the most effective cofactors [2]; <2> NADH is much more effective than NADPH [4]) [2, 3, 4, 8, 9, 10, 14, 16, 18, 20, 21, 22, 23]
NADPH <2, 5> (<5> NADPH and NADH are by far the most effective cofactors [2]; <2> NADH is much more effective than NADPH [4]) [2, 3, 4]
ferrocytochrome b_5 <2, 5, 7> (<7> evidence for the formation of a ternary complex of hydroxylase, CMP-Neu5Ac and cytochrome b_5 [21]; <2> binding of CMP-N-acetylneuraminate to CMP-N-acetylneuraminate hydroxylase changes conformation of the enzyme so as to construct a recognition site for cytochrome b_5, followed by the formation of a ternary complex through this domain. Then the transport of electrons from NAD(P)H to the enzyme through cytochrome b_5 takes place, CMP-N-acetylneuraminate is converted to CMP-N-glycoloylneuraminic acid and finally the ternary complex dissociates into its components to release CMP-N-glycoloylneuraminic acid [15]; <2> the electron carrier cytochrome b_5 is essential for activity [23]; <2,5> pure hydrophilic cytochrome b_5 interacts more effectively with the hydroxylase than isolated amphiphytic cytochrome b_5 [8]; <2> 2 electrons are donated by either NADH or NADPH are transported via cytochrome b_5 in the CMP-NeuAc hydroxylation system of mouse liver cytosol [4]) [4, 8, 15, 20, 21, 23]

Activating compounds

Ascorbate <7> (<7> activates [14]) [14]
Dithiothreitol <7> (<7> activates [14]) [14]
Glutathione <7> (<7> activates [14]) [14]
Nonidet P-40 <2> (<2> effective inhibitor [23]) [23]
Octanoic acid <2> (<2> 1 mM, modest activation [23]) [23]
Octyl glucoside <2> (<2> effective inhibitor [23]) [23]
SDS <2> (<2> 1 mM, modest activation [23]) [23]
Triton X-100 <2> (<2> effective inhibitor [23]) [23]
decyl glucoside <2> (<2> activation [23]) [23]
Additional information <2, 7> (<7> no activation by non-ionic detergents [14]; <2> highest activity in 50 mM Hepes buffer, significant inhibition at increasing concentrations [23]) [14, 23]

Metals, ions

$FeSO_4$ <7> (<7> activates [20]) [20]

Iron <2, 5, 7> (<5> iron-sulfur protein of the Rieske type [7]; <2> contains non-heme iron as an electron acceptor [10]; <7> the enzyme contains a non-haem iron cofactor [14]) [7, 10, 14]

NaCl <7, 9, 10, 11> (<7> maximal 3.5fold stimulation at 100 mM [20,21]; <9,10,11> 100 mM activates [20]) [20, 21]

Specific activity (U/mg)

0.00093 <7> [21]

0.126 <2> [16]

0.816 <5> [22]

6.8 <2> [10]

K_m-Value (mM)

0.0029 <7> (ferrocytochrome b_5, <7> pH 6.4, 25°C, amphiphilic form of ferrocytochrome b_5 [20]) [20]

0.003 <5> (CMP-N-acetylneuraminate, <5> pH 7.4, 37°C, reconstituted system consisting of purified hydroxylase, cytochrome b_5, cytochrome b_5 reductase and catalase [22]) [22]

0.0032 <7> (ferrocytochrome b_5, <7> pH 6.4, 25°C, soluble form of ferrocytochrome b_5 [20]) [20]

0.005 <2> (CMP-N-acetylneuraminate, <2> pH 7.5, 37°C [10]; <2> pH 7.4, 37°C, amphiphilic system [16]) [10, 16]

0.0072 <7> (CMP-N-acetylneuraminate, <7> pH 6.4, 25°C [20]) [20]

0.011 <5> (CMP-N-acetylneuraminate, <5> pH 7.4, 37°C, purified hydroxylase in presence of Triton X-100 solubilized microsomes [22]) [22]

0.013 <2> (CMP-N-acetylneuraminate, <2> pH 7.4, 37°C, soluble system [16]) [16]

0.018 <7> (CMP-N-acetylneuraminate, <7> 25°C [14]) [14]

0.6-0.9 <2> (CMP-N-acetylneuraminate, <2> pH 7.4, 37°C, enzyme from NS-1 cells [5]) [5]

2.5 <5> (CMP-N-acetylneuraminate, <5> pH 7.4, 37°C, enzyme from submaxillary gland [5]) [5]

Additional information <2> [3]

pH-Optimum

6-6.4 <7> [14]

6-6.6 <7> [20]

6.4-7.4 <2, 5> (<2,5> enzyme preparation [5]) [5]

6.8-7.4 <2> [23]

pH-Range

5.6-6.8 <7> (<7> about 60% of maximal activity at pH 5.6 and pH 6.8 [14]) [14]

Temperature optimum (°C)

22-27 <7> [20]

25-33 <7> [14]

Temperature range (°C)

15-43 <7> (<7> 15°C: about 55% of maximal activity, 43°C: about 30% of maximal activity [14]) [14]

4 Enzyme Structure

Molecular weight

17000 <2> (<2> gel filtration [23]) [23]
56000 <2> (<2> gel filtration [8]) [8]
58000 <2> (<2> gel filtration [10,16]) [10, 16]
60000 <5> (<5> gel filtration [22]) [22]

Subunits

monomer <2, 5> (<2> 1 * 64000, SDS-PAGE [10,16]; <5> 1 * 65000, SDS-PAGE [22]) [10, 16, 22]
Additional information <7> (<7> a main protein of 76000 da and a minor protein of 64000 Da correlates with CMP-neu5Ac hydroxylase activity [21]) [21]

5 Isolation/Preparation/Mutation/Application

Source/tissue

CHO cell <1> (<1> generation of a stable CHO cell line with significantly reduced levels of CMP-Neu5Ac hydroxylase activity and reduced amounts of N-glycolylneuraminic acid, using a rational antisense RNA strategy [25]) [25]
NS-1 cell <2> [5]
P3X63Ag8 cell <2> [5]
alimentary canal <7> [14]
body wall <7> [14]
colonic mucosa <4> [5]
gonad <7, 9, 10, 11> [14, 20, 21]
heart <5> (<5> weak activity [18]) [18]
kidney <5> (<5> weak activity [18]) [18]
liver <2, 4, 5> (<5> weak activity [18]) [1, 3, 4, 8, 10, 11, 15, 16, 17, 18, 23]
lung <5> [18]
lymph node <5> [18, 19]
lymphocyte <5> (<5> from thymus, spleen, lymph node and peripheral blood. Highest activity in peripheral blood lymphocytes [19]) [19]
serum <3> (<3> traces of Neu5Gc found in serum by paper chromatography and in serum glycoproteins by HPLC analysis of per-O-benzoylated sialic acids [26]) [26]
small intestine <5> (<5> no significant temporal alterations in the activity in foetal and newborn small intestine. Birth is followed by a 2-8fold decrease in activity, depending on the region of the small intestine. Increase in activity from duodenum to ileum [24]) [18, 24]

spleen <5> [18]
submandibular gland <5> [2, 7, 8, 18, 22]
submaxillary gland <5, 6> [5]
thymus <5> [18]

Localization

cytosol <2, 5> (<2> full-length enzyme with normal enzymatic activity [17]; <5> in the vicinity of the nuclear membrane and the outer membrane of the mitochondria [19]) [4, 5, 7, 10, 11, 17, 19]
endoplasmic reticulum <2> (<2> naturally occuring truncated protein lacking 46 amino acids in the middle of the normal full-length protein [17]) [17]
membrane <7> (<7> bound to [9,20]) [9, 20]
microsome <2> [16]
soluble <7> [6]

Purification

<2> [10, 11, 16]
<5> [22]
<7> [9, 21]

Cloning

<2> [13]
<2> (expression in COS-1 cells) [11, 17]
<5> [7]
<7> (expression in Escherichia coli) [9]

Engineering

Additional information <2> (<2> naturally occuring truncated protein, lacking 46 amino acids in the middle of the normal full-length protein, causes a change in intracellular distribution of the enzyme from cytosol to endoplasmic reticulum and a loss in activity [17]) [17]

6 Stability

Temperature stability

4 <7> (<7> 48 h, 10% loss of activity [21]) [21]
25 <7> (<7> 2 h, 50% loss of activity [21]) [21]

General stability information

<2>, enzyme is greatly stabilized by CMP-N-acetylneuraminate [10]
<7>, after each cycle of freezing and thawing, 25% loss of activity [21]
<2, 5>, 10 mM in the initial homogenization buffer is required for stability during 60-90 min incubation [5]

Storage stability

<2>, -80°C, 0.2 mM CMP-N-acetylneuraminate, stable for at least 6 months [10]
<5>, -70°C, stable in crude extract for several weeks [5]

<7>, -80°C, very stable [21]
<7>, 4°C, 48 h, 10% loss of activity [21]

References

[1] Lepers, A.; Shaw, L.; Schneckenburger, P.; Cacan, R.; Verbert, A.; Schauer, R.: A study on the regulation of N-glycoloylneuraminic acid biosynthesis and utilization in rat and mouse liver. Eur. J. Biochem., **193**, 715-723 (1990)

[2] Shaw, L.; Schauer, R.: The biosynthesis of N-glycoloylneuraminic acid occurs by hydroxylation of the CMP-glycoside of N-acetylneuraminic acid. Biol. Chem. Hoppe-Seyler, **369**, 477-486 (1988)

[3] Shaw, L.; Schauer, R.: Detection of CMP-N-acetylneuraminic acid hydroxylase activity in fractionated mouse liver. Biochem. J., **263**, 355-363 (1989)

[4] Kozutsumi, Y.; Kawano, T.; Yamakawa, T.; Suzuki, A.: Participation of cytochrome b_5 in CMP-N-acetylneuraminic acid hydroxylation in mouse liver cytosol. J. Biochem., **108**, 704-706 (1990)

[5] Muchmore, E.A.; Milewski, M.; Varki, A.; Diaz, S.: Biosynthesis of N-glycolyneuraminic acid. The primary site of hydroxylation of N-acetylneuraminic acid is the cytosolic sugar nucleotide pool. J. Biol. Chem., **264**, 20216-20223 (1989)

[6] Bergwerff, A.A.; Hulleman, S.H.D.; Kamerling, J.P.; Vliegenthart, J.F.G.; Shaw, L.; Reuter, G.; Schauer, R.: Nature and biosynthesis of sialic acids in the starfish Asterias rubens. Identification of sialo-oligomers and detection of S-adenosyl-L-methionine: N-acylneuraminate 8-O-methyltransferase and CMP-N-acetylneuraminate monooxygenase activities. Biochimie, **74**, 25-37 (1992)

[7] Schlenzka, W.; Shaw, L.; Kelm, S.; Schmidt, C.L.; Bill, E.; Trautwein, A.X.; Lottspeich, F.; Schauer, R.: CMP-N-acetylneuraminic acid hydroxylase: the first cytosolic Rieske iron-sulfur protein to be described in Eukarya. FEBS Lett., **385**, 197-200 (1996)

[8] Shaw, L.; Schneckenburger, P.; Schlenzka, W.; Carlsen, J.; Christiansen, K.; Juergensen, D.; Schauer, R.: CMP-N-acetylneuraminic acid hydroxylase from mouse liver and pig submandibular glands. Interaction with membrane-bound and soluble cytochrome b_5-dependent electron transport chains. Eur. J. Biochem., **219**, 1001-1011 (1994)

[9] Martensen, I.; Schauer, R.; Shaw, L.: Cloning and expression of a membrane-bound CMP-N-acetylneuraminic acid hydroxylase from the starfish Asterias rubens. Eur. J. Biochem., **268**, 5157-5166 (2001)

[10] Kawano, T.; Kozutsumi, Y.; Kawasaki, T.; Suzuki, A.: Biosynthesis of N-glycolylneuraminic acid-containing glycoconjugates. Purification and characterization of the key enzyme of the cytidine monophospho-N-acetylneuraminic acid hydroxylation system. J. Biol. Chem., **269**, 9024-9029 (1994)

[11] Kawano, T.; Koyama, S.; Takematsu, H.; Kozutsumi, Y.; Kawasaki, H.; Kawashima, S.; Kawasaki, T.; Suzuki, A.: Molecular cloning of cytidine monophospho-N-acetylneuraminic acid hydroxylase. Regulation of species- and

tissue-specific expression of N-glycolylneuraminic acid. J. Biol. Chem., **270**, 16458-16463 (1995)
[12] Chou, H.H.; Hayakawa, T.; Diaz, S.; Krings, M.; Indriati, E.; Leakey, M.; Paabo, S.; Satta, Y.; Takahata, N.; Varki, A.: Inactivation of CMP-N-acetylneuraminic acid hydroxylase occurred prior to brain expansion during human evolution. Proc. Natl. Acad. Sci. USA, **99**, 11736-11741 (2002)
[13] Irie, A.; Suzuki, A.: CMP-N-acetylneuraminic acid hydroxylase is exclusively inactive in humans. Biochem. Biophys. Res. Commun., **248**, 330-333 (1998)
[14] Schlenzka, W.; Shaw, L.; Schauer, R.: Catalytic properties of the CMP-N-acetylneuraminic acid hydroxylase from the starfish Asterias rubens: comparison with the mammalian enzyme. Biochim. Biophys. Acta, **1161**, 131-138 (1993)
[15] Takematsu, H.; Kawano, T.; Koyama, S.; Kozutsumi, Y.; Suzuki, A.; Kawasaki, T.: Reaction mechanism underlying CMP-N-acetylneuraminic acid hydroxylation in mouse liver: formation of a ternary complex of cytochrome b_5, CMP-N-acetylneuraminic acid, and a hydroxylation enzyme. J. Biochem., **115**, 381-386 (1994)
[16] Schneckenburger, P.; Shaw, L.; Schauer, R.: Purification, characterization and reconstitution of CMP-N-acetylneuraminate hydroxylase from mouse liver. Glycoconjugate J., **11**, 194-203 (1994)
[17] Koyama, S.; Yamaji, T.; Takematsu, H.; Kawano, T.; Kozutsumi, Y.; Suzuki, A.; Kawasaki, T.: A naturally occurring 46-amino acid deletion of cytidine monophospho-N-acetylneuraminic acid hydroxylase leads to a change in the intracellular distribution of the protein. Glycoconjugate J., **13**, 353-358 (1996)
[18] Malykh, Y.N.; Shaw, L.; Schauer, R.: The role of CMP-N-acetylneuraminic acid hydroxylase in determining the level of N-glycolylneuraminic acid in porcine tissues. Glycoconjugate J., **15**, 885-893 (1998)
[19] Malykh, Y.N.; Krisch, B.; Shaw, L.; Warner, T.G.; Sinicrop, D.; Smith, R.; Chang, J.; Schauer, R.: Distribution and localization of CMP-N-acetylneuraminic acid hydroxylase and N-glycolylneuraminic acid-containing glycoconjugates in porcine lymph node and peripheral blood lymphocytes. Eur. J. Cell Biol., **80**, 48-58 (2001)
[20] Gollub, M.; Schauer, R.; Shaw, L.: Cytidine monophosphate-N-acetylneuraminate hydroxylase in the starfish Asterias rubens and other echinoderms. Comp. Biochem. Physiol. B, **120**, 605-615 (1998)
[21] Gollub, M.; Shaw, L.: Isolation and characterization of cytidine-5'-monophosphate-N-acetylneuraminate hydroxylase from the starfish Asterias rubens. Comp. Biochem. Physiol. B, **134**, 89-101 (2003)
[22] Schlenzka, W.; Shaw, L.; Schneckenburger, P.; Schauer, R.: Purification and characterization of CMP-N-acetylneuraminic acid hydroxylase from pig submandibular glands. Glycobiology, **4**, 675-683 (1994)
[23] Shaw, L.; Schneckenburger, P.; Carlsen, J.; Christiansen, K.; Schauer, R.: Mouse liver cytidine-5-monophosphate-N-acetylneuraminic acid hydroxylase. Catalytic function and regulation. Eur. J. Biochem., **206**, 269-277 (1992)

[24] Malykh, Y.N.; King, T.P.; Logan, E.; Kelly, D.; Schauer, R.; Shaw, L.: Regulation of N-glycolylneuraminic acid biosynthesis in developing pig small intestine. Biochem. J., **22**, 2-29 (2002)

[25] Chenu, S.; Gregoire, A.; Malykh, Y.; Visvikis, A.; Monaco, L.; Shaw, L.; Schauer, R.; Marc, A.; Goergen, J.L.: Reduction of CMP-N-acetylneuraminic acid hydroxylase activity in engineered Chinese hamster ovary cells using an antisense-RNA strategy. Biochim. Biophys. Acta, **1622**, 133-144 (2003)

[26] Malykh, Y.N.; Schauer, R.; Shaw, L.: N-Glycolylneuraminic acid in human tumours. Biochimie, **83**, 623-634 (2001)

[27] Hayakawa, T.; Aki, I.; Varki, A.; Satta, Y.; Takahata, N.: Fixation of the human-specific CMP-N-acetylneuraminic acid hydroxylase pseudogene and implications of haplotype diversity for human evolution. Genetics, **172**, 1139-1146 (2006)

Lathosterol oxidase 1.14.21.6

1 Nomenclature

EC number
1.14.21.6

Systematic name
5α-cholest-7-en-3β-ol, NAD(P)H:oxygen 5-oxidoreductase

Recommended name
lathosterol oxidase

Synonyms
5-DES <7> [9]
5α-cholest-7-en-3β-ol:oxygen Δ^5-oxidoreductase
C-5 sterol desaturase
Δ^7-sterol 5-desaturase <7> [9]
Δ^7-sterol Δ^5-dehydrogenase <7> [9]
Δ^7-sterol-C5(6)-desaturase <7> [9]
Δ-7-sterol 5-desaturase
lathosterol 5-desaturase
Sc5d <1, 8> [13, 14]
sterol-C5-desaturase <8> [14]
lathosterol oxidase
oxidase, lathosterol

CAS registry number
37255-37-1

2 Source Organism

<1> *Mus musculus* (no sequence specified) [8, 13]
<2> *Homo sapiens* (no sequence specified) [8, 11]
<3> *Rattus norvegicus* (no sequence specified) [2, 3, 4, 5, 6]
<4> *Saccharomyces cerevisiae* (no sequence specified) [7]
<5> *Zea mays* (no sequence specified) [1, 7, 10, 12]
<6> *Nicotiana tabacum* (no sequence specified) [7]
<7> *Arabidopsis thaliana* (no sequence specified) [7,9,12]
<8> *Mus musculus* (UNIPROT accession number: O88822) [14]

3 Reaction and Specificity

Catalyzed reaction

5α-cholest-7-en-3β-ol + NAD(P)H + H^+ + O_2 = cholesta-5,7-dien-3β-ol + $NAD(P)^+$ + 2 H_2O (<5,7> enzyme initiates oxidation by cleavage of the chemically activated C6α-H bond, proposed mechanism [12])

Reaction type

oxidation
redox reaction
reduction

Natural substrates and products

S 5α-cholest-7-en-3β-ol + O_2 + NAD(P)H <3> (<3> cholesterol biosynthesis [2,3]) (Reversibility: ?) [2, 3]
P cholesta-5,7-dien-3β-ol + H_2O_2 + $NAD(P)^+$ <3> [2, 3]
S lathosterol + O_2 <1> (Reversibility: ?) [13]
P 7-dehydrocholesterol + H_2O_2

Substrates and products

S (24Z)-5α-stigmast-7,24(24^1)-dien-3β-ol + O_2 + NAD(P)H <5> (Reversibility: ?) [7]
P (24Z)-stigmasta-5,7,24(24^1)-trien-3-β-ol + H_2O_2 + $NAD(P)^+$ <5> [7]
S 5α-cholest-7-en-3β-ol + NADH + H^+ + O_2 <7> (<7> NADH is more efficient than NADPH [9]) (Reversibility: ?) [9]
P cholesta-5,7-dien-3β-ol + NAD^+ + 2 H_2O
S 5α-cholest-7-en-3β-ol + NADPH + H^+ + O_2 <7> (<7> NADH is more efficient than NADPH [9]) (Reversibility: ?) [9]
P cholesta-5,7-dien-3β-ol + $NADP^+$ + H_2O
S 5α-cholest-7-en-3β-ol + O_2 + NAD(P)H <3, 5> (<3> i.e. lathosterol [2,3,4]; <3> enzyme is rather a mixed function oxidase which requires O_2 and NAD(P)H for activity, than a dehydrogenase [2]; <3> NADPH, 40% of activity with NADH [2]; <3> cholesterol biosynthesis [2,3]) (Reversibility: ?) [2, 3, 4, 10]
P cholesta-5,7-dien-3β-ol + H_2O_2 + $NAD(P)^+$ <3, 5> [2, 3, 4, 10]
S 5α-ergost-7,24(241)-dien-3β-ol + O_2 + NAD(PH) <4, 7> (Reversibility: ?) [7]
P ergosta-5,7,24(241)-trien-3β-ol + H_2O_2 + $NAD(P)^+$ <4, 7> [7]
S lathosterol + O_2 <1> (Reversibility: ?) [13]
P 7-dehydrocholesterol + H_2O_2

Inhibitors

1,10-Phenanthroline <3, 5> (<3> 0.5 mM, 74% inhibition [2]) [2, 10]
4,5-dihydroxy-1,3-benzene-disulfonic acid <3> (<3> trivial name tiron, 0.5 mM, 80% inhibition [2]) [2]
Bathophenanthroline <3> (<3> 0.5 mM, 71% inhibition [2]) [2]

CN^- <3, 5, 7> (<3> 0.5 mM, 95% inhibition [2]; <7> 0.003 mM, 50% inhibition of recombinant enzyme [9]; <5> 0.003-0015 mM, 50% inhibition [10]) [2, 9, 10]
Dithiothreitol <3> (<3> 1.0 mM, 55% inhibition [2]) [2]
α-(2,4-dichlorophenyl)-α-phenyl-5-pyrimidine methanol <3> (<3> trivial name triarimol, weak inhibition [6]) [6]
anti-cytochrome b_5 antibodies <5> (<5> 0.5 mg/mg protein. 36% inhibition [1]) [1]
salicylic hydroxamic acid <5> [10]
Additional information <5> (<5> not inhibited by CO [10]) [10]

Cofactors/prosthetic groups

NAD(P)H <1, 2, 3, 4, 5, 7> (<5> NADH is more efficient [10]) [1, 2, 3, 4, 5, 6, 7, 8, 9, 10, 11, 12]
NADH <7> (<7> NADH is more efficient than NADPH [9]) [9]
NADPH <7> (<7> NADH is more efficient than NADPH [9]) [9]
cytochrome b_5 <3, 5, 7> (<3> required as electron carrier [2]; <5> suggested electron donor in plants [1]) [1, 2, 12]

Activating compounds

squalene and sterol carrier protein <3> (<3> 2fold stimulation of enzyme activity in microsomes, stimulation is lower if microsomes are pretreated with phospholipase A2, acetone powder, 0.05% sodium deoxycholate or 0.2% Triton X-100 [2]) [2, 5]

Metals, ions

Additional information <1, 2> (<1,2> the enzyme may hold non-heme irons in the catalytic center containing histidine residues [8]) [8]

Specific activity (U/mg)

0.0079 <3> [2]

K_m-Value (mM)

0.032 <7> (5α-Cholest-7-en-3β-ol, <7> 30°C, wild-type enzyme [9]) [9]
0.0357 <3> (5α-Cholest-7-en-3β-ol, <3> reconstituted enzyme system consisting of: cytochrome b_5, cytochrome b_5 reductase and egg lecithin liposomes [2]) [2]
0.07 <7> (5α-Cholest-7-en-3β-ol, <7> P175a mutant enzyme [9]; <7> 30°C, mutant enzyme P175A [9]) [9]
0.086 <7> (5α-Cholest-7-en-3β-ol, <7> P175V mutant enzyme [9]; <7> 30°C, mutant enzyme P175V [9]) [9]
0.09 <7> (5α-Cholest-7-en-3β-ol, <7> K115L mutant enzyme [9]; <7> 30°C, mutant enzyme K115L [9]) [9]
0.24 <7> (5α-Cholest-7-en-3β-ol, <7> T114I mutant enzyme [9]; <7> 30°C, mutant enzyme T114I [9]) [9]
0.244 <7> (5α-Cholest-7-en-3β-ol, <7> T114S mutant enzyme [9]; <7> 30°C, mutant enzyme T114S [9]) [9]

5 Isolation/Preparation/Mutation/Application

Source/tissue

brain <2> (<2> mRNA expression [11]) [11]
kidney <2> (<2> mRNA expression [11]) [11]
liver <2, 3> [2, 4, 6, 11]
lung <2> (<2> mRNA expression [11]) [11]
muscle <2> (<2> mRNA expression [11]) [11]
small intestine <2> (<2> mRNA expression [11]) [11]
stomach <2> (<2> mRNA expression [11]) [11]
testis <2> (<2> mRNA expression [11]) [11]

Localization

microsome <3, 5, 7> [2, 3, 4, 10, 12]

Purification

<3> [2]

Cloning

<1> (expression in Saccharomyces cerevisiae erg3 mutant) [8]
<2> [11]
<2> (expression in Saccharomyces cerevisiae erg3 mutant) [8]
<6> (expression in Saccharomyces cerevisiae erg3 mutant) [7]
<7> [9, 12]
<7> (expression in mutant erg3 of Saccharomyces cerevisiae) [9]
<7> (wild-type enzyme and ste1 mutant enzyme expressed in Sacharomyces cerevisiae erg3 mutant, wild-type enzyme overexpressed in Arabidopsis ste1 mutant) [7]

Engineering

G234A <7> (<7> no activity [9]; <7> inactive mutant enzyme [9]) [9]
G234D <7> (<7> 2% of wild-type activity [9]; <7> low activity [9]) [9]
H147E <7> (<7> no activity [9]; <7> inactive mutant enzyme [9]) [9]
H147L <7> (<7> no activity [9]; <7> inactive mutant enzyme [9]) [9]
H151E <7> (<7> no activity [9]; <7> inactive mutant enzyme [9]) [9]
H151l <7> (<7> no activity [9]; <7> inactive mutant enzyme [9]) [9]
H161E <7> (<7> no activity [9]; <7> inactive mutant enzyme [9]) [9]
H161L <7> (<7> no activity [9]; <7> inactive mutant enzyme [9]) [9]
H164L <7> (<7> no activity [9]; <7> inactive mutant enzyme [9]) [9]
H165L <7> (<7> no activity [9]; <7> inactive mutant enzyme [9]) [9]
H203L <7> (<7> reduced activity [9]; <7> 13% of wild-type activity [9]) [9]
H222E <7> (<7> reduced activity [9]; <7> 6% of wild-type activity [9]) [9]
H222L <7> (<7> no activity [9]; <7> inactive mutant enzyme [9]) [9]
H238E <7> (<7> no activity [9]; <7> inactive mutant enzyme [9]) [9]
H238L <7> (<7> no activity [9]; <7> inactive mutant enzyme [9]) [9]
H241L <7> (<7> no activity [9]; <7> inactive mutant enzyme [9]) [9]
H242L <7> (<7> no activity [9]; <7> inactive mutant enzyme [9]) [9]

K115L <7> (<7> 2times higher activity than wild-type [9]; <7> K_m-value for 5α-cholest-7-en-3β-ol is 2.8fold higher than that of wild-type activity [9]) [9]
P175A <7> (<7> wild-type activity [9]; <7> as active as wild-type enzyme [9]) [9]
P175V <7> (<7> 4times higher activity than wild-type [9]; <7> K_m-value for 5α-cholest-7-en-3β-ol is 2.7fold higher than that of wild-type activity [9]) [9]
P201A <7> (<7> 25% of wild-type activity [9]; <7> 20% of wild-type activity [9]) [9]
T114I <7> (<7> wild-type activity [9]) [9]
T114I | <7> (<7> K_m-value for 5α-cholest-7-en-3β-ol is 7.5fold higher than that of wild-type activity [9]) [9]
T114S <7> (<7> 28times higher activity than wild-type [9]; <7> K_m-value for 5α-cholest-7-en-3β-ol is 7.6fold higher than that of wild-type activity [9]) [9]

6 Stability

General stability information

<3>, extremely labile [2]

Storage stability

<3>, 0°C, 24 h, 30% loss of activity [2]

References

[1] Rahier, A.; Smith, M.; Taton, M.: The role of cytochrome b_5 in 4α-methyl-oxidation and C5(6) desaturation of plant sterol precursors. Biochem. Biophys. Res. Commun., **236**, 434-437 (1997)

[2] Kawata, S.; Traskos, J.M.; Gaylor, J.L.: Microsomal enzymes of cholesterol biosynthesis from lanosterol. Purification and characterization of Δ^7-sterol 5-desaturase of rat liver microsomes. J. Biol. Chem., **260**, 6609-6617 (1985)

[3] Grinstead, G.F.; Gaylor, J.L.: Total enzymic synthesis of cholesterol from 4,4,14α-trimethyl-5α-cholesta-8,24-dien-3β-ol. Solubilization, resolution, and reconstitution of Δ^7-sterol 5-desaturase. J. Biol. Chem., **257**, 13937-13944 (1982)

[4] Ishibashi, T.; Bloch, K.: Intermembrane transfer of 5α-cholest-7-en-3β-ol. Facilitation by supernatant protein (SCP). J. Biol. Chem., **256**, 12962-12967 (1981)

[5] Dempsey, M.E.; McCoy, K.E.; Barker, H.N.: Large scale purification and structural characterization of squalene and sterol carrier protein. J. Biol. Chem., **256**, 1867-1873 (1981)

[6] Mitropoulos, K.A.; Gibbons, G.F.: Effect of triarimol on cholesterol biosynthesis in rat-liver subcellular. Biochem. Biophys. Res. Commun., **71**, 892-900 (1976)

[7] Husselstein, T.; Schaller, H.; Gachotte, D.; Benveniste, P.: Δ^7-sterol-C5-desaturase: molecular characterization and functional expression of wild-type and mutant alleles. Plant Mol. Biol., **39**, 891-906 (1999)
[8] Nishi, S.; Nishino, H.; Ishibashi, T.: cDNA cloning of the mammalian sterol C5-desaturase and the expression in yeast mutant. Biochim. Biophys. Acta, **1490**, 106-108 (2000)
[9] Taton, M.; Husselstein, T.; Benveniste, P.; Rahier, A.: Role of highly conserved residues in the reaction catalyzed by recombinant Δ^7-sterol-C5(6)-desaturase studied by site-directed mutagenesis. Biochemistry, **39**, 701-711 (2000)
[10] Rahier, A.; Benveniste, P.; Husselstein, T.; Taton, M.: Biochemistry and site-directed mutational analysis of Δ^7-sterol-C5(6)-desaturase. Biochem. Soc. Trans., **28**, 799-803 (2000)
[11] Sugawara, T.; Fujimoto, Y.; Ishibashi, T.: Molecular cloning and structural analysis of human sterol C_5 desaturase. Biochim. Biophys. Acta, **1533**, 277-284 (2001)
[12] Rahier, A.: Deuterated Δ^7-cholestenol analogues as mechanistic probes for wild-type and mutated Δ^7-sterol-C5(6)-desaturase. Biochemistry, **40**, 256-267 (2001)
[13] Krakowiak, P.A.; Wassif, C.A.; Kratz, L.; Cozma, D.; Kovarova, M.; Harris, G.; Grinberg, A.; Yang, Y.; Hunter, A.G.; Tsokos, M.; Kelley, R.I.; Porter, F.D.: Lathosterolosis: an inborn error of human and murine cholesterol synthesis due to lathosterol 5-desaturase deficiency. Hum. Mol. Genet., **12**, 1631-1641 (2003)
[14] Lin, X.; Chen, Z.; Yue, P.; Averna, M.R.; Ostlund, R.E.; Watson, M.A.; Schonfeld, G.: A targeted apoB38.9 mutation in mice is associated with reduced hepatic cholesterol synthesis and enhanced lipid peroxidation. Am. J. Physiol. Gastrointest. Liver Physiol., **290**, G1170-G1176 (2006)

Cholesterol 25-hydroxylase 1.14.99.38

1 Nomenclature

EC number
1.14.99.38

Systematic name
cholesterol,hydrogen-donor:oxygen oxidoreductase (25-hydroxylating)

Recommended name
cholesterol 25-hydroxylase

Synonyms
cholesterol 25-hydroxylase <3, 4> [5]

CAS registry number
60202-07-5

2 Source Organism

<1> *Homo sapiens* (no sequence specified) [6, 7]
<2> *Rattus norvegicus* (no sequence specified) [2, 3, 4]
<3> *Mus musculus* (UNIPROT accession number: Q9Z0F5) [1, 5]
<4> *Homo sapiens* (UNIPROT accession number: O95992) [1, 5]

3 Reaction and Specificity

Catalyzed reaction
cholesterol + AH_2 + O_2 = 25-hydroxycholesterol + A + H_2O

Reaction type
oxidation
redox reaction
reduction

Natural substrates and products
S cholesterol + AH_2 + O_2 <1, 2, 3, 4> (<3,4> cholesterol 25-hydroxylase has the capacity to play an important role in regulating lipid metabolism by synthesizing a corepressor that blocks sterol regulatory element binding protein processing and ultimately leads to inhibition of gene transcription [5]; <1> genetic variability in the set of cholesterol metabolism genes: cholesterol 25-hydroxylase, cholesterol 24-hydroxylase and ATP-binding

cassette transporter A1 do not influence the development of Alzheimer's disease [6]; <2> testicular macrophagers play an important role in the differentiation of Leydig cells through the secretion of 25-hydroxycholesterol [3]; <3,4> the enzyme has the capacity to synthesize an active and potent regulator of the sterol response element binding protein pathway within the cell [1]) (Reversibility: ?) [1, 3, 5, 6]

P 25-hydroxycholesterol + A + H_2O

Substrates and products

S cholesterol + AH_2 + O_2 <1, 2, 3, 4> (<3,4> cholesterol 25-hydroxylase has the capacity to play an important role in regulating lipid metabolism by synthesizing a corepressor that blocks sterol regulatory element binding protein processing and ultimately leads to inhibition of gene transcription [5]; <1> genetic variability in the set of cholesterol metabolism genes: cholesterol 25-hydroxylase, cholesterol 24-hydroxylase and ATP-binding cassette transporter A1 do not influence the development of Alzheimer's disease [6]; <2> testicular macrophagers play an important role in the differentiation of Leydig cells through the secretion of 25-hydroxycholesterol [3]; <3,4> the enzyme has the capacity to synthesize an active and potent regulator of the sterol response element binding protein pathway within the cell [1]; <2> expression of 25-hydroxylase mRNA and production of 25-hydroxycholesterol by cultured testicular macrophages is significantly inhibited by testosterone at 0.01 mg/ml [4]) (Reversibility: ?) [1, 2, 3, 4, 5, 6]

P 25-hydroxycholesterol + A + H_2O

Metals, ions

Fe <3, 4> (<3,4> diiron cofactor [5]; <3,4> diiron-containing enzyme. The iron is present as Fe-O-Fe or Fe-OH-Fe and is coordinated with clusters of conserved histidine residues [1]) [1, 5]

4 Enzyme Structure

Subunits

? <3, 4> (<4> x * 31700 [1]; <3> x * 34700 [1]) [1]

5 Isolation/Preparation/Mutation/Application

Source/tissue

heart <3> [1]
kidney <3> [1]
lung <3> [1]
macrophage <2> (<2> testicular [2,4]; <2> testicular, 10-day-old animals have the highest steady-state levels of message [3]) [2, 3, 4]

Localization

Golgi apparatus <4> [1]
endoplasmic reticulum <4> [1]
membrane <3, 4> (<3,4> polytopic membrane protein [5]) [5]

Cloning

<3> [1]
<3> (expression in COS cells, expression of cholesterol 25-hydroxylase in transfected cells reduces the biosynthesis of cholesterol from acetate and suppresses the cleavage of sterol regulatory element binding protein-1 and -2) [5]
<4> [1]
<4> (expression in COS cells, expression of cholesterol 25-hydroxylase in transfected cells reduces the biosynthesis of cholesterol from acetate and suppresses the cleavage of sterol regulatory element binding protein-1 and -2) [5]

Application

medicine <1> (<1> genetic variability in the set of cholesterol metabolism genes: cholesterol 25-hydroxylase, cholesterol 24-hydroxylase and ATP-binding cassette transporter A1 do not influence the development of Alzheimer's disease [6]; <1> patients with Alzheimer's disease show high expression of cholesterol 25-hydroxylase gene in specifically vulnerable brain regions. The common enzyme haplotype CH25Hχ4 is a putative susceptibility factor for sporadic Alzheimer's disease [7]) [6, 7]

References

[1] Russell, D.W.: Oxysterol biosynthetic enzymes. Biochim. Biophys. Acta, **1529**, 126-135 (2000)

[2] Lukyanenko, Y.O.; Chen, J.J.; Hutson, J.C.: Production of 25-hydroxycholesterol by testicular macrophages and its effects on Leydig cells. Biol. Reprod., **64**, 790-796 (2001)

[3] Chen, J.J.; Lukyanenko, Y.; Hutson, J.C.: 25-hydroxycholesterol is produced by testicular macrophages during the early postnatal period and influences differentiation of Leydig cells in vitro. Biol. Reprod., **66**, 1336-1341 (2002)

[4] Lukyanenko, Y.; Chen, J.J.; Hutson, J.C.: Testosterone regulates 25-hydroxycholesterol production in testicular macrophages. Biol. Reprod., **67**, 1435-1438 (2002)

[5] Lund, E.G.; Kerr, T.A.; Sakai, J.; Li, W.P.; Russell, D.W.: cDNA cloning of mouse and human cholesterol 25-hydroxylases, polytopic membrane proteins that synthesize a potent oxysterol regulator of lipid metabolism. J. Biol. Chem., **273**, 34316-34327 (1998)

[6] Shibata, N.; Kawarai, T.; Lee, J.H.; Lee, H.S.; Shibata, E.; Sato, C.; Liang, Y.; Duara, R.; Mayeux, R.P.; St George-Hyslop, P.H.; Rogaeva, E.: Association studies of cholesterol metabolism genes (CH25H, ABCA1 and CH24H) in Alzheimer's disease. Neurosci. Lett., **391**, 142-146 (2006)

[7] Papassotiropoulos, A.; Lambert, J.C.; Wavrant-De Vrieze, F.; Wollmer, M.A.; von der Kammer, H.; Streffer, J.R.; Maddalena, A.; Huynh, K.D.; Wolleb, S.; Lutjohann, D.; Schneider, B.; Thal, D.R.; Grimaldi, L.M.; Tsolaki, M.; Kapaki, E.; Ravid, R.; Konietzko, U.; Hegi, T.; Pasch, T.; Jung, H.; Braak, H.: Cholesterol 25-hydroxylase on chromosome 10q is a susceptibility gene for sporadic Alzheimers disease. Neurodegener. Dis., **2**, 233-241 (2005)

Cob(II)yrinic acid a,c-diamide reductase 1.16.8.1

1 Nomenclature

EC number
1.16.8.1

Systematic name
cob(II)yrinic acid-a,c-diamide:FMN oxidoreductase

Recommended name
cob(II)yrinic acid a,c-diamide reductase

Synonyms
cob(II)yrinic acid a,c-diamide reductase <1> [1]

CAS registry number
145539-93-1

2 Source Organism

<1> *Pseudomonas denitrificans* (no sequence specified) [1]

3 Reaction and Specificity

Catalyzed reaction
2 cob(II)yrinic acid a,c-diamide + FMN = 2 cob(I)yrinic acid a,c-diamide + $FMNH_2$

Reaction type
oxidation
redox reaction
reduction

Natural substrates and products
S 2 cob(II)yrinic acid a,c-diamide + FMN <1> (Reversibility: ?) [1]
P cob(I)yrinic acid a,c-diamide + $FMNH_2$
S cob(II)yrinic acid a,c-diamide + FMN <1> (Reversibility: ?) [1]
P ? <1> [1]

Substrates and products
S 2 cob(II)yrinic acid a,c-diamide + FMN <1> (Reversibility: ?) [1]
P cob(I)yrinic acid a,c-diamide + $FMNH_2$

S GDP-cob(II)inamide + FMN <1> (<1> 71% activity compared to cob(II)yric acid [1]) (Reversibility: ?) [1]
P ? <1> [1]
S cob(II)alamin + FMN <1> (<1> 41% activity compared to cob(II)yric acid [1]) (Reversibility: ?) [1]
P ? <1> [1]
S cob(II)inamide + FMN <1> (<1> 67-48% activity compared to cob(II)yric acid, percentage depends on method of activity detection [1]) (Reversibility: ?) [1]
P ? <1> [1]
S cob(II)inamide phosphate + FMN <1> (<1> 66% activity compared to cob(II)yric acid [1]) (Reversibility: ?) [1]
P ? <1> [1]
S cob(II)yriC acid + FMN <1> (Reversibility: ?) [1]
P ? <1> [1]
S cob(II)yrinic acid a,c-diamide + FMN <1> (Reversibility: ?) [1]
P ? <1> [1]
S cob(II)yrinic acid a,c-diamide + FMN <1> (<1> 52-75% activity compared to cob(II)yric acid, percentage depends on method of activity detection [1]) (Reversibility: ?) [1]
P cob(I)yrinic acid a,c-diamide + $FMNH_2$ <1> [1]

K_m-Value (mM)

0.0059 <1> (FMN, <1> 0.2 M Tris-HCl buffer, pH 8.0, 30°C [1]) [1]

4 Enzyme Structure

Molecular weight

30000 <1> (<1> gel permeation HPLC [1]) [1]

Subunits

homodimer <1> (<1> 2 * 16000, SDS-PAGE [1]) [1]

5 Isolation/Preparation/Mutation/Application

Purification

<1> (6,300-fold purification including Mono Q and FMN agarose chromatography) [1]

References

[1] Blanche F.; Maton, L.; Debussche, L.; Thibaut, D.: Purification and characterization of cob(II)yrinic acid a,c-diamide reductase from Pseudomonas denitrificans. J. Bacteriol., **174**, 7452-7454 (1992)

Xanthine dehydrogenase 1.17.1.4

1 Nomenclature

EC number
1.17.1.4

Systematic name
xanthine:NAD^+ oxidoreductase

Recommended name
xanthine dehydrogenase

Synonyms
EC 1.1.1.204 (formerly)
EC 1.2.1.37 (formerly)
NAD-xanthine dehydrogenase
Rosy locus protein
XDH/XO
XOR
xanthine oxidoreductase
xanthine-NAD oxidoreductase
xanthine/NAD^+ oxidoreductase

CAS registry number
9054-84-6

2 Source Organism

<1> *Gallus gallus* (no sequence specified) [3, 19, 23, 26, 29, 53, 59]
<2> *Meleagris gallopavo* (no sequence specified) [21, 61]
<3> *Drosophila melanogaster* (no sequence specified) [7, 25, 31, 42, 43, 45, 52, 57]
<4> *Chlamydomonas reinhardtii* (no sequence specified) [20, 37, 58]
<5> *Bacillus subtilis* (no sequence specified) [16]
<6> *Mus musculus* (no sequence specified) [12,83,85,93]
<7> *Homo sapiens* (no sequence specified) [4,5,32,76,84,88]
<8> *Rattus norvegicus* (no sequence specified) [66,70,75,86,89,92]
<9> *Bos taurus* (no sequence specified) [2,11,14,34,35,42,64,71,72,76,79,84]
<10> *Triticum aestivum* (no sequence specified) (<10> α_1 subunit [15]) [15]
<11> *Oryctolagus cuniculus* (no sequence specified) (<11> fragment of dihydropteroate synthase [4,5]) [4, 5]
<12> *Neurospora crassa* (no sequence specified) [51]

<13> *Ovis aries* (no sequence specified) [76]
<14> *Pisum sativum* (no sequence specified) [36, 78]
<15> *Glycine max* (no sequence specified) (<15> alkB4 gene from Rhodococcus NRRL B-16531 [49,56]) [49, 56]
<16> *Arabidopsis thaliana* (no sequence specified) [77]
<17> *Pseudomonas aeruginosa* (no sequence specified) [81, 82]
<18> *Pseudomonas sp.* (no sequence specified) [41]
<19> *Pseudomonas putida* (no sequence specified) [8, 9, 40, 41]
<20> *Rhodobacter capsulatus* (no sequence specified) (<20> BCA_2 [72]) [50, 72, 74, 90, 91]
<21> *Clostridium cylindrosporum* (no sequence specified) [60]
<22> *Bacillus sp.* (no sequence specified) [41]
<23> *Lycopersicon esculentum* (no sequence specified) [77]
<24> *Arthrobacter sp.* (no sequence specified) [41]
<25> *Lactobacillus sp.* (no sequence specified) [41]
<26> *Capra hircus* (no sequence specified) [76, 87]
<27> *Alcaligenes sp.* (no sequence specified) [41]
<28> *Nocardia sp.* (no sequence specified) [41]
<29> *Pseudomonas acidovorans* (no sequence specified) [28]
<30> *Streptomyces sp.* (no sequence specified) [41]
<31> *Clostridium sp.* (no sequence specified) [41]
<32> *Persea americana* (no sequence specified) [65]
<33> *Calliphora vicina* (no sequence specified) [22]
<34> *Rhodopseudomonas capsulata* (no sequence specified) [55]
<35> *Veillonella sp.* (no sequence specified) [41]
<36> *Penicillium sp.* (no sequence specified) [41]
<37> *Camelus dromedarius* (no sequence specified) [67]
<38> *Serratia sp.* (no sequence specified) [41]
<39> *Rattus sp.* (no sequence specified) [1, 6, 10, 18, 27, 29, 34, 38, 46, 47, 48]
<40> *Pseudomonas synxantha* (no sequence specified) [63]
<41> *Comamonas acidovorans* (no sequence specified) [30, 80]
<42> *Clostridium acidiurici* (no sequence specified) [39, 60]
<43> *Escherichia sp.* (no sequence specified) [41]
<44> *Nicotiana plumbaginifolia* (no sequence specified) [33]
<45> *Streptomyces cyanogenus* (no sequence specified) [62]
<46> *Clostridium purinolyticum* (no sequence specified) [24, 69]
<47> no activity in *Klebsiella sp.* [41]
<48> *Colias eurytheme* (no sequence specified) [54]
<49> *Veillonella atypica* (no sequence specified) [17, 44]
<50> *Peptococcus sp.* (no sequence specified) [41]
<51> *Eubacterium barkeri* (no sequence specified) [13]
<52> no activity in *Columba domestica* [29]
<53> *Arabidopsis thaliana* (UNIPROT accession number: Q8GUQ8) [73]
<54> *Comamonas acidovorans* (UNIPROT accession number: Q8RLC1) [68]
<55> *Comamonas acidovorans* (UNIPROT accession number: Q8RLC0) [68]

3 Reaction and Specificity

Catalyzed reaction

xanthine + NAD^+ + H_2O = urate + NADH + H^+ (<9> catalytically labile Mo-OH oxygen forms a bond with a carbon atom of substrate, the Mo=S group of the oxidized enzyme becomes protonated to afford Mo-SH on reduction of the molybdenum center [79]; <8> mechanism of conversion of dehydrogenase form to oxidase form [75])

Reaction type

oxidation
redox reaction
reduction

Natural substrates and products

S 2-amino-4-hydroxy-pterine + NAD^+ + H_2O <3, 9, 10, 12, 33, 39, 48> (<3,9,10,12,33,39,48> i.e. pterin [15,22,25,31,38,42,43,45,51,54]; <39> conversion of xanthine dehydrogenase to xanthine oxidase is strongly influenced by in vitro cell culture of alveolar epithelial cells [38]; <10> 11% of the activity compared to xanthine [15]) (Reversibility: r) [15, 22, 25, 31, 38, 42, 43, 45, 51, 54]

P isoxanthopterin + NADH <3, 9, 10, 12, 33, 39, 48> (<33> precursor of the eye pigment drosopterin [22]) [15, 22, 25, 31, 38, 42, 43, 45, 51, 54]

S 2-hydroxypurine + NAD^+ + H_2O <45> (<45> considerable activity [62]) (Reversibility: ?) [62]

P xanthine + 2,8-dihydroxypurine + NADH <45> (<45> 84% xanthine, 8% 2,8-dihydroxypurine formed [62]) [62]

S 6,8-dihydroxypurine + NAD^+ + H_2O <45> (Reversibility: ?) [62]

P urate + NADH <45> (<45> 18% urate formed [62]) [62]

S 6,8-dihydroxypurine + O_2 <49> (Reversibility: ?) [44]

P ? + H_2O_2 <49> [44]

S NADPH + O_2 <51> (<51> high NADPH oxidase activity [13]) (Reversibility: ?) [13]

P $NADP^+$ + O_2^- + H^+ <51> [13]

S guanine + NAD^+ + H_2O <45> (<45> 81.3% of the activity compared to hypoxanthine [62]) (Reversibility: ?) [62]

P ? + NADH <45> [62]

S hypoxanthine + NAD^+ + H^+ + O_2^- <1, 2, 3, 4, 12, 15, 19, 21, 33, 34, 39, 42> (<34> preferred substrate [55]; <2> subtilisin treatment leads to an active component I of 120000 kDa [21]; <4,33> NAD^+-O_2- dependent xanthine oxidase activity [20,22]; <15> predominant reaction [56]; <19> more rapidly oxidized than xanthine [8]) (Reversibility: r) [8, 20, 21, 22, 46, 51, 53, 55, 56, 57, 58, 60]

P xanthine + NADH + H_2O_2 <1, 2, 3, 4, 12, 15, 19, 21, 33, 34, 39, 42> [8, 20, 21, 22, 46, 51, 53, 55, 56, 57, 58, 60]

S hypoxanthine + NAD^+ + H_2O <3, 4, 5, 9, 14, 19, 20, 29, 39, 40, 44, 48> (<40> preferred substrate [63]) (Reversibility: ?) [2, 9, 10, 11, 16, 28, 33, 36, 37, 50, 52, 54, 63]

P xanthine + NADH <3, 4, 5, 9, 14, 19, 20, 29, 39, 40, 44, 48> [2, 9, 10, 11, 16, 28, 33, 36, 37, 50, 52, 54, 63]

S hypoxanthine + NAD^+ + H_2O <45> (<45> preferred substrate [62]) (Reversibility: ?) [62]

P xanthine + 6,8-dihydroxypurine + NADH <45> (<45> 100% xanthine, 51% 6,8-dihydroxypurine formed [62]) [62]

S hypoxanthine + NADH <15> (<15> 0.1% of the xanthine oxidation rate [49]) (Reversibility: ?) [49]

P ? + NO_2^- + NAD^+ <15> [49]

S hypoxanthine + $NADP^+$ + H_2O <4, 14, 45, 51> (<45> 2.4% of the activity compared to NAD^+ [62]; <14> 40% of the activity compared to NAD^+ [36]; <51> strict specificity for $NADP^+$ [13]) (Reversibility: ?) [13, 36, 37, 62]

P xanthine + NADPH <4, 14, 45, 51> [13, 36, 37, 62]

S hypoxanthine + O_2 <45, 49> (<45> very low activity [62]) (Reversibility: ?) [44, 62]

P xanthine + H_2O_2 <45, 49> [44, 62]

S hypoxanthine + urate <49> (<49> oxygen-free assay [44]) (Reversibility: r) [44]

P xanthine + 6,8-dihydroxypurine <49> [44]

S purine + NAD^+ + H_2O <10, 19, 29, 40, 48> (<10,19,29,40> poor substrate [8,15,28,63]) (Reversibility: ?) [8, 15, 28, 54, 63]

P ? + NADH <10, 19, 29, 40, 48> [8, 15, 28, 54, 63]

S purine + NAD^+ + H_2O <45> (<45> poor substrate [62]) (Reversibility: ?) [62]

P hypoxanthine + 8-hydroxypurine + 2-hydroxypurine + NADH <45> (<45> 2.3% hypoxanthine, 2.3% 8-hydroxypurine and traces of 2-hydroxypurine formed [62]) [62]

S purine + $NADP^+$ + H_2O <51> (<51> 60% of the activity compared to hypoxanthine [13]) (Reversibility: ?) [13]

P ? + NADPH <51> [13]

S purine + O_2 + H_2O <49> (Reversibility: ?) [44]

P 8-hydroxypurine + H_2O_2 <49> [44]

S xanthine + NAD^+ + H_2O <1, 2, 3, 4, 5, 6, 7, 9, 10, 11, 12, 14, 15, 18, 19, 20, 21, 22, 27, 29, 30, 33, 34, 38, 39, 40, 41, 42, 45, 48> (<9> xanthine dehydrogenase form has distinct xanthine/oxygen activity, 35-42% of electrons transferred to O_2 to form O_2^- [35]; <9> conversion of dehydrogenase to oxidase type due to oxidation of sulfhydryl groups by molecular oxygen, dehydrogenase activity recovered by treatment with dithiothreitol [64]; <10> strict dehydrogenase activity, no utilization of O_2 [15]; <45> 61% of the activity compared to hypoxanthine [62]; <1> only dehydrogenase type D present [29]; <1> ping-pong reaction mechanism [26]; <39> NAD^+-dependent form is postulated to play a regulatory role in purine metabolism [47]; <19> NAD^+-linked activity, very low activity towards molecular oxygen [40]; <40> 75% of the activity compared to hypoxanthine [63]; <12> degradative pathway of conversion of purines to ammonia [51]; <14> only present as stable dehydrogenase from, no conversion to the oxidase form [36]; <6> xanthine oxidase form is the principle major form in fresh mouse milk, dehydrogenase form is the major form in mammary gland, conversion to the dehydrogenase form by thiol active compounds [12]; <3> involved in pteridine metabolism, 40% of activity compared to hypoxanthine [52]; <39> conversion of xanthine dehydrogenase to the oxidase type by thiol-disulfide oxidoreductase, thiol reagents or oxidized glutathione [1]; <34> 67% of the activity compared to hypoxanthine [55]; <5> regulation of xanthine dehydrogenase expression is subjected to nitrogen catabolite repression mediated through the GlnA-dependent signaling pathway [16]; <39> trypsin treatment leads to a complete conversion of xanthine dehydrogenase to xanthine oxidase activity [6]; <1> minimum degree of 1 : 1 for xanthine, 2 : 2 for NAD, 1 : 1 for urate and 1 : 2 for NADH in the xanthine/NAD^+ oxidoreductase reaction required [3]; <1> xanthine dehydrogenase can be partially reduced in a triphasic reaction by either xanthine or NADH, oxidation of fully, 6-electron-reduced xanthine dehydrogenase by either urate or NAD^+ is monophasic and depends on the oxidant concentration [23]; <2> subtilisin treatment leads to an active component I of 120000 kDa [21]; <9,39> NAD^+-dependent dehydrogenase type D [14,27,34]) (Reversibility: r) [1, 2, 3, 4, 5, 6, 7, 8, 9, 10, 11, 12, 14, 15, 16, 18, 19, 21, 22, 23, 25, 26, 27, 28, 29, 30, 31, 34, 35, 36, 37, 40, 41, 42, 45, 46, 47, 48, 50, 51, 52, 53, 54, 55, 56, 57, 59, 60, 62, 63, 64]

P urate + NADH <1, 2, 3, 4, 5, 6, 7, 9, 10, 11, 12, 14, 15, 18, 19, 20, 21, 22, 27, 29, 30, 33, 34, 38, 39, 40, 41, 42, 45, 48> (<1> NADH-binding to the 2-electron reduced enzyme is implicated in fixing end-point position in reactions involving pyridine nucleotides, urate-binding is involved in fixing end-point reactions involving xanthine and urate [23]; <45> 91% urate formed [62]) [1, 2, 3, 4, 5, 6, 7, 8, 9, 10, 11, 12, 14, 15, 16, 18, 19, 21, 22, 23, 25, 26, 27, 28, 29, 30, 31, 34, 35, 36, 37, 40, 41, 42, 45, 46, 47, 48, 50, 51, 52, 53, 54, 55, 56, 57, 59, 60, 62, 63, 64]

S xanthine + NAD^+ + O_2 <39> (<39> heat-treated intermediate dehydrogenase/oxidase type O [34]; <39> intermediate form of dehydrogenase/oxidase type D/O [29]) (Reversibility: ?) [27, 29, 34]

P urate + NADH + H_2O_2 <39> [27, 29, 34]

S xanthine + $NADP^+$ + H_2O <29, 51> (<51> strict specificity for $NADP^+$ [13]; <29> 11% of the activity compared to NAD^+ [28]) (Reversibility: ?) [13, 28]

P urate + NADPH <29, 51> [13, 28]

S xanthine + O_2 <1, 2, 3, 9, 14, 18, 19, 20, 24, 28, 29, 38, 39, 42, 43, 44, 49> (<19, 42, 49> very low activity [17, 39, 40]; <1, 14, 29> low activity [26, 28, 36]; <2> subtilisin treatment leads to an active component I of 120000 kDa [21]; <19> 4% of the activity compared to xanthine-NAD^+ [9]; <49> presence of ferredoxin enhances rate of oxygen reduction [44]; <39> toxic reactions of xanthine oxidase-derived radicals are critical factors in several mechanisms of tissue pathology [38]; <39> NAD^+-independent trypsin-treated oxidase type O [34]; <39> NAD^+-independent xanthine oxidase activity, low activity present in the enzyme preparation, conversion of the NAD^+-dependent to NAD^+-independent activity by some thiol reagents [47]; <9> xanthine oxidase form transfers 22% electrons to oxygen to form superoxide [35]) (Reversibility: r) [9, 14, 17, 21, 25, 26, 27, 28, 29, 31, 33, 34, 35, 36, 38, 39, 40, 41, 44, 45, 47, 50, 61]

P urate + O_2^- + H^+ <1, 2, 3, 9, 14, 18, 19, 20, 24, 28, 29, 38, 39, 42, 43, 44, 49> [9, 14, 17, 21, 25, 26, 27, 28, 29, 31, 33, 34, 35, 36, 38, 39, 40, 41, 44, 45, 47, 50, 61]

S xanthine + O_2 <49> (<49> dismutation reaction [44]) (Reversibility: ?) [44]

P hypoxanthine + ? <49> [44]

S xanthopterin + NAD^+ + H_2O <48> (<48> regulation of the pteridine pathway by competitive inhibition of reaction products and the precursor of xanthopterin, 7,8-dihydroxanthopterin [54]) (Reversibility: ?) [54]

P leucopterin + NADH <48> [54]

S Additional information <7, 8, 9> (<8> major function of enzyme in liver parenchymal and sinusoidal cells is the production of uric acid as a antioxidant [66]; <8> NADH oxidation by xanthine oxidoreductase may constitute an important pathway for reactive oxygen species-mediated tissue injuries. Xanthine oxidoreductase and xanthine oxidase catalyze the NADH oxidation, generating O_2^- radicals and inducing the peroxidation of liposomes, in a NADH and enzyme dependent manner [86]; <7,9> xanthine oxidoreductase plays a physiological role in milk equal in importance to its catalytic function as an enzyme [84]) (Reversibility: ?) [66, 84, 86]

P ?

Substrates and products

S 1-methylhypoxanthine + NAD^+ + H_2O <34> (<34> 10% of the activity compared to hypoxanthine [55]) (Reversibility: ?) [55]

P 1-methylxanthine + NADH <34> [55]

S 1-methylxanthine + NAD^+ + H_2O <20> (<20> 10fold reduced k_{red}-value compared to xanthine [91]) (Reversibility: ?) [91]

P 1-methylurate + NADH + H^+

S 1-methylxanthine + NAD^+ + H_2O <18, 19, 22, 27, 30, 34, 38, 43> (<34> 10% of the activity compared to hypoxanthine [55]) (Reversibility: ?) [41, 55]

P 1-methylurate + NADH <18, 19, 22, 27, 30, 34, 38, 43> [41, 55]

S 1-methylxanthine + ferricyanide + H_2O <18, 19, 22, 24, 25, 27, 28, 30, 36, 46, 50> (Reversibility: ?) [24, 41]

P 1-methylurate + ferrocyanide <18, 19, 22, 24, 25, 27, 28, 30, 36, 46, 50> [24, 41]

S 1-naphthaldehyde + NAD^+ + ? <53> (<53> 27.5% of the activity with xanthine [73]) (Reversibility: ?) [73]

P ? + NADH

S 2,6-diaminopurine + NAD$^+$ + H_2O <20> (<20> poor substrate [91]) (Reversibility: ?) [91]

P ? + NADH + H$^+$

S 2,6-dithiopurine + NAD$^+$ + H_2O <40> (<40> 26% of the activity compared to hypoxanthine [63]) (Reversibility: ?) [63]

P ? + NADH <40> [63]

S 2-OH-purine + ferricyanide + H_2O <46> (Reversibility: ?) [24]

P ? + ferrocyanide <46> [24]

S 2-amino-4-hydroxy-pterin + methylene blue + H_2O <3, 39> (Reversibility: ?) [31, 38, 42]

P isoxanthopterin + reduced methylene blue <3, 39> [31, 38, 42]

S 2-amino-4-hydroxy-pterin + NAD$^+$ + H_2O <3, 9, 10, 12, 33, 39, 48> (<3, 9, 10, 12, 33, 39, 48> i.e. pterin [15, 22, 25, 31, 38, 42, 43, 45, 51, 54]; <39> conversion of xanthine dehydrogenase to xanthine oxidase is strongly influenced by in vitro cell culture of alveolar epithelial cells [38]; <10> 11% of the activity compared to xanthine [15]) (Reversibility: r) [15, 22, 25, 31, 38, 42, 43, 45, 51, 54]

P isoxanthopterin + NADH <3, 9, 10, 12, 33, 39, 48> (<33> precursor of the eye pigment drosopterin [22]) [15, 22, 25, 31, 38, 42, 43, 45, 51, 54]

S 2-amino-4-hydroxypterin + NAD$^+$ + H_2O <3, 9, 10, 12, 33, 39, 48> (<3, 9, 10, 12, 33, 39, 48> i.e. pterin [15, 22, 25, 31, 38, 42, 43, 45, 51, 54]; <39> conversion of xanthine dehydrogenase to xanthine oxidase is strongly influenced by in vitro cell culture of alveolar epithelial cells [38]; <10> 11% of the activity compared to xanthine [15]) (Reversibility: r) [15, 22, 25, 31, 38, 42, 43, 45, 51, 54]

P isoxanthopterin + NADH <3, 9, 10, 12, 33, 39, 48> (<33> precursor of the eye pigment drosopterin [22]) [15, 22, 25, 31, 38, 42, 43, 45, 51, 54]

S 2-amino-4-hydroxypterin + nitroblue tetrazolium + H_2O <10> (<10> very low activity [15]; <10> i.e. pterin [15]) (Reversibility: ?) [15]

P isoxanthopterin + reduced nitroblue tetrazolium <10> [15]

S 2-hydroxy-6-methylpurine + NAD$^+$ + H_2O <20> (<20> poor substrate [91]) (Reversibility: ?) [91]

P ? + NADH + H$^+$

S 2-hydroxypurine + NAD$^+$ + H_2O <34> (<34> 35% of the activity compared to hypoxanthine, purine not oxidized [55]) (Reversibility: ?) [55]

P ? + NADH <34> [55]

S 2-hydroxypurine + NAD$^+$ + H_2O <45> (<45> considerable activity [62]) (Reversibility: ?) [62]

P xanthine + 2,8-dihydroxypurine + NADH <45> (<45> 84% xanthine, 8% 2,8-dihydroxypurine formed [62]) [62]

S 2-thioxanthine + NAD$^+$ + H_2O <20> (<20> good substrate [91]) (Reversibility: ?) [91]

P 2-thiourate + NADH + H$^+$

S 2-thioxanthine + NAD$^+$ + H_2O <40> (<40> 57% of the activity compared to hypoxanthine [63]) (Reversibility: ?) [63]

P 2-thiourate + NADH <40> [63]

S 3-methylxanthine + ferricyanide + H_2O <18, 19, 27> (Reversibility: ?) [41]

P 3-methylurate + ferrocyanide <18, 19, 27> [41]

S 4-aminoimidazole-5-carboxamide + $NADP^+$ + H_2O <51> (<51> 38% of the activity compared to hypoxanthine-$NADP^+$ [13]) (Reversibility: ?) [13]

P ? + NADPH <51> [13]

S 4-hydroxypyrazolo(3,4-d)pyrimidine + NAD^+ + H_2O <39, 48> (<48> best substrate tested [54]; <39,48> i.e. allopurinol [38,54]) (Reversibility: ?) [38, 54]

P 4,6-dihydroxypyrazolo(3,4-d)pyrimidine + NADH <39, 48> [38, 54]

S 4-hydroxypyrazolo(3,4-d)pyrimidine + ferricyanide + H_2O <46> (<46> i.e. allopurinol [24]) (Reversibility: ?) [24]

P 4,6-dihydroxypyrazolo(3,4-d)pyrimidine + ferrocyanide <46> [24]

S 4-hydroxypyrazolo(3,4-d)pyrimidine + methyl viologen + H_2O <42> (<42> i.e. allopurinol [39]; <42> 8% of the activity compared to xanthine [39]) (Reversibility: ?) [39]

P 4,6-dihydroxypyrazolo(3,4-d)pyrimidine + reduced methyl viologen <42> [39]

S 4-hydroxypyrazolo(3,4-d)pyrimidine + nitroblue tetrazolium + H_2O <10, 2> (<10> very low activity [15]; <10,20> i.e. allopurinol [15,50]) (Reversibility: ?) [15, 5]

P 4,6-dihydroxypyrazolo(3,4-d)pyrimidine + reduced nitroblue tetrazolium <10, 2> [15, 5]

S 5-aminoimidazol-4-carboxamide + methyl viologen + H_2O <42> (<42> 1.1% of the activity compared to xanthine [39]) (Reversibility: ?) [39]

P ? + reduced methyl viologen <42> [39]

S 6,8-dihydropurine + NAD^+ + H_2O <34> (<34> 50% of the activity compared to hypoxanthine [55]) (Reversibility: ?) [55]

P ? + NADH <34> [55]

S 6,8-dihydroxypurine + NAD^+ + H_2O <45> (Reversibility: ?) [62]

P urate + NADH <45> (<45> 18% urate formed [62]) [62]

S 6,8-dihydroxypurine + O_2 <49> (Reversibility: ?) [44]

P ? + H_2O_2 <49> [44]

S 6,8-dihydroxypurine + O_2 + H_2O <49> (Reversibility: ?) [44]

P ? + H_2O_2 <49> [44]

S 6-mercaptopurine + methyl viologen + H_2O <42> (<42> 9.5% of the activity compared to xanthine [39]) (Reversibility: ?) [39]

P ? + reduced methyl viologen <42> [39]

S 6-thioxanthine + NAD^+ + H_2O <20> (<20> good substrate [91]) (Reversibility: ?) [91]

P ? + NADH + H^+

S 6-thioxanthine + NAD^+ + H_2O <40> (<40> 63% of the activity compared to hypoxanthine [63]) (Reversibility: ?) [63]

P 6-thiourate + NADH <40> [63]

S 8-azahypoxanthine + NAD^+ + H_2O <34, 4> (<34> 39% 0f the activity compared to hypoxanthine [55]; <40> 42% of the activity compared to hypoxanthine [63]) (Reversibility: ?) [55, 63]

P 8-azaxanthine + NADH <34, 4> [55, 63]

S 8-azaxanthine + NAD^+ + H_2O <4> (<4> very low activity [37]) (Reversibility: ?) [37]

P 8-aza-urate + NADH <4> [37]

S 9-methylhypoxanthine + O_2 <49> (Reversibility: ?) [44]

P 9-methylxanthine + H_2O_2 <49> [44]

S N-methylnicontinamide + $NADP^+$ + H_2O <51> (<51> low activity, only a substrate at pH values above 8.0 [13]) (Reversibility: ?) [13]

P ? + NADPH <51> [13]

S NAD(P)H + 2,6-dichlorophenolindophenol + H_2O <34> (Reversibility: r) [55]

P $NAD(P)^+$ + ? <34> [55]

S NADH + electron acceptor + H_2O <1, 2, 3, 4, 10, 12, 29, 39, 41> (<39> conversion to the oxidase type O by trypsinization leads to 80-100% decrease in the oxidation rate of NADH, conversion to the oxidase type O by heat-treatment leads to a diminution of NADH oxidation [34]; <29> extremely slow reoxidation rate [28]; <4,10> electron acceptor: nitroblue tetrazolium [15,37]; <2> subtilisin treatment leads to an active component of 120000 kDa with enhanced activity [21]; <39> only dehydrogenase type D shows considerable activities, not oxidase type O [27]; <1,2,41> NADH diaphorase activity with several acceptors [21,26,30]; <12> 2,6-dichloroindophenol as electron acceptor [51]; <1> 2,6-dichloroindophenol, 3-acetylpyridine-adenine dinucleotide, methylene blue, phenazine methosulfate or trinitrobenzene sulfonate [26]; <3> electron acceptors: 2,6-dichlorphenolindophenol or methyl viologen [25]; <3> electron acceptor: 2,6-dichlorophenolindophenol, no activity for mutant E89K [45]) (Reversibility: ?) [15, 21, 25, 26, 27, 28, 30, 31, 34, 37, 45, 51]

P NAD^+ + reduced electron acceptor <1, 2, 3, 4, 10, 12, 29, 39, 41> [15, 21, 25, 26, 27, 28, 30, 31, 34, 37, 45, 51]

S NADH + reduced phenazine methosulfate + cytochrome c <3> (<3> no activity for the mutant E89K [45]) (Reversibility: ?) [45]

P NAD^+ + ? <3> [45]

S NADPH + O_2 <51> (<51> high NADPH oxidase activity [13]) (Reversibility: ?) [13]

P $NADP^+$ + O_2- + H^+ <51> [13]

S NADPH + electron acceptor + H_2O <51> (<51> electron acceptors: 2,6-dichlorophenolindophenol, methyl viologen, benzyl viologen, methylene blue [13]) (Reversibility: ?) [13]

P $NADP^+$ + reduced electron acceptor <51> [13]

S NADPH + nitroblue tetrazolium + H_2O <4, 1> (<4,10> diaphorase activity [15,37]) (Reversibility: ?) [15, 37]

P $NADP^+$ + reduced nitroblue tetrazolium <4, 1> [15, 37]

S NADPH + phenazine methosulfate + cytochrome c <3> (Reversibility: ?) [25]

P $NADP^+$ + ? <3> [25]

S abscisic aldehyde + NAD^+ + ? <53> (<53> 28.9% of the activity with xanthine [73]) (Reversibility: ?) [73]

P ? + NADH

S acetaldehyde + 2,6-dichloroindophenol + H_2O <9, 2> (<20> 0.1% of activity with xanthine [72]; <9> 1.2% of activity with xanthine [72]) (Reversibility: ?) [72]

P ?

S acetaldehyde + NAD^+ + H_2O <3, 4, 10, 45> (<4,10,45> very low activity [15,37,62]; <3> considerable activity for the recombinant enzyme [7]) (Reversibility: ?) [7, 15, 37, 62]

P acetic acid + NADH <3, 4, 10, 45> [7, 15, 37, 62]

S acetaldehyde + ferricyanide + H_2O <49> (<49> 5% of the activity compared to xanthine [17]) (Reversibility: ?) [17]

P acetic acid + ferrocyanide <49> [17]

S acetaldehyde + nitroblue tetrazolium + H_2O <10> (<10> very low activity [15]) (Reversibility: ?) [15]

P acetic acid + reduced nitroblue tetrazolium <10> [15]

S adenine + NAD^+ + H_2O <4, 48> (<4,48> low activity [37,54]) (Reversibility: ?) [37, 54]

P ? + NADH <4, 48> [37, 54]

S adenine + ferricyanide + H_2O <49> (<49> 12% of the activity compared to xanthine [17]) (Reversibility: ?) [17]

P urate + ferrocyanide <49> [17]

S adenine-N^1-oxide + NAD^+ + H_2O <40> (<40> 3.4% of the activity compared to hypoxanthine [63]) (Reversibility: ?) [63]

P ? + NADH <40> [63]

S benzaldehyde + O_2 <49> (Reversibility: ?) [44]

P benzoate + H_2O_2 <49> [44]

S benzaldehyde + ferricyanide + H_2O <49> (<49> 4% of the activity compared to xanthine [17]) (Reversibility: ?) [17]

P benzoate + ferrocyanide <49> [17]

S benzaldehyde + nitroblue tetrazolium + H_2O <10> (<10> very low activity [15]) (Reversibility: ?) [15]

P benzoate + reduced nitroblue tetrazolium <10> [15]

S glyceraldehyde + 2,6-dichloroindophenol + H_2O <9, 2> (<20> 0.3% of activity with xanthine [72]; <9> 12.1% of activity with xanthine [72]) (Reversibility: ?) [72]

P ?

S guanine + NAD^+ + H_2O <45> (<45> 81.3% of the activity compared to hypoxanthine [62]) (Reversibility: ?) [62]

P ? + NADH <45> [62]

S heptaldehyde + NAD^+ + ? <53> (<53> 12.5% of the activity with xanthine [73]) (Reversibility: ?) [73]

P ? + NADH

S hypoxanthine + 2,6-dichorophenolindophenol + H_2O <40, 45> (<40> considerable activity [63]; <45> 12.2% of the activity compared to NAD^+ [62]) (Reversibility: ?) [62, 63]

P xanthine + ? <40, 45> [62, 63]

S hypoxanthine + NAD^+ + H^+ + O_2- <1, 2, 3, 4, 7, 9, 12, 15, 19, 20, 21, 33, 34, 39, 42, 53> (<34> preferred substrate [55]; <2> subtilisin treatment leads to an active component I of 120000 kDa [21]; <4,33> NAD^+-O_2-dependent xanthine oxidase activity [20,22]; <15> predominant reaction [56]; <19> more rapidly oxidized than xanthine [8]; <9> 12.4% of activity with xanthine [72]; <20> 19% of activity with xanthine [72]; <53> 94.7% of the activity with xanthine [73]) (Reversibility: r) [8, 20, 21, 22, 46, 51, 53, 55, 56, 57, 58, 60, 72, 73, 88]

P xanthine + NADH + H_2O_2 <1, 2, 3, 4, 12, 15, 19, 21, 33, 34, 39, 42> [8, 20, 21, 22, 46, 51, 53, 55, 56, 57, 58, 60]

S hypoxanthine + NAD^+ + H_2O <3, 4, 5, 9, 14, 19, 20, 29, 39, 40, 44, 48> (<40> preferred substrate [63]; <14> 44% of the activity with xanthine [78]) (Reversibility: ?) [2, 9, 10, 11, 16, 28, 33, 36, 37, 50, 52, 54, 63, 78]

P xanthine + NADH <3, 4, 5, 9, 14, 19, 20, 29, 39, 40, 44, 48> [2, 9, 10, 11, 16, 28, 33, 36, 37, 50, 52, 54, 63]

S hypoxanthine + NAD^+ + H_2O <45> (<45> preferred substrate [62]) (Reversibility: ?) [62]

P xanthine + 6,8-dihydroxypurine + NADH <45> (<45> 100% xanthine, 51% 6,8-dihydroxypurine formed [62]) [62]

S hypoxanthine + NADH <15> (<15> 0.1% of the xanthine oxidation rate [49]) (Reversibility: ?) [49]

P ? + NO_2^- + NAD^+ <15> [49]

S hypoxanthine + $NADP^+$ + H_2O <4, 14, 45, 51> (<45> 2.4% of the activity compared to NAD^+ [62]; <14> 40% of the activity compared to NAD^+ [36]; <51> strict specificity for $NADP^+$ [13]) (Reversibility: ?) [13, 36, 37, 62]

P xanthine + NADPH <4, 14, 45, 51> [13, 36, 37, 62]

S hypoxanthine + O_2 <45, 49> (<45> very low activity [62]) (Reversibility: ?) [44, 62]

P xanthine + H_2O_2 <45, 49> [44, 62]

S hypoxanthine + O_2 + H_2O <16, 23> (Reversibility: ?) [77]

P xanthine + O_2- (<16,23> no production of H_2O_2 [77])

S hypoxanthine + cytochrome c + H_2O <45> (<45> 2.1% of the activity compared to NAD^+ [62]) (Reversibility: ?) [62]

P xanthine + reduced cytochrome c <45> [62]

S hypoxanthine + ferricyanide <46> (Reversibility: ?) [24]

P xanthine + ferrocyanide <46> [24]

S hypoxanthine + ferricyanide + H_2O <45, 49> (<45> 17.2% of the activity compared to NAD^+ [62]; <49> 98% of the activity compared to xanthine [17]) (Reversibility: ?) [17, 62]

P xanthine + ferrocyanide <45, 49> [17, 62]

S hypoxanthine + methyl viologen + H_2O <21, 42> (<42> 7% of the activity compared to xanthine [39]) (Reversibility: ?) [39, 6]

P xanthine + reduced methyl viologen <21, 42> [39, 6]

S hypoxanthine + methylene blue + H_2O <14> (<14> 39% of the activity compared to NAD^+ as electron acceptor [36]) (Reversibility: ?) [36]

P xanthine + reduced methylene blue <14> [36]

S hypoxanthine + nitroblue tetrazolium <10, 45> (<45> 18% of the activity compared to NAD^+ [62]) (Reversibility: ?) [15, 62]
P xanthine + ? <10, 45> [15, 62]
S hypoxanthine + phenazine methosulfate + H_2O <14, 4> (<40> low activity [63]; <14> effective electron acceptor [36]) (Reversibility: ?) [36, 63]
P urate + ? <14, 4> [36, 63]
S hypoxanthine + thio-NAD^+ + H_2O <3> (Reversibility: ?) [52]
P xanthine + thio-NADH <3> [52]
S hypoxanthine + urate <49> (<49> oxygen-free assay [44]) (Reversibility: r) [44]
P xanthine + 6,8-dihydroxypurine <49> [44]
S indole-3-acetaldehyde + NAD^+ + H_2O <14> (<14> and similar aldehydes, 2-3% of the activity with xanthine [78]) (Reversibility: ?) [78]
P ? + NADH
S indole-3-carboxaldehyde + NAD^+ + ? <53> (<53> 31.3% of the activity with xanthine [73]) (Reversibility: ?) [73]
P ? + NADH
S inosine + NAD^+ + H_2O <45> (<45> low activity [62]) (Reversibility: ?) [62]
P ? + NADH <45> [62]
S isoguanine + NAD^+ + H_2O <40> (<40> 7.3% of the activity compared to hypoxanthine [63]) (Reversibility: ?) [63]
P ? + NADH <40> [63]
S propionaldehyde + NAD^+ + H_2O <45> (<45> low activity [62]) (Reversibility: ?) [62]
P propionic acid + NADH <45> [62]
S pterin + 2,6-dichloroindophenol + H_2O <9, 2> (<9> 22.7% of activity with xanthine [72]; <20> 9.7% of activity with xanthine [72]) (Reversibility: ?) [72]
P ?
S pterin + NAD^+ + H_2O <37> (Reversibility: ?) [67]
P ? + NADH
S purine + 2,6-dichloroindophenol + H_2O <9, 2> (<9> 18.7% of activity with xanthine [72]; <20> 8.5% of activity with xanthine [72]) (Reversibility: ?) [72]
P ?
S purine + NAD^+ + ? <53> (<53> 10.3% of the activity with xanthine [73]) (Reversibility: ?) [73]
P ? + NADH
S purine + NAD^+ + H_2O <10, 19, 29, 40, 48> (<10,19,29,40> poor substrate [8,15,28,63]) (Reversibility: ?) [8, 15, 28, 54, 63]
P ? + NADH <10, 19, 29, 40, 48> [8, 15, 28, 54, 63]
S purine + NAD^+ + H_2O <45> (<45> poor substrate [62]) (Reversibility: ?) [62]
P hypoxanthine + 8-hydroxypurine + 2-hydroxypurine + NADH <45> (<45> 2.3% hypoxanthine, 2.3% 8-hydroxypurine and traces of 2-hydroxypurine formed [62]) [62]

S purine + $NADP^+$ + H_2O <51> (<51> 60% of the activity compared to hypoxanthine [13]) (Reversibility: ?) [13]

P ? + NADPH <51> [13]

S purine + O_2 + H_2O <49> (Reversibility: ?) [44]

P 8-hydroxypurine + H_2O_2 <49> [44]

S purine + ferricyanide + H_2O <49> (<49> 108% of the activity compared to xanthine [17]) (Reversibility: ?) [17]

P urate + ferrocyanide <49> [17]

S purine + methyl viologen + H_2O <42> (<42> 2% of the activity compared to xanthine [39]) (Reversibility: ?) [39]

P ? + reduced methyl viologen <42> [39]

S purine + nitroblue tetrazolium <10> (<10> low activity [15]) (Reversibility: ?) [15]

P ?

S salicylaldehyde + O_2 <49> (Reversibility: r) [44]

P salicylate + H_2O_2 <49> [44]

S salicylaldehyde + ferricyanide + H_2O <49> (<49> 2% of the activity compared to xanthine [17]) (Reversibility: ?) [17]

P salicylate + ferrocyanide <49> [17]

S theobromine + NAD^+ + H_2O <45> (<45> very low activity [62]) (Reversibility: ?) [62]

P ? + NADH <45> [62]

S theophylline + NAD^+ + H_2O <45> (<45> low activity [62]) (Reversibility: ?) [62]

P ? + NADH <45> [62]

S urate + NADH <41> (Reversibility: ?) [80]

P xanthine + NAD^+ + H_2O

S xanthine + 2,6-dichlorophenolindophenol + H_2O <1, 2, 3, 6, 9, 10, 19, 39, 41, 42, 46, 49> (<2> subtilisin treatment leads to an active component of 120000 kDa [21]; <42> 2.5% of the activity compared to methyl viologen as electron acceptor [39]; <49> 13% of the activity compared to methylene blue as electron acceptor [17]; <3> very low activity for the mutant E89K [45]; <39> greatly enhanced activity for dehydrogenase type D and trypsin- or heat-treated oxidase types O [34]; <1> same activity compared to NAD^+ as electron acceptor [26]; <19> 11% of the activity compared to ferricyanide as electron acceptor [40]) (Reversibility: r) [12, 15, 17, 21, 24, 25, 26, 30, 31, 34, 35, 39, 40, 45]

P urate + ? <1, 2, 3, 6, 9, 10, 19, 39, 41, 42, 46, 49> [12, 15, 17, 21, 24, 25, 26, 30, 31, 34, 35, 39, 40, 45]

S xanthine + 3-acetylpyridine-adenine dinucleotide + + H_2O <1, 15> (<1,15> same activity compared to NAD^+ [26,56]; <1> low reverse activity [26]) (Reversibility: r) [26, 56]

P urate + 3-acetylpyridine-adenine dinucleotide(H) <1, 15> [26, 56]

S xanthine + FAD + H_2O <49> (<49> 35% of the activity compared to methylene blue as electron acceptor [17]) (Reversibility: ?) [17]

P urate + FADH <49> [17]

S xanthine + FMN + H_2O <49> (<49> 44% of the activity compared to methylene blue as electron acceptor [17]) (Reversibility: ?) [17]

P urate + reduced FMN <49> [17]

S xanthine + NAD^+ + H_2O <7, 8, 20> (Reversibility: ?) [88, 89, 91]

P urate + NADH + H^+

S xanthine + NAD^+ + H_2O <1, 2, 3, 4, 5, 6, 7, 9, 10, 11, 12, 13, 14, 15, 18, 19, 20, 21, 22, 26, 27, 29, 30, 33, 34, 37, 38, 39, 40, 41, 42, 45, 48, 53, 54, 55> (<9> xanthine dehydrogenase form has distinct xanthine/oxygen activity, 35-42% of electrons transferred to O_2 to form O_2^- [35]; <9> conversion of dehydrogenase to oxidase type due to oxidation of sulfhydryl groups by molecular oxygen, dehydrogenase activity recovered by treatment with dithiothreitol [64]; <10> strict dehydrogenase activity, no utilization of O_2 [15]; <45> 61% of the activity compared to hypoxanthine [62]; <1> only dehydrogenase type D present [29]; <1> ping-pong reaction mechanism [26]; <39> NAD^+-dependent form is postulated to play a regulatory role in purine metabolism [47]; <19> NAD^+-linked activity, very low activity towards molecular oxygen [40]; <40> 75% of the activity compared to hypoxanthine [63]; <12> degradative pathway of conversion of purines to ammonia [51]; <14> only present as stable dehydrogenase from, no conversion to the oxidase form [36]; <6> xanthine oxidase form is the principle major form in fresh mouse milk, dehydrogenase form is the major form in mammary gland, conversion to the dehydrogenase form by thiol active compounds [12]; <3> involved in pteridine metabolism, 40% of activity compared to hypoxanthine [52]; <39> conversion of xanthine dehydrogenase to the oxidase type by thiol-disulfide oxidoreductase, thiol reagents or oxidized glutathione [1]; <34> 67% of the activity compared to hypoxanthine [55]; <5> regulation of xanthine dehydrogenase expression is subjected to nitrogen catabolite repression mediated through the GlnA-dependent signaling pathway [16]; <39> trypsin treatment leads to a complete conversion of xanthine dehydrogenase to xanthine oxidase activity [6]; <1> minimum degree of 1 : 1 for xanthine, 2 : 2 for NAD, 1 : 1 for urate and 1 : 2 for NADH in the xanthine/NAD^+ oxidoreductase reaction required [3]; <1> xanthine dehydrogenase can be partially reduced in a triphasic reaction by either xanthine or NADH, oxidation of fully, 6-electron-reduced xanthine dehydrogenase by either urate or NAD^+ is monophasic and depends on the oxidant concentration [23]; <2> subtilisin treatment leads to an active component I of 120000 kDa [21]; <9,39> NAD^+-dependent dehydrogenase type D [14,27,34]) (Reversibility: r) [1, 2, 3, 4, 5, 6, 7, 8, 9, 10, 11, 12, 14, 15, 16, 18, 19, 21, 22, 23, 25, 26, 27, 28, 29, 30, 31, 34, 35, 36, 37, 40, 41, 42, 45, 46, 47, 48, 50, 51, 52, 53, 54, 55, 56, 57, 59, 60, 62, 63, 64, 67, 68, 72, 73, 74, 76, 78]

P urate + NADH <1, 2, 3, 4, 5, 6, 7, 9, 10, 11, 12, 14, 15, 18, 19, 20, 21, 22, 27, 29, 30, 33, 34, 38, 39, 40, 41, 42, 45, 48> (<1> NADH-binding to the 2-electron reduced enzyme is implicated in fixing end-point position in reactions involving pyridine nucleotides, urate-binding is involved in fixing end-point reactions involving xanthine and urate [23]; <45> 91% urate formed [62]) [1, 2, 3, 4, 5, 6, 7, 8, 9, 10, 11, 12, 14, 15, 16, 18, 19, 21, 22,

23, 25, 26, 27, 28, 29, 30, 31, 34, 35, 36, 37, 40, 41, 42, 45, 46, 47, 48, 50, 51, 52, 53, 54, 55, 56, 57, 59, 60, 62, 63, 64]

S xanthine + NAD^+ + O_2 <39> (<39> heat-treated intermediate dehydrogenase/oxidase type O [34]; <39> intermediate form of dehydrogenase/oxidase type D/O [29]) (Reversibility: ?) [27, 29, 34]

P urate + NADH + H_2O_2 <39> [27, 29, 34]

S xanthine + $NADP^+$ + H_2O <29, 51> (<51> strict specificity for $NADP^+$ [13]; <29> 11% of the activity compared to NAD^+ [28]) (Reversibility: ?) [13, 28]

P urate + NADPH <29, 51> [13, 28]

S xanthine + O_2 <1, 2, 3, 9, 14, 16, 18, 19, 20, 23, 24, 28, 29, 38, 39, 42, 43, 44, 49> (<19, 42, 49> very low activity [17, 39, 40]; <1, 14, 29> low activity [26, 28, 36]; <2> subtilisin treatment leads to an active component I of 120000 kDa [21]; <19> 4% of the activity compared to xanthine-NAD^+ [9]; <49> presence of ferredoxin enhances rate of oxygen reduction [44]; <39> toxic reactions of xanthine oxidase-derived radicals are critical factors in several mechanisms of tissue pathology [38]; <39> NAD^+-independent trypsin-treated oxidase type O [34]; <39> NAD^+-independent xanthine oxidase activity, low activity present in the enzyme preparation, conversion of the NAD^+-dependent to NAD^+-independent activity by some thiol reagents [47]; <9> xanthine oxidase form transfers 22% electrons to oxygen to form superoxide [35]) (Reversibility: r) [9, 14, 17, 21, 25, 26, 27, 28, 29, 31, 33, 34, 35, 36, 38, 39, 40, 41, 44, 45, 47, 50, 61, 77]

P urate + O_2- + H^+ <1, 2, 3, 9, 14, 18, 19, 20, 24, 28, 29, 38, 39, 42, 43, 44, 49> (<16,23> no production of H_2O_2 [77]) [9, 14, 17, 21, 25, 26, 27, 28, 29, 31, 33, 34, 35, 36, 38, 39, 40, 41, 44, 45, 47, 50, 61]

S xanthine + O_2 <49> (<49> dismutation reaction [44]) (Reversibility: ?) [44]

P hypoxanthine + ? <49> [44]

S xanthine + benzyl viologen + H_2O <42, 49> (<42> 52% of the activity compared to methyl viologen as electron acceptor [39]; <49> 18% of the activity compared to methylene blue as electron acceptor [17]) (Reversibility: ?) [17, 39]

P urate + reduced benzyl viologen <42, 49> [17, 39]

S xanthine + cytochrome c + H_2O <39, 49> (<49> low activity [17]; <39> enhanced activity for heat- and trypsin-treated oxidase types O [34]; <49> presence of ferredoxin enhances cytochrom c reduction [44]) (Reversibility: ?) [17, 27, 34, 44]

P urate + reduced cytochrome c <39, 49> [17, 27, 34, 44]

S xanthine + ferredoxin + H_2O <49> (Reversibility: ?) [44]

P urate + reduced ferredoxin <49> [44]

S xanthine + ferricyanide + H_2O <18, 19, 22, 24, 25, 27, 28, 30, 31, 35, 36, 39, 41, 42, 43, 46, 49, 50> (<42> 99% of the activity compared to methyl viologen as electron acceptor [39]; <19> specific ferricyanide-dependent activity, no activity with NAD^+, $NADP^+$, oxygen, cytochrome c, FAD or FMN [40]; <19> 58% of the activity compared to xanthine-NAD^+ [9]; <46> preferred substrates, does not act with NAD^+ or $NADP^+$ [24];

<39> low activity for both dehydrogenase type D and oxidase type O [27,34]) (Reversibility: ?) [9, 17, 24, 27, 30, 34, 39, 40, 41, 44]

P urate + ferrocyanide <18, 19, 22, 24, 25, 27, 28, 30, 31, 35, 36, 39, 41, 42, 43, 46, 49, 50> [9, 17, 24, 27, 30, 34, 39, 40, 41, 44]

S xanthine + iodonitrotetrazolium + H_2O <49> (Reversibility: ?) [17]

P urate + reduced iodonitrotetrazolium <49> [17]

S xanthine + methyl viologen + H_2O <10, 21, 42, 49> (<49> low activity [17]; <42> best substrates tested, no activity with NAD^+ or $NADP^+$, 40% of the activity in the reverse reaction [39]) (Reversibility: r) [15, 17, 39, 60]

P urate + reduced methyl viologen <10, 21, 42, 49> [15, 17, 39, 60]

S xanthine + methylene blue + H_2O <1, 2, 3, 10, 39, 42, 49> (<2> subtilisin treatment leads to an active component of 120000 kDa [21]; <39> same activity for dehydrogenase type D and oxidase type O [27]; <39> enhanced oxidation of xanthine for dehydrogenase type D and trypsin- or heat-treated oxidase type O [34]; <10> 3fold higher activity compared to NAD^+ as electron acceptor [15]; <42> 87% of the activity compared to methyl viologen as electron acceptor [39]) (Reversibility: r) [15, 17, 21, 26, 27, 31, 34, 39, 44, 61]

P urate + reduced methylene blue <1, 2, 3, 10, 39, 42, 49> [15, 17, 21, 26, 27, 31, 34, 39, 44, 61]

S xanthine + myoglobin + H_2O <49> (Reversibility: ?) [44]

P urate + reduced myoglobin <49> [44]

S xanthine + nitroblue tetrazolium + H_2O <10, 41, 42, 49> (<42> 32% of the activity compared to methyl viologen as electron acceptor [39]) (Reversibility: ?) [15, 17, 30, 39, 44]

P urate + ? <10, 41, 42, 49> [15, 17, 30, 39, 44]

S xanthine + *p*-benzoquinone + H_2O <9> (<9> electron acceptor *p*-benzoquinone for both dehydrogenase and oxidase types [64]) (Reversibility: ?) [64]

P *p*-benzosemiquinone + urate <9> [64]

S xanthine + *p*-benzoquinone + H_2O <9> (<9> electron donor only for oxidase type [64]) (Reversibility: ?) [64]

P hypoxanthine + hydroquinone <9> [64]

S xanthine + phenazine methosulfate + H_2O <1, 10, 49> (<10> 5fold higher activity compared to NAD^+ as electron acceptor [15]) (Reversibility: r) [15, 17, 26, 44]

P urate + ? <1, 10, 49> [15, 17, 26, 44]

S xanthine + phenazine methosulfate + cytochrome c + H_2O <3> (<3> very low activity for the mutant E89K [45]) (Reversibility: ?) [7, 25, 31, 45]

P urate + ? <3> [7, 25, 31, 45]

S xanthine + pyridinealdehyde-NAD^+ + H_2O <1> (<1> 53% of the activity compared to NAD^+, low reverse activity [26]) (Reversibility: r) [26]

P urate + pyridinealdehyde-NADH <1> [26]

S xanthine + riboflavin + H_2O <49> (<49> 41% of the activity compared to methylene blue as electron acceptor [17]) (Reversibility: ?) [17]

P urate + reduced riboflavin <49> [17]

S xanthine + thio-NAD^+ + H_2O <1, 3> (<1> same activity compared to NAD^+ [26]) (Reversibility: r) [26, 52]
P urate + thio-NADH <1, 3> [26, 52]
S xanthine + trinitrobenzenesulfonate + H_2O <1, 2> (<2> subtilisin treatment leads to an active component of 120000 kDa [21]) (Reversibility: r) [21, 26]
P urate + ? <1, 2> [21, 26]
S xanthopterin + NAD^+ + H_2O <48> (<48> regulation of the pteridine pathway by competitive inhibition of reaction products and the precursor of xanthopterin, 7,8-dihydroxyxanthopterin [54]; <48> regulation of the pteridine pathway by competitive inhibition of reaction products and the precursor of xanthopterin, 7,8-dihydroxanthopterin [54]) (Reversibility: ?) [54]
P leucopterin + NADH <48> [54]
S xanthosine + NAD^+ + H_2O <45> (<45> 15.1% of the activity compared to hypoxanthine [62]) (Reversibility: ?) [62]
P ? + NADH <45> [62]
S Additional information <7, 8, 9, 26> (<8> major function of enzyme in liver parenchymal and sinusoidal cells is production of uric acid as a antioxidant [66]; <8> NADH oxidation by xanthine oxidoreductase may constitute an important pathway for reactive oxygen species-mediated tissue injuries. Xanthine oxidoreductase and xanthine oxidase catalyze the NADH oxidation, generating O_2^- radicals and inducing the peroxidation of liposomes, in a NADH and enzyme dependent manner [86]; <7,9> xanthine oxidoreductase plays a physiological role in milk equal in importance to its catalytic function as an enzyme [84]; <26> the dehydrogenase form of enzyme reacts significantly faster than the oxidase form [87]) (Reversibility: ?) [66, 84, 86, 87]
P ?

Inhibitors

1-methylhypoxanthine <34> (<34> 17% inhibition of xanthine dehydrogenase at 0.25 mM [55]) [55]

2,6-dihydroxanthopterin <48> (<48> competitive inhibition of pterin oxidation, K_i: 0.0084 mM [54]) [54]

2-(3-cyano-4-isobutoxyphenyl)-4-methyl-5-thiazolecarboxylic acid <9> (<9> i.e. TEI-6720, mixed type inhibitor, binds very tightly to active and inactive desulfo-form of enzyme [71]) [71]

2-iodosobenzoic acid <12, 39> (<12> effective inhibition of xanthine and pterine oxidation [51]; <39> 50% inhibition at 0.0025 mM of NAD^+-dependent activity by enzyme inactivation, not by conversion to the O_2-dependent activity [47]) [47, 51]

2-amino-4-hydroxypteridine-6-carboxyaldehyde <3, 9> (<3> competitive inhibition of 2-amino-4-hydroxy-pterine oxidation, K_i: 0.00025 mM, non-competitive inhibition of xanthine oxidation, K_i: 0.00051 mM [42]; <9> competitive inhibition of 2-amino-4-hydoxy-pterine oxidation, K_i: 0.000016 mM [42]) [42]

2-amino-4-hydroxypterin <12> (<12> substrate inhibition above 0.01 mM [51]) [51]
4-(5-pyridin-4-yl-1H-1,2,4-triazol-3-yl)pyridine-2-carbonitrile <9> (<9> i.e. FYX-051, strong, in absence of xanthine slow hydroxylation of inhibitor [79]) [79]
4-amino-2,6-dihydroxypyrimidine <48> (<48> competitive inhibition of xanthine oxidation, K_i: 0.106 mM [54]) [54]
4-chloromercuribenzoate <39, 45> (<39> decreases NAD^+-dependent activity from 0.01 up to 0.05 mM with simultaneous inactivation of the enzyme [47]; <45> 89.2% inhibition of hypoxanthine oxidation at 1 mM [62]) [47, 62]
4-hydroxymercuribenzoate <10, 12> (<12> effective inhibition of xanthine and pterine oxidation [51]; <10> 76% inhibition of xanthine-NAD^+-activity at 0.002 mM [15]) [15, 51]
4-chloromercuriphenyl sulfonic acid <40> (<40> 98% inhibition at 0.001 mM [63]) [63]
4-hydroxypyrazolo(3,4-d)pyrimidine <12, 14, 20, 41, 51> (<12,20,41,51> i.e. allopurinol [13,30,50,51]; <12> inhibition of pterine oxidation at 0.0003 mM [51]; <51> 95% inhibition of hypoxanthine-$NADP^+$-activity at 1 mM [13]; <41> rapid inactivation under anaerobic conditions at 0.1 mM [30]; <14> 51% inhibition of hypoxanthine oxidation at 0.0001 mM [36]; <20> inhibition by direct coordination of the reaction product alloxanthine, to the molybdenum via a nitrogen atom [50]) [13, 30, 36, 50, 51]
5,5-dithiobis-(2-nitrobenzoate) <15, 39, 45> (<39> conversion of dehydrogenase type D to oxidase type O, can be prevented and reversed by dithioerythritol [29]; <15> 30% inhibition at 1 mM, presence of NAD^+, no conversion from dehydrogenase to oxidase activity detectable [56]; <39> conversion from of NAD^+-dependent to O_2-dependent activity without any effect on the total activity [47]; <45> 45.4% inhibition of hypoxanthine oxidation at 1 mM [62]; <39> conversion of dehydrogenase type D to oxidase type O due to modification of a limited number of critical sulfhydryl groups [34]) [29, 34, 47, 56, 62]
6-mercaptopurine <51> (<51> 90% inhibition of hypoxanthine-$NADP^+$-activity at 1 mM [13]) [13]
8-azaadenine <4, 34> (<34> complete inhibition of xanthine dehydrogenase at 0.2 mM [55]; <4> competitive inhibition of xanthine oxidation, K_i: 0.25 mM [37]) [37, 55]
8-azaguanine <3, 9, 34> (<34> complete inhibition of xanthine dehydrogenase at 0.2 mM [55]; <3> competitive inhibition of xanthine oxidation, K_i: 0.037 mM, non-competitive inhibition of 2-amino-4-hydroxy-pterine oxidation, K_i: 0.071 mM [42]; <9> competitive inhibition of 2-amino-4-hydroxypterine oxidation, K_i: 0.0012 mM [42]) [42, 55]
8-azaxanthine <49> (<49> 50% inhibition of ferricyanide reduction in xanthine oxidation assay at 5 mM [44]) [44]
8-azohypoxanthine <34> (<34> 40% inhibition of xanthine dehydrogenase at 0.25 mM [55]) [55]
acetaldehyde <4> (<4> inactivation [37]) [37]

adenine <4, 12, 14, 29, 32, 34, 45, 51> (<12> effective inhibition of xanthine oxidation at 0.001 mM [51]; <34> competitive inhibition, K_i: 0.05 mM [55]; <29> some non-competitive inhibition [28]; <4> competitive inhibition of xanthine oxidation, K_i: 0.13 mM [37]; <51> 96% inhibition of hypoxanthine-$NADP^+$-activity at 1 mM [13]; <45> 62.8% inhibition of hypoxanthine oxidation at 0.5 mM [62]; <14> 0.1 mM, 21% residual activity [78]; <32> treatment of normal fruit in linear phase of growth arrests fruit growth [65]) [13, 28, 37, 51, 55, 62, 65, 78]

Ag^{2+} <40> (<40> complete inhibition of hypoxanthine oxidation at 0.01 mM [63]) [63]

allopurinol <14, 26, 32, 53> (<53> strong [73]; <14> 0.1 mM, complete inhibition [78]; <32> treatment of normal fruit in linear phase of growth arrests fruit growth [65]; <26> blocks xanthine dehydrogenase activity, without influencing xanthine oxidase activity [87]) [65, 73, 78, 87]

ammeline <1, 3, 9> (<1> competitive inhibition of xanthine oxidation K_i: 0.083 mM, uncompetitive inhibition of NADH oxidation K_i: 0.063 mM [26]; <3> competitive inhibition of xanthine oxidation, K_i: 0.021 mM, non-competitive inhibition of 2-amino-4-hydoxypterine oxidation, K_i: 0.045 mM [42]; <9> competitive inhibition of 2-amino-4-hydroxy-pterine oxidation, K_i: 0.016 mM [42]) [26, 42]

arsenite <1, 12, 19, 51> (<1> gradual inhibition of xanthine-2,6-dichloroindophenol-activity, paralleled by a corresponding increase of NADH-2,6-dichloroindophenol-activity [26]; <12> inhibition of xanthine or pterine oxidation at 0.3 mM, diaphorase activity unaffected [51]; <51> 50% inhibition of hypoxanthine-$NADP^+$-activity at 1 mM [13]; <19> 90% loss of the ferricyanide-linked activity in the presence of 1.78 mM [40]) [13, 26, 40, 51]

Cu^{2+} <40> (<40> 14% inhibition of hypoxanthine oxidation at 0.1 mM [63]) [63]

$CuSO_4$ <39> (<39> conversion of the dehydrogenase type D to oxidase type O, prolonged incubation leads to complete inactivation, conversion can be reversed and prevented by dithioerythritol [29]) [29]

diethyl dicarbonate <15> (<15> 90% loss of NAD^+ dependent activity at 1 mM, retains more than 90% of oxygen-dependent and 3-acetylpyridine adenine dinucleotide+-dependent NADH oxidation activity [56]) [56]

diethyldithiocarbamate <10> (<10> 72% inhibition of xanthine-NAD^+-activity at 10 mM [15]) [15]

diphenyleneiodonium <26> (<26> powerful inhibition of NADH oxidation [87]) [87]

EDTA <45> (<45> 19.8% inhibition of hypoxanthine oxidation at 10 mM [62]) [62]

GSH <9> (<9> 45% of the oxidase activity converted to dehydrogenase activity at 10 mM [35]) [35]

GSSG <9> (<9> 75% of the dehydrogenase activity converted to oxidase activity at 0.5 mM [35]) [35]

guanine <12, 29> (<29> some inhibition [28]; <12> effective inhibition of xanthine oxidation at 0.001 mM [51]) [28, 51]

Hg^{2+} <40> (<40> complete inhibition of hypoxanthine oxidation at 0.1 mM [63]) [63]
hypoxanthine <4> (<4> inactivation of xanthine oxidase activity, not in the presence of NAD^+ [37]) [37]
iodoacetic acid <45> (<45> 21% inhibition at 1 mM [62]) [62]
KCN <1, 12, 14, 19, 40, 41, 45, 46, 49, 51> (<41> irreversible inactivation [30]; <40> 37% inhibition at 1 mM [63]; <1> complete inhibition of xanthine dehydrogenase activity, 70% reduction of diaphorase activity [26]; <12> 35% inhibition of pterine oxidation, 60% inhibition of diaphorase activity at 5 mM [51]; <45> complete inhibition of hypoxanthine oxidation at 1 mM [62]; <49> 10-50% inhibition of xanthine oxidation only in the presence of Tris or phosphate buffers from 0.01 to 0.1 M inhibitor concentration [44]; <14> 63% inhibition at 1 mM [36]; <51> addition of selenide in the presence of dithionite reactivates the inhibited enzyme [13]; <46> cyanolyzable selenium, 75% inhibition at 15 mM [24]; <19> complete inactivation of oxygen-linked activity in 15 min, decline of NAD^+-linked activity in 75 min, ferricyanide-linked activity completely stable [40]) [13, 24, 26, 30, 36, 40, 44, 51, 62, 63]
leucopterin <48> (<48> competitive inhibition of xanthopterin oxidation, K_i: 0.0109 mM [54]) [54]
methanol <1, 19, 49> (<49> develops inhibition during course of catalysis, enhanced inhibition in the presence of ferricyanide in the oxygen-dependent oxidation of xanthine [44]; <1> slight inhibition of NAD^+ reduction at 1.5 M, rapid inactivation if NAD^+ is replaced by 2,6-dichloroindophenol, enhanced NADH diaphorase activity [26]; <19> 50% inhibition in 3 min at 1.5 M [40]) [26, 40, 44]
N-ethylmaleimide <39> (<39> conversion of dehydrogenase type D to oxidase type O, prevented by dithioerythritol but no reversible conversion [29]) [29]
N^6-furfuryladenine <14> (<14> 0.1 mM, 59% residual activity [78]) [78]
NAD^+ <1> (<1> competitive inhibition of NADH oxidation, K_i: 0.0143 mM [26]) [26]
NADH <4, 9, 15, 39> (<39> product inhibition [47]; <15> varied substrate: xanthine, product inhibition, K_i 0.05 mM, varied substrate: NAD^+, dead-end inhibition type, K_i 0.022 mM [56]; <39> accumulation of produced NADH inhibits activity to 50% [46]; <4> inactivation closely related to associated diaphorase activity [37]; <9> partial reduction of dehydrogenase activity under anaeroboic conditions, oxidase activity more slowly reduced [35]) [35, 37, 46, 47, 56]
NADPH <4> (<4> inactivation closely related to diaphorase activity [37]) [37]
NO <9> (<9> dose-dependent inhibition of xanthine dehydrogenase and oxidase activity, reaction with an essential sulfur in the molybdenum center, that damages the molybdopterin [14]) [14]
NaCN <10> (<10> 69% inhibition of xanthine-NAD^+-activity at 3.3 mM [15]) [15]
NaN_3 <12> (<12> slight inhibition of dehydrogenase activity [51]) [51]
potassium cyanide <53> (<53> strong [73]) [73]

purine <29> (<29> some inhibition [28]) [28]
pyridoxal <3> (<3> K_i for xanthine oxidation: 0.05 mM at 30°C, 0.11 mM at 50°C, competitive [42]; <3> K_i for 2-amino-4-hydroxypterine oxidation: 0.08 mM at 30 and 50°C, competitive [42]) [42]
quinacrine <45> (<45> 29.7% inhibition of hypoxanthine oxidation at 1 mM [62]) [62]
salicylhydroxamic acid <10, 14> (<14> 82% inhibition of hypoxanthine oxidation at 2.5 mM [36]; <10> 71% inhibition of xanthine-NAD^+-activity at 5 mM [15]) [15, 36]
sodium dithionite <4, 41, 51> (<4> irreversible inactivation by reduction of xanthine dehydrogenase, no recovery after dithionite elimination [37]; <41,51> reduction of enzyme [13,30]) [13, 30, 37]
superoxide dismutase <39> (<39> complete inhibition of the xanthine-cytochrome c activity for oxidase type O, lesser inhibition for dehydrogenase type D [27]) [27]
tetraethylthiuram disulfide <39> (<39> conversion of dehydrogenase type D to oxidase type O, can be prevented and reversed by dithioerythritol [29]) [29]
thiourea <12> (<12> 65% inhibition of xanthine dehydrogenase activity at 20 mM [51]) [51]
Tiron <45> (<45> 36.8% inhibition of hypoxanthine oxidation at 10 mM [62]) [62]
urate <10, 12, 15, 34, 45, 48, 49> (<15> varied substrate: xanthine, dead-end inhibition type, K_i 0.18 mM, varied substrate: NAD^+, product inhibition, K_i 0.45 mM [56]; <12> inhibition of xanthine oxidation at 0.06 mM [51]; <10> 15% inhibition of xanthine-NAD^+-activity [15]; <45> 36.6% inhibition of hypoxanthine oxidation at 0.5 mM [62]; <34> competitive inhibition of xanthine dehydrogenase, K_i: 0.144 mM [55]; <48> competitive inhibition of xanthine oxidation, K_i: 0.064 mM [54]; <49> 50% inhibition of xanthine oxidation at 0.5 mM [44]) [15, 44, 51, 54, 55, 56, 62]
urea <1> (<1> competitive inhibition of xanthine oxidation K_i: 0.28 M, uncompetitive inhibition of NADH oxidation K_i: 1 M [26]) [26]
Xanthine <3, 4, 12, 34, 41> (<41> reduction of enzyme [30]; <12> substrate inhibition above 0.13 mM [51]; <34> 40% inhibition of xanthine dehydrogenase at 0.25 mM [55]; <4> irreversible inactivation by reduction of xanthine dehydrogenase [37]; <3> substrate inhibition above 0.05 mM, but in the presence of NAD^+ [42]; <4> irreversible inhibition of xanthine oxidase activity, adenine and 8-azaadenine protects against inactivation, ferricyanide partially protects against inactivation, no inactivation in the presence of NAD^+ [37]) [30, 37, 42, 51, 55]
amflutizole <26> (<26> blocks xanthine dehydrogenase activity, without influencing xanthine oxidase activity [87]) [87]
cassia oil <6> (<6> oral adminstration of cassia oil significantly reduces serum and hepatic urate levels in hyperuricemic mice. At 600 mg/kg, cassia oil is as potent as allopurinol. This hypouricemic effect is explained by inhibiting activities of liver xanthine oxidase and xanthine oxidoreductase [93]) [93]
diphenylen iodinium <16, 23> [77]

o-phenanthroline <10, 12, 45> (<12> 50% inhibition at 5-15 mM [51]; <45> 19.5% inhibition of hypoxanthine oxidation at 10 mM [62]; <10> 12% inhibition of xanthine-NAD^+-activity at 7.5 mM [15]) [15, 51, 62]
oxypurinol <26> (<26> blocks xanthine dehydrogenase activity, without influencing xanthine oxidase activity [87]) [87]
p-hydroxymercuribenzoate <53> (<53> strong [73]) [73]
tetraethyldithiodicarbonic diamide <39> (<39> i.e. disulfiram [47]; <39> transformation of the NAD^+- to the O_2-dependent activity up to 0.025 mM, up to 80% loss of NAD^+-dependent activity, modification of one thiol group in the active centre, NAD^+ protects against modification due to a single thiol group involved in NAD^+-binding within the active centre [47]) [47]
Additional information <20> (<20> no substrate inhibition [74]) [74]

Cofactors/prosthetic groups

FAD <1, 2, 3, 7, 9, 12, 19, 20, 39, 40, 41, 42, 45, 46, 49, 51> (<7,9> 1 mol per mol of subunit [84]; <12> 2 mol per mol enzyme [51]; <39> two FAD per enzyme molecule [27]; <39> trypsin-treated oxidase type O shows absence of FAD, heat-treated oxidase type O shows small measurable FAD [34]; <1> one FAD per hemimolecule [3]; <1> generally viewed as the site at which NADH reacts [23]; <3> substitutions G348E, G353D, S357F located within the flavin/ NAD^+/NADH-domain [25]; <9> one FAD per subunit, central 40 kDa FAD-domain [2,11]; <9> deflavo-enzyme completely loses xanthine/NAD^+ activity, no loss of xanthine/2,6-dichorophenolindophenol activity, xanthine dehydrogenase form stabilizes the neutral form of flavin, xanthine oxidase form does not [35]; <42> 1.69 mol FAD per mol enzyme [39]; <20> 0.9 mol per subunit [72]; <41> 1.0-1.1 molecules per $\alpha\beta$ protomer [80]) [2, 3, 7, 9, 11, 13, 14, 17, 21, 23, 25, 26, 27, 30, 34, 35, 39, 44, 45, 51, 61, 62, 63, 69, 72, 80, 84]
FMN <15> (<15> primarily flavin cofactor [49]) [49]
ferricyanide <19, 46> (<46> does not act with NAD^+ or $NADP^+$ [24]; <19> specific xanthine oxidizing activity [40]) [24,40]
flavin <10, 37> [15,67]
heme <19> (<19> ferricyanide-linked activity, no FAD [40]) [40]
molybdenum cofactor <17> (<17> the enzyme contains molybdenum cofactor comprising only molybdopterin and molybdenum [82]) [82]
molybdopterin <7, 9, 17, 20> (<17> enzyme contains molybdopterin [81]; <7> 0.09 molecules per subunit [84]; <9> 0.62 molecules per subunit [84]; <20> protein XdhC binds molybdenum cofactor in stoichiometric amounts, which subsequently can be inserted into molybdenum-free apoxanthine dehydrogenase. Protein XdhC is required for the stabilization of the sulfurated form of molybdenum cofactor [90]) [81,84,90]
NAD^+ <1, 2, 3, 4, 6, 7, 9, 10, 11, 12, 14, 15, 18, 19, 21, 22, 27, 29, 30, 33, 34, 38, 39, 40, 41, 42, 45, 48, 53> (<10, 12, 14, 19, 34, 39, 40, 45> cannot be replaced by $NADP^+$ [8, 15, 27, 36, 51, 55, 62, 63]; <41> physiological cofactor [30]; <19> activity different from ferricyanide-linked activity [40]) [1, 2, 3, 4, 5, 6, 7, 8, 11, 12, 14, 15, 18, 19, 20, 21, 22, 23, 25, 26, 27, 28, 29, 30, 34, 35, 36, 37, 38, 40, 41, 42, 43, 45, 46, 47, 51, 52, 53, 54, 55, 56, 57, 58, 59, 60, 62, 63, 64, 73, 78]

NADH <1, 7, 9, 11, 12, 14, 15, 34, 39> (<9> reduces xanthine oxidase to reductase activity [64]) [3,4,5,6,18,23,26,51,55,56,64,78]
$NADP^+$ <4, 14, 29> (<14> 40% of the activity compared to NAD^+ [36]; <29> 11% of the activity compared to NAD^+ [28]) [28,36,37]
Additional information <1, 3, 9, 10, 12, 15, 19, 20, 34, 39, 41, 42, 45, 46, 49, 51> (<49> molybdoironflavoprotein: molar ratio of molybdenum to iron to acid-labile-sulfur to FAD is 1 : 2 : 1.9 : 0.8 [17]; <39,45> iron-containing flavoprotein [18,62]; <51> molybdoironflavoprotein: 17.5 mol iron, 18.4 mol acid-labile sulfur, 2.3 mol molybdenum, 1.1 mol tungsten, 0.95 mol selenium [13]; <42> molybdoironflavoprotein: ratio of non-heme iron to acid-labile sulfur to FAD to molybdenum to tungsten to selenium is 7.7 : 7.5 : 1.7 : 1.8 : 0.12 : 0.13 [39]; <49> molybdoironflavoprotein: 8 : 8 : 2 : 1.5 ratio of iron to sulfide to flavin to molybdenum [44]; <45> molar ratio of FAD to iron to labile sulfide per mol enzyme is 2 : 14 : 2 [62]; <9> one molybdopterin-cofactor, two Fe2-S2-cluster, one FAD per subunit [2]; <1> molybdoironflavoprotein: 1 : 1 : 4 ratio of molybdenum to FAD to iron [26]; <41> molybdoironflavoprotein: ratio of iron to FAD to molybdenum is 4 : 1 : 1 [30]; <19> cofactor compostion similar to eukaryotic enzymes [9]; <12> non-heme flavoprotein [51]; <39> molybdoironflavoprotein: ratio of 2 : 1.4 : 7.6 of FAD to molybdenum to Fe-S [27]) [2, 7, 9, 11, 13, 15, 17, 18, 24, 25, 26, 27, 30, 35, 39, 44, 45, 49, 50, 51, 55, 62]

Activating compounds

4,4'-Dithiodipyridine <9> (<9> reverses conversion of dehydrogenase to oxidase form [35]) [35]
4-Chloromercuribenzoate <39> (<39> increases total activity up to 0.005 mM [47]) [47]
DTT <9, 39> (<9,39> recovery from oxidase to dehydrogenase type [34,35,64]) [34, 35, 64]
Ferredoxin <49> (<49> 26fold increase of cytochrome c reduction, 2fold increase of oxygen reduction [44]) [44]
KCl <45> (<45> strong activation at 50 mM [62]) [62]
NAD^+ <39> (<39> 6fold stimulation of aerobic rate of oxidation of xanthine [27]) [27]
NH_4Cl <45> (<45> 2.5fold activation at 100 mM [62]) [62]
NaCl <45> (<45> nearly 2fold activation at 100 mM [62]) [62]
PD98059 <8> (<8> inhibitor used to block MEK-1/2 kinase, activates the promoter of xanthine oxidoreductase and significantly enhances expression of enzyme induced by insulin, acute phase cytokines, or growth factors [92]) [92]
Urate <10> (<10> 30% activation of hypoxanthine-NAD^+ activity [15]) [15]
dithioerythritol <39> (<39> recovery from oxidase to dehydrogenase type [29]) [29]
methylene blue <39> (<39> 6fold activation of aerobic rate of oxidation of xanthine [27]) [27]
thiol active compounds <6, 9> (<9> reversible conversion of xanthine dehydrogenase to xanthine oxidase by modification of C535 and C992 and forma-

tion of a disulfide bond, that induces a conformational change [11]; <6> conversion of oxidase to dehydrogenase form [12]) [11, 12]

Metals, ions

Fe <1, 2, 3, 12, 15, 34, 39, 42, 45, 49> (<42> non-heme iron-sulfur groups, 7.7 g-atom Fe, 7.5 g- atom S^{2-} per mol enzyme [39]; <15> 8.2 mol iron per mol of enzyme [49]; <34> 8 atoms iron per mol enzyme [55]; <2,3> iron-sulfur groups [21,25]; <12> 12.1 mol iron per mol enzyme [51]) [21, 25, 26, 27, 39, 44, 49, 51, 55, 62]

iron <7, 20, 41, 46> (<41> 3.7-3.9 atoms per $\alpha\beta$ protomer [80]; <46> FeS center [69]; <20> FeS center, EPR spectra, wild type 3.7 mol per subunit, active form of mutant R135C 2.8 mol per subunit, inactive form of mutant R135C 1.2 mol per subunit [72]; <7> enzyme is about 30% deficient in iron-sulfur centers on basis of UV/vis and CD spectra [84]) [69, 72, 80, 84]

iron-sulfur-center <1, 3, 9, 19, 39, 41, 42, 49, 51> (<9> 2 Fe2-S^{2-}clusters, located at the N-terminal 20 kDa domain [2,11]; <9,41,42,51> 2 Fe2-S^{2-}clusters [13,14,30,35,39]; <3> substitutions G42E, E89K, L127F located within the iron-sulfur domain [25]; <49> 1 Fe2-S^{2-}cluster [17]) [2, 7, 9, 11, 13, 14, 17, 23, 25, 27, 30, 35, 39, 45]

Mo <1, 2, 3, 4, 12, 15, 19, 34, 39, 42, 49> (<34> 2 molybdenum per mol enzyme [55]; <12> 1 mol molybdenum per mol enzyme [51]; <15> 1.7 mol molybdenum per mol enzyme [49]) [21, 25, 26, 27, 34, 39, 40, 44, 49, 51, 55, 57, 58]

molybdenum <7, 9, 13, 16, 17, 20, 23, 26, 41, 46> (<13> 0.18 atoms per subunit [76]; <41> 1.1 atoms per $\alpha\beta$ protomer in wild-type, 0.89 atoms per $\alpha\beta$ protomer after expression of xdhABC genes, 0.24 atoms per $\alpha\beta$ protomer after expression of xdhAB genes [80]; <20> active form of mutant R135C Moco content 97%, inactive form of mutant R135C Moco content 3.8% [72]; <16,23> superoxide production depends on sulfuration of molybdenum cofactor [77]; <17> the enzyme contains molybdenum cofactor comprising only molybdopterin and molybdenum [82]; <7> 0.04 atoms per subunit [84]; <9> 0.61 atoms per subunit [84]; <26> in milk, more than 90% of enzyme exists in the inactive demolybdo form [87]; <20> protein XdhC binds molybdenum cofactor in stoichiometric amounts, which subsequently can be inserted into molybdenum-free apoxanthine dehydrogenase. Protein XdhC is required for the stabilization of the sulfurated form of molybdenum cofactor [90]) [69, 72, 76, 77, 80, 82, 84, 87, 90]

molybdopterin <17> (<17> enzyme contains molybdopterin [81]) [81]

Se <42, 46, 51> (<42> 0.13 g-atom selenium per native enzyme molecule [39]; <51> 3.4 g-atom per mol enzyme [13]) [13, 24, 39, 69]

tungsten <42, 51> (<42> 0.12 g-atom tungsten per native enzyme molecule [39]; <51> 3.4 g-atom per mol of enzyme [13]) [13, 39]

Zn <49> (<49> 1.54 mol Zn per mol enzyme [17]) [17]

molybdenum-center <1, 3, 9, 19, 41, 42, 44, 49, 51> (<49> molybdopterin cytosine dinucleotide cofactor [17]; <9> C-terminal 85 kDa molybdopterin-binding domain [2,11]; <3> substitutions G800E, G1011E, G1164R, G1266D, S1275F located within the pterin molybdenum cofactor domain [25]; <1> urate, xanthine and 8-bromoxanthine interact with the molybdenum-site of fully reduced xanthine oxidase [23]; <42> 1.8 g-atom molybdenum per native molecule [39]; <41> contains molybdopterin mononucleotide rather than molybdopterin dinucleotide [30]) [2, 7, 9, 11, 13, 14, 17, 23, 25, 30, 33, 39, 40, 45]

Turnover number (min^{-1})

0.1 <49> (salicylaldehyde, <49> electron acceptor: oxygen [44]) [44]
0.33 <7> (hypoxanthine, <7> mutant R881M, pH 8.5, 25°C [88]) [88]
0.4 <54, 55> (xanthine, <54,55> coexpression of genes xdhAB, conditions of low aeration, pH 7.8, 25°C [68]) [68]
0.7 <49> (purine, <49> electron acceptor: oxygen [44]) [44]
0.717 <49> (cytochrome c, <49> cosubstrate xanthine [44]) [44]
0.75 <49> (hypoxanthine, <49> electron acceptor: oxygen [44]) [44]
0.8 <49> (oxygen, <49> cosubstrate xanthine [44]) [44]
0.99 <7> (xanthine, <7> wild-type, pH 8.5, 25°C [88]) [88]
1.35 <7> (xanthine, <7> mutant E803V, pH 8.5, 25°C [88]) [88]
2.5 <49> (xanthine, <49> electron acceptor: oxygen [44]) [44]
4.4 <20> (xanthine, <20> mutant E232A, pH 7.8, 25°C [74]) [74]
5.67 <49> (dichloroindophenol, <49> cosubstrate xanthine [44]) [44]
6.33 <9> (xanthine, <9> electron acceptor: NAD^+ [35]) [35]
9.55 <49> (Nitro blue tetrazolium, <49> cosubstrate xanthine [44]) [44]
12.2 <3> (xanthine, <3> electron acceptor: NAD^+ [45]) [45]
15 <9> (xanthine, <9> electron acceptor: oxygen [35]) [35]
17.1 <3> (NADH, <3> activation of mutant G1011E activity after preincubation with phenazine methosulfate and cytochrome c, electron acceptor: methylene blue [31]) [31]
18.2 <49> (ferricyanide, <49> cosubstrate xanthine [44]) [44]
20.2 <49> (ferredoxin, <49> cosubstrate xanthine [44]) [44]
22.8 <3> (xanthine, <3> mutant E89K, electron acceptor: NAD^+ [31]) [31]
24.9 <3> (xanthine, <3> electron acceptor: NAD^+ [31]) [31]
27 <54, 55> (xanthine, <54,55> coexpression of genes xdhAB, conditions of high aeration, pH 7.8, 25°C [68]) [68]
29 <3> (NADH, <3> activation of mutant G1011E activity after preincubation with ferricyanide, electron acceptor: methylene blue [31]) [31]
30 <7> (hypoxanthine, <7> wild-type, pH 8.5, 25°C [88]) [88]
31 <3> (xanthine, <3> recombinant enzyme, electron acceptor: phenazine methosulfate-cytochrome c [7]) [7]
86 <54, 55> (xanthine, <54,55> coexpression of genes xdhABC, conditions of low aeration, pH 7.8, 25°C [68]) [68]
108 <20> (xanthine, <20> wild-type, pH 7.8, 25°C [74]) [74]
118 <54, 55> (xanthine, <54,55> coexpression of genes xdhABC, conditions of high aeration, pH 7.8, 25°C [68]) [68]
120 <41> (xanthine, <41> electron acceptor: NAD^+ [30]) [30]

Specific activity (U/mg)

0.0011 <37> (<37> substrate pterin, 22°C, pH 8.3 [67]) [67]
0.013 <37> (<37> substrate xanthine, 22°C, pH 8.3 [67]) [67]

0.06 <7> (<7> substrate xanthine, pH 7.2, 25°C [76]) [76]
0.08 <49> [44]
0.152 <12> [51]
0.16 <9> (<9> xanthine oxidase activity [35]) [35]
0.21 <13> (<13> substrate NAD^+, pH 7.2, 25°C [76]) [76]
0.23 <10> (<10> xanthine-NAD^+ assay [15]) [15]
0.25 <9> (<9> substrate NAD^+, pH 7.2, 25°C [76]) [76]
0.27 <26> (<26> substrate NAD^+, pH 7.2, 25°C [76]; <26> substrate xanthine, pH 7.2, 25°C [76]) [76]
0.29 <7> (<7> substrate NAD^+, pH 7.2, 25°C [76]) [76]
0.54 <3> [45]
0.69 <13> (<13> substrate xanthine, pH 7.2, 25°C [76]) [76]
1.17 <14> (<14> pH 8.5 [78]) [78]
1.56 <9> (<9> xanthine dehydrogenase activity [35]) [35]
1.8 <3> [43]
1.83 <9> (<9> substrate xanthine, pH 7.2, 25°C [76]) [76]
2.4 <15> [49]
2.47 <39> [18]
2.6 <46> (<46> xanthine-2,6-dichlorophenolindophenol assay [24]) [24]
2.8 <42> (<42> highest activity if growth medium contains urate and is supplemented with 0.0001 mM selenite and 0.0001 mM tungstate [60]) [60]
3.7 <21> (<21> highest activity if growth medium contains urate and is supplemented with 0.0001 mM selenite and 0.0001 mM molybdate [60]) [60]
5.69 <4> [37]
6.58 <41> (<41> pH 7.8, 25°C, expression of xdhAB genes [80]) [80]
10.6 <45> [62]
20 <40> [63]
26.7 <19> (<19> hypoxanthine-NAD^+ assay [9]) [9]
35.9 <34> [55]
50 <41> [30]
72.56 <49> (<49> xanthine-ferricyanide assay [17]) [17]
80 <41> (<41> pH 7.8, 25°C, expression of xdhABC genes [80]) [80]
80.2 <19> [8]
163 <19> (<19> ferricyanide-linked activity [40]) [40]
164 <51> (<51> hypoxanthine-$NADP^+$-assay [13]) [13]
385 <42> [39]
1328 <3> [52]
Additional information <37> (<37> enzymic activity is only detectable after resulfuration of purified enzyme [67]) [67]

K_m-Value (mM)

0.0003 <9> (xanthine, <9> electron acceptor: NAD^+ [35]) [35]
0.00083 <3> (methylene blue, <3> electron donor: 2-amino-4-hydroxy-pterine [42]) [42]
0.001 <9> (2-amino-4-hydroxypterin) [42]
0.0017 <39> (xanthine, <39> dehydrogenase type D, absence of NAD^+ [27]) [27]

0.0018 <48> (xanthopterin, <48> cosubstrate: NAD^+ [54]) [54]
0.002 <39> (xanthine, <39> oxidase type O, absence of NAD^+ [27]) [27]
0.00215 <9> (xanthine, <9> pH 7.2, 25°C [76]) [76]
0.00219 <13> (NAD^+, <13> pH 7.2, 25°C [76]) [76]
0.0022 <12> (NAD^+, <12> oxidation of 2-amino-4-hydroxy-pterin [51]) [51]
0.00252 <7> (NAD^+, <7> pH 7.2, 25°C [76]) [76]
0.0026 <39> (xanthine, <39> cosubstrate NAD^+ [27]) [27]
0.00274 <9> (NAD^+, <9> pH 7.2, 25°C [76]) [76]
0.0033 <3> (NAD^+, <3> electron donor: 2-amino-4-hydroxypterine [42]) [42]
0.0035 <12> (2-amino-4-hydroxypterin) [51]
0.004 <3> (2-amino-4-hydroxypterin, <3> pH 7.0, electron acceptor: methylene blue [42]) [42]
0.0041 <10> (hypoxanthine, <10> electron acceptor: NAD^+ [15]) [15]
0.00412 <26> (NAD^+, <26> pH 7.2, 25°C [76]) [76]
0.0045 <46> (xanthine, <46> electron acceptor: 2,6-dichlorophenolindophenol [24]) [24]
0.005 <15> (xanthine, <15> at pH 7.5, ping-pong-reaction mechanism, strongly pH-dependent [56]) [56]
0.0056 <10> (phenazine methosulfate, <10> electron donor: xanthine [15]) [15]
0.0057 <10> (NADH, <10> electron acceptor: nitroblue tetrazolium [15]) [15]
0.006 <3> (NAD^+, <3> electron donor: pterin [31]) [31]
0.0061 <48> (2-amino-4-hydroxypterin, <48> body preparation, cosubstrate: NAD^+ [54]) [54]
0.0063 <15> (NADH, <15> at pH 7.0, Tris-maleate buffer [56]) [56]
0.00633 <26> (xanthine, <26> pH 7.2, 25°C [76]) [76]
0.0067 <9> (NAD^+, <9> electron donor: xanthine [35]) [35]
0.0071 <3> (2-amino-4-hydroxypterin, <3> pH 8.0, electron acceptor: methylene blue [42]) [42]
0.0071 <48> (adenine, <48> cosubstrate: NAD^+ [54]) [54]
0.00714 <13> (xanthine, <13> pH 7.2, 25°C [76]) [76]
0.0072 <48> (2-amino-4-hydroxypterin, <48> wing preparation, co substrate: NAD^+ [54]) [54]
0.0075 <10> (hypoxanthine, <10> electron acceptor: nitroblue tetrazolium [15]) [15]
0.00774 <7> (xanthine, <7> pH 7.2, 25°C [76]) [76]
0.0082 <10, 12> (xanthine, <10> electron acceptor: NAD^+ [15]; <12> xanthine dehydrogenase activity [51]) [15, 51]
0.0086 <10> (xanthine, <10> electron acceptor: nitroblue tetrazolium [15]) [15]
0.0088 <7> (hypoxanthine, <7> wild-type, pH 8.5, 25°C [88]) [88]
0.0088 <7> (xanthine, <7> wild-type, pH 8.5, 25°C [88]) [88]
0.009 <3> (NADH, <3> electron acceptor: 2,6-dichlorophenolindophenol [31]) [31]
0.011 <3> (2-amino-4-hydroxypterin, <3> increased K_m in the presence of pyridoxal [42]) [42]

0.011 <15> (NADH, <15> at pH 8.0, Tris-maleate buffer [56]) [56]
0.0113 <10> (Nitroblue tetrazolium, <10> electron donor: xanthine [15]) [15]
0.012 <9> (xanthine, <9> electron acceptor: oxygen [35]) [35]
0.0125 <15> (NAD^+, <15> xanthine oxidation at pH 7.5 [56]) [56]
0.015 <45> (guanine) [62]
0.017 <1, 14, 49> (xanthine, <14> pH 8.5 [78]; <49> electron acceptor: ferricyanide [17]; <1> pH 7.9, increasing K_m with increasing pH [59]) [17, 59, 78]
0.018 <14> (hypoxanthine, <14> pH 8.5 [78]) [78]
0.018 <3> (xanthine, <3> electron acceptor: NAD^+, 30°C [42]) [42]
0.0182 <48> (purine, <48> cosubstrate: NAD^+ [54]) [54]
0.019 <1> (NAD^+, <1> pH 7.9, pH dependence, increasing K_m with increasing pH [59]) [59]
0.02 <3> (hypoxanthine, <3> cosubstrate: NAD^+ [52]) [52]
0.02 <15> (NAD^+, <15> at pH 7.5, hypoxanthine oxidation [56]) [56]
0.02 <3> (thio-NAD^+, <3> cosubstrate: hypoxanthine [52]) [52]
0.021 <12> (2,6-dichloroindophenol, <12> diaphorase activity [51]) [51]
0.021 <15> (3-acetylpyridine-adenine dinucleotide, <15> xanthine oxidation [56]) [56]
0.021 <12> (hypoxanthine, <12> xanthine dehydrogenase activity [51]) [51]
0.0236 <3> (xanthine, <3> cosubstrate: NAD^+ [52]) [52]
0.024 <3> (xanthine, <3> ping-pong reaction mechanism [57]) [57]
0.025 <3, 14> (NAD^+, <14> pH 8.5 [78]; <3> electron donor: xanthine [42]; <3> cosubstrate: hypoxanthine [52]) [42, 52, 78]
0.0258 <48> (4-hydroxypyrazolo(3,4-d)pyrimidine, <48> cosubstrate: NAD^+ [54]) [54]
0.028 <12> (NAD^+, <12> xanthine dehydrogenase activity [51]) [51]
0.029 <3> (xanthine, <3> electron acceptors: phenazine methosulfate/cytochrome c [31]) [31]
0.032 <3> (xanthine, <3> electron acceptor: NAD^+ [7]) [7]
0.0325 <34> (xanthine) [55]
0.033 <3> (NAD^+, <3> increased K_m for mutant G353D, electron acceptor: pterin [31]) [31]
0.035 <1> (xanthine, <1> immobilized enzyme preparation, pH 7.9, pH dependence, minimum K_m at pH 8.1, increasing values below and above [59]) [59]
0.036 <3> (xanthine, <3> increased K_m in the presence of pyridoxal at 30°C [42]) [42]
0.038 <51> ($NADP^+$, <51> electron donor: hypoxanthine [13]) [13]
0.04 <19> (hypoxanthine) [8]
0.04 <3> (NAD^+, <3> ping-pong reaction mechanism [57]) [57]
0.04 <3> (xanthine, <3> electron acceptor: NAD^+, 50°C [42]) [42]
0.0451 <48> (hypoxanthine, <48> co substrate: NAD^+ [54]) [54]
0.047-0.079 <15> (hypoxanthine, <15> measured in the pH range from 6 to 8.9, pH-independent K_m [56]) [56]
0.048 <3> (NADH, <3> increased K_m for mutant G353D, electron acceptor: 2,6-dichlorophenolindophenol [31]) [31]

0.052 <19> (NAD^+) [8]
0.0525 <34> (hypoxanthine) [55]
0.054 <1> (NAD^+, <1> immobilized enzyme preparation, pH 7.9, pH dependence, minimum K_m at pH 8.1, increasing values below and above [59]) [59]
0.055 <45> (hypoxanthine) [62]
0.059 <3> (xanthine, <3> increased K_m in the presence of pyridoxal at 50°C [42]) [42]
0.061 <48> (xanthine, <48> wing preparation, cosubstrate: NAD^+ [54]) [54]
0.0612 <34> (NAD^+) [55]
0.064 <19, 20> (xanthine, <20> wild-type, pH 7.8, 25°C [74]) [8, 74]
0.0647 <48> (xanthine, <48> body preparation, cosubstrate: NAD^+ [54]) [54]
0.066 <41, 54, 55> (xanthine, <41> electron acceptor: NAD^+ [30]; <54,55> coexpression of genes xdhAB, conditions of high aeration, pH 7.8, 25°C [68]) [30, 68]
0.067 <51, 54, 55> (xanthine, <51> electron acceptor: $NADP^+$ [13]; <54,55> coexpression of genes xdhAB, conditions of low aeration, pH 7.8, 25°C [68]) [13, 68]
0.07 <29> (xanthine, <29> NAD^+ [28]; <29> cosubstrate [28]) [28]
0.072 <7> (hypoxanthine, <7> mutant R881M, pH 8.5, 25°C [88]) [88]
0.072 <7> (xanthine, <7> mutant E803V, pH 8.5, 25°C [88]) [88]
0.085 <54, 55> (xanthine, <54,55> coexpression of genes xdhABC, conditions of high aeration, pH 7.8, 25°C [68]) [68]
0.1 <54, 55> (NAD^+, <54,55> coexpression of genes xdhABC, conditions of high aeration, pH 7.8, 25°C [68]) [68]
0.102 <15> (3-acetylpyridine adenine dinucleotide+, <15> NADH oxidation [56]) [56]
0.103 <20> (NAD^+, <20> wild-type, pH 7.8, 25°C [74]) [74]
0.1065 <48> (NAD^+, <48> cosubstrate: hypoxanthine [54]) [54]
0.11 <45> (NAD^+) [62]
0.113 <54, 55> (NAD^+, <54,55> coexpression of genes xdhABC, conditions of low aeration, pH 7.8, 25°C [68]) [68]
0.12 <29> (NAD^+, <29> cosubstrate: xanthine [28]) [28]
0.124 <48> (NAD^+, <48> cosubstrate: xanthine [54]) [54]
0.148 <54, 55> (NAD^+, <54,55> coexpression of genes xdhAB, conditions of high aeration, pH 7.8, 25°C [68]) [68]
0.15 <45> (xanthine) [62]
0.156 <54, 55> (NAD^+, <54,55> coexpression of genes xdhAB, conditions of low aeration, pH 7.8, 25°C [68]) [68]
0.16 <29, 41> (NAD^+, <41> electron donor: xanthine [30]; <29> cosubstrate: hypoxanthine [28]) [28, 30]
0.163 <20> (xanthine, <20> mutant E232A, pH 7.8, 25°C [74]) [74]
0.171 <10> (NAD^+, <10> electron donor: xanthine [15]) [15]
0.21 <51> (hypoxanthine, <51> electron acceptor: $NADP^+$ [13]) [13]
0.29 <29> (hypoxanthine, <29> cosubstrate: NAD^+ [28]) [28]
0.44 <42> (methyl viologen, <42> electron donor: xanthine [39]) [39]

0.5 <46> (hypoxanthine, <46> electron acceptor: 2,6-dichlorophenolindophenol [24]) [24]
1 <46> (4-hydroxypyrazolo(3,4-d)pyrimidine, <46> electron acceptor: 2,6-dichlorophenolindophenol [24]; <46> i.e. allopurinol [24]) [24]
1.35 <42> (xanthine, <42> electron acceptor: methyl viologen [39]) [39]
6 <12> (NADH, <12> diaphorase activity [51]) [51]
80.09 <54, 55> (xanthine, <54,55> coexpression of genes xdhABC, conditions of low aeration, pH 7.8, 25°C [68]) [68]
Additional information <20> (<20> rapid reaction kinetic parameters for substrates xanthine, 2-thioxanthine, 6-thioxanthine, 1-methylxanthine, 2-hydroxy-6-methylpurine, and 2,6-diaminopurine, in wild-type and mutants R310K and R310M [91]) [91]

pH-Optimum
6.5 <49> (<49> xanthine oxidation, presence of oxygen, two pH-optima [44]) [44]
7 <42> (<42> optimum value in the reverse reaction with urate and reduced methyl viologen [39]) [39]
7.8 <48> (<48> at 30°C [54]) [54]
7.9 <29> [28]
8 <3> [52]
8-8.6 <10> (<10> xanthine-NAD^+, NADH-nitroblue tetrazolium or xanthine-nitroblue tetrazolium assay [15]) [15]
8.2-8.8 <40> [63]
8.3 <34> [55]
8.5 <14, 42> (<42> optimum value for oxidation of xanthine and reduction of methyl viologen [39]) [39, 78]
8.7 <45> [62]
9.5 <49> (<49> xanthine oxidation, presence of oxygen, two pH-optima [44]) [44]
9.8 <10> (<10> second optimum for NADH-nitroblue tetrazolium assay [15]) [15]

pH-Range
6-9.2 <7> (<7> trends to increasing activities at higher values [5]) [5]
8-8.5 <53> [73]

Temperature optimum (°C)
30 <1> [59]
37 <34> (<34> highest activity, but unstable [55]) [55]
40 <40, 45> [62, 63]
55 <42> [39]

4 Enzyme Structure

Molecular weight

125000 <45> (<45> gel filtration, sedimentation equilibrium analysis [62]) [62]
129000 <49> (<49> native PAGE [17]) [17]
224000 <42> (<42> gel filtration [39]) [39]
230000-300000 <48> (<48> gel filtration [54]) [54]
250000 <49> (<49> sucrose density gradient centrifugation [44]) [44]
255000 <19> (<19> sucrose density gradient centrifugation, gel filtration [40]) [40]
270000-300000 <53> (<53> gel filtration [73]) [73]
275000 <29> (<29> gel filtration [28]) [28]
285000 <15> (<15> native PAGE [49]) [49]
287000 <41> (<41> gel filtration [30]) [30]
290000 <9> [2, 11]
300000 <1, 3, 14, 39> (<3,14> gel filtration [43,78]; <1> native gel electrophoresis [19]; <1> sedimentation equilibrium measurements [26]; <39> non-denaturating disc gel electrophoresis [18]) [18, 19, 26, 43, 78]
325000 <15> (<15> gel filtration [56]) [56]
345000 <34> (<34> gel filtration [55]) [55]
350000 <10, 19> (<19> gel filtration [8]; <10> native PAGE [15]) [8, 15]
357000 <12> (<12> sucrose density gradient centrifugation [51]) [51]
529000 <46> (<46> gel filtration [24]) [24]
530000 <51> (<51> gel filtration [13]) [13]
540000 <40> (<40> gel filtration [63]) [63]

Subunits

? <3, 6, 7, 9, 13, 19, 37, 39, 40, 41> (<19> x * 72000, SDS-PAGE [40]; <6,13> x * 150000, SDS-PAGE [12,76]; <41> α_1, x * 90000 + β_1, x * 60000, SDS-PAGE [30]; <39> x * 140000, SDS-PAGE, trypsin-treated: x * 90000, SDS-PAGE, conversion of dehydrogenase to oxidase activity [6]; <39> x * 150000, SDS-urea - PAGE [27,34]; <40> x * 54000 + x * 76000, SDS-PAGE [63]; <3> x * 150000, SDS-phosphate gel electrophoresis [57]; <7> x * 147782, calculation from sequence of cDNA [32]; <19> x * 92000 + x * 46000, SDS-PAGE [8]; <37> x * 145300, SDS-PAGE [67]; <7> x * 148000, MALDI-TOF [84]; <9> x * 148300, MALDI-TOF [84]) [6, 8, 12, 27, 30, 32, 34, 40, 57, 63, 67, 76, 84]
dimer <1, 3, 9, 12, 14, 15, 45, 53> (<45> 2 * 67000, SDS-PAGE [62]; <14,53> 2 * 150000, SDS-PAGE [73,78]; <12> 2 * 155000, SDS-PAGE [51]; <1> α_1, 1 * 155000 + β_1, 1 * 135000, SDS-PAGE [19]; <9> 2 * 145000, data from crystallization [2,11]; <3> 1 * 130000 + 1 * 140000, SDS-PAGE [43]; <15> 2 * 141000, SDS-PAGE [49]) [2, 3, 11, 19, 43, 49, 51, 62, 73, 78]
dodecamer <46, 51> (<51> tetramer of $\alpha/\beta/\gamma$-protomers, α_4, 4 * 81300 + β_4, 4 * 30000 + γ4, 4 * 17500, SDS-PAGE [13]; <46> α_4, 4 * 80000 + β_4, 4 * 35000 + γ4, 4 * 16000, SDS-PAGE [24]) [13, 24]
pentamer <42> (<42> 1 * 110000 + 1 * 83000 + 1 * 56000 + 1 * 53000 + 1 * 26000, SDS-PAGE [39]) [39]

tetramer <29, 34> (<34> 4 * 84200, SDS-PAGE [55]; <29> 2 * 81000 + 2 * 63000, SDS-PAGE [28]) [28, 55]
trimer <49> (<49> 1 * 82400 + 1 * 28500 + 1 * 18400, SDS-PAGE [17]) [17]

Posttranslational modification

phosphoprotein <41> (<41> 3.6 mol per α/β-protomer [30]) [30]
Additional information <8> (<8> conversion from dehydrogenase to oxidase by incubation with 4,4'-dithiodipyridine [75]) [75]

5 Isolation/Preparation/Mutation/Application

Source/tissue

adrenal gland <7> [4, 5]
cell culture <4, 5, 18, 19, 20, 21, 22, 24, 25, 27, 28, 29, 30, 31, 34, 35, 36, 38, 39, 40, 41, 42, 43, 45, 46, 49, 50, 51> (<34> inducible [55]; <4,49> inducible by xanthine [17,20]; <21> inducible by urate, selenite and molybdate [60]; <19> 0.0025 mg/ml sodium tungstate in the growth medium reduces level of ferricyanide-linked enzyme to 3% [40]; <42> selenium is needed for xanthine dehydrogenase formation [39]; <39> alveolar type II cells [38]; <5> inducible by xanthine, hypoxanthine or adenine [16]; <51> inducible by xanthine, selenite and molybdate [13]; <42> inducible by urate, selenite and tungstate [60]; <19> inducible by hypoxanthine [8,9]) [8, 9, 13, 16, 17, 20, 24, 28, 30, 37, 38, 39, 40, 41, 44, 50, 55, 58, 60, 62, 63]
cerebellum <7, 11> (<7> high rate of dehydrogenase activity [4,5]) [4, 5]
cerebral cortex <7, 11> (<7> high rate of dehydrogenase activity [4,5]) [4, 5]
epithelium <8> (<8> lung epithelial cell [92]) [92]
fat body <33> [22]
fruit <32> (<32> enzyme activity is maximal in late linear phase of fruit growth [65]) [65]
heart <7, 11, 39> (<39> only oxidase type O form [29]; <7> high rate of dehydrogenase activity [4,5]) [4, 5, 29]
intestinal mucosa <39> (<39> exists preliminary as oxidase form, promotes oxidative incorporation of iron into mucosal transferrin, facilitates dietary iron absorption [48]) [48]
intestine <39> (<39> only oxidase type O form, can be converted to dehydrogenase type D by dithioerythritol [29]) [29]
kidney <7, 39> (<39> only oxidase type O form [29]) [4, 5, 29]
leaf <10, 14, 44, 53> (<44> analysis of an aba1 gene deficient mutant [33]; <53> cold stress causes decrease, desiccation and senesence cause strong increase in activity [73]) [15, 33, 36, 73]
liver <1, 2, 6, 7, 8, 11, 39> (<7,11> high oxidase and reductase activities at nearly the same rates [4,5]; <39> dehydrogenase and oxidase forms [29]; <39> exists preliminary as dehydrogenase form, participates in mobilization of iron from ferritin stores [48]; <8> isolated cells of liver [66]) [1, 4, 5, 6, 18, 19, 21, 23, 26, 27, 29, 32, 34, 46, 47, 48, 53, 59, 61, 66, 70, 75, 86, 88, 93]

lung <6, 7, 39> (<39> only oxidase type O form [29]; <6> lung xanthine oxidoreductase activity is significantly increased after 2 h of mechanical ventilation without changes in enzyme expression. Increase occurs via activation of p38 MAP kinase and ERK and plays a critical role in the pathogenesis of pulmonary edema associated with ventilator-induced lung injury [83]) [29, 32, 83]
malpighian tubule <33> [22]
mammary gland <6> [12]
milk <6, 7, 9, 13, 26> (<9> only oxidase type in stored milk, 94% NAD^+-dependent dehydrogenase type in fresh milk [64]; <26> in milk, more than 90% of enzyme exists in the inactive demolybdo form [87]) [2, 11, 12, 14, 34, 35, 42, 64, 71, 72, 76, 79, 84, 87]
mycelium <12> (<12> inducible [51]) [51]
nodule <15> [49, 56]
pancreas <7, 39> (<39> only oxidase type O form [29]) [29, 32]
paneth cell <39> (<39> identification of zinc-binding protein identical to xanthine oxidoreductase of 85 kDa [10]) [10]
placenta <7> (<7> very low oxidase and dehydrogenase activities [4,5]) [4, 5]
pons <7, 11> (<7> high rate of dehydrogenase activity [4,5]) [4, 5]
seedling <14> (<14> etiolated [78]) [78]
skeletal muscle <7> (<7> low oxidase activity [4,5]) [4, 5, 32]
small intestine <7> [5]
spleen <7, 39> (<39> only oxidase type O form [29]) [4, 5, 29]
whole body <3> (<3> five different forms in electrophoretic mobility, but no differences in kinetic constants [42]; <3> selection of 11 rosy mutant strain, corresponding to single amino acid substitutions for detailed studies on the activities [25]; <3> analysis of a rosy mutant strain [45]) [25, 31, 42, 43, 45, 52]
wing <48> [54]

Localization

cytoplasm <8, 33> (<8> parenchymal cells, dehydrogenase activity [66]) [22, 66]
cytosol <1, 7, 14> (<1,14> subcellular fractionation [36,53]; <7> 100000 x g supernatant, cerebral cortex, liver [5]) [5, 36, 53]
lipid droplet <9> (<9> fat-globule, milk [64]) [64]
peroxisome <8> (<8> matrix and core, of parenchymal cells, both oxidase and dehydrogenase activity [66]) [66]

Purification

<1> (immobilized on n-octylamine-substituted Sepharose 4B) [59]
<1> (to homogeneity, chromatography techniques) [3, 19, 26, 59]
<2> (to homogeneity, immobilized preparation) [61]
<3> (529fold purification, several $(NH_4)_2SO_4$ precipitation steps) [52]
<3> (near homogeneity, chromatography techniques) [43]
<3> (to homogeneity, chromatography techniques) [31, 45]
<3> (to homogeneity, immunoaffinity chromatography) [57]
<3> (to homogeneity, recombinant enzyme) [7]

<4> (partial) [20]
<4> (to homogeneity, chromatography, preparative gel electrophoresis) [37]
<6> (to homogeneity, affinity chromatography) [12]
<7> (after expression in Escherichia coli) [88]
<9> [2, 11]
<9> (to homogeneity, 2 types: dehydrogenase type, chromatography techniques, oxidase type: affinity chromatography) [64]
<9> (to homogeneity, presence of DTT required for purification of the dehydrogenase form) [35]
<10> (to homogeneity, chromatography techniques) [15]
<12> (to homogeneity, immunoabsorption chromatography) [51]
<13> [76]
<14> [78]
<15> (to homogeneity, chromatography techniques) [49, 56]
<19> [9]
<19> (to homogeneity, chromatography techniques) [8]
<19> (to homogeneity, mainly ferricyanide- and NAD^+-linked activities, chromatography steps) [40]
<20> [74]
<29> (near homogeneity, chromatography techniques) [28]
<34> (to homogeneity, affinity chromatography) [55]
<37> (in presence of dithiothreitol) [67]
<39> (during purification conversion from dehydrogenase type D to oxidase type O, reversible by dithioerythritol) [29]
<39> (nearly homogenous, dehydrogenase type D, chromatography steps, oxidase type O purified by nearly the same procedure, but heated to 60°C after homogenization) [27]
<39> (purified as antibody complex) [34]
<39> (to homogeneity, affinity chromatography) [46]
<39> (to homogeneity, chromatography techniques) [6, 18]
<39> (to homogeneity, mainly NAD^+-dependent activity) [47]
<40> (to crystalline state, chromatography techniques) [63]
<41> (to homogeneity, chromatography steps) [30]
<42> (to homogeneity, chromatography techniques, preparative gel electrophoresis) [39]
<45> (to homogeneity, chromatography techniques) [62]
<46> (to homogeneity, chromatography techniques) [24]
<48> (partial) [54]
<49> (to homogeneity, chromatography techniques) [17]
<49> (to homogeneity, precipitation and chromatography techniques) [44]
<51> (to homogeneity, chromatography techniques) [13]

Crystallization

<7> (mutant E803V, decrease in activity towards purine substrates is not due to large conformational change in the mutant protein) [88]
<8> (mutant C535A/C992R/C1324S) [75]

<8> (mutant W335A/F336L, showing two similar, but not identical subunits. The cluster involved in conformation-switching is completely disrupted in one subunit, but remains partly associated in the other. Xanthine oxidase and oxidoreductase forms of the mutant are in equilibrium that greatly favors the oxidase form, but upon incubation with dithiothreitol equilibrium is partly shifted towards the oxidoreductase form) [89]
<9> (2.1 A resolution for both xanthine oxidase and dehydrogenase forms, irreversible pancreatin cleaved xanthine oxidase form results in blocking access of substrate NAD^+ to the FAD cofactor) [11]
<9> (batch method, 2.5 and 3.3 A resolution for irreversible proteolytically cleaved xanthine oxidase form, 2.1 A resolution for dehydrogenase form) [2]
<9> (in complex with TEI-6720) [71]
<9> (in complex with inhibitor FYX051, which is slowly hydroxylated by the enzyme) [79]
<20> (2.7 A resolution, direct coordination of alloxanthine to the molybdenum via a nitrogen atom) [50]

Cloning

<3> (expression in Aspergillus nidulans) [7]
<53> [73]

Engineering

C535A <8> (<8> resistant to conversion from dehydrogenase to oxidase by incubation with 4,4'-dithiodipyridine [75]) [75]
C535A/C992R <8> (<8> slow conversion from dehydrogenase to oxidase by incubation with 4,4'-dithiodipyridine, conversion is blocked by NADH [75]) [75]
C535A/C992R/C1316S <8> (<8> completely resistant to conversion from dehydrogenase to oxidase by incubation with 4,4'–dithiodipyridine [75]) [75]
C535A/C992R/C1324S <8> (<8> completely resistant to conversion from dehydrogenase to oxidase by incubation with 4,4'–dithiodipyridine [75]) [75]
C992R <8> (<8> resistant to conversion from dehydrogenase to oxidase by incubation with 4,4'-dithiodipyridine [75]) [75]
E232A <20> (<20> decrease in k_{cat} value, increase in K_M-value [74]) [74]
E730A <20> (<20> no enzymic activity [74]) [74]
E730D <20> (<20> no enzymic activity [74]) [74]
E730Q <20> (<20> no enzymic activity [74]) [74]
E730R <20> (<20> no enzymic activity [74]) [74]
E803V <7> (<7> almost complete loss of activity with hypoxanthine, weak activity with xanthine, significant aldehyde oxidase activity [88]) [88]
E89K <3> (<3> natural mutant strain, lacking iron-sulfur centers, activity to xanthine/NAD^+ or xanthine/pterin not affected, but xanthine/phenazine methosulfate activity abolished [45]) [31, 45]
G1011E <3> (<3> within the molybdenum domain, no activity without oxidative activation [31]) [31]
G353D <3> (<3> modifications to the NAD^+-NADH-binding sites [31]) [31]
R135C <20> (<20> mutation corresponding to human protein variant of a patient suffering from xanthinuria I. Mutation results in an active $(\alpha\beta)2$ he-

terotetrameric form besides an inactive $\alpha\beta$ heterodimeric form missing the FeSI center [72]) [72]
R310K <20> (<20> absorption spectra similar to wild-type. 20fold decrease of k_{red}-value [91]) [91]
R310M <20> (<20> absorption spectra similar to wild-type. 20000fold decrease of k_{red}-value [91]) [91]
R881M <7> (<7> almost complete loss of activity with xanthine, weak activity with hypoxanthine, significant aldehyde oxidase activity [88]) [88]
S357F <3> (<3> modifications to the NAD^+-NADH-binding sites [31]) [31]
W335A/F336L <8> (<8> mutant oxidoreductase displaying xanthine oxidase activity [89]) [89]
Additional information <6, 16> (<16> T-DNA insertion mutant, loss of superoxide producing activity [77]; <6> enzyme null mutant mice demonstrate 50% reduction in adipose mass compared to control, while obese mice exhibit increased concentrations of xanthine oxidoreductase mRNA and urate in adipose tissues. In vitro, knockdown of xanthine oxidoreductase inhibits adipogenesis and nuclear receptor PPARγ activity [85]) [77, 85]

Application

agriculture <32> (<32> treatment of normal fruit in linear phase of growth with enzyme inhibitors allopurinol or adenine arrests fruit growth [65]) [65]
analysis <3> (<3> use of rosy mutant strains to probe structure and function of xanthine dehydrogenase [25]) [25]
medicine <6, 39> (<39> modification of xanthine oxidoreductase by disulfiram that is used in the management of chronic alcolism [47]; <6> enzyme null mutant mice demonstrate 50% reduction in adipose mass compared to control, while obese mice exhibit increased concentrations of xanthine oxidoreductase mRNA and urate in adipose tissues. In vitro, knockdown of xanthine oxidoreductase inhibits adipogenesis and nuclear receptor PPARγ activity. Xanthine oxidoreductase is a potential therapeutic target for metabolic abnormalities beyond hyperuricemia [85]; <6> lung xanthine oxidoreductase activity is significantly increased after 2 h of mechanical ventilation without changes in enzyme expression. Increase occurs via activation of p38 MAP kinase and ERK and plays a critical role in the pathogenesis of pulmonary edema associated with ventilator-induced lung injury [83]; <6> oral administration of cassia oil significantly reduces serum and hepatic urate levels in hyperuricemic mice. At 600mg/kg, cassia oil is as potent as allopurinol. This hypouricemic effect is explained by inhibiting activities of liver xanthine oxidase and xanthine oxidoreductase [93]) [47, 83, 85, 93]
synthesis <1, 8, 41, 54, 55> (<1> immobilized xanthine dehydrogenase for use in organic synthesis [59]; <8> expression of enzyme in baculovirus-insect cell system, yields a mixture of native dimeric, demolydbo-dimeric and monomeric forms. All forms contain flavin, the monomeric forms lack molybdopterin and the iron-sulfur centers. Monomeric forms require only three electrons for complete reduction [70]; <54,55> expression of genes xdhAB encoding the two subunits of enzyme, in Escherichia coli, produces active enzyme with moleybdenum content of 0.11-0.16 mol per $\alpha\beta$ protomer and

iron and FAD levels at stoichiometries similar to native enzyme. Coexpression of xdhAB genes with Pseudomonas aeruginosa xdhC gene increases level of molybdenum incorporated to a 1:1 stoichiometry and results in high levels of functional protein, up to 2284 units per mg and 8039 mg per l [68]; <41> for maximal level of functional expression, co-expression of xdhC gene is required which increases level of molybdenum incorporation. Iron and FAD content of expressed enzymes are independent of xdhC expression [80]) [59, 68, 70, 80]

6 Stability

pH-Stability

7 <49> (<49> unstable below, stabilized and activated at higher pH values [44]) [44]
7-8.5 <48> (<48> stable within, loss of activity at more acidic pHs [54]) [54]
7-12 <45> (<45> stable for 24 h if stored at 4°C [62]) [62]
7.5 <9> (<9> stable at 4°C, one day, absence of DTT [35]) [35]
8.5 <9> (<9> 4°C, absence of DTT, marked decrease of activity within hours [35]) [35]
9.2 <3> (<3> recombinant enzyme, half-life: 10-15 h at 0-4°C [7]) [7]
11 <2> (<2> no effect on activity [21]) [21]

Temperature stability

-20 <7, 14, 39> (<7> stable up to 10 days [5]; <39> conversion of dehydrogenase type D to oxidase type O [29]; <14> stable dehydrogenase form, no conversion to the oxidase form [36]) [5, 29, 36]
2 <9> (<9> dehydrogenase type stable under anaerobic conditions [64]) [64]
4 <1, 7, 45> (<1> free enzyme, half-life: 18 days, immobilized enzyme, half-life: 88 days [59]; <45> up to 150% activity when kept 24-120 h in a pH range of 9-11 [62]; <7> liver extract: conversion of dehydrogenase to oxidase activity, some loss of activity after 7 days, extract from cerebrellum and cerebral cortex: stable for 7 days [5]) [5, 59, 62]
30 <2> (<2> enhanced stability of immobilized enzyme preparation if N_2 and hemoglobin is included, reaction of oxygen products causes instability of the working enzyme, half-life: 250-560 min, compared to immobilized enzyme without additions of about 145 min [61]) [61]
37 <14, 34, 39> (<34> stable for 4 min [55]; <39> conversion to an intermediate dehydrogenase-oxidase type O form in 30 min [34]; <14> stable dehydrogenase activity, no conversion to the oxidase form [36]) [34, 36, 55]
38 <9> (<9> dehydrogenase type unstable, change to oxidase type inhibited under anaerobic conditions [64]) [64]
40 <40> (<40> stable for 10 min, pH 8.5 [63]) [63]
55 <45> (<45> stable below [62]) [62]
60 <39> (<39> conversion of dehydrogenase type D to oxidase type O [27]) [27]
65 <45> (<45> complete loss of activity in 10 min [62]) [62]

Oxidation stability

<9>, conversion of dehydrogenase to oxidase type under aerobic conditions [64]

Organic solvent stability

methanol <19> (<19> sensitive to, half-life: 3 min at 1.5 M [40]) [40]

General stability information

<1>, subunit α more sensitive to SDS-heat treatment [19]
<3>, extremely labile [52]
<19>, oxygen-linked activity rather unstable during storage at 4°C, ferricyanide-linked activity stable to freezing and thawing [40]
<29>, crude and purified enzyme preparations form readily insoluble inactive aggregates, not stable to freezing and thawing [28]
<39>, treatment accelerating conversion of dehydrogenase type D to oxidase type O is lyophilization, storage at -20°C, exposure to 20% ammonium sulfate, 0.5 mM NAD^+ or 0.01 mM xanthine [27]
<42>, complete loss of activity when stored below 0°C [39]
<45>, enzyme from crude extract can be stabilized by addition of nicotinamide, cysteine and EDTA, purified enzyme more stable [62]

Storage stability

<9>, -70°C, 0.1 M sodium diphosphate, 0.3 mM EDTA, pH 7.5, 1 mM salicylate, 2.5 mM DTT, 3 months [35]
<10>, 4°C, 25 mM Tris-boric acid, pH 8.5, 2.5 mM EDTA, 2 mM 2-mercaptoethanol, 0.2 mM sodium molybdate, one year [15]
<19>, 4°C, 1 week [40]
<39>, -20°C, pH 8.6, 3 days [46]
<41>, -80°C, 6 months [30]
<46>, -70°C, 6 months [24]
<48>, 1°C, 1 mM DTT in conjunction with 0.1 M ammonium sulfate, 60-70% retaining activity after 8-12 days [54]

References

[1] Battelli, M.G.; Lorenzoni, E.: Purification and properties of a new glutathione-dependent thiol:disulphide oxidoreductase from rat liver. Biochem. J., **207**, 133-138 (1982)

[2] Eger, B.T.; Okamoto, K.; Enroth, C.; Sato, M.; Nishino, T.; Pai, E.F.; Nishino, T.: Purification, crystallization and preliminary X-ray diffraction studies of xanthine dehydrogenase and xanthine oxidase isolated from bovine milk. Acta Crystallogr. Sect. D, **56**, 1656-1658 (2000)

[3] Bruguera, P.; Lopez-Cabrera, A.; Canela, E.I.: Kinetic mechanism of chicken liver xanthine dehydrogenase. Biochem. J., **249**, 171-178 (1988)

[4] Wajner, M.; Harkness, R.A.: Distribution of xanthine dehydrogenase and oxidase activities in human and rabbit tissues. Biochem. Soc. Trans., **16**, 358-359 (1988)

[5] Wajner, M.; Harkness, R.A.: Distribution of xanthine dehydrogenase and oxidase activities in human and rabbit tissues. Biochim. Biophys. Acta, **991**, 79-84 (1989)
[6] Stark, K.; Seubert, P.; Lynch, G.; Baudry, M.: Proteolytic conversion of xanthine dehydrogenase to xanthine oxidase: Evidence against a role for calcium-activated protease (calpain). Biochem. Biophys. Res. Commun., **165**, 858-864 (1989)
[7] Adams, B.; Lowe, D.J.; Smith, A.T.; Scazzocchio, C.; Demais, S.; Bray, R.C.: Expression of Drosophila melanogaster xanthine dehydrogenase in Aspergillus nidulans and some properties of the recombinant enzyme. Biochem. J., **362**, 223-229 (2002)
[8] Kim, J.M.; Schmid, R.D.: Purification and characterization of a microbial xanthine dehydrogenase highly active towards hypoxanthine. GBF Monogr., Biosens. Appl. Med., Environ. Prot. Process Control, **13**, 421-424 (1989)
[9] Parschat, K.; Canne, C.; Huttermann, J.; Kappl, R.; Fetzner, S.: Xanthine dehydrogenase from Pseudomonas putida 86: Specificity, oxidation-reduction potentials of its redox-active centers, and first EPR characterization. Biochim. Biophys. Acta, **1544**, 151-165 (2001)
[10] Morita, Y.; Sawada, M.; Seno, H.; Takaishi, S.; Fukutawa, H.; Miyake, N.; Hiai, H.; Chiba, T.: Identification of xanthine dehydrogenase/xanthine oxidase as a rat paneth cell zinc-binding protein. Biochim. Biophys. Acta, **1540**, 43-49 (2001)
[11] Enroth, C.; Eger, B.T.; Okamoto, K.; Nishino, T.; Nishino, T.; Pai, E.F.: Crystal structures of bovine milk xanthine dehydrogenase and xanthine oxidase: Structure-based mechanism of conversion. Proc. Natl. Acad. Sci. USA, **97**, 10723-10728 (2000)
[12] McManaman, J.L.; Neville, M.C.; Wright, R.M.: Mouse mammary gland xanthine oxidoreductase: Purification, characterization, and regulation. Arch. Biochem. Biophys., **371**, 308-316 (1999)
[13] Schräder, T.; Rienhöfer, A.; Andreesen, J.R.: Selenium-containing xanthine dehydrogenase from Eubacterium barkeri. Eur. J. Biochem., **264**, 862-871 (1999)
[14] Ichimori, K.; Fukahori, M.; Nakazawa, H.; Okamoto, K.; Nishino, T.: Inhibition of xanthine oxidase and xanthine dehydrogenase by nitric oxid. J. Biol. Chem., **274**, 7763-7768 (1999)
[15] Montalbini, P.: Purification and some properties of xanthine dehydrogenase from wheat leaves. Plant Sci., **134**, 89-102 (1998)
[16] Christiansen, L.C.; Schou, S.; Nygaard, P.; Saxild, H.H.: Xanthine metabolism in Bacillus subtilis: Characterization of the xpt-pbuX operon and evidence for purine- and nitrogen-controlled expression of genes involved in the xanthine salvage and catabolism. J. Bacteriol., **179**, 2540-2550 (1997)
[17] Gremer, L.; Meyer, O.: Characterization of xanthine dehydrogenase from the anaerobic bacterium Veillonella atypica and identification of molybdopterin-cytosine-dinucleotide-containing molybdenum cofactor. Eur. J. Biochem., **238**, 862-866 (1996)

[18] Suleiman, S.A.; Stevens, J.B.: Purification of xanthine dehydrogenase from rat liver: A rapid procedure with high enzyme yields. Arch. Biochem. Biophys., **258**, 219-225 (1987)
[19] Irie, S.: Subunit constitution of electrophoretically purified xanthine dehydrogenase of avian liver. J. Biochem., **95**, 405-412 (1984)
[20] Prez-Vicente, R.; Pineda, M.; Cardenas, J.: Occurrence of an NADH diaphorase activity associated with xanthine dehydrogenase in Chlamydomonas reinhardtii. FEMS Microbiol. Lett., **43**, 321-325 (1987)
[21] Fhaolain, I.N.; Coughlan, M.P.: Effects of limited proteolysis on the structure and activity of turkey liver xanthine dehydrogenase. Biochem. Soc. Trans., **5**, 1705-1707 (1977)
[22] Rocher-Chambonnet, C.; Berreur, P.; Houde, M.; Tiveron, M.C.; Lepesant, J.A.; Bregegere, F.: Cloning and partial characterization of the xanthine dehydrogenase gene of Calliphora vicina, a distant relative of Drosophila melanogaster. Gene, **59**, 201-212 (1987)
[23] Schopfer, L.M.; Massey, V.; Nishino, T.: Rapid reaction studies on the reduction and oxidation of chicken liver xanthine dehydrogenase by the xanthine/urate and NAD/NADH couples. J. Biol. Chem., **263**, 13528-13538 (1988)
[24] Self, W.T.; Stadtman, T.C.: Selenium-dependent metabolism of purines: A selenium-dependent purine hydroxylase and xanthine dehydrogenase were purified from Clostridium purinolyticum and characterized. Proc. Natl. Acad. Sci. USA, **97**, 7208-7213 (2000)
[25] Hughes, R.K.; Doyle, W.A.; Chovnick, A.; Whittle, J.R.S.; Burke, J.F.; Bray, R.C.: Use of rosy mutant strains of Drosophila melanogaster to probe the structure and function of xanthine dehydrogenase. Biochem. J., **285**, 507-513 (1992)
[26] Rajagopalan, K.V.; Handler, P.: Purification and properties of chicken liver xanthine dehydrogenase. J. Biol. Chem., **242**, 4097-4107 (1967)
[27] Waud, W.R.; Rajagopalan, K.V.: Purification and properties of the NAD^+-dependent (type D) and O_2-dependent (type O) forms of rat liver xanthine dehydrogenase. Arch. Biochem. Biophys., **172**, 354-364 (1976)
[28] Sin, I.L.: Purification and properties of xanthine dehydrogenase from Pseudomonas acidovorans. Biochim. Biophys. Acta, **410**, 12-20 (1975)
[29] Della Corte, E.; Stirpe, F.: The regulation of rat liver xanthine oxidase. Involvement of thiol groups in the conversion of the enzyme activity from dehydrogenase (type D) into oxidase (type O) and purification of the enzyme. Biochem. J., **126**, 739-745 (1972)
[30] Xiang, Q.; Edmondson, D.E.: Purification and characterization of a prokaryotic xanthine dehydrogenase from Comamonas acidovorans. Biochemistry, **35**, 5441-5450 (1996)
[31] Doyle, W.A.; Burke, J.F.; Chovnick, A.; Dutton, F.L.; Whittle, J.R.S.; Bray, R.C.: Properties of xanthine dehydrogenase variants from rosy mutant strains of Drosophila melanogaster and their relevance to the enzyme's structure and mechanism. Eur. J. Biochem., **239**, 782-795 (1996)
[32] Wright, R.M.; Vaitaitis, G.M.; Wilson, C.M.; Repine, T.B.; Terada, L.S.; Repine, J.E.: cDNA cloning, characterization, and tissue-specific expression of

human xanthine dehydrogenase/xanthine oxidase. Proc. Natl. Acad. Sci. USA, **90**, 10690-10694 (1993)

[33] Leydecker, M.T.; Moureaux, T.; Kraepiel, Y.; Schnorr, K.; Caboche, M.: Molybdenum cofactor mutants, specifically impaired in xanthine dehydrogenase activity and abscisic acid biosynthesis, simultaneously overexpress nitrate reductase. Plant Physiol., **107**, 1427-1431 (1995)

[34] Waud, W.R.; Rajagopalan, K.V.: The mechanism of conversion of rat liver xanthine dehydrogenase from an NAD^+-dependent form (type D) to an O_2-dependent form (type O). Arch. Biochem. Biophys., **172**, 365-379 (1976)

[35] Hunt, J.; Massey, V.: Purification and properties of milk xanthine dehydrogenase. J. Biol. Chem., **267**, 21479-21485 (1992)

[36] Nguyen, J.; Feierabend, J.: Some properties and subcellular localization of xanthine dehydrogenase in pea leaves. Plant Sci. Lett., **13**, 125-132 (1978)

[37] Perez-Vicente, R.; Alamillo, J.M.; Cardenas, J.; Pineda, M.: Purification and substrate inactivation of xanthine dehydrogenase from Chlamydomonas reinhardtii. Biochim. Biophys. Acta, **1117**, 159-166 (1992)

[38] Panus, P.C.; Burgess, B.; Freeman, B.A.: Characterization of cultured alveolar epithelial cell xanthine dehydrogenase/oxidase. Biochim. Biophys. Acta, **1091**, 303-309 (1991)

[39] Wagner, R.; Cammack, R.; Andreesen, J.R.: Purification and characterization of xanthine dehydrogenase from Clostridium acidiurici grown in the presence of selenium. Biochim. Biophys. Acta, **791**, 63-74 (1984)

[40] Woolfolk, C.A.: Purification and properties of a novel ferricyanide-linked xanthine dehydrogenase from Pseudomonas putida 40. J. Bacteriol., **163**, 600-609 (1985)

[41] Woolfolk, C.A.; Downard, J.S.: Distribution of xanthine oxidase and xanthine dehydrogenase specificity types among bacteria. J. Bacteriol., **130**, 1175-1191 (1977)

[42] Yen, T.T.T.; Glassman, E.: Electrophoretic variants of xanthine dehydrogenase in Drosophila melanogaster: II. Enzyme kinetics. Biochim. Biophys. Acta, **146**, 35-44 (1967)

[43] Seybold, W.D.: Purification and partial characterization of xanthine dehydrogenase from Drosophila melanogaster. Biochim. Biophys. Acta, **334**, 266-271 (1974)

[44] Smith, S.T.; Rajagopalan, K.V.; Handler, P.: Purification and properties of xanthine dehydrogenase from Micrococcus lactilyticus. J. Biol. Chem., **242**, 4108-4117 (1967)

[45] Hughes, R.K.; Bennett, B.; Bray, R.C.: Xanthine dehydrogenase from Drosophila melanogaster: Purification and properties of the wild-type enzyme and of a variant lacking iron-sulfur centers. Biochemistry, **31**, 3073-3083 (1992)

[46] Kaminski, Z.W.; Jezewska, M.M.: Effect of NADH on hypoxanthine hydroxylation by native NAD^+-dependent xanthine oxidoreductase of rat liver, and the possible biological role of this effect. Biochem. J., **200**, 597-603 (1981)

[47] Kaminsky, Z.W.; Jezewska, M.M.: Involvement of a single thiol group in the conversion of the NAD$^+$-dependent activity of rat liver xanthine oxidoreductase to the O_2-dependent activity. Biochem. J., **207**, 341-346 (1982)
[48] Topham, R.W.; Walker, M.C.; Calisch, M.: Liver xanthine dehydrogenase and iron mobilization. Biochem. Biophys. Res. Commun., **109**, 1240-1246 (1982)
[49] Triplett, E. W.; Blevins, D.G.; Randall, D.D.: Purification and properties of soybean nodule xanthine dehydrogenase. Arch. Biochem. Biophys., **219**, 39-46 (1982)
[50] Truglio, J.J.; Theis, K.; Leimkuhler, S.; Rappa, R.; Rajagopalan, K.V.; Kisker, C.: Crystal structures of the active and alloxanthine-inhibited forms of xanthine dehydrogenase from Rhodobacter capsulatus. Structure, **10**, 115-125 (2002)
[51] Lyon, E.S.; Garrett, R.H.: Regulation, purification, and properties of xanthine dehydrogenase in Neurospora crassa. J. Biol. Chem., **253**, 2604-2614 (1978)
[52] Parzen, S.D.; Fox, A.S.: Purification of xanthine dehydrogenase from Drosophila melanogaster. Biochim. Biophys. Acta, **92**, 465-471 (1964)
[53] Coolbear, K.P.; Herzberg, G.R.; Brosnan, J.T.: Xanthine dehydrogenase in chicken liver: Its subcelluar localization ant its possible role on gluconeogenesis from amino acids. Biochem. Soc. Trans., **9**, 394-395 (1981)
[54] Watt, W.B.: Xanthine dehydrogenase and pteridine metabolism in Colias butterflies. J. Biol. Chem., **247**, 1445-1451 (1972)
[55] Aretz, W.; Kaspari, H.; Klemme, J.H.: Molecular and kinetic characterization of xanthine dehydrogenase from the phototrophic bacterium Rodopseudomonas capsulata. Z. Naturforsch. C, **36**, 933-941 (1981)
[56] Boland, M.J.; Blevins, D.G.; Randall, D.D.: Soybean nodule xanthine dehydrogenase: A kinetic study. Arch. Biochem. Biophys., **222**, 435-441 (1983)
[57] Edwards, T.C.R.; Candido, E.P.M.: Xanthine dehydrogenase from Drosophila melanogaster. A comparison of the kinetic parameters of the pure enzyme from two wild-type isoalleles differing at a putative regulatory site. Mol. Gen. Genet., **154**, 1-6 (1977)
[58] Fernandez, E.; Cardenas, J.: Occurrence of xanthine dehydrogenase in Chlamydomonas reinhardii: A common cofactor shared by xanthine dehydrogenase and nitrate reductase. Planta, **153**, 254-257 (1981)
[59] Tramper, J.; Angelino, S.A.G.F.; Muller, F.; van der Plas, H.C.: Kinetics and stability of immobilized chicken liver xanthine dehydrogenase. Biotechnol. Bioeng., **21**, 1767-1786 (1979)
[60] Wagner, R.; Andreesen, J.R.: Selenium requirement for active xanthine dehydrogenase from Clostridium acidiurici and Clostridium cylindrosporum. Arch. Microbiol., **121**, 255-260 (1979)
[61] Coughlan, M.P.; Johnson, D.B.: The inactivation of xanthine-oxidizing enzymes, native and deflavo forms, in the presence of oxygen. Biochem. Soc. Trans., **7**, 18-21 (1979)
[62] Ohe, T.; Watanabe, Y.: Purification and properties of xanthine dehydrogenase from Streptomyces cyanogenus. J. Biochem., **86**, 45-53 (1979)

[63] Sakai, T.; Jun, H.K.: Purification, crystallization, and some properties of xanthine dehydrogenase from Pseudomonas synxantha A 3. Agric. Biol. Chem., **43**, 753-760 (1979)

[64] Nakamura, M.; Yamazaki, I.: Preparation of bovine milk xanthine oxidase as a dehydrogenase form. J. Biochem., **92**, 1279-1286 (1982)

[65] Taylor, N.J.; Cowan, A.K.: Xanthine dehydrogenase and aldehyde oxidase impact plant hormone homeostasis and affect fruit size in 'Hass' avocado. J. Plant Res., **117**, 121-130 (2004)

[66] Frederiks, W.M.; Vreeling-Sindelarova, H.: Ultrastructural localization of xanthine oxidoreductase activity in isolated rat liver cells. Acta Histochem., **104**, 29-37 (2002)

[67] Baghiani, A.; Harrison, R.; Benboubetra, M.: Purification and partial characterisation of camel milk xanthine oxidoreductase. Arch. Physiol. Biochem., **111**, 407-414 (2003)

[68] Ivanov, N.V.; Hubalek, F.; Trani, M.; Edmondson, D.E.: Factors involved in the assembly of a functional molybdopyranopterin center in recombinant Comamonas acidovorans xanthine dehydrogenase. Eur. J. Biochem., **270**, 4744-4754 (2003)

[69] Self, W.T.: Regulation of purine hydroxylase and xanthine dehydrogenase from Clostridium purinolyticum in response to purines, selenium, and molybdenum. J. Bacteriol., **184**, 2039-2044 (2002)

[70] Nishino, T.; Amaya, Y.; Kawamoto, S.; Kashima, Y.; Okamoto, K.: Purification and characterization of multiple forms of rat liver xanthine oxidoreductase expressed in baculovirus-insect cell system. J. Biochem., **132**, 597-606 (2002)

[71] Okamoto, K.; Eger, B.T.; Nishino, T.; Kondo, S.; Pai, E.F.: An extremely potent inhibitor of xanthine oxidoreductase. Crystal structure of the enzyme-inhibitor complex and mechanism of inhibition. J. Biol. Chem., **278**, 1848-1855 (2003)

[72] Leimkuhler, S.; Hodson, R.; George, G.N.; Rajagopalan, K.V.: Recombinant Rhodobacter capsulatus xanthine dehydrogenase, a useful model system for the characterization of protein variants leading to xanthinuria I in humans. J. Biol. Chem., **278**, 20802-20811 (2003)

[73] Hesberg, C.; Hansch, R.; Mendel, R.R.; Bittner, F.: Tandem orientation of duplicated xanthine dehydrogenase genes from Arabidopsis thaliana: differential gene expression and enzyme activities. J. Biol. Chem., **279**, 13547-13554 (2004)

[74] Leimkuhler, S.; Stockert, A.L.; Igarashi, K.; Nishino, T.; Hille, R.: The role of active site glutamate residues in catalysis of Rhodobacter capsulatus xanthine dehydrogenase. J. Biol. Chem., **279**, 40437-40444 (2004)

[75] Nishino, T.; Okamoto, K.; Kawaguchi, Y.; Hori, H.; Matsumura, T.; Eger, B.T.; Pai, E.F.: Mechanism of the conversion of xanthine dehydrogenase to xanthine oxidase: identification of the two cysteine disulfide bonds and crystal structure of a non-convertible rat liver xanthine dehydrogenase mutant. J. Biol. Chem., **280**, 24888-24894 (2005)

[76] Benboubetra, M.; Baghiani, A.; Atmani, D.; Harrison, R.: Physicochemical and kinetic properties of purified sheep's milk xanthine oxidoreductase. J. Dairy Sci., **87**, 1580-1584 (2004)
[77] Yesbergenova, Z.; Yang, G.; Oron, E.; Soffer, D.; Fluhr, R.; Sagi, M.: The plant Mo-hydroxylases aldehyde oxidase and xanthine dehydrogenase have distinct reactive oxygen species signatures and are induced by drought and abscisic acid. Plant J., **42**, 862-876 (2005)
[78] Sauer, P.; Frebortova, J.; Sebela, M.; Galuszka, P.; Jacobsen, S.; Pec, P.; Frebort, I.: Xanthine dehydrogenase of pea seedlings: a member of the plant molybdenum oxidoreductase family. Plant Physiol. Biochem., **40**, 393-400 (2002)
[79] Okamoto, K.; Matsumoto, K.; Hille, R.; Eger, B.T.; Pai, E.F.; Nishino, T.: The crystal structure of xanthine oxidoreductase during catalysis: implications for reaction mechanism and enzyme inhibition. Proc. Natl. Acad. Sci. USA, **101**, 7931-7936 (2004)
[80] Ivanov, N.V.; Trani, M.; Edmondson, D.E.: High-level expression and characterization of a highly functional Comamonas acidovorans xanthine dehydrogenase in Pseudomonas aeruginosa. Protein Expr. Purif., **37**, 72-82 (2004)
[81] Joshi, M.S.; Rajagopalan, K.V.: Specific incorporation of molybdopterin in xanthine dehydrogenase of Pseudomonas aeruginosa. Arch. Biochem. Biophys., **308**, 331-334 (1994)
[82] Johnson, J.L.; Chaudhury, M.; Rajagopalan, K.V.: Identification of a molybdopterin-containing molybdenum cofactor in xanthine dehydrogenase from Pseudomonas aeruginosa. BioFactors, **3**, 103-107 (1991)
[83] Abdulnour, R.E.; Peng, X.; Finigan, J.H.; Han, E.J.; Hasan, E.J.; Birukov, K.G.; Reddy, S.P.; Watkins, J.E.; Kayyali, U.S.; Garcia, J.G.; Tuder, R.M.; Hassoun, P.M.: Mechanical stress activates xanthine oxidoreductase through MAP kinase-dependent pathways. Am. J. Physiol. Lung Cell Mol. Physiol., **291**, L345-L353 (2006)
[84] Godber, B.L.; Schwarz, G.; Mendel, R.R.; Lowe, D.J.; Bray, R.C.; Eisenthal, R.; Harrison, R.: Molecular characterization of human xanthine oxidoreductase: the enzyme is grossly deficient in molybdenum and substantially deficient in iron-sulphur centres. Biochem. J., **388**, 501-508 (2005)
[85] Cheung, K.J.; Tzameli, I.; Pissios, P.; Rovira, I.; Gavrilova, O.; Ohtsubo, T.; Chen, Z.; Finkel, T.; Flier, J.S.; Friedman, J.M.: Xanthine oxidoreductase is a regulator of adipogenesis and PPARγ activity. Cell Metab., **5**, 115-128 (2007)
[86] Maia, L.; Vala, A.; Mira, L.: NADH oxidase activity of rat liver xanthine dehydrogenase and xanthine oxidase-contribution for damage mechanisms. Free Radic. Res., **39**, 979-986 (2005)
[87] Atmani, D.; Baghiani, A.; Harrison, R.; Benboubetra, M.: NADH oxidation and superoxide production by caprine milk xanthine oxidoreductase. Int. Dairy J., **15**, 1113-1121 (2005)
[88] Yamaguchi, Y.; Matsumura, T.; Ichida, K.; Okamoto, K.; Nishino, T.: Human Xanthine oxidase changes its substrate specificity to aldehyde oxidase type upon mutation of amino acid residues in the active site: roles of active site

residues in binding and activation of purine substrate. J. Biochem., **141**, 513-524 (2007)

[89] Asai, R.; Nishino, T.; Matsumura, T.; Okamoto, K.; Igarashi, K.; Pai, E.F.; Nishino, T.: Two mutations convert mammalian xanthine oxidoreductase to highly superoxide-productive xanthine oxidase. J. Biochem., **141**, 525-534 (2007)

[90] Neumann, M.; Schulte, M.; Juenemann, N.; Stoecklein, W.; Leimkuehler, S.: Rhodobacter capsulatus XdhC is involved in molybdenum cofactor binding and insertion into xanthine dehydrogenase. J. Biol. Chem., **281**, 15701-15708 (2006)

[91] Pauff, J.M.; Hemann, C.F.; Juenemann, N.; Leimkuehler, S.; Hille, R.: The role of arginine 310 in catalysis and substrate specificity in xanthine dehydrogenase from Rhodobacter capsulatus. J. Biol. Chem., **282**, 12785-12790 (2007)

[92] Roberts, L.E.; Fini, M.A.; Derkash, N.; Wright, R.M.: PD98059 enhanced insulin, cytokine, and growth factor activation of xanthine oxidoreductase in epithelial cells involves STAT3 and the glucocorticoid receptor. J. Cell. Biochem., **101**, 1567-1587 (2007)

[93] Zhao, X.; Zhu, J.X.; Mo, S.F.; Pan, Y.; Kong, L.D.: Effects of cassia oil on serum and hepatic uric acid levels in oxonate-induced mice and xanthine dehydrogenase and xanthine oxidase activities in mouse liver. J. Ethnopharmacol., **103**, 357-365 (2006)

Nicotinate dehydrogenase 1.17.1.5

1 Nomenclature

EC number
1.17.1.5

Systematic name
nicotinate:$NADP^+$ 6-oxidoreductase (hydroxylating)

Recommended name
nicotinate dehydrogenase

Synonyms
EC 1.5.1.13 (formerly)
NAH
NDH
dehydrogenase, nicotinate
nicotinic acid hydroxylase

CAS registry number
9059-03-4

2 Source Organism

<1> *Alcaligenes faecalis* (no sequence specified) [10]
<2> *Pseudomonas fluorescens* (no sequence specified) [3]
<3> *Serratia marcescens* (no sequence specified) [3]
<4> *Bacillus sp.* (no sequence specified) [4]
<5> *Clostridium barkeri* (no sequence specified) [2,8]
<6> *Clostridium sp.* (no sequence specified) [7]
<7> *Comamonas acidovorans* (no sequence specified) [3]
<8> *Bacillus niacini* (no sequence specified) [1,5,6]
<9> *Achromobacter xylosoxidans* (no sequence specified) [10]
<10> *Agrobacterium sp. DSM 6336* (no sequence specified) [9]

3 Reaction and Specificity

Catalyzed reaction
nicotinate + H_2O + $NADP^+$ = 6-hydroxynicotinate + NADPH + H^+

Reaction type
hydrolysis of amide bond
hydroxylation

oxidation
redox reaction
reduction

Natural substrates and products

S nicotinate + H_2O + $NADP^+$ <6> (<6> first step in the anaerobic decomposition of nicotinate, $NADP^+$ may be the natural, primary electron acceptor [7]) (Reversibility: r) [7]
P 6-hydroxynicotinate + NADPH <6> [7]
S nicotinate + H_2O + electron acceptor <2, 4, 8> (<2,4,8> first step of nicotinate catabolism [3,4,6]; <8> nicotinate degradation [5]; <2> enzyme is linked to the cytochrome respiratory chain, phenazine methosulfate may accelerate the electron transfer from the electron respiratory chain to oxygen [3]) (Reversibility: r) [3, 4, 5, 6]
P 6-hydroxynicotinate + reduced electron acceptor <2, 4, 8> [3, 4, 5, 6]

Substrates and products

S 2,3-pyrazinedicarboxylate + H_2O + $NADP^+$ <5> (<5> 10-20% of the activity with nicotinate [8]) (Reversibility: ?) [8]
P ?
S 2-hydroxynicotinate + H_2O + electron acceptor <2> (<2> hydroxylation at a low rate [3]) (Reversibility: ?) [3]
P ?
S 2-pyrazinecarboxylate + H_2O + $NADP^+$ <5, 6> (<6> 30% of the activity with nicotinate [7]; <5> equally as effective as nicotinate [8]) (Reversibility: r) [7, 8]
P ?
S 2-pyrazinecarboxylate + H_2O + electron acceptor <2> (Reversibility: ?) [3]
P ?
S 3,5-pyridinedicarboxylate + H_2O + $NADP^+$ <5> (<5> 5-10% of the activity with nicotinate [8]) (Reversibility: ?) [8]
P ?
S 3-cyanopyridine + H_2O + electron acceptor <2> (<2> hydroxylation at a low rate [3]) (Reversibility: ?) [3]
P ?
S 3-pyridinesulfonate + H_2O + electron acceptor <2> (Reversibility: ?) [3]
P ?
S 6-methylnicotinate + H_2O + $NADP^+$ <5> (<5> 5-10% of the activity with nicotinate [8]) (Reversibility: ?) [8]
P ?
S nicotinate + H_2O + $NADP^+$ <5, 6> (<6> specific for $NADP^+$, but artificial electron acceptors, i.e. benzyl viologen and 2,3,5-triphenyltetrazolium dyes, can replace $NADP^+$, high substrate specificity with respect to nicotinate, requiring an unsubstituted pyridine or pyrazine ring nitrogen and a carboxyl group meta to the nitrogen [7]; <5> high substrate specificity toward electron donor substrates, unsubstituted nitrogen and a carboxyl group at position 3 are absolutely required for substrate hydroxylation

and unsubstituted carbon-5 is important for oxidation at carbon-6 of substrate [8]; <6> equilibrium lies far in direction of 6-hydroxynicotinate formation [7]; <5> $NADP^+$ or various dyes as electron acceptors [2]; <6> first step in the anaerobic decomposition of nicotinate, $NADP^+$ may be the natural, primary electron acceptor [7]) (Reversibility: r) [2, 7, 8]

P 6-hydroxynicotinate + NADPH <5, 6> (<5> with NADPH as substrate, enzyme exhibits NADPH oxidase activity in presence of oxygen [2]) [2, 7, 8]

S nicotinate + H_2O + electron acceptor <2, 3, 4, 7, 8> (<8> equilibrium lies far in direction of 6-hydroxynicotinate formation [6]; <8> high substrate specificity, mixture of phenazine ethosulfate and 2,6-dichlorophenolindophenol as electron acceptor [5]; <8> high substrate specificity, transfers electrons only to artificial electron acceptors of high redox potential, i.e. thiazolyl blue, nitroblue tetrazolium chloride, mixture of phenazine ethosulfate and 2,6-dichlorophenolindophenol, methylene blue with 40% of the activity with phenazine ethosulfate and 2,6-dichlorophenolindophenol [6]; <2> purified enzyme probably loses its own native electron acceptor during the course of solubilization and purification, electron acceptors: phenazine methosulfate, potassium ferricyanide and with lower activity nitro blue tetrazolium, negative polarity at the 3-position on the pyridine ring of the substrate is important for reaction [3]; <4> flavin may be the primary electron acceptor, tetrazolium dye as electron acceptor [4]; <2,4,8> first step of nicotinate catabolism [3,4,6]; <8> nicotinate degradation [5]; <2> enzyme is linked to the cytochrome respiratory chain, phenazine methosulfate may accelerate the electron transfer from the electron respiratory chain to oxygen [3]) (Reversibility: r) [3, 4, 5, 6]

P 6-hydroxynicotinate + reduced electron acceptor <2, 3, 4, 7, 8> [3, 4, 5, 6]

S picolinic acid + $NADP^+$ + H_2O <1> (Reversibility: ?) [10]

P 6-hydroxypicolinic acid + $NADP^+$

S pyridine-3-carboxylate + $NADP^+$ + H_2O <9, 1> (Reversibility: ?) [9, 1]

P 6-hydroxy-pyridine-3-carboxylate + NADPH + H^+

S trigonelline + H_2O + $NADP^+$ <5> (<5> 5-10% of the activity with nicotinate [8]) (Reversibility: ?) [8]

P ?

S Additional information <2, 6, 8> (<2> not as substrates: nicotinamide, methyl nicotinate, β-picoline, 3-pyridylcarbinol, quinolinic acid, 3-pyridylacetic acid, 3-hydroxypyridine, 3-chloropyridine, not as electron acceptors: methylene blue, FAD, FMN, NAD^+, $NADP^+$, horse heart cytochrome c, 2,3,5-triphenyl-2H-tetrazolium, neutral red, safranin O, menadione [3]; <8> no activity with Brilliant cresyl blue, cytochrome c, FAD, ferricyanide, FMN, menadione, $NAD(P)^+$, and both viologens as electron acceptors, not as substrates: 2- or 6-aminonicotinamide, 2-aminonicotinic acid, chelidamic acid, cinchomeronic acid, citrazinic acid, dinicotinic acid, dipicolinic acid, 3-hydroxypicolinic acid, hypoxanthine, isocinchomeronic acid, isonicotinic acid, lutidinic acid, nicotinamide, xanthine [6]; <6> not as substrates: nicotinamide, isonicotinate, picolinate, nicotinic acid N-oxide, N-methylnicotinamide, trigonelline, hypoxanthine, not as electron acceptor: NAD^+ [7]) (Reversibility: ?) [3, 6, 7]

P ?

Inhibitors

1,10-phenanthroline <4> (<4> 1 mM, 30-50% inhibition [4]) [4]

2,2'-dipyridyl <2, 4> (<4> 0.1 mM, 30-50% inhibition [4]; <2> 1 mM, 10 min at 35°C, 21% loss of activity [3]) [3, 4]

6-hydroxynicotinate <5> (<5> inhibits effectively [8]) [8]

acriflavin <2> (<2> 1 mM, 10 min at 35°C, 22% loss of activity [3]) [3]

$AgNO_3$ <2> (<2> 1 mM, 10 min at 35°C, complete loss of activity [3]) [3]

atabrine <4> (<4> 1 mM, 30-50% inhibition [4]) [4]

$CuCl_2$ <2> (<2> 1 mM, 10 min at 35°C, complete loss of activity [3]) [3]

$HgCl_2$ <2> (<2> 1 mM, 10 min at 35°C, complete loss of activity [3]) [3]

KCN <2, 8> (<8> after removal of cyanide under anaerobic conditions, enzyme activity can be reconstituted by anaerobic incubation with Na_2S [6]; <8> 2 mM, complete loss of enzyme activity in anaerobically incubated crude extract [5]; <2> strong inhibition, partially restored by addition of phenazine methosulfate [3]) [3, 5, 6]

N-ethylmaleimide <2> (<2> 1 mM, 10 min at 35°C, complete loss of activity [3]) [3]

NaCl <4> (<4> sufficient NaCl inhibits enzyme activity significantly [4]) [4]

NaN_3 <2> (<2> 1 mM, 10 min at 35°C, 17% loss of activity [3]) [3]

selenophosphate <5> (<5> 7 mM, 30 min, anaerobic conditions, reversible complete inactivation, time-dependent [8]) [8]

sulfide <5, 8> (<8> complete loss of activity without previous KCN-treatment [6]; <5> 1 mM, 10 min, reversible time-dependent inactivation [8]) [6, 8]

p-chloromercuribenzoate <2> (<2> 1 mM, 10 min at 35°C, complete loss of activity [3]) [3]

phenylhydrazine <2> (<2> 1 mM, 10 min at 35°C, 15% loss of activity [3]) [3]

sodium selenide <5> (<5> 7 mM, 10 min, anaerobic conditions, reversible complete inactivation, time-dependent [8]) [8]

Additional information <5> (<5> not inhibited by incubation for 1 h at room temperature with 100 mM KCN [8]) [8]

Cofactors/prosthetic groups

4Fe-4S-center <5, 8> (<8> enzyme contains Fe/S centers, 8.3 mol iron per mol of enzyme, 1.5 mol acid-labile sulfur per mol of enzyme [6]; <5> two [2Fe-2S] clusters, 5-7 atoms Fe per 160 kDa enzyme molecule [8]; <5> enzyme contains iron-sulfur clusters, [2Fe-2S] centers [2]) [2,6,8]

Bactopterin <8> (<8> enzyme contains bactopterin, an eubacterial modification of molybdopterin, the organic moiety of the molybdenum cofactor [6]; <8> molybdoenzyme with bactopterin, which differs from molybdopterin in MW, phosphate content and structure [1]) [1,6]

FAD <5, 6, 8> (<8> non-covalently bound FAD [1]; <6> FAD as prosthetic group, 1.5 mol flavin per mol of enzyme [7]; <8> enzyme contains FAD, 2

mol flavin per mol of enzyme [6]; <5> 1 FAD molecule per 160 kDa protein protomer [8]) [1,2,6,7,8]
flavin <4> (<4> primary electron acceptor, 1.5-2 mol tightly bound flavin per mol of enzyme [4]) [4]
molybdenum cofactor <5, 8> (<8> molybdoenzyme with eubacterial molybdenum cofactor contains bactopterin [1]; <8> molybdoenzyme with eubacterial molybdenum cofactor [5]; <8> enzyme contains 1.5 mol molybdenum per mol of enzyme and bactopterin [6]) [1,2,5,6,8]
molybdopterin <5> (<5> Mo is bound to a dinucleotide form of molybdopterin and is coordinated with selenium, 1 mol Mo per 160 kDa enzyme molecule, molybdenum is directly coordinated to selenium, Se-Mo center is required for enzymic oxidation of nicotinate [8]; <5> molybdenum in the molybdopterin cofactor is coordinated directly with a dissociable selenium moiety, molybdenum center plays a role in the reaction step involving hydroxylation of the heterocyclic ring with oxygen derived from water [2]) [2,8]
$NADP^+$ <5, 6> (<6> specific for $NADP^+$, artificial electron acceptors, i.e. benzyl viologen and 2,3,5-triphenyltetrazolium dyes, can replace $NADP^+$ [7]; <5> $NADP^+$ as ultimate electron acceptor, can be replaced by various dyes [2]) [2,7,8]
NADPH <5, 6> (<5> is able to reduce FAD cofactor, NADH can replace NADPH [8]; <5,6> NADPH oxidase activity [7,8]; <5> NADPH oxidase activity in the presence of oxygen [2]) [2,7,8]
benzyl viologen <6> (<6> can replace $NADP^+$ [7]) [7]
phenazine methosulfate <2> (<2> phenazine methosulfate-requiring enzyme [3]) [3]
Additional information <2, 4, 5, 6> (<5> NAD^+ can not replace $NADP^+$, but NADH can replace NADPH [8]; <2> no cytochromes or flavoprotein components [3]; <5> enzyme contains a dissociable selenium moiety, which is coordinated directly with molybdenum in the molybdopterin cofactor, selenium is essential for enzyme activity, molar ratio of selenium to enzyme is 0.4-0.8, enzyme contains 1 g selenium per mol of native active enzyme, selenium is completely dissociated from protein in the SDS-gel [2]; <5> enzyme contains labile selenium cofactor which is essential for hydroxylase activity of enzyme, Se is directly coordinated to Mo, up to 1 Se atom per enzyme molecule [8]; <4> approximately 8 molecules tightly bound iron per enzyme molecule [4]; <6> NAD^+/NADH can not replace $NADP^+$/NADPH [7]; <6> enzyme contains 11 mol non-heme iron and 6 mol acid-labile sulfide per mole of enzyme [7]) [2,3,4,7,8]

Activating compounds

arsenate <6> (<6> activates less than phosphate [7]) [7]
phosphate <6> (<6> slight activation, half-maximal activation at 50 mM [7]) [7]
methylphosphate <6> (<6> activates less than phosphate [7]) [7]
Additional information <6> (<6> not activated by high concentrations of citrate, sulfate, or diphosphate [7]) [7]

Metals, ions

Additional information <8> (<8> enzyme activity increases with increasing concentrations of molybdate in growth medium [5]) [5]

Specific activity (U/mg)

0.54 <8> [6]
0.83 <4> [4]
1 <2> (<2> 2-hydroxynicotinate [3]) [3]
2.2 <8> (<8> crude extract, high activity achieved if nicotinate is the only substrate [5]) [5]
4.4 <5> (<5> purified enzyme, Difco yeast extract without selenium as growth medium [2]) [2]
12 <2> (<2> 3-cyanopyridine [3]) [3]
18 <5> [8]
20 <5> (<5> purified enzyme, average value, Difco yeast extract with selenium as growth medium [2]) [2]
28.8 <6> [7]
104 <2> (<2> 3-pyridinesulfonate [3]) [3]
170 <2> (<2> 2-pyrazinecarboxylate [3]) [3]
672 <2> (<2> nicotinate [3]) [3]
Additional information <5, 8> (<5> more values for specific activities in crude extract [2]) [1, 2]

K_m-Value (mM)

0.028 <6> ($NADP^+$) [7]
0.07 <4> (Nicotinate) [4]
0.11 <2, 6> (Nicotinate) [3, 7]
0.7 <2> (3-pyridinesulfonate) [3]
1 <8> (Nicotinate) [6]
2.1 <2> (2-Hydroxynicotinate) [3]
9.8 <2> (2-pyrazinecarboxylate) [3]
150 <2> (3-Cyanopyridine) [3]

pH-Optimum

7 <1, 9> [10]
7.5 <8> [6]
7.5-8 <6> (<6> phosphate buffer [7]) [7]
8 <6> (<6> triethanolamine phosphate buffer [7]) [7]
8-8.5 <6> (<6> triethanolamine buffer [7]) [7]
8.3 <2> [3]
8.3-9 <6> (<6> Tris buffer or diphosphate buffer [7]) [7]
8.5 <6> (<6> dimethylglutarate-phosphate buffer [7]) [7]

pH-Range

4.5-9 <2> (<2> broad pH-range with maximum at pH 8.3 [3]) [3]

Temperature optimum (°C)

25 <4> (<4> assay at [4]) [4]
30 <1, 8, 9> (<8> assay at [5,6]) [5, 6, 10]

35 <2> (<2> assay at [3]) [3]
50 <2> [3]
Additional information <5> (<5> assay at room temperature [8]) [8]

Temperature range (°C)
15-45 <10> [9]

4 Enzyme Structure

Molecular weight
100000 <2> (<2> gel filtration [3]) [3]
160000 <5> (<5> native PAGE [2]; <5> major form, occurence of additional enzyme forms of 400 kDa and 120 kDa with same subunit composition, gel filtration and native PAGE [8]) [2, 8]
280000-340000 <8> (<8> gel filtration, non-denaturing PAGE [6]) [6]
300000 <6> (<6> analytical ultracentrifugation [7]) [7]
400000 <4> (<4> gel filtration [4]) [4]

Subunits
heterotetramer <5> (<5> 1 * 50000 + 1 * 37000 + 1 * 33000 + 1 * 23000, SDS-PAGE [2,8]; <5> 23 kDa protein is less stained and may be a degradation product [8]) [2, 8]
monomer <2> (<2> 1 * 80000, SDS-PAGE, in presence of 50 mM 2-mercaptoethanol [3]) [3]
oligomer <8> (<8> x * 85000 + x * 34000 + x * 20000, additional bands with increasing period of storage, subunits are loosely linked by electrostatic interactions, SDS-PAGE [6]) [6]

5 Isolation/Preparation/Mutation/Application

Localization
membrane <2> [3]
soluble <4, 6, 8> [4, 6, 7]

Purification
<2> (126fold purification) [3]
<4> (24.4fold purification) [4]
<5> [2]
<5> (112fold purification) [8]
<6> (24fold increase in specific activity) [7]
<8> [6]
<8> (partial purification) [1]

Renaturation
<4> (after denaturation by acid precipitation at pH 4.5 with HCl, recombining and neutralizing of precipitated protein and yellow supernatant results in a reconstitution of over 90% of the original activity) [4]

Crystallization

<5> [8]

Engineering

Additional information <8> (<8> 4 chlorate-resistant mutants are no longer able to degrade nicotinate [5]) [5]

6 Stability

pH-Stability

5-8.8 <2> (<2> 10 min at 50°C in various buffers, stable [3]) [3]
8 <5> (<5> most stable at alkaline pH [8]) [8]
Additional information <5> (<5> maximal stability during incubation for 14 h at room temperature, after 9 days storage at pH 8, 25% of the initial activity retained [8]) [8]

Temperature stability

50 <2> (<2> 10 min in 0.1 M potassium phosphate buffer, pH 7, stable below [3]) [3]
Additional information <6> (<6> hydroxylase activity is more sensitive to heat inactivation than are diaphorase and NADPH-oxidase activities [7]) [7]

Oxidation stability

<5>, exposure of substrate-reduced enzyme to air results in a complete loss of activity, enzyme before reduction is much less sensitive to oxygen inactivation [8]
<5>, oxygen-sensitive enzyme [2]

General stability information

<2>, membrane-bound enzyme is stable, but solubilized enzyme is very labile, 20 mM nicotinate stabilizes [3]
<5>, enzyme is most stable at alkaline pH in the presence of glycerol, 20% glycerol and 400 mM KCl stabilize [8]
<8>, lyophilization completely inactivates, high protein concentration may have a stabilizing effect on enzyme activity, sugar protects enzyme from dissociation [6]

Storage stability

<5>, room temperature, 50 mM Tris-HCl buffer, pH 8.2, after 1 day 40% loss of hydroxylase activity, after 7 days 62% loss of hydroxylase activity, NADPH oxidase and diaphorase activity of enzyme are more stable [8]
<5>, room temperature, pH 8, 9 days, 75% loss of hydroxylase activity [8]
<8>, -20°C, increase of activity in crude extract [6]

References

[1] Kruger, B.; Meyer, O.; Nagel, M.; Andreesen, J.R.; Meincke, M.; Bock, E.; Blumle, S.; Zumft, W.G.: Evidence for the presence of bactopterin in the eubacterial molybdoenzymes nicotinic acid dehydrogenase, nitrite oxidoreductase, and respiratory nitrate reductase. FEMS Microbiol. Lett., **48**, 225-227 (1987)

[2] Gladyshev, V.N.; Khangulov, S.V.; Stadtman, T.C.: Nicotinic acid hydroxylase from Clostridium barkeri: Electron paramagnetic resonance studies show that selenium is coordinated with molybdenum in the catalytically active selenium-dependent enzyme. Proc. Natl. Acad. Sci. USA, **91**, 232-236 (1994)

[3] Hurh, B.; Yamane, T.; Nagasawa, T.: Purification and characterization of nicotinic acid dehydrogenase from Pseudomonas fluorescens TN5. J. Ferment. Bioeng., **78**, 19-26 (1994)

[4] Hirschberg, R.; Ensign, J.C.: Oxidation of nicotinic acid by a Bacillus species: purification and properties of nicotinic acid and 6-hydroxynicotinic acid hydroxylases. J. Bacteriol., **108**, 751-756 (1971)

[5] Nagel, M.; Andreesen, J.R.: Molybdenum-dependent degradation of nicotinic acid by Bacillus sp. DSM 2923. FEMS Microbiol. Lett., **59**, 147-152 (1989)

[6] Nagel, M.; Andreesen, J.R.: Purification and characterization of the molybdoenzymes nicotinate dehydrogenase and 6-hydroxynicotinate dehydrogenase from Bacillus niacini. Arch. Microbiol., **154**, 605-613 (1990)

[7] Holcenberg, J.S.; Stadtman, E.R.: Nicotinic acid metabolism. III. Purification and properties of a nicotinic acid hydroxylase. J. Biol. Chem., **244**, 1194-1203 (1969)

[8] Gladyshev, V.N.; Khangulov, S.V.; Stadtman, T.C.: Properties of the selenium- and molybdenum-containing nicotinic acid hydroxylase from Clostridium barkeri. Biochemistry, **35**, 212-223 (1996)

[9] Wieser, M.; Heinzmann, K.; Kiener, A.: Bioconversion of 2-cyanopyrazine to 5-hydroxypyrazine-2-carboxylic acid with Agrobacterium sp. DSM 6336. Appl. Microbiol. Biotechnol., **48**, 174-180 (1997)

[10] Petersen, M.; Kiener, A.: Biocatalysis - preparation and functionalization of N-heterocycles. Green Chem., **2**, 99-106 (1999)

Bile-acid 7α-dehydroxylase 1.17.1.6

1 Nomenclature

EC number

1.17.1.6 (transferred to EC 1.17.99.5 (2006))

Recommended name

bile-acid 7α-dehydroxylase

Xanthine oxidase 1.17.3.2

1 Nomenclature

EC number
1.17.3.2

Systematic name
xanthine:oxygen oxidoreductase

Recommended name
xanthine oxidase

Synonyms
EC 1.1.3.22
EC 1.2.3.2.
Schardinger enzyme
XO <6, 7> [1, 43, 45, 51]
XOD <7> [46]
XOR <6, 7> (<6> xanthine oxidoreductase can exists in a dehydrogenase form, XD, and an oxidase form, XO [43]) [43, 45]
XnOx <7> [56]
hypoxanthine oxidase
hypoxanthine-xanthine oxidase
hypoxanthine:oxygen oxidoreductase
oxidase, xanthine
xanthine oxidoreductase <6> (<6> xanthine oxidoreductase can exists in a dehydrogenase form, XD, and an oxidase form, XO [43]) [43]
xanthine:O_2 oxidoreductase
xanthine:xanthine oxidase

CAS registry number
9002-17-9

2 Source Organism

<1> *Gallus gallus* (no sequence specified) [19, 32, 33]
<2> *Cavia porcellus* (no sequence specified) [25]
<3> *Drosophila melanogaster* (no sequence specified) [19, 32]
<4> *Mus musculus* (no sequence specified) [6, 23, 25, 32, 41, 55, 65]
<5> *Homo sapiens* (no sequence specified) [15,17,22,25,58,67]
<6> *Rattus norvegicus* (no sequence specified) [1,7,8,21,25,28,33,43,61,63]

<7> *Bos taurus* (no sequence specified) [2, 3, 4, 5, 12, 14, 16, 18, 19, 20, 22, 25, 26, 27, 28, 29, 30, 31, 33, 34, 35, 36, 37, 38, 39, 40, 42, 44, 45, 46, 47, 48, 49, 50, 51, 52, 54, 56, 57, 59, 60, 62, 64, 66, 68, 69, 70]
<8> *Oryctolagus cuniculus* (no sequence specified) [24,25]
<9> *Ovis aries* (no sequence specified) [25]
<10> *Canis familiaris* (no sequence specified) [25]
<11> *Felis catus* (no sequence specified) [25]
<12> *Arthrobacter sp.* (no sequence specified) [11]
<13> *Capra hircus* (no sequence specified) [25]
<14> *Equus caballus* (no sequence specified) [25]
<15> *Enterobacter cloacae* (no sequence specified) [13]
<16> *Equus asinus* (no sequence specified) [25]
<17> *Desulfovibrio gigas* (no sequence specified) [9]
<18> *Gonyaulax polyedra* (no sequence specified) [53]
<19> *Lens esculenta* (no sequence specified) (<19> AtNTRB [10]) [10]
<20> *Locusta sp.* (no sequence specified) [32]
<21> *Erythrocebus patas* (no sequence specified) [25]

3 Reaction and Specificity

Catalyzed reaction

xanthine + H_2O + O_2 = urate + H_2O_2 (<7> mechanism [19,30])

Reaction type

oxidation
redox reaction
reduction

Natural substrates and products

S carboxylic aldehyde + H_2O + O_2 <2, 4, 5, 6, 7, 8, 9, 10, 11, 13, 14, 16, 21> (<2, 4, 5, 6, 7, 8, 9, 10, 11, 13, 14, 16, 21> enzyme is implicated in the control of various redox reactions in the cell, in milk: assures absorption of iron from the gut, coupling antibacterial effect via the lactoperoxidase system [25]) (Reversibility: ?) [25]
P carboxylic acid + H_2O_2 <5, 7, 13> [25]
S nitrate + NADH <7> (<7> reaction can be an important source of NO production in ischemic tissues [45]) (Reversibility: ?) [45]
P nitrite + NAD^+ + H_2O
S nitrite + NADH <7> (<7> reaction can be an important source of NO production in ischemic tissues [45]) (Reversibility: ?) [45]
P NO + NAD^+ + H_2O
S organic nitrate + NADH <7> (<7> organic nitrite is the initial product in the process of xanthine oxidase mediated organic nitrate biotransformation and is the precursor of NO and nitrosothiols, serving as the link between organic nitrate and soluble guanylyl cyclase [51]) (Reversibility: ?) [51]
P organic nitrite + NAD^+ + H_2O

S pteridine + H_2O + O_2 <2, 4, 5, 6, 7, 8, 9, 10, 11, 13, 14, 16, 21> (Reversibility: ?) [25]
P ?
S purine + H_2O + O_2 <2, 4, 5, 6, 7, 8, 9, 10, 11, 13, 14, 16, 21> (Reversibility: ?) [25]
P ?
S xanthine + H_2O + O_2 <6> (<6> xanthine oxidoreductase, XOR, can exist in a dehydrogenase form, XD, and an oxidase form, XO. Part of total XOR activity in peroxisomes is XO activity. The major function of XOR activity in the cytoplasm of rat liver parenchymal cells and in sinusoidal cells is not the production of O_2 radicals, but rather the production of uric acid which can act as a potent antioxidant [43]) (Reversibility: ?) [43]
P urate + H_2O_2
S Additional information <6> (<6> xanthine oxidase is necessary during the physiological involution of tissues [61]) (Reversibility: ?) [61]
P ?

Substrates and products

S 1,3-dimethylxanthine + H_2O + O_2 <7> (Reversibility: ?) [52]
P 1,3-dimethylurate + H_2O_2
S 1,7-dimethylxanthine + H_2O + O_2 <7> (Reversibility: ?) [52]
P 1,7-dimethylurate + $H2O_2$
S 1-methyl-2-hydroxypurine + H_2O + O_2 <7> (Reversibility: ?) [19]
P 1-methyl-2-hydroxy-7,9-dihydropurin-8-one + H_2O_2 <7> [19]
S 1-methylxanthine + H_2O + O_2 <7, 12> (Reversibility: ?) [11, 19, 52]
P 1-methylurate + H_2O_2 <7, 12> [11, 19]
S 2,3-dihydroxybenzaldehyde + H_2O + O_2 <7> (Reversibility: ?) [29]
P 2,3-dihydroxybenzoate + H_2O_2 <7> [29]
S 2,5-dihydroxybenzaldehyde + H_2O + O_2 <7> (Reversibility: ?) [39]
P ?
S 2,6-diaminopurine + H_2O + O_2 <7> (Reversibility: ?) [52]
P 2,6-diamino-7,9-dihydro-8H-purin-8-one
S 2-amino-6-chloro-purine + H_2O + O_2 <7> (Reversibility: ?) [52]
P 2-amino-6-chloro-7,9-dihydro-purin-8-one + H_2O_2
S 2-hydroxybenzaldehyde + H_2O + O_2 <7> (Reversibility: ?) [42]
P 2-hydroxybenzoate + H_2O_2
S 2-methoxybenzaldehyde + H_2O + O_2 <7> (Reversibility: ?) [42]
P 2-methoxybenzoate + H_2O_2
S 2-methylbenzaldehyde + H_2O + O_2 <7> (Reversibility: ?) [60]
P 2-methylbenzoate + H_2O_2
S 2-nitrobenzaldehyde + H_2O + O_2 <7> (Reversibility: ?) [60]
P 2-nitrobenzoate + H_2O_2
S 2-thioxanthine + H_2O + O_2 <7> (Reversibility: ?) [52]
P 2-thiourate + H_2O_2
S 3,4-dihydroxybenzaldehyde + H_2O + O_2 <7> (Reversibility: ?) [42]
P 3,4-dihydroxybenzoate + H_2O_2
S 3,4-dimethoxybenzaldehyde + H_2O + O_2 <7> (Reversibility: ?) [42]

P 3,4-dimethoxybenzoate + H_2O_2
S 3-hydroxy-4-methoxybenzaldehyde + H_2O + O_2 <7> (Reversibility: ?) [42]
P 3-hydroxy-4-methoxybenzoate + H_2O_2
S 3-hydroxybenzaldehyde + H_2O + O_2 <7> (Reversibility: ?) [42]
P 3-hydroxybenzoate + H_2O_2
S 3-methoxybenzaldehyde + H_2O + O_2 <7> (Reversibility: ?) [42]
P 3-methoxybenzoate + H_2O_2
S 3-methylbenzaldehyde + H_2O + O_2 <7> (Reversibility: ?) [60]
P 3-methylbenzoate + H_2O_2
S 3-methylhypoxanthine + H_2O + O_2 <7> (Reversibility: ?) [19]
P 3-methylxanthine + H_2O_2 <7> [19]
S 3-methylxanthine + H_2O + O_2 <7> (Reversibility: ?) [52]
P 3-methylurate + H_2O_2
S 3-nitrobenzaldehyde + H_2O + O_2 <7> (Reversibility: ?) [60]
P 3-nitrobenzoate + H_2O_2
S 4-hydroxy-3-methoxybenzaldehyde + H_2O + O_2 <7> (Reversibility: ?) [42]
P 4-hydroxy-3-methoxybenzoate + H_2O_2
S 4-hydroxybenzaldehyde + H_2O + O_2 <7> (Reversibility: ?) [42]
P 4-hydroxybenzoate + H_2O_2
S 4-hydroxyphenylglycoaldehyde + H_2O + O_2 <7> (Reversibility: ?) [39]
P ?
S 4-methoxybenzaldehyde + H_2O + O_2 <7> (Reversibility: ?) [42]
P 4-methoxybenzoate + H_2O_2
S 4-methylbenzaldehyde + H_2O + O_2 <7> (Reversibility: ?) [60]
P 4-methylbenzoate + H_2O_2
S 4-nitrobenzaldehyde + H_2O + O_2 <7> (Reversibility: ?) [60]
P 4-nitrobenzoate + H_2O_2
S 6'-deoxyacyclovir + H_2O + O_2 <5> (<5> prodrug of the antiviral agent acyclovir [22]) (Reversibility: ?) [22]
P acyclovir + H_2O_2 <5> [22]
S 6,8-dihydroxypurine + H_2O + O_2 <7> (Reversibility: ?) [19]
P ?
S 6-cyanopurine + H_2O + O_2 <7> (Reversibility: ?) [19]
P 6-cyano-7,9-dihydropurine-8-one <7> [19]
S 6-formylpterin + H_2O + O_2 <7> (Reversibility: ?) [46]
P ?
S 6-mercaptopurine + H_2O + O_2 <15> (<15> 4.4% of activity with xanthin [13]) (Reversibility: ?) [13]
P 6-mercapto-7,9-dihydropurin-8-one + H_2O_2 <15> [13]
S 6-thioxanthine + H_2O + O_2 <7> (Reversibility: ?) [19, 52]
P 6-thiourate + H_2O_2 <7> [19]
S 7-phenylpteridin-4-one + H_2O + O_2 <7> (Reversibility: ?) [34]
P 7-phenyllumazine + H_2O_2 <7> [34]
S 7-alkylpteridin-4-one + H_2O + O_2 <7> (Reversibility: ?) [34]
P 7-alkyllumazine + H_2O_2 <7> [34]

S 7-methylxanthine + H_2O + O_2 <7> (Reversibility: ?) [52]
P 7-methylurate + H_2O_2
S N^1-methylnicotinamide + H_2O + O_2 <5, 7> (Reversibility: ?) [15, 19]
P ?
S NADH + H_2O + O_2 <7> (Reversibility: ?) [19, 22]
P NAD^+ + H_2O_2 <7> [19]
S acetaldehyde + H_2O + O_2 <7> (Reversibility: ?) [19, 39]
P acetic acid + H_2O_2 <7> [19, 39]
S adenine + H_2O + O_2 <5, 7, 15> (<15> no activity with adenine [13]) (Reversibility: ?) [13, 15, 19]
P 6-amino-7,9-dihydropurin-8-one + H_2O_2 <5, 7> [15, 19]
S allopurinol + H_2O + O_2 <5, 7> (Reversibility: ?) [15, 19]
P ?
S allopurinol + H_2O + O_2 <7> (Reversibility: ?) [57]
P oxypurinol + H_2O_2 (<7> allopurinol is a conventional substrate that generates superoxide radicals during its oxidation [57])
S benzaldehyde + H_2O + O_2 <7> (Reversibility: ?) [42]
P benzoate + H_2O_2
S butanal + H_2O + O_2 <7> (Reversibility: ?) [39]
P butanoate + H_2O_2 <7> [39]
S carboxylic aldehyde + H_2O + O_2 <2, 4, 5, 6, 7, 8, 9, 10, 11, 13, 14, 16, 21> (<2, 4, 5, 6, 7, 8, 9, 10, 11, 13, 14, 16, 21> enzyme is implicated in the control of various redox reactions in the cell, in milk: assures absorption of iron from the gut, coupling antibacterial effect via the lactoperoxidase system [25]) (Reversibility: ?) [25]
P carboxylic acid + H_2O_2 <5, 7, 13> [25]
S formycin B + H_2O + O_2 <5> (Reversibility: ?) [15]
P ?
S glyceraldehyde-3-phosphate + H_2O + O_2 <7> (Reversibility: ?) [39]
P ?
S glyceryl trinitrate + 2,3-dihydroxybenzaldehyde <7> (Reversibility: ?) [51]
P ?
S glyceryl trinitrate + NADH <7> (Reversibility: ?) [51]
P ? + NAD^+ + H_2O (<7> further reaction of organic nitrite with thiols or ascorbate leads to generation of NO or nitrosothiols [51])
S glyceryl trinitrate + xanthine <7> (Reversibility: ?) [51]
P urate + ?
S guanine + H_2O + O_2 <7, 15> (<15> 53.3% of activity with xanthin [13]) (Reversibility: ?) [13, 52]
P 2-amino-1,7,8-trihydro-6H-purin-8-one + H_2O_2 <15> [13]
S hypoxanthine + H_2O + O_2 <1, 2, 3, 4, 5, 6, 7, 8, 9, 10, 11, 12, 13, 14, 15, 16, 17, 19, 20, 21> (<15> 92.3% of activity with xanthin [13]) (Reversibility: ?) [2, 3, 4, 5, 6, 7, 8, 9, 10, 11, 12, 13, 14, 15, 16, 17, 18, 19, 20, 21, 22, 23, 24, 25, 26, 27, 28, 29, 30, 31, 32, 33, 34, 35, 36, 37, 38, 39, 40, 41]

P urate + H_2O_2 <1, 2, 3, 4, 5, 6, 7, 8, 9, 10, 11, 12, 13, 14, 15, 16, 17, 19, 20, 21> [2, 3, 4, 5, 6, 7, 8, 9, 10, 11, 12, 13, 14, 15, 16, 17, 18, 19, 20, 21, 22, 23, 24, 25, 26, 27, 28, 29, 30, 31, 32, 33, 34, 35, 36, 37, 38, 39, 40, 41]

S indole-3-acetaldehyde + H_2O + O_2 <7> (Reversibility: ?) [39]

P ?

S indole-3-aldehyde + H_2O + O_2 <7> (Reversibility: ?) [39]

P ?

S isosorbide dinitrate + 2,3-dihydroxybenzaldehyde <7> (Reversibility: ?) [51]

P ?

S isosorbide dinitrate + NADH <7> (Reversibility: ?) [51]

P ? + NAD^+ + H_2O (<7> further reaction of organic nitrite with thiols or ascorbate leads to generation of NO or nitrosothiols [51])

S isosorbide dinitrate + xanthine <7> (Reversibility: ?) [51]

P urate + ?

S nitrate + 2,3-dihydroxybenzaldehyde <7> (Reversibility: ?) [45]

P nitrite + ?

S nitrate + NADH <7> (<7> reaction can be an important source of NO production in ischemic tissues [45]) (Reversibility: ?) [45]

P nitrite + NAD^+ + H_2O

S nitrate + xanthine <7> (Reversibility: ?) [45]

P nitrite + urate + ?

S nitrite + 2,3-dihydroxybenzaldehyde <7> (<7> NO generation occurs under aerobic conditions and is regulated by O_2 tension, pH, nitrite, and reducing substrate concentrations [50]) (Reversibility: ?) [50]

P NO + ?

S nitrite + NADH <7> (<7> reaction can be an important source of NO production in ischemic tissues [45]; <7> NO generation occurs under aerobic conditions and is regulated by O_2 tension, pH, nitrite, and reducing substrate concentrations [50]) (Reversibility: ?) [45, 49, 50]

P NO + NAD^+ + H_2O

S nitrite + O_2 + hypoxanthine <7> (Reversibility: ?) [49]

P peroxynitrite + ?

S nitrite + O_2 + pterin <7> (Reversibility: ?) [49]

P peroxynitrite + ?

S nitrite + xanthine <7> (<7> NO generation occurs under aerobic conditions and is regulated by O_2 tension, pH, nitrite, and reducing substrate concentrations [50]) (Reversibility: ?) [49, 5]

P NO + ?

S *o*-hydroxybenzaldehyde + H_2O + O_2 <7> (Reversibility: ?) [39]

P *o*-hydroxybenzoate + H_2O_2 <7> [39]

S organic nitrate + NADH <7> (<7> organic nitrite is the initial product in the process of xanthine oxidase mediated organic nitrate biotransformation and is the precursor of NO and nitrosothiols, serving as the link between organic nitrate and soluble guanylyl cyclase [51]) (Reversibility: ?) [51]

P organic nitrite + NAD^+ + H_2O

S propanal + H_2O + O_2 <7> (Reversibility: ?) [39]
P propanoate + H_2O_2 <7> [39]
S pteridine + H_2O + O_2 <2, 4, 5, 6, 7, 8, 9, 10, 11, 13, 14, 16, 21> (<7> and derivatives, e.g.: 4-amino-7-hydroxy pteridine, 4-hydroxy-7-azapteridine [19]) (Reversibility: ?) [19, 25]
P ?
S pterin + H_2O + O_2 <7> (Reversibility: ?) [20]
P isoxanthopterin + H_2O_2 <7> [20]
S purine + H_2O + O_2 <2, 4, 5, 6, 7, 8, 9, 10, 11, 13, 14, 16, 21> (Reversibility: ?) [25]
P ?
S purine + H_2O + O_2 <5, 7, 15> (<5, 7, 15> purine and derivatives [13, 15, 19]; <15> 23.3% of activity with xanthine [13]) (Reversibility: ?) [13, 15, 19]
P 7,9-dihydropurin-8-one + H_2O_2 <5, 7, 15> [13, 15, 19]
S pyridine-2-aldehyde + H_2O + O_2 <7> (Reversibility: ?) [39]
P ?
S pyridine-4-aldehyde + H_2O + O_2 <7> (Reversibility: ?) [39]
P ?
S pyrimidine derivatives + H_2O + O_2 <1, 3, 7> (<7> e.g. 2-hydroxypyrimidine [19]; <7> 6-hydroxy-6-aminepyrimidine [19]; <1,3> 7-hydroxy-(1,2,5)-thiadiazolo(3,4-d)-pyrimidine [19]) (Reversibility: ?) [19]
P ?
S pyrimidine-3-aldehyde + H_2O + O_2 <7> (Reversibility: ?) [39]
P ?
S salicylaldehyde + H_2O + O_2 <7> (Reversibility: ?) [19]
P salicylic acid + H_2O_2 <7> [19]
S succinate semialdehyde + H_2O + O_2 <7> (Reversibility: ?) [39]
P succinate + H_2O_2
S xanthine + H_2O + O_2 <7> (Reversibility: ?) [46]
P ?
S xanthine + H_2O + O_2 <7> (Reversibility: ?) [47]
P superoxide + urate + ?
S xanthine + H_2O + O_2 <6, 7> (<6> xanthine oxidoreductase, XOR, can exist in a dehydrogenase form, XD, and an oxidase form, XO. Part of total XOR activity in peroxisomes is XO activity. The major function of XOR activity in the cytoplasm of rat liver parenchymal cells and in sinusoidal cells is not the production of O_2 radicals, but rather the production of uric acid which can act as a potent antioxidant [43]) (Reversibility: ?) [1, 43, 49, 54, 60, 63]
P urate + H_2O_2
S xanthine + H_2O + O_2 <1, 2, 3, 4, 5, 6, 7, 8, 9, 10, 11, 12, 13, 14, 15, 16, 17, 19, 20, 21> (<7> electron acceptor methylene blue [19, 30]; <7> electron acceptor NAD^+ [19, 30]; <1, 2, 3, 4, 5, 6, 7, 8, 9, 10, 11, 12, 13, 14, 15, 16, 17, 19, 20, 21> electron acceptor O_2 [2, 3, 4, 5, 6, 7, 8, 9, 10, 11, 12, 13, 14, 15, 16, 17, 18, 19, 20, 21, 22, 23, 24, 25, 26, 27, 28, 29, 30, 31, 32, 33, 34, 35, 36, 37, 38, 39, 40, 41]; <7> electron acceptor quinones [19]; <7> specifi-

city for electron acceptor is low [19,30]; <7> electron acceptor triphenyltetrazolium chloride, phenazine methosulfate, nitrate, cytochrome c, ferritin [30]; <12> Arthrobacter S-2 enzyme: relatively specific [11]; <7> enzyme also oxidizes hypoxanthine, some other purines, pterines and aldehydes, i.e. possesses the activity of EC 1.2.3.1, probably acts on the hydrated derivatives of these substrates [19]; <7> low specificity to substrate [19,30]; <7,12,15> electron acceptor ferricyanide [11,13,19,30]; <7,12,15> electron acceptor 2,6-dichlorophenolindophenol [11,13,19,30]; <7> the enzyme accelerates reaction rate via base-catalyzed chemistry in which a Mo-OH group undertakes nucleophilic attack on the carbon center to be hydroxylated, with concomitant hydride transfer to a catalytically essential Mo=S group in the molybdenum coordination sphere. This chemistry appears to proceed via obligate two-electron chemistry rather than in individual steps to yield a reduced enzyme product complex with product coordinmated to the active site molybdenum by means of the newly introduced hydroxyl group in a sinple end-on fashion. Product displacement by hydroxide and electron transfer to other redox-active centers in the enzyme complete the catalytic sequence [52]) (Reversibility: ?) [2, 3, 4, 5, 6, 7, 8, 9, 10, 11, 12, 13, 14, 15, 16, 17, 18, 19, 20, 21, 22, 23, 24, 25, 26, 27, 28, 29, 30, 31, 32, 33, 34, 35, 36, 37, 38, 39, 40, 41, 52]

P uric acid + H_2O_2 <1, 2, 3, 4, 5, 6, 7, 8, 9, 10, 11, 12, 13, 14, 15, 16, 17, 19, 20, 21> (<7> under some conditions the product is mainly superoxide rather than peroxide: RH + H_2O + 2 O_2 = ROH + 2 H^+ + 2 O_2^- [19]; <5> enzyme contributes to the oxidant stress component of ischemia-reperfusion injury to intestine and liver, O_2- production decreases with increasing substrate concentrations [17]) [2, 3, 4, 5, 6, 7, 8, 9, 10, 11, 12, 13, 14, 15, 16, 17, 18, 19, 20, 21, 22, 23, 24, 25, 26, 27, 28, 29, 30, 31, 32, 33, 34, 35, 36, 37, 38, 39, 40, 41]

S xanthine + NO_2^- <7> (<7> oxidation of the enzyme by NO_2^- or reduction by xanthine take place at the molybdenum site [29]) (Reversibility: ?) [22, 29]

P uric acid + NO <7> [22]

S xanthine + methylene blue + O_2 <7> (Reversibility: ?) [56]

P urate + reduced methylene blue

S xanthine + thionine + O_2 <7> (Reversibility: ?) [56]

P urate + reduced thionine

S xanthopterin + H_2O + O_2 <7> (Reversibility: ?) [46]

P ?

S Additional information <6, 7> (<7> addition of xanthine oxidase to a solution of acetaldehyde and ascorbate increases the rate of ascorbate oxidation, due to the action of superoxide radicals generated in the process [48]; <7> hydroxyl free radicals generated by the hypoxanthine/xanthine oxidase/Fe system are implicated in oxidation of dibromoacetonitrile to CN^- [44]; <6> xanthine oxidase is necessary during the physiological involution of tissues [61]) (Reversibility: ?) [44, 48, 61]

P ?

Inhibitors

1,2-dihydroxybenzene 3,5-disulfonic acid <7> (<7> inhibits reaction with cytochrome c [30]) [30]
1-O-(4"-O-caffeoyl)-β-glucopyranosyl-1,4-dihydroxy-2-(3',3'-dimethylallyl)-benzene <7> [54]
2,4-diamino-6-hydroxy-s-triazine <7> [19]
2,4-dinitrofluorobenzene <7> [19]
2-(3,4-dihydroxy-5-methoxyphenyl)-5,7-dihydroxy-4H-chromen-4-one <7> (<7> competitive, 50% inhibition at 0.00022 mM [70]) [70]
2-(3,4-dihydroxyphenyl)-5,7-dihydroxy-4H-chromen-4-one <7> (<7> competitive, 50% inhibition at 0.00124 mM [70]) [70]
2-(3,4-dihydroxyphenyl)-5,7-dihydroxy-6-methoxy-4H-chromen-4-one <7> (<7> competitive, 50% inhibition at 0.00019 mM [70]; <7> competitive, 50% inhibition at 0.00020 mM [70]) [70]
2-amino-4-hydroxy-6-formylpterine <1, 6> [33]
2-amino-4-hydroxypterine-6-aldehyde <7> [19]
2-[3-cyano-4-(2-methylpropoxy)phenyl]-4-methylthiazole-5-carboxylic acid <7> (<7> i.e. febuxostat, TEI-6720, or TMX-67, mixed-type inhibition of both the oxidized and reduced form of xanthine oxidase [66]) [66]
2-chloroadenine <7> (<7> substrate analogue [69]) [69]
3,3',4,4'-tetrahydroxychalcone <7> [30]
3,4-dihydroxybenzaldehyde <7> (<7> mixed type inhibition, 50% inhibition at 0.0568 mM [70]) [70]
3,4-di-O-caffeoylquinic acid methyl ester <7> (<7> reversible inhibition, IC50: 0.0036 mM [54]) [54]
3,5-di-O-caffeoylquinic acid <7> [54]
4,5-di-O-caffeoylquinic acid <7> [54]
4,5-di-O-caffeoylquinic acid methyl ester <7> (<7> reversible inhibition [54]) [54]
5,7-dihydroxy-2-(3,4,5-trimethoxyphenyl)-4H-chromen-4-one <7> (<7> competitive, 50% inhibition at 0.00051 mM [70]) [70]
5,7-dihydroxy-2-(3-hydroxy-4,5-dimethoxyphenyl)-6-methoxy-4H-chromen-4-one <7> (<7> competitive, 50% inhibition at 0.00133 mM [70]) [70]
5,7-dihydroxy-2-(3-hydroxy-4-methoxyphenyl)-4H-chromen-4-one <7> (<7> competitive, 50% inhibition at 0.00013 mM [70]) [70]
5,7-dihydroxy-2-(4-hydroxy-3-methoxyphenyl)-3,6-dimethoxy-4H-chromen-4-one <7> (<7> competitive, 50% inhibition at 0.00115 mM [70]) [70]
5,7-dihydroxy-2-(4-hydroxyphenyl)-4H-chromen-4-one <7> (<7> competitive, 50% inhibition at 0.00036 mM [70]) [70]
5,7-dihydroxy-2-(4-methoxyphenyl)-4H-chromen-4-one <7> (<7> competitive, 50% inhibition at 0.080016mM [70]) [70]
6-(N-benzoylamino)purine <7> (<7> competitive, 50% inhibition at 0.00045 mM. Hydrogen bonding interaction involves N7 of the purine ring and N-H of R880, the N-H of the purine ring and OH of T1010 [64]) [64]
6-formylpterin <7> [46]
8-hydroxyquinoline-7-sulfonic acid <1, 6> [33]
Ag^+ <15> [13]

aldehydes <7> (<7> e.g. formaldehyde, 4-pyridinecarboxaldehyde, propionaldehyde, glycolaldehyde [38]) [38]
allopurinol <5, 7> (<7> IC50: 0.0026 mM [54]; <7> used as anti-gout drug [4]; <7> inhibits substrate binding at the molybdenum site [29]; <7> competitive inhibition of oxidation of dibromoacetonitrile by the hypoxanthine/xanthine oxidase/Fe system [44]; <7> inhibits peroxynitrite generation, IC50: 0.007 mM [49]; <7> allopurinol is a conventional substrate that generates superoxide radicals during its oxidation to oxypurinol [57]; <7> competitive, 50% inhibition at 0.0025 mM [70]; <5> increased activity of xanthine oxidase in cells exposed to $CoCl_2$ and subsequent increase in reactive oxygen species derived from enzyme activity, which results in accumulation of hypoxia-inducible factor 1α. Blockade of enzyme activity by allopurinol, N-acetyl-L-cysteine or siRNA significantly attenuates expression of hypoxia-inducible factor 1α and thus the induction of genes such as erythropoietin and vascular endothelial growth factor [58]) [4, 19, 20, 29, 44, 49, 54, 57, 58, 62, 70]
alloxan <1, 6> [33]
apigenin <7> (<7> mixed type inhibition of xanthine, strong inhibitor of xanthine oxidase, weak inhibition of monoamine oxidase [59]) [59]
Arsenite <6, 7> [19, 27, 33]
ascorbic acid <6, 15> [13, 33]
azaguanine <20> [32]
borate <1, 6> [33]
chalcones <1, 6> [33]
Co^{2+} <15> [13]
Cu^{2+} <7, 15> [13, 30, 33]
cyanide <1, 6, 7> [19, 27, 33]
desferrioxamine <7> (<7> significant decrease in CN^- formation from dibromoacetonitrile by the hypoxanthine/xanthine oxidase/Fe system [44]) [44]
dinitrophenol quinimine <7> [30]
dopamine <7> (<7> 23% inhibition at 1 mM and at 0.1 mM [42]) [42]
Fe^{3+} <15> [13]
folic acid <7> (<7> uncompetitive inhibition of oxidation of dibromoacetonitrile by the hypoxanthine/xanthine oxidase/Fe system [44]) [44]
formaldehyde <7> [19]
gallic acid <7> (<7> 50% inhibition of urate formation above 0.2 mM, 50% inhibition of superoxide anion generation at 0.0026 mM [68]) [68]
guanine <7> [19]
H_2O_2 <15> [13]
hydroxylamine <1, 6, 7> [30, 33]
imidazotriazines <1, 6> [33]
luteolin <7> (<7> mixed type inhibition of xanthine, strong inhibitor of xanthine oxidase, weak inhibition of monoamine oxidase [59]) [59]
methanol <1, 7> [19, 27]
myoglobin <7> (<7> inhibits reaction with cytochrome c as acceptor [30]) [30]

N-acetyl-L-cysteine <5> (<5> increased activity of xanthine oxidase in cells exposed to $CoCl_2$ and subsequent increase in reactive oxygen species derived from enzyme activity, which results in accumulation of hypoxia-inducible factor 1α. Blockade of enzyme activity by allopurinol, N-acetyl-L-cysteine or siRNA significantly attenuates expression of hypoxia-inducible factor 1α and thus the induction of genes such as erythropoietin and vascular endothelial growth factor [58]) [58]
noradrenaline <7> (<7> 15% inhibition at 0.1 mM, 30% inhibition at 1 mM [42]) [42]
O_2 <7> (<7> competitive inhibitor of NO production [50]) [50]
pterines <7> (<7> or other heterocyclic compounds, which are either not oxidized or oxidized rather slowly [30]) [30]
purine-6-aldehyde <7> [19]
purines <7> [30]
salicylate <7> [19, 30]
semicarbazide <7> [30]
superoxide dismutase <7> (<7> inhibits peroxynitrite generation [49]) [49]
tetraethylthiuram disulfide <1, 6> [33]
thiocyanate <1, 3> [19]
urea <7> [19]
xanthine <7, 12, 19> (<7> at high concentrations [45]; <12> no substrate inhibition [11]; <19> substrate inhibition at high concentration [10]) [10, 11, 45]
acacetin 7-O-(3-O-acetyl-β-D-glucopyranoside) <7> (<7> flavone glucoside from Chrysanthemum sinense, 50% inhibition at 0.080 mM [70]) [70]
anacardic acid <7> (<7> inhibits generation of superoxide radicals by xanthine oxicasein a sigmoidal inhibition, binds to allosteric sites near the xanthine-binding domain in xanthine oxidase [47]) [47]
cassia oil <4> (<4> oral adminstration of cassia oil significantly reduces serum and hepatic urate levels in hyperuricemic mice. At 600 mg/kg, cassia oil is as potent as allopurinol. This hypouricemic effect is explained by inhibiting activities of liver xanthine oxidase and xanthine oxidoreductase [65]) [65]
decyl gallate <7> (<7> 50% inhibition of urate formation at0.097 mM, 50% inhibition of superoxide anion generation at 0.0039 mM [68]) [68]
dimethylthiourea <7> (<7> significant decrease in rate of oxidation of dibromoacetonitrile by the hypoxanthine/xanthine oxidase/Fe system [44]) [44]
diphenylene iodonium <7> (<7> inhibits peroxynitrite generation [49]) [49]
diphenyleneiodonium chloride <7> (<7> inhibits nitrate reduction only when NADH is used as reducing substrate and does not inhibit nitrite generation when xanthine is used [45]; <7> strongly inhibits xanthine oxidase mediated NO generation with NADH serving as reducing substrate, with xanthine or 2,3-dihydroxybenzaldehyde as reducing substrates, NO generation is increased more than six times [50]) [45, 50]
genkwanin <7> [59]
hesperetin <7> (<7> 50% inhibition at 0.039 mM [62]) [62]
hexyl gallate <7> (<7> 50% inhibition of urate formation above 0.2 mM, 50% inhibition of superoxide anion generation at 0.0052 mM [68]) [68]

isatin <1, 6> [33]
luteolin 7-methyl ether <7> [59]
mannitol <7> (<7> significant decrease in rate of oxidation of dibromoacetonitrile by the hypoxanthine/xanthine oxidase/Fe system [44]) [44]
menthyl gallate <7> (<7> 50% inhibition of urate formation above 0.2 mM, 50% inhibition of superoxide anion generation at 0.0049 mM [68]) [68]
n-dodecyl gallate <7> (<7> 50% inhibition of urate formation at 0.049 mM, 50% inhibition of superoxide anion generation at 0.0036 mM [68]) [68]
ninhydrin <1, 6> [33]
octyl gallate <7> (<7> 50% inhibition of urate formation at 0.262 mM, 50% inhibition of superoxide anion generation at 0.0045 mM [68]) [68]
oxypurinol <7> (<7> inhibits NO generation triggered by xanthine, NADH or 2,3-dihydroxybenzaldehyde [50]) [45, 50]
p-aminophenol quinimine <1, 6> [33]
p-chloromercuribenzoate <1, 6, 7, 15> [13, 33]
phenylhydrazine <7> [30]
propyl gallate <7> (<7> 50% inhibition of urate formation above 0.2 mM, 50% inhibition of superoxide anion generation at 0.0064 mM [68]) [68]
pycnogenol <6> (<6> extract from french maritime pine bark, contains 75% (w/w) procyanidins formed by catechin and epicatechin but also taxifolin and phenolcarbonic acids and their glycosides, uncompetitive inhibitor, 0.01 mg/ml, 35% inhibition, 0.1 mg/ml, 80% inhibition, enzyme recovers activity upon dissociation of pycnogenol from enzyme [21]) [21]
theaflavin-3,3'-gallate <7> (<7> 50% inhibition at 0.049 mM [62]) [62]
Additional information <7> (<7> study on the effect of food extracts on enzyme activity in vitro. Extract of black tea, extract of rooibus herbal tea, purple grape juice, extract of clove, and cranberry juice are inhibitory [62]) [62]

Cofactors/prosthetic groups

FAD <1, 2, 3, 4, 5, 6, 7, 8, 9, 10, 11, 12, 13, 14, 15, 16, 17, 19, 20, 21> (<1, 2, 3, 4, 5, 6, 7, 8, 9, 10, 11, 12, 13, 14, 15, 16, 17, 19, 20, 21> flavoprotein [2, 3, 4, 5, 6, 7, 8, 9, 10, 11, 12, 13, 14, 15, 16, 17, 18, 19, 20, 21, 22, 23, 24, 25, 26, 27, 28, 29, 30, 31, 32, 33, 34, 35, 36, 37, 38, 39, 40, 41]; <1, 3, 7> contains FAD, molybdenum and iron in the ratio 1 : 4 : 4 [19,30]; <7> bovine liver: 1 FAD per subunit (2 per molecule) [37]) [2, 3, 4, 5, 6, 7, 8, 9, 10, 11, 12, 13, 14, 15, 16, 17, 18, 19, 20, 21, 22, 23, 24, 25, 26, 27, 28, 29, 30, 31, 32, 33, 34, 35, 36, 37, 38, 39, 40, 41]
molybdopterin <7> (<7> one cofactor per subunit, oxidation of xanthine takes place at this center, electrons are rapidly distributed to the other centers by intramolecular electron transfer [5]) [5]
NADH <7> [45,51]

Activating compounds

ascorbate <7> (<7> in absence of thiols or ascorbate, no NO generation is detected from xanthine oxidase mediated organic nitrate reduction [51]) [51]
dithiothreitol <7> (<7> enhance oxidation of dibromoacetonitrile by the hypoxanthine/xanthine oxidase/Fe system [44]) [44]

glutathione <7> (<7> enhances oxidation of dibromoacetonitrile by the hypoxanthine/xanthine oxidase/Fe system [44]) [44]
L-cysteine <7> (<7> in absence of thiols or ascorbate, no NO generation is detected from xanthine oxidase mediated organic nitrate reduction [51]) [51]
N-acetyl-L-Cys <7> (<7> enhance oxidation of dibromoacetonitrile by the hypoxanthine/xanthine oxidase/Fe system [44]) [44]
thiol <7> (<7> in absence of thiols or ascorbate, no NO generation is detected from xanthine oxidase mediated organic nitrate reduction [51]) [51]
diphenyleneiodonium chloride <7> (<7> strongly inhibits xanthine oxidase mediated NO generation with NADH serving as reducing substrate, with xanthine or 2,3-dihydroxybenzaldehyde as reducing substrates, NO generation is increased more than six times [50]) [50]
Additional information <7> (<7> study on the effect of food extracts on enzyme activity in vitro. Orange juice and pink grapefruit juice are activating [62]) [62]

Metals, ions

$CoCl_2$ <5> (<5> increased activity of xanthine oxidase in cells exposed to $CoCl_2$ and subsequent increase in reactive oxygen species derived from enzyme activity, which results in accumulation of hypoxia-inducible factor 1α. Blockade of enzyme activity by allopurinol, N-acetyl-L-cysteine or siRNA significantly attenuates expression of hypoxia-inducible factor 1α and thus the induction of genes such as erythropoietin and vascular endothelial growth factor [58]) [58]
Fe^{2+} <7> (<7> xanthine oxidase dependent oxidation of ascorbate is markedly increased in presence of iron [48]) [48]
iron <1, 2, 3, 4, 5, 6, 7, 8, 9, 10, 11, 12, 13, 14, 15, 16, 17, 19, 20, 21> (<1, 2, 3, 4, 5, 6, 7, 8, 9, 10, 11, 12, 13, 14, 15, 16, 17, 19, 20, 21> iron-molybdenum protein [2, 3, 4, 5, 6, 7, 8, 9, 10, 11, 12, 13, 14, 15, 16, 17, 18, 19, 20, 21, 22, 23, 24, 25, 26, 27, 28, 29, 30, 31, 32, 33, 34, 35, 36, 37, 38, 39, 40, 41]; <7> bovine liver: 2 Fe-S centres per subunit, 4 per molecule [37]; <1,3,7> contains FAD, iron and molybdenum in the ratio 1/4/4 [19,30]) [2, 3, 4, 5, 6, 7, 8, 9, 10, 11, 12, 13, 14, 15, 16, 17, 18, 19, 20, 21, 22, 23, 24, 25, 26, 27, 28, 29, 30, 31, 32, 33, 34, 35, 36, 37, 38, 39, 40, 41]
molybdenum <1, 2, 3, 4, 5, 6, 7, 8, 9, 10, 11, 12, 13, 14, 15, 16, 17, 19, 20, 21> (<7> liver enzyme: 1 molybdenum atom per subunit, 2 per molecule [37]; <7> molybdenum cofactor is a complex between molybdenum and molybdopterin (a 6-alkylpterin with 4-carbon side chain which has an enedithiol at carbon 1 and 2, a hydroxyl at carbon 3, and a terminal phosphate group) [2,3]; <1,3,7> contains FAD, molybdenum and iron in the ratio 1/4/4 [19, 30]; <7> purification of the molybdenum cofactor from milk xanthine oxidase [36]; <1, 2, 3, 4, 5, 6, 7, 8, 9, 10, 11, 12, 13, 14, 15, 16, 17, 19, 20, 21> an iron-molybdenum protein [2, 3, 4, 5, 6, 7, 8, 9, 10, 11, 12, 13, 14, 15, 16, 17, 18, 19, 20, 21, 22, 23, 24, 25, 26, 27, 28, 29, 30, 31, 32, 33, 34, 35, 36, 37, 38, 39, 40, 41]; <7> nitrate reduction to nitrite as well as nitrite reduction to NO occurs at the molybdenum site [45]; <7> the enzyme accelerates reaction rate via base-catalyzed chemistry in which a Mo-OH group undertakes nucleophi-

lic attack on the carbon center to be hydroxylated, with concomitant hydride transfer to a catalytically essential Mo=S group in the molybdenum coordination sphere. This chemistry appears to proceed via obligate two-electron chemistry rather than in individual steps to yield a reduced enzyme product complex with product coordinated to the active site molybdenum by means of the newly introduced hydroxyl group in a sinple end-on fashion [52]) [2, 3, 4, 5, 6, 7, 8, 9, 10, 11, 12, 13, 14, 15, 16, 17, 18, 19, 20, 21, 22, 23, 24, 25, 26, 27, 28, 29, 30, 31, 32, 33, 34, 35, 36, 37, 38, 39, 40, 41, 45, 52]

Turnover number (min^{-1})

0.41 <6> (xanthine, <6> xanthine oxidoreductase mutant W335A/F336L, pH 7.8, 25°C [63]) [63]
1.28 <7> (NADPH, <7> cosubstrate cytochrome c [33]) [33]
2.57 <7> (NADPH, <7> cosubstrate indophenol [33]) [33]
4.35 <7> (hypoxanthine, <7> cofactor 2,6-dichlorophenolindophenol [33]) [33]
4.92 <7> (xanthine, <7> (+ O_2) [33]) [33]
5.25 <7> (xanthine, <7> cofactor 2,6-dichlorophenolindophenol [33]) [33]
9.73 <6> (xanthine, <6> xanthine oxidoreductase mutant W335A/F336L treated with dithiothreitol, pH 7.8, 25°C [63]) [63]
16.2 <7> (xanthine, <7> pH 8.2, 23.5°C [19]) [19]

Specific activity (U/mg)

0.0002 <4> (<4> enzyme activity in embryo cell lines [23]) [23]
0.278 <8> [24]
0.885 <6> [21]
1.67 <7> [31]
1.8 <5> [15]
2.04 <5> (<5> colostrum [25]) [25]
2.4 <7> [19]
2.963 <7> [37]
3.64 <6> [1]
5.2 <4> [6]
7.8 <7> (<7> milk [25]) [25]
8.571 <4> [41]
10.16 <7> [18]
15.2 <15> [13]
123 <13> (<13> milk [25]) [25]
189.6 <6> (<6> electron acceptor O_2 [8]) [8]
230 <12> [11]
800 <6> (<6> electron acceptors O_2 and NAD^+ [8]) [8]

K_m-Value (mM)

0.000077 <7> (methylene blue, <7> pH 6.5 [56]) [56]
0.000325 <7> (thionine, <7> pH 6.5 [56]) [56]
0.0005 <7> (NADPH, <7> cofactor 2,6-dichlorophenolindophenol [33]) [33]
0.001 <7> (3-nitrobenzaldehyde, <7> pH 7.0 [60]) [60]

0.00143 <20> (2-amino-4-hydroxypteridine, <20> + methylene blue [32]) [32]
0.00146 <7> (xanthine) [29]
0.0015 <7> (pterin, <7> cell-associated enzyme, determined in intact cell monolayers [20]) [20]
0.0016 <7> (pterin) [20]
0.0018 <6> (xanthine) [21]
0.002 <6> (xanthine) [7]
0.003 <5> (xanthine) [17]
0.0034 <7> (xanthine, <7> cell-associated enzyme, determined in intact cell monolayers [20]) [20]
0.0036 <7> (xanthine) [20]
0.004 <5> (allopurinol) [15]
0.0065 <7> (xanthine) [31]
0.0067 <3> (2-amino-4-hydroxypteridine, <3> + methylene blue [32]) [32]
0.00924 <5> (hypoxanthine, <5> membrane-bound enzyme [25]) [25]
0.01 <7> (2-nitrobenzaldehyde, <7> pH 7.0 [60]) [60]
0.011 <7> (4-nitrobenzaldehyde, <7> pH 7.0 [60]) [60]
0.0134 <5> (hypoxanthine, <5> free enzyme [25]) [25]
0.015 <5> (xanthine, <5> heparin-Sepharose 6B-bound enzyme [17]) [17]
0.022 <7> (hypoxanthine) [49]
0.024 <7> (3-methylbenzaldehyde, <7> pH 7.0 [60]) [60]
0.024 <1> (xanthine, <1> kidney enzyme [32]) [32]
0.025 <4> (xanthine) [32]
0.0253 <7> (2,3-dihydroxybenzaldehyde, <7> in presence of 1.0 mM nitrate [45]) [45]
0.026 <7> (O_2, <7> with xanthine as cosubstrate [49]) [49]
0.0264 <5> (xanthine, <5> membrane-bound enzyme [25]) [25]
0.0279 <5> (xanthine, <5> free enzyme [25]) [25]
0.03 <7> (2-methylbenzaldehyde, <7> pH 7.0 [60]) [60]
0.035 <7> (2,3-dihydroxybenzaldehyde) [29]
0.036 <7> (pterin) [49]
0.039 <7> (3,4-dihydroxybenzylaldehyde) [42]
0.04 <7> (4-methylbenzaldehyde, <7> pH 7.0 [60]) [60]
0.04 <7> (nitrite, <7> pterin as cosubstrate [49]) [49]
0.046 <7> (pyridine-3-aldehyde) [39]
0.0475 <7> (hypoxanthine, <7> cofactor 2,6-dichlorophenolindophenol [33]) [33]
0.05 <7> (O_2, <7> cosubstrate xanthine, pH 8.5 [19]) [19]
0.068 <7> (2,5-dihydroxybenzaldehyde) [39]
0.07 <7> (nitrite, <7> hypoxanthine as cosubstrate [49]) [49]
0.08 <7> (O_2, <7> cosubstrate xanthine, pH 10.0 [19]) [19]
0.085 <7> (indole-3-aldehyde) [39]
0.088 <7> (xanthine, <7> cofactor 2,6-dichlorophenolindophenol [33]) [33]
0.1 <7> (O_2, <7> with NADH as cosubstrate [49]) [49]
0.12 <7> (4-hydroxy-3-methoxybenzaldehyde) [42]

0.129 <6> (O_2, <6> xanthine oxidoreductase mutant W335A/F336L treated with dithiothreitol, pH 7.8, 25°C [63]; <6> xanthine oxidoreductase mutant W335A/F336L, pH 7.8, 25°C [63]) [63]
0.13 <12> (1-methylxanthine) [11]
0.13 <12, 20> (canthine) [11, 32]
0.17 <7> (3-methoxybenzylaldehyde) [42]
0.17 <7> (4-hydroxybenzylaldehyde) [42]
0.17 <7> (4-methoxybenzylaldehyde) [42]
0.18 <7> (3-hydroxy-4-methoxybenzaldehyde) [42]
0.18 <7> (3-hydroxybenzylaldehyde) [42]
0.19 <7> (2-methoxybenzaldehyde) [42]
0.19 <7> (2-hydroxybenzylaldehyde) [42]
0.26 <1> (xanthine, <1> liver enzyme [32]) [32]
0.3 <7> (benzaldehyde) [42]
0.3 <7> (indole-3-acetaldehyde) [39]
0.36 <7> (pyridine-2-aldehyde) [39]
0.49 <7> (glyceryl trinitrate, <7> xanthine as cosubstrate [51]) [51]
0.55 <5> (formycin B) [15]
0.56 <7> (4-hydroxyphenylglycolaldehyde) [39]
0.68 <7> (3,4-dimethoxybenzaldehyde) [42]
0.86 <7> (NADH, <7> in presence of 1.0 mM nitrate [45]) [45]
0.878 <7> (NADH) [29]
0.93 <5> (benzaldehyde) [15]
1 <7> (nitrite, <7> with xanthine as cosubstrate [49]) [49]
1 <7> (succinate semialdehyde) [39]
1.03 <7> (*o*-hydroxybenzaldehyde) [39]
1.2 <7> (nitrite, <7> with NADH as cosubstrate [49]) [49]
1.64 <7> (isosorbide dinitrate, <7> xanthine as cosubstrate [51]) [51]
1.7 <7> (pyridine-4-aldehyde) [39]
2 <7> (glyceraldehyde 3-phosphate) [39]
2.25 <7> (NO_2^-) [29]
6.6 <7> (xanthine, <7> in presence of 2 mM NADH [45]) [45]
6.8 <7> (xanthine, <7> in presence of 0.5 mM 2,3-dihydroxybenzaldehyde [45]) [45]
20 <7> (aldehyde, <7> cofactor 2,6-dichlorophenolindophenol [33]) [33]
48 <5> (N-methylnicotinamide) [15]
97 <7> (dibromoacetonitrile) [44]
130 <7> (acetaldehyde) [39]
142 <7> (butyraldehyde) [39]
161.5 <7> (formaldehyde) [39]
430 <7> (propionaldehyde) [39]
Additional information <2, 3, 4, 5, 6, 7, 8, 9, 10, 11, 12, 13, 14, 16, 19, 20, 21> (<7> enzyme kinetics at all pH values studied is non-hyperbolic, and the use of the Michaelis-Menten equation is not adequate [69]) [10, 11, 15, 19, 24, 25, 30, 32, 39, 41, 69]

K_i-Value (mM)

0.0000006 <7> (2-[3-cyano-4-(2-methylpropoxy)phenyl]-4-methylthiazole-5-carboxylic acid, <7> 25°C, pH 7.4, oxidized form of enzyme [66]) [66]
0.0000031 <7> (2-[3-cyano-4-(2-methylpropoxy)phenyl]-4-methylthiazole-5-carboxylic acid, <7> 25°C, pH 7.4, reduced form of enzyme [66]) [66]
0.0000475 <7> (6-(N-benzoylamino)purine, <7> 25°C, pH 7.5 [64]) [64]
0.000094 <7> (6-formylpterin) [46]
0.00011 <7> (2-(3,4-dihydroxyphenyl)-5,7-dihydroxy-6-methoxy-4H-chromen-4-one, <7> pH 7.5, 25°C [70]) [70]
0.00011 <7> (5,7-dihydroxy-2-(4-methoxyphenyl)-4H-chromen-4-one, <7> pH 7.5, 25°C [70]) [70]
0.00012 <7> (5,7-dihydroxy-2-(3-hydroxy-4-methoxyphenyl)-4H-chromen-4-one, <7> pH 7.5, 25°C [70]) [70]
0.00015 <7> (2-(3,4-dihydroxy-5-methoxyphenyl)-5,7-dihydroxy-4H-chromen-4-one, <7> pH 7.5, 25°C [70]) [70]
0.00015 <7> (2-(3,4-dihydroxyphenyl)-5,7-dihydroxy-6-methoxy-4H-chromen-4-one, <7> pH 7.5, 25°C [70]) [70]
0.00024 <7> (4,5-di-O-caffeoylquinic acid methyl este, <7> 25°C [54]) [54]
0.00028 <7> (5,7-dihydroxy-2-(4-hydroxyphenyl)-4H-chromen-4-one, <7> pH 7.5, 25°C [70]) [70]
0.00037 <7> (5,7-dihydroxy-2-(3,4,5-trimethoxyphenyl)-4H-chromen-4-one, <7> pH 7.5, 25°C [70]) [70]
0.00079 <7> (5,7-dihydroxy-2-(4-hydroxy-3-methoxyphenyl)-3,6-dimethoxy-4H-chromen-4-one, <7> pH 7.5, 25°C [70]) [70]
0.0009 <7> (2-(3,4-dihydroxyphenyl)-5,7-dihydroxy-4H-chromen-4-one, <7> pH 7.5, 25°C [70]) [70]
0.00102 <7> (5,7-dihydroxy-2-(3-hydroxy-4,5-dimethoxyphenyl)-6-methoxy-4H-chromen-4-one, <7> pH 7.5, 25°C [70]) [70]
0.0018 <7> (allopurinol, <7> pH 7.5, 25°C [70]) [70]
0.0031 <7> (3,4-di-O-caffeoylquinic acid methyl ester, <7> 25°C [54]) [54]
0.0399 <7> (3,4-dihydroxybenzaldehyde, <7> pH 7.5, 25°C [70]) [70]
0.0538 <7> (acacetin 7-O-(3-O-acetyl-β-D-glucopyranoside), <7> pH 7.5, 25°C [70]) [70]
0.08 <7> (xanthine) [45]
0.4 <7> (8-bromoxanthine, <7> 25°C, pH 7.5 [64]) [64]

pH-Optimum

5 <7> (<7> generation of nitrite or NO [45]) [45, 50]
7 <20> [32]
7.1 <15> [13]
8 <19> [10]
8.1 <8> [24]
8.2 <5> [25]
8.3 <7> [19]
8.35 <13> [25]
8.5 <7> (<7> conversion of dibromoacetonitrile to CN^- [44]) [44]

Additional information <7> (<7> the rate of xanthine oxidation is pH dependent, and the neutral form of xanthine binds stronger to the enzyme than the monoanion [69]) [69]

pH-Range

5-9 <15> (<15> pH 5.0: about 10% of activity maximum, pH 9.0: about 25% of activity maximum [13]) [13]
6-7.6 <7> (<7> under aerobic conditions, NO generation increases more than three times as the pH value decreases from pH 7.4 to 6.0 [50]) [50]
6-9 <19> (<19> pH 6.0: about 25% of activity maximum, pH 9.0: about 20% of activity maximum [10]) [10]
6.5-9.5 <7> (<7> pH 6.5: about 40% of maximal activity, pH 9.5: about 65% of maximal activity, conversion of dibromoacetonitrile to CN^- [44]) [44]
7.7-9.1 <8> (<8> at pH 7.7 and 9.1: about 30% of activity maximum [24]) [24]

Temperature optimum (°C)

23 <2, 4, 5, 6, 7, 8, 9, 10, 11, 13, 14, 16, 21> (<2,4,5,6,7,8,9,10,11,13,14,16,21> assay at [25]) [25]
25 <4> (<4> assay at [41]) [41]
35 <15> [13]
37 <7, 15> (<15> assay at [13]; <7> conversion of dibromoacetonitrile to CN^- [44]) [13, 44]

Temperature range (°C)

5 <15> (<15> 5°C: about 10% of activity maximum, 60°C: about 50% of activity maximum [13]) [13]
30-45 <7> (<7> 30°C: about 60% of maximal activity, 45°C: about 75% of maximal activity, conversion of dibromoacetonitrile to CN^- [44]) [44]

4 Enzyme Structure

Molecular weight

100000 <7> (<7> approximate value, bovine milk, ultrafiltration [12]) [12]
128000 <15> (<15> gel filtration [13]) [13]
146000 <12> (<12> gel filtration [11]) [11]
275000 <7> (<7> gel filtration [37]) [37]
280000 <8> (<8> gel filtration [24]) [24]
300000 <4, 5> (<4,5> gel filtration [15,41]) [15, 41]
303000 <7> (<7> sedimentation equilibrium [35]) [35]
310000 <5> (<5> gel filtration [25]) [25]
Additional information <1, 6, 7> [19, 33]

Subunits

dimer <1, 3, 4, 5, 7, 12, 15> (<12> 2 * 80000, SDS-PAGE [11]; <4,5,7> 2 * 150000, SDS-PAGE [15,19,35,41]; <15> 2 * 69000, SDS-PAGE [13]; <7> 1 * 152000, 1 * 131000, SDS-PAGE [18]; <3> 2 * 130000-140000 [19]; <1> 2 *

120000 [19]; <7> the two XOD subunits are strongly cooperative in both binding and catalysis [46]) [11, 13, 15, 16, 18, 19, 35, 41, 46]
oligomer <8> (<8> x * 52000 + x * 99000, SDS-PAGE [24]) [24]
Additional information <6> (<6> on polyacrylamide gel electrophoresis in presence of SDS and 2-mercaptoethanol the purified xanthine oxidase is dissociated into one major band of approximately 150000 Da and three bands of 100000 Da, 60000 Da and 40000 Da [1]) [1]

Posttranslational modification

proteolytic modification <1, 2, 3, 4, 5, 6, 7, 8, 9, 10, 11, 12, 13, 14, 15, 16, 17, 19, 20, 21> [2, 3, 4, 5, 6, 7, 8, 9, 10, 11, 12, 13, 14, 15, 16, 17, 18, 19, 20, 21, 22, 23, 24, 25, 26, 27, 28, 29, 30, 31, 32, 33, 34, 35, 36, 37, 38, 39, 40, 41]
Additional information <1, 2, 3, 4, 5, 6, 7, 8, 9, 10, 11, 12, 13, 14, 15, 16, 17, 19, 20, 21> (<2, 4, 5, 6, 7, 8, 9, 10, 11, 13, 14, 16, 21> the enzyme from animal tissues can be interconverted to xanthine dehydrogenase, EC 1.1.1.204, the liver enzyme exists in vivo mainly in its dehydrogenase form, but can be converted into the oxidase form by storage at -20°C, by treatment with proteolytic enzymes or with organic solvents, or by thiol reagents such as Cu^{2+}, N-ethylmaleimide or 4-hydroxymercuribenzoate, the effect of the thiol reagents can be reversed by thiols such as 1,4-dithioerythritol, in other animal tissues the enzyme exists almost entirely as EC 1.1.3.22 but can be converted into the dehydrogenase form by 1,4-dithioerythritol [8,25,28]; <7> bovine milk enzyme: conversion from oxidase into dehydrogenase by treatment with dithioerythritol or dihydrolipoic acid [31]; <6> liver enzyme is unstable as dehydrogenase and is gradually converted to oxidase [7]; <7> oxidase is converted into an irreversible oxidase form by pretreatment with chymotrypsin, papain or subtilisin, but only partially with trypsin [31]; <7> reversible conversion of xanthine dehydrogenase to xanthin oxidase can be achieved by modification of Cys535 and Cys992, tryptic proteolysis of xanthine dehydrogenase after Lys551 or pancreatin cleavage after Leu219 and Lys569 results in irreversible transformation to xanthine oxidase [5]) [2, 3, 4, 5, 6, 7, 8, 9, 10, 11, 12, 13, 14, 15, 16, 17, 18, 19, 20, 21, 22, 23, 24, 25, 26, 27, 28, 29, 30, 31, 32, 33, 34, 35, 36, 37, 38, 39, 40, 41]

5 Isolation/Preparation/Mutation/Application

Source/tissue

3T3 cell <4> [23]
3T6-Swiss albino cell <4> [23]
BALB/3T12-3 cell <4> (<4> mouse embryo cell line: 3T12 [23]) [23]
U-251MG cell <5> (<5> glioma cell [58]) [58]
U-87MG cell <5> (<5> glioma cell [58]) [58]
colostrum <2, 4, 5, 6, 7, 8, 9, 10, 11, 13, 14, 16, 21> [25]
duodenum <4, 20> [32]

heart <7> (<7> xanthine oxidase mediated nitrite reduction can be a source of NO in heart tissue under conditions of tissue normoxia, and it is further increased with mild hypoxia [50]) [50]
kidney <4, 20> [32]
liver <1, 2, 4, 5, 6, 7, 8, 9, 10, 11, 13, 14, 16, 20, 21> (<5> together with intestine highest enzyme activity of any tissue due to xanthine oxidase rich parenchyma cells of this tissues [22]; <4> xanthine oxidase activity falls for the first 11 days after infection with smooth type Salmonella typhimurium, coinciding with the period of bacterial growth in the liver. Mild infection with Pseudomonas aeruginosa stimulates xanthine oxidase [55]) [1, 7, 8, 14, 15, 19, 22, 24, 25, 28, 30, 32, 41, 43, 55, 65]
lung <4, 7, 20> [14, 32]
mammary gland <4, 6> (<6> after weaning, during the involution of the mammary gland, enzyme activity increases and high mitochondrial H_2O_2 production takes place. Inhibition of xanthine oxidase slows down the involution of the mammary gland due to the decrease in the number of apoptotic cells and prevents the production of H_2O_2 that occurs during apoptosis [61]) [6, 61]
milk <2, 4, 5, 6, 7, 8, 9, 10, 11, 13, 14, 16, 21> (<7> buttermilk [56]; <7> grade I from buttermilk [62]; <7> unpasteurized milk [57]) [2, 3, 12, 14, 19, 25, 26, 27, 30, 31, 33, 34, 35, 36, 38, 39, 40, 42, 45, 49, 50, 51, 52, 54, 56, 57, 60, 62, 66, 68, 69, 70]
mucosa <5> (<5> gastric mucosa of Helicobacter pylori positive and negative pediatric patients [67]) [67]
pancreas <4, 20> [32]
parenchymal cell <6> [43]
seedling <19> [10]
spleen <4, 20> [32]
Additional information <18> (<18> control of the enzyme under the circadian clock, activity is 15times higher in light than in darkness [53]) [53]

Localization

cytosol <6> [1]
lipid droplet <4, 7, 11> (<11> membrane-bound [25]; <7> enzyme is associated with bovine milk-fat-globule membrane [12,40]; <4> milk lipid globule membrane possesses enzyme activity capable of catalyzing the conversion of xanthine dehydrogenase to xanthine oxidase, kinetic of this process is consistent with the rapid appearence of xanthine oxidase in milk [6]) [6, 12, 25, 40]
lipid particle <7> [12, 14, 40]
microsome <7> (<7> enzyme from milk and mammary gland [14,30]) [14, 30]
mitochondrion <6> [61]
peroxisome <6> (<6> matrix and core of peroxisomes of liver parenchymal cells, xanthine oxidoreductase and xanthine oxidase detected [43]) [43]
Additional information <5, 6, 7> (<7> enzyme binds specifically and with high affinity to bovine aortic endothelial cells, increasing cell xanthine oxi-

dase activity up to 10fold, circulating xanthine oxidase may therefore bind to vascular cells, impairing cell function via oxidative mechanisms [20]; <7> distribution of enzyme in milk globules, mammary gland and liver [14]; <5> enzyme binds to endothelial cells in a partially heparin-reversible manner [17]; <7> liver enzyme: in supernatant fraction [30]; <6> only XOR activity is present in the cytoplasm of rat liver parenchymal cells. In Kupffer cells and sinusoidal endothelial cell xanthine oxidoreductase activity is demonstrated in vesicles and occasionally on granular endoplasmic reticulum. Xanthine oxidase activity is not found in Kupffer cells and sinusoidal cells [43]) [14, 17, 20, 30, 43]

Purification

<4> [41]
<4> (60% ammonium sulfate, benzamidine-Sepharose) [6]
<5> [15, 25]
<6> [7]
<6> (ammonium sulfate, hydroxyapatite, DEAE-cellulose) [8]
<6> (simultaneous purification of xanthine oxidase and aldehyde oxidase) [1]
<7> [4, 12, 14, 25, 31, 35, 37]
<7> (ammonium sulfate, DEAE-Sepaharose) [18]
<7> (purification of the molybdenum cofactor of milk xanthine oxidase) [36]
<8> [24]
<12> [11]
<13> [25]
<15> [13]

Crystallization

<6> (xanthine oxidoreductase mutant W335A/F336L, showing two similar, but not identical subunits. The cluster involved in conformation-switching is completely disrupted in one subunit, but remains partly associated in the other. Xanthine oxidase and oxidoreductase forms of the mutant are in equilibrium that greatly favors the oxidase form, but upon incubation with dithiothreitol equilibrium is partly shifted towards the oxidoreductase form) [63]
<7> [26]
<7> (2.5 A resolution, each enzyme subunit is composed of an N-terminal 20000 Da domain containing two iron sulfur centers, a central 40000 Da FAD domain and a C-terminal 85000 molybdopterin binding domain, the four redox centers are aligned in a linear fashion) [5]
<7> (batch method most suitable for crystallization) [4]
<17> (2.25 A resolution, enzyme contains a molybdoptererin cofactor and two different [2Fe-2S] centers, enzyme is folded into four domains, the first two bind the iron sulfur centers, the last two are involved in molybtopterin binding) [9]

Cloning

<7> (expression in Escherichia coli) [16]

Engineering

W335A/F336L <6> (<6> xanthine oxidoreductase mutant displaying xanthine oxidase activity [63]) [63]

Application

biotechnology <7> (<7> construction of amperometric biosensors based in xanthine oxidase which has been immobilized by covalent binding to gold electrodes modified with dithiobis-N-succinimidyl propionate. Redox dyes thionine and methylene blue work well as electron acceptors for reduced enzyme [56]) [56]

medicine <4, 5> (<5> in Helicobacter pylori positive and negative pediatric patients, the activities of xanthine oxidase, myeloperoxidase, and superoxide dismutase in gastric mucosa are not affected by presence/absence of Helicobacter pylori [67]; <5> increased activity of xanthine oxidase in cells exposed to $CoCl_2$ and subsequent increase in reactive oxygen species derived from enzyme activity, which results in accumulation of hypoxia-inducible factor 1α. Blockade of enzyme activity by allopurinol, N-acetyl-L-cysteine or siRNA significantly attenuates expression of hypoxia-inducible factor 1α and thus the induction of genes such as erythropoietin and vascular endothelial growth factor [58]; <4> oral adminstration of cassia oil significantly reduces serum and hepatic urate levels in hyperuricemic mice. At 600 mg/kg, cassia oil is as potent as allopurinol. This hypouricemic effect is explained by inhibiting activities of liver xanthine oxidase and xanthine oxidoreductase [65]) [58, 65, 67]

6 Stability

pH-Stability

6.2-6.8 <15> (<15> 60°C, 30 min [13]) [13]

7.8 <6> (<6> 5°C or 10°C, 6 months, no significant loss of activity [1]) [1]

Temperature stability

60 <15> (<15> pH 6.2-6.8, 30 min, stable [13]) [13]

70 <15> (<15> 30 min, complete loss of activity [13]) [13]

Oxidation stability

<7>, photooxidation, protection by competitive inhibitors [30]

General stability information

<5>, heparin-Sepharose 6B-bound enzyme, half-life: 120 h and 67 h at 4°C and 20°C respectively, free enzyme, 61 h and 45 h at 4°C and 67°C respectively [17]

<6>, rat liver enzyme is unstable as dehydrogenase and is gradually converted to oxidase [7]

<6>, the xanthine dehydrogenase form can be obtained through incubation of xanthine oxidase with sulfhydryl reducing reagents [1]

<7>, low operational stability by immobilization [34]

<7>, phosphate stabilizes [30]
<12>, stable for more than 1 year with alternate freezing and thawing [11]
<2, 4, 5, 6, 7, 8, 9, 10, 11, 13, 14, 16, 21>, the enzyme from animal tissues can be interconverted to EC 1.1.1.204, that from liver exists in vivo mainly as the dehydrogenase form, but can be converted into the oxidase form by storage at -20°C, by treatment with proteolytic enzymes or with organic solvents, or by thiol reagents such as Cu^{2+}, N-ethylmaleimide or 4-hydroxymercuribenzoate, the effect of the thiol reagents can be reversed by thiols such as 1,4-dithioerythritol, in other animal tissues the enzyme exists almost entirely as EC 1.1.3.22 but can be converted into the dehydrogenase form by 1,4-dithioerythritol [8, 25, 28]

Storage stability
<12>, -75°C, more than 1 year with alternate freezing and thawing, stable [11]
<13>, -20°C, 27% loss of activity after 2 weeks, 51% loss of activity after 4 weeks, 89% loss of activity after 12 weeks, goat enzyme [25]
<13>, 4°C, 31% loss of activity after 6 days, 54% loss of activity after 12 days, 72% loss of activity after 16 days, goat enzyme [25]

References

[1] Maia, L.; Mira, L.: Xanthine oxidase and aldehyde oxidase: A simple procedure for the simultaneous purification from rat liver. Arch. Biochem. Biophys., **400**, 48-53 (2002)
[2] Kramer, S.P.; Johnson, J.L.; Ribeiro, A.A.; Millington, D.S.; Rajagopalan, K.V.: The structure of the molybdenum cofactor. Characterization of di-(carboxamidomethyl)molybdopterin from sulfite oxidase and xanthine oxidase. J. Biol. Chem., **262**, 16357-16363 (1987)
[3] Johnson, J.L.; Hainline, B.E.; Rajagopalan, K.V.: Characterization of the molybdenum cofactor of sulfite oxidase, xanthine, oxidase, and nitrate reductase. Identification of a pteridine as a structural component. J. Biol. Chem., **255**, 1783-1786 (1980)
[4] Eger, B.T.; Okamoto, K.; Enroth, C.; Sato, M.; Nishino, T.; Pai, E.F.; Nishino, T.: Purification, crystallization and preliminary X-ray diffraction studies of xanthine dehydrogenase and xanthine oxidase isolated from bovine milk. Acta Crystallogr. Sect. D, **56**, 1656-1658 (2000)
[5] Enroth, C.; Eger, B.T.; Okamoto, K.; Nishino, T.; Nishino, T.; Pai, E.F.: Crystal structures of bovine milk xanthine dehydrogenase and xanthine oxidase: Structure-based mechanism of conversion. Proc. Natl. Acad. Sci. USA, **97**, 10723-10728 (2000)
[6] McManaman, J.L.; Neville, M.C.; Wright, R.M.: Mouse mammary gland xanthine oxidoreductase: Purification, characterization, and regulation. Arch. Biochem. Biophys., **371**, 308-316 (1999)

[7] Waud, W.R.; Rajagopalan, K.V.: Purification and properties of the NAD^+-dependent (type D) and O_2-dependent (type O) forms of rat liver xanthine dehydrogenase. Arch. Biochem. Biophys., **172**, 354-364 (1976)

[8] Della Corte, E.; Stirpe, F.: The regulation of rat liver xanthine oxidase. Involvement of thiol groups in the conversion of the enzyme activity from dehydrogenase (type D) into oxidase (type O) and purification of the enzyme. Biochem. J., **126**, 739-745 (1972)

[9] Romao, M.J.; Archer, M.; Moura, I.; Moura, J.J.G.; Legall, J.; Engh, R.; Schneider, M.; Hof, P.; Huber, R.: Crystal structure of the xanthine oxidase-related aldehyde oxido-reductase from D. gigas. Science, **270**, 1170-1176 (1995)

[10] Taneja, R.; Taneja, V.: Xanthine oxidase in lentil (Lens esculenta) seedlings. Biochim. Biophys. Acta, **485**, 489-491 (1977)

[11] Woolfolk, C.A.; Downard, J.S.: Bacterial xanthine oxidase from Arthrobacter S-2. J. Bacteriol., **135**, 422-428 (1978)

[12] Nathans, G.R.; Hade, E.P.K.: Bovine milk xanthine oxidase: purification by ultrafiltration and conventional methods which omit addition of proteases: some criteria for homogeneity of native xanthine oxidase. Biochim. Biophys. Acta, **526**, 328-344 (1978)

[13] Machida, Y.; Nakanishi, T.: Purification and properties of xanthine oxidase from Enterobacter cloacae. Agric. Biol. Chem., **45**, 425-432 (1981)

[14] Bruder, G.; Heid, H.; Jarasch, E.D.; Keenan, T.W.; Mather, I.H.: Characteristics of membrane-bound and soluble forms of xanthine oxidase from milk and endothelial cells of capillaries. Biochim. Biophys. Acta, **701**, 357-369 (1982)

[15] Krenitsky, T.A.; Spector, T.; Hall, W.W.: Xanthine oxidase from human liver: purification and characterization. Arch. Biochem. Biophys., **247**, 108-119 (1986)

[16] Berglund, L.; Rasmussen, J.T.; Andersen, M.D.; Rasmussen, M.S.; Petersen, T.E.: Purification of the bovine xanthine oxidoreductase from milk fat globule membranes and cloning of complementary deoxyribonucleic acid. J. Dairy Sci., **79**, 198-204 (1996)

[17] Radi, R.; Rubbo, H.; Bush, K.; Freeman, B.A.: Xanthine oxidase binding to glycosaminoglycans: kinetics and superoxide dismutase interactions of immobilized xanthine oxidase-heparin complexes. Arch. Biochem. Biophys., **339**, 125-135 (1997)

[18] Özer, N.; Muftuoglu, M.; Ataman, D.; Ercan, A.; Ögus, I.H.: Simple, high-yield purification of xanthine oxidase from bovine milk. J. Biochem. Biophys. Methods, **39**, 153-159 (1999)

[19] Bray, R.C.: Molybdenum iron-sulfur flavin hydroxylases and related enzymes. The Enzymes, 3rd Ed. (Boyer, P.D., ed.), **12**, 299-412 (1975)

[20] Houston, M.; Esteveze, A.; Chumley, P.; Aslan, M.; Marklund, S.; Parks, D.A.; Freeman, B.A.: Binding of xanthine oxidase to vascular endothelium. J. Biol. Chem., **274**, 4985-4994 (1999)

[21] Moini, H.; Guo, Q.; Packer, L.: Enzyme inhibiton and protein-binding action of the procyanidin-rich french maritime pine bark extract, pycnogenol: effect on xanthine oxidase. J. Agric. Food Chem., **48**, 5630-5639 (2000)
[22] Pritsos, C.A.: Cellular distribution, metabolism and regulation of the xanthine oxidoreductase enzyme system. Chem. Biol. Interact., **129**, 195-208 (2000)
[23] Clynes, M.M.; Hurley, M.P.; Shannon, M.F.: Xanthine oxidase activity in mammalian cell cultures. Biochem. Soc. Trans., **7**, 72-74 (1979)
[24] Catigani, G.L.; Chytil, F.; Darby, W.J.: Purification and characterization of xanthine oxidase from livers of vitamin E deficient rabbits. Biochim. Biophys. Acta, **377**, 34-41 (1975)
[25] Zikakis, J.P.; Dressel, M.A.; Silver, M.R.: Bovine, caprine, and human milk xanthine oxidases: isolation, purification, and characterization. Instrum. Anal. Foods, Recent Prog. (Proc. Symp. Int. Flavor Conf., 3rd Ed., Charalambous, G., Inglett, G., eds.), **2**, 243-303 (1983)
[26] Avis, P.G.; Bergel, F.; Bray, R.C.: Cellular constituents. The chemistry of xanthine oxidase. Part I. The preparation of a crystalline xanthine oxidase from cow's milk. J. Chem. Soc., **2**, 1100-1105 (1955)
[27] Coughlan, M.P.; Rajagopalan, K.V.; Handler, P.: The role of molybdenum in xanthine oxidase and related enzymes. Reactivity with cyanide, arsenite, and methanol. J. Biol. Chem., **244**, 2658-2663 (1969)
[28] Stirpe, F.; Della Corte, E.: The regulation of rat liver xanthine oxidase: conversion of type D (dehydrogenase) into type O (oxidase) by a thermolabile factor, and reversibility by dithioerythritol. Biochim. Biophys. Acta, **212**, 195-197 (1970)
[29] Li, H.; Samouilov, A.; Liu, X.; Zweier, J.L.: characterization of the magnitude and kinetics of xanthine oxidase-catalyzed nitrite reduction. J. Biol. Chem., **276**, 24482-24489 (2001)
[30] Bray, R.C.: Xanthine oxidase. The Enzymes, 2nd Ed.(Boyer, P.D., Lardy, H., Myrbäck, K., eds.), **7**, 533-556 (1963)
[31] Batelli, M.G.; Lorenzoni, E.; Stirpe, F.: Milk xanthine oxidase type D (dehydrogenase) and type O (oxidase). Purification, interconversion and some properties. Biochem. J., **131**, 191-198 (1973)
[32] Hayden, T.J.; Ryan, J.P.; Duke, E.J.: Comparative aspects of xanthine-oxygen oxidoreductase systems. Biochem. Soc. Trans., **1**, 247-250 (1973)
[33] De Renzo, E.C.: Chemistry and biochemistry of xanthine oxidase. Adv. Enzymol. Relat. Subj. Biochem., **17**, 293-328 (1956)
[34] Tramper, J.: Preparation of immobilized milk xanthine oxidase and application in organic synthesis. Methods Enzymol., **136**, 254-262 (1987)
[35] Waud, W.R.; Brady, F.O.; Wiley, R.D.; Rajagopalan, K. V.: A new purification procedure for bovine milk xanthine oxidase: effect of proteolysis on the subunit structure. Arch. Biochem. Biophys., **169**, 695-701 (1975)
[36] Van Spanning, R.J.M.; Wansell-Bettenhaussen, C.W.; Oltmann, L.F.; Stouthamer, A.H.: Extraction and purification of molybdenum cofactor from milk xanthine oxidase. Eur. J. Biochem., **169**, 349-352 (1987)
[37] Cabre, F.; Canela, E.I.: Purification properties and functional groups of bovine liver xanthine oxidase. Biochem. Soc. Trans., **15**, 511-512 (1987)

[38] Morpeth, F.F.; Bray, R.C.: Mechanism-based inactivation of mitochondrial monoamine oxidase by N-(1-methylcyclopropyl)benzylamine. Biochemistry, **23**, 1322-1338 (1984)
[39] Morpeth, F.F.: Studies on the specificity toward aldehyde substrates and steady-state kinetics of xanthine oxidase. Biochim. Biophys. Acta, **744**, 328-334 (1983)
[40] Briley, M.S.; Eisenthal, R.: Association of xanthine oxidase with the bovine milk-fat-globule membrane. Nature of the enzyme-membrane association. Biochem. J., **147**, 417-423 (1975)
[41] Carpani, G.; Racchi, M.; Ghezzi, P.; Terao, M.; Garattini, E.: Purification and characterization of mouse liver xanthine oxidase. Arch. Biochem. Biophys., **279**, 237-241 (1990)
[42] Panoutsopoulos, G.I.; Beedham, C.: Kinetics and specificity of guinea pig liver aldehyde oxidase and bovine milk xanthine oxidase towards substituted benzaldehydes. Acta Biochim. Pol., **51**, 649-663 (2004)
[43] Frederiks, W.M.; Vreeling-Sindelarova, H.: Ultrastructural localization of xanthine oxidoreductase activity in isolated rat liver cells. Acta Histochem., **104**, 29-37 (2002)
[44] Mohamadin, A.M.; Abdel-Naim, A.B.: In vitro activation of dibromoacetonitrile to cyanide: role of xanthine oxidase. Arch. Toxicol., **77**, 86-93 (2003)
[45] Li, H.; Samouilov, A.; Liu, X.; Zweier, J.L.: Characterization of the magnitude and kinetics of xanthine oxidase-catalyzed nitrate reduction: evaluation of its role in nitrite and nitric oxide generation in anoxic tissues. Biochemistry, **42**, 1150-1159 (2003)
[46] Tai, L.A.; Hwang, K.C.: Cooperative catalysis in the homodimer subunits of xanthine oxidase. Biochemistry, **43**, 4869-4876 (2004)
[47] Masuoka, N.; Kubo, I.: Characterization of xanthine oxidase inhibition by anacardic acids. Biochim. Biophys. Acta, **1688**, 245-249 (2004)
[48] Lovstad, R.A.: A kinetic study on iron stimulation of the xanthine oxidase dependent oxidation of ascorbate. BioMetals, **16**, 435-439 (2003)
[49] Millar, T.M.: Peroxynitrite formation from the simultaneous reduction of nitrite and oxygen by xanthine oxidase. FEBS Lett., **562**, 129-133 (2004)
[50] Li, H.; Samouilov, A.; Liu, X.; Zweier, J.L.: Characterization of the effects of oxygen on xanthine oxidase-mediated nitric oxide formation. J. Biol. Chem., **279**, 16939-16946 (2004)
[51] Li, H.; Cui, H.; Liu, X.; Zweier, J.L.: Xanthine oxidase catalyzes anaerobic transformation of organic nitrates to nitric oxide and nitrosothiols: characterization of this mechanism and the link between organic nitrate and guanylyl cyclase activation. J. Biol. Chem., **280**, 16594-16600 (2005)
[52] Choi, E.-Y.; Stockert, A.L.; Leimkuhler, S.; Hille, R.: Studies on the mechanism of action of xanthine oxidase. J. Inorg. Biochem., **98**, 841-848 (2004)
[53] Deng, T.-S.; Roenneberg, T.: The flavo-enzyme xanthine oxidase is under circadian control in the marine alga Gonyaulax. Naturwissenschaften, **89**, 171-175 (2002)
[54] Gongora, L.; Manez, S.; Giner, R.M.; Recio, M.d.C.; Schinella, G.; Rios, J.L.: Inhibition of xanthine oxidase by phenolic conjugates of methylated quinic acid. Planta Med., **69**, 396-401 (2003)

[55] Takahashi, M.; Ushijima, T.; Ozaki, Y.: Changes in hepatic superoxide dismutase and xanthine oxidase activity in mice infected with Salmonella typhimurium and Pseudomonas aeruginosa. J. Med. Microbiol., **26**, 281-284 (1988)

[56] Casero, E.; de Quesada, A.M.; Jin, J.; Quintana, M.C.; Pariente, F.; Abruna, H.D.; Vazquez, L.; Lorenzo, E.: Comprehensive study of bioanalytical platforms: xanthine oxidase. Anal. Chem., **78**, 530-537 (2006)

[57] Galbusera, C.; Orth, P.; Fedida, D.; Spector, T.: Superoxide radical production by allopurinol and xanthine oxidase. Biochem. Pharmacol., **71**, 1747-1752 (2006)

[58] Griguer, C.E.; Oliva, C.R.; Kelley, E.E.; Giles, G.I.; Lancaster, J.R.; Gillespie, G.Y.: Xanthine oxidase-dependent regulation of hypoxia-inducible factor in cancer cells. Cancer Res., **66**, 2257-2263 (2006)

[59] Noro, T.; Oda, Y.; Miyase, T.; Ueno, A.; Fukushima, S.: Inhibitors of xanthine oxidase from the flowers and buds of Daphne genkwa. Chem. Pharm. Bull., **31**, 3984-3987 (1983)

[60] Veskoukis, A.S.; Kouretas, D.; Panoutsopoulos, G.I.: Substrate specificity of guinea pig liver aldehyde oxidase and bovine milk xanthine oxidase for methyl- and nitrobenzaldehydes. Eur. J. Drug Metab. Pharmacokin., **31**, 11-16 (2005)

[61] Rus, D.A.; Sastre, J.; Vina, J.; Pallardo, F.V.: Induction of mitochondrial xanthine oxidase activity during apoptosis in the rat mammary gland. Front. Biosci., **12**, 1184-1189 (2007)

[62] Dew, T.P.; Day, A.J.; Morgan, M.R.: Xanthine oxidase activity in vitro: effects of food extracts and components. J. Agric. Food Chem., **53**, 6510-6515 (2005)

[63] Asai, R.; Nishino, T.; Matsumura, T.; Okamoto, K.; Igarashi, K.; Pai, E.F.; Nishino, T.: Two mutations convert mammalian xanthine oxidoreductase to highly superoxide-productive xanthine oxidase. J. Biochem., **141**, 525-534 (2007)

[64] Tamta, H.; Thilagavathi, R.; Chakraborti, A.K.; Mukhopadhyay, A.K.: 6-(N-benzoylamino)purine as a novel and potent inhibitor of xanthine oxidase: inhibition mechanism and molecular modeling studies. J. Enzyme Inhib. Med. Chem., **20**, 317-324 (2005)

[65] Zhao, X.; Zhu, J.X.; Mo, S.F.; Pan, Y.; Kong, L.D.: Effects of cassia oil on serum and hepatic uric acid levels in oxonate-induced mice and xanthine dehydrogenase and xanthine oxidase activities in mouse liver. J. Ethnopharmacol., **103**, 357-365 (2006)

[66] Takano, Y.; Hase-Aoki, K.; Horiuchi, H.; Zhao, L.; Kasahara, Y.; Kondo, S.; Becker, M.A.: Selectivity of febuxostat, a novel non-purine inhibitor of xanthine oxidase/xanthine dehydrogenase. Life Sci., **76**, 1835-1847 (2005)

[67] Akcam, M.; Elmas, O.; Yilmaz, A.; Caglar, S.; Artan, R.; Gelen, T.; Aliciguezel, Y.: Myeloperoxidase, xanthine oxidase and superoxide dismutase in the gastric mucosa of Helicobacter pylori positive and negative pediatric patients. Mol. Cell. Biochem., **290**, 125-130 (2006)

[68] Masuoka, N.; Nihei, K.; Kubo, I.: Xanthine oxidase inhibitory activity of alkyl gallates. Mol. Nutr. Food Res., **50**, 725-731 (2006)

[69] Banach, K.; Bojarska, E.; Kazimierczuk, Z.; Magnowska, L.; Bzowska, A.: Kinetic model of oxidation catalyzed by xanthine oxidase-the final enzyme in degradation of purine nucleosides and nucleotides. Nucleosides Nucleotides Nucleic Acids, **24**, 465-469 (2005)

[70] Nguyen, M.T.; Awale, S.; Tezuka, Y.; Ueda, J.Y.; Tran, Q.; Kadota, S.: Xanthine oxidase inhibitors from the flowers of Chrysanthemum sinense. Planta Med., **72**, 46-51 (2006)

6-Hydroxynicotinate dehydrogenase 1.17.3.3

1 Nomenclature

EC number
1.17.3.3

Systematic name
6-hydroxynicotinate:O_2 oxidoreductase

Recommended name
6-hydroxynicotinate dehydrogenase

Synonyms
6-HDH <2> [3]
6-hydroxynicotinate hydroxylase <2> [3]
6-hydroxynicotinic acid dehydrogenase <2> [3]
6-hydroxynicotinic acid hydroxylase <2> [3]
dehydrogenase, 6-hydroxynicotinate <2> [3]

CAS registry number
122191-32-6

2 Source Organism

<1> *Bacillus sp.* (no sequence specified) [2]
<2> *Bacillus niacini* (no sequence specified) [1, 3]
<3> *Eubacterium barkeri* (no sequence specified) [4]

3 Reaction and Specificity

Catalyzed reaction
6-hydroxynicotinate + H_2O + O_2 = 2,6-dihydroxynicotinate + H_2O_2 (<2> contains [2Fe-2S] iron-sulfur centres, FAD and molybdenum, it also has a catalytically essential, labile selenium that can be removed by reaction with cyanide, in Bacillus niacini, this enzyme is required for growth on nicotinic acid [3])

Reaction type
oxidation
redox reaction
reduction

Natural substrates and products

S Additional information <2> (<2> acts together with nicotinate dehydrogenase in nicotinate catabolism [3]; <2> catalyzes the second step in the degradation of nicotinate [1]) (Reversibility: ?) [1, 3]

P ? <2> [1, 3]

Substrates and products

S 6-hydroxynicotinate + H_2O + 2,6-dichlorophenol indophenol <2> (<2> highly specific for 6-hydroxynicotinate, transfers electrons only to artificial acceptors of high redox potential, the mixture of phenazine ethosulfate and 2,6-dichlorophenol indophenol is a good electron acceptor [3]) (Reversibility: ?) [3]

P 2,6-dihydroxynicotinate + H_2O_2 + reduced 2,6-dichlorophenol indophenol <2> [3]

S 6-hydroxynicotinate + H_2O + O_2 <3> (Reversibility: ?) [4]

P 1,4,5,6-tetrahydro-6-oxonicotinate + H_2O_2

S 6-hydroxynicotinate + H_2O + O_2 <1, 2> (<2> highly specific for 6-hydroxynicotinate [3]) (Reversibility: ?) [1, 2, 3]

P 2,6-dihydroxynicotinate + H_2O_2 <2> [1, 3]

S 6-hydroxynicotinate + H_2O + methylene blue <2> (<2> highly specific for 6-hydroxynicotinate, transfers electrons only to artificial acceptors of high redox potential, 40% of the activity with the mixture of phenazine ethosulfate and 2,6-dichlorophenol indophenol [3]) (Reversibility: ?) [3]

P 2,6-dihydroxynicotinate + H_2O_2 + reduced methylene blue <2> [3]

S 6-hydroxynicotinate + H_2O + nitroblue tetrazolium chloride <2> (<2> highly specific for 6-hydroxynicotinate, transfers electrons only to artificial acceptors of high redox potential, same activity as with the mixture of phenazine ethosulfate and 2,6-dichlorophenol indophenol [3]) (Reversibility: ?) [3]

P 2,6-dihydroxynicotinate + H_2O_2 + reduced nitroblue tetrazolium chloride <2> [3]

S 6-hydroxynicotinate + H_2O + phenazine ethosulfate <2> (<2> highly specific for 6-hydroxynicotinate, transfers electrons only to artificial acceptors of high redox potential, the mixture of phenazine ethosulfate and 2,6-dichlorophenol indophenol is a good electron acceptor [3]) (Reversibility: ?) [3]

P 2,6-dihydroxynicotinate + H_2O_2 + reduced phenazine ethosulfate <2> [3]

S 6-hydroxynicotinate + H_2O + thiazolyl blue <2> (<2> highly specific for 6-hydroxynicotinate, transfers electrons only to artificial acceptors of high redox potential, same activity as with the mixture of phenazine ethosulfate and 2,6-dichlorophenol indophenol [3]) (Reversibility: ?) [3]

P 2,6-dihydroxynicotinate + H_2O_2 + reduced thiazolyl blue <2> [3]

S 6-hydroxynicotinate + H_2O + thionin <2> (<2> highly specific for 6-hydroxynicotinate, transfers electrons only to artificial acceptors of high redox potential, 2% of the activity with the mixture of phenazine ethosulfate and 2,6-dichlorophenol indophenol [3]) (Reversibility: ?) [3]

P 2,6-dihydroxynicotinate + H_2O_2 + reduced thionin <2> [3]

S Additional information <2> (<2> no electron acceptors: Brilliant cresyl blue, cytochrome c, FAD, ferricyanide, FMN, menadione, NAD^+, $NADP^+$, and both viologens [3]; <2> acts together with nicotinate dehydrogenase in nicotinate catabolism [3]; <2> catalyzes the second step in the degradation of nicotinate [1]) (Reversibility: ?) [1, 3]

P ? <2> [1, 3]

Inhibitors

glycerol <2> (<2> at a final concentration 10% (v/v), inactivates to 30% of the initial activity [3]) [3]
KCN <2> [3]
Na_2S <2> (<2> partially purified 6-HDH, 90% loss of activity [3]) [3]
Additional information <2> (<2> not inhibited by NaCl [3]) [3]

Cofactors/prosthetic groups

bactopterin <2> (<2> component of the molybdenum cofactor [1]) [1]
FAD <2> (<2> contains FAD as cofactor [1]; <2> contains FAD [3]) [1,3]
flavin <3> (<3> covalently bound flavin [4]) [4]
molybdenum cofactor <2> (<2> contains molybdenum [3]; <2> with bactopterin as component [1]) [1,3]
iron-sulfur centre <2> (<2> contains iron-sulfur centres, iron and acid-labile sulfur [3]) [3]

Metals, ions

iron <2, 3> (<2> contains iron, iron-sulfur centres [3]; <3> enzyme contains one [2Fe-2S] and two [4Fe-4S] clusters [4]) [3, 4]
sulfur <2> (<2> contains acid-labile sulfur, iron-sulfur centres [3]) [3]

Specific activity (U/mg)

0.7 <2> (<2> pH 7.5, 30°C [3]) [3]
350 <3> [4]

K_m-Value (mM)

0.8 <2> (6-hydroxynicotinate, <2> pH 7.5, 30°C [3]) [3]

pH-Optimum

7.5 <2> [3]

Temperature optimum (°C)

30 <2> (<2> assay at [3]) [3]

4 Enzyme Structure

Molecular weight

100000-140000 <2> (<2> non-denaturing PAGE, gel filtration [3]) [3]

Subunits

? <2, 3> (<3> x * 53000, SDS-PAGE [4]; <2> x * 85000 + x * 34000 + x * 15000, SDS-PAGE [3]) [3, 4]

5 Isolation/Preparation/Mutation/Application

Localization

soluble <2, 3> [3, 4]

Purification

<2> [3]
<2> (free of nicotinate dehydrogenase) [1]
<3> (purification under strictly unaerobic conditions) [4]

6 Stability

Oxidation stability

<3>, activity is lost with a half-life of 90 min in air-saturated solutions [4]

General stability information

<2>, significant loss of activity during the purification procedure, lyophilization results in a complete loss of activity [3]

References

[1] Nagel, M.; Koenig, K.; Andreesen, J.R.: Bactopterin as component of eubacterial dehydrogenases involved in hydroxylation reactions initiating the degradation of nicotine, nicotinate, and 2-furancarboxylate. FEMS Microbiol. Lett., **60**, 323-326 (1989)
[2] Nagel, M.; Andreesen, J.R.: Molybdenum-dependent degradation of nicotinic acid by Bacillus sp. DSM 2923. FEMS Microbiol. Lett., **59**, 147-152 (1989)
[3] Nagel, M.; Andreesen, J.R.: Purification and characterization of the molybdoenzymes nicotinate dehydrogenase and 6-hydroxynicotinate dehydrogenase from Bacillus niacini. Arch. Microbiol., **154**, 605-613 (1990)
[4] Alhapel, A.; Darley, D.J.; Wagener, N.; Eckel, E.; Elsner, N.; Pierik, A.J.: Molecular and functional analysis of nicotinate catabolism in Eubacterium barkeri. Proc. Natl. Acad. Sci. USA, **103**, 12341-12346 (2006)

Phenylacetyl-CoA dehydrogenase 1.17.5.1

1 Nomenclature

EC number
1.17.5.1

Systematic name
phenylacetyl-CoA:quinone oxidoreductase

Recommended name
phenylacetyl-CoA dehydrogenase

Synonyms
oxidase, phenylacetyl coenzyme A (α-oxidizing) <1> [2]
phenylacetyl-CoA α-carbon-oxidizing oxidase <1> [2]
phenylacetyl-CoA-acceptor oxidoreductase <1> [2]
phenylacetyl-CoA:acceptor oxidoreductase <1> [2]
Additional information <1> (<1> catalyzes the first step in the conversion of phenylacetyl-CoA to phenylglyoxylate, the second step being carried out by EC 3.1.2.25 [2]) [2]

CAS registry number
210756-43-7 (cf. EC 3.1.2.25)

2 Source Organism

<1> *Thauera aromatica* (no sequence specified) [1, 2]

3 Reaction and Specificity

Catalyzed reaction
phenylacetyl-CoA + H_2O + 2 quinone = phenylglyoxylyl-CoA + 2 quinol (<1> the enzyme from Thauera aromatica is a membrane-bound molybdenum-iron-sulfur protein, the enzyme is specific for phenylacetyl-CoA as substrate, phenylacetate, acetyl-CoA, benzoyl-CoA, propionyl-CoA, crotonyl-CoA, succinyl-CoA and 3-hydroxybenzoyl-CoA cannot act as substrates, the oxygen atom introduced into the product, phenylglyoxylyl-CoA, is derived from water and not molecular oxygen, duroquinone, menaquinone and 2,6-dichlorophenolindophenol(DCPIP) can act as acceptor, but the likely physiological acceptor is ubiquinone, a second enzyme, EC 3.1.2.25, phenylglyoxy-

lyl-CoA hydrolase, converts the phenylglyoxylyl-CoA formed into phenylglyoxylate [2])

Reaction type

oxidation (<1> α-oxidation, anaerobic [2])
redox reaction
reduction

Natural substrates and products

S phenylacetyl-CoA + H_2O + ubiquinone <1> (<1> ubiquinone is most likely the natural electron acceptor, involved in anaerobic metabolism of phenylalanine, catalyzes the first step in the conversion of phenylacetyl-CoA to phenylglyoxylate, the second step being carried out by EC 3.1.2.25 [1]) (Reversibility: ?) [1]
P phenylglyoxylyl-CoA + ubiquinol <1> [1]
S Additional information <1> (<1> anaerobic metabolism of phenylacetate, catalyzes the first step in the conversion of phenylacetyl-CoA to phenylglyoxylate, the second step being carried out by EC 3.1.2.25, cytochrome c probably does not act as physiological electron acceptor [2]) (Reversibility: ?) [2]
P ? <1> [2]

Substrates and products

S phenylacetyl-CoA + H_2O + 2,6-dichlorophenolindophenol <1> (<1> four-electron oxidation, uses dichlorophenolindophenol as artificial electron acceptor, 2 mol dichlorophenolindophenol is reduced per mol phenylacetyl-CoA, catalyzes the anaerobic oxidation of the methylene group in the α-position to the CoA-activated carboxyl group, molecular oxygen is not required for the oxidation [2]; <1> uses 2,6-dichlorophenolindophenol as artificial electron acceptor [1]) (Reversibility: ?) [1, 2]
P phenylglyoxylyl-CoA + reduced 2,6-dichlorophenolindophenol <1> [1, 2]
S phenylacetyl-CoA + H_2O + duroquinone <1> (<1> at 0.3 mM duroquinone, 150% of the activity with 0.25 mM 2,6-dichlorophenolindophenol [1]) (Reversibility: ?) [1]
P phenylglyoxylyl-CoA + duroquinol <1> [1]
S phenylacetyl-CoA + H_2O + menadione <1> (<1> at 0.3 mM menadione, 60% of the activity with 0.25 mM 2,6-dichlorophenolindophenol [1]) (Reversibility: ?) [1]
P phenylglyoxylyl-CoA + menadiol <1> [1]
S phenylacetyl-CoA + H_2O + quinone <1> (<1> four-electron oxidation, catalyzes an α-oxidation without utilizing molecular oxygen, ubiquinone is most likely the natural electron acceptor, specific for phenylacetyl-CoA, enzyme preparation catalyzes the reaction phenylacetyl-CoA + 2 H_2O + 2 quinone = phenylglyoxylate + 2 quinone H_2 + CoASH with phenylglyoxylyl-CoA as free intermediate, which is subsequently hydrolyzed [1]) (Reversibility: ?) [1, 2]
P phenylglyoxylyl-CoA + quinol <1> [1]

S phenylacetyl-CoA + H_2O + ubiquinone <1> (<1> ubiquinone is most likely the natural electron acceptor, involved in anaerobic metabolism of phenylalanine, catalyzes the first step in the conversion of phenylacetyl-CoA to phenylglyoxylate, the second step being carried out by EC 3.1.2.25 [1]) (Reversibility: ?) [1]
P phenylglyoxylyl-CoA + ubiquinol <1> [1]
S Additional information <1> (<1> no substrate: phenylacetate, inactive electron acceptors: NAD^+, $NADP^+$, FAD, FMN, triphenyltetrazoliumchloride [2]; <1> no substrates: phenylacetate, acetyl-CoA, benzoyl-CoA, propionyl-CoA, crotonyl-CoA, succinyl-CoA, 3-hydroxybenzoyl-CoA [1]; <1> anaerobic metabolism of phenylacetate, catalyzes the first step in the conversion of phenylacetyl-CoA to phenylglyoxylate, the second step being carried out by EC 3.1.2.25, cytochrome c probably does not act as physiological electron acceptor [2]) (Reversibility: ?) [1, 2]
P ? <1> [1, 2]

Inhibitors

CN^- <1> (<1> irreversible inactivation at low concentrations, faster inactivation at more alkaline pH [2]; <1> reversible inactivation by a low concentration of cyanide [1]) [1, 2]
Additional information <1> (<1> low enzyme activity in cells grown with phenylglyoxylate, no activity in cells grown with benzoate and nitrate or after aerobic growth with phenylacetate, not inhibited by sodium salts of azide, thiocyanate, nitrate, nitrite, fluoride, cyanate, bipyridine, hydroxylamine, hydroxyurea, each at 10 mM [2]) [2]

Cofactors/prosthetic groups

molybdenum cofactor <1> (<1> molybdenum-iron-sulfur enzyme, contains 0.66 mol Mo per mol of native enzyme [1]) [1]
molybdopterin <1> (<1> presence of a molybdopterin cofactor, contains 0.66 mol Mo per mol of native enzyme [1]) [1]
iron-sulfur centre <1> (<1> molybdenum-iron-sulfur enzyme, contains 30 mol Fe and 25 mol acid-labile sulfur per mol of native enzyme [1]) [1]
Additional information <1> (<1> catalyzes an α-oxidation without utilizing molecular oxygen [1]; <1> no requirement of molecular oxygen for phenylacetyl-CoA oxidation [2]) [1,2]

Activating compounds

Additional information <1> (<1> enzyme activity is induced under denitrifying conditions with phenylalanine and phenylacetate [2]) [2]

Turnover number (min^{-1})

12 <1> (phenylacetyl-CoA, <1> pH 7.5, 30°C, duroquinone as electron acceptor [1]) [1]

Specific activity (U/mg)

0.03 <1> (<1> pH 7.5, 30°C [2]) [2]
0.18 <1> (<1> pH 7.5, 30°C, 2,6-dichlorophenolindophenol as electron acceptor [1]) [1]

K_m-Value (mM)
0.02 <1> (phenylacetyl-CoA, <1> pH 7.5, 30°C, 2,6-dichlorophenolindophenol as electron acceptor [1]) [1]
Additional information <1> [2]

pH-Optimum
7 <1> (<1> 50 mM potassium phosphate buffer [2]; <1> 2,6-dichlorophenolindophenol as electron acceptor [1]) [1, 2]

pH-Range
6.5-7.5 <1> (<1> with 50 mM potassium phosphate buffer: 60% of optimum activity at pH 6.5, 87% of optimum activity at pH 7.5, with 50 mM Tris-HCl buffer: 60% of optimum activity at pH 7.5 [2]) [2]

Temperature optimum (°C)
30 <1> (<1> assay at [1,2]) [1, 2]

4 Enzyme Structure

Molecular weight
280000 <1> (<1> gel filtration [1]) [1]

Subunits
hexamer <1> (<1> $\alpha_2\beta_2\gamma_2$, 2 * 93000 + 2 * 27000 + 2 * 26000, SDS-PAGE [1]) [1]

5 Isolation/Preparation/Mutation/Application

Localization
membrane <1> (<1> bound [1,2]) [1, 2]

Purification
<1> (12fold, partial) [1]

6 Stability

Oxidation stability
<1>, oxygen-insensitive, 73% of the initial activity is recovered after 3 h incubation in air on ice [2]
<1>, remarkably stable with respect to oxygen [1]

Storage stability
<1>, -20°C, 20% (w/v) glycerol, 0.1% (w/v) lauryldimethylamine oxide, partially purified enzyme, several weeks, stable [1]

References

[1] Rhee, S.K.; Fuchs, G.: Phenylacetyl-CoA:acceptor oxidoreductase, a membrane-bound molybdenum-iron-sulfur enzyme involved in anaerobic metabolism of phenylalanine in the denitrifying bacterium Thauera aromatica. Eur. J. Biochem., **262**, 507-515 (1999)
[2] Schneider, S.; Fuchs, G.: Phenylacetyl-CoA:acceptor oxidoreductase, a new a-oxidizing enzyme that produces phenylglyoxylate. Assay, membrane localization, and differential production in Thauera aromatica. Arch. Microbiol., **169**, 509-516 (1998)

$3\alpha,7\alpha,12\alpha$-Trihydroxy-5β-cholestanoyl-CoA 24-hydroxylase 1.17.99.3

1 Nomenclature

EC number
1.17.99.3

Systematic name
(25R)-$3\alpha,7\alpha,12\alpha$-trihydroxy-5β-cholestan-26-oyl-CoA:acceptor 24-oxidoreductase (24R-hydroxylating)

Recommended name
$3\alpha,7\alpha,12\alpha$-trihydroxy-5β-cholestanoyl-CoA 24-hydroxylase

Synonyms
$3\alpha,7\alpha,12\alpha$-trihydroxy-5β-cholestan-26-oate 24-hydroxylase <4> [4]
$3\alpha,7\alpha,12\alpha$-trihydroxy-5β-cholestanoyl-CoA oxidase <3, 4> [2, 4]
THC-CoA oxidase <1, 2, 4> [1, 4, 11]
THCA-CoA oxidase <2, 4> [4, 12]
THCCox <2> [3]
Trihydroxycoprostanoyl-CoA oxidase <1, 2, 4> [3, 4, 6, 9, 11]
$\alpha,7\alpha,12\alpha$-trihydroxy-5β-cholestanoyl-CoA oxidase <5> [10]

CAS registry number
119799-47-2
152787-68-3

2 Source Organism

<1> *Homo sapiens* (no sequence specified) [11]
<2> *Rattus norvegicus* (no sequence specified) [1, 3, 5, 6, 7, 8, 9, 12]
<3> *Oryctolagus cuniculus* (no sequence specified) [2]
<4> *Rattus norvegicus* (UNIPROT accession number: P97562) [4]
<5> *Oryctolagus cuniculus* (UNIPROT accession number: O02767) [10]

3 Reaction and Specificity

Catalyzed reaction
(25R)-$3\alpha,7\alpha,12\alpha$-trihydroxy-5β-cholestan-26-oyl-CoA + H_2O + acceptor = (24R,25R)-$3\alpha,7\alpha,12\alpha$,24-tetrahydroxy-5β-cholestan-26-oyl-CoA + reduced acceptor

Reaction type

oxidation
redox reaction
reduction

Natural substrates and products

- **S** Additional information <2> (<2> noninducible enzyme [1]) (Reversibility: ?) [1]
- **P** ?

Substrates and products

- **S** (25S)-$3\alpha,7\alpha,12\alpha$-trihydroxy-5β-cholestan-26-oyl-CoA + H_2O + acceptor <2> (Reversibility: ?) [6]
- **P** $3\alpha,7\alpha,12\alpha,24$-tetrahydroxy-$5\beta$-cholestan-26-oyl-CoA + reduced acceptor
- **S** (25S)-$3\alpha,7\alpha,12\alpha$-trihydroxy-5β-cholestan-26-oyl-CoA + H_2O + acceptor <2> (<2> the 25S epimer is a preferential substrate [12]) (Reversibility: ?) [12]
- **P** (24R)-$3\alpha,7\alpha,12\alpha,24$-tetrahydroxy-$5\beta$-cholestan-26-oyl-CoA + reduced acceptor
- **S** (25S)-$3\alpha,7\alpha,12\alpha$-trihydroxy-5β-cholestanoyl-CoA + H_2O + acceptor <3> (<3> no activity with (25R)-$3\alpha,7\alpha,12\alpha$-trihydroxy-5β-cholestanoyl-CoA [2]) (Reversibility: ?) [2]
- **P** (24E)-$3\alpha,7\alpha,12\alpha$-trihydroxy-5β-cholest-24-en-27-oic acid + reduced acceptor
- **S** 2-methylhexanoyl-CoA + H_2O + acceptor <2> (Reversibility: ?) [3]
- **P** 3-hydroxy-2-methylhexanoyl-CoA + reduced acceptor
- **S** 2-methylpalmitoyl-CoA + H_2O + acceptor <2> (Reversibility: ?) [3]
- **P** 3-hydroxy-2-methylhexandecaoyl-CoA + reduced acceptor
- **S** $3\alpha,7\alpha,12\alpha$-trihydroxy-5β-cholestan-26-oyl-CoA + H_2O + acceptor <2> (<2> conversion of the coenzyme A ester is lower than with the free acid as substrate [7]) (Reversibility: ?) [7]
- **P** $3\alpha,7\alpha,12\alpha,24$-tetrahydroxy-$5\beta$-cholestan-26-oyl-CoA + reduced acceptor
- **S** $3\alpha,7\alpha,12\alpha$-trihydroxy-5β-cholestanoic acid + H_2O + acceptor <2> (Reversibility: ?) [7]
- **P** $3\alpha,7\alpha,12\alpha,24$-tetrahydroxy-$5\beta$-cholestanoic acid + reduced acceptor
- **S** pristanoyl-CoA + H_2O + acceptor <2> (Reversibility: ?) [3]
- **P** 3-hydroxy-1,6,10,14-tetramethylpentadecanoyl-CoA + reduced acceptor
- **S** trihydroxycoprostanoyl-CoA + H_2O + acceptor <2> (Reversibility: ?) [3]
- **P** $3\alpha,7\alpha,12\alpha,24$-tetrahydroxycoprostanoyl-CoA + reduced acceptor
- **S** Additional information <2> (<2> noninducible enzyme [1]) (Reversibility: ?) [1]
- **P** ?

Inhibitors

(25R)-$3\alpha,7\alpha,12\alpha$-trihydroxy-5β-cholestan-26-oyl-CoA <2> [6]
palmitoyl-CoA <2> [8]

Additional information <2> (<2> oxygen may be obligatory since there is almost complete inhibition when the reaction is performed in an atmosphere consisting of nitrogen [7]) [7]

Cofactors/prosthetic groups

coenzyme A <2> (<2> marked inhibition of reaction with 3α,7α,12α-trihydroxy-5β-cholestanoic acid [7]) [7]
FAD <2> (<2> contains most likely 2 mol of loosely bound FAD per mol of enzyme [3]) [3]
FMN <2> (<2> slight [7]) [7]

Activating compounds

1,2-butanediol <2> (<2> activates [3]) [3]
1,2-propanediol <2> (<2> activates [3]) [3]
1,3-propanediol <2> (<2> activates [3]) [3]
2,3-butanediol <2> (<2> activates [3]) [3]
ATP <2> (<2> required [7]) [7]
albumin <2> (<2> required [3]) [3]
GTP <2> (<2> stimulates to a lesser degree than ATP [7]) [7]
ITP <2> (<2> stimulates to a lesser degree than ATP [7]) [7]

Specific activity (U/mg)

0.0453 <5> [10]
Additional information <2> [3]

pH-Optimum

7.5 <2> [7]

pH-Range

6.5-8.5 <2> (<2> pH 6.5: about 30% of maximal activity, pH 8.5: about 60% of maximal activity [7]) [7]

4 Enzyme Structure

Molecular weight

120000 <2> (<2> gel filtration [3]) [3]
139000 <2> (<2> gel filtration [1]) [1]
175000 <2> (<2> non-denaturing PAGE [3]) [3]

Subunits

dimer <2> (<2> 2 * 69000, SDS-PAGE [1]; <2> 2 * 68800, SDS-PAGE [3]) [1, 3]

5 Isolation/Preparation/Mutation/Application

Source/tissue

liver <1, 2, 3, 5> [1, 2, 3, 5, 6, 7, 8, 9, 10, 11, 12]

Localization

mitochondrion <2> [7]
peroxisome <2, 3, 4> [1, 2, 3, 4, 5, 6, 9]

Purification

<2> [3]
<2> (partial) [1]
<5> (partial) [10]

Cloning

<4> [4]
<5> (expression in COS cells) [10]

6 Stability

Organic solvent stability

1,2-butanediol <2> (<2> 40-60 v/v, in presence of dithiothreitol, the enzyme is very stable [3]) [3]
1,2-propanediol <2> (<2> 40-60 v/v, in presence of dithiothreitol, the enzyme is very stable [3]) [3]
1,3-butanediol <2> (<2> 40-60 v/v, in presence of dithiothreitol, the enzyme is very stable [3]) [3]
1,3-propanediol <2> (<2> 40-60 v/v, in presence of dithiothreitol, the enzyme is very stable [3]) [3]
2,3-butanediol <2> (<2> 40-60 v/v, in presence of dithiothreitol, the enzyme is very stable [3]) [3]
ethylene glycol <2> (<2> 40-60 v/v, in presence of dithiothreitol, the enzyme is very stable [3]) [3]

General stability information

<2>, in presence of dithiothreitol, the enzyme is very stable in high concentrations of ethylene glycol,1,2-propanediol, 1,3-propanediol, 1,2-butanediol, 2,3-butanediol and 1,3-butanediol, 40-60 v/v [3]

References

[1] Schepers, L.; Van Veldhoven, P.P.; Casteels, M.; Eyssen, H.J.; Mannaerts, G.P.: Presence of three acyl-CoA oxidases in rat liver peroxisomes. An inducible fatty acyl-CoA oxidase, a noninducible fatty acyl-CoA oxidase, and a noninducible trihydroxycoprostanoyl-CoA oxidase. J. Biol. Chem., **265**, 5242-5246 (1990)

[2] Pedersen, J.I.; Veggan, T.; Bjorkhem, I.: Substrate stereospecificity in oxidation of (25S)-$3\alpha,7\alpha,12\alpha$-trihydroxy-5β-cholestanoyl-CoA by peroxisomal trihydroxy-5β-cholestanoyl-CoA oxidase. Biochem. Biophys. Res. Commun., **224**, 37-42 (1996)

[3] van Veldhoven, P.P.; van Rompuy, P.; Vanhooren, J.C.; Mannaerts, G.P.: Purification and further characterization of peroxisomal trihydroxycoprostanoyl-CoA oxidase from rat liver. Biochem. J., **304**, 195-200 (1994)
[4] Baumgart, E.; Vanhooren, J.C.; Fransen, M.; van Leuven, F.; Fahimi, H.D.; van Veldhoven, P.P.; Mannaerts, G.P.: Molecular cloning and further characterization of rat peroxisomal trihydroxycoprostanoyl-CoA oxidase. Biochem. J., **320**, 115-121 (1996)
[5] Dieuaide-Noubhani, M.; Novikov, D.; Baumgart, E.; Vanhooren, J.C.T.; Fransen, M.; Goethals, M.; Vanderkerckhove, J.; van Veldhoven, P.P.; Mannaerts, G.P.: Further characterization of the peroxisomal 3-hydroxyacyl-CoA dehydrogenases from rat liver. Relationship between the different dehydrogenases and evidence that fatty acids and the C27 bile acids di- and trihydroxycoprostanic acids are metabolized by separate multifunctional proteins. Eur. J. Biochem., **240**, 660-666 (1996)
[6] van Veldhoven, P.P.; Croes, K.; Asselberghs, S.; Herdewijn, P.; Mannaerts, G.P.: Peroxisomal β-oxidation of 2-methyl-branched acyl-CoA esters: stereospecific recognition of the 2S-methyl compounds by trihydroxycoprostanoyl-CoA oxidase and pristanoyl-CoA oxidase. FEBS Lett., **388**, 80-84 (1996)
[7] Gustafsson, J.: Biosynthesis of cholic acid in rat liver. 24-Hydroxylation of 3α,7α,12α-trihydroxy-5β-cholestanoic acid. J. Biol. Chem., **250**, 8243-8247 (1975)
[8] Casteels, M.; Schepers, L.; Van Eldere, J.; Eyssen, H.J.; Mannaerts, G.P.: Inhibition of 3α,7α,12α-trihydroxy-5β-cholestanoic acid oxidation and of bile acid secretion in rat liver by fatty acids. J. Biol. Chem., **263**, 4654-4661 (1988)
[9] Van Veldhoven, P.P.; Vanhove, G.; Assselberghs, S.; Eyssen, H.J.; Mannaerts, G.P.: Substrate specificities of rat liver peroxisomal acyl-CoA oxidases: palmitoyl-CoA oxidase (inducible acyl-CoA oxidase), pristanoyl-CoA oxidase (non-inducible acyl-CoA oxidase), and trihydroxycoprostanoyl-CoA oxidase. J. Biol. Chem., **267**, 20065-20074 (1992)
[10] Pedersen, J.I.; Eggertsen, G.; Hellman, U.; Andersson, U.; Bjorkhem, I.: Molecular cloning and expression of cDNA encoding 3α,7α,12α-trihydroxy-5β-cholestanoyl-CoA oxidase from rabbit liver. J. Biol. Chem., **272**, 18481-18489 (1997)
[11] Casteels, M.; Schepers, L.; Van Veldhoven, P.P.; Eyssen, H.J.; Mannaerts, G.P.: Separate peroxisomal oxidases for fatty acyl-CoAs and trihydroxycoprostanoyl-CoA in human liver. J. Lipid Res., **31**, 1865-1872 (1990)
[12] Ikegawa, S.; Goto, T.; Mano, N.; Goto, J.: Substrate specificity of THCA-CoA oxidases from rat liver light mitochondrial fractions on dehydrogenation of 3α,7α,12α-trihydroxy-5β-cholestanoic acid CoA thioester. Steroids, **63**, 603-607 (1998)

Uracil/thymine dehydrogenase 1.17.99.4

1 Nomenclature

EC number
1.17.99.4

Systematic name
uracil:acceptor oxidoreductase

Recommended name
uracil/thymine dehydrogenase

Synonyms
EC 1.1.99.19 <4> (<4> formerly [4]) [4]
EC 1.2.99.1 (formerly)
dehydrogenase, uracil
uracil dehydrogenase <4> [4]
uracil oxidase <4> [4]
uracil-thymidine oxidase <4> [4]

CAS registry number
9029-00-9

2 Source Organism

<1> *Corynebacterium sp.* (no sequence specified) [2]
<2> *Enterobacter aerogenes* (no sequence specified) [1]
<3> *Mycobacterium sp.* (no sequence specified) [2]
<4> *Rhodococcus erythropolis* (no sequence specified) [4]
<5> *Aerobic soil bacterium* (no sequence specified) [3]

3 Reaction and Specificity

Catalyzed reaction
thymine + H_2O + acceptor = 5-methylbarbiturate + reduced acceptor
uracil + H_2O + acceptor = barbiturate + reduced acceptor

Reaction type
oxidation
redox reaction
reduction

Natural substrates and products

S thymine + H_2O + aceptor <4> (<4> involved in oxidative pyrimidine metabolism [4]) (Reversibility: ?) [4]
P 5-methylbarbiturate + reduced acceptor
S uracil + H_2O + acceptor <4> (<4> involved in oxidative pyrimidine metabolism [4]) (Reversibility: ?) [4]
P barbiturate + reduced acceptor
S uracil + acceptor <2> (<2> reaction in two-step oxidative catabolism of uracil in prokaryotes [1]) (Reversibility: ?) [1]
P barbiturate + reduced acceptor <2> [1]

Substrates and products

S 5-aminouracil + acceptor <5> (Reversibility: ?) [3]
P ?
S thymine + H_2O + aceptor <4> (<4> involved in oxidative pyrimidine metabolism [4]) (Reversibility: ?) [4]
P 5-methylbarbiturate + reduced acceptor
S thymine + acceptor <1, 3, 5> (Reversibility: ?) [2, 3]
P 5-methylbarbiturate + reduced acceptor <1, 3, 5> [2, 3]
S uracil + H_2O + acceptor <4> (<4> involved in oxidative pyrimidine metabolism [4]) (Reversibility: ?) [4]
P barbiturate + reduced acceptor
S uracil + acceptor <1, 2, 3, 5> (<1, 2, 3, 5> barbiturate, isobarbiturate, 5-methylbarbiturate, 6-methyluracil, dihydrothymine, dihydrouracil, 2-thiouracil, 2-thiothymine, 2-thio-5-methyluracil and cytosine are no substrates [1,2,3]; <5> acceptor: 2,6-dichlorophenolindophenol or methylene blue act as artificial electron acceptors, O_2 only in the presence of methylene blue [3]; <2> reaction in two-step oxidative catabolism of uracil in prokaryotes [1]) (Reversibility: ?) [1, 2, 3]
P barbiturate + reduced acceptor <1, 2, 3> [1, 2]

Cofactors/prosthetic groups

2,6-dichlorophenolindophenol <2> (<2> required as electron acceptor, promotes barbiturate formation better than methylene blue [1]) [1]
Additional information <1, 3, 5> (<1,3,5> no oxidation in the presence of O_2 without methylene blue as electron acceptor [2,3]) [2, 3]

Activating compounds

methylene blue <1, 2, 3, 5> (<1, 2, 3, 5> required as electron acceptor [1,2,3]) [1, 2, 3]

Specific activity (U/mg)

0.0115 <2> [1]

K_m-Value (mM)

0.035 <3> (thymine) [2]
0.131 <3> (uracil) [2]

pH-Optimum

8.5 <3> [2]

pH-Range

8-9.7 <3> (<3> about half-maximal activity at pH 8.0 and 9.7 with uracil as substrate [2]) [2]
8-10 <3> (<3> about half-maximal activity at pH 8.0 and 10 with thymine as substrate [2]) [2]

Temperature optimum (°C)

22-25 <1, 3> (<1,3> assay at [2]) [2]
30 <2, 5> (<2,5> assay at [1,3]) [1, 3]

5 Isolation/Preparation/Mutation/Application

Localization

cytoplasm <1, 2, 3, 5> [1, 2, 3]

6 Stability

pH-Stability

7 <3> (<3> 90% loss of activity [2]) [2]
9 <3> (<3> most stable at -10°C, no appreciable loss of activity for several months [2]) [2]

General stability information

<4>, enzyme in cell-free extract is unstable, activity totally diappears within 3 days [4]

Storage stability

<3>, -10°C, most stable at pH 9.0, no appreciable loss of activity for several months [2]
<5>, -20°C, 80% loss of activity in crude extracts within two months [3]

References

[1] Patel, B.N.; West, T.P.: Oxidative catabolism of uracil by Enterobacter aerogenes. FEMS Microbiol. Lett., **40**, 33-36 (1987)
[2] Hayaishi, O.; Kornberg, A.: Metabolism of cytosine, thymine, uracil, and barbituric acid by bacterial enzymes. J. Biol. Chem., **197**, 717-732 (1952)
[3] Wang, T.P.; Lampen, J.O.: Uracil oxidase and the isolation of barbituric acid from uracil oxidation. J. Biol. Chem., **194**, 785-791 (1952)
[4] Soong, C.L.; Ogawa, J.; Shimizu, S.: Novel amidohydrolytic reactions in oxidative pyrimidine metabolism: analysis of the barbiturase reaction and discovery of a novel enzyme, ureidomalonase. Biochem. Biophys. Res. Commun., **286**, 222-226 (2001)

Bile-acid 7α-dehydroxylase 1.17.99.5

1 Nomenclature

EC number
1.17.99.5

Systematic name
deoxycholate:FAD oxidoreductase (7α-dehydroxylating)

Recommended name
bile-acid 7α-dehydroxylase

Synonyms
EC 1.17.1.6

CAS registry number
85130-33-2

2 Source Organism

<1> *Eubacterium sp.* (no sequence specified) [1, 2, 3]

3 Reaction and Specificity

Catalyzed reaction
deoxycholate + FAD + H_2O = cholate + $FADH_2$
lithocholate + FAD + H_2O = chenodeoxycholate + $FADH_2$

Substrates and products

S 3α-hydroxy-5β-6-cholen-24-oic acid + reduced acceptor <1> (Reversibility: ?) [2]
P lithocholic acid + ?
S chenodeoxycholate + reduced acceptor <1> (Reversibility: ?) [2]
P ?
S chenodeoxycholate + reduced acceptor <1> (Reversibility: ?) [3]
P lithocholate + ?
S cholate + reduced acceptor <1> (Reversibility: ?) [2]
P ?
S cholate + reduced acceptor <1> (<1> it is hypothesized that a highly polar bile acid metabolite is an intermediate in 7-dehydroxylation [1]) (Reversibility: ?) [1, 3]
P deoxycholate + ?

S ursodeoxycholate + reduced acceptor <1> (Reversibility: ?) [2]
P ?

Inhibitors

chenodeoxycholate <1> (<1> substrate inhibition above 0.05 mM [3]) [3]
NADH <1> (<1> 0.5 mM, 62% inhibition of 7α-dehydroxylation of cholic acid, 75% inhibition of 7α-dehydroxylation of deoxycholic acid [3]; <1> 0.5 mM, more than 50% inhibition when added to reaction mixtures containing NAD^+ (0.5 mM). Inhibitory effect of NADH is reduced (50%) by the presence of 0.1 M NaCl. NADH inhibition is consistent with negative cooperativity. 0.5 mM NADH inhibits reduction of Δ^6-intermediates to lithocholic acid 78% [2]) [2, 3]
NADPH <1> (<1> 0.5 mM, 62% inhibition of 7α-dehydroxylation of cholic acid, 42% inhibition of 7α-dehydroxylation of deoxycholic acid [3]) [3]

Cofactors/prosthetic groups

NAD^+ <1> (<1> specific activity increases 4-fold to 6-fold with either cholic acid or chenodeoxycholic acid as substrate in presence of NAD^+. NAD^+ stimulates 7α-dehydroxylation activity without any detectable reduction. K_m: 0.13 mM (rection with cholate), 0.06 mM (reaction with chenodexycholate) [3]) [3]

Activating compounds

$FADH_2$ <1> (<1> stimulates reduction when added to reaction mixtures containing NAD^+ [2]) [2]
$FMNH_2$ <1> (<1> stimulates reduction when added to reaction mixtures containing NAD^+ [2]) [2]
NADH <1> (<1> stimulates 30-50% at low concentrations (below 0.15 mM). 7-Dehydroxylase activity is modulated by the molar ratio of NAD^+ to NADH with maximal activity at a NAD^+ mole fraction of 0.75 to 0.85 [2]) [2]

K_m-Value (mM)

0.025 <1> (cholate, <1> pH 7.5, 37°C [3]) [3]

4 Enzyme Structure

Molecular weight

114000 <1> (<1> gel filtration [3]) [3]

References

[1] Coleman, J.P.; White, W.B.; Egestad, B.; Sjoevall, J.; Hylemon, P.B.: Biosynthesis of a novel bile acid nucleotide and mechanism of 7α-dehydroxylation by an intestinal Eubacterium species. J. Biol. Chem., **262**, 4701-4707 (1987)
[2] White, B.A.; Paone, D.A.; Cacciapuoti, A.F.; Fricke, R.J.; Mosbach, E.H.; Hylemon, P.B.: Regulation of bile acid 7-dehydroxylase activity by NAD^+ and

NADH in cell extracts of Eubacterium species V.P.I. 12708. J. Lipid Res., **24**, 20-27 (2003)

[3] White, B.A.; Cacciapuoti, A.F.; Fricke, R.J.; Whitehead, T.R.; Mosbach, E.H.; Hylemon, P.B.: Cofactor requirements for 7α-dehydroxylation of cholic and chenodeoxycholic acid in cell extracts of the intestinal anaerobic bacterium, Eubacterium species V.P.I. 12708. J. Lipid Res., **22**, 891-989 (1981)

Aureusidin synthase 1.21.3.6

1 Nomenclature

EC number
1.21.3.6

Systematic name
2',4,4',6'-tetrahydroxychalcone:oxygen oxidoreductase

Recommended name
aureusidin synthase

Synonyms
AmAS1 <1> [6]
synthase, aureusidin
Additional information [1, 2]

CAS registry number
320784-48-3

2 Source Organism

<1> *Antirrhinum majus* (no sequence specified) [1, 2, 3, 4, 5, 6]

3 Reaction and Specificity

Catalyzed reaction
2',3,4,4',6'-pentahydroxychalcone + O_2 = 2 aureusidin + 2 H_2O (<1> mechanism [2,3])
2',4,4',6'-tetrahydroxychalcone + O_2 = aureusidin + H_2O (<1> mechanism [2,3])

Reaction type
dehydrogenation
hydroxylation
oxidation
oxidative cyclization
oxygenation
redox reaction
reduction

Natural substrates and products

S 2',3,4,4',6'-pentahydroxychalcone + 0.5 O_2 <1> (Reversibility: ir) [1, 2, 3]
P aureusidin + bracteatin + H_2O <1> [1, 2, 3]
S 2',4,4',6'-tetrahydroxychalcone + O_2 <1> (<1> key intermediate in the general flonoid metabolism [3]) (Reversibility: ir) [1, 2, 3]
P aureusidin + H_2O <1> [1, 2, 3]
S Additional information <1> (<1> pathway, overview [1]; <1> chalcone-specific polyphenol oxidase specialized for aurone biosynthesis [2]; <1> enzyme plays a key role in the yellow coloration of flowers, enzyme is a homologue of plant polyphenol oxidase [1,2,3]) (Reversibility: ?) [1, 2, 3]
P ?

Substrates and products

S 2',3,4,4',6'-pentahydroxychalcone + O_2 <1> (<1> best substrate [1,3]; <1> 2210% activity compared to 2,4,4,6-tetrahydroxychalcone [1]; <1> 2-6fold higher activity than with 2,4,4,6-tetrahydroxychalcone [3]) (Reversibility: ir) [1, 2, 3]
P aureusidin + bracteatin + H_2O <1> (<1> products are formed in a 6:1 ratio [1,2,3]) [1, 2, 3]
S 2',3,4,4',6'-pentahydroxychalcone 4'-β-D-glucopyranoside + O_2 <1> (Reversibility: ir) [1, 2, 3]
P aureusidin 6-β-D-glucopyranoside + bracteatin 6-β-D-glucopyranoside + H_2O <1> (<1> products are formed in a 5:1 ratio [2]; <1> products provide the yellow colour of flowers [3]) [2, 3]
S 2',4,4',6'-tetrahydroxychalcone + O_2 <1> (<1> no formation of 2-(α-hydroxybenzyl)coumaranone intermediate [3]; <1> key intermediate in the general flonoid metabolism [3]) (Reversibility: ir) [1, 2, 3]
P aureusidin + H_2O <1> [1, 2, 3]
S 2',4,4',6'-tetrahydroxychalcone 4'-β-D-glucopyranoside + O_2 <1> (Reversibility: ir) [1, 2, 3]
P aureusidin 6-β-D-glucopyranoside + H_2O <1> [2, 3]
S butein + O_2 <1> (Reversibility: ir) [2]
P sulfuretin + 3',4',5',6'-tetrahydroxyaurone + H_2O <1> (<1> products are formed in a 23:1 ratio [2]) [2]
S isoliquiritigenin + O_2 <1> (Reversibility: ir) [2]
P sulfuretin + H_2O <1> [2]
S Additional information <1> (<1> no activity with L-tyrosine, L-DOPA, 4-coumaric acid, caffeic acid, naringenin, eriodictyol, 4,4,6-trihydroxyaurone, and aureusidin [2]; <1> absolutely specific for chalcones [2]; <1> no activity with 2-hydroxychalcone, 4-hydroxychalcone, 2,3,4,4,6-pentahydroxychalcone 3-glucoside, and 2,6-dihydroxy-4,4-dimethoxychalcone [2]; <1> enzyme shows no 3,4-dehydrogenase activity towards aureusidin [1,2]; <1> pathway, overview [1]; <1> chalcone-specific polyphenol oxidase specialized for aurone biosynthesis [2]; <1> enzyme plays a key role in the yellow coloration of flowers, enzyme is a homologue of plant polyphenol oxidase [1,2,3]) (Reversibility: ?) [1, 2, 3]
P ?

Inhibitors

H_2O_2 <1> (<1> inhibits oxidation of 2,3,4,4,6-pentahydroxychalcone [1,3]) [1, 3]
phenylthiourea <1> (<1> competitive inhibition [2]) [2]

Activating compounds

H_2O_2 <1> (<1> absolutely required [3]; <1> activates oxidation of 2,4,4,6-tetrahydroxychalcone [1,3]; <1> optimal at 5 mM [3]) [1, 3]

Metals, ions

Cu^{2+} <1> (<1> copper-containing enzyme, bound via histidine residues in the active site [1]) [1]

Specific activity (U/mg)

578 <1> (<1> purified enzyme [1]) [1]

K_m-Value (mM)

0.0025 <1> (isoliquiritigenin, <1> pH 6.6 [2]) [2]
0.0039 <1> (2',4,4',6'-tetrahydroxychalcone 4'-β-D-glucopyranoside, <1> pH 6.6 [2]) [2]
0.0043 <1> (2',4,4',6'-tetrahydroxychalcone, <1> pH 6.6 [2]) [2]
0.0081 <1> (2',3,4,4',6'-pentahydroxychalcone 4'-β-D-glucopyranoside, <1> pH 6.6 [2]) [2]
0.0147 <1> (butein, <1> pH 6.6 [2]) [2]
0.0157 <1> (2',3,4,4',6'-pentahydroxychalcone, <1> pH 6.6 [2]) [2]

K_i-Value (mM)

0.001 <1> (phenylthiourea, <1> pH 6.6 [2]) [2]

pH-Optimum

5-7 <1> (<1> broad, formation of aureusidin from 2,3,4,4,6-pentahydroxychalcone [1]) [1]
5.4 <1> (<1> 3-hydroxylation and cyclization of 2,4,4,6-tetrahydroxychalcone [1,3]) [1, 3]
Additional information <1> (<1> at pH values above pH 7.0, 2,4,4,6-tetrahydroxychalcone undergoes a very rapid isomerization to the inactive naringenin [3]) [3]

Temperature optimum (°C)

30 <1> (<1> assay at [3]) [3]

4 Enzyme Structure

Molecular weight

40000 <1> (<1> mature enzyme, gel filtration [1]) [1]
Additional information <1> (<1> enzyme is processed to the mature from a 65 kDa precursor protein by cleavage of the N-terminal part [1]) [1]

Subunits

? <1> (<1> x * 39000, mature form, SDS-PAGE [5]) [5]
monomer <1> (<1> 1 * 39000, SDS-PAGE [1]) [1]

Posttranslational modification

glycoprotein <1> [1]
proteolytic modification <1> (<1> vacuolar targeting sequence is encoded within a 53-residue N-terminal sequence, but not in the C-terminal sequence of the precursor [5]) [5]

5 Isolation/Preparation/Mutation/Application

Source/tissue

flower <1> [1, 2, 3]
petal <1> (<1> enzyme expression [1]; <1> inner epidermis of the face and throat of lower petal [4]) [1, 4, 5]
Additional information <1> (<1> analysis of spatial and temporal expression [1]; <1> not in stem and leaf [1]) [1]

Localization

vacuole <1> (<1> chalcones are 4'-O-glucosylated in the cytoplasm and thereafter transported to the vacuole, where the enzyme converts them to aurone 6-O-glucosides [6]) [1, 6]

Purification

<1> (107fold) [1]
<1> (to homogeneity) [1, 3]

Cloning

<1> (gene AmAS1, DNA and amino acid sequence determination and analysis) [1]

Engineering

Additional information <1> (<1> fusion of putative enzyme propepetide to green fluorescent protein, localization within the vacuole [5]; <1> fusion protein of enzyme N-terminal 60 amino acids with red fluorescence protein localizes within the vacuole [6]; <1> natural enzyme mutants sulfurea and violacea showing increased aurone production in petals and reduced aurone production, resp. Enzyme and aureusidin 7-O-glucosyltransferase transcript abundance and spatial pattern is similar in wild-type and mutants. Recessive mutant line CFR1011 with greatly reduced aurone production also shows no change in transcript abundance or any point mutantions in the coding sequences of enzyme or aureusidin 7-O-glucosyltransferase [4]) [4, 5, 6]

Application

agriculture <1> (<1> coexpression of enzyme and chalcone 4'-O-glucosyltransferase is sufficient for accumulation of aureusidin 6-O-glucoside in transgenic flowers. Additional down-regulation of anthocyanin biosynthesis by RNAi results in yellow flowers [6]) [6]

References

[1] Nakayama, T.; Yonekura-Sakakibara, K.; Sato, T.; Kikuchi, S.; Fukui, Y.; Fukuchi-Mizutani, M.; Ueda, T.; Nakao, M.; Tanaka, Y.; Kusumi, T.; Nishino, T.: Aureusidin synthase: a polyphenol oxidase homolog responsible for flower coloration. Science, **290**, 1163-1166 (2000)

[2] Nakayama, T.; Sato, T.; Fukui, Y.; Yonekura-Sakakibara, K.; Hayashi, H.; Tanaka, Y.; Kusumi, T.; Nishino, T.: Specificity analysis and mechanism of aurone synthesis catalyzed by aureusidin synthase, a polyphenol oxidase homolog responsible for flower coloration. FEBS Lett., **499**, 107-111 (2001)

[3] Sato, T.; Nakayama, T.; Kikuchi, S.; Fukui, Y.; Yonekura-Sakakibara, K.; Ueda, T.; Nishino, T.; Tanaka, Y.; Kusumi, T.: Enzymatic formation of aurones in the extracts of yellow snapdragon flowers. Plant Sci., **160**, 229-236 (2001)

[4] Davies, K.M.; Marshall, G.B.; Bradley, J.M.; Schwinn, K.E.; Bloor, S.J.; Winefield, C.S.; Martin, C.R.: Characterization of aurone biosynthesis in Antirrhinum majus. Physiol. Plant., **128**, 593-603 (2006)

[5] Ono, E.; Hatayama, M.; Isono, Y.; Sato, T.; Watanabe, R.; Yonekura-Sakakibara, K.; Fukuchi-Mizutani, M.; Tanaka, Y.; Kusumi, T.; Nishino, T.; Nakayama, T.: Localization of a flavonoid biosynthetic polyphenol oxidase in vacuoles. Plant J., **45**, 133-143 (2006)

[6] Ono, E.; Fukuchi-Mizutani, M.; Nakamura, N.; Fukui, Y.; Yonekura-Sakakibara, K.; Yamaguchi, M.; Nakayama, T.; Tanaka, T.; Kusumi, T.; Tanaka, Y.: Yellow flowers generated by expression of the aurone biosynthetic pathway. Proc. Natl. Acad. Sci. USA, **103**, 11075-11080 (2006)

Selenate reductase 1.97.1.9

1 Nomenclature

EC number
1.97.1.9

Systematic name
selenite:reduced acceptor oxidoreductase

Recommended name
selenate reductase

Synonyms
SR <3> [1]

Additional information <4> (<4> enzyme probably belongs to the DMSO reductase family of mononuclear molybdenum enzymes [3]) [3]

CAS registry number
146359-71-9

2 Source Organism

<1> *Escherichia coli* (no sequence specified) [6]
<2> *Enterobacter cloacae* (no sequence specified) [7, 8, 10, 11]
<3> *Thauera selenatis* (no sequence specified) [1, 2, 3, 5, 8]
<4> *Sulfurospirillum barnesii* (no sequence specified) [3, 8]
<5> *Bacillus arsenicoselenatis* (no sequence specified) [3]
<6> *Thauera selenatis* (UNIPROT accession number: Q9S1H0, Q9S1G9, Q9S1G7, 3 subunits) [4]
<7> *Thaurea selenatis* (no sequence specified) [9]

3 Reaction and Specificity

Catalyzed reaction
selenite + H_2O + acceptor = selenate + reduced acceptor (<4> active site molybdenum [3]; <3> electron transport mechanism [1]; <1> selenate transport mechanism [6])

Reaction type
oxidation
redox reaction
reduction

Natural substrates and products

S selenate + H_2 <7> (Reversibility: ?) [9]
P selenite + H_2O
S selenate + electron donor <3, 4, 5, 6> (<3, 4, 5, 6> selenate is the terminal electron acceptor in dissimilatory selenate reduction [1,2,3,4]) (Reversibility: ?) [1, 2, 3, 4]
P selenite + H_2O + oxidized electron donor <3, 4, 5, 6> [1, 2, 3, 4]
S selenate + reduced acceptor <2, 3, 4> (<2> detoxification of selenate [7]) (Reversibility: ?) [7, 8]
P selenite + H_2O + acceptor

Substrates and products

S selenate + H_2 <7> (Reversibility: ?) [9]
P selenite + H_2O
S selenate + acetate <3> (Reversibility: ?) [3]
P selenite + H_2O + CO_2 <3> [3]
S selenate + electron donor <2> (Reversibility: ?) [11]
P selenite + H_2O + oxidized donor
S selenate + electron donor <1, 2, 3, 4, 5, 6> (<3> specific for selenate [2,3]; <4> enzyme also reduces nitrate, thiosulfate, and fumarate [3]; <3,4,5,6> selenate is the terminal electron acceptor in dissimilatory selenate reduction [1,2,3,4]) (Reversibility: ?) [1, 2, 3, 4, 5, 6, 7]
P selenite + H_2O + oxidized electron donor <2, 3, 4, 5, 6> [1, 2, 3, 4, 5, 7]
S selenate + glucose <3> (Reversibility: ?) [3]
P ?
S selenate + lactate <4, 5, 6> (Reversibility: ?) [3, 4]
P selenite + H_2O + acetate + HCO_3^- <4, 5> [3]
S selenate + malate <5> (Reversibility: ?) [3]
P ?
S selenate + pyruvate <4> (Reversibility: ?) [3]
P ?
S selenate + reduced acceptor <2, 3, 4, 7> (<2> detoxification of selenate [7]) (Reversibility: ?) [7, 8, 9]
P selenite + H_2O + acceptor
S selenate + reduced benzyl viologen <2> (Reversibility: ?) [7]
P selenite + H_2O + benzyl viologen
S selenate + reduced benzyl viologen <2, 3> (<3> best electron donor [2]) (Reversibility: ?) [2, 3, 7]
P selenite + H_2O + oxidized benzyl viologen
S selenate + reduced methyl viologen <2> (Reversibility: ?) [7]
P selenite + H_2O + methyl viologen
S selenate + reduced methyl viologen <2, 3, 4> (<2> preferred electron donor [7]; <3> 11% of the activity with benzyl viologen [2]) (Reversibility: ?) [2, 3, 7]
P selenite + H_2O + oxidized methyl viologen
S Additional information <2, 3, 4> (<3> NADH, succinate, and lactate are no electron donors [2]; <3> no activity with nitrate, nitrite, sulfate, and

chlorate [2,3]; <4> also active on nitrate, thiosulfate and fumarate [8]; <2> no substrate: nitrate, sulfate, perchlorate, thiosulfate [11]) (Reversibility: ?) [2, 3, 8, 11]

P ?

Inhibitors

thiocyanate <2> (<2> mixed-type inhibition [10]) [10]
selenate <2> (<2> mixed-type inhibition [10]) [10]
tungstate <2, 4> (<2> in vitro and in vivo [7]; <4> highly inhibitory [3]) [3, 7, 8]
Additional information <2, 3> (<2> no inhibition by sodium azide [7]; <3> no inhibition by nitrate [1]) [1, 7]

Cofactors/prosthetic groups

cytochrome b <2, 3, 7> (<3> 1 cytochrome b per $\alpha\beta\gamma$-trimer [2]) [2,3,8,9]
heme <2> (<2> 0.9 mol per mol of enzyme [11]) [11]
molybdopterin <7> [9]
iron-sulfur centre <3> (<3> at least 2 [Fe-S]-centre as prosthetic groups per enzyme molecule [2]) [2]

Activating compounds

molybdate <2> [8]
selenate <3> (<3> induction of enzyme activity when included in growth medium [1]) [1]

Metals, ions

iron <1, 2, 3, 7> (<3> part of at least 2 iron-sulfur centres per enzyme molecule and of the heme group of cytochrome b [2]; <1> in FeS-centres [6]; <3> 12.9 mol per mol of enzyme, based on MW 159000 [2]; <3,7> iron-sulfur centre [8,9]; <2> non-heme iron, 18 mol per mol of enzyme [11]) [2, 6, 8, 9, 11]
molybdenum <1, 2, 3, 4, 7> (<2> required [7]; <2> enhances the activity 10fold in vivo when added to the growth medium [7]; <3> 1 molybdenum per mol of trimer [2]; <4> in the active site [3]; <7> Mo-S, Mo=O and Mo-O bound to enzyme [9]; <3> present at the active site [8]; <2> 0.6 mol per mol of enzyme [11]) [2, 3, 6, 7, 8, 9, 11]
selenium <7> (<7> isolated enzyme contains a reduced form of selenium, probably as selenocysteine [9]) [9]
sulfur <3, 7> (<3,7> iron-sulfur centre [8,9]) [8, 9]

Turnover number (min^{-1})

387 <3> (selenate, <3> pH 6.0 [2]) [2]

Specific activity (U/mg)

0.29 <2> (<2> membrane fraction [7]) [7]
0.42 <6> (<6> wild-type [4]) [4]
0.76 <3> (<3> wild-type, cells grown on nitrate [1]) [1]
1.6 <3> (<3> wild-type, cells grown on selenate + nitrate [1]) [1]
3.84 <3> (<3> wild-type, cells grown on selenate [1]) [1]
41.4 <3> (<3> purified enzyme [2]) [2]

500 <2> (<2> pH 7.2, 30°C [11]) [11]
Additional information <2, 3> (<3> activity in nirate reductase deficient mutants under different growth conditions [1]; <2> microtiter plate assay method based on enzyme-dependent reoxidation of reduced methyl viologen detected at 600 nm. Assay is fast and allows for simultaneous testing of a range of alternative substrates and multiple samples [10]) [1, 10]

K_m-Value (mM)
0.012 <4> (selenate) [3]
0.016 <3> (selenate, <3> pH 6.0 [2]) [2, 3]
2.1 <2> (selenate, <2> pH 7.2, 30°C, holenzyme complex [11]) [11]
5.5 <2> (selenate, <2> pH 7.2, 30°C, isolated α subunit [11]) [11]
6.25 <2> (benzyl viologen, <2> 20°C [7]) [7]
6.25 <2> (selenate, <2> with reduced benzyl viologen, 20°C [7]) [7]

K_i-Value (mM)
2.9 <2> (Thiocyanate, <2> 30°C [10]) [10]

pH-Optimum
6 <3> [1, 2]
8 <2> [10]

Temperature optimum (°C)
20 <2> (<2> assay at [7]) [7]

Temperature range (°C)
Additional information <5> (<5> maximal growth at a pH range of pH 9.0-11.0 [3]) [3]

4 Enzyme Structure

Molecular weight
160000 <7> [9]
180000 <3> (<3> gel filtration [2,3]) [2, 3, 8]
600000 <2> (<2> gel filtration [11]) [11]
600000 <2> (<2> gel filtration [8]) [8]

Subunits
nonamer <2> (<2> 3* 100, α-subunit, 3 * 55000, β-subunit, 3 * 36000, γ-subunit, SDS-PAGE [11]) [11]
tetramer <4> (<4> 1 * 82000 + 1 * 53000 + 1 *34000 + 1* 21000 [3]; <4> $\alpha\beta\gamma\delta$, 1 * 82000 + 1 * 53000 + 1 * 34000 + 1 * 21000 [8]) [3, 8]
trimer <3, 7> (<3> 1 * 99000, α, + 1 * 37000, β, + 1 *23000, γ, SDS-PAGE [5]; <3> 1 * 96000, α, + 1 * 40000, β, + 1 * 23000, γ, SDS-PAGE [2,3]; <7> α, β, γ [9]; <3> $\alpha\beta\gamma$, 1 * 96000 + 1 * 40000 + 1 * 23000 [8]) [2, 3, 5, 8, 9]
Additional information <3> (<3> the γ subunit may be identical with the cytochrome b [3]; <3> 1 cytochrome b per $\alpha\beta\gamma$-complex [2]) [2, 3]

5 Isolation/Preparation/Mutation/Application

Localization

cytoplasmic membrane <2> (<2> with localisation of the catalytic site to the periplasmic side of the membrane [7]) [7]
membrane <1, 2, 4> [3, 6, 7, 8, 10, 11]
periplasm <2, 3, 6, 7> (<3> loosely associated with the cytoplasmic membrane [1]) [1, 2, 3, 4, 7, 8, 9]
soluble <7> [9]
Additional information <2> (<2> not in the cytosol [7]) [7]

Purification

<2> [11]
<3> [5]
<3> (57fold, to near homogeneity) [2]

Crystallization

<3> (hanging-drop vapour diffusion method, precipitation by ammonium sulfate, protein solution: 10 mg/ml, 0.3-0.5 M ammonium sulfate, 50 mM piperazine, pH 6.0, reservoir solution: 1.8-2.2 M ammonium sulfate, 100 mM Tris-HCl, pH 8.0-8.9, 293 K, 2-4 weeks, cryoprotection by 25% glycerol, X-ray structure determination and analysis) [5]

Cloning

<6> (DNA sequence determination and analysis, contruction of gene bank, genomic organisationand potential function of: genes serA, serB, and serC, additional overlapping serD) [4]

Engineering

Additional information <1, 6> (<1> construction of diverse Tn5 insertion mutants by transposon mutagenesis, overview [6]; <6> construction of Tn5 insertion mutants by transposon mutagenesis using Escherichia coli strain S17-1 as partner, loss of activity [4]) [4, 6]

6 Stability

Storage stability

<2>, -20°C, enzyme extracted with Thesit retains 50% activity after being frozen for prolonged periods [11]
<2>, 4°C, enzyme extracted with Thesit remains active for 24 h [11]

References

[1] Rech, S.A.; Macy, J.M.: The terminal reductases for selenate and nitrate respiration in Thauera selenatis are two distinct enzymes. J. Bacteriol., **174**, 7316-7320 (1992)

[2] Schroder, I.; Rech, S.; Krafft, T.; Macy, J.M.: Purification and characterization of the selenate reductase from Thauera selenatis. J. Biol. Chem., **272**, 23765-23768 (1997)
[3] Stolz, J.F.; Oremland, R.S.: Bacterial respiration of arsenic and selenium. FEMS Microbiol. Rev., **23**, 615-627 (1999)
[4] Krafft, T.; Bowen, A.; Theis, F.; Macy, J.M.: Cloning and sequencing of the genes encoding the periplasmic-cytochrome B-containing selenate reductase of Thauera selenatis. DNA Seq., **10**, 365-377 (2000)
[5] Maher, M.J.; Macy, J.M.: Crystallization and preliminary X-ray analysis of the selenate reductase from Thauera selenatis. Acta Crystallogr. Sect. D, **58**, 706-708 (2002)
[6] Bebien, M.; Kirsch, J.; Mejean, V.; Vermeglio, A.: Involvement of a putative molybdenum enzyme in the reduction of selenate by Escherichia coli. Microbiology, **148**, 3865-3872 (2002)
[7] Watts, C.A.; Ridley, H.; Condie, K.L.; Leaver, J.T.; Richardson, D.J.; Butler, C.S.: Selenate reduction by Enterobacter cloacae SLD1a-1 is catalysed by a molybdenum-dependent membrane-bound enzyme that is distinct from the membrane-bound nitrate reductase. FEMS Microbiol. Lett., **228**, 273-279 (2003)
[8] Watts, C.A.; Ridley, H.; Dridge, E.J.; Leaver, J.T.; Reilly, A.J.; Richardson, D.J.; Butler, C.S.: Microbial reduction of selenate and nitrate: common themes and variations. Biochem. Soc. Trans., **33**, 173-175 (2005)
[9] Maher, M.J.; Santini, J.; Pickering, I.J.; Prince, R.C.; Macy, J.M.; George, G.N.: X-ray absorption spectroscopy of selenate reductase. Inorg. Chem., **43**, 402-404 (2004)
[10] Ridley, H.; Watts, C.A.; Richardson, D.J.; Butler, C.S.: Development of a viologen-based microtiter plate assay for the analysis of oxyanion reductase activity: application to the membrane-bound selenate reductase from Enterobacter cloacae SLD1a-1. Anal. Biochem., **358**, 289-294 (2006)
[11] Ridley, H.; Watts, C.A.; Richardson, D.J.; Butler, C.S.: Resolution of distinct membrane-bound enzymes from Enterobacter cloacae SLD1a-1 that are responsible for selective reduction of nitrate and selenate oxyanions. Appl. Environ. Microbiol., **72**, 5173-5180 (2006)

Thyroxine 5'-deiodinase 1.97.1.10

1 Nomenclature

EC number

1.97.1.10

Systematic name

acceptor:3,5,3'-triiodo-L-thyronine oxidoreductase (iodinating)

Recommended name

thyroxine 5'-deiodinase

Synonyms

5DI
5DII
5DIII
DIO1 <16> [31]
DIO2 <14> [31]
DIOI
DIOII
DIOIII
EC 3.8.1.4
Iodothyronine 5'-monodeiodinase <13> (<13> isoenzyme type I [18]) [18]
L-thyroxine iodohydrolase (reducing)
Thyroxine deiodinase
type 1 DI
type 2 DI
type 3 DI
type-I 5'deiodinase
type-II 5'deiodinase
type-III 5'deiodinase
XL-15
diiodothyronine 5'-deiodinase
iodothyronine 5'-deiodinase <12, 13> (<13> type II, no selenoprotein, induced by bt2cAMP and hydrocortisone (ED50 4 nM) [9,10,12,15]; <13> iodothyronine 5-deiodinase type [6]; <13> type I, selenoprotein, inducible by T3 (higher synthesis) and angiotensin II (positive chronotropic effect in myocard) [10,11,12,14,20]; <13> isoenzymes type I (D1) and type II (D2) [10,12,13]) [6, 9, 10, 11, 12, 13, 14, 15, 16, 20]
iodothyronine 5-deiodinase <13> (<13> iodothyronine 5-deiodinase type [6]) [6]
iodothyronine inner ring monodeiodinase <9, 13> (<9> isoenzyme type III [16]) [16]

iodothyronine outer ring monodeiodinase
thyroxine 5-deiodinase
type 2 deiodinase <14> [27]
type 2 iodothyronine deiodinase <14> [28, 35]
type I iodothyronine deiodinase <1, 15, 17> [34, 36, 37]
type I-like deiodinase <19> [32]
type II iodothyronine deiodinase <14, 18> [29, 38]
type II-like deiodinase <19> [32]
types 1 iodothyronine selenodeiodinase <16> [31]
types 2 iodothyronine selenodeiodinase <14> [31]

CAS registry number
70712-46-8

2 Source Organism

<1> *Gallus gallus* (no sequence specified) [16, 18, 24, 36]
<2> *Cavia porcellus* (no sequence specified) [18]
<3> *Mus musculus* (no sequence specified) [11, 18, 26]
<4> *Homo sapiens* (no sequence specified) [11, 15, 18, 26]
<5> *Rattus norvegicus* (no sequence specified) [2,3,7,19,25,26]
<6> *Sus scrofa* (no sequence specified) [18,33]
<7> *Bos taurus* (no sequence specified) [18,26]
<8> *Ovis aries* (no sequence specified) [18]
<9> *Rana catesbeiana* (no sequence specified) [16]
<10> *Felis catus* (no sequence specified) [30]
<11> *vertebrata* (no sequence specified) [26]
<12> *Salmo gairdneri* (no sequence specified) [16]
<13> *Rattus sp.* (no sequence specified) (<13> enzyme contains MsrA and MsrB domains [18]) [1, 4, 5, 6, 8, 9, 10, 11, 12, 13, 14, 15, 16, 17, 18, 20, 21, 22, 23]
<14> *Homo sapiens* (UNIPROT accession number: Q92813) [27, 28, 29, 31, 35]
<15> *Gallus gallus* (UNIPROT accession number: O42411) [36]
<16> *Homo sapiens* (UNIPROT accession number: P49895) [31]
<17> *Rattus norvegicus* (UNIPROT accession number: P24389) [34, 37]
<18> *Mus musculus* (UNIPROT accession number: Q9Z1Y9) [38]
<19> *Crocodylus porosus* (no sequence specified) [32]

3 Reaction and Specificity

Catalyzed reaction
3,5,3'-triiodo-L-thyronine + iodide + A + H^+ = L-thyroxine + AH_2 (<13> reaction mechanism of thyronine 5-deiodinase [6]; <13> type III, inner (tyrosyl) ring deiodination [6,10]; <13> type II, outer ring (phenolic) deiodination [6,9,10,12]; <13> type I, outer ring (phenolic) deiodination [6,10,12];

<3,5,7,11> isozyme type I, reaction mechanism, both 5'-deiodinase isoforms and 5-deiodinase contain a selenocysteine residue at the active site [26])

Reaction type

hydrolysis of C-halide

Natural substrates and products

- **S** 3,3',5'-triiodo-L-thyronine + AH_2 <5> (Reversibility: ?) [2]
- **P** 3,3'-diiodothyronine + iodide + A + H^+
- **S** 3,3',5'-triiodo-L-thyronine + AH_2 <3, 4, 5, 11> (<3,4,5,11> isozyme type I [26]) (Reversibility: ?) [26]
- **P** 3,3'-diiodo-L-thyronine + iodide + A + H^+
- **S** 3,3',5'-triiodothyronine + AH_2 <1, 2, 3, 4, 6, 7, 8, 12, 13> (<12, 13> isoenzyme type I [6,9,10,11,12,13,14,15,16,18]) (Reversibility: ?) [6, 9, 10, 11, 12, 13, 14, 15, 16, 18, 20]
- **P** 3,3'-diiodothyronine + iodide + A + H^+
- **S** 3,3',5-triiodothyronine + AH_2 <13> (Reversibility: ?) [6]
- **P** 3,5-diiodothyronine + iodide + A + H^+
- **S** 3,3',5-triiodothyronine + AH_2 <13> (<13> isoenzyme type II and type III [6,14]) (Reversibility: ?) [6, 14]
- **P** 3,3'-diiodothyronine + iodide + A + H^+
- **S** 3,5'-diiodo-L-thyronine + AH_2 <5> (Reversibility: ?) [2]
- **P** 3-iodothyronine + iodide + A + H^+
- **S** L-thyroxine + AH_2 <3, 4, 5, 11> (Reversibility: ?) [26]
- **P** 3,3',5-triiodo-L-thyronine + iodide + A + H^+ (<3, 4, 5, 11> from microsomal enzyme complex, 3,3,5-triiodo-L-thyronine is thyromimetically active [26])
- **S** L-thyroxine + AH_2 <12, 13> (<12,13> idothyronine 5-deiodinase type I [6,16]) (Reversibility: ?) [6, 16, 20]
- **P** 3,3',5'-triiodothyronine + iodide + A + H^+
- **S** L-thyroxine + AH_2 <12, 13, 17> (<12,13> thyronine 5-deiodinase type II [6,16,21,22]; <17> in kidney and liver cells the enzyme can serve as an enzymic barrier to the entry of the prohormone thyroxine and the source of bioactive 3,3,5-triiodothyronine for the cell [34]) (Reversibility: ?) [6, 16, 17, 21, 22, 34]
- **P** 3,3',5-triiodothyronine + iodide + A + H^+
- **S** L-thyroxine + AH_2 <3, 4, 5> (<3,4> isozymes type II and I, production of 3,3,5-triiodo-L-thyronine [26]) (Reversibility: ?) [2, 26]
- **P** 3,5,3'-triiodo-L-thyronine + iodide + A + H^+ (<3,4> hormonally inactive [26])
- **S** L-thyroxine + AH_2 <5, 11> (<5,11> isozymes type II and I, production of 3,3,5-triiodo-L-thyronine [26]) (Reversibility: ?) [26]
- **P** 3,3',5'-triiodo-L-thyronine + iodide + A + H^+ (<5,11> hormonally inactive [26])
- **S** L-thyroxine + electron acceptor <18> (<18> the enzyme is regulated at the pretranslational level [38]) (Reversibility: ?) [38]
- **P** 3,3',5-triiodothyronine + iodide + reduced electron acceptor

S Additional information <5> (<5> L-hormone analogues are preferentially deiodinated via the T4-5-deiodination pathway, whereas D-analogues produce products via the T4-5-deiodination pathway [2]) (Reversibility: ?) [2]
P ?

Substrates and products

S 3,3',5'-triiodo-L-thyronine + AH_2 <5> (Reversibility: ?) [2, 3]
P 3,3'-diiodothyronine + iodide + A + H^+
S 3,3',5'-triiodo-L-thyronine + AH_2 <3, 4, 5, 7, 11> (<3, 4, 5, 11> isozyme type I [26]; <3,5> isozyme type I: degradation of 3,3,5-triiodo-L-thyronine [26]; <4,7,11> isozyme type I, degradation of 3,3,5-triiodo-L-thyronine [26]) (Reversibility: ?) [26]
P 3,3'-diiodo-L-thyronine + iodide + A + H^+ (<3, 4, 5, 7, 11> hormonally inactive [26])
S 3,3',5'-triiodothyronine + AH_2 <1, 2, 3, 4, 6, 7, 8, 12, 13> (<12, 13> isoenzyme type I [6,9,10,11,12,13,14,15,16,18]) (Reversibility: ?) [6, 9, 10, 11, 12, 13, 14, 15, 16, 18, 20]
P 3,3'-diiodothyronine + iodide + A + H^+
S 3,3',5-triiodo-L-thyronine sulfate + reduced electron acceptor <17> (Reversibility: ?) [37]
P diiodothyronine + iodide + electron acceptor
S 3,3',5-triiodothyronine + AH_2 <13> (Reversibility: ?) [6]
P 3,5-diiodothyronine + iodide + A + H^+
S 3,3',5-triiodothyronine + AH_2 <13, 19> (<13> isoenzyme type II and type III [6,14]; <19> rT3 ORD activity [32]) (Reversibility: ?) [6, 14, 32]
P 3,3'-diiodothyronine + iodide + A + H^+
S 3,3',5-triiodothyronine + AH_2 <14> (<14> rT3 ORD activity [29]) (Reversibility: ?) [29]
P diiodothyronine + iodide + A + H^+
S 3,5'-diiodo-L-thyronine + AH_2 <5> (Reversibility: ?) [2]
P 3-iodothyronine + iodide + A + H^+
S L-thyroxine + AH_2 <12, 13, 14, 17, 19> (<12,13> thyronine 5-deiodinase type II [6,16,21,22]; <17> in kidney and liver cells the enzyme can serve as an enzymic barrier to the entry of the prohormone thyroxine and the source of bioactive 3,3,5-triiodothyronine for the cell [34]; <19> outer ring iodothyronine deiodinase activity, T4 ORD activity [32]; <14> reaction with dithiothreitol [29]) (Reversibility: ?) [6, 16, 17, 21, 22, 27, 28, 29, 32, 34, 35]
P 3,3',5-triiodothyronine + iodide + A + H^+
S L-thyroxine + AH_2 <12, 13> (<13> type II [22]; <13> thyronine 5-deiodinase type [6]; <12,13> idothyronine 5-deiodinase type I [6,16]) (Reversibility: ?) [6, 16, 20, 22]
P 3,3',5'-triiodothyronine + iodide + A + H^+
S L-thyroxine + AH_2 <3, 4, 5, 7, 11> (<3,4,7,11> isozyme type II [26]) (Reversibility: ?) [26]

P 3,3',5-triiodo-L-thyronine + iodide + A + H^+ (<3,4,7,11> hormonally active [26]; <3,4,5,7,11> from microsomal enzyme complex, 3,3,5-triiodo-L-thyronine is thyromimetically active [26])
S L-thyroxine + AH_2 <3, 4, 5> (<5> isozyme type II [26]; <3,4> isozymes type II and I, production of 3,3,5-triiodo-L-thyronine [26]) (Reversibility: ?) [2, 3, 7, 26]
P 3,5,3'-triiodo-L-thyronine + iodide + A + H^+ (<3,4> hormonally inactive [26]; <5> hormonally active [26])
S L-thyroxine + AH_2 <5, 11> (<5,11> isozymes type II and I, production of 3,3,5-triiodo-L-thyronine [26]) (Reversibility: ?) [26]
P 3,3',5'-triiodo-L-thyronine + iodide + A + H^+ (<5,11> hormonally inactive [26])
S L-thyroxine + electron acceptor <17, 18> (<18> the enzyme is regulated at the pretranslational level [38]) (Reversibility: ?) [34, 38]
P 3,3',5-triiodothyronine + iodide + reduced electron acceptor
S reverse triiodothyronine + AH_2 <6, 1> (Reversibility: ?) [30, 33]
P diiodothyronine + iodide + A + H^+
S sulfated reverse triiodothyronine + AH_2 <10> (Reversibility: ?) [30]
P diiodothyronine + iodide + A + H^+ + ?
S Additional information <4, 5> (<5> L-hormone analogues are preferentially deiodinated via the T4-5-deiodination pathway, whereas D-analogues produce products via the T4-5-deiodination pathway [2]; <4> selenium-level of cells influences the enzyme activity [26]) (Reversibility: ?) [2, 26]
P ?

Inhibitors

(2E)-2-(3,4-dihydroxybenzylidene)-4,6-dihydroxy-1-benzofuran-3(2H)-one <5> (<5> aurone derivative isolated from plant extract. Comparison with inhibition of 5-deiodinase EC 1.97.1.11 [3]) [3]
(2E)-2-(3,4-dihydroxybenzylidene)-6-hydroxy-1-benzofuran-3(2H)-one <5> (<5> aurone derivative isolated from plant extract. Comparison with inhibition of 5-deiodinase EC 1.97.1.11 [3]) [3]
(2E)-4,6-dihydroxy-2-(4-hydroxy-3-iodobenzylidene)-1-benzofuran-3(2H)-one <5> (<5> aurone derivative isolated from plant extract. Comparison with inhibition of 5-deiodinase EC 1.97.1.11 [3]) [3]
(2E)-4,6-dihydroxy-2-(4-hydroxybenzylidene)-1-benzofuran-3(2H)-one <5> (<5> aurone derivative isolated from plant extract. Comparison with inhibition of 5-deiodinase EC 1.97.1.11 [3]) [3]
(4-[(E)-(4,6-dihydroxy-3-oxo-1-benzofuran-2(^{3}H)-ylidene)methyl]phenoxy)acetic acid <5> (<5> aurone derivative isolated from plant extract. Comparison with inhibition of 5-deiodinase EC 1.97.1.11 [3]) [3]
2-(3,4-dihydroxybenzyl)-4,6-dihydroxy-1-benzofuran-3(2H)-one <5> (<5> aurone derivative isolated from plant extract. Comparison with inhibition of 5-deiodinase EC 1.97.1.11 [3]) [3]
2-thiouracil <13> [6, 8]

2-amino-3-[4-[(4-hydroxy-3,5-diiodophenyl)thio]-3,5-diiodophenyl]propanoic acid <5> (<5> comparison with 5-deiodinase EC 1.97.1.11 [2]) [2]
3',5'-diiodothyronine <5> (<5> comparison with 5-deiodinase EC 1.97.1.11 [2]) [2]
3,3',5'-triiodo-L-thyronine <4, 5, 11> (<5> isozyme type II, rapid inactivation in brain [26]; <11> inhibits expression of isozyme type II [26]) [26]
3,3',5'-triiodothyronine <5> (<5> comparison with 5-deiodinase EC 1.97.1.11 [2]) [2]
3,5,3'-triiodo-5'-nitrothyronine <5> (<5> comparison with 5-deiodinase EC 1.97.1.11 [2]) [2]
3,5,5'-triiodo-2'-methylthyronine <5> (<5> comparison with 5-deiodinase EC 1.97.1.11 [2]) [2]
3,5-diiodo-2'-hydroxythyronine <5> (<5> comparison with 5-deiodinase EC 1.97.1.11 [2]) [2]
3,5-diiodo-3',5'-dinitrothyronine <5> (<5> comparison with 5-deiodinase EC 1.97.1.11 [2]) [2]
3,5-diiodo-3'-hydroxythyronine <5> (<5> comparison with 5-deiodinase EC 1.97.1.11 [2]) [2]
3,5-diiodo-4'-amino-3',5'-dimethylthyronine <5> (<5> comparison with 5-deiodinase EC 1.97.1.11 [2]) [2]
6-propyl-2-thiouracil <12, 13> (<13> inhibits selectively only thyronine 5-deiodinase type I activity not type II activity [6,12,15,20,21]; <13> competes with thiol cofactor [14]; <13> type I mutant with exchange of selenocysteine by cystein is 300fold less sensitive [14]; <12> only type II sensitive to inhibition [16]) [6, 12, 14, 15, 16, 20, 21]
Ca^{2+} <11> (<11> inhibits expression of isozyme type I in thyroid gland [26]) [26]
coumarin <13> (<13> anticoagulants [1]) [1]
dexamethasone <11> (<11> inhibits expression of isozyme type I [26]) [26]
dicoumarol <13> [1]
insulin <13> (<13> inhibits enzyme expression in glial cell culture [9]) [9]
iodoacetate <11, 14> (<11> isozyme type I [26]) [26, 29]
L-thyroxine <11> (<11> isozyme type II [26]) [26]
N-acetyl-3,5,3'-triiodo-5'-nitrothyronine <5> (<5> comparison with 5-deiodinase EC 1.97.1.11 [2]) [2]
N-acetyl-3,5-diiodo-3',5'-dinitrothyronine <5> (<5> comparison with 5-deiodinase EC 1.97.1.11 [2]) [2]
N-acetyl-3,5-diiodo-3'-bromo-5'-nitrothyronine <5> (<5> comparison with 5-deiodinase EC 1.97.1.11 [2]) [2]
N-bromoacetyl-L-thyroxine <13> (<13> i.e. BrAc-3,3,5-triiodothyronine, all isoenzymes [13]) [13]
propanolol <13> (<13> type II [15]) [15]
SH-group blocking reagents <13> [6]
salicylate <13> [1]
soybean phospholipids <13> (<13> competition with microsomal lipids [23]) [23]

tetraiodothyroacetic acid <5> (<5> comparison with 5-deiodinase EC 1.97.1.11 [2]) [2]

thyroxine <5> (<5> isozyme type II, rapid inactivation in brain [26]) [26]

warfarin <13> [1]

[4-(4-hydroxy-3,5-diiodophenoxy)phenyl]acetic acid <5> (<5> comparison with 5-deiodinase EC 1.97.1.11 [2]) [2]

[4-(4-hydroxy-3,5-dinitrophenoxy)-3,5-diiodophenyl]acetic acid <5> (<5> comparison with 5-deiodinase EC 1.97.1.11 [2]) [2]

[4-(4-hydroxy-3-iodo-5-nitrophenoxy)-3,5-diiodophenyl]acetic acid <5> (<5> comparison with 5-deiodinase EC 1.97.1.11 [2]) [2]

[4-(4-hydroxy-3-iodo-5-nitrophenoxy)phenyl]acetic acid <5> (<5> comparison with 5-deiodinase EC 1.97.1.11 [2]) [2]

aurothioglucose <1, 2, 3, 4, 5, 6, 7, 8, 11, 13, 18> (<5> competitive [19]; <1,2,3,4,6,7,8,13> type I, selenium-containing enzyme, competitive [18]; <11> isoyzme type I [26]) [18, 19, 26, 38]

ethyl 4-(4-hydroxy-3,5-diiodobenzoyl)phenyl carbonate <5> (<5> comparison with 5-deiodinase EC 1.97.1.11 [2]) [2]

gold thioglucose <14> [29]

interleukin 1 and interleukin 6 <13> (<13> interleukins are competing for transcriptional coactivators and inhibit thyronine 5-deiodinase induction by 3,3,5-triiodothyronine [11]) [11]

iopanoic acid <4, 11> (<4,11> both isozyme type I and type II [26]) [26]

ipanoic acid <13> (<13> type II [15,21]) [12, 15, 21]

propylthiouracil <6, 10, 11, 14, 19> (<11> isozyme type I [26]; <10> native and recombinant enzyme, IC50: 0.005-0.01 mM [30]; <6> partially, uncompetitively [33]; <14> wild-type and mutant enzymes A131C and A131S are insensitive to propylthiouracil in presence ogf 20 mM dithiothreitol. When tested in presence of 0.2 mM dithiothreitol the Ic50 value for propylthiouracil is reduced to about 0.1 mM [29]) [26, 29, 30, 32, 33]

Additional information <4, 13, 14, 18, 19> (<13> L-thyroxine inhibited competitively 3,3,5-triiodothyronine utilization of isoenzyme type II [10]; <13> 3,3,5-triiodothyronine blocks 5-deiodination of L-thyroxine blocks 5-deiodination of 3,3,5-triiodothyronine [6,14,20]; <18> no inhibition by propylthiouracil [38]; <19> no inhibition by propylthiouracil, type II-like enzyme [32]; <14> substrate-induced loss of activity of the type 2 deiodinase is due to proteasomal degradation of the enzyme and requires interaction with the catalytic center [27]; <4> expression and function of the 5-deiodinase and 5'-deiodinase isozymes are sensitive to thyroid hormone status, various cytokines and growth factors, severe illness, reactive oxygen species, a variety of hormones and signaling compounds, circadian rythm, and pharmacological agents, and therefore might be useful in sensor function of physiology and pathophysiology [26]) [6, 10, 14, 20, 26, 27, 32, 38]

Cofactors/prosthetic groups

Dithiothreitol <5, 11> (<5,11> can be substituted by dithioerythritol, isozyme type II, 5-20 mM, isozyme type I, 2-5 mM [26]) [26]

dithioerythritol <4, 5, 11> (<5,11> can be substituted by dithiothreitol, isozyme type II, 5-20 mM, isozyme type I, 2-5 mM [26]) [26]
Additional information <5, 11> (<5,11> no activity with glutathione or thioredoxin in vivo, isozyme type I [26]) [26]

Activating compounds

3,3',5-triiodo-L-thyronine <11> (<11> induces isozyme type I [26]) [26]
C-type natriuretic protein <11> (<11> induces isozyme type II via cGMP [26]) [26]
dithiothreitol <13, 19> (<13> high concentrations of cofactor increases type I activity [10,12,14,20]; <13> high concentrations increase type II activity [21]; <19> 10 mM required [32]; <19> type II-like enzyme from liver microsomes: 15 mM required. Type II-like enzyme from gut and kidney microsomes: 20 mM required [32]) [10, 12, 14, 20, 21, 32]
EDTA <13> [6]
epidermal growth factor <5> [26]
isoproterenol <13> (<13> type II [15,21]) [15, 21]
nicotine <11> (<11> stimulates isozyme type II [26]) [26]
phorbol esters <11> (<11> induce isozyme type II via protein kinase C [26]) [26]
retinoic acid <5, 11> (<11> stimulates isozyme type I [26]) [26]
Se^{2+} <4, 5, 11> (<4> severe selenium deficiency leads to a decrease in activity of isozyme type I in liver, kidney, and several other organs, but not in the thyroid gland, the central nervous system, and several other endrocrine organs [26]; <11> stimulates isozyme type I, absolutely required, isozymes of the deiodinase enzyme family are selenoproteins [26]) [26]
testosterone <11> (<11> stimulates isozyme type I in liver [26]) [26]
atrial natriuretic protein <11> (<11> induces isozyme type II via cGMP [26]) [26]
β-adrenergic agonists <11> (<11> stimulates isozyme type II [26]) [26]
cAMP <5, 11> (<11> induces isozyme type I in thyroid gland only, induces isozyme type II [26]) [26]
fibroblast growth factor <11> (<11> induces isozymes type II, and I [26]) [26]
reduced thiols <12, 13> (<12,13> e.g. dithiothreitol [6,10,16,20]; <13> e.g. ethanethiol, glutathione [6]) [1, 6, 10, 12, 16, 20]
thyroid hormone L-thyroxine and 3,3',5'-triiodothyronine <13> (<13> extranuclear site of action [9]; <13> type I [17]) [9, 17]
thyroid stimulating hormone <11> (<11> induces isozyme type I [26]) [26]
Additional information <1, 4, 13> (<13> phospholipid required [4]; <13> hypothyroidism enhances type II activity and decreases type I activity [12, 15, 17, 21, 22]; <13> no effect of insulin and dexamethasone on type I activity [20]; <13> thiol cofactor reaction stimulation reduced in type I mutant with exchange of selenocysteine by cysteine [14]; <4> expression and function of the deiodinase isozymes are sensitive to thyroid hormone status, various cytokines and growth factors, severe illness, reactive oxygen species, a variety of hormones and signaling compounds, circadian rythm, and pharmacological

agents [26]; <1> growth hormone increases plasma T3 and decreases plasma T4 levels in 8-day old chicken embryos, in newly hatched chicks and in adult chickens within 2 h after injection. Growth hormone has no effect at all on the amount of hepatic type I enzyme catalyzing T4 deiodination to T3 but acutely decreases the amount of type III enzyme catalyzing T3 deiodination [24]) [4, 12, 14, 15, 17, 20, 21, 22, 24, 26]

Metals, ions

Se <14, 17> (<14> contains selenocysteine in the active center [29]; <17> enzyme contains selenocysteine [37]; <14> enzyme contains selenocysteine in the highly conserved active senter at position 133. Selenium plays a critical role in deiodination [28]; <14> selenoprotein [35]) [28, 29, 35, 37]

Turnover number (min^{-1})

0.000467 <13> (3,3',5'-triiodothyronine) [9]
0.00107 <13> (3,3',5'-triiodothyronine, <13> in hypothyroid medium [9]) [9]
Additional information <14> [28]

Specific activity (U/mg)

0.0001 <4, 8> (<4,8> isoenzyme type I [18]) [18]
0.0002 <1, 7> (<1,7> isoenzyme type I [18]) [18]
0.0003 <6> (<6> isoenzyme type I [18]) [18]
0.0004 <3> (<3> isoenzyme type I [18]) [18]
0.002 <2> (<2> isoenzyme type I [18]) [18]
0.2 <13> (<13> isoenzyme D2 (type II) in cerebral cortex [10]) [10]
2 <13> (<13> isoenzyme D2 (type II) induced in cultured astrocytes [10]) [10]
Additional information <3, 5, 7, 11> (<3,5,7,11> overview on assay methods [26]) [26]

K_m-Value (mM)

0.00000101 <14> (reverse triiodothyronine, <14> mutant enzyme A131S, in presence of 20 mM dithiothreitol [29]) [29]
0.0000014 <14> (L-thyroxine, <14> wild-type enzyme [28]) [28]
0.0000044 <18> (L-thyroxine) [38]
0.0000045 <14> (L-thyroxine, <14> mutant enzyme A131C, in presence of 20 mM dithiothreitol [29]) [29]
0.0000049 <14> (L-thyroxine, <14> wild-type enzyme, in presence of 20 mM dithiothreitol [29]) [29]
0.0000054 <14> (L-thyroxine, <14> mutant enzyme A131S, in presence of 20 mM dithiothreitol [29]) [29]
0.0000089 <14> (reverse triiodothyronine, <14> wild-type enzyme, in presence of 20 mM dithiothreitol [29]) [29]
0.0000098 <14> (reverse triiodothyronine, <14> mutant enzyme A131C, in presence of 20 mM dithiothreitol [29]) [29]
0.00003 <14> (thyroxine) [35]
0.0002-0.0006 <5> (L-thyroxine, <5> isoform III, 37°C, pH 6.0 [7]) [7]
0.00036 <6> (reverse triiodothyronine) [33]

0.0007 <10> (sulfated reverse triiodothyronine, <10> wild-type enzyme [30]) [30]
0.001 <5> (L-thyroxine, <5> isoform II, 37°C, pH 6.0 [7]) [7]
0.001-0.005 <13> (L-thyroxine, <13> type II [15]) [15]
0.0015 <17> (3,3',5-triiodo-L-thyronine sulfate, <17> mutant enzyme G2A, duplicate determination give the values: 0.0015 mM and 0.0024 mM [37]) [37]
0.0018 <17> (3,3',5-triiodo-L-thyronine sulfate, <17> mutant enzyme N94Q [37]) [37]
0.002 <17> (3,3',5-triiodo-L-thyronine sulfate, <17> mutant enzyme S176A, duplicate determination give the values: 0.002 mM and 0.0024 mM [37]; <17> mutant enzyme S176A, duplicate determination give the values: 0.0024 mM and 0.0024 mM [37]; <17> mutant enzymes, in which selenocysteine residues in the core catalytic center is replaced by cysteine [37]) [37]
0.002 <5> (L-thyroxine, <5> pH 7.4, 37°C [2]) [2]
0.002-0.0035 <5> (L-thyroxine, <5> isoform I, 37°C, pH 6.8 [7]) [7]
0.0021 <14> (L-thyroxine, <14> mutant enzyme SeC133C [28]) [28]
0.0022 <17> (3,3',5-triiodo-L-thyronine sulfate, <17> mutant enzyme S101A, duplicate determination give the values: 0.0022 mM and 3.2 mM [37]) [37]
0.0024 <17> (3,3',5-triiodo-L-thyronine sulfate, <17> mutant enzyme G2A, duplicate determination give the values: 0.0015 mM and 0.0024 mM [37]) [37]
0.00256 <19> (3,3',5-triiodothyronine, <19> type I-like activity in kidney microsomes [32]) [32]
0.0028 <17> (3,3',5-triiodo-L-thyronine sulfate, <17> native enzyme from rat liver microsomes [37]) [37]
0.0029 <17> (3,3',5-triiodo-L-thyronine sulfate, <17> mutant enzymeN94Q/N203Q [37]) [37]
0.003 <13> (thyroxine) [8]
0.0032 <17> (3,3',5-triiodo-L-thyronine sulfate, <17> mutant enzyme S101A, duplicate determination give the values: 0.0022 mM and 0.0032 mM [37]) [37]
0.0045 <17> (3,3',5-triiodo-L-thyronine sulfate, <17> mutant enzyme N203Q [37]) [37]
0.005 <19> (3,3',5-triiodothyronine, <19> type I-like activity in liver microsomes [32]) [32]
0.005 <14> (L-thyroxine, <14> mutant enzyme SeC133C, in presence of 20 mM dithiothreitol [29]) [29]
0.0066 <14> (L-thyroxine, <14> mutant enzyme SeC133C/A131C, in presence of 20 mM dithiothreitol [29]) [29]
0.0067 <13> (3,3',5'-triiodothyronine, <13> recombinant isoenzyme D2 (type II) from induced brain astrocytes with cosubstrate dithiothreitol K_M: 18 mM [10]) [10]
0.009 <13> (L-thyroxine, <13> DTT 20 mM [21]; <13> type II, thymus [21]; <13> 0.0105 mM during hypothyroidism [21]) [21]
0.01 <5> (3,5'-diiodo-L-thyronine, <5> pH 7.4, 37°C [2]) [2]
0.015 <10> (reverse triiodothyronine, <10> wild-type enzyme [30]) [30]
0.02 <5> (3,3',5'-triiodo-L-thyronine, <5> pH 7.4, 37°C [2]) [2]
0.065 <13> (3,3',5'-triiodothyronine, <13> DTT 10 mM [20]; <13> type I, myocardial [20]) [20]

0.1 <13> (3,3',5'-triiodothyronine, <13> type I [12]; <13> DTT 20 mM [12]) [12]
0.18 <12> (3,3',5'-triiodothyronine, <12> type I, liver [16]) [16]
0.21 <13> (3,3',5'-triiodothyronine, <13> type I [18]; <13> DTT 10 mM [18]) [18]
0.24 <7> (3,3',5'-triiodothyronine, <7> DTT 10 mM [18]; <7> type I, liver [18]) [18]
0.25 <13> (3,3',5'-triiodothyronine, <13> type I [14]; <13> DTT 30 mM [14]) [14]
0.28 <3> (3,3',5'-triiodothyronine, <3> DTT 10 mM [18]; <3> type I, liver [18]) [18]
0.33 <6> (3,3',5'-triiodothyronine, <6> DTT 10 mM [18]; <6> type I, liver [18]) [18]
0.51 <2> (3,3',5'-triiodothyronine, <2> DTT 10 mM [18]; <2> type I, liver [18]) [18]
0.66 <1> (3,3',5'-triiodothyronine, <1> DTT 10 mM [18]; <1> type I,l iver [18]) [18]
0.69 <8> (3,3',5'-triiodothyronine, <8> DTT 10 mM [18]; <8> type I, liver [18]) [18]
0.7 <4> (3,3',5'-triiodothyronine, <4> DTT 10 mM [18]; <4> type I, liver [18]) [18]
2.7 <13> (3,3',5'-triiodothyronine, <13> type I mutant, exchange of selenocysteine by cysteine [14]) [14]
Additional information <4, 5, 7, 11, 17, 19> (<17> mutant enzymes S101A/S176A and Y209F/Y217F show K_M-values above 0.005 mM [37]; <4,5,7,11> review on enzyme isoforms type I and II and 5-deiodinase, type III [26]) [26, 32, 37]

pH-Optimum

5-6 <5> (<5> isoform III [7]) [7]
5.8-6.3 <5> (<5> isoform II [7]) [7]
6.5 <13> [8]
6.8 <5> (<5> isoform I [7]) [7]
7 <12, 13> [16, 21]
7-7.4 <13> [20]
7.2 <13> [5]
8 <4, 13> (<4> assay at [26]) [6, 26]

pH-Range

6.5-8 <13> [6]

Temperature optimum (°C)

23-25 <12> [16]
25-30 <19> (<19> T4 ORD activity of type II-like enzyme from gut and kidney microsomes [32]) [32]
30-35 <19> (<19> rT3 ORD activity of type I-like enzyme [32]; <19> T4 ORD activity of type II-like enzyme from liver microsomes [32]) [32]
37 <4, 5, 7, 11, 13> (<4,5,7,11> assay at [26]) [4, 21, 26]

4 Enzyme Structure

Molecular weight

49900 <13> (<13> gel filtration [4]) [4]
55400 <13> (<13> type I, sucrose density gradient centrifugation and gel filtration [13]) [13]
56000 <13> (<13> SDS-PAGE [23]) [23]
198700 <13> (<13> type II, sucrose density gradient centrifugation and gel filtration [13]) [13]
200000 <13> [10]

Subunits

? <5> (<5> x * 29000, SDS-PAGE [19]) [19]
dimer <5, 14, 16, 17> (<17> 2 * 27000 [34]; <5> 2 * 27000, isozyme type I [26]; <16> overexpressed enzyme can homodimerize probably through disulfide bridges, type 1 iodothyronine selenodeiodinase [31]; <14> overexpressed enzyme can homodimerize probably through disulfide bridges, type 2 iodothyronine selenodeiodinase [31]; <16> x * 55000, SDS-PAGE, overexpressed enzyme can homodimerize probably through disulfide bridges. Monomeric form is also catalytically active [31]; <14> x * 62000, SDS-PAGE, overexpressed enzyme can homodimerize probably through disulfide bridges. Monomeric form is also catalytically active [31]) [26, 31, 34]
Additional information <1, 2, 3, 4, 6, 7, 8, 13> (<6> substrate binding unit type I [18]; <13> several subunits, thereof 1 * 29000 substrate binding subunit + 1 * 60000 cAMP-activation subunit + unknown number of other subunits, sequence analysis [10]; <4> substrate binding unit type I 31 kD [18]; <13> substrate binding subunit of multimeric isoenzyme type I has 27 kD and of multimeric isoenzyme type II 29 kD, SDS-PAGE [13]; <1,3,7,8,13> substrate binding unit type I 29 kD, SDS-PAGE [18]; <2> substrate binding unit type I 33 kD [18]) [10, 13, 18]

Posttranslational modification

glycoprotein <17> (<17> heterogeneity is not caused by N-linked glycosylation, but probably by a combination of O-linked glycosylation and phosphorylation [37]) [37]
no glycoprotein <13> (<13> no glycosylation of substrate binding subunits of both isoenzymes [13]) [13]
phosphoprotein <17> (<17> heterogeneity is not caused by N-linked glycosylation, but probably by a combination of O-linked glycosylation and phosphorylation [37]) [37]
Additional information <5> (<5> both type I and type II enzyme as well as type III enzyme EC 1.97.1.11 are selenoproteins. All tissues studied maintain more than 50% deiodinase activity during prolonged selenium-deficiency. Only when selenium levels decrease by more than 80%, deiodinase activity markedly decreases [25]) [25]

5 Isolation/Preparation/Mutation/Application

Source/tissue

MSTO-211H cell <14> (<14> mesothelioma cell line, high expression of type 2 iodothyronine deiodinase [35]) [35]
adipocyte <3, 5, 11> (<3,11> isozyme type II [26]) [26]
brain <3, 11, 13> (<13> thyronine 5-deiodinase type II [12,15,17,22]; <13> isoenzyme D2 [10,17]; <3,11> isozyme type II [26]) [10, 12, 15, 17, 22, 26]
brown adipose tissue <3, 5, 11, 13> (<13> thyronine 5-deiodinase type II [12,15,17,22]; <3,5,11> isozyme type II [26]) [12, 15, 17, 22, 26]
central nervous system <4> (<4> isozyme type I [26]) [26]
colon <18> (<18> low level of activity [38]) [38]
duodenum <18> (<18> low level of activity [38]) [38]
epithelium <3, 11> (<3,11> of kidney, isozyme type I [26]) [26]
glial cell <3, 11, 13> (<13> thyronine 5-deiodinase type II [12]; <13> i.e. brain astrocytes in culture [9,10]; <13> measurable activity only during induction with bt2cAMP [9]; <13> isoenzyme type II (D2) [10]; <3,11> isozyme type II [26]) [9, 10, 12, 13, 26]
gut <19> (<19> low type I-like activity, type II-like activity is the predominant activity [32]; <19> type II-like activity is the predominant activity [32]) [32]
harderian gland <3, 11, 13> (<13> thyronine 5-deiodinase type II [15,22]; <3,11> isozyme type II vertebrata [26]) [15, 22, 26]
heart <3, 11, 18> (<18> low level of activity [38]; <3,11> isozyme type I [26]) [26, 38]
hepatocyte <4> [26]
hypophysis <13> (<13> thyronine 5-deiodinase type II [12,15,17,22]) [12, 15, 17, 22]
keratinocyte <13> (<13> thyronine 5-deiodinase type II [15]) [15]
kidney <1, 3, 4, 11, 13, 15, 17, 18, 19> (<18> low level of activity [38]; <13> thyronine 5-deiodinase type I [12,14]; <13> isoenzyme type I (D1) [10]; <3,4,11> isozyme type I [26]; <1> expression is associated with the tubular epithelial cells and with the transitional epithelium, and the inner longotudinal and outer circular muscle layers of the ureter [36]; <15> maximal expression in a thin layer of hepatocates bordering the blood veins [36]; <19> type I-like activity [32]) [1, 4, 5, 6, 10, 12, 14, 15, 26, 32, 34, 36, 38]
liver <1, 2, 3, 4, 5, 6, 7, 8, 9, 11, 12, 13, 15, 17, 18, 19> (<18> low level of activity [38]; <13> isoenzyme typ I (D1) [10]; <1> type I and II [16]; <9> type II and III [16]; <12, 13> thyronine 5-deiodinase type I [12, 14, 15, 16]; <3, 4, 5, 11> isozyme type I [26]; <15> expressionis associated with tubular epithelial cells and with the transitional epithelium, and the inner longitudinal and outer circular muscle layers of the ureter [36]; <19> low type II-like activity, type I-like activity is the predominant activity [32]; <1> maximum protein expression is shown in a thin layer of hepatocytes bordering the blood veins [36]; <19> type I-like activity is the predominant activity [32]) [3, 5, 6, 7, 10, 11, 12, 14, 15, 16, 18, 19, 23, 24, 26, 32, 34, 36, 37, 38]

lung <18> (<18> low level of activity [38]) [38]
mammary gland <18> [38]
myocardium <13> (<13> thyronine 5-deiodinase type I [20]) [20]
pineal gland <3, 11, 13> (<13> thyronine 5-deiodinase type II [15,22]; <3,11> isozyme type II [26]) [15, 22, 26]
pituitary gland <3, 5, 7, 11> (<5> isozyme type I [26]; <7> anterior, isozyme type I [26]; <3,11> anterior, isozyme type II, isozyme type I [26]) [26]
placenta <3, 5, 11> (<3,5,11> isozyme type II [26]) [26]
prostate <18> (<18> low level of activity [38]) [38]
seminal plasma <6> [33]
skin <3, 5, 11> (<3,11> isozyme type II [26]) [26]
tanycyte <3, 11> (<3,11> isozyme type II [26]) [26]
testis <6, 18> (<18> low level of activity [38]; <6> increase in the prepubertal testis [33]) [33, 38]
thymus <3, 11, 13> (<13> thyronine 5-deiodinase type II [21]; <3,11> isozyme type II [26]) [21, 26]
thyroid gland <3, 4, 5, 11, 13> (<13> thyronine 5-deiodinase type I [12,14]; <5> isozyme type I [26]; <3,11> isozyme types I and II [26]) [8, 12, 14, 26]
Additional information <4, 13, 19> (<13> no type II activity in spleen [21]; <4> severe selenium deficiency leads to a decrease in activity of isozyme type I in liver, kidney, and several other organs, but not in the thyroid gland, the central nervous system, and several other endrocrine organs [26]; <19> no type-II-like activity in kidney [32]) [21, 26, 32]

Localization

endoplasmic reticulum <7, 11> (<11> liver, smooth and rough, isozyme type I, at the cytosolic surface [26]; <7> anterior pituitary gland, isozyme type I [26]) [26]
membrane <13, 17> (<17> bound to [37]; <13> bound, type II integral in neurolemmal membranes [9,10]; <13> isoenzyme type I and type II are both integral membrane proteins [13]; <13> type II in membranes of stromal cells and thymocytes [21]; <17> integral membrane protein. Once assembled the D1 holoenzyme is sorted to the plasma membrane in both kidney and liver cells, where it can serve as an enzymic barrier to the entry of the prohormone thyroxine and the source of bioactive 3,3,5-triiodothyronine for the cell [34]) [4, 9, 10, 13, 21, 34, 37]
microsome <3, 4, 5, 11, 13, 19> (<13> type I [14]; <5> isozyme type I, liver [26]; <5> enzyme isoforms I, II [7]) [1, 2, 3, 7, 12, 14, 19, 26, 32]
mitochondrion <5, 11> [26]
plasma membrane <11> (<11> at the inner leaflet of the basolateral plasma membrane, isozyme type I, kidney [26]) [26]
Additional information <5, 11> (<5> membrane association is not essential for maintenance of functional activity [26]; <11> subcellular localization of the different isozyme types in different tissues, overview [26]) [26]

Purification

<13> [4, 21]
<13> (isoenzyme type I and type II) [13]

Renaturation

<13> (partly reconstitution of activity of delipinated purified thyronine 5'-deiodinase from liver microsomes in soybean phospholipids) [23]

Cloning

<3> (isozyme type I, mapping on chromosome 4) [26]
<4> (isozyme type I, mapping on chromosome 1p32-p33, DNA sequence analysis) [26]
<10> (expression in COS cells) [30]
<13> (expression in Escherichia coli via phage lambda infection) [10]
<13> (expression in JEG-3 cells, transient transfection) [14]
<13> (overexpression in rat astrocytes of 29 kD subunit with and without GFP-tag via adenovirus infection) [10]
<13> (transcription in vitro, translation in Xenopus laevis oocytes) [12]
<13> (translation of purified rat RNA in Xenopus laevis oocytes) [15]
<14> (expression in COS cells) [29]
<14> (expression in HEK-293 cells) [27, 31]
<16> (expression in HEK-293 cells) [31]
<17> (expression of mutant enzymes, in which selenocysteine residues in the core catalytic center is replaced by cysteine, in Saccharomyces cerevisiae) [37]
<17> (overexpression in LLC-PK1 cells) [34]
<18> [38]

Engineering

A131C <14> (<14> similar K_m-values for L-thyroxine and 3,3,5-triiodothyronine as the wild-type enzyme. Mutation improves the interaction with the reducing cofactor dithiothreitol [29]) [29]
A131S <14> (<14> similar K_m-values for L-thyroxine and 3,3,5-triiodothyronine as the wild-type enzyme. Mutation improves the interaction with the reducing cofactor dithiothreitol [29]) [29]
N203Q <17> (<17> decrease in K_m-value for 3,3,5-triiodo-L-thyronine sulfate [37]) [37]
N94Q <17> (<17> decrease in K_m-value for 3,3,5-triiodo-L-thyronine sulfate [37]) [37]
N94Q/N203Q <17> (<17> increase in K_m-value for 3,3,5-triiodo-L-thyronine sulfate [37]) [37]
S101A <17> (<17> K_m-value for 3,3,5-triiodo-L-thyronine sulfate is nearly identical [37]) [37]
S101A/S176A <17> (<17> increase in K_m-value for 3,3,5-triiodo-L-thyronine sulfate [37]) [37]
SeC133A <14> (<14> inactive mutant enzyme [28]) [28]
SeC133C <14> (<14> mutant enzyme with a 1000fold increase in K_m-value for L-thyroxine [29]; <14> the K_m-value for L-thyroxine is 1500fold higher than the wild-type value, the turnover-number is 10fold lower than the wild-type value [28]) [28, 29]
SeC133C/A131C <14> (<14> mutant enzyme with more than 1000fold increase in K_m-value for L-thyroxine [29]) [29]

Y209F/Y217F <17> (<17> increase in K_m-value for 3,3,5-triiodo-L-thyronine sulfate [37]) [37]
Additional information <13, 17> (<13> exchange selenocysteine to cysteine by site-directed mutagenesis of isoenzyme type I [14]; <17> mutant enzymes, in which selenocysteine residues in the core catalytic center is replaced by cysteine, shows a 10fold increase in K_m-value for 3,3,5-triiodo-L-thyronine sulfate. Deletion of the endoplasmic reticulum (ER)-signal sequence and the membrane-spanning domain, amino acids 2-35, does not result in the production of a soluble type I¯like enzyme. This mutant protein is inactive but is still membrane-bound [37]) [14, 37]

Application

medicine <4> (<4> expression and function of the 5-deiodinase and 5'-deiodinase isozymes are sensitive to thyroid hormone status, various cytokines and growth factors, severe illness, reactive oxygen species, a variety of hormones and signaling compounds, circadian rythm, and pharmacological agents, and therefore might be useful in sensor function of physiology and pathophysiology [26]) [26]

6 Stability

Temperature stability

4 <13> (<13> inactivation after 30 min [20]) [20]
56 <13> (<13> inactivation after 30 min [20]) [20]
60 <13> (<13> inactivation after 30 min [8]) [8]

General stability information

<13>, activity mostly lost during purification procedures, catalytic activity lost in detergent solution [13, 23]
<13>, very short lifetime in vivo of isoenzyme type II (D2) from brain [9, 10]

Storage stability

<5>, -20°C, presence of 3 mM dithiothreitol, stable for at least 6 weeks, after 3 months, loss of55% of activity [7]

References

[1] Goswami, A.; Leonard, J.L.; Rosenberg, I.N.: Inhibition by coumardin anticoagulants of enzymatic outer ring monodeiodination of iodothyronines. Biochem. Biophys. Res. Commun., **104**, 1231-1238 (1982)

[2] Koehrle, J.; Auf'mkolk, M.; Rokos, H.; Hesch, R.D.; Cody, V.: Rat liver iodothyronine monodeiodinase. Evaluation of the iodothyronine ligand-binding site. J. Biol. Chem., **261**, 11613-11622 (1986)

[3] Auf'mkolk, M.; Koehrle, J.; Hesch, R.D.; Cody, V.: Inhibition of rat liver iodothyronine deiodinase. Interaction of aurones with the iodothyronine ligand-binding site. J. Biol. Chem., **261**, 11623-11630 (1986)

[4] Leonard, J.L.; Rosenberg, I.N.: Solubilization of a phospholipid-requiring enzyme, iodothyronine 5-deiodinase, from rat kidney membranes. Biochim. Biophys. Acta, **659**, 205-218 (1981)
[5] Colquhoun, E.Q.; Thomson, R.M.: Lysosomal thyroid hormone 5-deiodinase. FEBS Lett., **177**, 221-226 (1984)
[6] Visser, T.J.: Deiodination of thyroid hormone and the role of glutathione. Trends Biochem. Sci., **5**, 222-224 (1980)
[7] Auf dem Brinke, D.; Hesch, R.D.; Köhrle, J.: Re-examination of the subcellular localization of thyroxine 5-deiodination in rat liver. Biochem. J., **180**, 273-279 (1979)
[8] Ericksen, V.J.; Cavalieri, R.R.; Rosenberg, L.L.: Phenolic and nonphenolic ring iodothyronine deiodinases from rat thyroid gland. Endocrinology, **108**, 1257-1264 (1981)
[9] Leonard, J.L.; Siegrist-Kaiser, C.A.; Zuckerman, C.J.: Regulation of type II iodothyronine 5'-deiodinase by thyroid hormone. J. Biol. Chem., **265**, 940-946 (1990)
[10] Leonard, D.M.; Stachelek, S.J.; Safran, M.; Farwell, A.P.; Kowalik, T.F.; Leonard, J.L.: Cloning, expression, and functional characterization of the substrate binding subunit of rat type II iodothyroine 5'-deiodinase. J. Biol. Chem., **275**, 25194-25201 (2000)
[11] Yu, J.; Koenig, R.J.: Regulation of hepatocyte thyroxine 5'-deiodinase by T3 and nuclear receptor coactivators as a model of the sick euthyroid syndrome. J. Biol. Chem., **275**, 38296-38301 (2000)
[12] Sharifi, J.; St.Germain, D.L.: The cDNA for the type I iodothyroine 5'-deiodinase encodes an enzyme manifesting both high K_M and low K_M activity. J. Biol. Chem., **267**, 12539-12544 (1992)
[13] Safran, M.; Leonard, J.L.: Comparison of the physicochemical properties of type I and type II iodothyroine 5'-deiodinase. J. Biol. Chem., **266**, 3233-3238 (1991)
[14] Berry, M.J.; Kieffer, J.D.; Harney, J.W.; Larsen, P.R.: Selenocysteine confers the biochemical properties charateristic of the type I iodothyronine deiodinase. J. Biol. Chem., **266**, 14155-14158 (1991)
[15] Garcia-Macias, J.F.; Molinero, P.; Guerrero, J.M.; Osuna, C.: Expression of type II thyroxine 5'-deiodinase from rat harderian gland in Xenopus laevis oocytes. FEBS Lett., **354**, 110-112 (1994)
[16] Orozco, A.; Silva; J.E.; Valverde-R; C.: Rainbow trout liver expresses two iodothyronine phenolic ring deiodinase pathways with the characteristics of mammalian types I and II 5'-deiodinases. Endocrinology, **138**, 254-258 (1997)
[17] Burmeister, L.A.; Pachucki, J.; St.Germain, D.L.: Thyroid hormones inhibit type 2 iodothyronine deiodinase in the rat cerebral cortex by both pre- and posttranslational mechanisms. Endocrinology, **138**, 5231-5237 (1997)
[18] Santini, F.; Chopra, I.J.; Hurd, R.E.; Chua Teco, G.N.: A study of the characteristics of hepatic iodothyronine 5'-monodeiodinase in various vertebrate species. Endocrinology, **131**, 830-834 (1992)
[19] Santini, F.; Chopra, I.J.; Hurd, R.E.; Solomon, D.H.; Chua Teco, G.N.: A study of the characteristics of the rat placental iodothyronine 5'-monodeio-

dinase: evidence that it is distinct from the rat hepatic iodothyronine 5'-monoiodinase. Endocrinology, **130**, 2325-2332 (1992)

[20] Mori, Y.; Nishikawa, M.; Toyoda, N.; Yonemoto, T.; Matsubara, H.; Inada, M.: Iodothyronine 5'-deiodinase activity in cultured rat myocardial cells: characteristics and effects of triiodothyronine and angiotensin II. Endocrinology, **128**, 3105-3112 (1991)

[21] Molinero, P.; Osuna, C.; Guerrero, J.M.: Type II thyroxine 5'-deiodinase in the rat thymus. J. Endocrinol., **146**, 105-111 (1995)

[22] Osuna, C.; Orta, J.M.; Rubio, A.; Molinero, P.; Guerrero, J.M.: Thyroxine type II 5'-deiodinase activity in pineal and harderian glands is enhanced by hypothyroidism but is independent of serum thyroxine concentrations during hyperthyroidism. Int. J. Biochem., **25**, 1041-1046 (1993)

[23] Goswami, A.; Rosenberg, I.N.: Reconstitution of a purified and partially active low K_M iodothyronine 5'-deiodinase with phospholipids. Biochem. Int., **27**, 257-263 (1992)

[24] Darras, V.M.; Berghman, L.R.; Vanderpooten, A.; Kuhn, E.R.: Growth hormone acutely decreases type III iodothyronine deiodinase in chicken liver. FEBS Lett., **310**, 5-8 (1992)

[25] Bates, J.M.; Spate, V.L.; Morris, J.S.; St. Germain, D.L.; Galton, V.A.: Effects of selenium deficiency on tissue selenium content, deiodinase activity, and thyroid hormone economy in the rat during development. Endocrinology, **141**, 2490-2500 (2000)

[26] Koehrle, J.: Iodothyronine deiodinases. Methods Enzymol., **347**, 125-167 (2002)

[27] Steinsapir, J.; Bianco, A.C.; Buettner, C.; Harney, J.; Larsen, P.R.: Substrate-induced down-regulation of human type 2 deiodinase (hD2) is mediated through proteasomal degradation and requires interaction with the enzyme's active center. Endocrinology, **141**, 1127-1135 (2000)

[28] Buettner, C.; Harney, J.W.; Larsen, P.R.: The role of selenocysteine 133 in catalysis by the human type 2 iodothyronine deiodinase. Endocrinology, **141**, 4606-4612 (2000)

[29] Kuiper, G.G.; Klootwijk, W.; Visser, T.J.: Substitution of cysteine for a conserved alanine residue in the catalytic center of type II iodothyronine deiodinase alters interaction with reducing cofactor. Endocrinology, **143**, 1190-1198 (2002)

[30] Kuiper, G.G.; Wassen, F.; Klootwijk, W.; Van Toor, H.; Kaptein, E.; Visser, T.J.: Molecular basis for the substrate selectivity of cat type I iodothyronine deiodinase. Endocrinology, **144**, 5411-5421 (2003)

[31] Curcio-Morelli, C.; Gereben, B.; Zavacki, A.M.; Kim, B.W.; Huang, S.; Harney, J.W.; Larsen, P.R.; Bianco, A.C.: In vivo dimerization of types 1, 2, and 3 iodothyronine selenodeiodinases. Endocrinology, **144**, 937-946 (2003)

[32] Shepherdley, C.A.; Richardson, S.J.; Evans, B.K.; Kuehn, E.R.; Darras, V.M.: Characterization of outer ring iodothyronine deiodinases in tissues of the saltwater crocodile (Crocodylus porosus). Gen. Comp. Endocrinol., **125**, 387-398 (2002)

[33] Brzezinska-Slebodzinska, E.; Slebodzinski, A.B.; Kowalska, K.: Evidence for the presence of 5'-deiodinase in mammalian seminal plasma and for the

increase in enzyme activity in the prepubertal testis. Int. J. Androl., **23**, 218-224 (2000)

[34] Leonard, J.L.; Visser, T.J.; Leonard, D.M.: Characterization of the subunit structure of the catalytically active type I iodothyronine deiodinase. J. Biol. Chem., **276**, 2600-2607 (2001)

[35] Curcio, C.; Baqui, M.M.; Salvatore, D.; Rihn, B.H.; Mohr, S.; Harney, J.W.; Larsen, P.R.; Bianco, A.C.: The human type 2 iodothyronine deiodinase is a selenoprotein highly expressed in a mesothelioma cell line. J. Biol. Chem., **276**, 30183-30187 (2001)

[36] Verhoelst, C.H.; Van Der Geyten, S.; Darras, V.M.: Renal and hepatic distribution of type I and type III iodothyronine deiodinase protein in chicken. J. Endocrinol., **181**, 85-90 (2004)

[37] Kuiper, G.G.; Klootwijk, W.; Visser, T.J.: Expression of recombinant membrane-bound type I iodothyronine deiodinase in yeast. J. Mol. Endocrinol., **34**, 865-878 (2005)

[38] Song, S.; Sorimachi, K.; Adachi, K.; Oka, T.: Biochemical and molecular biological evidence for the presence of type II iodothyronine deiodinase in mouse mammary gland. Mol. Cell. Endocrinol., **160**, 173-181 (2000)

Thyroxine 5-deiodinase 1.97.1.11

1 Nomenclature

EC number
1.97.1.11

Systematic name
acceptor:3,3',5'-triiodo-L-thyronine oxidoreductase (iodinating)

Recommended name
thyroxine 5-deiodinase

Synonyms
ITHD <4> [2]
T4-5'-deiodinase <4> [4]
T4-5-deiodinase <4> [4]
deiodinase, thyroxine 5-
diiodothyronine 5'-deiodinase
iodothyronine 5-deiodinase
iodothyronine inner ring monodeiodinase
iodothyronine monodeiodinase <4> [2]
type 3 iodothyronine deiodinase <3> [24]
type III iodothyronine deiodinase <14> [25]

CAS registry number
74506-30-2

2 Source Organism

<1> *Gallus gallus* (no sequence specified) [7, 9, 20, 22]
<2> *Mus musculus* (no sequence specified) [19]
<3> *Homo sapiens* (no sequence specified) [9, 10, 17, 19, 21, 23, 24]
<4> *Rattus norvegicus* (no sequence specified) [1, 2, 3, 4, 5, 8, 9, 11, 12, 14, 16, 19, 22]
<5> *Bos taurus* (no sequence specified) [19]
<6> *Xenopus laevis* (no sequence specified) [18]
<7> *Rana catesbeiana* (no sequence specified) [6,9]
<8> *Monkey* (no sequence specified) [9]
<9> *vertebrata* (no sequence specified) [19]
<10> *Rattus norvegicus* (UNIPROT accession number: P49897) [13]
<11> *Oreochromis niloticus* (UNIPROT accession number: Q9I966) [15]
<12> *Xenopus laevis* (UNIPROT accession number: Q90Z53) [18]

<13> *Homo sapiens* (GenBank accession number: XM007250) [21]
<14> *Gallus gallus* (UNIPROT accession number: 042412) [25]

3 Reaction and Specificity

Catalyzed reaction

3,3',5'-triiodo-L-thyronine + iodide + A + H^+ = L-thyroxine + AH_2 (<3,13> all 3 isozyme types contain a selenocysteine residue at the active site [21]; <4> enzyme has both thyroxine 5-deiodinase and thyroxine 5-deiodinase activities [2,4,5]; <3,13> inner ring deiodination [21])
3,3?,5?-triiodo-L-thyronine + iodide + A + H^+ = L-thyroxine + AH_2 (<9> both 5-deiodinase isoforms and 5-deiodinase contain a selenocysteine residue at the active site [19])

Reaction type

oxidation
redox reaction
reduction

Natural substrates and products

S 3,3',5-triiodo-L-thyronine + AH_2 <1, 2, 3, 4, 7, 9, 13> (<1,4> isozyme type III [20,22]; <13> selenocysteine residue in the active site is essential for efficient inner ring deiodination [21]; <1,3,13> enzyme plays a critical role in regulating and maintaining the local 3,3,5-triiodo-L-thyronine content [17,20,21]; <4> no detectable activity in vivo, isozyme type III [12]) (Reversibility: ?) [6, 12, 17, 19, 20, 21, 22]

P 3,3'-diiodo-L-thyronine + iodide + A + H^+ <1, 3, 4, 7, 13> (<1, 2, 3, 4, 9> hormonally inactive [17, 19, 22]; <1, 3, 4, 7, 13> i.e. 3,3-diiodo-L-thyronine [6,17,21,22]) [6, 17, 21, 22]

S L-thyroxine + AH_2 <1, 2, 3, 4, 7, 9, 13> (<1, 3, 4, 7, 13> 5-deiodinase activity [2, 4, 6, 8, 17, 21, 22]; <7> in pre- and prometamorphic tadpoles, undetectable during metamorphic climax [6]; <13> selenocysteine residue in the active site is absolutely essential for efficient inner ring deiodination [21]; <2,3,9> enzyme III performs the inactivation of 3,3,5,5-tetraiodo-L-thyronine [19]; <4> enzyme performs the inactivation of 3,3,5,5-tetraiodo-L-thyronine [19]) (Reversibility: ?) [1, 2, 4, 6, 8, 17, 19, 21, 22]

P 3,3',5'-triiodo-L-thyronine + iodide + A + H^+ <1, 3, 4, 7, 13> (<1, 2, 3, 4, 9> hormonally inactive [8, 17, 19, 22]; <3,13> enzyme plays a critical role in regulating and maintaining the local 3,3,5-triiodo-L-thyronine content [17,21]; <4> from microsomal enzyme complex, 3,3,5-triiodo-L-thyronine is regulatory active [4]) [2, 4, 6, 8, 17, 21, 22]

S L-thyroxine + AH_2 <3, 4, 7> (<3> isozyme type III [17]; <4,7> 5-deiodinase activity [2,4,6]; <7> in vivo after entering metamorphic climax, low 5-deiodinase activity [6]) (Reversibility: ?) [2, 4, 6, 17]

P 3,3',5-triiodo-L-thyronine + iodide + A + H^+ <3, 4, 7> (<4> from microsomal enzyme complex, 3,3,5-triiodo-L-thyronine is thyromimetically active [4]) [2, 4, 6, 17]

S Additional information <1, 2, 3, 4, 7, 8, 9> (<4> enzyme plays a critical role in regulating and maintaining the local 3,3,5-triiodo-L-thyronine content [19]; <7> enzyme activity is developmentally regulated, overview [6]; <4> 5-deiodinase activity in brain is slightly decreased in hypo- and slightly increased in hyperthyroidism [9]; <4> placental isozyme type III is only slightly increased in hyperthyroidism, while brain isozyme type III is strongly induced [11]; <1,4> enzyme performs inactivation of thyroid hormones [22]; <1,3,4,7,8> isozyme type III is a oncofetal enzyme, regulation [9]; <3> thyroid hormone is essential for maintaining normal neurological functions both during development and in adult life [17]; <3> the enzyme is the major physiological inactivator of thyroid hormone. The presence of the enzyme at maternal-fetal interfaces is consistent with its role in modulating the thyroid status of the human fetus and its expression in endometrium suggests that local regulation of thyroid status is important in implantation [24]; <2,3,9> isozyme type III performs the inactivation of 3,3,5,5-tetraiodo-L-thyronine and enzyme plays a critical role in regulating and maintaining the local 3,3,5-triiodo-L-thyronine content [19]; <4> L-hormone analogues are preferentially deiodinated via the T4-5-deiodination pathway, whereas D-analogues produce products via the T4-5-deiodination pathway [1]) (Reversibility: ?) [1, 6, 9, 11, 17, 19, 22, 24]

P ?

Substrates and products

S 3,3',5'-triido-L-thyronine + AH <3> (Reversibility: ?) [24]

P 3,3'-diiodothyronine + iodide + A + H^+

S 3,3',5'-triiodo-L-thyronine + AH_2 <4, 7, 10, 12> (<10> no activity [13]; <4> preferred substrate [2]; <4,12> type I isozyme [2,18]) (Reversibility: ?) [2, 6, 13, 18]

P 3,3'-diiodo-L-thyronine + iodide + A + H^+ <4> [2]

S 3,3',5-triiodo-L-thyronine + AH_2 <1, 2, 3, 4, 5, 6, 7, 8, 9, 10, 11, 13> (<1, 3, 4, 7, 10, 11> best substrate [6, 8, 9, 11, 13, 15]; <1, 3, 4, 6, 7, 10, 11, 13> isozyme type III [6, 8, 9, 11, 13, 14, 15, 17, 18, 20, 21, 22]; <13> selenocysteine residue in the active site is essential for efficient inner ring deiodination [21]; <1, 3, 13> enzyme plays a critical role in regulating and maintaining the local 3,3,5-triiodo-L-thyronine content [17, 20, 21]; <4> no detectable activity in vivo, isozyme type III [12]) (Reversibility: ?) [2, 6, 8, 9, 11, 12, 13, 14, 15, 17, 18, 19, 20, 21, 22]

P 3,3'-diiodo-L-thyronine + iodide + A + H^+ <1, 3, 4, 7, 8, 10, 11, 13> (<1, 2, 3, 4, 5, 9, 10> hormonally inactive [13, 17, 19, 22]; <1, 3, 4, 7, 13> i.e. 3,3-diiodo-L-thyronine [6, 17, 21, 22]) [2, 6, 8, 9, 11, 13, 14, 15, 17, 20, 21, 22]

S 3,3',5-triiodo-L-thyronine sulfate + AH_2 <11> (<11> isozyme type III [15]; <11> about 14% of the activity with 3,3,5-triiodo-L-thyronine [15]) (Reversibility: ?) [15]

P 3,3'-diiodo-L-thyronine sulfate + iodide + A + H^+

S L-thyroxine + AH_2 <3> (Reversibility: ?) [24]

P 3,3',5'-triido-L-thyronine + iodide + A + H^+

S L-thyroxine + AH_2 <3, 4, 7, 10> (<10> no activity [13]; <7> low activity [6]; <4> preferred substrate [2,8]; <3> isozyme type III [17]; <4,7> 5-deiodinase activity [2,4,5,6,8]; <4> isozyme type II [8]; <4> isozyme type I [2,8]; <7> in vivo after entering metamorphic climax, low 5-deiodinase activity [6]) (Reversibility: ?) [2, 4, 5, 6, 8, 13, 17]

P 3,3',5-triiodo-L-thyronine + iodide + A + H^+ <3, 4, 7> (<4> from microsomal enzyme complex, 3,3,5-triiodo-L-thyronine is thyromimetically active [4]) [2, 4, 5, 6, 8, 17]

S L-thyroxine + AH_2 <1, 2, 3, 4, 5, 7, 8, 9, 10, 11, 13> (<11> very low activity [15]; <1, 3, 4, 7, 8, 10> preferred substrate [8, 9, 11, 13, 14]; <4> type I isozyme [2]; <1, 3, 4, 7, 8, 10, 11, 13> isozyme type III [8, 9, 11, 13, 14, 15, 17, 21, 22]; <1, 3, 4, 7, 8, 10, 11, 13> 5-deiodinase activity [2, 4, 5, 6, 8, 9, 10, 11, 13, 14, 15, 17, 21, 22]; <4> i.e. 3,3,5,5-tetraiodo-L-thyronine [2]; <7> in pre- and prometamorphic tadpoles, undetectable during metamorphic climax [6]; <13> selenocysteine residue in the active site is absolutely essential for efficient inner ring deiodination [21]; <2, 3, 9> enzyme III performs the inactivation of 3,3,5,5-tetraiodo-L-thyronine [19]; <4> enzyme performs the inactivation of 3,3,5,5-tetraiodo-L-thyronine [19]; <2, 3, 4, 5, 9> enzyme performs the inactivation of 3,3,5,5-tetraiodo-L-thyronine and plays a critical role in regulating and maintaining the local 3,3,5-triiodo-L-thyronine content [19]) (Reversibility: ?) [1, 2, 4, 5, 6, 8, 9, 10, 11, 13, 14, 15, 17, 19, 21, 22]

P 3,3',5'-triiodo-L-thyronine + iodide + A + H^+ <1, 3, 4, 7, 8, 10, 11, 13> (<1, 2, 3, 4, 9, 10> hormonally inactive [8, 13, 17, 19, 22]; <3, 13> enzyme plays a critical role in regulating and maintaining the local 3,3,5-triiodo-L-thyronine content [17, 21]; <4> from microsomal enzyme complex, 3,3,5-triiodo-L-thyronine is regulatory active [4]) [2, 4, 5, 6, 8, 9, 10, 13, 14, 15, 17, 21, 22]

S Additional information <1, 2, 3, 4, 7, 8, 9> (<1, 3, 4, 7, 8> isozyme type III predominantly performs 5-deiodination reactions, while isozymes type I and type II perform 5-deiodination reactions [8,9]; <4> enzyme plays a critical role in regulating and maintaining the local 3,3,5-triiodo-L-thyronine content [19]; <7> enzyme activity is developmentally regulated, overview [6]; <4> 5-deiodinase activity in brain is slightly decreased in hypo- and slightly increased in hyperthyroidism [9]; <4> placental isozyme type III is only slightly increased in hyperthyroidism, while brain isozyme type III is strongly induced [11]; <1,4> enzyme performs inactivation of thyroid hormones [22]; <1,3,4,7,8> isozyme type III is a oncofetal enzyme, regulation [9]; <3> thyroid hormone is essential for maintaining normal neurological functions both during development and in adult life [17]; <3> the enzyme is the major physiological inactivator of thyroid hormone. The presence of the enzyme at maternal-fetal interfaces is consistent with its role in modulating the thyroid status of the human fetus and its expression in endometrium suggests that local regulation of thyroid status is important in implantation [24]; <2,3,9> isozyme type III performs the inactivation of 3,3,5,5-tetraiodo-L-thyronine and enzyme plays a critical role in regulating and maintaining the local 3,3,5-triiodo-L-thyronine content [19]; <4> L-hormone analogues are preferentially deiodinated via the T4-5-deiodination pathway, whereas D-analogues produce products via the T4-5-deiodination pathway [1]; <3> selenium-level of cells influences the enzyme activity [19]) (Reversibility: ?) [1, 6, 8, 9, 11, 17, 19, 22, 24]

P ?

Inhibitors

(2E)-2-(3,4-dihydroxybenzylidene)-4,6-dihydroxy-1-benzofuran-3(2H)-one <4> (<4> aurone derivative isolated from plant extract. Comparison with inhibition of 5-deiodinase EC 1.97.1.10 [2]) [2]
(2E)-2-(3,4-dihydroxybenzylidene)-6-hydroxy-1-benzofuran-3(2H)-one <4> (<4> aurone derivative isolated from plant extract. Comparison with inhibition of 5-deiodinase EC 1.97.1.10 [2]) [2]
(2E)-4,6-dihydroxy-2-(4-hydroxy-3-iodobenzylidene)-1-benzofuran-3(2H)-one <4> (<4> aurone derivative isolated from plant extract. Comparison with inhibition of 5-deiodinase EC 1.97.1.10 [2]) [2]
(2E)-4,6-dihydroxy-2-(4-hydroxybenzylidene)-1-benzofuran-3(2H)-one <4> (<4> aurone derivative isolated from plant extract. Comparison with inhibition of 5-deiodinase EC 1.97.1.10 [2]) [2]
(4-[(E)-(4,6-dihydroxy-3-oxo-1-benzofuran-2(^{3}H)-ylidene)methyl]phenoxy)acetic acid <4> (<4> aurone derivative isolated from plant extract. Comparison with inhibition of 5-deiodinase EC 1.97.1.10 [2]) [2]
2-(3,4-dihydroxybenzyl)-4,6-dihydroxy-1-benzofuran-3(2H)-one <4> (<4> aurone derivative isolated from plant extract. Comparison with inhibition of 5-deiodinase EC 1.97.1.10 [2]) [2]
3',4',4,6-tetrahydroxyaurone <4> (<4> inhibition kinetics [2]; <4> very potent naturally occuring plant-derived inhibitor, conformation study [2]) [2]
3',5'-diiodothyronine <4> (<4> comparison with 5-deiodinase EC 1.97.1.10 [1]) [1]
3,3',5'-triiodo-L-thyronine <3, 4, 7> (<4,7> substrate inhibition [2,6]; <3> no effect up to 100 nM [17]; <7> 75% inhibition at 0.01 mM [6]) [2, 6, 17]
3,3',5'-triiodothyronine <4> (<4> comparison with 5-deiodinase EC 1.97.1.10 [1]) [1]
3,3',5,5'-tetraiodo-L-thyronine <3, 7> (<3> substrate inhibition of isozyme type III [17]; <7> 95% substrate inhibition at 0.001 mM [6]) [6, 17]
3,3',5-triiodo-L-thyronine <3, 6, 7> (<3> substrate inhibition of isozyme type III, complete inhibition at 25 nM [17]; <6> isozyme type III, inhibition of tail growth [18]; <7> substrate inhibition, completely at 0.001 mM [6]) [6, 17, 18]
3,3'-diiodo-L-thyronine <7> (<7> product inhibition, 82% at 0.001 mM [6]) [6]
3,5,3'-triiodo-5'-nitrothyronine <4> (<4> comparison with 5-deiodinase EC 1.97.1.10 [1]) [1]
3,5,5'-triiodo-2'-methylthyronine <4> (<4> comparison with 5-deiodinase EC 1.97.1.10 [1]) [1]

3,5-diiodo-2'-hydroxythyronine <4> (<4> comparison with 5-deiodinase EC 1.97.1.10 [1]) [1]
3,5-diiodo-3',5'-dinitrothyronine <4> (<4> comparison with 5-deiodinase EC 1.97.1.10 [1]) [1]
3,5-diiodo-3'-hydroxythyronine <4> (<4> comparison with 5-deiodinase EC 1.97.1.10 [1]) [1]
3,5-diiodo-4'-amino-3',5'-dimethylthyronine <4> (<4> comparison with 5-deiodinase EC 1.97.1.10 [1]) [1]
3,5-diiodo-L-thyronine <3, 7> (<3> substrate inhibition of isozyme type III [17]; <7> 75% inhibition of 3,3-diiodo-L-thyronine formation at 0.001 mM [6]) [6, 17]
4',4,6-trihydroxyaurone <4> (<4> inhibition kinetics [2]; <4> cofactor competitive mechanism, displacement of thyroxine from the binding site of thyroxine-binding prealbumin [2]; <4> very potent naturally occuring plant-derived inhibitor [2]) [2]
cycloheximide <4> (<4> isozyme type III, slight decrease of activity in vivo [14]) [14]
ellagic acid <4> [2]
iodoacetate <11> (<11> recombinant and native isozyme type III [15]) [15]
N-acetyl-3,5,3'-triiodo-5'-nitrothyronine <4> (<4> comparison with 5-deiodinase EC 1.97.1.10 [1]) [1]
N-acetyl-3,5-diiodo-3',5'-dinitrothyronine <4> (<4> comparison with 5-deiodinase EC 1.97.1.10 [1]) [1]
N-acetyl-3,5-diiodo-3'-bromo-5'-nitrothyronine <4> (<4> comparison with 5-deiodinase EC 1.97.1.10 [1]) [1]
N-bromoacetyl-3,3',5-triiodo-L-thyronine <1, 4> (<1,4> irreversible inactivation [22]; <1,4> reacts with the selenocysteine residue in the active site [22]; <1,4> isozyme type III [22]) [22]
phospholipids <4> (<4> no inhibition of placental isozyme type III [11]; <4> possible regulatory role [5]) [5, 11]
tetraiodothyroacetic acid <4> (<4> comparison with 5-deiodinase EC 1.97.1.10 [1]) [1]
[4-(4-hydroxy-3,5-diiodophenoxy)phenyl]acetic acid <4> (<4> comparison with 5-deiodinase EC 1.97.1.10 [1]) [1]
[4-(4-hydroxy-3,5-dinitrophenoxy)-3,5-diiodophenyl]acetic acid <4> (<4> comparison with 5-deiodinase EC 1.97.1.10 [1]) [1]
[4-(4-hydroxy-3-iodo-5-nitrophenoxy)-3,5-diiodophenyl]acetic acid <4> (<4> comparison with 5-deiodinase EC 1.97.1.10 [1]) [1]
[4-(4-hydroxy-3-iodo-5-nitrophenoxy)phenyl]acetic acid <4> (<4> comparison with 5-deiodinase EC 1.97.1.10 [1]) [1]
aurothioglucose <3, 4, 10, 11> (<4> noncompetitive [3]; <3> 3,3,5,5-tetraiodo-L-thyronine 5-deiodinase activity [10]; <3> isozyme type III, 90% inhibition at 0.001 mM [17]; <11> recombinant and native isozyme type III [15]; <4,10> no inhibition, isozyme type III [13,14]) [3, 10, 13, 14, 15, 17]
bromoacetyl-triiodothyronine <4> [3]
growth hormone <6> (<6> upregulates the 5-deiodinase mRNA expression in the tail, but downregulates it in the liver [18]) [18]

iopanoic acid <3, 6, 9> (<3,6> isozyme type III [17,18]; <6> tail growth inhibition [18]; <3> 90% inhibition at 0.002 mM [17]; <3> 3,3,5,5-tetraiodo-L-thyronine 5-deiodinase activity [10]) [10, 17, 18, 19]
luteolin-7-β-glucoside <4> [2]
prolactin <6> (<6> upregulates the 5-deiodinase mRNA expression in the tail, but downregulates it in the liver [18]; <6> inhibits the regression of the tail fin [18]; <6> inhibits the 3,3,5-triiodo-L-thyronine-dependent upregulation of thyroid hormone receptor β mRNA expression in the tail [18]) [18]
propylthiouracil <11> (<11> slight inhibition of recombinant and native isozyme type III [15]) [15]
rosmarinic acid <4> [2]
sodium iodate <7> (<7> 95% inhibition at 1 mM [6]) [6]
Additional information <3, 10> (<3> 3,3,5-triiodo-L-thyronine sulfate has no effect on isozyme type III in hippocampus up to 100 nM [17]; <3,10> no inhibition of 5-deiodinase activity by propylthiouracil [10,13]; <3> expression and function of the 5-deiodinase and 5'-deiodinase isozymes are sensitive to thyroid hormone status, various cytokines and growth factors, severe illness, reactive oxygen species, a variety of hormones and signaling compounds, circadian rythm, and pharmacological agents, and therefore might be useful in sensor function of physiology and pathophysiology [19]) [10, 13, 17, 19]

Cofactors/prosthetic groups

dithiothreitol <1, 3, 4, 7, 8, 9, 11, 13> (<3> absolutely required [10]; <1,3,4,7,8> cofactor, influences the kinetic parameters [9]; <1,3,4,7,8,11> i.e. DTT [9,15,22]; <13> reducing cofactor [21]; <1,3,4,7,8> optimal at 50-100 mM [9,10]; <4,9> 5-20 mM, can be substituted by dithioerythritol [19]) [9,10,15,17,19,21,22]
dithioerythritol <4, 9> (<4> i.e. DTE [2]; <4> cofactor, influences the kinetic parameters [2]; <9> 5-20 mM, can be substituted by dithiothreititol [19]; <4> 5-20 mM, can be substituted by dithiothreitol [19]) [2,19]

Activating compounds

12-O-tetradecanoylphorbol-13-acetate <4> (<4> induction of isozyme type III from brain astrocytes [14]) [14]
3,3',5-triiodo-L-thyronine <9> (<9> induces enzyme [19]) [19]
actinomycin D <4> (<4> induction of isozyme type III from brain astrocytes [14]) [14]
epidermal growth factor <4, 9> (<9> induces enzyme [19]; <4> 25fold induction of isozyme type III 5-deiodinase activity in vitro [12]) [12, 19]
phospholipids <4> (<4> no influence on the kinetic parameters of 5-deiodinase activity of isozyme type III in placenta, but essential for activity [11]) [11]
retinoic acid <4> (<4> induction of isozyme type III from brain astrocytes [14]) [14]
Se^{2+} <4> (<4> required, depletion reversibly reduces the activity of isozyme type III by 3-10fold in cultured brain astrocytes [14]) [14]

vasopressin <4> (<4> slight induction of isozyme type III 5-deiodinase activity in vitro [12]) [12]
acidic fibroblast growth factor <4> (<4> 50fold induction of isozyme type III 5-deiodinase activity in vitro [12]; <4> slight increase of 3,3,5,5-tetraiodo-L-thyronine 5-deiodinase activity by isozyme type III [14]; <4> induction of isozyme type III 3,3,5-triiodo-L-thyronine 5-deiodinase activity from brain astrocytes, 6fold higher in Se-repleted than in depleted cells [14]) [12, 14]
basic fibroblast growth factor <4> (<4> 45fold induction of isozyme type III 5-deiodinase activity in vitro [12]) [12]
cAMP <4> (<4> induction of isozyme type III from brain astrocytes [14]) [14]
fibroblast growth factor <9> (<9> induces enzyme [19]) [19]
growth hormone <6> (<6> slightly upregulates the 5-deiodinase mRNA expression in the tail, but strongly downregulates it in the liver [18]) [18]
insulin-like growth factor I <4> (<4> slight induction of isozyme type III 5-deiodinase activity in vitro [12]) [12]
platelet-derived growth factor <4> (<4> slight induction of isozyme type III 5-deiodinase activity in vitro [12]) [12]
prolactin <6> (<6> upregulates the 5-deiodinase mRNA expression in the tail, but downregulates it in the liver [18]) [18]
Additional information <1, 3> (<3> expression and function of 5-deiodinase are sensitive to thyroid hormone status, various cytokines and growth factors, severe illness, reactive oxygen species, a variety of hormones and signaling compounds, circadian rythm, and pharmacological agents [19]; <1> growth hormone increases plasma T3 and decreases plasma T4 levels in 8-day old chicken embryos, in newly hatched chicks and in adult chickens within 2 h after injection. Growth hormone has no effect at all on the amount of hepatic type I enzyme catalyzing T4 deiodination to T3 but acutely decreases the amount of type III enzyme catalyzing T3 deiodination [7]) [7, 19]

Metals, ions

Se <3> (<3> selenoenzyme [24]) [24]

Specific activity (U/mg)

0.00000024 <4> (<4> 3,3,5-triiodo-L-thyronine 5-deiodination activity, isozyme type III, posterior pituitary gland [8]) [8]
0.000042 <4> (<4> isozyme type III, brown adipocyte cell culture, in presence of 10% newborn calf serum, substrate 3,3,5-triiodo-L-thyronine [12]) [12]
Additional information <1, 2, 3, 4, 9> (<4> effect of growth factors on isozyme type III [12]; <3> 3,3,5,5-tetraiodo-L-thyronine 5-deiodinase activity in brain carcinoma types [10]; <4> activity in hyperthyroid and control rats at different substrate concentrations [11]; <2,4,9> overview on assay methods [19]) [10, 11, 12, 17, 19, 20]

K_m-Value (mM)

0.000001 <4> (3,3',5-triiodo-L-thyronine, <4> pH 7.4, 37°C [14]; <4> not influenced by selenium-deficiency [14]) [14]

0.0000028 <13> (3,3',5-triiodo-L-thyronine, <13> native wild-type, pH 7.2, 37°C, in presence of 1 mM DTT [21]) [21]
0.0000031 <13> (3,3',5-triiodo-L-thyronine, <13> recombinant wild-type, pH 7.2, 37°C, in presence of 0.3 mM DTT [21]) [21]
0.0000035 <13> (L-thyroxine, <13> recombinant wild-type, pH 7.2, 37°C, in presence of 3 mM DTT [21]) [21]
0.0000036 <7> (3,3',5-triiodo-L-thyronine, <7> pH 7.0 [6]) [6]
0.0000042 <13> (3,3',5-triiodo-L-thyronine, <13> recombinant mutant SeC$_{14}$4C, pH 7.2, 37°C, in presence of 0.3 mM DTT [21]) [21]
0.000005 <1, 4> (3,3',5-triiodo-L-thyronine, <1,4> pH 7.2, 37°C [22]; <1,4> isozyme type III [22]) [22]
0.0000055 <4> (3,3',5-triiodo-L-thyronine) [9]
0.000008 <13> (L-thyroxine, <13> native wild-type, pH 7.2, 37°C, in presence of 10 mM DTT [21]) [21]
0.000011 <3> (3,3',5'-triiodo-L-thyronine, <3> pH 7.4, 37°C [17]) [17]
0.00002 <11> (3,3',5-triiodo-L-thyronine, <11> pH 7.2, 37°C [15]) [15]
0.0000225 <4> (3,3',5-triiodo-L-thyronine, <4> pH 7.5, 37°C, isozyme type III [12]) [12]
0.000037 <4> (L-thyroxine) [9]
0.000107 <3> (L-thyroxine, <3> pH 8.0, 37°C [10]) [10]
0.000265 <13> (L-thyroxine, <13> recombinant mutant SeC144C, pH 7.2, 37°C, in presence of 1 mM DTT [21]) [21]
Additional information <3, 4, 8, 9> (<4> kinetics, isozyme type III [11]; <3,4,9> review on enzyme and 5'-deiodinase isoforms type I and II [19]) [5, 9, 11, 19]

K_i-Value (mM)

0.000005 <4> (N-bromoacetyl-3,3',5-triiodo-L-thyronine, <4> pH 7.2, 37°C [22]; <4> isozyme type III [22]) [22]
0.00014 <4> (4',4,6-trihydroxyaurone, <4> at 2-20 mM dithioerythreitol, pH 7.4, 37°C [2]) [2]
0.00017 <4> (4',4,6-trihydroxyaurone, <4> at 1.8 mM dithioerythreitol, pH 7.4, 37°C [2]) [2]
Additional information <4> [2]

pH-Optimum

7 <7> (<7> assay at [6]) [6]
7-7.5 <4> (<4> isozyme type III, 3,3,5-triiodo-L-thyronine 5-deiodinase activity [8]) [8]
7.2 <1, 4, 11, 13> (<1,4,11,13> assay at [15,21,22]) [15, 21, 22]
7.4 <3, 4> (<3,4> assay at [2,14,17]) [2, 14, 17]
7.5 <4> (<4> assay at [11,12]) [11, 12]
8 <3, 5, 9> (<3,5,9> assay at [19]; <3> 3,3,5,5-tetraiodo-L-thyronine 5-deiodinase activity [10]) [10, 19]

Temperature optimum (°C)

37 <1, 3, 4, 5, 9, 11, 13> (<1,3,4,5,9,11,13> assay at [2,11,12,14,15,17,19,21,22]; <3> 3,3,5,5-tetraiodo-L-thyronine 5-deiodinase activity [10]) [2, 10, 11, 12, 14, 15, 17, 19, 21, 22]

4 Enzyme Structure

Subunits

? <1, 4, 13> (<4> x * 31000, SDS-PAGE [3]; <13> x * 32000, isozyme type III, SDS-PAGE [21]; <1> x * 30000, isozyme type III, SDS-PAGE [20]) [3, 20, 21]
dimer <3> (<3> x * 65000, SDS-PAGE, overexpressed enzyme can homodimerize probably through disulfide bridges [23]) [23]
Additional information <1, 4> (<1,4> the N-bromoacetyl-3,3,5-triiodo-L-thyronine labeled 32 kDa protein is not isozyme type III, labeling and activity are not corresponding [22]) [22]

Posttranslational modification

Additional information <4> (<4> both type I and type II 5-deiodinase enzymes of EC 1.97.1.10 as well as type III enzyme are selenoproteins. All tissues studied maintain more than 50% deiodinase activity during prolonged selenium-deficiency. Only when selenium levels decrease by more than 80%, deiodinase activity markedly decreases [16]; <4> enzyme is not a selenoprotein [3]) [3, 16]

5 Isolation/Preparation/Mutation/Application

Source/tissue

adipocyte <4> (<4> cell culture [12]; <4> isozyme type III [12]) [12]
astrocyte <4> (<4> cell culture [14]; <4> isozyme type III [14]) [14]
astrocytoma cell <3> [10]
brain <1, 2, 3, 4, 6, 9, 11> (<1, 3, 4, 6, 11> isozyme type III [9, 11, 15, 17, 18, 20, 22]; <4> more abundant in fetal than in adult rat brain [22]) [9, 11, 15, 17, 18, 19, 20, 22]
brain cancer cell <3> (<3> 5-deiodinase activity [10]) [10]
brown adipose tissue <4> (<4> cell culture [12]; <4> isozyme type III [12]) [12]
cell culture <4> (<4> astrocytes [14]; <4> brown adipocytes [12]) [12, 14]
central nervous system <3> (<3> isozyme type III [17]) [17]
cerebellar Purkinje cell <1> (<1> isozyme type III [20]) [20]
cerebellum <1> (<1> Purkinje cells [20]; <1> isozyme type III [20]) [20]
cerebral cortex <4> (<4> isozyme type III [9,11]) [9, 11]
colon carcinoma cell <3> (<3> isozyme type III [9]) [9]
colonic epithelium <3> (<3> fetal [24]) [24]
cytotrophoblast <3> [24]
decidua <3> [24]

embryo <1> [20, 22]
endothelium <3> (<3> of fetal vessels [24]) [24]
gill <11> (<11> gene expression [15]) [15]
glioblastoma cell <3> [10]
gut <11> (<11> very low activity [15]; <11> isozyme type III [15]) [15]
hepatocarcinoma cell <8> (<8> isozyme type III [9]) [9]
hepatocyte <6> (<6> larval, primary [18]) [18]
hippocampus <3> (<3> isozyme type III [17]) [17]
intestine <4, 6> (<4> fetal [9]; <4> isozyme type III [9]; <6> no measurable expression of isozyme type III [18]) [9, 18]
kidney <6, 11, 14> (<11> low activity [15]; <11> isozyme type III [15]; <6> no measurable expression of isozyme type III [18]; <14> expression is associated with the tubular epithelial cells and with the transitional epithelium, and the inner longotudinal and outer circular muscle layers of the ureter [25]) [15, 18, 25]
larva <6> (<6> premetamorphic [18]) [18]
liver <1, 3, 4, 6, 7, 11, 14> (<11> low activity [15]; <1> embryonic [9,20,22]; <1,6,11> isozyme type III [9,15,18,20,22]; <4> not in fetal liver microsomes [22]; <7> tadpole [6,9]; <14> maximum protein expression is shown in a thin layer of hepatocytes bordering the blood veins [25]) [2, 4, 5, 6, 7, 9, 15, 18, 20, 22, 25]
lymphoma cell <3> (<3> malignant [10]) [10]
medulloblastoma cell <3> [10]
meningioma cell <3> [10]
muscle <11> (<11> very low activity [15]; <11> isozyme type III [15]) [15]
oligodendroglioma cell <3> [10]
pituitary gland <4> (<4> posterior [8]) [8]
placenta <2, 3, 4, 9, 13> (<3,4,13> isozyme type III [9,11,21,22]) [3, 9, 11, 19, 21, 22]
respiratory epithelium <3> (<3> fetal [24]) [24]
skin <4, 10> (<4> isozyme type III [9]) [9, 13]
syncytiotrophoblast <3> [24]
tail <6> (<6> organ culture of tadpole tail tips [18]) [18]
temporal cortex <3> (<3> isozyme type III [17]) [17]
tissue culture <6> (<6> of tadpole tail tips [18]) [18]
umbilical artery <3> [24]
umbilical vein <3> [24]
urinary tract <3> (<3> epithelium [24]) [24]
uterine endometrium <3> (<3> of nonpregnant human uteri [24]) [24]
uterus <3> (<3> endometrial gland [24]) [24]
Additional information <2, 3, 4, 7, 9> (<7> 5-deiodinase and 5'-deiodinase activity are developmentally regulated [6]; <3> isozyme type III activity in brain areas, overview [17]; <4> no enzymic activity in testes and spleen [3]; <2,3,9> not: kidney, healthy adult liver, pituitary gland, thyroid gland [19]) [3, 6, 17, 19]

Localization

microsome <1, 3, 4, 7, 11, 13> (<4> 5-deiodinase and 5'-deiodinase activity [2]; <4> 3,3,5,5-tetraiodo-L-thyronine-5-deiodinase and 3,3,5,5-tetraiodo-L-thyronine-5-deiodinase are solely localized in liver microsomes [4]; <1, 3, 4, 11, 13> isozyme type III [9,11,15,17,20,21,22]) [1, 2, 3, 4, 5, 6, 9, 11, 15, 17, 19, 20, 21, 22]

mitochondrion <4> (<4> isozyme type III [11]) [11]

Additional information <4, 9> (<4> membrane association is not essential for maintenance of functional activity [19]; <9> subcellular localization of the different isozyme types in different tissues, overview [19]) [19]

Purification

<4> [3]

<4> (partial, 3fold) [5]

<11> (recombinant from COS-1 cells) [15]

Renaturation

<4> (delipidation of microsomes during purification results in loss of enzyme activity. Full recovering of activity is achieved by recombining phospholipids and protein, partial recovery may be achieved by addition of exogenous phospholipids to protein) [3]

Cloning

<1> (isozyme type III, via RT-PCR from cerebellum mRNA, expression in bacteria) [20]

<10> (screening of neonatal skin cDNA library, DNA sequence determination and analysis, isozyme type III, an in-frame TGA codon encodes a selenocysteine residue, functional translation in Xenopus laevis oocytes, catalytis properties of recombinant and native enzyme are identical) [13]

<11> (DNA sequence determination and analysis, isozyme type III, an in-frame TGA codon encodes a selenocysteine residue, contains a putative Sec insertion sequence element at 3' UTR, functional expression in COS-1 cells) [15]

<13> (isozyme type III, expression of wild-type and active site mutants in COS cells, 50fold higher expression level for the mutants than for the wild-type enzyme) [21]

Engineering

SeC144A <13> (<13> site-directed mutagenesis, exchange of the selenocysteine residue in the highly conserved active site sequence, enzymatically inactive [21]) [21]

SeC144C <13> (<13> site-directed mutagenesis, exchange of the selenocysteine residue in the highly conserved active site sequence, 5fold increased K_m for 3,3,5-triiodo-L-thyronine and 100fold increased K_m for 3,3,5,5-tetraiodo-L-thyronine compared to the wild-type, 2-6fold reduced turnover [21]) [21]

Application

medicine <3> (<3> expression and function of the 5-deiodinase and 5'-deiodinase isozymes are sensitive to thyroid hormone status, various cytokines and growth factors, severe illness, reactive oxygen species, a variety of hormones and signaling compounds, circadian rythm, and pharmacological agents, and therefore might be useful in sensor function of physiology and pathophysiology [19]) [19]

6 Stability

Temperature stability

70 <3> (<3> 30 min, inactivation of 5-deiodinase [10]) [10]

References

[1] Koehrle, J.; Auf'mkolk, M.; Rokos, H.; Hesch, R.D.; Cody, V.: Rat liver iodothyronine monodeiodinase. Evaluation of the iodothyronine ligand-binding site. J. Biol. Chem., **261**, 11613-11622 (1986)

[2] Auf'mkolk, M.; Koehrle, J.; Hesch, R.D.; Cody, V.: Inhibition of rat liver iodothyronine deiodinase. Interaction of aurones with the iodothyronine ligand-binding site. J. Biol. Chem., **261**, 11623-11630 (1986)

[3] Santini, F.; Chopra, I.J.; Hurd, R.E.; Solomon, D.H.; Chua Teco, G.N.: A study of the characteristics of the rat placental iodothyronine 5'-monodeiodinase: evidence that it is distinct from the rat hepatic iodothyronine 5'-monoiodinase. Endocrinology, **130**, 2325-2332 (1992)

[4] Auf dem Brinke, D.; Kohrle, J.; Kodding, R.; Hesch, R.D.: Subcellular localization of thyroxine-5-deiodinase in rat liver. J. Endocrinol. Invest., **3**, 73-76 (1980)

[5] Fekkes, D.; Hennemann, G.; Visser, T.J.: Properties of detergent-dispersed iodothyronine 5- and 5'-deiodinase activities from rat liver. Biochim. Biophys. Acta, **742**, 324-333 (1983)

[6] Galton, V.A.; Hiebert, A.: Hepatic iodothyronine 5-deiodinase activity in Rana catesbeiana tadpoles at different stages of the life cycle. Endocrinology, **121**, 42-47 (1987)

[7] Darras, V.M.; Berghman, L.R.; Vanderpooten, A.; Kuhn, E.R.: Growth hormone acutely decreases type III iodothyronine deiodinase in chicken liver. FEBS Lett., **310**, 5-8 (1992)

[8] Tanaka, K.; Shimatsu, A.; Imura, H.: Iodothyronine 5-deiodinase in rat posterior pituitary. Biochem. Biophys. Res. Commun., **188**, 272-277 (1992)

[9] Visser, T.J.; Schoenmakers, C.H.: Characteristics of type III iodothyronine deiodinase. Acta Med. Austriaca, **19 Suppl 1**, 18-21 (1992)

[10] Mori, K.; Yoshida, K.; Kayama, T.; Kaise, N.; Fukazawa, H.; Kiso, Y.; Kikuchi, K.; Aizawa, Y.; Abe, K.: Thyroxine 5-deiodinase in human brain tumors. J. Clin. Endocrinol. Metab., **77**, 1198-1202 (1993)

[11] Mori, K.; Yoshida, K.; Fukazawa, H.; Kiso, Y.; Sayama, N.; Kikuchi, K.; Aizawa, Y.; Abe, K.: Thyroid hormone regulates rat placental type III iodothyronine deiodinase activity by inducing kinetic changes different from those in the same isozyme in rat brain. Endocr. J., **42**, 753-760 (1995)
[12] Hernandez, A.; Obregon, M.J.: Presence of growth factors-induced type III iodothyronine 5-deiodinase in cultured rat brown adipocytes. Endocrinology, **136**, 4543-4550 (1995)
[13] Croteau, W.; Whittemore, S.L.; Schneider, M.J.; St Germain, D.L.: Cloning and expression of a cDNA for a mammalian type III iodothyronine deiodinase. J. Biol. Chem., **270**, 16569-16575 (1995)
[14] Ramauge, M.; Pallud, S.; Esfandiari, A.; Gavaret, J.; Lennon, A.; Pierre, M.; Courtin, F.: Evidence that type III iodothyronine deiodinase in rat astrocyte is a selenoprotein. Endocrinology, **137**, 3021-3025 (1996)
[15] Sanders, J.P.; Van der Geyten, S.; Kaptein, E.; Darras, V.M.; Kuhn, E.R.; Leonard, J.L.; Visser, T.J.: Cloning and characterization of type III iodothyronine deiodinase from the fish Oreochromis niloticus. Endocrinology, **140**, 3666-3673 (1999)
[16] Bates, J.M.; Spate, V.L.; Morris, J.S.; St. Germain, D.L.; Galton, V.A.: Effects of selenium deficiency on tissue selenium content, deiodinase activity, and thyroid hormone economy in the rat during development. Endocrinology, **141**, 2490-2500 (2000)
[17] Santini, F.; Pinchera, A.; Ceccarini, G.; Castagna, M.; Rosellini, V.; Mammoli, C.; Montanelli, L.; Zucchi, V.; Chopra, I.J.; Chiovato, L.: Evidence for a role of the type III-iodothyronine deiodinase in the regulation of 3,5,3'-triiodothyronine content in the human central nervous system. Eur. J. Endocrinol., **144**, 577-583 (2001)
[18] Shintani, N.; Nohira, T.; Hikosaka, A.; Kawahara, A.: Tissue-specific regulation of type III iodothyronine 5-deiodinase gene expression mediates the effects of prolactin and growth hormone in Xenopus metamorphosis. Dev. Growth Differ., **44**, 327-335 (2002)
[19] Koehrle, J.: Iodothyronine deiodinases. Methods Enzymol., **347**, 125-167 (2002)
[20] Verhoelst, C.H.; Vandenborne, K.; Severi, T.; Bakker, O.; Zandieh Doulabi, B.; Leonard, J.L.; Kuhn, E.R.; van der Geyten, S.; Darras, V.M.: Specific detection of type III iodothyronine deiodinase protein in chicken cerebellar purkinje cells. Endocrinology, **143**, 2700-2707 (2002)
[21] Kuiper, G.G.; Klootwijk, W.; Visser, T.J.: Substitution of cysteine for selenocysteine in the catalytic center of type III iodothyronine deiodinase reduces catalytic efficiency and alters substrate preference. Endocrinology, **144**, 2505-2513 (2003)
[22] Schoenmakers, C.H.; Pigmans, I.G.; Kaptein, E.; Darras, V.M.; Visser, T.J.: Reaction of the type III iodothyronine deiodinase with the affinity label N-bromoacetyl-triiodothyronine. FEBS Lett., **335**, 104-108 (1993)
[23] Curcio-Morelli, C.; Gereben, B.; Zavacki, A.M.; Kim, B.W.; Huang, S.; Harney, J.W.; Larsen, P.R.; Bianco, A.C.: In vivo dimerization of types 1, 2, and 3 iodothyronine selenodeiodinases. Endocrinology, **144**, 937-946 (2003)

[24] Huang, S.A.; Dorfman, D.M.; Genest, D.R.; Salvatore, D.; Larsen, P.R.: Type 3 iodothyronine deiodinase is highly expressed in the human uteroplacental unit and in fetal epithelium. J. Clin. Endocrinol. Metab., **88**, 1384-1388 (2003)
[25] Verhoelst, C.H.; Van Der Geyten, S.; Darras, V.M.: Renal and hepatic distribution of type I and type III iodothyronine deiodinase protein in chicken. J. Endocrinol., **181**, 85-90 (2004)

Printing: Krips bv, Meppel, The Netherlands
Binding: Stürtz, Würzburg, Germany